ELEMENTARY STATISTICS:
From Discovery to Decision

MARILYN K. PELOSI, PH.D.

Western New England College
Springfield, MA

THERESA M. SANDIFER, PH.D.

Southern Connecticut State University
New Haven, CT

JOHN WILEY & SONS, INC.

ASSOCIATE PUBLISHER	Laurie Rosatone
ACQUISITION EDITOR	Angela Battle
DEVELOPMENT EDITOR	Caroline Ryan
MARKETING MANAGER	Julie Lindstrom
PRODUCTION SERVICES MANAGER	Jeanine Furino
SENIOR DESIGNER	Karin Gerdes Kincheloe
BOOK DESIGNER	Nancy B. Field
ASSISTANT EDITOR	Jennifer Battista
ILLUSTRATION EDITOR	Anna Melhorn
PHOTO MANAGER	Hilary Newman
MEDIA EDITOR	Lisa Schlettner
PRODUCTION MANAGEMENT SERVICES	Susan L. Reiland
COVER PHOTO	© Tom Wilkins

Photo Credits for Chapter Openers: Introduction, Chapters 1–7, 9, 12, 14–16: ©PhotoDisc; Chapters 8 and 10: ©Corbis Digital Stock; Chapters 11 and 13: ©Eyewire. **Cartoon credit**: Page 3, ©J. B. Handelsman/The Cartoon Bank.

This book was set in Times Ten by Progressive Information Technologies, and printed and bound by Von Hoffmann. The cover was printed by Von Hoffmann.

This book is printed on acid-free paper.

0-471-40142-0
WIE: 0-471-42903-1

Printed in the United States of America

10 9 8 7 6 5 4 3 2 1

About the Authors

Dr. Marilyn Pelosi is a Professor of Quantitative Methods at Western New England College in Springfield, MA. Dr. Pelosi, a Rhode Island native, graduated from Brown University, where she met her husband. She received her Ph.D. in Industrial Engineering from the University of Massachusetts, where she met co-author Terry Sandifer. Dr. Pelosi has worked as a statistician for the government in Washington, DC, and taught Industrial Engineering at Western New England College before moving to her current position in the School of Business. Dr. Pelosi is known for her enthusiastic teaching style and for using real data from consulting projects in the classroom. Dr. Pelosi has two children and enjoys reading, traveling, and watching movies.

Dr. Terry Sandifer is a Professor of Mathematics at Southern Connecticut State University in New Haven. She graduated from Iona College and received her Ph.D. in Industrial Engineering and Operations Research from the University of Massachusetts, where she met both her husband and Dr. Pelosi. She continued her career teaching Industrial Engineering at Western New England College with Dr. Pelosi. Dr. Sandifer then spent five years working as a statistician for the Kimberly Clark Corporation before returning to teaching. A Brooklyn, NY, native, Dr. Sandifer has two children and enjoys reading, shopping, and playing video games.

Dr. Pelosi and Dr. Sandifer have been best friends since graduate school and have co-authored three statistics workbooks and a business statistics textbook.

Preface

"Statistical thinking will one day be as necessary for efficient citizenship as the ability to read and write." — H.G. Wells

Our objective in writing *Elementary Statistics: From Discovery to Decision* is to teach students to use statistics as a tool for making informed decisions about the data that surround them in their everyday lives. The pedagogical focus of this text is on statistical reasoning, interpretation, and decision making. We want students to perceive themselves as consumers of data by learning how to turn data into useful, inference-based information. Therefore, the stress is on conceptual understanding rather than rote calculation. To support this approach, we emphasize practice as well as theory.

Elementary Statistics: From Discovery to Decision is intended for use in a one- or two-semester, algebra-based introductory statistics course. The first ten chapters cover topics usually found in a one-semester course, while the remaining chapters on advanced topics are designed for those courses that emphasize topics after regression. Due to the wide range of applications included in the examples and exercises, this book should be meaningful to students majoring in a variety of disciplines including allied health, liberal arts, humanities, and the social sciences.

GENERAL FEATURES

Enjoyable to Read As you read through the text, you'll notice that we employ a truly conversational tone. Students find the book approachable and are not intimidated by the subject matter. Our goal is that students will understand how and why a technique works, while also learning to discern which technique to use and when to use it.

Real-Life Relevance Data are all around us. This text enables students to draw connections between real data and the techniques for analysis and interpretation. All of the cases are based on real-life situations and are designed to help students relate their personal experiences with data to the course. The data used in the examples and exercises are based on our consulting and teaching experiences. In order to protect the confidentiality of our clients, we've rescaled or altered some of the data, but only in a way that does not hinder its usefulness from a pedagogical perspective, and is still true to life.

Activity-Based Approach The design of this text is an integral part of the learning experience. After reflecting on our years of teaching experience and examining how students learn new things, we adopted a step-by-step, activity-based model. This model allows students to gain a basic understanding of a concept in terms they can grasp, pushes them to apply their knowledge, and requires them to make decisions based on what they've learned. The following features comprise the activity-based approach: First, there are activities that encourage *discovery*, such as the "Try It Now" and "Discovery Exercises." Second, we incorporate features that encourage students to *learn by doing*, such as the "Learning It," "Thinking About It," and "Doing It" exercises. And third, there is pedagogy that stimulates *conceptual understanding* in the "Get It in Writing" and "Making Your Case" features. Overall, students are required to take an active role throughout the learning process.

Appropriate Use of Technology Although we assume that most students have access to some form of technology, we understand that not every school chooses to use it. Only a scientific calculator is necessary. Each chapter concludes with an *optional* technology section for those using a statistical software package. These sections include statistical output screen displays and brief instruction for using Minitab, the TI-83 graphing calculator, and Excel. Four text-specific technology manuals (for Minitab, Excel, the TI-83, and SPSS) are available from the publisher and may be packaged with the text. Additionally, the student versions of Minitab and SPSS are also available.

FLEXIBLE SYLLABUS

The chapter organization will suit most introductory courses. In keeping with the goals of the text, we offer several innovations that work well with students:

Introductory Chapter This distinctive chapter sets the tone for the text and gives students an overarching introduction to the field of statistics. It introduces students to the idea of statistical thinking and problem solving. Our students like this chapter, but coverage of this material is not essential; in other words, an instructor can begin immediately with Chapter 1.

Coverage of Descriptive Statistics Descriptive statistics is covered in Chapters 2–4 of the text. We believe that students need to fully understand how to describe data before they learn probability in Chapters 5 and 6. Chapter 4 offers students a preview of topics such as regression, qualitative data analysis, and nonparametrics, discussed later in Chapters 11, 14, and 15. Those instructors who prefer an accelerated treatment of descriptive statistics can skip Chapter 4 and move directly to Chapter 5 on probability.

Advanced Probability Topics Coverage of Bayes' Rule and the Poisson distribution is provided in Appendices C and D.

Coverage of Hypothesis Testing Beginning with Chapter 8, we offer a unique parallel treatment of approaches to hypothesis testing. The critical value and *p*-value approaches are presented in tandem with examples, so that students understand and can use either.

Case Studies Chapter 16 entitled "Making Your Case" is provided as an option for those instructors who wish to incorporate more detailed real-life cases into the course. Designed to be assigned individually or in groups, these cases ask students to apply their knowledge, draw conclusions, and make decisions about data. Each case references the chapters that introduced the tools necessary for completing it. Therefore, these cases can be used before the end of the book is reached.

CHAPTER FEATURES

The chapter openers provide real-world motivation for the topics that follow, and also encourage students to consider the information they will use in analysis. Each describes a real-life situation, and will prompt students to think about how statistics are involved in many aspects of the world around them.

Once students have been briefly introduced to a new topic, they can enjoy the challenge of the *Try It Now!* exercises. These self-test checkpoints allow students to assess their own understanding before moving on. Repeated use of these exercises helps to promote student confidence and prepares them to complete exercises. The answers to the *Try It Now!* exercises are provided at the bottom of the page, upside down, for immediate feedback.

Hundreds of *Worked Examples* are included throughout the text to reinforce each chapter topic and to demonstrate techniques in action. Examples include real data and a *Four-Step Problem-Solving Method.* Each problem-solving step is noted with an icon. Often, the worked examples will apply the techniques presented to the chapter-opening problem.

The *Discovery Exercises* highlight another important goal of introductory statistics: the leap from discovery to decision. These exercises function as expanded case studies and can be used for group work. They are written to achieve a high level of student understanding—an "aha" discovery of a key concept—and prompt students to make an informed decision.

A blend of computational, conceptual, and activity-based *Exercises* has been carefully crafted to help students check their understanding of key topics. Chapter exercises are divided into three graded categories: "Learning It" exercises allow students to practice the basics; "Thinking About It" exercises require students to interpret results and make decisions based on their analysis; "Doing It" exercises use large real-world data sets and ask students to perform and interpret statistical analyses in order to solve a problem. The complete data sets are provided on the free CD-ROM that accompanies the text.

Every chapter features a *Using Technology* section that includes screen displays of statistical output and brief instruction for Minitab, Excel, and the TI-83 graphing calculator. The Wiley Spreadsheet Toolkit, KADDSTAT, is provided free with the accompanying student CD-ROM. This Microsoft® Excel add-in helps students to perform statistical analyses more quickly and easily than with Excel alone, and provides the functionality of more powerful packages for free. Detailed instruction is included in the accompanying technology manuals.

At the end of each chapter, the *Get It in Writing* feature highlights an overarching goal that is sometimes overlooked: the importance of communicating your results. We want students to understand that if you fail to effectively communicate the results of your analysis, then your work can be for naught. *Get It in Writing* summarizes the answer to the chapter-opening problem in memo or presentation form. Thus, each chapter opens by posing a problem, continues by analyzing this problem throughout the chapter in the worked examples, and ends with a written summary of the solution.

EXERCISE APPENDIX

Successfully completing exercises is a fundamental part of any statistics course. Carefully crafted exercises can reinforce basic concepts and test students' overall understanding of ideas. Exercises can also teach students to analyze, interpret, and make decisions about data. Throughout the text, we have provided a variety of types of exercises to achieve both goals.

However, we understand that depending on individual needs, some instructors may want more emphasis on skill-building exercises. As a result, we have included an additional bank of skill-building exercises in Appendix E. This convenient placement will allow instructors greater flexibility when assigning exercises.

CHAPTER COMMENTARY

Introductory Chapter: The Role of Statistics in Life This chapter motivates the study of statistics as a necessary tool to allow the average citizen to read the newspaper responsibly. It also directly addresses fears and/or anxieties students may have about statistics. If the instructor chooses to forego lecturing on this chapter, it can be assigned as reading and the instructor can start immediately with Chapter 1.

Chapter 1: The Language of Statistics Most students find this chapter easy to read and helpful. Just like learning any new language, learning statistics starts with the vocabulary. It begins with the discussion of sampling error, a concept that is repeatedly emphasized throughout the text. An early discussion of sample design and ethics sets up the discussion for all subsequent chapters. The chapter also uses examples from actual surveys to teach the different types of data.

The instructor can ask each student to bring in a questionnaire that they find either on-line or in a newspaper or magazine. These can be used to reinforce all the concepts in this chapter and make for a fun and applied class discussion.

Chapters 2 and 3: Graphical and Numerical Descriptors of Data These two chapters teach the main tools of descriptive statistics/exploratory data analysis. The techniques are presented in a way that forces the students to *think* before manipulating numbers or graphs. A good deal of discussion is provided regarding what the summaries really tell us and which graphs and statistics are appropriate for a particular data set.

These chapters rely heavily on technology and focus on helping students discern which tool to use and how to interpret it, rather than on the creation and calculation aspects of the tools. The daily newspaper is filled with statistics and graphs that can be employed with these chapters. These are typically very popular and useful chapters for students.

Chapter 4: Analyzing Bivariate Data This is the last chapter that covers descriptive statistics/exploratory data analysis (EDA). This chapter looks at the relationship between two variables and builds from ideas introduced in Chapters 2 and 3, which cover the EDA tools to describe a single variable. Some interesting discussions can occur as a result. Cause and effect can be raised as a discussion point, for example.

Those instructors who wish to cover more inferential statistics can certainly skip Chapter 4. However, it is a very valuable chapter for those students who will mainly be doing descriptive statistics/EDA. It is best taught with the use of technology.

Chapter 5: Probability Chapters 5 and 6 present all the probability necessary in order to teach inferential statistics. The basic rules and definitions of probability are presented first. These are linked to the corresponding concepts that the students learned in the descriptive statistics chapters.

The emphasis is on learning how to calculate and interpret probabilities so that confidence intervals and hypothesis testing can be understood. For this reason, Bayes' Rule is presented in Appendix C so that it can be easily omitted if the instructor chooses to move quickly to the inferential statistical tools. This chapter should not be skipped.

Chapter 6: Random Variables and Probability Distributions The concepts of a random variable and a distribution are presented in this chapter. The link is made between histograms and distributions. The most commonly used distributions are presented: the binomial and the normal distributions. For those instructors who would like to cover additional distributions, Appendix D presents the Poisson distribution.

Chapter 7: Sampling Distributions and Confidence Intervals This chapter links the probability chapters to the inferential statistics chapters. It covers the concept of an estimator and its properties in a "big-picture way" linking back to the EDA chapters. If the instructor prefers to begin directly with the Central Limit Theorem, then Sections 7.1–7.4 can be skipped. These first few sections talk about estimators of any parameter and include discussion of the mean, proportion, and variance. The idea is to help the students see that all estimators serve the same purpose but the specifics are different depending, of course, on what you are estimating.

The properties of the estimator $\bar{x}$ are then covered in detail as the Central Limit Theorem. The Central Limit Theorem is explained in detail to help the students understand the concept of a sampling distribution in addition to understanding the details of the theorem. Applications of the CLT are presented, including confidence intervals and sample size calculations.

Chapter 8: Hypothesis Testing: An Introduction This chapter introduces the language and steps of hypothesis testing that are used for many of the chapters that follow. It teaches the students that all hypothesis tests follow the same set of steps but that the calculations are different depending on what hypothesis is being tested. Students often see hypothesis tests as a collection of independent and unrelated tools. This chapter sets up the "big picture" and then presents large-sample, one-population tests of the mean. The chapter uses both p values and critical values in tandem and therefore can be taught using either or both approaches. The instructor who prefers to jump right into the test of the mean can skip Sections 8.2 and 8.3.

Chapter 9: Inferences: More One-Population Tests This chapter should be taught immediately after Chapter 8. It covers small-sample tests of the mean, tests of proportion, and test of the variance for one population. For each test, the same hypothesis-testing steps are utilized and reinforced to help the students see the "big picture" in addition to the details. The chapter concludes by linking the tool of hypothesis testing to that of confidence intervals. The Discovery Exercise in this chapter can be used in class to help students draw this link themselves.

Chapter 10: Comparing Two Populations Moving from estimating parameters for a single population to comparing parameters from two populations makes for interesting discussion. The chapter reminds a student that oftentimes a qualitative variable can be used to split a group into two populations for comparison purposes.

The hypothesis test to compare two means is covered, including an introduction to designed experiments through the paired t test. Hypothesis tests to compare two proportions and two variances are also presented. This chapter can be skipped if the instructor wants to move directly on to regression.

Chapters 11 and 12: Simple Linear Regression and Multiple Linear Regression These two chapters present the tool of regression as a model-building technique. Both chapters rely heavily on the use of technology. The emphasis is on understanding the relationship between two or more variables and developing a model to predict one variable from one or more other variables. If the instructor has covered Chapter 4, then the student has seen the concept of a regression line but not done any of the inferential statistics connected with the model testing. In this case, some of the material in Chapter 11 will be a nice review of the technique. If Chapter 4 has been skipped, then Chapter 11 presents simple linear regression from the ground up.

We have found that the tool of regression is often the "tool of choice" simply because most statistical packages do regression and people who have had an introduction to statistics remember it. These chapters point out the power of the tool of regression but also its potential misuse. The chapters teach how to do residual analysis and assumption checking in order to check the validity of the use of the tool of regression. If the instructor prefers to take an ANOVA approach to regression, then Chapter 13 should be presented first. Otherwise, Chapter 12 provides a nice step into the technique of ANOVA presented in the next chapter.

Chapter 13: Experimental Design and ANOVA   This chapter continues the discussion of getting the most information for your time and money, which starts in Chapter 1. It also follows nicely after the discussion of a paired t test in Chapter 10. This chapter needs to be taught with some technology, as the calculations are long and tedious and not the focus of the chapter. The focus is on the output of ANOVA and what it tells you.

Chapter 14: The Analysis of Qualitative Data   This chapter can be taught any time after Chapter 8 on hypothesis testing has been presented. Typically it would be presented after Chapter 10 has been covered. It covers the standard hypothesis tests relating to qualitative data including the goodness-of-fit test, equality of proportions from more than two populations, and the test for independence of two qualitative variables. The chapter points out that the goodness-of-fit test can be used to determine whether the assumption of normality is met and can thus be used as in introduction to the nonparametric chapter.

Chapter 15: Nonparametric Statistics   This chapter provides an introduction to the tools available when the data are not normally distributed or the data are not truly quantitative (i.e., ratings or rankings). The same hypothesis-testing steps are followed, allowing students to see the parallel with parametric tests.

This chapter can be done at the end of the semester (when the students are tired) because the computations are fairly easy and the concepts have all been introduced in Chapter 8. Alternatively, for disciplines where most of the data are ranked data, this chapter can be done after Chapters 8 and 9 have been covered. If Chapter 14 has been covered, then students could test to see whether the normal assumption has been violated and, if so, then use the nonparametric tests. This makes a nice connection between the chapters and is a very practical approach to many data sets.

Chapter 16: Making Your Case   Very often you need to use a combination of techniques to analyze a large data set, and deciding on the right tool is part of the challenge. This chapter provides the student with a summary of the tools that he or she has learned and a framework for tackling the analysis of large, unstructured, real-life data sets. This is very much a capstone chapter. The student is walked through the analysis of one data set from this perspective and then given a variety of large cases to analyze.

This chapter can clearly be used at the end of the course to assess students' ability to actually use the statistical tools they have been taught. The cases can also be used as student projects. Some of the cases can be used after Chapter 4 has been covered, as they focus primarily on EDA; some of the cases can be used after the tool of hypothesis testing has been taught, and these are so identified.

Supplements

INSTRUCTOR SUPPLEMENTS

Instructor's Manual

0-471-26707-4
This print manual includes full solutions to the "Thinking About It," "Learning It," and "Discovery Exercises" throughout the text. It also includes answers to the exercises in Appendix E as well as additional teaching materials.

Printed Test Bank

0-471-26708-2
Written by Larry Stephens of the University of Nebraska-Omaha, the test bank includes hundreds of additional exercises in true–false, sentence completion, multiple-choice, and problem/essay formats.

Computerized Test Bank

0-471-42061-1
All questions from the printed test bank will be available in an electronic format. Please contact your local sales representative for further information.

eGrade

eGrade is an on-line assessment system that contains a large bank of skill-building problems and solutions. Instructors can now automate the process of assigning, delivering, grading, and routing all kinds of homework, quizzes, and tests while providing students with immediate scoring and feedback on their work. Wiley *eGrade* "does the math"… and much more. For more information and an on-line demonstration, visit www.wiley.com/college/egrade.

Book Companion Website – Instructor

This password-protected site located at www.wiley.com/college/pelosi contains the electronic files for the Instructor's Manual and Printed Test Bank, as well as the text's data sets. Please contact your local sales representative for further information.

STUDENT SUPPLEMENTS

Student Resource CD-ROM

The student resource CD-ROM is packaged free with every new copy of the text and includes the following resources:

> *PowerPoint Slides*
> Prepared by Nancy Proyect of Storm King Publishing, this resource includes chapter-by-chapter slides to illustrate key concepts. It is designed for use in class or for review.

> *Data Sets*
> Provided in Minitab, Excel, and SPSS formats, the data sets are linked to the "Doing It" and "Learning It" exercises in the text. The name of the appropriate data file is listed next to problems requiring use of the files.

Excel Interactive Tutorial: Tutorials for Statistics
Developed by Barbara Miller of Indiana University, this tutorial provides hands-on training in the basic functions of Microsoft® Excel.

Wiley Spreadsheet Tool Kit, KADDSTAT
Brian Kilmer created this menu-driven, Microsoft® Excel add-in statistical analysis program in collaboration with Donald L. Harnett. The programmed statistical routines provide "point and click" computational capability to perform graphical analysis, descriptive statistics, and inferential model analysis activities.

Technology Manuals

Prepared by the authors, the following text-specific manuals work seamlessly with the text. Each manual provides screen displays and step-by-step keystroke level instruction to help students learn how to utilize the technology. Individual manuals can be packaged with the text for a reduced price.

Minitab Manual	**Excel Manual**
0-471-26722-8	0-471-26721-X
TI-83 Calculator Manual	**SPSS Manual**
0-471-26724-4	0-471-26725-2

Book Companion Website – Student

This site located at www.wiley.com/college/pelosi contains all of the resources available on the Student Resource CD-ROM. Some of the resources are password-protected. Please visit the site for further information.

ADDITIONAL SOFTWARE OPTIONS

Student versions of Minitab and SPSS can be packaged with the text for a small additional fee. Consult your local Wiley sales representative for further details.

Acknowledgments

No work of this magnitude can happen without a lot of help. We would like to take this opportunity to thank those people who have made this book a reality. First and most important, we want to thank Brad Wiley II, our original editor, who has always believed in us and what we wanted to accomplish. We could not have done this without the help of Mary O'Sullivan, who kept us sane during a very difficult process, and Susan Reiland, who carried us through the production process without losing her sense of humor. There are many other people at Wiley who deserve our thanks, in particular, Bruce Spatz, Laurie Rosatone, Angela Y. Battle, Julie Lindstrom, Jennifer Battista, Stacy French, Jeanine Furino, and Martin Batey.

Part of what makes this book unique is the amount of student input in the development process. We thank all of our students who read and used the book in draft form, finding our errors and making suggestions.

LIST OF REVIEWERS

We are indebted to the many reviewers who offered their time and expertise to help hone and craft this text. The suggestions and advice of the following individuals have proved invaluable:

Marla M. Bell, Kennesaw State University

Ann Marie Harris, Brigham Young University

Robert Herrman, Elmhurst College

Thomas P. Kline, University of Northern Iowa

Christopher Lacke, Rowan University

Ben Lev, University of Michigan–Dearborn

Alexander Levin, Catholic University of America

John Loase, Westchester Community College

Rhonda Magel, North Dakota State University

Joyce McQuade, Westchester Community College

Rene Leo E. Ordonez, Southern Oregon University

Lindsay Packer, College of Charleston

Scott Perkins, Lake Sumpter Community College

Larry Ringer, Texas A & M University

Neal Rogness, Grand Valley State University

Ronald Schwartz, Florida Atlantic University

Lynn Smith, Gloucester County College

Sandra Spain, Thomas Nelson Community College

CONTENT CONTRIBUTORS

Larry Stephens, University of Nebraska, Omaha

Tammiee Dickenson, University of South Carolina

ACCURACY CHECKER

Paul Lorczak

Although we have made every effort to ensure that this text is both current and accurate, users may have feedback to offer. We welcome your suggestions. Please feel free to forward them to us at the email addresses listed below. We hope that you and your students enjoy teaching and learning from the text as much as we have.

Marilyn Pelosi
mpelosi@wnec.edu

Theresa Sandifer
sandifer@southernct.edu

Features Walkthrough

Chapter Openers

An opening story motivates each chapter by discussing a real-world situation. This opening problem is referenced in examples throughout the remainder of the chapter.

CHAPTER

3

Numerical Descriptors of Data

THE GOLF BALL DILEMMA

Have you ever gone to the store to buy something and been completely overwhelmed by the number of variations that a company has for a single product? Is there really that much difference between basic and deluxe? What do premium and XL really mean? One example of this problem is golf balls. A single manufacturer can have as many as six variations in a single product line. The question is, Are all the variations really different and, if they are, is the difference worth the increase in cost? A golf pro at a country club decided to run a test of the golf balls at the top and bottom of a particular manufacturer's product line to see how they differ. In preparation for the study, he collected sample data on different variables for the two types of balls. You are asked to summarize the data and report on what you find. Here is a portion of the data:

Ball number	Model number	S1	S2	S3	Weight (g)	Dimple width (mm)	Dimple depth (mm)	Head	Temperature (°F)	Carry (yd)	Total distance (yd)	Date	Time
1	M1	81	81	82	45.3	0.1450	0.0110	686	77	257	270	8/20	8:15
2	M1	83	83	84	45.2	0.1510	0.0111	688	77	255	267	8/20	8:15
3	M1	81	82	84	45.2	0.1450	0.0105	687	77	256	267	8/20	8:15
4	M1	81	81	83	45.3	0.1440	0.0117	688	77	255	271	8/20	8:15
5	M1	83	83	82	45.5	0.1460	0.0108	687	77	255	268	8/20	8:15
6	M1	83	83	82	45.3	0.1560	0.0111	687	77	256	267	8/20	8:15
7	M1	81	81	82	45.2	0.1495	0.0111	687	77	255	264	8/20	8:15
8	M1	83	81	82	45.1	0.1505	0.0110	690	78	258	269	8/20	8:15

The golf pro wants to know how the two products compare on factors such as how far most of the balls go when they are hit, how much variation there is in the distance, and what percentage of the balls go beyond a certain distance. You need to figure out which numerical measures will provide the most information.

Chapter Objectives

The first section of every chapter draws connections between previous chapters and the topics in the current chapter. Additionally, these sections outline the major topics in the chapter.

3.1 Chapter Objectives

Remember that when you looked at graphical displays of numerical data, you were interested in three characteristics: center, spread or dispersion, and shape. In this chapter we look at numerical measures that can be used to describe the same features of the data. The chapter covers the following material:

- Numerical measures of center: the mean, the median, and the mode
- Numerical measures of variability: the range and the standard deviation
- Descriptions of a set of data: the empirical rule and boxplots
- Measures of relative standing: percentiles and percentile rank
- Identification of outliers: z-scores and boxplots

3.2 Describing Data Numerically

Although we say that a picture is worth a thousand words, numerical quantities are also useful to describe the data. You may wonder why we need numerical descriptors when the graphs let us *see* the data. There are two reasons. First, although we can certainly see the data using histograms and bar charts, it is difficult to *talk* or *write* about pictures. We often need other references to describe the data. Second, we may want to make inferences based on the sample data. To make statistical inferences we need to use *numerical measures*.

When we collect data, we may have either a *population* or a *sample* from the population. Numerical measures calculated from the data are known as either **statistics** or **parameters.**

A *statistic* is a numerical descriptor that is calculated from sample data and is used to describe the sample. Statistics are usually represented by Roman letters.

A *parameter* is a numerical descriptor that is used to describe a population. Parameters are usually represented by Greek letters.

Most of the time in statistics we work with sample data, but sometimes we may have the entire population available for study. We usually use Greek letters to denote parameters and Roman letters to describe statistics.

3.3 Measures of Central Tendency

When we look at numerical descriptors for a set of data, we want to describe the same properties of the data that we described from the graphical displays. You will find, however, that several different statistics can be used to describe each property and that the choice of the statistic is dependent on the problem you are trying to solve.

Definitions

The definitions of key terms are highlighted in green boxes for easy reference. Any term that is in a definition box is also listed in the chapter summary's "Key Terms."

Worked Examples

Hundreds of worked examples are included throughout to reinforce each chapter topic and to demonstrate techniques in action. Examples include real data and the four-step problem-solving method

Problem-Solving Steps

Each worked example is accompanied by problem-solving steps using yellow marginal icons. These steps provide a template for students to model for exercises.

Marginal Notes

These helpful italicized notes contain key-words, hints, and tips. They are designed to help students review the material and locate key formulas, definitions, and concepts.

EXAMPLE 3.3

THE GOLF BALL DILEMMA

Using the Sample Mean

The golf pro is interested in describing the way the two different types of golf balls behave so that he can see if there really is a difference between the two. One way to describe the distance that the balls travel is to use the variable *Carry*, which measures the distance (in yards) from the point where the ball was hit to the point where it hit the ground. Since the pro is interested in *comparing* the two different designs, you will want to look at the designs separately. Using a computer package, you can calculate a set of descriptive statistics for the two different ball designs. The output provides the following information:

Understand the problem.

Collect the data.

Analyze the data.

	Type M1	Type M2
Sample size	36	36
Sum of data	9627	9244
Sample mean	257.4	256.8

It appears from these values that there is not much difference in the way a "typical" ball of the two types behaves, but at this point that is just conjecture.

To better understand what the numbers really mean, you can locate the values obtained for the mean on a histogram of the data. Perhaps the numbers and graphs together will provide more information.

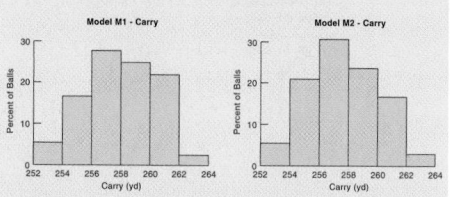

When you locate the mean *Carry* for each ball type on the appropriate histogram, you see that the mean appears to be a good measure of the center of the data and that it does not appear to be influenced by extreme values. However, you notice that the sample mean does not provide any information about the number of golf balls that went more or less than that number of yards. This could be useful information, and so you need another measure of center.

3.3.2 THE SAMPLE MEDIAN

Although the sample mean measures the center of the data, its value might be influenced by unusually high or low values in the sample and might not present a true picture of the sample data. For this reason we often look at other measures

---

Remember! Since 100 is even, you average the two middle values to find the median.

Step 1: Since 100 observations are in the sample, the median must be the average of the 50th and 51st data values:

$$Q_2 = \frac{26,100 + 26,200}{2} = \$26,150$$

The lower half of the data set, everything *below* the median, consists of the first 50 values, and the upper half of the data, everything above the median, consists of the second 50 values.

Analyze the data.

Step 2: To find the first quartile, we find the median of the first 50 values, which is the average of the 25th and 26th observations:

$$Q_1 = \frac{25,700 + 25,700}{2} = \$25,700$$

Step 3: To find the median of the 51st through 100th values, we average the 75th and 76th observations:

$$Q_3 = \frac{26,600 + 26,700}{2} = \$26,650$$

So, the college administrator finds that 25% of the graduates earn less than $25,700 and 25% of the graduates earn more than $26,650.

✓ TRY IT NOW!

Aptitude Test Scores
Finding the Quartiles

The manufacturing facility that is looking at training aptitude wants to give employees who scored in the top 25% on the test the opportunity to attend a seminar on training. The test scores are reprinted here:

185	227	241	257	281	299	314	329
195	228	243	261	283	304	318	333
196	234	248	269	283	307	319	335
199	238	250	271	291	309	322	349
223	241	253	272	297	310	328	353

In the sample, what is the cutoff score for those people who will be able to attend the seminar?

Hint: The value that defines the top 25% is the same as the value that defines the bottom 75%.

Suppose the company decides that the employees who scored in the bottom 25% need some additional classes on team building. What is the cutoff score for those employees who need the classes on team building?

Ans. $Q_1 = 312; Q_3 = 241$

Try It Now

These self-test checkpoints are designed for students to assess their own understanding before moving on to the end-of-section exercises. The answers to the Try It Now! exercises are provided at the bottom of the page, upside down, so that students will have immediate feedback.

CHAPTER 3 EXERCISES

Learning It!

3.25 A manufacturer of pain relievers is interested in studying the amount of time it takes a person to be relieved of headache pain after taking the medication. The manufacturer selects a random sample of 12 people and conducts a study. The data are in minutes:

13.0 12.9 13.2 12.7 13.1 13.0 13.1 13.0 12.6 13.1 13.0 13.1

(a) Find the mean, median, and mode of the relief times.
(b) Find the range and standard deviation of the relief times.
(c) What is the percentile rank of the person who took 12.9 minutes to be relieved of headache pain?
(d) Find the quartiles for the data set and interpret them.

3.26 A group of elderly people in a town filed a grievance with the town council, saying that the length of the walk signals was inadequate for many elderly people to cross safely. In an attempt to investigate the claim, the town collected data on street crossing times for 12 elderly persons. The times are to the nearest tenth of a second:

21.4 15.1 13.6 16.0 15.0 19.1 21.0 14.2 15.6 20.1 21.1 22.2

(a) Find the mean, median, and mode of the crossing times.
(b) Find the range and standard deviation of the crossing times.
(c) Compare the mean and the median. Do you think the data are skewed?
(d) What is the percentile rank of the person who took 20.1 seconds to cross the street?
(e) Find the quartiles and interpret them.
(f) Make a boxplot of the data.
(g) Describe the distribution of the crossing times.

3.27 In a study of the price variation for a 24-pack of a popular cold relief tablet, a random sample of the product at ten different stores yields the following prices in dollars:

4.74 4.62 4.76 4.72 4.99 5.29 4.98 4.79 4.75 4.75

(a) Find the mean selling price of the product.
(b) Find the median selling price of the product.
(c) Compare the median and the mean. Is the distribution of the prices symmetric? Why or why not?
(d) Would it make sense to use the mode as a measure of the center for this sample? Why or why not?
(e) Find the range of the prices.
(f) Find the standard deviation of the prices.

3.28 The vice president of marketing at a large corporation is wondering how many of the company's employees arrive at work before he does. He decides to collect some data by counting the numbers of cars that are in the parking lot when he arrives at 6:30 A.M. on ten different days:

23 24 24 25 25 25 25 26 26 37

(a) Find the sample mean of the data.
(b) Find the median of the data.
(c) Find the range of the data.
(d) Find the standard deviation of the data.

3.29 A company that sells air purifiers via mail order and telephone wants to know the amount of time that a customer spends on the phone with a sales agent during a call.

Doing It!

3.46 The golf pro wants to expand the analysis. He recognizes that you could collect additional data, and he wants to look at other aspects of the balls' performance. In addition to measures of the balls' performance, such as the variable *Carry*, he knows that other factors, both internal (ball related) and external (environment related), could affect performance. He tests 36 of each model of ball at three different times using a machine to launch the balls. Data are recorded on 14 different variables. A portion of the data is shown here:

Datafile:
GOLFBALL.XXX

Ball	Model	S1	S2	S3	Wgt	Dw	Dd	Head	Temp	Carry	TotDist	Date	Time
1	M1	81	81	82	45.3	0.145	0.0110	686	77	257	270	8/20	8:15
2	M1	83	83	84	45.2	0.151	0.0111	688	77	255	267	8/20	8:15
3	M1	81	82	84	45.2	0.145	0.0105	687	77	256	267	8/20	8:15
4	M1	81	81	83	45.3	0.144	0.0117	688	77	255	271	8/20	8:15
5	M1	83	81	82	45.5	0.146	0.0108	687	77	255	268	8/20	8:15

Thinking About It!

Requires Exercise 3.28

3.34 Consider the data on the numbers of cars that are in the parking lot when the vice president arrives:

23 24 24 25 25 25 25 26 26 37

(a) If the vice president wants an estimate of the typical number of employees who arrive at work before he does, is the mean or the median a better statistic to use? Why?
(b) Is the standard deviation or the range more suitable for these data? Why?
(c) Find the z-score of the data value of 37 cars. Does it confirm your suspicion that the data value is unusual?

3.35 Grading homework is a real problem and takes an enormous amount of time. Many students do not do a good job on their assignments or they copy answers from other students or the back of the book. A teacher of elementary statistics decided to conduct a study to determine what effect grading homework had on her students' exam scores. She taught three sections of elementary statistics and randomly assigned each class to one of three conditions: (1) no homework given, (2) homework given but not collected, and (3) homework given, collected, and graded. After the first exam, she collected the data (exam scores) and made histograms and calculated some numerical measures.

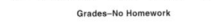 $\overline{X} = 76.6$; median = 77.5; $s = 10.1$; $R = 45$ 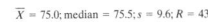 $\overline{X} = 75.0$; median = 75.5; $s = 9.6$; $R = 43$

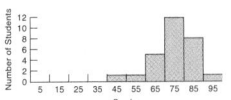

No homework   **Homework, not collected**

Discovery Exercise 3.1

THE TRIMMED MEAN

Part I. Investigating the Data

In a report to the administration of a large university, the Psychology Department states that the average class size is larger than the 35 students allowed by the university charter. The report indicates that the mean class size is 39.4.

No data are appended to the report, but you can obtain the current enrollments easily. You find these data:

3	14	22	26	42
3	15	23	27	45
5	15	24	28	45
9	17	24	28	190
11	21	25	36	193
13	22	26	38	193

1. Do you think that the mean is a good measure of center for these data? Why or why not?

2. By simply studying the data, what do you think is a typical class size for the Psychology Department?

3. What is the median of the data? Is this close to what you thought?

4. Compare the mean and the median. What does the comparison lead you to believe about the data?

5. Display the data graphically. Does your opinion change?

Part II. Solving the Problem

Neither the mean nor the median gives a very good measure of a typical class size. In addition, a comparison of the mean and the median leads you to believe that the data are skewed right! From the histogram you can see that is not the case.

Discovery Exercises

The Discovery Exercises help students to transition from discovery to decision making and function as expanded case studies. They are written to achieve a high level of student understanding by directing student thinking, resulting in an "aha" discovery of a key concept, and then prompting the student to make an informed decision. Most of the Discovery Exercises require the students to write a memo or short report to explain their decision and recommendations.

Get It In Writing

Get It In Writing summarizes the answer to the chapter-opening problem in memo or presentation form. This provides an example for the students to follow when completing the writing assignments that are included in the end-of-chapter Exercises and the Discovery Exercises. Additionally, they model the type of communication students may be asked to create in real-life settings.

GET IT IN WRITING

Report on Comparison of Two Golf Ball Models

We were asked to compare two different models of golf balls in order to determine what differences, if any, exist between the top and bottom of the product line. The study was conducted as a blind study; that is, we did not know the golf ball types at the time we collected and analyzed the data. This procedure ensures that our results are as unbiased as possible.

Using a mechanical hitting device, we hit 72 different golf balls, 36 of each type. The balls were hit in batches of 12, alternating brands. To measure distance traveled we used carry, which is the distance from the point of impact with the club to the place where the ball first hit the ground, and total distance, which is the distance from the point of impact to the final position of the ball. This report focuses on the carries of the two ball types.

Table 1 contains summary statistics for the two ball types. As you can see, the two models are very similar. The average carry for model 1 is 257.4 yards, and for model 2 it is 256.8 yards. This difference is less than 1 yard and might not be distinguishable to most golfers. The medians for models 1 and 2 are 257.5 and 257.0 yards, respectively. This means that for each model, 50% of the balls hit carried farther than this value and 50% carried less. The standard deviation for the carry is 2.4 yards for model 1 and 2.3 yards for model 2. This, along with the ranges, indicates that the variation in carry for the two balls is about the same.

TABLE 1

	Model 1	Model 2
Mean	257.4	256.8
Standard error	0.4	0.4
Median	257.5	257.0
Mode	258.0	255.0
Standard deviation	2.4	2.3
Range	10.0	11.0
Minimum	252.0	251.0
Maximum	262.0	262.0
Count	36.0	36.0

Although these summary statistics indicate that the balls are not very different, we decided to look at some other measures. Table 2 lists quartile values for the two models. You can see that for model 1 25% of the balls went farther than 259 yards, whereas for model 2 the top 25% cutoff was 258 yards. The bottom 25% of the model 1 balls flew less than 256 yards, whereas for model 2 the value was 255 yards.

Minitab

1. From the **Stat** menu, choose **Basic Statistics** and then **Display Descriptive Statistics.**
2. Select the variable for which you want the statistics calculated (in this case, "Carry"). Since you want summary statistics for the two balls separately, click **By Variable,** then **Model,** and then **OK.**
3. The output from the **Display Descriptive Statistics** command appears in the **Session** window, since the output is all text, as shown in Figure 3.10. You can see

```
 Session

        9/26/01 10:53:18 AM

Welcome to Minitab, press F1 for help.
MTB > Retrieving worksheet from file: D:\Textbook Business 2e\Data Sets\GOLFBAL
# Worksheet was saved on Mon Mar 22 1999
MTB > Describe 'Carry'.

Descriptive Statistics: Carry by Model _

Variable   Model _        N       Mean     Median     TrMean     StDev
Carry      M1            36     257.42     257.50     257.47      2.36
           M2            36     256.78     257.00     256.78      2.29

Variable   Model _   SE Mean    Minimum    Maximum         Q1         Q3
Carry      M1           0.39     252.00     262.00     256.00     259.75
           M2           0.38     251.00     262.00     255.00     258.00
```

FIGURE 3.10   Descriptive Statistics output from Minitab

TI-83 Graphing Calculator

Here we create a set of summary statistics for the aptitude test scores from Try It Now! In Section 3.5.1.

1. Press $\boxed{\text{STAT}}$ and use the $\boxed{\blacktriangleright}$ key to move over to the [CALC] selection. The screen should look like the one shown in Figure 3.11.

```
EDIT CALC TESTS
1:1-Var Stats
2:2-Var Stats
3:Med-Med
4:LinReg(ax+b)
5:QuadReg
6:CubicReg
7↓QuartReg
```

FIGURE 3.11   Calc > 1-Var Stats screen

2. Choose option 1: **1-Var Stats** and press $\boxed{\text{ENTER}}$. The words **1-Var Stats** appear on the screen. If your data are in L_1, then just press $\boxed{\text{ENTER}}$. If your data are in another list, then press $\boxed{\text{2nd}}$ and the key that has the correct label (L_2 through L_6) above it.
3. The screen should display a set of summary statistics as shown in Figure 3.12.

Key Terms & Key Formulas

Each chapter concludes with a quick reference table detailing key terms and definitions. Key formulas are also summarized in a separate table. Page references are provided for both tables so that students can easily turn back to the appropriate chapter page for more details.

CHAPTER 3 SUMMARY

There are many ways to describe a set of data using sample statistics. No single number will do the job, nor is there any standard way to proceed. The measures that you choose must reflect the characteristics of the data themselves. Most of the time the best descriptions come from using multiple measures and the conclusions that can be reached by comparing them.

Rather than summarize a sample with a list of numbers it is often useful to create images of the data using combinations of different statistics. An example of this is the *empirical rule*, which gives a picture of the distribution of the data. Another example of using summary statistics to get a picture of the distribution is the *boxplot*.

Data analysis is not a static tool. You need to look at a set of data in every way possible to obtain all of the information that it contains. Sometimes different methods all lead to the same conclusions, and sometimes one method yields an insight that is hidden in every other method.

KEY TERMS

Term	Definition	Page reference
Boxplot or box and whisker diagram	A **boxplot** is a graphical display that summarizes the distribution of a sample using the quartiles and the median.	145
Empirical rule	The **empirical rule** states that, for a symmetric distribution: • About 68% of all observations are within one standard deviation of the mean. • About 95% of all observations are within two standard deviations of the mean. • Almost all (more than 99%) of the observations are within three standard deviations of the mean.	134
First quartile, Q_1	The **first quartile** is the value in the sample that has 25% of the data at or below it.	142
Interquartile range, IQR	The **interquartile range** is the difference between the third and first quartiles, $Q_3 - Q_1$.	145
Modal class	The **modal class** is the class interval in a frequency distribution or histogram that has the highest frequency.	121
Parameter	A **parameter** is a numerical descriptor that is used to describe a population.	111

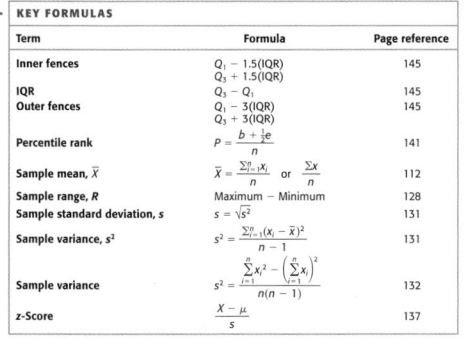

KEY FORMULAS

Term	Formula	Page reference
Inner fences	$Q_1 - 1.5(\text{IQR})$ $Q_3 + 1.5(\text{IQR})$	145
IQR	$Q_3 - Q_1$	145
Outer fences	$Q_1 - 3(\text{IQR})$ $Q_3 + 3(\text{IQR})$	145
Percentile rank	$P = \dfrac{b + \frac{1}{2}e}{n}$	141
Sample mean, $\bar{X}$	$\bar{X} = \dfrac{\sum_{i=1}^{n} x_i}{n}$ or $\dfrac{\sum x}{n}$	112
Sample range, R	Maximum − Minimum	128
Sample standard deviation, s	$s = \sqrt{s^2}$	131
Sample variance, s^2	$s^2 = \dfrac{\sum_{i=1}^{n}(x_i - \bar{x})^2}{n - 1}$	131
Sample variance	$s^2 = \dfrac{\sum_{i=1}^{n} x_i^2 - \left(\sum_{i=1}^{n} x_i\right)^2}{n(n-1)}$	132
z-Score	$\dfrac{X - \mu}{s}$	137

Making Your Case

These cases take a broad look at the techniques learned throughout the course, and pose scenarios that require students to use a variety of the statistical tools. Each case references the chapters that introduced the tools necessary for completing it. Designed to be assigned individually or in groups, these cases ask students to apply their knowledge, draw conclusions, and make decisions about data.

Now that you have gotten the idea, you understand what you need to do for each of the rest of the categories in order to have your report on the President's desk by the end of the week. Oh, by the way, the President is wondering where the best place in the country to live is.

MAKING YOUR CASE

What's on the Road? Who's Driving It?: Exploratory Data Analysis

Before you can make decisions about the future, you need to understand the current situation. The transportation industry controls the infrastructure of the country, and the decisions it makes affect everyone.

One method the industry uses to understand current trends is the Nationwide Personal Transportation Survey (NPTS). This survey is done approximately every 7 years under the sponsorship of the U.S. Department of Transportation. The NPTS compiles national data on the nature and characteristics of personal travel. NPTS data may be used to describe current travel patterns and, given projections of demographic change, can provide a valuable tool to forecast future travel demand. One aspect of transportation planning is knowing what is currently on the road. The NPTS data include motor vehicle information such as year, make, model, and other vehicle-related information.

You are asked to describe the current vehicle population of the United States. To assist in this, you have been given a portion of the NPTS data from the 1990 survey, in which data were collected on 26,172 households using computer-assisted telephone interviewing (CATI). Specifically, you have been given data from all survey interviews conducted in April 1990. A portion of the data is shown here:

CMSA	HHFAMINC	HHLOC	HHMSA	HHSIZE	HH_RACE	HOUSEID	LIF_CYC	MAKECODE
5602	05	1	5640	01	01	00038	01	014
5602	04	1	5640	06	01	00039	08	038
5602	04	1	5640	06	01	00039	08	022
5602	04	1	5640	06	01	00039	08	021
5602	98	1	5640	05	01	00041	06	021
5602	98	1	5640	05	01	00041	06	021
5602	98	1	5640	05	01	00041	06	035
5602	14	2	5190	02	01	00045	10	009
5602	14	2	5190	02	01	00045	10	006
5602	99	2	5190	03	01	00047	06	019

The data are found in a datafile called *CARS.XXX*. As is the case with much publicly available data, the variable names are cryptic and the qualitative variables are coded. Clearly, you will need to know what each variable and code mean. This information is detailed in Appendix B.

Of particular interest to automobile manufacturers is information about the ages and types of vehicles on the road.

1. Use graphical and numerical summary tools to describe the distribution of vehicles by make and year.

2. Are the distributions affected by various demographics? Who drives foreign cars? Who drives old cars?

3. Is the distance driven annually affected by where people live, gender, age, or type of vehicle?

Contents

Introduction: The Role of Statistics in Life

I.1 Objectives   1

I.2 Dispelling the Myths About Statistics   1

I.3 What You Should Know About Statistics   2

I.4 Statistical Thinking—A New Paradigm   2

I.5 Problem-Solving Steps   3

I.6 Situations That Call for Statistical Thinking   4

Discovery Exercise I.1:
STARTING TO THINK STATISTICALLY   6

I.7 Key Components of Statistical Thinking   8

I.8 Use of Minitab, Excel, TI-83, and Other Statistical Software   8

SUMMARY   9

EXERCISES   9

CHAPTER 1

The Language of Statistics   10

1.1 Chapter Objectives   11

1.2 The Difference Between the Population and a Sample of the Population   11

Discovery Exercise 1.1:
INTRODUCTION TO SAMPLING AND VARIABILITY   16

1.3 The Difference Between a Parameter and a Statistic   18

Discovery Exercise 1.2:
POPULATIONS, PARAMETERS, SAMPLES, AND STATISTICS   22

1.4 Factors That Influence Sample Size: Some Sampling and Sample Size Considerations   24

1.5 Selecting the Sample   26

Discovery Exercise 1.3:
INTRODUCTION TO SAMPLING   30

1.6 Types of Data   32

1.7 The Difference Between Descriptive Statistics and Inferential Statistics   38

1.8 Ethical Issues in Data Analysis   41

1.9 Communicating the Results   42

1.10 Basic Summation Notation   43

1.11 Selecting a Sample with Excel or Minitab   46

CHAPTER 1 SUMMARY   48

CHAPTER 1 EXERCISES   49

CHAPTER 2

Graphical Displays of Data   56

2.1 Chapter Objectives   57

2.2 Organizing Data   57

2.3 Graphical Displays of Data   72

2.4 Describing and Comparing Data   86

Discovery Exercise 2.1:
THINKING ABOUT VARIABILITY   93

2.5 Creating Graphical Displays Using Technology   96

CHAPTER 2 SUMMARY   99

CHAPTER 2 EXERCISES   100

CHAPTER 3

Numerical Descriptors of Data   110

3.1 Chapter Objectives   111

3.2 Describing Data Numerically   111

3.3 Measures of Central Tendency   111

Discovery Exercise 3.1:
THE TRIMMED MEAN   120

Discovery Exercise 3.2:
INVESTIGATING VARIABILITY   125

3.4 Measures of Dispersion or Spread   128

3.5 Measures of Relative Standing   140

3.6 Numerical Descriptors and Technology   153

    CHAPTER 3 SUMMARY   156

    CHAPTER 3 EXERCISES   158

CHAPTER 4

Analyzing Bivariate Data   166

4.1 Chapter Objectives   167

4.2 Qualitative Bivariate Data   167

4.3 Quantitative Bivariate Data   178

    Discovery Exercise 4.1:
    DISCOVERING RELATIONSHIPS   184

4.4 Investigating Bivariate Data
with Technology   194

    CHAPTER 4 SUMMARY   197

    CHAPTER 4 EXERCISES   198

CHAPTER 5

Probability   206

5.1 Chapter Objectives   207

5.2 The Language of Probability   207

    Discovery Exercise 5.1:
    LAW OF LARGE NUMBERS   214

5.3 Laws of Probability: OR and AND   217

5.4 Conditional Probability
and Independence   226

5.5 Generating Random Data
with Technology   232

    CHAPTER 5 SUMMARY   235

    CHAPTER 5 EXERCISES   237

CHAPTER 6

Random Variables and Probability Distributions   242

6.1 Chapter Objectives   243

6.2 Random Variables   243

6.3 The Binomial Probability Distribution   249

    Discovery Exercise 6.1:
    EXPLORING THE BINOMIAL DISTRIBUTION   260

6.4 Continuous Random Variables   264

6.5 The Normal Probability Distribution   266

6.6 Generating Probability Distributions with
Technology   282

    CHAPTER 6 SUMMARY   284

    CHAPTER 6 EXERCISES   285

CHAPTER 7

Sampling Distributions and Confidence Intervals   292

7.1 Chapter Objectives   293

7.2 Motivation for Point Estimators   294

7.3 Common Point Estimators   295

7.4 Desirable Properties of Point Estimators   300

7.5 Distribution of the Sample Mean, $\bar{X}$: The Central
Limit Theorem   305

7.6 The Central Limit Theorem—A More
Detailed Look   307

    Discovery Exercise 7.1:
    THE CENTRAL LIMIT THEOREM IN ACTION   315

7.7 Drawing Inferences by Using the Central
Limit Theorem   316

7.8 Large-Sample Confidence Intervals
for the Mean   319

    Discovery Exercise 7.2:
    EXPLORING CONFIDENCE INTERVALS FOR μ 330

7.9 Distribution of the Sample Mean: Small Sample
and Unknown σ 331

7.10 Small-Sample Confidence Intervals
for the Mean   333

7.11 Confidence Intervals for Qualitative Data   336

7.12 Sample Size Calculations   339

7.13 Using Technology to Find Confidence
Intervals   345

    CHAPTER 7 SUMMARY   347

    CHAPTER 7 EXERCISES   349

CHAPTER 8

Hypothesis Testing: An Introduction   356

8.1 Chapter Objectives   357

8.2 What Is a Hypothesis Test?   358

8.3 Designing Hypotheses to Be
Tested—An Overview   359

8.4 The Steps in a Hypothesis Test   362

    Discovery Exercise 8.1:
    FORMULATING HYPOTHESES   365

8.5 Large-Sample vs Small-Sample Tests   368

8.6 Large-Sample Test of the Mean:
Two-Tail Tests   369

    Discovery Exercise 8.2:
    EXPLORING THE IMPACT OF VARYING THE
    VALUE OF α 376

8.7 What Error Could You Be Making?   382

8.8 Which Theory Should Go into the Null Hypothesis? 387

8.9 One-Tail Tests of the Mean: Large Sample 393

8.10 Using Technology in Hypothesis Testing 404

CHAPTER 8 SUMMARY 404

CHAPTER 8 EXERCISES 406

CHAPTER **9**

Inferences: More One-Population Tests 410

9.1 Chapter Objectives 411

9.2 Hypothesis Test of the Mean: Small Sample 412

9.3 Hypothesis Test of a Single Variance 421

9.4 Hypothesis Test of a Single Proportion 429

9.5 Summary of One-Population Hypothesis Tests 435

Discovery Exercise 9.1:
THE CONNECTION BETWEEN CONFIDENCE INTERVALS AND HYPOTHESIS TESTING 436

9.6 Connection Between Hypothesis Testing and Confidence Intervals 437

9.7 Using Technology for Hypothesis Testing 439

CHAPTER 9 SUMMARY 444

CHAPTER 9 EXERCISES 445

CHAPTER **10**

Comparing Two Populations 450

10.1 Chapter Objectives 451

10.2 Collecting Data from Two Populations 451

10.3 Hypothesis Test of the Difference in Two Population Means—Overview 452

10.4 Large-Sample Tests of the Difference in Two Population Means 453

10.5 Small-Sample Tests of the Difference in Two Population Means 462

10.6 Summary of Tests of Two Population Means: Independent Samples 469

10.7 Tests of the Difference in Two Population Means: Dependent Samples 470

Discovery Exercise 10.1:
INTRODUCTION TO EXPERIMENTAL DESIGN 471

10.8 Test of the Difference in Two Population Proportions 479

10.9 Test of the Difference in Two Population Variances 485

10.10 Technology and Two-Population Hypothesis Tests 492

CHAPTER 10 SUMMARY 496

CHAPTER 10 EXERCISES 497

CHAPTER **11**

Regression Analysis 502

11.1 Chapter Objectives 503

11.2 The Simple Linear Regression Model 503

11.3 Inferences About the Linear Regression Model 524

11.4 Confidence Intervals and Prediction Intervals 532

11.5 Correlation Analysis 537

11.6 Regression Assumptions and Residual Analysis 540

11.7 Using Technology for Simple Linear Regression 547

CHAPTER 11 SUMMARY 558

CHAPTER 11 EXERCISES 560

CHAPTER **12**

Multiple Regression Models 568

12.1 Chapter Objectives 569

12.2 The Multiple Regression Model 569

12.3 Assessing the Multiple Regression Model 580

Discovery Exercise 12.1:
FINDING THE BEST MODEL 595

12.4 Building a Multiple Regression Model 596

12.5 Checking Model Adequacy 605

12.6 Using Technology for Multiple Regression Models 612

CHAPTER 12 SUMMARY 616

CHAPTER 12 EXERCISES 617

CHAPTER **13**

Experimental Design and ANOVA 624

13.1 Chapter Objectives 625

13.2 Motivation for Using a Designed Experiment 625

13.3 Analysis of Data from One-Way Designs 627

13.4 Assumptions of ANOVA 648

13.5 Analysis of Data from Blocked Designs   651

Discovery Exercise 13.1:
THE BENEFITS OF BLOCKING   660

13.6 Analysis of Data from Two-Way Designs   662

13.7 Other Types of Experimental Designs   671

13.8 Using Technology for Analysis of Variance   674

CHAPTER 13 SUMMARY   682

CHAPTER 13 EXERCISES   685

CHAPTER 14

The Analysis of Qualitative Data   692

14.1 Chapter Objectives   693

14.2 Test for Goodness of Fit   693

14.3 Test for Equality of Proportions   711

14.4 Chi-Square Test for Independence   722

14.5 Using Technology for the Chi-Square Test   733

CHAPTER 14 SUMMARY   737

CHAPTER 14 EXERCISES   738

CHAPTER 15

Nonparametric Statistics   748

15.1 Chapter Objectives   749

15.2 Method for Comparing Two Populations   750

15.3 Method for Comparing More Than Two Populations   760

15.4 Using Nonparametric Tests   768

15.5 Using Technology for Nonparametric Tests   769

CHAPTER 15 SUMMARY   771

CHAPTER 15 EXERCISES   773

CHAPTER 16

Making Your Case   778

16.1 Chapter Objectives   779

16.2 Summary of the Techniques Covered in This Book   780

16.3 How to Analyze the "Great Divide" Data Set   781

CHAPTER 16 SUMMARY   783

CHAPTER 16 EXERCISES   783

MAKING YOUR CASE: ADVISING THE PRESIDENT: EXPLORATORY DATA ANALYSIS   784

MAKING YOUR CASE: WHAT'S ON THE ROAD? WHO'S DRIVING IT?: EXPLORATORY DATA ANALYSIS   785

MAKING YOUR CASE: WHO SPENDS MONEY? WHAT DO THEY BUY?: INFERENTIAL STATISTICS   786

MAKING YOUR CASE: ADVISING THE PRESIDENT: INFERENTIAL STATISTICS   787

MAKING YOUR CASE: RESPONDING TO THE CUSTOMER: PULLING IT ALL TOGETHER   788

MAKING YOUR CASE: NEEDS ASSESSMENT: PULLING IT ALL TOGETHER   789

APPENDIX A   Statistical Tables   A1

APPENDIX B   Explanation of Large Data Sets   B1

APPENDIX C   Conditional Probability and Bayes' Rule   C1

APPENDIX D   The Poisson Probability Distribution   D1

APPENDIX E   Exercises Appendix   E1

INDEX   I1

Introduction: The Role of Statistics in Life

DATA ARE EVERYWHERE!

Every time you pick up the newspaper you see statistics! There is no escape. People quote statistics to support whatever it is they wish you to believe. We live in an era of increasing amounts of data. Data are collected constantly—sometimes with our knowledge and sometimes without it. There are databases of information about our families, about our spending habits, about the books we read (or at least order online), about the places we travel, about our jobs, about our health, and on and on. When collected and used properly, data and the statistics calculated from them can help us to understand situations and make well-informed decisions.

To give you some initial idea of the variety of topics that use data, here are some headlines from recent newspaper articles:

- Study finds weekend care in hospitals inferior
- Professor tells how to make the grade
- SAT scores steady, but reveal gender, racial gaps
- Accidents kill more people on less traveled rural roads
- Study: Tobacco ads still targeting teens

Each of these articles quotes statistics calculated from the data to support a particular theory or idea. Even if you are not destined to become a world-renowned statistician (although that might be cool!), you must still live in today's society and be able to make sense of data in order to evaluate your options and make wise decisions.

You may be wondering why you have to take this course. Students often think that taking a course in statistics causes unnecessary pain and that they will never again use the statistical techniques they learn. The objective of this book is to refute both of these ideas: The course need not be painful, and you certainly will be asked to collect and analyze data and use the information to make informed decisions.

The objective of this introduction is to start you down this path. Specifically, this introduction covers the following topics:

- Dispelling the myths about statistics
- What you should know about statistics
- Statistical thinking—a new paradigm
- Problem-solving steps
- Situations that call for statistical thinking
- Key components of statistical thinking
- Use of Minitab, Excel, TI-83, and other statistical software

I.2　Dispelling the Myths About Statistics

We should first dispel some of the common myths about statistics. Here are three myths we run across regularly:

Myth 1: "If I had one hour left to live, I would choose to live it in statistics class because it would seem to last forever!"

A STUDENT'S LAMENT

Myth 2: "There are three kinds of lies—lies, damned lies, and statistics."

BENJAMIN DISRAELI

Myth 3: "If it moves, it's biology; if it changes color, it's chemistry; if it breaks, it's physics; if it puts you to sleep, it's statistics."

BOB HOGG, UNIVERSITY OF IOWA

If you are like most people, you can relate to one of these myths about the "S-word," *statistics*, or *sadistics* as some people refer to it. Statistics is boring and not useful! This book will lead you to another view of the dreaded S-word—one that sees statistics not as a sleeping pill but as a way to view all sorts of exciting, amazing, and valuable things.

✔ TRY IT NOW!

Myths and Fears
Identifying Some of Your Own

Be honest. Write down your myths and fears about statistics right now. Get them out in the open so you can deal with them directly and put them behind you. To get you started, here is a common student fear:

- I am worried about all of the math in this course.

You have taken the first step. It is easier to combat the myths about statistics when you acknowledge them. Good work. Next we will show you why it is important to learn about the tools of statistics.

1.3 What You Should Know About Statistics

The use of statistical techniques has long played an important role in quality control and quality improvement in business and industry. Before high-speed computers sat on everyone's desks, data analysis and statistics were the tools of business and research. Today, however, we all need to know how to analyze data and interpret results simply to read the newspaper or decide what car to buy.

One reason people prefer not to think about data analysis is that the analysis requires the use of statistical techniques, which are often viewed as difficult. But statistical thinking is not difficult to comprehend. Three steps are involved in using statistics effectively in your personal and professional life:

Life—both professional and personal—and statistics go hand in hand.

1. You must understand *why* you need statistical knowledge.
2. You must know *how* to get statistical knowledge.
3. You must *use* the statistical knowledge.

To integrate statistics into your life, you are going to have to change the way you see the world. You need a new pair of glasses.

1.4 Statistical Thinking—A New Paradigm

Look at Figure I.1. What do you see? Ask the person next to you what he or she sees. Some of you will see two faces and some will see a vase. You probably can see both of these things, but which one did you see first? There is no right or wrong answer here. What you see first depends on your viewpoint, your lens, your glasses. This is your **paradigm**—how you *see* things. Each of us has a paradigm or view of the world that has developed based on our individual experiences. The hard part is changing our lens. Many times you must change your paradigm to see the same picture in a new way. This book is designed to help you shift your paradigm so that you can see the world through the lens of statistical thinking.

> A *paradigm* is commonly used today to mean a model, theory, perception, assumption, or frame of reference. It was originally a scientific term.

FIGURE I.1 Two faces or a vase?

"It's nice, but I never smile like that. I smile like this."

FIGURE I.2 **Cartoon depicting Mona Lisa and da Vinci** (©The New Yorker Collection, 1994. J.B. Handelsman from cartoonbank.com. All rights reserved.)

Did you ever take a class on the great artists of the Italian Renaissance? Even if you did not, you might have heard of Michelangelo, Leonardo da Vinci, Raphael, and Donatello. If you study Michelangelo, you would probably study his statue of David, and if you study Leonardo, you would certainly learn about his painting of the Mona Lisa. Then you would begin to see David and Mona in all sorts of places including cartoons, advertisements, and logos. The cartoon shown in Figure I.2 shows Mona Lisa commenting on the painting of herself by da Vinci.

Are these ads new? No, you just became more aware of their existence after you studied about them. This is similar to what happened when you looked at the faces/vase illustration in Figure I.1. Once you become aware of the second way of looking at the picture, you see both things each time you look at the picture. Similarly, once you study Michelangelo and Leonardo, your view is different because you are now aware of their works and begin to see them everywhere. You have a new paradigm or view of the world. Your paradigm has *shifted*.

This book is about changing your paradigm to one that allows you to *see* things using statistical thinking. Statistical thinking is public property. Everyone owns it and everyone must use it—not just statisticians. Over the short term, statistical thinking can improve the quality of decisions; over the long term, it can turn people into leaders.

You will learn to see data everywhere you look and to see the information hidden in the data as well. This information will help you make informed professional and personal decisions. How is this possible? By learning and understanding the tools of statistics, you will be able to see the world differently. The tools of statistics are actually logical and simple to use, yet many people do not use them because they lack exposure to statistics or they dislike math. Hence, these people often do not make well-informed decisions. You will be different because you will learn to use statistics as decision-making tools and not as mathematical manipulation. Think about this book as a handbook on decision making, not a math text. You will soon see what we mean.

I.5 Problem-Solving Steps

As you learn how to analyze data to make well-informed decisions, you will note three recurring themes:

- We must use data to *continually learn* about our world.
- We must *be ethical* in our data analysis to maintain our standards.
- We must be able to *communicate* our analysis effectively to make our case.

These themes are woven throughout this book. They provide a context for our data analysis. It is not possible to isolate these topics in a single chapter because they pervade virtually every situation that calls for statistical thinking.

The next section contains three examples. In each situation the data were collected and must be analyzed to *understand* the environment. You will also see that there are ways to "*lie* with statistics," as Disraeli warned us in myth 2. We present discussions of what not to do and things to watch out for. Finally, you will see that you can be the best data analyst in the world, but if you cannot effectively *communicate* the results of your analysis, then it is as though you never did the analysis. So, you will learn how to synthesize and summarize your analysis so it can be used.

Typically, communicating the results of your analysis is the last step. Sometimes this last step becomes the first step of another analysis, but clearly some systematic steps are generally followed in any type of problem solving. There are many different models to follow in problem solving, but we use a simple, four-step model in this book:

1. Understand the problem.
2. Collect and analyze the data.
3. Draw conclusions and make recommendations.
4. Communicate the results.

The goal of this book is to teach you the tools to do the second step. Most of your work in this text will be analyzing data. However, the second step does not take place without an understanding of a problem to solve or a situation to improve, and the analysis is worthless if it does not lead to conclusions and recommendations that are properly communicated.

The organization of every chapter illustrates the four steps and helps you recognize them in several different ways. First, every chapter opens with a situation that calls for some data analysis; this is step 1. The examples worked in the chapter carry out some of the analysis for the opening problem and draw some conclusions based on the analysis; these are steps 2 and 3. Every chapter concludes with a summary of the analysis of the opening problem that goes over the results of the worked examples. This is called "Get It in Writing" and is one part of step 4. Some chapters also include a PowerPoint presentation as another way of communicating the results. To reinforce the communication theme, you will practice writing memos and summaries as part of the exercises.

In addition to written communication, you are often asked to communicate your analysis orally. This might take the form of a PowerPoint presentation. Section 1.9 guides you in preparing such a presentation by addressing questions about what type of visuals to use and how much detail to provide. For some of the case assignments in this book you are asked to prepare a short presentation of your analysis.

1.6 Situations That Call for Statistical Thinking

After finishing this course and graduating from college, you may find yourself facing decisions similar to any one of the following scenarios. In each case you must make decisions based on what you see in the data. We provide a short discussion of possible approaches you might use to investigate the situations. This should convince you that the tools of statistics make logical sense.

EXAMPLE 1.1

WAITING FOR A PRESCRIPTION

Understand the problem.

Deciding on an Approach

In June 2001 Opinion Research for AmeriSource Health reported that nearly three in ten Americans wait more than 20 minutes to have a prescription filled. Only 13% say they typically wait 5 minutes or less. You are trying to select a pharmacy to use in your community. Each one, of course, claims to be the best and the fastest. In order to decide you talk to people who have been using each of the two closest pharmacies. What is your decision?

Here is one possible approach along with some things to think about in making a decision:

- Make a table listing how many minutes it takes people to get their prescriptions filled.
- Draw a graph from this table.
- Find the average wait time for each pharmacy.
- Look to see how these pharmacies compare to the national data reported by AmeriSource Health.

Let's consider another situation.

EXAMPLE 1.2

GOLF BALL DESIGN

Understand the problem.

Deciding on an Approach

Your golf game is an important part of your personal and professional life. You want to buy the golf balls that will fly the farthest! You decide to test two ball designs. You go out to the course and collect data on how far the balls carry (in yards) for 12 balls of each design.

Design 1						Design 2					
257	259	255	256	260	258	254	252	256	255	255	257
260	259	259	257	255	260	253	255	254	254	256	255

You must recommend which design to buy. What is your decision?

Here is one possible approach along with some things to think about in making a decision:

- For each design, find the average distance the balls carry.
- Compare the two average distances.
- Think about whether the difference between these averages is large.
- Think about whether the difference between the averages could be caused by something other than the ball design.

Here is one more example.

EXAMPLE 1.3

Understand the problem.

DRESS DOWN DAY

Deciding on an Approach

Many companies have designated one day a week as dress down day. This idea has been implemented even in traditionally conservative companies such as IBM. What reason(s) does management have for adopting such a policy? In April 1992, Levi Strauss and Co. conducted the first national survey on business casual dress issues. A phone survey gathered data and opinions from managers in a wide range of industries.

At the company you work for, your boss has conducted a survey of employees. Thirty employees were asked whether they strongly agree (SA), agree (A), disagree (D), or strongly disagree (SD) with each of the following two statements:

Statement 1: Casual dress improves morale.

A SA A A A D SD SA A A D D A A SA
A A D D A A SA SD A D A A D A SA

Statement 2: I do my best work when dressed casually.

D SD A A A A SA A A A D D A A SA
A A A D A D SD D A D A A D A SA

Your boss wants your recommendation regarding the likely increase in productivity if a dress down day is adopted by your company. What is your decision?

Here is one possible approach along with some things to think about in making a decision:

- Summarize the results in a table.
- Examine what percentage of people strongly agreed, agreed, disagreed, or strongly disagreed with the two statements presented to them.
- Consider how productivity is defined in this survey.
- Look to see whether the responses are different in different industries.

Discovery Exercise 1.1

STARTING TO THINK STATISTICALLY

This exercise allows you to begin to think statistically. Although you have not yet learned any statistical techniques, you will be surprised to discover that many of the ideas you suggest are the basis for some of the techniques you will learn about later in this book. Do not try to solve the problem, but rather focus on what you might do with the data to analyze the situation and what additional information you would like to know about the data or the situation. Remember that statistical thinking is logical, uses data, recognizes the interdependence of activities, and looks at how things vary.

Part I. Living Together Now a Fact of Life in the United States

The following excerpt is taken from an article with this headline from the *Springfield Union News* (September 2001):

The latest statistical soup being served up is data from the 2000 census, heralding either the disintegration of the American family or the resurgence of traditional values, depending on how you read the numbers.

> *Last week it was widely reported that the number of U.S. unmarried partner households spiked 72% in the 1990's. That is a striking number. Too bad it seems to strike different people different ways. Some say cohabitation is replacing marriage, an institution that provides increased health, longevity, and family stability. Others say living together before marriage enhances a couple's ability to forge a long-term union.*

The article ends:

> *As for encouraging marriage with government financial incentives, this plan seems foolhardy. If Dr. Phil and Oprah and vast armies of therapists and clergymen can't fix the problem, are a few shekels going to do it?*

Has cohabitation replaced the institution of marriage? Which side is right?

1. What else would you like to know about these data before using them?
2. How could you find out if marriage provides increased health, longevity, and family stability, as one side claims?
3. How could you find out if living together prior to marriage decreases the chances of divorce?
4. How should lawmakers decide if the government should provide financial incentives for marriage?

Part II. Are We There Yet?

In this hurry-up world people are always concerned about how long it will take to get somewhere. The amount of time it takes to travel is clearly dependent on how far you travel and the type of vehicle you are driving. The data shown here give information about how long it took to travel a certain distance in either a BMW Z3 or a Dodge Caravan.

Type of vehicle	Distance (miles)	Travel time (minutes)
Caravan	55	35
Caravan	102	56
Z3	33	23
Caravan	20	12
Z3	55	65
Caravan	48	34
Z3	53	23
Caravan	22	12
Z3	45	35
Z3	44	46
Z3	12	14
Caravan	45	34

Think about how you would analyze the impact that distance and type of vehicle have on travel time.

1. What else would you like to know about the data?
2. How could you display the relationship between travel time and distance in a graph?
3. How could you predict the time it would take a BMW Z3 to travel a distance of 90 miles?

Your eye can process and understand millions of tiny dots of paint when they are all put together in a meaningful fashion. In this way you see a beautiful picture. Well, that is exactly what the tools of statistics do for you. They put together millions of bits of data in a meaningful fashion. In this way you can make the best decision based on what you see in the data.

I.7 Key Components of Statistical Thinking

It is important to understand the components of this new paradigm that we have called statistical thinking. There are three key parts:

- Use data whenever possible to guide the analysis.
- Look for connections and relationships.
- Understand why data values differ from one another.

Now it is time to introduce an official definition of **statistics:**

Statistics is a branch of mathematics dealing with the analysis and interpretation of masses of data.

With statistical thinking as the new paradigm, your way of thinking about numbers has begun to change. You must discipline yourself to think systematically and to collect data systematically. Then you must learn how to see the information contained in the data you have collected. Finally, you must use the information to make informed decisions. The rest of this book is designed to increase your inventory of specific statistical skills to help you see the information contained in the data and strengthen your decision-making skills. In highly successful organizations, the statistical-thinking paradigm is linked to other important paradigms, such as providing leadership, promoting teamwork, working toward continuous improvement, creating innovative channels of communication, and delighting customers.

I.8 Use of Minitab, Excel, TI-83, and Other Statistical Software

Many statistical techniques require long and involved calculations. Often you want to look at the data in a variety of ways to get a clear picture of what the data are telling you. Clearly, you want to use software to assist you. Many different statistical packages are available, including Minitab, SPSS, SAS, and Statgraphics. We will show you how to do statistical computations with Minitab, Microsoft Excel, and the TI-83 calculator. In your professional and personal life you will be more likely to use statistical techniques if you know how to use the technology to crunch the numbers.

Minitab is clearly a statistical package. Although Minitab was once mainly a tool for teaching statistics, today it is more widely used by industry and business. So you may have access to Minitab not only as a student but also as a working professional.

As a student you may have a TI-83 calculator that you know how to use, so learning how to do statistical analysis on that calculator is helpful to you now and perhaps later.

Finally, you may have noticed that Excel is not on the list of statistical packages we presented. This is not a mistake. Excel is a spreadsheet tool that also has the ability to perform many statistical functions, but it is not first and foremost a statistics package. Excel is used widely in the business world, and this is precisely why we have chosen to include Excel whenever possible. We do not want you to think of Excel as a statistical package. But if you learn how to do data analysis using Excel, which you will most likely have on your desktop at work, then you will be able to apply the statistical tools you learn in this book.

To assist you in using Minitab, the TI-83, and Excel to analyze data, the last section of each chapter shows how to perform the statistical techniques using these technologies.

You will see Minitab and Excel output throughout this book. However, if you choose not to use these packages, this will not be a problem because the output is there to show the results of the "number crunching." If you do not use Minitab or Excel, you will use some other tool that will give you the same results.

Perhaps the most important point to make about using software to support your data analysis is that you should, whenever possible, choose the tool that does the job easily and correctly. Many of the statistical features of Excel are found in the **Tools** menu under an option called **Data Analysis**. If this option does not appear on your **Tools** pull-down menu, you will be shown how to add it when you first need this feature in Chapter 2. There are some statistical techniques that Excel does not perform. Some of these can be added easily to Excel via an add-in feature called KADDSTAT found on the accompanying CD. The first time you need KADDSTAT you will be instructed on how to add it to Excel. Even with the add-ins, however, Excel is not adequate in all cases, and then we will provide output from Minitab and SPSS. Remember that the software is there only to save you the drudgery of the calculations in step 2 of the problem-solving process. You must interpret the data, draw the conclusions, and communicate the results!

You will first need KADDSTAT in Chapter 3 to do boxplots.

SUMMARY

In this introduction you learned how important it is for you as a consumer of the news and a professional to know how to use the tools of statistics. Our world is changing rapidly and more data are available than ever before. To make informed decisions you need the tools of statistics. You need to see the world through a different lens or paradigm.

KEY TERMS		
Term	**Definition**	**Page reference**
Paradigm	A **paradigm** is commonly used today to mean a model, theory, perception, assumption, or frame of reference.	2
Statistics	**Statistics** is a branch of mathematics dealing with the analysis and interpretation of masses of data.	8

EXERCISES

Thinking About It!

I.1 We have said that statistics has invaded virtually all aspects of our lives. List areas of your life that could be affected by the collection and analysis of data. Keep this list and review it again at the end of the course.

I.2 Describe a situation in which you had a paradigm shift or change.

I.3 There has been an increased use of cell phones. Certainly cell phones have changed the way we do business and in fact the way we lead our lives. Is there a difference in the number of minutes that men speak on cell phones compared to women? You collect these data on the numbers of minutes ten men and ten women spent on their cell phones:

Women: 100 150 200 210 140 350 190 120 180 190
Men: 259 275 220 248 231 215 252 260 296 191

(a) What can you do with these data to investigate whether there is a difference in cell phone use by gender?

(b) What factors, other than gender, might influence cell phone use?

I.4 You have to get your car repaired. One thing that concerns you is how long you will be without your car. You decide to examine the service times for all the cars worked on last Friday at the dealership you use. The service time is the number of minutes between the time the car arrived at the dealership and the time the service was completed. The data for last Friday are

15 125 45 65 35 50 20

(a) What additional data would you like to examine to get a more complete picture of the level of service?

(b) How can you collect the data identified in part (a)?

CHAPTER 1

The Language of Statistics

THE COLLEGE SNACK BAR

Would you like fries with that? Have you been to the snack bar on your campus recently? Most college campuses have a cafeteria for students who are on a meal plan, but typically there is also a snack bar or cafe for other students and for the faculty and staff. A group of students were not happy with the quality of service they were receiving at their school's snack bar. To understand how to improve the service at the snack bar, the students set about collecting data to correctly identify the problems.

The students complained about two problem areas. First was the amount of time it took to get through the line, place the order, and of course get the food. Sometimes just getting a bagel and cream cheese for breakfast made you late for your 9:00 A.M. class! Second was the frequent mistakes in the orders, such as a tomato on your sandwich when you asked for no tomato or a large french fry order when you wanted a small order.

In the introduction you learned that data are everywhere! It is easier than ever to collect data as a result of advances in information technology and especially the increased use of the Internet. You need to understand data and statistics in order to make informed decisions in both your professional and personal life. As a person of the 21st century, you are a consumer of statistics. But just like "consuming" fine food or high-tech toys, you need to understand the language.

In order to communicate in another language, you must first learn the basics of that language. Likewise, the first step to understanding what to do with data and statistics is to learn the language of statistics. People sometimes joke about talking in "statistical-eze." They ask you to say something in statistics and then wait for you to say "standard deviation"! The concept of a standard deviation is indeed covered in Chapter 3, but you will need to learn some more fundamental words before then. Fortunately, many statistical terms are familiar words that have a statistical meaning similar to their everyday meaning. This chapter develops the basic language that you will need. The following material is covered:

- The difference between the population and a sample of the population
- The difference between a parameter and a statistic
- Factors that influence sample size: some sampling and sample size considerations
- Selecting the sample
- Collecting the data
- Types of data
- The difference between descriptive statistics and inferential statistics
- Ethical issues in data analysis
- Communicating the results
- Basic summation notation

1.2   The Difference Between the Population and a Sample of the Population

1.2.1   THE POPULATION OF INTEREST

Section I.6 presented situations that used statistics. You can see that regardless of the subject matter, there is always a group of people or things that we must study and understand in order to make the necessary decision. For the consumer, it was the times it took pharmacy customers to have their prescriptions filled; for the golfer, it was the distances the golf balls traveled; and for the business manager, it was all the responses to the survey on casual dress. These are examples of what is called the **population** in the language of statistics. Thus, all the waiting times for people who fill their prescriptions is the population of interest to the person who wants to choose a pharmacy, all the distances traveled is the population of interest to the golfer who wants to choose a golf ball design, and all the survey responses is the population of interest to the manager who must make a recommendation about dress down day. In each case, the decision maker must learn about the population of interest to reach a conclusion.

The *population* is everyone or everything you wish to study.

EXAMPLE 1.1

Understand the problem.

THE COLLEGE SNACK BAR

Identifying the Population

The students who want to study the service at the college snack bar are interested in all customers of the snack bar. Clearly, all students, faculty, staff, and guests are potential customers of the snack bar; however, some of these people may never eat in the snack bar. For example, the faculty member who eats her lunch at her desk and the student who is a strict vegetarian and cooks all his own meals may not be of interest in the study. Should these people be included in the population? Probably not, although if you were interested in attracting new customers, then you might want to include these individuals. Some way of identifying the customers of the snack bar is needed. One possibility is to identify as the population all members of the college community who have eaten in the snack bar at least once in the past year.

Often, when we study a population, we are really interested in knowing about different characteristics of each member of the population. These characteristics are known as **variables.** For each member of the population, we may be interested in knowing about one, two, or even more different variables.

A *variable* is a characteristic of each member of the population.

EXAMPLE 1.2

Understand the problem.

THE COLLEGE SNACK BAR

Identifying Variables of Interest

In the study of the snack bar, many variables could be of interest depending on what the students are trying to accomplish. They may wish to know how long it takes to get an order, the customers' sense of the quality of the food, the appeal of the menu, and the convenience of the hours. They may wish to know how many times a week customers eat in the snack bar and what time of day they come in. The students in our particular case wish to know how long it takes to get the food and what kind of mistake, if any, is made in the order. They wish to study two variables or characteristics of the members of the population.

✔ TRY IT NOW!

The Scooter Company
Identifying Possible Variables to Study

Despite the popularity of the push or kick scooters, concerns have been raised about their safety. These sleek, fold-up scooters have sent nearly 9500 Americans to emergency rooms with injuries. Before you purchase one for your nephew, you wish to check out the safety issues. What is the population of interest?

Name two variables or characteristics that you might wish to study.

ANS. ALL SCOOTER ACCIDENTS, LOCATION OF ACCIDENT, AGE OF USER

1.2.2 THE SAMPLE

Now that you understand what a population is, we need to formally define a **sample.** The statistical definition of the word *sample* is much like the general use of the word. If someone tells you to sample the apple pies made by a bakery, then you take a small piece of one pie and eat it, or you might eat one whole pie out of the many pies made by the bakery. It is the same in statistics. If you take a sample, you take a small piece of the population and look at it or test it. Thus, we have our next definition.

> A *sample* is a piece of the population.

If we think about the population as the big oval (all the pies) shown in Figure 1.1, then a sample from this population is the small oval. This is clearly not the only sample that could be taken from this population; many different samples may be picked. Some of the samples may overlap and some may be bigger or smaller than the one shown.

1.2.3 WHY PICK A SAMPLE AT ALL?

We have said that the population is everything we wish to study and that the sample is only a piece of the population. A natural question to ask is, Why should we bother to examine a sample when what we really want to know about is the population? Most of the time we cannot study the entire population, so we must use a sample as a guide. The main reasons are fairly clear when you think about it for a minute:

- It would take too much time to study the entire population.
- It would cost too much money to study the entire population.
- It might not be possible to identify all the members of the population.
- If we test the entire population, we might not have anything left.

Reasons we can't study the entire population

Let's consider a couple of populations that illustrate these reasons.

EXAMPLE 1.3

THE COLLEGE SNACK BAR

Understanding Why the Population Cannot Be Studied

To study the entire population the students must observe every customer. Let's say they define a customer as someone who has eaten in the snack bar at least once in the past year. So, the students would have to collect data over an entire year. It would clearly take a great deal of time and money to collect data on every customer.

Understand the problem.

FIGURE 1.1 **The population and a sample**

Here's another example based on a situation described in the introduction.

Understand the problem.

EXAMPLE 1.4

DRESS DOWN DAY

Understanding Why the Population Cannot Be Studied

A company wants to study the productivity of its employees before and after casual dress is permitted in the workplace. If the company has 5000 employees all doing different jobs, it would be virtually impossible to measure the productivity of all of them.

1.2.4   AN INTRODUCTION TO SAMPLING ERROR

At this point it might look like we should throw up our hands and go home because we will never have the time, money, or ability to study an entire population. The good news is that we don't need to study the entire population. By studying the sample we can get a good understanding of the population. The picture of the population will not be perfect, but it will be good enough to guide our decision making. The situation is just like eating a piece of the apple pie. You don't know precisely how the whole pie tastes, but you have a pretty good idea that it is delicious!

Remember, it is really the population that we wish to understand and study. The sample is a means to this end. By understanding and studying a sample, we gain insight and knowledge about the population. The amount of information in the sample is not perfect but adequate.

Imagine that you took a Polaroid picture of the population but let it develop for only a few seconds. You would be able to get an overall sense of what the population looked like, but you would not have all of the details. This is like taking a sample and using the sample to determine how the population behaves. If you let this picture develop a few seconds longer, you would get a little clearer view of the population. This is equivalent to taking a larger sample. Finally, if you let the picture develop completely, then you would have complete information. This is equivalent to studying the entire population. If you study the entire population, then you have taken what is called a **census.**

A *census* is a study of the entire population.

In addition to sampling error, other nonsampling errors include the respondent's lying, measurement error, and error due to people not responding.

Since the sample is an imperfect snapshot of the population, you know that there will be differences between the sample and the population. This problem is disconcerting at first. Unless we study the entire population, which we usually cannot do, we will always have incomplete information about the population. That is, unless we study the entire population, we cannot eliminate what is known as **sampling error.**

Sampling error is the difference between a characteristic of the entire population and a sample of that population.

The size of the sample influences the size of the sampling error.

The size of the sampling error is determined by two factors. The first factor is the size of the sample. Clearly, the larger the sample you take, the more similar the sample will be to the population, thus decreasing the sampling error. The second factor is the amount of **variation** that exists in the population. *Variation* in statistics has the same meaning that it has in general language usage.

> The amount of *variation* refers to how different the members of the population are from one another with regard to the variable being studied.

The amount of variability in the population influences the size of the sampling error.

For example, suppose your population of interest is all attendees at a concert. The variable of interest is the age of these people. If all attendees were exactly the same age, then you would say that there was no variation in the age of the members of the population. In this extreme case you would need to take a sample of only one student to have perfect information about the age of the concertgoers.

You will almost never study a characteristic that has no variability. Suppose that the ages of the concertgoers have a small amount of variability. Let's say the ages range from 18 to 22. Clearly, you have to take a sample of more than one attendee to understand how the ages vary in this population. Now suppose the ages of the concertgoers range from 17 to 60. You have to take a much larger sample to understand how the ages vary in this population. As the amount of variability in the population increases, the sampling error also increases.

1.2.5   EXERCISES—LEARNING IT!

1.1   "I'm so stressed" is a common cry among college students. What is stressing you out?

(a) What is the population of interest?

(b) Identify which of the reasons for taking a sample (listed on page 13) apply in this case. (There are more than one.)

(c) Identify two variables or characteristics of the members of this population that you may wish to study.

1.2   A university president wishes to see what types of activities and jobs graduates of the university are doing 5 years after graduation.

(a) What is the population of interest?

(b) Identify which of the reasons for taking a sample (listed on page 13) apply in this case. (There are more than one.)

(c) Identify two variables or characteristics of the members of this population that you may wish to study.

1.3   A Division III school is having trouble attracting and retaining football players. The school wishes to retain a higher percentage of athletes.

(a) What is the population of interest?

(b) What are some reasons there would be varying views about why the school is having trouble attracting and retaining athletes?

(c) Why might the athletic director be interested in these data?

1.4   Does heading a ball in soccer lead to head injuries later in life? This headline for a news article gets your attention.

(a) What is the population of interest?

(b) Suppose you had enough resources to do a census. Why might it be difficult to do a census?

1.5   The illustration at the top of page 16 ranks the top ten international airports in passenger satisfaction.

(a) What is the population of interest?

(b) What variable is being studied?

(c) Speculate on how the data were obtained.

Cincinnati among satisfying airports	1. Copenhagen, Denmark
The top ten international airports in passenger satisfaction, according to 1999's annual survey of international travelers:	2. Changi, Singapore
	3. Helsinki, Finland
	4. Vancouver, Canada
	5. Manchester, England
	6. Kuala Lumpur, Malaysia
	7. Cincinnati, USA
	8. Perth, Australia
	9. Amsterdam, Netherlands
	10. Hong Kong, China

Source: International Air Transport Association

1.6 Nearly half (48%) of all moms say the best Mother's Day gift would be hearing from their adult children more often during the year. Why don't kids call mom?

(a) What is the population of interest?

(b) What variable is being studied?

(c) Do you think a census was taken to arrive at the figure of 48%? Explain why or why not.

1.7 Before selecting a major, a student decides to study the salaries of nurses. The student is thinking that the choice of a major is related to salary. What other factors might cause variation in the salaries of nurses?

Discovery Exercise 1.1

INTRODUCTION TO SAMPLING AND VARIABILITY

Suppose that each set of data in this exercise represents an entire population. Since we don't yet have a way to quantify the amount of variability in a population, the data sets are labeled as having a small amount of variability or a large amount of variability. The first data set gives the numbers of people in 50 families who live in a small college town in New England. The second set of data lists the numbers of people in 50 families who live in a large city in the South.

New England families: Large amount of variability (average number of people in 50 families: 4.50)					Southern families: Small amount of variability (average number of people in 50 families: 4.36)				
1	4	5	7	8	1	3	4	3	5
3	9	8	8	8	3	4	6	6	3
4	9	9	1	6	6	4	3	4	6
4	1	3	9	7	7	5	4	5	5
8	2	3	1	9	4	4	5	5	6
1	7	5	1	1	3	4	4	7	4
1	6	8	2	9	6	3	5	5	5
4	1	1	1	3	5	4	4	4	5
4	2	4	9	4	4	5	4	3	4
1	3	8	1	1	5	4	2	4	4

Step 1. Select a sample of five numbers from each population.

	Sample from New England families	Sample from southern families
Selection 1		
Selection 2		
Selection 3		
Selection 4		
Selection 5		
Average		

Step 2. Calculate the sample average of the five numbers by adding them together and dividing the sum by 5.

Step 3. Calculate how far away the sample average is from the true population average by subtracting the sample average from the population average that is provided for you.

Step 4. Which data set has the greater error—the one with a small amount of variability (southern families) or the one with a large amount of variability (New England families)?

Repeat Steps 1–4 with a sample of ten numbers.

Step 1. Select a sample of ten numbers from each population.

	Sample from New England families	Sample from southern families
Selection 1		
Selection 2		
Selection 3		
Selection 4		
Selection 5		
Selection 6		
Selection 7		
Selection 8		
Selection 9		
Selection 10		
Average		

Step 2. Calculate the sample average of the ten numbers by adding them together and dividing the sum by 10.

Step 3. Calculate how far away the sample average is from the true population average by subtracting the sample average from the population average that is provided for you.

Step 4. Which data set has the greater error—the one with a small amount of variability or the one with a large amount of variability?

Step 5. Record the errors in the following table:

	Sample of size 5	**Sample of size 10**
New England families (large variability)		
Southern families (small variability)		

What happened to the error when you increased the sample size?

1.3 The Difference Between a Parameter and a Statistic

In the preceding section we explained that by understanding and studying the sample, we can gain insight and knowledge about the population. Next we must think about how to describe the population to someone else. Suppose you were asked to describe your instructor in the classroom. You might say that he or she is informative, humorous, organized, and helpful (we hope!). These are verbal descriptors of your instructor. If you were asked to describe the appearance of your instructor, you might say that she is about 5 feet 2 inches tall, weighs approximately 110 pounds, and is about 40 years old. These are numerical descriptors of the person. To paint a picture of a population and a sample, we use descriptors that are both verbal and numerical, just like the ones we used to describe your instructor.

1.3.1 PARAMETERS: NUMERICAL DESCRIPTORS OF THE POPULATION

There are many different ways to describe a population. One is to use numerical values to paint a picture of the population. For instance, if you are told that all the values of a variable of the population fall between 0 and 10, you form a mental image of the population that is quite different from the picture that is conjured up if you learn that all the values fall between 0 and 1000. Chapter 3 presents the traditional numerical measures that are used to describe the population. They are all examples of what are known as **parameters.**

A *parameter* is a number that describes a characteristic of the population.

The following example gives three different parameters that might be of interest to the students studying the college snack bar.

EXAMPLE 1.5

THE COLLEGE SNACK BAR

Identifying Parameters of Interest

The students may wish to know the average number of minutes it takes a customer to receive an order at the snack bar. In this case the parameter of interest is an average value. The students may also wish to know the longest time it takes a customer to get food. In this case the parameter of interest is the maximum time. Likewise, the students might wish to know what time of day is the least busy at the snack bar. So the least busy time of day is also a parameter of interest.

Understand the problem.

The next example shows that the parameter of interest might be a percentage.

EXAMPLE 1.6

DRESS DOWN DAY

Identifying a Parameter of Interest

A company that is thinking of adopting casual dress in the workplace would like to know the percentage of employees whose productivity increased as a result of this policy change. The parameter of interest here is a percentage or a proportion.

Understand the problem.

Remember that generally we do not know much about the population. If we did, we would not need to take a sample and we would not need the tools of statistics (and we could all go home now!). We are usually trying to discover information about parameters. In particular, we are often trying to estimate the value of a parameter. This is the job of statistics.

1.3.2 STATISTICS: NUMERICAL DESCRIPTORS OF THE SAMPLE

Since we most likely will not know the value of the parameters needed to describe the population, we must resort to using the information contained in the sample. It seems logical that a numerical descriptor of the sample might somehow be used to estimate the corresponding measure for the population. This is the right idea.

So we need numbers to describe the sample for two reasons: (1) to paint a picture of the sample and (2) to help us estimate the corresponding population parameters. There is nothing difficult here—in fact, it is quite simple. A **statistic** is nothing more than a number that describes a characteristic of the sample. Let's put that into a definition box.

A *statistic* is a number that describes a characteristic of a sample.

According to this definition we could dream up any formula to describe the sample and it would count as a statistic. Surprisingly enough, this is exactly right. There are, however, a few measures that are typically used because they convey some fairly standard information. These measures are the subject of Chapter 3.

Let's look at two examples: the college snack bar and the company that is considering adding dress down day to the workplace. In each case we will see what statistics might be calculated.

EXAMPLE 1.7

THE COLLEGE SNACK BAR AND DRESS DOWN DAY

Statistics That Could Be Calculated

The students who are examining the service at the snack bar might calculate the average service time from the sample data. This is one statistic. They might also record the shortest and the longest service times in the sample. These are two more statistics.

Let's assume the company that is thinking of adopting dress down day in the workplace has instituted the policy in one branch of the company. A manager has taken a sample of employees in this branch and measured the productivity of each employee in the sample both before and after the casual dress policy was adopted. Should the policy be adopted companywide? The statistic to calculate is the percentage of employees in the sample whose productivity increased after the casual dress policy was adopted.

Notice that the statistics that can be calculated from the sample data are closely related to the parameters of interest. This is often the case. Remember that when you calculate the average service time for the sample, it is exactly that: a description of the sample. Is the average service time for the population of all customers the same as the average you found in the sample? Probably not. Is it close? That depends on how well your sample reflects the population. This is the subject of the next two sections.

1.3.3 TELLING THE DIFFERENCE BETWEEN PARAMETERS AND STATISTICS

Sometimes it is difficult to tell whether the numbers reported in the newspaper or on television are parameter values based on a census or statistics based on a sample. To make this distinction, you must ask whether the data are based on complete population information. Sometimes this is the case, but often data are not based on complete information even though they are presented as such. Sometimes you cannot tell from the information provided, and sometimes the writer doesn't want you to think that much about it. This is actually an ethical issue.

For example, newspapers often present lists of "top tens" such as the top ten vacation spots, the top ten car features desired by car owners, the top ten busiest airports, and the top ten movies. In the case of the top ten vacation spots, it is possible that a survey was conducted and respondents were asked about their favorite vacation spots. Then the data reported would be sample statistics. It is also possible that a count was taken of the number of people who stayed at hotels in a number of different vacation spots. If hotel registration data were used, the result could be census data for the hotels.

Consider the following list of the box office revenues for the top ten movies in the weekend ending November 30, 2001.

Movie	Gross
1. *Harry Potter and the Sorcerer's Stone*	$23,642,327
2. *Behind Enemy Lines*	$18,736,133
3. *Spy Game*	$11,013,350
4. *Monsters, Inc.*	$9,105,664
5. *Black Knight*	$5,522,248
6. *Shallow Hal*	$4,525,000
7. *Out Cold*	$2,718,839

(continued)

Movie	Gross
8. *Domestic Disturbance*	$1,912,678
9. *Amelie*	$1,358,649
10. *Heist*	$1,182,497

Are these revenues based on parameter values or statistics? You need to know how the researchers got the data. Most likely they took a sample of movie theaters across the nation; thus, the data are statistics.

Now consider the average of the top ten fastest times for completing the Boston Marathon. This average is based on the actual top ten fastest times and therefore is a parameter for the population of the ten fastest race times. Many sports statistics are based on complete historical information. As such they are parameters. There is no sampling error because the numbers were calculated from census data.

EXAMPLE 1.8

AFFORDABLE HOUSING SHORTAGE

Distinguishing Parameters from Statistics

Because wages are not keeping up with skyrocketing housing prices and because there is not enough affordable housing being built, a growing number of moderate-income families face a housing crisis. Look at the accompanying data about the numbers of families facing a housing crisis. Are these population parameters or sample statistics?

Draw conclusions and make recommendations.

Metro area	Number of families in crisis	Percentage of all working families
San Jose, CA	45,278	27%
San Francisco	49,609	26%
Oakland	63,952	22%
Tampa	56,206	21%
Boston	88,573	20%
Providence	20,008	17%
Washington	90,280	16%
Norfolk, VA	31,965	15%
Rochester, NY	17,919	13%
Baltimore	44,628	13%
Salt Lake City	20,600	12%
New York	513,649	12%
Houston	55,870	10%
Minneapolis–St. Paul	46,496	10%
Cincinnati	19,972	10%
Los Angeles	401,402	9%
Birmingham, AL	7,974	7%
United States	**3,046,000**	**10%**

Source: The Center for Housing Policy, June 2, 2000.

The data in the table were gathered by the Center for Housing Policy and are most likely census data for the cities listed. But what about the values for the United States shown at the bottom of the table? Are these the totals of the values in the table? Check for yourself. The values are not totals, although the layout of the table suggests that. So the United States figures are sample statistics.

Discovery Exercise 1.2

POPULATIONS, PARAMETERS, SAMPLES, AND STATISTICS

Read the following article about couples who live together before marriage.
Web site: http://www.psu.edu/ur/2000/co-habit.html

Married Couples Who Lived Together Before Wedding
Show Poorer Communication Skills

July 17, 2000.

University Park, PA—A Penn State study that involved videotaping 92 recently married couples talking in their living rooms showed that those that had lived together before tying the knot had poorer communication skills when trying to solve a marital or personal problem than couples who didn't previously cohabit.

The study director, Dr. Catherine Cohan, assistant professor of human development and family studies, says, "These data suggest that couples who choose to live together before marrying might need to work on these skills to counteract the risk of divorce that poorer communication competence could pose. However, we believe it is premature to draw firm conclusions about intervention or policy since this is the first study of its type."

Cohan presented details of the study June 29 at the International Conference on Personal Relationships in Brisbane, Australia. Her paper, "Toward a Greater Understanding of the Cohabitation Effect: Premarital Cohabitation and Marital Communication," was co-authored by Stacey Kleinbaum, a master's degree candidate in human development and family studies.

Contrary to the popular belief that living together will improve a person's ability to choose a marriage partner and stay married, the opposite is actually the case, Cohan notes.

"It has been consistently shown that, compared to spouses who did not cohabit, spouses who cohabited before marriage have a higher perceived likelihood of divorce and higher rates of marital separation and divorce," she says. However, why cohabitation and divorce are related remains unknown.

Cohan's study is the first to compare the problem-solving and support behavior skills of married couples who did not previously cohabit with those that did. In the study, the researchers visited the couples at home and talked with both the husband and wife separately. During these separate interviews, both the husband and wife had an opportunity to review a list of common marital problems and then select one that was important in their own marriage. The list included problems with sex, money, in-laws, career decisions, religion, the timing of children, who should do what chores around the house, and other typical sources of marital strife.

Later, the husband and wife sat together on the sofa in their living room in front of a video camera and were instructed by the researcher to spend 15 minutes trying to solve the problem that one of them had selected as important in their marriage. The researcher then left the room and the video camera recorded the interaction between husband and wife. Two conversations were videotaped: one in which the couple discussed the problem selected by the husband and one in which they discussed the problem selected by the wife.

On viewing the videotapes and analyzing them using standard psychological techniques, the researchers found that the couples who had lived together before marriage, even for as little as a month, displayed more negative and less positive problem-solving behaviors compared with the spouses who had not cohabited. In particular, when the husband had picked the problem the couple was trying to solve, the spouses who had cohabited expressed more negative behaviors such as coerciveness and attempts to control. In general, wives who had lived with their husbands before marriage expressed more verbal aggression than wives in couples with no prior cohabitation.

In a similar fashion the couples were asked to select and discuss a personal problem from a list, such as trying to quit smoking or to lose weight, to develop a hobby, to improve a relationship with a family member, or other common personal improvement goals. Then, in front of the video camera in their living room, they were instructed to talk about the personal problem with their spouse. Again, the videotaped conversations showed that couples who had lived together before wedding demonstrated more negative and less positive support behavior.

Currently, about half of U.S. couples live together before marrying. Cohan says this fact makes the need for further research evident. The results of this study combined with results from past cohabitation research and marital communication research suggest that problems with communication may be the link between cohabitation and the high rates of divorce.

"An important next step is to follow newlywed couples with and without cohabitation experience over four to five years to test whether communication skills link living together before marriage with a higher likelihood of divorce after marriage," she adds.

Now consider these questions:

1. What population is being studied?
2. Was a sample selected? If so, can you tell how it was selected and what size it was?
3. What are the variables of interest?
4. Does the article give information about parameters? If so, what are they?
5. Does the article give sample statistics? If so, what are they?
6. What theory, hypothesis, or idea is being promoted by the article?
7. What additional questions do you have about the data?

1.3.4   EXERCISES—LEARNING IT!

1.8   Consider the illustration shown here.

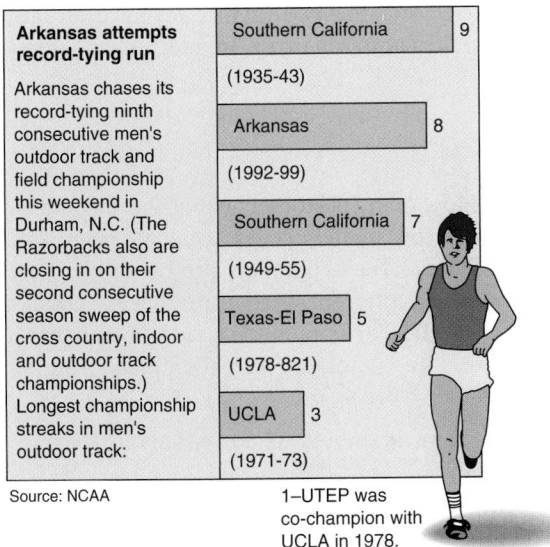

Arkansas attempts record-tying run

Arkansas chases its record-tying ninth consecutive men's outdoor track and field championship this weekend in Durham, N.C. (The Razorbacks also are closing in on their second consecutive season sweep of the cross country, indoor and outdoor track championships.) Longest championship streaks in men's outdoor track:

Southern California   9
(1935-43)
Arkansas   8
(1992-99)
Southern California   7
(1949-55)
Texas-El Paso   5
(1978-821)
UCLA   3
(1971-73)

Source: NCAA

1–UTEP was co-champion with UCLA in 1978.

Are the numbers cited parameter values or statistics? Explain your reasoning.

1.9   Penn State researchers have concluded that eating tree nuts or peanuts can have a strong protective effect against coronary heart disease. The researchers' review of the existing published epidemiologic studies found that consuming 1 ounce of nuts more than five times a week can result in a 25%–39% reduction in coronary heart disease risk among people whose characteristics match those of the general adult U.S. population.

(a) Are the values a statistic or a parameter?

(b) What else would you like to know about the data before deciding to add nuts to your diet?

1.10 J. K. Rowling's best-selling Harry Potter books are popular with adults; 59% of the series' books sold last year were purchased for readers age 14 and older.

(a) Is 59% a statistic or a parameter?

(b) What else would you like to know about the data before running out to buy the book?

1.11 Most brides are willing to pay top dollar for bridal dresses, up to $1000 in some cases.

(a) What parameter is of most interest to the bride-to-be?

(b) What is the $1000 quoted in the statement—a parameter or a statistic?

(c) Is $1000 a helpful value if you were engaged to be married?

1.12 The U.S. government wishes to know how many people are unemployed. The unemployment rate for the United States in 1998 was 4.5%. Is this a statistic or a parameter value? Explain your answer.

1.13 Before selecting a major, a student decides to study the salaries of individuals who work as high school teachers.

(a) What parameter is of interest?

(b) What statistic might you calculate from the sample?

1.4 Factors That Influence Sample Size: Some Sampling and Sample Size Considerations

The next question is, How big must the sample be? Do we need to eat half of one pie to know how all the pies in a batch taste? What factors will influence our decision in this matter? We touched on this question when we observed the impact of the sample size and the amount of variability on the size of the sampling error. This section reinforces the connection between the sample size and the two factors we have already studied: the variability in the population and the sampling error. In addition, we introduce two other factors.

1.4.1 SIZE OF THE POPULATION

The size of the population affects the sample size.

The first factor is the **size of the population.** How many customers eat at the snack bar? How many employees work in the company? How many scooter accidents occur in a year? How many adults live in the United States? These numbers are the sizes of the populations. Intuitively it seems like it would take a different sample size to learn about a population of 1000 compared to a population of 100,000. Population size is a factor, but it turns out to be less important than some of the other factors we have identified.

> The *size of the population* is the number of members of the population. Its symbol is N.

1.4.2 RESOURCES AVAILABLE

The amount of resources available affects the sample size.

The next factor that influences the sample size is the amount of time, money, and other resources you have available. Depending on whether you need a decision in 6 months or next week, the amount of data you can possibly collect and analyze is different. Also, remember that it is expensive to collect and process data, so cost must be a factor. We still have not identified the two most important factors in dictating sample size.

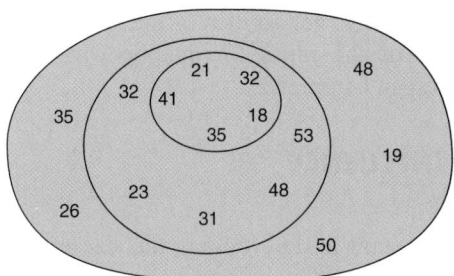

FIGURE 1.2   **Bigger and bigger samples**

1.4.3 ERROR THAT CAN BE TOLERATED

Section 1.2.4 explained that whenever you use a sample to draw conclusions about the population, you will have some sampling error. And the bigger the sample we pick, the less error we will have in our conclusions. Returning to the ovals in Figure 1.1, you can see that if you take a bigger sample, then you capture more of the population. Therefore, the conclusions you draw about the population based on the sample will be more accurate. Consider the increasing sample sizes in Figure 1.2. The numbers are the ages of members of the population being studied. You can see that the smallest oval represents a sample size of five because it contains five members of the population. The next larger oval represents a sample of size ten. In addition to the five members contained in the small oval, it contains five more members of the population.

Remember! Sampling error does not imply that you did anything wrong. It results simply because you have an incomplete picture of the population.

If we continue taking bigger sample sizes, eventually we will end up taking a census, or studying the entire population. This situation is represented by the largest oval in Figure 1.2. When this happens, the sample size is the same as the population size, *N*, and there is no sampling error.

It seems, then, that if we can live with a little more error, we can get by with a smaller sample size. The more costly the error, the larger our sample size must be. Thus, the amount of sampling error (dictated by the cost of the error) we can live with is a factor in determining sample size. We return to this matter in greater detail in Chapters 7–9.

The size of the error that we can afford is an important factor in determining the sample size.

1.4.4 VARIATION IN THE POPULATION

Even if we can tolerate only a small amount of error, we still may not need a really large sample size. One other factor is very important: the variability that exists in the population. Suppose that I wish to study all the students in your college. I want to be very accurate in my conclusions, but all of the students feel exactly the same about the issue I am studying. How many students do I need to talk to in order to get very precise information about the population? The correct answer is that only one student is needed. A sample size of one is adequate despite the need to be very accurate and despite the large size of the population. This is because there is no variation in the population; that is, everyone feels the same way about the issue. Granted, this is not likely to happen, but you can see the impact this factor has on the sample size needed. The more similar the members of the population are to one another, the less variation there is within the population. So the sample size can be smaller. If the population is highly diverse, then I need to talk to more people to get a sense of how the population feels about the issue.

The variation in the population is a key factor in determining the sample size.

A natural question to ask is, How can we know the amount of variability in the population to determine the sample size? Given that we are taking a sample to estimate such parameters as variability in the population, it is a problem if we need to know the variability to determine the sample size. We are locked in a loop at this point, a classic "catch-22" situation. We could estimate the amount of variability using information about the variability in a population similar to the one we are studying, we could use previous studies of the same population to get an idea of the variability, or we could do a pilot study. These are some of the ideas for dealing with this dilemma, and they are addressed more completely in Chapter 7.

1.4.5 SUMMARY OF FACTORS THAT INFLUENCE SAMPLE SIZE

We have identified the factors that are important in determining the size of the sample needed to draw conclusions about a population:

- The variation in the population
- The amount of error that can be tolerated
- The resources available for the project
- The size of the population, N

You will see that the first two factors show up in the formula developed in Chapter 7 for determining the size of the sample. For now it is sufficient for you to have a general understanding of the impact that each of these factors has on the sample size determination.

As a final note in this section, we should agree on a label for the sample size. No matter what statistics book you pick up, you will always see the **sample size** symbolized by n. Remember that the size of the population is N.

The *size of the sample* is symbolized by n.

1.5 Selecting the Sample

Now that you have a general idea of the factors that influence the size of the sample to be selected, we are ready to discuss how to select a sample. Let's decide what qualities we would like our sample to have.

1.5.1 SELECTING AN UNBIASED SAMPLE

Ideally, we would like the sample to be a miniversion of the population. Remember that we will be using the sample to understand the population. The sample should thus contain all the key features found in the population. It is easiest to understand what this means by looking at some examples of samples that may not be a miniversion of the population.

Collect and analyze the data.

EXAMPLE 1.9

THE COLLEGE SNACK BAR

Selecting a Sample

Suppose the students who are studying the snack bar decided to use the first 50 customers on a Monday morning as the sample.

Why is this not a good idea? It is possible that the service changes during the day. Different people may work at different times of the day and on different days of the week. It is possible that the service times and mistakes made vary with both which employees are working and how busy the snack bar is. If our sample is the first 50 customers on a Monday morning, we may not see the impact of these other factors on the service time or accuracy of the order.

EXAMPLE 1.10

DRESS DOWN DAY

Selecting a Sample

Suppose the company that is considering allowing casual dress in the workplace decides to use the recently hired women employees as their sample.

Again the problem is that this sample may not truly represent the population of employees of the company. Why not? First of all, the people in this sample are all women. Maybe women in this company are consistently more or less productive than men. Second, the sample is all new hires, who are likely to be putting their best efforts into the job. At the same time, they are in training and may make more than the average number of errors. For many reasons this group is not a miniversion of the company's employees.

> Collect and analyze the data.

In both examples the proposed sample may not let us see all of the variation that exists in the population. In other words, these samples might be **biased.** The word *bias* in statistics has the same meaning that it has in general usage. It means that the sample is somehow not a fair reflection of the reality we would see in the population. A biased sample gives an unfair or prejudiced view of the population. This is precisely what we wish to avoid. Let's capture this in a definition.

> A *biased sample* is a sample that does not fairly represent the population.

✔ TRY IT NOW!

Stress Relief
Identifying Possible Biases in a Sample

You are studying the methods that students at your school use to relieve stress. You decide to use your statistics class as your sample. Why might this be a biased sample?

1.5.2 SELECTING A SIMPLE RANDOM SAMPLE

In selecting any sample, we wish to pick an unbiased or fair sample. We also want to pick a sample that contains as much information as possible. There are many ways to pick a sample, but for the purposes of an introductory course we will use what is known as a **simple random sample (srs).** This means that each member of the population has an equal chance of being selected as a member of the sample. In addition, every sample of size n has the same chance of becoming the sample. For example, we could place the names of all the members of the population in a hat and then reach in and pick out members to be in the sample.

> A *simple random sample* is a sample that has been selected in such a way that all members of the population have an equal chance of being chosen. In addition, every sample of size n has the same chance of becoming the sample.

ANS. THE STUDENTS IN THIS CLASS MAY BE MOSTLY SOPHOMORES WHO HAVE THE SAME MAJOR.

Simple random sampling is the most obvious way to select a sample. In fact, you might wonder how else you could do it. Suppose you wished to learn about the prices charged for advertising in newspapers. Your population would be all companies in the United States that publish newspapers. If you picked a simple random sample of newspaper companies, you might end up with all small companies purely by chance. You might not get any of the "biggies" such as the *New York Times*, the *Boston Globe*, the *Chicago Tribune*, or the *Los Angeles Sun* in the sample. Is this bad? Well, that depends on what variable you are studying. We are interested in the prices for advertising in all papers, and the larger companies might set the pace for the smaller companies with regard to pricing. In this case you might choose to use something other than simple random sampling to guarantee that you get some "biggies" in your sample.

As you learn more about statistics, you will see that simple random sampling does not always provide the most information for your money. All of the concepts that you will learn in this book carry over to the other kinds of selection methods, but the formulas might be altered if you use a procedure other than simple random sampling. In Chapter 13 you will learn how to select a sample by designing an experiment.

1.5.3 CREATING A SAMPLING FRAME

Now that we have decided that we want to select a simple random sample, what do we need to pick the sample? Returning to the example of the newspapers, we need a list of all the newspapers in the United States. Such a list is known as the **sampling frame.**

> A *sampling frame* is a list of all members of the population.

Some government agencies and private companies, such as Dun & Bradstreet, collect data for sampling frames. However, creating a sampling frame may take a fair amount of time, energy, and money. The list from which you pick your sample must be accurate and complete. The difficulty in creating these lists is not the focus of this book. However, you should know what a sampling frame is and know enough to watch out for potential problems.

1.5.4 USING A TABLE OF RANDOM NUMBERS TO SELECT THE SAMPLE

Once you have the sampling frame and you have determined your sample size (using the formula developed in Chapter 7), you are ready to select your sample. Although it is easy to understand the process of placing the name of each member of the population in a hat and selecting the sample by picking names from the hat, this is not a practical way to select the sample. To imitate this process you can use a **table of random numbers.**

> A *table of random numbers* is a list of numbers randomly generated and listed in the order in which they are generated.

A random number table is provided in Table 1 in Appendix A. A portion of this table is shown in Table 1.1.

A random number table should contain roughly as many 0s as 1s as 2s and so forth. There is a way to check that such a table does in fact contain random numbers, but for our purposes we will assume that the random number table has been generated correctly.

TABLE 1.1

PORTION OF A RANDOM NUMBER TABLE

Row	Column 1	2	3	4	5	6	7
1	094632795	711501513	537971597	562758635	410398128	182794408	773761503
2	033413186	653475420	289063704	485441982	460744361	328703833	289612212
3	297556368	658953044	738968017	414437050	296126017	075254187	702140315
4	472960570	785645638	574817322	817883255	976076280	843373358	118284363
5	256883707	716249997	378236162	467694224	193707682	380141891	605807481
6	179451522	878902420	602450872	987686989	686677180	242196303	517640224
7	894964682	704841116	241902107	750429362	794778197	693242123	316755091
8	738120861	744470405	873393138	758824215	394004646	496696605	006936567
9	803156944	653387115	716335974	835667154	066959782	908783760	165946696
10	187636922	321421098	638210137	055734541	493193305	566923120	435549770

To use the random number table to identify which members of our population will be selected for the sample, we must first assign each member of the population an identification (ID) number. Suppose that you are studying the students at your college or university. Each student probably has a student ID number. Since many schools use the student's Social Security number as the ID number, let's suppose that the ID number is a nine-digit number. To use a table of random numbers to select a sample of 30 students from this population, you first select where in the random number table you will start reading. One way to do this is to close your eyes and point to a spot in the table. Suppose that you do this and you select row 10, column 2 as your starting point. Then you read the next nine digits from Table 1.1 as 321421098. This is the ID number of the first member of your sample. The next nine-digit number from the table is 638210137 and this is the ID number of the second member of your sample. You continue this process until you have selected 30 ID numbers.

The sample obtained in this manner is a simple random sample. However, you certainly need to read more than 30 ID numbers from the table to get 30 usable Social Security numbers. Most of the ID numbers you read from the table of random numbers will not correspond to anyone at your school. Although you will eventually find 30 usable ID numbers, you will waste an incredible amount of time.

This method works a little better when you can assign ID codes to the members of your population in such a way that they are sequential, so that you will not select any nonusable ID numbers. Let's see how this works in the following example.

EXAMPLE 1.11

STUDENT SAMPLE

Selecting a Sample Using a Table of Random Numbers

Each student at your college has a mailbox on campus. The mailboxes are numbered from 0000 to 9000. To select a simple random sample of ten students, we can select ten mailbox numbers at random using the random number table. Suppose we choose to start at row 7, column 3 of Table 1.1. The first student selected has mailbox 2419, which is a valid number. If we continue to read off four-digit

Collect and analyze the data.

numbers from this table, the second number selected is for mailbox 0210. The list of all ten mailbox numbers selected is:

2419   0210   7750   4293   6279   4778   1976   2123   3167   5509

Since the numbers in the table are organized in nine-digit blocks and we need only four-digit mailbox numbers, we just keep reading the numbers sequentially, wrapping down to the next row upon reaching the end of column 7. Notice that the selection after 1976 would have been 9324, which is not a valid mailbox number, so we simply skipped it.

The students who have these mailboxes are the students in the sample.

For the most part, this book assumes that the sample has been selected and the data have been collected. Your job is to see the information in the data and make some recommendations or decisions based on the data. We do not focus on the actual selection of the sample or the job of collecting the data. This is the subject of another course.

TRY IT NOW!

The College Snack Bar
Selecting a Simple Random Sample

Use Table 1.1 to select a sample of five customers of the snack bar. You can assume that each customer has a five-digit ID number on the ID card, which can be used for the point system.

Discovery Exercise 1.3

INTRODUCTION TO SAMPLING

Suppose the data shown here represent an entire population. They are the numbers of people in 50 families who live in a small college town in New England. (If you did Discovery Exercise 1.1, then you will recognize this as the same data set.) Note that a two-digit ID number has also been included.

New England families: Large amount of variability
(average number of people in 50 families: 4.50)

ID: 01	1	ID: 02	4	ID: 03	5	ID: 04	7	ID: 05	8
ID: 06	3	ID: 07	9	ID: 08	8	ID: 09	8	ID: 10	8
ID: 11	4	ID: 12	9	ID: 13	9	ID: 14	1	ID: 15	6
ID: 16	4	ID: 17	1	ID: 18	3	ID: 19	9	ID: 20	7
ID: 21	8	ID: 22	2	ID: 23	3	ID: 24	1	ID: 25	9

New England families: Large amount of variability
(average number of people in 50 families: 4.50)

ID: 26	1	ID: 27	7	ID: 28	5	ID: 29	1	ID: 30	1
ID: 31	1	ID: 32	6	ID: 33	8	ID: 34	2	ID: 35	9
ID: 36	4	ID: 37	1	ID: 38	1	ID: 39	1	ID: 40	3
ID: 41	4	ID: 42	2	ID: 43	4	ID: 44	9	ID: 45	4
ID: 46	1	ID: 47	3	ID: 48	8	ID: 49	1	ID: 50	1

Step 1: Select a sample of five numbers from this population using the table of random numbers. Record your sample in the table.

Selection 1	
Selection 2	
Selection 3	
Selection 4	
Selection 5	
Average	

Step 2: Calculate the sample average of the five numbers by adding them together and dividing the sum by 5.

Step 3: Calculate how far away the sample average is from the true population average by subtracting the sample average from the population average of 4.50.

Step 4: If you did Discovery Exercise 1.1, compare the sample you selected using the random number table with the sample you selected without the use of the table. Which sample average is closer to the population average?

Repeat Steps 1–4 with a sample of ten numbers.

Step 1: Select a sample of ten numbers from the population using the table of random numbers.

Selection 1		Selection 7	
Selection 2		Selection 8	
Selection 3		Selection 9	
Selection 4		Selection 10	
Selection 5		**Average**	
Selection 6			

Step 2: Calculate the sample average of the ten numbers by adding them together and dividing the sum by 10.

Step 3: Calculate how far away the sample average is from the true population average by subtracting the sample average from the population average of 4.50.

Step 4: If you did Discovery Exercise 1.1, compare the sample you selected using the random number table with the sample you selected without the use of the table. Which sample average is closer to the population average?

1.5.5 EXERCISES—LEARNING IT!

1.14 The President of the United States wishes to see how popular he or she is after 2 years in office. A sample of 1000 voters is taken from the state of California.

(a) Why might this sample be biased?

(b) How could you get a simple random sample?

1.15 Tasty Ice Cream Corporation wishes to be sure that all of the half-gallon ice cream cartons actually contain one-half gallon of ice cream. A sample of 30 cartons are measured. All of the cartons in the sample were filled on Friday.

(a) Why might this sample be biased?

(b) How could you get a simple random sample?

1.16 A skier wants to know the average price of lift tickets. He goes online and gets the prices at five ski resorts in Vermont.

(a) Why might this sample be biased?

(b) How could you get a simple random sample?

1.17 The university wishes to know how students view the new athletic center on campus. A questionnaire is distributed to people as they enter the building.

(a) Why might this sample be biased?

(b) How could you get a simple random sample?

1.18 The U.S. government wishes to know how many people are unemployed. A sample of 1000 individuals over age 18 is selected from a national listing. Explain why this is probably an unbiased sample.

1.19 You are in the market for a new car. You decide that gas mileage is an important characteristic of the car. You find one friend who owns each of the cars you are considering and use their mileages as your data. Explain why this is probably not a very good way to assess the gas mileage of these cars.

1.20 The Coca-Cola Company wishes to know the proportion of people who prefer Coke over Pepsi. A sample of 100 people is taken at a county fair.

(a) Why might this sample be biased?

(b) How could you get a simple random sample?

1.21 Before selecting a major, a student decides to study the salaries of individuals who work as nurses. A sample of nurses is selected from the list of alumni of the school that the student attends.

(a) Why might this sample be biased?

(b) How could you get a simple random sample?

1.6 Types of Data

NOTE: The words **data** *and* **variable** *are used interchangeably.*

Although we will not study the whole topic of data collection, we need to think about the different kinds of data or variables that we might encounter. The kind of statistical analysis that we do depends on the type of data we have. There are two major types of variables: **qualitative** and **quantitative**. In this section we concentrate on identifying the different types of data. Since different statistical techniques are used with different types of data, it is important to identify the data you have before you analyze them so that you don't use the wrong technique.

1.6.1 QUALITATIVE DATA

Qualitative data, also known as **nominal** or **categorical data,** are the simplest form of data. Examples of qualitative data are variables such as gender (male or female) or the grade in a course (A, B, C, D, or F). Each item in the sample falls into one of a finite number of possible categories.

> *Qualitative data* describe a particular characteristic of a sample item. They are most often nonnumerical.

Suppose you are interested in determining the length of time it takes to get food at the snack bar. You collect the data from a sample of customers and after analyzing them, you find that there appear to be two subgroups in the data. Each subgroup contains service times that cluster around a different value. Why does this happen? When you ask some questions about how the data were collected, you learn that some of the customers were in the snack bar for breakfast and some were there for lunch. When you return to the service times, you find that all of the breakfast values are clustered at the lower number, whereas the lunch service times are clustered at the higher number. The variable that tells you what meal is being served is an example of qualitative data. Knowing the meal helps to explain the differences in the service times that you see in the data.

✔ TRY IT NOW!

Dress Down Day
Identifying Qualitative Variables

A company that is considering instituting a dress down day in the workplace has a sample of employees wear casual clothes to work on Fridays. After the policy has been in effect for some time, the company measures the changes in productivity for the employees. It appears that for some employees productivity has increased, whereas for others it has decreased. What qualitative data might be collected to help the company understand the differences observed?

The statistical techniques that are used to analyze qualitative data are limited but are often critical to our understanding of the results of statistical analyses. When you are collecting data, it is important to think of qualitative data that may be relevant to the problem. Qualitative data are easily collected at the time but almost impossible to reconstruct after the fact. When in doubt, qualitative data should always be collected.

Sometimes numbers are used to classify qualitative data. For example, in surveys that ask for gender, a 0 is often used to denote "male" and a 1 is used to denote "female." These data are referred to as **nominal data.** The numbers are used simply to represent different categories and have no real meaning as numbers.

ANS. GENDER, DEPARTMENT, LEVEL WITHIN THE ORGANIZATION

> Data that are created by assigning numbers to different categories when the numbers have no real meaning are called *nominal data.*

Nominal data are treated the same way as ordinary qualitative data.

The order in which numbers are assigned to qualitative data may have some meaning. When numbers are used to name ordered categories, the data are called **ordinal.** In the case of gender, the numbers 0 and 1 could very easily be reversed and so they have no intrinsic meaning. But suppose you asked a group of people to rank five different versions of a new soft drink. The resulting data might look like this:

Version	3	2	5	1	4
Ranking	1	2	3	4	5

In this case the numbers are not entirely meaningless because they indicate a relative position on a scale for each version of the product. However, you cannot tell from this scale whether the person doing the ranking liked version 3 a lot and really hated the other four or whether versions 3 and 2 were similar and much superior to the remaining three versions. The distances between the assigned numbers are not necessarily equal.

Another example of ordinal data is when a characteristic of a sample item, such as income, is classified as 1 = low, 2 = medium, and 3 = high. Here, the numbers have a relative ordering, but there is no way to compare 1 to 2 to 3 numerically.

> Data that are created by assigning numbers to categories where the order of assignment has meaning are called *ordinal data.*

Another common type of survey question that yields ordinal data uses what is known as a Likert scale. There are many different ways to design a survey question using a Likert scale. Some of the most popular designs are shown in the tables that follow. The shaded sets are the most popular.

> *Likert scales* are used to collect information on attitudes, including degree of agreement with a statement, frequency of use, importance of an issue, quality, and likelihood.

Agreement

Strongly Agree Agree Undecided Disagree Strongly Disagree	Agree Strongly Agree Moderately Agree Slightly Disagree Slightly Disagree Moderately Disagree Strongly	Agree Disagree	Agree Undecided Disagree
Agree Very Strongly Agree Strongly Agree Disagree Disagree Strongly Disagree Very Strongly	Yes No	Completely Agree Mostly Agree Slightly Agree Slightly Disagree Mostly Disagree Completely Disagree	Disagree Strongly Disagree Tend to Disagree Tend to Agree Agree Agree Strongly

Frequency

Very Frequently Frequently Occasionally Rarely Very Rarely Never	Always Very Frequently Occasionally Rarely Very Rarely Never	Always Usually About Half the Time Seldom Never	Almost Always To a Considerable Degree Occasionally Seldom
A Great Deal Much Somewhat Little Never	Often Sometimes Seldom Never	Always Very Often Sometimes Rarely Never	

Importance

Very Important Important Moderately Important Of Little Importance Unimportant	Very Important Moderately Important Unimportant

Quality

Very Good Good Barely Acceptable Poor Very Poor	Extremely Poor Below Average Average Above Average Excellent	Good Fair Poor

Likelihood

Like Me Unlike Me	To a Great Extent Somewhat Very Little Not at All	True False
Definitely Very Probably Probably Possibly Probably Not Very Probably Not	Almost Always True Usually True Often True Occasionally True Sometimes But Infrequently True Usually Not True Almost Never True	True of Myself Mostly True of Myself About Halfway True of Myself Slightly True of Myself Not at All True of Myself

EXAMPLE 1.12

WORKING IN TEAMS

Using a Likert Scale

Many instructors use team assignments in their classes to prepare students for working in teams on the job. However, students have differing opinions about the use of teams, particularly when their grade depends on the work of the other team members. Here is one question in a survey on group assignments:

Collect and analyze the data.

I believe that I could have received a better grade on group assignment(s) working by myself than in a group.

1 Strongly Disagree	2 Disagree	3 Somewhat Disagree	4 Neither Agree Nor Disagree	5 Somewhat Agree	6 Agree	7 Strongly Agree

Clearly ordinal data are generated because the responses are ordered.

If you use an odd number of categories in a Likert scale, then you allow the respondent to be neutral. In Example 1.12 the response of 4 is neutral. If you use an even number of categories, then you force the respondent to take a position, positive or negative. These design decisions are tied to what you are trying to learn about the population and whether or not a neutral position is really an option.

Most often, ordinal data are analyzed using the same graphical techniques that apply to other qualitative data. Some limited statistical techniques can be used to further analyze ordinal data. These techniques are called nonparametric analyses and are covered in Chapter 15. The techniques provide less information about the population than when we can collect meaningful numerical data. However, often you can consider Likert scale data to be numerical, as discussed in the next section.

1.6.2 QUANTITATIVE DATA

The next type of data to be considered is quantitative data. This type of data falls into two categories: **discrete** and **continuous.**

> Data that are inherently numerical are called *quantitative data.*

Discrete data usually result from counting or enumerating. The only possible values are positive integers and whole numbers. Examples of discrete data are the number of prior convictions for a person who has been arrested and the number of defective items in a sample.

> *Discrete data* can take on only certain values. These values are often integers or whole numbers.

Discrete data that result from counting the number of times something occurs are important. This type of data is used in analyzing many public opinion polls and other surveys. You will encounter this kind of data many times in the text.

In addition, sometimes Likert scale data are considered discrete data. Certainly, the values assigned to the categories of agreement in Example 1.12 appear to be discrete quantitative data. However, remember that the question you need to ask yourself is, Do the numbers carry any intrinsic meaning? Would it make any difference if you changed the scale to 10, 20, 30, 40, 50, 60, 70 instead of 1 through 7? Doing so would change the average, but the average is meaningful only relative to the original scale. The numbers themselves do carry some meaning because it would not make sense to use the scale 1, 3, 5, 7, 2, 4, 6! A higher number tells you that a person is more in agreement with the statement. You must decide if it makes sense to do mathematical functions with these numbers. If so, then it is fine to consider them as discrete data. If not, as is the case with data on gender (nominal data), then you cannot consider them to be discrete data.

When you treat Likert scale data as discrete data, you are also making some assumptions about the differences between categories. You are saying that the difference between the opinion of *Agree* and *Somewhat Agree* is the difference between the assigned numerical values of 6 and 5, which is $6 - 5 = 1$. This difference alone is not critical, but it is important whether the difference between any two other adjacent categories is somehow the same difference in attitude. For instance, the difference between the person who said *Strongly Agree* and the person who said *Agree* is also 1 $(7 - 6 = 1)$. Making the Likert responses discrete quantitative data means that the difference between the attitudes of *Agree* and *Somewhat Agree* is the same as the difference between the attitudes *Strongly Agree* and *Agree*. Often this is a legitimate assumption and you can in fact treat Likert data as discrete quantitative data.

Numerical data that are not discrete are called **continuous**. Continuous data most often result from measuring and are sometimes referred to as measurement data. Continuous data usually consist of real numbers, which, in this book, will be in decimal form.

> *Continuous data* can take on any one of an infinite number of possible values over an interval on the number line. These values are most often the result of measurement.

EXAMPLE 1.13

STUDY TIMES

Collect and analyze the data.

Using Continuous Data

Suppose you are interested in determining the length of time a typical student spends studying statistics on any given night. Initially, the number of possibilities for values is infinite, ranging from 0 minutes up. These data are continuous. Once you decide on a measuring device, the number of possibilities is limited, but the limitation is caused by the measuring device and not the variable itself. For example, if you measure the time spent studying to the nearest hour, the possible values are 0, 1, 2, ... If, however, you decide to measure to the nearest one-tenth of an hour (6 minutes), then the possible values are 0.0, 0.1, 0.2, . . . , 3.0, 3.1, ... The more precise your measuring device, the more possible values your data can assume.

1.6.3 EXERCISES–LEARNING IT!

1.22 The President of the United States wishes to see how popular he or she is after 2 years in office. A sample of voting-age adults are asked if they would reelect him or her today.

(a) Are the data qualitative or quantitative?

(b) If the data are qualitative, are they nominal or ordinal? If the data are quantitative, are they discrete or continuous?

1.23 Tasty Ice Cream Corporation wishes to be sure that all of the half-gallon ice cream cartons actually contain one-half gallon of ice cream.

(a) What is one variable of interest?

(b) Are the data qualitative or quantitative?

(c) If the data are qualitative, are they nominal or ordinal? If the data are quantitative, are they discrete or continuous?

1.24 The university wishes to know how students view the new athletic center on campus. Students are asked to rank the new center against four other schools' athletic centers.

(a) Are the data qualitative or quantitative?

(b) If the data are qualitative, are they nominal or ordinal? If the data are quantitative, are they discrete or continuous?

1.25 The U.S. government wishes to know how many people are unemployed. A sample of people are asked whether or not they are employed full time.

(a) Are the data qualitative or quantitative?

(b) If the data are qualitative, are they nominal or ordinal? If the data are quantitative, are they discrete or continuous?

1.26 Before selecting a major, a student decides to study the salaries of individuals who work as counselors. A sample of counselors are asked their salaries.

(a) Are the data qualitative or quantitative?

(b) If the data are qualitative, are they nominal or ordinal? If the data are quantitative, are they discrete or continuous?

1.27 Here are a few questions from a survey about the "I Love You" computer virus. For each question, identify what type of data will be collected.

(a) Was your computer infected by the "I Love You" virus?

- Yes, one or more of my computers was infected by it.
- No, none of my computers was infected by it.
- I'm really not sure whether it was.

(b) Do you know anyone else whose computers were infected by the "I Love You" virus?

- Yes
- No

(c) How often do you carefully examine the subject and body of your e-mail messages before you open their attachments?

- Always
- Often
- Occasionally
- Seldom or never

1.7 The Difference Between Descriptive Statistics and Inferential Statistics

So far, we have looked at the concept of population and recognized the usefulness of the sample in providing information about the population. We have discussed factors that influence how large the sample needs to be and how we go about picking the sample. And we have looked at the types of data that might be collected from the sample. Now we need an answer to the question, What do I do with the data? The right answer is, It depends. It depends on what type of data you have and on what questions you want answered about the behavior of the population. The rest of this book is devoted to providing tools to answer the question, What do I do with the data? These tools fall into two main categories: the tools of descriptive statistics and the tools of inferential statistics. Although these tools support each other, the jobs they do are quite different.

Descriptive statistics is often referred to as exploratory data analysis (EDA).

1.7.1 DESCRIPTIVE STATISTICS

The tools of descriptive statistics are usually the first ones used in any data analysis. They allow you to describe the sample; however, that is all they do. Description is generally the first step in any data analysis, after the data have been collected, of course. If you were given the results of 1000 surveys of parents about the DARE (Drug Abuse Resistance Education) program in your city, the first thing you would want to do is summarize the data. This is precisely what the **tools of descriptive statistics** do for you: summarize the data.

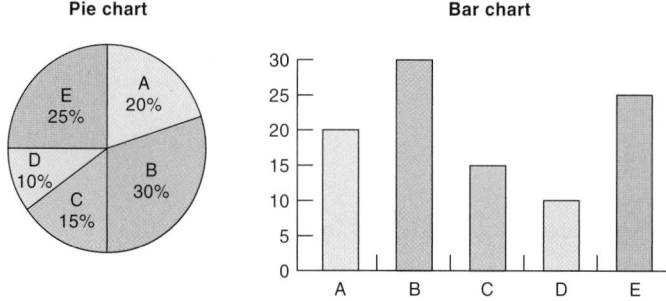

FIGURE 1.3 **Pie chart and bar chart**

The *tools of descriptive statistics* allow you to summarize the data.

Graphical and visual descriptive tools include bar charts, pie charts, and histograms. These tools are discussed in Chapter 2. Graphical tools help you to see how the data behave and to summarize the data visually. These tools are used all the time and appear frequently in newspapers and in reports. Two examples are shown in Figure 1.3.

Numerical descriptive tools allow you to summarize the data numerically. Typically, a numerical summary provides such statistics as the average, the median, the mode, and the largest and smallest data values. We used the word *statistics* in the last sentence because that is precisely what these values are. Remember that a statistic is simply a number that describes the sample. These numbers are explained in more detail in Chapter 3.

Descriptive tools would be adequate if all we wanted to do was describe the sample. But remember that the sample is a means to an end. In the final analysis, it is not the sample that we care about but the population. Somehow we must make the leap from the information contained in the sample to the behavior of the population. To do this we need the tools of inferential statistics.

1.7.2 INFERENTIAL STATISTICS

Let's start by defining the word **inference.** If your friend tells you not to infer that all teenagers are disrespectful, then your friend is telling you not to reach that conclusion!

An *inference* is a deduction or a conclusion.

We wish to draw conclusions about a population based on the information in the sample. We have already seen that the sample is only a piece of the population, so we do not have complete information. We have also seen that if the sample is properly selected, then the information contained in the sample will give us reliable information about the population.

How are we going to make this leap from describing the sample data to drawing an inference or conclusion about the behavior of the population? The answer is found in the study of probability. Indeed, without the tools of probability, we could not do anything more than describe the sample data. We could do no inferential statistics.

Think about the population and the sample as the circles shown in Figure 1.4 (page 40). The larger circle represents the population, and the smaller circle represents the sample. This figure illustrates how probability is used to draw inferences about the population from the sample. In reality, the sample circle should sit inside the population circle because it is a piece of it. But for now let us continue with this picture. The line that moves from the sample to the population is labeled "Inferential

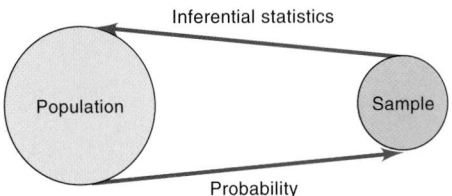

FIGURE 1.4 **Relationship between probability and inferential statistics**

statistics." The tools of inferential statistics are used to move from sample information to population information. Let's save that thought as a definition.

> The *techniques of inferential statistics* allow us to draw inferences or conclusions about the population from the sample.

To understand the line that moves from the population to the sample, consider the following situation. You and I have decided to play poker and I have dealt each of us five cards. You have a full house and are wildly excited. However, your excitement is quickly contained as I reveal to you that I have not one, not two, not three, not four, but five aces! Immediately, you accuse me of cheating (imagine that!). You conclude that I have not used a standard deck.

You reach a conclusion based on a sample. Did I cheat?

Let's take a closer look at how you reached that conclusion. The deck of cards is the population. You had a sample of five cards selected from that population, and I had a sample of five cards from that population. On the basis of the information in my sample, you calculated the probability of observing five aces if I was using a standard deck and found it to be zero. There is no chance of getting five aces if the population follows the behavior of a standard deck. So you quickly rejected the notion that I was using a standard deck. This is precisely what we need to do in general. We need to draw conclusions about a population based on the observed sample and the theory of probability. The conclusions will not always be so obvious as they were in this example. So we have a definition for the line in Figure 1.4 labeled "Probability," which leads from the population to the sample.

> We use *probability* theory to calculate the likelihood of observing or selecting a particular sample from a population.

Although the branch of mathematics known as probability theory is interesting in its own right, we present only those particular parts of probability that help us in our ultimate goal: to make inferences about the population from the sample.

Let's summarize Figure 1.4. As we make the trip across the bottom line, we make some assumptions about the population behavior and determine the likelihood of various samples that might come from this population. As we make the trip across the top line, we take the information in the sample and draw conclusions about the population from which it came. The tools of probability ultimately allow us to accomplish what we want, which is to do inferential statistics, the trip across the top line.

1.7.3 UPCOMING CHAPTERS

Based on our discussions in this section, it is clear that the first thing you should do is learn the tools of descriptive statistics. These are covered in Chapters 2, 3, and 4. Chapter 6 presents the necessary probability theory that helps us move beyond simply describing the sample data. Chapters 7–15 present various techniques of inferential statistics. The last chapter (Chapter 16) provides a synthesis of the book and a set of cases.

1.8 Ethical Issues in Data Analysis

In this text we caution you numerous times to present the data fairly and accurately. You can distort the message of the data by altering the scale of the graphs, by changing the level of α at which you perform the hypothesis test, by withholding data, and by other means.

The American Statistical Association (ASA) is one of the professional associations for statisticians. The ASA is concerned about the misuse of data and has published the following "Ethical Guidelines for Statistical Practice":

1. Statisticians have a public duty to maintain integrity in their professional work, particularly in the application of statistical skills to problems where private interests may inappropriately affect the development or application of statistical knowledge. For these reasons, statisticians should:

- *present their findings and interpretations honestly and objectively*
- *avoid untrue, deceptive, or undocumented statements*
- *disclose any financial or other interests that may affect, or appear to affect, their professional statements*

2. Recognizing that collecting data for a statistical inquiry may impose a burden on respondents, that it may be viewed by some as an invasion of privacy, and that it often involves legitimate confidentiality considerations, statisticians should:

- *collect only the data needed for the purpose of their inquiry*
- *inform each potential respondent about the general nature and sponsorship of the inquiry and the intended uses of the data*
- *establish their intentions, where pertinent, to protect the confidentiality of information collected from respondents and strive to ensure that these intentions realistically reflect their pledges of confidentiality and their limitations to the respondents*
- *ensure that the means are adequate to protect confidentiality to the extent pledged or intended, that processing and use of data conform with the pledges made, that appropriate care is taken with directly identifying information, that appropriate techniques are applied to control statistical disclosure*
- *ensure that, whenever data are transferred to other persons or organizations, this transfer conforms with the established confidentiality pledges, and require written assurance from the recipients of the data that the measures employed to protect confidentiality will be at least equal to those originally pledged*

3. Recognizing that statistical work must be visible and open to assessment with respect to quality and appropriateness to advance knowledge and that such assessment may involve an explanation of the assumptions, methodology, and data processing used, statisticians should:

- *delineate the boundaries of the inquiry as well as the boundaries of the statistical inferences that can be derived from it*
- *emphasize that statistical analysis may be an essential component of an inquiry and should be acknowledged in the same manner as other essential components*
- *be prepared to document data sources used in an inquiry, known inaccuracies in the data, and steps taken to correct or refine the data, statistical procedures applied to the data, and the assumptions required for their application*

- *make the data available for analysis by other responsible parties with appropriate safeguards for privacy concerns*
- *recognize that the selection of a statistical procedure may to some extent be a matter of judgment and that other statisticians may select alternative procedures*
- *direct any criticism of a statistical inquiry to the inquiry itself and not to the individuals conducting it*

4. Recognizing that a client or employer may be unfamiliar with statistical practice and be dependent upon the statistician for expert advice, statisticians should:
- *make clear their qualifications to undertake the statistical inquiry at hand*
- *inform a client or employer of all factors that may affect or conflict with their impartiality*
- *accept no contingency fee arrangements*
- *fulfill all commitments in any inquiry undertaken*
- *apply statistical procedures without concern for a favorable outcome*
- *state clearly, accurately, and completely to a client the characteristics of alternate statistical procedures along with the recommended methodology and the usefulness and implications of all possible approaches*
- *disclose no private information about or belonging to any present or former client without the client's approval*

1.9 Communicating the Results

Before concluding this chapter on the language of statistics, we should talk about the language you use to communicate your results. Remember that the last step of the problem-solving process is to communicate your results. In a sense you need to make your case based on your analysis. How should you do this? Should you show all of your analysis, or should you show only a summary of your findings? The best data analysis in the world is not worth much if it is not presented and explained clearly so that the person who must make decisions based on the data understands them. This section provides you with some general guidelines for clear communication.

First, understand your audience. Consider the technical background of the people you are talking to, and make your decisions about level of detail and degree of technical sophistication accordingly. Use language that is understandable.

Second, provide your audience with a written summary of your analysis. This is called an executive summary, and the first example is presented next. Each chapter of this book concludes with an executive summary. Use these as models. A summary should be at most one page long.

Third, use an electronic slide show tool, such as PowerPoint, to present the results. Be sure you use only the graphs and results necessary to make your case. Be sure all your graphs have titles and all the axes are properly labeled. Do not show every graph you created but only those that led you to your conclusions and recommendations. Likewise, show only relevant calculations. Begin your slide show with a slide that gives your name, the date, and the project title. Follow this with a slide that outlines your presentation. The results of your analysis should follow next, with summary slides after each point is made. Complete your presentation with a slide that shows your conclusions and recommendations.

GET IT IN WRITING

The College Snack Bar

TO: **Snack Bar Manager**
FROM: **Student Senate Committee**
RE: **Service at the campus snack bar**

Many students and faculty eat at the snack bar on campus. In the past semester people who regularly eat in the snack bar have noticed that it takes a long time to get their orders and often there are mistakes in the orders. Before making a big fuss about this based on only anecdotal data, a few people's stories, we would like permission to survey snack bar customers and collect data on service times and actual errors.

In order to get a fair understanding of the service at the snack bar, we propose to randomly sample ten customers at breakfast, lunch, and dinner on each day of the week (M–F) for one week. We will choose a typical week in the school calendar and not one with a holiday. This will help us to identify whether there is a problem with one particular time of the day or day of the week. We will sit in the snack bar and randomly select customers as they enter the line. We will have to record what they are wearing (unless we happen to know them) in order to be able to record their service time—from the time they place their order until the time their number is called to go back and pick up their food. We will also record their order number so that we can collect the slips from you later to determine what they ordered. Finally, we will ask the customers if they received the correct orders. It is recommended that we collect the following data:

Qualitative variables: Day of the week data are taken
Name of the person recording the data
Time of the day data are taken
What was ordered (grill vs. nongrill item)
Type of error

Quantitative variable: Time from order being placed to order received

Graphs and numerical summaries will be provided. Thank you for your cooperation in our efforts to improve the snack bar service.

1.10 Basic Summation Notation

In all of the material that follows this chapter, you will need to use basic summation notation. If you are familiar with this notation, you may skip this section. But you may wish to use it as a quick review and refer to it as you proceed through the rest of the book. A shorthand notation useful for writing statistical formulas is called **sigma notation**.

> *Sigma notation* is shorthand notation used to write formulas. It is so named because it uses the Greek capital letter sigma, written Σ.

1.10.1 SUMMING DATA VALUES

If we have a sample of size n, we can refer to the individual data values as $x_1, x_2, x_3, \ldots, x_n$, where x_1 represents the first data value, x_2 the second data value, and so on. Many of the formulas you encounter require that you sum the data values. We could write the sum as $x_1 + x_2 + x_3 + \cdots + x_n$, but this gets cumbersome and is awkward to use.

Remember that we use n to represent the size of the sample.

Using the sigma notation, we can write the sum $x_1 + x_2 + x_3 + \cdots + x_n$ as

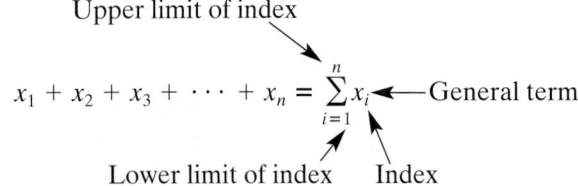

Upper limit of index

$$x_1 + x_2 + x_3 + \cdots + x_n = \sum_{i=1}^{n} x_i \longleftarrow \text{General term}$$

Lower limit of index   Index

The sigma notation on the right of the equation is read "the sum of x_i as i goes from 1 to n." The letter under the Σ is called the index; it really doesn't matter what letter you use because you use the same letter in describing the general term. The value of the index indicated under the Σ is the starting value for the index and is usually 1, meaning that we are going to start with x_1. The value above the Σ is the ending value for the index and is usually n, the sample size. When the index goes from $i = 1$ to n, all of the data values in the sample are being used.

Let's try some examples.

EXAMPLE 1.14

Collect and analyze the data.

THE COLLEGE SNACK BAR

Using Summation Notation

The students who are studying the service at the snack bar conducted a pilot study to finalize their data collection procedure. These are the service times (in minutes) they collected for ten customers: 23, 16, 5, 8, 25, 19, 10, 5, 4, 13. Thus, x_1 is 23, the first member of the sample. In a similar fashion we can label the rest of the data values: $x_2 = 16$, $x_3 = 5$, $x_4 = 8$, $x_5 = 25$, $x_6 = 19$, $x_7 = 10$, $x_8 = 5$, $x_9 = 4$, and $x_{10} = 13$. The value of n is 10, the size of the sample.

To find $x_1 + x_2 + x_3 + \cdots + x_{10}$ we first write it as $\sum_{i=1}^{10} x_i$. This means that we start the value of i at 1 and start with x_1, which we know is 23. We check to see whether the value for i is the upper limit for i, which in this case is 10. Since i is only 1, we continue. Next, we put a plus sign, change i to 2, and add x_2, which we know is 16. So far we have $23 + 16$. Again, we check to see whether the value for i is the upper limit of 10. It is not, so we continue by putting a plus sign, changing i to 3, and adding x_3, which we know is 5. Now we have $23 + 16 + 5$. This process continues until the value of i is 10 and we have written $23 + 16 + 5 + 8 + 25 + 19 + 10 + 5 + 4 + 13$ to get 128.

✔ TRY IT NOW!

Stats on the Slopes
Using Sigma Notation

A skier is looking at the wait time for a lift. She records the amount of time it takes from when she gets in line until she gets on the lift seat. For eight ski runs down the mountain, she collects the following times (in minutes):

31   28   29   32   30   31   29   30

Use sigma notation to write the expression that adds all the data values.

ANS. $\sum_{i=1}^{8} x_i$

1.10.2 SUMMING DIFFERENCES

When we analyze data, we often need to look at how far away data values are from some number. Clearly, we can determine how far away any one of the data values is from a number by simply subtracting the two numbers. For the snack bar data in Example 1.14, we can say that x_1 is 13 minutes away from the target service time of 10 minutes. Remember that x_1 is 23. In addition to being interested in how far away x_1 is from 10, we often have to state how far away the whole data set is from 10. To do this we have to subtract 10 from each value in the data set, each x_i, and then add up all these differences.

EXAMPLE 1.15

THE COLLEGE SNACK BAR

Using Sigma Notation to Calculate Differences

Suppose we wish to quantify how far away the whole snack bar data set is from 10 minutes. Here is the calculation we want:

$$(23 - 10) + (16 - 10) + (5 - 10) + (8 - 10) + (25 - 10) + (19 - 10) + (10 - 10)$$
$$+ (5 - 10) + (4 - 10) + (13 - 10)$$
$$= (13) + (6) + (-5) + (-2) + (15) + (9) + (0) + (-5) + (-6) + (3)$$
$$= 28$$

Using the notation x_1 to represent the first data value, x_2 to represent the second data value, and so on, we can write this expression more generally as

$$(x_1 - 10) + (x_2 - 10) + (x_3 - 10) + (x_4 - 10) + (x_5 - 10) + (x_6 - 10) + (x_7 - 10)$$
$$+ (x_8 - 10) + (x_9 - 10) + (x_{10} - 10)$$

Taking this one step further, we can use sigma notation to make this expression less cumbersome:

$$\sum_{i=1}^{10} (x_i - 10)$$

> **Collect and analyze the data.**

This shorthand way of writing the long summation will help us when we get to Chapter 3. Now you try one of these.

✔ TRY IT NOW!

Stats on the Slopes
Using Sigma Notation to Sum Differences

The ski resort advertises that the typical wait time for a chair lift is 30 minutes. Use sigma notation to write the expression that adds all the differences between the number 30 and the data values: 31, 28, 29, 32, 30, 31, 29, and 30.

ANS. $\sum_{i=1}^{8} (x_i - 30)$

1.10.3 EXERCISES—LEARNING IT!

1.28 From a sample of 15 towns in Massachusetts researchers recorded the numbers of reported burglaries in 2000:

305   61   8   2   77   72   13   355   104   1628   86   25   2   4   99

Write the sigma notation to represent the sum of the 15 data values.

1.29 A recent study by the Economic Poll Institute tells us that Americans are working more hours than they did in 1989. You decide to ask some of your friends how many hours a week they work, and you get these data:

42   34   45   22   62   30   45   46   40   37

(a) Write the sigma notation to represent the sum of these data values.

(b) Government figures show that the average workweek is 43 hours. Write the sigma notation to find the sum of the difference between each observation and 43.

(c) Compute the expression found in part (b).

1.30 Before selecting a major, a student decides to study the salaries of individuals who work as teachers. A sample of 15 kindergarten teachers is selected from the list of alumni of the school that the student attends. These are their salaries:

$25,100 $33,000 $27,500 $29,000 $35,000 $26,400 $29,100
$31,050 $32,100 $29,100 $40,000 $21,000 $25,000 $26,500 $33,000

(a) Write the sigma notation for the sum of all the salaries in the sample.

(b) Write the sigma notation for the sum of the first five salaries in the sample.

(c) Write the sigma notation for the sum of the last five salaries in the sample.

1.11 Selecting a Sample with Excel or Minitab

When selecting a random sample from a population, you can use either Microsoft Excel or Minitab. Refer to the manual that accompanies this book for more detailed instructions on how to use these software packages. An example of their capabilities follows.

Excel 2000

There are two ways to use Excel to select a simple random sample from a population. One method assumes you already have population values in a worksheet. Suppose you have a population with 50 two-digit ID numbers.

1. Use the **Data Analysis** tools in Excel. Select **Sampling** from the **Analysis Tools** list as in Figure 1.5. You must tell Excel three things: (1) the location of the population, (2) the type of sampling method and number of samples, and (3) where you want the sample placed.

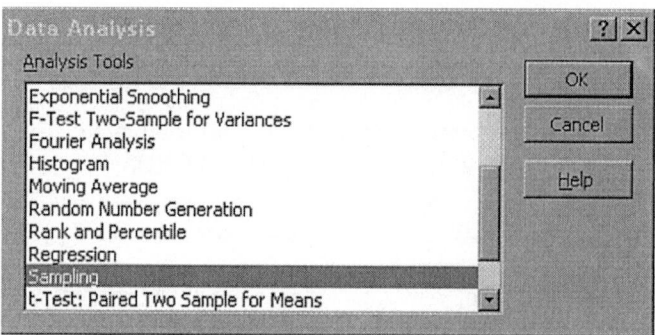

FIGURE 1.5   **Excel Data Analysis dialog box**

2. Position the cursor in the **Input Range** box and then highlight the range in the worksheet that contains the data.

3. In the sampling method section, click for the **Random** sampling method and type 5 in the text box for the **Number of Samples.**

4. In the **Output Options** section, you can specify that the sample be located in a section of the current worksheet, a new worksheet in the same workbook, or a new workbook. See Figure 1.6.

	A	B	C
1	ID Codes		42
2	1		11
3	2		10
4	3		40
5	4		46
6	5		
7	6		
8	7		

↓	C1
	ID Codes
1	1
2	2
3	3
4	4
5	5
6	6
7	7

FIGURE 1.6 Random sample of five ID codes in Excel

FIGURE 1.7 ID code data in Minitab

Minitab

You can easily use Minitab to select a random sample from a population. Suppose a Minitab worksheet contains a set of 50 two-digit ID numbers as shown in Figure 1.7.

1. From the Minitab menu, select ID **Calc>Random Data>Sample from columns.**

2. In the **Sample _____ rows from column(s):** box, enter the size of the sample you want to select—in this case, 5 (see Figure 1.8).

3. In the next box, select the column(s) that contain the population values.

4. Tell Minitab where to store the sample. You can type a column heading, like **C2,** or give the column a name, like **ID Sample.**

5. Click OK.

6. In this example, we didn't choose to sample with replacement. In most real cases you do not want to do that. What new information can you gain by sampling the same person or object twice?

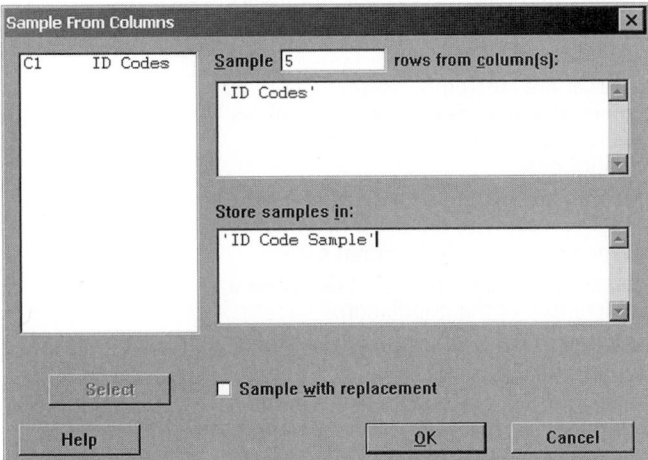

FIGURE 1.8 Completed Sample From Columns dialog box in Minitab

CHAPTER 1 SUMMARY

In this chapter you learned the basic language of statistics. You learned that the complete group you wish to study is called the population. Typically you want to know about several characteristics of each member of the population. These are called variables. The two types of variables are quantitative and qualitative. Ideally you would like to study the entire population or take what is known as a census, but most of the time a census is too expensive and too time-consuming. For these reasons and others you select a sample from a list of the members of the population. In doing so, you end up with sampling error because you are not studying every member of the population. The only way to eliminate sampling error is to study the entire population.

We always want our sample to be a fair representation of the population, or an unbiased sample. There are many ways to select the sample, but we use simple random samples in this book. You learned how to select a simple random sample using a table of random numbers or a software tool such as Excel. You were also introduced to a shorthand notation known as sigma notation, which is used to write equations throughout the remainder of this book.

KEY TERMS		
Term	**Definition**	**Page reference**
Biased sample	A **biased sample** is a sample that does not fairly represent the population.	27
Census	A **census** is a study of the entire population.	14
Continuous data	**Continuous data** can take on any one of an infinite number of possible values over an interval on the number line.	37
Descriptive statistics	The **tools of descriptive statistics** allow you to summarize the data.	39
Discrete data	**Discrete data** can take on only certain values. These values are often integers or whole numbers.	36
Inference	An **inference** is a deduction or a conclusion.	39
Inferential statistics	The **techniques of inferential statistics** allow us to draw inferences or conclusions about the population from the sample.	40
Likert scale	A **Likert scale** is a scale used to collect data on attitudes, including degree of agreement with a statement, frequency of use, importance of an issue, quality, and likelihood.	34
Nominal data	Data that are created by assigning numbers to different categories when the numbers have no real meaning are called **nominal data.**	34
Ordinal data	Data that are created by assigning numbers to categories where the order of assignment has meaning are called **ordinal data.**	34
Parameter	A **parameter** is a number that describes a characteristic of the population.	18
Population	The **population** is everyone or everything you wish to study.	12
Population size	The **size of the population** is the number of members of the population. Its symbol is N.	24
Probability	We use **probability** theory to calculate the likelihood of observing or selecting a particular sample from a population.	40

Term	Definition	Page reference
Qualitative data	**Qualitative data** describe a particular characteristic of a sample item. They are most often nonnumerical.	33
Quantitative data	Data that are inherently numerical in form are called **quantitative data**.	36
Sample	A **sample** is a piece of the population.	13
Sample size	The **size of the sample** is symbolized by n.	26
Sampling error	**Sampling error** is the difference between a characteristic of the entire population and a sample of that population.	14
Sampling frame	A **sampling frame** is a list of all members of the population.	28
Sigma notation	**Sigma notation** is shorthand notation used to write formulas. It is so named because it uses the Greek capital letter sigma, written as Σ.	43
Simple random sample	A **simple random sample** is a sample that has been selected in such a way that all members of the population have an equal chance of being chosen. In addition, every sample of size n has the same chance of becoming the sample.	27
Statistic	A **statistic** is a number that describes a characteristic of a sample.	19
Table of random numbers	A **table of random numbers** is a list of numbers randomly generated and listed in the order they are generated.	28
Variable	A **variable** is a characteristic of each member of the population	12
Variation	The amount of **variation** refers to how different the members of the population are from one another with regard to the variable being studied.	15

CHAPTER 1 EXERCISES

Learning It!

1.31 State senators and representatives are evaluating the effectiveness of the drug education program in schools called DARE (Drug Abuse Resistance Education).

(a) What is the population of interest?

(b) Identify which of the reasons for taking a sample (listed on page 13) apply in this case. (There are more than one.)

(c) What type of data would you collect?

1.32 A bank currently offers its customers online banking but is concerned about security in this electronic environment. The bank wants to determine how much time to allow an account to be idle (logged on but no activity registered) before timing out the account.

(a) What is the population of interest?

(b) Identify which of the reasons for taking a sample (listed on page 13) apply in this case.

(c) What type of data would you collect?

(d) How would you get a sample to address this issue?

1.33 You decide to take a sample of gas prices. You visit five stations in your town and price a gallon of unleaded gas at each station.

(a) Why might this sample be biased?

(b) How could you get a simple random sample?

1.34 Continuing with Exercise 1.28, you learn that the researchers chose the sample of 15 towns in Massachusetts by using an alphabetical list of towns and choosing every other one from the first 30 towns listed. What do you think of this sampling procedure? Is this a representative sample of the towns in the state of Massachusetts? Explain.

1.35 You are trying to decide where to live after you graduate from college. You would like to live somewhere where there are a lot of young people. You collect the following data on the percentages of adults between ages 20 and 24 in seven states across the nation:

Vermont	6.2%	Maryland	5.9%
Illinois	6.9%	Connecticut	5.5%
Virginia	6.8%	Utah	10.1%
Nevada	6.5%		

(a) Write the sigma notation for the sum of the data.

(b) For the entire country, 6.7% of the population is between ages 20 and 24. Write the summation notation to find the sum of the differences of the percentages listed from the national value of 6.7%.

(c) Find the numerical value of the expression in part (b).

1.36 The survey reprinted on pages 51–54 was used to collect some information on online shopping. For each question, indicate what type of data will be collected: qualitative (nominal or ordinal) or quantitative (discrete or continuous).

Thinking About It!

1.37 Will your information be accurate if you study every member of a population? Why or why not?

1.38 In trying to understand a population, what is the only way to eliminate the possibility of reaching the wrong conclusion?

1.39 Is it likely that the value of the statistic calculated from the sample is exactly the same as the value of the population parameter?

1.40 Remembering that the percentage of voters who support the President of the United States is an unknown number, what might you suggest using as a best guess for this percentage?

1.41 This classic story illustrates how the wrong inferences can be reached if the sample is biased. Just before the 1948 presidential election between Harry Truman and Thomas E. Dewey, the *Chicago Daily Tribune* selected a national sample of voters from the telephone book. These voters overwhelmingly supported Dewey in the presidential race, and so the newspaper confidently printed the headline "Dewey Defeats Truman" before the results were finally tallied. As we know, the headline was wrong: Truman became the 33rd president in 1945. What mistake did the *Chicago Daily Tribune* make?

1.42 Recall that you want your sample to be a miniversion of the population. Candidates for president think they have found a sample that is a miniversion of the population. Republican and Democratic leaders believe that whoever wins in Michigan, a state with a demographic makeup that closely mirrors the nation, will be the next president. Examine the following evidence.

1998 data	**Michigan**	**United States**
Per capita income	$14,154	$18,685
Unemployment rate	3.9%	4.5%
Median age	34.6	34.6
Population over age 65	12.5%	12.8%
Minority populations		
Blacks	13.9%	12.1%
Hispanics	2%	9%
Asians	1.1%	2.9%
American Indians	0.6%	0.8%
Portion of workforce in unions	21.6%	13.9%*
Median house value	$60,600	$84,209
Married-couple households	55.1%	55.9%

*Sources: The Almanac of American Politics; *USA Today research; 1999 data.

(a) What do you think about this claim?

(b) What other information would you like to know to evaluate the claim?

(*Exercises continue on page 54.*)

Online Shopping Survey

Information about consumer online shopping experiences is being collected as an exercise for my electronic marketing course at Western New England College. Thank you for your participation in this project.

Section 1: Internet Usage

1. **What is the most important reason you use the Internet and World Wide Web?**

   _____

   _____

2. **From where do you <u>most often</u> access the Internet and World Wide Web?**

   ☐ Work   ☐ Home   ☐ Other:_____

3. **How much time do you spend on the Internet per week?**

   ☐ Less than 1 hour   ☐ 8 to 10 hours
   ☐ 1 to 3 hours   ☐ 10 to 15 hours
   ☐ 4 to 7 hours   ☐ More than 15 hours

4. **Have you ever purchased products or services from an Internet retailer?**

   ☐ Yes (*please proceed to Section 2*)   ☐ No (*please answer the question below and SKIP to Section 3*)

   If no, what is your most important reason for not shopping online?

   ☐ Concerns about security   ☐ Inconvenience
   ☐ Shipping charges   ☐ Concerns about privacy
   ☐ No opportunity   ☐ Other:_____

Section 2:   Online Shopping Experiences

5. **What is your primary reason for purchasing a product or service online? (check one)**

   ☐ Reputation of the company   ☐ Price
   ☐ Reliability   ☐ Service
   ☐ Security of your information   ☐ Product or service not available offline

   ☐ Other:_____

6. **What types of products do you purchase online? (please check all that apply)**

   ☐ Consumer electronics   ☐ Sporting goods   ☐ Software
   ☐ Tickets for travel   ☐ Apparel   ☐ Toys
   ☐ Gifts   ☐ Books   ☐ Food
   ☐ Music   ☐ Other: _____

7. **Listed below are factors that may be important to you when deciding to purchase from an online retailer. Please rank order the top five factors in order of importance from 1 (most important) to 5 (least important)**

   ___ Product selection   ___ Shipping and handling charges
   ___ Quality of product information   ___ Delivery options
   ___ Quantity of product information   ___ Customer support
   ___ Prices relative to similar online stores   ___ Friend or family recommendation
   ___ Ease of website navigation   ___ Privacy policy
   ___ Trust

8. **Do you comparison shop the price of the product or service at different websites offering the same, or a similar, product or service?**

☐ Yes ☐ No

If yes, how many sites did you look at? _____

How did you find these sites?_____

9. **Have you ever purchased a product or service online as a result of the following factors? (check all that apply)**

☐ Web advertisement you saw online
☐ Promotional advertisement via e-mail
☐ Friend or relative recommendation
☐ Advertisement for a website on TV or radio
☐ Advertisement for a website in newspapers or magazines
☐ Other:_____

10. **For what occasions are you likely to purchase products or services from online retailers? (check all that apply)**

☐ Personal use
☐ Family or household use
☐ Gift-giving occasions, nonholiday (birthdays, anniversaries, etc.)
☐ Holiday gift-giving occasions
☐ Other:_____

11. **How much do you spend with online retailers per month?**

☐ Less than $50   ☐ $50–$100   ☐ $101–$200   ☐ More than $200

12. **How long, on average, does it take you to complete an online purchase?**

_____

13. **Overall, how would you evaluate your online shopping experiences?**

☐ Delightful
☐ Very satisfactory
☐ Satisfactory
☐ Unsatisfactory
☐ Very unsatisfactory

14. **What can online retailers do to increase your online shopping activities?**

_____

_____

_____

15. **Please rate your general satisfaction with online shopping using the factors listed below:**

	Very Satisfied	Moderately Satisfied	Somewhat Satisfied	Somewhat Dissatisfied	Moderately Dissatisfied	Very Dissatisfied
Ease of ordering.............................. (convenience and speed of ordering)	☐	☐	☐	☐	☐	☐
Product selection.............................. (breadth and depth of products offered)	☐	☐	☐	☐	☐	☐
Product information......................... (information quality, quantity, and relevance)	☐	☐	☐	☐	☐	☐
Website navigation and looks........... (speed of site, quality of layout, etc.)	☐	☐	☐	☐	☐	☐
On-time delivery.............................. (expected versus actual delivery date)	☐	☐	☐	☐	☐	☐

	Very Satisfied	Moderately Satisfied	Somewhat Satisfied	Somewhat Dissatisfied	Moderately Dissatisfied	Very Dissatisfied
Product representation..................................... (product description versus what was actually received)	☐	☐	☐	☐	☐	☐
Quality of customer support........................... (status updates, complaint and question handling)	☐	☐	☐	☐	☐	☐
Posted privacy policies..................................... (efforts to inform you of policies)	☐	☐	☐	☐	☐	☐
Product shipping and handling....................... (appropriateness and condition of packaging)	☐	☐	☐	☐	☐	☐

16. **How secure do you feel about providing the following personal information online?**

    Name and address   Very Secure   ____   ____   ____   ____   ____   Not At All Secure

    Credit card information   Very Secure   ____   ____   ____   ____   ____   Not At All Secure

17. **What can businesses that sell products and services online do to earn your trust?**

    _____

    _____

    _____

18. **Indicate the extent to which you agree or disagree with the following statements:**

	Strongly Disagree	Disagree	Neither Agree Nor Disagree	Agree	Strongly Agree
Online shopping allows you to obtain greater value than shopping at offline, physical retail outlets..	1	2	3	4	5
In terms of satisfaction, shopping at offline, physical retail outlets is more satisfying than online shopping......................................	1	2	3	4	5
I am more confident when shopping online than I am when shopping at offline, physical retail outlets..	1	2	3	4	5

Section 3: Classification Information

---

19. **Do you plan to purchase products or services from online retailers in the future?**

    ☐ Yes   ☐ No

    Why? _____

    _____

20. **Highest level of education you have completed is:**

    ☐ High school   ☐ Attended college but did not graduate
    ☐ Post-secondary technical school   ☐ 4-year college degree
    ☐ 2-year college degree   ☐ Graduate study

21. **Occupation**_____

22 **Are You:** ☐ Single ☐ Married ☐ Other:_____

23. **Do you have children?** ☐ Yes ☐ No

      If yes, do you have children: ☐ Under age 5 ☐ Ages 14 to 18
                                              ☐ Ages 5 to 9 ☐ Over 18
                                              ☐ Ages 10 to 13

24. **Are you:** ☐ Male ☐ Female

25. **Your age is:** ☐ 18 to 24 ☐ 35 to 44 ☐ 55 to 64
                              ☐ 25 to 34 ☐ 45 to 54 ☐ 65 and over

26. **Please express any other comments, observations, or feelings you have about online shopping in the space below.**

_____

_____

_____

_____

_____

THANK YOU FOR COMPLETING THIS SURVEY

(Continued from page 50.)

Doing It!

1.43 Three different student teams have developed surveys to study the snack bar service at their university. The surveys are shown here.

TEAM A

1. How do you rate the speed of food service?

Poor	Decent	Good	Very Good	Excellent
1	2	3	4	5

2. How do you rate the training of the order taker?

Poor	Decent	Good	Very Good	Excellent
1	2	3	4	5

3. How do you rate the training of the grill chefs?

Poor	Decent	Good	Very Good	Excellent
1	2	3	4	5

4. How do you rate the training of the cashiers?

Poor	Decent	Good	Very Good	Excellent
1	2	3	4	5

5. Should snack bar workers have a regular position until they become proficient at that position?
Yes   No

TEAM B

Please answer the following questions about the last time you ordered food in the snack bar during dinner hours (5–7 p.m.)

1. What time did you order?

2. How busy was the snack bar at that time?

    a. Not at all busy (0–4 people in line)   c. Pretty busy (10–15 people in line)
    b. Somewhat busy (5–10 people in line)   d. Very busy (more than 15 people in line)

3. What did you order?

    a. Grilled item   c. Soup/salad or other prepared item
    b. Nongrilled sandwich   d. Other:_____

4. How long did it take from the time you entered the line until the time you paid for and got your food?
   a. Under 5 minutes   c. 10–15 minutes
   b. 5–10 minutes   d. Over 15 minutes

5. What would you suggest be done to speed up the service?

TEAM C

The information collected in this survey will be used to try and correct the issues dealing with the inadequacy of orders at the snack bar. If you do not eat any meals at the snack bar, please do not fill out this survey.

Scale used to measure inadequacies:

1–Never   2–Sometimes   3–Often   4–Always

Circle the appropriate number.

How many times has this happened with your order?
Paid a different price for an identical order

1	2	3	4
N	ST	O	A

Asked for a special request that was not granted
(Example: no cheese, add tomato, toast the roll)

1	2	3	4
N	ST	O	A

Received the wrong order

1	2	3	4
N	ST	O	A

Got an incorrect order
(Example: ordered a large fry and got a small)

1	2	3	4
N	ST	O	A

Got a meal prepared improperly
(Example: over- or undercooked, turkey club not sliced in fours, soup not hot)

1	2	3	4
N	ST	O	A

(a) Each student team seems to be focusing on a different aspect of the problems at the snack bar. Summarize the main area of study for each questionnaire.

(b) For each question in each of the survey instruments, identify the type of data that will result.

(c) Each of the questionnaires has some room for improvement. Critique each one and suggest changes to the existing questions.

(d) Suggest a way to select a simple random sample of snack bar customers to survey for each team.

(e) Prepare a memo to go to the director of food services with a suggested sampling plan and one survey instrument that represents the combined concerns of all three of these student teams.

1.44 More and more surveys are being conducted online as more and more consumers become Internet users. In this exercise you will take an online survey. Go to the web address http://www.survey.net/ and scroll down until you see the words Society and Entertainment. Click on it and take one of the surveys. Then answer these questions:

(a) What is the population of interest?

(b) Do you think this online survey will yield an unbiased sample of this population? Why or why not?

(c) What are the variables of interest?

(d) What kind of data will be generated from each question in this survey?

(e) Do you think there is any need to be concerned about people lying in their responses to this survey?

CHAPTER 2

Graphical
Displays of Data

AM I PAID FAIRLY?

Are women really paid less than men for comparable work? Many newspaper and magazine articles suggest that gender bias exists in different fields. One recent focus has been on the field of higher education. At a small public university, ABC College, a committee was formed to study whether or not the claim of gender bias was true. The committee decided that one area they wanted to study was salary, so they determined that their first step must be to collect some data on faculty salaries.

Knowing that faculty salary can depend on several different variables, the committee asked the university administration to gather data for each faculty member on current rank (Professor, Associate Professor, Assistant Professor, and Instructor), number of years of service to the college, and current salary. The data, which were provided in the form of a list, are not very enlightening. A portion of the data follows:

Rank	Years of service	1996–97 Salary
ASST	22	$53,316
PROF	11	64,375
ASSO	7	63,501
ASSO	6	59,426
ASSO	20	49,058
PROF	4	94,969
ASST	21	54,762
ASSO	9	55,516

To gain any information from the data, the committee will need to summarize them, probably with some graphical displays.

In Chapter 1 you learned about the different types of data and about how data can and should be collected. You learned that people collect data to provide information about something they are interested in understanding, such as why it takes so long to fill an order at the snack bar or why different scooters have accidents. Usually there is some problem that needs to be solved, like a new scooter design, as a result of this understanding.

In this chapter we look at ways to display different types of data. The chapter covers the following material:

- Graphical methods for qualitative data: frequency tables, bar charts, and pie charts
- Graphical methods for quantitative data: frequency tables and histograms
- Other graphical methods: dotplots

2.2 Organizing Data

If one objective of statistics is to obtain information about a set of data, then we need to organize the data in some way. When data are collected, the initial result is usually a list of the observations for each variable. This is referred to as the *raw data*. Raw data, such as those given to the university committee that is studying gender bias, provide little information. Statistics gives us some tools or techniques for turning raw data into useful information.

2.2.1 THE FREQUENCY DISTRIBUTION

One way to organize data is to consolidate them by determining how many times each value occurs in the data and making a table that summarizes this information. The result is called a **frequency table** or **frequency distribution.** A frequency distribution divides the data into categories or classes. These terms are used interchangeably, although *categories* usually refer to qualitative data and *classes* to quantitative data.

> A *frequency table* or *frequency distribution* is a table that records each category, value, or class of values that a variable might have and the number of times that each one occurs in the data. The frequency of the ith class is denoted by f_i.

2.2.2 FREQUENCY TABLES FOR QUALITATIVE DATA

Creating a frequency table for qualitative data is not difficult. A basic frequency table has two columns. In the first column, each row lists one of the values for the variable of interest. The second column lists the corresponding number of times that the value occurs in the set of data. Later on we will add some additional columns to the table, but right now we will work with only two. Table 2.1 (page 58) is a typical frequency table.

TABLE 2.1

A FREQUENCY TABLE

Category	Frequency
Category 1	f_1
Category 2	f_2
Total	n

EXAMPLE 2.1

ABC FACULTY SALARIES

Creating a Frequency Table

Understand the problem.

To see whether women are indeed paid less than men, the committee at ABC College needs to understand how the current faculty is made up. They know that several variables, or factors, affect salary. One of these factors is the current rank of the faculty member. They decide to look at the entire faculty by rank to see how it is made up. They put the data in the form of a frequency table:

Analyze the data.

Rank	Frequency
Professor	55
Associate Professor	67
Assistant Professor	77
Instructor	8
Total	**207**

From the frequency table the committee sees that, although Assistant Professor has the highest frequency, the three highest ranks do not differ much. Not many faculty have the rank of Instructor.

The order for the categories in the frequency table is not important. If there is a logical order, or if you know what the classifications will be before you fill out the table, then you might use a particular order. If you are creating the categories as you make the table, you might list them in the order in which they first appear in the raw data.

EXAMPLE 2.2

STUDENT DISTRIBUTION

Creating a Frequency Table

The dean of the School of Arts and Sciences at a university is concerned that many students delay taking the Introductory Statistics course required for many of the majors. Delays in taking this course mean that students cannot take other required courses in their majors, which in turn causes scheduling problems. The course can be taken as early as the freshman year, but traditionally it is taken by sophomores. To determine whether his concerns are justified, the dean selects a section of Introductory Statistics at random and classifies each student as a freshman (F), sophomore (S), junior (J), or senior (Sr) according to the number of credits completed to date. Here are the raw data for the 28 students:

Understand the problem.

Sr	Sr	S	S	S	Sr	Sr
S	S	S	S	S	F	F
Sr	S	Sr	Sr	Sr	F	F
J	S	J	S	S	Sr	Sr

Collect the data.

To use the data, the dean organizes them in the form of a frequency table:

Class	Frequency
Freshman	4
Sophomore	12
Junior	2
Senior	10
Total	**28**

Analyze the data.

The table shows that very few students who take the course are freshmen or juniors and the rest are about evenly divided between sophomores and seniors.

Draw conclusions.

Although the information we get from a frequency table is certainly better than the information from raw data, it is still not as informative as we would like. The values in a frequency table are influenced by the sample size. For large samples the individual frequencies are much higher numbers than for small samples. This makes comparisons of different samples difficult. To solve this problem we must find a way to express the frequency so that the sample size does not matter. One solution is to use the **relative frequency.**

The *relative frequency* of a classification is the number of times an observation falls into that classification represented as a portion of the total number of observations. Relative frequency can be expressed as a fraction, decimal, or percentage.

Thus, to find the relative frequency for the ith classification, rf_i, we use the equation

$$rf_i = \frac{\text{Frequency of the } i\text{th classification}}{\text{Sample size}} = \frac{f_i}{n}$$

EXAMPLE 2.3

ABC FACULTY SALARIES

Finding Relative Frequencies

To get a better description of how the faculty are distributed among the ranks, the committee decides to calculate the relative frequency for each rank. The calculations and results are shown in the table:

Analyze the data.

Rank	Frequency	Relative frequency	Relative frequency (%)
Professor	55	55/207	26.6
Associate Professor	67	67/207	32.4

(continued)

Note: If you add the percentages for relative frequency, you do not get 100% here because the percentages have been rounded.

Note: Use the actual variable name for the classification to provide the most information.

Draw conclusions.

Assistant Professor	77	77/207	37.2
Instructor	8	8/207	3.9
Total	**207**	**207/207**	**100.0**

It appears from the relative frequencies that the three largest classes are not very different and no single class constitutes a majority.

It does not matter whether you use a fraction, decimal, or percentage to express the relative frequency. Percentages are easiest for most people to understand. When using percentages, you need to be careful about rounding. When you report relative frequency to the nearest percent or even one decimal place, the values may not sum to exactly 100% because of rounding. This is not a problem.

EXAMPLE 2.4

STUDENT DISTRIBUTION

Calculating Relative Frequencies

To get a better picture of the students in the Introductory Statistics course, the dean of the School of Arts and Sciences who is studying the data decides to look at the relative frequency for each class.

If these calculations are not easy for you, then you should get out a calculator and practice them until they are!

Freshmen: $rf_1 = 4/28 = 0.143$ or as a percentage $0.143 \times 100 = 14.3\%$

Sophomores: $rf_2 = 12/28 = 0.429$ or as a percentage $0.429 \times 100 = 42.9\%$

Juniors: $rf_3 = 2/28 = 0.071$ or as a percentage $0.071 \times 100 = 7.1\%$

Seniors: $rf_4 = 10/28 = 0.357$ or as a percentage $0.357 \times 100 = 35.7\%$

Adding a new column to the frequency table gives this table:

Analyze the data.

Class	Frequency	Relative frequency (%)
Freshman	4	14.3
Sophomore	12	42.9
Junior	2	7.1
Senior	10	35.7
Total	**28**	**100.0**

Draw conclusions.

Now the dean can see that 35.7% of the students in the class are seniors and that 35.7% + 7.1% = 42.8% of the students delayed the course beyond their sophomore year.

✔ TRY IT NOW!

Student Grades
Creating a Frequency Table

A professor in an Introductory Statistics course knows that, although students dread taking the course, they have unusually high expectations for their performance. She surveys (anonymously, of course) her students and asks them what grade they expect to get in the course. Here are the raw data:

A	C	B	A	A	B	B
B	A	B	A	A	C	B
B	A	F	C	B	D	C
B	B	D	B	A	B	A

Make a frequency table for the data, including both frequencies and relative frequencies.

2.2.3 FREQUENCY TABLES FOR QUANTITATIVE DATA

Frequency Tables for Integer Data

Remember from Chapter 1 that one type of quantitative data you might collect results from *counting* the number of times that something occurs in a set of data, or from *ranking* or *rating* objects. This type of data takes on integer values.

Creating a frequency table for integer data is exactly the same as creating one for qualitative data. Each value of the variable represents one category or classification in the table. The only real difference is that because the numbers in integer data have some meaning, the categories must be in numerical order. For qualitative data, the order in which you list the categories is not rigid.

EXAMPLE 2.5

STUDENT ATTITUDES

Creating a Frequency Table for Integer Data

The statistics professor who has been trying to learn about her class is also interested in her students' attitudes toward the study of statistics. She is aware that most of the students are enrolled in the course because it is required, and she wonders if this is reflected in their attitude toward the subject. As part of her anonymous survey on grade expectations, she asks the class to rank "the importance of statistics in my life" on a scale of 0 to 10, where 0 = of no importance whatsoever and 10 = the most important thing I will ever study. These are the raw data from the study:

0	10	8	0	0	8	6
5	1	6	1	4	10	8
7	0	10	9	4	10	10
7	6	10	9	1	5	1

Since the raw data are not very informative, she decides to create a frequency table for the data:

Rating	Frequency	Relative frequency (%)
0	4	4/28 = 14.3
1	4	4/28 = 14.3
2	0	0/28 = 0.0
3	0	0/28 = 0.0
4	2	2/28 = 7.1

(continued)

Understand the problem.

Collect the data.

Analyze the data.

5	2	2/28 = 7.1
6	3	3/28 = 10.7
7	2	2/28 = 7.1
8	3	3/28 = 10.7
9	2	2/28 = 7.1
10	6	6/28 = 21.4
Total	**28**	**28/28 = 100**

Draw conclusions.

From the frequency table the professor sees that most students have attitudes on either end of the scale, with fewer in the center neutral zone.

Most data that are truly integer in nature result from counting, rankings, or ratings. Sometimes data that are really measurements are recorded as integers. In this case it is important to think about where the data came from and the number of possible values before you pick the classes for your frequency table. Since the data are numbers, you cannot leave out a class (value) if there are no observations in it. If the number of possible values is much larger than 15 or 20, you probably do not want to treat the data as integer data. Instead you should use the methods for continuous data shown later in this section.

EXAMPLE 2.6

ABC FACULTY SALARIES

Looking at Integer Data

The committee that is looking at faculty salaries knows that salary also depends on the length of time a faculty member has been at the college. They decide to summarize these data using a frequency table, and they notice that the data are integer data.

Analyze the data.

They also notice that the shortest time of service is 0 years and the longest is 41 years. That means that 42 different values could appear in the data. The committee realizes that time is actually a measurement and it is integer data because it is recorded to the nearest whole year. They see that treating these numbers as integer data will not be appropriate.

Since quantitative data are ordered by nature, it is interesting to ask questions like, What percentage of the class rates statistics on the bottom half of the scale? Neutral? The top half? This is a good way to analyze data that come from the Likert scale (Strongly Agree to Strongly Disagree) that you learned about in Chapter 1. You can answer such questions by summing the relative frequencies or by using the **cumulative relative frequency.**

The *cumulative relative frequency* of a class is the sum of the relative frequencies of all classes at or below that class represented as a portion of the total number of observations. It can be expressed as a fraction, decimal, or percentage.

EXAMPLE 2.7

STUDENT ATTITUDES

Calculating Cumulative Frequencies

For her data on student attitudes toward statistics, the professor calculated the cumulative frequency for each class:

Rating	Frequency	Cumulative relative frequency (%)
0	4	4/28 = 14.3
1	4	8/28 = 28.6
2	0	8/28 = 28.6
3	0	8/28 = 28.6
4	2	10/28 = 35.7
5	2	12/28 = 42.9
6	3	15/28 = 53.6
7	2	17/28 = 60.7
8	3	20/28 = 71.4
9	2	22/28 = 78.6
10	6	28/28 = 100.00
Total	**28**	

Analyze the data.

From the table, the professor can easily see that 35.7% of the students have negative attitudes toward statistics. To find the percentage of students who have neutral attitudes, she must do some arithmetic. From the table she can see that 42.9% of the students gave ratings from 5 down. Since she knows that 35.7% gave ratings from 4 down, she can find that 42.9% − 35.7% = 7.2% have a neutral attitude toward statistics. This agrees with the relative frequency from Example 2.5.

Draw conclusions.

To find the percentage of students who have positive attitudes, again a little arithmetic is needed. Since the professor knows that 42.9% gave ratings from 5 down, the remainder must have given ratings from 6 up. She finds that 100% − 42.9% = 57.1% have positive attitudes toward statistics. Teaching this class may be an uphill battle!

At this point you must be thinking that it would be simpler to use the relative frequencies and just add up the ones you are interested in, like you did for the qualitative data. Sometimes this is true, but when there are a lot of classes, the procedure becomes very tedious. There are two other reasons for learning how to deal with cumulative relative frequencies: (1) When you are using statistics or spreadsheet software to analyze data, the cumulative frequencies are part of the output, which makes it a much easier job; and (2) certain statistical tables such as probability tables are cumulative and there is just no other way to use them.

✔ TRY IT NOW!

New Product Survey
Using Cumulative Relative Frequencies

A marketing research firm conducted a survey of consumers who invariably use a particular brand of bath soap. The consumers were given a competitor's version of the bath soap with nonallergenic enhancements and asked whether they would consider buying the new product. Their answers were ratings on a scale of 1 to 5, where 1 = would not ever buy this product and 5 = will buy this product immediately. These are the raw data from the survey:

5	4	1	4	2	2
3	1	4	4	2	4
3	1	3	3	5	4
4	4	2	3	2	5
4	3	4	4	1	5

Create a frequency table for the data that includes both frequency and cumulative relative frequency.

Which ratings indicate a negative attitude toward the new product? What percentage of the consumers surveyed had a negative attitude?

Which rating indicates a neutral attitude? What percentage of the consumers surveyed had a neutral attitude?

Which ratings indicate a positive attitude? What percentage had a positive attitude?

Frequency Tables for Continuous Data

Another type of quantitative data that is frequently encountered in statistics is measurement, or continuous, data. This type of data can take on many different values in a sample. For this reason, we cannot use each value in the sample as a category for the frequency table. It is possible that each value will occur only once in the sample! The result would be entirely too many classes for the frequency table and no real information. Thus, before we can create a frequency table for measurement data, we must figure out how to define the categories or classes for the data.

Two questions must be answered to create the frequency table for continuous (measurement) data:

Question 1: How many classes should be in the table?

Question 2: How large (wide) should each class be?

Although there are no absolute rules for creating the frequency classes for continuous data, you can apply some guidelines to answer the questions.

Question 1: There are two basic rules for determining the number of classes to include in the frequency table.

Rule 1: The number of classes in the frequency table should be approximately equal to the square root of the sample size, n; that is,

$$\text{Number of classes} = \sqrt{n}$$

You can see already that this is just a guideline because the calculation may not result in a whole number. When that happens, we usually just take the integer part of the answer. For example, for a sample size of $n = 30$, we get $\sqrt{30} = 5.477$ so we would use five classes.

Rule 2: The number of classes should not be less than 5 or more than 20. The reasons for this rule have to do with the kind of information that we are looking for in the frequency table. The reasons will become more obvious when we discuss the graphical methods for displaying data.

Once we have decided on the number of classes to use in the frequency table, we can answer the second question.

ANS. RF: 13.3%, 16.7%, 20.0%, 36.7%, 13.3%; CRF: 13.3%, 30.0%, 50.0%, 86.7%, 100.0%; 1 AND 2, 30.0%; 3, 20.0%; 4 AND 5, 50%

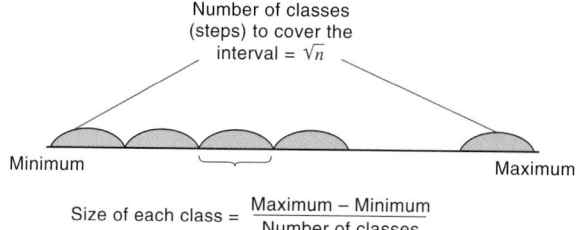

FIGURE 2.1 **Number of classes and class interval for continuous data**

Question 2: To determine how large each class should be, we need to think about what the table is supposed to accomplish. Once the classes have been determined, each piece of data in the sample must be counted in one and only one of the classes. Therefore:

1. The classes must span the entire data set.
2. The classes must not overlap.

Think about the number of classes as the number of steps you can take to get from the smallest data value (minimum) to the largest data value (maximum). Then the size of each class should equal

$$\frac{\text{Maximum} - \text{Minimum}}{\text{Number of classes}}$$

Figure 2.1 illustrates the preceding rules.

After the number of classes and the size of each class are determined, we can define each class precisely. That is, we can specify the smallest and largest data values that will fall into each class. Now is probably a good time for an example.

EXAMPLE 2.8

ABC FACULTY SALARIES

Setting Up Class Intervals for Continuous Data

The committee that is studying gender equity in salaries wants to know about the current faculty as a whole. The first variable they choose to look at is the number of years of service that a faculty member has at the university, so they create a frequency table for those data. Although they will probably use a computer to do the final analysis, they also know that some software asks the user for input on the number of classes to use as well as the width or starting points.

The committee knows that the sample contains 207 faculty members, so they calculate $\sqrt{207} = 14.39 \cong 14$ and decide to use about 14 classes. Since the highest data point is 41 years and the lowest is 0, the width of each class should be

$$\text{Width} = \frac{14 - 0}{14} = 2.93 \text{ years}$$

Analyze the data.

Note: Round results to a reasonable number of decimal places, usually one or two more than the data have.

EXAMPLE 2.9

FLIGHT DELAYS

Setting Up Class Intervals for Continuous Data

As more and more people take to the skies in airline travel, it seems that more and more flights do not leave on schedule. In an effort to understand how bad the problem really is, a consumer group decides to look at the percentages of delayed flights on 30 days for a major airline. The data for the month of April 2001 are listed here:

Understand the problem.

Collect the data.

Analyze the data.

27.4	18.1	26.1	17.6	11.1
17.1	23.8	11.5	15.0	10.5
41.9	21.0	24.6	10.7	8.5
20.2	20.1	15.6	19.7	5.1
22.4	23.7	14.5	10.9	8.6
34.7	39.2	16.8	13.5	18.6

To make some sense of the data, the consumer group decides to use a frequency table. Before they can create the table, however, they need to know how many classes to use. The group's first calculation is

$$\sqrt{30} = 5.48$$

and so they decide to use five classes.

Next they must decide how large each class should be. The data show that the maximum value is 41.9% and the minimum value is 5.1%. The total distance that needs to be covered by the classes is

$$41.9 - 5.1 = 36.8$$

so the size of each class must be

$$\frac{36.8}{5} = 7.36$$

To determine the starting and ending points (boundaries) for each class, we must remember that the classes cannot overlap and there can be no gaps in the interval. There are several ways to accomplish this, but we define a class as containing all observations from the lower boundary up to *but not including* the upper boundary:

$$\text{Lower boundary} \leq x < \text{Upper boundary}$$

where x represents the value of the variable being studied.

Different texts use different methods for defining intervals, but they all yield similar results. The interval definition used here is consistent with most major statistical packages such as Minitab and SPSS. Some spreadsheets, such as Excel, define the classes so that they do not include the lower boundary and go up to and include the upper boundary:

$$\text{Lower boundary} < x \leq \text{Upper boundary}$$

This particular definition causes problems when the data include the number 0 because they force the lowest boundary to be negative so that 0 gets included. This often results in awkward looking intervals.

Remember that the intervals are defined so that they include the lower boundary. To create each class, we can start at the minimum data value and add the class width. The lower boundary for the next class is the upper boundary for the preceding class. In the next example we use the exact values that we obtained from our previous calculations. This procedure does not always lead to intervals that "look nice," but it is the simplest way to learn.

EXAMPLE 2.10

FLIGHT DELAYS

Creating Class Intervals for Continuous Data

In Example 2.9 the consumer group found that the width of each class should be 7.36. The data are measured only to the nearest tenth, so the group uses increments of 7.4. Since the smallest value in the data is 5.1, the first interval starts at that value and adds 7.4. That is, the first class interval contains all values, x, of the variable such that

$$5.1 \leqslant x < 12.5$$

The next class is formed by starting at 12.5 and again adding 7.4:

$$12.5 \leqslant x < 9.9$$

The remainder of the classes are

$$19.9 \leqslant x < 27.3$$
$$27.3 \leqslant x < 34.7$$
$$34.7 \leqslant x < 42.1$$

Since the last class includes the highest value, we can end here.

To create the frequency table for the data, the consumer group counts the number of data values that fall into each interval and completes the table.

Percentage of delayed flights	Frequency	Relative frequency (%)	Cumulative relative frequency (%)
$5.1 \leqslant x < 12.5$	8	26.7	26.7
$12.5 \leqslant x < 19.9$	10	33.3	60.0
$19.9 \leqslant x < 27.3$	8	26.7	86.7
$27.3 \leqslant x < 34.7$	1	3.3	90.0
$34.7 \leqslant x < 42.1$	3	10.3	100.0

From the frequency table the consumer group can obtain information such as: On 60% of the days, less than 20% of the flights were delayed. They get a good picture of the airline's performance during April 2001.

Analyze the data.

It is okay to wind up with one more class than you planned on. Remember that the rules are only guidelines.

Notice that the data value of 34.7 goes in the last interval because the fourth interval goes up to but does not include 34.7.

The preceding example showed exactly how to use the rules for creating class intervals for continuous data. However, you can see that by following the rules exactly you sometimes have to work with awkward numbers. Such numbers are difficult for you to work with and also difficult for the people who will use the information you give them. Most people tend to think in "nice" or "round" numbers that end in 0 or 5, not decimals. If decimals are necessary, it is easier to deal with numbers like 0.1 and 0.2 than with 0.08 and 0.16.

Remember that the rules we gave you are just *guidelines*. You can adjust the results, within reason, to make your results more appealing. You might change the starting value. Instead of using the actual lowest value in the data, you might pick a "nice" number just below that. For example, with the airline delay data, instead of starting at 5.1, you might start at 5.0. Or, you might want to change the class width. Certainly most people do not think in intervals of 7.4. It would be reasonable to use interval widths of 8 or even 10.

You may think, "Well, I will just use the computer to create the intervals," but software packages follow rules very similar to the ones outlined here and the results are very often not what you want. You can usually adjust the output, but without knowing where you are, it is difficult to figure out where you are going!

EXAMPLE 2.11

ABC FACULTY SALARIES

Creating Class Intervals

The committee that is studying gender equity in salaries needs to set up class intervals that have a width of 2.93 years. They recognize that this is an awkward number to work with, so they decide to try a different number. They would like to use a class width of 5, but that seems too much larger than 2.93. The closest

Analyze the data.

whole number is 3, which isn't very round but is certainly better than 2.93. Finally, they decide to try both class widths and see what difference it makes. Since the lowest data value is 0, which is a nice, round number, they start there. They obtain the following class intervals:

Class	Frequency	Class	Frequency
$0 \leqslant x < 3$	34	$0 \leqslant x < 5$	49
$3 \leqslant x < 6$	19	$5 \leqslant x < 10$	43
$6 \leqslant x < 9$	34	$10 \leqslant x < 15$	25
$9 \leqslant x < 12$	16	$15 \leqslant x < 20$	24
$12 \leqslant x < 15$	14	$20 \leqslant x < 25$	48
$15 \leqslant x < 18$	13	$25 \leqslant x < 30$	15
$18 \leqslant x < 21$	17	$30 \leqslant x < 35$	2
$21 \leqslant x < 24$	33	$35 \leqslant x < 40$	0
$24 \leqslant x < 27$	20	$40 \leqslant x < 45$	1
$27 \leqslant x < 30$	4		
$30 \leqslant x < 33$	2		
$33 \leqslant x < 36$	0		
$36 \leqslant x < 39$	0		
$39 \leqslant x < 42$	1		

With a class width of 3, there are 14 classes and the end classes are fairly empty. With a class width of 5, the number of classes is reduced to 9 and there are fewer empty or nearly empty classes. The committee is not sure whether this matters, but they will keep both tables for the moment.

EXAMPLE 2.12

FLIGHT DELAYS

Adjusting Class Boundaries

After examining the frequency table for the flight delay data in Example 2.10, the consumer group decides that they would prefer to see the data in rounder, more understandable numbers. They choose to start the first interval at 5%, which is just below the smallest data value, and use a width of 10%. The new frequency table looks like this:

Analyze the data.

Percentage of delayed flights	Frequency	Relative frequency (%)	Cumulative relative frequency (%)
$0.0 \leqslant x < 10.0$	3	10.0	10.0
$10.0 \leqslant x < 20.0$	15	50.0	60.0
$20.0 \leqslant x < 30.0$	9	30.0	90.0
$30.0 \leqslant x < 40.0$	2	6.7	96.7
$40.0 \leqslant x < 50.0$	1	3.3	100.0

The changes still result in five classes; even though the first interval starts at a lower number, the intervals are wider.

Assignment Times
Creating a Frequency Table for Continuous Data

The instructor for an Introductory Statistics class wonders about the complaints that she is hearing about the time it takes students to complete a computer assignment. The assignments are designed to be done in about 25 minutes. She asks the members of the class to time how long it takes to do the next assignment and to hand the data in with the assignment. The times (in minutes) are listed here:

22.8	27.0	27.9	30.4	33.4
24.8	27.2	27.9	31.1	33.9
24.8	27.4	28.2	31.4	35.3
26.0	27.4	29.4	32.4	35.7
26.0	27.4	29.6	33.1	36.3
26.1	27.6	29.8	33.2	40.4

Approximately how many classes should the frequency table have?

What should the class width be?

Create a frequency table for the data.

If you were going to adjust the class widths to use "nicer" numbers, what would you recommend?

2.2.4 EXERCISES—LEARNING IT!

2.1 The administrators of a local university are trying to determine whether they have an adequate amount of parking. They take a random sample of 25 entering freshmen and ask them what mode of transportation they will use to travel to classes. Here are the responses:

Car	Car	Car	Car	Car
Car	Walk	Car	Car	Car
Public	Car	Walk	Car	Car
Other	Car	Bicycle	Car	Car
Public	Car	Car	Public	Walk

(a) Create a frequency table for the responses.

(b) What mode of transportation is the most popular? The least popular?

(c) Does any class in the frequency table constitute a majority? If so, which one?

2.2 To assess the dominance of Microsoft for college campus networks, a local computer society randomly sampled 30 colleges in the state and asked which operating system they were currently running for their main computer network. The society got these data:

Mac OS	Mac OS	Mac OS	W NT	W 2000
W NT	W NT	Linux	W NT	W 2000
W 2000	W 2000	W NT	Mac OS	W NT
W NT	Mac OS	Mac OS	W NT	Mac OS
W 2000	Mac OS	Linux	W 2000	Linux
Mac OS	W NT	W NT	W 2000	W 2000

ANS. 6; 3.52; 22.70 ≤ x < 26.22, 26.22 ≤ x < 29.74, 29.74 ≤ x < 33.26, 33.26 ≤ x < 36.78, 36.78 ≤ x < 40.30, 40.30 ≤ x < 43.82; FREQUENCIES 6, 11, 7, 5, 0, 1. YOU COULD USE A CLASS WIDTH OF 5 MINUTES AND START AT 20.

(a) Create a frequency table for the data.

(b) What percentage of the colleges surveyed used W 2000?

(c) What percentage of the colleges surveyed did not use a Microsoft product?

(d) Was any operating system used by a majority of those surveyed?

2.3 A local supermarket has been receiving many complaints about the condition of cardboard cereal boxes. Customers are refusing to buy boxes that are dented and crushed, and the store is losing sales. Since the supermarket manager is fairly sure the damage is not occurring when the boxes are put on the shelves, she goes to the warehouse to check the cases as they arrive from the distributor. The manager takes a random sample of the cases as they arrive and examines them for various defects. The sample provides these data:

Unsealed	Crushed	No defect	Dented	Crushed
No defect	Crushed	No defect	No defect	No defect
Crushed	No defect	No defect	Dented	Crushed
Crushed	Unsealed	No defect	Dented	Dented

(a) Create a frequency table for the data.

(b) Based on the data, does it seem that the manager is correct that the cases are arriving damaged?

2.4 A survey in a campus newsletter asked students how many times per week they logged onto the Internet using computers in the main computing labs. The responses from 50 students are listed here:

1	7	3	4	1
0	7	0	1	8
3	3	4	2	4
3	0	3	3	5
3	3	7	4	1
7	5	3	3	4
4	3	3	3	7
6	8	0	7	2
4	0	3	9	3
4	4	4	4	3

(a) Create a frequency table for the data.

(b) What percentage of students surveyed logged onto the Internet exactly five times per week? More than five times per week?

(c) What percentage of students surveyed logged on four, five, or six times per week? Less than twice per week?

2.5 Sometimes it seems as though the people who repair photocopy machines should be the busiest people on earth! A large hospital surveyed all departments with copy machines and asked how many times per week their machines needed to be serviced. These data were obtained:

3	2	3	0	3
0	2	2	2	1
3	2	2	1	0
2	3	2	1	2
3	4	2	7	1
2	2	1	3	1
0	3	1	1	2

(a) Create a frequency table for the data, including relative frequency and cumulative relative frequency.

(b) What percentage of the copiers needed repairs more than twice per week?

(c) What percentage needed repairs at least four times per week?

(d) Did a majority of the copiers require repairs more than three times per week?

2.6 Technology is playing a larger and larger role in elementary schools. To determine whether it was reasonable to request that students use computers to write reports, a school administrator surveyed a fifth-grade class and asked the students how many computers they had at home. Their responses are listed here:

1	0	1	1	0
2	2	0	1	0
0	0	0	0	3
0	1	1	2	1
1	2	0	0	1

(a) Create a frequency table for the data, including relative frequency and cumulative relative frequency.

(b) What percentage of the students did not have access to computers at home?

(c) What percentage of the students had easy access (two or more) to computers at home?

(d) Based on the data collected, do you think that asking students to use computers to write reports is reasonable? Why or why not?

2.7   As part of a study to decide whether to renew the contract of a food service company at a local university, the administration surveyed a group of students who use the service on a regular basis. The students were asked to rate the food quality on a scale of 1 to 5, where 1 was extremely bad and 5 was extremely good. These are the results:

2	4	3	4	1
3	2	3	1	3
1	1	4	1	4
1	3	3	4	1
4	2	3	4	3
3	3	3	1	3

(a) Create a frequency table for the data.

(b) What rating had the highest frequency? The lowest?

(c) Based on the data, do you think that the students like the food provided by the company? Why or why not?

2.8   A local computer group surveyed high school students to find out how many hours per week the students spent instant messaging with friends. The responses follow:

0.0	4.1	11.9	15.5	17.8
0.0	4.2	13.0	15.5	17.9
1.0	6.9	13.4	15.5	18.4
1.4	8.0	13.5	15.5	18.7
2.5	8.2	14.9	16.7	19.5
2.9	8.6	15.1	16.8	19.7

Create a frequency table for the amount of time students spend instant messaging. Include relative frequency and cumulative relative frequency.

2.9   Most people hate housecleaning. A recent study asked people how many hours they spend per week cleaning their homes. The following data were obtained:

0.6	2.6	3.3	4.2	5.0
1.2	2.6	3.3	4.4	5.0
1.3	2.7	3.3	4.4	5.6
1.9	2.9	3.4	4.4	5.7
2.0	3.0	3.6	4.5	5.8
2.1	3.0	3.8	4.7	6.0
2.2	3.1	3.8	4.7	6.0
2.4	3.1	4.1	4.9	7.4

(a) Create a frequency table for the data, including relative frequency and cumulative relative frequency.

(b) What percentage of the people spend more than 3 hours per week cleaning their homes? (If this and the following questions are not easy to answer with your frequency table, modify the class intervals so that they are.)

(c) What percentage of those surveyed clean 1 hour or less? Between 2 and 5 hours?

(d) Approximately what percentage clean more than 6 hours?

(e) The person who conducted the study asked for only the time spent cleaning. Are there other factors that might be important? If you were conducting the study, what additional data would you collect? Why?

2.10 The Bureau of Weights and Measures conducts random checks of products in supermarkets. The bureau decides to check half-gallon containers of a particular brand of orange juice. These are the weights of each of the containers tested (measured to the nearest 0.1 ounce):

64.8	65.2	65.6	65.7	65.9
64.8	65.3	65.7	65.7	66.0
64.9	65.3	65.7	65.7	66.1
65.1	65.4	65.7	65.8	66.1
65.2	65.4	65.7	65.8	66.3

(a) Create a frequency table for the data.

(b) If the containers are subject to bursting when filled with more than 66 ounces of juice, what percentage of the containers are in danger?

2.11 The local refuse company in charge of recycling is receiving complaints from the people at the end of the pickup route. Those people report that their recycling is not being picked up because the drivers say that the trucks are too full. In an effort to determine how much capacity is really needed for the route, the refuse company randomly samples the newspaper recycling that is put out and measures each pile to the nearest 0.1 inch. These data are obtained:

8.3	10.2	12.2	13.1	13.9
8.5	10.4	12.3	13.3	14.1
8.8	11.0	12.3	13.3	14.3
9.1	11.0	12.3	13.4	14.3
9.7	11.2	12.3	13.4	14.4
9.8	11.4	12.5	13.4	14.4
9.9	12.0	12.7	13.6	14.4
9.9	12.0	12.8	13.6	14.8
10.1	12.2	12.9	13.8	15.2

(a) Create a frequency table for the data, including relative frequency and cumulative relative frequency.

(b) If necessary, modify the frequency table you created in part (a) so that you can use it easily answer the questions in parts (c)–(e).

(c) What percentage of the newspaper recycling piles are more than 1 foot tall?

(d) What percentage of the recycling piles are 10 inches or shorter?

(e) Approximately what percentage of the recycling piles are between 10 and 14 inches?

2.3 Graphical Displays of Data

Although the frequency distribution does a good job of organizing and summarizing a set of data, it does not have much visual impact. For a data set to make a more immediate impression we need to create a pictorial (graphical) representation of the information in the frequency distribution. Most of the work is already done in creating the frequency table. The displays have the same basic structures with some changes to accommodate the type of data being displayed.

2.3.1 GRAPHICAL DISPLAYS FOR QUALITATIVE DATA

You can use two types of graphs to display qualitative data: a **bar chart** and a **pie chart.**

Chapter 12: Multiple Regression Models

$$F = \frac{MSR}{MSE} \quad \text{F test statistic}$$

$$F_{\alpha, k, n-k-1} \quad \text{F critical value}$$

$$t = \frac{b_i}{s_{b_i}} \quad \text{Test statistic for individual coefficients}$$

$$C_p = \frac{SSE_p}{MSE_{ALL}}(n - 2p) \quad \text{Bias for model with p terms}$$

Chapter 13: Experimental Design and ANOVA

$$SST = \sum_{j=1}^{c}\sum_{i=1}^{n_i}(x_{ij} - \bar{\bar{x}})^2 \left.\vphantom{\sum}\right\}$$

$$SSA = \sum_{j=1}^{c} n_j(\bar{x}_j - \bar{\bar{x}})^2 \left.\vphantom{\sum}\right\} \quad \text{Sums of squares—One-way}$$

$$SSE = \sum_{j=1}^{c}\sum_{i=1}^{n_i}(x_{ij} - \bar{x}_j)^2 \left.\vphantom{\sum}\right\}$$

$$MST = \frac{SST}{n-1} \left.\vphantom{\frac{a}{b}}\right\}$$

$$MSA = \frac{SSA}{c-1} \left.\vphantom{\frac{a}{b}}\right\} \quad \text{Mean squares—One-way}$$

$$MSE = \frac{SSE}{n-c} \left.\vphantom{\frac{a}{b}}\right\}$$

$$F = \frac{MSA}{MSE} \quad \text{F statistic—One-way}$$

$$SST = SSA + SSBL + SSE \quad \text{Sum of squares—Block}$$

$$RE = \frac{(r-1)MSBL + r(c-1)MSA}{(n-1)MSE} \quad \begin{array}{l}\text{Relative efficiency}\\\text{of block design}\end{array}$$

$$SST = SSA + SSB + SSE \quad \text{Sum of squares—Two-way}$$

Chapter 14: The Analysis of Qualitative Data

$$\chi^2 = \sum \frac{(o - e)^2}{e} \quad \text{Chi-square test statistic}$$

$$k - p - 1 \quad \text{Degrees of freedom—goodness of fit}$$
$$(r - 1)(c - 1) \quad \text{Degrees of freedom—independence}$$

Expected frequencies:

$$e_i = np_i \quad \text{Goodness of fit}$$

$$e_i = \frac{s}{n} n_i = pn_i \quad \text{Test for proportions}$$

$$e_{ij} = \left(\frac{r_i}{n}\frac{c_j}{n}\right)n = \frac{r_i c_j}{n} \quad \text{Test for independence}$$

Chapter 15: Nonparametric Statistics

$$\frac{(n)(n+1)}{2} \quad \text{Sum of integers from 1 to n}$$

Mean and standard deviation of Wilcoxon rank sum test:

$$\mu = \frac{n_1(n_1 + n_2 + 1)}{2}$$

$$\sigma = \sqrt{\frac{n_1 n_2(n_1 + n_2 + 1)}{12}}$$

Test statistic for Kruskal–Wallis test:

$$H = \frac{SSA}{SST/(n-1)}$$

$$SST = \sum_{j=1}^{c}\sum_{i=1}^{n_i}(r_{ij} - \bar{\bar{r}})^2$$

$$SSA = \sum_{j=1}^{c} n_j(\bar{r}_j - \bar{\bar{r}})^2$$

Approximate Kruskal–Wallis test statistic:

$$H = \frac{12}{n(n+1)}\sum_{j=1}^{c}\frac{R_j^2}{n_j} - 3(n+1)$$

Steps for Hypothesis Testing

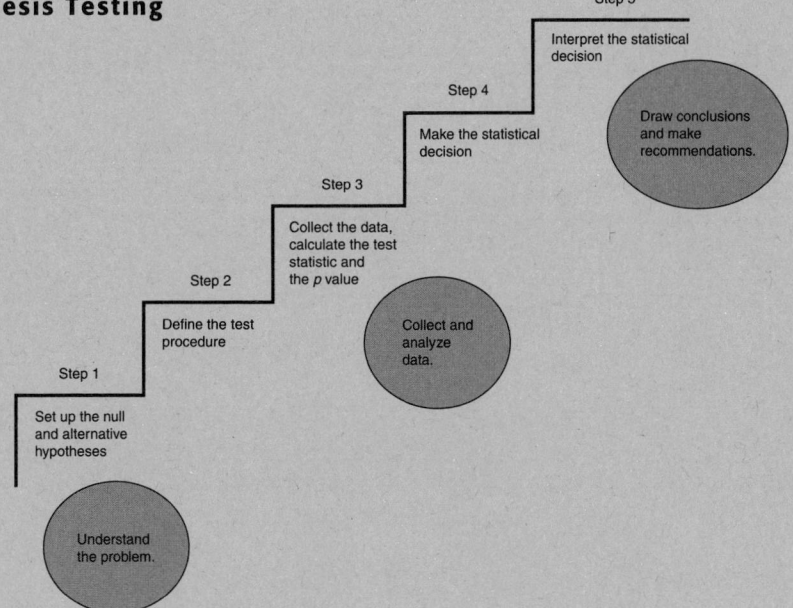

TABLE 4 t Critical Values

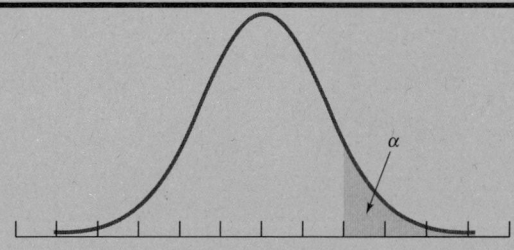

Degrees of Freedom	Upper Tail Probability (α)								
	0.15	0.10	0.05	0.025	0.015	0.01	0.005	0.001	0.0005
1	1.963	3.078	6.314	12.706	21.205	31.821	63.657	318.309	1273.155
2	1.386	1.886	2.920	4.303	5.643	6.965	9.925	22.327	44.703
3	1.250	1.638	2.353	3.182	3.896	4.541	5.841	10.215	16.326
4	1.190	1.533	2.132	2.776	3.298	3.747	4.604	7.173	10.305
5	1.156	1.476	2.015	2.571	3.003	3.365	4.032	5.893	7.976
6	1.134	1.440	1.943	2.447	2.829	3.143	3.707	5.208	6.788
7	1.119	1.415	1.895	2.365	2.715	2.998	3.499	4.785	6.082
8	1.108	1.397	1.860	2.306	2.634	2.896	3.355	4.501	5.617
9	1.100	1.383	1.833	2.262	2.574	2.821	3.250	4.297	5.291
10	1.093	1.372	1.812	2.228	2.527	2.764	3.169	4.144	5.049
11	1.088	1.363	1.796	2.201	2.491	2.718	3.106	4.025	4.863
12	1.083	1.356	1.782	2.179	2.461	2.681	3.055	3.930	4.717
13	1.079	1.350	1.771	2.160	2.436	2.650	3.012	3.852	4.597
14	1.076	1.345	1.761	2.145	2.415	2.625	2.977	3.787	4.499
15	1.074	1.341	1.753	2.131	2.397	2.602	2.947	3.733	4.417
16	1.071	1.337	1.746	2.120	2.382	2.583	2.921	3.686	4.346
17	1.069	1.333	1.740	2.110	2.368	2.567	2.898	3.646	4.286
18	1.067	1.330	1.734	2.101	2.356	2.552	2.878	3.611	4.233
19	1.066	1.328	1.729	2.093	2.346	2.539	2.861	3.579	4.187
20	1.064	1.325	1.725	2.086	2.336	2.528	2.845	3.552	4.146
21	1.063	1.323	1.721	2.080	2.328	2.518	2.831	3.527	4.109
22	1.061	1.321	1.717	2.074	2.320	2.508	2.819	3.505	4.077
23	1.060	1.319	1.714	2.069	2.313	2.500	2.807	3.485	4.047
24	1.059	1.318	1.711	2.064	2.307	2.492	2.797	3.467	4.021
25	1.058	1.316	1.708	2.060	2.301	2.485	2.787	3.450	3.997
26	1.058	1.315	1.706	2.056	2.296	2.479	2.779	3.435	3.974
27	1.057	1.314	1.703	2.052	2.291	2.473	2.771	3.421	3.954
28	1.056	1.313	1.701	2.048	2.286	2.467	2.763	3.408	3.935
29	1.055	1.311	1.699	2.045	2.282	2.462	2.756	3.396	3.918
30	1.055	1.310	1.697	2.042	2.278	2.457	2.750	3.385	3.902
40	1.050	1.303	1.684	2.021	2.250	2.423	2.704	3.307	3.788
50	1.047	1.299	1.676	2.009	2.234	2.403	2.678	3.261	3.723
60	1.045	1.296	1.671	2.000	2.223	2.390	2.660	3.232	3.681
120	1.041	1.289	1.658	1.980	2.196	2.358	2.617	3.160	3.578
Z critical value									
	1.036	1.282	1.645	1.960	2.170	2.326	2.576	3.090	3.290
Level of Significance for a one-tailed test	0.15	0.10	0.05	0.025	0.015	0.01	0.005	0.001	0.0005
Level of Significance for a two-tailed test	0.30	0.20	0.10	0.05	0.03	0.02	0.01	0.002	0.001

TABLE 3   Standard Normal Table

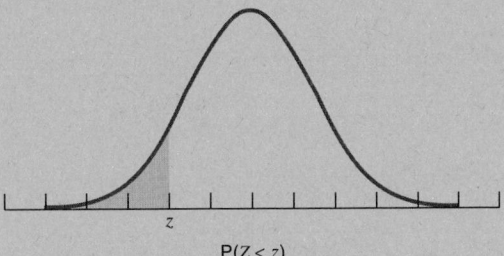

$P(Z < z)$

	Second Decimal Place									
z	0.00	0.01	0.02	0.03	0.04	0.05	0.06	0.07	0.08	0.09
-3.9	0.0000	0.0000	0.0000	0.0000	0.0000	0.0000	0.0000	0.0000	0.0000	0.0000
-3.8	0.0001	0.0001	0.0001	0.0001	0.0001	0.0001	0.0001	0.0001	0.0001	0.0001
-3.7	0.0001	0.0001	0.0001	0.0001	0.0001	0.0001	0.0001	0.0001	0.0001	0.0001
-3.6	0.0002	0.0002	0.0001	0.0001	0.0001	0.0001	0.0001	0.0001	0.0001	0.0001
-3.5	0.0002	0.0002	0.0002	0.0002	0.0002	0.0002	0.0002	0.0002	0.0002	0.0002
-3.4	0.0003	0.0003	0.0003	0.0003	0.0003	0.0003	0.0003	0.0003	0.0003	0.0002
-3.3	0.0005	0.0005	0.0005	0.0004	0.0004	0.0004	0.0004	0.0004	0.0004	0.0003
-3.2	0.0007	0.0007	0.0006	0.0006	0.0006	0.0006	0.0006	0.0005	0.0005	0.0005
-3.1	0.0010	0.0009	0.0009	0.0009	0.0008	0.0008	0.0008	0.0008	0.0007	0.0007
-3.0	0.0013	0.0013	0.0013	0.0012	0.0012	0.0011	0.0011	0.0011	0.0010	0.0010
-2.9	0.0019	0.0018	0.0018	0.0017	0.0016	0.0016	0.0015	0.0015	0.0014	0.0014
-2.8	0.0026	0.0025	0.0024	0.0023	0.0023	0.0022	0.0021	0.0021	0.0020	0.0019
-2.7	0.0035	0.0034	0.0033	0.0032	0.0031	0.0030	0.0029	0.0028	0.0027	0.0026
-2.6	0.0047	0.0045	0.0044	0.0043	0.0041	0.0040	0.0039	0.0038	0.0037	0.0036
-2.5	0.0062	0.0060	0.0059	0.0057	0.0055	0.0054	0.0052	0.0051	0.0049	0.0048
-2.4	0.0082	0.0080	0.0078	0.0075	0.0073	0.0071	0.0069	0.0068	0.0066	0.0064
-2.3	0.0107	0.0104	0.0102	0.0099	0.0096	0.0094	0.0091	0.0089	0.0087	0.0084
-2.2	0.0139	0.0136	0.0132	0.0129	0.0125	0.0122	0.0119	0.0116	0.0113	0.0110
-2.1	0.0179	0.0174	0.0170	0.0166	0.0162	0.0158	0.0154	0.0150	0.0146	0.0143
-2.0	0.0228	0.0222	0.0217	0.0212	0.0207	0.0202	0.0197	0.0192	0.0188	0.0183
-1.9	0.0287	0.0281	0.0274	0.0268	0.0262	0.0256	0.0250	0.0244	0.0239	0.0233
-1.8	0.0359	0.0351	0.0344	0.0336	0.0329	0.0322	0.0314	0.0307	0.0301	0.0294
-1.7	0.0446	0.0436	0.0427	0.0418	0.0409	0.0401	0.0392	0.0384	0.0375	0.0367
-1.6	0.0548	0.0537	0.0526	0.0516	0.0505	0.0495	0.0485	0.0475	0.0465	0.0455
-1.5	0.0668	0.0655	0.0643	0.0630	0.0618	0.0606	0.0594	0.0582	0.0571	0.0559
-1.4	0.0808	0.0793	0.0778	0.0764	0.0749	0.0735	0.0721	0.0708	0.0694	0.0681
-1.3	0.0968	0.0951	0.0934	0.0918	0.0901	0.0885	0.0869	0.0853	0.0838	0.0823
-1.2	0.1151	0.1131	0.1112	0.1093	0.1075	0.1056	0.1038	0.1020	0.1003	0.0985
-1.1	0.1357	0.1335	0.1314	0.1292	0.1271	0.1251	0.1230	0.1210	0.1190	0.1170
-1.0	0.1587	0.1562	0.1539	0.1515	0.1492	0.1469	0.1446	0.1423	0.1401	0.1379
-0.9	0.1841	0.1814	0.1788	0.1762	0.1736	0.1711	0.1685	0.1660	0.1635	0.1611
-0.8	0.2119	0.2090	0.2061	0.2033	0.2005	0.1977	0.1949	0.1922	0.1894	0.1867
-0.7	0.2420	0.2389	0.2358	0.2327	0.2296	0.2266	0.2236	0.2206	0.2177	0.2148
-0.6	0.2743	0.2709	0.2676	0.2643	0.2611	0.2578	0.2546	0.2514	0.2483	0.2451
-0.5	0.3085	0.3050	0.3015	0.2981	0.2946	0.2912	0.2877	0.2843	0.2810	0.2776
-0.4	0.3446	0.3409	0.3372	0.3336	0.3300	0.3264	0.3228	0.3192	0.3156	0.3121
-0.3	0.3821	0.3783	0.3745	0.3707	0.3669	0.3632	0.3594	0.3557	0.3520	0.3483
-0.2	0.4207	0.4168	0.4129	0.4090	0.4052	0.4013	0.3974	0.3936	0.3897	0.3859
-0.1	0.4602	0.4562	0.4522	0.4483	0.4443	0.4404	0.4364	0.4325	0.4286	0.4247
0.0	0.5000	0.4960	0.4920	0.4880	0.4840	0.4801	0.4761	0.4721	0.4681	0.4641

Chapter 8: Hypothesis Testing: An Introduction

$Z = \dfrac{\overline{X} - \mu}{\sigma/\sqrt{n}}$ Z test statistic

Chapter 9: Inferences: More One-Population Tests

$t = \dfrac{\overline{X} - \mu}{s/\sqrt{n}}$ t test statistic

$\chi^2 = \dfrac{(n-1)s^2}{\sigma^2}$ χ^2 test statistic

$Z = \dfrac{\hat{p} - p}{\sqrt{p(1-p)/n}}$ Z test statistic for proportion

Chapter 10: Comparing Two Populations

Tests for Two Population Means

$Z = \dfrac{(\overline{X}_1 - \overline{X}_2)}{\sqrt{\sigma_1^2/n_1 + \sigma_2^2/n_2}}$ Variances known

$Z = \dfrac{(\overline{X}_1 - \overline{X}_2)}{\sqrt{s_1^2/n_1 + s_2^2/n_2}}$ Variances unknown; $n_1, n_2 \geq 30$

$t = \dfrac{(\overline{X}_1 - \overline{X}_2)}{s_p\sqrt{1/n_1 + 1/n_2}}$ Variances unknown but equal; n_1, $n_2 < 30$; populations normal

$t = \dfrac{(\overline{X}_1 - \overline{X}_2)}{\sqrt{s_1^2/n_1 + s_2^2/n_2}}$ Variances unknown, not equal; n_1, $n_2 < 30$; degrees of freedom adjusted

$s_p^2 = \dfrac{(n_1 - 1)s_1^2 + (n_2 - 1)s_2^2}{n_1 + n_2 - 2}$ Pooled variance

$v = \dfrac{\left(\dfrac{s_1^2}{n_1} + \dfrac{s_2^2}{n_2}\right)^2}{\dfrac{(s_1^2/n_1)^2}{n_1 - 1} + \dfrac{(s_2^2/n_2)^2}{n_2 - 1}}$ Adjusted degrees of freedom

$t = \dfrac{\overline{d}}{s_d/\sqrt{n}}$ Test for paired difference

$\overline{d} = \dfrac{\sum d}{n}$ Average difference

$Z = \dfrac{\hat{p}_1 - \hat{p}_2}{\sqrt{\overline{p}(1 - \overline{p})\left(\dfrac{1}{n_1} + \dfrac{1}{n_2}\right)}}$ Test for two population proportions

$\overline{p} = \dfrac{x_1 + x_2}{n_1 + n_2}$ Common population proportion

$F = \dfrac{s_1^2}{s_2^2}$ Test for two population variances

$F_{\text{lower, df1, df2}} = \dfrac{1}{F_{\text{upper, df2,df1}}}$ Lower critical value for F test

Chapter 11: Regression Analysis

Slope and intercept of least squares line: see Chapter 5

$s_{y|x} = \sqrt{\dfrac{\sum(y - \hat{y})^2}{n - 2}}$ Standard error of estimate

$r = \dfrac{\sum xy - \sum x \sum y/n}{\sqrt{\sum x^2 - (\sum x)^2/n}\sqrt{\sum y^2 - (\sum y)^2/n}}$ Correlation coefficient

$t = \dfrac{b_1 - \beta_1}{s_{b_1}}$ Test statistic for the slope

$s_{b_1} = \dfrac{s_{y|x}}{\sqrt{\sum x^2 - (\sum x)^2/n}}$ Standard error of the slope

$\hat{y}_i \pm t_{\alpha/2, n-2}s_{y|x}\sqrt{\dfrac{1}{n} + \dfrac{(x_i - \overline{x})^2}{\sum x^2 - (\sum x)^2/n}}$ Confidence interval

$\hat{y}_i \pm t_{\alpha/2, n-2}s_{y|x}\sqrt{1 + \dfrac{1}{n} + \dfrac{(x_i - \overline{x})^2}{\sum x^2 - (\sum x)^2/n}}$ Prediction interval

$\left.\begin{array}{l} \text{SST} = \sum(y - \overline{y})^2 \\ \text{SSR} = \sum(\hat{y} - \overline{y})^2 \\ \text{SSE} = \sum(y - \hat{y})^2 \\ \text{SST} = \text{SSR} + \text{SSE} \end{array}\right\}$ Sums of squares

$\left.\begin{array}{l} \text{MSR} = \dfrac{\text{SSR}}{1} \\[2mm] \text{MSE} = \dfrac{\text{SSE}}{n - 2} \end{array}\right\}$ Mean squares

$R^2 = \dfrac{\text{SSR}}{\text{SST}}100\%$ or r^2 Coefficient of determination

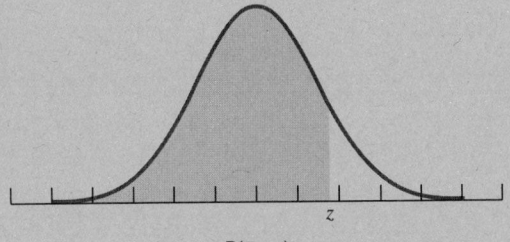

$P(Z < z)$

Second Decimal Place

z	0.00	0.01	0.02	0.03	0.04	0.05	0.06	0.07	0.08	0.09
0.0	0.5000	0.5040	0.5080	0.5120	0.5160	0.5199	0.5239	0.5279	0.5319	0.5359
0.1	0.5398	0.5438	0.5478	0.5517	0.5557	0.5596	0.5636	0.5675	0.5714	0.5753
0.2	0.5793	0.5832	0.5871	0.5910	0.5948	0.5987	0.6026	0.6064	0.6103	0.6141
0.3	0.6179	0.6217	0.6255	0.6293	0.6331	0.6368	0.6406	0.6443	0.6480	0.6517
0.4	0.6554	0.6591	0.6628	0.6664	0.6700	0.6736	0.6772	0.6808	0.6844	0.6879
0.5	0.6915	0.6950	0.6985	0.7019	0.7054	0.7088	0.7123	0.7157	0.7190	0.7224
0.6	0.7257	0.7291	0.7324	0.7357	0.7389	0.7422	0.7454	0.7486	0.7517	0.7549
0.7	0.7580	0.7611	0.7642	0.7673	0.7704	0.7734	0.7764	0.7794	0.7823	0.7852
0.8	0.7881	0.7910	0.7939	0.7967	0.7995	0.8023	0.8051	0.8078	0.8106	0.8133
0.9	0.8159	0.8186	0.8212	0.8238	0.8264	0.8289	0.8315	0.8340	0.8365	0.8389
1.0	0.8413	0.8438	0.8461	0.8485	0.8508	0.8531	0.8554	0.8577	0.8599	0.8621
1.1	0.8643	0.8665	0.8686	0.8708	0.8729	0.8749	0.8770	0.8790	0.8810	0.8830
1.2	0.8849	0.8869	0.8888	0.8907	0.8925	0.8944	0.8962	0.8980	0.8997	0.9015
1.3	0.9032	0.9049	0.9066	0.9082	0.9099	0.9115	0.9131	0.9147	0.9162	0.9177
1.4	0.9192	0.9207	0.9222	0.9236	0.9251	0.9265	0.9279	0.9292	0.9306	0.9319
1.5	0.9332	0.9345	0.9357	0.9370	0.9382	0.9394	0.9406	0.9418	0.9429	0.9441
1.6	0.9452	0.9463	0.9474	0.9484	0.9495	0.9505	0.9515	0.9525	0.9535	0.9545
1.7	0.9554	0.9564	0.9573	0.9582	0.9591	0.9599	0.9608	0.9616	0.9625	0.9633
1.8	0.9641	0.9649	0.9656	0.9664	0.9671	0.9678	0.9686	0.9693	0.9699	0.9706
1.9	0.9713	0.9719	0.9726	0.9732	0.9738	0.9744	0.9750	0.9756	0.9761	0.9767
2.0	0.9772	0.9778	0.9783	0.9788	0.9793	0.9798	0.9803	0.9808	0.9812	0.9817
2.1	0.9821	0.9826	0.9830	0.9834	0.9838	0.9842	0.9846	0.9850	0.9854	0.9857
2.2	0.9861	0.9864	0.9868	0.9871	0.9875	0.9878	0.9881	0.9884	0.9887	0.9890
2.3	0.9893	0.9896	0.9898	0.9901	0.9904	0.9906	0.9909	0.9911	0.9913	0.9916
2.4	0.9918	0.9920	0.9922	0.9925	0.9927	0.9929	0.9931	0.9932	0.9934	0.9936
2.5	0.9938	0.9940	0.9941	0.9943	0.9945	0.9946	0.9948	0.9949	0.9951	0.9952
2.6	0.9953	0.9955	0.9956	0.9957	0.9959	0.9960	0.9961	0.9962	0.9963	0.9964
2.7	0.9965	0.9966	0.9967	0.9968	0.9969	0.9970	0.9971	0.9972	0.9973	0.9974
2.8	0.9974	0.9975	0.9976	0.9977	0.9977	0.9978	0.9979	0.9979	0.9980	0.9981
2.9	0.9981	0.9982	0.9982	0.9983	0.9984	0.9984	0.9985	0.9985	0.9986	0.9986
3.0	0.9987	0.9987	0.9987	0.9988	0.9988	0.9989	0.9989	0.9989	0.9990	0.9990
3.1	0.9990	0.9991	0.9991	0.9991	0.9992	0.9992	0.9992	0.9992	0.9993	0.9993
3.2	0.9993	0.9993	0.9994	0.9994	0.9994	0.9994	0.9994	0.9995	0.9995	0.9995
3.3	0.9995	0.9995	0.9995	0.9996	0.9996	0.9996	0.9996	0.9996	0.9996	0.9997
3.4	0.9997	0.9997	0.9997	0.9997	0.9997	0.9997	0.9997	0.9997	0.9997	0.9998
3.5	0.9998	0.9998	0.9998	0.9998	0.9998	0.9998	0.9998	0.9998	0.9998	0.9998
3.6	0.9998	0.9998	0.9999	0.9999	0.9999	0.9999	0.9999	0.9999	0.9999	0.9999
3.7	0.9999	0.9999	0.9999	0.9999	0.9999	0.9999	0.9999	0.9999	0.9999	0.9999
3.8	0.9999	0.9999	0.9999	0.9999	0.9999	0.9999	0.9999	0.9999	0.9999	0.9999
3.9	1.0000	1.0000	1.0000	1.0000	1.0000	1.0000	1.0000	1.0000	1.0000	1.0000

Formulas and Tables

for *Elementary Statistics: From Discovery to Decision, 1/e*
©2003 John Wiley & Sons, Inc.

Chapter 3: Numerical Descriptors of Data

$\overline{X} = \dfrac{\Sigma x}{n}$ Sample mean

$s^2 = \dfrac{\Sigma(x - \overline{x})^2}{n - 1}$ Sample variance (definition)

$s^2 = \dfrac{\Sigma x^2 - (\Sigma x)^2}{n(n - 1)}$ Sample variance (shortcut)

$s = \sqrt{s^2}$ Sample standard deviation

$R = \text{Max} - \text{Min}$ Sample range

$P = \dfrac{b + \frac{1}{2}e}{n}$ Percentile rank

$z = \dfrac{x - \overline{x}}{s}$ z-Score

$IQR = Q_3 - Q_1$ Interquartile range

$\left.\begin{array}{l} Q_1 - 1.5(IQR) \\ Q_3 + 1.5(IQR) \end{array}\right\}$ Inner fences

$\left.\begin{array}{l} Q_1 - 3(IQR) \\ Q_3 + 3(IQR) \end{array}\right\}$ Outer fences

Chapter 4: Analyzing Bivariate Data

$\hat{y} = a + bx$ Least squares line

$b = \dfrac{n \sum XY - \sum X \sum Y}{n \sum X^2 - (\sum X)^2}$ Slope estimate

$a = \dfrac{\sum y}{n} - b\dfrac{\sum x}{n}$ Y-intercept estimate

$e = \hat{y} - y$ Deviation

Chapter 5: Probability

$P(A) = \dfrac{\text{Number of ways A can occur}}{\text{Total number of outcomes}} = \dfrac{n_A}{N}$

$P(A') = 1 - P(A)$ Probability of complement

$P(A \text{ OR } B) = P(A) + P(B)$ Simple Addition Rule

$P(A \text{ OR } B) = P(A) + P(B) - P(A \text{ AND } B)$ General Addition Rule

$P(A \mid B) = \dfrac{P(A \text{ AND } B)}{P(B)}$ Conditional probability

$P(A \text{ AND } B) = P(A) \times P(B)$ Independent events

Chapter 6: Random Variables and Probability Distributions

$\left.\begin{array}{l} \mu = np \\ \sigma = \sqrt{np(1 - p)} \end{array}\right\}$ Mean and standard deviation of a binomial random variable

$p(x) = \dfrac{n!}{x!(n - x)!} p^x (1 - p)^{n-x}$ Binomial probability distribution

$Z = \dfrac{X - \mu}{\sigma}$ Z-score

Chapter 7: Sampling Distributions and Confidence Intervals

$\mu_{\overline{X}} = \mu$ Mean of $\overline{X}$

$\sigma_{\overline{X}} = \dfrac{\sigma}{\sqrt{n}}$ Standard error of $\overline{X}$

$Z = \dfrac{\overline{X} - \mu_{\overline{X}}}{\sigma_{\overline{X}}} = \dfrac{\overline{X} - \mu}{\sigma/\sqrt{n}}$ Z-score for $\overline{X}$

$\overline{X} \pm Z_{\alpha/2} \dfrac{\sigma}{\sqrt{n}}$ Confidence interval for μ when σ is known

$\overline{X} \pm Z_{\alpha/2} \dfrac{s}{\sqrt{n}}$ Large-sample confidence interval for μ when σ is unknown

$\overline{X} \pm t_{\alpha/2} \dfrac{s}{\sqrt{n}}$ Small-sample confidence interval for μ, σ unknown, population normal

$\hat{p} \pm Z_{\alpha/2} \sqrt{\dfrac{\hat{p}(1 - \hat{p})}{n}}$ Confidence interval for population proportion

$n = \dfrac{Z_{\alpha/2}^2 \sigma^2}{e^2}$ Sample size for estimating μ with error e

$n = \dfrac{Z_{\alpha/2}^2 p(1 - p)}{e^2}$ Sample size for estimating p with error e

A *bar chart* represents the frequency or relative frequency from a frequency table in the form of a rectangle or bar.

Creating a Bar Chart for Qualitative Data

In a bar chart, one of the axes is used to represent the categories in the frequency table and the other axis is used to represent the frequencies or relative frequencies for the categories. For qualitative data, assignment of the axes is a matter of preference. We use the x axis for the categories and the y axis for the frequencies and relative frequencies to be consistent with the graphs for the other types of data. Here are the steps to use in drawing a bar chart:

Step 1: Draw a pair of axes, x and y.

Step 2: At evenly spaced intervals on the x axis put tick marks and label them with the categories from the frequency table.

Step 3: Scale the y axis so that it includes the category with the highest frequency or relative frequency. Choose the scale so that you can distinguish different frequencies or relative frequencies.

Step 4: At each category on the x axis draw a rectangle (bar) with its height equal to the frequency or relative frequency for the category. The bases of the rectangles must be the same width, and the bars should not touch each other.

Step 5: Label the axes and give the graph an appropriate title.

Often there is no predetermined order for the categories for qualitative data, but you might want to think about the order in which you place them because it may affect the impact of the graph. Some common choices are by descending or ascending frequency or in alphabetical order.

EXAMPLE 2.13

STUDENT DISTRIBUTION

Creating a Bar Chart

The dean of the School of Arts and Sciences wants to create a bar chart from the data that he collected on the classes of students who are taking the Introductory Statistics course. The frequency table created earlier is reprinted here:

Class	Frequency
Freshman	4
Sophomore	12
Junior	2
Senior	10
Total	**28**

Since in this case there is a natural order for the categories, the dean uses that order when labeling the x axis. The y axis must accommodate the highest frequency, 12, so the scale for that axis goes from 0 to 12 by 2s. The completed bar chart is shown in the figure.

Analyze the data.

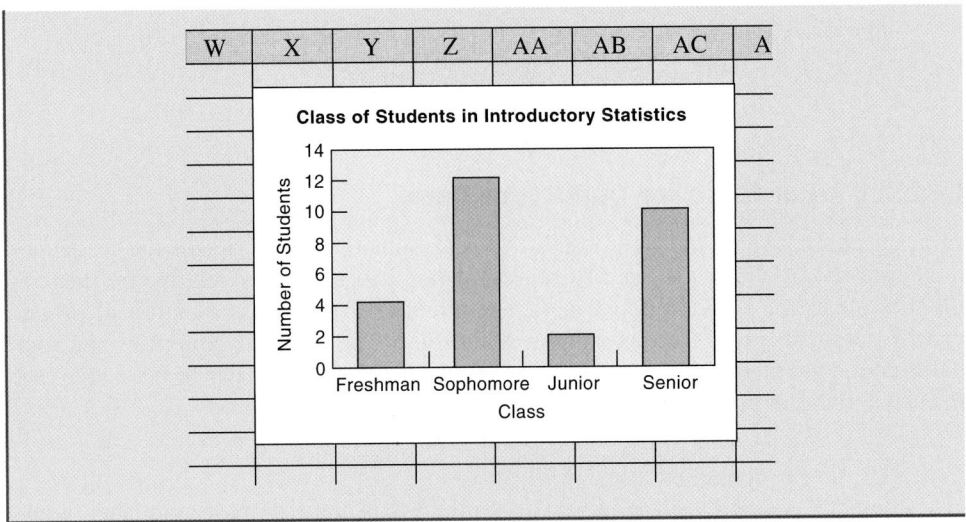

When you make a bar chart, remember that there are no absolute rules. The idea is to create a chart that conveys clear information to the viewer. In general, it is desirable to keep the number of categories in a bar chart to no more than ten. When there are too many categories, the axis becomes crowded and difficult to read, which distracts the viewer. To keep the number of categories down, categories with very low frequencies are often consolidated into a single category labeled "Other." Even though the "Other" category might have a higher frequency than some of the categories, it must appear last in the bar chart.

EXAMPLE 2.14

ABC FACULTY SALARIES

Creating a Bar Chart

The ABC College committee that is investigating gender equity in salaries decides to create a bar chart for the data they have collected on faculty rank. The frequency table for the data is reprinted here:

Analyze the data.

Rank	Frequency	Relative frequency (%)
Professor	55	26.6
Associate Professor	67	32.4
Assistant Professor	77	37.2
Instructor	8	3.9
Total	**207**	**100.0**

The committee decides to use a computer and see what kind of graph it produces. The output from Minitab is shown here:

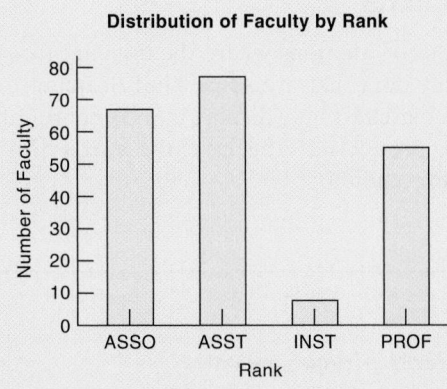

Distribution of Faculty by Rank

After examining the output, the committee decides that rank order might be better than alphabetical order for the *x* axis.

✔ **TRY IT NOW!**

Student Grades
Creating a Bar Chart

The professor who surveyed her students about their grade expectations wants to create a bar chart from the data. Here is the frequency table for the data:

Grade	Frequency	Relative frequency (%)
A	9	32.1
B	12	42.9
C	4	14.3
D	2	7.1
F	1	3.6
Total	28	100.0

Create a bar chart for the data with relative frequency on the *y* axis. Be sure to label the axes and add an appropriate title.

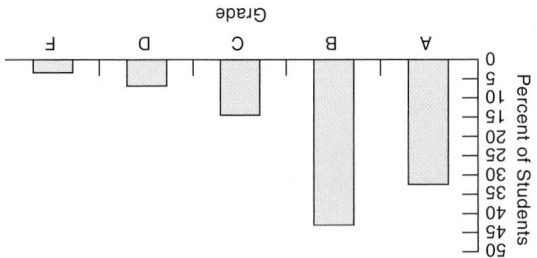

Ans. Expected Grades—Introductory Statistics

Other Uses for Bar Charts

Bar charts are not limited to situations where the data are frequencies. Very often bar charts are used to display data that are some kind of quantitative measure for each category. For example, a bar chart may display and compare sales revenues for a sample of different software products. In this case, the *y* axis of the chart represents the value of the variable being studied.

EXAMPLE 2.15

FOOD SALES

Creating a Bar Chart for Nonfrequency Data

A company studied the sales of a group of foods that are advertised as having "added nutrients." The company looked at supermarket sales for a 52-week period and gathered these data:

Product	Manufacturer	Sales (millions)
Hawaiian Punch	Procter & Gamble	$125.7
Yoo Hoo Chocolate-Flavored Drink	Austin Nichols	32.9
Life Savers Flavor Pops	Agway	9.4
Wonder Calcium Enriched White Bread	Ralston-Continental Baking	1.0
Vicks Vitamin C Lemon Drops	Procter & Gamble	0.5

Collect the data.

Analyze the data.

The company created a bar chart for the data using the product for the *x* axis and the sales (in millions of dollars) for the *y* axis.

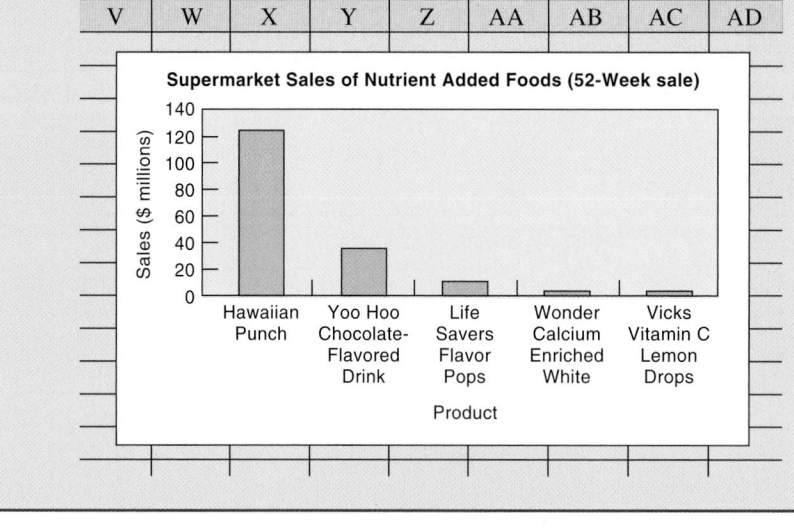

Creating a Pie Chart for Qualitative Data

Qualitative data can also be displayed on a pie chart. A pie chart is often used when the categories of the data represent some part or portion of a whole.

> A *pie chart* represents data in the form of slices or sections of a circle. Each slice represents a category, and the size of the slice is proportional to the relative frequency of the category.

Drawing a pie chart by hand requires the use of a protractor to accurately measure the sizes of the slices. It is fairly tedious. Since a circle has 360 degrees, you can use the relative frequency to determine how many degrees should be allocated to each slice. Because of these calculations it is usually best to use computer software to create a pie chart.

EXAMPLE 2.16

STUDENT DISTRIBUTION

Creating a Pie Chart

The dean of the School of Arts and Sciences wants to use the data on students who are taking Introductory Statistics, for a presentation to the faculty. He thinks that a pie chart might have the greatest impact. He has this relative frequency table:

Class	Relative frequency (%)
Freshman	14.3
Sophomore	42.9
Junior	7.1
Senior	35.7
Total	**100.0**

The dean uses Microsoft Excel to create this pie chart from the relative frequency table:

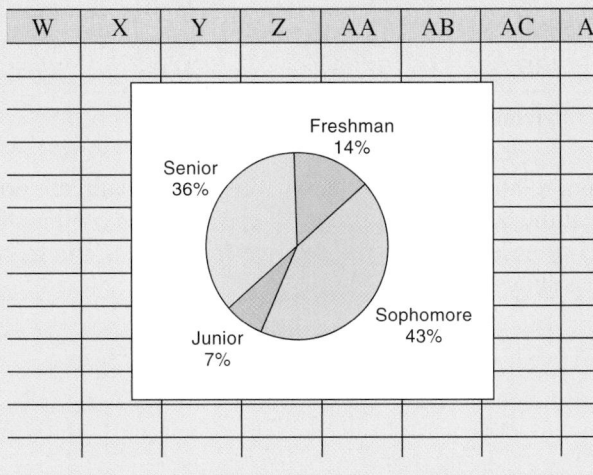

2.3.2 GRAPHICAL DISPLAYS FOR QUANTITATIVE DATA

The tool used to display quantitative data is called a **histogram.**

A *histogram* is very similar to a bar chart, but because the data are naturally ordered, the *x* axis of the graph must be scaled to reflect this.

There are slight differences between histograms for dealing with integer and continuous data.

Histograms for Integer Data

As in a bar chart, we use a rectangle to represent each possible data value, with the height of the bar corresponding to the frequency or relative frequency for that value. Remember that integers are numbers and they have a definite ordering. The x axis must accommodate all of the possible values, whether or not any observations had the value. The rectangles are centered on the data values as in a bar chart, but the bars are contiguous; that is, they touch each other.

EXAMPLE 2.17

STUDENT ATTITUDES

Creating a Histogram for Integer Data

Understand the problem.

The professor who is interested in student attitudes toward statistics wants to create a graphical display of the data. She thinks it might provide another view of the situation. Here is the frequency table for the data:

Rating	Frequency	Relative frequency (%)
0	4	14.3
1	4	14.3
2	0	0.0
3	0	0.0
4	2	7.1
5	2	7.1
6	3	10.7
7	2	7.1
8	3	10.7
9	2	7.1
10	6	21.4
Total	28	100.0

Collect the data.

Analyze the data.

To create a relative frequency histogram for the data, the professor scales the x axis so that the numbers 0 through 10 appear at the centers of the bars. The y axis is scaled to accommodate the relative frequencies; the highest is 21.4%. The histogram is shown here:

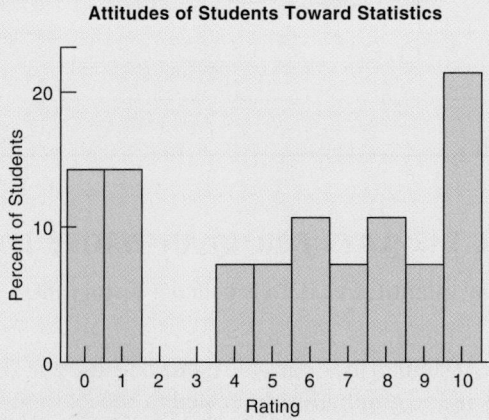

Attitudes of Students Toward Statistics

Draw conclusions.

From the histogram the professor sees that the strong negative attitudes are separate from the rest of the group and that the most typical response is a

10. The remainder of the responses are evenly spread out over ratings 4–9. Although there is no new information in the graph, the visual display adds another dimension to the information.

 TRY IT NOW!

New Product Survey
Drawing Histograms for Discrete Data

The marketing research firm that is conducting a survey about consumer attitudes toward a new brand of bath soap wants to present its data graphically. The frequency table for the data is shown here:

Rating	Frequency	Relative frequency (%)
1	4	13.3
2	5	16.7
3	6	20.0
4	11	36.7
5	4	13.3
Total	30	100.0

Create a relative frequency histogram for the data.

Histograms for Continuous Data

A histogram for continuous data differs from a histogram for integer data in that each rectangle represents a class interval, which is a range of values. For this reason, the rectangles are not centered on values but begin and end at the class boundaries.

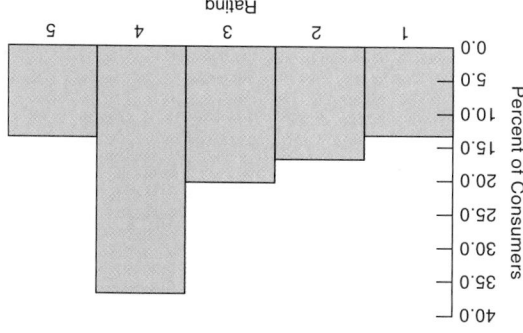

Ans.

EXAMPLE 2.18

ABC FACULTY SALARIES

Creating a Histogram for Continuous Data

The committee that is studying gender equity in salaries wants to draw a histogram of the data on number of years of service. They decide to use the frequency distribution that has 5-year intervals:

Analyze the data.

Years of service	Frequency
$0 \leq x < 5$	49
$5 \leq x < 10$	43
$10 \leq x < 15$	25
$15 \leq x < 20$	24
$20 \leq x < 25$	48
$25 \leq x < 30$	15
$30 \leq x < 35$	2
$35 \leq x < 40$	0
$40 \leq x < 45$	1

The histogram is shown here:

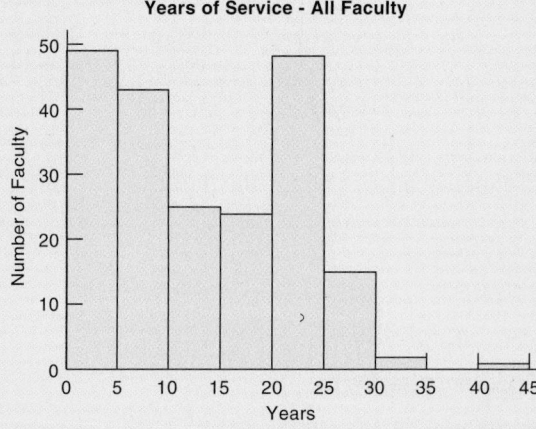

Draw conclusions.

The committee sees at first glance that the data are heavily concentrated at the bottom half of the class intervals and that there is also a large concentration at 20–25 years.

Sometimes you will see histograms that do not use the endpoints of the intervals to label the *x* axis. The bars of the histogram are centered on the axis tick marks and the midpoint of each interval is used instead. Minitab uses this convention as the default for histograms.

EXAMPLE 2.19

FLIGHT DELAYS

Creating a Histogram for Continuous Data

The consumer group that is looking at airline delays wants to make a graphical display of the data they have collected. Here is the frequency table for the data:

Percentage of delayed flights	Frequency	Relative frequency (%)	Cumulative relative frequency (%)
$0.0 \leqslant x < 10.0$	3	10.0	10.0
$10.0 \leqslant x < 20.0$	15	50.0	60.0
$20.0 \leqslant x < 30.0$	9	30.0	90.0
$30.0 \leqslant x < 40.0$	2	6.7	96.7
$40.0 \leqslant x < 50.0$	1	3.3	100.0

Analyze the data.

The group decides to create a relative frequency histogram. They must scale the y axis so that it can accommodate percentages from 0% to 50%. The group decides to go from 0% to 50% in increments of 10%. The tick marks on the x axis are the values at the beginning of each class interval. The last tick mark is the end of the last interval. The histogram is shown here:

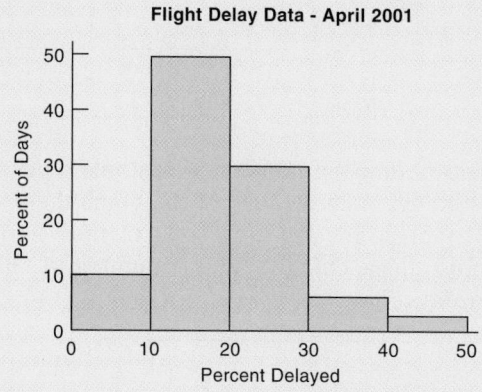

Flight Delay Data - April 2001

From the histogram the consumer group sees that on about half the days between 10% and 20% of flights were delayed, and that on 80% of the days there were 10% to 30% delayed flights.

Draw conclusions.

2.3.3 DISPLAYING SMALL DATA SETS

The rules for creating histograms are not suitable for data sets that contain less than 25 observations. This is because the number of classes should not be less than 5, and we determine the number of classes by taking the square root of the sample size. Often when we collect data, we do not have more than 25 observations. Is there a way to display such data sets graphically? The solution is a graphical method called a **dotplot.**

> A **dotplot** is a graph used for small data sets, in which each observation is plotted as a point on a single horizontal axis. The dotplot's axis is scaled so that each of the data points can be located uniquely on the axis. When more than one observation have the same value, the points are "stacked" on top of each other.

A dotplot can show many of the same features of the data as a histogram.

BANK CUSTOMERS

Creating a Dotplot for a Data Set

Understand the problem.

After hearing complaints and noticing long lines, a bank manager is interested in looking at the number of customers who arrive hourly on the Fridays preceding holiday weekends. She collects data for 8 hours on two different Fridays:

Collect the data.

| 10 | 14 | 15 | 14 | 19 | 12 | 11 | 14 |
| 15 | 14 | 20 | 19 | 11 | 12 | 17 | 17 |

After looking at the data, the bank manager decides to use a dotplot to display them because there are not very many observations. An axis is scaled from 10 (the smallest value) to 20 (the largest value) in increments of 1 because the data are integers:

Analyze the data.

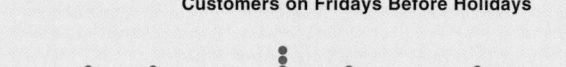

Customers on Fridays Before Holidays

Number of Customers per Hour

Draw conclusions.

The dotplot shows that the number of customers per hour varies widely over the interval, although in 25% (4/16) of the hours there were 14 customers, and in 69% (11/16) there were at least 14 (14 or more) customers. The bank manager decides to use these data to plan the number of tellers who should be working on those Fridays.

2.3.4 USING THE COMPUTER TO CREATE GRAPHICAL DISPLAYS

You may be wondering why it is necessary to learn how to create bar charts, pie charts, and histograms by hand, when you will almost certainly be using a computer to create them. The graphs produced by computer software are superior to those produced by hand, but a good deal of critical thought goes into creating a graph that conveys *information* to the viewer. It is not so important that you create the charts by hand, as that you know *how* they are created.

Most computer software packages ask the user for input when creating graphs. Knowing how the charts are created will help you make decisions about how the charts look. In addition, not all software allows the user to make the same decisions. The default graphs produced by computer software offer great starting points for the final graphical display, but they are not usually the end product that you are looking for.

When you create a bar chart by hand, you have the freedom to put the bars in any order you want and to combine categories as you wish. Some software allows you to choose ascending or descending order of frequency for the bars; others require you to sort the frequency table first, by either frequency or alphabetical order of categories. The same is true about combining categories and using frequency or relative frequency. If you do not know what the chart should contain, you will not know how to process the raw data that you collect.

Creating histograms differs among software packages. There are differences in the way you go about defining and creating the chart. Almost all software packages create a histogram using a default set of class intervals, but they differ widely in what they allow the user to specify. You will see in the next section that when

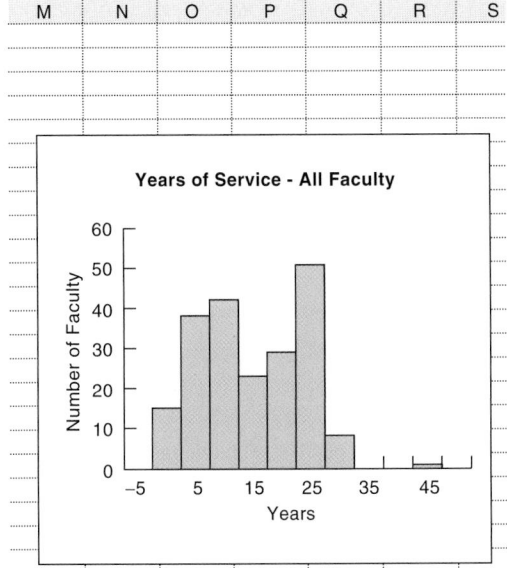

Note that regardless of the convention, Excel puts the labels at the midpoints of the bars. They should actually all shift to the right and line up with the tick marks.

FIGURE 2.2 Excel histogram for faculty years of service

you are trying to make graphical displays to compare different samples, you need to be able to control the x axis of the graph. Microsoft Excel allows you to specify the ending value for each of the classes, but when it creates the histogram from the data, it puts the endpoints in the centers of the bars. Minitab and SPSS allow you to specify endpoint values or midpoint values *or* the number of classes you want to use, but not both.

There are also differences in the way the final chart looks. Remember that Minitab and SPSS define their intervals to include the lower value and go up to but not include the upper value. Microsoft Excel, on the other hand, uses the opposite convention. The differences in processing cause the histograms from various software packages to look a little different. Figure 2.2 shows the histogram produced by Microsoft Excel for the data on faculty years of service.

If you compare this histogram to the one in Example 2.18, you see that the major difference is in the first class. In the Excel histogram the first class contains only those people with 0 years of service ($-5 < x \le 0$), whereas in Example 2.20 the first class contains all people with 0, 1, 2, 3, or 4 years of service ($0 \le x < 5$). This difference does change the lower end of the histogram, but it is not very different. Knowing the basics of creating histograms enables you to make smarter, more informed decisions, no matter what software you use.

2.3.5 EXERCISES—LEARNING IT!

2.12   The data that the computer society collected on college operating systems are given here again: *Requires Exercise 2.2*

Mac OS	Mac OS	Mac OS	W NT	W 2000
W NT	W NT	Linux	W NT	W 2000
W 2000	W 2000	W NT	Mac OS	W NT
W NT	Mac OS	Mac OS	W NT	Mac OS
W 2000	Mac OS	Linux	W 2000	Linux
Mac OS	W NT	W NT	W 2000	W 2000

(a) Suppose you were asked to create a bar chart for the data. How would you order the categories? Why?

(b) Make the bar chart for the data.

2.13 Every month the Department of Transportation (DOT) releases the number of mishandled baggage reports filed by passengers on U.S. airlines. The data for April 2000 follow:

Airline	Reports per 10,000 passengers
Alaska	277
America West	581
American	502
Continental	397
Delta	381
Northwest	424
Southwest	401
TWA	452
United	587
US Airways	429

(a) Create a bar chart for the numbers of reported problems per 10,000 passengers.

(b) Comment on any interesting features of the data.

2.14 Columbia University conducted a survey of 637 members of the Authors Guild and the Dramatists Guild, all of whom have published at least one book, one play, or three magazine articles. As part of the survey, the members were asked to classify the amount of money they made from writing. This frequency table gives the results of the survey:

Revenue from writing	Number of members
More than $50,000	57
More than $20,000 but less than $50,000	127
More than $1000 but less than $20,000	192
More than $0 but less than $1000	102
No money at all	159
Total	637

(a) Create a relative frequency bar chart for the data.

(b) Would you conclude that a majority of those surveyed can or cannot earn a living solely from writing?

(c) Although the data are numerical, why did you have to make a bar chart rather than a histogram?

(d) What problem do you see with the way the classes are defined?

Requires Exercise 2.5 **2.15** The hospital that collected data on the number of weekly repairs needed by copy machines decided that it wanted a graphical display of the data to support a request for more funds for repairs. The data are reprinted here:

```
3  2  3  0  3
0  2  2  2  1
3  2  2  1  0
2  3  2  1  2
3  4  2  7  1
2  2  1  3  1
0  3  1  1  2
```

Create a histogram for the data.

Requires Exercise 2.3 **2.16** After collecting the following data on the conditions of cereal box cases, the supermarket manager wants a graph of the data to use in a brainstorming session with the employees.

Unsealed	Crushed	No defect	Dented	Crushed
No defect	Crushed	No defect	No defect	No defect
Crushed	No defect	No defect	Dented	Crushed
Crushed	Unsealed	No defect	Dented	Dented

(a) What type of graph is appropriate for these data?

(b) Create a graphical display of the data.

(c) What order did you select for the categories? Why?

2.17 Retaining employees is a problem for many industries. Just about the time the employee is at the height of productivity, he or she often leaves the company. In an attempt to figure out why it is losing a large number of salespeople, a large pharmaceutical company decides to look at the reasons the employees gave for leaving. Managers examine the exit interviews of 35 salespeople who left the company in the past two years. The table lists the reason codes:

Location	L
Raise in pay	P
Creative differences	C
Relationship with manager	M
More responsibilities	R
Leaving the workforce	Q
Too much travel	T

The data follow:

```
T   T   L   L   T   L   L
L   L   L   Q   R   L   L
R   M   T   P   C   L   R
T   P   L   L   L   T   M
M   L   Q   T   T   L   T
```

(a) Create a bar chart for the data, including relative frequency.

(b) What is the major reason salespeople left their jobs?

(c) Does it appear that the pharmaceutical company is losing salespeople for one or two very specific reasons, or are there a wide variety of reasons?

(d) If you were a manager of this company, what would you do about the problem? Why?

2.18 After the cable television service changed the channels in a town, subscribers were asked to rate the new service on a scale of 1 = worst service I have ever had to 10 = best service I have ever had. The responses from 40 random subscribers follow:

```
1   4   5   7   8
2   4   5   7   9
2   4   5   7   9
3   4   6   7   9
3   4   6   7   9
3   4   6   7   9
3   4   6   8   9
3   5   7   8   10
```

(a) Create a relative frequency histogram of the ratings.

(b) Are the majority of the ratings favorable, unfavorable, or neither?

2.19 The company in Exercise 2.17 that is worried about losing its salespeople also looked at the lengths of time (in months) that the people in the sample worked for the company:

```
18.5   19.9   22.1   23.1   25.5   26.3   27.4
18.6   20.5   22.4   23.3   25.6   26.6   27.6
18.7   20.8   22.6   24.7   25.8   26.8   29.8
19.8   20.9   22.9   25.1   26.1   27.0   31.4
19.8   21.7   23.0   25.5   26.3   27.0   47.5
```

Create a relative frequency histogram for the data.

2.20 The computer group that is looking at the hours per week high school students spend instant messaging decides to use the data it collected in a presentation to parents: *Requires Exercise 2.8*

```
0.0   4.1   11.9   15.5   17.8
0.0   4.2   13.0   15.5   17.9
1.0   6.9   13.4   15.5   18.4
1.4   8.0   13.5   15.5   18.7
2.5   8.2   14.9   16.7   19.5
2.9   8.6   15.1   16.8   19.7
```

(a) Create a relative frequency histogram for the data.

(b) From the histogram, does it appear that a majority of the students spend more than 14 hours per week instant messaging?

Requires Exercise 2.10   **2.21**   The Bureau of Weight and Measures, which looked at the contents of orange juice containers from supermarkets, wants to create a graphical display of the following weights (in ounces):

64.8	65.2	65.6	65.7	65.9
64.8	65.3	65.7	65.7	66.0
64.9	65.3	65.7	65.7	66.1
65.1	65.4	65.7	65.7	66.1
65.2	65.4	65.7	65.8	66.3

Draw an appropriate graphical display of the data.

2.4   Describing and Comparing Data

Since the reason for displaying data is to gain understanding, it is important to know what we can learn from the graph of a data set. For quantitative data we usually want to know what a typical observation is and how the actual observations differ from the typical values. We also like to be able to compare different data sets and make some decisions based on the comparisons.

2.4.1   DESCRIBING QUANTITATIVE DATA

In statistics, the features of interest for a set of numerical data can be classified as **center, shape,** and **variability.**

> The *center* of a set of data is where, numerically, the data are centered or concentrated.

> The *shape* of a set of data describes how the data are spread out around the center with respect to the symmetry of skewness of the data.

> The *variability* of a set of data describes how the data are spread out around the center with respect to the smoothness and magnitude of the variation.

Together these three features describe the distribution of the data. It is useful to picture the distribution as a smooth curve that captures the "shape" of the histogram. One way to create this curve is to plot a point at the top of each bar of the histogram and then connect the dots as shown in Figure 2.3. This is sometimes known as a

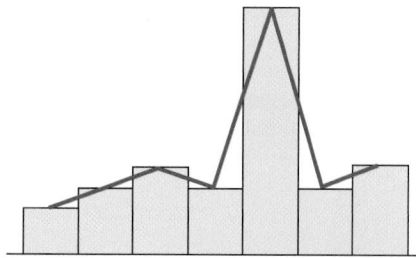

FIGURE 2.3   Histogram with a line showing the shape of the distribution

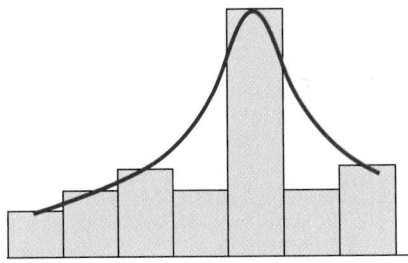

FIGURE 2.4 Histogram with a smoothed curve representing the distribution

frequency polygon. If we smooth out the plotted line, we get a curve that looks like the one shown in Figure 2.4. The curve shows the bump or high point of the curve. Since the bump in the distribution is the location of the class that has the highest frequency, it locates the most typical data points. In general, a data set will have one center. We refer to such a data set as **unimodal.** Sometimes a set of data will appear to have two centers. We call such a data set **bimodal.** More than two centers is very unlikely. Often, when there is more than one center, it indicates that another important variable is affecting the results.

The shape of a distribution illustrates how the data are spread out on either side of the center—that is, whether they are **symmetric** or **skewed.**

When the data are evenly spread out on both sides of the center, we describe the distribution of the data as *symmetric.*

When the data are not evenly spread out on either side of the center, we refer to the distribution as being *skewed.*

A typical symmetric distribution is shown in Figure 2.5.

Skewness has a direction associated with it, either left (negative) or right (positive). The direction of the skew describes the side on which the distribution of the data covers a larger distance or has more variability. It is the direction in which the distribution "tails off" more slowly. Figure 2.6 shows both right and left skewed distributions.

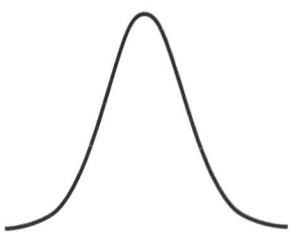

FIGURE 2.5 A typical symmetric distribution

 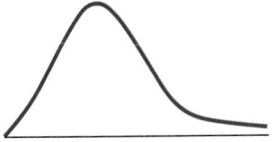

FIGURE 2.6 Typical skewed distributions

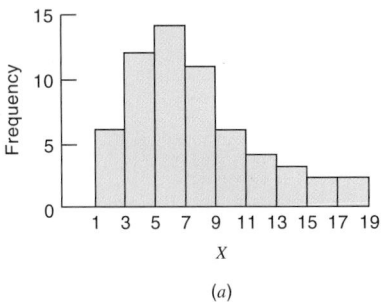 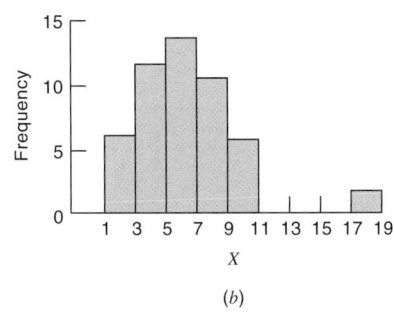

FIGURE 2.7 **Histograms with (*a*) skewed data and (*b*) extreme values**

When data are skewed, either left or right, the tailing off of the data is continuous and gradual as shown in Figure 2.7(*a*). When the tailing off includes a gap in the data—a place where classes in the frequency histogram have no observations—as shown in Figure 2.7(*b*), the data are not really skewed. More likely the data contain some extreme or unusual observations. We will talk about these extreme values more in Chapter 3 when we discuss outliers.

In addition to center and shape, we are interested in how much the data differ from the center or typical values. We need some way to measure the differences. We will examine measures of variability in Chapter 3. At this point we can describe the variability of the distribution in two ways: the "smoothness" of the curve and the total spread of the data.

When data are not very variable, the frequency of the observations decreases steadily as we move away from the center. Sometimes when data are highly variable, the distribution is jagged; that is, the frequency of the data values does not decrease steadily as we move away from the center. Another way to understand variability is to describe the distance from one end of the data to the other. Figures 2.8 and 2.9 show histograms with different degrees of variability.

Dotplots can also be used to describe data distributions, but they typically represent only small data sets.

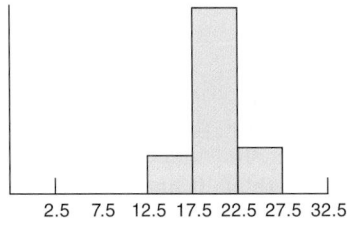 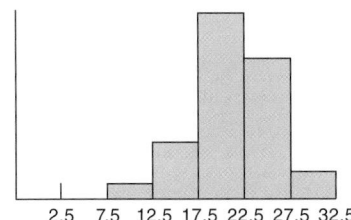

FIGURE 2.8 **Histograms with different variability showing spread of data**

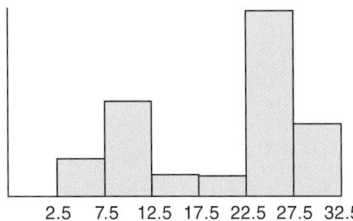 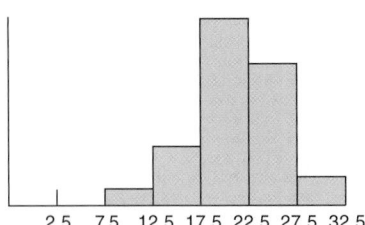

FIGURE 2.9 **Histograms with different variability showing smoothness**

EXAMPLE 2.21

ABC FACULTY SALARIES

Describing Data Distributions

The committee at ABC College looks at the distribution of years of faculty service in Figure 2.2 to see whether they can determine how the data are distributed. They find that the data are highly variable and the distribution does not have a smooth shape. In addition, there does not seem to be a unique center or shape. The committee wonders why this might be. Perhaps another variable is affecting the picture. Since rank is often determined by years of service, it might make more sense to look at the years of service for each rank separately.

Analyze the data.

Sometimes when we collect data and make a histogram, there is no cohesive picture to be seen. This often happens when other underlying variables have an effect on the variable we are studying. When this happens it is useful to separate the data into groups on the basis of another variable and look at them again.

EXAMPLE 2.22

FLIGHT DELAYS

Describing Data Distributions

The consumer group that is looking at flight delays obtained this histogram:

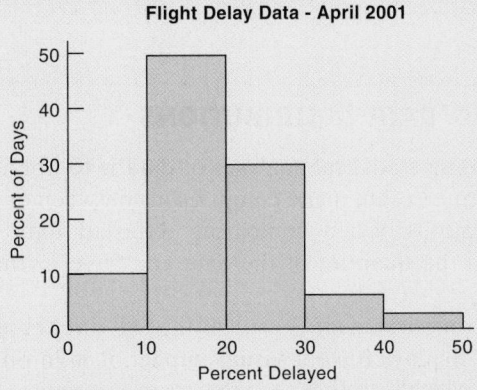

Analyze the data.

From the histogram the group can see that the data have a distinct center and that, typically, between 10% and 20% of the flights were delayed. The frequency of the data drops off rapidly from the center. The data are relatively smooth and not too variable. The distribution is skewed to the right, which means that the data are more spread out on the high side (three classes) than on the low side (one class).

The choice of the number of class intervals for a histogram is very important. If a histogram has too many intervals for the number of data values, then the data might appear to have a lot of variability when, in fact, classes are empty or have few observations because of the lack of data. If a histogram has too few intervals, then the distribution looks like a lump and important features of the data might be hidden. It is helpful to look at the data in several different ways to make sure that you are getting a true and consistent picture. You will see later on that it is possible to distort graphical displays and bias an analysis by playing with the scale.

Assignment Times
Creating a Frequency Table for Continuous Data

The instructor for the Introductory Statistics class wants to see, in graphical form, the data she collected on the times (in minutes) it took the students to do the assignment. Here is the frequency distribution for the data:

Time	Frequency	Relative frequency (%)
$22.70 \leq x < 26.22$	6	20.0
$26.22 \leq x < 29.74$	11	36.7
$29.74 \leq x < 33.26$	7	23.3
$33.26 \leq x < 36.78$	5	16.7
$36.78 \leq x < 40.30$	0	0.0
$40.30 \leq x < 43.82$	1	3.3

Create a relative frequency histogram for the data.

Use the histogram to describe the distribution of the times the students took to complete the assignment.

2.4.2 COMPARING DATA DISTRIBUTIONS

A major reason for doing statistical analyses of data is to obtain facts for making informed decisions. We must often make comparisons between or among samples taken from different populations. When comparing different data sets, we make those comparisons based on the qualities of the data you have learned: center, shape, and variability.

To make valid comparisons it is critical that all data be displayed in the same way. Since graphical displays have a visual impact, it is important that the graphs used are comparable in scale so that the viewers are not misled.

To permit accurate visual comparisons about center, shape, and variability, the class intervals in different graphs should be the same. When this is not the case, it is

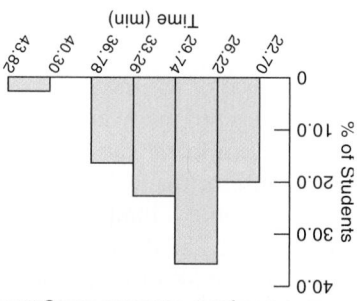

ANS. **Time to Complete Statistics Assignment** TYPICALLY BETWEEN 26.22 AND 29.74 MINUTES, SKEWED RIGHT, NOT VERY VARIABLE.

difficult if not impossible to compare these qualities in a meaningful way. The following example shows how using different scales can mask the visual impact of the display.

EXAMPLE 2.23

FINANCIAL AID APPLICATION TIMES

Comparing Data Distributions

Several colleges and universities have been developing new procedures for filling out financial aid forms. The new procedures have one universal form that students use to apply for all financial aid, whether public (state and federal) or private (scholarships and loans from various sources). After the procedures were in place for several semesters, the group that designed the new form collected data on the amounts of time (in minutes) it took students to complete the financial aid application. The times were recorded at two different colleges that were similar in size and student population. One used the new method, and the other retained the old method.

New method					Old method			
39.2	41.2	44.1	46.2	49.9	45.1	47.6	48.3	49.7
39.2	42.2	44.3	46.3	51.2	45.7	47.7	48.4	49.4
40.0	43.4	44.4	47.4	51.7	46.9	47.6	48.4	49.6
40.0	43.5	44.9	47.7	52.3	46.8	48.1	48.7	49.9
40.8	43.5	45.3	48.1	53.7	46.6	48.1	49.0	50.0
41.0	44.0	45.6	48.3	56.5	47.2	48.4	49.0	50.6

To better understand the data and to compare the application times for the two methods, a computer software package is used to create a histogram for each of the data sets. The histograms show that the application times for the new method are typically between 44 and 46 minutes and that the data are slightly skewed to the right. The loan application times for the old method are typically between 47.5 and 48.5 minutes and the data are also skewed right. On first glance, it appears that the new method is better. Or is it?

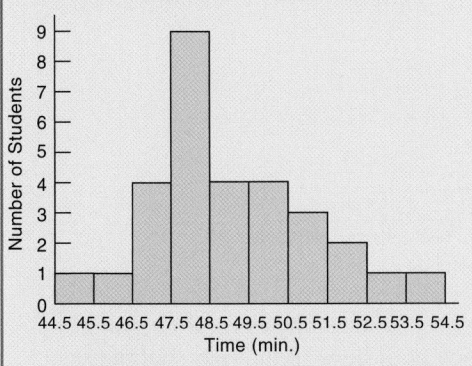

Financial Aid Application Times - Old Method

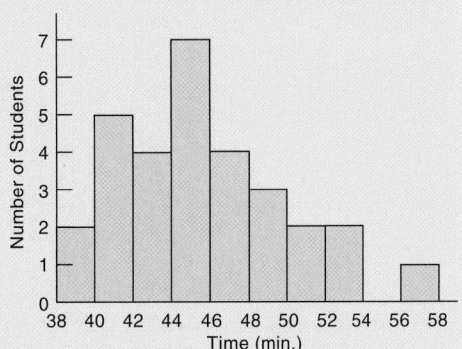

Financial Aid Application Times - New Method

The group decides to recreate the histograms, this time using a common scale for the *x* axis. The results are shown in the following histograms.

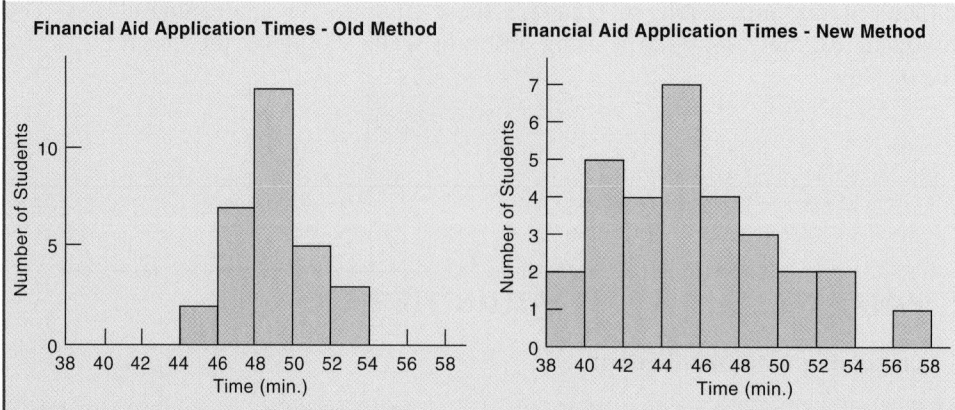

From the second set of histograms it appears that the times for the new form are much more variable than those for the old form. The typical times to complete the new application form are really not that much different from those for the old form. The second set of graphs shows that the old application form has more consistent times than the new form, and it has comparable completion times.

In the preceding example it does not really matter that the *y* axis of the histograms is frequency rather than relative frequency because the number of observations is the same in each sample. If this is not the case, you should use relative frequency to make valid and meaningful comparisons.

ABC FACULTY SALARIES

Drawing Conclusions from Graphs

Now that the committee understands a little more about the characteristics of the faculty at ABC College, they finally decide to look at salaries. It would seem that a faculty member's salary should be closely related to years of service. The committee decides to make a histogram of faculty salaries and compare it with the histogram for years of service. The histogram for faculty salary is shown here:

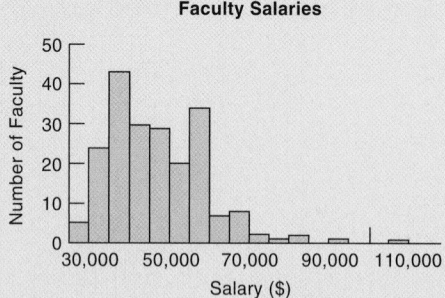

Comparing this graph with Figure 2.2, they see some similarities. Both graphs appear to be bimodal and highly variable. They both skew to the right, but the salary histogram shows more unusually high values. It appears that there might be other factors to consider in this analysis. Perhaps gender does make a difference!

Discovery Exercise 2.1

THINKING ABOUT VARIABILITY

A manufacturer of compact discs uses two different suppliers for the jewel boxes used to hold the discs. There have been problems with these boxes in the past. The inside width of the jewel box has critical specifications of 119.0 ± 0.2 mm. If the case is too narrow, the disc will not fit in it; if it is too wide, the front label insert slips around, making the box hard to close. This is the number one cause of customer complaints. Because it is time to renew the purchasing contracts for the jewel boxes, the CD manufacturer decides to look at a sample of the boxes from each supplier. The data for each source are in millimeters:

	Supplier A					Supplier B			
118.7	118.9	119.0	119.1	119.2	118.8	118.8	118.9	118.9	118.9
118.8	119.0	119.0	119.1	119.2	118.8	118.8	118.9	118.9	119.0
118.8	119.0	119.0	119.1	119.2	118.8	118.8	118.9	118.9	119.0
118.9	119.0	119.1	119.2	119.2	118.8	118.8	118.9	118.9	119.0
118.9	119.0	119.1	119.2	119.3	118.8	118.9	118.9	118.9	119.1

1. Make a relative frequency histogram of the data for each supplier.

2. Describe the distribution of jewel box widths for each supplier and compare them.

3. Your company has decided to buy its jewel boxes from a single source. The purchasing agent in charge of the accounts argues that supplier B should not get a renewed contract because the jewel boxes from that source are not centered at the target specification of 119.0 mm, whereas the jewel boxes from supplier A are right on target. Can you explain to him why, although his observation is true, his decision to use supplier A is not necessarily correct? What factor has he failed to consider?

4. Which supplier would you recommend that the company use? Write a short memo to the manager with your recommendation and supporting reasons.

2.4.3 EXERCISES—LEARNING IT!

2.22 A university collected some data on the amount of money that students spend on textbooks in a typical semester. Here are the dollar amounts:

239	289	304	323	336
256	290	307	324	394
280	295	310	326	397
284	295	314	330	415
284	298	315	331	429
287	298	319	332	445
287	299	321	334	447

(a) Create a histogram of the data.

(b) Where is the center of the distribution?

(c) Are the data symmetric or skewed? If they are skewed, are they left or right skewed?

(d) Describe the variability of the data.

2.23 The pharmaceutical company that is looking at retaining employees wants to summarize its data. The lengths of time (in months) that salespeople stayed at the company are listed at the top of page 94.

Requires Exercise 2.17

18.5	19.9	22.1	23.1	25.5	26.3	27.4
18.6	20.5	22.4	23.3	25.6	26.6	27.6
18.7	20.8	22.6	24.7	25.8	26.8	29.8
19.8	20.9	22.9	25.1	26.1	27.0	31.4
19.8	21.7	23.0	25.5	26.3	27.0	47.5

(a) Describe the center of the data.

(b) Describe the shape of the data.

(c) Describe the variability of the data.

Requires Exercise 2.11 **2.24** The refuse company decides to use the data it collected in a presentation to the town. These are the heights (in inches) of the newspaper piles put out for recycling:

8.3	9.1	9.9	10.2	11.0
8.5	9.7	9.9	10.4	11.2
8.8	9.8	10.1	11.0	11.4
12.0	12.3	13.1	13.6	14.3
12.0	12.3	13.3	13.6	14.4
12.2	12.5	13.3	13.8	14.4
12.2	12.7	13.4	13.9	14.4
12.3	12.8	13.4	14.1	14.8
12.3	12.9	13.4	14.3	15.2

Describe the distribution of the data. Remember to discuss center, shape, and variability.

Requires Exercise 2.9 **2.25** The people doing the study on how many hours people spend cleaning their houses are planning to write a newspaper feature on the topic. They want to be able to describe the data they collected:

0.6	2.7	3.6	4.7
1.2	2.9	3.8	4.9
1.3	3.0	3.8	5.0
1.9	3.0	4.1	5.0
2.0	3.1	4.2	5.6
2.1	3.1	4.4	5.7
2.2	3.3	4.4	5.8
2.4	3.3	4.4	6.0
2.6	3.3	4.5	6.0
2.6	3.4	4.7	7.4

(a) How long does a typical person spend cleaning the house?

(b) What shape is the distribution of cleaning times?

(c) Describe the variability of cleaning times.

GET IT IN WRITING

ABC College

TO: **President of the College**
FROM: Committee on Gender Equity
RE: **Current Faculty Salaries**

Before we look at the issue of gender equity, it will be helpful to understand the current faculty salary situation. This will give us a frame of reference to use in responding to the issue of gender equity. The data we collected were for

207 current faculty members. The variables chosen were rank, years of service, and salary. The results of the analysis are presented in summary form and recommendations follow.

Figure 1 shows the distribution of faculty by rank. No single rank constitutes a majority of the faculty. With the exception of the rank of Instructor, which makes up only 4% of the total faculty, the distribution by rank is fairly uniform. The low percentage in the Instructor rank is not unusual, since faculty at this rank do not have a doctorate; therefore, the category consists of special terminal appointments. It is interesting to note that the percentage of faculty at each of the other ranks (Assistant, 37%; Associate, 32%; Professor, 27%) decreases. This might turn out to be an important phenomenon.

Distribution of Faculty by Rank

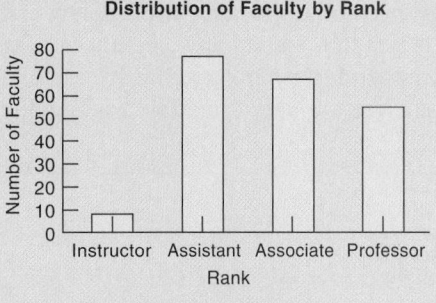

FIGURE 1

The graph that shows the distribution of faculty by years of service (Figure 2) shows that the data are bimodal; that is, they center in two different places. A typical faculty member has between 0 and 10 years or between 20 and 25 years of service. The number of faculty with between 10 and 20 years of service is considerably smaller. This drop might indicate that faculty members at this point in their careers are leaving the college, which is cause for some concern. The data are highly variable, indicating that perhaps other factors affect years of service.

Faculty Years of Service

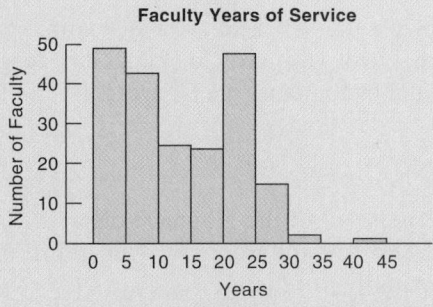

FIGURE 2

The distribution of faculty by salaries is shown in Figure 3. It seems reasonable that salary and years of service would be related, so we might expect the distributions in Figures 2 and 3 to be similar, which they are in several ways. First, the salary histogram is also bimodal, indicating that typical salaries are in the $35,000–$40,000 range and the $55,000–$60,000 range. This is consistent with the data on years of service. The salary data are also highly variable, particularly on the high side. The sample contains several unusually high salaries. Again, this indicates that other factors should be considered.

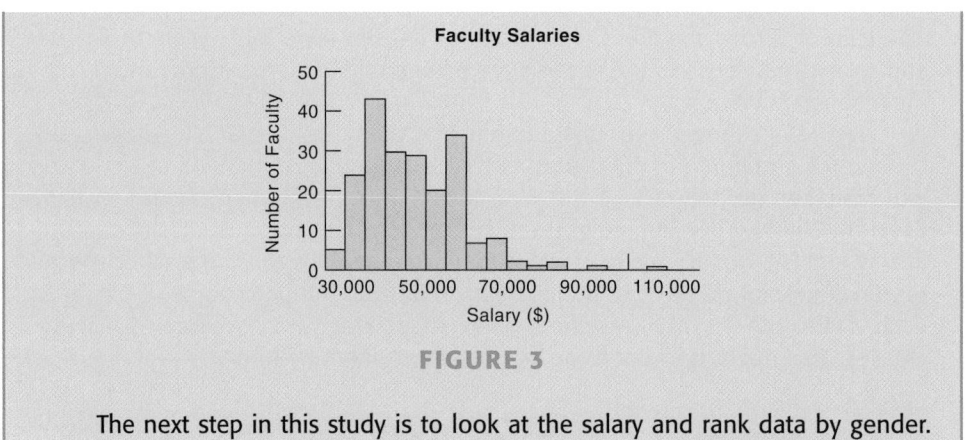

FIGURE 3

The next step in this study is to look at the salary and rank data by gender. This will help us to see whether indeed any differences exist. We are going to generate additional data on gender, tenure, and the school in which the faculty member teaches. We will analyze the data using this additional information and report back within 2 weeks.

2.5   Creating Graphical Displays
Using Technology

The graphical displays that you learned about in this chapter can be created by using one of the computer packages available for statistics. Microsoft Excel 2000, Minitab, and the TI-83 graphing calculator are three of the most popular and effective packages. For detailed instructions on how to use each particular computer package, refer to the Technology Manual that accompanies this text. For an overview of each computer package's capabilities, read on.

Excel 2000

Using a computer package such as Excel to create graphs gives you the power to modify them easily. With Excel you have the capability to generate bar charts, pie charts, frequency tables, and histograms (see Figures 2.10–2.12).

Minitab

Minitab can be used to create all of the graphical displays you learned about in this chapter. Charts in Minitab are created from raw data, not from frequency tables. With Minitab you have the capability to generate bar charts, pie charts, and histograms (see Figures 2.13–2.15).

	A	B	C
1			
2	Rank	Total	
3	Associate	67	
4	Assistant	77	
5	Instructor	8	
6	Professor	55	
7	Grand Total	207	

FIGURE 2.10   **Excel frequency table**

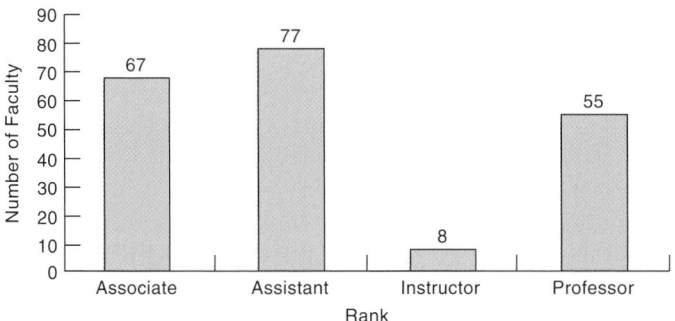

FIGURE 2.11   **Excel column chart**

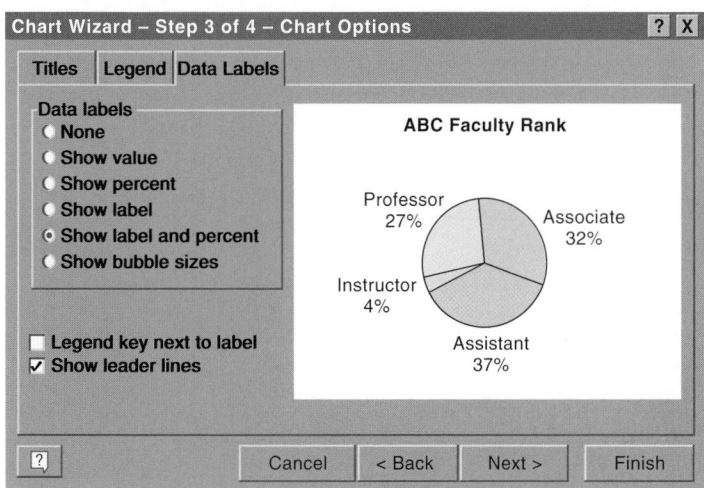

FIGURE 2.12   **Formatting the pie chart in Excel**

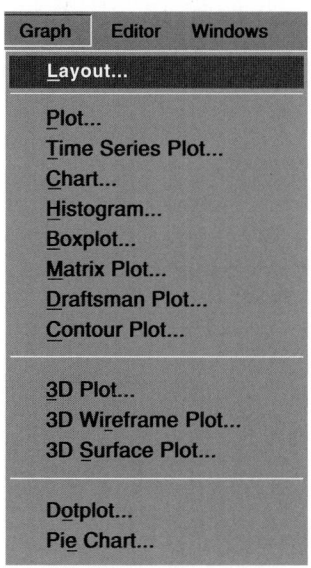

FIGURE 2.13   **The Graph menu in Minitab**

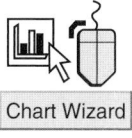

Chart Wizard belongs to the Standard Toolbar.

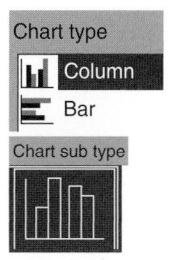

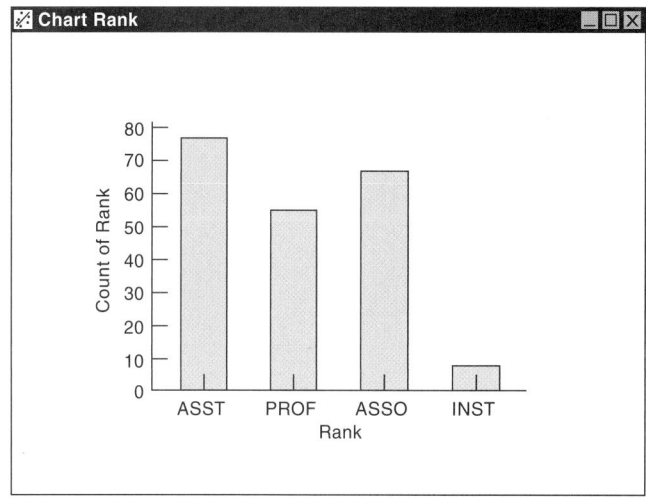

FIGURE 2.14 **Minitab bar chart of rank**

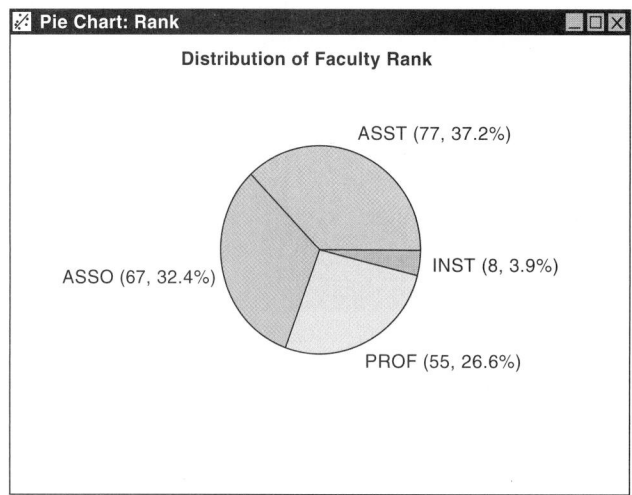

FIGURE 2.15 **Complete pie chart in Minitab**

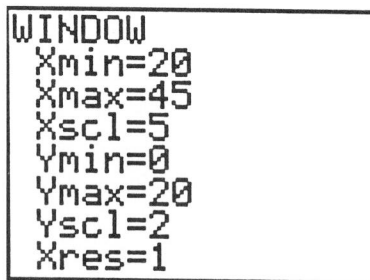

FIGURE 2.16 **TI-83 window screen with values to create a histogram**

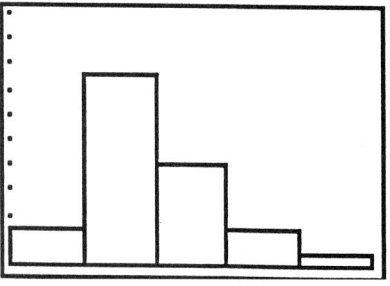

FIGURE 2.17 **Completed histogram on TI-83**

TI-83 Graphing Calculator

You cannot create bar charts or pie charts using the TI-83 graphing calculator, but you can create histograms (see Figures 2.16 and 2.17).

CHAPTER 2 SUMMARY

Raw data are nothing more than a list of words or numbers. The purpose of descriptive statistics is to turn *data* into *information*. One way to do this is by summarizing the data with a frequency distribution and then using a graphical display that is appropriate for the type of data. Qualitative data can be displayed in *bar charts* or *pie charts*, whereas quantitative data are usually displayed in *histograms* or *dotplots*.

Graphical displays can be used to describe data in terms of their *center*, *shape*, and *variability*. They can also be used to compare different data sets.

Although it is possible to create graphs by hand, using a computer software package such as Minitab or Excel greatly enhances the quality of the finished product. It is important, however, to know how each graphical technique works so that you can use the software intelligently and make useful (and not just flashy) graphs.

KEY TERMS		
Term	**Definition**	**Page reference**
Bar chart	A **bar chart** is a graph that represents the frequency or relative frequency from a frequency table in the form of a rectangle or bar.	73
Center	The **center** of a distribution is where, numerically, the data are centered or concentrated.	86
Cumulative relative frequency	The **cumulative relative frequency** of a class is the sum of the relative frequencies of all classes at or below that class represented as a portion of the total number of observations.	62
Dotplot	A **dotplot** is a graph used for small data sets, in which each observation is plotted as a point on a single horizontal axis.	81
Frequency table	A **frequency table** or **frequency distribution** is a table that records each category, value, or class of values that a variable might have and the number of times that each one occurs in the data.	57
Histogram	A **histogram** is very similar to a bar chart, but because the data are naturally ordered, the x axis of the graph must be scaled to reflect this.	77
Pie chart	A **pie chart** is a graph in which a circle is used to represent the whole, and each "slice" is used to represent one of the categories. The size of the slice is proportional to the relative frequency of the category.	76
Relative frequency	The **relative frequency** of a classification is the number of times an observation falls into that classification represented as a portion of the total number of observations.	59

(continued)

Term	Definition	Page reference
Shape	The **shape** of a distribution describes how the data are spread out around the center with respect to symmetry or skewness.	86
Skewed	When the data are not evenly spread out on either side of the center, we refer to the distribution as being **skewed.**	87
Symmetric	When the data are evenly spread out on both sides of the center, we describe the distribution of the data as **symmetric.**	87
Variability	The **variability** of a distribution describes how the data are spread out around the center with respect to the smoothness and magnitude of the variation.	86

CHAPTER 2 EXERCISES

Learning It!

2.26 In a study on housing costs in major international cities, data were collected on the construction costs (dollars per square meter) for housing in 23 capital cities. Here are the data:

Jakarta	65	Melbourne	383
Dar Es Salaam	67	Algiers	500
Karachi	87	Washington, DC	500
Beijing	90	Madrid	510
New Delhi	94	London	560
Istanbul	110	Toronto	608
Manila	148	Seoul	617
Bangkok	156	Hong Kong	641
Kingston	157	Singapore	749
Bogota	171	Paris	990
Johannesburg	192	Tokyo	2604
Rio de Janeiro	214		

(a) What graphical display is appropriate for these data? Why?

(b) Create a dotplot of the data.

(c) Use the data to describe housing construction costs for these cities.

(d) Do any of the data values appear to be unusual? If so, which one(s)?

2.27 In an ongoing study of trends in higher education, an article in *Dartmouth Life* (September 15, 1996) reported on the number of students from different classes who enrolled in various majors as freshmen. The data for the class of 1986 are:

Major	1986
Anthropology	10
Art History	28
Asia	11
Biology	31
Chemistry	28
Classics	2
Comparative Literature	5
Drama	13
Earth Science	8
Economics	90
English	156

(continued)

Major	1986
French	23
Geography	23
Government	147
History	102
Mathematics	35
Music	10
Philosophy	21
Physics	13
Psychology	70
Religion	22
Sociology	17

(a) Create a bar chart for these data.

(b) What percentage of the students majored in the sciences?

(c) What order did you choose for the categories on the x axis? Why did you choose this order? Can you think of any other orders that might also be appropriate? If so, what are they?

2.28 The *Chronicle of Higher Education* (June 2000) reported on the distribution of college and university faculty by rank:

Academic rank	Relative frequency (%)
Professor	28.9
Associate Professor	22.7
Assistant Professor	23.5
Instructor	12.1
Lecturer	2.3
No rank	1

(a) Create a bar chart for these data.

(b) What do the data tell about the distribution of faculty over the various ranks?

(c) Compare the distribution reported in the *Chronicle of Higher Education* with the distribution of faculty ranks at ABC College in Example 2.1.

2.29 A report by the Bureau of Labor Statistics said that in 1998, the last year for which such data are available, 709 homicides occurred while people were on the job. These homicides were classified by the type of job that the victim had:

Managerial and professional	131
Sales	239
Service industry	145
Drivers/factory workers	129
Other	65

(a) Create a bar chart for these data.

(b) Describe any interesting features of the data.

2.30 The publishing industry has long felt that the food magazine sector is in a recession. Researchers looked at the paid circulations for all of the magazines in the field and obtained these data:

Magazine	Paid circulations
Bon Appetit	$1,294,945
Cooking Light	1,119,811
Gourmet	906,299
Food & Wine	734,831
Eating Well	648,697
Cook's Illustrated	160,000
Fine Cooking	100,000
Saveur	100,000

(a) Create a bar chart for the data.

(b) Does one of the magazines account for a majority of the paid circulations in food magazines? If so, which one?

(c) Do you think that *Bon Appetit, Food & Wine,* and *Gourmet* deserve their reputation as the Big Three? Why or why not?

2.31 As part of a survey conducted by the Bureau of Transportation in 1996, drivers were asked about the age of the vehicle that they normally drive. The survey recorded the year that the vehicle was manufactured. Data for 50 drivers surveyed follow:

1955	1983	1989	1991	1993
1967	1984	1989	1992	1994
1970	1984	1989	1993	1994
1974	1984	1989	1993	1994
1977	1986	1990	1993	1994
1978	1988	1990	1993	1994
1978	1988	1990	1993	1995
1979	1989	1990	1993	1995
1981	1989	1991	1993	1995
1982	1989	1991	1993	1995

(a) Create a histogram for these data.

(b) Use the histogram to describe the distribution of the age of vehicles driven by the people in the sample.

(c) These data are really integer data. Why would a histogram using each year as a separate class not work for these data?

2.32 As part of the 2000 census of the United States, 700,000 households were asked to fill out an additional survey called the American Community Survey. One question that was asked was how long (in minutes) a person traveled to get to work each day. Here are the averages for the 50 states plus the District of Columbia:

Alabama	22.4	Kentucky	22.5	North Dakota	15.4
Alaska	18.2	Louisiana	23.7	Ohio	22.1
Arizona	23.7	Maine	21.2	Oklahoma	19.6
Arkansas	19.8	Maryland	29.2	Oregon	21.4
California	26.7	Massachusetts	26.1	Pennsylvania	23.8
Colorado	23.4	Michigan	22.7	Rhode Island	21.9
Connecticut	23.5	Minnesota	21.6	South Carolina	21.9
Delaware	22.5	Mississippi	21.9	South Dakota	15.6
District of Columbia	28.5	Missouri	23.2	Tennessee	22.7
Florida	24.3	Montana	16	Texas	23.6
Georgia	26.7	Nebraska	16.1	Utah	20.8
Hawaii	24.2	Nevada	22.3	Vermont	21.1
Idaho	19.7	New Hampshire	24.4	Virginia	25.4
Illinois	27.0	New Jersey	28.7	Washington	24.9
Indiana	21.7	New Mexico	19.9	West Virginia	25.5
Iowa	17.6	New York	31.2	Wisconsin	20.1
Kansas	17.7	North Carolina	22.6	Wyoming	17.1

(a) Create a graphical display of people's travel time to work.

(b) Use the graph to describe the distribution of the data.

(c) Are any of the data values unusual? If so, which one(s)?

2.33 *The Panel Study on Income Dynamics* (PSID) is a longitudinal survey of individuals and families that has been ongoing since 1968. A random sample of 40 respondents from Maryland to the 1997 Family Survey—Public Release I was surveyed. One question in the survey is how many rooms are in each family's place of residence. The data follow:

1	5	6	7	8
2	5	6	7	8
3	5	6	7	8
3	5	6	8	8
4	5	6	8	8
4	5	6	8	8
4	5	7	8	9
5	6	7	8	10

(a) Create a histogram of the data. Treat the data as integer data.

(b) Describe the distribution of the data.

(c) The *American Community Survey*, which is part of the 2000 census of the United States, is a detailed survey of 700,000 households. Results from this survey released in August 2001 indicate that 25% of the homes in Maryland have 8 or more rooms. How does the result from the PSID compare to this statement?

2.34 On May 1, 2000, United Airlines had 294 scheduled flights that departed from Denver International Airport. A random sample of 50 of those flights was taken and the departure delays (in minutes) were recorded. Negative numbers indicate that the flight left earlier than scheduled.

−7	−2	−1	0	8
−5	−2	−1	0	8
−5	−2	−1	1	9
−4	−2	−1	1	11
−4	−2	−1	2	15
−4	−2	−1	2	15
−3	−2	−1	3	54
−3	−1	0	5	56
−2	−1	0	6	63
−2	−1	0	8	111

Source: Bureau of Transportation Statistics (BTS)

(a) Create a relative frequency histogram for the data.

(b) What percentage of the flights were early or on time? What percentage were more than 30 minutes late?

(c) If you were preparing a report on flight delays for the management of United Airlines, what other variables would you include in the data?

(d) The sampling scheme described here is obviously not adequate for such a report. If you had to collect the data, how would you select the sample?

2.35 A survey of 40 households asked each head of the household how much was spent on clothing during the previous 3 months. The data are in dollars:

0.00	0.00	0.00	35.00	210.00
0.00	0.00	0.00	36.75	221.00
0.00	0.00	0.00	60.89	224.00
0.00	0.00	0.00	64.66	303.33
0.00	0.00	10.62	96.00	310.55
0.00	0.00	15.00	114.61	365.95
0.00	0.00	17.21	149.27	365.95
0.00	0.00	21.33	165.37	702.60

(a) Create a frequency table for the data.

(b) Use the frequency table to draw a histogram of the data.

(c) Describe the distribution of the amounts of money that households spend on clothing in a 3-month period.

Thinking About It!

2.36 Picking a health plan can be very confusing to new employees. The four major plan types have different costs and rules. Here are the major types of plans:

HMO—health maintenance organization
PPO—preferred provider organization
POS—point of service program
FFS—fee for service plan

The table lists enrollment figures for employees at 4200 different employers:

	Enrollment (%)		
Type of plan	**1993**	**1995**	**1999**
HMO	19	27	30
PPO	27	29	43
POS	7	14	16
FFS	48	15	11

Source: William M. Mercer Inc.

(a) Create a bar chart for each year for these data.

(b) Create a pie chart for each year for these data.

(c) Which chart does a better job of displaying the data and what they mean? Why?

(d) Are there any changes or trends in which health plans employees select?

2.37 Consumers are increasingly complaining that it is no longer worth the effort to clip coupons. One reason they cite is that the life of the coupons is decreasing and it takes more time to sort and discard expired coupons. In an effort to substantiate these claims, a consumer group looked at samples of 500 coupons from both 1995 and 1996.

	Number of coupons	
Coupon life	**1995**	**1996**
Shorter than 1 month	28	123
1–4 months	88	140
5–8 months	204	122
9–12 months	141	88
Longer than 1 year	39	27

(a) Create bar charts for 1995 and 1996.

(b) Compare and describe the data for the two years. Has there been a change in coupon life? Why or why not?

2.38 The problem of workplace violence is growing. Following are the results of a survey of 600 full-time American workers who were the victims of violence while working:

Type of violence	**Cases reported (%)**
Harassment	19
Threat of physical harm	7
Physical attack	3

	Type of violence (%)		
Major effect on worker	**Harassment**	**Threat of harm**	**Physical attack**
Psychological	49	53	49
Disrupted work life	34	25	25
Physically injured or sick	13	9	17
No negative effect	4	13	9

Create graphical displays of these data. Use any techniques you consider appropriate to convey the information in the data.

Requires Exercise 2.27   **2.39** The article that looked at the majors of students at Dartmouth College also reported data for the classes of 1991 and 1996:

Major	1991	1996
Anthropology	19	17
Art History	20	15
Asia	15	22
Biology	44	108
Chemistry	15	55
Classics	1	8
Comparative Literature	6	6
Drama	8	5
Earth Science	15	7
Economics	66	93
English	143	92
French	15	19
Geography	28	28
Government	172	142
History	125	115
Mathematics	23	14
Music	3	11
Philosophy	28	25
Physics	10	16
Psychology	74	87
Religion	31	8
Sociology	26	15

(a) Create bar charts for the 1991 and 1996 data. Design the charts so that you will be able to compare the two classes and also compare them with the data for the class of 1986 in Exercise 2.27.

(b) Did the fact that you needed to make comparisons change the way you might have arranged the categories? Why or why not?

(c) Did the fact that you needed to make comparisons affect the way you scaled the y axis?

(d) Is there any basis for the claim that there has been an increase in the number of students who are interested in the sciences? Why or why not?

(e) Make comparative bar charts for only those majors that are sciences. Do these charts make trends in the sciences easier to see? Why or why not?

(f) Compare the distribution of majors in general for each of the classes. Do you see any other changes or trends?

2.40 The computer center at a college is receiving many student complaints about disk errors. *Datafile: DISKS.XXX* The students say that when they try to save data to their removable disks, the files do not save properly. Students often have to spend countless hours on their work and they are reasonably frustrated. Since the director of the computer center is relatively sure it is not a hardware error, he decides to look at the two different brands of removable disks that the college bookstore stocks. The director selects 45 diskettes from each supplier and distributes them to a random sample of students. The students are asked to use the diskettes as they normally would but to check them regularly for signs of wear using a disk-checking utility. The students are to record how long (in minutes of use) their disks last before they exhibit signs of wear. Here are the data for each supplier:

	Supplier A					Supplier B			
474	492	498	504	511	487	492	495	497	499
486	492	500	505	511	488	492	495	497	499
489	494	501	506	512	489	492	495	497	499
490	494	501	507	513	489	492	495	497	501
490	494	501	508	514	489	493	495	498	502
490	495	502	508	515	491	493	496	498	503
491	496	502	509	517	491	494	496	498	503
491	498	504	509	519	491	494	496	498	505
491	498	504	510	528	492	494	496	499	506

(a) Create histograms for the lifetimes of the diskettes from each supplier. Be sure to choose your scales so that you can make meaningful comparisons.

(b) Describe the distribution of the diskette lifetimes for each supplier.

(c) Compare the lifetimes of the diskettes from supplier B with those from the company's current supplier, supplier A. Which diskettes would you recommend for use? Why?

(d) Write a memo to the director of the computer center with your results and recommendations.

2.41 Recent data from the Department of Justice (June 1995) give a state-by-state report on the number of people incarcerated per 100,000 people in the population.

District of Columbia	1151	New York	382	Tennessee	260
Delaware	890	North Carolina	356	Wyoming	239
Texas	643	Kentucky	354	Iowa	235
South Carolina	512	Virginia	354	Idaho	231
Alaska	505	Missouri	352	New Mexico	229
Arizona	501	Mississippi	346	Wisconsin	217
Connecticut	481	Kansas	332	West Virginia	211
Oklahoma	479	Illinois	329	Washington	197
Alabama	468	South Dakota	328	Montana	191
Florida	444	Arkansas	326	New Hampshire	183
Georgia	439	Pennsylvania	326	Utah	178
Nevada	431	Rhode Island	310	Massachusetts	174
Maryland	424	New Jersey	294	Nebraska	173
Michigan	424	Colorado	289	Vermont	173
Lousiana	399	Hawaii	279	Minnesota	139
California	398	Oregon	275	Maine	117
Ohio	390	Indiana	267	North Dakota	107

(a) Create a histogram for the data.

(b) Describe the number of people incarcerated (per 100,000) in the United States.

(c) Does the observation from the District of Columbia have an effect on the way your histogram turned out? If so, what is it?

(d) Recreate the histogram without the data from the District of Columbia.

(e) Use the histogram in part (d) to describe the number of people incarcerated (per 100,000) in the United States.

(f) Does your description in part (e) differ from that in part (b)? If so, how?

Requires Exercise 2.18

2.42 The cable television service that is looking at its ratings after changing the channels wonders whether customer feelings about its service have changed. The company looks back at the responses from a survey done 6 months prior to the change and finds the following data:

```
2   3   3   4   5
2   3   3   4   6
2   3   3   4   6
2   3   3   4   6
2   3   4   5   6
2   3   4   5   7
3   3   4   5   8
```

(a) Create a graphical display for the ratings data before the channels were changed.

(b) By comparing the ratings before and after the changes, can the company conclude that the ratings have improved? Why or why not?

2.43 A large hospital recently did a study on the treatment of patients who were admitted with pancreatitis. The hospital wanted to know, among other things, how long a patient spent in the hospital. Since patients were admitted to either the medical or surgical service, hospital administrators decided to look at the data for both services separately.

Medical

```
2   2   2   3   3   4   4   5   5   6
7   7   7   7   8   10  12  13  15
```

Surgical

3	4	4	5	5	6	6
6	7	8	9	10	11	13

(a) Create a graphical display for each of the samples.

(b) Use the graphical display to describe the distribution of the lengths of stay for each service.

(c) Based on the data, do you think there is a difference in the lengths of stay for the two services? Why or why not?

(d) On the basis of these data, can the hospital conclude that patients who are admitted to the medical service are cured more quickly than those admitted to the surgical service? Why or why not?

(e) Write a memo to the director of the hospital summarizing the analysis and make recommendations for improving the study.

Doing It!

2.44 On May 2, 2000, the National Highway Traffic Safety Administration (NHTSA) opened a defect investigation into tires manufactured by Bridgestone/Firestone Inc. Many of the tires in question were sold as original equipment on vehicles manufactured by Ford Motor Company. On August 9, 2000, Firestone recalled 14.4 million tires, including the ATX and ATXII models and the Wilderness AT. The NHTSA set up a website to record complaints about Firestone tires. As of June 14, 2001, there were 6184 reports of problems with Firestone tires. The data file was cleaned up so that it could be used, and there are 2173 usable records. A portion of the data file is shown here:

Datafile: TIRES.XXX

B	C	D	E	F	G	H	I
TIRE_MODEL	TIRE_SIZE	Actual Tire Info	STATE	INJURIES	FATALITIES	VEH_MFR	VEH_MODEL
WILDERNESS	14	14	TN	0	0	FORD	EXPLORER
ATX	15	15	TX	0	0	FORD	EXPLORER XLT
ATX	15	15	AZ	0	0	FORD	EXPLORER EDDIE BAUER
FIRESTONE	15	15	TX	0	0	FORD	EXPLORER
WILDERNESS	15	15	FL	0	0	FORD	EXPLORER
WILDERNESS	15	15	FL	1	1	FORD	EXPLORER

J	K	L	M	N	O	P	Q	R
VEH_MODEL_YEAR	FAILURE_DATE	BLOWOUT	TREAD_SEPARATION	ROLLOVER	LOST_CONTROL	CRASH	SPEED	TIRE_POSITION
1993	4/10/99	N	Y	N	N	N	65	PR
1994	10/22/96	Y	UNK	Y	Y	Y	UNK	DR
1993	4/10/98	Y	Y	Y	Y	Y	70	DR
1993	6/4/00	UNK	P	N	N	N	65	TWO
1994	7/21/99	Y	N	Y	Y	Y	65	PR
1996	7/4/00	N	Y	N	N	N	UNK	DR

Here is a brief description of the variables:

 TIRE_MODEL—the specific model of Firestone tire

 TIRE_SIZE—the tire size (in inches)

 Actual Tire Info—the tire size as entered in the original data

 STATE—the state from which the complaint originated

 INJURIES—the number of injuries involved

 FATALITIES—the number of fatalities involved

 VEH_MFR—the manufacturer of the vehicle involved

 VEH_MODEL—the specific model of the car

 VEH_MODEL_YEAR—the model year of the car

 FAILURE_DATE—the date of the tire failure

 BLOWOUT—whether or not the tire blew out

 TREAD_SEPARATION—whether the tread separated from the tire

ROLLOVER—whether or not the vehicle involved rolled over

LOST_CONTROL—whether or not the driver lost control of the vehicle

CRASH—whether or not the vehicle was involved in a crash

SPEED—the speed of the vehicle at the time of the tire failure

TIRE_POSITION—the location of the tires that failed

For the variables *BLOWOUT, ROLLOVER, LOST_CONTROL*, and *CRASH*, the coded values are:

Y = yes

N = no

UNK = unknown

For the variable *TREAD_SEPARATION*, the coded values are:

Y = yes

N = no

C = complete tread separation

P = partial tread separation

UNK = unknown

For the variable *TIRE_POSITION*, the coded values are:

DF = driver front

DR = driver rear

PF = passenger front

PR = passenger rear

ONE, TWO, or THREE = number of tires (not specific) that failed

(a) Create a bar chart for the tire model. Does any one model constitute a majority of the complaints?

(b) Create a bar chart for the vehicle manufacturer. Do you notice anything striking in the data?

(c) Create bar charts for the vehicle model for each of the vehicle manufacturers that have more than one complaint. Do any of the models have a particularly large number of complaints associated with them?

(d) For any vehicle models that appear to have a large number of complaints, make bar charts for tire model, vehicle year, and all the types of problems that occurred (blowout, tread separation). What do you learn from these data?

(e) Create similar bar charts for the types of problems that the vehicle had (rollover, loss of control, crash). Does any one problem constitute a majority?

(f) Redraw the bar charts as pie charts. Which type of chart is more effective? Why?

(g) Create histograms of the numbers of injuries and fatalities involved in all of the complaints. Do the same thing for the tire type and vehicle model that were associated with the most complaints. Is there any new information?

(h) For the vehicle/tire combination that had the most complaints, create a bar chart or pie chart of the position of the tires that failed. What do you learn?

(i) The NHSA started its investigation in May 2000. How many failures were there prior to this date? How many more failures occurred between the time the NHSA started its investigation and Firestone issued the recall?

(j) Create any additional graphs that you think might be informative. Do you think that the people who had these data should have started the investigation long before they did? If not, why not? If so, when do you think that the data became compelling enough to warrant investigation.

(k) Write a newspaper article about your findings in the data.

Note: The NHSA reports that 2910 complaints were verified by telephone callbacks. The dataset itself is full of conflicting coding (using both UNK and UKN as unknown), incomplete coding, and missing data. Some variables from the original file had so much missing data that they were excluded from the file. It was impossible to obtain the exact tire size. It was actually the P235/75R15 size that was first recalled, but inconsistent and missing information dictated that only the actual tire size (in inches) could be used. The original dataset is included on the CD-ROM so that you can see what kind of data you might encounter in reality.

2.45 The committee at ABC College that is studying gender equity in salaries is ready to look at the data they have collected in a different way. They now want to look at salary, rank, school, gender, and tenure. A sample of the data is shown here:

Datafile:
FACULTY.XXX

Salary	Years of service	Rank	School	Gender (M/F)	Tenure? (Y/N)
$53,316	22	ASST	BUSINESS	F	Y
64,375	11	PROF	BUSINESS	M	Y
63,501	7	ASSO	BUSINESS	M	Y
59,426	6	ASSO	BUSINESS	M	N
49,058	20	ASSO	BUSINESS	M	Y
94,969	4	PROF	BUSINESS	M	N
54,762	21	ASST	BUSINESS	M	Y
55,516	9	ASSO	BUSINESS	M	Y

(a) Create a histogram of the salaries for each rank separately. Compare the distribution of salaries for the four ranks.

(b) Repeat part (a) for each gender. Compare the salary distributions by gender.

(c) Create bar charts or pie charts of rank and tenure for each gender. Compare them.

(d) Create a bar chart or pie chart of the school for each gender.

(e) Create a bar chart of the years of service for each gender separately.

(f) Create any additional graphs that you think might be important.

(g) Write a memo of your findings to the college president. Attach the answers to parts (a)–(e) as an appendix to the memo.

3

Numerical Descriptors of Data

THE GOLF BALL DILEMMA

Have you ever gone to the store to buy something and been completely overwhelmed by the number of variations that a company has for a single product? Is there really that much difference between basic and deluxe? What do premium and XL really mean? One example of this problem is golf balls. A single manufacturer can have as many as six variations in a single product line. The question is, Are all the variations really different and, if they are, is the difference worth the increase in cost? A golf pro at a country club decided to run a test of the golf balls at the top and bottom of a particular manufacturer's product line to see how they differ. In preparation for the study, he collected sample data on different variables for the two types of balls. You are asked to summarize the data and report on what you find. Here is a portion of the data:

Ball number	Model number	S1	S2	S3	Weight (g)	Dimple width (mm)	Dimple depth (mm)	Head	Temperature (°F)	Carry (yd)	Total distance (yd)	Date	Time
1	M1	81	81	82	45.3	0.1450	0.0110	686	77	257	270	8/20	8:15
2	M1	83	83	84	45.2	0.1510	0.0111	688	77	255	267	8/20	8:15
3	M1	81	82	84	45.2	0.1450	0.0105	687	77	256	267	8/20	8:15
4	M1	81	81	83	45.3	0.1440	0.0117	688	77	255	271	8/20	8:15
5	M1	83	81	82	45.5	0.1460	0.0108	687	77	255	268	8/20	8:15
6	M1	83	83	82	45.3	0.1560	0.0111	687	77	256	267	8/20	8:15
7	M1	81	81	82	45.2	0.1495	0.0111	687	77	255	264	8/20	8:15
8	M1	83	81	82	45.1	0.1505	0.0110	690	78	258	269	8/20	8:15

The golf pro wants to know how the two products compare on factors such as how far most of the balls go when they are hit, how much variation there is in the distance, and what percentage of the balls go beyond a certain distance. You need to figure out which numerical measures will provide the most information.

3.1 Chapter Objectives

Remember that when you looked at graphical displays of numerical data, you were interested in three characteristics: center, spread or dispersion, and shape. In this chapter we look at numerical measures that can be used to describe the same features of the data. The chapter covers the following material:

- Numerical measures of center: the mean, the median, and the mode
- Numerical measures of variability: the range and the standard deviation
- Descriptions of a set of data: the empirical rule and boxplots
- Measures of relative standing: percentiles and percentile rank
- Identification of outliers: z-scores and boxplots

3.2 Describing Data Numerically

Although we say that a picture is worth a thousand words, numerical quantities are also useful to describe the data. You may wonder why we need numerical descriptors when the graphs let us *see* the data. There are two reasons. First, although we can certainly see the data using histograms and bar charts, it is difficult to *talk* or *write* about pictures. We often need other references to describe the data. Second, we may want to make inferences based on the sample data. To make statistical inferences we need to use *numerical measures*.

When we collect data, we may have either a *population* or a *sample* from the population. Numerical measures calculated from the data are known as either **statistics** or **parameters.**

> A *statistic* is a numerical descriptor that is calculated from sample data and is used to describe the sample. Statistics are usually represented by Roman letters.

> A *parameter* is a numerical descriptor that is used to describe a population. Parameters are usually represented by Greek letters.

Most of the time in statistics we work with sample data, but sometimes we may have the entire population available for study. We usually use Greek letters to denote parameters and Roman letters to describe statistics.

3.3 Measures of Central Tendency

When we look at numerical descriptors for a set of data, we want to describe the same properties of the data that we described from the graphical displays. You will find, however, that several different statistics can be used to describe each property and that the choice of the statistic is dependent on the problem you are trying to solve.

The golf pro would like to know what values represent a "typical" golf ball. He wants to measure the *center* of the data. We look at three different statistics that measure central tendency: the *sample mean*, the *median*, and the *mode*.

3.3.1 THE ARITHMETIC MEAN

You are probably already familiar with the most common measure of center, the **sample mean.** The mean, or average as it is commonly known, is calculated by adding all of the data values in the sample and then dividing the sum by the number of values. The symbol for the sample mean is $\overline{X}$ (this is read as "X-bar"):

$$\overline{X} = \frac{\text{Sum of all the values in the sample}}{\text{Total number of observations}}$$

The population parameter that corresponds to the sample mean is the **population mean, μ** (mu).

> The *sample mean* is the center of balance of a set of data. It is found by adding up all of the data values and dividing by the number of observations.

> The *population mean* is represented by the Greek letter μ (mu).

Using the Σ notation from Chapter 1, we can write the formula for the sample mean as

Sample mean

$$\overline{X} = \frac{\sum_{i=1}^{n} x_i}{n} \quad \text{or} \quad \frac{\Sigma x}{n}$$

Most of the time in statistics, it is understood that the sum is over the entire sample and so we can leave out the index on the summation sign.

When we talk about a variable or a statistic in general, we use *capital* letters, such as X or $\overline{X}$. When we talk about a *specific value* of a variable, we use *lowercase* letters, such as x. For example, we write, "The sample mean, $\overline{X}$, is calculated by . . ." or "The third value in the sample is $x = 27.2$."

Understand the problem.

Collect the data.

Analyze the data.

EXAMPLE 3.1

THE MAIL-ORDER COMPANY

Calculating the Sample Mean

A mail-order company wants some information about the daily demand for a product that has been heavily advertised. The company wants a measure of what it might typically expect the demand to be. The company looks at the number of orders over a 10-day period and obtains these data: 29, 28, 29, 31, 30, 31, 27, 29, 30, and 32. Since it is interested in a *typical* value for the demand, the company decides to calculate the sample mean, $\overline{X}$:

$$\overline{X} = \frac{29 + 28 + 29 + 31 + 30 + 31 + 27 + 29 + 30 + 32}{10} = \frac{296}{10} = 29.6 \text{ orders}$$

What Does the Sample Mean Really Measure?

You can think of the sample mean as the *balance point* of the data. The value of $\overline{X}$ balances the higher values against the lower ones. It is easiest to see this when you look at the mean on a dotplot of the data. In Figure 3.1 you can see that the mean, 29.6, is in the center of the data and that the data are fairly evenly spread out on both sides of the mean.

Suppose that the data on the right (high) side are more spread out than those on the left. What will happen to the value of the sample mean? Remember, we said that the sample mean is the balance point of the data values. When a few data

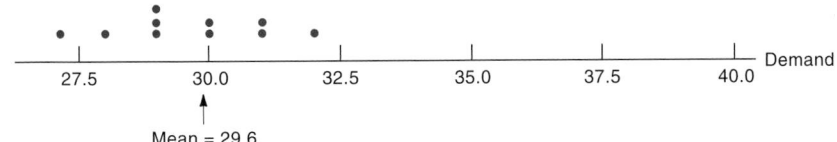

FIGURE 3.1 Dotplot of the data for the mail-order company

points on one side are far from the bulk of the data (the bump), the sample mean moves toward them to balance with the data on the other side. The next example shows what happens.

EXAMPLE 3.2

THE MAIL-ORDER COMPANY

Looking at the Sample Mean As a Balance Point

Suppose that the mail-order company found these data: 40, 28, 29, 31, 30, 31, 27, 29, 39, and 36. Most of the data values are still around 30, but three of them are quite a bit higher.

If you calculate the sample mean from the data, you get

$$\overline{X} = \frac{40 + 28 + 29 + 31 + 30 + 31 + 27 + 29 + 39 + 36}{10} = \frac{320}{10} = 32.0 \text{ items}$$

You see that the sample mean has changed from 29.6 to 32.0.

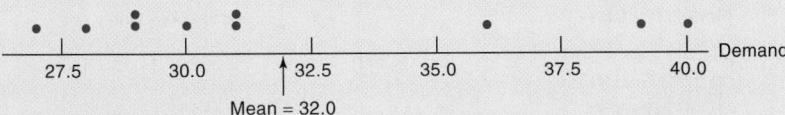

Although the bulk of the data are still located around 30, the value of the sample mean has changed to 32.0 to balance the three data values at the high end.

✓ TRY IT NOW!

Restaurant Table Times
Calculating the Sample Mean

A restaurant is trying to decide whether it has an adequate number of tables available. The restaurant owner would like some information on the amount of time a table is occupied by a customer. She collects data on the length of time a customer occupies a table for a random sample of ten customers:

Customer	1	2	3	4	5	6	7	8	9	10
Time (min)	59.3	58.6	62.7	65.4	59.0	67.3	62.8	68.1	59.4	63.7

Calculate the sample mean for the length of time a table is occupied.

EXAMPLE 3.3

THE GOLF BALL DILEMMA

Using the Sample Mean

Understand the problem.

The golf pro is interested in describing the way the two different types of golf balls behave so that he can see if there really is a difference between the two. One way to describe the distance that the balls travel is to use the variable *Carry*, which measures the distance (in yards) from the point where the ball was hit to the point where it hit the ground. Since the pro is interested in *comparing* the two different designs, you will want to look at the designs separately. Using a computer package, you can calculate a set of descriptive statistics for the two different ball designs. The output provides the following information:

	Type M1	Type M2
Sample size	36	36
Sum of data	9627	9244
Sample mean	257.4	256.8

Collect the data.

It appears from these values that there is not much difference in the way a "typical" ball of the two types behaves, but at this point that is just conjecture.

To better understand what the numbers really mean, you can locate the values obtained for the mean on a histogram of the data. Perhaps the numbers and graphs together will provide more information.

Analyze the data.

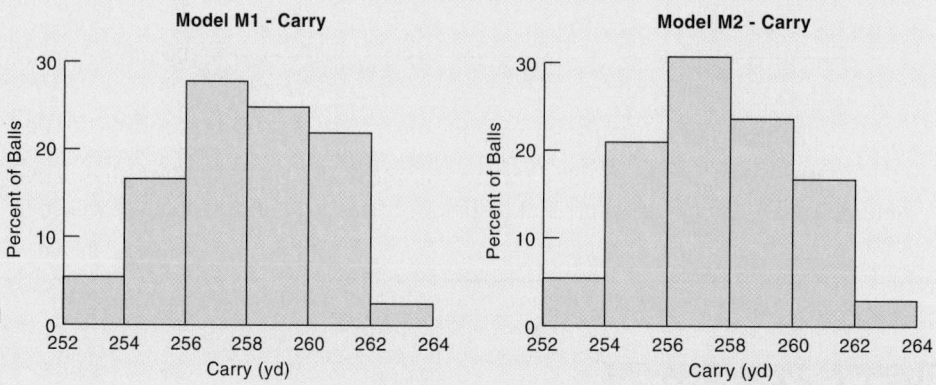

When you locate the mean *Carry* for each ball type on the appropriate histogram, you see that the mean appears to be a good measure of the center of the data and that it does not appear to be influenced by extreme values. However, you notice that the sample mean does not provide any information about the number of golf balls that went more or less than that number of yards. This could be useful information, and so you need another measure of center.

3.3.2 THE SAMPLE MEDIAN

Although the sample mean measures the center of the data, its value might be influenced by unusually high or low values in the sample and might not present a true picture of the sample data. For this reason we often look at other measures of center in addition to the sample mean so that we can see a better picture of the data.

Another measure of central tendency that is often used is the **sample median.**

> The *sample median* is the value of the middle observation in an ordered set of data.

The sample mean is a measure of the center of balance of the data and is sensitive to the actual values, whereas the sample median is a measure of the middle of the data set after it is sorted from lowest to highest.

Finding the sample median requires sorting the data set first. Then the sample median is the value of the observation that is in the middle of the data. The exact location of the middle depends on whether the number of observations in the sample is even or odd.

Step 1: If the number of observations in the sample, n, is odd, then the median is the value of the observation in the $(n + 1)/2$ position.

Step 2: If n is even, then the median is the average of the values in the $n/2$ and $n/2 + 1$ positions.

Steps for locating the median

EXAMPLE 3.4

MORTGAGE WAITING TIMES

Calculating the Sample Median

Understand the problem.

A bank has been receiving complaints from real estate agents that their customers have been waiting too long for mortgage confirmations. The bank prides itself on its mortgage application processing and decides to investigate the claims. The bank manager takes a random sample of 15 customers whose mortgage applications have been processed in the last 6 months and finds the following wait times (in days):

$$8 \quad 10 \quad 12 \quad 12 \quad 6 \quad 10 \quad 6 \quad 15 \quad 8 \quad 7 \quad 13 \quad 9 \quad 6 \quad 12 \quad 14$$

What is the median number of days required for mortgage confirmation?

Step 1: Put the data in numerical order from lowest to highest.

Collect and analyze the data.

Position	1	2	3	4	5	6	7	8	9	10	11	12	13	14	15
Wait	6	6	6	7	8	8	9	10	10	12	12	12	13	14	15

Step 2: Locate the middle observation.

Since $n = 15$ is odd, the middle position is the $(15 + 1)/2 = 16/2 = $ 8th observation. The median of the sample is the *value* of the eighth observation, 10 days.

Since the median represents the middle of the data set, it tells the manager that about half of the customers in the sample waited less than 10 days for a mortgage confirmation and about half of the customers waited more than 10 days.

Draw conclusions.

EXAMPLE 3.5

THE MAIL-ORDER COMPANY

Calculating the Sample Median

Look at the data from the mail-order company in Example 3.1. We can sort the data in numerical order from highest to lowest:

Position	1	2	3	4	5	6	7	8	9	10
Demand	27	28	29	29	29	30	30	31	31	32

Analyze the data.

Since $n = 10$ is even, the sample median is the average of the observations in positions $10/2 = 5$ and $10/2 + 1 = 6$. So we find that the median is $(29 + 30)/2 = 29.5$ orders.

✔ ## TRY IT NOW!

Town Hall Traffic
Calculating the Sample Median

In the past few years the town council of a small town has received complaints that it has become increasingly difficult to cross the main street near the library. The council decides to look at traffic flow on the street. It selects a site directly in front of the library where most people try to cross the street and records the number of cars that pass the point in a 2-minute period. Counts are taken in ten 2-minute periods at 3:00 P.M. over several weeks. The following numbers of cars are obtained:

$$20 \quad 27 \quad 29 \quad 28 \quad 37 \quad 23 \quad 21 \quad 28 \quad 29 \quad 28$$

Find the median number of cars that pass the site in 2 minutes.

Remember to SORT the data before you locate the median!

Why Use Two Different Measures?

At this point you still may be wondering why we need to have two different measures of center. For most sets of sample data, the mean and the median are very close to each other in value. If you need to decide on a single measure, then the choice will depend on the problem you are trying to solve and what information you need. The median tells you that half of the observations in the sample are above that value and half of the observations are below it. Because it is a measure of *location*, however, it ignores the actual values of the observations and may not fully reflect the sample data. The mean uses all of the data values in its calculation and measures the center of balance of the data. Although it can be shifted by extreme values, it does reflect all of the data values equally.

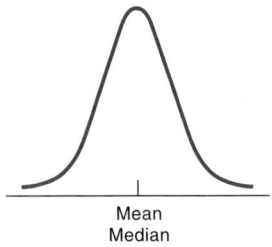

FIGURE 3.2 **Mean and median for a symmetric distribution**

3.3.3 COMPARING THE MEAN AND THE MEDIAN

We know that the mean and the median provide different information about the center of a set of sample data, but can anything additional be gained from knowing both of them?

If we start out with a symmetric, mound-shaped distribution, then the mean and the median are both located at the center of the distribution, at the bump. This is illustrated in Figure 3.2. When the data values are more spread out in one direction (that is, when the data are skewed), the mean is pulled toward these values, in the direction of the skew. This is illustrated in Figure 3.3. If we compare the mean and the median, then we can learn about the shape of the distribution. In particular,

If	Distribution is	Illustration
Mean = Median	Symmetric	Figure 3.2
Mean < Median	Skewed left	Figure 3.3(*a*)
Mean > Median	Skewed right	Figure 3.3(*b*)

At this point you might be saying, "Hey, hold on here! The mean and the median will almost never be exactly equal. How different do they have to be to say that the data are skewed?" This is a really good question and is often ignored. There is no exact answer, but several rules of thumb can be used. One rule is that the mean and the median should differ by at least the width of a class in a histogram of the data. For small data sets or data values that do not have large magnitudes, another rule is that they should differ by at least 10% of either measure. Because these are rules of thumb, they do not work in all situations. Knowledge of the data, intuition, and experience are always important in interpreting statistics. Sometimes, as you will see in the next example, you really have to draw a graph and *look* at the data.

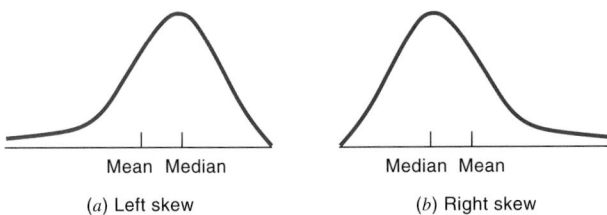

(*a*) Left skew   (*b*) Right skew

FIGURE 3.3 **Mean and median for skewed distributions**

EXAMPLE 3.6

STARTING SALARIES

Comparing the Mean and the Median

Understand the problem.

A college administrator wonders about the numbers reported by the placement office on graduates' salaries. She thinks that the mean seems inflated and wonders whether it gives an accurate picture of what is really happening. She decides to collect some data on the salaries earned by students who graduate from the School of Business and to calculate both the mean and the median. The college takes a random sample of 100 graduates from the past year and obtains these data:

$25,000	$25,400	$25,600	$25,800	$25,900	$26,200	$26,400	$26,600	$27,000	$28,000
25,100	25,400	25,600	25,800	25,900	26,200	26,400	26,600	27,100	28,200
25,100	25,400	25,600	25,800	26,000	26,200	26,400	26,600	27,100	28,300
25,200	25,500	25,700	25,800	26,000	26,200	26,400	26,600	27,100	28,300
25,200	25,500	25,700	25,800	26,000	26,200	26,400	26,600	27,200	28,400
25,200	25,500	25,700	25,800	26,100	26,200	26,400	26,700	27,300	28,500
25,200	25,500	25,700	25,900	26,100	26,200	26,500	26,700	27,400	28,600
25,300	25,600	25,700	25,900	26,100	26,300	26,500	26,800	27,600	29,400
25,300	25,600	25,700	25,900	26,100	26,300	26,500	26,900	27,700	30,700
25,300	25,600	25,800	25,900	26,100	26,300	26,500	26,900	27,700	30,800

The mean of the salaries is

$$\overline{X} = \frac{2,638,500}{100} = \$26,385$$

The median is the average of the 50th and 51st observations in the data set:

$$\text{Median} = \frac{26,100 + 26,200}{2} = \$26,150$$

In this case, the actual difference between the mean and the median is

$$\$26,385 - \$26,150 = \$235$$

Analyze the data.

Clearly the mean is larger than the median, but are they different enough to indicate that the data are skewed? If we apply the 10% rule, we find

$$10\% \text{ of } \$26,385 = \$2638.50$$

Remember! No one statistical tool will provide all of the information in a sample.

The actual difference of $235 is much less than $2638.50, so if we use the 10% rule, the salaries do not appear to be skewed.

The college administrator decides to use a histogram from a statistical software package to decide whether the conclusion of no skew is consistent with what she sees. From the histogram, we see that, in fact, the data are skewed right. This is not what the administrator expected.

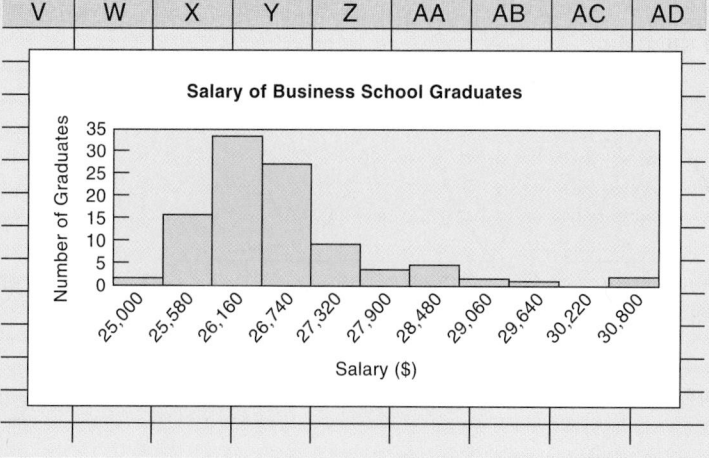

The width of the classes in the histogram is $800, and the difference between the mean and median, $235, is not more than this either. This is why it is important to look at data in more than one way.

Knowing that the data are skewed, the college administrator decides that she should look at other variables to see why this might be true. Then the placement office will be able to prepare a report that better represents the true situation.

Draw conclusions.

✔ TRY IT NOW!

Airline Cancellations
Comparing the Mean and the Median

An airline company is wondering about the number of cancellations it receives for a particular business commuter flight. The airline takes a random sample of 15 days from the first quarter of the year and obtains the following numbers of cancellations:

| 4 | 9 | 9 | 12 | 12 | 13 | 14 | 14 | 15 | 15 | 16 | 16 | 17 | 17 | 24 |

Find the mean and median for the number of cancellations for the commuter flight.

Note: The data have been sorted for you.

When compared, do the data appear symmetric or skewed?

Make a dotplot of the data.

From the dotplot, do the data appear symmetric or skewed?

EXAMPLE 3.7

THE GOLF BALL DILEMMA

Comparing the Mean and the Median

The golf pro needs another number to describe a "typical" golf ball. Although the mean is a good statistic, you know that he can get more information if he looks at the median, too. You know that the value of the sample median represents the distance that has half of the golf balls above it and half below it. The median will tell the pro a little more about how the golf balls behave.

From the same computer output that gave us the mean, we learn that the median for the M1 balls is 257.5 yards and the median for the M2 balls is 257.0 yards. This information tells the pro that 50% of the M1 balls in the sample went farther than 257.5 yards and 50% of the M2 balls in the sample went farther than 257.0 yards.

Understand the problem.

Analyze the data.

ANS. MEAN = 13.8; MEDIAN = 14; SYMMETRIC

Analyze the data.

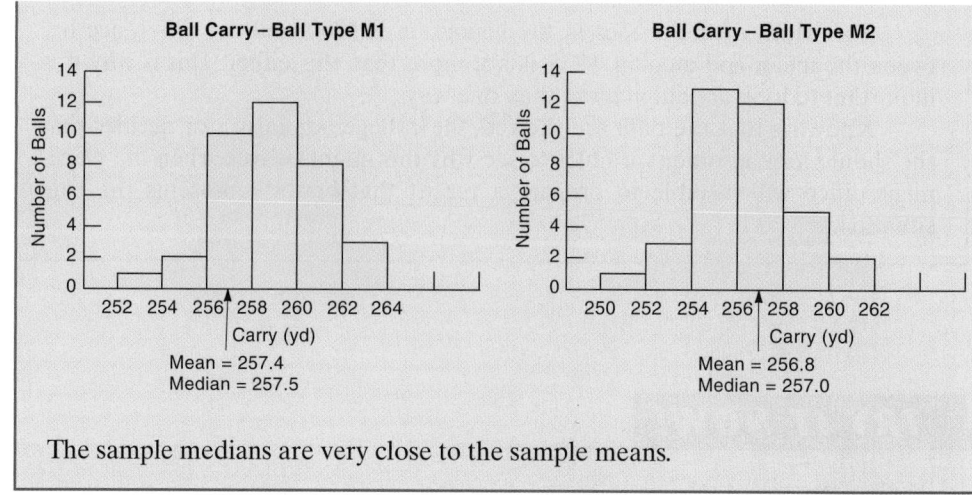

The sample medians are very close to the sample means.

Discovery Exercise 3.1

THE TRIMMED MEAN

Part I. Investigating the Data

In a report to the administration of a large university, the Psychology Department states that the average class size is larger than the 35 students allowed by the university charter. The report indicates that the mean class size is 39.4.

No data are appended to the report, but you can obtain the current enrollments easily. You find these data:

3	14	22	26	42
3	15	23	27	45
5	15	24	28	45
9	17	24	28	190
11	21	25	36	193
13	22	26	38	193

1. Do you think that the mean is a good measure of center for these data? Why or why not?

2. By simply studying the data, what do you think is a typical class size for the Psychology Department?

3. What is the median of the data? Is this close to what you thought?

4. Compare the mean and the median. What does the comparison lead you to believe about the data?

5. Display the data graphically. Does your opinion change?

Part II. Solving the Problem

Neither the mean nor the median gives a very good measure of a typical class size. In addition, a comparison of the mean and the median leads you to believe that the data are skewed right! From the histogram you can see that is not the case.

How can we measure the center of the data when we have extreme values that influence the statistics we usually rely on?

You really can't just discard the extremes without a careful investigation of their causes. One way to do this is to use a measure other than the median that removes the effect of the extremes. The *trimmed mean* is such a statistic. The trimmed mean allows you to drop a specified percentage of the observations in the data set from *each end.* By doing this, you are not simply dropping extreme values; you are looking at the center of the data, which may prove to be more reliable. Typically the trimmed mean is used with a percentage of 10%, but the percentage can be varied to suit the specific circumstances.

Computing the trimmed mean

To compute a 10% trimmed mean you must first determine how many values will be dropped from each end of the data set.

1. Find 10% of the sample size.

2. Do you think that dropping this many values from each end will be effective? Why or why not? If not, how many values would be effective?

3. Drop the top and bottom three observations from the data and recalculate the mean.

4. Compare the trimmed mean and the median. What do you think about the data? Is this a more accurate representation?

By using the trimmed mean together with the mean and the median, you find that the mean was not influenced by the *skewness* of the data but by extreme values on the high end.

5. As an administrator at this university, do you think that the Psychology Department can claim that it exceeds the class size specification of 35 students? Use all of the information you have gathered to write a memo explaining your decision.

3.3.4 THE SAMPLE MODE

There is another value that is used in statistics to measure the center of the data: the **mode.** In a bar chart, the mode is analogous to the bar with the highest frequency.

The *sample mode* is the data value that has the highest frequency of occurrence in the sample.

It appears that the mode would be a very good measure of a typical value, but there are some obvious reasons it does not always provide useful information. Depending on the size of the sample and the number of possible data values, there may not be any repeated values in the sample. *That is, for some samples, the mode may not exist.* For continuous data, which have many different possible values, we do not usually talk about the mode because of that problem. In these cases, we often refer to the **modal class** in a frequency distribution or histogram.

The *modal class* is the class interval in a frequency distribution or histogram that has the highest frequency.

Another problem with the mode is that a sample may appear to have more than one mode. This frequently happens with small samples. When this occurs, there may not be two or three values that occur much more frequently than any others. Rather, most of the values happen to occur more than once.

For these reasons the mode is considered by many people to be an unreliable measure of central tendency. However, sometimes a sample does have more than one distinct mode. A sample that is *bimodal* (two modes) or *multimodal* (many modes) should raise questions in your mind.

1. Is it likely that these data could have two or more distinct centers? What would cause such a phenomenon?

2. Is it possible that the sample represents two or more different populations that were not understood when the data were collected?

EXAMPLE 3.8

THE CLOTHING STORE

Calculating the Sample Mode

Understand the problem.

A large retailer of women's clothing is trying to obtain some information that will help it formulate an ordering policy for clothing sizes. The retailer decides to look at a single line of apparel and collect data on the sizes of the items sold in a 2-week period. These are the data after sorting in size order:

6	10	10	12	12	14
8	10	10	12	12	14
8	10	10	12	12	14
10	10	10	12	12	14
10	10	10	12	12	16

A frequency table of the data follows:

Collect and analyze the data.

Size	Frequency
6	1
8	2
10	12
12	10
14	4
16	1

From the frequency table you can see that the mode of the sample is 10. That is, size 10 was sold the most often during the period of study.

What are the sample mean and median for these data?

Sample mean: $\overline{X} = \dfrac{334}{30} = 11.1$

Sample median: Average of the $\dfrac{30}{2} = 15$th and $\dfrac{30}{2} + 1 = 16$th observations

$$= \dfrac{10 + 12}{2} = 11$$

Neither the mean nor the median is a real size. The mean measures the balance point of the sizes and the median is simply the size that had half the sales below it and half above. Neither supplies information that helps the retailer develop an ordering policy.

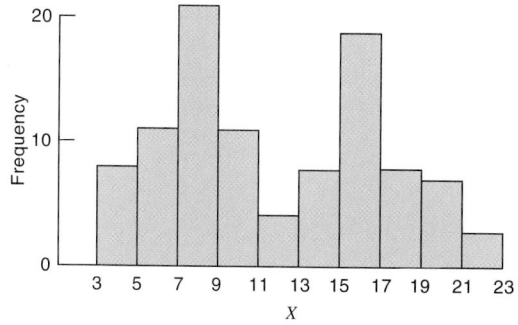

FIGURE 3.4 **Histogram of bimodal data**

You might wonder why, in the preceding example, we did not say that the data are bimodal, since size 12 has a frequency that is very close to the frequency for size 10. Usually when we refer to bimodal or multimodal data, we are talking about data with very distinct different centers. In a histogram, this means that the two modal classes are not adjacent classes. Figure 3.4 shows a histogram that is bimodal.

3.3.5 EXERCISES—LEARNING IT!

3.1 The management of Disney World would like to know the amount of time that visitors spend waiting for the monorail at one of the hotels at the resort. They take some sample data of times (in minutes):

In many exercises, the data are already sorted for you.

   5.5   9.6   5.1   13.6   6.5   8.6   9.3   9.1   9.5   15.0   9.7   14.1

(a) What is the mean time spent waiting for a monorail?

(b) What is the median time spent waiting for a monorail?

3.2 As part of a study to decide whether to renew the contract of a food service company at a local university, the administration surveyed a group of students who use the service on a regular basis. The students were asked to rate the food quality on a scale of 1 to 5, where 1 was extremely bad and 5 was extremely good. The results are listed here:

$$
\begin{array}{ccccc}
1 & 1 & 3 & 3 & 4 \\
1 & 1 & 3 & 3 & 4 \\
1 & 2 & 3 & 3 & 4 \\
1 & 2 & 3 & 3 & 4 \\
1 & 2 & 3 & 3 & 4 \\
1 & 3 & 3 & 4 & 4 \\
\end{array}
$$

(a) What is the average rating of the food service?

(b) What is the median rating of the food service?

(c) What is the mode of the ratings?

(d) Based on the data, would you say that the students surveyed have a favorable, neutral, or unfavorable opinion of the food service? Why?

3.3 A group of elementary school children gave these responses about how many hours per week they spend playing outdoors:

   6   2   9   13   13   14   0   6   13   2   4   1

(a) What is the average amount of time a child spent playing outdoors?

(b) What is the median amount of time a child spent playing outdoors?

3.4 How much do NBA basketball players really get paid? The 2000/2001 salaries for the Boston Celtics are listed in the table at the top of page 124.

Player	2000/2001 salary (millions)	Player	2000/2001 salary (millions)
Chris Herren	$0.420	Walter McCarty	$2.670
Mark Blount	0.420	Tony Battie	3.200
Adrian Griffin	0.500	Greg Minor	3.240
Doug Overton	0.550	Eric Williams	3.890
Chris Carr	1.200	Vitaly Potapenko	4.290
Jerome Moiso	1.460	Bryant Stith	5.920
Paul Pierce	1.610	Kenny Anderson	7.520
John Williams	2.110	Antoine Walker	10.130
Randy Brown	2.250		

(a) What is the average salary for a Boston Celtics player?

(b) What is the median salary for a Boston Celtics player?

(c) Which statistic does a better job of describing the salary of a typical player? Why?

3.5 Does a newer NBA team pay its players salaries that are comparable to those of older teams? The 2000/2001 salaries for the L.A. Clippers are listed in the table:

Player	2000/2001 salary (millions)	Player	2000/2001 salary (millions)
Zendon Hamilton	$0.320	Keith Closs	$1.920
Earl Boykins	0.500	Cherokee Parks	2.100
Jeff McInnis	0.520	Sean Rooks	2.160
Etdrick Bohannon	0.520	Eric Piatkowski	2.200
Brian Skinner	0.900	Lamar Odom	2.630
Quentin Richardson	1.020	Darius Miles	2.840
Corey Maggette	1.350	Derek Strong	3.510
Keyon Dooling	1.540	Michael Olowokandi	3.700
James Robinson	1.870		

(a) What is the mean salary for an L.A. Clippers player?

(b) What is the median salary for an L.A. Clippers player?

(c) Compare the typical salaries for the Boston Celtics and the L.A. Clippers.

3.6 The *2001 Kids Count Data Book* published by the Annie E. Casey Foundation has data on the percentages of children who live in working-poor families (a family whose income is less than twice the federal poverty level and who has at least one parent working 50 or more weeks per year) without a telephone at home. Here are the data for the western United States:

Alaska	26%	Nevada	27%
Arizona	30	New Mexico	38
California	28	Oregon	38
Colorado	35	Utah	44
Idaho	38	Washington	42
Montana	48	Wyoming	48

(a) What is the mean percentage of children without a telephone at home?

(b) What is the median percentage of children without a telephone at home?

(c) Do you think that either measure is better than the other at measuring typical? Why or why not?

3.7 The transportation department of a small city has received numerous complaints about the length of the light cycle at a busy intersection. Drivers complain that so many cars are backed up at the light that they cannot all get through the intersection when the light turns green. In an effort to determine the validity of the complaints, the transportation engineer decides to collect some data on the numbers of cars that are waiting at the light when it turns green. These data are for ten randomly selected light cycles:

4   4   5   6   7   8   9   10   11   12

(a) Find the mean and median of the numbers of cars waiting at the light.

(b) Are the data skewed or symmetric?

3.8   A study on the effects of television on behavior in adolescents uses, as part of the data, the number of hours per day that the television set is turned on in a household. Twenty-six households are randomly selected:

3.6   3.7   3.7   3.8   3.9   3.9   3.9   3.9   3.9   3.9   4.2   4.3   4.6
5.0   5.3   5.6   5.7   5.8   6.0   6.0   6.0   6.0   6.0   6.0   6.3   6.9

(a) Find the mean and median for the numbers of hours that the television is turned on in a household.

(b) Do you think that these statistics are good measures for a typical household?

(c) What is the mode or modal class for the data? Does this statistic provide any information that is not given by the mean or median? Why or why not?

Discovery Exercise 3.2

INVESTIGATING VARIABILITY

The table contains air-quality data collected by the Environmental Protection Agency. The data are the numbers of unhealthy days in which the ozone level was dangerous for 14 major U.S. cities in 2000:

Atlanta	18	Los Angeles	1
Boston	0	New York	13
Chicago	0	Philadelphia	2
Dallas	5	Pittsburgh	3
Denver	0	San Francisco	0
Houston	94	Seattle	0
Kansas City	0	Washington, DC	0

Part I

1. Display these data using a dotplot.

2. Find the typical number of unhealthy days by calculating the average value.

3. Can you expect every observation to be typical? Why not?

Part II

1. What you noticed in Part I is that although we know how to measure a typical value, we don't have any way to describe the differences from this typical value that we see in the data. The difference is called *variability*. Using the dotplot, decide whether you think this data set has a lot of variability, a little variability, or somewhere between a lot and a little.

2. How can you measure this variability?

3. You might have tried subtracting the smallest value from the largest value in an attempt to measure variability. If you did so, you calculated what is known as the *range*. If you did not do this already, calculate the range.

4. In this case the range is a large number, which presumably tells you that there is a large amount of variability in the data. Do you agree? Based on this, do you think that the range is a good measure of variability?

Part III

1. We hope you concluded that the range can, in fact, give a very misleading picture of variability in the data. The range is quite large for this data set. Why?

2. What if you try to measure how far away each data point is from the middle (i.e., the average) of the data? Fill in the following table. Remember that the average is 9.71 days. The first measurement has been done for you.

City	Number of unhealthy days	Distance from middle
Atlanta	18	$18 - 9.7 = 8.3$
Boston	0	
Chicago	0	
Dallas	5	
Denver	0	
Houston	94	
Kansas City	0	
Los Angeles	1	
New York	13	
Philadelphia	2	
Pittsburgh	3	
San Francisco	0	
Seattle	0	
Washington, DC	0	
Average or typical	**9.7**	

3. The results are still not informative because you want a single number that will give the typical deviation from the middle. What is one way to measure "typical"?

Now, calculate the typical deviation.

4. Does this value give a good idea of how much variation there is in the data? Why not?

5. What caused the typical variation just calculated to be such a small number? How can you fix this?

PART IV

1. There are two ways to handle the problem. One way is to convert all the numbers to positive values by squaring them. The other way is to take the absolute values of the numbers. Fill in the following table. Then calculate the typical or average of the values in the fourth column and the average of the values in the fifth column.

City	Number of unhealthy days	Distance from the middle	Absolute value of distance from the middle	Distance from the middle squared
Atlanta	18	$18 - 9.7 = 8.3$	8.3	$(8.3)^2 = 68.89$
Boston	0			
Chicago	0			
Dallas	5			
Denver	0			
Houston	94			
Kansas City	0			
Los Angeles	1			
New York	13			
Philadelphia	2			
Pittsburgh	3			
San Francisco	0			
Seattle	0			
Washington, DC	0			
Average	**9.7**			

2. Why are these two averages not close in magnitude?

3. The average of the absolute values of the distances (fourth column) is called the *mean absolute deviation* (MAD), and the average of the squared distances (fifth column) is called the *variance*. What are the units for the MAD? What are the units for the variance?

4. The MAD is actually easier to interpret but is not often used because absolute values do not "behave well." What can you do to the variance to get the magnitude to be about the same as the MAD? Do it.

What you have calculated is called the *standard deviation*.

3.4 Measures of Dispersion or Spread

In Chapter 2 you saw that simply describing the center of the data or a typical data value does not provide complete information about the data set. In addition to knowing what a typical value for the sample is, it is important to know how diverse the values in the sample can be. That is, we need to know how *spread out* or *dispersed* the data values are relative to the typical values. Understanding the *variation* in a set of data is of critical importance in statistics. When you use statistics to make decisions, it is important to understand not only a typical outcome, but all possibilities as well. We look at two different measures of dispersion: the *sample range* and the *sample standard deviation.*

3.4.1 THE SAMPLE RANGE

The simplest measure of dispersion, the **sample range,** involves looking at the two extreme values in the sample: the highest (maximum) and the lowest (minimum) values.

> The *sample range, R,* is the difference between the maximum and minimum observations in the sample.
>
> $$\text{Range} = \text{Max} - \text{Min}$$

The sample range is very easy to calculate and understand. It gives information about the distance from one end of an ordered data set to the other. If the sample data are symmetric, then the range also gives information about the spread of the data relative to the measures of central tendency.

EXAMPLE 3.9

THE MAIL-ORDER COMPANY

Calculating the Sample Range

Look at the sample data on demand from the mail-order company from Example 3.5:

Position	1	2	3	4	5	6	7	8	9	10
Demand	27	28	29	29	29	30	30	31	31	32

The sample range for these data is

$$R = 32 - 27 = 5$$

The range tells the company that the demand for the product has a range or spread of 5 units around its center. In this case the range is a reliable measure of how spread out the demands are.

Information on the range *along with a measure of central tendency* gives us a mental image of the data. If the data are symmetric, then the company in Example 3.9 would expect that a typical demand for the product is 30 units and that the demand is evenly spread out on either side of the center. The actual demand might be as low as 27.1 (29.6 − 2.5) or as high as 32.1 (29.6 + 2.5). In Example 3.9 these bounds agree very well with the actual data.

The range seems to be a good statistic because it gives a clear picture of the spread and is easy to calculate and understand. The next example illustrates why this is not always the case.

EXAMPLE 3.10

CUSTOMER HOLD TIMES

Understanding the Sample Range

Understand the problem.

A company is wondering whether complaints about the amount of time customers spend on hold for customer service are justified or whether the complaints are a result of the "squeaky wheel" phenomenon. Company managers take a sample of 15 customer calls to the technical service phone line and record the amount of time each customer spends on hold.

Collect the data.

Customer	1	2	3	4	5	6	7	8	9	10	11	12	13	14	15
Wait (minutes)	5.6	10.2	6.6	6.9	9.4	6.7	0.6	9.2	7.6	10.7	9.6	2.9	6.0	8.6	4.6

To find the range, find the minimum and maximum observations and subtract:

$$R = 10.7 - 0.6 = 10.1 \text{ minutes}$$

Analyze the data.

What does the range tell the company?

At first glance, the company would think that the length of time a customer spends on hold has a spread of 10.1 minutes. If the company considers that the largest value in the sample is 10.7 minutes, it could conclude that the data are quite variable. Or are they?

Waiting Time on Hold (min.)

A dotplot of the data shows that the value of 0.6 minute is a large distance from the next nearest value, whereas quite a few observations are in the 9- to 10-minute range. The picture that the company receives from the data is actually distorted. If it calculates the average hold time

$$\overline{X} = \frac{105.2}{15} = 7.0 \text{ minutes}$$

Draw conclusions.

then the company would think that a typical caller is on hold for 7.0 minutes and that the time a customer spends on hold might be as short as 1.95 (7.0 − 5.05) minutes or as long as 12.05 (7.0 + 5.05) minutes. This is certainly not an accurate picture of the data because the mean is biased downward by the few extremely low values and the range is biased upward by the same values!

✔ TRY IT NOW!

Restaurant Table Times
Calculating the Sample Range

A restaurant that is looking at the turnaround time for its tables wonders how variable the occupation time for a table really is. The restaurant collected these turnaround times in minutes:

59.3   58.6   62.7   65.4   59.0   67.3   62.8   68.1   59.4   63.7

What is the range of the turnaround times?

Previously (page 113) you calculated the mean turnaround time to be 62.6 minutes. Using this information and the value for the range, what can the restaurant expect as its shortest turnaround time? Its longest turnaround time?

The range is heavily influenced by unusual or extreme values in the sample. You saw that the mean is also influenced by extremes, but much less bias is introduced because when you calculate the sample mean, you use all of the values in the sample. When you calculate the range, you use only the extreme values, so any unusual values have a large impact on the statistic. In fact, the sample range is one of the only statistics that gets more *unreliable* as the sample size gets larger. As a rule, when the sample size is larger than 25, the sample range should not be used as a measure of variability.

EXAMPLE 3.11

THE GOLF BALL DILEMMA

Looking at Variation

Understand the problem.

The golf pro who is investigating the behavior of the two golf ball designs needs to know about the variation in the characteristics as well as the measure of typical distance. Unless the distance that the golf balls carry is fairly consistent, the conclusions the pro draws might not make sense for all consumers. From the histograms in Example 3.3, we see that the data do not appear to have any unusual values, and so we decide to use the sample range as a measure of variation. We look at the computer printout and find the following information:

Analyze the data.

	Type M1	Type M2
Minimum	252	251
Maximum	262	262
Range	10	11
Average	257.4	256.8

From the data it appears that the range is about the same for both ball types.

Draw conclusions.

Since we also know that the data are symmetric, by using the sample ranges and sample means together, we can provide the pro with a better picture of how the golf balls behave. The M1 balls appear to carry between a low of 252.4 yards and a high of 262.4 yards. The M2 balls have a low carry of 251.3 yards and a high of 262.3 yards. The fact that these numbers agree with the observed minimums and maximums for the data support the range as a good measure for these data.

Still, the range does not answer all of the questions. Although we know that an M2 golf ball *could* travel between 251.3 and 262.3 yards, we do not have a measure of how the actual values are spread out within that range. What we would like to know, really, is the typical variation in travel.

ANS. range = 9.5 min; shortest = 57.9 min; longest = 67.4 min

3.4.2 THE SAMPLE STANDARD DEVIATION

You have seen that the sample range proves unreliable in the presence of extreme data values. As the size of the sample increases, the chance increases that the sample will contain an extreme value. So for large sample sizes another measure of dispersion or spread must often be used.

If you think about what we are really trying to describe when we measure variability, it seems logical to try to define a measure of how far away from the center of the data a value might be. In fact, what we would probably like to know is, On the average, or typically, how far do the values *vary* (differ) from the center? This measure is the **sample standard deviation.** The standard deviation is most often defined relative to another measure of dispersion called the **sample variance.** In practice, the measure that is used is the standard deviation because its units and order of magnitude are the same as those of the actual data.

> The **sample variance, s^2,** is the average of the squared deviations of the data values from the sample mean.

> The **sample standard deviation, s,** is the positive square root of the sample variance.

By their definitions alone the sample variance and standard deviation seem overly complicated, especially when compared to the sample range. To calculate the sample standard deviation, we calculate the sample variance first, using the formula

$$s^2 = \frac{\sum_{i=1}^{n}(x_i - \bar{x})^2}{n - 1}$$

Sample variance

Note: The sum for the variance still includes all n sample values even though we divide by n − 1.

To obtain the sample standard deviation, s, we take the positive square root of the sample variance:

$$s = \sqrt{s^2}$$

Sample standard deviation

> The population variance and standard deviation are represented by the Greek letter σ (sigma), where σ^2 is the **population variance** and σ is the **population standard deviation.**

EXAMPLE 3.12

CUSTOMER HOLD TIMES

Calculating the Sample Standard Deviation with the Definition

The company that is looking at customer hold times knows that a measure of a typical hold time does not provide complete information about what its customers might encounter. The managers of the company need to know how much the time of a call may vary from the center. They decide to calculate the sample standard deviation for their data on customer hold times:

Understand the problem.

Customer	1	2	3	4	5	6	7	8	9	10	11	12	13	14	15
Wait (minutes)	5.6	10.2	6.6	6.9	9.4	6.7	0.6	9.2	7.6	10.7	9.6	2.9	6.0	8.6	4.6

To calculate the sample standard deviation the managers first need to calculate the sample mean because the standard deviation measures the average of

Analyze the data.

the squared distances from the sample mean. For this set of data we found in Example 3.10 that the mean hold time is 7.0 minutes.

The next part of the calculation for the sample variance finds the distance of each data value, X, from the sample mean, $\overline{X}$, squares each of them, and then adds them. The accompanying table presents the details of each part of the calculation. Each column of the table represents one part of the calculation. The last row is the sum of the rows above.

Customer	Wait (minutes)	$(X - \overline{X})$	$(X - \overline{X})^2$
1	5.6	−1.4	1.96
2	10.2	3.2	10.24
3	6.6	−0.4	0.16
4	6.9	−0.1	0.01
5	9.4	2.4	5.76
6	6.7	−0.3	0.09
7	0.6	−6.4	40.96
8	9.2	2.2	4.84
9	7.6	0.6	0.36
10	10.7	3.7	13.69
11	9.6	2.6	6.76
12	2.9	−4.1	16.81
13	6.0	−1.0	1.00
14	8.6	1.6	2.56
15	4.6	−2.4	5.76
Sum	**105.2**	**0.2**	**110.96**

In the last part of the calculation, we divide the sum of the squared deviations by $n - 1$—in this case, 14. So we have the sample variance

$$s^2 = \frac{110.96}{14} = 7.93 \text{ minutes}^2$$

and the sample standard deviation

$$s = \sqrt{7.93} = 2.82 \text{ minutes}$$

The results tell the company that although an average call lasts 7.0 minutes, the actual call times vary from that value. Typically, the variations or differences from the average are about 2.82 minutes.

By now you must be convinced that there has to be a really good reason to use the standard deviation instead of the range! In fact, almost nobody actually *calculates* the sample variance and standard deviation using the definitions. A shortcut formula reduces considerably the number of calculations you have to perform:

Shortcut formula for sample variance

$$s^2 = \frac{n\Sigma x^2 - (\Sigma x)^2}{n(n - 1)}$$

To calculate the sample variance and sample standard deviation this way, you need to sum all of the data values and also to square each value and sum the squares. The next example recalculates both the sample variance and standard deviation from Example 3.12 using the new, shorter method.

EXAMPLE 3.13

CUSTOMER HOLD TIMES

Calculating *s* with the Shortcut Formula

Using the same data, we need to sum both the values and the values squared. The table gives the details of the calculations:

Customer	Wait, X (minutes)	X^2
1	5.6	31.36
2	10.2	104.04
3	6.6	43.56
4	6.9	47.61
5	9.4	88.36
6	6.7	44.89
7	0.6	0.36
8	9.2	84.64
9	7.6	57.76
10	10.7	114.49
11	9.6	92.16
12	2.9	8.41
13	6.0	36.00
14	8.6	73.96
15	4.6	21.16
Sum	105.2	848.76

Substituting the values in the last row of the table into the formula, we get

$$s^2 = \frac{(15)(848.76) - (105.2)^2}{(15)(14)} = \frac{1664.36}{210} = 7.93 \text{ minutes}^2$$

and

$$s = \sqrt{7.93} = 2.82 \text{ minutes}$$

which is the same answer we got in Example 3.12 using the original formula.

It is probably hard to convince yourself that the shortcut formula is actually any better than the first set of calculations. Actually it is, but unless you do it several times, it is hard to see. In fact, there is no real reason to calculate the sample variance and sample standard deviation by hand more than once or twice in your lifetime! Many calculators include statistical calculations like the sample mean and sample standard deviation, and these functions are also built into every spreadsheet package. What is important is understanding what the standard deviation measures and how it can be used to interpret sample data.

Caution! Calculators and spreadsheets can calculate more than one type of standard deviation. Be sure that you are finding the **sample** *standard deviation.*

 **TRY IT NOW!**

Town Hall Traffic

Calculating the Sample Variance and Standard Deviation

The town council that is looking at the traffic flow problem has seen reports that use the standard deviation and they want to use it to describe the variability of traffic flow. The data are the numbers of cars that pass a certain point in a 2-minute period:

20	27	29	28	37	23	21	28	29	28

What is the sample standard deviation of the traffic flow?

Use whatever method you feel most comfortable with. If you have a statistical calculator, learn how to use it NOW!

3.4.3 INTERPRETING THE STANDARD DEVIATION— THE EMPIRICAL RULE

Admittedly, the standard deviation is not so intuitive or appealing as the sample range. The sample range gives an immediate (although sometimes false) picture of how far the data spread out around the center. The sample standard deviation does not have the same intuitive appeal.

One way to understand what information the standard deviation gives is to use the **empirical rule.**

The *empirical rule* says that for a mound-shaped, symmetric distribution:

- About 68% of all observations are within one standard deviation of the mean.
- About 95% of all observations are within two standard deviations of the mean.
- Almost all (more than 99%) of the observations are within three standard deviations of the mean.

The empirical rule is defined for large data sets and distributions that are symmetric and mound-shaped, often called bell-shaped or normal curves (see Figure 3.5). For distributions that are only slightly skewed, the empirical rule is surprisingly accurate as well. It provides bounds for data values from a given population.

In general, if we are talking about sample data, we do not know the population mean and standard deviation, so it is necessary to substitute $\overline{X}$ and s for μ and σ. To really understand the empirical rule, an example is necessary.

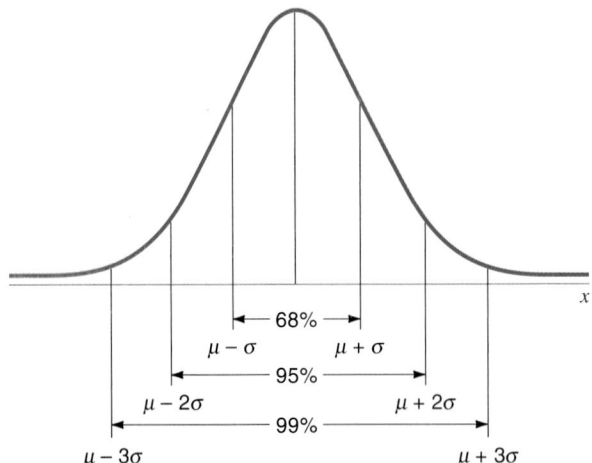

FIGURE 3.5 The empirical rule

EXAMPLE 3.14

Analyze the data.

CUSTOMER HOLD TIMES

Using the Empirical Rule

Suppose that the company that is looking at customer hold times had collected a total of 50 observations. The sorted hold times are shown here:

0.6	4.6	5.6	6.3	6.8	7.5	7.8	8.3	8.9	9.6
2.9	4.7	6.0	6.3	6.9	7.5	7.9	8.4	9.2	10.1
3.4	5.2	6.0	6.6	6.9	7.6	8.0	8.4	9.2	10.2
3.8	5.5	6.1	6.6	7.0	7.6	8.1	8.6	9.4	10.7
4.5	5.5	6.1	6.7	7.2	7.8	8.2	8.6	9.4	11.1

To investigate the empirical rule we need to find the sample mean and sample standard deviation of the data. For this data set $\bar{x} = 7.12$ minutes and $s = 2.08$ minutes. We should also verify that the empirical rule applies—that is, that the data follow a normal curve. To do this we can make a histogram of the data:

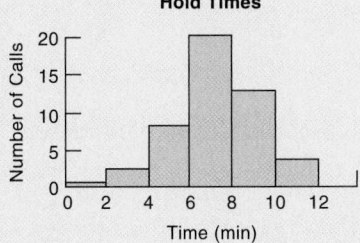

The histogram shows that the data are reasonably symmetric and bell-shaped, so we can use the empirical rule:

$\bar{x} + s = 7.12 + 2.08 = 9.20$ minutes — *About 68% of the data values should be*
$\bar{x} - s = 7.12 - 2.08 = 5.04$ minutes — *between 5.04 and 9.20 minutes.*
$\bar{x} + 2s = 7.12 + 2(2.08) = 11.28$ minutes — *About 95% of the data values should be*
$\bar{x} - 2s = 7.12 - 2(2.08) = 2.96$ minutes — *between 2.96 and 11.28 minutes.*
$\bar{x} + 3s = 7.12 + 3(2.08) = 13.36$ minutes — *Almost all of the data (more than 99%)*
$\bar{x} - 3s = 7.12 - 3(2.08) = 0.88$ minute — *should be between 0.88 and 13.36 minutes.*

For a data set this large, you should use your calculator or a computer software package to find $\overline{X}$ and s.

To see how well the data agree with the empirical rule, we can calculate the actual percentage of data values that fall within each interval. The table lists the number of data values and the percentage that fall within each of the three intervals.

Interval	Number of data values	Percentage of data values	Empirical rule (%)
5.04 to 9.20	36	72	68
2.96 to 11.28	48	96	95
0.88 to 13.36	49	98	99

Draw conclusions.

Looking at the table, we see that the actual percentages are slightly different from those predicted by the empirical rule. Remember that variation in data is to be expected. In statistics it is important to understand when the variation is within the limits of what is expected or typical. When the variation from what is expected is too large, something may be happening to cause it.

In this case the actual percentages are fairly close to those predicted by the empirical rule. The largest difference occurs in the first interval. If the percentages are very different from what is expected, then the data are probably from a distribution that is not symmetric. The more skewed a data set is, the more it deviates from the empirical rule in the first two intervals.

✔ *TRY IT NOW!*

Loan Processing
Using the Empirical Rule

Errors in filling out loan applications can lead to delays in the loan approval. Bank employees must contact applicants to correct the errors, and this sometimes requires multiple contacts. To understand the extent to which errors affect the application process, a bank collected data on the number of follow-up contacts required before a loan could be processed. The bank looked at the numbers of contacts for 25 different applications:

0	1	2	3	4
0	2	2	4	4
1	2	3	4	5
1	2	3	4	5
1	2	3	4	7

Make a dotplot of the data.

From the dotplot, is the assumption that the data have a symmetric, bell-shaped distribution reasonable?

Find the mean and standard deviation of the data.

According to the empirical rule, between what two values should 68% of the observations fall?

Between what two values should 95% of the observations fall?

Between what two values should more than 99% of the observations fall?

3.4.4 *z*-SCORES

In the preceding section you saw that the standard deviation can be used to measure how likely it is for a data value to occur. Since, for a symmetric, bell-shaped distribution, 68% of the data values fall within one standard deviation of the mean, it is reasonable to assume that data values in that interval make up about 68% of the sample values. Certainly, most (more than 99%) of the data values in the sample should be within three standard deviations of the mean! We can use this rule as a measure of how "usual" a data value is. This measure is called the **z-score.**

> A **z-score** measures the number of standard deviations that a data value is from the mean.

To calculate the z-score of a data value, we first find the *distance* that the data value is from the mean and then divide that by the standard deviation:

ANS. DATA ARE SOMEWHAT SYMMETRIC AND BELL-SHAPED. MEAN = 2.8, s = 1.7; 68%, (1.1, 4.5);
95%, (−0.6, 6.2); > 99%, (−2.3, 7.9)

$$z = \frac{\text{Distance between the data value and the mean}}{\text{Standard deviation}} = \frac{X - \mu}{s}$$

z-Score

As in the empirical rule, for sample data we substitute $\overline{X}$ and s for μ and σ, respectively. A positive *z*-score indicates that the data value is *above* the mean, whereas a negative *z*-score indicates that the data value is *below* the mean.

Remember! When we substitute $\overline{X}$ and s for μ and σ, we are relying on large sample sizes to ensure that they are good estimates.

If you think about the empirical rule together with the *z*-score, you can begin to make some inferences about how data values compare to what is expected from a random sample. Table 3.1 gives you an idea of how we do this.

TABLE 3.1

USING THE EMPIRICAL RULE AND *z*-SCORES TOGETHER

If the *z*-score is . . .	The empirical rule says it will occur . . .	You can conclude that . . .
Less than −2 or more than 2	About 5% of the time	It is unusual and possibly an outlier
Less than −3 or more than 3	Less than 1% of the time	It is very unusual and probably an outlier.

It is possible to use the *z*-score of a data value to identify *outliers*. The problem with doing this is that everyone's definition of unusual is not the same and so inconsistencies in data analysis can result. In Chapters 6 and 8, *z*-scores will be discussed in much more detail.

Just because a data value is identified as an outlier does not mean you can discard it from the sample!

An **outlier** is a data value that has a very low probability of occurrence—that is, it is unusual.

EXAMPLE 3.15

THE GOLF BALL DILEMMA

Calculating *z*-scores

From the histograms in Example 3.7, the golf pro who is investigating the two designs of golf balls does not think that any of the data are potential outliers. Still, it is better to be sure, so he asks you to calculate *z*-scores. Rather than calculate the *z*-score for every data value, it is easier to look at just the maximum and minimum values first. If neither of these has an unusual *z*-score, then you know that none of the other values will either. From the tables of summary statistics in Example 3.11, you can find the information you need and compute the *z*-scores shown in the table. The *z*-scores are indeed in the 2–3 range, and they do not indicate that the extreme sample values are outliers.

Analyze the data.

Ball Type	Carry (yards)	Carry (yards)	z-score
M1	Max 262	$\overline{x} = 257.4$	1.92
	Min 252	$s = 2.4$	−2.25
M2	Max 262	$\overline{x} = 256.8$	2.26
	Min 251	$s = 2.3$	−2.52

TRY IT NOW!

Town Hall Traffic
Calculating z-Scores

The town council that is looking at traffic flow in front of the town hall wonders whether the observation of 37 cars in 2 minutes is unusual. Although the town officials know that their sample size of ten cars is not large enough to ensure accuracy, they want to use z-scores to look at the data on numbers of cars:

20   27   29   28   37   23   21   28   29   28

What is the z-score for the observation of 37 cars?

Comparing the z-score to the empirical rule, do you think that the value of 37 is unusual?

3.4.5 EXERCISES–LEARNING IT!

Requires Exercise 3.1

3.9 In addition to knowing the typical time visitors to Disney World wait for a monorail, management wants to know about the variation in waiting times. These are the waiting times in minutes:

5.5   9.6   5.1   13.6   6.5   8.6   9.3   9.1   9.5   15.0   9.7   14.1

(a) What is the range of the waiting times?

(b) What are the variance and the standard deviation of the waiting times?

3.10 To understand its customer base, a mail-order book club must look at the frequency with which customers make subsequent purchases. Managers collect data on the numbers of purchases that customers make during their initial 3-year membership. From a sample of 15 randomly selected customers they obtain these data:

12   12   14   15   14   11   13   10   14   14   13   14   14   11   15

(a) Find the range of the numbers of purchases in a 3-year period.

(b) Find the variance and the standard deviation of the numbers of purchases.

(c) Is the range or the standard deviation a better measure of variation for these data? Why?

3.11 A professor in an introductory statistics course is interested in the numbers of hours that students spend doing homework during the week. A random selection of 12 students yields these values:

4.1   2.8   6.1   4.9   4.2   5.5   3.2   5.9   2.7   5.4   6.9   3.7

(a) Find the range and the standard deviation of the numbers of hours spent doing homework.

(b) According to the empirical rule, between what two amounts of time will 68% of the students spend on homework? 95%? More than 99%?

(c) How do the actual data compare to the predictions from the empirical rule?

3.12 A manager at XYZ Corporation thinks that the numbers of travel miles claimed in the expense reports of its sales personnel have increased in recent months. To substantiate his idea he collects historical data on travel miles from expense reports filed in the previous year:

1558   1617   1655   1709   1716   1733   1758   1908   1963   2062   2075   2171

(a) Find the range of the numbers of miles claimed.

(b) Find the standard deviation of the numbers of miles claimed.

(c) What is the z-score of the smallest data value? Is this value unusual? Why or why not?

(d) What is the z-score of the largest data value? Is this value unusual? Why or why not?

3.13 Look at the data from the elementary school children who were surveyed about how *Requires Exercise 3.3* many hours per week they spend playing outdoors:

$$6 \quad 2 \quad 9 \quad 13 \quad 13 \quad 14 \quad 0 \quad 6 \quad 13 \quad 2 \quad 4 \quad 1$$

(a) What are the range and standard deviation of the hours spent outdoors?

(b) According to the empirical rule, between what two values will 68% of the times fall? 95%? More than 99%?

(c) Do you think that the empirical rule is appropriate for these data? Why or why not?

(d) What is the z-score for the child who played outdoors for 4 hours? What does the z-score say about that observation?

3.14 To make a decision about replacing the cars in its current fleet, a car rental company looks at the amount of money (in dollars) spent on repairs in the past 12 months for a random sample of ten cars:

$$472 \quad 472 \quad 603 \quad 459 \quad 538 \quad 601 \quad 449 \quad 588 \quad 539 \quad 521$$

(a) Find the range, variance, and standard deviation of the amounts spent on repairs.

(b) Use the empirical rule to describe the distribution of the amounts spent on repairs.

(c) Is the use of the empirical rule appropriate here? Why or why not?

(d) Find the z-score for each data value. Are any of the values unusual? If so, why?

3.15 Look at the data on the percentages of material loss per day in manufacturing:

$$10 \quad 12 \quad 12 \quad 13 \quad 14 \quad 14 \quad 18 \quad 19 \quad 19 \quad 20$$

(a) What is the range of percentages of material loss for the process?

(b) What is the standard deviation of the percentages of material loss for the process?

(c) Find the z-score for each data value. Are any of the values unusual? Why or why not?

3.16 Look again at the salary data for the Boston Celtics in 2000/2001. *Requires Exercise 3.4*

Player	2000/2001 salary (millions)	Player	2000/2001 salary (millions)
Chris Herren	$0.420	Walter McCarty	$2.670
Mark Blount	0.420	Tony Battie	3.200
Adrian Griffin	0.500	Greg Minor	3.240
Doug Overton	0.550	Eric Williams	3.890
Chris Carr	1.200	Vitaly Potapenko	4.290
Jerome Moiso	1.460	Bryant Stith	5.920
Paul Pierce	1.610	Kenny Anderson	7.520
John Williams	2.110	Antoine Walker	10.130
Randy Brown	2.250		

(a) What is the range of the actual salaries for the players? What are the variance and standard deviation of the salaries?

(b) Make a dotplot or a histogram of the data.

(c) Use the empirical rule to find the intervals $\overline{X} \pm 1s$, $\overline{X} \pm 2s$, and $\overline{X} \pm 3s$, and mark them on the graphical display.

(d) Find the actual number of data points that fall in each interval. Does this agree with the predictions of the empirical rule? Why or why not?

3.17 In the study on the effects of television on behavior in adolescents, the number of hours *Requires Exercise 3.8* per day that the television set is turned on is one variable considered. Twenty-six households are randomly selected and the data are in hours per day:

$$3.6 \quad 3.7 \quad 3.7 \quad 3.8 \quad 3.9 \quad 3.9 \quad 3.9 \quad 3.9 \quad 3.9 \quad 3.9 \quad 4.2 \quad 4.3 \quad 4.6$$
$$5.0 \quad 5.3 \quad 5.6 \quad 5.7 \quad 5.8 \quad 6.0 \quad 6.0 \quad 6.0 \quad 6.0 \quad 6.0 \quad 6.0 \quad 6.3 \quad 6.9$$

(a) Find the range and standard deviation of the hours per day that the television is turned on.

(b) Make a dotplot of the data. Do any of the data values appear unusual?

(c) Is the range or the standard deviation a more reliable measure for this set of data? Why?

(d) Find the z-scores of the two largest and the two smallest values. What do the z-scores tell about the observations?

3.18 Data taken from the latest Department of Justice (June 1995) survey of prisons includes information on inmates involved in educational programs. A random sample was taken of 40 of the institutions surveyed and the percentages of inmates in educational programs were recorded:

0	4	13	19	43
0	4	14	19	44
0	4	14	22	46
0	7	15	28	51
0	7	16	29	51
0	8	17	33	56
0	9	19	35	84
3	13	19	36	89

(a) Find the range and standard deviation of the percentages of inmates enrolled in educational programs.

(b) Use the empirical rule to give a picture of how the percentages vary.

(c) Find the z-scores of the maximum and minimum percentages. Is either of these two values unusual? Why or why not?

3.5   Measures of Relative Standing

Measures of center and dispersion are certainly important, but they are not the only numerical measures that can be used to obtain information about a set of data. Other measures, called measures of relative standing or *order statistics*, give information about the position of an observation in the sample. We looked at one measure of relative standing, the median, when we presented measures of center. Now we look at some additional measures.

The methods used for calculating percentiles vary slightly among software packages, and so the values you get might differ.

3.5.1   PERCENTILES

It is useful in some real situations to know what data value in a sample has a certain percentage of the sample above or below it. This measure is known as the **percentile** of the data.

> The pth **percentile** of a data set is the value that has $p\%$ of the data at or below it.

Two questions involve percentiles: What value has $p\%$ of the data at or below it? and What is the percentile rank of a particular data value? The first question involves finding either a particular percentile or a set of percentiles, such as the *deciles* (10%, 20%, . . . , 90%). It can be tedious to do this by hand but easy using most statistical software packages. Percentiles are particularly useful when the observations in the data set are to be compared with one another or with some other norm.

EXAMPLE 3.16

THE GOLF BALL DILEMMA

Understand the problem.

Finding Percentiles

The golf pro's statistics so far tell a lot about how the golf balls behave when they are hit, but these statistics are not useful for drawing conclusions like "Ninety percent of model M2 balls go *x* yards when they are hit!" No number that we have looked at so far provides that information, although the median carry comes close. The golf pro wonders whether there is a more general statistical measure similar to the median.

The measures the golf pro wants are called *percentiles*, which are easily obtained from a statistical software package:

Analyze the data.

Remember! The median gives the data value that has 50% of the values above or below it.

Type M1		Type M2	
Percentage	Percentile	Percentage	Percentile
10	255	10	254
20	255	20	255
30	256	30	255
40	257	40	256
50	257.5	50	257
60	258	60	257
70	259	70	258
80	260	80	258
90	260	90	260

From this information the golf pro sees, for example, that 20% of the M1 balls went farther than 260 yards, compared to only 10% of the M2 balls. This is really the first difference we have seen in the way the two balls behave, but it might be just part of the natural variation.

The second question about percentiles involves finding the **percentile rank** of a particular value in a data set.

> The *percentile rank* of a value is the percentage of the data in the sample that are at or below the value of interest.

This measure allows you to determine the *relative standing* of an observation in a set of data. To find the percentile rank of an observation, the data must be put in numerical order. The percentile rank, *P*, is then found by

$$P = \frac{b + \frac{1}{2}e}{n}$$

Percentile rank

where

b = the number of data values *below* the value of interest

e = the number of data values *equal* to the value of interest

n = the sample size

EXAMPLE 3.17

STARTING SALARIES

Finding the Percentile Rank

Suppose that in the example where we looked at starting salaries (see Example 3.6), a person in the group who had a starting salary of $26,200 wanted to know how her salary ranked relative to her peers. Looking at the data, she sees that

$$b = 50 \quad \text{(50 salaries are below \$26,200)}$$

$$e = 7 \quad \text{(7 salaries are equal to \$26,200)}$$

$$n = 100$$

Analyze the data.

so

$$P = \frac{50 + \frac{1}{2}(7)}{100} = 0.535 = 53.5$$

This tells her that 53.5% of the starting salaries were at or below hers.

✔ TRY IT NOW!

Aptitude Test Scores
Calculating the Percentile Rank

A group of employees at a manufacturing facility take a test to determine their aptitude for training. The tests are scored on a 400-point scale and the scores are listed in increasing order:

185	227	241	257	281	299	314	329
195	228	243	261	283	304	318	333
196	234	248	269	283	307	319	335
199	238	250	271	291	309	322	349
223	241	253	272	297	310	328	353

One of the employees who scored 283 wants to know how he stands relative to the other employees who took the exam. What is the percentile rank for the employee's score?

What is the percentile rank of the employee who scored 319?

3.5.2 QUARTILES

Although calculating percentiles can be tedious, certain percentiles are used frequently. These percentiles are the 25th percentile and the 75th percentile, also known as the **first** and **third quartiles.**

> The *first quartile, Q_1,* is the value in the sample that has 25% of the data at or below it.

Ans. 55%; 81.25%

The *third quartile, Q_3,* is the value in the sample that has 75% of the data at or below it.

You may be wondering what happened to the second quartile, but we have already seen it! If you think about the definition of the quartile, you recognize that the second quartile, Q_2, must be the median.

Just like finding percentiles, finding the quartiles for a set of data can be tedious, depending on how you choose to do it. Actually, several methods can be used to find the quartiles of a set of data. As with the percentiles, different software packages use different methods, and so it is not unusual to get two slightly different answers when you use two different packages on the same data set. The method that we use estimates the quartiles and is not really exact. It gives values that are quite close to the more exact methods and it is much simpler.

If you think about the quartiles, you see that they divide the data set into fourths. The median divides the data set in half, and if you take half of one-half you get a quarter. So, it would appear that if we find the median of each of the two halves of the data, we have our quartiles! It may sound complicated, but it really isn't.

Since percentiles and quartiles are order statistics, finding them requires that the data set be sorted from lowest to highest.

Step 1: Put the data set in order and find the median of the data.

Step 2: Take the lower half of the data (all of the values that are below the median) and find the median of the lower half of the data. This value is the first quartile, Q_1.

Step 3: Take the upper half of the data (all of the values that are above the median) and find the median of the upper half of the data. This value is the third quartile, Q_3.

Steps for finding the quartiles

When the median is actually one of the data values (when *n* is odd), do not include the median in either half of the data.

EXAMPLE 3.18

STARTING SALARIES

Finding the Quartiles

A college administrator is studying the graduates' starting salaries. Among the graduates sampled, what salary has 25% of the students earning less than that salary, and what salary has 25% of the students earning more than that salary? In other words, the administrator wants to know the quartiles of the data. To make this problem easier, we repeat the data set from Example 3.6:

$25,000	$25,400	$25,600	$25,800	$25,900	$26,200	$26,400	$26,600	$27,000	$28,000
25,100	25,400	25,600	25,800	25,900	26,200	26,400	26,600	27,100	28,200
25,100	25,400	25,600	25,800	26,000	26,200	26,400	26,600	27,100	28,300
25,200	25,500	25,700	25,800	26,000	26,200	26,400	26,600	27,100	28,300
25,200	25,500	25,700	25,800	26,000	26,200	26,400	26,600	27,200	28,400
25,200	25,500	25,700	25,800	26,100	26,200	26,400	26,700	27,300	28,500
25,200	25,500	25,700	25,900	26,100	26,200	26,500	26,700	27,400	28,600
25,300	25,600	25,700	25,900	26,100	26,300	26,500	26,800	27,600	29,400
25,300	25,600	25,700	25,900	26,100	26,300	26,500	26,900	27,700	30,700
25,300	25,600	25,800	25,900	26,100	26,300	26,500	26,900	27,700	30,800

Remember! Since 100 is even, you average the two middle values to find the median.

Step 1: Since 100 observations are in the sample, the median must be the average of the 50th and 51st data values:

$$Q_2 = \frac{26{,}100 + 26{,}200}{2} = \$26{,}150$$

The lower half of the data set, everything *below* the median, consists of the first 50 values, and the upper half of the data, everything above the median, consists of the second 50 values.

Analyze the data.

Step 2: To find the first quartile, we find the median of the first 50 values, which is the average of the 25th and 26th observations:

$$Q_1 = \frac{25{,}700 + 25{,}700}{2} = \$25{,}700$$

Step 3: To find the median of the 51st through 100th values, we average the 75th and 76th observations:

$$Q_3 = \frac{26{,}600 + 26{,}700}{2} = \$26{,}650$$

So, the college administrator finds that 25% of the graduates earn less than $25,700 and 25% of the graduates earn more than $26,650.

✔ TRY IT NOW!

Aptitude Test Scores
Finding the Quartiles

The manufacturing facility that is looking at training aptitude wants to give employees who scored in the top 25% on the test the opportunity to attend a seminar on training. The test scores are reprinted here:

185	227	241	257	281	299	314	329
195	228	243	261	283	304	318	333
196	234	248	269	283	307	319	335
199	238	250	271	291	309	322	349
223	241	253	272	297	310	328	353

In the sample, what is the cutoff score for those people who will be able to attend the seminar?

Hint: The value that defines the top 25% is the same as the value that defines the bottom 75%.

Suppose the company decides that the employees who scored in the bottom 25% need some additional classes on team building. What is the cutoff score for those employees who need the classes on team building?

Ans. $Q_3 = 312$; $Q_1 = 241$

3.5.3 DISPLAYING THE DATA USING BOXPLOTS

In Chapter 2 you saw several different methods for displaying quantitative data. These methods allow you to look at the data and describe central tendency, shape, variability, and outliers. Both histograms and dotplots are displays of the actual data in the data set.

Another method for displaying a set of data uses not the individual data values, but rather a set of summary statistics taken from the data. The plot is called a **boxplot** or a **box and whisker diagram.**

A *boxplot* or *box and whisker diagram* is a graphical display that uses summary statistics to display the distribution of a set of data.

A boxplot summarizes a sample using the quartiles and the median. You know that the median is a measure of center, but how can these statistics be used to show shape and variability? How can they identify outliers?

If you look at the first and third quartiles of a sample, Q_1 and Q_3, you see that 50% of the data in the sample fall between these two values. The distance between these two values is called the **interquartile range (IQR).**

The *interquartile range* **(IQR)** is the difference between the third and first quartiles, $Q_3 - Q_1$.

You learned that one way to describe the variability of a set of data is to use the empirical rule. The empirical rule says that approximately 68% of the data in the sample should be within one standard deviation of the mean (in the interval $\mu - 1\sigma, \mu + 1\sigma$). The interval from Q_1 to Q_3 is similar to the first interval of the empirical rule.

The main part of the boxplot is a rectangle (box) constructed from the median and the quartiles, as shown in Figure 3.6. The box provides a partial picture of the data set. To complete the description with the empirical rule we use two additional intervals, $\mu \pm 2\sigma$ and $\mu \pm 3\sigma$. The boxplot uses multiples of the IQR instead of the standard deviation. The second interval in a boxplot is described by the **inner fences.** This is similar to the 2σ interval of the empirical rule. The last interval is described by the **outer fences.**

Note: The parts of a boxplot are defined from the quartiles, not the median.

The *inner fences* of a boxplot are located at $Q_1 - 1.5(\text{IQR})$ and $Q_3 + 1.5(\text{IQR})$.

The *outer fences* of a boxplot are located at $Q_1 - 3(\text{IQR})$ and $Q_3 + 3(\text{IQR})$.

The inner and outer fences are not drawn on the boxplot but are used to define the "whisker" portions of the plot. The whiskers are constructed by drawing horizontal lines from the quartiles to the smallest and largest observations in the sample that are *within* the inner fences. The whiskers are illustrated in Figure 3.7 (page 146).

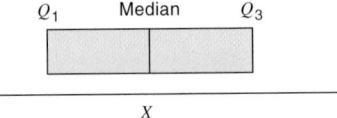

FIGURE 3.6   Box portion of boxplot

Note: The height of the vertical lines in the box is arbitrary. The scale on the x axis should span the values of the data similar to a histogram or dotplot.

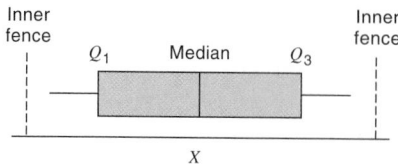

FIGURE 3.7 Boxplot with whiskers

It is easy to see that the median gives us a picture of the center of the data, but you may be wondering how the boxplot shows the shape of the distribution. When data are symmetric, the mean and the median are both located at the bump in the distribution and the distribution tails off at the same rate in both directions. In a boxplot, if the data are symmetric, then the median should be located halfway between the two quartiles and the whiskers should be the same length.

When the data are skewed, the median stays near the bump in the distribution and one side of the distribution tails off at a slower rate than the other. In a boxplot, the median is closer to one of the quartiles than to the other and/or the whisker on the tail side of the distribution is longer than the other whisker. Figure 3.8 shows boxplots for data that are skewed.

Although the boxplot may seem complicated, it really is not. It is actually *easier* to make a boxplot than a histogram because the rules are standard and no judgment is involved in setting it up. Making a boxplot can be separated into two parts: the calculations and the construction.

Calculations

Creating a boxplot

Step 1: Find the median and the first and third quartiles for the data.

Step 2: Calculate the interquartile range (IQR) by finding $Q_3 - Q_1$.

Step 3: Find the values for locating the inner and outer fences:
Lower inner fence (LIF): $Q_1 - 1.5(IQR)$
Upper inner fence (UIF): $Q_3 + 1.5(IQR)$
Lower outer fence (LOF): $Q_1 - 3(IQR)$
Upper outer fence (UOF): $Q_3 + 3(IQR)$

Construction

Step 1: Draw the *x* axis and select a scale that will accommodate the largest and smallest values in the sample.

Step 2: Draw three vertical lines, all the same height, one at Q_1, one at Q_3, and one at the median.

Step 3: Construct a box using the quartiles as the vertical edges and enclosing the median.

Step 4: Locate the inner fences on the axis. Draw a horizontal line from Q_1 to the smallest value in the sample that is within (larger than) the lower inner fence. Draw a similar horizontal line from Q_3 to the largest sample value that is within (less than) the upper inner fence.

FIGURE 3.8 Boxplots for skewed data

Now it is probably time for an example!

EXAMPLE 3.19

STARTING SALARIES

Constructing a Boxplot

Consider the data on starting salaries in Examples 3.6 and 3.18. We have summarized the data by constructing a histogram and by looking at sample statistics. What can we learn from a boxplot?

The first step in the calculations is done because in Example 3.18 we found the median and the quartiles for the data:

$$\text{Median} = \$26,150$$
$$Q_1 = \$25,700$$
$$Q_3 = \$26,650$$

The interquartile range is

$$Q_3 - Q_1 = \$26,650 - \$25,700 = \$950$$

For the inner fences we find

$$\text{Lower inner fence} = 25,700 - 1.5(950) = \$24,275$$
$$\text{Upper inner fence} = 26,650 + 1.5(950) = \$28,075$$

and for the outer fences we find

$$\text{Lower outer fence} = 25,700 - 3(950) = \$22,850$$
$$\text{Upper outer fence} = 26,650 + 3(950) = \$29,500$$

We now have all of the information we need to draw the boxplot.

We select the scale for the x axis so that it will accommodate the smallest and largest values in the sample—in this case, $25,000 and $30,800. We draw the axis from $24,000 to $31,000 and make tick marks at $1000 increments.

Locating the box portion of the plot is not difficult. To determine how far to extend the whiskers, we compare the minimum and maximum values in the data set to the lower and upper inner fences:

Minimum: $25,000   Lower inner fence: $24,275
Maximum: $30,800   Upper inner fence: $28,075

We see that the lower whisker can extend all the way to the minimum data value, but the upper whisker cannot go to the maximum because it is beyond the inner fence.

Looking back at the data, we see that the largest data value that is less than $28,075 is $28,000. The boxplot is shown in the figure:

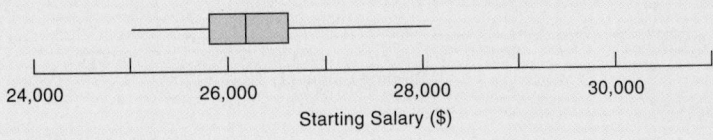

From the boxplot it appears that the data are only slightly skewed to the right because the median is located only a little closer to the lower quartile and the right whisker is only slightly longer than the left.

3.5.4 USING A BOXPLOT TO IDENTIFY OUTLIERS

At this point you may have questions: Why did we find the outer fences if we don't use them? What happens to the values in the sample that are beyond the inner fences? Both of these questions are reasonable and we are ready to answer them.

When we started talking about boxplots, we said that they could be used to identify outliers. The intervals defined by the fences are similar to the 2σ and 3σ intervals from the empirical rule. Sample data that fall between the inner and outer fences are called *possible outliers*, whereas data values that fall beyond the outer fences are called *probable outliers*. If you are having trouble figuring out the difference between *probable* and *possible*, think about the difference in your reaction when your instructor tells you, "It is *possible* that you will pass this course" versus "It is *probable* that you will pass this course."

The observations that fall between the inner and outer fences, the possible outliers, correspond to those that fall between two and three standard deviations from the mean using the empirical rule. Remember that less than 5% of the data should fall in this region, so data values there are fairly unlikely events.

The observations that fall beyond the outer fences, the probable outliers, correspond to those that are more than three standard deviations from the mean in the empirical rule. Less than 1% of the data should fall in this region, so data values there are very unlikely events.

The possible and probable outliers are plotted individually on the boxplot using special symbols, such as an open dot for possible outliers and a closed dot for probable outliers.

EXAMPLE 3.20

STARTING SALARIES

Locating Outliers on a Boxplot

Continuing on from Example 3.19, let's finish the boxplot for the starting salaries data by identifying the possible and probable outliers. To do this we need to compare the data values that are not included in the whiskers with the inner and outer fences. Nine observations are beyond the inner fence on the upper side of the boxplot. We can compare these values with the values of the inner and outer fences:

Upper inner fence:	$28,075
Possible outliers:	$28,200; 28,300; 28,300; 28,400; 28,500; 28,600; 29,400
Upper outer fence:	$29,500
Probable outliers:	$30,700; 30,800

The completed boxplot is shown in the figure:

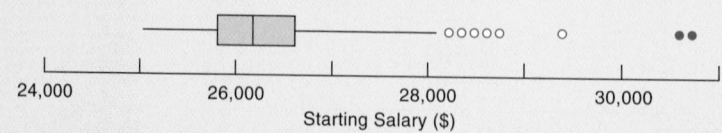

When the outliers are added to the boxplot, the extreme skewness shown in the histogram is evident. Now we know that it is not just a case of skewed data but that in fact at least two of the data points are very different from the rest of the data.

Now that you know that some of the data might be outliers, you may wonder what we do about it. Remember that when we sample from populations, we want our sample to represent the population of interest. An outlier is a data value that has a large variation from the center. The first thing to do when we have identified outliers is to search for the *cause* of the variation. If some identifiable, assignable cause makes the data value *not* representative of the data, then it is reasonable to drop the observation from the sample and redo the analysis. We cannot drop the observation without cause, but we can note in our report or analysis that an observation might have influenced the results.

Why do we need *another* method for displaying sample data and identifying outliers? One important reason for using the boxplot is that it relies on statistics that are *invariant* (do not change) to the outliers themselves. When we use z-scores to identify outliers, we use the mean and the standard deviation, both of which are sensitive to extreme values. If we have cause to actually delete the extreme observations, then the mean and standard deviation will change and *so will the definition of an outlier!* In a way, using z-scores to identify outliers is like trying to hit a moving target. The median and quartiles, on the other hand, remain relatively stable. Removing outliers does not affect the locations of the inner and outer fences, and so the definition of an outlier stays the same.

Another reason for using boxplots to display data is that they give an excellent picture of the distribution and they make it easy to compare samples from the same or different populations. Multiple boxplots may be put on the same axes and thus make comparisons easier than multiple histograms, each of which requires a separate graph.

EXAMPLE 3.21

THE GOLF BALL DILEMMA

Comparing Data Using Boxplots

As the statistical consultant, you decide to give the golf pro a final picture by showing him the boxplots for each sample. You use a statistical software package to create the plots; the results are shown here:

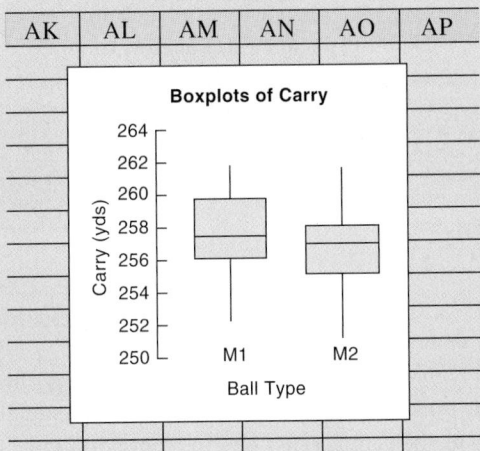

This is the first time the two ball types have actually been compared side by side, and the boxplots provide new information. It appears that although the middle 50% of the model 1 balls carry between 256 and 259 yards, the middle 50% of the model 2 balls carry only between 255 and 258 yards. In addition, based on the size of the plot, it appears that the model 2 balls are slightly more variable, but, again, this may be due entirely to the samples selected.

TRY IT NOW!

Aptitude Test Scores
Drawing a Boxplot

The manufacturing facility that administered the training aptitude test to its employees wants a better picture of how the employees performed on the test. We reprint the test scores here:

185	227	241	257	281	299	314	329
195	228	243	261	283	304	318	333
196	234	248	269	283	307	319	335
199	238	250	271	291	309	322	349
223	241	253	272	297	310	328	353

In Section 3.5.2 you found the first and third quartiles of the data set. Use these values to complete the calculations needed for a boxplot.

Draw a complete boxplot of the data.

Are there any outliers? If so, which data values are they?

3.5.5 EXERCISES—LEARNING IT!

Requires Exercise 3.1

3.19 The management of Disney World thinks that between 5 and 10 minutes is about the right amount of time for someone to wait for a monorail. Here are the sample wait times in minutes:

$$5.5 \quad 9.6 \quad 5.1 \quad 13.6 \quad 6.5 \quad 8.6 \quad 9.3 \quad 9.1 \quad 9.5 \quad 15.0 \quad 9.7 \quad 14.1$$

(a) What is the percentile rank of the person who waited 9.7 minutes?

(b) What is the percentile rank of the person who waited 5.1 minutes?

(c) Approximately what percentage of the people in the sample waited between 5 and 10 minutes?

Requires Exercise 3.7

3.20 The transportation department of the small city wants to know what percentage of the time ten cars were waiting at the light. Use the following data on the numbers of cars waiting at ten light cycles:

$$4 \quad 4 \quad 5 \quad 6 \quad 7 \quad 8 \quad 9 \quad 10 \quad 11 \quad 12$$

(a) What is the percentile rank of the observation of ten cars waiting at the light?

(b) What is the percentile rank of the observation of seven cars waiting at the light?

(c) What information do the percentile ranks give the transportation department?

Requires Exercise 3.10

3.21 The mail-order book club wants to know about the frequency with which customers make subsequent purchases. Use the following sample of the numbers of purchases made by 15 randomly selected customers:

ANS. IQR = 71, LIF = 134.5, UIF = 418.5, LOF = 28, UOF = 525; NO OUTLIERS

12 12 14 15 14 11 13 10 14 14 13 14 14 11 15

(a) What is the percentile rank of a person who makes 13 subsequent purchases?

(b) What are the first and third quartiles of the data?

(c) What information do the quartiles give the book club managers?

3.22 The manager who is looking at the travel expenses of sales personnel wants to look at the data in terms of percentiles. Here are the travel miles claimed in expense reports: *Requires Exercise 3.12*

1558   1617   1655   1709   1716   1733   1758   1908   1963   2062   2075   2171

(a) Find the first and third quartiles of the data and explain what they mean.

(b) What is the interquartile range? What does it tell the manager?

(c) Make a boxplot of the data.

(d) Use the boxplot to describe the distribution of travel miles claimed.

(e) Are any of the observations unusual? Why or why not?

3.23 The university that surveyed students about the food service wants to look at the ratings of food quality using measures of relative standing. Here are the data: *Requires Exercise 3.2*

```
1   1   3   3   4
1   1   3   3   4
1   2   3   3   4
1   2   3   3   4
1   2   3   3   4
1   3   3   4   4
```

(a) Find the first and third quartiles of the data and explain what they mean.

(b) What is the interquartile range? What does it mean?

(c) Make a boxplot of the data.

(d) Use the boxplot to describe the distribution of food ratings.

(e) Are any of the observations unusual? Why or why not?

3.24 Recent data from the Department of Justice (June 1995) give a state-by-state report on the number of people incarcerated per 100,000 people in the population.

District of Columbia	1151	New York	382	Tennessee	260
Delaware	890	North Carolina	356	Wyoming	239
Texas	643	Kentucky	354	Iowa	235
South Carolina	512	Virginia	354	Idaho	231
Alaska	505	Missouri	352	New Mexico	229
Arizona	501	Mississippi	346	Wisconsin	217
Connecticut	481	Kansas	332	West Virginia	211
Oklahoma	479	Illinois	329	Washington	197
Alabama	468	South Dakota	328	Montana	191
Florida	444	Arkansas	326	New Hampshire	183
Georgia	439	Pennsylvania	326	Utah	178
Nevada	431	Rhode Island	310	Massachusetts	174
Maryland	424	New Jersey	294	Nebraska	173
Michigan	424	Colorado	289	Vermont	173
Louisiana	399	Hawaii	279	Minnesota	139
California	398	Oregon	275	Maine	117
Ohio	390	Indiana	267	North Dakota	107

(a) Make a boxplot of the data.

(b) Use the boxplot to describe the number of people incarcerated in the United States.

(c) Are any of the data values unusual? Why or why not?

GET IT IN WRITING

Report on Comparison of Two Golf Ball Models

We were asked to compare two different models of golf balls in order to determine what differences, if any, exist between the top and bottom of the product line. The study was conducted as a blind study; that is, we did not know the golf ball types at the time we collected and analyzed the data. This procedure ensures that our results are as unbiased as possible.

Using a mechanical hitting device, we hit 72 different golf balls, 36 of each type. The balls were hit in batches of 12, alternating brands. To measure distance traveled we used carry, which is the distance from the point of impact with the club to the place where the ball first hit the ground, and total distance, which is the distance from the point of impact to the final position of the ball. This report focuses on the carries of the two ball types.

Table 1 contains summary statistics for the two ball types. As you can see, the two models are very similar. The average carry for model 1 is 257.4 yards, and for model 2 it is 256.8 yards. This difference is less than 1 yard and might not be distinguishable to most golfers. The medians for models 1 and 2 are 257.5 and 257.0 yards, respectively. This means that for each model, 50% of the balls hit carried farther than this value and 50% carried less. The standard deviation for the carry is 2.4 yards for model 1 and 2.3 yards for model 2. This, along with the ranges, indicates that the variation in carry for the two balls is about the same.

TABLE 1

	Model 1	Model 2
Mean	257.4	256.8
Standard error	0.4	0.4
Median	257.5	257.0
Mode	258.0	255.0
Standard deviation	2.4	2.3
Range	10.0	11.0
Minimum	252.0	251.0
Maximum	262.0	262.0
Count	36.0	36.0

Although these summary statistics indicate that the balls are not very different, we decided to look at some other measures. Table 2 lists quartile values for the two models. You can see that for model 1 25% of the balls went farther than 259 yards, whereas for model 2 the top 25% cutoff was 258 yards. The bottom 25% of the model 1 balls flew less than 256 yards, whereas for model 2 the value was 255 yards.

TABLE 2

	Model 1	Model 2
First quartile	256.0	255.0
Third quartile	259.3	258.0

These results are best illustrated with the boxplots in Figure 1. You can see that the middle 50% of the model 2 balls (indicated by the box) carry less far than those for model 1. In addition, there is more variation in the carry for model 2.

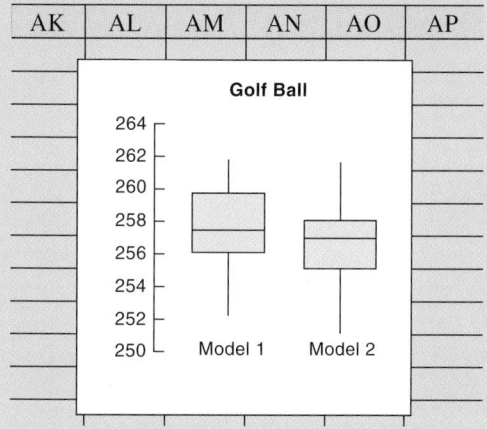

At this point, it is difficult to say that there is any real difference between the two balls. However, we suggest that a similar analysis be done using total distance instead of carry, since this is what matters to most golfers. Also, we recommend that golf balls be hit under a wider variety of conditions to better simulate use by consumers.

3.6 Numerical Descriptors and Technology

To calculate summary statistics and to create boxplots, you can use Excel, Minitab, or the TI-83 Plus graphing calculator. In some cases, you need to download Kaddstat, the add-in feature on the CD that accompanies this book.

We present some of the procedures for creating summary statistics for the golf ball and the aptitude test scores examples discussed in this chapter. For more detailed instructions on how to use these packages for all of the concepts in this chapter, refer to the technology manual that accompanies this text.

Excel 2000

1. Select **Data Analysis** from the **Tools** menu, and choose **Descriptive Statistics** from the list of analysis tools.
2. Highlight the range of data you want to calculate in the **Input Range** box.

3. Click the radio button to specify a location for output. If you select **Output Range,** you must specify a location on the worksheet.

4. Specify what kind of descriptive statistics you want. Click on **Summary Statistics** and finally click on **OK.** The output will appear in the location you specified and should look like Figure 3.9. When a statistic cannot be computed, the output will read N/A (not available).

Note: You will have to adjust the column widths to be able to read the first column of the output.

AE	AF
Carry	
Mean	257.0972
Standard Error	0.274902
Median	257
Mode	258
Standard Deviation	2.332621
Sample Variance	5.441119
Kurtosis	-0.18199
Skewness	-0.04536
Range	11
Minimum	251
Maximum	262
Sum	18511
Count	72

FIGURE 3.9 Excel output from Tools > Data Analysis > Descriptive Statistics

Minitab

1. From the **Stat** menu, choose **Basic Statistics** and then **Display Descriptive Statistics.**

2. Select the variable for which you want the statistics calculated (in this case, "Carry"). Since you want summary statistics for the two balls separately, click **By Variable,** then **Model,** and then **OK.**

3. The output from the **Display Descriptive Statistics** command appears in the **Session** window, since the output is all text, as shown in Figure 3.10. You can see

```
          —— 9/26/01 10:53:18 AM ——

Welcome to Minitab, press F1 for help.
MTB > Retrieving worksheet from file: D:\Textbook Business 2e\Data Sets\GOLFBAL
# Worksheet was saved on Mon Mar 22 1999
MTB > Describe 'Carry'.

Descriptive Statistics: Carry by Model _

Variable   Model _        N      Mean    Median    TrMean     StDev
Carry      M1            36    257.42    257.50    257.47      2.36
           M2            36    256.78    257.00    256.78      2.29

Variable   Model _   SE Mean   Minimum   Maximum        Q1        Q3
Carry      M1           0.39    252.00    262.00    256.00    259.75
           M2           0.38    251.00    262.00    255.00    258.00
```

FIGURE 3.10 Descriptive Statistics output from Minitab

that the output includes all of the summary statistics you learned about in this chapter. It also includes the trimmed mean and the standard error of the mean (SE Mean). You will learn more about the SE Mean in Chapter 7.

TI-83 Graphing Calculator

Here we create a set of summary statistics for the aptitude test scores from Try It Now! In Section 3.5.1.

1. Press [STAT] and use the ▶ key to move over to the [CALC] selection. The screen should look like the one shown in Figure 3.11.

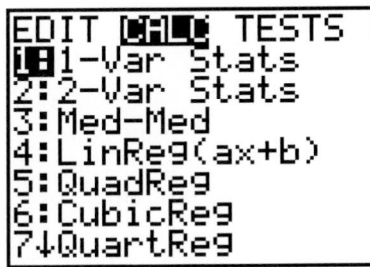

FIGURE 3.11 **Calc > 1-Var Stats screen**

2. Choose option 1: **1-Var Stats** and press [ENTER]. The words **1-Var Stats** appear on the screen. If your data are in L_1, then just press [ENTER]. If your data are in another list, then press [2nd] and the key that has the correct label (L_2 through L_6) above it.

3. The screen should display a set of summary statistics as shown in Figure 3.12.

```
1-Var Stats
 x̄=274.875
 Σx=10995
 Σx²=3102279
 Sx=45.29911274
 σx=44.7292899
↓n=40
█
```

FIGURE 3.12 **1-Var Stats**

The first value is the mean, $\overline{X}$, which is 274.875. The second and third values are the sum of the x values and the sum of the x^2 values, which you know from the formula for the standard deviation. The fourth value is the sample standard deviation, s, and the fifth value is the population standard deviation, σ, which is obtained by dividing by n instead of $n - 1$ in the formula. The last value, n, is important for checking your work. It is very easy to forget to clear out old data or to make mistakes when entering data. Make sure this value is correct before you use the results.

Next to the value of n is a down arrow indicating that there is additional output. Use the ▼ button to scroll down to see the output. You should see values for Min, Q_1, median, Q_3, and Max, as shown in Figure 3.13 on page 156.

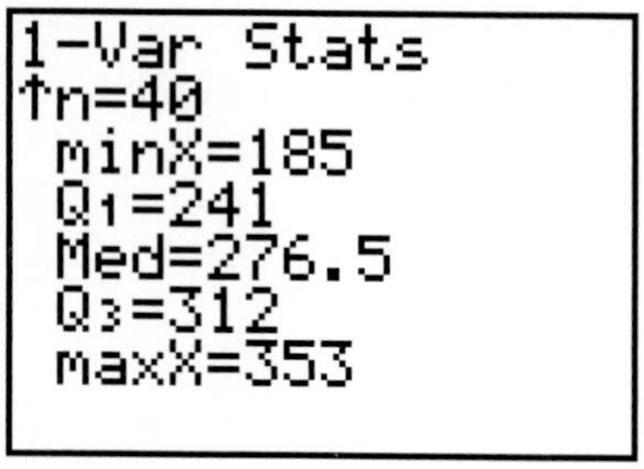

```
1-Var Stats
↑n=40
 minX=185
 Q₁=241
 Med=276.5
 Q₃=312
 maxX=353
```

**FIGURE 3.13 Additional
1-Var output**

CHAPTER 3 SUMMARY

There are many ways to describe a set of data using sample statistics. No single number will do the job, nor is there any standard way to proceed. The measures that you choose must reflect the characteristics of the data themselves. Most of the time the best descriptions come from using multiple measures and the conclusions that can be reached by comparing them.

Rather than summarize a sample with a list of numbers it is often useful to create images of the data using combinations of different statistics. An example of this is the *empirical rule*, which gives a picture of the distribution of the data. Another example of using summary statistics to get a picture of the distribution is the *boxplot*.

Data analysis is not a static tool. You need to look at a set of data in every way possible to obtain all of the information that it contains. Sometimes different methods all lead to the same conclusions, and sometimes one method yields an insight that is hidden in every other method.

KEY TERMS

Term	Definition	Page reference
Boxplot or box and whisker diagram	A **boxplot** is a graphical display that summarizes the distribution of a sample using the quartiles and the median.	145
Empirical rule	The **empirical rule** states that, for a symmetric distribution:	134
	• About 68% of all observations are within one standard deviation of the mean.	
	• About 95% of all observations are within two standard deviations of the mean.	
	• Almost all (more than 99%) of the observations are within three standard deviations of the mean.	
First quartile, Q_1	The **first quartile** is the value in the sample that has 25% of the data at or below it.	142
Interquartile range, IQR	The **interquartile range** is the difference between the third and first quartiles, $Q_3 - Q_1$.	145
Modal class	The **modal class** is the class interval in a frequency distribution or histogram that has the highest frequency.	121
Parameter	A **parameter** is a numerical descriptor that is used to describe a population.	111

Term	Definition	Page reference
Possible outlier	A **possible outlier** is an observation that falls between the inner and outer fences.	148
Probable outlier	A **probable outlier** is an observation that falls beyond the outer fences.	148
pth percentile	The **p**th percentile is the value in the data that has $p\%$ of the data at or below it.	140
Sample mean	The **sample mean** is the center of balance of a set of data, found by adding all of the data values and dividing the sum by the number of observations.	112
Sample median	The **sample median** is the value of the middle observation in an ordered set of data.	115
Sample mode	The **sample mode** is the data value that has the highest frequency of occurrence in the sample.	121
Sample range, **R**	The **sample range** is difference between the maximum and minimum observations in the sample.	128
Sample standard deviation, **s**	The **sample standard deviation** is positive square root of the sample variance.	131
Sample variance, **s^2**	The **sample variance** is the average of the squared deviations of the data values from the sample mean.	131
Statistic	A **statistic** is a numerical descriptor that is calculated from sample data and is used to describe the sample.	111
Third quartile, **Q_3**	The **third quartile** is the value in the sample that has 75% of the data at or below it.	143
z-Score	A **z-score** measures the number of standard deviations that a data value is from the mean.	136

KEY FORMULAS

Term	Formula	Page reference
Inner fences	$Q_1 - 1.5(IQR)$ $Q_3 + 1.5(IQR)$	145
IQR	$Q_3 - Q_1$	145
Outer fences	$Q_1 - 3(IQR)$ $Q_3 + 3(IQR)$	145
Percentile rank	$P = \dfrac{b + \frac{1}{2}e}{n}$	141
Sample mean, $\overline{X}$	$\overline{X} = \dfrac{\sum_{i=1}^{n} x_i}{n}$ or $\dfrac{\sum x}{n}$	112
Sample range, R	Maximum $-$ Minimum	128
Sample standard deviation, s	$s = \sqrt{s^2}$	131
Sample variance, s^2	$s^2 = \dfrac{\sum_{i=1}^{n}(x_i - \overline{x})^2}{n - 1}$	131
Sample variance	$s^2 = \dfrac{\sum_{i=1}^{n} x_i^2 - \left(\sum_{i=1}^{n} x_i\right)^2}{n(n-1)}$	132
z-Score	$\dfrac{X - \mu}{s}$	137

CHAPTER 3 EXERCISES

Learning It!

3.25 A manufacturer of pain relievers is interested in studying the amount of time it takes a person to be relieved of headache pain after taking the medication. The manufacturer selects a random sample of 12 people and conducts a study. The data are in minutes:

13.0   12.9   13.2   12.7   13.1   13.0   13.1   13.0   12.6   13.1   13.0   13.1

(a) Find the mean, median, and mode of the relief times.

(b) Find the range and standard deviation of the relief times.

(c) What is the percentile rank of the person who took 12.9 minutes to be relieved of headache pain?

(d) Find the quartiles for the data set and interpret them.

3.26 A group of elderly people in a town filed a grievance with the town council, saying that the length of the walk signals was inadequate for many elderly people to cross safely. In an attempt to investigate the claim, the town collected data on street crossing times for 12 elderly persons. The times are to the nearest tenth of a second:

21.4   15.1   13.6   16.0   15.0   19.1   21.0   14.2   15.6   20.1   21.1   22.2

(a) Find the mean, median, and mode of the crossing times.

(b) Find the range and standard deviation of the crossing times.

(c) Compare the mean and the median. Do you think the data are skewed?

(d) What is the percentile rank of the person who took 20.1 seconds to cross the street?

(e) Find the quartiles and interpret them.

(f) Make a boxplot of the data.

(g) Describe the distribution of the crossing times.

3.27 In a study of the price variation for a 24-pack of a popular cold relief tablet, a random sample of the product at ten different stores yields the following prices in dollars:

4.74   4.62   4.76   4.72   4.99   5.29   4.98   4.79   4.75   4.75

(a) Find the mean selling price of the product.

(b) Find the median selling price of the product.

(c) Compare the median and the mean. Is the distribution of the prices symmetric? Why or why not?

(d) Would it make sense to use the mode as a measure of the center for this sample? Why or why not?

(e) Find the range of the prices.

(f) Find the standard deviation of the prices.

3.28 The vice president of marketing at a large corporation is wondering how many of the company's employees arrive at work before he does. He decides to collect some data by counting the numbers of cars that are in the parking lot when he arrives at 6:30 A.M. on ten different days:

23   24   24   25   25   25   25   26   26   37

(a) Find the sample mean of the data.

(b) Find the median of the data.

(c) Find the range of the data.

(d) Find the standard deviation of the data.

3.29 A company that sells air purifiers via mail order and telephone wants to know the amount of time that a customer spends on the phone with a sales agent during a call.

The company collects information on the lengths of calls (in minutes) for 20 different customers:

15	23	34	38	41
16	25	37	40	43
19	25	38	40	43
20	28	38	41	44

(a) What is the average length of a call to a sales agent? What is the median length of a call?

(b) What is the range of the call lengths? What is the standard deviation of the call lengths?

(c) Compare the mean and the median. Do you think the distribution of call times is symmetric or skewed? If it is skewed, which way is it skewed?

(d) Use one of the graphical methods you learned in Chapter 2 to display the data. Does the display agree with your answer to part (c)? If not, why not?

3.30 A manufacturer of breakfast cereals packs the cereal in 15-ounce boxes. Periodic quality checks are made to ensure that the boxes are filled adequately. A random sample of 35 boxes of the cereal is selected and their weights (in ounces) are recorded:

14.91	15.23	15.34	15.40	15.48	15.59	15.62
14.96	15.24	15.34	15.42	15.49	15.60	15.62
15.11	15.28	15.36	15.45	15.50	15.60	15.63
15.21	15.30	15.38	15.46	15.52	15.61	15.64
15.22	15.30	15.39	15.48	15.58	15.61	15.67

(a) Find the mean, median, and mode for the weights of the cereal boxes.

(b) Find the range and the standard deviation of the cereal box weights.

(c) Describe the distribution of the data using the information from parts (a) and (b).

(d) Find the quartiles of the data and interpret them.

(e) Draw a boxplot of the data.

3.31 A university collected some data on the amount of money (in dollars) that students spend on textbooks in a typical semester:

239	289	304	323	336
256	290	307	324	394
280	295	310	326	397
284	295	314	330	415
284	298	315	331	429
287	298	319	332	445
287	299	321	334	447

(a) Find the mean and the standard deviation of the amounts spent on textbooks.

(b) Find the intervals $\overline{X} \pm 1s$, $\overline{X} \pm 2s$, and $\overline{X} \pm 3s$.

(c) Make a histogram of the data and mark the intervals on the histogram.

(d) Calculate the actual percentage of the data that fall within each interval.

(e) Compare the actual percentages with those predicted by the empirical rule.

3.32 A pharmaceutical company is looking at the lengths of time that salespeople stayed at the company. The data (in months) are shown here:

18.5	19.9	22.1	23.1	25.5	26.3	27.4
18.6	20.5	22.4	23.3	25.6	26.6	27.6
18.7	20.8	22.6	24.7	25.8	26.8	29.8
19.8	20.9	22.9	25.1	26.1	27.0	31.4
19.8	21.7	23.0	25.5	26.3	27.0	47.5

(a) Find the mean and the standard deviation of the lengths of time that salespeople stayed at the company.

(b) Find the intervals $\overline{X} \pm 1s$, $\overline{X} \pm 2s$, and $\overline{X} \pm 3s$.

(c) Calculate the actual percentage of the data that fall within each interval.

(d) Compare the actual percentages with those predicted by the empirical rule.

(e) Create a boxplot of the times salespeople stayed at the company.

(f) Use the boxplot to describe the time that salespeople stayed at the company.

3.33 A large hospital recently did a study on the treatment of patients who were admitted with pancreatitis. Administrators wanted to know, among other things, how long patients spent in the hospital. Since patients were admitted to either the medical or surgical service, the administrators decided to look at the numbers of days in the hospital for both services separately.

Medical				Surgical		
2	4	7	10	3	6	9
2	4	7	12	4	6	10
2	5	7	13	4	6	11
3	5	7	15	5	7	13
3	6	8		5	8	

(a) Find the mean, median, and mode for the numbers of hospital days for each service.

(b) Find the range and standard deviation for the two sets of data.

(c) Make a boxplot for each set of data.

Thinking About It!

Requires Exercise 3.28

3.34 Consider the data on the numbers of cars that are in the parking lot when the vice president arrives:

$$23 \quad 24 \quad 24 \quad 25 \quad 25 \quad 25 \quad 25 \quad 26 \quad 26 \quad 37$$

(a) If the vice president wants an estimate of the typical number of employees who arrive at work before he does, is the mean or the median a better statistic to use? Why?

(b) Is the standard deviation or the range more suitable for these data? Why?

(c) Find the z-score of the data value of 37 cars. Does it confirm your suspicion that the data value is unusual?

3.35 Grading homework is a real problem and takes an enormous amount of time. Many students do not do a good job on their assignments or they copy answers from other students or the back of the book. A teacher of elementary statistics decided to conduct a study to determine what effect grading homework had on her students' exam scores. She taught three sections of elementary statistics and randomly assigned each class to one of three conditions: (1) no homework given, (2) homework given but not collected, and (3) homework given, collected, and graded. After the first exam, she collected the data (exam scores) and made histograms and calculated some numerical measures.

$\overline{X} = 76.6$; median $= 77.5$; $s = 10.1$; $R = 45$ $\overline{X} = 75.0$; median $= 75.5$; $s = 9.6$; $R = 43$

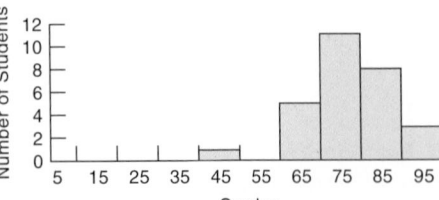

No homework

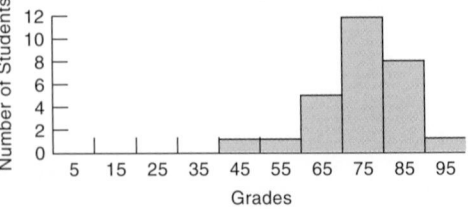

Homework, not collected

$\overline{X} = 83.5$; median $= 83.0$; $s = 8.0$; $R = 32$

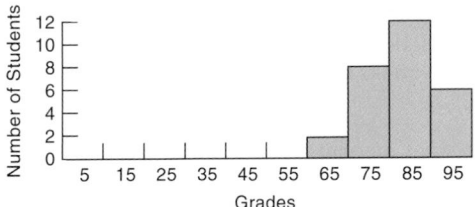

Homework collected and graded

(a) Describe the distribution of the data for each of the groups and use the information to compare the three data sets. What conclusions can you draw about the effects of homework on exam grades?

(b) Do you think that this was a valid way for the instructor to examine the effects of homework on exam scores? Why or why not? What other factors might need to be considered? Based on the data, if you were the instructor, what would you do about assigning homework?

3.36 The air purifier company that collected data on the lengths of customer calls to the sales agents is unsure about which sample statistics to use to summarize the data and what the statistics really mean. The company calls on you to be its expert. What can you tell them about the lengths of phone calls to the sales agent? Write a memo summarizing your findings.

Requires Exercise 3.29

3.37 A company has been recording data on the numbers of defective items found in the daily production lots after 100% inspection. The data recorded for the last five days showed an average of 20 defectives. One of the people on the inspection team thought this seemed strange and realized that on one of the days the number of defectives was really 10 and had been miscopied in the records. When the correct value was written, the average number of defectives changed from 20 to 15. What value had been written down instead of the 10?

3.38 Create a boxplot for the data on the numbers of cars that are in the parking lot when the vice president arrives:

Requires Exercises 3.28 and 3.34

$$23 \quad 24 \quad 24 \quad 25 \quad 25 \quad 25 \quad 25 \quad 26 \quad 26 \quad 37$$

(a) What does the boxplot indicate about outliers?

(b) Does part (a) agree with the information you obtained from the z-score? Why or why not?

(c) Is the boxplot or the z-score a more reliable indicator of outliers in this case?

3.39 A college computer center is looking at the times to failure of the diskettes that students buy in the college bookstore. The manager decides to look at two different suppliers of the product. The times to failure (in hours) for its current supplier and for the new supplier are listed here:

Supplier A					**Supplier B**				
474	492	498	504	511	487	492	495	497	499
486	492	500	505	511	488	492	495	497	499
489	494	501	506	512	489	492	495	497	499
490	494	501	507	513	489	492	495	497	501
490	494	501	508	514	489	493	495	498	502
490	495	502	508	515	491	493	496	498	503
491	496	502	509	517	491	494	496	498	503
491	498	504	509	519	491	494	496	498	505
491	498	504	510	528	492	494	496	499	506

(a) Find the mean, median, range, and standard deviation of the times for each supplier.

(b) Make a boxplot of the data for each supplier.

(c) Based on these statistics, describe the diskettes for each supplier and write a recommendation to the computer center on which supplier to choose. Be sure to support your recommendation with references to the data.

Requires Exercise 3.31 **3.40** The university that collected data on the amount of money that students spend on textbooks wants to include the information in their catalog on typical semester expenses. The current catalog estimates that students spend $320 on books in a semester.

(a) Do you think the catalog should be changed?

(b) If you think the catalog should be changed, what amount should be listed? Justify your choice. If you think the catalog does not need to be changed, explain why.

(c) Suppose that changing catalog copy is costly considering the number of catalogs that are currently on hand. Does this change your answer to part (b)? If so, why? If not, why not?

3.41 The news reports that the average yearly salary of entry-level social workers in the country is $32,500, with a standard deviation of $930. A state agency collects data on recent hires and asks you to interpret the data and to show how the state's social workers' salaries compare with those of the country in general. The salaries of 30 social workers are given here:

26,633	29,045	30,192	30,628	33,397
26,764	29,453	30,260	30,829	33,553
27,306	29,620	30,269	31,270	33,685
27,619	29,690	30,346	31,489	34,466
28,445	29,864	30,400	32,731	34,944
28,827	29,974	30,532	33,190	35,751

Analyze the data using any techniques you think are appropriate, and prepare a report for the state agency on your findings.

3.42 As part of an annual program to calibrate quality inspectors, each inspector in a company is asked to do a 100% inspection of a lot of 500 items. These data are the numbers of defectives found by each of the 35 inspectors:

8	10	13	14	15	16	17
9	11	13	15	15	16	18
9	12	13	15	15	17	18
9	12	14	15	15	17	19
10	12	14	15	15	17	20

(a) If you did not know the actual number of defectives in a lot, would you use the sample mean, median, or mode as your estimate of the number? Justify your decision.

(b) Suppose that you were the training coordinator for the quality inspectors. How would you interpret the data? What would you do as a result?

Requires Exercise 3.33 **3.43** The hospital that is looking at the data on the numbers of days that patients with pancreatitis spend in the hospital asks you to examine the summary statistics and compare the lengths of stay for the two services.

Medical				**Surgical**		
2	4	7	10	3	6	9
2	4	7	12	4	6	10
2	5	7	13	4	6	11
3	5	7	15	5	7	13
3	6	8		5	8	

(a) Based on the data, what would you tell the hospital?

(b) Are the data sufficient for you to conclude that admission to the medical service reduces the length of stay for patients?

(c) If your answer to part (b) is yes, why can you reach this conclusion? If your answer to part (b) is no, why do you think this conclusion is not valid?

(d) As a consultant write a memo to the hospital. Be sure to include suggestions for additional data collection.

3.44 A study was recently completed by an insurance company concerning a particular surgical procedure. The study looked at the hospital records of 40 patients at two hospitals and compared the numbers of days the patients stayed in the hospital. The data were analyzed using Excel. The accompanying output shows the descriptive statistics for the data and boxplots of the two samples.

E	F	G	H	I	J	K	L
Boxplot Output	Hospital 1	Hospital 2					
First Quartile	6.0000	8.0000					
Median	7.5000	10.0000					
Third Quartile	9.0000	12.5000					
Interquartile Range	3.0000	4.5000					
Moderate Outliers (△)	1	0					
Extreme Outliers (▲)	0	0					

(a) Using the computer output, describe each of the data sets. Be sure to include information about central tendency, variability, shape, and outliers.

(b) Compare the lengths of stay for the two hospitals.

(c) Proponents of hospital 1 say that this hospital is better than hospital 2 at the surgical procedure because the patients from hospital 1 recover faster than the patients from hospital 2. Can you conclude, on the basis of these data, that hospital 1 is better at the surgical procedure than hospital 2?

(d) If you think that the conclusion described in part (c) is reasonable, justify it. If you think this conclusion is not justifiable, what can you conclude? What additional information would you, as the insurance company president, find useful?

3.45 The town that is studying crossing times consults you about what the data mean. Currently the walk signals are set for 15 seconds. Write a report telling the town council what you have found.   *Requires Exercise 3.26*

Doing It!

3.46 The golf pro wants to expand the analysis. He recognizes that you could collect additional data, and he wants to look at other aspects of the balls' performance. In addition to measures of the balls' performance, such as the variable *Carry*, he knows that other factors, both internal (ball related) and external (environment related), could affect performance. He tests 36 of each model of ball at three different times using a machine to launch the balls. Data are recorded on 14 different variables. A portion of the data is shown here: *Datafile: GOLFBALL.XXX*

Ball	Model	S1	S2	S3	Wgt	Dw	Dd	Head	Temp	Carry	TotDist	Date	Time
1	M1	81	81	82	45.3	0.145	0.0110	686	77	257	270	8/20	8:15
2	M1	83	83	84	45.2	0.151	0.0111	688	77	255	267	8/20	8:15
3	M1	81	82	84	45.2	0.145	0.0105	687	77	256	267	8/20	8:15
4	M1	81	81	83	45.3	0.144	0.0117	688	77	255	271	8/20	8:15
5	M1	83	81	82	45.5	0.146	0.0108	687	77	255	268	8/20	8:15

Here is a brief description of the variables:

Ball—the observation number; goes from 1 to 72
Model—the ball type, M1 and M2
S1, S2, S3—measurement of the ball's circumference at three different points
Wgt—weight of the ball
Dw, Dd—dimple width and depth
Head—head speed of the ball when it is hit
Temp—environmental temperature, in degrees Fahrenheit
Carry—distance from the point where the ball was hit to the point where it first hit the ground, in yards
TotDist—total distance the ball travels from the point where it was hit to its final position
Date—date of the testing
Time—time of day when the observation took place; there were three different time periods

(a) For each model number, calculate the mean, median, mode, range, and standard deviation of the variable *TotDist*.

(b) Describe the distribution of *TotDist* for each ball. How does *TotDist* compare with *Carry*? Which measure do you think is better from a statistical point of view?

(c) What can you tell the golf pro about the effects of temperature during the trials?

(d) What can you say about the variable *Head* over the entire trial? Is it consistent for both ball models? Should the head speed be considered as a variable in this trial or is it controlled well enough?

(e) Would you expect there to be much difference in performance from one time period to the next? Examine the effects of *Time* for each model separately and describe what you see. Make sure you look at both *Carry* and *TotDist*.

(f) Look at the variables *S1, S2,* and *S3*. The closer these measures are to being equal, the closer to spherical the ball is. Do you think the balls are spherical? Why or why not?

(g) Look at *Dw* and *Dd*. How do these compare for each ball type?

(h) Think about the problem and analyze the data in any other way that you think might be interesting or important.

(i) Write a summary memo to the golf pro with your analysis. Include the answers to parts (a)–(g) as appendix material.

Datafile: UNITED.XXX **3.47** Because more and more consumers travel by air, the Bureau of Transportation Statistics publishes data on airline performance. One variable of interest to consumers is the delay time for a flight—that is, the difference in time between when a flight was scheduled to depart and when it actually did depart. Here is a portion of the flight delay data for United Airlines flights that departed from Denver, Colorado, in May 2000.

Carrier	Date	Flight No.	Tail No.	Desti-nation	Scheduled departure time	Actual departure time	Departure delay (minutes)	Wheels-off time	Taxi-out time (minutes)
UA	05/01/2000	59	N670UA	SFO	10:15	10:16	1	10:28	12
UA	05/01/2000	61	N161UA	SFO	14:47	14:55	8	15:07	12
UA	05/01/2000	240	N209UA	ORD	6:40	6:43	3	6:56	13
UA	05/01/2000	241	N7453U	PDX	8:06	8:06	0	8:26	20
UA	05/01/2000	248	N674UA	ORD	10:20	10:19	−1	10:35	16
UA	05/01/2000	252	N768UA	ORD	12:30	12:29	−1	12:47	18
UA	05/01/2000	256	N342UA	ORD	13:42	13:38	−4	13:52	14

(a) Calculate summary statistics for departure delay (negative numbers mean the flight departed early) for the whole 30-day period.

(b) Use the summary statistics to describe the departure data for United Airlines in May 2000.

(c) Display the data for the whole month using a boxplot. What does this tell you about the departure times?

(d) Look at the departure data for LaGuardia Airport (LGA) in New York City. Calculate summary statistics and create a boxplot for departure delay for LaGuardia. How do they compare with the overall performance?

(e) Repeat part (d) for O'Hare Airport (ORD) in Chicago. How do the data for O'Hare compare with those for LaGuardia?

(f) Look at the data for LaGuardia and O'Hare for days around holiday weekends (Mother's Day, Memorial Day). How do the holiday weekends compare with the overall performance?

(g) Calculate summary statistics and create boxplots for the early-morning departures (before 8:00 A.M.) from LaGuardia and O'Hare. Do the same thing for midafternoon departures (from 12:00 noon to 2:00 P.M.) and early evening departures (4:00–7:00 P.M.). How do these compare with each other and with the overall statistics?

(h) Write a newspaper article about airline delays using the data. Be sure to include any appropriate statistics and data. Include the answers to parts (a)–(g) as an appendix.

4 Analyzing Bivariate Data

AM I PAID FAIRLY?

The ABC College committee that is studying gender equity in salaries decides that it is time to look at more data than they currently have available. They want to consider some additional variables that may help them to understand the problem. The committee wonders whether, in addition to gender, they should look at the department or school in which the faculty member teaches and whether or not the faculty member has tenure. Perhaps these additional variables will give them some insight into the issue of gender equity. A portion of the data is shown here:

Rank	Years of service	Salary	Gender (M/F)	Tenure? (Y/N)	School
ASST	22	$53,316	F	Y	Business
PROF	11	64,375	M	Y	Business
ASSO	7	63,501	M	Y	Business
ASSO	6	59,426	M	N	Business
ASSO	20	49,058	M	Y	Business
PROF	4	94,969	M	N	Business
ASST	21	54,762	M	Y	Business

4.1   Chapter Objectives

In Chapters 2 and 3 you learned how to summarize data using both graphical and numerical measures. You learned how to use these techniques to obtain *information* about the data and to use that information to help in making decisions. In this chapter we look at some additional methods for summarizing data and also at how to display and analyze *bivariate* data. The chapter covers these topics:

- Qualitative bivariate data: contingency tables and clustered and stacked bar charts
- Quantitative bivariate data: scatter plots and the least-squares line

4.2   Qualitative Bivariate Data

In earlier chapters of this book we looked at ways to organize, summarize, and describe data that represent a *single* variable or characteristic of the population under study. Most often when data are collected, however, multiple variables are involved. Phenomena that occur in one variable usually are connected to or can be explained by the values of other related variables. In this section we discuss methods for summarizing and displaying data for two variables, known as *bivariate data*.

4.2.1   SUMMARIZING TWO QUALITATIVE VARIABLES

In Chapter 2 you learned that the first step in organizing and summarizing data for a single variable is to create a frequency table. The frequency table lists the number of occurrences of each value of the variable expressed as a count, fraction, decimal, or percentage. When data are collected on two *related* variables, they are organized by means of a **cross-classification table** or **contingency table.**

> A *contingency table* is a table with rows that represent the possible values of one variable and columns that represent the possible values for a second variable. The entries in the table are the numbers of times that each *pair* of values occurs.

Creating a contingency table for a set of bivariate data is similar to creating a frequency table for a single variable. The basic layout of a contingency table is shown in Table 4.1 on page 168. Each category for one of the variables is represented by a column, whereas each category for the other variable is represented by a row. The numbers in the table, the values of f_{ij}, represent the number of observations in the data set that are in category i of variable 1 *and* category j of variable 2.

By using a contingency table, we can look at the number, proportion, or percentage of the data that fall into each *pair* of values for variable 1 and variable 2. Adding a second variable to an analysis gives us a different, and sometimes valuable, perspective on the data we have collected.

TABLE 4.1

LAYOUT OF A CONTINGENCY TABLE

		Variable 2				
		Category I	Category 2	. . .	Category n	Total
Variable 1	Category 1	f_{11}	f_{12}	. . .	f_{1n}	
	Category 2	f_{21}	f_{22}	. . .	f_{2n}	
	.	.	.	.	.	
	.	.	.	.	.	
	.	.	.	.	.	
	Category m	f_{m1}	f_{m2}	. . .	f_{mn}	
	Total					

EXAMPLE 4.1

ABC FACULTY SALARIES

Creating a Contingency Table

Understand the problem.

Since the committee at ABC College knows that other variables should be factors in determining faculty salaries, they decide to look at the gender of faculty members in addition to the school in which they teach. The college has four schools: Business, Liberal Studies, Professional Studies, and Sciences. They create a contingency table for the data:

	School				
Gender	Business	Liberal Studies	Professional Studies	Sciences	Total
Female	6	22	15	27	70
Male	21	58	19	39	137
Total	27	80	34	66	207

Analyze the data.

Making comparisons with the actual frequencies is difficult, so the committee decides to use the relative frequency for each cell. The relative frequency table is shown here:

	School (%)				
Gender	Business	Liberal Studies	Professional Studies	Sciences	Total (%)
Female	3	11	7	13	34
Male	10	28	9	19	66
Total	13	39	16	32	100

Draw conclusions.

The committee sees that the distribution of faculty over the schools is not as uniform as the distribution over ranks. The Schools of Liberal Studies and Sciences have 71% of the faculty, whereas Business and Professional Studies have only 29%. They also notice that the distribution of gender differs in each of the schools. Although the percentage of females is always lower, in the School of Professional Studies the percentages of males and females are almost the same, whereas in the Schools of Business, Liberal Studies, and Sciences they are not. They wonder whether this has any impact on the distribution of faculty salaries.

EXAMPLE 4.2

STUDENT DISTRIBUTION

Creating a Contingency Table

When the dean of the School of Arts and Sciences looked at the data on students who were taking the Introductory Statistics course, he wondered whether he could identify the reason so many seniors were taking the course. He went back and asked again about their class but also asked a few more questions to help classify them. One of the questions was whether the student was a transfer student. He guessed that students who transferred into the college might not have had the course previously. The data looked like this:

Fr Transfer	So Nontransfer	So Nontransfer	Sr Transfer
Fr Nontransfer	So Nontransfer	So Transfer	Sr Nontransfer
Fr Nontransfer	So Nontransfer	Jr Nontransfer	Sr Transfer
Fr Nontransfer	So Nontransfer	Jr Nontransfer	Sr Transfer
So Nontransfer	So Transfer	Sr Transfer	Sr Transfer
So Nontransfer	So Nontransfer	Sr Transfer	Sr Transfer
So Nontransfer	So Nontransfer	Sr Transfer	Sr Transfer

To make sense of the data he decided to create a contingency table using the students' year as the row variable and the transfer status as the column variable.

	Status		
Year	**Nontransfer**	**Transfer**	**Total**
Fr	3	1	4
So	10	2	12
Jr	2	0	2
Sr	1	9	10
Total	16	12	28

From the table the dean saw that nine of the seniors who were taking the course were indeed transfer students. To make the table more general, he decided to use the same table but express each frequency as a percentage of the total:

	Status (%)		
Year	**Nontransfer**	**Transfer**	**Total (%)**
Fr	11	4	14
So	36	7	43
Jr	7	0	7
Sr	3	32	36
Total	57	43	100

From the second table the dean learned that, in fact, 43% of the students in the class were sophomores, which is supposed to be the norm. The dean also learned that senior transfer students made up 32% of the total students enrolled in the course. Since the table also indicated that 43% of the students in the course were transfer students, he decided that establishing a task force to study transfer students was probably a good idea.

Understand the problem.

Analyze the data.

Draw conclusions.

Creating contingency tables from large data sets can be tedious and time-consuming. Often, when data are collected, many more than two variables are considered and the person doing the analysis wants to look at different pairs of variables. Almost any software package that can be used for statistical analyses can create contingency tables easily. It is the *information* obtained from the tables that is important.

✓ **TRY IT NOW!**

Quality Problems
Creating a Contingency Table

A company that manufactures cardboard boxes is trying to understand some of its quality problems. The company has analyzed some data and determined that the major defects in the boxes are printing, color, and skewness (how square the box is). Further attempts to pinpoint the problems have resulted in many opinions and finger-pointing about responsibility. The company decides to collect some additional data on the type of defect and the shift during which the defective box was produced. The data on defect type and shift follow:

Color	1	Color	1	Color	1	Printing	2	Printing	2
Color	2	Color	1	Color	3	Printing	1	Printing	1
Color	1	Color	1	Color	1	Printing	3	Printing	2
Color	3	Color	2	Printing	3	Printing	1	Printing	1
Color	3	Color	2	Printing	2	Printing	2	Skewness	2
Color	2	Color	1	Printing	1	Printing	1	Skewness	2

Create a relative frequency contingency table for the data.

What percentage of the defects are color?

Does there appear to be any evidence for the claim that the majority of the defects occur on the third shift?

4.2.2 DISPLAYING CONTINGENCY TABLES

Although a great deal of information can be obtained from a contingency table, you learned in Chapter 2 that a visual representation of the data can increase your under-

Ans.

Defect	Shift (%)			
	1	2	3	Total (%)
Color	27	13	10	50
Printing	20	17	7	43
Skewness	0	7	0	7
Total	47	37	17	100

50%. NO, THE THIRD SHIFT HAS THE SMALL-EST PERCENTAGE OF DEFECTS.

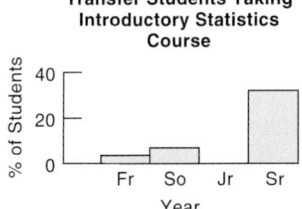
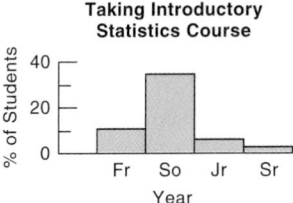

FIGURE 4.1 **Two bar charts representing a contingency table**

standing of what is happening. The question then is, How can we best represent bivariate data?

If we select one of the variables, it is possible to create a bar or pie chart for each of the possible categories of that variable. For the student distribution data in Example 4.2, the corresponding bar charts are shown in Figure 4.1. The pictures accomplish what we wanted, but our task will certainly be cumbersome if the number of possible categories increases! Also, as you learned in Chapter 2, when you are making comparisons on separate charts, the scales of the charts must be comparable.

One way to display contingency table data on the same chart is to use what are called *clustered* or *stacked* bar charts. Both of these charts use a different type of bar (color, shading) to represent one of the values of a selected qualitative variable.

Clustered Bar Charts

Clustered bar charts can be used in place of multiple bar charts when we want to compare the data for the values of selected qualitative variables. An example of a clustered bar chart is shown in Figure 4.2.

> In a ***clustered bar chart,*** the bars for one variable are grouped according to the values of the other qualitative variable.

You can see that a clustered bar chart really *combines* the results of several bar charts into a single chart. A legend is included to identify each of the categories.

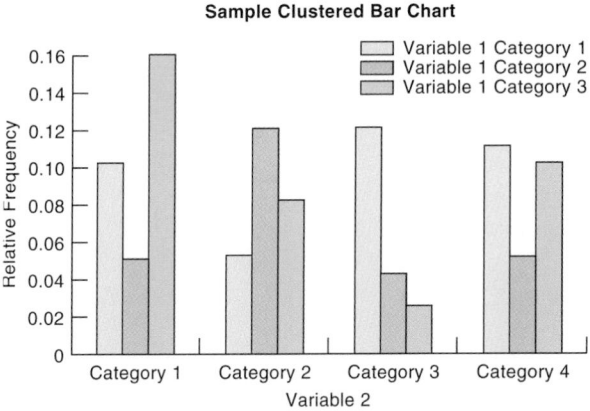

FIGURE 4.2 **Example of a clustered bar chart**

EXAMPLE 4.3

ABC FACULTY SALARIES

Creating a Clustered Bar Chart

The provost would like to see how the genders of the faculty are distributed over the schools. She decides to create a clustered bar chart using school as the category on the *x* axis and a different bar for each gender:

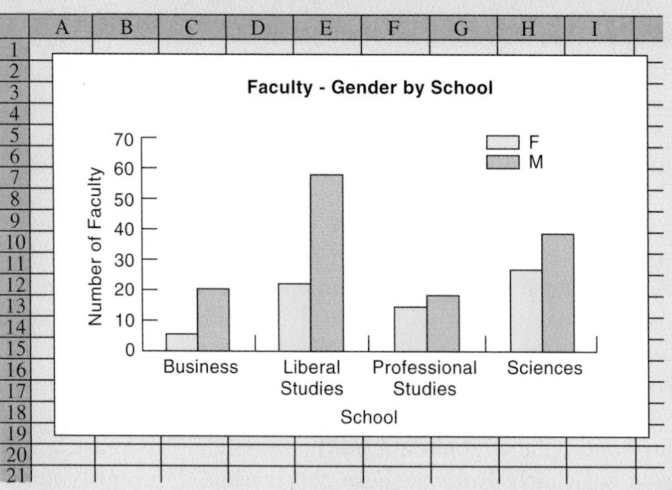

Draw conclusions.

Looking at the bar chart, the provost sees that the largest gender differences are in the Schools of Business and Liberal Studies, whereas the genders in the Schools of Professional Studies and Sciences are closer to being equal.

EXAMPLE 4.4

STUDENT DISTRIBUTION

Creating a Clustered Bar Chart

The dean wanted to compare transfer students to nontransfer students. To get a good picture of the data he decided to use a clustered bar chart to display the data. Since the variable he wanted to compare was transfer status, he made a bar for each of the two student types for each of the four years. There are really no new calculations to do. The chart that resulted is shown here:

Understand the problem.

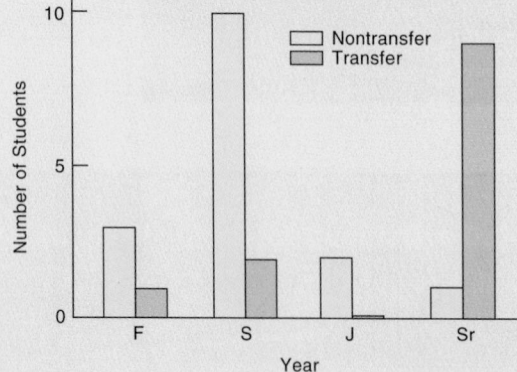

Analyze the data.

Draw conclusions.

From the chart the dean could see that transfer students take the statistics course primarily as seniors, whereas nontransfer students take the course primarily as sophomores.

TRY IT NOW!

Quality Problems
Creating a Clustered Bar Chart

The management of the company that manufactures cardboard boxes decides to have a meeting with the crews about the quality problems. The managers decide that a graphical display of the data is the best way to convey the information they have obtained. They want to display the shift data for each of the quality defects.

Defect	Shift (%) 1	2	3	Total (%)
Color	27	13	10	50
Printing	20	17	7	43
Skewness	0	7	0	7
Total	47	37	17	100

Create a clustered bar chart that displays the percentage of defects for each shift using the quality problems as the categories for the *x* axis.

Which shift contributes the largest percentage of color defects? The smallest percentage?

What do you notice about skewness defects?

Stacked Bar Charts

In Chapter 2 you learned that when you want to look at different categories as part of a whole, a pie chart can be more informative than a bar chart. For bivariate data, instead of using multiple pie charts you can use a **stacked bar chart.**

In a *stacked bar chart* the data for the selected variable are represented as a percentage of the total for each category of the second variable. Each value of the selected variable is represented in a different way, and the bars are "stacked" to total 100%.

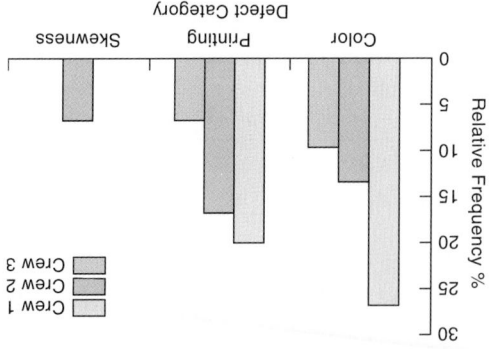

Defect Data by Crew

Stacked Bar Chart for Bivariate Data

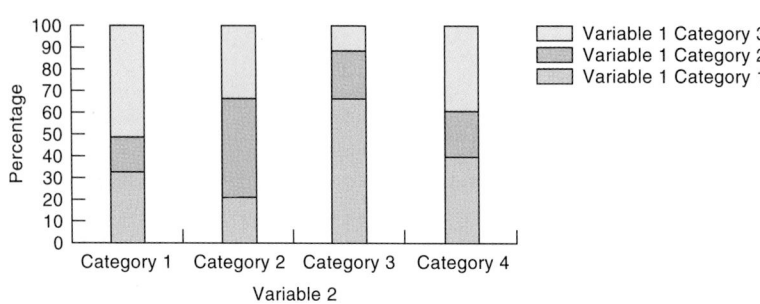

FIGURE 4.3 **Stacked bar chart**

An example of a stacked bar chart is shown in Figure 4.3.

EXAMPLE 4.5

ABC FACULTY SALARIES

Creating a Stacked Bar Chart

After looking at the clustered bar charts, the committee is interested in how the different schools are made up by gender of faculty. They decide to use a stacked bar chart to look at this, and so they modify the original contingency table shown here:

| | | School | | | |
Gender	Business	Liberal Studies	Professional Studies	Sciences	Total
Female	6	22	15	27	70
Male	21	58	19	39	137
Total	27	80	34	66	207

Analyze the data.

Since they want to know, for example, what percentage of the faculty in the School of Business is female, they need to calculate each frequency as a percentage of the column total. This calculation is 6/27 = 22.2%, or approximately 22%. The next calculation is for females in the School of Liberal Arts: 22/80 = 27.5%, or approximately 28%. The completed table is shown here:

| | | School (%) | | | |
Gender	Business	Liberal Studies	Professional Studies	Sciences	Total (%)
Female	22	28	44	41	34
Male	78	73	56	59	66
Total	100	100	100	100	100

Here is the stacked bar chart for this table:

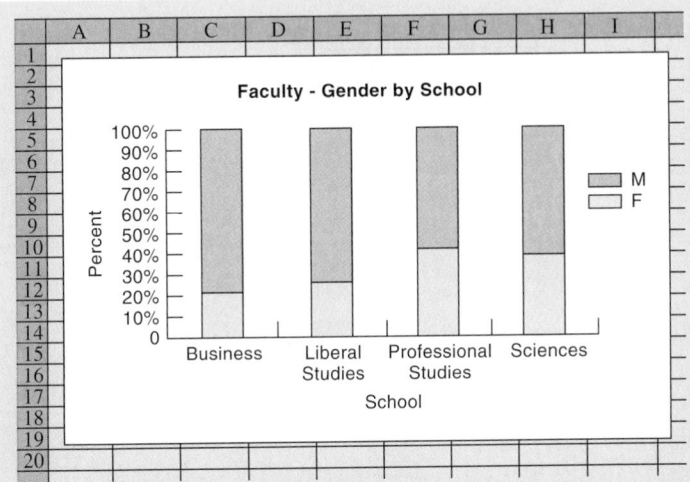

They see from the graph that the Schools of Business and Liberal Studies have a much smaller percentage of females than the other schools and that Professional Studies has the highest percentage of females. To determine whether this is important, they should now compare these figures with the gender breakdown of the faculty as a whole.

Draw conclusions.

EXAMPLE 4.6

STUDENT DISTRIBUTION

Creating a Stacked Bar Chart

The dean of the School of Arts and Sciences decides to look at the data one more time. He wants to see whether he can determine the proportion of transfer students enrolled in the course who are freshmen, sophomores, juniors, and seniors. The first task is to figure out the percentages in each column of the contingency table. The dean wants to express the year as a percentage of the type of student rather than as a percentage of the total, so he uses the column totals instead of the grand total to calculate them. Here is the original contingency table:

Understand the problem.

Analyze the data.

	Status		
Year	**Nontransfer**	**Transfer**	**Total**
Fr	3	1	4
So	10	2	12
Jr	2	0	2
Sr	1	9	10
Total	16	12	28

The number of nontransfer students who are freshmen is 3. To express this as a percentage of the nontransfer students, he calculates $3/16 = 18.75\%$, or approximately 19%. In the same way, to find the percentage of transfer students who are freshmen, he calculates $1/12 = 8.33\%$, or approximately 8%. The completed table follows:

	Status (%)	
Year	**Nontransfer**	**Transfer**
Fr	19	8
So	63	17
Jr	13	0
Sr	6	75
Total	100	100

Here is the stacked bar chart for the data:

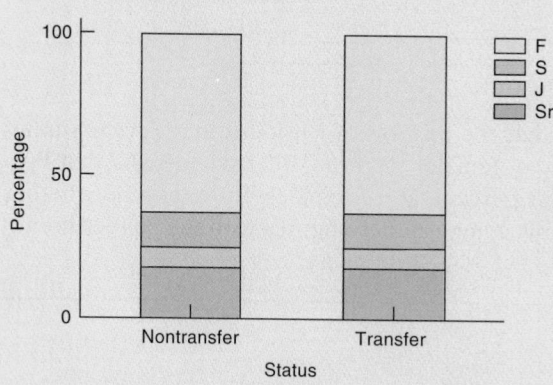

This chart shows that the majority of the nontransfer students enrolled in the course are freshmen and sophomores, which is appropriate. The majority of the transfer students are seniors, and there are no juniors. This finding reinforces the earlier idea that better advising of transfer students might alleviate the problem.

> **Draw conclusions.**

✓ TRY IT NOW!

Quality Problems
Creating a Stacked Bar Chart

Managers of the company that manufactures cardboard boxes decide to create a graphical display of the data to show to the employees. The managers want to display the type of defect as a percentage of total defects for each shift. It is hoped that these data will help each shift concentrate on its own problems. The contingency table for the defect data is shown here:

	Shift			
Defect	**1**	**2**	**3**	**Total**
Color	8	4	3	15
Printing	6	5	2	13
Skewness	0	2	0	2
Total	14	11	5	30

Modify the contingency table to display the defect types as percentages of total defects for each shift.

Create a stacked bar chart for the data.

Which defect type should each crew concentrate on? Why?

4.2.3 EXERCISES—LEARNING IT!

4.1 In an article about Internet access and minorities (*The Truth About the Digital Divide*, www.forrester.com, April 11, 2000) Forrester Research looked at data obtained from surveying 103,200 households. Data on percentages of people online for two time periods by race are listed here:

	Time (%)	
Race/Ethnicity	January 2000	January 1999
Caucasian American	43	34
African American	33	23
Hispanic American	47	36
Asian American	69	64
All households	43	35

(a) Create a clustered bar chart for the data using the variable *Race/Ethnicity* as the category for the *x* axis.

(b) Create a clustered bar chart using the variable *Time* on the *x* axis.

(c) Which graph do you think presents this information better? Why?

4.2 The Bureau of Labor Statistics (BLS) reports that the second leading cause of on-the-job deaths fell to its lowest level in seven years in 1998. The accompanying table lists the numbers of workplace homicides in the retail trade industry, broken down into three subcategories, for 1994 through 1998.

Retail industry	1994	1995	1996	1997	1998
Grocery stores	196	152	146	141	95
Eating and drinking places	135	121	135	109	69
Gasoline service stations	41	36	23	34	24

(a) Create a clustered bar chart using the year as the *x*-axis category.

(b) Create a clustered bar chart using the type of retail industry on the *x* axis.

(c) Which chart does a better job of illustrating the claim made by the BLS? Why?

(d) Use the data in the table to create a stacked bar chart with the year as the *x*-axis category.

(e) Are homicides more prevalent in any one type of retail industry? What conclusions can you draw about the numbers of homicides in the different retail industries? What other information would you want to know to better interpret the data?

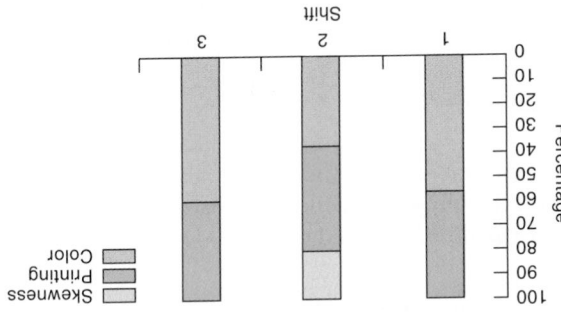

Defect Data for Cardboard Box Company

Defect	1	2	3	Total (%)
Color	57	36	60	50
Printing	43	45	40	43
Skewness	0	18	0	7
Total	100	100	100	100

Shift (%)

ANS. CREW 1 COLOR; CREW 2 PRINTING; CREW 3 COLOR

4.3 The Fatal Accident Reporting System (FARS), which is part of the National Highway Traffic Safety Administration, maintains a database of all traffic accidents in which fatalities are recorded. In 2000 there were 357 accidents in Wyoming that involved fatalities. In 136 of those accidents the results of blood alcohol testing of the driver are available. The table gives data on the blood alcohol levels and the ages of the drivers.

Blood alcohol level (%)	Age									
	Under 16	16–20	21–24	25–34	35–44	45–54	55–64	65–69	Over 69	Total
0.00	7	12	6	12	8	20	10	2	8	85
0.01–0.09	0	6	2	0	0	0	1	0	0	9
0.10+	0	8	4	11	11	6	1	1	0	42
Total	7	26	12	23	19	26	12	3	8	136

(a) Create a stacked bar chart of the data with *Age* as the x-axis category. What conclusions, if any, can you draw?

(b) Create a stacked bar chart of the data with *Blood alcohol level* as the x-axis category. Can you draw any new conclusions from this presentation?

(c) Which graph, if any, do you think displays the data better? Why?

(d) Do you want any additional information? If so, what would you like to know?

4.3   Quantitative Bivariate Data

In the preceding section you learned ways to investigate the relationships among *qualitative* variables. It is also helpful to understand how *quantitative* variables are related.

When we say that two variables are *related*, we mean that they vary in some systematic way. For example, as one variable increases, the other might also increase. This is true for variables such as height and weight, and it might or might not be true for variables such as the amount of time you study and the grade you earn on an exam. In another type of relationship, one variable increases as the other decreases. This might be true for the price of a product and sales of that same product.

In statistics it is often important to investigate the relationships between variables to learn how to predict the value of one variable based on the value of the other related variable. This is called *regression analysis* and will be covered in Chapter 11. But, just as in all statistical studies, before any analytical tool is used, it is important to look at the data to get an understanding of what is happening. Viewing the data with different graphical tools helps a statistician to decide what statistical techniques are appropriate. The most important tool for exploring relationships between quantitative variables is a graphical tool called a scatter plot.

4.3.1   SCATTER PLOTS

Each observation of bivariate quantitative data can be thought of as a data pair of the form (x, y), where x represents the value of the first variable and y represents the value of the second variable. The x variable is often called the **independent** variable, and the y variable is often called the **dependent** variable. The type of graph used to display this kind of data is called a **scatter plot.**

> In a *scatter plot* an axis is used to represent each of the variables, and the data are plotted as points on the graph. Typically, the *independent variable* is plotted on the x axis and the *dependent variable* is plotted on the y axis.

An example of a scatter plot is shown in Figure 4.4. The plot gives information about whether a relationship exists between the two variables and what the

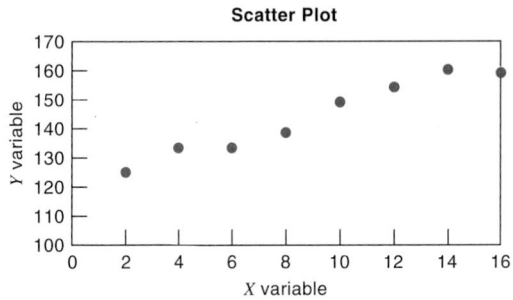

FIGURE 4.4 **Scatter plot**

form of that relationship might be. From the plot it appears that as the value of X increases, so does the value of Y. Of course, what we know about variation in data tells us that the relationship will not be perfect. If there is a strong enough relationship between the two variables, then it should be apparent even in the presence of normal variation.

Creating a scatter plot is not difficult. Once you have selected which variable is the *independent* variable and which is the *dependent* variable, all you do is draw a set of axes and scale them so that you can plot all of the data. You might have learned in a mathematics class that when you create a graph on a pair of x–y axes, the scales should both start at 0. Although this is often desirable in mathematics, it may not be suitable for a scatter plot in statistics. If your data values are very large, then starting both axes at 0 might squash the graph into the top portion and actually hide the relationship. For this reason, we usually pick the starting points on the axes to be some reasonable numbers that are smaller than the smallest x and y values. When the axes are drawn, we plot the data pairs as points on the graph.

EXAMPLE 4.7

ABC FACULTY SALARIES

Looking for Relationships

The Committee that is looking at faculty salaries knows that there should be a relationship between a faculty member's number of years of service to the university and salary. They want to look at these two variables together. Since both variables are quantitative, the first step in understanding the relationship is to make a scatter plot of salary versus number of years of service. Salary depends on years of service, so they pick salary for the y (dependent) variable and years of service for the x (independent) variable. The plot is shown here:

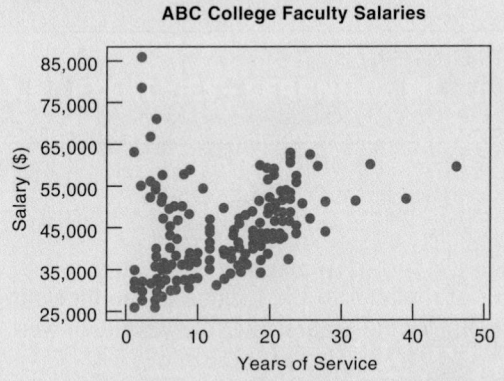

From the plot they know that they are basically correct—salary and years of service are related and the relationship is, for the most part, linear. They see,

Understand the problem.

Analyze the data.

however, that many data points do not quite fit the picture. Some of the other qualitative variables that they have been looking at might be of interest. Perhaps this is where gender becomes an issue.

EXAMPLE 4.8

SALES AND ADVERTISING

Creating a Scatter Plot

A pharmaceutical company is interested in determining whether there is a relationship between the amount of money it spends on advertising and factory sales for a group of over-the-counter products. It seems reasonable that spending more on advertising will increase sales, but the company is not sure. The managers collect some data:

Advertising ($ millions)	143.8	91.7	43.8	26.7
Factory sales ($ millions)	855	360	170	130

Understand the problem.

They decide to create a scatter plot to look at the data. Since they are expecting the amount of sales to depend on the amount spent on advertising, they pick sales to be the dependent (y) variable and advertising expenditures to be the independent (x) variable. After looking at the data, they choose to scale the x axis from $0 to $150 million and the y axis from $100 to $900 million. They then plot the four data points on the graph.

Collect the data.

Analyze the data.

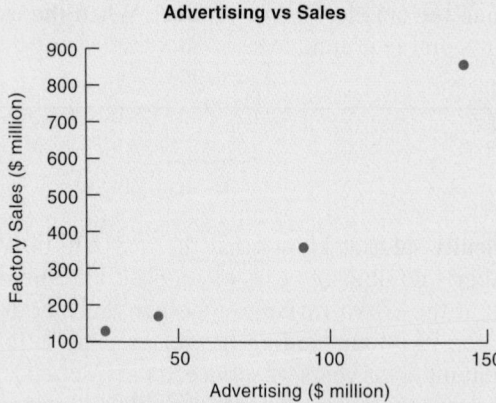

From the graph they see that their idea was correct and that as advertising expenditures increase so do sales.

✔ TRY IT NOW!

Airline Fares
Creating a Scatter Plot

A travel agency was interested in knowing how airline fares are related to the lengths of the flights in miles. The agency hypothesized that the longer the flight, the higher the airfare. The following data were collected:

Flight length (miles)	2375	1400	1250	2325	985	2025
Airfare ($)	430	272	252	422	207	373

Which variable is the dependent variable and which is the independent variable?

Make a scatter plot of the data.

From your plot, do you think that the travel agency's hypothesis is correct? Why or why not?

4.3.2 TYPES OF BIVARIATE RELATIONSHIPS

After we determine that two variables are related, we often want to find an equation that describes the relationship. The equation shows how the variables are related and can then be used to predict values of the dependent variable for different values of the independent variable. The purpose of a scatter plot is to give a visual clue about whether a relationship exists between two quantitative variables. A scatter plot also gives information about the type of relationship, which can be used to determine the correct method for finding the equation that relates the variables.

When two variables are related, the scatter plot forms a pattern that characterizes the form of the relationship. Figure 4.5 shows some of the different types of relationships that can exist between variables.

When two variables are not related, the scatter plot forms no pattern, as is seen in Figure 4.6 on page 182.

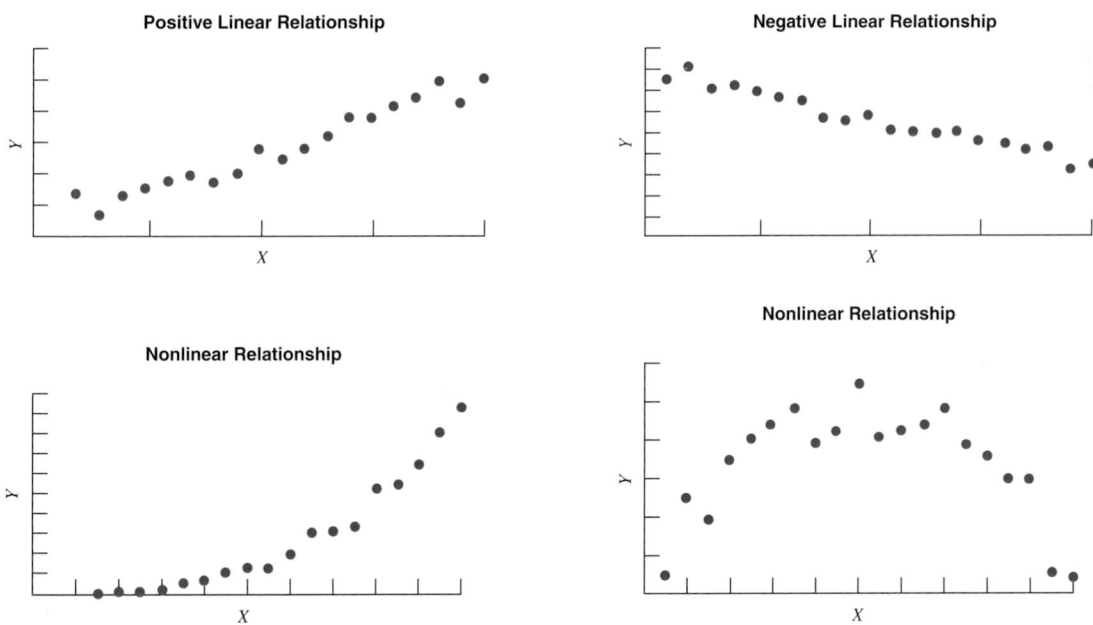

FIGURE 4.5 **Types of relationships between two variables**

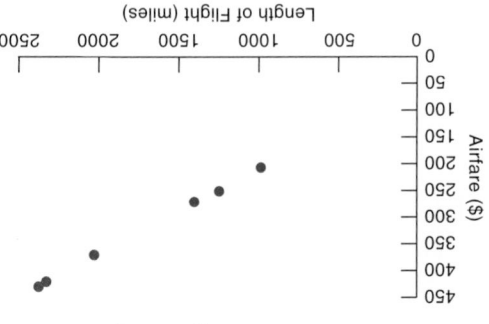

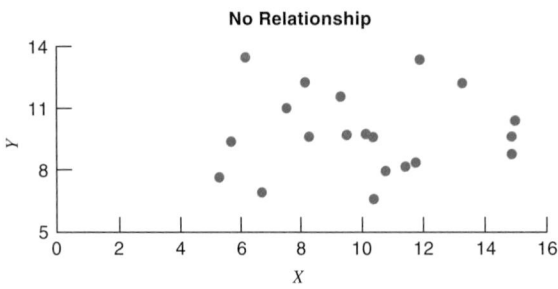

FIGURE 4.6 **No relationship between two variables**

TRAINING AND PERFORMANCE

Determining the Type of Relationship

Understand the problem.

A company is interested in determining whether there is a relationship between the number of days that it spends training employees at a particular job and the employees' performance as measured on a standardized test. The human resources office collects these data:

Training days	1.0	1.5	2.0	2.5	3.0
Test score	41	60	72	91	99

The first thing they do is display the data on a scatter plot to determine whether the two variables are related. Since the company thinks that the score depends on the number of training days, they plot the test score on the *y* axis and the number of training days on the *x* axis:

Collect the data.

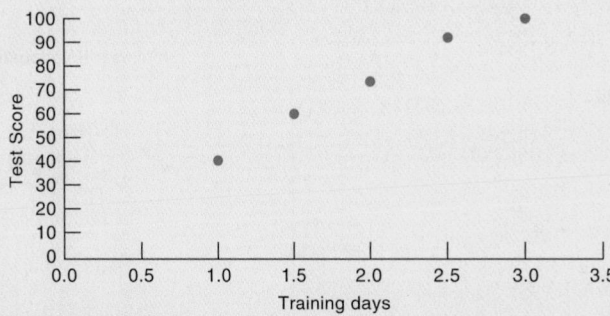

Analyze the data.

The plot confirms the company's hypothesis and indicates that a linear relationship is appropriate.

✔ TRY IT NOW!

Starting Salaries and Math Courses
Determining the Type of Relationship

The Career Planning Office of a large university is interested in knowing whether there is a relationship between the starting salary of a graduate and the number of mathematics courses the graduate had taken as a student. An employee goes through the records for the last year and finds the following data:

Number of math courses	Starting salary
1	$26,284
1	25,470
2	26,777
3	27,269
4	28,553
6	30,054

Which variable is the independent variable? The dependent variable?

Use the grid to create a scatter plot of the data.

Is there a linear relationship between the number of math courses taken and the starting salary? If so, describe the relationship.

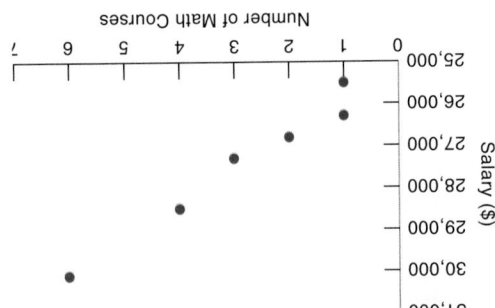

ANS. INDEPENDENT—NUMBER OF MATH COURSES; DEPENDENT—SALARY; YES, AS NUMBER OF MATH COURSES INCREASES, SO DOES SALARY.

Discovery Exercise 4.1

DISCOVERING RELATIONSHIPS

The following data are values for the number of weeks a student was enrolled in a speed reading program and the speed the student gained in words per minute:

Weeks	2	3	4	6	8
Speed gain	49	86	109	164	193

Part I

1. Plot the data on the grid. Be sure to label the axes and construct the graph so that it uses most of the grid.

2. Draw a straight line through the points that you think best represent the relationship between x and y. What criteria did you use for drawing the line you selected?

Part II

Use your **_line_** and the **_graph_** to answer the following questions. Do NOT use the data given in the table.

1. What is the speed gain for a person who spent 4 weeks in the program as predicted by the line?

2. If you selected a person at random who had spent 3 weeks in the program, what would you predict for that person's speed gain?

3. By how much would you predict the speed gain will change for a 1-week increase in time in the program?

4. Draw the line with the equation $y = 10.1 + 23.9x$ on the graph.

5. Answer the prediction questions 1–3 using the new "best" line.

6. Agree that my line is better than yours (if it is different).

7. What do you think would happen if you used the new line to predict the speed gain of a person who had been in the program for 1 week? 0 weeks? Is this reasonable?

8. What problems can you see with what we have done?

4.3.3 LEAST-SQUARES LINE

If the relationship between the variables looks linear (like a straight line), then we can use a curve-fitting technique called *least squares* to find the equation of the line that best fits the data. In Chapters 11 and 12 you will use least squares in more depth in a statistical technique called regression analysis. Now we use least squares to find the equation that relates X and Y.

In drawing a line to fit a set of points, we want to satisfy some criteria. First, we want the line that is drawn to be unique. It will certainly not be useful if everyone who sees the same set of data uses a technique that produces a *different* line! Second, we want the line to be as close as possible to all of the points. In Figure 4.7 you see some data points and a line drawn to represent the relationship between the variables. The distances from each point to the line are called the *deviations* or *errors*. We can use the **least-squares technique** to find the equation of the line that best represents the relationship between the two variables.

The ***least-squares technique*** finds the equation of the line that minimizes the sum of the squared errors between the actual data points and the line.

The reasons for minimizing the sum of the squared deviations rather than just the deviations or the absolute deviations are mathematical and are discussed in more detail in Chapter 11.

The ***least-squares line*** is defined by

$$\hat{y} = a + bx$$

Least-squares line

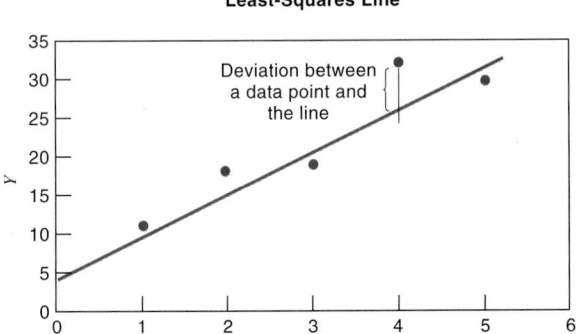

FIGURE 4.7 Deviations between the data points and the line

TABLE 4.2

TABLE FOR CALCULATING THE LEAST-SQUARES LINE

Observation number	X	Y	XY	X²
1	x_1	y_1	$x_1 \times y_1$	$x_1 \times x_1$
2	x_2	y_2	$x_2 \times y_2$	$x_2 \times x_2$
.	.	.	.	.
.	.	.	.	.
.	.	.	.	.
n	x_n	y_n	$x_n \times y_n$	$x_n \times x_n$
Total	ΣX	ΣY	ΣXY	ΣX^2

The value $\hat{y}$ (y-hat) is the **predicted value** of y for a selected value of x. We are also interested in the **deviation** of the data from the least-squares line.

The distance between the predicted value of Y, $\hat{y}$, and the actual value of Y, y, is called the **deviation** or **error**. The **deviation, e,** is equal to $\hat{y} - y$.

Calculating the Least-Squares Line

The actual method of least squares finds the *slope* and the *y intercept* of the line $y = a + bx$. The least-squares estimates for the slope, b, and the y intercept, a, are given by

Slope and intercept of the least-squares line

$$b = \frac{n \sum XY - \sum X \sum Y}{n \sum X^2 - (\sum X)^2} \quad \text{and} \quad a = \frac{\sum Y}{n} - b \frac{\sum X}{n}$$

Although these equations might seem too complicated to use, if you look at them carefully you see that only five quantities have to be calculated. Four of the quantities are sums involving the data values x and y, and the fifth is the number of observations in the sample. As an example, look at the quantity $\sum XY$. This quantity is the sum of the products of the x and y values for each observation. The easiest way to calculate all of the quantities involved is to make a table with a column for each sum needed (see Table 4.2).

In the table each x_i represents the value of X for the ith data point, and each y_i represents the Y value of the ith data point. The first column keeps track of the observation number, the next two columns are the values of X and Y for each observation, the next column is the product of the X and Y values for each observation, and the last column is the square of the X value of the observation.

EXAMPLE 4.10

TRAINING AND PERFORMANCE

Finding the Least-Squares Line

The company that is looking at the relationship between training and performance decides to use the least-squares method to find the equation of the line that describes the relationship. The first thing the human resources people do is to create this table from the data in Example 4.9:

Observation number	Training days, X	Test score, Y	XY	X²
1	1.0	41	41.0	1.00
2	1.5	60	90.0	2.25
3	2.0	72	144.0	4.00
4	2.5	91	227.5	6.25
5	3.0	99	297.0	9.00
Total	10.0	363	799.5	22.50

Analyze the data.

Substituting the values in the last row of the table into the formulas, the company obtains

$$b = \frac{(5)(799.5) - (10)(363)}{(5)(22.5) - (10^2)} = \frac{367.5}{12.5} = 29.4$$

$$a = \frac{363}{5} - (29.4)\frac{10}{5} = 13.8$$

When you are substituting the value for b into the equation for a, do NOT round it. Round all of the values AFTER the calculations are done.

Using these values, they find the least-squares line to be

$$\hat{y} = 13.8 + 29.4x$$

The equation tells them that the base score on the test is 13.8 points and that every day of training increases the score by 29.4 points.

✓ TRY IT NOW!

Starting Salaries and Math Courses
Finding the Least-Squares Line

The Career Planning Office wants to use the least-squares technique to find the equation that relates the number of math courses taken and starting salary. They use these data:

Number of math courses	Starting salary
1	$ 26,284
1	25,470
2	26,777
3	27,269
4	28,553
6	30,054

Many calculators that do statistical calculations also find the least-squares line. If you have such a calculator, it is probably worthwhile to learn how to do this.

Find the equation for the least-squares line for the data. (*Hint:* Create a table like the one in Example 4.10 or use a computer software package.)

Explain what the least-squares line tells the Career Planning Office about math courses and starting salary.

ANS. $\hat{y} = 25,019 + 840.6x$ THE BASE SALARY IS $25,019 AND EVERY MATH CLASS INCREASES IT BY $840.60.

When you obtain the least-squares line, it is often useful to plot it on the same graph as the original data. Then you can see how well the line fits the data values. Since two points determine a straight line, the easiest way to plot the line is to substitute two different values for X into the least-squares equation and solve for $\hat{y}$. Once you have two pairs of points, you can plot them and draw the line.

EXAMPLE 4.11

TRAINING AND PERFORMANCE

Plotting the Least-Squares Line

The company that is looking at performance test scores and training days decides to plot the data and the line it had found on the same graph to see how well the line represents the data. To do this the company chooses two different values for X, 1 and 3 training days, and uses the equation to find the predicted scores. The following equations are obtained:

$$y = 13.8 + 29.4(1) = 43.2 \quad \text{and} \quad y = 13.8 + 29.4(3) = 102$$

Thus, the least-squares line predicts that a person with 1 day of training will score 43.2 points and a person with 3 days of training will score 102 points. The company plots these points on the graph with the original data and draws the line. Here is the resulting plot:

Analyze the data.

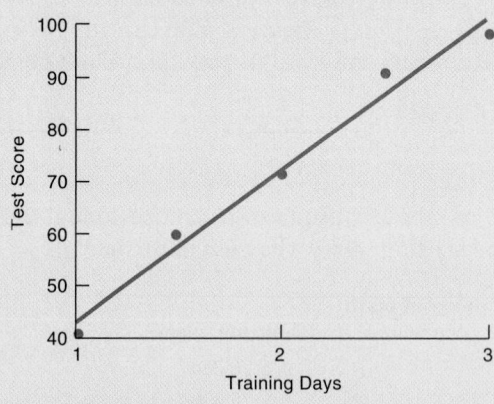

Days of Training vs. Performance Score
Score = 13.8 + 29.4 Training Day

The plot shows that the line fits the data very well.

✔ **TRY IT NOW!**

Starting Salaries and Math Courses
Finding the Least-Squares Line

After the least-squares equation is found, the Career Planning Office wants to see whether the equation does a good job of predicting the starting salary for a given number of math courses. Plot the data and the least-squares line on the same grid.

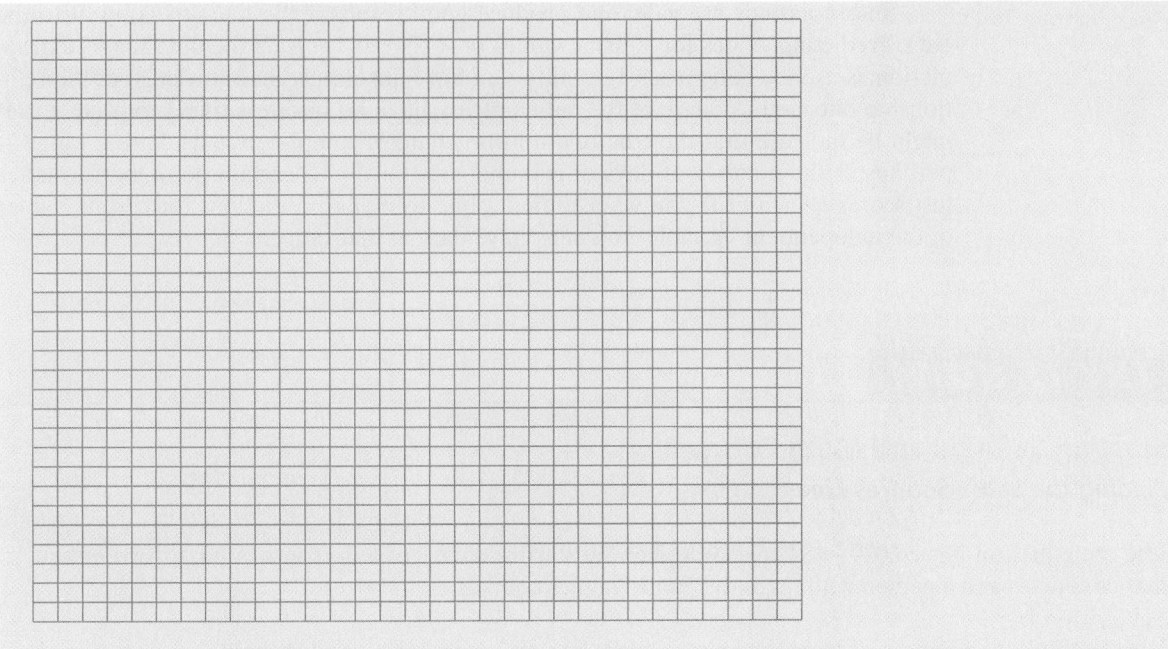

Interpolation and Extrapolation

After we find the least-squares line, it is reasonable to wonder when we can use it. When are the predictions useful? Are there any times when the predictions are not useful?

In the example on training days and performance test scores, the company used the equation of the least-squares line to predict scores for a person who was trained for 1 day and a person who was trained for 3 days. Notice that both of the values they chose for X were *within* the range of X from the data they collected. This is called ***interpolation.*** Another kind of prediction is called ***extrapolation.***

Predicting values for Y using values of X that are outside the data range is called ***extrapolation.*** Finding values for Y using X values that are within the data range is called ***interpolation.***

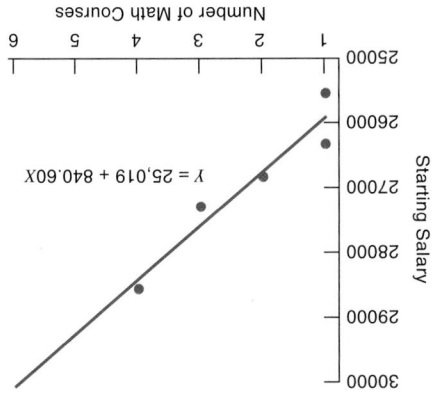

Ans. **Least-squares Line for Math Courses vs. Salary**

Interpolations are valid and produce good results if the line is a good fit to the data. Predicting values for *Y* using values of *X* that are *outside* the data range, extrapolation, is usually dangerous. In reality, you have absolutely no knowledge of the relationship outside the range of the data and you have no business trying to predict it. It might be the case that the true relationship changes completely outside that range—perhaps as the number of days of training goes beyond a certain point, performance test scores *decrease!* If you want to find a line to make predictions for certain values of the independent variable, you must have data in that range.

✔ **TRY IT NOW!**

Starting Salaries and Math Courses
Finding the Least-Squares Line

Use the equation $\hat{y} = 25{,}019 + 840.6x$ to predict the starting salary of a person who takes two mathematics courses and a person who takes five mathematics courses.

Do you think the predictions from the least-squares line are useful? Why or why not?

Is what you just did interpolation or extrapolation?

Now use the equation of the least-squares line to predict the starting salary of a person who takes no math courses and a person who takes ten math courses.

Do you think these predictions are valid? Why or why not?

4.3.4 EXERCISES—LEARNING IT!

4.4 In a study about postage and revenue, the European Economic Community wanted to look at the relationship between the number of postal employees in a country and the amount of domestic mail that was processed. They collected data for six different countries:

Country	Number of employees	Domestic mail (billions of pieces)
Germany	342,413	18.32
France	289,156	23.87
Italy	221,534	6.62
Britain	189,000	16.75
Spain	65,355	4.06
Sweden	52,251	4.21

Source: The Economist, June 15, 1996

(a) Create a scatter plot of the data and determine whether a linear relationship is appropriate.

(b) Find the equation of the least-squares line for the data.

(c) The Netherlands employs 53,560 postal workers. Use the least-squares equation to predict its domestic mail traffic.

4.5   In looking at the effects of shopping center expansion, the Commerce Department decided to examine the relationship between the number of shopping centers and the retail sales for different states in the same region. The department collected these data for the North Central states:

State	Number of shopping centers	Retail sales (billions)
Illinois	2096	$41.8
Indiana	905	21.4
Iowa	308	7.5
Kansas	481	11.6
Michigan	1018	25.3
Minnesota	471	13.9
Missouri	887	22.7
Nebraska	264	5.7
North Dakota	87	2.1
Ohio	1704	41.6
South Dakota	58	1.3
Wisconsin	625	14.6

Source: Statistical Abstract of the United States 1999

(a) Create a scatter plot of the data.

(b) Find the equation of the least-squares line that relates retail sales and number of shopping centers.

(c) Plot the least-squares line and the data on the same plot. Do you think the line fits the data well? Why or why not?

(d) Use the least-squares equation to predict retail sales for each state.

4.6   As part of an international study on energy consumption, data were collected on the number of cars in a country and the total travel. The data for 12 of the countries are shown here:

Country	Total cars (millions)	Travel (billion km)
United States	142,352	3140
Finland	1,823	35
Denmark	1,664	31
Britain	21,316	353
Australia	8,534	138
Sweden	3,321	53
Netherlands	5,527	84
France	23,268	348
Norway	1,592	24
Italy	26,117	368
Germany	43,752	609
Japan	40,246	439

Source: The Economist, June 22, 1996

(a) Create a scatter plot of the data. Do you think there is a linear relationship between the number of kilometers traveled and the number of cars?

(b) Find the least-squares line for the data. Interpret the value of the slope.

(c) Does the intercept make sense for these data? Why or why not?

(d) Plot the least-squares line on the same plot with the data. Does the line make you feel confident about predicting travel from the number of cars?

(e) Use the least-squares equation to predict the number of kilometers traveled for Sweden and Japan. How well do the predictions agree with the original data?

4.7 In a study of why retail stores fail, data were collected on the type of retailer, rental charges, and sales for super regional shopping malls in the United States in 1993. The data are listed in the table:

Type of retailer	Rental charges ($/sq ft)	Sales ($/sq ft)
Department	2.0	131
Clothing and accessories	19.0	237
Gift/specialty	22.5	250
Shoes	22.5	259
Food service	32.0	342
Jewelry	40.0	555

Source: *The Economist,* March 1, 1997

(a) Create a scatter plot for the data. Do you think the relationship between rental charges and sales is linear?

(b) Find the equation of the least-squares line for the data and interpret the value of the slope and the intercept.

(c) Does the intercept make sense for these data? Why or why not?

(d) Plot the least-squares line on the same plot with the data. Do you think the line does a good job of predicting sales from rental charges? Why or why not?

GET IT IN WRITING

TO: President, ABC College
FROM: Gender Equity Committee
RE: Study on Gender Equity

To increase our knowledge of faculty salaries, we decided to look at the variables of rank, gender, school, and tenure. These variables should provide additional insight into the makeup of the current faculty. The first phase of the analysis deals with the school and gender of each faculty member. The other variables will be examined in the next phase of the analysis.

When you look at faculty gender by school as shown in Figures 1 and 2, you can see that there are differences in the percentages of male and female faculty in each school. The biggest differences are in the Schools of Business and Liberal Studies. From Figure 2 you can see that 73% and 78% of the faculty of the Schools of Liberal Studies and Business are male, while the Schools of Sciences and Professional Studies have 59% and 56% males, respectively. It seems unlikely that the professions included in each school are that different in terms of gender makeup.

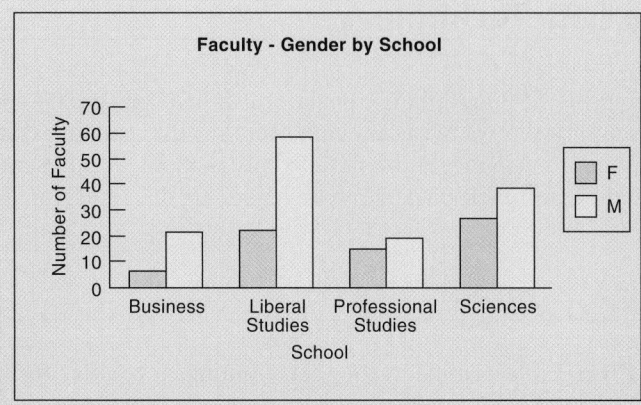

FIGURE 1

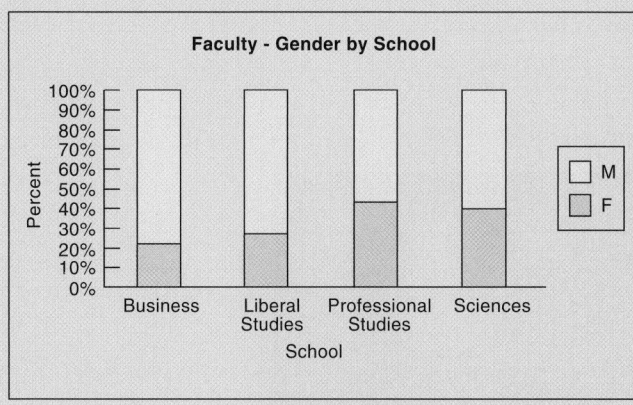

FIGURE 2

Based on the past, we expect to see a relationship between years of service and salary. Figure 3 shows that although such a relationship does appear to exist, it is not clear. From the graph of salary versus years of service, you can see that there is a great deal of variability in the data as well as some anomalies that must be explained before we can understand the salary structure. For example, the faculty member with the highest salary has less than 5 years of service to the college.

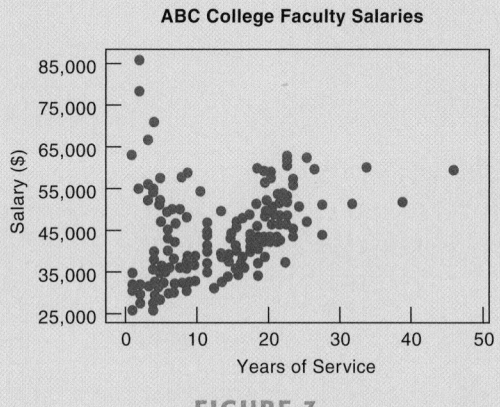

FIGURE 3

It appears that other variables are influencing salary. Our next step will be to compare the gender distribution for each school with the overall distribution at the college. We will also look at gender by rank and tenure and at salary as it relates to these other variables. We will get back to you with our findings in 2 weeks.

4.4 Investigating Bivariate Data with Technology

The use of Excel, Minitab, and the TI-83 graphing calculator can help you create contingency tables, charts, and scatter plots for bivariate data. Refer to the manual associated with the software that you are using for more detailed instructions. We present a brief overview of the capabilities of each package with regard to the topics of this chapter.

Excel 2000

For Excel you can use the **Pivot Table** command to create contingency tables for bivariate data. After you have created a contingency table of data, you can use the data to create clustered bar charts, stacked bar charts, and scatter plots (see Figures 4.8–4.11).

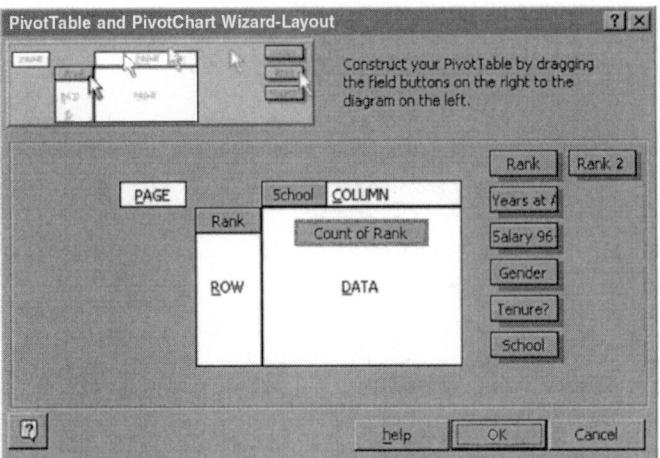

FIGURE 4.8   **Creating the contingency table in Excel**

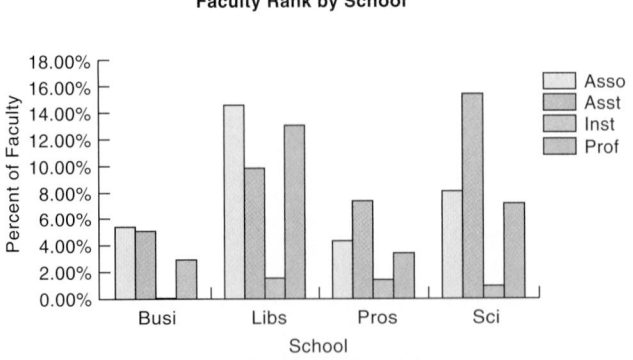

FIGURE 4.9   **Excel clustered bar chart**

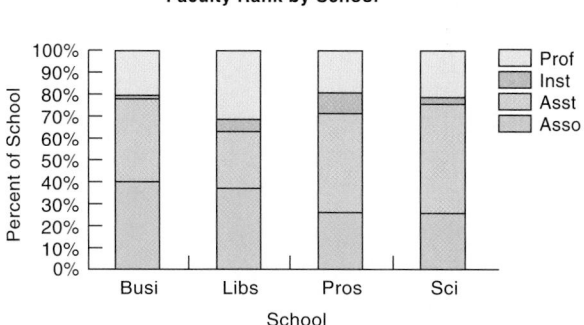

FIGURE 4.10 Excel stacked bar chart

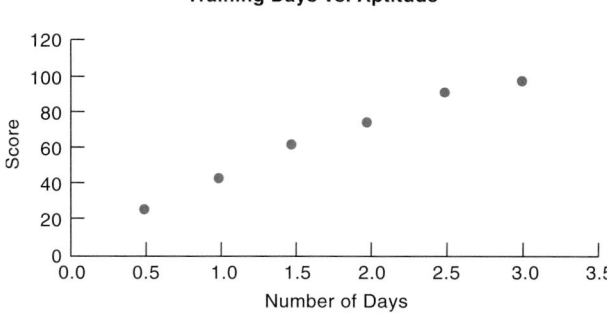

FIGURE 4.11 Excel scatter plot

Minitab

For Minitab it is unnecessary to create the contingency table before you create the charts associated with it. To create a stacked or clustered bar chart, use the same **Chart** dialog box that you would use to create a bar chart. You can create clustered bar charts, stacked bar charts, and scatter plots (see Figures 4.12–4.14).

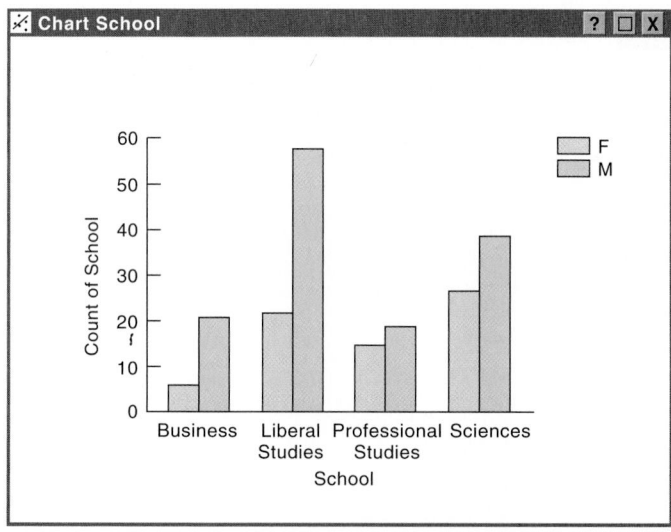

FIGURE 4.12 Clustered bar chart in Minitab

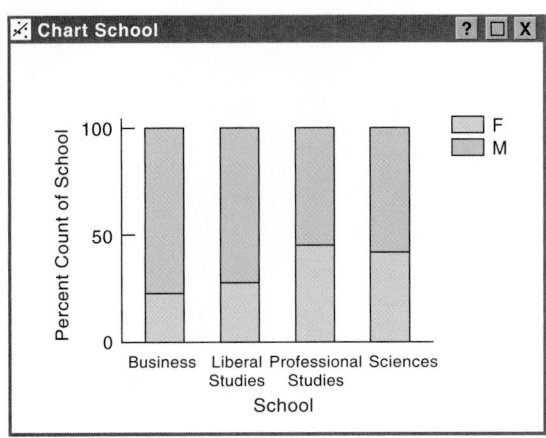

FIGURE 4.13 **Stacked bar chart in Minitab**

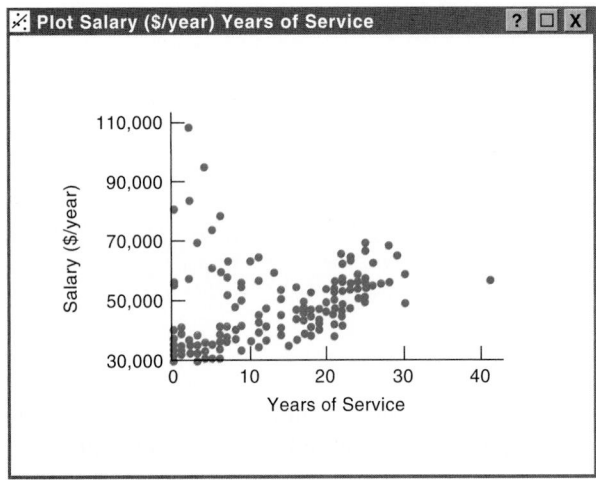

FIGURE 4.14 **Scatter plot in Minitab**

TI-83 Graphing Calculator

It is not possible to create contingency tables or clustered or stacked bar charts using the TI-83 graphing calculator. You can, however, use the TI-83 to create a scatter plot of a set of quantitative bivariate data and find and plot the least-squares line (see Figures 4.15 and 4.16).

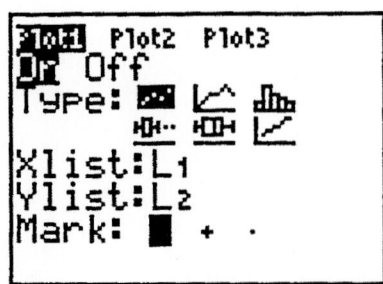

FIGURE 4.15 **TI-83 screen for scatter plot**

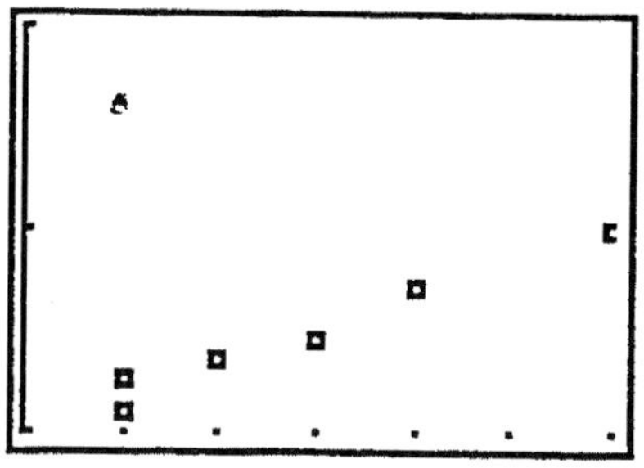

FIGURE 4.16 **TI-83 scatter plot of math courses and salary**

CHAPTER 4 SUMMARY

In this chapter you learned some methods for looking at how two variables are related. Very often you are interested not in a single variable, but rather in how variables are *related*. There are several different methods for looking at these relationships. Graphical methods, such as *stacked bar charts, clustered bar charts*, and *scatter plots*, allow you to see the nature of the relationships. Quantitative methods, such as *least-squares analysis*, allow you to describe the relationship numerically.

By now you should be more convinced than ever that there is no standard way to analyze a set of data. Rather, various statistical tools are used together to produce the most complete *picture* of the data and the information they contain.

KEY TERMS		
Term	**Definition**	**Page reference**
Clustered bar chart	A **clustered bar chart** is used in place of multiple bar charts to make comparisons for one of the qualitative variables. The bars are grouped according to the values of the other variable.	171
Contingency table	A contingency table is a table with rows that represent the possible values of one variable and columns that represent the possible values for a second variable. The entries in the table are the numbers of times that each *pair* of values occurs.	167
Deviation	The difference between the data point and the value predicted by the least-squares equation is the **deviation, e.**	186
Extrapolation	Predicting values for Y using values of X that are outside the data range is called **extrapolation.**	189
Interpolation	Finding values for Y using X values that are within the data range is called **interpolation.**	189
Least-squares technique	The **least-squares technique** finds the equation of the line that minimizes the sum of the squared errors between the actual data points and the line.	185
Scatter plot	A **scatter plot** is a graph used to represent bivariate quantitative data. An axis is used to represent each of the variables, and the data are plotted as points on the graph.	178

(continued)

Term	Definition	Page reference
Stacked bar chart	In a **stacked bar chart** the data for the selected variable are represented as a percentage of the total for each category of the second variable. Each value of the selected variable is represented in a different way, and the bars are "stacked" to total 100%.	173
$\hat{y}$	$\hat{y}$ (y-hat) is the predicted value of y for a selected value of x.	186

KEY FORMULAS

Term	Formula	Page reference
Least-squares equation	$\hat{y} = a + bx$	185
$\hat{y}$ (predicted value)	$\hat{y}$	186
e (deviation)	$\hat{y} - y$	186
a (y intercept estimate)	$a = \dfrac{\sum Y}{n} - b\dfrac{\sum X}{n}$	186
b (slope estimate)	$b = \dfrac{n\sum XY - \sum X \sum Y}{n\sum X^2 - (\sum X)^2}$	186

CHAPTER 4 EXERCISES

Learning It!

4.8 Consider the data on the numbers of students in the different majors at Dartmouth College that you looked at in Chapter 2.

Major	1986	1991	1996
Anthropology	10	19	17
Art History	28	20	15
Asia	11	15	22
Biology	31	44	108
Chemistry	28	15	55
Classics	2	1	8
Comparative Literature	5	6	6
Drama	13	8	5
Earth Science	8	15	7
Economics	90	66	93
English	156	143	92
French	23	15	19
Geography	23	28	28
Government	147	172	142
History	102	125	115
Mathematics	35	23	14
Music	10	3	11
Philosophy	21	28	25
Physics	13	10	16
Psychology	70	74	87
Religion	22	31	8
Sociology	17	26	15

(a) What type of bivariate chart is appropriate for displaying these data?

(b) What variable do you choose for the *x*-axis categories?

(c) Create a clustered bar chart for the data.

4.9 In an effort to learn whether customers are correct that the lifetimes of coupons are getting shorter, the following numbers of coupons are recorded:

Coupon life	1995	1996
Shorter than 1 month	28	123
1–4 months	88	140
5–8 months	204	122
9–12 months	141	88
Longer than 1 year	39	27

(a) Use a clustered bar chart to display the data.

(b) Is the perception that coupon life is getting shorter justified? Why or why not?

4.10 Look again at the data presented in Chapter 2 on the percentages of people who reported workplace violence:

Type of violence	Cases reported (%)
Harassment	19
Threat of physical harm	7
Physical attack	3

	Type of violence (%)		
Major effect on worker	Harassment	Threat of harm	Physical attack
Psychological	49	53	49
Disrupted work life	34	25	25
Physically injured or sick	13	9	17
No negative effect	4	13	9

Display these data on a single graph.

4.11 Large companies are always looking for expanding markets. Media companies in particular are looking to foreign markets for expansion. To obtain some information about expansion possibilities in Central Europe, a major communications company collected data from seven countries on revenues from television advertising and percentages of households with cable or satellite television:

Country	Percent of households	TV advertising revenues (billions)
Hungary	55	$170
Slovakia	47	28
Poland	33	369
Czech Republic	28	105
Romania	27	40
Russia	15	250
Bulgaria	9	23

Source: The Economist, 6 July 1996

(a) Create a scatter plot of the data. Does it appear that advertising revenues are related to the percentage of households that have cable or satellite TV?

(b) Find the equation of the least-squares line for the data.

(c) Plot the least-squares line on the same graph as the data. Does the line do a good job of predicting TV advertising revenues? Why or why not?

(d) In the Ukraine 8% of the households have cable or satellite television. Use the least-squares line to predict TV advertising revenues for the Ukraine.

4.12 For many countries, tourism contributes a large part of revenues. In attempts to predict tourism revenues, one of the independent variables that is considered important is the number of foreign visitors to the country. Data for six different countries were collected:

Country	Foreign visitors (millions)	Tourism receipts (billions)
France	60	$ 27.3
Spain	48	25.1
United States	45	58.4
Italy	30	27.1
Britain	23	17.5
Germany	15	11.9

Source: The Economist, July 1996

(a) Create a scatter plot of the data and find the equation of the least-squares line.

(b) Use the least-squares line to predict tourism revenues for Italy and the United States.

(c) For which country does the least-squares line do the best job of predicting tourism receipts?

4.13 As part of the anticrime bill passed in 1995, the U.S. government granted money to cities to hire new beat-patrol officers. Nine cities in New Jersey were given grants and hired officers. In trying to assess the program, government analysts wanted to look at the relationship between the numbers of officers hired and the sizes of the grant. They used these data:

Grant	Officers hired
$150,000	2
375,000	5
471,125	6
70,967	1
450,000	6
525,000	7
375,370	7
750,000	10
1,000,000	12

Source: U.S. Justice Department

(a) Create a scatter plot of the data.

(b) Find the equation of the least-squares line.

(c) Plot the least-squares line on the same graph as the data.

(d) How well does the line fit the data?

(e) Use the equation to predict the number of officers hired for each city.

(f) Compare the values from the least-squares line with the actual data.

4.14 How well does performance partway through the baseball season predict the outcome at the end of the season? For each of the record-holders for homeruns during a season in which the record was broken, data were recorded on the number of runs after 76 games were played and the number of runs at the end of the season:

Player	Runs after game 76	Total runs	Year
Barry Bonds	39*		2001
Luis Gonzalez	32*		2001
Mark McGwire	34	70	1998
Sammy Sosa	30	66	1998
Roger Maris	30	61	1961
Babe Ruth	26	60	1927
Mark McGwire	23	65	1999
Sammy Sosa	30	63	1999

* At the time these data were collected, the 2001 season had not yet ended.

(a) Create a scatter plot of the data. Do you think there is a relationship between the number of homeruns hit after 76 games and the number hit at the end of the season?

(b) Find the least-squares line for the data.

(c) Use the equation of the least-squares line to predict the number of homeruns hit in 2001 by Barry Bonds and Luis Gonzalez.

(d) Find out how many homeruns Barry Bonds and Luis Gonzalez actually hit during the 2001 season. How well did the least-squares line predict the outcomes?

Thinking About It!

4.15 The summer of 1995 was one of the hottest on record. The *New York Times* reported the amount of electricity used and the temperature on ten record-breaking days:

Date	Temperature (°F)	Electricity usage (MW)
August 2, 1995	95	10,805
July 23, 1991	99	10,752
July 8, 1993	100	10,667
July 27, 1993	93	10,654
July 27, 1995	90	10,567
June 20, 1995	95	10,551
July 9, 1995	101	10,398
July 26, 1995	90	10,391
July 8, 1994	90	10,368
July 19, 1991	96	10,349

(a) Is it likely that the amount of electricity used in New York City depends on the temperature?

(b) Create a scatter plot showing the relationship between electricity usage and temperature.

(c) Do you think the data indicate a relationship between the two variables? Why or why not?

(d) What other variables or factors might be hiding a relationship? If so, what might that variable or factor be?

(e) Eliminate the days from years other than 1995 and replot the data. Now what do you see?

(f) Do any of the data points appear to be unusual? Can you offer any explanation for this?

(g) July 9, 1995, was a Sunday (which means that businesses in New York City were closed). Eliminate that point and replot the data. Now what do you see?

(h) Find the least-squares line for the data from the plot in part (g).

4.16 The pharmaceutical company that is investigating the relationship between advertising expenditures and sales of over-the-counter drugs collects some additional data:

Drug	Advertising expenditures (millions)	Sales (millions)
Tylenol	$143.8	$855
Advil	91.7	360
Vicks	26.6	350
One Touch	2.0	220
Robitussin	37.7	205
Bayer Aspirin	43.8	170
Alka-Seltzer	52.2	160
Centrum	16.5	150
Mylanta	32.8	135
Tums	27.6	135
Excedrin	26.7	130
Benadryl	30.9	130
Halls	17.4	130
Metamucil	12.1	125
Sudafed	28.6	115

(a) Make a scatter plot of the data.

(b) What explanation can you offer for the fact that there does not appear to be a relationship between advertising expenditures and sales? What additional information would help put the data in perspective?

(c) Separate the drugs listed into four categories: pain relievers, digestive, cold/allergy, and other.

(d) Create four different scatter plots based on part (c). Does this change anything? If so, how?

(e) For each of the categories that shows a relationship between sales and advertising expenditures, find the equation of the least-squares line.

(f) Do the equations in part (e) have any similarities? If so, what are they?

Datafile:
ATHLETES.XXX

4.17 What is the relationship between the graduation rate for all students and the graduation rate for athletes? Data on these two variables were collected for NCAA Division I schools for students who entered schools in 1994–1995:

Institution	Graduation rate (%) All students	Athletes	Institution	Graduation rate (%) All students	Athletes
Arizona State U.	47	44	San Jose State U.	34	41
Arkansas State U.	32	36	Southern Methodist U.	71	63
Auburn U.	67	55	Stanford U.	93	86
Ball State U.	46	68	SUNY—Buffalo	56	60
Baylor U.	67	67	Syracuse U.	74	81
Boise State U.	25	53	Temple U.	44	57
Boston College	86	83	Texas A&M	69	52
Bowling Green State U.	58	59	Texas Christian U.	63	57
Brigham Young U.	71	55	Texas Tech U.	48	51
California State U.—Fresno	40	47	Tulane U.	70	75
Central Michigan U.	50	63	U. of Akron	34	57
Clemson U.	72	63	U. of Alabama—Birmingham	33	42
Colorado State U.	62	56	U. of Alabama—Tuscaloosa	61	59
Duke U.	93	90	U. of Arizona	54	53
East Carolina U.	51	58	U. of Arkansas—Fayetteville	46	51
Eastern Michigan U.	34	31	U. of California—Berkeley	83	59
Florida State U.	61	54	U. of California—Los Angeles	80	58
Georgia Inst. of Tech.	69	54	U. of Central Florida	49	50
Indiana U.—Bloomington	65	63	U. of Cincinnati	31	48
Iowa State U.	62	58	U. of Colorado—Boulder	64	52
Kansas State U.	54	61	U. of Florida	69	41
Kent State U.	42	58	U.of Georgia	66	65
Louisiana State U.	52	50	U. of Hawaii—Manoa	54	64
Louisiana Tech. U.	45	52	U. of Houston	35	27
Marshall U.	34	47	U. of Idaho	52	36
Miami U. (Ohio)	80	72	U. of Illinois—Urbana–Champ.	76	72
Michigan State U.	66	62	U. of Iowa	63	74
Middle Tennessee State U.	37	50	U. of Kansas	55	62
Mississippi State U.	48	67	U. of Kentucky	55	45
New Mexico State U.	45	63	U. of Louisiana—Lafayette	25	43
North Carolina State U.	60	60	U. of Louisiana—Monroe	30	41
Northern Illinois U.	47	53	U. of Louisville	30	35
Northwestern U.	92	90	U. of Maryland—College Park	64	63
Ohio State U.	55	62	U. of Memphis	33	52
Ohio U.	68	64	U. of Miami	63	58
Oklahoma State U.	49	24	U. of Michigan—Ann Arbor	82	71
Oregon State U.	57	43	U. of Minnesota—Twin Cities	50	56
Pennsylvania State U.	80	75	U. of Mississippi	51	72
Purdue U.	64	67	U. of Missouri—Columbia	60	57
Rice U.	89	76	U. of Nebraska—Lincoln	51	49
Rutgers U.—New Brunswick	75	68	U. of Nevada—Las Vegas	23	46
San Diego State U.	36	55	U. of Nevada—Reno	48	47

U. of New Mexico	40	47	U. of Toledo	34	55
U. of N. Carolina—Chapel Hill	79	71	U. of Tulsa	53	57
U. of North Texas	36	46	U. of Utah	52	62
U. of Notre Dame	94	74	U. of Virginia	91	84
U. of Oklahoma—Norman	50	49	U. of Washington	68	64
U. of Oregon	58	66	U. of Wisconsin—Madison	76	74
U. of Pittsburgh	60	66	U. of Wyoming	52	51
U. of S. Carolina—Columbia	55	62	Utah State U.	41	64
U. of South Florida	44	48	Vanderbilt U.	84	74
U. of Southern California	73	66	Virginia Tech.	72	67
U. of Southern Mississippi	54	47	Wake Forest U.	87	73
U. of Tennessee—Knoxville	56	59	Washington State U.	57	62
U. of Texas—Austin	69	63	West Virginia U.	54	54
U. of Texas—El Paso	23	41	Western Michigan U.	55	60

(a) Find the least-squares line to predict the athlete graduation rate from the overall graduation rate.

(b) Interpret the slope and intercept of the line. Do they make sense? Why or why not?

(c) Does the least-squares line do a good job of predicting the graduation rate for athletes from the overall graduation rate? Why or why not?

4.18 The Commerce Department has data on the number of shopping centers and retail sales for the South Central states: *Requires Exercise 4.5*

State	Number of shopping centers	Retail sales (billions)
Alabama	630	$15.5
Arkansas	370	7.5
Kentucky	616	13.9
Louisiana	700	18.7
Mississippi	430	8.2
Oklahoma	568	13.2
Tennessee	1200	23.0
Texas	2976	87.3

Source: Statistical Abstract of the United States 1999

(a) Create a scatter plot of the data. Do you think that the variables are linearly related?

(b) Find the equation of the least-squares line for the data.

(c) How does the equation for the South Central states compare with the one you found in Exercise 4.5 for the North Central states?

(d) Would you expect the least-squares lines for the South Central and North Central states to be exactly the same? Why or why not?

(e) What similarities would you expect the two lines to have? What differences?

4.19 Combine the data on retail sales and numbers of shopping centers for the North Central and South Central states. The data for the North Central states are repeated here for convenience. *Requires Exercises 4.5, 4.18*

State	Number of shopping centers	Retail sales (billions)
Illinois	2096	$41.8
Indiana	905	21.4
Iowa	308	7.5
Kansas	481	11.6
Michigan	1018	25.3
Minnesota	471	13.9
Missouri	887	22.7
Nebraska	264	5.7
North Dakota	87	2.1
Ohio	1704	41.6
South Dakota	58	1.3
Wisconsin	625	14.6

(a) Create a scatter plot of the data and find the equation of the least-squares line.

(b) Graph the least-squares line on the scatter plot.

(c) Your boss says that the model for the combined data will be better than the model for the two regions separately because more data go into the combined equation. Write a memo to your boss explaining your results and which method you think is better. Be sure to back up your opinions with the analyses.

Doing It!

Datafile:
FACULTY.XXX

4.20 The committee that is studying gender equity at ABC College wants to look at the faculty data in some other ways. A sample of the data is shown here:

Salary	Years of service	Rank	School	Gender (M/F)	Tenure? (Y/N)
$53,316	22	ASST	BUSINESS	F	Y
64,375	11	PROF	BUSINESS	M	Y
63,501	7	ASSO	BUSINESS	M	Y
59,426	6	ASSO	BUSINESS	M	N
49,058	20	ASSO	BUSINESS	M	Y
94,969	4	PROF	BUSINESS	M	N
54,762	21	ASST	BUSINESS	M	Y
55,516	9	ASSO	BUSINESS	M	Y

(a) Create graphical displays that analyze the relationships between gender and rank and between gender and tenure. Is there anything that the committee might find interesting? Why or why not?

(b) Create graphical displays and look at the overall distribution of rank, gender, and tenure at the school. How do these compare with the distribution by school and rank?

(c) Look at salary versus years of service for each rank separately. Do you think the relationship between the two variables is still valid? Is it easier to see the relationship when the data are separated like this? Why or why not?

(d) Find least-squares equations that will enable the committee to predict salary based on years of service for each rank. Do the equations do a good job of predicting salary? Why or why not?

(e) Are all the equations you found in part (d) the same? Would you have expected them to be? How do they compare?

(f) For each rank, find the least-squares line for males and females separately. How do these compare with each other and with the overall equation?

(g) Repeat part (f) for tenure and for school. What can you conclude?

(h) Write a memo to the president of the college summarizing your results. Include suggestions for the next phase of the analysis.

Datafile:
CELTICS.XXX

4.21 Why are some athletes paid such high salaries? To look at this phenomenon, data were collected on the Boston Celtics, a team in the National Basketball Association. The data give the salary for each player on the roster in the 2000–2001 season as well as other variables about that player, including personal data and season statistics. A portion of the data are shown here:

Player	2000–2001 salary (millions)	Position	Height (inches)	Weight	Birthday	College	Year	Years of Experience
Chris Herren	$0.42	G	74	190	9/27/75	Fresno State	1999	2
Mark Blount	0.42	C	84	230	1/30/75	Pittsburgh	1999	1
Adrian Griffin	0.5	G-F	77	215	7/4/74	Seton Hall	1996	2
Doug Overton	0.55	G	75	190	8/3/69	LaSalle	1991	9
Chris Carr	1.2	G	78	220	3/12/74	Southern Illinois	1995	8

Player	G	GS	MPG	FG%	3P%	FT%	OFF	DEF	TOT	APG	SPG	BPG	TO	PF	PPG
Chris Herren	25	7	16.3	0.302	0.291	0.75	0.2	0.7	0.8	2.2	0.56	0	0.8	1.7	3.3
Mark Blount	64	50	17.2	0.505	0	0.697	1.5	2.1	3.6	0.5	0.61	1.19	0.97	2.9	3.9
Adrian Griffin	44	0	8.6	0.34	0.346	0.75	0.6	1.4	2	0.6	0.41	0.11	0.41	1	2.1
Doug Overton	7	1	20.6	0.25	0.4	2.1	2.7	0.6	0	1.86	2.1	5.4	1.9	2	5.4
Chris Carr	35	0	8.8	0.47	0.46	0.77	0.3	0.9	1.3	0.3	0.1	0.09	0.5	1	4.8

Here is an explanation of the variables:

G—games played

GS—games started

MPG—minutes per game (average)

FG%—field goal percentage

3P%—3-point shot percentage

FT%—free throw percentage

OFF—offensive rebounds

DEF—defensive rebounds

TOT—total rebounds

APG—assists per game (average)

SPG—steals per game (average)

BPG—blocks per game (average)

TO—turnovers (average per game)

PF—personal fouls (average per game)

PPG—points per game (average)

(a) Find a least-squares line that relates the height and weight for the players. How well does the line predict a player's weight from his height?

(b) Create plots of salary against each of the other variables. From the plots, pick the three variables that you think are most closely related to salary.

(c) For each variable you selected in part (b) find the least-squares line.

(d) Graph the least-squares lines from part (c) on the same axes as the data for each variable. Which line do you think does the best job of predicting salary? Why?

(e) Can you see any problem with the variables that are used here? What are they?

(f) Which variables might be affected by the number of minutes per game that a player was on the court? Do you think that this might be a problem?

(g) Write a newspaper article about NBA player salaries using the results you found.

5

Probability

FIRESTONE TIRE DEFECTS

Should Ford and Firestone have known there was a problem with the Wilderness AT tires long before the government forced them to act? On May 2, 2000, the National Highway Traffic Safety Administration (NHTSA) opened an investigation into defective tires manufactured by Bridgestone/Firestone Inc. Many of the tires in question were sold as original equipment on vehicles manufactured by Ford Motor Company. On August 9, 2000, Firestone recalled 14.4 million tires, including the ATX and ATX II models and the Wilderness AT. The NHTSA set up a website to record complaints about Firestone Tires. As of June 14, 2001, there were 6184 reports of problems with Firestone Tires. The data file was cleaned up so that it could be used, and there are 2173 usable records. Here is a portion of the data file:

Tire model	Tire size	Veh mfr	Veh model	Veh year	Blow out	Tread separ	Roll over	Lost control	Crash	Tire position
WILDERNESS	14	FORD	EXPLORER	1993	N		N	N	N	PR
ATX	15	FORD	EXPLORER	1994	Y	UNK	Y	Y	Y	DR
ATX	15	FORD	EXPLORER	1993	Y	Y	Y	Y	Y	DR
FIRESTONE	15	FORD	EXPLORER	1993	UNK	P	N	N	N	TWO
WILDERNESS	15	FORD	EXPLORER	1994	Y	N	Y	Y	Y	PR

5.1   Chapter Objectives

In Chapters 2–4 you learned different ways to summarize and describe data that were collected. The data were usually random samples from different populations, but the techniques used really described only the sample data.

In this chapter we look at how *probability theory* can be used to measure and predict what is *likely* to happen when data are collected from different populations. You will see how the relative frequency of an event that you learned in Chapter 2 is used to define probabilities. You will also see how probability can be used to analyze data from the contingency tables of Chapter 4.

This chapter covers the following topics:

- Basic probability rules
- The law of large numbers
- Conditional probability and independence

5.2   The Language of Probability

Probability is a concept that you are most likely familiar with, though perhaps not in a formal sense. Whenever you talk about whether or not some event is likely to occur, such as whether it will rain tomorrow, you are using the concepts of **probability.**

> *Probability* is a measure of how likely it is that something will occur.

To talk about probability using the language of probability, it is necessary to define some terms. In probability and statistics we often speak of **experiments.** These are not experiments in the laboratory sense, but rather the actions we perform to collect data. As an experiment we might count the number of students who miss statistics class each day or record the color of the car parked next to ours in the parking lot.

> An *experiment* is any action with outcomes that are recordable data.

When we perform an experiment, or collect data, we must think about what outcomes might occur so that we know what form our data will take. The possible outcomes are called the **sample space** of the experiment. You remember from earlier chapters that data can be qualitative or quantitative and that we use different techniques when we analyze different data.

> The *sample space,* **S,** is the set of all possible outcomes of an experiment.

As a very simple experiment, consider rolling a single, six-sided die and recording the number of spots on the top side. You know that the experiment has six possible outcomes, so the sample space is

The brackets { } are used to indicate a sample space.

$$S = \left\{ \;\begin{array}{cccccc} \boxdot & \boxdot & \boxdot & \boxdot & \boxdot & \boxdot \end{array}\; \right\}$$

or   S = {1,2,3,4,5,6}

Another example is the sample space for the experiment of tossing a coin. In this case the sample space is

$$S = \left\{ \;\begin{array}{cc} \text{} \end{array}\; \right\}$$

or   S = {H, T}, where H represents a head and T represents a tail

Depending on the experiment and the type of data collected, a sample space might have a *finite* number of elements such as the ones we just described. The word *finite* means that we can count the number of elements. This most often happens when the experiment results in qualitative data or in quantitative data that are integers.

When the data from the experiment are quantitative and continuous, the sample space contains an *infinite* number of possible values and must be described mathematically or in words. For example, suppose that as our experiment we decide to choose college students at random and measure their heights. Heights are an example of *continuous* quantitative data. The number of possible values that we get is *bounded* (there are probably no college students shorter than 1 foot or taller than 8 feet) but *infinite* (since height is a measurement and there are an infinite number of measurements between 1 foot and 8 feet). To describe this sample space we might write

S = {all numbers greater than or equal to 1 and less than or equal to 8}

or, using mathematical notation,

$$S = \{x : 1 \leq x \leq 8\}$$

Figuring out the sample space for an experiment is important in data collection, too. It helps us to think about all the possible results we might obtain and to plan for them.

EXAMPLE 5.1

FLIPPING TWO COINS

Writing Out a Sample Space

An experiment consists of flipping two standard coins and writing down what is on the side facing up on each of the coins. If we let H represent a head and T represent a tail, then the sample space for this experiment is

$$S = \{HH, HT, TH, TT\}$$

If you are thinking that HT and TH are the same outcome, then try to think of tossing the coins one after another or think of the problem as if the coins are different, such as a penny and a quarter. In writing out sample spaces, it is best not to combine outcomes but to write each one separately.

HT and TH are different even if the coins are both the same, but this is another way to think about it.

✔ TRY IT NOW!

The Spinner Problem
Writing Out the Sample Space

An experiment consists of spinning the two different spinners pictured here:

 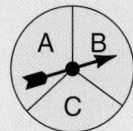

Write down the sample space for this experiment.

You may be wondering what rolling dice and spinning spinners have to do with collecting data and you are right to wonder. In fact, they have little to do with the types of problems you will encounter in statistics, but they are an easy tool to use for discussing and understanding the general rules of probability.

5.2.1 PROBABILITY OF AN EVENT

Several of the sample spaces that you looked at have something in common that makes calculating probabilities easy: Each of the outcomes in the sample space is equally likely to occur on any given trial of the experiment. For example, when you roll a single die, each of the six possible outcomes in S is equally likely to occur. The same is true about flipping coins, both one coin and two coins.

Often we are interested in knowing how likely it is that a certain outcome or outcomes of the experiment will occur. We call these outcomes the **events** of interest.

> An *event,* **A,** is an outcome or a set of outcomes that are of interest to the experimenter.

The **probability** that an event A will occur is written as P(A) and is read "the probability of A."

> The *probability of an event* **A, P(A),** is a measure of the likelihood that an event A will occur.

When each of the outcomes in a sample space *is equally likely*, the probability of event A can be calculated using the formula:

$$P(A) = \frac{\text{Number of ways that A can occur}}{\text{Total number of possible outcomes}}$$

Probability of event A

or

$$P(A) = \frac{n_A}{N}$$

Ans. {1A, 1B, 1C, 2A, 2B, 2C, 3A, 3B, 3C}

where n_A is the number of outcomes that correspond to the event A, and N is the total number of outcomes in the sample space S.

Although you have been taught to reduce fractions and use decimals, we will not do that when we are calculating probabilities so you can see the actual number of outcomes of interest.

In the simple experiment of flipping a coin, if S = {H, T} and you look at the event A = coin turns up heads, then $P(A) = \frac{1}{2}$ because there is only one way that the event A can occur and there are two possible outcomes of the experiment. This should not be surprising—it is one of the ways you understand probability intuitively.

Similarly if you look at the experiment of rolling a die, where S = {1, 2, 3, 4, 5, 6} and A is the event that the number that comes up is even, then

$$P(A) = \frac{3}{6} = \frac{1}{2}$$

since 3 even numbers are possible out of the 6 possible outcomes of the experiment. Again, this is not surprising because it is somewhat intuitive.

There are some rules about probabilities that must hold true:

*These facts may seem trivial, but try to keep them in mind when you calculate probabilities. If you get an answer that is not a number between 0 and 1, then **it cannot be correct!***

1. $0 \le P(A) \le 1$. The probability of an event must be a number between 0 and 1 inclusive. Since you know that probabilities are formed by taking the ratio of the number of ways that A can happen to the total number of outcomes, the numerator is a subset of (smaller than or equal to) the denominator.

2. $P(S) = 1$. The sum of the probabilities for the entire sample space must be equal to 1, or essentially, when you perform an experiment, something must happen!

3. If an event A *must* happen, then $P(A) = 1$, and if the event *cannot* happen, then $P(A) = 0$.

EXAMPLE 5.2

FLIPPING TWO COINS

Finding Probabilities

Joe and Tom decide to flip coins to determine who will pay for dinner. They agree that if the coins match Joe will pay, and if they do not match Tom will pay. What is the probability that Tom pays for dinner?

Remember from Example 5.1 that the sample space for this experiment is S = {HH, HT, TH, TT}. If we let A be the event that Tom pays, then we can write

A = the two coins do not match

It is important to list HT and TH as different outcomes so that all outcomes in S are equally likely.

Looking at S, we see that all the outcomes in the sample space are equally likely to occur and that two outcomes correspond to the event A: HT and TH. Thus, the probability that Tom pays for dinner is

$$P(A) = \frac{2}{4}$$

Similarly, if we let B be the event that Joe pays for dinner, or

B = the two coins match

then two outcomes in S correspond to the event B, HH and TT, so

$$P(B) = \frac{2}{4}$$

We see that the arrangement is fair to both people.

Finding probabilities when the events are equally likely is not hard if you can write down the sample space, S, without much difficulty. These types of problems are known as "classical" probability problems. Although the situations where the ideas of classical probability apply are limited, they do make it easy to illustrate the rules of probability.

✔ *TRY IT NOW!*

The Spinner Problem
Applying Classical Probability

Earlier you found this sample space for the example of the two spinners:

$$S = \{1A, 1B, 1C, 2A, 2B, 2C, 3A, 3B, 3C\}$$

Let A be the event that the first spinner lands on an odd number. Find P(A).

Let B be the event that the second spinner is a vowel. Find P(B).

When the number of possible outcomes of an experiment is large, calculating probabilities can be tedious. At times it is easier to solve the opposite or **complement** of the problem you are interested in! You may wonder how this can be possible. It is because of one of the three rules of probabilities: the probabilities must sum to 1.

The complement of an event A is often referred to as the event "not A."

> The *complement* of an event A, denoted **A′,** is the set of all outcomes in the sample space, S, that do not correspond to the event A.

Since the event A is a set of outcomes in S and the complement of the event, A′, is the set of all outcomes in S that do *not* correspond to A, we can see that

$$P(A) + P(A') = 1$$

Probability of the complement of event A

We can find P(A) by calculating P(A′) and subtracting it from 1; that is,

$$P(A) = 1 - P(A')$$

For the experiment of rolling a single die, the sample space is

$$S = \{1, 2, 3, 4, 5, 6\}$$

If we define the event A to be that the number that comes up is greater than 2, we can calculate $P(A) = \frac{4}{6}$ directly because there are 4 outcomes that are greater than 2 out of the 6 possible outcomes. We notice that the complement of A, A′, is defined as all of the numbers that are *not* larger than 2. Since two outcomes correspond to A′, we know that $P(A') = \frac{2}{6}$ and we can find $P(A) = 1 - P(A') = 1 - \frac{2}{6} = \frac{4}{6}$.

Although this calculation seems to complicate something that was not very complicated, there are times when using the complement of an event to find a probability is the simplest approach. Usually this is when the number of possible outcomes that correspond to the event of interest is a large part of the sample space, S.

Despite what it may look like, the idea of equally likely outcomes is not limited to dice, coins, and spinners! In fact, the ideas of classical probability apply to random sampling as well. Whenever an experiment involves selecting an item at random from a group of items, the sample space consists of all of the possible items. If the sample is truly random, then each item in the group is just as likely to be selected as any other item.

ANS. $P(A) = \frac{6}{9}$; $P(B) = \frac{3}{6}$

EXAMPLE 5.3

TIRE QUALITY

Relating Probabilities and Random Sampling

Collect and analyze the data.

The NHTSA data on accidents involving Firestone tires include the manufacturer of each vehicle involved in an accident. These data can tell us whether it is more likely that one manufacturer's vehicles will be involved in an accident than another. A frequency table lists the following manufacturers and numbers of accidents:

Ford	1810
Mercury	108
Chevrolet	60
Toyota	53
Mazda	26
GMC	24
Nissan	22
Jeep	20
Dodge	16
Isuzu	7
Cadillac	4
Lincoln	4
Honda	3
Oldsmobile	3
Mitsubishi	2
Subaru	2
Other	9
Total	2173

If a vehicle is chosen at random, what is the probability that the vehicle is a Ford?

In this problem, the sample space is the 2173 vehicles with Firestone tires involved in accidents. That is, S = {V1, V2, . . . , V2173} and each element of the sample space is equally likely to be chosen. We are interested in the event that the vehicle is a Ford (call the event A), so there are 1810 elements (vehicles) in the sample space that correspond to the event A and

$$P(A) = \tfrac{1810}{2173}$$

EXAMPLE 5.4

PRODUCT PREFERENCE

Using Random Sampling

A market research firm has conducted a product preference survey at a large manufacturing company. A group of 250 production workers were asked to use three different pairs of safety glasses and to select the one that they preferred. The results of the survey showed that 120 preferred product A, 85 preferred product B, and 45 preferred product C. The marketing research firm decides to select a worker at random and interview him more thoroughly about his choice. They want to determine the probability that the worker interviewed prefers product B.

Collect the data.

If you think about the problem, you will see that the experiment consists of selecting a worker at random. Thus, the sample space of the experiment consists of the 250 workers:

$$S = \{W1, W2, W3, \ldots, W248, W249, W250\}$$

Let B be the event that the worker chosen prefers product B. Then $P(B) = \frac{85}{250}$ because 85 outcomes (workers) in S correspond to the event B and there are 250 possible outcomes.

5.2.2 ESTIMATING PROBABILITIES

At this point you may wonder why we need to study probability at all. That is a reasonable question if you think of probability only in terms of coins, dice, and spinners, or as random samples taken from known populations. These contexts are good for explaining how probabilities are calculated. The situations serve as models for examining probabilities for other situations.

Earlier in this book you looked at methods for describing sample data taken from some population. You learned that the descriptions are exact for the sample but are only estimates when applied to the whole population. How good those estimates are depends on many factors, such as the size of the sample and how well the sample represents the population of interest.

The same is true of probability models. When you collect sample data and calculate relative frequencies, you can find the exact probability that an item taken from the sample will have some characteristic(s) of interest. If the data are a good representation of the population, then you can also use the relative frequencies as estimates of the true probabilities for the population. Probabilities calculated in this way are often called **empirical probabilities.** The same rules of probability that you have already learned apply to these problems as well.

An *empirical probability* is one that is calculated from sample data and is an estimate for the true probability.

EXAMPLE 5.5

RESTAURANT SURVEY

Thinking of Probabilities As Relative Frequencies

A local marketing firm took a random sample of people in a large city to learn what kind of restaurants they think the city needs. The firm asked the people to choose from this list:

- Fast food
- Family
- Adult economical
- Adult moderate
- Adult upscale

As a result of their research, the firm obtained the following data:

Type of restaurant	Number of people	Relative frequency (%)
Fast food	127	20
Family	234	36
Adult economical	158	24
Adult moderate	72	11
Adult upscale	56	9
Total	647	100.00

The marketing firm assumes that the sample is a good representation of the adult residents of the city and wants to determine the probability that a person selected at random favors some type of adult restaurant.

This problem is really asking for the probability that a person favors an adult economical (AE), an adult moderate (AM), or an adult upscale (AU) restaurant. Since the events are mutually exclusive, we can use the simple addition rule to find the answer. What has changed is that the sample space is no longer the 647 original respondents to the survey. Instead, the sample space of the experiment is the possible responses that a person can give:

S = {fast food, family, adult economical, adult moderate, adult upscale}

and the outcomes are no longer equally likely. We can, however, use the relative frequencies as estimates of the probabilities for the outcomes:

P(AE OR AM OR AU) = 24% + 11% + 9% = 44%

What makes one empirical probability better than another? You have already learned that one important element is how well the sample represents the population. In Chapter 1 we discussed many different factors that affect the bias of a sample. Another factor is the size of the sample. How does the sample size affect the accuracy of the probability estimate?

Let's look at the experiment of flipping a coin. Suppose we want to estimate the probability that the coin will come up "heads" and we toss the coin once. If the coin does land heads up, our estimate of the probability will be 1/1 = 1, and if it does not, the estimate will be 0/1 = 0. These estimates could not be farther from correct! Clearly, as we increase the sample size we will get a better estimate. If we try it with two tosses, we know the sample space is S = {HH, HT, TH, TT}, which gives estimates of 2/2 = 1, 1/2, 1/2, and 0/2 = 0. At least two of these estimates are better (in fact, they are correct).

TRY IT NOW!

Estimating Probabilities
How Sample Size Affects Probability Estimates

Write down the sample space for the experiment of tossing a coin three times.

Use the sample space to figure out the estimate of the probability of getting a head for each outcome.

How do these estimates compare to the estimates obtained by using one and two tosses?

Discovery Exercise 5.1

LAW OF LARGE NUMBERS

After being in your introductory statistics course for a few weeks, you begin to see data everywhere in your life, even at the Department of Motor Vehicles (DMV). You are at the DMV waiting to get your

ANS. S = {HHH, HHT, HTH, HTT, THH, THT, TTH, TTT}; 1, 2/3, 2/3, 1/3, 2/3, 1/3, 1/3, 0; although none are exactly correct, more of them (6/8) are not 0 or 1.

license renewed, and you begin to observe the gender of the people as they are called up to be served. Working under the assumption that there is a 50–50 split between men and women who drive, you wonder how many people you would have to observe in order to get a decent estimate of the probability that the next person called up is a male.

Being bored, you begin to write down the gender as the DMV worker calls out names. Ten people are called before they get to you. The genders of these 10 people are F F M F F M M F F M, where M's are males and F's are females. Based on your sample data thus far, your estimate is that there is a 4/10 or 40% chance that the next person called up is male.

Your name is called next and you don't want to hang around the DMV all day collecting data, so you decide to simulate the gender of people at the DMV by using a coin toss. If you toss a head, record that as F (female), and if you toss a tail, record that as M (male). How many "people" will you have to observe before your estimate of the probability comes close to the true 50–50 split? Will more observations ever make the estimate better?

Toss the coin 10 times and then another 10 times, and so forth. After each additional 10 tosses, compute the probability of a male being called up next using ALL the data up to that point. Use the following table to record your data and your calculations. [*Note: Alternatively, you can use the computer to generate the samples using the information for this chapter in the technology manual.*]

Column 1	Column 2	Column 3	Column 4
Sample Data	Cumulative Sample Size	Probability of a Male Based on the Cumulative Data	Error–Difference Between Your Estimate in Column 3 and the Actual Probability of 0.50
F F M F F M M F F M	10	4/10 = 0.40	0.40 − 0.50 = −0.10
	20		
	30		
	40		
	50		
	60		
	70		
	80		
	90		
	100		

1. What happens to the probabilities in column 3 as the sample size increases?

2. What happens to the error in column 4 as the sample size increases?

3. If the sign of the error is negative, what does this tell you?

4. If the sign of the error is positive, what does this tell you?

5. Display the estimates on the following graph:

(*continued*)

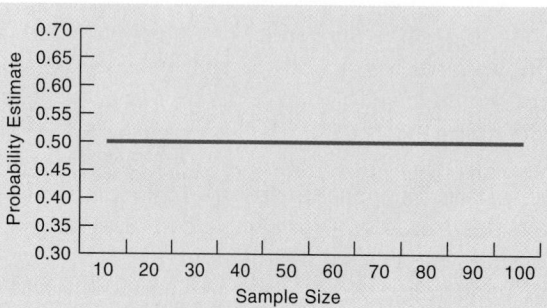

6. After how many tosses would you say you have a "good" estimate of the true probability of 0.50? Why?

7. Do you think that the estimate will ever reach 0.50 and stay there? Why or why not?

There is a law in probability and statistics called the **law of large numbers,** which is sometimes known informally as the *law of averages.* The actual law is beyond the scope of this text, but it states mathematically what you have already observed. Stated informally:

> The **law of large numbers** says that as the number of replications of an experiment increases, the estimate of the probability of an event gets closer to the true or real probability.

In Chapter 7 you will see how to determine how large a sample is necessary to assure that an estimated value is within a certain margin of error from the true value.

5.2.3 EXERCISES—LEARNING IT!

5.1 A poll regarding the amount of time that President Bush spent on vacation was conducted by the Gallup Organization in 2001. Each of 1017 adults was asked whether 30 days was too long for President Bush to be away from the White House. Of the 1017 respondents, 559 said yes.

(a) What is the probability that a person polled thought that 30 days was too long?

(b) What is the probability that a person polled did not think that 30 days was too long?

5.2 A poll of 1167 people was taken to see what kinds of bills people pay electronically. Of the 1167 people, 537 said that they paid their credit card bills electronically.

(a) What is the probability that a person selected at random from the people polled pay their credit card bills electronically?

(b) What is the probability that a person selected at random from those polled does not pay their credit card bills electronically?

5.3 A random sample of 450 students at an urban high school are polled and asked about the legalization of marijuana use. Of the students polled, 280 say that they do favor legalization of marijuana use.

(a) What is the probability that a student selected at random from the 450 will favor the legalization of marijuana?

(b) What is the probability that a randomly selected student will not favor the legalization of marijuana?

(c) Are these probabilities estimated or exact?

5.4 An article published in the journal *Pediatrics* (August 2001) looked at suicide among children and adoption. The study look at 6,577 seventh- to twelfth-graders across the nation. Of those students, 214 were adopted. Of the 213 youngsters who said that that they had attempted suicide, 16 were adopted and 197 were not.

(a) What is the probability that a student in the study was not adopted?

(b) What is the probability that a student in the study who attempted suicide was not adopted?

(c) What is the probability that a student in the study was adopted?

(d) What is the probability that a student in the study who attempted suicide was adopted?

5.5 A board in Connecticut looked at data collected by the Department of Education on the number of student misdeeds that lead to suspension and expulsion. For the towns in a particular area, the data for the 1998–1999 school year are given in the table.

Town/Region	Number of Students Suspended or Expelled
Bethel	420
Brookfield	302
Danbury	3427
New Fairfield	233
New Milford	740
Newtown	202
Redding/Region 9	258
Region 15	410
Ridgefield	500
Total	6492

(a) What is the probability that a student who was suspended or expelled in 1998–1999 was from New Milford?

(b) What is the probability that he or she was from Danbury?

(c) What is the probability that he or she was from Redding/Region 9?

(d) What is the probability that he or she was not from Brookfield?

5.3   Laws of Probability: OR and AND

Calculating the probability of a single event is not difficult. Much of the time, however, we are interested in looking at more than one event. For example, if an experiment consisted of selecting employees at random from a company, we might be interested in the event A = the employee is an hourly worker or the event B = the employee participates in the company's stock purchase program. From what you have learned, if we know the number of employees in the company and the number that are in each category, then we can calculate P(A) and P(B) without much trouble. But suppose that the company is interested in how these two events behave *together*? We must look at the events **A OR B** and **A AND B.** Often the symbols $\cup$ and $\cap$ are used to represent OR and AND, respectively.

> The event **A OR B (A $\cup$ B)** is the event that either A happens or B happens or they both happen. It is often referred to as the union of the two events.

> The event **A AND B (A $\cap$ B)** is the event that A and B both occur.

Let's first look at one of the simpler problems we studied earlier. When you roll a single die, the sample space is S = {1, 2, 3, 4, 5, 6}. If you let the event A = the number that comes up is even, and the event B = the number that comes up is a 3, then you can easily see that $P(A) = \frac{3}{6}$ and $P(B) = \frac{1}{6}$. What if you are interested in the event that the number that comes up is even OR a 3? The same rules that we used to calculate simple probabilities apply here.

If we look at the sample space S, we can count how many of the possible outcomes correspond to the event A OR B. We see that 2, 4, and 6 correspond to A and that 3 corresponds to B. Thus, using the rules for probability when events are equally likely, we find that P(A OR B) = $\frac{4}{6}$ because 4 outcomes out of the 6 possible outcomes correspond to what we are interested in.

You may have noticed two things about this problem. First, the answer, $\frac{4}{6}$, is simply the sum of the two individual probabilities, $\frac{3}{6}$ and $\frac{1}{6}$. This is not a coincidence. Second, the events A and B have no outcomes in common. In probability we refer to these kinds of events as **mutually exclusive.**

> Two events, A and B, are said to be ***mutually exclusive*** if they have no outcomes in common.

When two events are mutually exclusive, then the probability that A occurs or B occurs, P(A OR B), is the sum of the individual probabilities. This is known as the **simple addition rule** and is expressed as

Simple addition rule:
P(A OR B)

$$P(A \ OR \ B) = P(A) + P(B)$$

The simple addition rule extends easily to any number of mutually exclusive events. For example, if A, B, C, and D are four mutually exclusive events, then

$$P(A \ OR \ B \ OR \ C \ OR \ D) = P(A) + P(B) + P(C) + P(D)$$

EXAMPLE 5.6

FLIPPING TWO COINS

Finding the Probability of A OR B

Remember that Joe and Tom are matching coin flips to see who will pay for dinner. The sample space for the experiment is S = {HH, HT, TH, TT}. If A is the event that the two coins both come up tails, and B is the event that the two coins do not match, what is P(A OR B)?

You can see that one outcome, TT, corresponds to the event A and that two outcomes, HT and TH, correspond to the event B, so that

$$P(A \ OR \ B) = \tfrac{3}{4}$$

You can also see that A and B have no outcomes in common. That is, A and B are mutually exclusive, so you should be able to find P(A OR B) by adding P(A) and P(B). Since P(A) = $\frac{1}{4}$ and P(B) = $\frac{2}{4}$ (these should be easy for you by now),

$$P(A \ OR \ B) = \tfrac{1}{4} + \tfrac{2}{4} = \tfrac{3}{4}$$

In probability, as in any kind of mathematics, we must be very precise in our use of words. The word OR in probability is *inclusive*. That means when we talk about the event A OR B occurring, we mean that A occurs or B occurs or BOTH occur. This is not a problem when we are talking about rolling a die because there is no way that a number can be both even and a 3!

EXAMPLE 5.7

TIRE QUALITY

Applying the Simple Addition Rule

Since Ford and Mercury are essentially the same company, we would like to know the probability that an accident involved a Ford or a Mercury. In Example 5.3 we defined A to be the event that the reported accident involved a Ford and found that P(A) was $\frac{1810}{2173}$. We can define the event B as the event that a reported accident involved a Mercury. Now are trying to find P(A OR B) or P(A ∪ B).

It is easy to see that the events A and B are mutually exclusive; that is, they have no outcomes in common. This is the same as saying that no vehicle is both a Ford and a Mercury. We can use the simple addition rule here because the sample space is much too large to work with comfortably:

$$P(A\ OR\ B) = P(A) + P(B)$$

Now we need to find P(B). The results of the survey showed that 108 accidents involved Mercurys, so $P(B) = \frac{108}{2173}$. Thus,

$$P(A\ OR\ B) = \tfrac{1810}{2173} + \tfrac{108}{2173} = \tfrac{1918}{2173}$$

Analyze the data.

EXAMPLE 5.8

PRODUCT PREFERENCE

Finding the Probability of A OR B

The random sample of 250 production workers were asked to use three different pairs of safety glasses and to select the one that they preferred. The results of the survey found that 120 preferred product A, 85 preferred product B, and 45 preferred product C.

In Example 5.4 we defined an experiment where a worker was selected at random for further interviewing. We let the event B = the worker preferred product B, and we found that $P(B) = \frac{85}{250}$. We can also let the event C = the worker preferred product C.

To find the probability that the worker selected for further interviewing preferred product B or C, we must use the simple addition rule because the sample space for this experiment does not help us. We need to find P(C). Since 45 outcomes (workers) in the sample space preferred product C, we know that

$$P(C) = \tfrac{45}{250}$$

To find P(B OR C) we use the simple addition rule:

$$P(B\ OR\ C) = \tfrac{85}{250} + \tfrac{45}{250} = \tfrac{130}{250}$$

Up to this point we have been expressing probabilities as fractions without reducing them in order to illustrate what numbers we use to calculate the probabilities. In fact, probabilities are often expressed as decimals or percentages. In the preceding example, we could also say that the probability that the worker selected preferred product B or C is 0.52 or 52%. Looking at the probability as a percentage is illuminating in this case. We see that although product A was preferred by the largest number of people, a majority of the workers preferred something else.

✔ **TRY IT NOW!**

The Spinner Problem
Calculating the Probability of A OR B

The sample space for the experiment of spinning the two spinners is

$$S = \{1A, 1B, 1C, 2A, 2B, 2C, 3A, 3B, 3C\}$$

Let A be the event that the first spinner points to 1 and let B be the event that it points to 3. Find the probability that A OR B occurs using the sample space.

Now find the same probability using the simple addition rule.

Why are the two answers the same?

The simple addition rule for probability applies only when the events of interest have nothing in common—that is, when they are mutually exclusive. In the product preference example, it is not physically possible for a single worker to prefer more than one of the products, just as it is not possible for a single roll of a die to result in both a 2 and a 3 at the same time. What happens when this is *not* the case—when it is possible for both of the events. A and B, to occur?

Look at the example of rolling a single die. The sample space for the experiment is

$$S = \{1, 2, 3, 4, 5, 6\}$$

When we let A be the event that the number that comes up is even and B be the event that the number that comes up is a 3, these events have nothing in common. It is impossible for them both to occur on a *single trial* of the experiment. The law of probability works so that $P(A \text{ OR } B) = \frac{3}{6} + \frac{1}{6} = \frac{4}{6}$.

Now suppose we define a new event C, where

$$C = \text{the number that comes up is exactly divisible by 3}$$

This is the same as asking for the probability that A OR C will occur.

We want to know the probability that the number that comes up is even *and* is exactly divisible by 3. Remember that our definition of the word AND is that *both* of the events occur. This means that we are looking at any outcomes in S that are both even AND exactly divisible by 3. Looking at the sample space, you can see that there is only one outcome that fits this description, 6. Thus, the probability that both events occur is $P(A \text{ AND } C) = \frac{1}{6}$. The probability that both events occur is often referred to as the probability of the *intersection* of the events.

EXAMPLE 5.9

TIRE QUALITY

Finding the Probability of Intersections

Just because so many of the accidents reported for Firestone tires involved Ford vehicles doesn't mean that the NHTSA can immediately start pointing fingers at any particular vehicle or tire. The contingency table lists the tire models and sizes for the accidents that involved a Ford vehicle.

Suppose we want to know how likely it was that a reported accident involved a Ford vehicle that had tires that were the ATX model (event A) and 16 inches (event B). We want to find $P(A \text{ AND } B)$, or $P(A \cap B)$. Remember that the event A AND B occurs when both of the events occur and that we often refer to this as the *intersection* of the two events. If we look up both of the events in the table and find where they intersect, we see that 17 vehicles are included in both events. Therefore,

$$P(A \text{ AND } B) = \frac{17}{1810}$$

Now suppose that we want to know the probability that the tires in an accident were Wilderness AT (event C) and 15 inches (event D). Again since we want to find the probability that both of these events occur, we look for their intersection and find that

$$P(C \text{ AND } D) = \frac{605}{1810}$$

It is much more likely that a vehicle reported in an accident had 15-inch Wilderness AT tires.

	Tire size					
Tire model	**Less than 15 inch**	**15 inch**	**16 inch**	**17 inch**	**18 inch**	**Total**
ATX	0	432	17	0	0	449
ATX II	0	3	0	0	0	3
Daytona Radial	0	2	0	0	0	2
Firehawk	2	12	4	0	0	18
Firehawk ATX	2	7	3	0	0	12
Other	0	6	1	0	0	7
Steel Tex	0	0	3	0	0	3
UNK	0	5	3	0	0	8
Widetrack Radial Baja	1	1	0	1	0	3
Wilderness	6	209	70	4	1	290
Wilderness AT	9	605	222	15	0	851
Wilderness HT	21	69	62	0	0	152
Wilderness LE	0	0	12	0	0	12
Total	41	1351	397	20	1	1810

Does the fact that the events A and C have something in common change the way we look at the event A OR C? Remember that our definition of the word OR is that one or the other or *both* of the events can occur. In the example of rolling a single die, we are looking at any outcomes in S that are even (event A), or exactly divisible by 3 (event C), or both even AND exactly divisible by 3. Four outcomes in the sample space fit this description: 2, 3, 4, and 6. Thus, from the formula for probability, $P(A \text{ OR } C) = \frac{4}{6}$.

What happens if we try to use the simple addition rule in this example? From the sample space we see that $P(A) = \frac{3}{6}$ (three outcomes are even) and $P(C) = \frac{3}{6}$ (two outcomes are exactly divisible by 3). If we add these two probabilities, we get $\frac{3}{6} + \frac{2}{6} = \frac{5}{6}$. Wait! This is *not* the answer we got using the formula for probability. What went wrong?

What went wrong is what usually goes wrong when mathematics leads to an incorrect answer: We violated the rules—in this case, the rules for using the simple addition rule. The simple addition rule is valid only if the events of interest are *mutually exclusive*. In this case, events A and C are *not* mutually exclusive; they have an outcome in common, a 6. When we calculated P(A), the outcome that a 6 comes up was included in that probability. Then, when we calculated P(C), the outcome of a 6 was included again. When we added the two probabilities together, the outcome of a 6 was included twice, once for each event. Thus, the answer we obtained was too large by $\frac{1}{6}$, which is the probability that a 6 comes up.

How can we adjust the simple addition rule of probability to work in situations when the events are *not* mutually exclusive? You can see from the example that the problem occurs when an outcome is included in *both* events—that is, in both A and C. The answer we got by adding the individual probabilities is too large because the probability that both A and C occur, P(A AND C), is included in both individual probabilities. The **general addition rule** for probability gives the probability of A OR B when the events are not mutually exclusive:

General addition rule:
P(A OR B)

$$P(A \text{ OR } B) = P(A) + P(B) - P(A \text{ AND } B)$$

EXAMPLE 5.10

FLIPPING TWO COINS

Calculating P(A OR B)

Joe and Tom are flipping coins and the sample space is

$$S = \{HH, HT, TH, TT\}$$

We let A be the event that both coins come up tails, so that $P(A) = \frac{1}{4}$. Now we define the event B = the coins match and get $P(B) = \frac{2}{4}$. We want to find the probability that the coins match or they both come up tails.

From the sample space we can see that two outcomes (HH and TT) correspond to the event A OR B, so $P(A \text{ OR } B) = \frac{2}{4}$. To use the general addition rule we need to look at events A and B and decide whether they have any elements in common. They do because the outcome TT satisfies both the definition of event A and the definition of event B. Thus, $P(A \text{ AND } B) = \frac{1}{4}$. We can use the general addition rule and remember to subtract P(A AND B):

$$P(A \text{ OR } B) = P(A) + P(B) - P(A \text{ AND } B)$$

or

$$P(A \text{ OR } B) = \frac{1}{4} + \frac{2}{4} - \frac{1}{4} = \frac{2}{4}$$

which we know is the correct answer.

In Chapter 4 you learned about collecting data that involve two qualitative variables. You learned how to organize that data into a contingency table and to use the data to describe the sample. Data of this type are also important in probability problems.

EXAMPLE 5.11

TIRE QUALITY

Using the General Addition Rule

Suppose the NHTSA wants to find the probability that a vehicle reported in an accident had the ATX tire model (event A) or had 15-inch tires (event C). They must find the probability P(A OR C) from the data on the next page.

From the table we see that $P(A) = \frac{449}{1810}$ and $P(C) = \frac{1351}{1810}$. Now, because we are going to *add* the probabilities, we have to determine whether the two events have any outcomes in common. We need to know whether any vehicles had tires that are ATX *and* 15 inches; that is, we need to find P(A AND C). Looking at the table, we find the intersection $P(A \text{ AND } C) = \frac{432}{1810}$. We can use the general addition rule:

$$P(A \text{ OR } C) = P(A) + P(C) - P(A \text{ AND } C)$$

or

$$P(A \text{ OR } C) = \frac{449}{1810} + \frac{1351}{1810} - \frac{432}{1810} = \frac{1368}{1810}$$

| Tire model | Tire size | | | | | |
	Less than 15 inch	15 inch	16 inch	17 inch	18 inch	Total
ATX	0	432	17	0	0	449
ATX II	0	3	0	0	0	3
Daytona Radial	0	2	0	0	0	2
Firehawk	2	12	4	0	0	18
Firehawk ATX	2	7	3	0	0	12
Other	0	6	1	0	0	7
Steel Tex	0	0	3	0	0	3
UNK	0	5	3	0	0	8
Widetrack Radial Baja	1	1	0	1	0	3
Wilderness	6	209	70	4	1	290
Wilderness AT	9	605	222	15	0	851
Wilderness HT	21	69	62	0	0	152
Wilderness LE	0	0	12	0	0	12
Total	41	1351	397	20	1	1810

When you are looking for the probability of the event A OR B in a contingency table problem, it is not difficult to recognize when you need to use the general addition rule and when the simple addition rule will work. You can think about the outcomes that the events might have in common as the *intersection* of the two simple events in the table. If one event is represented by a row and the other by a column, then there is a cell where they intersect or overlap and you need to use the general addition rule. If both of the simple events A and B are rows or both are columns, then they cannot intersect and the simple addition rule applies.

EXAMPLE 5.12

STUDENT DISTRIBUTION

Using the General Addition Rule

The dean of the School of Arts and Sciences is concerned about the number of upper-level students who are enrolled in the Introductory Statistics course. The school collected data on 28 students about their year and whether they had transferred to the university. The data are shown in this contingency table:

| Year | Status | | |
	Nontransfer	Transfer	Total
Freshman	3	1	4
Sophomore	10	2	12
Junior	2	0	2
Senior	1	9	10
Total	16	12	28

Understand the problem.

Suppose the dean decides to select a student at random and look at that student's records more closely. The sample space of the experiment is the 28

Analyze the data.

students. Since every student is equally likely to be selected, all the outcomes in S are equally likely. We are looking for the probability that the student whose records are examined is a sophomore or a transfer student.

In general, it is helpful to represent events with letters that are coded to the actual data rather than A or B.

Let S represent the event that the student selected is a sophomore and T be the event that the student selected is a transfer student. We are looking for P(S OR T). To start, we need to find P(S) and P(T). From the table we see that $P(S) = \frac{12}{28}$ and $P(T) = \frac{12}{28}$. Next, we need to think about whether these events have any outcomes in common. Are any students both sophomores AND transfer students? The answer is yes; in the table two students are classified as both sophomores and transfer students. These two students are included in both P(S) and P(T), so we must use the general addition rule. Thus,

$$P(S \text{ OR } T) = \frac{12}{28} + \frac{12}{28} - \frac{2}{28} = \frac{22}{28}$$

If we look at the cells in the table that correspond to S or T instead of the totals, we get

$$P(S \text{ OR } T) = \frac{10 + 2 + 1 + 0 + 9}{28} = \frac{22}{28}$$

which is the same answer.

✔ TRY IT NOW!

Quality Problems
Using the General Addition Rule

A company that manufactures cardboard boxes is trying to understand some of its quality problems. Managers have collected data on defect types and production shifts. The data are summarized in the contingency table:

Defect	Shift 1	2	3	Total
Color	8	4	3	15
Printing	6	5	2	13
Skewness	0	2	0	2
Total	14	11	5	30

If a box has more than one defect, then it is classified by the more serious of the defects only. Suppose that a box from the sample is selected at random and examined more closely. What is the probability that the box has a color defect?

What is the probability that the box was produced during the second shift?

Is it possible for the selected box to have a color defect and to have been produced on the second shift? If so, what is the probability?

What is the probability that the selected box has a color defect or was produced on the second shift?

Ans. $P(C) = \frac{15}{30}$; $P(2) = \frac{11}{30}$; yes, $\frac{4}{30}$; $\frac{22}{30}$

5.3.1 EXERCISES—LEARNING IT!

5.6 In 1997 the competition for local phone service increased significantly as a result of changes in federal law. A survey of more than 10,000 people in a large metropolitan area asked whether they rated their current local phone service provider as Excellent, Very Good–Good, Satisfactory, or Poor–Very Poor. The results are shown in the table:

Rating	Excellent	Very Good–Good	Satisfactory	Poor–Very Poor
% Responding	12	21	25	42

(a) What is the probability that a person selected at random from the area rates the current phone service as Excellent or Very Good–Good?

(b) What is the probability that a person selected at random from the area does not rate the service as Poor–Very Poor?

5.7 In a 1999 study on attitudes toward crime and punishment in Vermont respondents were asked to compare crime in Vermont at the current time with crime 5 years ago. The table lists the responses by the gender of the respondent:

	Compared to 5 years ago, crime in Vermont is . . .				
Gender	**Increasing**	**About the same**	**Decreasing**	**Not sure/don't know**	**Total**
Male	125	135	26	14	300
Female	159	104	13	25	301
Total	284	239	39	39	601

Source: National Archive of Criminal Justice Data

(a) What is the probability that a respondent said crime was increasing?

(b) What is the probability that a respondent was female and said crime was decreasing?

(c) What is the probability that a respondent said crime was about the same or decreasing?

(d) What is the probability that a respondent was male or said crime was increasing?

5.8 One survey conducted by the National Center for Health Statistics is the Immunization Survey. In the survey conducted in 1999, respondents were asked whether or not they had immunization records for the child in the survey and whether or not the parents were present in the family. A random sample of 3000 of the surveys provided the following table:

	Have shot records?		
Parents in family	**Yes**	**No**	**Total**
Mother, no father	189	614	803
Father, no mother	32	119	151
Mother and father	564	1386	1950
Neither mother nor father	19	77	96
Total	804	2196	3000

(a) What is the probability that the household had shot records for the child?

(b) What is the probability that both mother and father were present and they had shot records?

(c) What is the probability that exactly one parent was present and they did not have shot records?

(d) What is the probability that neither mother nor father was present or they did not have shot records?

5.9 A pharmaceutical firm is looking at the type of medication that people with allergies took during the autumn allergy season. In particular, the firm wants to know whether the person took medication daily and whether that medication was prescribed by a physician or purchased over the counter. Here are the results of the survey:

Frequency of taking medicine	Type of medicine	
	Prescription	Over the counter
Daily	86	43
As needed (sporadic)	23	156

(a) What is the probability that an allergy sufferer took prescription medication daily?

(b) What is the probability that an allergy sufferer took over-the-counter medication or took medication daily?

(c) What is the probability that an allergy sufferer took medication daily and used over-the-counter medication?

5.10 The Department of Justice reported on interviews with 3721 rape victims. The attacks were classified by age of the victim and the relationship of the victim to the rapist.

Age of victim	Relationship of rapist		
	Family	Acquaintance or friend	Stranger
Under 12	153	167	13
12–17	230	746	172
Over 17	269	1232	739

(a) What is the probability that a victim was under 12 years of age?

(b) What is the probability that a victim was between 12 and 17 years old and the rapist was a member of the family?

(c) What is the probability that a victim was under 12 or the rapist was an acquaintance or friend?

(d) What is the probability that the victim was not under 12 years of age?

(e) What is the probability that the rapist was not a family member, acquaintance, or friend?

5.4   Conditional Probability and Independence

In Chapter 4 you learned about relationships between two variables. Now we discuss that same topic with respect to probabilities.

Suppose we are interested in marketing a particular product. If we define the event A to be that a person chosen at random buys the product, then, with the relative frequency approach, P(A) is simply the percentage of people in the sample space who buy the product. Suppose, however, we decide to look at a second event, B, which is that a person in the sample space has seen an advertisement for the product. Now, as marketers, we might be interested in whether the fact that a person sees the ad affects the probability that the person buys the product. That is, we might be interested in these questions: *If* a person has seen the advertisement, what is the probability that he/she will buy the product? *If* a person has not seen the advertisement, what is the probability that he/she will buy the product? As marketers, we certainly hope that the first probability is higher than the second!

5.4.1 CONDITIONAL PROBABILITY

Up to now we have been considering problems in which we sample from the entire sample space. Now, our knowledge of the outcome of event B reduces the number of sample space elements from which we choose. This type of probability is called a **conditional probability.** Conditional probabilities are written as **P(A|B)** and read "the probability that A will occur *given that* B has occurred" or "the probability of A given B."

Several phrases mean the same thing as "given that." Some are "of the," "if," and "when."

Conditional probability is usually defined by a formula that makes it seem more complicated than it really is. Keep in mind that a conditional probability is simply one that is calculated from a reduced sample space.

The ***conditional probability*** of an event A given an event B is

$$P(A \mid B) = \frac{P(A \text{ AND } B)}{P(B)}$$

Conditional probability

EXAMPLE 5.13

PRODUCT MARKETING

Using Conditional Probability

Does advertising influence buying? A random sample of 500 people were asked whether they bought a new product and whether they saw an advertisement for the product before the purchase. The results are listed in the table:

	Saw advertisement	Did not see advertisement	Total
Purchased product	175	45	220
Did not purchase	100	180	280
Total	275	225	500

We are interested in finding out whether seeing the advertisement (event B) affects the probability that a person buys the product (event A). To do this we need to find P(A) and P(A|B).

We already know how to do the first calculation and so, from the table, we have

$$P(A) = \tfrac{220}{500} = 44\%$$

Now, what does the second probability mean? We want to know the probability that a person purchases the product *given that* he/she saw the advertisement. If we know that the person saw the advertisement, then our sample space no longer has 500 possible outcomes in it. It has been reduced to only the outcomes in the first column of the table, 275 possibilities. From our new sample space, we see that 175 of those people purchased the product. Thus,

$$P(A \mid B) = \tfrac{175}{275} = 63.6\%$$

This probability is different (much higher) than just P(A). It appears that the two events might be related.

The formula

$$P(B|A) = \frac{P(B \cap A)}{P(A)}$$

is also used to calculate probabilities in more complex situations where the denominator event, A, is a complex event that can happen in more than one way. We solve these problems using Bayes' theorem. To learn more about this refer to the extended discussion of this topic in Appendix C of this book.

In the preceding example the two probabilities, P(A) and P(A|B), are not the same. However, this does not mean that the two events, A and B, are *definitely* related. Because we calculated the probabilities using the relative frequency method, we know that another sample might produce different results. To be able to say with some certainty that A and B are related, we need to use the methods of inferential statistics that will be presented in Chapter 15. Right now, however, the conditional probability leads us to believe that the events might be related.

EXAMPLE 5.14

TIRE QUALITY

Using Conditional Probability

Consider again the Firestone tire contingency table:

Tire model	Tire size					Total
	Less than 15 inch	**15 inch**	**16 inch**	**17 inch**	**18 inch**	
ATX	0	432	17	0	0	449
ATX II	0	3	0	0	0	3
Daytona Radial	0	2	0	0	0	2
Firehawk	2	12	4	0	0	18
Firehawk ATX	2	7	3	0	0	12
Other	0	6	1	0	0	7
Steel Tex	0	0	3	0	0	3
UNK	0	5	3	0	0	8
Widetrack Radial Baja	1	1	0	1	0	3
Wilderness	6	209	70	4	1	290
Wilderness AT	9	605	222	15	0	851
Wilderness HT	21	69	62	0	0	152
Wilderness LE	0	0	12	0	0	12
Total	41	1351	397	20	1	1810

Suppose we want to know if there is a relationship between the model of tire involved in the accident and the size of the tire. In particular, we want to know what percentage of the ATX tires (event A) are 15 inches (event B)—in other words, the probability that a tire involved in an accident is 15 inches **given that** it is an ATX tire, or P(B|A).

We must first determine how the given information reduces the sample space. Since we know that the tire was an ATX tire, our sample space is limited to the row of the table labeled "ATX." This reduces the sample space to 449 possibilities. Now, of those companies we see that 432 were 15-inch tires. Thus

$$P(B|A) = \tfrac{432}{449} \quad \text{or} \quad 96.2\%$$

Note! The strategy in solving these problems is to identify the "given" information first. This becomes the denominator of the probability.

5.4.2 INDEPENDENT EVENTS

When you looked at bivariate relationships in Chapter 4, you were trying to decide whether the value of one quantitative variable depended on another or whether they were **independent.**

> Two events are *independent* if the probability that one event occurs on any given trial of an experiment is not affected or changed by the occurrence of the other event.

We can also use the word *independent* in relation to events and probability. If two events are independent, then $P(A|B) = P(A)$. In probability, two events, A and B, are independent *exactly when*

$$P(A \text{ AND } B) = P(A) \times P(B)$$

Independent events

The phrase *exactly when* means that the statement applies in both directions. That is, if we know that two events are independent, then to find the probability that both will occur, we can multiply the individual probabilities together. Also, if we know that the probability that both will occur is equal to the product of the individual probabilities, then we can conclude that the events are independent.

EXAMPLE 5.15

PRODUCT MARKETING

Looking at Independence and Probability

We can check to see whether purchasing the product (event A) and seeing the advertisement (event B) are independent by comparing the quantity $P(A \text{ AND } B)$ with the quantity $P(A) \times P(B)$. The data on advertising and buying are repeated here:

	Saw advertisement	Did not see advertisement	Total
Purchased product	175	45	220
Did not purchase	100	180	280
Total	275	225	500

From the table we see that

$$P(A) = \tfrac{222}{500} \qquad P(B) = \tfrac{275}{500} \qquad P(A \text{ AND } B) = \tfrac{175}{500}$$

If we calculate $P(A) \times P(B)$, we get $\tfrac{220}{500} \times \tfrac{275}{500} = 0.242$ and $P(A \text{ AND } B) = \tfrac{175}{500} = 0.35$.

Thus, the events A and B are not independent *in this sample*. We know that when we use sample data, the probabilities are *exactly correct* only for the sample. Does this mean that the two events A and B are not independent in the population? Not really. We need the tools of inferential statistics to answer that question.

In reality, knowing absolutely that two events are independent is very difficult. Most of the time, that is what we are really trying to determine from the sample data.

Most of the examples that allow us to use the formula $P(A \text{ AND } B) = P(A) \times P(B)$ for independent events are limited to the classical probability examples using coins and dice that we discussed in the beginning of this chapter. To calculate

P(A AND B) in other situations we use the definition of conditional probability with the terms rearranged:

$$P(A \text{ AND } B) = P(A|B)P(B)$$

That is, the probability that both events A and B happen is the probability that A happens *given that* B has happened, weighted by the percentage of time that B happens. You can see that if A and B are independent, then the term $P(A|B)$ reduces to $P(A)$ and we have the formula for independent events.

5.4.3 EXERCISES – LEARNING IT!

Requires Exercise 5.7

5.11 Consider the survey on crime in Vermont. The data are shown here in a contingency table:

	Compared to 5 years ago, crime in Vermont is . . .				
Gender	**Increasing**	**About the same**	**Decreasing**	**Not sure/don't know**	**Total**
Male	125	135	26	14	300
Female	159	104	13	25	301
Total	284	239	39	39	601

(a) What is the probability that a person said that crime was increasing given that he/she is a male?

(b) Of the people who said that crime was decreasing, what is the probability of being female?

(c) What is the probability that if the person was female, she said that crime was increasing or about the same?

(d) Are the events that a person was female and was not sure or didn't know independent in this sample? Why or why not?

Requires Exercise 5.8

5.12 The data from the 1999 survey on immunization are reprinted here:

	Have shot records?		
Parents in family	**Yes**	**No**	**Total**
Mother, no father	189	614	803
Father, no mother	32	119	151
Mother and father	564	1386	1950
Neither mother nor father	19	77	96
Total	804	2196	3000

(a) What is the probability that the household had shot records for the child given that neither mother nor father is present?

(b) Of the households where the mother but no father is present, what is the probability that they did have shot records?

(c) What is the probability that exactly one parent was present given that they did not have shot records?

(d) What is the probability that only one parent was present given that they had shot records?

Requires Exercise 5.9

5.13 A pharmaceutical company is looking at the type of medication that people with allergies took during the autumn allergy season. Here are their survey results:

Frequency of taking medicine	**Type of medicine**	
	Prescription	**Over the counter**
Daily	86	43
As needed (sporadic)	23	156

(a) What is the probability that allergy sufferers took medication daily given that they took over-the-counter medicine?

(b) Of the allergy sufferers who took prescription medicine, what is the probability that they took the medication daily?

(c) If allergy sufferers take over-the-counter medicine, what is the probability that they take it only as needed?

(d) Are the events that a person takes over-the-counter medicine and that he/she takes it sporadically independent in this sample? Why or why not?

5.14 The Department of Justice reported on interviews with 3721 rape victims. The attacks were classified by age of the victim and the relationship of the victim to the rapist.   *Requires Exercise 5.10*

	Relationship of rapist		
Age of victim	Family	Acquaintance or friend	Stranger
Under 12	153	167	13
12–17	230	746	172
Over 17	269	1232	739

(a) What is the probability that a victim was under age 12 given that the rapist was a family member?

(b) If the victim was under age 12, what is the probability that the rapist was a stranger?

(c) Of the victims who were raped by a family member, what is the probability that a victim was between 12 and 17 years old?

(d) Given that the victim was not under age 12, what is the probability that she was raped by a stranger?

GET IT IN WRITING

Chamber of Commerce

TO: **Secretary of Transportation**
FROM: **NHTSA**
RE: **Firestone Tire Complaints**

We have done a preliminary analysis of the accidents that involved Firestone tires. There were actually 6184 accidents reported. Data from many of the reports were not consistent. After double checking the data and taking only those accidents with usable data, we had 2173 accidents to analyze. Based on our data on the vehicle manufacturer and the number of accidents shown in Table 1, we have determined that if you select a single response to follow up, you will most likely select an accident that involves a vehicle manufactured by Ford Motor Company.

TABLE 1

Ford	1810
Mercury	108
Chevrolet	60
Toyota	53
Mazda	26
GMC	24

(continued)

TABLE 1 *(continued)*

Nissan	22
Jeep	20
Dodge	16
Isuzu	7
Cadillac	4
Lincoln	4
Honda	3
Oldsmobile	3
Mitsubishi	2
Subaru	2
OTHER	9
Total	2173

Furthermore if you look at just those accidents that involved a Ford vehicle, listed in Table 2, it is highly likely that the model of tire involved was a Wilderness AT (47%). Of the Wilderness AT tires involved, 71% $\left(\frac{605}{1810}\right)$ were 15-inch tires.

TABLE 2

	Tire size					
Tire model	**Less than 15 inch**	**15 inch**	**16 inch**	**17 inch**	**18 inch**	**Total**
ATX	0	432	17	0	0	449
ATX II	0	3	0	0	0	3
Daytona Radial	0	2	0	0	0	2
Firehawk	2	12	4	0	0	18
Firehawk ATX	2	7	3	0	0	12
Other	0	6	1	0	0	7
Steel Tex	0	0	3	0	0	3
UNK	0	5	3	0	0	8
Widetrack Radial Baja	1	1	0	1	0	3
Wilderness	6	209	70	4	1	290
Wilderness AT	9	605	222	15	0	851
Wilderness HT	21	69	62	0	0	152
Wilderness LE	0	0	12	0	0	12
Total	41	1351	397	20	1	1810

At this point further investigation certainly is warranted. We must follow up on the remainder of the accidents reported to get the final set of data. If the data in the other accidents (approximately 4000) are consistent with the data we have analyzed, a recall of the tires seems more than appropriate.

5.5 Generating Random Data with Technology

You can use Minitab, Excel, or the TI-83 to generate random data for simulating experiments.

Excel 2000

Before you generate random data in Excel, you must specify the data that you want to generate and enter the associated probabilities into the Excel spreadsheet. We will look at generating coin tosses with Excel. The data values we would like are heads and tails, but Excel, like almost every other random data generator, insists on generating only numerical data. So, we will use a 1 for a head and a 2 for a tail. We know that the probabilities are both 0.50.

Once you have entered the probabilities, you are ready to begin. Open the Data Analysis tool, scroll down to **Random Number Generation,** and click **OK.** A dialog box will open.

1. The first step is to determine what kind of random numbers you want to generate since the dialog box input will differ. In this case we will select **Discrete.** In Chapter 6 you will learn how to generate data from several different probability distributions.

   You must give Excel some general information about the data you want to generate. The first is the number of variables or samples you want to generate. The second is how many random numbers you want in each sample.

2. Position the cursor in the box labeled **Number of <u>V</u>ariables:** and type in 1. This will generate one sample.

3. Move the cursor to the box labeled **Number of Random Num<u>b</u>ers:** and type in 10. This means that we want to generate 10 coin tosses. Obviously, this will change depending on what you are trying to accomplish.

We have already selected the type of data we want to generate, so we will need to tell Excel the location of our specific information—the values and probabilities.

4. Position the cursor in the box labeled **Value and Probability <u>I</u>nput Range:** Move the cursor to the spreadsheet and highlight the range that contains the information.

5. The last step is to tell Excel where you want the output located. Position the cursor in the box labeled **<u>O</u>utput Range** and then click on the cell location where you want the first sample value to be placed. Alternatively, you can select to put the output in a separate worksheet.

6. Click **OK,** and data similar to that shown in Figure 5.1 should appear. *Note:* The data are supposed to be random, so don't expect yours to look exactly the same as ours!

   From the output, you see that in this set of 10 coin tosses, you had 5 heads and 5 tails.

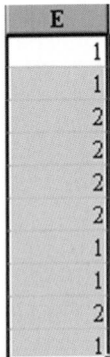

FIGURE 5.1 **Data from generating random coin tosses**

If you need to generate more than one sample of data, you can do so by inputting the number of samples you want in the box labeled **Number of Variables** as long as all the samples are the same size.

Minitab

Before you start, you must specify the data that you want to generate and enter the associated probabilities into two columns in the worksheet. We will look at generating coin tosses with Minitab. The data values we would like are heads and tails, but Minitab, like almost every other random data generator, insists on generating only numerical data. So, we will use a 1 for a head and a 2 for a tail. We know that the probabilities are both 0.50.

From the main menu select **Calc > Random Data > Discrete....** A dialog box will open.

To generate a random sample of data, you have to indicate how many random data values you want (the sample size), the column to use for storing the data, and the location of the random data and probabilities that you inputted to the worksheet.

1. Position the cursor in the box labeled **Generate _____ rows of data** and type in 10.

2. Move the cursor to the box labeled **Store in column(s):** and enter the label of an empty column in the worksheet—for example, C5.

3. Place the cursor in the box labeled **Values in:** and enter the label of the column that contains the set of possible data values that you want. You can also double click on the column from the box on the left.

4. Place the cursor in the box labeled **Probabilities in:** and enter the column that contains the probabilities for the data values.

5. Click **OK** and output similar to Figure 5.2 should appear. *Note:* Your output will not likely be the same output, since it is supposed to be a random sample.

    From the output, you can see that the ten coin tosses resulted in 6 heads and 4 tails.

If you want to generate more than one sample of the same size, you can do so easily by indicating more columns in the **Store in column(s):** box. For example, for 10 samples of 10 tosses each, you would enter C5-C14.

TI-83 Graphing Calculator

It is possible to generate random data with the TI-83. You are somewhat limited in that you can generate data from only a few distributions, but it can be a useful tool.

C5
1
2
1
2
1
2
1
1
2
1

FIGURE 5.2 **Output from random coin toss simulation**

Suppose you want to simulate tossing a coin 10 times. That is, you want to generate 10 random data values that are either heads or tails, each of which occurs with probability 0.5. This is possible with the **IntRand** function. The limitation of this function compared to using Excel or Minitab is that you can generate only consecutive-integer data and they have to be equally likely. This fits with the coin toss, though, so we can proceed. Because the data generated must be integer-valued, we will use a 1 instead of a head and a 2 instead of a tail.

1. Press the ⟨MATH⟩ key and use the ▶ to move the cursor to the **PRB** menu.

2. Select option 5, **randInt(.** The function will appear on the screen.

    The **randInt** function expects two or three input parameters: the lowest integer value you want, the highest integer value you want, and the number of random values you wish to generate. If you leave the third parameter out, you will get exactly one random value.

3. Complete the **randInt** function by entering 1, 2, and 10 separated by commas.

4. Press the ⟨ENTER⟩ key, and the random data will appear as a list as shown in Figure 5.3. You will have to use the arrow keys to see all of the values.

    It is difficult to see because of the overlap, but this sample of 10 tosses had 3 heads and 7 tails.

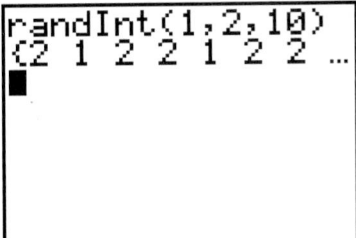 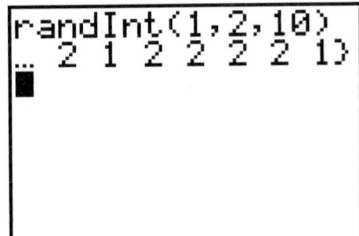

FIGURE 5.3 Output from generating 10 coin tosses

CHAPTER 5 SUMMARY

Probability is an important and interesting subject in its own right, but the study of probability is also important in the study and development of statistics.

Probability is the bridge between what we studied in the first four chapters of this book, *descriptive statistics,* and what is to follow in subsequent chapters, *inferential statistics.* In this chapter we saw the connection between the relative frequency that we learned about in descriptive statistics and how probabilities are estimated. We also saw how the contingency tables we learned about in Chapter 4 can be used to determine whether events are independent in a particular sample. In inferential statistics, we will learn more formal techniques that allow us to make general statements—hypotheses—that extend beyond the sample to the populations from which these samples came. Probability is the tool that allows us to reconcile what happened (descriptive) with what we think is true by determining how likely are the outcomes of the experiment we perform.

KEY TERMS

Term	Definition	Page reference
A AND B	The event **A AND B** is the event that A and B both occur.	217
A OR B	The event **A OR B** is the event that either A happens or B happens or they both happen.	217
Complement	The **complement** of an event A, denoted **A′**, is the set of all outcomes in the sample space, S, that do not correspond to the event A.	211
Empirical probability	An **empirical probability** is one that is calculated from sample data and is an estimate for the true probability.	213
Event	An **event, A,** is an outcome or a set of outcomes that are of interest to the experimenter.	209
Experiment	An **experiment** is any action with outcomes that are recordable data.	207
Independent events	Two events are **independent** if the probability that one event occurs on any given trial of an experiment is not affected or changed by the occurrence of the other event.	229
Law of large numbers	The **law of large numbers** says that as the number of replications of an experiment increases, the estimate of the probability of an event gets closer to the true or real probability.	216
Mutually exclusive	Two events, A and B, are said to be **mutually exclusive** if they have no outcomes in common.	218
Probability	**Probability** is a measure of how likely it is that something will occur.	207
Probability of an event, P(A)	The **probability of an event A, P(A),** is a measure of the likelihood that an event A will occur.	209
Sample space, S	The **sample space, S,** is the set of all possible outcomes of an experiment.	207

KEY FORMULAS

Term	Formula	Page reference
Conditional probability of an event A given an event B	$P(A\mid B) = \dfrac{P(A \text{ AND } B)}{P(B)}$	227
General addition rule	$P(A \text{ OR } B) = P(A) + P(B) - P(A \text{ AND } B)$	222
Independent events	$P(A \text{ AND } B) = P(A) \times P(B)$	229
Probability of an event A	$P(A) = \dfrac{\text{Number of ways that A can occur}}{\text{Total number of possible outcomes}}$ $= \dfrac{n_A}{N}$	209
Probability of the complement, P(A′)	$P(A) + P(A') = 1$	211
Simple addition rule	$P(A \text{ OR } B) = P(A) + P(B)$	218

CHAPTER 5 EXERCISES

Learning It!

5.15 In 1953 the Gallup Organization conducted a poll asking 1730 Americans whether they favored inserting the words "under God" into the Pledge of Allegiance. Of those polled, 1105 said yes, 465 said no, and 160 had no opinion.

(a) What is the probability that a person polled in 1953 thought that the words "under God" should be inserted into the Pledge of Allegiance?

(b) What is the probability that a person did not have an opinion?

5.16 A poll conducted by *The Washington Post* on January 15–19, 1998, asked 1206 adults whether they have ever considered themselves a fan of professional baseball. Of those polled, 169 said yes and 1433 said no.

(a) What is the probability that a person polled considered himself or herself a fan of professional baseball?

(b) Based on this poll, do you think the image of baseball as "the American pastime" is reasonable? Why or why not?

5.17 A poll was conducted by CNN/Gallup on the two days following the June 26, 2002, decision by the Ninth Circuit Court of Appeals of San Francisco that said the words "under God" in the United States Pledge of Allegiance were constitutional. Of 1000 adults surveyed, 868 said that they support the phrase while 132 did not.

Requires Exercise 5.15

(a) What is the probability that a person polled favored the words "under God" in the Pledge of Allegiance?

(b) How does this probability compare to the 1953 poll conducted by Gallup?

5.18 A survey of 120 elementary school children who attend camp during the summer asked what activity they most liked to do during summer camp. The results were:

Activity	Number of children
Arts and crafts	18
Swimming	77
Nature	11
Snack	14
Total	120

(a) What is the probability that a child surveyed said arts and crafts?

(b) What is the probability that a child surveyed selected nature or snack?

(c) What is the probability that a child surveyed did not select swimming?

5.19 In a recent survey, people were asked what they do while they are stuck at the airport. Of the 2200 people surveyed, the results were:

What they do	Number of People
Have a drink	1125
East fast food	436
Read	231
Eat local cuisine	215
Eat junk food	193
Total	2200

(a) What is the probability that a person selected at random from those surveyed eats fast food when stuck at the airport?

(b) What is the probability that a person selected at random from those surveyed reads when stuck at the airport?

(c) What is the probability that a person selected at random from those surveyed does not have a drink?

(d) What is the probability that a person selected at random from those surveyed does not eat or drink?

Requires Exercise 5.6 **5.20** In the survey about satisfaction with the local phone service, those respondents who rated their current service Excellent and those who rated it Poor–Very Poor were asked what type of company their current local service provider was. The results are listed here:

Current service source	Excellent	Poor–Very Poor
Long-distance company	264	1394
Local phone company	444	1318
Power company	131	485
Cable TV company	215	431
Cellular phone company	198	572

(a) What is the probability that a person selected from this group will rate current service Excellent and have a long-distance company as the current provider?

(b) What is the probability that a person selected from this group will use a power company or a cable TV company as the current provider?

(c) What is the probability that a person selected from this group will rate current service as Poor–Very Poor or use a local phone company?

(d) What is the probability that a person selected from this group will rate current service as Poor–Very Poor and will use a cellular phone company?

5.21 In a study on college students and binge drinking, researchers are interested in looking at binge drinking and gender. They asked questions about gender and the number of times a student had five or more drinks in a row in the last 2 weeks.

Gender	\multicolumn{6}{c}{Number of times had five or more drinks in a row in last 2 weeks}

Gender	0	1	2	3	4	5
Male	55	20	18	19	8	4
Female	75	14	12	24	2	0

(a) What is the probability that a student is a male and had five or more drinks in a row two times in the last 2 weeks?

(b) What is the probability that a student is a female or had five or more drinks in a row three times in the last 2 weeks?

(c) What is the probability that a student had five or more drinks in a row in the last 2 weeks three or four times?

(d) What is the probability that a student did not have five or more drinks in a row in the last 2 weeks?

Requires Exercise 5.21 **5.22** In the same study of binge drinking and college students, the researchers were also interested in the number of times that a student had a hangover during the semester. The data they collected are given here:

Gender	\multicolumn{3}{c}{Hangover since beginning of semester}

Gender	Not at all	Once	Twice or more
Male	61	23	40
Female	66	25	36

(a) What is the probability that a student is a female and had a hangover twice or more during the semester?

(b) What is the probability that a student is a male and did not have a hangover during the semester?

(c) What is the probability that a student had a hangover once or more during the semester?

Thinking About It

5.23 Consider again the poll conducted by CNN/Gallup on the two days following the June 26, 2002, decision by the Ninth Circuit Court of Appeals of San Francisco that said the words "under God" in the United States Pledge of Allegiance were unconstitutional. Of 1000 adults surveyed, 868 said that they support the phrase while 132 did not.

Requires Exercises 5.15 and 5.17

(a) Do you think that it would be reasonable to compare this poll to the one taken in 1953? Why or why not?

(b) Do you think that the new poll is fair or might it have bias? Why or why not?

(c) Would you use 0.868 as an estimate of the probability that an American adult selected at random would be in favor of the phrase being in the Pledge of Allegiance? Why or why not?

5.24 During the poll conducted by *The Washington Post* concerning professional baseball fans, data were also recorded on gender. The expanded data are given here.

Requires Exercise 5.16

Have You Ever Considered Yourself a Professional Baseball Fan, or Not?

| Response | Gender | | Total |
	Male	Female	
Yes	128	53	181
No	672	353	1025
Total	800	406	1206

(a) What is the probability that a person polled was male? Female?

(b) If a person polled was female, what is the probability that she was a professional baseball fan?

(c) If a person polled was a professional baseball fan, what is the probability that the person was male?

(d) What can you conclude about professional baseball fans from these data?

5.25 An article published in the journal *Pediatrics* (August 2001) looked at suicide among children and adoption. The study looked at 6,577 seventh- to twelfth-graders across the nation. Of those students, 214 were adopted. Of the 213 youngsters who said that they had attempted suicide, 16 were adopted and 197 were not.

(a) Create a contingency table showing these data.

(b) What is the probability that an adopted child attempted suicide?

(c) If a child selected at random from the study was not adopted, what is the probability that he or she had attempted suicide?

(d) Are the events adopted and attempted suicide independent in this sample? Why or why not?

5.26 Look again at the data about student suspensions and expulsions.

Requires Exercise 5.5

Town/Region	Number of Students Suspended or Expelled
Bethel	420
Brookfield	302
Danbury	3427
New Fairfield	233
New Milford	740
Newtown	202
Redding/Region 9	258
Region 15	410
Ridgefield	500
Total	6492

(a) Based on the data, what is the probability that a student who was expelled or suspended came from Danbury schools?

(b) Based on these data, do you think that it would be reasonable to say that Danbury students are more likely to be suspended or expelled? If so, why? If not, what other information do you think you would want to know?

5.27 The Fatal Accident Reporting System (FARS) collects data on fatal traffic accidents in the United States. For 281 reported fatalities in Connecticut in 2000 in which alcohol tests were administered, data were collected on blood alcohol level and number of previous convictions for driving while impaired.

Blood alcohol	Previous DWI convictions				
level (%)	None	1	2	3	Total
0.00	159	4	0	0	163
0.01–0.09	26	1	0	0	27
0.10+	83	4	1	3	91
Total	268	9	1	3	281

(a) What is the probability that a fatal accident victim had a blood alcohol level higher than 0.10%?

(b) Of the people with previous DWI convictions, what is the probability that the victim had a blood alcohol level higher than 0.10%?

(c) How do the probabilities in parts (a) and (b) compare?

(d) What is the probability that the victim had a blood alcohol level of 0.00% given that he/she had previous DWI convictions?

(e) Can you draw any reasonable conclusions about blood alcohol levels and DWI convictions from the data? If so, what are they? Why?

5.28 The Centers for Disease Control maintains a database on leading causes of death. In 1998 in the western United States, there were 904 deaths of teenagers in which firearms were the cause of death. A contingency table of age versus race for the 904 deaths follows:

Race	Age (years)							
	13	14	15	16	17	18	19	Total
American Indian/Alaska native	1	1	1	7	7	6	9	32
Asian/Pacific Islander	3	3	8	3	9	15	8	49
Black	2	4	7	17	23	45	39	137
White	32	46	64	85	130	174	155	686
Total	38	54	80	112	169	240	211	904

(a) What is the probability that the person was over 16 years old?

(b) What is the probability that the person was white given that he/she was over 16?

(c) What is the probability that the person was under 16 given that he/she was neither white nor black?

Requires Exercise 5.22 **5.29** Consider the data from the study of binge drinking and college students:

	Hangover since beginning of semester		
Gender	Not at all	Once	Twice or more
Male	61	23	40
Female	66	25	36

(a) Given that the student is a female, what is the probability that she had a hangover twice or more during the semester?

(b) What is the probability that a student is a male given that he had a hangover once or less during the semester?

(c) What is the probability that a student had two or more hangovers in a semester given that the student is a male?

(d) Compare your answers to parts (a) and (c) and interpret the results for the researchers.

Random Variables and Probability Distributions

WHAT HAPPENS WHEN THEY FAIL?

It seems likely that Ford and Firestone should have known about problems with Firestone tires, particularly the ATX and ATX II models. But questions remain about what exactly was happening? Were there accidents involved? How serious were they? Did all of the vehicles have similar accidents, or were they all different? These questions need to be addressed so that particular problems can be identified and fixed. When the NHTSA set up the website to record complaints about Firestone tires, it collected information about the nature of the incident as well as vehicle information. In particular, it asked whether the tire failure was a blowout or a tread separation, whether the vehicle rolled over, whether the driver lost control, and whether the car was involved in a collision. A portion of the data file is shown here:

TIRE MODEL	TIRE SIZE	VEH MFR	VEH MODEL	VEH YEAR	BLOW OUT	TREAD SEPAR	ROLL OVER	LOST CONTROL	CRASH	TIRE POSITION
WILDERNESS	14	FORD	EXPLORER	1993	N		N	N	N	PR
ATX	15	FORD	EXPLORER	1994	Y	UNK	Y	Y	Y	DR
ATX	15	FORD	EXPLORER	1993	Y	Y	Y	Y	Y	DR
FIRESTONE	15	FORD	EXPLORER	1993	UNK	P	N	N	N	TWO
WILDERNESS	15	FORD	EXPLORER	1994	Y	N	Y	Y	Y	PR

6.1 Chapter Objectives

In Chapters 2–4 you learned different ways to summarize and describe data that were collected. The data were usually random samples from different populations, but the techniques used really described only the sample data.

In Chapter 5 you learned some basic laws of probability and you saw how probability relates to some of the topics you have studied, such as relative frequencies and contingency tables. In this chapter we will continue our study of probability by looking at random variables and studying some probability distributions. The characteristics of these probability distributions relate back to the way you learned to describe the distribution of sample data and to the empirical rule and Z-scores. We will continue to look at how *probability theory* can be used to measure and predict what is *likely* to happen when data are collected from different populations. This chapter covers the following topics:

- Random Variables and Probability Distributions
- The Binomial Probability Distribution
- The Normal Probability Distribution

6.2 Random Variables

When you studied the different techniques for summarizing and analyzing data, you learned that although qualitative data are important for understanding and interpreting the information obtained from a set of data, there is not much you can do with the raw data. Most statistical analyses use numerical or quantitative data as their basis. In much the same way, we prefer to discuss probabilities for experiments that have numerical outcomes.

You know from algebra that a variable is a quantity that has values that can change or vary. The exact value that the variable takes on depends on the equation that includes it. We have referred to the different characteristics of data that can be collected from a population as *variables*. The exact value that a statistical variable takes on depends on the laws of chance or probability. When the variables in question are quantitative, they are known as **random variables.**

> A *random variable, X,* is a quantitative variable that has values that vary according to the rules of probability.

When the outcome of an experiment is a random variable, the elements of the sample space are all the possible values that the variable can have. Random variables, like data, can be discrete (integers) or continuous (real numbers). We first consider only the discrete case.

In the experiment of rolling a single die and recording the number of spots that are on the top face, the sample space for the experiment can be written as

$$S = \{1, 2, 3, 4, 5, 6\}$$

The outcomes are numerical and the exact value that will turn up varies. We know from the rules of probability that each of the values is equally likely, or has an equal probability of happening. Thus, X = the number of spots on the top face of a die is a random variable.

EXAMPLE 6.1

FLIPPING TWO COINS

Defining a Random Variable

In the experiment of tossing two coins, the sample space is

$$S = \{HH, HT, TH, TT\}$$

We always use capital letters, like X, to represent the random variable and lowercase letters, such as x, to represent the values of the random variable.

This experiment does not involve a random variable because the possible outcomes are not numerical. It is possible, however, to describe the outcomes of the experiment numerically by defining a random variable for the experiment. We let X = the number of heads that appear on the two coins, and then the possible values of X are $x = 0, 1$, and 2. Every time we conduct the experiment, we do not know what will happen, but the laws of probability give us some insight into what is likely to occur.

6.2.1 PROBABILITY DISTRIBUTION OF A DISCRETE RANDOM VARIABLE

The rules of probability that describe the way a random variable behaves are known as the **probability distribution** of the random variable. The probability distribution of a discrete random variable assigns a probability to each of the possible values that can occur.

Read p(x) as "p of x."

> The **probability distribution** of a random variable, X, written as **p(x),** gives the probability that the random variable will take on each of its possible values.

The notation that we use for the probability distribution of a random variable is similar to the notation for the probability of an event. For a random variable,

$$p(x) = P(X = x) \quad \text{for all possible values of } X$$

Most often the probability distribution is written in the form of a table.

EXAMPLE 6.2

FLIPPING TWO COINS

Writing the Probability Distribution

In the experiment of tossing two coins, X = the number of heads is a random variable that can take on three values: $0, 1$, and 2. Using the rules of probability, we can find the probability distribution of X, the number of heads in two tosses of a coin:

$p(0) = P(X = 0)$ corresponds to only one outcome, TT, so $p(0) = \frac{1}{4}$

$p(1) = P(X = 1)$ corresponds to two outcomes, HT and TH, so $p(1) = \frac{2}{4}$

$p(2) = P(X = 2)$ corresponds to one outcome, HH, so $p(2) = \frac{1}{4}$

We display the results in a table:

x	0	1	2
p(x)	$\frac{1}{4}$	$\frac{2}{4}$	$\frac{1}{4}$

The rules for probability distributions are the same as the rules for probabilities. For each value of the random variable, X:

1. $0 \leq p(x) \leq 1$. The probability must be a number between 0 and 1 inclusive.
2. $\Sigma p(x) = 1$ for all values of X. The probabilities of all values of X must sum to 1.

Notice that the outcomes of a random variable are *mutually exclusive*. This means that when we are interested in finding the probability that the random variable takes on one of its values OR another of its values, we can use the simple addition rule.

EXAMPLE 6.3

NEWSPAPER SALES

Finding Probabilities from a Probability Distribution

The number of copies of *USA Today* that are sold daily by a convenience store is a random variable X that has the following probability distribution:

x	0	1	2	3	4	5
$p(x)$	0.10	0.12	0.25	0.30	0.20	0.03

The store manager wants to know the probability that on any given day he will sell exactly two copies of *USA Today*. We can find this answer directly from the table:

$$p(2) = 0.25$$

Now what is the probability that two or three copies are sold on any given day? It is first important to recognize that we are trying to find $P(X = 2$ OR $X = 3)$. Since the values of a random variable are mutually exclusive, we can simply add the probabilities to obtain the answer:

$$P(X = 2 \text{ OR } X = 3) = p(2) + p(3) = 0.25 + 0.30 = 0.55$$

TABLE 6.1

PROBABILITY NOTATION SUMMARY

Find the probability that X takes on a value that is . . .	What it means	Notation
at least x	All the values of the random variable that are the value x or greater (up to n)	$P(X \geq x)$
more than x	All the values of the random variable that are greater than the value x (up to n)	$P(X > x)$
at most x	All the values of the random variable that are the value x or less	$P(X \leq x)$
less than x	All the values of the random variable that are less than the value x	$P(X < x)$
between x_1 and x_2	All the values of the random variable that are greater than the value x_1 and less than the value x_2	$P(x_1 < X < x_2)$
between x_1 and x_2 inclusive	All the values of the random variable that start with the value x_1 and go up to and include the value x_2	$P(x_1 \leq X \leq x_2)$

Notation and Interval Probabilities

Very often we are interested in the probability that a random variable takes on any one of a set of its possible values. We might be interested in finding the probability that the random variable takes on a value that is **"at least x," "more than x," "at most x," "less than x," "between x_1 and x_2,"** or **"between x_1 and x_2 inclusive."** There is nothing new that you need to know to find the probabilities. It just takes a little practice to be able to recognize each of the different problems and to write down what you are looking for. Since the words can get cumbersome, we use a standard notation. For example, we use a random variable X that can take on values of $x = 0, 1, 2, 3, \ldots, n$. Table 6.1 provides a summary of each problem, what it means, and the correct notation.

EXAMPLE 6.4

NEWSPAPER SALES

Finding Interval Probabilities

The manager of the convenience store that sells *USA Today* wants to know more about the probability that he will sell newspapers. The probability distribution of X, the number of copies of *USA Today* sold per day, is reprinted here:

x	0	1	2	3	4	5
$p(x)$	0.10	0.12	0.25	0.30	0.20	0.03

In particular, the store needs to sell at least three copies per day to make a profit from the sales. The manager is looking for the probability that the store sells at least three copies, or $P(X \geq 3)$. This means $P(X = 3$ OR $X = 4$ OR $X = 5)$, which is calculated as

$$p(3) + p(4) + p(5) = 0.30 + 0.20 + 0.03 = 0.53$$

Now, what is the probability that the store will not make a profit? The manager can answer this question directly by determining that the store will not make a profit if it sells less than three copies of the newspaper and finding $P(X < 3)$. We can also do this problem by considering the complement of an event. The events "make a profit" or "$X \geq 3$" and "do not make a profit" or "$X < 3$" are complements of each other. This means that their probabilities must sum to 1. Since we already know that $P(X \geq 3) = 0.53$, we can find $P(X < 3)$ from

$$1 - P(X \geq 3) = 1 - 0.53 = 0.47$$

 TRY IT NOW!

Defective Pens
Finding Interval Probabilities

A company that sells ballpoint pens in bulk packages to a warehouse club knows that the number of defective pens in a package is a random variable with the probability distribution given here:

x	0	1	2	3	4	5	6
$p(x)$	0.30	0.21	0.12	0.10	0.10	0.09	0.08

Find the probability that a package of pens will contain at least three defective pens.

Find the probability that the package will contain between two and five defective pens.

Find the probability that the number of defective pens will be at most two.

6.2.2 PROBABILITY HISTOGRAMS

Random variables and their probability distributions are models for the populations from which sample data are taken. You learned in Chapter 2 that you can display quantitative data on a relative frequency table or a relative frequency histogram. In much the same way, a random variable can be displayed on a probability distribution table or a probability distribution histogram.

NEWSPAPER SALES

Creating a Probability Histogram

The manager of the convenience store that sells *USA Today* wants to see what the distribution of newspaper sales looks like. Here is a probability histogram for the random variable:

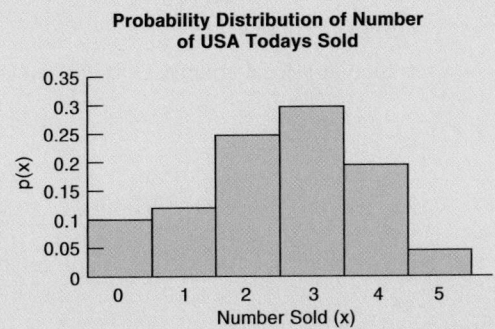

The histogram shows that the probability distribution of sales is not very variable, approximately symmetric, and centered at about three newspapers per day.

In a probability histogram the area of each rectangle is equal to the probability that the random variable takes on the given value. This may not seem important right now, but we use this fact later in the chapter when we move from discrete to continuous random variables.

TRY IT NOW!

Defective Pens
Creating a Probability Histogram

The company that sells ballpoint pens in bulk packages to a warehouse club wants to have a picture of how the number of defective pens in a package behaves. The probability distribution is reprinted here:

x	0	1	2	3	4	5	6
$p(x)$	0.30	0.21	0.12	0.10	0.10	0.09	0.08

Create a probability histogram for the number of defective pens.

Use the probability histogram to describe the distribution of the number of defective pens in a package.

6.2.3 EXERCISES—LEARNING IT!

6.1 The number of employees who call in sick on any given day in a small business is a random variable with this probability distribution:

x	0	1	2	3	4	5	6
$p(x)$	0.10	0.23	0.18	0.16	0.13	0.10	0.10

(a) What is the probability that on any given day at most four employees call in sick?

(b) What is the probability that between two and four employees call in sick?

(c) What is the probability that more than four employees call in sick?

6.2 The number of members who cannot get a tee time at a local country club is a random variable with this probability distribution:

x	0	1	2	3	4	5	6	7	8
$p(x)$	0.11	0.12	0.13	0.19	0.12	0.09	0.09	0.08	0.07

(a) What is the probability that between two and five members inclusive cannot get tee times?

(b) What is the probability that fewer than three cannot get tee times?

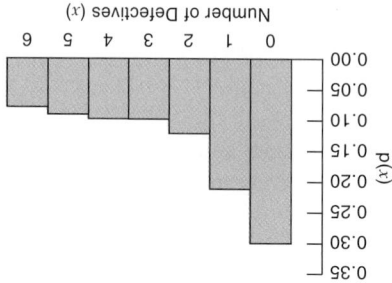

Probability Distribution of Number of Defective Pens

ANS. SKEWED RIGHT, MUCH MORE LIKELY TO FIND SMALL NUMBER OF DEFECTIVES.

(c) What is the probability that at least six cannot get tee times?

(d) What is the probability that at most four cannot get tee times?

6.3 The number of times a person gets a busy signal when calling the customer service number of a local cable television office is a random variable with the following probability distribution:

x	0	1	2	3	4
$p(x)$	0.23	0.34	0.17	0.15	?

(a) What is p(4)?

(b) What is the probability that a person does not get a busy signal?

(c) What is the probability that a person gets at least one busy signal?

(d) What is the probability that a person gets more than two busy signals?

6.4 A company that packages small items for resale is looking at the problems of incorrect packaging. In a box that is supposed to contain two dozen items, the number of missing items is a random variable with this probability distribution:

x	0	1	2	3	4	5	6
$p(x)$	0.13	0.17	0.26	0.30	0.07	0.05	0.02

(a) What is the probability that in a box of two dozen, exactly three are missing?

(b) What is the probability that the number of missing items is less than four?

(c) What is the probability that exactly 20 items are in the box?

(d) What is the probability that more than 20 items are in the box?

6.5 A large airline keeps track of the number of no-shows for one of its most important commuter flights. Over time the airline has found that the number of ticketed passengers who do not show up is a random variable with this probability distribution:

x	0	1	2	3	4	5	6	7	8
$p(x)$	0.05	0.08	0.13	0.23	0.18	0.13	0.08	0.06	0.06

(a) What is the probability that at least three ticketed passengers do not show up for the flight?

(b) What is the probability that between two and five passengers do not show up for the flight?

(c) What is the probability that not more than six passengers do not show up for the flight?

(d) The aircraft used for the flight has 35 seats. If the airline routinely overbooks the flight by four passengers, what is the probability that on any given day every ticketed passenger who shows up will get a seat?

6.3 The Binomial Probability Distribution

In the preceding section you learned the definitions of a random variable and a probability distribution. Each of the random variables you looked at was described by a probability distribution presented in a table. The random variables represented many different types of data that might be collected in a statistical study. In fact, most of the random variables that we see in the real world fall into specific categories and can be described by a set of special models or probability distributions. We now look at one of these models, the binomial probability distribution, in detail.

6.3.1 THE BINOMIAL MODEL

One of the most common types of data that people collect is the numbers of times that some phenomenon occurs in a sample of given size. For example, you may be interested in the number of people in a sample who are in favor of certain legislation or in the number of people who like a new flavor of ice cream. The sample does not have to consist of people; it may be the number of defective diskettes in a box of ten or the number of times a coin turns up heads in a certain number of tosses. The random variable in each case is the *number of times the phenomenon occurs* in the sample. These types of data are examples of a **binomial random variable.**

> A *binomial random variable* is the number of successes in *n* trials or in a sample of size *n*.

Five characteristics define binomial random variables:

1. An experiment has a fixed number of identical trials. *This is the same as taking a sample of size n. Just think of each trial as selecting the next item for the sample.*

2. The outcome in each trial of the experiment can be classified in one of two ways: a *success*, S (when the phenomenon of interest happens), or a *failure*, F (when the phenomenon of interest does not happen).

3. The probability that a success occurs in any sample element or on any trial of the experiment, *p*, is the same for each element or trial. This means that the probability of a failure, which is $1 - p$, is also constant. *Nothing happens over the course of the experiment to change the probability of a success, such as a change in the population.*

4. The trials of the experiment are independent. *The outcome from one trial does not affect the outcomes of subsequent trials.*

5. The random variable is the number of successes that occur in the *n* trials of the experiment.

In looking for real-world situations that meet the criteria for a binomial random variable when we are sampling from a population, it is difficult to ensure that the trials are independent of each other and that *p*, the probability of a success on any trial or the proportion of successes in the population, remains constant. Keep in mind that we want the trials to be independent because if they are not, then the probability of a success changes from trial to trial. This is a violation of the third characteristic. In fact, the only way to ensure independence when sampling is to sample from a finite population with replacement or from an infinite population. Sampling *with replacement* means that after we sample an item from the population, we return it to the population, which allows for the possibility that we might sample the exact same item again.

The first method, sampling with replacement, is not very appealing from a practical point of view. Since the purpose of statistics is to obtain information, sampling the same item or person repeatedly does not add to the information contained in the sample. The second method, using an infinite population, is more philosophical. Although no populations are truly infinite, many populations are infinite for all practical purposes, such as the number of people in the world (or even in a large country for that matter) or all of the production, past and present, of a machine or factory.

What happens when the population from which we sample is finite and we do not sample with replacement?

EXAMPLE 6.6

CD JEWEL CASES

Sampling Without Replacement

Suppose that you are inspecting a shipment of CD jewel cases for cracks in the cover. The box from which you are sampling contains 30 cases, of which 5 have cracked covers. The first time you select an item from the population, the probability of getting a case with a cracked cover (the proportion of successes in the population) is $\frac{5}{30} = 0.167$. Now, you certainly are not going to sample with replacement in this situation. What would you gain by throwing a case you have already inspected back into the box, particularly if it is defective?

The next time you sample an item from the box, what is the probability of a success? If you are thinking, "It depends on what happened the first time," then you are absolutely correct! Look at the diagram shown here. You can see that as you continue to sample from the box, the probability of obtaining a case with a cracked cover on *that trial* is different from the probability on the previous trial; that is, p changes from trial to trial. The trials are not independent because the outcome of one trial directly affects the outcome of the next!

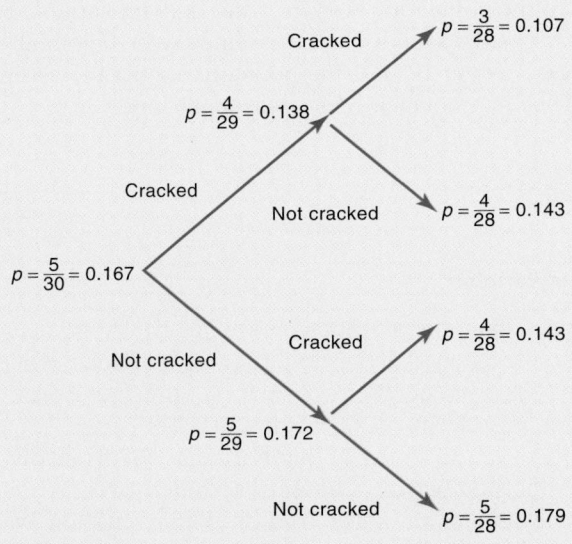

What could we do to fix the problem? The real issue is that when the denominator in the proportion is small and changing, the value of the fraction changes considerably. What happens if the denominator is much larger? Will the change still be noticeable? Suppose that the box contains 3000 jewel cases, of which 500 have cracked covers. In this case, p, the probability or proportion of cracked cases, is still 0.167 when we begin sampling. The next figure illustrates what happens to p as items are sampled and not replaced in the population.

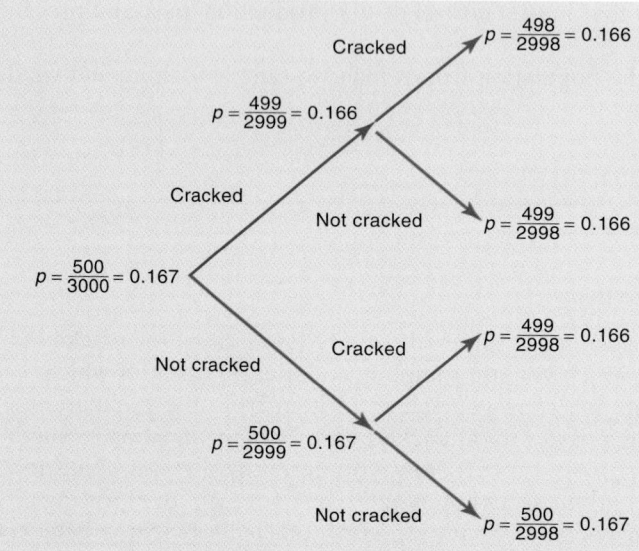

You see that although π changes, the changes are much less noticeable. In fact, if you were writing π to the nearest percent, you would not see the change at all.

Usually the population size must be at least 1000 to ensure that the changes do not occur before the third decimal place. The sample size should be no more than 10% of the population.

Clearly, as the population gets larger, the problem gets smaller. The magnitude of the change in p depends on two things: the size of the population and the size of the sample that is drawn from the population. In general, if the population is large and the sample is small relative to the population size, then the random variable represented by the number of successes in the sample can be assumed to be binomial.

In general, if the sample size is about 10% of the population, then the assumption that the probability of a success does not change is a reasonable one.

EXAMPLE 6.7

TIRE QUALITY

Identifying a Binomial Random Variable

In addition to identifying variables about the vehicles and tires involved in the accidents, the National Highway Traffic Safety Administration asked questions about the nature of the accidents. One yes-or-no question was, Did the vehicle roll over? In general when a bad tire causes an accident, about 6% of vehicles roll over.

We take a random sample of 100 from the 2173 reported accidents involving Firestone tires and count the number of vehicles that rolled over. Is the number of rollovers a binomial random variable? To answer this question, we need to evaluate the variable in light of the five characteristics of the binomial random variable.

1. Yes. There is a fixed number of trials, in this case 100, and the selection is random.
2. Yes. Each trial can result in only one of two outcomes: a success (the vehicle did roll over) or a failure (the vehicle did not roll over).
3. Yes. Because most of the accidents reported occurred between 1998 and 2000, the percentage of rollovers in the actual population did not change according to the national statistics.

4. Yes. Here we have to be careful. The sample of 100 is taken from a population of 2173 reported accidents. If 6% of the accidents involved rollovers, then that would correspond to about 130 successes in the population. In the worst case, the probability of a success could change to 1.3% (if all of the sample winds up being successes) but that is nearly impossible. (If it did happen, then either the sample was not random or else the estimate of 6% was not at all realistic.) Most likely the probability would change from $\frac{130}{2173}$ (5.98%) at the beginning of the sampling to about $\frac{124}{2073}$ (5.98%) when the entire sample of 100 was completed.

5. Yes. The data collected are the numbers of rollovers in the accidents reported.

The number of rollovers is a binomial random variable.

EXAMPLE 6.8

WEARING SEATBELTS

Identifying a Binomial Random Variable

The state of Connecticut has seatbelt laws for drivers and passengers. It is known that 70% of the drivers in Connecticut wear their seatbelts when they drive. A police trap selects a random sample of five drivers and counts the number who are wearing seatbelts. We want to know if the number of drivers who wear seatbelts in the sample of five is a binomial random variable.

Again, we need to relate the situation to the five characteristics of a binomial random variable.

1. Yes. There is a fixed number of identical trials if we assume that each driver is selected randomly.

2. Yes. Each trial can result in only one of two outcomes: a success (the driver is wearing a seatbelt) or a failure (the driver is not wearing a seatbelt).

3. Yes. If we assume that the sample is taken over a short enough period of time that driving habits will not change drastically, then the 70% remains constant.

4. Not necessarily. Usually when people see a police trap, they immediately do things like fasten their seatbelts. It is possible that if the road block is easily visible, then the trials are not independent and the proportion of people who wear seatbelts does not stay constant. However, if police at the road block were checking for some other violation like an expired emissions sticker, then the driver could do nothing to change the outcome. The population of the drivers in Connecticut is very large and the sample size is definitely small relative to the population, so the trials are independent and the proportion of successes in the population remains essentially constant.

5. Yes. The data collected are the number of people who are wearing seatbelts in the sample of five selected at random.

The number of people who wear seatbelts might not be a binomial random variable.

It is important that you learn to recognize problems to which the binomial distribution applies. It is easy enough to do in a textbook (after some practice), but when you are actually practicing statistics, you will need to know what to look for. Keep in mind the five characteristics of a binomial random variable when you are looking at both problems in this book and data that have been collected in real statistical studies.

✓ TRY IT NOW!

Kids Count
Recognizing a Binomial Random Variable

The Annie E. Casey Foundation is a private charitable organization dedicated to helping disadvantaged children. Through a program called Kids Count they conduct yearly surveys of the status of children in the United States. According to the *2001 Kids Count Data Book* about 80% of children under the age of 2 in the United States are immunized. Suppose that a public health agency in Indiana wanted to see how Indiana children compare with the national average. They select a random sample of 25 children under age 2 and count the number who are immunized. Does the number of immunized children qualify as a binomial probability distribution?

6.3.2 CALCULATING THE BINOMIAL PROBABILITY DISTRIBUTION

Remember! We use parameters for populations, and a probability distribution is a model for a population.

Now that you can identify binomial random variables, you need to know how to find their probability distributions. The probability distribution for a binomial random variable depends on two quantities or parameters: (1) n, the number of trials or the sample size, and (2) p, the proportion of successes in the population. The random variable X is the number of successes in n trials of the experiment. The probability distribution of X is determined by this formula:

Binomial probability distribution

$$p(x) = \frac{n!}{(x!)(n-x)!} p^x (1-p)^{n-x} \quad \text{for } x = 0, 1, 2, \ldots, n$$

The terms that contain ! are known as factorials; n! is equal to $n \times (n-1) \times \cdots \times 1$. The other factorial terms are calculated in a similar manner.

The formula considers the sample size, n, the probability of a success, p, and the value of X that you are interested in, x. The number of trials in the experiment defines the possible values that the random variable can have. If there are n trials of the experiment, then the smallest number of successes that can occur is 0 and the largest number of successes is n. Thus, a binomial random variable, X, can have values $x = 0$, 1, 2, . . . , n. The probability of a success on any trial, p, determines how the probabilities are distributed over the values of X. When p is large, you expect to see a lot of successes in the sample, so higher values of X have higher probabilities. When p is small, lower values of X have higher probabilities.

You do not really need to know how to use the formula to find the probability distribution of a binomial random variable. There are tables for this purpose for many different combinations of n and p. A good set of these tables is found in Appendix A at the end of this book.

TABLE 6.2

PORTION OF THE BINOMIAL TABLE FOR $n = 5$

$n = 5$						p						
x	0.05	0.10	0.20	0.25	. . .	0.50	0.60	0.70	0.75	. . .	0.90	0.95
0	0.774	0.590	0.328	0.237	. . .	0.031	0.010	0.002	0.001	. . .	0.000	0.000
1	0.204	0.328	0.410	0.396	. . .	0.156	0.077	0.028	0.015	. . .	0.000	0.000
2	0.021	0.073	0.205	0.264	. . .	0.313	0.230	0.132	0.088	. . .	0.008	0.001
3	0.001	0.008	0.051	0.088	. . .	0.313	0.346	0.309	0.264	. . .	0.073	0.021
4	0.000	0.000	0.006	0.015	. . .	0.156	0.259	0.360	0.396	. . .	0.328	0.204
5	0.000	0.000	0.000	0.001	. . .	0.031	0.078	0.168	0.237	. . .	0.590	0.774

Binomial Probability Tables

The probability distribution tables for the binomial distribution are classified according to the value of n, the number of trials of the experiment. There are tables for values of n from 5 to 30, and each table covers a range of values for p from 0.05 to 0.95. A portion of the table for $n = 5$ is shown in Table 6.2. The two columns that are shaded make up the probability distribution table for $n = 5$ and $p = 0.70$, or 70%. If you just take those columns and transpose them as shown in Table 6.3, the result looks like every probability distribution table you have seen before!

Except for recognizing binomial random variables and using the tables, you have nothing new to learn in order to answer questions involving binomial random variables.

Most statistical software packages, such as Minitab and SPSS, and many spreadsheets, such as Excel, find binomial probabilities.

Don't forget that decimals and percentages are equivalent!

EXAMPLE 6.9

WEARING SEATBELTS

Using Binomial Probability Tables

In Example 6.8 about Connecticut drivers wearing seatbelts, we identified the random variable as a binomial random variable with $n = 5$ and $p = 0.70$. We want to find the probability that, in the sample of five drivers, exactly three are wearing seatbelts. From Table 6.2 or 6.3, we see that p(3) = 0.309.

Now, what is the probability that, in the sample of five drivers, at most two are wearing seatbelts? Again, from Table 6.2 or 6.3,

$$P(X \leq 2) = p(0) + p(1) + p(2) = 0.002 + 0.028 + 0.132 = 0.162$$

TABLE 6.3

BINOMIAL PROBABILITY DISTRIBUTION FOR $n = 5$ AND $p = 0.70$

x	0	1	2	3	4	5
$p(x)$	0.002	0.028	0.132	0.309	0.360	0.168

You can see that nothing new is involved in solving problems that involve the binomial distribution. The steps that require some thinking and work are recognizing the problem as binomial, identifying the parameters n and p for the particular problem, and finding the correct table to use.

It is important to recognize that the definition of a success is critical to problem solving with the binomial distribution. A success is not necessarily always something good, nor does it have to stay the same in a given problem. To determine what a success is, you must look at the question to be answered, define a success, and use the appropriate value of p.

EXAMPLE 6.10

PARKING TICKETS

Calculating Binomial Probabilities

A local watchdog agency has been looking at parking problems at the city court building. It estimates that 40% of all cars parked in the metered lot receive parking tickets for meter violations. The agency decides to take a random sample of ten cars from the lot and see whether they have a parking ticket on the windshield. We want to find the probability that, of the ten cars sampled, exactly six have parking tickets. For this problem a success is having a parking ticket, and so we use the table with $n = 10$ and $p = 0.40$ to find that $P(X = 6) = 0.111$.

Now, what is the probability that between four and seven cars inclusive do not have parking tickets? For this problem a success is *not* having a parking ticket. Since 40% of the cars have parking tickets, we can use the definition of the complement to find that $100\% - 40\% = 60\%$ do not have parking tickets. We use the table with $n = 10$ and $p = 0.60$ to get $P(4 \leqslant X \leqslant 7) = 0.111 + 0.201 + 0.251 + 0.215 = 0.778$.

If the watchdog group found in its sample of ten cars that none of the cars had parking tickets, would you consider the estimate of 40% reasonable? With $n = 10$ and $p = 0.40$, the probability that no cars have tickets, $P(X = 0)$, is 0.006, which is very small. Thus, it is highly unlikely that, if 40% of the cars get tickets, the sample of ten would have no cars with tickets. The estimate seems to be high.

Caution! No matter what you may think, it is NOT easier to keep using p = 0.40 and try to change the question to be in terms of cars having parking tickets!

The preceding example is a preview of the next topic you will study in statistics. You first learned how to describe data that are collected and how to calculate different sample statistics. Now you are learning about probability models and how they can be used to determine the likelihood that different events will occur when we assume some parameters for our population data. In Example 6.10 we looked at how well what we observe (the data) fit the probability model (binomial with $p = 40\%$ and $n = 10$). We used the probability that such an event would happen to come to the conclusion that the model did not seem appropriate. In the remainder of this book you will learn the more formal methods of hypothesis testing to accomplish this same thing.

TRY IT NOW!

Kids Count

Solving Binomial Probability Problems

The data from Kids Count estimate that 80% of all children under the age of 2 are immunized. A public health agency in Indiana takes a random sample of 25 children under 2 and counts the number who have been immunized. Define a success for this problem.

Describe the random variable, X, in words.

What are the parameters of the binomial distribution?

Find the probability that, in the sample of 25 children, more than 19 have been immunized.

What is the probability that between 16 and 21 inclusive have been immunized.

6.3.3   THE MEAN AND STANDARD DEVIATION OF THE BINOMIAL DISTRIBUTION

Because probability distributions are models of populations and random variables are numerical, it makes sense that, just like quantitative sample data, they have means and standard deviations. There are some general formulas for calculating the mean and standard deviation of a random variable, but we concentrate on probability distributions.

Since probability distributions are population models, their means and standard deviations are parameters represented by the Greek letters μ and σ. In particular, for a binomial random variable, X, the **mean, μ,** and the **standard deviation, σ,** are found using the following formulas:

$$\mu = np \quad \text{and} \quad \sigma = \sqrt{np(1 - p)}$$

Binomial mean and standard deviation

You see that the mean and standard deviation depend on the parameters of the probability distribution, n and p. The formula for the mean is actually intuitive. If you knew, for example, that 40% of a certain population wore eyeglasses, and you took a sample of ten people from that population, how many of the ten would you *expect* to wear glasses? Naturally you would take 40% of the ten to get four people who wear glasses. From the formula for μ, we get

$$\mu = np = (10)(0.40) = 4 \text{ people}$$

The formula for the standard deviation is not at all intuitive and deriving it is beyond the scope of this text. To see how it is used, we can calculate the standard deviation of our example and find that

$$\sigma = \sqrt{np(1 - p)} = \sqrt{(10)(0.40)(0.60)} = 1.55 \text{ people}$$

Thus, we know that if $n = 10$ and $p = 0.40$, then the number of people who wear glasses is a random variable with a mean of 4 and a standard deviation of 1.55.

EXAMPLE 6.11

TIRE QUALITY

Calculating the Mean and Standard Deviation of a Binomial Random Variable

With a sample of $n = 100$ we cannot use the tables to determine the probabilities for the number of vehicles that rolled over. Most statistical software packages calculate probability distributions for binomial random variables for any values of n and p, but we can find the mean and standard deviation of this random variable without any problems.

ANS, IMMUNIZED: X = THE NUMBER OF CHILDREN IN THE 25 WHO HAVE BEEN IMMUNIZED; $n = 25$, $p = 0.80$; 0.618; 0.748

The mean represents the expected number of vehicles that rolled over in the sample of 100. Remember that the national percentage is 6%. That is, with $n = 100$ and $p = 0.06$, the mean is

$$\mu = np = (100)(0.06) = 6.0 \text{ vehicles}$$

The standard deviation of the number of vehicles in 100 that rolled over is

$$\sigma = \sqrt{np(1 - p)} = \sqrt{(100)(0.06)(0.94)} = 2.37 \text{ vehicles}$$

EXAMPLE 6.12

WEARING SEATBELTS

Finding the Mean and Standard Deviation of a Binomial Random Variable

The Connecticut Department of Transportation would like to know how many drivers the state troopers should expect to find wearing their seatbelts at a checkpoint. It decides to calculate the mean of the binomial random variable. In this case, $n = 5$ and $p = 0.70$, so it calculates

$$\mu = np = (5)(0.70) = 3.5 \text{ people}$$

The department would also like to know the standard deviation of the number that would be wearing seatbelts. In this case,

$$\sigma = \sqrt{np(1 - p)} = \sqrt{(5)(0.70)(1 - 0.70)} = \sqrt{(5)(0.70)(0.30)} = \sqrt{1.05} = 1.02 \text{ people}$$

Remember that the mean and standard deviation tell us something about how random variables behave. In particular, they tell us where the center of the probability distribution is located and how much the random variable varies around that center.

✔ TRY IT NOW!

Kids Count

Calculating the Mean and Standard Deviation of a Binomial Random Variable

The public health agency in Indiana that was looking at immunization in children under 2 years of age wants to know the mean and standard deviation for the binomial random variable with $n = 25$ and $p = 0.80$. Find the mean and standard deviation of the number of children under age 2 in 25 who have been immunized.

How can we use our knowledge of the mean and standard deviation? We can compare our knowledge of what *should* happen with the reality of what *did* happen to get some idea of how well the theory fits reality. This is the basis for the work you will do in the remainder of this book—*inferential statistics*. Right now, we look at this work in an informal way.

Ans. $\mu = 20$, $\sigma = 2$

EXAMPLE 6.13

TIRE QUALITY

Using the Mean and the Standard Deviation

The NHTSA wonders whether the random sample of 100 vehicles from the Firestone tire complaints is comparable to the national figures in terms of rollovers. It already has the mean and standard deviation for its model, so it decides to compare the model with the actual data.

The mean of the binomial random variable with $n = 100$ and $p = 0.006$ is 6.0 and the standard deviation is 2.37. From the actual data, the NHTSA analysts find that the number of vehicles that rolled over was 7. Clearly, 7 is not equal to 6, but how different is it?

If the analysts consider the standard deviation as a first check, they can calculate the number of standard deviations that their data value is from the mean:

$$Z = \frac{7 - 6}{2.37} = 0.42$$

They know that the empirical rule holds only for symmetric distributions and they have no idea whether that applies here, but they do know that 0.42 standard deviation away from the mean is not unusual.

As a second check the analysts use the data to estimate the probability that a vehicle rolled over. They found 7 rollovers in their sample of 100, so they estimate the probability that a vehicle will roll over to be $\frac{7}{100} = 0.07$, or approximately 7%. This is different from the 6% they were expecting, but the difference may just be a result of sampling.

As a third check they use a statistical software package to calculate the probability of having 7 successes in a sample of 100 when $p = 0.06$. They find that the probability is 0.142, which is about 14%. It seems that this sample fits the national data.

EXAMPLE 6.14

PARKING TICKETS

Using the Mean and Standard Deviation

The watchdog agency that is looking at the problem of parking tickets wants to know what it should expect to see in the data it collects. The analysts in the agency decide to calculate the mean and standard deviation of the binomial random variable:

$$\mu = (10)(0.40) = 4 \text{ cars}$$

$$\sigma = \sqrt{np(1 - p)} = \sqrt{(10)(0.40)(0.60)} = \sqrt{2.4} = 1.55 \text{ cars}$$

This tells the analysts that they should expect to find four cars with parking tickets in a sample of ten cars. Wondering how much the number of cars might vary, they decide to find the probability that the number of tickets will be within two standard deviations of the mean. To do this they calculate $\mu \pm 2\sigma = 4 \pm (2)(1.55) = 4 \pm 3.1 = (0.9, 7.1)$. They want to find the probability that the random variable is between these two values, or $P(0.90 < X < 7.1)$.

How will they do this? The binomial tables certainly do not include numbers like 0.90 and 7.1! The analysts realize that they will have to convert the problem to the nearest integer values that satisfy the probability expression, so they need to find $P(1 \leq X \leq 7)$. From the tables for $n = 10$ and $p = 0.40$, they find that the answer is 0.981. They interpret this to mean that 98.1% of the time, the number of tickets they find will be between 1 and 7 inclusive. This gives them a good idea of what to expect in their data.

Comparing what *should* happen with what *does* happen is a critical idea in statistics. In the next few chapters we will develop a formal method for doing this called *hypothesis testing*.

Discovery Exercise 6.1

EXPLORING THE BINOMIAL DISTRIBUTION

Dear Mom and Dad: Send cash
According to USA Today, *70% of college students receive spending money from their parents when at school.*

For this exercise you need to simulate selecting 30 samples of 5 students from this population of college students and observe whether they receive spending money from their parents. Consider the successful outcome to be "receives money" with $p = 0.70$ and the failure outcome to be "does not receive money." If your instructor does not provide you with a method, you can take ten (small) pieces of paper and write an S on seven of them and an F on three of them. Put the papers in a bag or other container and select one at random to simulate an observation. *Note*: Be sure to replace the paper each time or p will not always be 0.70.

1. Record an S when you select a student who receives money from his/her parents and an F when you select a student who does not receive spending money from his/her parents. For each sample, record the number of successes you sampled.

   In the last column compute a running estimate of p. Remember that p is the probability of a successful outcome. In this case, p is known to be 0.70. Let's see how close the estimate gets to 0.70 as the sample size increases. So, after the first sample is selected, your estimate of p is simply the number of successes divided by 5. After the second sample is selected, your estimate of p is the number of successes in both samples divided by 10, and so forth.

Sample number	Observation number 1	2	3	4	5	X = Total number of successes in n = 5 trials	Running total number of successes	Running total of number sampled	Estimate of p
1								5	
2								10	
3								15	
4								20	
5								25	
6								30	
7								35	
8								40	
9								45	

10							50	
11							55	
12							60	
13							65	
14							70	
15							75	
16							80	
17							85	
18							90	
19							95	
20							100	
21							105	
22							110	
23							115	
24							120	
25							125	
26							130	
27							135	
28							140	
29							145	
30							150	

2. Using the results of the seventh column of the table, construct a relative frequency distribution for the number of successes in five trials.

X = number of successes	Number of samples that had X successes	Observed relative frequency = $X/30$	Theoretical relative frequency
0			
1			
2			
3			
4			
5			
Total	**30**	**1.00**	**1.00**

(continued)

3. Display the relative frequency as a bar chart.

4. Complete the theoretical relative frequency column in the table by using the binomial table with $n = 5$ and $p = 0.70$.

5. Display the binomial distribution in a bar chart.

6. Compare the bar chart from step 3 with the binomial distribution displayed in step 5. How do they compare? Why are they different?

7. The graph below has a line at the theoretical value of p, 0.70. Graph your estimates of p for each sample on the same graph. What happens to your estimate as the sample size increases?

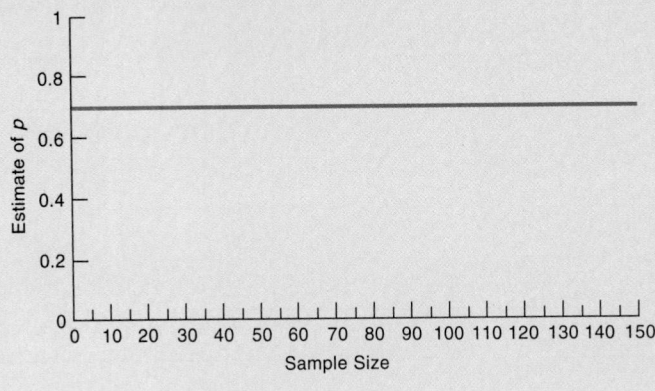

6.3.4 EXPLORING THE BINOMIAL DISTRIBUTION

You are becoming familiar with using the binomial probability tables and solving binomial probability problems, but you still may not understand what role the parameters n and p play in determining what the probability distribution of a binomial random variable looks like.

In Figure 6.1 you see the effects of changing the value of the parameter p for a fixed value of n. The graphs illustrate that when p is small, it is more likely that the

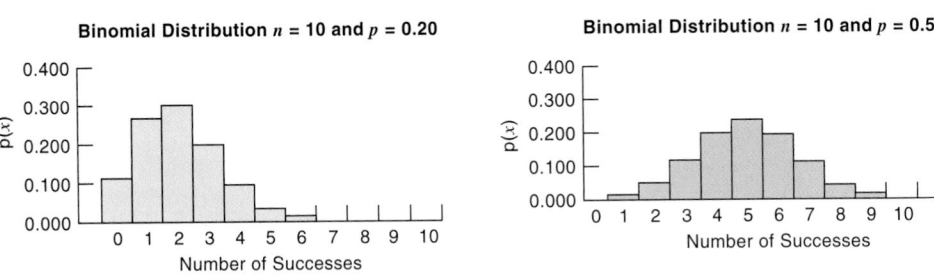

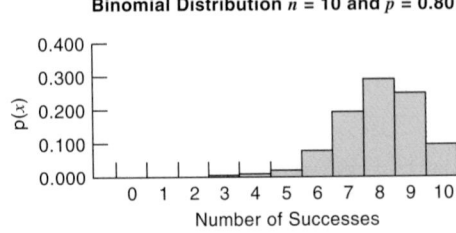

FIGURE 6.1 **Effects of changing p when n is fixed**

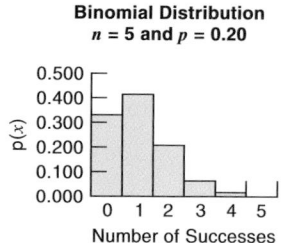

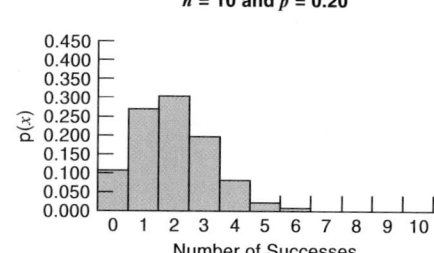

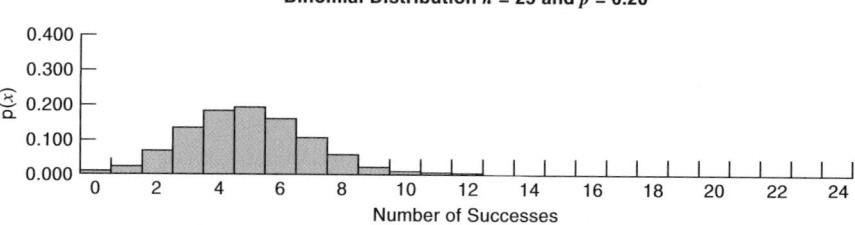

FIGURE 6.2 **Effects of changing *n* for a fixed value of *p***

number of successes will be small and the distribution skews to the right. When *p* is large, it is more likely that the number of successes will be large and the distribution is skewed to the left. You can also see that when *p* is equal to 0.50, the distribution is symmetric. From the binomial distributions for *n* = 10, you will notice, not surprisingly, that the probability distributions for *p* = 0.20 and *p* = 0.80 are mirror images of each other.

Where does the mean fit into the picture? The mean is the value that you expect to occur; that is, it is the most likely outcome. In Figure 6.1 you see that in the first probability histogram the highest bar is for *X* = 2 and the mean of a binomial random variable with *n* = 10 and *p* = 0.20 is $\mu = (10)(0.20) = 2$.

Figure 6.2 shows the effects of changing the value of the parameter *n* for a fixed value of *p*. The *y*-axis scale for each histogram is approximately the same. You can see that, for the most part, the shape of the distribution remains the same as *n* varies, but the number of possible values and the probability of each value of *X* (bars in the histogram) change. In fact, as *n* gets larger, the individual probabilities get smaller. This makes sense because we know that the probabilities must sum to 1, and if there are more values of *X*, then each value gets a smaller share of the total.

The binomial distribution is not the only discrete probability distribution that deals with the number of successes. The Poisson distribution is used when we are interested in the number of successes in a time interval or region. To find out more about this, refer to the extended discussion of this topic in Appendix D of the text.

6.3.5 EXERCISES – LEARNING IT!

6.6 The board of realtors of a small city reports that 80% of the homes that are sold have been on the market for more than 6 months. The board takes a random sample of 15 homes that have recently been sold and counts the number that were on the market longer than 6 months. What is the probability that of the 15 houses in the sample:

(a) Less than 12 have been on the market longer than 6 months?

(b) Between 8 and 13 have been on the market longer than 6 months?

(c) At least 10 have been on the market longer than 6 months?

(d) At most 4 have been on the market for less than 6 months?

6.7 The Department of Transportation for a city has found that 25% of all parking tickets issued are not paid within 1 month of issue. The department takes a random sample of 20 parking tickets that were issued 1 month ago and counts the number that have not been paid. What is the probability that:

(a) At most 5 have not been paid?

(b) Between 4 and 8 inclusive have not been paid?

(c) More than 7 have not been paid?

(d) At least 6 have been paid?

6.8 It is estimated that in 1998 10% of teens between the ages of 16 and 19 were high school dropouts. A random sample of 5 people between ages 16 and 19 is taken.

(a) What is the probability that 4 of the 5 people were high school dropouts?

(b) What is the mean number of high school dropouts in the sample of 5?

(c) What is the standard deviation of the number of high school dropouts in a sample of size 5?

6.9 A poll taken by the Gallup organization in June/July 2001 estimated that 70% of adults thought their state government should pass a law making it illegal to talk on handheld cell phones while driving. Several months later another poll of 25 adults was taken and asked the same question. Nothing happened to change public opinion between the polls.

(a) What is the probability that at most 5 adults did not think that such a law should be passed?

(b) What is the probability that between 10 and 17 approved of such a law?

(c) What is the mean number of adults in 25 that approved a law banning cell phone use while driving?

(d) What is the standard deviation of the number of adults in 25 that approved banning cell phone use while driving?

(e) What is the probability that the number of adults in 25 that approved of the law is within three standard deviations of the mean?

6.10 Companies are having problems with employees playing computer games at work. As the size and complexity of such games increase, computer system administrators find that network resources are being drained and that the games are using more and more hard disk space. A recent survey across various industries revealed that 30% of workers said that the last computer game they played had been played at work. A random sample of 15 employees is taken.

(a) What is the probability that fewer than six said they played their last computer game at work?

(b) What is the probability that at least four said they played their last computer game at work?

(c) What is the probability that at most ten said they did not play their last computer game at work?

(d) What is the expected number of employees in a sample of size 15 who played their last computer game at work? What is the standard deviation of the number of employees in 15 who played their last computer game at work?

(e) What is the probability that the number of employees in 15 who played their last computer game at work is within two standard deviations of the mean?

6.4 Continuous Random Variables

In Chapter 1 you learned about different types of data. In particular, you learned that the two types of quantitative data are discrete and continuous. Discrete data are integer and are often a count of the number of times that something happens. The binomial distribution that you just studied is a good probability model for discrete data.

For continuous data the variable of interest can take on any one of an infinite number of values over some *interval* on the real number line. Continuous data usually result from taking data on a *measurement*. Examples are the heights of students and grade point averages (GPAs). The actual number of values that we can obtain is limited by the measuring instrument (we choose to measure height to the nearest inch or report GPA to two decimal places), but any value in the interval is valid.

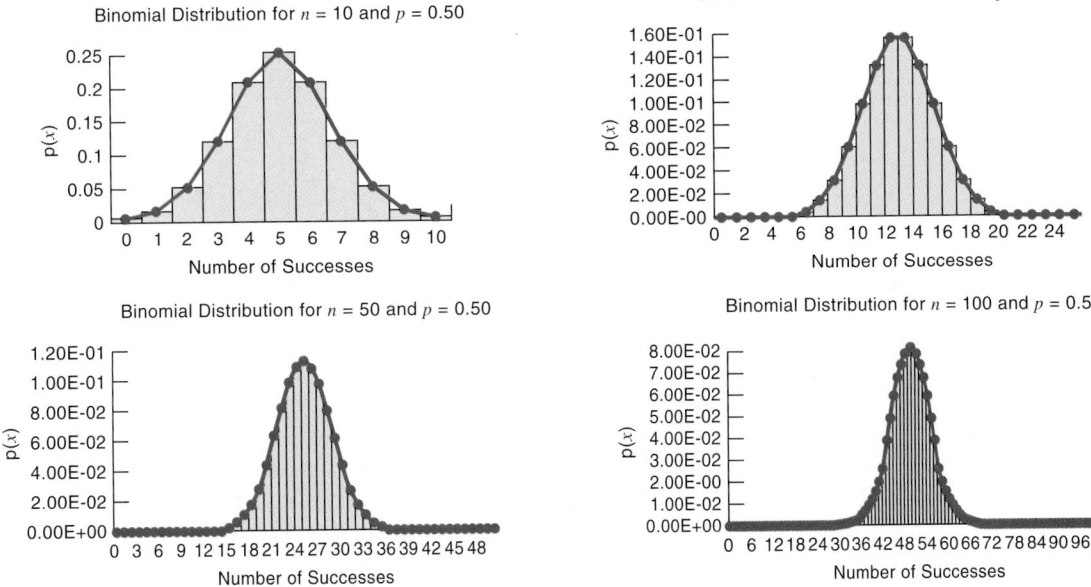

FIGURE 6.3 **Binomial distributions for large values of *n***

For discrete random variables you learned that to find the probabilities associated with the random variable, you simply add the relevant individual probabilities. For continuous random variables you have to shift your thinking a little bit to see how probabilities are calculated.

If you look at the effects of changing the value of *n* in the binomial distribution, you can see that as *n* increased, the number of possible values of *X* increased and the individual probabilities got smaller. When *n* = 10 it is not that tedious to calculate $P(X \geq 4)$, but when *n* = 20 or *n* = 25 it is considerably more tedious to calculate it directly. You have to add many more terms and they are all very small numbers. If you let *n* get very large—say, *n* = 100—it would be tiring indeed! Figure 6.3 shows the binomial distributions for $\pi = 0.50$ and *n* = 10, 25, 50, and 100.

If we connect the tops of all of the histogram bars in the graphs, then as *n* gets larger, the curve becomes smooth. For continuous random variables, which can take on many more than 100 possible values, the probability distribution is called a **probability density function, *f*(*x*)**, and is represented by a smooth curve like the one in Figure 6.4.

> A *probability density function, f(x),* is a smooth curve that represents the probability distribution of a continuous random variable.

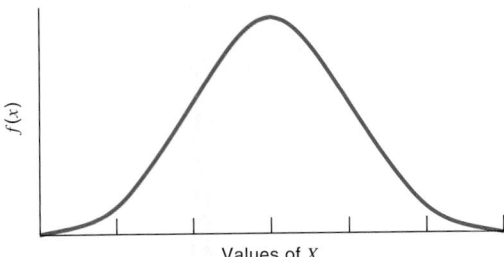

FIGURE 6.4 **Probability distribution for a continuous random variable**

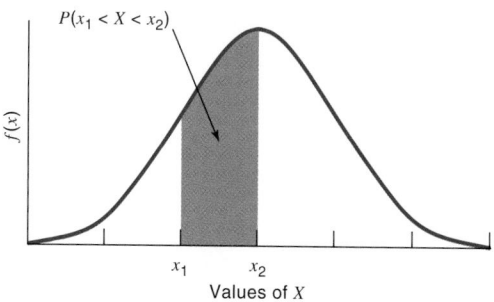

FIGURE 6.5 **Probability represented by an area under the curve**

You might be wondering how this curve relates to probabilities because no corresponding probability table with numbers is involved. How do you find the probabilities that you are interested in? These are good questions.

When we increased the value of n for the binomial distribution, the individual probabilities got much smaller. In fact, from the scales on the y axes in Figure 6.3, you can see that for $n = 100$, the *largest* number on the graph is 0.08! Again, this makes sense because the total probability is always equal to 1, and when you have to divide it up over more and more possible values, each individual probability gets closer and closer to 0.

For continuous random variables, $P(x_1 < X < x_2)$ is exactly the same as $P(x_1 \leq X \leq x_2)$ because $P(X = x) = 0$.

In fact, for continuous random variables we can no longer talk about the probability that the random variable will assume one particular value, $P(X = x)$, because this is equal to 0. Instead, we talk about the probability that the random variable will take on any one of the values over an *interval* of interest; that is, we can find $P(x_1 < X < x_2)$. This probability is represented by the area under the probability density curve as shown in Figure 6.5. Finding probabilities for a particular probability density involves calculus, but for certain distributions the probabilities have already been calculated and tabulated.

In the next section we look at one particular continuous probability distribution that serves as a model for many natural phenomena. It is known as the *normal distribution*.

6.5 The Normal Probability Distribution

6.5.1 THE NORMAL CURVE

The normal probability distribution is the symmetric, bell-shaped curve shown in Figure 6.6. It is usually called the *normal curve*. This is the same curve you saw when we discussed the empirical rule.

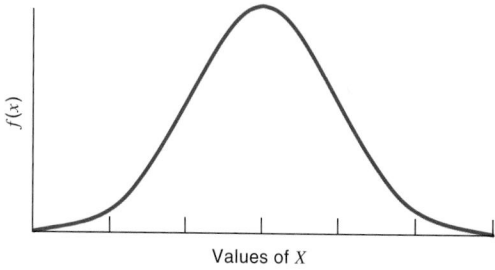

FIGURE 6.6 **Normal probability curve**

In reality, most measurement data are modeled very well by this distribution. The actual formula for the normal probability density is

$$f(x) = \frac{1}{\sigma\sqrt{2\pi}}\, e^{-(x-\mu)^2/2\sigma^2} \quad \text{for } -\infty < x < \infty$$

*Normal probability
density*

where $e = 2.71828\ldots$ and $\pi = 3.14159\ldots$. You can see that the probability density for a value, x, relies on two **parameters,** μ and σ. The normal distribution is the ultimate symmetric, bell-shaped distribution.

> For a normal random variable, the parameter μ is the ***mean*** of the normal random variable, X, and σ is the ***standard deviation.***

When we refer to a normally distributed random variable, we often use a special notation:

$$X \sim N(\mu, \sigma)$$

This shorthand is equivalent to saying that the random variable X is normally distributed with a mean of μ and a standard deviation of σ.

In our discussion of the binomial probability distribution, you learned that the parameters, in that case n and p, define how the probability is distributed over the set of possible values of X. That is, the parameters define the shape of the probability histogram. In a similar way, μ and σ define the way the normal distribution looks.

You have come across the mean in two other situations. In the section on descriptive statistics, you learned that the mean is a measure of central tendency, or an estimate of the most typical data value. Then, in the section on the binomial distribution, we said that the mean is the most likely value, or the one that we expect to happen most often. For a normally distributed random variable, the mean is the center of the distribution. It is the value that determines the location of the probability density curve on the number line.

You have also learned about the standard deviation. You know that the standard deviation measures how far the data are spread out around the mean. For a normally distributed random variable, σ determines how spread out the probability density is around its center. You may remember that in Chapter 3 we used the mean and the standard deviation to create a mental image of the distribution of the data. The empirical rule told us that for a symmetric distribution, virtually all of the data fall within three standard deviations of the mean. Since the normal distribution is the fundamental symmetric curve, we can use this same idea to get a picture of a normal probability distribution.

Figure 6.7 shows the effect that changes in μ and σ have on the way the normal curve looks. As the standard deviation increases, the distribution gets more spread out around its center. For the curve with a mean of 30 and a standard deviation of 5, the

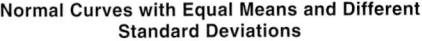

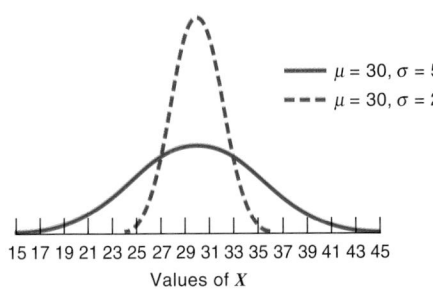

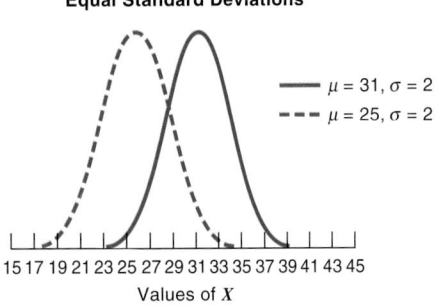

FIGURE 6.7   Effects of changing μ and σ on the normal curve

ends of the curve are at around 15 and 45, three standard deviations away from the mean! As the mean increases, the distribution moves to the right (in the direction of increasing values of X), and as the mean decreases, the distribution moves to the left. Notice that although the appearance of the normal curve changes when μ and σ are varied, the basic shape does not change. The curve is always bell-shaped and symmetric. This is an important feature of the normal probability distribution.

✔ **TRY IT NOW!**

Food Expenditures
Looking at the Normal Curve

The amount of money that a person working in a large city spends each week for lunch is a normally distributed random variable. For professional and management personnel, the random variable has a mean of \$35 and a standard deviation of \$5. For hourly employees, the mean is \$30 with a standard deviation of \$2. Sketch the normal curves for each of the two random variables on the same graph.

6.5.2 THE STANDARD NORMAL CURVE

In our discussion of continuous random variables we said that the probability that the random variable takes on any value in an interval is equal to the corresponding area under the probability density curve. Finding these probabilities, however, requires the use of integral calculus, which is beyond the scope of this book. How do we find the probabilities we are interested in?

The formula for finding probabilities with the binomial probability distribution was also complicated, and we solved the problem by using tables for different sets of values for n and p. Because p is constrained to be between 0 and 1, it is not hard to get tables for many different values of n.

Using probability tables for the normal probability distribution seems like a good idea, but there is a practical limitation. Unlike n and p, the values of μ and σ are not at all constrained. Depending on the variable of interest, there are an infinite number of pairs of μ and σ that we might be interested in. It is not feasible to have a table for every possible pair.

The problem is solved by a method that transforms every normally distributed random variable into a single, standard normal random variable. This standard normal random variable is known as **Z** and has a mean of 0 and a standard deviation of 1. We can use the shorthand notation to write that as $Z \sim N(0, 1)$. Once we transform the problem, we can use a single normal probability table, a **standard normal table,** to solve it.

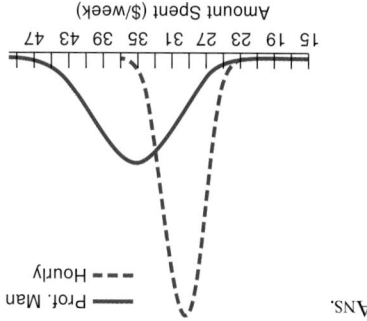

Ans.

> A *standard normal random variable, Z,* is normally distributed with a mean of 0 and a standard deviation of 1; $Z \sim N(0, 1)$.

> A *standard normal table* is a table of probabilities for a Z random variable.

How does the transform work? Remember that the mean of a normally distributed random variable locates the normal curve on the number line. What we want to do is move any arbitrary normal random variable that is centered at μ down the number line to center at 0. It is not hard to see that this can be accomplished by subtracting the value of μ from the random variable X.

This manipulation makes sense. For example, if you think about it for a minute, if the class average on an exam were a 70 and you wanted to make it a 75, you would *add* 5 points to everyone's grade. Similarly, to move the mean from μ to 0, you *subtract* μ from the values.

The change from a standard deviation of σ to a standard deviation of 1 is not difficult either. Essentially, we need to scale the probability distribution so that the total area under the curve stays equal to 1 (one of the rules of probability). A value of X that is two standard deviations away from μ for the original random variable must stay two standard deviations away from 0 when it is transformed to Z. To accomplish this we describe the distance that X is from μ, $X - \mu$, in terms of the number of standard deviations it is away from μ. The transform from X to Z is then given by

$$Z = \frac{X - \mu}{\sigma}$$

where $X \sim N(\mu, \sigma)$ and $Z \sim N(0, 1)$. Now an example will help you see what is happening.

Z-score
This is the population version of the Z-scores you calculated in Chapter 3.

EXAMPLE 6.15

APTITUDE SCORES

Transforming from *X* to *Z*

Scores on an aptitude test given by a community job training and placement agency are normally distributed with a mean of 75 points and a standard deviation of 5 points. The directors of the agency are interested in knowing what proportion of the people who take the test score between 65 and 85 points.

Our first step is to *draw a picture* to represent the problem we are trying to solve.

Note: It may seem like a waste of time to draw the pictures, but a little time spent here will avoid a LOT of time wasted later on.

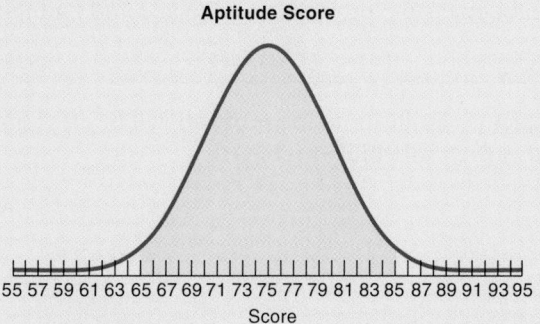

Aptitude Score

55 57 59 61 63 65 67 69 71 73 75 77 79 81 83 85 87 89 91 93 95
Score

The picture shows the distribution of the scores and the area under the curve that corresponds to the problem the directors want to solve. We know that to solve the problem we must transform it into a standard normal problem. We have to transform each of the values of X into the corresponding values of Z. In this case,

$$\text{for } X = 65: \quad Z = \frac{65 - 75}{5} = -2$$

$$\text{for } X = 85: \quad Z = \frac{85 - 75}{5} = 2$$

Caution! It is easy to mix up X and μ when you use the formula for Z. Use the sign of the answer you get to check for errors. Make sure that the sign of your Z agrees with the location of X relative to μ.

Neither of these values of Z should be a surprise because you know that a Z-score measures the number of standard deviations that a value is from its mean. The score of 65 is 10 points, or exactly two standard deviations, *below* the mean of 75, and the score of 85 is exactly two standard deviations *above* the mean of 75. The sign on the value of Z indicates whether the value of X is above (+) or below (−) the mean.

For the picture of the transformed problem, you can see that it looks the same as the original picture except for the scale on the x axis. It is not necessary to draw both X and Z pictures.

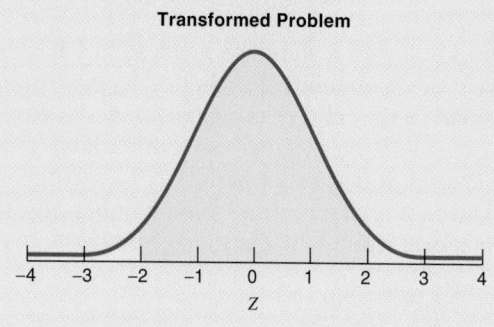

Transforming problems about normally distributed random variables into problems about standard normals is not difficult. Just as in all probability problems, the hardest part is understanding what problems you need to solve.

✔ TRY IT NOW!

Speed Reading
Transforming from X to Z

The number of pages of a statistics textbook that a student can read in a given hour is a normally distributed random variable with a mean of 7 pages and a standard deviation of 1.5 pages. One professor who uses the book wants to know the probability that a randomly selected student can read more than 8.5 pages of the textbook in an hour. Draw a picture that depicts the problem to be solved and find the Z value necessary to solve the problem.

The importance of drawing a picture to illustrate the problem will be clear in the next section when we look at how to solve probability problems.

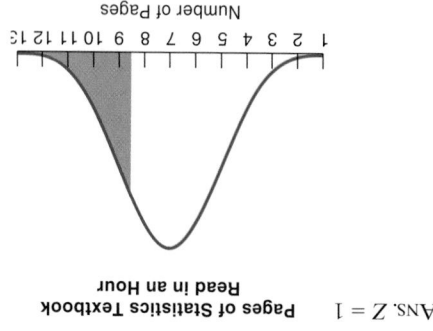

Ans. $Z = 1$

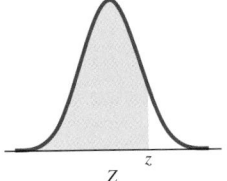

FIGURE 6.8 **Probability given by the standard normal table**

6.5.3 THE STANDARD NORMAL TABLES

The tables we used for the binomial distribution listed $P(X = x)$ for different combinations of n and p. To find quantities like $P(x_1 \leq X \leq x_2)$ we summed the probabilities for all the values of X that were included in the interval. We know that $P(X = x) = 0$ for continuous random variables, so the probability tables for the standard normal probability distribution must contain some kind of **interval probability.**

> An *interval probability* gives the probability that a random variable takes on a value *between* two given values $P(x_1 \leq X \leq x_2)$.

Several different types of standard normal tables can be used. Each gives probabilities for different types of intervals. The tables in this book give $P(Z < z)$; that is, each entry in the table is the probability that the random variable Z takes on a value that is less than some value z. Figure 6.8 corresponds to this situation.

The table in Appendix A has two parts, for negative and positive values of Z, to make solving problems less cumbersome. All tables for the standard normal probabilities are set up the same way. The values of Z are defined to two decimal places, and the corresponding probability is located at the intersection of the appropriate row and column of the table. If you think of the Z value as always being in the form $X.XX$, then the rows contain the first two digits $X.X$ and the columns contain the last digit $0.0X$. It sounds complicated but it is easy once you are familiar with it. A portion of the table is shown in Table 6.4.

TABLE 6.4

THE STANDARD NORMAL TABLE

z	Second decimal place									
	0.00	**0.01**	**0.02**	**0.03**	**0.04**	**0.05**	**0.06**	**0.07**	**0.08**	**0.09**
0.0	0.5000	0.5040	0.5080	0.5120	0.5160	0.5199	0.5239	0.5279	0.5319	0.5359
0.1	0.5398	0.5438	0.5478	0.5517	0.5557	0.5596	0.5636	0.5675	0.5714	0.5753
0.2	0.5793	0.5832	0.5871	0.5910	0.5948	0.5987	0.6026	0.6064	0.6103	0.6141
0.3	0.6179	0.6217	0.6255	0.6293	0.6331	0.6368	0.6406	0.6443	0.6480	0.6517
0.4	0.6554	0.6591	0.6628	0.6664	0.6700	0.6736	0.6772	0.6808	0.6844	0.6879
0.5	0.6915	0.6950	0.6985	0.7019	0.7054	0.7088	0.7123	0.7157	0.7190	0.7224
0.6	0.7257	0.7291	0.7324	0.7357	0.7389	0.7422	0.7454	0.7486	0.7517	0.7549
0.7	0.7580	0.7611	0.7642	0.7673	0.7704	0.7734	0.7764	0.7794	0.7823	0.7852
0.8	0.7881	0.7910	0.7939	0.7967	0.7995	0.8023	0.8051	0.8078	0.8106	0.8133
0.9	0.8159	0.8186	0.8212	0.8238	0.8264	0.8289	0.8315	0.8340	0.8365	0.8389
1.0	0.8413	0.8438	0.8461	0.8485	0.8508	0.8531	0.8554	0.8577	0.8599	0.8621
1.1	0.8643	0.8665	0.8686	0.8708	0.8729	0.8749	0.8770	0.8790	0.8810	0.8830
1.2	0.8849	0.8869	0.8888	0.8907	0.8925	0.8944	0.8962	0.8980	0.8997	0.9015

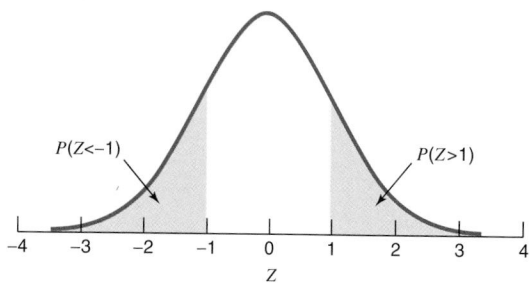

FIGURE 6.9 **Comparison of upper and lower probabilities**

To look up a value of Z—say, $Z = 1.14$—we first locate the row that corresponds to the first two digits, 1.1. This is the shaded row in Table 6.4. The column we need is the one that corresponds to the last digit, 0.04, which is also shaded. The entry at the intersection of the row and column, 0.8729, gives the probability of obtaining a value of the random variable Z that is less than 1.14, or the area under the curve to the left of $Z = 1.14$.

Caution! When looking up Z values such as 0.08, the first two digits are 0.0 not 0.8.

The half of the table for negative values of Z is used in exactly the same way. It is important to be consistent about the way you calculate and look up Z values. Because the tables give the values of Z to two decimal places, you need to calculate Z values to two decimal places, rounding correctly, and look them up that way.

You now know how to use the table to find $P(Z < z)$. How does the procedure change if you need $P(Z > z)$ or $P(z_1 < Z < z_2)$? The first question is not difficult to answer. Since the tables give $P(Z < z)$ and the total area under the curve must always equal 1, you can find $P(Z > z)$ by using the definition of the complement of an event; that is,

$$P(Z > z) = 1 - P(Z < z)$$

For continuous random variables, probability is often referred to as the area under the curve.

So, to find the area under the curve above a given value of Z, you look up the Z value, find the probability below Z, and subtract it from 1. Thus, from the earlier example,

$$P(Z > 1.14) = 1 - 0.8729 = 0.1271$$

Once you are comfortable using the normal probability table you can find upper-area probabilities easily by relying on the symmetry of the normal distribution. Since the normal probability distribution is symmetric about the mean, the area to the right of a Z value must be equal to the area to the left of the negative of that Z value, or

$$P(Z > z) = P(Z < -z)$$

This symmetry is illustrated in Figure 6.9.

Using the tables to find $P(z_1 < Z < z_2)$ is not difficult either. Figure 6.10 illustrates the process. When you look up the larger Z value—in this case, $Z = +2.00$—

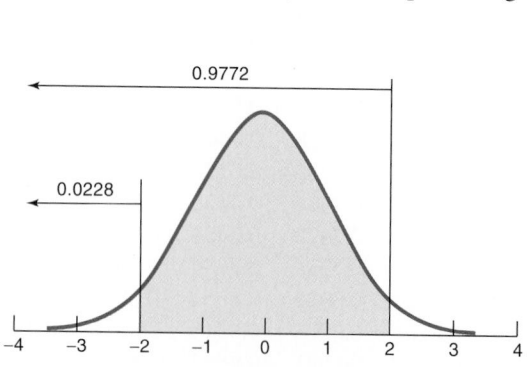

FIGURE 6.10 **Finding the area between two Z values**

TABLE 6.5

RULES FOR SOLVING NORMAL PROBABILITY PROBLEMS

To find . . .	Area under the curve . . .	Look up . . .
$P(Z < z)$	below a value of Z	the Z value and use the table directly
$P(Z > z)$	above a value of Z	1. the Z value and subtract the value in the table from 1
		OR
		2. the negative of the Z value and use the table directly
$P(z_1 < Z < z_2)$	between two values of Z	both Z values and subtract the lower value from the higher value

the table gives the entire area under the curve to the left of +2.00, which is 0.9772. Looking at the picture, you see that this includes the portion of the curve to the left of the lower value—in this case, $Z = -2.00$—which is 0.0228. Since you do not want to include this lower portion, you must subtract it from the first value. Thus, in general,

$$P(z_1 < Z < z_2) = P(Z < z_2) - P(Z < z_1)$$

and for this example,

$$P(-2 < Z < +2) = P(Z < +2) - P(Z < -2) = 0.9772 - 0.0228 = 0.9544$$

If you think about it, you already knew this! Remember that the Z random variable is $N(0, 1)$, so finding $P(-2 < Z < +2)$ is the same as finding the probability that the random variable is within two standard deviations of the mean. The answer, not surprisingly, is 0.9544, or 95.44%. Remember the empirical rule? It told us that for a symmetric distribution, approximately 95% of the data fall within two standard deviations of the mean. This is where the empirical rule comes from.

Remember that Z can also be thought of as the number of standard deviations that a value is from its mean.

Table 6.5 summarizes the rules for using the standard normal probability tables. The technique of finding probabilities using the Z-score is important not only in solving probability problems. It is a *very* important skill used to find p values in hypothesis testing. This topic is covered in Chapters 8–10.

TRY IT NOW!

The Standard Normal Table
Using the Table to Find Probabilities

For doing each problem *draw a picture* of what you are trying to find *before* you use the table to find it.
   Find the probability that a Z random variable takes on a value that is less than 2.74.

Find the probability that a Z random variable is greater than 0.85.

Drawing a picture will help you use the table correctly.

Find the probability that Z is between −1.36 and 1.87.

ANS. 0.9969; 0.1977; 0.8824

6.5.4 SOLVING NORMAL PROBABILITY PROBLEMS

Just as you found with the binomial distribution, once you understand how the random variable and the probability distribution work, solving problems is really not hard. The hardest part is understanding what the problem really means.

You can use the following steps to solve problems that involve normally distributed random variables with mean μ and standard deviation σ.

Step 1: Write down the information about the random variable—that is, $X \sim N(\mu, \sigma)$.

Step 2: Draw a picture that represents the problem and write down the probability statement—for example, $P(X > 30)$.

Step 3: Transform the values of X in the problem into Z values.

Step 4: Look up the Z values on the standard normal tables.

Step 5: Perform any additional calculations that need to be done, as described in Table 6.5.

EXAMPLE 6.16

APTITUDE SCORES

Solving Normal Probability Problems

The job placement agency that administered the aptitude test to its clients wants to know what proportion of the people who take the test score between 65 and 85 points. From Example 6.15 we know that $X \sim N(75, 5)$ and that we are looking for $P(65 < X < 85)$, or

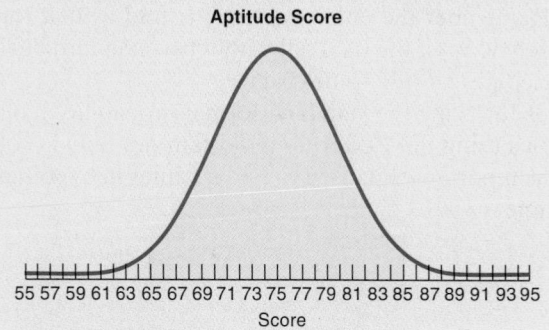

Aptitude Score

55 57 59 61 63 65 67 69 71 73 75 77 79 81 83 85 87 89 91 93 95
Score

We found the Z values of interest to be -2 and $+2$. Looking these values up in the standard normal tables, we find that the probability for $Z = -2$ is 0.0228 and for $Z = +2$ it is 0.9772. From the picture we see that we are trying to find the area *between* these two values, and so we subtract the probability values:

$$0.9772 - 0.0228 = 0.9544$$

The agency concludes that 95.44% of its clients score between 65 and 85 points on the test.

The agency is also interested in knowing what proportion of the clients who take the test score in what is considered to be the superior range, above 87 points. We know that $X \sim N(75, 5)$ and we need to find $P(X > 87)$. This case is represented by the graph:

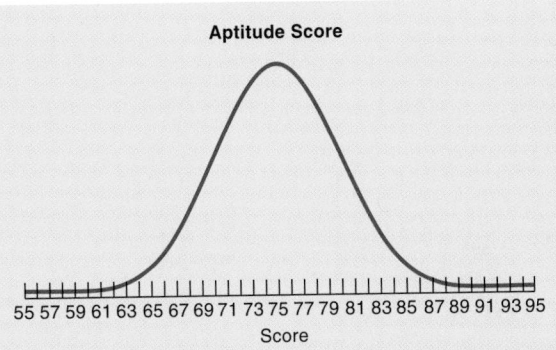

Aptitude Score

55 57 59 61 63 65 67 69 71 73 75 77 79 81 83 85 87 89 91 93 95
Score

Transforming the X value of 87 to Z, we get

$$Z = \frac{87 - 75}{5} = \frac{12}{5} = 2.40$$

Notice that this answer makes sense because it is positive and the X value of interest is to the right of the mean. When we look up the Z value of 2.40 in the table, we get a probability of 0.9918.

Is there anything left to do? The answer is most certainly *yes!* Common sense tells us that 0.9918 cannot possibly be the answer. Look at how much of the area in the picture is shaded—it is very small, hardly more than 99%! We are looking for the area *above* a value and the table gives the opposite area. We must subtract the value in the table from 1 to obtain

$$1 - 0.9918 = 0.0082, \text{ or } 0.82\%$$

Thus, the agency sees that less than 1% of the people who take the test score in the superior range.

When doing problems, keep in mind that probabilities, proportions, and percentages are all ways of expressing the same quantity. Do not be misled by the way a question is asked. It is best to answer the question using the quantity type specified, but you are not wrong in using either decimals or percentages.

EXAMPLE 6.17

ON-TIME FLIGHTS

Calculating Normal Probabilities

The number of minutes late that a flight arrives at Chicago's O'Hare Airport from Dallas/Ft. Worth Airport is normally distributed with a mean of 5.4 minutes and a standard deviation of 1.8 minutes. The airline is interested in finding out what percentage of the flights on this route are more than 10 minutes late.

Steps 1 and 2: Write down the information and draw a picture. The first step in solving the problem is to understand what we need to find. We know that $X \sim N(5.4, 1.8)$ and we are trying to find $P(X > 10)$. The drawing illustrates the problem. Since we are looking for the probability that the flight is more than 10 minutes late, we shade the area to the right of the value 10.

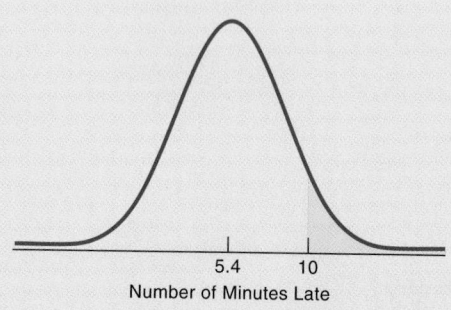

Number of Minutes Late

Step 3. Transform the values of *X* into *Z* values. To transform the problem into a standard normal problem, we need to find the *Z*-score for the value of 10:

$$Z = \frac{10 - 5.4}{1.8} = 2.56$$

Step 4. Look up the *Z* values on the standard normal table. From the standard normal table, we find that the probability associated with this *Z*-score is 0.9948.

Step 5. Perform any additional calculations. The last step is to reconcile the number in the table to the problem. Since the table always gives the area to the left of *Z* and we are looking for the area to the right of *Z*, we need to subtract the table value from 1. Thus, the answer to the problem is

$$1 - 0.9948 = 0.0052$$

That is, 0.52% of the flights are more than 10 minutes late.

Suppose the airline also wants to know the probability that flights on this route are not late. Now we are looking for $P(X \leq 0)$.

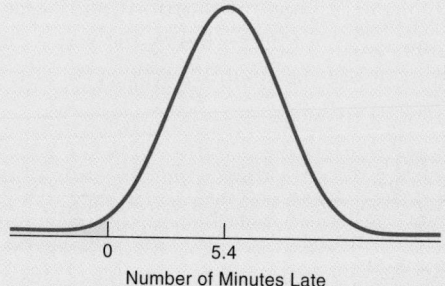

Number of Minutes Late

Transforming the problem into a standard normal problem, we calculate

$$Z = \frac{0 - 5.4}{1.8} = -3.00$$

When we look up the *Z* value in the standard normal table, we find that the probability is 0.0013. The table and the problem we are trying to solve agree, so this is the answer to the question.

The airline has learned that although a very small percentage (0.52%) of its flights on the Chicago–Dallas/Ft. Worth route are more than 10 minutes late, an even smaller percentage (0.13%) are on time.

TRY IT NOW!

Speed Reading
Solving a Normal Probability Problem

The professor who is interested in how many pages of the statistics textbook a student can read in an hour knows that the random variable is $N(7, 1.5)$. Find the probability that a student can read more than 11.5 pages in an hour.

The instructor is concerned about the percentage of students who read less than a five-page section in the given hour. What percentage of the students is this?

6.5.5 FINDING VALUES THAT CORRESPOND TO KNOWN PROBABILITIES

In the previous section you learned how to use the standard normal probability tables to solve problems involving any normally distributed random variable. You were interested in finding the probability or the area under the curve that corresponds to a specific value of the random variable, X.

Another interesting type of problem uses the normal distribution and the standard normal tables. The probabilities of the normal distribution can be used to make decisions about what values of a variable define certain outcomes.

Suppose a company wants to reward with an incentive bonus those salespeople whose sales are in the top 10%, or suppose a city decides that only those people who score in the top 5% on a civil service exam will get job interviews. In many cases it does not make sense simply to take the top 5% of each group that takes the exam or the top 10% of each month's sales. If a particularly ill-suited group of people take the exam, the city could wind up interviewing people who are in no way qualified for the positions available. Similarly, the company could reward people in one month and deny people in the next who have the same sales figures. Instead, these decision makers rely on the normal probability model to define the cutoff points for the variables of interest. How do they find the numbers that they are looking for?

You know how to use the normal probability tables to find the probability that corresponds to a given value of X or Z. The problem we just described is really the exact opposite of that problem. The decision makers know the percentage or area under the curve that they want to define. They do not know the value of Z (and subsequently X) that it corresponds to. This problem is illustrated in Figure 6.11 on page 278.

Solving this type of problem involves using the normal probability tables "inside out." Earlier you located the Z value on the *outside* of the table and determined the answer to the question by looking on the *inside* of the table. For this type of problem we start by looking on the *inside* of the table for the known probability and then

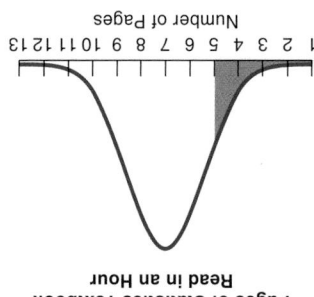

ANS. 0.0013; 0.0918 OR 9.18%

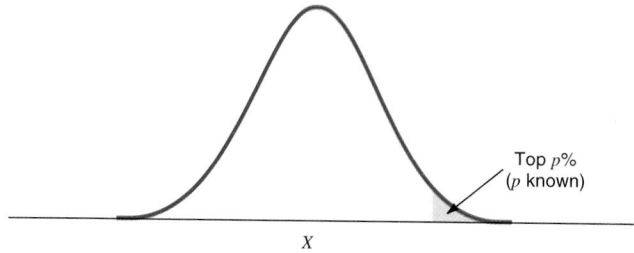

FIGURE 6.11 **The "inverse" normal probability problem**

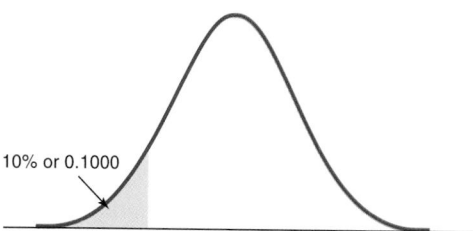

FIGURE 6.12 **Bottom 10% of the normal distribution**

move to the *outside* to find the answer. It sounds more complicated than it really is. Again, you simply need to be systematic in your approach.

To get started, we concentrate on the technique. Suppose we want to find the value of Z that has 10% of the area below it (see Figure 6.12). Since we know that the normal tables give the area below the value of Z and that is what we are looking for, we can look the probability up directly. At first glance it may seem like you are looking for a needle in a haystack, but with a little bit of common sense and insight you will see that it is really not difficult.

From the picture (it is still important to draw them) we see immediately that the value of Z we want must be negative. Now we have already halved our search! Also, we know that the numbers in the table are in numerical order. Once we find a suitable starting place for the search, we simply have to search systematically for the given number. The section of the normal probability table that contains the value we are looking for is shown in Table 6.6.

TABLE 6.6

NORMAL PROBABILITY TABLE CONTAINING THE PROBABILITY 0.1000

z	0.00	0.01	0.02	0.03	0.04	0.05	0.06	0.07	0.08	0.09
−1.9	0.0287	0.0281	0.0274	0.0268	0.0262	0.0256	0.0250	0.0244	0.0239	0.0233
−1.8	0.0359	0.0351	0.0344	0.0336	0.0329	0.0322	0.0314	0.0307	0.0301	0.0294
−1.7	0.0446	0.0436	0.0427	0.0418	0.0409	0.0401	0.0392	0.0384	0.0375	0.0367
−1.6	0.0548	0.0537	0.0526	0.0516	0.0505	0.0495	0.0485	0.0475	0.0465	0.0455
−1.5	0.0668	0.0655	0.0643	0.0630	0.0618	0.0606	0.0594	0.0582	0.0571	0.0559
−1.4	0.0808	0.0793	0.0778	0.0764	0.0749	0.0735	0.0721	0.0708	0.0694	0.0681
−1.3	0.0968	0.0951	0.0934	0.0918	0.0901	0.0885	0.0869	0.0853	0.0838	0.0823
−1.2	0.1151	0.1131	0.1112	0.1093	0.1075	0.1056	0.1038	0.1020	0.1003	0.0985
−1.1	0.1357	0.1335	0.1314	0.1292	0.1271	0.1251	0.1230	0.1210	0.1190	0.1170
−1.0	0.1587	0.1562	0.1539	0.1515	0.1492	0.1469	0.1446	0.1423	0.1401	0.1379
−0.9	0.1841	0.1814	0.1788	0.1762	0.1736	0.1711	0.1685	0.1660	0.1635	0.1611
−0.8	0.2119	0.2090	0.2061	0.2033	0.2005	0.1977	0.1949	0.1922	0.1894	0.1867

This is probably a good place to point out that since the tables give probabilities to four decimal places, we are not likely to find the exact value we are looking for. If the value we want is not listed, we can always locate two adjacent values, one smaller and one larger than the value we are looking for. From Table 6.6 we see that the numbers are increasing from top to bottom and that at the row for $z = -1.3$ they are too small and at the $z = -1.2$ row they are too large. The value we want must be in there somewhere! A more careful search leads to the two cells that are shaded in the table, 0.0985 and 0.1003.

As we guessed, one of these is a bit too small and the other is too large. The value of 0.0985 corresponds to a value of Z of -1.29, whereas 0.1003 corresponds to $z = -1.28$. Thus, the value of Z that we are looking for must be in between these two.

How do we determine which Z value to use? Some people suggest performing a linear interpolation between the two values to get an estimate of the correct value of Z. Even though the normal curve is certainly not linear, for small increments in Z this is not a bad approximation. The only problem is, it is not worth the trouble.

For our purposes we use the value of the probability that is closest to the value we are looking for unless some information in the problem tells us to use the one that is smaller or larger. The information might be some directional description of the percentage we are looking for, like "at least 10%," in which case we would choose 0.1003, or "at most 10%," which would lead us to 0.0985. When we present an actual application of this technique, you will see how little difference the exact answer makes.

Now on to solving real problems. We start with a simple example that uses what we have already figured out.

EXAMPLE 6.18

APTITUDE SCORES

Solving the Inverse Problem

Training and retooling people are expensive. Job training and placement agencies must find people who are capable of doing the jobs they are hired for. The agency that gives clients the aptitude test has decided that people who score in the bottom 10% of the test scores will receive less job training than those who score higher. If jobs that require additional skills are available, those people will be the last to receive training. The question is, What cutoff score on the test should the agency use?

We already know that the bottom 10% corresponds to a Z-score of -1.28. But clearly this cannot be the answer to the question. How do we translate the Z-score back into the realm of the variable of interest, the test scores?

The Z-score represents the number of standard deviations away from the mean that a value is. The negative sign on the Z value tells us to move down from the mean (subtract the number of standard deviations). Thus, we know that the cutoff score must be 1.28 standard deviations *below* the mean; that is,

$$X = 75 - (1.28)(5) = 68.6 \text{ points}$$

The agency has to decide whether to use 68 points (and affect less than 10% of the people) or 69 points (and affect more than 10%).

From this example you see why it is not necessary to interpolate to get the exact value of Z. If the answer is to be used to make a decision, the extra precision gained by getting the exact Z value does not usually translate into any practical information.

The test is not scored to the nearest tenth of a point. Even if it were, the change in the Z-score affects only the hundredths place of the score.

We used our understanding of the Z value to determine the cutoff score, but the formula we used really comes from algebraically solving the definition of the Z value for X:

$$Z = \frac{X - \mu}{\sigma} \longrightarrow X = \mu + Z\sigma$$

When Z is positive we add to μ, and when Z is negative we subtract from μ.

For problems that define the probability and look for the value of the random variable, it is important to know whether you are specifying an area above the value of interest or an area below the value. When the area is below the value you are looking for, you can look up the specified probability directly. When the area is above the value you are looking for, you have to either subtract the specified probability from 1 and look that up, or look up the given probability and rely on the symmetry of the table. The technique of solving the inverse normal problem is used in finding critical values for tests of hypotheses.

EXAMPLE 6.19

APTITUDE SCORES

Specifying Upper-Area Probabilities

The job placement agency plans to give extra training to clients who score in the top 2% of people who take the aptitude test. The agency wants to identify the score to use as the cutoff point. We draw a picture that represents the problem and see that the Z value we are looking for will most certainly be positive.

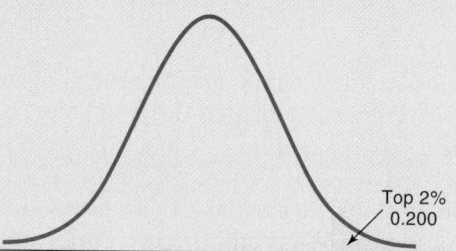

Top 2%
0.200

If we want to look up the value directly, we have to look up from the bottom area or 0.9800. It is just as easy to use common sense and rely on the symmetry of the normal distribution. We do not find 0.0200 in the table, but we can choose from 0.0197 and 0.0202. Because 0.0202 is closer to 2% (and slightly more generous to the clients) we use the corresponding Z value of -2.05.

But we said that the answer has to be positive. This is where common sense, symmetry, and the picture come together. We know the answer should be positive from the picture and common sense, but we used the symmetry feature to make the lookup easier. We use $+2.05$ to find the answer.

The cutoff score for the top 2% is then

$$75 + (2.05)(5) = 85.25 \text{ points}$$

The agency can give extra training to 85 people (and train more than 2%) or 86 people (and train less than 2%).

✓ TRY IT NOW!

Speed Reading
Solving the Inverse Problem

The professor who is interested in how fast students can read the statistics textbook wants to identify the bottom 25% of the class in terms of the number of pages they can read in an hour. Find the number of pages per hour that defines the bottom 25% of the students.

Inverse normal probability problems are an important application of the normal distribution. These types of problems occur often in real life when standards or cutoff points must be determined.

6.5.6 EXERCISES – LEARNING IT!

6.11 The amount of money spent by students for textbooks in a semester is a normally distributed random variable with a mean of $235 and a standard deviation of $15.

(a) Sketch the normal distribution that describes the amount of money spent on textbooks in a semester.

(b) What is the probability that a student spends between $220 and $250 in any semester?

(c) What percentage of students spend more than $270 on textbooks in any semester?

(d) What percentage of students spend less than $225 in a semester?

6.12 On any given day, the number of leasable square feet of office space available in a small city is a normally distributed random variable with a mean of 850,000 square feet and a standard deviation of 25,000 square feet. The number of leasable square feet available in another small city is a normally distributed variable with a mean of 900,000 square feet and a standard deviation of 25,000 square feet.

(a) Sketch the distribution of leasable office space for both cities on the same graph.

(b) What is the probability that the number of leasable square feet in the first city is less than 925,000 square feet?

(c) What is the probability that the amount available in the second city is less than 925,000?

6.13 The actual amount of a certain brand of orange juice in a container marked half-gallon is a normally distributed random variable with a mean of 65 ounces and a standard deviation of 0.35 ounce.

(a) What percentage of the containers contain more than 64.5 ounces of juice?

(b) What percentage of the containers contain between 64 and 66 ounces?

(c) If federal law says that 98% of all containers must be at or above the labeled weight, does this brand of orange juice meet the requirement?

6.14 The amount of money per month earned by a nurse practitioner is a normally distributed random variable with mean $5000 and standard deviation $1200.

(a) What percentage of nurse practitioners earn more than $4000 per month?

(b) What percentage of nurse practitioners earn less than $3000 per month?

(c) What is the probability that a randomly selected nurse practitioner earns between $3250 and $3800 per month?

(d) What monthly income cuts off the top 10% of all nurse practitioners?

GET IT IN WRITING

Firestone Tires

TO: **Secretary of Transportation**
FROM: **NHTSA**
RE: **Firestone Tire Accident Information – Preliminary Results**

After the the investigation concerning vehicles and tire models of Firestone tires, we decided to begin an initial investigation into the types of accidents that occurred. As you recall, data from many of the reports were not consistent. After taking only those accidents with usable data, we had 2173 accidents to analyze. We decided to look at accidents in which the vehicle involved rolled over since we had information on the overall national rate of rollovers.

For the preliminary work, we took a random sample of 100 from the 2173 accidents for which we had usable data and counted the number of accidents that involved a rollover. We did this so that we could use the binomial distribution as the probability model. We carefully checked to ensure that the criteria of the model were satisfied.

In the sample of 100 accidents involving Firestone tires, we found that there were 7 rollovers. The national rate is 6%, which would mean that we expect to find 6 rollovers in a sample of size 100. The question is, then, how different from the expected value of 6 is our sample value of 7? We calculated the standard deviation of the national norm to be 2.37 cars and so it would seem that the sample value is well within 2 standard deviations of the mean. In fact, it is within 0.42 standard deviation of the mean. Although a binomial distribution with $p = 0.06$ is not really symmetric, it would appear that the sample data conform very well to the national norm. The probability of having 7 rollovers in a sample of size 100 when $p = 0.06$ is 0.142 or 14%. Thus the sample results seem consistent with the overall national rate.

We believe that further investigation is still warranted. It would be interesting to look at the data for different types of vehicles and different types of tires and see whether the results are still consistent. We will begin that analysis immediately and inform you of our progress within a week.

6.6 Generating Probability Distributions with Technology

Excel, Minitab, and the TI-83 graphing calculator can help you create probability distributions. Refer to the manual associated with the software that you are using for more detailed instructions. A brief overview of the capabilities of each package follows.

Excel 2000

Excel allows you to calculate probabilities and generate random data from many probability distributions. One of the advantages of Excel is that it has built-in functions that allow you to solve a problem for values that are not in tables (see Figure 6.13). You can use the Function Wizard to find out about Excel functions you

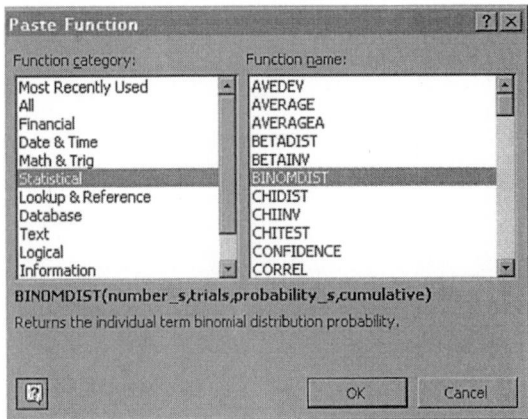

FIGURE 6.13 **Choosing an Excel function**

have never used. You can also use the macro included on the KADDSTAT disk that accompanies this text (see Figure 6.14).

Minitab

Minitab also can help you calculate probabilities and generate random data from different probability distributions. See Figures 6.15 and 6.16 (page 284).

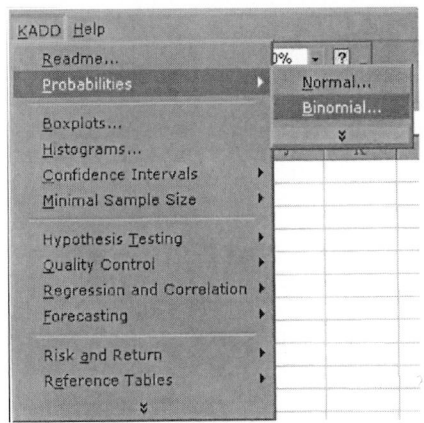

FIGURE 6.14 **KADD menu**

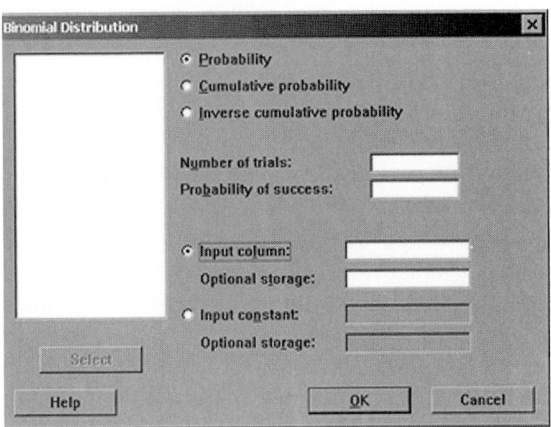

FIGURE 6.15 **Binomial Distribution dialog box in Minitab**

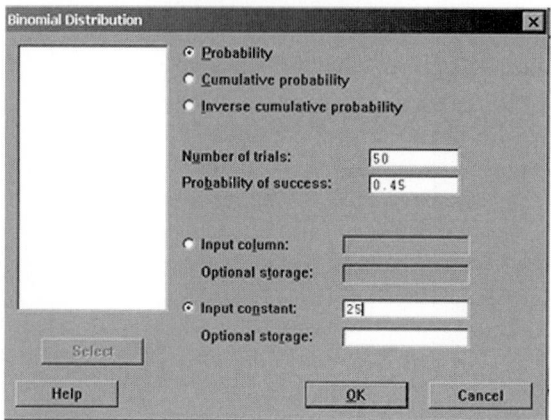

FIGURE 6.16 **Completed dialog box for creating a set of numbers in Minitab**

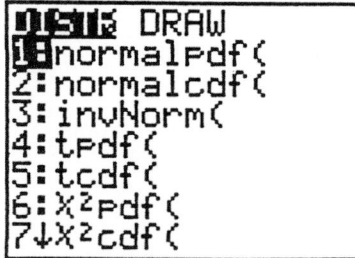

FIGURE 6.17 **Probability distributions screen on the TI-83**

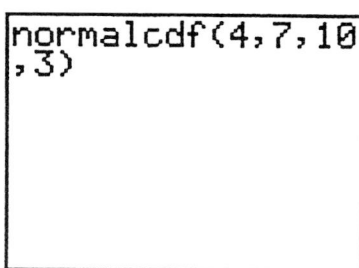

FIGURE 6.18 **Normal probability function on the TI-83**

TI-83 Graphing Calculator

You can use the TI-83 graphing calculator to calculate probabilities from the binomial and normal distributions: See Figures 6.17 and 6.18.

CHAPTER 6 SUMMARY

Probability is an important and interesting subject in its own right, but the study of random variables and probability distributions is important in the study and development of statistics. Most important, probability is the bridge between what we studied in the first four chapters of this book, *descriptive statistics*, and what is to follow in subsequent chapters, *inferential statistics*. In descriptive statistics we use different techniques to describe sample data. In inferential statistics we will test hypotheses about the populations from which these samples came. Probability is the tool that allows us to reconcile what happened (descriptive) with what we think is true by determining how likely the outcomes of the experiment we perform are.

In this chapter you saw how the way you learned to describe the distribution of sample data relates to the characteristics of probability distributions. You also saw how Z-scores and the empirical rule come from the normal distribution. The techniques of this chapter will be important in the chapters that follow.

KEY TERMS		
Term	**Definition**	**Page reference**
Binomial random variable	A **binomial random variable** is the number of successes in n trials or in a sample of size n.	250
Interval probability	An **interval probability** gives the probability that a random variable takes on a value *between* two given values $P(x_1 \leq X \leq x_2)$.	271
Probability density function, $f(x)$	A **probability density function, $f(x)$,** is a smooth curve that represents the probability distribution of a continuous random variable.	265
Probability distribution, $p(x)$	The **probability distribution** of a random variable X, written as $p(x)$, gives the probability that the random variable will take on each of its possible values.	244
Random variable	A **random variable, X,** is a quantitative variable that has values that vary according to the rules of probability.	243
Standard normal random variable, Z	A **standard normal random variable** is normally distributed with a mean of 0 and a standard deviation of 1; $Z \sim N(0, 1)$.	269
Standard normal table	A **standard normal table** is a table of probabilities for a Z random variable.	269

KEY FORMULAS		
Term	**Formula**	**Page reference**
Binomial mean and standard deviation	$\mu = np$ and $\sigma = \sqrt{np(1-p)}$	257
Binomial probability distribution	$p(x) = \dfrac{n!}{(x!)(n-x)!} p^x (1-p)^{n-x}$ for $x = 0, 1, \ldots, n$	254
Z-score	$Z = \dfrac{X - \mu}{\sigma}$	269

CHAPTER 6 EXERCISES

Learning It!

6.15 A recent survey reported that 30% of people pay their car loans electronically. If a random sample of 15 people is taken, what is the probability that in the sample of 15

(a) at least 5 pay their car loans electronically?

(b) between 6 and 12 inclusive pay their car loans electronically?

(c) fewer than 10 do not pay their car loans electronically?

6.16 In a large study conducted by the Horatio Alger Association, it was reported that 75% of all teens expected to go to either a two-year or four-year college. If a random sample of 25 teens is taken, what is the probability that

(a) at most 20 plan to go to a two-year or four-year college?

(b) at least 7 do not plan to go to a two-year or four-year college?

(c) Find the mean and standard deviation of the number of teens in the sample who would expect to go to a two-year or four-year college.

6.17 According to the *Journal of the American Medical Association* published in August 2001, 20% of adolescent girls endure either sexual or physical violence while dating. If a sample of 20 adolescent girls is taken at random, find the probability that

(a) fewer than 8 endure violence while dating.

(b) at most 10 do not endure violence while dating.

(c) between 7 and 19 endure violence while dating.

(d) Find the mean and standard deviation of the number of adolescent girls in a sample of 20 who endure violence while dating.

6.18 A study conducted by *GamePro* magazine, a magazine about video and computer games, reports that subscribers to the magazine spend an average of 12 hours per week playing such games. Suppose that the time spent playing is normally distributed with a standard deviation of 2.3 hours.

(a) What is the probability that a subscriber to the magazine spends more than 20 hours a week playing video and computer games?

(b) What percent of the subscribers spend between 10 and 20 hours per week playing video and computer games?

(c) What percent of the subscribers spend less than 5 hours per week playing video and computer games?

6.19 The amount of money that a driver in Boston, Massachusetts, pays for unreserved monthly parking is normally distributed with a mean of $350 and a standard deviation of $28.

(a) What percent of Boston drivers pay between $300 and $400 for unreserved monthly parking?

(b) What is the probability that a randomly selected driver pays more than $375 for unreserved monthly parking?

(c) What is the probability that a randomly selected driver pays at most $425 for unreserved monthly parking?

6.20 Surveys indicate that 40% of all small businesses do not provide any kind of health insurance options for their employees. The Chamber of Commerce in a large city took a random sample of 20 small businesses. Calculate the probabilities of these events:

(a) At least 12 do not provide health insurance options for employees.

(b) At most 4 do not provide health insurance options for employees.

(c) Between 8 and 13 do not provide health insurance options for employees.

(d) At least 5 provide health insurance options for employees.

6.21 Recent studies have shown that 30% of employees in the insurance industry telecommute 4 days a week. For a random sample of 15 insurance industry employees find these probabilities:

(a) Between 4 and 7 inclusive telecommute 4 days a week.

(b) Fewer than 6 telecommute 4 days a week.

(c) At least 9 telecommute 4 days a week.

(d) Between 5 and 10 do not telecommute 4 days a week.

6.22 A survey done at a state university in New England found that 40% of all seniors have encountered academic problems related to binge drinking. If a random sample of 25 seniors at the university is taken, what are these probabilities?

(a) At most 6 have encountered academic problems related to binge drinking.

(b) Between 4 and 9 have encountered academic problems related to binge drinking.

(c) More than 15 have encountered academic problems related to binge drinking.

(d) Find the mean and standard deviation of the number of seniors in a sample of 25 who have encountered problems related to binge drinking.

6.23 The amount of office space allocated to social workers in state agencies is a normally distributed random variable with a mean of 90 square feet and a standard deviation of 6 sq ft.

(a) What percentage of social workers' offices are larger than 105 sq ft?

(b) What percentage of social workers' offices have between 85 and 105 sq ft?

(c) What percentage of social workers' offices are smaller than 110 sq ft?

6.24 The size of a gift/specialty store in a regional supermall is a normally distributed random variable with a mean of 8500 sq ft and a standard deviation of 260 sq ft. What is the probability of the following sizes for a randomly selected gift/specialty store in a regional supermall?

(a) Larger than 8000 sq ft

(b) Between 8300 and 9000 sq ft

(c) Smaller than 9500 sq ft

6.25 A recent study done at a New England university found that 60% of all students have missed class during the semester because of drinking. A random sample of 20 students is taken. Find these probabilities:

(a) At least 15 students have missed class because of drinking.

(b) Between 12 and 17 inclusive have missed class because of drinking.

(c) Fewer than 5 have not missed class because of drinking.

(d) At most 13 have not missed class because of drinking.

Thinking About It!

6.26 Data from the National Highway Traffic Safety Administration (NHTSA) survey in 2000 found that 50% of drivers in West Virginia wear seatbelts.

(a) If a random sample of drivers in West Virginia is taken, how many drivers would you expect to find wearing a seatbelt?

(b) What is the standard deviation of the number of drivers in the sample who are wearing seatbelts?

(c) What is the probability that the number of drivers in the sample who are wearing seatbelts is within 2 standard deviations of the mean?

(d) Do you expect that this will compare well with what the empirical rule predicts? Why or why not?

6.27 The same survey by NHTSA about drivers wearing seatbelts reported that 90% of the drivers in California wear seatbelts.

(a) If a random sample of drivers in California is taken, how many drivers would you expect to find wearing a seatbelt?

(b) What is the standard deviation of the number of drivers in the sample who are wearing seatbelts?

(c) What is the probability that the number of drivers in the sample who are wearing seatbelts is within 2 standard deviations of the mean?

(d) Do you expect that this will compare well with what the empirical rule predicts? Why or why not?

6.28 Suppose that you were given the results of a random sample of 25 drivers and were told that there were 17 drivers in the sample who were wearing seatbelts. Would you be more likely to think that these data came from West Virginia or California? Be sure to support your answer with facts.

Requires Exercises 6.26 and 6.27

6.29 A study done at the University of Michigan reported that young adults who grow up in homes that were rated as very clean to clean earn wages that are normally distributed with a mean of $14.17 per hour and a standard deviation of $0.60 per hour. Further, the number of years of school that they complete is normally distributed with an average of 13.6 years and a standard deviation of 0.9 year. On the other hand, young adults who grew up in dirtier homes earn wages that are normally distributed with a mean of $12.60 and a standard deviation of $0.75. The number of years of school completed for these young adults is normally distributed with a mean of 12 years and a standard deviation of 1.2 years.

(a) Draw the distributions for wages for each group on the same graph and discuss them.

(b) Draw the distributions for years of school for each group on the same graph and discuss them.

(c) What is the probability that a person who grows up in a clean home completes college?

(d) What is the probability that a person who grows up in a dirty home completes college?

(e) Do you think there is really a difference based on the numbers in the report? Why or why not?

6.30 A recent study done on college drinking reported that the number of drinks consumed weekly by a male freshman was normally distributed with a mean of 8.5 drinks and a standard deviation of 0.7 drink. The same information for male seniors indicated that the mean was 10.1 drinks per week with a standard deviation of 0.9. For a female freshman, the number of drinks per week was normally distributed with a mean of 3.7 drinks and a standard deviation of 0.4, while for a female senior the average was 2.3 drinks per week and the standard deviation was 0.4.

(a) Draw the distributions for number of drinks for both male populations on the same graph and discuss them.

(b) Draw the distributions for number of drinks for both female populations on the same graph and discuss them.

(c) Draw a graph relating drinking among female and male freshmen and discuss it.

(d) Draw a graph relating drinking among female and male seniors and discuss it.

(e) Do you think that males and females are very different with respect to college drinking? Why or why not?

(f) Which other comparisons do you think are very different? Why?

6.31 A study reported by *USA Today* in July 2001 found that Americans spend an average of 3.4 hours per week cleaning their houses. Suppose that the amount of time spent cleaning a house per week is normally distributed and that the standard deviation is 0.9 hours.

(a) If the top 10% of the houses with the longest times spent cleaning are to be rated as very clean, what is the smallest number of hours that a house can be cleaned per week and still be rated as very clean?

(b) Suppose that the houses with cleaning times in the bottom 2% are to be rated as uninhabitable. Below what cleaning time will a house be rated as uninhabitable?

Requires Exercise 6.20 **6.32** The Chamber of Commerce that is interested in small businesses providing health insurance options for employees wants a few more questions answered. The surveys indicated that 40% of employers do not provide health insurance options for employees.

(a) In the sample of 20 small businesses, how many should the Chamber of Commerce expect to provide health insurance options for employees?

(b) What is the standard deviation of the number of small businesses in the 20 that provide health insurance options?

(c) Find the probability that the number of small businesses in a sample of 20 will be within two standard deviations of the mean.

(d) How does the probability in part (c) compare with the percentage predicted by the empirical rule? If it does not agree, why not?

Requires Exercises 6.20, 6.32 **6.33** The Chamber of Commerce that is looking at health insurance options finds four small businesses in 20 that provide health insurance options for employees and thinks this is unusual.

(a) Do you agree that the finding is unusual? Why or why not?

(b) What will you tell the Chamber of Commerce this might mean?

Requires Exercise 6.24 **6.34** The manager of a regional supermall wants to compare the size of the gift/specialty stores in her mall with the size of the smallest 20% of the gift/specialty shops in similar malls. She knows that the size of such stores is normally distributed with a mean of 8500 square feet and a standard deviation of 260 square feet.

(a) What square footage defines the smallest 20% of such stores?

(b) Harriet's Gift Boutique has complained that it is much smaller than similar stores in other malls. If this store is 7800 sq ft, is the complaint reasonable?

Requires Exercise 6.25 **6.35** The dean of students at another New England university read the report on binge drinking and decided to conduct a small survey on her own campus. She took a random sample of 20 students and found that 15 of them had missed class because of drinking. She decides that this indicates that her university is within the norm for this problem. Do you agree with her conclusion? Why or why not?

Doing It!

6.36   The data from the NHTSA file on accidents involving Firestone tires include five different variables that relate to the nature of the accident itself.   *Datafile: TIRES.XXX*

B	C	D	E	F	G	H	I
TIRE_MODEL	TIRE_SIZE	Actual Tire Info	STATE	INJURIES	FATALITIES	VEH_MFR	VEH_MODEL
WILDERNESS	14	14	TN	0	0	FORD	EXPLORER
ATX	15	15	TX	0	0	FORD	EXPLORER XLT
ATX	15	15	AZ	0	0	FORD	EXPLORER EDDIE BAUER
FIRESTONE	15	15	TX	0	0	FORD	EXPLORER
WILDERNESS	15	15	FL	0	0	FORD	EXPLORER
WILDERNESS	15	15	FL	1	1	FORD	EXPLORER

J	K	L	M	N	O	P	Q	R
VEH_MODEL_YEAR	FAILURE_DATE	BLOWOUT	TREAD_SEPARATION	ROLLOVER	LOST_CONTROL	CRASH	SPEED	TIRE_POSITION
1993	4/10/99	N	Y	N	N	N	65	PR
1994	10/22/96	Y	UNK	Y	Y	Y	UNK	DR
1993	4/10/98	Y	Y	Y	Y	Y	70	DR
1993	6/4/00	UNK	P	N	N	N	65	TWO
1994	7/21/99	Y	N	Y	Y	Y	65	PR
1996	7/4/00	N	Y	N	N	N	UNK	DR

A brief description of the variables follows:

TIRE_MODEL—the specific model of Firestone tire

TIRE_SIZE—the tire size, in inches

Actual Tire Info—the tire size as entered in the original data

STATE—the state from which the complaint originated

INJURIES—the number of injuries involved

FATALITIES—the number of fatalities involved

VEH_MFR—the manufacturer of the vehicle involved

VEH_MODEL—the specific model of the car

VEH_MODEL_YEAR—the model year of the car

FAILURE_DATE—the date of the tire failure

BLOWOUT—whether or not the tire blew out

TREAD_SEPARATION—whether or not the tread separated from the tire

ROLLOVER—whether or not the vehicle involved rolled over

LOST_CONTROL—whether or not the driver lost control of the vehicle

CRASH—whether or not the vehicle was involved in a crash

SPEED—the speed of the vehicle at the time of the tire failure

TIRE_POSITION—the location of the tires that failed

For the variables *BLOWOUT, ROLLOVER, LOST_CONTROL,* and *CRASH,* the coded values are:

Y = yes

N = no

UNK = unknown

For the variable *TREAD_SEPARATION,* the coded values are:

Y = yes

N = no

C = complete tread separation

P = partial tread separation

UNK = unknown

For the variable *TIRE_POSITION*, the coded values are:

DF = driver front

DR = driver rear

PF = passenger front

PR = passenger rear

ONE, TWO, or THREE = number of tires (not specific) that failed

(a) Consider the variable *ROLLOVER*, which was discussed in Example 6.7. Separate the accidents according to the manufacturer of the vehicle. For each manufacturer generate a binomial probability distribution with $p = 0.06$ and $n =$ the number of vehicles from that manufacturer.

(b) For each manufacturer calculate the probability that the number of rollovers reported is the actual number found in that sample.

(c) For each manufacturer calculate an estimate of the value of p. Compare these estimates with one another and with the overall value of 6% reported by the NHTSA.

(d) The NHTSA does not give overall population estimates for the likelihood of blowouts, tread separations, crashes, or loss of control for accidents involving tires. Use the entire data set to estimate the probability that a vehicle involved in an accident had each of these problems. How does the response UNK affect your estimate? Do you think it is better to count the unknowns in the estimate or to just use the accidents that reported Yes or No? Why do you think this?

Use the estimates from part (d) as the values of p from a binomial distribution to answer the following questions:

(e) Look at just those accidents that involved Ford vehicles. Generate binomial probability distributions for each value of p for blowouts, tread separations, crashes, and loss of control. Use the distributions to find the probabilities that the number of accidents reporting each problem is the actual number in the sample. How do the Ford data compare with your estimates for the entire population?

(f) Separate the Ford vehicles by model and do the same analysis as in part (e). What conclusions can you draw?

(g) Separate the Ford vehicles by tire model and repeat the analysis. What did you learn?

(h) Write a newspaper article about your findings from the data.

Datafile:
TISSUES.XXX

6.37 In the current competitive market, most companies pay close attention to customer complaints. Sometimes customer complaints seem to be about trivial matters, but they can have a large impact on sales for the company. For a company that manufactures tissues, one of the problems that make up a large percentage of complaints is dispensing—that is, how the tissues come out of the box. In that category, the most common complaint is that the sheets tear as they are removed from the box.

The managers of the tissue company have decided to address this problem. They know that tensile strength (the amount of force the tissue can withstand before it tears) determines when a tissue will tear, and they have decided that to solve the problem they need to investigate the tensile strength of the tissues. As part of the quality control program at the company, facial tissue has certain product specifications—that is, criteria that must be met for the product to be acceptable to consumers. One of the characteristics that is specified is tensile strength. The managers have decided to look at the current levels of tissue strength. They know the target values for the process and the parameters that should be met, and they want to check whether the process is meeting the current specifications. If it is not, then changes need to be made to see that it does. If it is, then perhaps the process specifications need to be changed. The managers will collect data on two variables:

- **Machine-direction (MD) strength:** This is the strength in the direction that the machine pulls on the tissue during manufacture. It has to be high enough that the tissues do not break, which causes machine downtime.

- **Cross-direction (CD) strength:** This is the strength in the direction that the tissue is pulled out of the box. It is the variable that determines whether the sheets tear when customers remove them from the box.

Samples are taken from tissues produced on a single tissue machine. The samples are taken over three different days and the results recorded. A portion of the datafile and an explanation of the variables are presented here:

Day	MDStrength	CDStrength
1	1006	422
1	994	448
1	1032	423
1	875	435
1	1043	445
1	962	464
1	973	472

Day—the day on which the sample was taken, goes from 1 to 3

MDStrength—the machine-direction strength, measured in lb/ream

CDStrength—the cross-direction strength, measured in lb/ream

According to the specifications, *MDStrength* is supposed to be normally distributed with a mean of 1000 and a standard deviation of 50 lb/ream. *CDStrength* should be normally distributed with a mean of 400 and a standard deviation of 25 lb/ream.

(a) Use a computer software package to create normal probability tables for the specified distributions of *MDStrength*. Use increments of 50 and go from 800 to 1200.

(b) Use the tables from part (a) to determine the probability that a tissue manufactured according to specifications will have an *MDStrength* less than 850 lb/ream.

(c) Create a relative frequency histogram for the variable *MDStrength* that goes from 800 to 1200 in class intervals of 50 units.

(d) Do the data in the histogram from part (c) appear to have the shape of a normal distribution? What is the center?

(e) Use the relative frequencies on the histogram from part (c) to calculate what percentage of the data are within one standard deviation of the mean value of 1000. According to the empirical rule what percentage should be within one standard deviation? How do the actual data compare with the empirical rule prediction?

(f) Use the procedure from part (e) to determine the percentages of *MDStrength* data that are within two and three standard deviations of the mean.

(g) The critical specifications for *MDStrength* are 850 on the low side and 1075 on the high side. Use the frequency distribution to determine the percentage of the tissues that actually do not meet these specifications.

(h) How does the percentage from part (g) compare with the percentage defective expected by the product specifications?

(i) Do you think that the company should be concerned about the difference in part (h)? Why or why not?

(j) Create a set of normal probabilities for the theoretical distribution of *CDStrength*.

(k) The critical values for *CDStrength* are 480 on the high side and 390 on the low side. *CDStrength* that is too high creates a stiff tissue. Since the cross-direction is the one in which tissues are pulled from the box, a value of *CDStrength* that is too low can cause sheets to tear as they are removed from the box. According to the specifications, what percentage of the tissues should have *CDStrength* values that are too high? Too low?

(l) Create a graph of the theoretical distribution of *CDStrength*.

(m) Generate a set of descriptive statistics for the variable *CDStrength*. Compare the mean and the median. Do you think that the distribution of *CDStrength* is symmetric?

(n) Create a relative frequency histogram for the variable *CDStrength*. Does it support the assumption of normality?

(o) Compare the percentages of *CDStrength* values that are within one, two, and three standard deviations of the mean with the predictions of the empirical rule. Is the assumption of normality still reasonable?

(p) Prepare a report to management that indicates whether the process appears to be meeting the product specifications. Include any changes that need to be made (in terms of mean and standard deviation) to bring the process back to target values.

Sampling Distributions and Confidence Intervals

CHANGING JOBS AND MOVING ON!

If you don't know where you're going, any road will get you there.
—LEWIS CARROLL, *THROUGH THE LOOKING GLASS*

Often entrenched in our thinking is the belief that career planning is logical, linear, and indeed planned. People who are reluctant to answer questions like "What do you want to do?" (". . . when you grow up" is, of course, implied) are often described as undecided or indecisive. True, there was always good old George who knew from the start that being a radiologist was his true calling. But for most of us, answering requires us to wrestle with the complex question about our role in an increasingly complicated workplace. So in this world of downsizing, rightsizing, and capsizing, there are things you should think about and take action on—assuming you're in a situation somewhat more stable than Alice's in *Through the Looking Glass* and somewhat less stable than good old George's.

This paragraph from the infotoday website *(http://www.infotoday.com/it/feb00/ream.htm)* suggests that most of you will be changing jobs, if not careers, every 6 to 8 years. Clearly you will be moving many times as well.

So are you ready to consider (or do circumstances require you to consider) getting a new job? Let's assume that you are in one of those times of transition and you need to sell your house in order to move on to your new position. You decide to use your statistical tools to determine which real estate agency to use, how long it will take to sell your house, and how much you can sell it for.

You do a little research on the web and find that a typical house for sale stays on the market for 50 days. But you know that this is a regional average and so you go in search of local data. You find the following information about houses that have sold in your area in a recent week:

List price	Sale price	DOM (days on market)
$89,900	$90,000	76
96,500	90,500	48
70,000	68,000	42
79,900	73,500	9
94,900	92,400	27

List price is the original asking price, whereas sale price is the amount the house sold for.

You find that the mean number of days on the market is 40.4 for this sample of five houses, but you remember that the sample mean is only an estimate of the population mean and you wonder if the population mean is 50 days in your local area. You must decide which realtor to use, when to place your home on the market, and of course the list price.

7.1   Chapter Objectives

To properly make inferences about the population mean, you must understand the behavior of the sample mean, $\overline{X}$. Gaining an understanding of $\overline{X}$ is the focus of this chapter. In Chapter 6 you saw that the behavior of a quantitative variable can be described by its probability distribution, and this distribution in turn can be described by parameters such as the mean and the variance. In this chapter we make the link from the probability concepts you learned in Chapters 5–6 to the inferential statistics techniques covered in the remaining chapters in this book.

You first saw Figure 7.1 in Chapter 1. The larger circle represents the population you are studying, and the smaller circle represents the sample you have taken from this population. In reality, the sample circle should sit inside the population circle since it is a portion of it. This figure helps you see the relationship between probability and inferential statistics.

Chapters 2, 3, and 4 taught you the techniques of *descriptive statistics*, which you use to describe the sample. These techniques work on the smaller circle and are very helpful in giving an understanding of the sample data, but they do not let you use the information in the sample to draw conclusions or inferences about the population. In

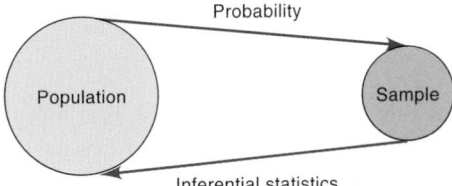

FIGURE 7.1   Relationship between probability and inferential statistics

293

other words, you do not yet have the ability to link that smaller circle to the larger one. In Chapters 5 and 6 you learned some tools of probability. In particular, you learned how to calculate the likelihood of a particular sample being selected from a population. You were thus making the trip across the top arrow in Figure 7.1, from the population to the sample.

But this is not really our desired goal. We wish ultimately to make the trip from the sample to the population along the bottom arrow in Figure 7.1. We want to use the information in the sample to make probabilistic statements about the behavior of the population. This is the purpose of *inferential statistics*. The key to this process is found in this chapter, and the rest of the chapters in this book can be opened only with the key.

The key to inferential statistics is found in this chapter.

More specifically, this chapter covers the following material:

- Point estimators—motivation and properties
- The Central Limit Theorem
- Large-sample confidence intervals for the mean
- Small-sample confidence intervals for the mean
- Confidence intervals for qualitative data
- Sample size calculations

7.2 Motivation for Point Estimators

Remember! A parameter is a numerical descriptor of the population. Parameters are typically unknown.

Since most likely we will not know the values of the parameters that describe the population, we must resort to using the information contained in the sample. It seems logical that if we can identify a numerical descriptor for the sample, then this statistic, called a point estimate, can be used to estimate the corresponding measure for the population. This is the right idea and leads us to the definitions of a **point estimate** and a **point estimator.**

> A *point estimate* is a single number calculated from sample data. It is used to estimate a parameter of the population.

> A *point estimator* is the formula or rule that is used to calculate the point estimate for a particular set of data.

Sample statistics become point estimates.

You have been working with point estimates without really knowing it. Up to this point, values calculated from the sample have been called sample statistics. Now we establish the link between sample statistics and the corresponding population parameters. In doing so we create point estimates out of these sample statistics. Let's take a look at an example.

EXAMPLE 7.1

THE HOUSING MARKET

Looking at Some Point Estimates

Suppose you have sample data on 20 houses that recently sold in your area, including the numbers of days on the market. You decide to use the tools of descriptive statistics to describe this sample. The results of your analysis in Minitab are printed here. Each of the values in the table meets the definition of a point estimate. Each is a single number calculated from the sample, and it can be used to estimate an unknown population parameter. The sample mean, median, and trimmed mean (TrMean) can be used to estimate the true unknown mean days on the market for the population of all houses in this area.

Collect and analyze the data.

Descriptive Statistics: DOM (days on market)

Variable	N	Mean	Median	TrMean
DOM	20	52.9	45.0	47.1

Variable	Minimum	Maximum	Q1	Q3
DOM	2.0	208.0	15.5	83.3

Before we go any further, we must consider what parameters of the population we wish to estimate.

7.3 Common Point Estimators

In Chapter 1 you learned that two major types of data can be used to classify variables: *qualitative* and *quantitative*. You also learned that the tools that can be used to analyze data depend on the type of data. Statisticians must know what kind of data they are analyzing before they select analysis tools and make inferences.

Remember! A quantitative variable is numerical and a qualitative variable is descriptive.

7.3.1 POINT ESTIMATORS FOR QUANTITATIVE VARIABLES

If we are studying a *single quantitative variable*, then we typically wish to know the values of the *population mean* and the *population standard deviation*. That is, we wish to know the center and the variability in the population. Remember that we want to know the values of those parameters that describe the behavior of the population.

EXAMPLE 7.2

THE HOUSING MARKET

Identifying Parameters of Interest

For the housing market, the population consists of all houses that are for sale in the area. One of the variables is the number of days on the market. The number is a quantitative variable, and so we are interested in knowing the mean number of days on the market and the standard deviation of the number of days on the market. Knowing the mean and standard deviation of the number of days will allow you to determine how to plan your move and when you should put your house on the market to be relatively sure of selling it in time to move. These are the population parameters of interest.

Understand the problem.

The population mean is always labeled with the Greek letter μ (pronounced *mu*), and the population standard deviation is labeled with the Greek letter σ (pronounced *sigma*). These values are typically unknown, so we use point estimates to estimate these values. In particular, we use the sample mean, $\overline{X}$, to estimate μ, and we use the sample standard deviation, s, to estimate σ. It is logical to use the sample mean to estimate the population mean and the sample standard deviation to estimate the population standard deviation. In the next section we discuss in more detail why these are the point estimates of choice.

This notation was introduced in Chapter 3.

Now suppose we wish to compare two quantitative variables. For example, we may want to compare the following pairs:

- The times to sell a house in two different areas
- The mile splits of runners under age 25 and those of runners between ages 25 and 30
- The times it takes to get your burger at McDonalds and at Burger King
- The salaries for men and women in the same occupation
- The lengths of patient stays in the hospital for two different teams of doctors

In many situations we study two quantitative variables and wish to compare them. Often we want to know how the true population means compare and how the amount of variation in one population compares with the amount of variation in the other population. In order to keep track of the different populations, we arbitrarily label one population as population 1 and the other as population 2.

Use subtraction to compare two population means.

To compare two numbers, what computation would you do? For example, if you want to compare your salary to with your friend's salary, what do you do with the two salary figures? If you said *subtract* one number from the other, then you are correct. By convention we always subtract the second mean from the first, and thus we need to estimate the *true difference* between μ_1 and μ_2, or $\mu_1 - \mu_2$. It makes sense to estimate this true difference by using the actual difference in the two sample means, or $\overline{X}_1 - \overline{X}_2$. If the two population means are the same, then the difference in the sample means should be close to 0. But remember that even if the two population means are the same and we select a sample from each population, we will most likely get two slightly different sample means. This is because we are looking at only a piece of each of the populations. This is what we have called *sampling error*. So just because the sample means are different, this does not at all imply that the population means are different.

Use a ratio to compare variability.

Another way to compare two numbers is to find their *ratio*. If the numbers are the same, then their ratio is 1. We use ratios to compare the amount of variation in the first population with the amount of variation in the second population. So far in this book we have been using the standard deviation as a measure of variation. Unfortunately, for mathematical reasons, to compare the variations in two populations, we must compare the variances. Recall that the variance is the standard deviation squared. So if the standard deviation is 3 grams, then the variance is 9 grams squared. The variance is not usually quoted because the units of measure are so odd—no one thinks in terms of grams squared. In this case what we wish to estimate is σ_1^2/σ_2^2. It makes sense to estimate this by calculating the ratio of the sample variances, or s_1^2/s_2^2.

✔ TRY IT NOW!

Trial Days

Comparing Point Estimators Calculated from Samples Selected from Different Populations

Use the samples given here. Suppose the first variable is the number of days trials last in Bigcity, and the second variable is the number of days trials last in Smalltown. Select a sample of size $n = 10$ trials from both locations. Assume that the trial lengths in both locations are normally distributed with a mean of 5 days and a standard deviation of 1 day. Find $\overline{X}_1$, $\overline{X}_2$, and $\overline{X}_1 - \overline{X}_2$.

Bigcity	Smalltown
4	3
5	6
6	6
2	4
6	3
4	3
3	4
4	2
4	4
5	4

What do you notice about the difference in the sample means even though the population means are the same?

Find s_1^2 (Bigcity standard deviation squared), s_2^2, and s_1^2/s_2^2.

What do you notice about the ratio of the two sample variances?

7.3.2 POINT ESTIMATORS FOR QUALITATIVE VARIABLES

If the variable we are studying is a *qualitative variable*, then we typically wish to know what *proportion* or *percentage* of the population has that particular characteristic. For example, we may wish to know the percentage of people who change jobs every 6 years, the percentage of people who graduate from college in 4 years, or the percentage of people who favor the death penalty. The true unknown *population percentage* is labeled p. We use the sample proportion $\hat{p}$ (read p-hat) to estimate p.

p is the notation for population proportion.

We have used Greek letters (μ and σ) to represent population parameters, but we do not follow the same convention for population proportions. We want to avoid the use of π, which, of course, is traditionally reserved for the transcendental number 3.1415. . . .

EXAMPLE 7.3

TIME MAKES MEMORIES FONDER — BUT FALSE

Collect and analyze the data.

Using the Sample Proportion to Estimate the Population Proportion

Ah, memories—like the time you saw that car accident, or the spill in the grocery store, or Bugs Bunny at Disneyland. Only you really didn't! A novel set of experiments suggests that people unconsciously tamper with their own memories, inventing causes for events they see around them to help make sense of things. Researchers from the University of Puget Sound and Boston University showed slides of scenes from settings such as a grocery store or a restaurant to 144 student study participants. Of the 144 students in the study, 97 "filled in the blanks" in their memories, claiming to have seen slides that were not originally presented but that matched a scene's narrative. For example, to explain a spill in the grocery store scene, many claimed to have seen a picture of someone pulling an orange from a pile of fruit. We find $\hat{p} = \frac{97}{144} = 0.6736$ or 67.36% as the estimate for the population proportion of people who do this.

ANS. SAMPLE MEANS ARE 4.3 AND 3.9. THEY ARE NOT THE SAME, SO THE DIFFERENCE IS NOT 0. SAMPLE VARIANCES ARE 1.57 AND 1.66. THEY ARE NOT THE SAME, SO THE RATIO IS NOT EXACTLY 1.

As with quantitative variables, we often wish to compare two qualitative variables. Here are some example comparisons:

- The proportion of men who "make memories fonder" compared with the proportion of women who do this
- The proportion of those younger than 18 living in Chicago compared with the corresponding proportion in San Francisco
- The proportion of single-parent households in 2000 compared with the proportion in 1990
- The proportion of unsafe bridges in Massachusetts compared with the corresponding proportion in Wyoming

When we are comparing two different population proportions, we must be sure that the proportions use the same variable in both populations. In the first example in this list, the two populations are men and women, and in both populations we are studying the proportion of the population that "make memories fonder." We would not, for example, compare the proportion of men who tamper with their memories with the proportion of men who do not do this. That would be silly. These proportions would add up to 1 and so we would not learn anything.

Use subtraction to compare two population proportions.

As we did with population means, we want to compare the true population proportions π_1 to π_2 and we do this with subtraction. We wish to estimate the true difference in the population proportions, or $p_1 - p_2$. It makes sense to estimate this true difference by using the difference between the two sample proportions, or $\hat{p}_1 - \hat{p}_2$. If the two population proportions are the same, then the difference between the sample proportions should be close to 0. Remember that even if the two population proportions are both 0.20 and we select a sample from each population, we will most likely get two slightly different sample proportions. This is because we are looking at only a piece of each of the populations and we have what is called sampling error.

7.3.3 SUMMARY OF COMMONLY USED POINT ESTIMATORS

Table 7.1 summarizes the most commonly used point estimators. In the next section you will see that the estimators listed in the third column are not the only estimators for the corresponding population parameters, but they are the most common ones

TABLE 7.1

SUMMARY OF COMMON POINT ESTIMATORS

Type of variable(s)	Population parameter (unknown, what we wish to estimate)	Point estimator or sample statistic (calculated from sample data)
A single quantitative variable	Population mean, μ	Sample mean, $\overline{X}$
	Population standard deviation, σ	Sample standard deviation, s
Two quantitative variables	Difference in population means, $\mu_1 - \mu_2$	Difference in sample means, $\overline{X}_1 - \overline{X}_2$
	Ratio of population variances, σ_1^2/σ_2^2	Ratio of sample variances, s_1^2/s_2^2
A single qualitative variable	Population proportion, p	Sample proportion, $\hat{p}$
Two qualitative variables	Difference in population proportions, $p_1 - p_2$	Difference in sample proportions, $\hat{p}_1 - \hat{p}_2$

because, in a sense, they are the "best" estimators. In the next section we look at what makes them best.

7.3.4 EXERCISES – LEARNING IT!

7.1 Samples of ten homerun distances for Sammy Sosa and Mark McGwire are listed in the tables:

Sammy Sosa: 1998 season

Date	Against	Pitcher	Runners on	Inning	Distance (ft)
Apr 4	Montreal	R	0	3	371
Apr 11	Montreal	R	0	7	350
Apr 15	N.Y. Mets	L	0	8	430
Apr 23	San Diego	R	0	9	420
Apr 24	L.A.	R	0	1	430
Apr 27	San Diego	R	1	1	434
May 3	St. Louis	R	0	1	370
May 16	Cincinnati	R	2	3	420
May 22	Atlanta	R	0	1	440
May 25	Atlanta	R	0	4	410

Mark McGwire: 1998 season

Date	Against	Pitcher	Runners on	Inning	Distance (ft)
Mar 31	L.A.	R	3	5	364
Apr 2	L.A.	R	2	12	368
Apr 3	San Diego	L	1	5	364
Apr 4	San Diego	R	2	6	419
Apr 14	Arizona	R	1	3	424
Apr 14	Arizona	R	0	5	347
Apr 14	Arizona	R	1	8	462
Apr 17	Philadelphia	R	1	4	419
Apr 21	Montreal	L	1	3	437
Apr 25	Philadelphia	R	0	7	419

(a) What is your estimate of the average homerun distance for Sosa?

(b) What is your estimate of the average homerun distance for McGwire?

(c) What is the difference between the two estimates you found in parts (a) and (b)?

(d) These samples were both selected from populations with a normal population mean of $\mu = 425$ feet. Explain why they have different sample averages.

7.2 Using the data for the homerun distances in Exercise 7.1, estimate the variability in the home run distances for the two players.

(a) What is your estimate of the variability of the homerun distances for Sosa?

(b) What is your estimate of the variability of the homerun distances for McGwire?

(c) What is the ratio of the two estimates you found in parts (a) and (b)?

(d) These samples were both selected from populations with a population variance of $\sigma^2 = 2500$ feet^2. Explain why they have different sample variances.

7.3 Airlines keep a record of the number of minutes each flight is late (or early). Here are the data for 20 flights into Ronald Reagan National Airport on September 1, 2000:

Scheduled arrival time	Arrival delay (minutes)
9:14	−16
9:12	14
10:08	12
11:06	9
15:08	0
17:00	5
19:12	3

(continued)

21:12	7
23:26	7
22:11	0
20:21	0
18:12	91
16:02	8
14:07	15
13:13	−8
12:06	−4
9:14	−14
9:12	−4
12:06	1
14:07	0

(a) Estimate the average delay for flights.

(b) What is the difference between your estimate and a target of 10 minutes?

7.4   The Gallup organization conducted a telephone survey of 1004 U.S. adults and asked the following question: "How satisfied are you with your job, or the work you do?" Here's how the sample responded:

Very satisfied	502
Somewhat satisfied	392
Somewhat dissatisfied	70
Very dissatisfied	40

Source: Survey conducted by the Gallup organization, June 11– June 17, 2001

(a) What is the type of data?

(b) What is your estimate of the proportion of adults who are very satisfied with their jobs?

(c) What other information would you like to know about these respondents?

7.5   A company that buys blank VHS tapes for video recording is concerned about the amount of time the tapes are able to record. The tapes are rated at 120 minutes, but the company knows that the actual recording times vary. Data are collected on 25 randomly selected tapes and their actual recording times (in minutes) are listed here:

116	118	119	119	120
117	118	119	119	121
117	118	119	120	121
117	119	119	120	121
117	119	119	120	121

(a) What is your estimate of the average time available on the tapes?

(b) Does your estimate differ from 120 minutes? If so, by how much?

7.4   Desirable Properties of Point Estimators

In this section we develop the properties of point estimators. We focus on estimators for the population mean, μ, but the resulting properties apply to any point estimator.

In Chapter 3 you learned how to calculate measures of the middle of the sample data—particularly the sample mean, $\overline{X}$, the sample median, and the sample mode. The trimmed mean was also examined as a measure of the middle of the sample data. It makes sense to consider using one or more of these statistics as the point estimator of the unknown value μ.

Collect and analyze the data.

EXAMPLE 7.4

THE HOUSING MARKET

Considering Possible Point Estimates of μ

Suppose we take a sample of five homes and record the number of days each was on the market before it sold. The data are 76, 48, 42, 9, and 27. The sample mean is 40.4 days and the sample median (the middle score) is 42 days. This data set has no mode because none of the values occurs more than once. We could use the value of either 40.4 or 42 as our point estimate for the unknown true mean number of days to sell a home.

It appears that the mean, the median, and the mode all fit the definition of a point estimator. In fact, we could dream up many other formulas that would fit the definition of a point estimator. For example, we could decide to average the minimum and maximum values in the sample and call this a point estimator. It would be a legitimate point estimator because the calculation is based on only sample data and yields a single number. Clearly, we need some criteria to judge which of these point estimators is the best one to use to estimate μ.

Let's think about the criteria we could use. If the point estimate is to be our best guess of the value of the unknown population parameter—in this case, μ—then we want the point estimate to be close to μ. Ideally, we want our point estimate to hit μ on the nose. But we know that the chance of that happening is pretty slim. In fact, we know from Chapter 1 that the sample is only a part of the population and if we were to take a different sample, we would get a different sample mean, a different sample median, and a different sample mode.

Collect and analyze the data.

EXAMPLE 7.5

THE HOUSING MARKET

Looking at Point Estimates from a Second Sample

We take a second sample of five houses that sold the next week and we find they were on the market these numbers of days: 140, 10, 75, 24, and 4. Some quick calculations yield

Sample mean = 50.6 days   Sample median = 24 days   No sample mode

The point estimates are different now. We used the same point estimators (the formulas to calculate the sample mean and sample median), but we got different numbers because the data in the second sample are different from those in the first sample. This tells us that the point estimate depends on the particular sample we happen to choose and it changes from sample to sample. When we combine this fact with the criterion of getting close to the unknown μ, we see that we want the point estimator that comes closest to μ for most samples. That is, we want a point estimator that does not yield radically different numbers from sample to sample.

Suppose we dreamed up a formula for a point estimator for μ and we used it with two different samples from the same population. For the first sample our point estimate was calculated to be 24, and for the second sample it was calculated to be 245. How much faith should we have in our estimate? Not much. Why? The feature of this point estimator that is troubling is that it yields wildly different estimates from sample to sample. The estimator has too much variability; that is, it jumps around too much from sample to sample.

In summary, what we really want is a point estimator with these three properties:

1. The point estimator should estimate fairly the unknown population parameter.

2. The point estimator should yield a number close to the unknown population parameter as the sample size increases.

3. The point estimator should not have a great deal of variability.

Even though we have been considering point estimators for the unknown population mean, μ, these properties are desirable for any point estimator. We state these three properties more precisely:

Properties of point estimators

1. The point estimator should be unbiased.

2. The point estimator should be consistent.

3. The point estimator should be efficient.

Unbiased estimators are fair. They do not systematically overestimate or underestimate the parameter.

The word *unbiased* in statistics has basically the same meaning that it has in general language. When the word is used as a character trait for somebody, it implies that the person is fair and does not favor something or someone. We want our point estimator to behave in this fashion as well. The point estimator should not always overestimate μ or always underestimate μ. That estimator would be a biased or an unfair estimate. Sometimes the estimator $\overline{X}$ will be smaller than μ and sometimes it will be larger than μ, but we know that it will be close to μ most of the time.

This is basically the same definition you saw in Chapter 1 when you learned about bias in a sample.

> An **unbiased estimator** yields an estimate that is fair. It neither systematically overestimates the parameter nor systematically underestimates the parameter.

Proving that an estimator is unbiased is the job of the mathematical statisticians. They have shown that $\overline{X}$ is an unbiased estimator and we believe them! They have also shown that the median is an unbiased estimator only if the population has a normal distribution. If a proposed point estimator is wildly biased, the bias may be obvious simply from the formula. Let's look at one such point estimator for μ, one that is clearly ridiculous.

Collect and analyze the data.

EXAMPLE 7.6

THE HOUSING MARKET

Using a Biased Estimator

Someone in the real estate company proposes using the smallest observation in the sample as the estimate of μ. For the sample in Example 7.4 this yields an estimate of 9 days, and for the sample in Example 7.5 this gives an estimate of 4 days. Clearly, this rule of selecting the minimum value as the estimate of μ is silly. It will always yield a number that is too small and is clearly an unfair estimate of the middle.

Mathematical statisticians have also shown that $\overline{X}$ does not fluctuate a great deal. $\overline{X}$ has a smaller standard deviation than the median or the mode; that is, it jumps around less from sample to sample. If you think about the way the sample mean, the sample median, and the sample mode are found, you can see that this makes sense. The sample mean incorporates all of the data in the sample and thus washes out the effect of any single number in the sample that is unusual. Both the median and the mode are based on only a portion of the sample and thus tend to vary more from sample to sample. The sample mean, $\overline{X}$, is clearly the best estimator of the unknown population mean, μ. The following example demonstrates this fact.

EXAMPLE 7.7

THE HOUSING MARKET

Collect and analyze the data.

Looking at Three More Samples

Let's follow the housing market for three more weeks. Each week a sample of five observations is collected on the numbers of days houses were on the market before they sold. The observations are listed here along with the samples from Examples 7.4 and 7.5:

						Sample mean	Sample median	Sample mode
Week 1	76	48	42	9	27	40.4	42	None
Week 2	140	10	75	24	4	50.6	24	None
Week 3	14	87	68	94	120	76.6	87	None
Week 4	3	73	24	39	160	59.8	39	None
Week 5	52	13	4	30	13	22.4	13	13

You can quickly see that the sample mode is not a very useful point estimator. Only one of the samples has a mode.

As we look at the sample median and the sample mean, let's think about the two properties of point estimators that we have been discussing. We must look at where the values of the point estimator are centered and how much the values vary from sample to sample. Which numbers have more variation: the sample means or the sample medians? How can we tell? We can calculate the average and the standard deviation of the sample means as well as the average and the standard deviation of the sample medians.

The average of the sample means is 49.96 days and the standard deviation is 20.36 days. The average of the sample medians is 41 days and the standard deviation is 28.26 days. The sample medians jump around more than the sample means; that is, they have more dispersion.

You can understand that the point estimator yields values that are clustered around the true population parameter and do not vary much from sample to sample. Now consider the next five weeks of data.

✔ TRY IT NOW!

The Housing Market
Comparing the Variability of Two Point Estimators

The numbers of days homes are on the market in the next five weeks of sales are listed here:

Week 6	84	57	41	38	33
Week 7	2	55	2	163	33
Week 8	89	22	108	130	15
Week 9	125	3	93	22	27
Week 10	108	9	62	49	14

For each sample, calculate the sample mean and the sample median.

Find the average of the sample means and the average of the sample medians.

Find the standard deviation of the sample means and the standard deviation of the sample medians.

Which point estimator has less variability?

Here is a different variable.

✓ TRY IT NOW!

The Housing Market
Comparing the Variability of Two Point Estimators

In addition to being concerned about how long it will take to sell your house, you are clearly interested in the price you can sell it for. The selling prices for samples of homes sold over the past five weeks are listed:

Week 1	$75,000	$90,500	$68,000	$73,500	$92,400
Week 2	84,000	99,000	123,000	123,000	109,900
Week 3	118,000	109,000	89,900	99,000	112,000
Week 4	95,000	100,000	105,000	97,900	127,500
Week 5	89,900	99,900	149,500	79,900	96,000

For each sample, calculate the sample mean and the sample median.

Find the average of the sample means and the average of the sample medians.

Find the standard deviation of the sample means and the standard deviation of the sample medians.

Which point estimator has less variability?

ANS.

Sample mean	Sample median		
$79,880	$75,000		
107,780	109,900		
105,580	109,000		
105,080	100,000		
103,040	96,000		
Average $100,272	$97,980		
Standard deviation $11,523	$14,139		

THE MEAN HAS LESS VARIABILITY.

ANS.

Sample mean	Sample median		
50.6	41		
51	33		
72.8	69		
54	27		
48.4	49		
Average 55.36	47.8		
Standard deviation 9.95	24.48		

THE MEAN HAS LESS VARIABILITY.

7.5 Distribution of the Sample Mean, $\overline{X}$: The Central Limit Theorem

From your work in the preceding section you decide to use $\overline{X}$ as the point estimator for the true unknown population mean, μ. You are trying to decide on a real estate agency, and you know that the average time a house is on the market is 50 days. The sample mean $\overline{X}$ will not equal 50 days, but it should be close to 50 days if the agency is doing its job. You take a sample and find a sample mean of 94.4 days. By looking at the last ten values of $\overline{X}$, you realize that a sample mean of 94.4 is high, but you are not sure whether this is caused by simple sampling error or whether the agency is not aggressively marketing the homes. How can you decide?

7.5.1 PUTTING Z-SCORES AND THE EMPIRICAL RULE TO USE

In Chapter 3 you learned about the *empirical rule* and *Z-scores*. These tools allow us to decide whether an individual observation is an outlier or unusual. The formula for the z-score is

$$Z = \frac{\text{Distance between the data value and the mean}}{\text{Standard deviation}}$$

A Z-score measures the number of standard deviations that a data value is from the mean.

Since we are trying to decide whether an $\overline{X}$ value of 94.4 days is unusual, it seems like a good idea to calculate the Z-score for the sample mean of 94.4 days. If 94.4 days is more than three standard deviations away from the regional mean of 50 days, then we are pretty sure there is a problem. Why? Because the empirical rule tells that Z-scores beyond ± 3 occur less than 1% of the time. A sample mean of 94.4 days would be *extremely unlikely* (less than 1% chance) *if* $\mu = 50$ days.

Now our data value is the observed sample mean, $\overline{X}$. This is a bit different from our earlier use of Z-scores, but the idea is exactly the same. We are trying to decide whether a particular $\overline{X}$ value is too unusual, whereas in Chapter 3 we were trying to decide whether a particular individual observation was too unusual. We can rewrite the formula for Z to reflect this fact:

$$Z = \frac{\overline{X} - \mu_{\overline{X}}}{\sigma_{\overline{X}}}$$

Z-score for $\overline{X}$

Notice that the average is now the average of all of the possible $\overline{X}$ values and is therefore labeled $\mu_{\overline{X}}$. The standard deviation is the standard deviation of the $\overline{X}$ values and is labeled $\sigma_{\overline{X}}$. Although this calculation for the Z-score looks different, it is really the same beast we used in Chapter 3. It simply measures the number of standard deviations that a particular $\overline{X}$ value is from the mean of all possible $\overline{X}$ values.

So, we should just calculate the Z-score for the sample mean of 94.4 days. But if we look at the formula, we see that to calculate the Z-score we need to know the *average of the sample means* and the *standard deviation of the sample means*. In Try It Now! in Section 7.4, you calculated the average and the standard deviation of five sample means. We need to expand these calculations to include all possible sample means. In other words, what we need now is the average of all possible sample means and the standard deviation of all possible sample means.

Clearly, we do not want to select all possible samples from the population. This would defeat the whole idea of using one sample to learn about the population. Fortunately, we can once again rely on our friends the mathematical statisticians who have proved a theorem that gives us the results we need. This Central Limit Theorem is the foundation for virtually all of inferential statistics. It is the key that unlocks the rest of the course.

7.5.2 THE CENTRAL LIMIT THEOREM

Let's look at the housing data again. In Section 7.4 we looked at ten weekly samples. Each sample consisted of $n = 5$ observations of houses sold and the numbers of days they were on the market. The value of the sample mean, $\overline{X}$, varied from sample to sample. We called $\overline{X}$ a point estimator, but it is also a *random variable*. In Chapter 6 you learned that a random variable has a mean, a standard deviation, and a probability distribution. Thus, the random variable $\overline{X}$ has a mean, a standard deviation, and a probability distribution.

Remember! A random variable is a quantitative variable with a value that varies according to the rules of probability.

When the random variable is a point estimator, then the standard deviation and the probability distribution are labeled with slightly different words.

> The probability distribution of a point estimator or a sample statistic is called a **sampling distribution**.

> The **standard error** is the standard deviation of the sampling distribution of a point estimator. It measures how much the point estimator or sample statistic varies from sample to sample.

EXAMPLE 7.8

Collect and analyze the data.

THE HOUSING MARKET

Looking at $\overline{X}$ As a Random Variable with a Mean, Standard Error, and Sampling Distribution

Each of our samples from ten weeks had $n = 5$ observations (remember that sample size is always labeled n). These are the ten sample means of days on the market:

40.4 50.6 76.6 59.8 22.4 50.6 51 72.8 54 48.4

These ten numbers are some of the possible values for $\overline{X}$ but not all of them. To list all of the possible sample means, we would have to take all possible samples of size five from the population of all houses sold. This is not possible and, we will see, not necessary.

We can get a handle on how $\overline{X}$ behaves by looking at just these ten values: The average of the ten $\overline{X}$ values is 52.66 days, and the standard error of the ten $\overline{X}$ values is 15.37 days. The dotplot (done in Minitab) of these ten $\overline{X}$ values looks like this:

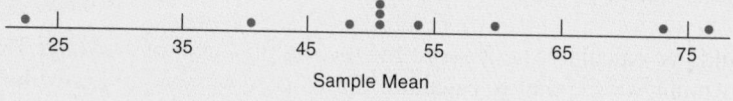

Sample Mean

Before stating the theorem let's see if we can anticipate the results by looking at Example 7.8. It looks like the mean of the $\overline{X}$ values is close to the population mean, $\mu = 50$ days. The standard error of the $\overline{X}$ values is less than the population standard deviation, $\sigma = 60$ days. It makes sense that the sample means are more similar (have a smaller standard error) than the individual days on the market because the averaging process washes out the effect of individual values. Finally, the sampling distribution or the histogram of the $\overline{X}$ values doesn't have much of a shape now, but remember that we have only ten $\overline{X}$ values in this example.

Central Limit Theorem (CLT): In random sampling from a population with mean μ and standard deviation σ, when n is large enough, the distribution of $\overline{X}$ is

- approximately normal with
- a mean, $\mu_{\overline{X}}$, equal to μ and
- a standard error, $\sigma_{\overline{X}}$, equal to $\sigma/\sqrt{n}$

The CLT holds regardless of the shape of the population distribution if n > 30.

The Central Limit Theorem applies for large enough sample sizes. A "large enough" sample size depends on how much the population distribution deviates from a normal distribution. Typically, if the sample size is greater than 30, then it is considered large enough. The larger the sample size, the better the normal approximation is. If the population we are sampling from has a normal distribution, then the results of the CLT hold even for small samples ($n \leq 30$).

The CLT holds for small samples when the popultion has a normal or symmetric distribution.

7.6 The Central Limit Theorem—A More Detailed Look

Now we restate the Central Limit Theorem and examine each of its three points in terms of the housing data.

7.6.1 THE SHAPE OF THE SAMPLING DISTRIBUTION OF $\overline{X}$

The *first point* of the Central Limit Theorem (CLT) is that if we take all possible random samples from an arbitrary population and calculate all the possible sample means, then the distribution of the sample means will be approximately normal. Remember that this point applies when the sample size is large enough or when the underlying population distribution is normally distributed.

The first point of the CLT: The distribution of all possible sample means is approximately normal.

The housing market data are samples of size $n = 5$ for each week. Because each individual sample size is small ($n = 5$), to apply the CLT we need to assume that the underlying distribution of days on the market is normal or approximates a normal shape. We can get a sense of the shape of the population distribution by examining a histogram of sample observations. The histogram of the days on the market for 150 houses (representing 30 weeks of sampling) is shown in Figure 7.2 on page 308.

This histogram gives a picture of the sample, not the population. But remember that we can learn about the population by studying a representative sample like this one. If the sample looks reasonably bell-shaped, the population is probably also bell-shaped. In Chapter 15 you will learn a more precise way to make an inference about the shape of the population.

From the distribution of a sample of 150 values of days on the market, based on the histogram of the sample, we conclude that the population is not reasonably close to a normal distribution. Thus, we don't expect the distribution of the sample means to be approximately normal even though each sample size is small ($n = 5$). This conclusion is supported by the histogram of the 50 sample means shown in Figure 7.3.

The distribution of the sample means does not look normal, as we expected. Yet, even though the underlying population mean is not normal (see Figure 7.2) and the sample size is small ($n = 5$), the sampling distribution of the sample mean of days on the market is not that far from a normal curve. The CLT states that the sampling distribution of $\overline{X}$ is approximately normal if we are sampling from a normal distribution. Although this is not the case for the variable *Days on market*, it is true for the variable *Selling price* for the housing data. The histogram of the selling

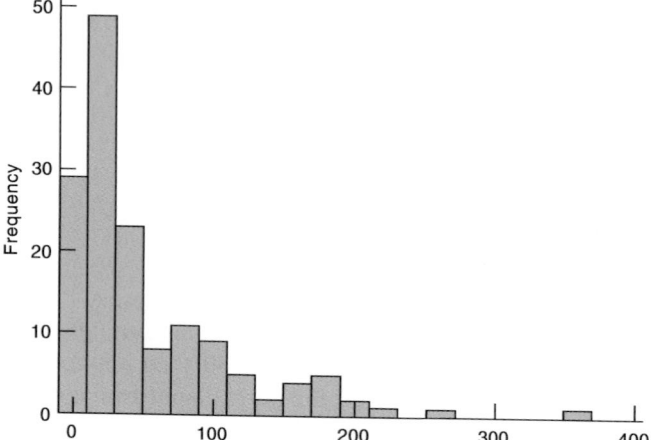

Remember! You can alter the shape of the histogram by adjusting the class interval width.

FIGURE 7.2 **Minitab histogram of the days on the market**

prices for all 150 values is shown in Figure 7.4, and it is approximately normal. Figure 7.5 is the histogram of 50 sample mean selling prices, and it does indeed look approximately normal.

If the sample size is sufficiently large, the CLT states that the sampling distribution of $\overline{X}$ will be approximately normal regardless of the shape of the underlying population distribution. This is great news because most of the time we will not know the shape of the population from which we are sampling.

7.6.2 THE MEAN OF THE SAMPLING DISTRIBUTION OF $\overline{X}$

The second point of the CLT: The average of all possible sample means is μ.

The *second point* of the Central Limit Theorem is that the mean of the sampling distributions of $\overline{X}$ equals the mean of the original population. This means that the center of the histogram of the $\overline{X}$ values should be μ. Consider the selling prices of houses. Figure 7.5 illustrates that most of the $\overline{X}$ values are close to $\mu = \$105,000$. The CLT states that if we take all possible samples of size n, with n sufficiently large, calculate

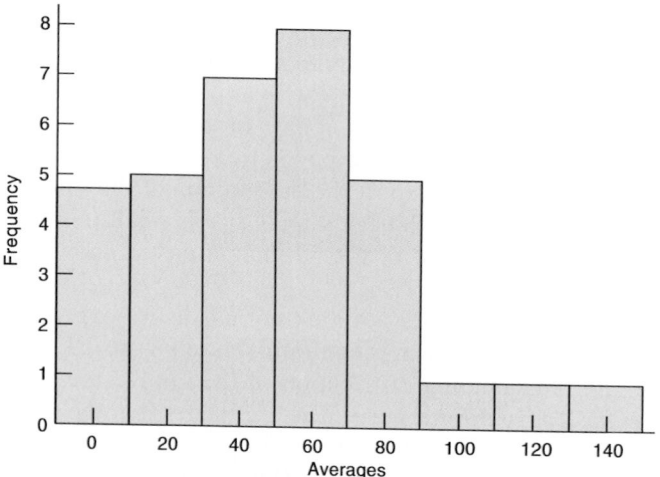

FIGURE 7.3 **Minitab histogram of 50 sample means of the days on the market**

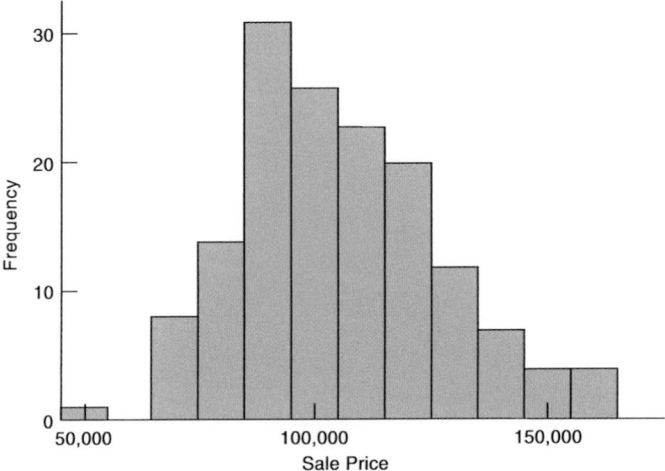

FIGURE 7.4   Minitab histogram of 150 selling prices

the sample mean for each of these samples, and average the sample means, then we will get μ. In terms of our notation, this means $\mu_{\overline{X}} = \mu$. For the housing market the average of all the $\overline{X}$ values should be close to the population mean of $\mu = \$105,000$. In fact, the average of the 50 sample means is $\$104,884$.

Why is this important? Clearly, we are not going to take more than one sample. We will take one sample and have one sample mean, $\overline{X}$, which will be our estimate of μ. The CLT states that we can be reasonably sure that the sample mean that we get will be close to the true population mean because the sample means cluster around the true mean, μ. Later in this chapter we quantify the words *reasonably sure* and *close* by using what is known as a *confidence interval*.

7.6.3 THE STANDARD ERROR OF THE SAMPLING DISTRIBUTION OF $\overline{X}$

The *third point* of the Central Limit Theorem says that the standard deviation of the $\overline{X}$ values (also called the standard error) depends on two factors: the amount of variability in the population, σ, and the sample size, n.

The third point of the CLT: The standard error of $\overline{X}$ is $\sigma/\sqrt{n}$.

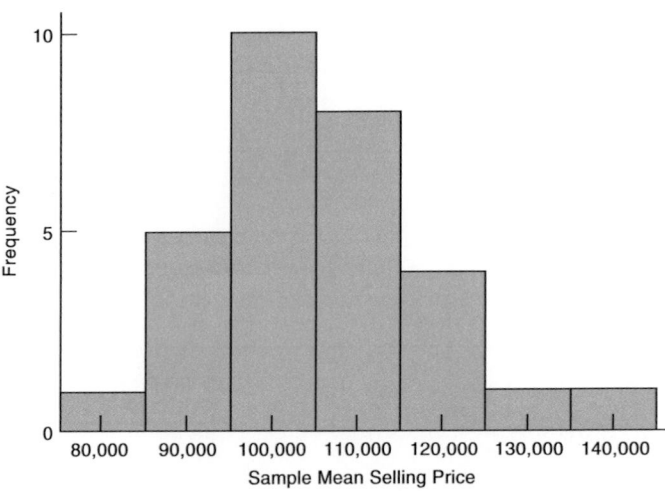

FIGURE 7.5   Minitab histogram of the mean selling prices

Two Populations with Normal Distributions

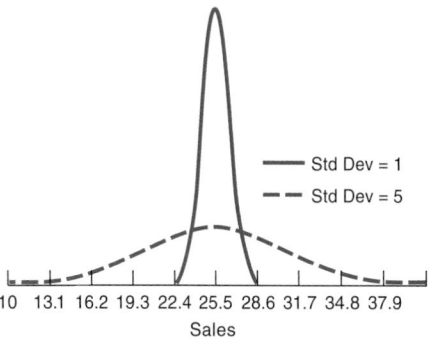

FIGURE 7.6 **Graphs of two populations with the same mean but different standard deviations**

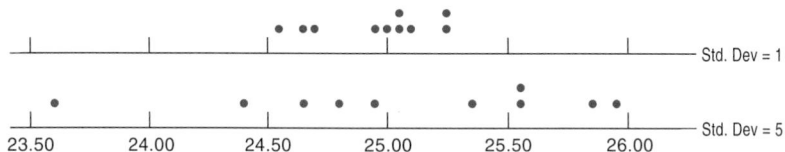

FIGURE 7.7 **Dotplots of sample means**

Let's think about what this means. Suppose we have a population of students who are all the *same age*, we pick ten different samples from this population, and we calculate the average for each of the ten samples. How different will the sample means be from one another? They will all be the same because everyone in the population is the same age! There is no variability in the population, $\sigma = 0$, and so there is no variability in the $\overline{X}$ values. Clearly, this is an extreme example.

Now consider one population that has a mean of 25 and a standard deviation of 1 and another population that has a mean of 25 and a standard deviation of 5. Graphs of these distributions are shown in Figure 7.6. We take ten samples of size $n = 30$ from each population and calculate the ten sample means. How do the ten sample means from the first population behave compared with the ten sample means from the second population? The dotplots of the sample means are shown in Figure 7.7. In both cases, the sample means cluster around $\mu = 25$, but the sample means from the first population appear to be more similar (less dispersion) simply because the values in the first population are more similar. We can capture this idea in a statement:

> As the standard deviation in the population increases, the standard error of $\overline{X}$ also increases.

✓ **TRY IT NOW!**

The Central Limit Theorem
Investigating the Impact of Variability on Standard Error

Use a software package such as Excel or Minitab to simulate picking ten samples each of size $n = 35$ from two different populations:

Population 1: Monthly sales of a leading online bookstore, normally distributed with mean $\mu = \$25{,}000$ and standard deviation $\sigma = \$1000$

Population 2: Monthly sales of a competing online bookstore, normally distributed with mean $\mu = \$25{,}000$ and standard deviation $\sigma = \$3000$.

Alternatively, you can use these samples:

Samples from population 1

1	2	3	4	5	6	7	8	9	10
24.7	23.7	25.2	26.3	26.2	26.7	22.8	24.8	26.1	23.9
24.3	23.3	23.2	24	24.2	22.9	24.4	24.6	25.1	24.6
24.7	24.6	26.3	24.9	24.8	24.5	27	25.9	27.4	24.3
26.7	23.4	25.5	25.9	26.9	24.9	24.5	25.7	24.6	25.8
23.6	24.2	23.5	24.6	25	25	24.7	27.2	23.3	24.3
22.4	26.4	23.7	24.3	25.8	25.5	25.9	25.6	23.6	23.9
25.7	25.3	24.1	24.8	25.1	25.6	25.1	24.1	26.9	25.5
25.1	25.8	25.9	24.4	24.1	26.1	23.8	23.4	25.7	25.6
27.2	26.4	26.3	25.1	25	25.5	25	23.9	23.2	25.8
25.4	25.6	25.2	24	26.2	24.7	24.2	24.2	24.6	24.5
24.5	25.8	25.5	24.4	26.3	23.2	25.6	24.9	25	24.3
24.5	25.8	25.8	25.5	25.7	26.6	25.3	25.6	26.9	24.7
26	25.1	26.1	24.9	24.2	26.1	24.4	25.5	25.5	24.7
24.6	23.6	27	24.4	25.1	24.8	27.8	26.3	25.9	26.3
25.2	25.5	24.8	23.8	26.3	24.7	23.7	25.8	25.8	25.4
25.4	24.4	25.7	25.5	26.1	22.2	25.5	26.5	23.3	25
26.4	26.7	25.2	25.1	26.9	25	24.3	25.9	25	26
25.4	26.5	24.9	24	23.5	22.4	25.1	25.3	25	25.2
25.3	26.5	25.1	23.9	23.6	25.2	26	23.5	24.4	25.5
24.8	24.7	24.8	24.6	24.2	25.6	25.1	26.3	25.3	25.1
25.2	25.2	25	23.8	25.2	25.1	25.8	24.3	25.6	25.2
25.1	25.9	26.8	23.7	23.8	25.6	26.2	24.1	22.9	24.1
25.2	25.4	24	24.1	25.5	25.1	24.5	25.1	24.1	24.1
26	26.6	24.3	24.8	25.1	26.8	25.6	26.7	25.2	26.6
24.4	24	26.2	25.9	23.3	26.3	24.8	23.9	23.1	24.2
25.5	25.9	24.9	24.9	25.2	24.6	25.9	24.5	27.3	25.7
25.6	24	25.9	26.4	24.3	24.7	23.7	26.4	26	25.2
22.8	25.5	25.5	26.2	24	26.2	24.4	24.3	23.3	26.4
24	26.5	23.8	24.6	23.7	26.6	25	25.3	26	26.9
24.9	24	24.4	25.8	25.9	25.5	23.5	23.6	24.9	24.4
24.4	25.9	24.2	24.9	24.8	26.5	23.7	24.1	24.8	23.9
23.6	22.8	25.6	24.7	25.4	23	26.2	23.1	24.5	24
24.9	25.1	26.7	24.5	24.2	24.3	25.6	26.3	23.6	26
26.6	24.3	25.1	24.2	25.7	25.2	25	24	24.7	25.7
25	26	24	25.8	26.2	23.8	25.4	23.2	25.1	26

Sample averages

1	2	3	4	5	6	7	8	9	10
25	25.2	25.1	24.8	25.1	25	25	25	25	25.1

Samples from population 2

1	2	3	4	5	6	7	8	9	10
21.7	19.2	29.3	23	17.9	15.9	26.3	23.2	31.2	29.7
18.5	25.2	27.8	29.3	29.1	25.5	19.9	22.4	28.9	25.3
24.1	27.7	19.9	28	25.6	19.2	29.6	23.1	26.7	23.9
31.8	23.7	16.1	18.7	26.7	21.5	18.6	19.5	29.2	20.6

(*continued*)

21.8	27.5	29.9	25.2	24.5	19.8	25.8	22.3	22.8	25.5
25.5	25.6	22.9	23	19.7	34.9	21.8	20.1	23.4	29.9
19.1	14.2	24.3	20.8	17.4	28.9	17.8	20.8	22.3	27.5
28.4	24.1	24	20.6	27.3	19.7	24.9	19.9	28.6	24.9
9.4	30.2	23.7	22.1	18.6	25.2	21.9	26.6	23.9	16.1
18.2	32.6	25.3	25	23.7	26.2	28.4	20.3	22.9	29.9
16.9	28.4	25.8	10.8	29.5	21.2	20.7	33.6	23.4	23.5
35.2	26.6	31.6	21.9	25.3	24.7	27.5	25.5	29.8	37.1
21.9	19.9	33.9	30.3	25	20.4	17	35.7	20	22.7
25.4	16.2	32.9	31.1	28.3	26.7	25.7	27.9	30.2	29.2
29.7	31.9	20.6	37.3	24.2	30.1	35.7	20.3	21.6	21.8
25.6	20.6	28.6	33.2	32.3	18.1	30	31.7	18.8	25.4
14.7	22.4	27.5	38.8	27.3	21.1	21.1	29.3	23.8	26.6
14.6	29	28.1	23.4	29.2	27	27	27.4	28.5	27.6
17.2	16.6	25.5	18.6	25.1	31.6	29.4	23.5	25.1	17.8
27	20.8	27.8	21	28.2	20.2	19	23.4	22.4	29.3
28.9	33	35.8	23	19.1	31.8	26.3	24.4	21.9	31.5
28	26.8	25.9	29.4	22.9	23.6	28	25.3	23	29.1
26.3	26.1	20.9	28	22	20.9	25.2	22.2	22.3	25.5
28.1	27.4	28.6	27.5	32.5	19.6	20.4	20.9	29.4	14.7
26.8	26.9	27.4	31	23.7	36.9	30.7	25.1	23	23
20.2	27	23.7	31.5	27.1	26.6	23.6	24.3	28.5	23.6
23.6	24	26.8	22.2	18.1	31.9	33.9	19.8	21.8	24.6
15.9	28.6	29	19	22.6	26.2	22.1	25.7	24.8	27.5
31.5	32.6	29.9	25.8	22.7	36.5	23	20.9	24.1	19.3
19.1	28.6	28.2	21.5	26	21.4	19.4	27.7	17.6	26.5
30.4	24.3	22.3	30.9	23.7	25.1	21.4	21.2	19.9	13.9
32.1	32.7	26.7	31.1	29.2	24	33.4	23.2	35	25.5
23.5	24.8	20	13.2	29.9	28.8	24.8	32.1	33.3	20.2
20	18.2	26.8	23.8	14.4	22.9	29.1	18.2	32.9	24.9
18.1	23.6	22.8	16.7	23.8	30	23.9	23.6	15.3	16.9

Sample averages

23.4	25.3	26.3	25	24.6	25.3	25	24.3	25	24.6

If you are using your own samples, calculate the sample mean, $\overline{X}$, for each sample.

Find the average and standard deviation of the ten $\overline{X}$ values from the population 1 samples.

Find the average and standard deviation of the ten $\overline{X}$ values from the population 2 samples.

Compare the results for the two populations.

The second number that influences the size of the standard error of $\overline{X}$ is the sample size, n. This should also make sense. Consider the smallest possible sample size: $n = 1$. Clearly, the sample mean of one number is just the single value, and any other sample means have the same variation as the population since $\sigma/\sqrt{n} = \sigma/\sqrt{1} = \sigma$. Typically, though, a sample is more than a single observation

ANS. THE AVERAGE OF THE $\overline{X}$'s SHOULD BE CLOSE TO $\mu = \$25,000$ IN BOTH CASES. THE STANDARD DEVIATION OF THE $\overline{X}$'s FROM POPULATION 1 SAMPLES SHOULD BE LESS THAN THE STANDARD DEVIATION OF THE $\overline{X}$'s FROM POPULATION 2 SAMPLES.

and thus n is usually 2 or greater. No matter what the sample size is, then, the *amount of variation in the $\overline{X}$ values is less than the amount of variation in the original population.*

Collect and analyze the data.

EXAMPLE 7.9

THE HOUSING MARKET

Creating a Frequency Histogram of Selling Prices and Average Selling Prices, $\overline{X}$

We have taken 30 samples of houses sold over the last 30 weeks. For each weekly sample of five houses, a sample mean selling price, $\overline{X}$, has been calculated. The frequency histogram displays the distribution of the selling prices ($5 \times 30 = 150$ observations) (solid) and the distribution of the 30 $\overline{X}$ values (cross-hatch). Clearly, the distribution of the sample means has less variability.

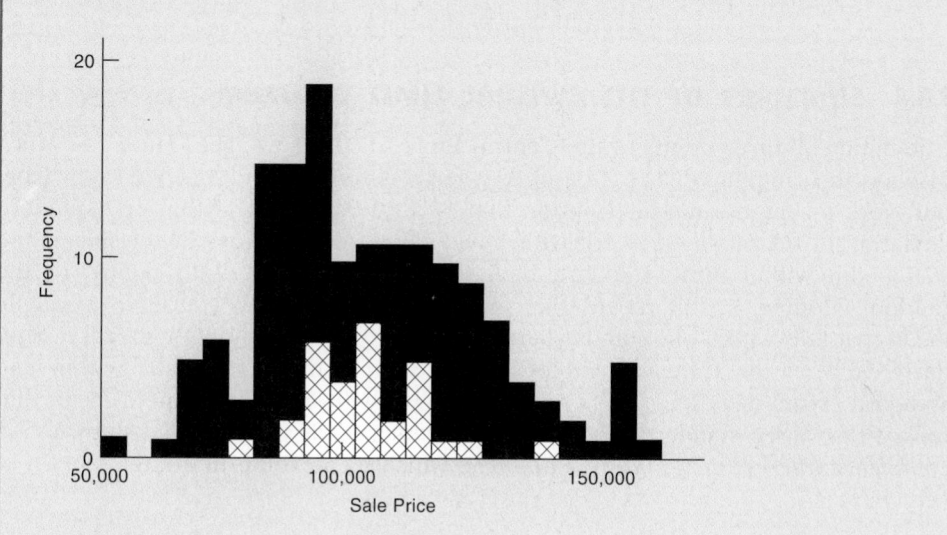

What is the impact of increasing the sample size? Suppose we return to the population with a mean of 25 and a standard deviation of 1, take ten samples of size five, and calculate the ten sample means. Then we take ten samples of size 35 from this same population and calculate ten more sample means. How should we expect the ten sample means based on the samples with $n = 5$ to compare with the ten sample means based on the samples with $n = 35$? The sample means calculated from the bigger ($n = 35$) samples are more similar (and thus have a smaller standard deviation) to one another because each sample mean contains more information and is thus a better estimate of μ.

The CLT states that the standard error of $\overline{X}, \sigma_{\overline{X}}$, equals $\sigma/\sqrt{n}$. So as n increases from 5 to 35, the denominator of the standard error gets larger, which means we are dividing by a bigger number, yielding a smaller standard error. Let's follow this example and see what the numbers should be:

$\overline{X}$ values based on samples of size 5: The standard error should be $1/\sqrt{5} = 1/2.24 = 0.45$.

$\overline{X}$ values based on samples of size 35: The standard error should be $1/\sqrt{35} = 1/5.92 = 0.17$.

This is the second general result:

As the sample size increases, the standard error of $\overline{X}$ decreases.

Thus, $\overline{X}$ is a consistent estimator.

✓ TRY IT NOW!

Central Limit Theorem
Investigating the Impact of Sample Size on Standard Error

Using the random number table in Appendix A and the 350 values shown in the previous Try It Now section, select ten samples of size five each from population 1. Review Section 1.5 if you need a refresher on how to use the random number table. For each sample, calculate a sample mean, $\overline{X}$.

Find the average and standard deviation of the ten $\overline{X}$ values. Compare these values with the corresponding values you found in the previous Try It Now section for population 1. In that case the sample size was $n = 35$.

7.6.4 SUMMARY OF THE CENTRAL LIMIT THEOREM

Remember! The standard deviation of $\overline{X}$ is called the standard error. It is equal to $\sigma/\sqrt{n}$.

Combining all three points of the Central Limit Theorem, we get Figure 7.8, which displays the sampling distribution of $\overline{X}$ when n is sufficiently large. We know from our work on the normal distribution that 68% of values fall within one standard deviation of the mean, 95% fall within two standard deviations of the mean, and 99.7% fall within three standard deviations of the mean. With regard to the random variable $\overline{X}$, this means that 68% of the time we will observe a sample mean that falls within one standard error of the unknown population mean, μ. Similarly, 95% of the time we will observe a sample mean that falls within two standard errors of μ, and 99.7% of the time we will observe a sample mean that falls within three standard errors of μ. This idea leads to a concept known as a *confidence interval* or an *interval estimate*, which we develop in a later section of this chapter.

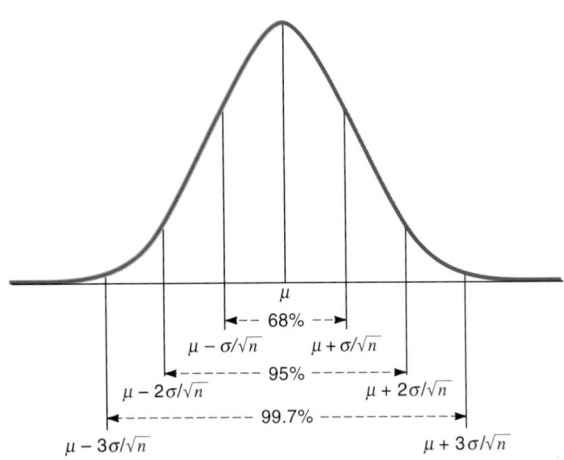

FIGURE 7.8   Sampling distribution of $\overline{X}$

Discovery Exercise 7.1

THE CENTRAL LIMIT THEOREM IN ACTION

Part I. Draw a picture of a normal distribution with mean 80 and standard deviation 5.

Perhaps this is the distribution of exam scores in your statistics class!

Part II. Generate and examine 100 random samples.

Each sample should consist of 30 values selected from a normal distribution with mean 80 and standard deviation 5. If you do not know how to generate this data, your instructor will provide you with data.

1. Plot the values from one of the samples. What shape does the histogram have?

2. What value do you expect the average of the 30 values from sample 1 to be close to? Find the average of the first sample of 30. Is it close to the number you thought it would be?

3. Do you expect the average of the 30 values in sample 2 to be close to the same number? Why or why not?

4. What value should the standard deviation of the first sample of 30 values be close to? Find the standard deviation of the first sample of 30 values. Were you right?

5. Do you expect the standard deviation of the 30 values in sample 2 to be close to the same number? How about sample 3?

Part III. Create a distribution of $\overline{X}$ for the 100 samples of size $n = 30$.

For each of the 100 samples find the sample mean, $\overline{X}_1$, $\overline{X}_2$, and so on. You should have a column of 100 $\overline{X}$ values.

1. Examine these $\overline{X}$ values. Are they all the same? Why or why not?

2. Are they all equal to 80? Why or why not? What is the reason for this?

Construct a histogram of the 100 $\overline{X}$ values.

3. What shape does the distribution of $\overline{X}$ values have?

4. Is the shape of the distribution of $\overline{X}$ values always normal, or is it just normal because we sample from a normal population?

Find the average of the 100 $\overline{X}$ values and the standard deviation of the 100 $\overline{X}$ values. The CLT states that the average of the $\overline{X}$ values should be close to the mean of the distribution of the population.

5. What is the mean of the population we sampled from?

6. What is the average of the 100 $\overline{X}$ values?

7. Are the two means close?

8. How could we get them to be closer?

The CLT states that the standard deviation of the $\overline{X}$ values should be about $\sigma/\sqrt{n}$.

9. What is σ?

10. What is n?

11. What value should the standard deviation of the $\overline{X}$ values be close to?

12. What is the standard deviation of the $\overline{X}$ values? Is it close to the value you suggested?

13. How could we get the two standard deviations to be closer?

7.7 Drawing Inferences by Using the Central Limit Theorem

Now we can return to the problem of choosing a real estate agency. We want to know whether an $\overline{X}$ value of 94.4 days on the market is indeed unusual or whether it is just reasonable variation caused by sampling error. Recall that the population mean number of days on the market is 50 and the population standard deviation is 60. Also recall that the distribution of days on the market does not appear to be normally distributed. Therefore, to use the Central Limit Theorem we need a sample size of at least 30. Let's say that when we take a sample of size 30, we find a sample mean of 90 days. Ultimately you want to know if you should list your house with this agency and when you should put your house on the market so that you are reasonably sure that your house will sell in that time frame.

7.7.1 USING THE CENTRAL LIMIT THEOREM

We draw a histogram of the days on the market to check the assumption of normality. We want create a picture similar to Figure 7.8 for the number of days a house is on the market. With a sample of 30 houses we do not need the assumption of normality. The $\overline{X}$ values should be centered at $\mu_{\overline{X}} = 50$ days and the standard error should be $\sigma_{\overline{X}} = \sigma/\sqrt{n} = 60/\sqrt{30} = 10.95$. Figure 7.9(a) combines these results.

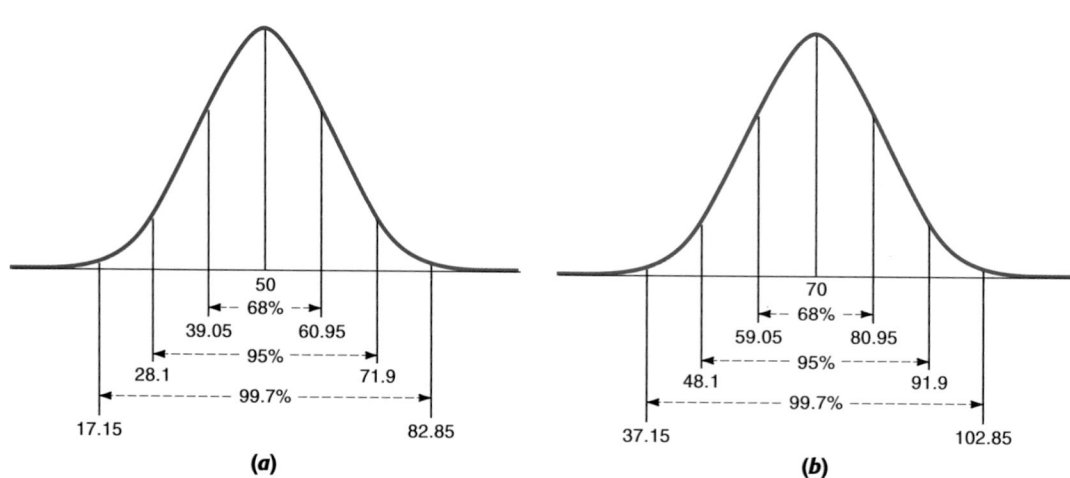

FIGURE 7.9 **Sampling distribution of $\overline{X}$ when (a) $\mu = 50$ days and (b) $\mu = 70$ days**

A sample mean of 90 days clearly falls in the right tail of this distribution. This conclusion takes into account that the numbers of days on the market themselves have some variability as measured by σ and the sample size is $n = 30$. Thus, we must conclude that this agency is taking too long to sell homes. We reach the same conclusion by calculating the Z-score for the observed $\overline{X}$ value of 90 days:

$$Z = \frac{\overline{X} - \mu_{\overline{X}}}{\sigma_{\overline{X}}} = \frac{90 - 50}{10.95} = 3.65$$

Use the Z-score to decide whether $\overline{X} = 90$ is unusual.

We know that a Z-score greater than 3 is extremely unusual. Thus, we conclude that you should find a different agency.

Now suppose the market shifts and the true population mean time on the market has increased to 70 days. In this case the sampling distribution looks like Figure 7.9(b). A sample mean of $\overline{X} = 90$ days is not quite so unlikely now. The CLT has provided a basis for judging whether the observed sample mean value of 90 days is too unusual and, if so, what could have happened to result in such a high sample mean.

We can begin to tackle some other problems now that we have learned about the Central Limit Theorem.

EXAMPLE 7.10

CAR BATTERIES

Collect and analyze the data.

Using the Central Limit Theorem

A company that produces car batteries guarantees them for 36 months. The battery lives have a standard deviation of 2 months. You and eight of your friends purchased these batteries and collected these data on the length of time each of them lasted (in months):

> 35.38   35.87   35.49   34.38   34.55   39.99   36.18   35.09   34.57

You calculate the average length of life of these batteries and find it to be 35.72 months. On the basis of your data, do you have reason to be suspicious of the company's claim of 36 months?

The Z-score for the sample mean of 35.72 is

$$z = \frac{\overline{X} - \mu_{\overline{X}}}{\sigma_{\overline{X}}} = \frac{35.72 - 36}{2/\sqrt{9}} = -0.42$$

This Z-score indicates that the sample mean value of 35.72 months is within one standard deviation (Z-score between -1 and 1) of the warranty value of 36 months. You know that 68% of Z-scores are between -1 and 1. Thus, your sample average is *not* unusually low and you have no reason to be suspicious of this company's claim.

In this example we used the value of the population standard deviation claimed by the company. What happens if we do not know the population standard deviation? This is common. If we need to estimate the mean, why would we *know* the standard deviation? If the sample is sufficiently large, $n > 30$, then we can use the point estimate s instead of σ in the calculation of the Z-score. The next example demonstrates this situation.

Collect and
analyze the data.

EXAMPLE 7.11

RUNNING TIMES

Using the Central Limit Theorem

You have been training for the Boston Marathon and you are trying to decide how your times stack up against those of the average runner. You record your times for 5 miles on your next 35 runs and you find an average of 35.2 minutes with a standard deviation of 1.6 minutes. The average male runner in the age category 18–25 can run 5 miles in 36.2 minutes. You can see that your average is faster than the population mean of 36.2 minutes. Did you just have some good times or are you really pretty fast?

The Z-score for the sample average of 35.2 minutes is

Note! ≈ means approximate. In this case the approximation is due to the estimate of σ.

$$ Z = \frac{\overline{X} - \mu_{\overline{X}}}{\sigma_{\overline{X}}} \approx \frac{35.2 - 36.2}{1.6/\sqrt{35}} = \frac{1}{0.27} = -3.7 $$

This Z-score is a large negative number. In fact, your average time is more than three standard errors less than the population mean time. You really are faster than most of the other runners in this age group.

In the preceding example we used s, the sample standard deviation, as the estimate of σ in finding the z-score because σ was not given. This procedure isn't precisely correct but it is close enough given the large sample size. In this case, the formula to compute the approximate Z-score is

Z-score when σ is unknown and n is large

$$ Z \approx \frac{\overline{X} - \mu_{\overline{X}}}{s_{\overline{X}}} = \frac{\overline{X} - \mu}{s/\sqrt{n}} $$

This issue is discussed more thoroughly later in this chapter along with what to do if the standard deviation is unknown and the sample size is small ($n \leq 30$).

✓ TRY IT NOW!

Cost of Books

Comparing the Sample Mean with the Claimed Population Mean

A university states that students spend an average of $325 per semester on books. Based on your own experience you think that underestimates the true expenditure. You ask 30 of your friends how much they spent on textbooks last semester and get these amounts (in dollars):

344.00	324.45	354.89	375.53	373.97	384.66	306.33	345.32	371.90	328.27
336.20	316.19	313.06	330.45	334.53	307.64	338.64	341.92	352.70	342.69
343.46	342.60	376.85	348.29	346.28	339.74	389.44	367.31	397.51	336.90

Based on these data, do you have reason to tell the university that its statement is inaccurate?

ANS. $\overline{X}$ = $347.06, s = 23.75, Z = 5.48; THE UNIVERSITY'S PUBLISHED AVERAGE IS INACCURATE.

7.7.2 EXERCISES–LEARNING IT!

7.6 A manufacturer of pain relievers claims that it takes an average of 12.75 minutes for a person to be relieved of headache pain after taking its pain reliever. The time it takes to get relief is normally distributed with a standard deviation of 0.5 minute. A sample of 12 people is taken and their times to relief are listed here:

12.9 13.2 12.7 13.1 13.0 13.1 13.0 12.6 13.1 13.0 13.1 12.8

(a) Find the sample mean.

(b) Find the standard error of $\overline{X}$.

(c) If the manufacturer claims that the mean is 12.75 minutes, find the Z-score of the sample mean.

(d) What do you think about the manufacturer's claim based on the Z-score?

7.7 It seems like it has been raining forever! You think that there has been more rainfall than usual. You measure the rainfall for the 30 days in May and get these amounts (in inches):

0	0	0.8	0	0.1	0	0	0.2	0.1	0
0	0.5	0.2	0.1	0	0.5	0	0.2	0	0
1.2	0	0.1	0	0	0.4	0.1	0	0	0.5

(a) Find the sample mean rainfall for this May.

(b) From the National Weather Service, you find that the standard deviation for the rainfall is 0.1 inch. Find the standard error of $\overline{X}$.

(c) If the May total rainfall is typically 4.88 inches, what is the typical May average daily rainfall?

(d) Using your answer to part (c) as the population mean, find the Z-score for the sample mean.

(e) Based on the Z-score you found in part (d), was it an unusually wet May?

7.8 You are trying to decide what teacher you should have for your next statistics course. You have access to teacher evaluation results. From a sample of 100 student responses, one professor scored an average of 4.1 with a standard deviation of 0.2 on the question about overall performance of the instructor. The scale went from 1 to 5, with 5 being the best. The national average for statistics teachers is 3.5.

(a) Find the Z-score for 4.1.

(b) Based on this Z-score, do you believe this professor is truly superior?

7.9 Scores on a national test are normally distributed with a mean of 75 points and a standard deviation of 5 points. At a certain high school 50 students took the test and their average was 80. On the basis of these data, can this high school be considered unusually strong?

7.10 Starting salaries nationally for business majors follow a normal distribution with a mean of $28,900 and a standard deviation of $1200. The career and human resource director at College XYZ has taken a sample of 25 of the most recent graduates of the Business School and found that their average salary is $28,500.

(a) Between what two values will 99.7% of the salary averages fall?

(b) Based on your answer to part (a), what can you conclude about the average starting salary of the most recent graduates of the Business School?

7.8 Large-Sample Confidence Intervals for the Mean

In this chapter you have learned about point estimators, which give you a single number to be used as an estimate of the population parameter. A confidence interval (also called an interval estimate) takes the point estimate a step further and gives a range of values and a probability. The probability value is the likelihood that an interval actually includes the value of the unknown population parameter.

7.8.1 THE BASICS OF CONFIDENCE INTERVALS

Let's return to the problem at the beginning of this chapter. You are trying to decide which realtor to use, when to put your home on the market, and how much you can expect to sell your house for. You base your decisions on a sample of 30 observations that give a mean selling price of $\overline{X} = \$105,000$. The sample mean, $\overline{X}$, is a reliable point estimate of μ, but it is not likely to hit μ on the nose. Initially, it looks like $\overline{X}$ is not very helpful—it is an estimate of μ that is pretty likely to be wrong! We would like to use $\overline{X}$ to find a range of values that definitely contains μ, but this is not possible because of sampling error. Instead, we can specify a high probability—say, 0.90 or 0.95—that a particular range or interval covers the true mean, μ. For example, we would like to be able to say that, based on the sample data, the interval from $100,000 to $110,000 covers μ with a probability of 0.95.

This is an example of a **confidence interval.** Let's examine the components of the confidence interval. First of all, it has a lower bound for μ; call this L. In the example, L is $100,000. The confidence interval also has an upper bound for μ; call this U. In the example, U is $110,000. Finally, the confidence interval has a probability value, which is called the **confidence level** and is labeled $1 - \alpha$. In the example, α is 0.05 and $1 - \alpha$ is 0.95. We are constructing a $100(1 - 0.05\%) = 95\%$ confidence interval. This means that if we constructed 20 intervals based on 20 different sample means, we would expect 95% of them or 19 intervals to actually contain μ. For any *individual* interval, either μ is in the interval or it is not. Due to sampling error, we do not know for sure whether or not we have an interval that includes the value of μ. Thus, the probability refers to the chance that we have one of the intervals that does indeed contain μ.

In general, a confidence interval for the population mean has the form

$$P(L \leq \mu \leq U) = 1 - \alpha$$

> The *confidence interval* or an *interval estimate* is a range of values with an associated probability or *confidence level* $1 - \alpha$. The probability quantifies the chance that the interval contains the true population parameter.

Think about the information a confidence interval gives you. Instead of being able to say the true μ is somewhere close to $\overline{X} = \$105,000$, now you can say the probability that the interval from $100,000 to $110,000 covers μ is 0.95. You have a way to quantify the error you know is inherent in $\overline{X}$.

The next few sections provide the details for calculating the lower and upper bounds for confidence intervals for μ. These sections are followed by a section that focuses on the interpretation of confidence intervals.

7.8.2 CONFIDENCE INTERVAL FOR μ: NORMALLY DISTRIBUTED POPULATION AND KNOWN STANDARD DEVIATION

Remember! An unbiased estimator of μ is a fair estimator that does not systematically overestimate or underestimate μ.

We start by developing a confidence interval for the mean of a population that is normally distributed and for which the population standard deviation, σ, is known. In the next section we will explain what happens as we relax both the assumption of normality and the assumption of a known standard deviation.

As you consider how to calculate the lower and upper bounds for the confidence interval, think about where $\overline{X}$ should fall in the interval. Remember that $\overline{X}$ is

TABLE 7.2

A PORTION OF THE Z TABLE

Z	0.00	0.01	0.02	0.03	0.04	0.05	0.06	0.07	0.08	0.09
−2.1	0.0179	0.0174	0.0170	0.0166	0.0162	0.0158	0.0154	0.0150	0.0146	0.0143
−2.0	0.0228	0.0222	0.0217	0.0212	0.0207	0.0202	0.0197	0.0192	0.0188	0.0183
−1.9	0.0287	0.0281	0.0274	0.0268	0.0262	0.0256	0.0250	0.0244	0.0239	0.0233
−1.8	0.0359	0.0351	0.0344	0.0336	0.0329	0.0322	0.0314	0.0307	0.0301	0.0294
−1.7	0.0446	0.0436	0.0427	0.0418	0.0409	0.0401	0.0392	0.0384	0.0375	0.0367

an *unbiased* estimator of μ and so it makes sense to put $\overline{X}$ right in the middle of the interval as shown here:

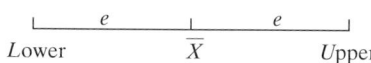

The value of $\overline{X}$ should always go in the center of a confidence interval for μ.

Notice that the distance from the lower bound to the middle of the interval is labeled e. The distance from the middle of the interval to the upper bound is also labeled e. Thus, the width of the interval is $2e$. To find the lower bound we take $\overline{X}$ and subtract e, and to find the upper bound we take $\overline{X}$ and add e. Now, all we need is the value of e, which stands for *error*.

Suppose we want to construct a 95% confidence interval ($\alpha = 0.05$) for the population mean, μ. In a sense, we wish to "cover" the value of μ 95% of the time. We know that $\overline{X}$ is the best point estimate we can use for μ and that $\overline{X}$ has a normal distribution when n is large enough. Remember that to cover 95% of the values in a normal distribution, we need to go out about two standard deviations from the mean in both directions. Finally, we recall that the standard deviation of the sample mean, also known as the standard error, is equal to $\sigma/\sqrt{n}$. These three pieces of information together tell us that e should equal $2\sigma/\sqrt{n}$. This is just about right.

The CLT states that the sampling distribution of $\overline{X}$ is a normal distribution.

In Chapter 6 you learned how to find Z values given tail area probabilities. To get the correct value for Z, we must use that procedure with a tail area probability equal to $\alpha/2$. We label this value as $Z_{\alpha/2}$. For a 95% confidence interval we divide $\alpha = 0.05$ by 2 and find that the area in one of the tails is 0.025. This means we are given a probability and we need the corresponding Z value. A portion of the Z table is reprinted in Table 7.2.

We want to find the Z value that has 0.025 in the lower tail. We look inside the table for the number closest to 0.025 and see that it corresponds to a Z value of −1.96. This gives us the value of $-Z_{\alpha/2}$. To get the positive critical value we simply drop the negative sign because of the symmetry of the normal distribution. So $Z_{\alpha/2} = 1.96$. Figure 7.10 shows the normal distribution with 5% in the tail area.

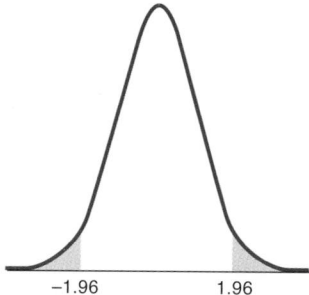

FIGURE 7.10 Normal distribution with 0.05 in the tails

So, the value of e is calculated as $1.96\sigma/\sqrt{n}$. This gives us the following formulas for the lower and upper bounds for a 95% confidence interval for μ:

Lower and upper bounds for a 95% confidence interval for μ

$$L = \overline{X} - e = \overline{X} - \frac{1.96\sigma}{\sqrt{n}}$$

$$U = \overline{X} + e = \overline{X} + \frac{1.96\sigma}{\sqrt{n}}$$

Let's use these equations to find a 95% confidence interval for the housing problem.

Collect and analyze the data.

EXAMPLE 7.12

THE HOUSING MARKET

Finding the 95% Confidence Interval for the Mean Selling Price

A sample of 30 homes yielded a sample average selling price of $99,223.30. Based on this sample, you want to find a 95% confidence interval for μ. You are given:

$$\overline{X} = \$99,223.30$$
$$n = 30$$
$$\sigma = \$21,000$$

We perform these calculations:

$$\text{Standard error} = \frac{\sigma}{\sqrt{n}} = \frac{21,000}{\sqrt{30}} = 3832.12$$

$$\text{Error} = e = \frac{1.96\sigma}{\sqrt{n}} = (1.96)(3832.12) = 7510.96$$

$$\text{Lower bound} = \overline{X} - e = 99,223.30 - 7510.96 = \$91,712.34$$

$$\text{Upper bound} = \overline{X} + e = 99,223.30 + 7510.96 = \$106,734.26$$

We can state that we are 95% confident that the true mean selling price is between $91,712.34 and $106,734.26. There is a 95% chance that we have constructed an interval that does indeed contain the true mean selling price.

The formulas used in the example are for a 95% confidence interval. Suppose instead that we wish to find a 90% confidence interval or a 99% confidence interval. The only adjustment that must be made to the formula is the value of Z that we multiply the standard error by to get e. For the 95% confidence level we used the value of Z that cut off a probability of 0.05 in the two tails of the normal distribution. The procedure for figuring out the correct Z value for any confidence level follows:

Steps for finding the value of Z for a confidence interval

Step 1. Take the value of α and divide it by 2.

Step 2. Look that value up in the body of the Z table (find the entry closest to it).

Step 3. Read off the corresponding Z value to get the value of $-Z_{\alpha/2}$.

Step 4. Drop the negative sign to get the value of $Z_{\alpha/2}$.

Remember that Z tells how many standard errors away from the mean we need to go to get the coverage we desire.

The formulas for a $100(1 - \alpha)\%$ confidence interval for μ when the population standard deviation, σ, is known are given here:

TABLE 7.3

COMMON VALUES OF α AND THE CORRESPONDING Z VALUES

	Confidence level $(1 - \alpha)$				
	0.90	**0.95**	**0.98**	**0.99**	**0.995**
α	0.10	0.05	0.02	0.01	0.005
$Z_{\alpha/2}$	1.645	1.96	2.33	2.58	2.81

Error: $$e = \frac{Z_{\alpha/2}\sigma}{\sqrt{n}}$$

Width of the interval: $w = 2e$

Lower bound: $L = \overline{X} - e$

Upper bound: $U = \overline{X} + e$

Confidence interval for μ when σ is known

where $Z_{\alpha/2}$ is the value that cuts off $\alpha/2$ in the upper tail of the standard normal distribution and $\alpha/2$ in the lower tail of the standard normal distribution.

Some common values of α and the corresponding Z values are listed in Table 7.3.

Notice that as the level of confidence increases, so does the value of Z used in the formula. Since the value of Z affects the size of e, the interval gets wider as the confidence level increases. Are wider intervals better? Suppose you had a choice between two intervals with the same level of confidence: one that stated the average age, μ, of the target population was between 2 and 95 years and one that stated the average age, μ, of the target population was between 25 and 35 years. Which interval gives better information? Clearly, the interval that is narrower, the second one, does. This simple example demonstrates that the narrower the interval, the more precise our estimate. Thus, as we *increase* the level of confidence, we *decrease* the precision of the estimate.

As you increase the level of confidence, you lose precision.

TRY IT NOW!

Bridge Traffic
Finding a Confidence Interval for μ

Many states are worried about the condition of road bridges. One bridge in Massachusetts has been rated a 3, which means "a bridge is basically intolerable, requiring a high priority of corrective action" (source: *http://www.townonline.com/tolhome/bridges/database_index.html*).

A sample of 30 days of traffic found that a particular bridge in Framingham, Massachusetts, which was built in 1937, sees average daily traffic of 15,800 cars. If it is known that the population standard deviation, σ, is 500 cars, find a 95% confidence interval for μ.

How wide is the interval?

Now find a 98% confidence interval for μ.

Which interval is wider?

ANS. 95% INTERVAL: 15,621.08 TO 15,978.92; 98% INTERVAL: 15,587.64 TO 16,012.36; THE 98% INTERVAL IS WIDER.

Assumptions of confidence interval formulas

The procedure just described depends on our knowing that $\overline{X}$ has a normal distribution and the population standard deviation is known. The Central Limit Theorem states that $\overline{X}$ has a normal distribution when $n > 30$ or when the underlying population distribution is normally distributed (or close to normal). Therefore, for a *large sample* ($n > 30$), we can use this procedure to find confidence intervals for μ regardless of the shape of the underlying distribution.

For a *small sample* ($n \leq 30$), we need to visually check the shape of the distribution of the sample data to get a rough idea of the population shape. If the histogram or other plot of the data looks reasonably symmetric, then we can use the formulas we just developed.

7.8.3 WHAT HAPPENS IF σ IS UNKNOWN OR THE DISTRIBUTION IS NOT NORMAL?

In many cases we do not know the population standard deviation, σ, and/or we do not have an underlying normally distributed population. First, consider what happens to the calculation of the confidence interval for μ if we do not know σ but we still have a normally distributed population. There are two cases: large sample size and small sample size.

Use s to estimate σ when n > 30.

If we have a large sample size, then we can use s, the sample standard deviation, to estimate σ. Using s in the formulas for the upper and lower bounds of the confidence interval does not appreciably affect the confidence level. This is the same approach we took in computing the Z-score when we did not know σ but the sample size was sufficiently large.

The formulas for a $100(1 - \alpha)\%$ confidence interval for the mean, μ, of a normally distributed population when the population standard deviation, σ, is unknown and n is sufficiently large are given here:

Confidence interval for μ when σ is unknown

$$\text{Error:} \qquad e = \frac{Z_{\alpha/2}\sigma}{\sqrt{n}}$$

Width of the interval: $w = 2e$

Lower bound: $L = \overline{X} - e$

Upper bound: $U = \overline{X} + e$

If the sample size is small and the standard deviation is unknown, then the sampling distribution of $\overline{X}$ is not normally distributed. The sampling distribution is called a t distribution and is covered in the next section.

Transform the data or use a nonparametric procedure when $n \leq 30$ and the distribution is not normal.

Finally, if the sample size is small, the histogram of the data does not look roughly normal in shape, and the population standard deviation is unknown, then we cannot use this procedure for finding a confidence interval for μ. It may be possible to use a transformation to get normally distributed data or it may be necessary to use a different distribution or a nonparametric procedure. Nonparametric procedures are typically not as powerful, but they do not require the assumption of an underlying normal population. They are the subject of Chapter 16. Table 7.4 summarizes the different scenarios.

7.8.4 INTERPRETING THE CONFIDENCE INTERVAL

The last major point of discussion for confidence intervals has to do with what the words "I am 95% confident" imply. Many people think this means that there is a 95% chance that the population mean is in the interval we have constructed. This is wrong! After all, μ is a number even if it is unknown to us. Therefore, there is no probability associated with it being between two other numbers; either it is in the interval or it is not.

TABLE 7.4

SUMMARY OF CONFIDENCE INTERVALS FOR μ

Population	Standard deviation	Sample size	Confidence interval for μ
Normal	Known	$n \geq 1$	$L = \bar{X} - Z\sigma/\sqrt{n}$ $U = \bar{X} + Z\sigma/\sqrt{n}$
Normal	Unknown	$n > 30$	$L = \bar{X} - Zs/\sqrt{n}$ $U = \bar{X} + Zs/\sqrt{n}$
Normal	Unknown	$n \leq 30$	Use the t distribution covered in Section 7.9
Nonnormal	Known	$n > 30$	$L = \bar{X} - Z\sigma/\sqrt{n}$ $U = \bar{X} + Z\sigma/\sqrt{n}$
Nonnormal	Unknown	$n \leq 30$	Use transformation or a nonparametric procedure covered in Chapter 16

So, then, what is the proper interpretation of those words? The correct interpretation of the 95% confidence level has to do with the chance that we have an interval that does in fact contain the population parameter. Remember that if we take 100 different samples from the same population, we get 100 different sample means and therefore 100 different confidence intervals. If each of them is a 95% confidence interval, then theoretically 95 out of the 100 intervals will contain μ and 5 of them will not.

Suppose that ten weekly samples are selected from the population of homes for sale. For each sample, an average sale price is calculated and a 90% confidence interval is determined based on this value of $\bar{X}$. These ten confidence intervals are listed in Table 7.5. Let's assume that the true population mean selling price is $110,000 and the population standard deviation is $21,000. According to the theory, nine out of ten of these intervals should include the value of $\mu = \$110,000$.

The last column of Table 7.5 indicates whether the interval includes the value of $110,000. Figure 7.11 on page 326 illustrates that only one interval does not include the value of $110,000 and nine do. This is just what we expected, since 90% of ten intervals is nine intervals that cover μ. The problem, of course, is that we don't know the

TABLE 7.5

TEN 90% CONFIDENCE INTERVALS

Sample mean	Error	Upper bound	Lower bound	Covers μ?
$83,580	$15,448	$99,028	$68,132	No
103,500	15,448	118,948	88,052	Yes
120,780	15,448	136,227	105,332	Yes
110,400	15,448	125,848	94,952	Yes
125,200	15,448	140,648	109,752	Yes
107,500	15,448	122,948	92,052	Yes
97,960	15,448	113,408	82,512	Yes
98,860	15,448	114,308	83,412	Yes
116,740	15,448	132,188	101,292	Yes
113,400	15,448	128,848	97,952	Yes

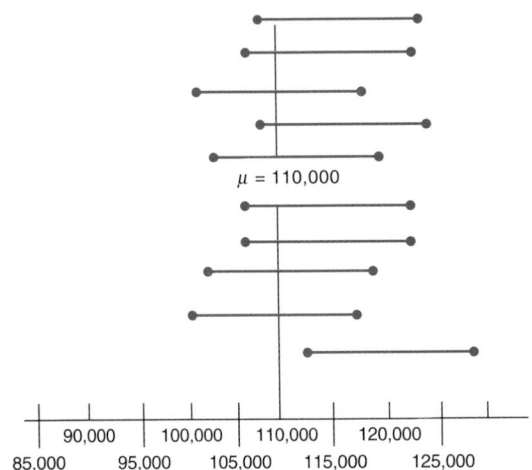

FIGURE 7.11 **Comparison of confidence intervals and μ**

actual value of μ and therefore we don't know whether or not we have an interval that actually contains μ. The confidence level gives us the probability of our having an interval that does in fact contain the population parameter.

To increase the probability that we have a "good" interval, we could widen the intervals. This is precisely what happens when we increase the confidence level. When we increase the confidence level from 90% to 95%, we are increasing the chance that we have a good interval, but we are losing precision by widening the interval. It is a tradeoff.

7.8.5 EXERCISES–LEARNING IT!

7.11 A university wants to estimate the average amount of money that students spend on textbooks in a semester. It takes a random sample of 45 students and finds that the average amount of money they spent was $382 with a standard deviation of $21. Find a 98% confidence interval estimate for the true mean amount of money spent on textbooks in a semester.

7.12 Birth weights of newborns are normally distributed with a standard deviation of 14 ounces. Here are data for a random sample of 16 babies born at Baystate Health Center in Springfield, Massachusetts, in 2001:

Date	Weight Pounds	Ounces	Length (inches)	Gender
June 1	9	6	22	F
	6	12	21.5	F
	8	4	20.5	F
	6	11	19.5	F
June 2	6	13	20.25	F
	6	15	20	M
June 3	6	13	20.5	F
	8	4	19	F
June 4	9	6	22.5	M
	8	1	21	M
	7	12	21.5	F
	6	14	20.5	M
June 5	7	15	20.5	M
	8	9	22	F
	7	7	21.5	M
	7	6	19.5	M

(a) Find a 95% confidence interval for the true mean weight of newborns. Be careful to convert the weights to ounces first.

(b) What other factors might influence the newborns' weights?

7.13 The average January snowfall in Rochester, New York, is 22.7 inches with a standard deviation of 1.5 inches. These figures are based on a sample of 53 years. Syracuse, New York, has an average of 29.6 inches with a standard deviation of 1.8 inches based on 44 years of data.

(a) Assuming that the amount of snowfall in January is normally distributed, calculate a 95% confidence interval for the average January snowfall in Rochester.

(b) Assuming that the amount of snowfall in January is normally distributed, calculate a 95% confidence interval for the average January snowfall in Syracuse.

7.14 Is it a lottery jackpot or a bonus? A salary or a life's savings? No, it's the annual compensation paid out to football players. The table lists the 2000 salaries and bonuses for 50 AFC players.

Datafile:
AFCSALARIES.XXX

Player	Position	Base salary	Bonus	Total pay
Joe Cummings	LB	$325,000	$1,300	$326,300
Jay Foreman	LB	175,000	31,000	206,000
Chris Mohr	P/K	400,000	200,000	600,000
Marlo Perry	LB	400,000	103,000	503,000
Jay Riemersma	TE	934,000	3,600	937,600
Alex Van Pelt	QB	400,000	101,900	501,900
Obafemi Ayanbadejo	RB	250,000	0	250,000
Greg DeLong	TE	375,000	108,600	483,600
Fernando Smith	DE	700,000	600	700,600
Corey Fuller	S	500,000	1,125,000	1,625,000
Chris Gardocki	P/K	400,000	842,000	1,242,000
Madre Hill	RB	175,000	0	175,000
Darius Holland	DT	375,000	51,800	426,800
Raymond Jackson	CB	429,000	4,100	433,100
Kevin Johnson	WR	300,000	185,000	485,000
Daylon McCutcheon	CB	175,000	170,000	345,000
Jeff Blake	QB	2,150,000	501,400	2,651,400
Corey Dillon	RB	503,000	152,200	655,200
Charles Fisher	CB	274,000	300,000	574,000
James Hundon	WR	350,000	2,000	352,000
Damian Vaughn	TE	62,000	0	62,000
Kimo Von Oelhoffen	DT	750,000	416,700	1,166,700
Nick Williams	RB	175,000	35,000	210,000
Desmond Clark	TE	175,000	20,000	195,000
Matt Lepsis	OL	325,000	37,700	362,700
Trey Teague	OL	250,000	17,700	267,700
Shane Bonham	DT	400,000	0	400,000
Peyton Manning	QB	1,430,000	1,935,200	3,365,200
Spencer Reid	LB	250,000	0	250,000
Tavian Banks	RB	189,000	87,500	276,500
Lenzie Jackson	WR	175,000	0	175,000
Damon Jones	TE	325,000	34,500	359,500
Seth Payne	DT	325,000	77,500	402,500
John Wade	OL	250,000	34,500	284,500
Joe Horn	WR	429,000	1,100	430,100
Ty Parten	DE	429,000	2,300	431,300
John Bock	OL	400,000	54,100	454,100
O'Lester Pope	OL	62,000	0	62,000
Terry Allen	RB	400,000	50,000	450,000
Chad Eaton	DT	775,000	106,000	881,000
Max Lane	OL	400,000	1,013,400	1,413,400
John Elliott	OL	2,200,000	844,000	3,044,000
Bobby Hamilton	DE	350,000	51,000	401,000

(continued)

Ian Rafferty	OL	175,000	5,000	180,000
Bobby Brooks	LB	175,000	0	175,000
Steve Wisniewski	OL	500,000	1,552,000	2,052,000
Wayne Gandy	OL	400,000	962,600	1,362,600
Chris Oldham	CB	400,000	32,100	432,100
Ryan Thelwell	WR	250,000	0	250,000
Charles Dimry	CB	400,000	250,300	650,300

Assume these players represent a random sample of football players.

(a) Find a 90% confidence interval for the mean base salary.

(b) Find a 90% confidence interval for the mean bonus.

(c) Find a 90% confidence interval for the mean total pay.

Database:
CHIPS.XXX

7.15 The U.S. Department of Labor publishes a great deal of monthly data. The following data on the average price of a 16-ounce bag of potato chips were extracted from the World Wide Web:

	1991	**1992**	**1993**	**1994**	**1995**	**1996**	**1997**	**1998**	**1999**	**2000**
Jan.	$2.99	$2.942	$2.832	$2.872	$3.079	$3.004	$3.125	$3.149	$3.217	$3.386
Feb.	2.982	2.914	2.935	3.018	3.014	2.96	3.099	3.136	3.223	3.448
Mar.	2.954	2.943	2.839	2.97	2.975	3.035	3.145	3.107	3.249	3.354
Apr.	2.96	2.857	2.836	3.012	3.048	3.075	3.184	3.175	3.264	3.409
May	2.937	2.871	2.898	2.999	2.952	2.989	3.056	3.161	3.212	3.345
June	3.062	2.906	2.827	2.968	2.987	2.981	3.118	3.153	3.235	3.302
July	2.929	2.911	2.893	2.902	2.971	3.106	3.117	3.128	3.255	3.31
Aug.	2.939	2.974	2.966	2.992	3.035	3.086	3.128	3.21	3.279	3.302
Sep.	2.901	2.856	2.895	2.965	2.97	3.069	3.176	3.121	3.237	3.416
Oct.	3.021	2.937	2.911	2.924	3.02	3.155	3.152	3.219	3.289	3.341
Nov.	2.93	2.836	2.843	3.02	2.979	3.102	3.133	3.207	3.299	3.276
Dec.	2.954	2.84	2.918	3.012	3.026	3.118	3.166	3.177	3.33	3.437

Treat these 10 years of data as a sample, and find a 90% confidence interval for the average price. Assume that the population standard deviation is $0.15.

7.16 A national grocery chain is considering opening a new store. To be sure that enough traffic goes by the new location, the grocery chain took a sample of vehicles crossing the intersection on 40 days. The numbers of cars per day at the intersection are shown in the table.

1431	1302	1255	1377	1450	1483	1529	1588	1535	1533
1540	1700	1840	1642	1139	1227	1684	1782	1491	1513
1293	1533	1272	1572	1520	1227	1257	1238	1276	1420
1340	1402	1467	1220	1477	1515	1242	1350	1367	1375

(a) Find a 95% confidence interval for the average number of cars per day that pass this location. The standard deviation is assumed to be 165 cars.

(b) The company has decided to open a store at the new location only if there is a daily average of at least 1400 cars at the intersection. Based on your confidence interval from part (a), would you advise the company to open a store at this location? Explain why or why not.

7.17 Is that basketball player really worth that much money? The number of points per game is a key way to evaluate the contribution of an NBA player. The following data are the numbers of points scored per game for the first 40 games of the 2000—2001 season by Shaquille O'Neal.

Points per game	Game date	Points per game	Game date
36	Oct. 31	39	Nov. 5
34	Nov. 1	24	Nov. 7
27	Nov. 4	13	Nov. 8

14	Nov. 12	19	Dec. 13
34	Nov. 14	0	Dec. 15
33	Nov. 16	28	Dec. 17
0	Nov. 18	22	Dec. 19
14	Nov. 19	25	Dec. 21
15	Nov. 22	28	Dec. 22
28	Nov. 24	32	Dec. 25
16	Nov. 27	18	Dec. 28
27	Nov. 28	29	Dec. 30
23	Nov. 30	24	Jan. 3
36	Dec. 1	33	Jan. 7
25	Dec. 3	34	Jan. 12
27	Dec. 5	30	Jan. 13
25	Dec. 6	31	Jan. 15
26	Dec. 8	41	Jan. 19
26	Dec. 10	31	Jan. 21
26	Dec. 12	29	Jan. 23

Using these data as a sample:

(a) Find a 95% confidence interval for the average number of points scored per game by O'Neal.

(b) What do you think happened to O'Neal on November 18 and December 15?

(c) Drop the November 18 and December 15 data values, and find a 95% confidence interval for the average number of points scored by Shaq per game.

(d) If a "typical" NBA player scores an average of 25 points per game, is Shaq much different? Use the confidence interval you found in part (c) to support your answer.

7.18 Most companies have increased their dependence on computers and software. As a result, more employee time is spent on the telephone with technical support for the software. A sample of 22 times spent on hold (in minutes) for technical support is shown here:

8.6	12.7	8.7	12.2	11.7	7.3	9.8	14.5	13.0	12.9	11.4
12.9	11.4	10.3	7.9	9.2	10.5	11.8	10.9	11.5	10.6	11.7

Assume that the standard deviation is 2 minutes.

(a) Because the sample size is less than 30 and you are not told that the population of times on hold is normally distributed, display the data in a histogram and comment on the shape of the data. Is it reasonable to assume that the data come from a population that has a normal distribution? Why or why not?

(b) Find a 99% confidence interval for the average amount of time spent on hold per call.

(c) Find a 95% confidence interval for the average amount of time spent on hold per call.

(d) If you are the manager of these employees and you are trying to argue for additional staff, which of the two confidence intervals would you use and why?

7.19 Managers of the symphony in a medium-sized New England city are surveying the community to determine the average number of times during a year that a person would attend a concert at a particular price. The responses from 50 adults are listed here:

1	4	3	2	2	2	3	2	4	3
3	4	2	1	3	3	5	4	3	3
2	4	4	4	3	3	5	4	2	2
2	4	4	4	3	3	5	4	2	2
2	4	4	4	3	3	5	4	2	2

(a) Find a 90% confidence interval for the average number of times a year a person would attend a concert at the given price.

(b) How could this confidence interval be useful to the managers of the symphony?

Discovery Exercise 7.2

EXPLORING CONFIDENCE INTERVALS FOR μ

From a population of college students across the United States, a sample was selected to find out how many hours per week a typical student spends playing sports.

Part I

A random sample of 2500 students was selected. The sample mean, $\overline{X}$, was found to be 12.5 hours. The population standard deviation, σ, is known to be 1.05 hours.

   1. Find a 90% confidence interval for μ.

   2. Find a 92% confidence interval for μ.

   3. Find a 94% confidence interval for μ.

   4. Find a 96% confidence interval for μ.

   5. Find a 98% confidence interval for μ.

   6. Discuss what happens to the size of the confidence interval as the level of confidence increases.

Part II

A random sample of 2500 students was selected. The sample mean, $\overline{X}$, was found to be 10.5 hours. The population standard deviation, σ, is known to be 1.05 hours.

   1. Find a 90% confidence interval for μ.

   2. Find a 92% confidence interval for μ.

   3. Find a 94% confidence interval for μ.

   4. Find a 96% confidence interval for μ.

   5. Find a 98% confidence interval for μ.

   6. Compare the confidence intervals you found in Part I with those in Part II. Discuss what happened to the confidence interval as a result of the change in the value of the sample mean, $\overline{X}$.

Part III

A random sample of 2500 students was selected. The sample mean, $\overline{X}$, was found to be 12.5 hours. Suppose you learn that the population standard deviation, σ, is actually 2.05 hours.

   1. Find a 90% confidence interval for μ.

   2. Find a 92% confidence interval for μ.

   3. Find a 94% confidence interval for μ.

   4. Find a 96% confidence interval for μ.

5. Find a 98% confidence interval for μ.

6. Compare the confidence intervals you found in Part I with those in Part III. Discuss what happened to the confidence intervals as a result of the change in the value of the population standard deviation, σ.

Part IV

A random sample of 2000 students was selected. The sample mean, $\overline{X}$, was found to be 12.5 hours. The population standard deviation, σ, is known to be 1.05 hours.

1. Find a 90% confidence interval for μ.

2. Find a 92% confidence interval for μ.

3. Find a 94% confidence interval for μ.

4. Find a 96% confidence interval for μ.

5. Find a 98% confidence interval for μ.

6. Compare the confidence intervals you found in Part I with those in Part IV. Discuss what happened to the confidence intervals as a result of the change in the sample size, n.

7.9 Distribution of the Sample Mean: Small Sample and Unknown σ

In the preceding section we noted that if the standard deviation of a normally distributed population is unknown and the sample size is small ($n \le 30$), then the sampling distribution of $\overline{X}$ does not follow a normal distribution. As in the large-sample situation, logic tells us to try using s instead of σ in the formula for the Z-score. When we do this, we get a t-score instead of a Z-score. That is, the sampling distribution of $\overline{X}$ for small samples, when σ is unknown and the population is normally distributed, follows what is called *Student's t distribution*. The t-score is calculated as

$$t = \frac{\overline{X} - \mu}{s/\sqrt{n}}$$

t-score

Notice that the calculation for t is the same as for Z with σ replaced with s.

The t distribution was developed in 1908 by William S. Gosset, who worked for Guinness Breweries in Ireland. The company did not allow employees to publish their research results, so Gosset published his results under a pen name: *Student*.

This distribution is not named for you, the student!

The graph of the t distribution looks very much like the standard normal distribution. It is symmetric, it has a bell shape, and it is centered at 0 just like the Z distribution. The major difference between the Z and the t distributions has to do with the spread or variability of the distributions. The standard deviation of the t distribution is not 1 as it is for the Z distribution. Instead, the variability of the t distribution is related to a number that is called the *degrees of freedom*. Thus, there are many different t distributions, but all have the same general shape. A t distribution with 5 degrees of freedom is shown in Figure 7.12 on page 332.

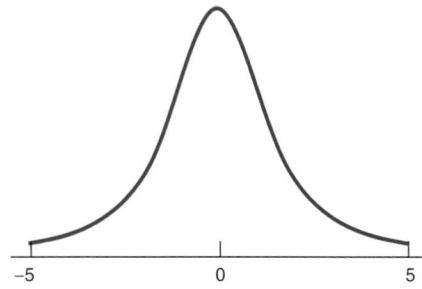

FIGURE 7.12 *t* **distribution with 5 degrees of freedom**

Let's examine the Z and t calculations, shown side by side:

$$Z = \frac{\overline{X} - \mu}{\sigma/\sqrt{n}} \qquad t = \frac{\overline{X} - \mu}{s/\sqrt{n}}$$

Each sample yields a different sample mean, $\overline{X}$, and a different sample standard deviation, s.

Think about what causes the variability in the Z statistic. For a given sample size—say, $n = 26$—the only thing that causes the Z value to change from sample to sample is $\overline{X}$. The other elements of the calculation stay the same. Now look at the t distribution. Both $\overline{X}$ and s change from sample to sample, so t has more variability than Z. We can conclude that t will therefore have "fatter" tails than Z and indeed it does. The t distribution for a sample size of $n = 26$ is shown in Figure 7.13.

You should ask, Why do I need to know the value of n to draw the graph? You might remember that as we take larger and larger samples (n gets larger), our estimate of the unknown population standard deviation gets better and better. Why is this so? Well, think of the whole population as a big pie. When you take a small sample, it is like tasting a small piece of the pie and trying to decide how the whole pie tastes based on that piece. As the piece you taste gets bigger, you have a much better idea of how the whole pie tastes. Eventually, if you take a big enough sample, you will eat the whole pie and you will know exactly how the whole pie tastes! This is the same as sampling the whole population. If the whole population were sampled, you wouldn't need to estimate σ because you would know it.

Better estimates of σ make t less variable.

So as we take a larger and larger sample, we get a better and better estimate of σ, which, in turn, causes t to be less and less variable. This means that the graph of t depends on the size of the sample. In particular, the degrees of freedom associated with the t distribution is related to the sample size. The value for the degrees of freedom is $n - 1$, one less than the sample size. If we have a sample of size 26, then the t distribution has 25 degrees of freedom. This is the graph shown in Figure 7.13. Examine the t distributions in Figure 7.14 for 1 and 50 degrees of freedom (df).

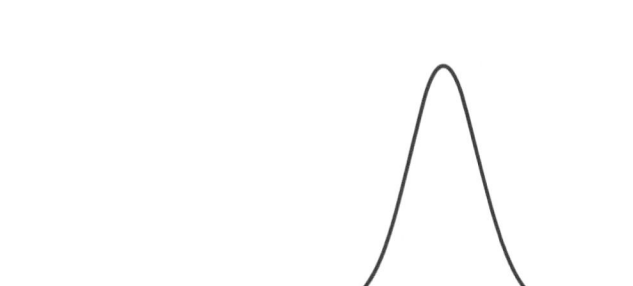

FIGURE 7.13 *t* **distribution with 25 degrees of freedom**

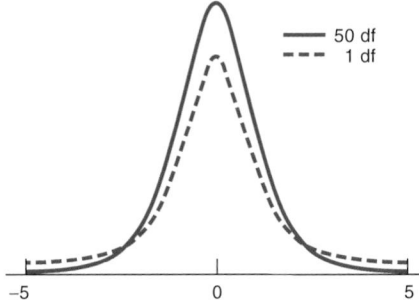

FIGURE 7.14 *t* **distributions with 1 and 50 degrees of freedom**

What do you notice about the graph of *t* as the number of degrees of freedom gets larger (i.e., as the sample size gets larger)? The *t* distribution gets tighter, meaning that the variability of the distribution gets smaller. This is precisely what we expect to happen based on our discussion of our estimate of *s* as we sample bigger pieces of the pie. In fact, eventually, if we use a large enough value for *n*, the graph of *t* will be indistinguishable from the standard normal distribution, *Z*. Typically, the value of *n* that is considered large enough is $n > 30$.

7.10 Small-Sample Confidence Intervals for the Mean

Now that you have an understanding of the *t* distribution, you need to understand how the confidence interval for μ changes for small samples. Replacing the *Z* table value with a *t* value in the formulas for the upper and lower bounds of the confidence interval for μ gives us the following equations:

Error: $$e = \frac{t_{\alpha/2, n-1} s}{\sqrt{n}}$$

Width of the interval: $w = 2e$

Lower bound: $L = \overline{X} - e$

Upper bound: $U = \overline{X} + e$

Confidence interval for μ, small sample and σ unknown

In addition, you must now assume that the underlying population of the variable you are estimating is normally distributed.

7.10.1 USING THE *t* TABLE

The procedure for figuring out the correct *t* value for any confidence level is similar to the method we used to find the correct *Z* value. We need to find the *t* value so that there is a total probability of α in the tails of the distribution. Since the *t* distribution is symmetric, there will be $\alpha/2$ in each of the tail areas. Unlike with the *Z* table, there is no need to work the table backward. Typically, the *t* table is used to find the *t* value that corresponds to a certain probability in one of the tails of the distribution rather than to find probabilities under the *t* distribution. A portion of the *t* table is shown in Table 7.6 on page 334.

The complete *t* table is found in Table 4 in Appendix A. Notice that the table has several columns and rows. The columns correspond to the tail area probability and the rows correspond to the number of degrees of freedom, which is 1 less than the sample size. The procedure for using the table follows:

TABLE 7.6

A PORTION OF THE t TABLE

Degrees of freedom	Upper-Tail Areas					
	0.25	0.1	0.05	0.025	0.01	0.005
20	0.687	1.325	1.725	2.086	2.528	2.845
21	0.686	1.323	1.721	2.080	2.518	2.831
22	0.686	1.321	1.717	2.074	2.508	2.819
23	0.685	1.320	1.714	2.069	2.500	2.807
24	0.685	1.318	1.711	2.064	2.492	2.797
25	0.684	1.316	1.708	2.060	2.485	2.787
26	0.684	1.315	1.706	2.056	2.479	2.779

Steps for finding the value of t for a confidence interval

Step 1: Take the value of α and divide it by 2.

Step 2: Use that column of the t table.

Step 3: Find the number of degrees of freedom by calculating $n - 1$. Use that row of the t table.

Step 4: Read the t value at the intersection of the row and column you identified. Label this value $t_{\alpha/2, n-1}$ to indicate that the tail area probability is $\alpha/2$ and there are $n - 1$ degrees of freedom.

Let's look at an example to see how to use the t table.

Collect and analyze the data.

EXAMPLE 7.13

Using the t Table

We can follow the steps to find the t value for a 95% confidence interval for a sample size of 25.

Step 1. Since a 95% confidence interval is required, α is 0.05 and so $\alpha/2$ is 0.025.

Step 2. Use the 0.025 column of the t table.

Step 3. Since the sample size is $n = 25$, use the row corresponding to $n - 1 = 25 - 1 = 24$ degrees of freedom.

Step 4. The t value 2.064 is found on Table 7.6. Label it $t_{0.025, 24}$.

Now we can use this t value to find a confidence interval for μ.

Collect and analyze the data.

EXAMPLE 7.14

GAS MILEAGE

Finding the Small-Sample Confidence Interval for μ

You have been recording the gas mileage each time you fill up your Honda Civic, and after 25 fillups, you have these data (in miles per gallon):

29.1	26.17	30.73	33.83	33.6
35.2	23.45	29.3	33.29	26.74
27.93	29.44	24.93	24.46	27.07
27.68	23.65	28.3	28.79	30.4
28.9	29.02	28.89	34.03	29.74

To find the confidence interval for μ, we need to calculate $\overline{X}$ and s and find the appropriate t value from Table 7.6. We get

$$\overline{X} = 28.99$$

$$s = 3.22$$

The t value for 24 degrees of freedom and a tail area of 0.025 is $t_{0.025,24} = 2.064$. The upper and lower 95% confidence bounds for μ are calculated using the following formulas:

Error: $\qquad e = \dfrac{t_{\alpha/2,n-1}s}{\sqrt{n}} = \dfrac{(2.064)(3.22)}{\sqrt{25}} = 1.33$

Width of the interval: $\quad w = 2e = 2(1.33) = 2.66$

Lower bound: $\qquad L = \overline{X} - e = 28.99 - 2.66 = 26.33$

Upper bound: $\qquad U = \overline{X} + e = 28.99 + 2.66 = 31.65$

Notice that the t value for the 95% confidence interval with 24 degrees of freedom is 2.064. If we had known the value of σ, then we would have used a Z value of 1.96 instead of the t value. This would have made the value of e smaller and the interval narrower. Confidence intervals constructed using t values are always wider than the corresponding intervals constructed from Z values. This is a direct result of the increased variability caused by having to estimate σ.

The t distribution is the correct sampling distribution to use whenever σ is unknown. However, when the sample size is large, $n > 30$, it is common practice to use the Z distribution instead of the t distribution because the graphs are virtually identical. The original reason for using Z instead of t for large samples has all but vanished, but the practice continues. Before the use of statistical software, analysts had to rely on printed t tables. Since the t distribution depends on the sample size, very large t tables were needed to accommodate many different sample sizes. Thus, it was generally agreed that Z could be used for sample sizes greater than 30, which eliminated the need for t tables beyond 30 degrees of freedom. Now software packages have eliminated the need for printed tables. Nevertheless, the convention of using Z instead of t for large samples has become entrenched, and in this book when the sample size is larger than 30 we use the Z distribution. Note that you are certainly correct to use t whenever the standard deviation is unknown; some software packages may take that approach.

7.10.2 EXERCISES—LEARNING IT!

7.20 The police department is concerned about the ability of officers to identify drunk drivers on the road. Before instituting a new training program, the department takes a sample of 28 arrests and records the level of alcohol in the blood at the time of the arrest. Assume that the level of alcohol in the blood is normally distributed. The levels of alcohol (in %) are listed here:

92	93	108	173	194	133	207	127	256	252
184	253	159	101	133	204	182	173	105	
153	150	180	209	141	151	133	147	209	

(a) Find a 90% confidence interval for the average blood alcohol level at the time of arrest.

(b) Find a 95% confidence interval for the average blood alcohol level at the time of arrest.

7.21 A large amusement park has recently added five new rides, including a large roller coaster called the Mind Eraser. Management is concerned about the waiting times for the new roller coaster. A random sample of ten people is selected and the time (in minutes) each person waits to ride the Mind Eraser is recorded:

43   80   48   61   74   66   54   72   58   68

(a) Find a 95% confidence interval for the average waiting time for the Mind Eraser, assuming that the waiting times are normally distributed.

(b) The park management thinks that if customers have to wait longer than 60 minutes for a ride, then the park should increase the staff to reduce the waiting time. Based on your confidence interval, does the park need to increase the staff? Explain why or why not.

7.22 Hospital administrators are paying increasing attention to the lengths of stays for patients. The stays (in days) from a sample of 14 patients are listed:

7   2   6   7   8   8   3
6   5   4   4   2   3   7

(a) Find a 95% confidence interval for the length of patient stay, assuming that the lengths of stay are normally distributed.

(b) Find a 90% confidence interval for the length of patient stay, assuming that the lengths of stay are normally distributed.

(c) Which of these intervals should the administrator use in negotiating with insurance companies and why?

7.23 The numbers of workers per vehicle at the top ten car assembly plants are shown here:

Nissan Smyrna	2.22	Honda Marysville	2.57
Toyota Cambridge	2.35	Ford Atlanta	2.63
Honda East Liberty	2.38	Ford Chicago	2.66
Toyota Georgetown #1	2.50	Chrysler Belvidere	2.68
Chrysler Bramalea	2.54	GM Oshawa #1	2.68

Source: *Manufacturing Engineering*, August 1997

Assuming that the number of workers per vehicle has a normal distribution, find a 95% confidence interval for the average number of workers per vehicle.

7.24 The annual number of deaths per 100,000 people in the population is called the death rate. Here are the death rates for a sample of 14 states for 1999:

Alabama	1025.3	New Hampshire	794.0
Alaska	437.1	North Dakota	963.1
California	692.1	Ohio	964.0
Kansas	922.1	Rhode Island	979.8
Kentucky	992.8	Texas	733.4
Massachusetts	904.3	Utah	566.1
Minnesota	807.0	Wyoming	842.8

Find a 95% confidence interval for the average death rate, assuming that the death rates are normally distributed.

7.11   Confidence Intervals for Qualitative Data

Remember! Qualitative data describe a particular characteristic of a sample item. They are most often nonnumerical.

Often we are interested in estimating what proportion of the population has a particular characteristic or opinion. In fact, most of the data printed in newspapers and magazines are the result of surveys and often report proportions or percentages. This is particularly true at election times. For example, the candidate for mayor wishes to know what proportion or percentage of voters in the city favor him.

The increasing focus on quality and customer needs has led to increased data collection, and much of the data are qualitative. A manufacturing company is clearly interested in the proportion of products that are defective. All businesses, both manufacturing and service industries, are interested in knowing whether their products/services are meeting the needs of the customer. We are often asked to fill out questionnaires about how we like a product or service. The data from such questionnaires are often qualitative data.

If the data we are analyzing are qualitative data, then we are most likely interested in estimating the proportion, p, of population members that have a certain characteristic (one of the categories of the nominal variable). Confidence intervals for the population proportion, p, have the same basic structure as those for μ. Remember that the sample proportion, $\hat{p}$, is the best point estimate for p, and so we center the confidence interval at the value of $\hat{p}$.

7.11.1 FINDING THE CONFIDENCE INTERVAL FOR p

To develop the formulas for the lower and upper bounds of a $100(1 - \alpha)\%$ confidence interval for p we need to know that the sampling distribution of $\hat{p}$ is a normal distribution. Again drawing on the properties of the normal distribution, we must extend the interval a certain number of standard errors to get the coverage we desire. The standard error of the point estimator, $\hat{p}$ is $\sqrt{p(1-p)/n}$. Notice that to calculate the standard error of $\hat{p}$, we must know the value of p. But we are trying to estimate p, so, clearly, we do not know it. It makes the most sense to use $\hat{p}$ as an estimate of p in the formula for the standard error.

The point estimate, $\hat{p}$, should be at the center of any confidence intervals of p.

Combining these pieces of information with the knowledge that we wish to place $\hat{p}$ at the center of the confidence interval, we arrive at the following formulas for the lower and upper bounds:

Confidence interval for p

Error:
$$e = Z_{\alpha/2}\sqrt{\frac{p(1-p)}{n}} \approx Z_{\alpha/2}\sqrt{\frac{\hat{p}(1-\hat{p})}{n}}$$

Width of the interval: $w = 2e$

Lower bound: $L = \hat{p} - e$

Upper bound: $U = \hat{p} + e$

where $Z_{\alpha/2}$ is the Z value that cuts off $\alpha/2$ in the upper tail of the standard normal distribution and $\alpha/2$ in the lower tail of the standard normal distribution.

Let's look at an example.

EXAMPLE 7.15

UNDERAGE DRINKING

Collect and analyze the data.

Calculating the Confidence Interval for p

There is much discussion in state legislatures about the appropriate drinking age. A Fox News opinion poll from June 8, 2001, posed this question to a sample of 900 registered voters:

"Did you ever drink alcohol before you were legally old enough to do so?" Here are the results:

Yes	63%
No	35%
Not sure	2%

To find a 95% confidence interval for the proportion of registered voters who did drink underage, we are given

$$\hat{p} = 0.63$$
$$n = 900$$

Now we can perform the calculations:

Standard error: $\sqrt{\dfrac{\hat{p}(1 - \hat{p})}{n}} = \sqrt{\dfrac{0.47(1 - 0.47)}{2000}} = 0.016$

Error: $e = z_{\alpha/2}\sqrt{\dfrac{\hat{p}(1 - \hat{p})}{n}} = (1.96)(0.016) = 0.031$

Lower bound: $\hat{p} - e = 0.63 - 0.031 = 0.599$

Upper bound: $\hat{p} + e = 0.63 + 0.031 = 0.661$

So we can state that we are 95% confident that the percentage of all registered voters who drank underage is between 59.9% and 66.1%.

✔ *TRY IT NOW!*

Retirement Years
Finding a Confidence Interval for p

A survey shows that a growing number of Americans are willing to make sacrifices to become home-owners despite increasing job and financial worries. The Federal National Mortgage Association surveyed 1857 Americans and found that 67% would put off retirement for 10 years to own a home. Find a 90% confidence interval for the proportion of all Americans who would put off retirement for 10 years to own a home.

7.11.2 EXERCISES—LEARNING IT!

7.25 In a survey 100 friends were asked the question, "Do you regularly watch *The Weakest Link*?" Of the 100, 35 answered yes.

(a) Calculate a 95% confidence interval for the "viewership" of this show.

(b) The network is considering canceling the show if less than one-third of the population regularly watches it. Based on this information, what will the network do?

7.26 Americans are often viewed as being impatient. A recent survey of 2000 drivers found that 52% of them are irritated by slow drivers and 60% report that they are pressed for time.

(a) Find a 95% confidence interval for the proportion of drivers who are irritated by slow drivers.

(b) Based on your confidence interval, what recommendations might you make to the Department of Motor Vehicles?

7.27 A poll of 450 registered Massachusetts voters found 46% opposed to allowing casino gambling in the state.

(a) Find a 95% confidence interval for the proportion of all Bay State voters opposed to allowing casino gambling in the state.

(b) Is 50% in the confidence interval? If so, what does this tell you? If not, what does this tell you?

(c) What would you recommend to the governor of Massachusetts, who is pushing for expanded gambling?

7.28 Thousands of years ago people hung the yellow blossoms of Saint-John's-wort over their doorways, hoping to ward off evil spirits. Today German physicians write nearly 3 million prescriptions a year for pills made from extracts of the plant, meant to relieve depression. But doctors on the other side of the Atlantic haven't followed suit. U.S. scientists say there is no reliable evidence that the herb can help. To find out more, researchers at the University of Munich and the University of Texas in San Antonio studied 1500 people with mild to moderate

ANS. $L = 0.652, U = 0.688$

depression. One-third were treated with Saint-John's-wort, one-third were treated with antidepressant drugs, and one-third were treated with a placebo. The researchers found that after 6 weeks of treatment, 64% of the people taking Saint-John's-wort pills felt markedly better compared to 59% of those who received synthetic antidepressant drugs.

(a) Construct a 95% confidence interval for the proportion of people who would feel markedly better using Saint-John's-wort pills.

(b) Construct a 95% confidence interval for the proportion of people who would feel markedly better using synthetic antidepressant drugs.

(c) Compare the two confidence intervals. Do they overlap at all? What is your recommendation to doctors?

7.29 In designing a new dormitory, a progressive university wishes to determine where students prefer to study in order to provide appropriate space. A survey of 100 randomly selected undergraduate students found that 33% prefer to study in their rooms.

(a) Construct a 95% confidence interval for the proportion of students who prefer to study in their rooms.

(b) If the university has the option of making various room sizes, what proportion of the rooms should be made larger to accommodate students who wish to study in their rooms?

7.30 A medium-sized city hospital is concerned about the number of ventilator-acquired respiratory infections in each of the past three years. A random sample of infections were studied and the numbers of infections that were vent-related are given here:

Year	Number of infections	Number of vent-related infections
1998	19	13
1999	14	11
2000	13	10

(a) Find the sample proportion of vent-related infections for each of the three years.

(b) Construct a confidence interval for the proportion of vent-related infections for each of the three years.

(c) Should the hospital be concerned that the proportion of vent-related infections is increasing? Justify your answer using the confidence intervals you found in part (b).

7.12 Sample Size Calculations

Confidence intervals are easy to calculate and are commonly used to provide interval estimates for either the population mean or the population proportion. In fact, newspaper articles often report the sample mean or sample proportion and a number called the sampling error or *margin of error*.

EXAMPLE 7.16

SPOUSAL SECRETS

Illustrating Sampling Error

According to an article written by the Associated Press and published in the *Springfield Union News* on July 21, 2001, 40% of married Americans admit keeping a secret from their spouses. Most of the secrets have nothing to do with an affair or fantasy. The most common secret is how much money a spouse spends. The poll was conducted by Ipsos-NPD, an Illinois-based research group, which surveyed 1000 husbands and wives by telephone in March 2001. The poll has a margin of error of plus or minus 3 percentage points.

Collect and analyze the data.

This sampling error or margin of error is what we have labeled *error*. Thus, if we want to construct a confidence interval, we simply add and subtract the sampling error to the point estimate.

Collect and analyze the data.

EXAMPLE 7.17

SPOUSAL SECRETS

Looking at the Margin of Error

From the results of the Ipsos-NPD study, the proportion of married Americans who admit to keeping a secret is 40% with a 3% margin of error. Thus, the percentage is between 37% and 43%. The article in the newspaper did not report the confidence level, but typically 95% confidence intervals are calculated.

7.12.1 UNDERSTANDING THE SAMPLING ERROR

Sampling error results because only a piece of the population has been studied.

We have not really examined why the distance between the center of the confidence interval and either endpoint is called an *error*. As you have learned, the term *sampling error* does not imply that we have made an error. It does indicate that we have imperfect information about the population because we have studied only a sample of that population. There is a different reason this value is called an error. Consider the confidence interval calculated in Example 7.12 for the housing market. We found this 95% confidence interval for the mean selling price:

$99,223.30

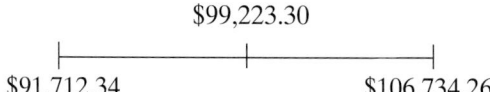

$91,712.34 $106,734.26

Remember that the value $99,223.30 is the sample mean, $\overline{X}$. Suppose for the moment that we have found an interval that does contain μ but we don't know where in the interval μ is located. If we use $\overline{X}$ to estimate μ, what is the largest amount by which we could miss the value of μ? This *worst-case scenario* happens if μ is at either endpoint of the interval. Then our point estimate $\overline{X}$ is in error by the amount $e = 7510.96$. If μ is anywhere else in the interval, then our point estimate $\overline{X}$ is off by some amount less than 7510.96. So we are 95% confident that the error is at most 7510.96. Remember that a 95% confidence level indicates that 19 out of 20 intervals of this width contain the true population parameter. This is why we have labeled the distance from the middle of the interval to either endpoint as e, to stand for *error*. It is precisely this value that is often quoted in newspaper or research articles, although the newspaper rarely gives the corresponding level of confidence. Typically, a 95% level of confidence is used.

You might wonder if it is possible to specify the size of this error to achieve a certain level of accuracy. This is indeed often desirable. By specifying the amount of error that we can tolerate in a particular situation, we are determining the sample size needed. Often we cannot achieve a particular level of accuracy because to do so would require a sample size that we cannot afford. This is our next subject.

7.12.2 DETERMINING THE SAMPLE SIZE

In Chapter 1 we identified the following factors that are important in determining the size of the sample needed:

- The variation in the population
- The error that can be tolerated

- The resources available
- The size of the population

We now know enough statistics to develop a formula for the sample size that incorporates the first two of these factors. You should remember that these two factors were identified as most important to the sample size determination.

Recall that the expression for e is one of the following, depending on whether you are estimating the mean, μ, or the proportion, p:

$$\text{Estimating } \mu: \quad e = \frac{Z_{\alpha/2}\sigma}{\sqrt{n}}$$

$$\text{Estimating } p: \quad e = Z_{\alpha/2}\sqrt{\frac{p(1-p)}{n}}$$

Each of these equations can be rewritten and solved for n, the sample size, by using some basic algebra. We get two equations for n that are algebraically equivalent to the preceding two equations:

$$\text{Estimating } \mu: \quad n = \frac{Z_{\alpha/2}^2\sigma^2}{e^2} = \left(\frac{Z_{\alpha/2}\sigma}{e}\right)^2 \qquad \text{\textit{Sample size for estimating } μ}$$

$$\text{Estimating } p: \quad n = \frac{Z_{\alpha/2}^2\, p(1-p)}{e^2} \qquad \text{\textit{Sample size for estimating } p}$$

Let's examine the equation for n if we are trying to estimate the population mean, μ. If we specify a certain level of confidence and a value for the maximum difference between $\overline{X}$ and μ, then we can calculate how large a sample we must select.

EXAMPLE 7.18

THE HOUSING MARKET

Collect and analyze the data.

Determining the Sample Size

You, the seller, want to be 95% confident that the estimate $\overline{X}$ of the number of days it will take to sell your house is not in error by more than 10 days. That is, you want to be sure that 19 out of 20 possible confidence intervals of width 10 days contain the true mean, μ. Recall that the number of days a house is on the market has a standard deviation of 60. Thus, we have

$$e = 5 \text{ days}$$
$$Z_{\alpha/2} = 1.96$$
$$\sigma = 160$$

Using this information to calculate n yields

$$n = \frac{Z_{\alpha/2}^2\sigma^2}{e^2} = \frac{(1.96^2)(60^2)}{5^2} = 553.19$$

Since we can't sample a fractional house, we must round this value up to 554 houses.

In sample size calculations you should always round up to guarantee the level of confidence and error you have specified.

Always round up in sample size calculations.

✓ TRY IT NOW!

Bridge Traffic
Finding the Sample Size

How many days does the Department of Transportation need to sample to be 98% confident that the mean number of cars passing over the bridge in Framingham has an error of at most 1000 cars? Remember that the population standard deviation is 500 cars.

Ways to proceed when standard deviation is unknown

Notice that in the preceding problem we needed a value for the population standard deviation to calculate n. If this value is unknown, then we must estimate it. Unfortunately, often we need to take a sample to estimate σ, but we need σ to figure out how large our sample needs to be. This is a classic "catch-22" situation. There appears to be no way out.

Fortunately, there are a couple of ways to handle this dilemma. The first way is to take a small sample to get a rough estimate of the standard deviation and use this estimate in the formula for n. Suppose that we take a sample of size 10 and use it to calculate the sample standard deviation, s. We can then use this value of s as an estimate for σ and plug it into the formula for n. Suppose when we do this the formula tells us that we need a sample of size $n = 55$. Well, we already have data on 10 observations so we need an additional 45 to complete the sample. The second way is to use information about the variability of a similar product or process. This may not be perfect but it gives a rough idea of the value of σ that can, in turn, be used in the sample size calculation.

To be more confident, the sample size must be larger.

Very often we cannot use the sample size determined by the formula because of limited resources (i.e., cost). Then we must be willing to accept a lower level of confidence or tolerate a higher value of e or both. Let's consider the impact on n of varying the level of confidence. We know that as the level of confidence increases, the Z value also increases. Since Z is in the numerator of the formula for n, we would expect the sample size needed to increase as the level of confidence increases. This should make intuitive sense as well. For us to be more confident, the sample size must be larger.

Collect and analyze the data.

EXAMPLE 7.19

THE HOUSING MARKET

Increasing the Sample Sizes for Increasing Levels of Confidence

Suppose we consider several different levels of confidence for the average number of days the house needs to be on the market in order to sell it. The sample sizes needed to achieve an error of at most 10 days for various levels of confidence are calculated here:

90% confidence level: $\quad n = \dfrac{Z_{\alpha/2}^2 \sigma^2}{e^2} = \dfrac{(1.645^2)(60^2)}{10^2} = 97.42$ rounded up to 98

95% confidence level: $\quad n = \dfrac{Z_{\alpha/2}^2 \sigma^2}{e^2} = \dfrac{(1.96^2)(60^2)}{10^2} = 138.30$ rounded up to 139

98% confidence level: $\quad n = \dfrac{Z_{\alpha/2}^2 \sigma^2}{e^2} = \dfrac{(2.33^2)(60^2)}{10^2} = 195.44$ rounded up to 196

99% confidence level: $\quad n = \dfrac{Z_{\alpha/2}^2 \sigma^2}{e^2} = \dfrac{(2.58^2)(60^2)}{10^2} = 239.63$ rounded up to 240

Ans. $n = 22$

If the sample size we calculate is too expensive, the other way we can cut costs is to increase the size of the error we can tolerate. Let's see what happens to the sample size as the maximum error is increased. Again, the formula for *n* has the value for *e* in the denominator. So by increasing *e* we are dividing by a larger number and hence the sample size needed will be smaller. Intuitively, this too makes sense. If we can tolerate a larger error, then we can take a smaller sample.

As the maximum error increases, the sample size decreases.

EXAMPLE 7.20

THE HOUSING MARKET

Looking at the Impact of Larger Errors

Consider the following values for the maximum error that we are willing to tolerate in our estimate of μ. In all cases we wish to be 95% confident that the error is at most the specified value.

Error = 5 days: $n = \dfrac{Z_{\alpha/2}^2 \sigma^2}{e^2} = \dfrac{(1.96^2)(60^2)}{5^2} = 553.19$ rounded up to 554

Error = 10 days: $n = \dfrac{Z_{\alpha/2}^2 \sigma^2}{e^2} = \dfrac{(1.96^2)(60^2)}{10^2} = 138.29$ rounded up to 139

Error = 15 days: $n = \dfrac{Z_{\alpha/2}^2 \sigma^2}{e^2} = \dfrac{(1.96^2)(60^2)}{15^2} = 61.47$ rounded up to 62

Error = 20 days: $n = \dfrac{Z_{\alpha/2}^2 \sigma^2}{e^2} = \dfrac{(1.96^2)(60^2)}{20^2} = 34.57$ rounded up to 35

Collect and analyze the data.

Calculations of the sample size when we wish to estimate the population proportion, *p*, are done in a similar manner. The only difference is the particular formula used. Consider the poll on underage drinking described in Example 7.15.

EXAMPLE 7.21

UNDERAGE DRINKING

Calculating the Sample Size for Proportions

How many registered voters must be sampled to be 95% confident that our estimate of *p* is off by at most 5%? Remember that the point estimate is $\hat{p}$, the sample proportion or percentage. If we want the sample proportion to be in error by at most 5%, then $e = 0.05$. The formula for *n* is

$$n = \frac{Z_{\alpha/2}^2 \, p(1 - p)}{e^2}$$

Plugging in the values we know gives us

$$n = \frac{(1.96^2)p(1 - p)}{0.05^2}$$

At this point we realize that we can't continue without a value for *p*, but we need a sample to get a sample proportion.

Collect and analyze the data.

This is again a circular problem. We need *p* to find *n*, but we need *n* to get *p*! Two approaches are possible here. First, if we have any information about the value of *p* from other samples or experience, then we may use that information as an estimate of p. Second, if we have no information at all, then we may use a value of $p = 0.50$ in the formula for the sample size calculation. When we do this, the sample size that is calculated is as large as it can be for the specified confidence level and error. It is the most conservative approach we can take and often leads to a sample size larger than we really need. Let's continue Example 7.21 by using a value of $p = 0.50$.

Collect and analyze the data.

EXAMPLE 7.22

UNDERAGE DRINKING

Calculating the Sample Size for Proportions

Using a value $p = 0.50$ in the formula for n gives us

$$n = \frac{(1.96^2)p(1 - p)}{0.05^2} = \frac{(1.96^2)(0.5)(1 - 0.5)}{0.05^2} = 384.16$$

which we round up to 385 registered voters.

✔ TRY IT NOW!

Retirement Years
Calculating the Sample Size for p

How many Americans must be sampled to determine the percentage who would put off retirement for 10 years in order to own a home? The estimate should not differ from the actual population proportion by more than 3% with a confidence of 90%.

The conclusions that we reached about what happens to the required sample size as the confidence level and the error vary are the same for proportions as for means. As the confidence level increases, the sample size needed to estimate the population proportion also increases. As the error that can be tolerated increases, the sample size needed decreases.

7.12.3 EXERCISES—LEARNING IT!

Requires Exercise 7.13

7.31 How many years must be sampled to be 95% confident that the error in estimating the average January snowfall in Rochester, New York, is at most 0.25 inch? The standard deviation of the amount of snowfall is known to be 1.5 inches.

Requires Exercise 7.15

7.32 How many months must be sampled for analysts to be 99% confident that the error in estimating the average monthly price of a 16-ounce bag of potato chips is at most $0.02? Assume the standard deviation is $0.15.

Requires Exercise 7.16

7.33 How many days must be observed for the grocery store chain to be 90% confident that the error in estimating the average number of vehicles that pass a certain location is at most ten?

Requires Exercise 7.25

7.34 How many friends must be surveyed to be 98% confident that the estimate of the viewership is in error by at most 3%?

Requires Exercise 7.26

7.35 How many drivers must be sampled to be 96% confident that the error in estimating the proportion of drivers who are pressed for time is at most 0.02?

GET IT IN WRITING

The Housing Market

TO: **Manager, Real Estate Agency**
FROM: **Erica Q. Homeowner**
RE: **Changing Jobs and Moving On**

As you can see by looking at the website I cited in the beginning of this chapter, people change jobs more frequently in today's marketplace, thereby necessitating more housing moves. In trying to estimate the average selling price of houses in this market as well as the average number of days a house is on the market, I have analyzed the data you sent me.

The data gave the details of 150 home sales by your agency in my area. Based on these data, the average number of days houses sold by your agency are on the market is 52.5. After calculating a confidence interval, I feel that this is quite typical of the national trends. In addition, it seems that the average selling price of homes sold by your agency is close to $105,000. Based on this analysis, I plan to list my home with your agency and I would like to put it on the market 75 days before I plan to move. This will give me a high probability of selling the home before I need to relocate. In order to determine the list price for my home, I need to do some additional analysis. I will calculate the average price of homes that have similar characteristics to my home and get back to you when I have completed this analysis.

Thank you for sharing these data with me, and I look forward to doing business with your agency.

7.13   Using Technology to Find Confidence Intervals

Excel, Minitab, and the TI-83 graphing calculator can be used to help you find confidence intervals. Refer to the manual associated with the software that you are using for more detailed instructions. A brief overview of the capabilities of each package with regard to the topics of this chapter follows.

Excel 2000

Excel does not have a tool that automatically calculates confidence intervals. Instead, you will have to use some of the functions you already learned about in a formula. To find a confidence interval for a population parameter, you need a point estimate and a formula for the error of the estimate. To find a confidence interval for the mean, you need a point estimate, $\overline{X}$. You can obtain this from the Descriptive Statistics in the Data Analysis Tools or by using the **AVERAGE** function. You also need a value for σ or else a sample estimate, s, based on a large sample. The last thing you need is the Z value, which you obtain using the **NORMSINV** function discussed in Chapter 6.

Minitab

Minitab allows you to calculate confidence intervals for many different population parameters. The functions that calculate the intervals are part of hypothesis testing, which is the subject of Chapters 8–10.

TI-83 Graphing Calculator

You can use the TI-83 to calculate any of the confidence intervals you learned about in this chapter. One convenient feature of the TI-83 is that it allows you to do the calculations either from raw data or using summary statistics (as in some of the exercises at the end of the chapter).

KADDSTAT

KADDSTAT provides statistical functions that allow you to find confidence intervals for both means and proportions. Following are detailed directions for the small-sample confidence interval for the mean.

From the KADD menu, select **Confidence intervals > One sample > Population Mean using t.** The dialog box shown in Figure 7.15 opens.

1. First, indicate the level of confidence, as a percentage, that you want to use.

2. You have the option of calculating a confidence interval either from raw data or from summary statistics (like a problem from a textbook). Select **User Input** if you already have the summary statistics or **Input Range** if you have raw data. As you can see from Figure 7.15, for **User Input,** you must supply Excel with the mean, the standard deviation, and the sample size. If you select **Input Range,** the dialog box will change as shown in Figure 7.16, and you will input the location of the data. For **Input Range** you must also indicate whether the data have a header row.

3. You must indicate how the sampling was done: from a large (infinite) population or from a finite population without replacement.

4. Indicate where you want the output to appear and click **OK.**

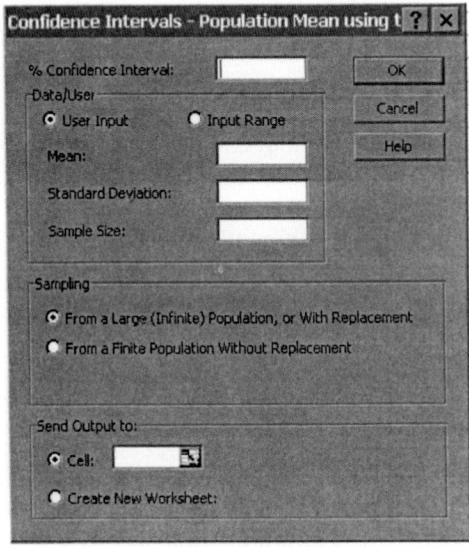

FIGURE 7.15 Dialog box for Confidence Intervals (User Input)

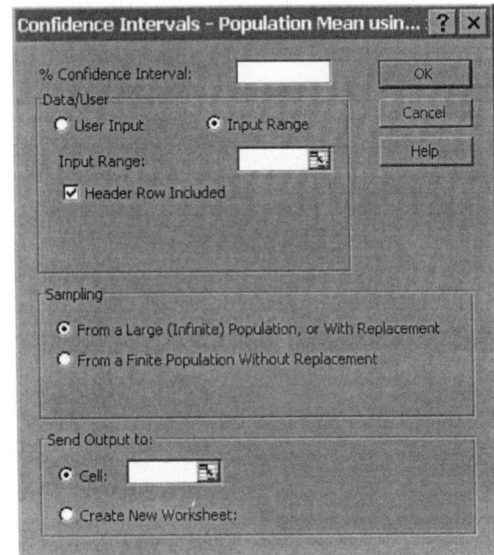

FIGURE 7.16 Dialog box for Confidence Intervals (Input Range)

CHAPTER 7 SUMMARY

This chapter covered the basics of estimating population parameters. In particular, you learned how to estimate the average of a numerical characteristic of a population, μ, and the proportion of a population that has a certain characteristic, p. The estimates are calculated from a sample selected from the population. Each sample yields a slightly different estimate of the population parameter, and thus the estimators are themselves random variables. Just like the random variables you studied in Chapter 6, the estimators have a distribution. When the random variable is an estimator, the distribution is called a sampling distribution. The sampling distribution has a mean and a standard deviation, which is called the standard error.

You learned how to use the sampling distribution of $\overline{X}$ to calculate probabilities and make inferences about μ. You also learned how to create confidence intervals for μ and for p. Finally, you learned how to calculate the sample size required to achieve a certain level of precision with a specified confidence.

KEY TERMS

Key Term	Definition	Page Reference
Central Limit Theorem	The **Central Limit Theorem** states that in random sampling from a population with mean μ and standard deviation σ, when n is large enough, the distribution of $\overline{X}$ is approximately normal with a mean, $\mu_{\bar{x}}$, equal to μ and a standard deviation, $\sigma_{\bar{x}}$, equal to $\sigma/\sqrt{n}$.	307
Confidence interval	A **confidence interval** or an **interval estimate** is a range of values with an associated probability or confidence level, $1 - \alpha$. The probability quantifies the chance that the interval contains the true population parameter.	320
Point estimate	A **point estimate** is a single number calculated from sample data. It is used to estimate a parameter of the population.	294
Point estimator	A **point estimator** is the formula or rule that is used to calculate the point estimate for a particular set of data.	294
Sampling distribution	The distribution of a point estimator or a sample statistic is called a **sampling distribution**.	306
Standard error	The **standard error** is the standard deviation of the sampling distribution of a point estimator. It measures how much the point estimator or sample statistic varies from sample to sample.	306
Unbiased estimator	An **unbiased estimator** yields an estimate that is fair. It neither systematically overestimates the parameter nor systematically underestimates the parameter.	302

KEY FORMULAS

Description	Formula	Page Reference
Z-score for $\bar{X}$	$Z = \dfrac{\bar{X} - \mu_{\bar{X}}}{\sigma_{\bar{X}}} = \dfrac{\bar{X} - \mu}{\sigma/\sqrt{n}}$	305
Mean of $\bar{X}$	$\mu_{\bar{X}} = \mu$	308
Standard error of $\bar{X}$	$\sigma_{\bar{X}} = \dfrac{\sigma}{\sqrt{n}}$	309
Confidence interval for μ when σ is known	Error: $e = \dfrac{Z_{\alpha/2}\sigma}{\sqrt{n}}$ Width of the interval: $w = 2e$ Lower bound: $L = \bar{X} - e$ Upper bound: $U = \bar{X} + e$	323
Large-sample confidence interval for μ when σ is unknown	Error: $e = \dfrac{Z_{\alpha/2}s}{\sqrt{n}}$ Width of the interval: $w = 2e$ Lower bound: $L = \bar{X} - e$ Upper bound: $U = \bar{X} + e$	324
t	$t = \dfrac{\bar{X} - \mu}{s/\sqrt{n}}$	331
Small-sample confidence interval for μ when σ is unknown, normal population	Error: $e = \dfrac{t_{\alpha/2,n-1}s}{\sqrt{n}}$ Width of the interval: $w = 2e$ Lower bound: $L = \bar{X} - e$ Upper bound: $U = \bar{X} + e$	333
Confidence interval for p	Error: $e = Z_{\alpha/2}\sqrt{\dfrac{p(1-p)}{n}} \approx Z_{\alpha/2}\sqrt{\dfrac{\hat{p}(1-\hat{p})}{n}}$ Width of the interval: $w = 2e$ Lower bound: $L = \hat{p} - e$ Upper bound: $U = \hat{p} + e$	337
Sample size for estimating μ	$n = \dfrac{Z_{\alpha/2}^2 \sigma^2}{e^2}$	341
Sample size for estimating p	$n = \dfrac{Z_{\alpha/2}^2 p(1-p)}{e^2}$	341

CHAPTER 7 EXERCISES

Learning It!

7.36 Does it sometimes seem like the cereal box you bought is not quite full? According to the information on most cereal boxes, the "contents may have settled" and so it appears less than full. Still, you would like to be sure that you are getting your money's worth of Fruit Loops. You decide to sample 35 15-ounce cereal boxes, and you find these weights (in ounces):

14.91	15.23	15.34	15.40	15.48	15.59	15.62
14.96	15.24	15.34	15.42	15.49	15.60	15.62
15.11	15.28	15.36	15.45	15.50	15.60	15.63
15.21	15.30	15.38	15.46	15.52	15.61	15.64
15.22	15.30	15.39	15.48	15.58	15.61	15.67

(a) Find the upper and lower confidence bounds for the average weight of the cereal boxes. Use $\alpha = 0.05$. Is the value 15.00 contained in the interval?

(b) Based on your answer to part (a), are you getting your money's worth of cereal?

7.37 A photographer who transfers old pictures to video and sets them to music uses blank VHS tapes for video recording. He is concerned about the actual amount of time that the tapes are able to record. The tapes are rated at 120 minutes, but he knows from experience there is variation in the actual times that the tapes can be used. He collects data on 25 randomly selected tapes and finds these recording times (in minutes):

116	118	119	119	120
117	118	119	119	121
117	118	119	120	121
117	119	119	120	121
117	119	119	120	121

(a) Find the average and the standard deviation of these data.

(b) Find a 95% confidence interval for the average recording time, assuming that the recording times are normally distributed.

(c) What do you conclude about the claim that the tapes have an average of 120 minutes of usable recording time? Use the confidence interval to support your conclusion.

7.38 A cross-country runner is evaluating his times against a team average of 18.5 minutes. The standard deviation of his times is 0.25 minute. This season the runner's average is 17.0 minutes for ten races.

(a) Assuming that race times are normally distributed, calculate the Z-score for the runner's average time.

(b) Using the Z-score you found in part (a), decide whether this runner is unusually fast compared with the team.

7.39 One of the concerns you have in choosing an insurance company is the amount of time it takes before you speak to a real person, or the response time. After all, you don't call your insurance agent just to chat; you usually have a problem that requires immediate attention. You are considering a company that claims its average response time is 30 seconds. A sample of 40 response times (in minutes) are recorded. The results of generating descriptive statistics in Minitab are shown here:

Descriptive Statistics: Time

Variable	N	Mean	Median	TrMean	StDev	SE Mean
Time	40	39.425	38.000	39.417	5.773	0.913

Variable	Minimum	Maximum	Q1	Q3
Time	29.000	51.000	36.000	44.000

(a) What are the average and standard deviation of the response times?

(b) Calculate the standard error of $\overline{X}$ using the standard deviation and n.

(c) Do you see the number calculated in part (b) anywhere in the output?

(d) Calculate the Z-score for the average response time.

(e) Based on the Z-score, do you think this company's claim is correct?

7.40 A company concerned about the health of its employees has offered them free membership in a local health club. One year after this benefit was adopted, 50 employees were surveyed to determine the average number of hours per week they exercise. The 50 employees surveyed exercised an average of 3.5 hours a week with a standard deviation of 0.5 hour. The national average exercise is 3 hours a week.

(a) Find the Z-score for the average of 3.5 hours.

(b) Display the sampling distribution of $\overline{X}$.

(c) Based on the Z-score, do you think the employees of this company exercise more than the national average?

7.41 The Bureau of Labor Statistics publishes a great deal of monthly data. The following data on the average price of bread were extracted from the World Wide Web:

Month/Year	Price	Month/Year	Price
Jan-98	$0.855	Jan-00	$0.907
Feb-98	0.860	Feb-00	0.924
Mar-98	0.853	Mar-00	0.924
Apr-98	0.863	Apr-00	0.927
May-98	0.866	May-00	0.915
Jun-98	0.859	Jun-00	0.915
Jul-98	0.867	Jul-00	0.935
Aug-98	0.869	Aug-00	0.923
Sep-98	0.860	Sep-00	0.918
Oct-98	0.849	Oct-00	0.934
Nov-98	0.855	Nov-00	0.953
Dec-98	0.866	Dec-00	0.987
Jan-99	0.872	Jan-01	0.982
Feb-99	0.880	Feb-01	0.994
Mar-99	0.883	Mar-01	1.020
Apr-99	0.897	Apr-01	1.008
May-99	0.886	May-01	0.995
Jun-99	0.885	Jun-01	0.989
Jul-99	0.893	Jul-01	0.987
Aug-99	0.884	Aug-01	0.991
Sep-99	0.878	Sep-01	0.996
Oct-99	0.889	Oct-01	1.010
Nov-99	0.899	Nov-01	1.012
Dec-99	0.899	Dec-01	1.012

Treating these three years of data as a sample, find a 90% confidence interval for the average price of bread. Assume that the standard deviation is $0.02 per pound.

7.42 You have heard the statistic that one in two marriages ends in divorce. You decide to check out your own family. Of the 20 marriages among your family members (cousins and siblings), only three have ended in divorce.

(a) Using your family as a sample, find a 90% confidence interval for the proportion of divorces in the population.

(b) Explain why your family might not be an unbiased sample. What might be a better way to check this statistic?

7.43 A nationwide survey of practicing physicians will be taken to estimate μ, the true mean number of prescriptions written per day. The desired margin of sampling error is 0.75. A pilot study revealed that a reasonable planning value for the population standard deviation is 5.

(a) If the desired level of confidence is 99%, how many physicians should be contacted in the survey to estimate μ?

(b) If the desired confidence level is lowered to 95%, what should the sample size be?

(c) Discuss the results of parts (a) and (b).

7.44 You wish to evaluate hotels in a particular chain to determine the proportion of rooms that are not ready when customers check in to the hotel.

(a) How many rooms must be in the sample for the hotel to be 95% confident that the margin of error is at most 1%?

(b) How many rooms must be in the sample for the hotel to be 95% confident that the margin of error is at most 3%?

Thinking About It!

7.45 Suppose you are studying the lengths of stays at a hospital and the population has a mean of 8 days with a standard deviation of 2.5 days. The lengths of hospital stays have a normal distribution.

(a) Display the population distribution of hospital stays.

(b) If you observed the hospital stays of 20 patients, find the standard error of $\overline{X}$. Display the sampling distribution of $\overline{X}$ on the same graph as part (a) using a different color pen or pencil.

(c) If you observed the hospital stays of 30 patients, find the standard error of $\overline{X}$. Display the sampling distribution of $\overline{X}$ on the same graph as (a) and (b) using a different color.

(d) If you observed the hospital stays of 80 patients, find the standard error of $\overline{X}$. Display the sampling distribution of $\overline{X}$ on the same graph as (a) and (b) using a different color.

(e) What happens to the sampling distribution of $\overline{X}$ as the sample size increases? What does this tell you about the accuracy of $\overline{X}$ as an estimate for μ?

7.46 Suppose you are studying the salaries of three different populations that all have the same average salary: $\mu = \$25,000$. The three populations have different standard deviations: criminal justice major, $\sigma = 1000$; social work major, $\sigma = 2000$; and nursing major, $\sigma = 3000$. You select a sample of 30 from each population.

(a) Find the standard error of $\overline{X}$ for each of the three groups.

(b) What happens to the standard error as the population variability increases?

(c) What does this tell you about the accuracy of $\overline{X}$ as an estimate for μ?

7.47 The resident populations of six states are listed for the year 2000. You think you can use this information to estimate the average state population. You are trying to decide where to relocate on your next move and you wish to avoid crowded locations.

State	Resident population (April 1, 2000)
Connecticut	3,405,565
Massachusetts	6,349,097
New York	18,976,457
Rhode Island	1,048,319
New Hampshire	1,235,786
Vermont	608,827

(a) What do you notice about your sample of states?

(b) How does your answer to part (a) affect the usefulness of the data?

(c) Compute a 90% confidence interval for the mean population. What assumption do you need to make to do this?

(d) Why is the average state size not a particularly useful statistic? What might be a better measure to help you identify "crowded locations"?

7.48 Are women well represented in politics in Europe today? The data shown here give the percentages of women in government as of 2000 for some of the countries that belonged to the Council of Europe.

Country	Number of Members	Number of Women	Women %
Sweden	20	10	50.0
Denmark	20	9	45.0
Finland	18	7	38.9
United Kingdom	85	30	35.3
Germany	40	14	35.0
Austria	16	5	31.3
Netherlands	29	9	31.0
France	33	10	30.3
Luxembourg	14	4	28.6
Belgium	18	4	22.2
Ireland	32	6	18.8
Spain	17	3	17.6
Italy	78	11	14.1
Greece	83	10	12.0
Portugal	60	7	11.7

(a) Leaving Sweden out of the data set, construct a 95% confidence interval for the average percentage of seats held by women in international government.

(b) Does Sweden appear to be atypical? Explain your answer using the confidence interval calculated in part (a).

7.49 A CBS News poll taken on May 13, 2001, found that 47% of those surveyed did not think capital punishment is a deterrent to murder. The margin of error was 5%. Your friend, who is in favor of capital punishment, argues that more than half of the population agrees with him. What is wrong with his argument?

Requires Exercise 7.24 **7.50** The state of Virginia had a death rate of 804.9 in 1999. Using the confidence interval you found in Exercise 7.24, what can you conclude about Virginia's death rate in relation to the national average?

Doing It!

Datafile: *WINDSPEEDS.XXX* **7.51** The data in this problem are the hourly wind speeds at 60 feet on the island of Oahu in Hawaii for almost one year. Here is a brief description of the variable names presented on each of the four sheets:

Popdata—population data for all hours (there are some missing values)

Sample30—thirty samples of $n = 30$ randomly selected weeks from the population

Sample40—thirty samples of $n = 40$ randomly selected weeks from the population

Sample50—thirty samples of $n = 50$ randomly selected weeks from the population

(a) Using the 30 samples of size $n = 30$ (sheet *Sample30*), calculate a 95% confidence interval for each of the 30 samples.

(b) Count how many of the 30 confidence intervals calculated in part (a) actually include the true population mean, μ. Is your count consistent with what you expected? Why or why not?

(c) Calculate 90%, 95%, and 99% confidence intervals for the first sample on the sheet *Sample40*. How does the width of the confidence interval vary with the level of confidence?

(d) Calculate a 95% confidence interval for each of the following: (i) the first sample on the sheet *Sample30*, (ii) the first sample on the sheet *Sample40* (note: you did this one in part (c) above), and (iii) the first sample on the sheet *Sample50*. How does the width of the confidence interval vary with the sample size?

(e) Write a memo summarizing your findings.

7.52 The real estate agency you have been working with has collected data on 150 *Datafile:* houses sold. They have hired you to help them analyze these data. The first lines of the data *HOUSES.XXX* follow:

List Price	Sale Price	DOM	# Rooms	# Bed- rooms	Full Baths	Half Baths	Master Bath	Lot Size (sq ft)	Style	Color	Vinyl Siding	Above Ground Pool	In Ground Pool	Heating	Closing Date	Area
89900	75000	76	5	3	1	0	N	21932	Ranch	Beige	Y	N	N	Gas	1/6/01	East Forest Park
96500	90500	48	7	4	1	1	N	4792	Cape	White	N	N	N	Gas	1/22/01	East Forest Park
70000	68000	42	5	2	1	1	N	6700	Cape	Almond	Y	N	N	Gas	1/26/01	East Forest Park
79900	73500	9	5	3	1	0	N	8626	Ranch	Yellow	N	N	N	Gas	1/26/01	East Forest Park
94900	92400	27	6	2	1	1	N	5184	Colonial	White	N	N	N	Gas	1/29/01	East Forest Park
84900	84000	140	6	4	1	0	N	7045	Cape	Brown	N	N	N	Gas	1/30/01	East Forest Park
99000	99000	10	6	3	2	0	N	5000	Cape	Blue	Y	N	N	Oil	1/31/01	East Forest Park
124900	123000	75	8	4	1	1	N	5700	Tudor	Brick	N	N	N	Gas	2/6/01	East Forest Park

Here is an explanation of the variables:

List Price—the price the house was originally listed at when it came on the market

Sale Price—the actual selling price of the house

DOM—the number of days the house was on the market before it sold

 # Rooms—the number of rooms in the house

Full Baths—the number of full bathrooms (including a shower)

Half Baths—the number of half bathrooms (no shower) in the house

Master Bath—a yes/no variable to indicate whether or not there is a master bathroom in the house

Lot Size—the size of the lot in square feet

Style—the architectural style of the house

Color—the color of the exterior of the house

Vinyl Siding—whether or not the house has vinyl siding

Above Ground Pool—whether or not the house has an above-ground pool

In Ground Pool—whether or not the house has an in-ground pool

Heating—the type of heating in the house

Closing Date—the date the house changed ownership

Area—the section of the city where the house is located

In the chapter, Figure 7.3 is the sampling distribution for the average sale price when each sample has $n = 5$ observations. If you combine the data from two samples and assume that a sample of size $n = 10$ was taken, you can examine the effects of a change in the sample size on the sampling distribution. Similarly, if you combine the data from three samples, you can see the impact if the sample size had been $n = 15$. Finally if you combine the data from four samples, you can see the impact if the sample size had been $n = 20$.

(a) Do these combinations and display the histograms of the average sale price. You will have three histograms in addition to Figure 7.3.

(b) For each set of average prices, find the mean and standard deviation and fill out the first three columns of the table:

Sample size	Average of the $\overline{X}$ values	Standard error of the $\overline{X}$ values	Theoretical standard deviation of the $\overline{X}$ values
5			
10			
15			
20			

(c) What do you notice about the values for the average of the $\overline{X}$'s as n increases?

(d) What do you notice about the values for the standard error of the $\overline{X}$'s as n increases?

(e) Complete the last column in the table. How do the standard errors that you found compare with what the Central Limit Theorem says they should be?

(f) Repeat this analysis for the number of days on the market.

(g) Now consider the 150 houses that you have to be a sample of houses sold. Construct 90%, 95%, and 99% confidence intervals for the average sale price.

(h) Select one of the qualitative variables in the data set and separate the sample into two groups based on this qualitative variable (i.e., Capes vs. all others). Construct a 95% confidence interval for the average sale price in each group. Do they overlap?

(i) Construct 90%, 95%, and 99% confidence intervals for the numbers of days on the market using the full data set as your sample.

(j) Using the same qualitative variable that you selected in part (h), construct a 95% confidence interval for the average number of days on the market for each group. Do they overlap?

(k) If you are going to sell your home in this area what do you recommend as the selling price and how many days should you expect your home to be on the market? Summarize your findings in a report to the real estate agency.

7.53 Go to this website:

http://davidmlane.com/hyperstat/Instructional_Demos.html

Choose **Confidence Intervals** and then **Confidence Interval by Rice Virtual Lab**.

You will see a **Begin** button. When you click on the **Begin** button, you will be able to choose a sample size of 10, 15, or 20. Once you do, the applet will generate 100 samples and the corresponding confidence intervals for the mean. The population mean is 50 and the population standard deviation is 10.

First set $n = 10$ and record what percentage of the 95% confidence intervals actually contain the true population mean of 50. Record what percentage of the 99% confidence intervals actually contain the value of 50.

What percentage of the 95% confidence intervals would you expect to contain 50?

What percentage of the 99% confidence intervals would you expect to contain 50?

Did what you expect match what happened? Why or why not?

Which is wider, the 95% or 99% confidence interval? Why? Will this always be the case?

Now set $n = 15$ and repeat the experiment.

Now set $n = 20$ and repeat the experiment.

How does sample size affect the number of intervals that contain 50? Explain.

How does sample size affect the width of the intervals?

If you already knew that the population mean was 50, what value would there be in computing a confidence interval?

There is an important determinant of the width of confidence intervals that you cannot modify in this simulation. What is it?

8

Hypothesis Testing: An Introduction

WILL YOUR PLANE DEPART AND LAND ON TIME?

The Travel Industry Association of America (TIA) is forecasting that the U.S. domestic travel volume will total more than 1.0 billion person-trips for the full year 2001, up 1% from the year-end projection in 2000 of 997.6 million person-trips. A person-trip is one person traveling on a trip, 50 miles or more one way, away from home. The growth in domestic travel volume is expected to be greater in 2002 and 2003. Americans are expected to take 1.03 billion person-trips in 2002 (up 2.1%) and 1.05 billion person-trips in 2003 (up 2.2%) (source: *http://www.tia.org/Press/pressrec.asp?Item=128*).

Based on this forecast, it is likely that you will be traveling by airplane for business or pleasure sometime in the next year. You know that problems, often nightmarish, arise when your plane departs or lands late. You might miss a connecting flight, you might miss an important client or meeting, or you might just keep a special friend waiting longer than you would like.

Before deciding on an airline to use for your business travel, you decide to have a look at airline track records. Fortunately for you, the airline industry keeps a great deal of data and much of them are public. You are considering making American Airlines your primary carrier, and since you live in New York City, you land at LaGuardia Airport. You are able to download all of the American Airlines flights into LaGuardia for the month of December 2000. A portion of the data set is shown here:

Carrier code	Date	Flight no.	Tail no.	Origin airport	Scheduled arrival time	Actual arrival time	Arrival delay (minutes)
AA	12/01/2000	158	N628AA	DFW	22:03	22:03	0
AA	12/01/2000	338	N3BGAA	ORD	0:54	0:56	2
AA	12/01/2000	340	N3BFAA	ORD	23:31	23:45	14
AA	12/01/2000	342	N3BEAA	ORD	22:54	23:04	10
AA	12/01/2000	362	N3BUAA	ORD	18:06	19:24	78
AA	12/01/2000	370	N502AA	ORD	16:00	17:07	67
AA	12/01/2000	378	N453AA	ORD	13:55	14:31	36
AA	12/01/2000	386	N425AA	ORD	12:01	12:18	17

8.1 Chapter Objectives

In Chapter 7 you learned that the sample mean is a good point estimate of the population mean. Suppose we use the sample of flights and the sample average delay in minutes as a way to evaluate the average delay of *all* American Airlines flights. You would like to use an airline that on average is no more than 30 minutes late. So you figure you can just find the sample mean for the sample of flights and, if the delay is more than 30 minutes, you should not use that airline. Simple, right? Well, not quite.

After a bit more thought you remember that although the sample mean, $\overline{X}$, is a single number for any particular sample, if you pick a different sample you will probably get a different sample mean. In fact, you could get many different possible values for the sample mean, and virtually none of them would actually equal the true population mean, μ.

Remember! A point estimate is a single number, calculated from sample data, that is used to estimate a population parameter.

EXAMPLE 8.1

AIRLINE ARRIVALS

Understanding Sampling Error

You decide to calculate the sample mean, $\overline{X}$, for the number of minutes that American Airlines flights are late. You use the first 200 flights in December as your sample and find that the sample mean delay into LaGuardia Airport is 40 minutes. Clearly the sample mean is higher than the acceptable mean of 30 minutes.

But the sample average will be different each time you take a sample and will almost never be equal to 30 minutes even if the true average is 30 minutes! Remember that you are examining only a sample of the population of flights and no two flights have exactly the same delay. The variability in the flight delay times combined with the fact that you have data on only a piece of the population leads to variability in the sample averages, the $\overline{X}$ values. Recall that the variability of $\overline{X}$ is called the *standard error* and it is based on the amount of variability in the population, σ, and the sample size, n. Based on many years of data, the population standard deviation is 50 minutes.

Understand the problem.

This example shows that you should not simply compare the $\overline{X}$ value you get from the sample with the desired target mean because even if the airlines have an average delay of 30 minutes, the $\overline{X}$ that you find will almost never equal 30. So, a simple point estimate does not do the job.

This chapter introduces the major concepts and philosophy of a technique called **hypothesis testing,** which does do the job. When we talked about the need for inferential statistics, we said that most of the techniques of inferential statistics can be classified as either estimation tools or hypothesis-testing tools. This chapter lays the groundwork for hypothesis testing. The remaining chapters of this text rely heavily on the ideas developed in this chapter.

In this chapter we look at the following topics:

- Overview and steps of hypothesis tests
- Two-tail tests of the mean: large sample
- One-tail tests of the mean: large sample
- Type I and Type II errors

8.2   What Is a Hypothesis Test?

The word **hypothesis** has the same meaning in statistics as it does in everyday use. What does this word mean to you? Here are some possibilities:

- An idea
- An assumption
- A guess
- A theory

Often the hypothesis is about the value of a parameter, such as the population mean.

> In statistics, a **hypothesis** is an idea, an assumption, or a theory about the characteristics of one or more variables in one or more populations.

We actually work with many hypotheses every day without even realizing it. For example, we might think that regular exercise improves our ability to study. The hypothesis in this case is that there is a relationship between exercising and effective studying. To decide whether the hypothesis is correct, we might use our recollection of past experiences.

The sample data are the evidence, and on the basis of these data you must make a decision.

Once a hypothesis is formed, we must *test* it: We must decide whether to believe the hypothesis. The only information we have to help us decide is contained in the sample. Thus, to do a **hypothesis test,** we use the information in the sample data to decide whether or not to believe the hypothesis. Basically, we are trying to decide whether the sample is consistent with the hypothesis (in which case we believe the hypothesis) or the sample is inconsistent with the hypothesis (in which case we choose not to believe it or to reject it).

> A **hypothesis test** is a statistical procedure that involves formulating a hypothesis and using sample data to decide on the validity of the hypothesis.

The framework for all hypothesis tests is the same.

We can test many different types of hypotheses and these are the subject of the next section. The purpose of this chapter is to explain the framework of the hypothesis-testing procedure. The details of the test depend on the particular hypothesis that we are testing, but the purpose and general approach are the same for all tests. In fact, virtually all of the remaining chapters in this book are devoted to spelling out the details for the various tests we use for analyzing sample data and making informed decisions. It is important that you not view these as separate and isolated chapters. They add flesh to the skeleton we will develop in this chapter. Think of each chapter as a variation on the same theme.

8.3 Designing Hypotheses to Be Tested—An Overview

We can start by thinking about what types of variables we might examine. In Chapter 1 you learned that variables can be classified in two major types: *qualitative* and *quantitative*. We made the distinction because different techniques are used for different types of data. In Chapters 2–4 and 7 we explained that different *descriptive* tools and point estimators are used, depending on what type of data we wish to evaluate. Now you will see that there are different *inferential* tools for different types of data.

Different types of data require different analysis tools.

8.3.1 HYPOTHESES ABOUT QUANTITATIVE VARIABLES

Sample data that are inherently numerical are called *quantitative* data. Recall from Chapter 6 that if we are analyzing a quantitative variable, then we can describe it by its distribution. The distribution in turn can be described by parameters. Thus, if we are constructing a theory or hypothesis about a quantitative variable, it might be a statement about one of these characteristics:

- The shape of the distribution of the variable in one population
- The mean value, μ, of the variable in one population
- How the mean value of the variable in one population, μ_1, compares with the mean value of the variable in a second population, μ_2
- The equality of the mean values of the variable in more than two populations
- The amount of variability, σ^2, of the variable in one population
- How the amount of variability of the variable in one population, σ_1^2, compares with the amount of variability of the variable in a second population, σ_2^2

Hypotheses about quantitative variables

All but the first item are generally referred to as tests of means and tests of variances. The first item requires a "goodness of fit" test, which is introduced in Chapter 14.

EXAMPLE 8.2

AIRLINE ARRIVALS

Looking at Possible Hypotheses

For the airline problem that you are investigating, delay time is a quantitative variable. We present an example of each of the hypotheses just discussed for this variable:

Understand the problem.

Type of hypothesis	Specific hypothesis
The shape of the distribution of the variable in one population	The variable *Arrival Delay* has a normal distribution.
The mean value, μ, of the variable in one population	The population mean *Arrival Delay* is 30 minutes.
How the mean value of the variable in one population compares with the mean value of the variable in a second population	The mean *Arrival Delay* of morning flights is less than the mean *Arrival Delay* of afternoon flights.

The equality of the mean values of the variable in more than two populations	The mean *Arrival Delays* of flights on Monday, Tuesday, and Wednesday are equal.
The amount of variability, σ^2, of the variable in one population	The variability of *Arrival Delay* is less than 50 minutes[2]. (Recall that the variance is the standard deviation squared.)
How the amount of variability of the variable in one population compares with the amount of variability of the variable in a second population	The *Arrival Delays* on Friday have more variability than the delays of flights on Tuesday

Remember! Qualitative data describe a particular characteristic of a sample item.

Notice that we used a qualitative variable, in this case the time of day or the day of the week, to divide the data into two or more populations. This is often the major use of a qualitative variable, particularly nominal data. Recall that nominal data are one type of qualitative data. Nominal data are created by assigning numbers to different categories when the numbers have no real meaning.

TRY IT NOW!

Smog Days
Looking at Possible Hypotheses

The Environmental Protection Agency (EPA) is studying the number of smog days per year in different states. Develop a specific hypothesis for each of the hypotheses we have presented. We've listed one example to get you started.

Type of hypothesis	Specific hypothesis
The shape of the distribution of the variable in one population	The numbers of smog days per year in California are normally distributed.
The mean value, μ, of the variable in one population	
How the mean value of the variable in one population compares with the mean value of the variable in a second population	
The equality of the mean values of the variable in more than two populations	
The amount of variability, σ^2, of the variable in one population	
How the amount of variability of the variable in one population compares with the amount of variability of the variable in a second population	

ANS. (MAY VARY) THE AVERAGE NUMBER OF SMOG DAYS NATIONWIDE IS 15. THE AVERAGE NUMBER OF SMOG DAYS IN TEXAS IS LESS THAN IN CALIFORNIA. THE AVERAGE NUMBERS OF SMOG DAYS IN ALL NEW ENGLAND STATES ARE THE SAME. THE VARIABILITY OF THE NUMBER OF SMOG DAYS IN NEW YORK IS 2 DAYS[2]. THE VARIABILITY OF THE NUMBER OF SMOG DAYS IN NEW MEXICO IS GREATER THAN IN OREGON.

8.3.2 HYPOTHESES ABOUT NOMINAL VARIABLES

In addition to using nominal variables as a way to divide quantitative data into groups, we are often interested in what percentage or proportion of a population has a particular characteristic. For example, we might be interested in what proportion of Firestone tires have caused accidents due to defects. In this case, the nominal variable is the quality status of the product, nondefective or defective, and we are interested in the percentage or proportion of the population that has the quality status "defective." If we are analyzing *nominal* data, then the hypothesis might be a statement about one of these:

- The proportion, p, of population members that have a certain characteristic (one of the categories of the nominal variable)
- How the proportion that have a certain characteristic in one population, p_1, compares with the corresponding proportion in a second population, p_2

EXAMPLE 8.3

SURFING THE NET

Looking at Possible Hypotheses

A recent study reported that many parents are concerned about the websites their teens are visiting. Here are examples of the hypotheses we presented for nominal variables:

Type of hypothesis	Specific hypothesis
The value of the proportion, p, of the population members that have a certain characteristic	At least 60% of the parents reported they had checked to see which websites their teens had visited.
How the proportion that have a certain characteristic in one population compares with the corresponding proportion in a second population	A greater proportion of parents reported checking on younger teens (under 16) than on older teens.

Understand the problem.

In this example, we are interested in the proportion of the population of parents that monitor the websites used by their teenage children. The nominal variable is whether a parent monitors net surfing. We used a second variable (age) to divide the parents into two populations and then constructed a hypothesis about the proportion of parents who monitor the websites of younger teens compared with the proportion who monitor older teens.

 **TRY IT NOW!**

The Sports Complex
Looking at Possible Hypotheses

A particular university is considering building a new sports complex and wants to know whether it will be widely used by students. Develop a specific hypothesis for each type of hypothesis we have discussed.

Type of hypothesis	Specific hypothesis
The value of the proportion, p, of the population members that have a certain characteristic	
How the proportion that have a certain characteristic in one population compares with the corresponding proportion in a second population	

8.3.3 SUMMARY OF KINDS OF HYPOTHESES

Table 8.1 summarizes the kinds of hypotheses you are likely to use to analyze sample data and make informed decisions.

TABLE 8.1

SUMMARY OF HYPOTHESES TO BE TESTED

Type of data	One population (Chapters 8, 9, and 14)	Two populations (a qualitative variable may be used to divide the data into populations) (Chapter 10)	More than two populations (Chapters 13 and 14)
Quantitative data	The shape of the distribution. The mean value, μ, of the variable	Compare the mean value of the variable in one population, μ_1, with the mean value of the variable in a second population, μ_2	The equality of the mean values of the variable
	The amount of variability, σ^2, of the variable	Compare the amount of variability of the variable in one population, σ_1^2, with the amount of variability of the variable in a second population, σ_2^2	
Nominal data	The value of the proportion, p, of the population members that have a certain characteristic (one of the categories of the nominal variable)	Compare the proportion that have a certain characteristic in one population, p_1, with the corresponding proportion in a second population, p_2	Test for independence

8.4 The Steps in a Hypothesis Test

To carry out any hypothesis test, we set it up according to some general guidelines. We use this five-step procedure to do any hypothesis test:

Steps for any hypothesis test

Step 1: Set up the ***null and alternative hypotheses.***

Step 2: Define the test procedure. This includes selecting the right ***test statistic*** and the value of ***α***, the significance level.

Step 3: Collect the data and calculate the ***test statistic*** and the ***p* value.**

Step 4: Make a statistical decision. Decide whether to reject the ***null hypothesis.***

Step 5: ***Interpret*** the statistical decision in terms of the stated problem.

Let's see how this procedure fits into our problem-solving model. Remember the problem-solving steps we identified in Chapter 1: (1) Understand the problem. (2) Collect and analyze the data. (3) Draw conclusions and make recommendations. We have been identifying which of the step(s) is involved for every example in the book. Most of the examples illustrate one step in this procedure. The five-step hypothesis-testing procedure encompasses all of the basic problem-solving steps. Step 1 is "understand the problem"; Steps 2 and 3 require us to "collect and analyze the data"; Steps 4 and 5 involve "drawing conclusions and making recommendations."

The staircase below shows the relationship between the five-step procedure used to do any hypothesis test and the three problem-solving steps we have been using. We will use the staircase as a visual cue as we look at the details of each of these steps.

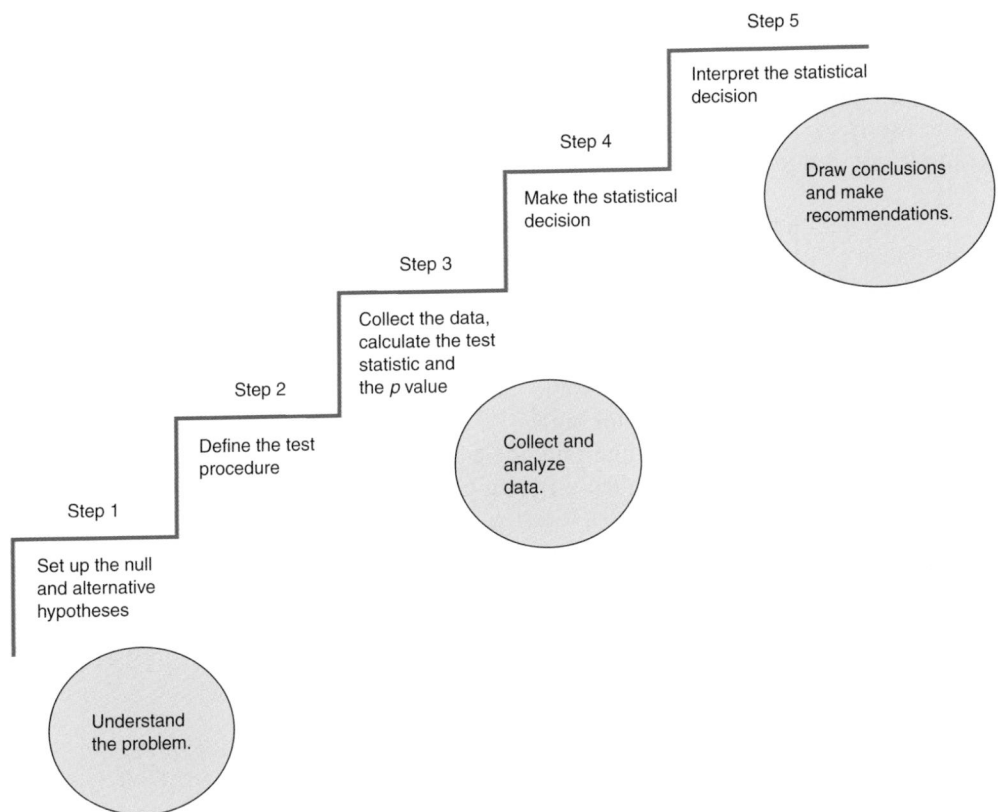

As you can see, hypothesis testing has a language of its own. The good news is that all hypothesis tests can be done using this five-step procedure. The details change from one test to another, but the overall approach is always the same. Each of the next subsections defines the terms printed in bold and explains a step of this procedure.

8.4.1 STEP 1: SET UP THE NULL AND ALTERNATIVE HYPOTHESES

The first step is to take your idea or hypothesis and construct two opposing views. One of these is called the **null hypothesis** and the other is the **alternative hypothesis.**

> The ***null hypothesis*** is a statement about a parameter of the population(s). It is labeled H_0. The ***alternative hypothesis*** is a statement about a parameter of the population(s) that is opposite to the null hypothesis. It is labeled H_A.

Understand the problem.

EXAMPLE 8.4

AIRLINE ARRIVALS

Step 1: Set Up the Null and Alternative Hypotheses—Two-Tail Test

You may choose to set up the airline delay problem this way:

Null hypothesis: The true mean *Arrival Delay* is equal to 30 minutes.
Alternative hypothesis: The true mean *Arrival Delay* is not equal to 30 minutes.

Remember! You do not know μ! You are checking to see if μ is 30 by using the sample mean as your evidence.

Clearly, if you believe the null hypothesis, then you cannot believe the alternative hypothesis. You must choose between them on the basis of the evidence in the sample. Using the notation of hypothesis testing, you can rewrite the hypotheses as

$$H_0: \quad \mu = 30 \text{ minutes}$$
$$H_A: \quad \mu \neq 30 \text{ minutes}$$

Sometimes the alternative hypothesis is labeled H_1. It doesn't matter whether you use H_1 or H_A as long as you are consistent. In this book we use H_A. In the example we rewrote the hypotheses using the standard symbol for the population mean, μ.

H_0 and H_A must be mutually exclusive and exhaustive.

We said that the null and alternative hypotheses are opposing views. What does this mean? Notice that no value of μ is part of both the null and the alternative hypotheses. The null and alternative hypotheses cannot overlap; that is, the two hypotheses are **mutually exclusive.** In addition, when considered together, the null and alternative hypotheses must cover all the possibilities; this means they are **exhaustive.** In the airline example, all the possible values of μ are included in either the null or the alternative hypothesis but not both.

Example 8.4 shows the null and alternative hypotheses for a **two-tail test.** It is difficult to explain why it is called a two-tail test until we define the rejection region. However, we can say that a two-tail test is used to test the hypothesis that a population parameter is actually equal to a particular value. Thus, the = sign is in the null hypothesis and the ≠ is in the alternative hypothesis.

The next example shows the null and alternative hypotheses for a **one-tail test.** Section 8.8 gives you some tips on how to decide whether you should use a two-tail test or a one-tail test.

Understand the problem.

EXAMPLE 8.5

PURCHASING A KAYAK

Step 1: Set Up the Null and Alternative Hypotheses—One-Tail Test

You have decided to purchase a kayak to use on the river near your home. In reading about kayaks you learn that the rear cockpit height should be at most 10 inches. This is the height measured from the bottom of the boat to the top rim of the kayak where the rower sits. If the cockpit is too high, then the seat will not fit and other accessories that go into the boat cannot be put in correctly—thus, the person cannot use the boat properly. The same problem results if the cockpit is too low, and if it is low enough, some people cannot fit into the boat! You wish to set up a hypothesis test to decide whether a certain manufacturer is producing kayaks that have an average height of at most 10 inches:

$$H_0: \quad \mu \geq 10 \text{ inches}$$
$$H_A: \quad \mu < 10 \text{ inches}$$

A one-tail test uses an inequality sign, which includes the equal portion ($\leq$ or $\geq$) in the null hypothesis and a strict inequality sign ($<$ or $>$) in the alternative hypothesis. Notice that the $=$ sign is still in the null hypothesis. This will always be the case. Again, more details are given in Section 8.8 to help you decide which way to set up the null and alternative hypotheses.

Now let's look at a situation in which the data to be analyzed are nominal data.

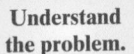
Understand
the problem.

EXAMPLE 8.6

U.S. UNEMPLOYMENT

Step 1: Set Up the Null and Alternative Hypotheses—Test of Proportion

The U.S. government wishes to know what proportion of the workforce is unemployed. Last month the estimated proportion was 6%, and the government wants to set up a hypothesis test to decide whether the proportion for this month has changed from 6%. Here are the two opposing views:

$$H_0: \quad p = 0.06$$
$$H_A: \quad p \neq 0.06$$

Now it is time for you to try your hand at setting up the null and alternative hypotheses.

✔ TRY IT NOW!

Gas Prices
Setting Up Null and Alternative Hypotheses

Gas prices soared during the summer of 2001. You wish to test whether the average price per gallon nationwide has changed from $1.50. What null and alternative hypotheses will you use?

Discovery Exercise 8.1

FORMULATING HYPOTHESES

Consider the population of all M&M packages like the one you have in your hand.

1. Identify as many different variables as you can. Be sure you have some quantitative and some nominal variables. Record the values of these variables for your package. (*Hint*: You should carefully examine the package before you rip it open.)

2. Select one of the quantitative variables and set up null and alternative hypotheses for a parameter of this variable.

3. Select one of the nominal variables and set up null and alternative hypotheses for a parameter of this variable.

4. As a class, agree on several quantitative and nominal variables that are important. Record all the data on each of these variables.

5. Use the tools of descriptive statistics to display these data.

6. Enjoy the M&M's!

ANS. $H_0: \mu = \$1.50; H_A: \mu \neq \$1.50$

8.4.2 STEP 2: DEFINE THE TEST PROCEDURE

The second step of any hypothesis test is to **define the test procedure,** which includes selecting the appropriate test statistic and the value of α, the significance level of the test.

Much of the rest of this textbook is devoted to various types of hypothesis tests. So, we must decide on the right test for the particular problem or situation. Table 8.1 is a starting point for this task. We must decide whether the data are quantitative or nominal and whether we are looking at one, two, or more than two populations. For the moment, we consider situations that involve hypothesis tests for the mean, μ, of a quantitative variable for one population. After we get through the basics of hypothesis testing, we will return to this issue and do a better job of determining which test to use.

To perform the test, we need to choose between the null and the alternative hypotheses. More specifically we must decide to reject or not to reject the null hypothesis. We always phrase the decision in terms of the null hypothesis. If we choose not to reject the null hypothesis, this means that the sample data are consistent with the null hypothesis and we have no reason to reject it. If we choose to reject the null hypothesis, this means that the sample data are sufficiently inconsistent with the null hypothesis and we have reason to reject it.

Remember! In general, a statistic is a number calculated from sample data.

So it seems that the information in the sample is the evidence that we use to decide between H_0 and H_A. However, the sample consists of n observations. Somehow we must find a single number that captures the information in the sample. This number is called the **test statistic.** The formula for calculating the test statistic depends on the particular test we choose, but its function is always the same.

> A **test statistic** is a number that captures the information in the sample data. It is used to decide between the null and alternative hypotheses.

How will we know if we should reject the null hypothesis? This is precisely the question that we must answer next. There are two approaches to answering this question, but both involve specifying the **significance level of the test,** called **alpha (α).** For now, think of α as the greatest probability (or chance) of rejecting a true null hypothesis that you are willing to tolerate. In other words, α is the greatest chance of this type of error that you are willing to live with. This error is called a **Type I error** and it will be further discussed in Section 8.7.

> The **significance level, α,** is the maximum probability tolerated for rejecting a true null hypothesis.

One method for deciding if we should reject the null hypothesis is based on the concept of a **p value** and a comparison of the p value with the significance level, α. Remember that α is a probability.

> The **p value** is the actual probability of rejecting a true null hypothesis based on the evidence in the test statistic.

If the p value is "very small," then we know that the actual probability of rejecting a true null hypothesis is less than the maximum we are willing to tolerate and so we reject the null hypothesis, H_0. "Very small" is thus defined to be less than α, the maximum probability imposed by the significance level.

The second and more traditional approach to deciding if we should reject the null hypothesis is called the **critical value** or **rejection region** method. It may be a bit easier to grasp this approach as you learn the material the first time, but eventually you will want to understand both this approach and the p value approach. Prior to the prevalence of statistical software, the rejection region approach was used almost exclusively as p values are sometimes difficult to calculate. However, today, most statistical software packages report p values for most tests. Because there are some tests where the p value is not reported by the software, it is important for you to learn both approaches.

> The *rejection region* is the range of values of the test statistic that will lead us to reject the null hypothesis. It is defined by the *critical value(s).* The area of the rejection region is α, the significance level.

Remember that you learned in Chapter 6 how to find the Z values that cut off specified areas in the tails.

Notice that α is needed for this approach as well. Remember from Chapter 6 that areas correspond to probabilities. Alpha is the area of the rejection region. If the test statistic falls in the rejection region, then we will reject the null hypothesis.

Once the significance level is specified, you will reach the same conclusion with regard to rejecting the null hypothesis regardless of which approach you use. That is why it is important to specify the significance level at this step of the procedure. Otherwise, we run the risk of waiting to see the results before deciding on the maximum chance of error we can tolerate. This is not ethical. It is like waiting to see whether we get an 85 on an exam and then declaring that we were hoping to get an 85!

EXAMPLE 8.7

AIRLINE ARRIVALS

Understand the problem.

Step 2. Define the Test Procedure

In evaluating American Airlines flights into LaGuardia, you are concerned about the average arrival delay. The null and alternative hypotheses were set up in Example 8.4. The appropriate test is a one-population test of the mean. The test statistic is Z (you would not know this yet). You decide that you can tolerate a certain kind of error 5% of the time, so $\alpha = 0.05$.

8.4.3 STEP 3: COLLECT THE DATA, CALCULATE THE TEST STATISTIC AND THE p VALUE

The third step of any hypothesis test requires that we use the data we have collected and calculate the test statistic and the p value for the sample data. We will use the value of the test statistic and the p value from this step to do the fourth step. The details for calculating the test statistic and p value depend on the specific hypothesis test you are doing.

8.4.4 STEP 4: MAKE THE STATISTICAL DECISION

Use the p value you calculated in Step 3. Using p values frees us from having to actually find the rejection region. If the p value is smaller than the specified value of α, then we reject the null hypothesis.

Alternatively, we can use the rejection region approach. If the test statistic calculated in Step 3 falls in the rejection region, then we reject the null hypothesis. If it does not fall in the rejection region, then we do not reject the null hypothesis. We say that we *fail to reject* the null hypothesis. It is technically incorrect to say that we *accept* the null hypothesis because the word *accept* implies that we have, in a sense, proven this hypothesis. The only thing we have done is shown that the data are not inconsistent with the null hypothesis. This point is discussed further in Section 8.8.2.

Most software packages report a p value, but for some tests the p value is not reported, so it is important that you understand both approaches.

8.4.5 STEP 5: INTERPRET THE STATISTICAL DECISION

The final step in any hypothesis test is to interpret the statistical decision from Step 4 in terms of the stated problem. In other words, we translate the decision to either reject the null hypothesis or fail to reject the null hypothesis into a business decision. What are our recommendations and conclusions based on this test? This step is clearly very important and goes along with the third step in problem solving: "Draw conclusions and make recommendations."

8.4.6 LOOKING AHEAD

As noted in Section 8.3, you can conduct a hypothesis test about many different population characteristics. Section 8.4 has introduced the five-step hypothesis-testing procedure and its associated vocabulary. Although the five-step procedure is the same for any hypothesis test, the details of which test statistic to use, how to calculate the p value, and how to find the critical values vary from test to test. So, it is time to get more specific and get some practice with these steps.

Very often you are interested in the population mean. Therefore, we will start by getting practice with a hypothesis test of the mean. The Central Limit Theorem (CLT) from Chapter 7 was said to be the key to unlocking the door to inferential statistics. Section 8.5 will explain what constitutes a *large-sample test* of the mean, and Section 8.6 will explain how to conduct a two-sided large-sample test of the mean. This test relies on the Central Limit Theorem to give us our test statistic.

Then we will return to a discussion of the types of errors you could be making as well as a discussion of how to actually decide which hypothesis to put in the null and which one to put in the alternative hypothesis. We will finish the chapter with some more large-sample tests of the mean.

Both Chapters 9 and 10 are also hypothesis-testing chapters to cover small-sample tests of the mean, tests of the variance and the population proportion, and two-population tests. Although the details for each test are different, remember that they all follow the same five-step procedure and they all have the same objective: to choose between two opposing viewpoints on the basis of sample evidence.

8.5 Large-Sample vs Small-Sample Tests

One of the most important characteristics of a population is the mean. When testing the population mean, there are two different cases you might encounter:

- When you know the value of the population standard deviation, σ
- When you don't know the value of the population standard deviation, σ

Now you might be thinking that it is unlikely that you would know the population standard deviation, σ, when you don't know the population mean. After all, why would you be doing a hypothesis test of the mean if you knew it! If you were thinking this, then you have a sharp understanding of this material. Most of the time, if you don't know the mean (which is a measure of the middle of the population) then you probably don't know the standard deviation (which is a measure of the spread of the values in the population). So why should we even bother with the first case?

This is an excellent question. Let's think about what makes these two cases different. In the first case, when we know the value of σ, the Central Limit Theorem tells

Remember we said that the CLT in Chapter 7 was the key to inferential statistics.

us that $\dfrac{\overline{X} - \mu}{\sigma/\sqrt{n}}$ has a normal distribution. However, you might also remember from Chapter 7 that when the population standard deviation is unknown AND the sample size is sufficiently large enough, $n > 30$, you can use the sample standard deviation, s, as an estimate of σ and still claim the results of the Central Limit Theorem. That is, you can claim that the distribution of $\frac{\overline{X} - \mu}{s/\sqrt{n}}$ is *approximately* normal or *approximately* a Z test statistic. So, when we know the population standard deviation, σ, OR when $n > 30$ we can use the results of the Central Limit Theorem. These are the two situations when we can use what are known as **large-sample tests of the mean.**

> A *large-sample test of the mean* is conducted when the characteristic of interest is the population mean, μ, and either of these situations exists:
>
> **(a)** the population standard deviation is known (regardless of the sample size).
>
> OR
>
> **(b)** the population standard deviation is unknown but the sample size, n, is greater than 30.

Clearly, when the population standard deviation is unknown and $n > 30$, then we will need to use what is known as a *small-sample test* of the mean. In this case we will not be able to rely on the results of the Central Limit Theorem to get our test statistic. Instead, the test statistic will be what is known as a t test statistic. This is covered in Chapter 9.

> A *small-sample test of the mean* is conducted when the characteristic of interest is the population mean, μ, and the population standard deviation is unknown but the sample size, n, is less than or equal to 30.

Many folks have suggested that we should do away with the large-sample versus small-sample distinction and simply use the Z test statistic whenever σ, the population standard deviation, is known, and we should use the t test statistic whenever σ is unknown. Their argument is that since we have technology to assist us in calculating p values and critical values, why should we use an approximate Z test statistic when we can use an exact t test statistic? This part of their argument is indeed valid, since one of the reasons for using the Z approximation was that it was difficult to find p values and critical values for the t test (a fact that you will understand when we get to Chapter 9). However, this argument forgets that one of the benefits of using the Central Limit Theorem is that the results are true regardless of the shape of the underlying population distribution. The t test statistic, as we will see in Chapter 9, requires that the underlying population distribution be bell shaped. Using the Z approximation frees us of having to worry about and check the assumption of normality. Believe it or not, this paragraph is an ongoing debate among statisticians!

An ongoing debate!

For our purposes in learning elementary statistics, we will use the Z approximation and thus a large-sample test whenever the sample size is large enough, $n > 30$. You will see in Chapter 9 that the difference in the p values and the critical values between Z and t are small when $n > 30$, and if using one test statistic instead of another changes your conclusions, then perhaps you should collect some more data! Now on to the large-sample tests of the mean!

Large-sample tests are used whenever we know σ, or $n > 30$.

8.6 Large-Sample Test of the Mean: Two-Tail Tests

Once you have decided that you need to do a large-sample test of the mean, there is yet another decision to be made. This decision is between a *two-tail test of the mean* and a *one-tail test of the mean*. In this section we will consider only large-sample two-tail tests of the mean. This test should be used when you are interested in testing whether the population mean is equal to a certain value or not. Another way to say this is that you are testing whether the population mean is *different from* a certain value. The reason it is called a two-tail test will become clear when we look at how to find the *p* value and the rejection region for this test.

This is the easiest place to begin our discussion of the details of hypothesis testing because the null and alternative hypotheses are always set up in the following way.

Set-up for a two-tail test of the mean

> A ***two-tail test of the population mean*** has these null and alternative hypotheses:
>
> $$H_0: \quad \mu = [\text{a specific number}]$$
> $$H_A: \quad \mu \neq [\text{a specific number}]$$

Notice that the "=" hypothesis is in the null hypothesis (H_0), and the "≠" theory is in the alternative hypothesis (H_A). These cannot be reversed.

The next section walks you through a detailed example.

8.6.1 A DETAILED EXAMPLE

The first step is to take our idea or hypothesis about the mean and construct the null and alternative hypotheses.

Understand the problem.

EXAMPLE 8.8

AIRLINE ARRIVALS

Step 1: Set Up the Null and Alternative Hypotheses

Suppose you are interested in whether the average *Arrival Delay* is different from the acceptable value of 30 minutes. Remember that if a two-tail test is used, the = sign is in the null hypothesis. We complete the first step of the procedure by writing the null and alternative hypotheses:

$$H_0: \quad \mu = 30$$
$$H_A: \quad \mu \neq 30$$

Remember that $\overline{X}$ is the best point estimate for μ.

The second step is to define the test procedure. Because we are talking about a hypothesis test of μ, it makes sense to use the value of $\overline{X}$ as the evidence from the sample. Remember that according to the Central Limit Theorem (see Chapter 7), the sample mean, $\overline{X}$, tends to have a normal distribution with a mean equal to the true population mean, μ. The theorem also states that the distribution of the sample mean, $\overline{X}$, has a standard deviation or standard error equal to $\sigma/\sqrt{n}$. We can therefore convert $\overline{X}$ to the standard normal variable, Z, by the formula

$$Z = \frac{\overline{X} - \mu}{\sigma/\sqrt{n}}$$

This is the test statistic for a large-sample test of the mean when the population standard deviation is known. However, the Central Limit Theorem also tells us that when the sample size is large, $n > 30$, we can substitute the sample standard deviation for σ and the test statistic will be approximately normal.

So the test statistic for a large-sample test of the mean is

$$Z = \frac{\overline{X} - \mu}{\sigma/\sqrt{n}} \quad \text{when } \sigma \text{ is known}$$

and

$$Z = \frac{\overline{X} - \mu}{s/\sqrt{n}} \quad \text{when } \sigma \text{ is unknown but } n > 30$$

Test statistic for large-sample test of the mean

EXAMPLE 8.9

AIRLINE ARRIVALS

Step 2: Define the Test Procedure

Since we know σ, the test statistic is $Z = \frac{\overline{X} - \mu}{\sigma/\sqrt{n}}$. You decide that you do not want to reject a true null hypothesis more than 5% of the time. Thus, $\alpha = 0.05$.

Understand the problem.

The third step is to capture the information in the sample into a single number called the test statistic and then calculate the p value. It is the sample mean, $\overline{X}$, that is particularly relevant because we are testing an idea about the true population mean, μ. Suppose the sample mean *Arrival Delay* of 200 flights is found to be 40 minutes. On the surface it looks like you should not use this airline. After all, 40 is 10 minutes more than the acceptable value of 30 minutes. You must remember that this $\overline{X}$ value of 40 is based on a sample, though. If you looked at another 200 flights, you might get a sample mean of 20. You cannot decide whether a difference of 10 is really big until you compare it with the standard error! This is precisely what the Z-score calculation does. The p value tells you the likelihood of observing this Z-score or one more extreme.

We can calculate Z using the following formula for standarding $\overline{X}$:

$$Z = \frac{\overline{X} - \mu}{\sigma/\sqrt{n}}$$

It seems like we have a problem: We need to know μ to calculate the Z statistic, but we don't know μ. This is easily resolved here because μ is the **target** value. In general, for the value of μ we use the number that we are testing the sample evidence against—the value in the null hypothesis.

Use the value of μ specified in the null hypothesis.

Once the test statistic is calculated, the p value can be found. Remember that the p value is the probability of obtaining the observed value of the test statistic or a more extreme value if the null hypothesis is true. Remember also that probabilities correspond to areas, and we know that for a large-sample test of the mean the appropriate distribution function is the normal distribution. The area that corresponds to the p value is thus shown in Figure 8.1.

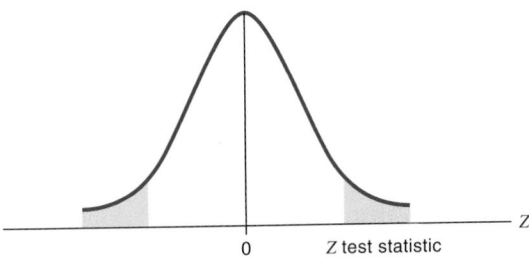

FIGURE 8.1   Area corresponding to p value when Z test statistic is greater than 0

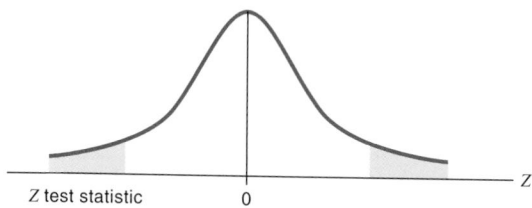

FIGURE 8.2 **Area corresponding to *p* value when *Z* test statistic is less than 0**

Notice that the area we wish to find corresponds to those values of *Z* that are "more extreme" than the observed test statistic. In this graph, the calculated test statistic, *Z*, is a positive number. This is the correct graph when the *Z* value is positive. But what if the test statistic value, *Z*, is negative? Then the area is shown in Figure 8.2.

Notice that these two areas are precisely the same—in both cases they correspond to values of *Z* that are farther away from the mean than the calculated test statistic. In addition, remember that the normal distribution is symmetric. So regardless of the sign of *Z*, we need to find only one of the tail area probabilities and then double it to get the *p* value.

These steps are summarized here. To find the *p* value for a large-sample two-sided test of the mean:

Steps for finding the p value for a two-tail test of the mean

1. Calculate the test statistic, *Z*.
2. If *Z* is negative, drop the sign.
3. Find the area to the right of *Z*. That is, find P(*Z* > test statistic).
4. Double this probability to get the *p* value.

Let's see how this test statistic and *p* value are calculated for the airline arrival example.

Collect and analyze the data.

EXAMPLE 8.10

AIRLINE ARRIVALS

Step 3: Collect the Data, Calculate the Test Statistic and the *p* Value

The data have been collected, and a portion were listed at the beginning of this chapter. Recall that the standard deviation is 50 minutes and the sample mean was found to be 40 minutes. The sample size is 200. Thus

$$n = 200$$
$$\sigma = 50 \text{ minutes}$$
$$\overline{X} = 40 \text{ minutes}$$

and

$$Z = \frac{\overline{X} - \mu}{\sigma/\sqrt{n}} = \frac{40 - 30}{50/\sqrt{200}} = \frac{10}{3.54} = 2.82$$

Notice that we use the target value of 30 as the value for μ in this calculation. You will always use the value in the null hypothesis for the value of μ in the test statistic.

To find the *p* value, we must find the area of the region shown below:

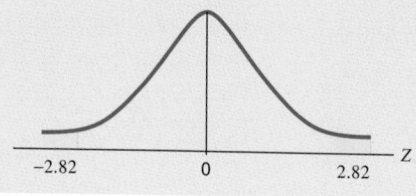

Follow the steps:

1. Calculate the test statistic, Z.	$Z = 2.82$, from above.
2. If Z is negative, drop the sign.	Z is not negative.
3. Find the area to the right of Z. That is find $P(Z > \text{test statistic})$.	From Table 3 in Appendix A we get $P(Z > 2.82) = 0.0024$.
4. Double this probability to get the p value.	p value $= 2(0.0024) = 0.0048$

The fourth step of the hypothesis-testing procedure is to decide whether to reject the null hypothesis. All of your data are evidence and are now summarized in the test statistic. There are two ways to decide—using the p value method or using the rejection region method.

Using the p value method, we will reject the null hypothesis if the p value is less than the value of α. Let's see if we can understand why this makes sense. Recall that α is the largest chance of making an error that you are willing to tolerate. This error is the mistake of rejecting the null hypothesis when it is actually true. Remember that we used the value of μ from the null hypothesis to calculate the Z test statistic. So the p value is the chance that you would observe the test statistic or a value more extreme IF the null hypothesis is true. When this chance, captured in the p value, is less than the maximum chance you are willing to live with, α, then it makes sense to reject the null hypothesis. So the rejection rule using the p value is quite simple.

Finding the p value is quite easy for a large-sample test of the mean because the appropriate distribution is the normal distribution, thanks to the Central Limit Theorem. However, you will see that this is not always the case, and sometimes it is more difficult to find the p value. Fortunately, most statistical software packages will calculate and print the p value for you for most hypothesis tests so this is the easier approach. However, it is also useful to know how to use the rejection region approach in case you don't have access to statistical software or if the p value is not printed out for you.

Using the rejection region approach, we will reject the null hypothesis if the test statistic falls in the rejection region. If the test statistic does not fall in the rejection region, then we fail to reject the null hypothesis. Let's see if we can understand why this makes sense.

Let's think about what Z really tells us and what values of the Z test statistic will lead us to reject H_0. We are interested in detecting whether the true mean, μ, is different from 30 minutes. Values of $\overline{X}$ that are far from 30 will lead us to be suspicious of and reject $H_0: \mu = 30$.

If the $\overline{X}$ value is considerably less than 30, what happens to the numerator of the Z test statistic? The numerator is $(\overline{X} - \mu)$. If $\overline{X}$ is much less than μ, then the difference will be a negative number of large magnitude. For instance, if $\overline{X} = 20$, then the difference is -10. As a result of the "large" negative value in the numerator, Z will be a "large" negative number. Keep in mind that the difference of -10 must be deemed "large" relative to the size of the standard error. Similarly, if $\overline{X}$ is a number much greater than μ—say, 40, then the difference $(\overline{X} - \mu)$ will be a large positive number and Z will be a large positive number. It seems then that "large" negative values of Z (resulting from $\overline{X}$ values much less than 30) and large positive values of Z (resulting from $\overline{X}$ values much greater than 30) are the ones that will lead us to reject the null hypothesis, H_0. We need a cutoff or **critical value** of Z to help us decide what constitutes a "large" Z value.

The specification of these cutoff or critical values for Z is completely determined by the size of the rejection region, α. We want the total area of the rejection region to be α, and because the rejection region consists of two tails of the Z distribution (large positive Z values and "large" negative Z values), the area in each tail must

Note: The expression "large" negative values means negative values of large magnitude. The term "large" negative is not technically correct but using it makes the explanations easier to follow.

We determined the rejection region by logically thinking about what Z values should lead us to reject the null hypothesis.

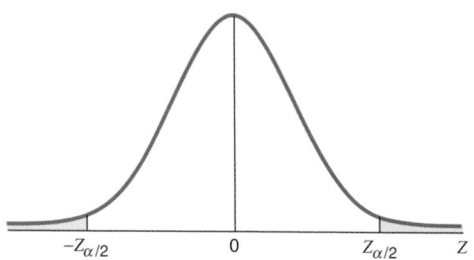

FIGURE 8.3 **Rejection region for the two-tail test of** μ

be $\alpha/2$. Combining all these ideas results in the rejection region shaded in Figure 8.3. The rejection region is those values of the test statistic (in this case, Z) that lead us to reject the null hypothesis. In this case, the rejection region is all Z values that are greater than $Z_{\alpha/2}$ or less than $-Z_{\alpha/2}$. These cutoff values are so labeled because they are Z values (from the Z table) and they cut off a tail area of $\alpha/2$. They are sometimes referred to as **critical values.**

In Chapters 6 and 7 you learned how to find Z values for given tail area probabilities. This is the procedure we use to get $Z_{\alpha/2}$. We repeat the steps from Chapter 7 here:

Steps for finding $Z_{\alpha/2}$ and $-Z_{\alpha/2}$

Step 1: Take the value of α and divide it by 2.

Step 2: Look up that value in the body of the Z table (find the value closest to it).

Step 3: Read off the corresponding Z value. This gives the value of $-Z_{\alpha/2}$.

Step 4: Drop the negative sign to get the value $Z_{\alpha/2}$.

You should have been able to guess that the critical value of Z is close to 2. Why? Remember the empirical rule from Chapter 3 states that about 95% of the values fall within two standard deviations of the mean. For Z the mean is 0 and the standard deviation is 1, so about 95% of the Z values should fall between $0 - 2(1) = -2$ and $0 + 2(1) = 2$. We found a critical Z value of 1.96, which is indeed very close to 2!

Use the bottom of the t table to quickly find the Z critical value.

There is a shorter way to find the values of $Z_{\alpha/2}$ and $-Z_{\alpha/2}$. At the bottom of Table 4 in Appendix A, you see the Z critical values for different values of α for both one-tail and two-tail tests. The value of 1.96 is found in the column that corresponds to a level of significance of 0.05 for a two-tail test. You might be wondering why these Z values are at the bottom of the t table. Recall from Chapter 7 that the t distribution approaches the Z distribution as the sample size (and hence the number of degrees of freedom) gets larger. To demonstrate this, the t value for 120 degrees of freedom is shown and then just below it is the corresponding Z value. So you can use the row labeled Z critical value to quickly find the values of $Z_{\alpha/2}$ and $-Z_{\alpha/2}$.

✔ **TRY IT NOW!**

Airline Arrivals
Finding the Rejection Region

Suppose you decide to set $\alpha = 0.10$ in the airline arrival example. Find the rejection region.

ANS. REJECT H_0 IF $Z < -1.64$ OR $Z > 1.64$.

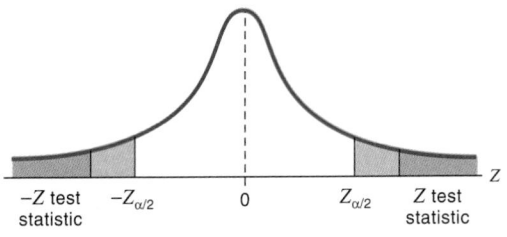

FIGURE 8.4

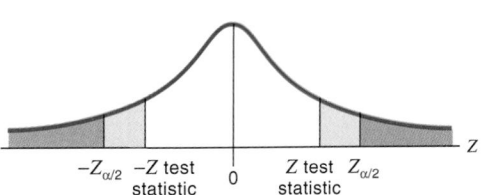

FIGURE 8.5

It is interesting to see how the p value method and the rejection region method are connected. The easiest way to do this is visually. First consider the case where the p value is less than α. We would reject the null hypothesis, and so the test statistic will clearly lie in the rejection region. The relationship between the p value and α is shown in Figure 8.4.

Now consider the case where the p value is greater than α. We would fail to reject the null hypothesis, and so the test statistic is not in the rejection region. This situation is shown in Figure 8.5.

Clearly, both the p value method and the rejection method lead to the same conclusion. Let's do the fourth step for the airline data.

EXAMPLE 8.11

AIRLINE ARRIVALS

Analyze the data.

Step 4: Make the Statistical Decision

We will make the statistical decision using both the rejection region method and the p value method. The rejection region and p value for this two-tail test are shown here:

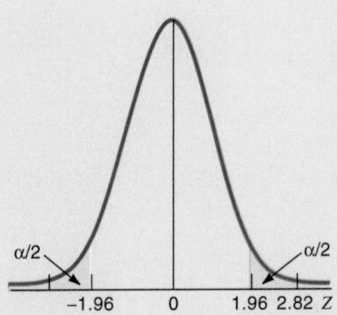

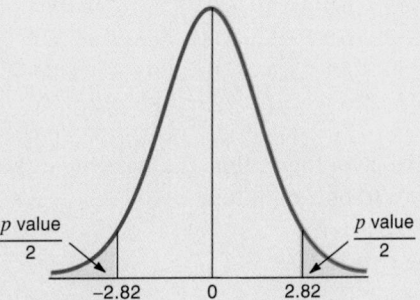

The calculated Z value of 2.82 clearly falls in the rejection region. Thus, we reject H_0 and believe the alternative hypothesis.

The p value of 0.0048 is less than the specified value of $\alpha = 0.05$, so we reject the null hypothesis. This of course is the same conclusion we reached using the rejection region method. The two approaches will always give the same results.

$Z = 2.82$ is greater than $+1.96$; it is a "large" positive value.

The fifth step is to interpret the statistical decision in terms of the stated problem. At this step we "translate" the result of the statistical test into a recommended action, decision, or conclusion depending on the nature of the problem.

> **Draw conclusions and make recommendations.**

EXAMPLE 8.12

AIRLINE ARRIVALS

Step 5: Interpret the Statistical Decision

In Step 4 we rejected the null hypothesis, so we must believe the alternative hypothesis. The alternative hypothesis is that the true mean is not equal to 30 minutes. Therefore, on the basis of the evidence in the sample that is captured in the Z value of 2.82, you decide that American Airlines has an unacceptable average arrival delay.

Your decision is based on the fact that it is unlikely that you would get such a large Z value ($Z = 2.82$) if the true mean arrival delay was really 30 minutes. You know that this is an unlikely Z value because the p value is so small (p value = 0.0048). But is it possible that the population mean really is 30 minutes? You bet! Maybe you got a sample with a large $\overline{X}$, which gave the large Z value. How likely is this? The chance of it happening is the p value, which is 0.0048.

Discovery Exercise 8.2

EXPLORING THE IMPACT OF VARYING THE VALUE OF α

In Chapter 3 we examined the data collected by a company that was concerned about the length of time its customers were on hold on the phone. The data shown here were first presented in Example 3.14. They are 50 observations (in minutes) on customer hold times:

0.6	4.6	5.6	6.3	6.8	7.5	7.8	8.3	8.9	9.6
2.9	4.7	6.0	6.3	6.9	7.5	7.9	8.4	9.2	10.1
3.4	5.2	6.0	6.6	6.9	7.6	8.0	8.4	9.2	10.2
3.8	5.5	6.1	6.6	7.0	7.6	8.1	8.6	9.4	10.7
4.5	5.5	6.1	6.7	7.2	7.8	8.2	8.6	9.4	11.1

The company wishes to test the hypothesis that the true mean hold time is 7 minutes. The standard deviation of hold times is known to be 2 minutes.

1. Test this hypothesis using $\alpha = 0.10$.

    Step 1: Set up the null and alternative hypotheses.

    Step 2: Define the test procedure.

    Step 3: Collect the data and calculate the test statistic and the p value.

    Step 4: Decide whether to reject the null hypothesis.

    Step 5: Interpret the statistical decision in terms of the stated problem.

2. Now vary the value of α and complete the table.

α	Critical value	Decision
0.10 (from part 1)		
0.05		
0.02		
0.01		

3. What happened to your decision as the value of α changed?

8.6.2 UNDERSTANDING THE RELATIONSHIP BETWEEN p VALUES AND α

You may have noticed that the rejection region is tied to the value of α. You may therefore be wondering whether selecting a different value of α would lead to a different conclusion. This is indeed a good point and worthy of consideration. Let's follow the details of the hypothesis-testing procedure through another example and then examine the impact of selecting a different value for α.

EXAMPLE 8.13

THE VENDING MACHINE

Understand the problem.

Step 1: Set Up the Null and Alternative Hypotheses

On your break at work you usually get a soda from the vending machine. Each cup should contain 12 ounces of soda. You have been getting soda at this machine for a while and you think it is cheating you by either underfilling your cup or spilling soda over the cup (which of course would have been extra for you!). You want to know whether the machine is underfilling or overfilling the cups. Since you are interested in detecting differences from 12 ounces in either direction, you use a two-tail test. These are the null and alternative hypotheses:

$$H_0: \quad \mu = 12 \text{ ounces}$$
$$H_A: \quad \mu \neq 12 \text{ ounces}$$

Continuing with this example, we complete Step 2 by selecting the value of α and finding the rejection region.

EXAMPLE 8.14

THE VENDING MACHINE

Understand the problem.

Step 2: Define the Test Procedure

You know from the machine specifications that the population standard deviation, σ, is supposed to be 0.6 ounce. Since you are doing a hypothesis test of μ and you know σ, the appropriate test statistic is $Z = \dfrac{\overline{X} - \mu}{\sigma/\sqrt{n}}$. Set $\alpha = 0.05$.

Collect and
analyze the data.

EXAMPLE 8.15

THE VENDING MACHINE

Step 3: Collect the Data, Calculate the Test Statistic and p Value

Suppose you observe the machine filling 30 cups over a period of a month and you collect these data (in ounces):

11.6	12.62	11.4	10.69	11.59	11.49	11.11	10.66	10.9	11.47
11.56	12.74	11.87	11.63	10.83	11.55	12.42	12.18	12.39	12.3
11.26	11.82	11.71	12.94	12.35	11.21	11.36	11.57	11.66	11.42

For these data, the sample mean is $\overline{X} = 11.68$ and there are $n = 30$ observations. You know that $\sigma = 0.6$ ounce.

To find the Z test statistic and the p value, follow the steps:

1. Calculate the test statistic, Z. $Z = \dfrac{\overline{X} - \mu}{\sigma/\sqrt{n}} = \dfrac{11.68 - 12}{0.6/\sqrt{30}} = -2.92$

2. If Z is negative, drop the sign. Z is negative, so use 2.92.
3. Find the area to the right of Z. That is, $P(Z > 2.92) = 0.0018$
   find $P(Z > $ test statistic$)$.
4. Double this probability to get the p value. p value $= 2(0.0018) = 0.0036$

Now we are ready to use the evidence in the sample, which has been captured in the Z test statistic, and the p value to decide whether to reject the null hypothesis.

Collect and
analyze the data.

EXAMPLE 8.16

THE VENDING MACHINE

Step 4: Make the Statistical Decision

Using the p value method: The p value of 0.0036 is less than $\alpha = 0.05$. So your statistical decision is to reject the null hypothesis and conclude that the true mean is not equal to 12 ounces.

Using the rejection region method: The rejection region is the same one used in the airline arrival example. Reject H_0 if $Z > 1.96$ or $Z < -1.96$. The value of $Z = -2.92$ clearly falls in the rejection region.

Step 5 requires that we interpret this statistical decision in terms of the problem statement.

Draw
conclusions and
make recommen-
dations.

EXAMPLE 8.17

THE VENDING MACHINE

Step 5: Interpret the Statistical Decision

You concluded that the true mean of the soda-filling process is not 12 ounces. Something is wrong with the filling process. Possibly the true mean has shifted to a value different from 12 ounces.

Table 8.2 lists five different α values for the bottle-filling example and shows the impact on the rejection region and ultimately on your decision.

Notice that when α changes from 0.005 to 0.001 the decision switches from rejecting H_0 to failing to reject H_0. Initially, this should make you a bit uncomfortable. After all, you need to know whether you have a legitimate gripe with the vending

TABLE 8.2

THE IMPACT OF VARYING THE VALUE OF α

α	Critical value	Decision
0.10	1.64 or 1.65	$-2.92 < -1.64$, so reject H_0
0.05	1.96	$-2.92 < -1.96$, so reject H_0
0.01	2.58	$-2.92 < -2.58$, so reject H_0
0.005	2.81	$-2.92 < -2.81$, so reject H_0
0.001	3.29	$-2.92 > -3.29$, so fail to reject H_0

Find the critical values at the bottom of the table.

machine company and you get a different answer depending on how you set α! This phenomenon is not unique to this problem. There will always be some value for α where your decision switches—called the *p* **value.**

The *p value* is the smallest value of α for which we can reject H_0.

Another definition for p value

For some value of α, the value of the test statistic (based on sample data) is the same as the critical value. When that happens, we have the *p* value.

In the vending machine example, the *p* value is very small, less than 0.005. This tells us that the smallest value of α for which we can reject H_0 is 0.0036. It also tells us that there is a chance of 36 in 10,000 that we would observe a sample mean of $\overline{X} = 11.68$ ounces if the population mean was really 12 ounces. Clearly, the data are inconsistent with the null hypothesis and you have a "legitimate beef" with the vending machine company. They cannot claim that you just got a "bad sample" because the data are telling you to reject the null hypothesis for virtually every possible value of α.

Consider another two-tail test where the calculated Z test statistic is 1.80. To calculate the *p* value for this situation, we find

$$p \text{ value} = 2P(Z > 1.8) = (2)(0.0359) = 0.0718$$

A *p* value between 0.05 and 0.10 indicates that the data are not strongly consistent with the null hypothesis, but neither are they strongly inconsistent with the null hypothesis. In such cases the data are "on the fence."

Finally, consider an example where the data are extremely consistent with H_0. Suppose we are doing a two-tail test of μ and the calculated Z statistic is 0.95. The *p* value calculation is

$$p \text{ value} = 2P(Z > 0.95) = (2)(0.1711) = 0.3422$$

You should have realized that a Z test statistic of less than 1 means that the observed sample mean is less than one standard error away from the hypothesized value and we know that such Z values are likely. This points to the same conclusion. So in this case the *p* value is large (> 0.10), which means we will almost always fail to reject the null hypothesis. The data are *highly consistent* with the null hypothesis.

In summary, we present the following table:

p Value	Data are . . .	Decision
Less than 0.05	Highly inconsistent with H_0	Strong rejection of H_0
Between 0.05 and 0.10	"On the fence"	Collect more data (if possible)
Greater than 0.10	Highly consistent with H_0	Strong failure to reject H_0

This table is just a *guideline*. There are no absolute rules about *p* values. The *p* value provides another way to do Step 4 of the hypothesis-testing procedure, when we decide

p values give an alternative way to do Step 4.

whether or not to reject the null hypothesis. We have also used the rejection region approach to make this decision. But now we see that by picking a value of α, the rejection region is determined, which of course determines whether we reject the null hypothesis.

A better and more widely used approach is to report the p value for the hypothesis test. This gives the decision maker much more information than simply the results of the hypothesis test for one particular value of α. In a sense the p value provides the information to do Step 4 for any value of α.

> If the p value is less than the value of α, then we reject H_0. If the p value is greater than or equal to the value of α, then we fail to reject H_0.

This rule is summarized in the next table, which is a generalization of the earlier table.

p Value	Decision	Implications
Less than α	Reject H_0	Unlikely that we would observe this test statistic value if the null hypothesis were true. The data are sufficiently inconsistent with the null hypothesis, so we reject it.
Greater than α	Fail to reject H_0	Likely that we would observe this test statistic value if the null hypothesis were true. The data are sufficiently consistent with the null hypothesis, so we fail to reject it.

Because α is the chance of incorrectly rejecting a true null hypothesis (read more about this in Section 8.7), the appropriate value of α depends very much on the cost of making this kind of error. For instance, if making this kind of wrong decision involves loss of human life, then we must be very sure of the decision. We would require an extremely small p value (< 0.0001) before we are willing to reject H_0. If, however, the cost of making this kind of error is that we adjust a machine that was actually working well, then we would require less evidence and reject H_0 for p values less than 0.05.

You might be thinking that you should always use a "tiny" value for α so that you almost never reject a true null hypothesis. The problem with this way of thinking is that you could be making another kind of error: to continue to believe the null hypothesis (fail to reject it) when it is false. This problem is thoroughly discussed in the next section.

8.6.3 EXERCISES—LEARNING IT!

8.1 The school committee members of a midsize New England city agree that a strict discipline code has led to an increase in the number of student suspensions. Here are the numbers of suspensions for a sample of schools in the city:

Central	245	Putnam	1024
MCDI	1	Kiley	56
Chestnut	65	Central Academy	254
Duggan	133	Commerce	114
Kennedy	97	Bridge	7
Forest Park	149		

The average number of suspensions in the past year was 130.5 with a population standard deviation of 158.2.

(a) Set up the null and the alternative hypotheses to test whether the average number of suspensions has changed.

(b) Complete the remaining hypothesis-testing steps using $\alpha = 0.05$.

(c) Find the p value.

(d) Display the data to see whether the committee can reasonably assume that the underlying population distribution is normal.

(e) Based on the p value, what can the committee conclude about the average number of suspensions?

8.2 The Educational Testing Service (ETS) designs and administers the SATs. Recently the format of the exam changed and the claim has been made that the new exam can be completed in an average time of 120 minutes. A sample of 50 new exam times yielded an average of 115 minutes. The standard deviation is assumed to be 2 minutes.

(a) Set up the null and the alternative hypotheses to test whether the average time to complete the exam has changed from 120 minutes.

(b) Complete the remaining hypothesis-testing steps using $\alpha = 0.05$.

(c) Find the p value.

(d) Based on the p value, what can the ETS conclude about the average time to complete the new exam?

8.3 A vending machine that dispenses coffee into cups must fill the cups with 7.8 ounces of liquid. Before selling the vending machine to a college or business, the company tests the machine to be sure it is dispensing an average of 7.8 ounces. A sample of 20 amounts (in ounces) is listed here. The amount of coffee dispensed is assumed to be normally distributed with a standard deviation of 0.05 ounce.

| 7.78 | 7.79 | 7.82 | 7.82 | 7.87 | 7.84 | 7.80 | 7.82 | 7.80 | 7.78 |
| 7.83 | 7.75 | 7.85 | 7.83 | 7.84 | 7.73 | 7.82 | 7.87 | 7.81 | 7.88 |

(a) Set up the null and the alternative hypotheses to test whether the average amount of coffee dispensed is different from 7.8 ounces.

(b) Complete the remaining hypothesis-testing steps using $\alpha = 0.05$.

(c) Find the p value.

(d) Based on the p value, what can the company conclude about the average amount of fluid dispensed by the machine?

8.4 According to data released by the U.S. Department of Labor, the average price of a half-gallon of ice cream in 1996 was $2.86. The monthly averages (in dollars) for the price of a half-gallon of ice cream at 12 stores in your town in 1996 are shown here:

| 2.87 | 2.93 | 3.13 | 2.92 | 2.99 | 3.01 |
| 3.02 | 3.01 | 2.93 | 3.08 | 2.67 | 2.89 |

Assume that the ice cream prices are normally distributed with a standard deviation of $0.10.

(a) Set up the null and the alternative hypotheses to test whether the average price of a half-gallon of ice cream in your town is different from the 1996 national average.

(b) Complete the remaining hypothesis-testing steps using $\alpha = 0.05$.

(c) Find the p value.

(d) What can you conclude about the average price of a half-gallon of ice cream in your town?

8.5 You have just read an article stating that, on the average, employees spend 1 hour a day playing video games at work. You decide to check whether your fellow employees are really spending this much work time playing games. A random sample of 35 employees played games an average of 55 minutes per day with a standard deviation of 5 minutes. You report this to your manager, and she is relieved to learn that her employees are playing less than the national average. You quickly remind her that she has information on only a sample of employees and suggest that she test the hypothesis that the population mean is 60 minutes. She is impressed with your knowledge of statistics and asks you to do the hypothesis test and report back to her.

(a) Set up the null and the alternative hypotheses to test whether the average amount of time spent playing games at work is different from 60 minutes.

(b) Complete the remaining hypothesis-testing steps using $\alpha = 0.05$.

(c) Find the p value.

(d) Based on the p value, are the employees at your company different from the average employee?

8.6 The U.S. government keeps track of tourism in the United States. Consider the following data for 1997:

Month	Length of stay (days)	Tourist arrivals
January	9.2	19,941
February	8.5	22,481
March	8.1	24,748
April	8.2	20,681
May	7.9	23,179
June	9	17,730
July	9.8	22,635
August	9	21,470
September	9.3	14,366
October	9	18,647
November	8.4	19,552
December	9.4	22,976
Average	**8.8**	**20,701**

(a) Set up the null and the alternative hypotheses to test whether 1997 had a significantly different number of tourists than the population mean of 20,000.

(b) Complete the remaining hypothesis-testing steps using $\alpha = 0.10$.

(c) Find the p value.

(d) Based on the p value, what can you conclude about the average number of tourists in 1997?

8.7 What Error Could You Be Making?

At the end of the last section, we noted that we can make two kinds of errors in any hypothesis test. Those errors can occur because we use only *sample* information to decide about the validity of a hypothesis about the *population*. We discussed the error of rejecting a true null hypothesis. The maximum chance of this kind of error is set to be α.

Let's look in more detail at the two ways we can be wrong.

8.7.1 TWO TYPES OF ERRORS

Four outcomes can result from the decisions made in any hypothesis test. Look at Table 8.3 and think carefully about how each of the outcomes could occur.

As you can see from Table 8.3, we can make the right decision in two ways and we can make a mistake in two ways. Unfortunately, we will not know whether we made the right decision or made an error. This is because to know which category we are in requires that we know what the population actually looks like and thus if H_0 or H_A is true. If we knew that, we wouldn't be taking a sample!

Consider the four possible outcomes in the example of the airline arrival times. They are presented in Table 8.4.

To be clear which of the two possible errors we are talking about, we give them names. Table 8.5 shows the same four situations with each error named. You can see

TABLE 8.3

POSSIBLE OUTCOMES IN A HYPOTHESIS TEST

Sample indicates we should	In actuality:	
	H_0 is true	H_A is true
Believe H_0	We made the right decision—no error	Error
Believe H_A	Error	We made the right decision—no error

that statisticians are not very creative at naming errors; they are simply called **Type I** and **Type II errors.**

> A *Type I error* is made when we reject the null hypothesis and the null hypothesis is actually true. In other words, we incorrectly reject a true null hypothesis.

> A *Type II error* is made when we fail to reject the null hypothesis and the null hypothesis is false. In other words, we continue to believe a false null hypothesis.

TABLE 8.4

FOUR POSSIBLE OUTCOMES FOR THE AIRLINE ARRIVAL EXAMPLE

Sample indicates you should	In actuality:	
	H_0 is true $\mu = 30$	H_A is true $\mu \neq 30$
Believe H_0: You believe that $\mu = 30$ and so the airline meets your needs.	You **correctly** decide that the airline averages 30-minute delays.	You **incorrectly** decide that the airline can meet your needs when it cannot.
Believe H_A: You believe that $\mu \neq 30$ and so it does not meet your needs.	You **incorrectly** decide that the airline does not meet your needs when it actually does.	You **correctly** decide that the airline cannot meet your needs.

TABLE 8.5

NAME THOSE ERRORS: TYPE I AND TYPE II

Sample indicates you should	In actuality:	
	H_0 is true	H_A is true
Believe H_0	You made the right decision—no error	Type II error
Believe H_A	Type I error	You made the right decision—no error

EXAMPLE 8.18

THE VENDING MACHINE

Identifying Type I and Type II Errors

In Example 8.13 we set up these null and alternative hypotheses:

$$H_0: \quad \mu = 12 \text{ ounces}$$

$$H_A: \quad \mu \neq 12 \text{ ounces}$$

Remember that a Type I error is made when we believe H_A but H_0 is true. For this problem that means you believe the true mean is not 12 ounces (H_A) and therefore you believe the cups are not being filled properly. You complain to the vending machine company when in fact the true mean really is 12 ounces (H_0) and the machine did not need to be adjusted.

Draw conclusions and make recommendations.

A Type II error is made when we believe H_0 is true but H_A is true. In this case, you believe that the vending machine is properly filling the cups when, in fact, it is either overfilling or underfilling them. So, you do not complain when in fact the machine needs to be adjusted.

✔ TRY IT NOW!

Gas Prices
Examining Type I and Type II Errors

Find the Type I and Type II errors for the hypothesis test you set up to test whether the nationwide average price per gallon of gas changed from $1.50 during the summer of 2001.

8.7.2 PROBABILITY OF MAKING AN ERROR

Since we can't eliminate sampling error unless we sample the entire population, the next best thing we can do is try to control the chances that we are in error. The standard labels for the chances of making these errors are α and β.

This is the same α that we have been specifying in Step 2 of the hypothesis-testing procedure.

> The probability of making a Type I error is labeled α *(alpha)*.

> The probability of making a Type II error is labeled β *(beta)*.

Which error should be avoided?

Clearly, α and β must be numbers between 0 and 1 because they are probabilities. As the investigator, you decide the value of α. This means you can specify the chances of making a Type I error to be anything you wish. Once you set α, the value of β is completely determined. Your first thought might be to set α as small as possible so there is hardly any chance of making a Type I error. But, there is a price to pay for making α very small. As α gets smaller, the value of β gets larger. So, just like most things in life, there is a tradeoff. You can force the chance of making a Type I error to be very small, but then you have to live with a greater chance of making a Type II error.

Figure 8.6 shows the rejection region for a two-tail test of μ for $\alpha = 0.05$. Remember that the rejection region corresponds to those values of the Z test statistic that lead us to reject the null hypothesis. Also recall that the Z test statistic is

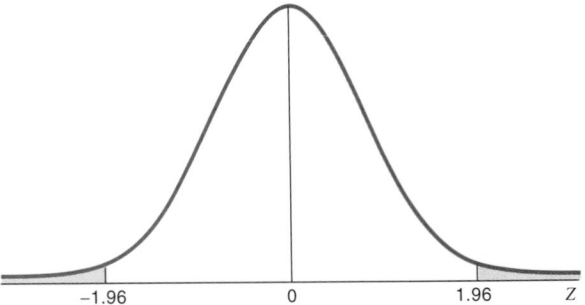

FIGURE 8.6 **Rejection region for a two-tail test of μ with $\alpha = 0.05$**

ANS. TYPE I ERROR—YOU CONCLUDE THAT THE GAS PRICES ARE DIFFERENT FROM $1.50 WHEN THEY ARE ON AVERAGE EQUAL TO $1.50. TYPE II ERROR—YOU CONCLUDE THAT PRICES HAVE NOT CHANGED FROM $1.50 BUT THEY REALLY HAVE.

calculated using the value of μ from the null hypothesis. So the rejection region is drawn based on the assumption that the null hypothesis is *true*. The shaded region then corresponds to the chance of getting a "large" negative Z test statistic or a large positive Z test statistic when the null hypothesis is true. If we make this region smaller by making α smaller, then we reduce the likelihood of rejecting a true null hypothesis, which means we fail to reject the null hypothesis more often. For some of those times that we fail to reject the null hypothesis, we should have rejected it. This is a Type II error. Thus, by decreasing the chance of a Type I error, α, we increase the chance of making a Type II error, β.

Let's examine this tradeoff in the example of airline arrival delays.

EXAMPLE 8.19

AIRLINE ARRIVALS

Looking at Tradeoffs in Errors

Suppose you decide to set the chance of making a Type I error, α, very low — say, 0.01. This means that only 1% of the time you will incorrectly decide that the airline cannot meet your requirements when it really can. This is the good news. The tradeoff is that the chance of making a Type II error increases. This means that the likelihood of deciding that the airline is acceptable when the average arrival delay is not really 30 minutes increases. How large the likelihood gets depends on the real value of μ.

Understand the problem.

The standard value of α used to test hypotheses is 0.05. However, if making a Type I error is much more costly than making a Type II error, then we consider testing at a value of $\alpha = 0.01$. This of course causes the value of β to increase. If, on the other hand, making a Type II error is particularly costly, then we consider testing at a value of $\alpha = 0.10$. If both errors are equally costly, then we generally use the standard value of $\alpha = 0.05$.

EXAMPLE 8.20

THE AIRPLANE BOLTS PROBLEM

Considering the Costs of Errors

Suppose you are inspecting bolts that are used to fasten engines to airplanes. The mean strength of these bolts must be at least 100 pounds per square inch (psi). You need to decide whether a shipment of these bolts is acceptable — that is, whether they are strong enough. You cannot test them all because once you test a bolt, it cannot be used in the airplane. This is called destructive sampling. If you test all the bolts, you will have none to use. You decide to take a sample of bolts and, on the basis of the mean strength of the sample, you will decide whether the shipment is acceptable.

Here is one possible way to set up the hypothesis test:

$$H_0: \quad \mu \geq 100 \text{ psi} \quad \text{(The shipment is acceptable.)}$$
$$H_A: \quad \mu < 100 \text{ psi} \quad \text{(The shipment is unacceptable.)}$$

For this hypothesis test, what is a Type I error? You believe that the shipment is unacceptable when in fact it is fine. So, you would send the bolts back for rework when they did not need to be reworked. Now what is a Type II error? You keep the bolts even though they are not strong enough.

Which of these errors is more costly?

- Send bolts back for rework when they do not need to be reworked.
- Keep bolts that are not strong enough.

Understand the problem.

Remember! A Type I error is that we reject the null hypothesis when, in fact, the null hypothesis is true.

A Type II error is that we fail to reject the null hypothesis when, in fact, the alternative hypothesis is true.

We would probably all agree that keeping bolts that are not strong enough is the more costly error (in terms of human life). This is a Type II error. Thus, you should try to minimize the value of β by testing at $\alpha = 0.10$.

 **TRY IT NOW!**

New Approach to Rehabilitating Criminals
Setting the Value of α

You have seen that a one-tail test is often used to investigate whether a new method is better than an existing one. Consider a prison system that is trying a new approach to rehabilitating criminals. With the current approach the average number of months until the next arrest is 3.2. To test whether the new approach is an improvement, the null and alternative hypotheses are

$$H_0: \quad \mu \le 3.2 \text{ months}$$
$$H_A: \quad \mu > 3.2 \text{ months}$$

In the warden's decision to adopt or not adopt this new approach, what are the Type I and Type II errors?

What value of α do you suggest be used to conduct the test?

8.7.3   EXERCISES—LEARNING IT!

8.7   Administrators at a small college are concerned that part-time evening students may not be familiar with all the services of the college. They wish to offer an orientation program to these students but recognize that most of the part-time students work during the day and are very busy. The administrators do not want to prepare an elaborate presentation if only a handful of part-time students will attend. They decide to conduct the orientation if more than 25% of the part-time students are interested in attending.

(a) State the consequence of a Type I error.

(b) State the consequence of a Type II error.

(c) Suggest a value for α and justify your choice.

8.8   A company is thinking about setting up an on-site day-care program for its employees. The CEO has stated that she will do so only if more than 80% of the employees favor such a decision.

(a) State the consequence of a Type I error.

(b) State the consequence of a Type II error.

(c) Suggest a value for α and justify your choice.

8.9   When Aretha Franklin crooned the words "I'll say a little prayer for you" in the hit 1960s song, she probably didn't imagine that the soulful pledge would become the stuff of serious science. But increasingly scientists are studying the power of prayer and in particular its role in healing people who are sick. Most research in the field looks at how sick people are affected by their own spiritual beliefs and practices. These studies have suggested that people who are religious seem to heal faster or cope with illness more effectively than do the nondevout. Suppose that the average healing time for all people with a particular illness is 1 week. You wish to test the hypothesis.

ANS. TYPE I ERROR—CONCLUDE THE NEW METHOD IS BETTER WHEN IT IS NOT. TYPE II ERROR—CONCLUDE THE NEW METHOD IS NOT BETTER WHEN IT IS. ERRORS ARE EQUALLY COSTLY SO USE $\alpha = 0.05$.

(a) State the consequence of a Type I error.

(b) State the consequence of a Type II error.

(c) Suggest a value for α and justify your choice.

8.10 You are a connoisseur of chocolate chip cookies and you do not think that Nabisco's claim that every bag of Chips Ahoy cookies has 1000 chocolate morsels is correct.

(a) State the consequence of a Type I error.

(b) State the consequence of a Type II error.

(c) Suggest a value for α and justify your choice.

8.11 Antilock brake systems (ABS) have been hailed as a revolutionary safety feature. A study by the National Highway Traffic Safety Administration looked at fatal accidents. The claim is that cars with ABS are involved in fewer fatal crashes than those without ABS.

(a) State the consequence of a Type I error.

(b) State the consequence of a Type II error.

(c) Suggest a value for α and justify your choice.

8.12 A college placement office wonders whether there is a difference between the average salaries of criminal justice graduates and social work graduates.

(a) State the consequence of a Type I error.

(b) State the consequence of a Type II error.

(c) Suggest a value for α and justify your choice.

8.13 Your new television has a 1-year warranty. You are given the option to buy a 3-year warranty and you wonder whether it is worth it. You wish to test the hypothesis that the average time before a problem occurs is more than 3 years.

(a) State the consequence of a Type I error.

(b) State the consequence of a Type II error.

(c) Suggest a value for α and justify your choice.

8.14 M&M/Mars claims that at least 20% of the M&M's in each package are the new blue color.

(a) State the consequence of a Type I error.

(b) State the consequence of a Type II error.

(c) Suggest a value for α and justify your choice.

8.15 A computer center is arguing for more computers in the lab for students at a midsize college. The computer center at a university claims that the average amount of time students spend online has increased from last year's average of 1 hour per day.

(a) State the consequence of a Type I error.

(b) State the consequence of a Type II error.

(c) Suggest a value for α and justify your choice.

8.8 Which Theory Should Go into the Null Hypothesis?

The first step in hypothesis testing is to construct two opposing views: the null hypothesis and the alternative hypothesis. Because of the way the hypothesis-testing procedure works, it is important to consider carefully which of the views is called the null hypothesis and which is called the alternative hypothesis.

Several different approaches may be taken to determine how to set up the null and alternative hypotheses. These approaches are our next subject. We first develop the concept of two-tail and one-tail tests and then illustrate the conservative nature of the hypothesis-testing procedure. Depending on the specifics of each problem, the approach we choose sheds some light on answering the question, Which theory should go into the null hypothesis?

8.8.1 TWO-TAIL TESTS AND ONE-TAIL TESTS

Much of our discussion so far has been about the null and alternative hypotheses in **two-tail tests.** Their format is familiar to you by now.

> A *two-tail test* of the population mean has these null and alternative hypotheses:
>
> $$H_0: \quad \mu = \text{[a specific number]}$$
> $$H_A: \quad \mu \neq \text{[a specific number]}$$

Remember! You could be testing a variance or a proportion instead.

The null hypothesis of a two-tail test claims that the mean (or whatever parameter we are testing) is *equal* to the specific number stated. The opposing view is clearly that the mean is *not equal* to that particular value; this is the alternative hypothesis.

When we use a two-tail test, we are interested in seeing whether the true mean is *different* from the number specified. We want to know if the true mean is *higher than* the number or *lower than* the number. In other words, we want to test for deviations from the number in either direction—on the high side or on the low side. When we are doing a two-tail test, no decision has to be made about how the null and alternative hypotheses are set up. The hypothesis-testing procedure requires that the view with the = sign be in the null hypothesis.

Understand the problem.

EXAMPLE 8.21

AIRLINE ARRIVALS

Setting Up a Two-Tail Test of the Mean

For the evaluation of airline arrival delays, we set up a two-tail test:

$$H_0: \quad \mu = 30 \text{ minutes}$$
$$H_A: \quad \mu \neq 30 \text{ minutes}$$

Here the specific number is 30.

Both the vending machine problem and the gas prices problem discussed earlier are also two-tail tests. The view that the true mean, μ, is equal to 12 ounces (for the soda) or \$1.50 (for the gasoline) is the null hypothesis.

✔ TRY IT NOW!

Are We Working More Hours?
Setting Up a Two-Tail Test of the Mean

A recent study by the Economic Policy Institute found that an average middle-class family income rose by 9.2% from 1989 to 1998. But workers also spent more time on the job. Set up the null and alternative hypotheses to test whether the average full-time worker's workweek changed from 43 hours.

Sometimes we want to test only whether the population mean (or whatever parameter we are testing) is less than the stated value. Then we use what is called a **lower-tail test,** which is one of two types of one-tail tests.

ANS: $H_0: \mu = 43$ HOURS; $H_A: \mu \neq 43$ HOURS

A *lower-tail test* of a population mean has these null and alternative hypotheses:

$$H_0: \quad \mu \geq \text{[a specific number]}$$
$$H_A: \quad \mu < \text{[a specific number]}$$

As with the two-tail test, we may be testing a proportion, a variance, or the difference between two means or proportions.

EXAMPLE 8.22

AIRLINE ARRIVALS

Illustrating a Lower-Tail Test of the Mean

You really only want to be sure that the airline is on the average less than 30 minutes late. The test is set up this way:

$$H_0: \quad \mu \geq 30 \text{ minutes}$$
$$H_A: \quad \mu < 30 \text{ minutes}$$

Understand the problem.

In the example notice that the viewpoint you are interested in testing became the alternative hypothesis. You should also notice that the inequality with the equal sign attached to it (in this case $\geq$) is in the null hypothesis.

EXAMPLE 8.23

THE VENDING MACHINE

Illustrating a Lower-Tail Test of the Mean

Recall that you are suspicious that the vending machine is cheating you out of soda. What you really want to know is whether the average amount of soda is less than 12 ounces. It makes more sense to set this test up as a lower-tail test:

$$H_0: \quad \mu \geq 12 \text{ ounces}$$
$$H_A: \quad \mu < 12 \text{ ounces}$$

Understand the problem.

We made the theory that we wish to test the alternative hypothesis, and again the inequality with the equal sign (in this case, $\geq$) is placed in the null hypothesis.

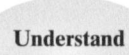

TRY IT NOW!

Bank Waiting Times
Setting Up a Lower-Tail Test of the Mean

Suppose a bank knows that its customers are waiting in line an average of 10.2 minutes during the lunch hour. The branch manager has decided to add an additional teller during the 12–2 P.M. period and wishes to test the hypothesis that the average wait has decreased because of the additional teller. Set up the null and alternative hypotheses for the bank manager.

Ans: $H_0: \mu \geq 10.2$ MIN; $H_A: \mu < 10.2$ MIN

We may also wish to see whether the true mean is greater than a specific value. In this case, we are interested only in testing whether the true mean is *greater than* some number. This is called an **upper-tail test,** which is the other kind of one-tail test.

An *upper-tail test* of a population mean has these null and alternative hypotheses:

$$H_0: \quad \mu \leq \text{[a specific number]}$$
$$H_A: \quad \mu > \text{[a specific number]}$$

Understand the problem.

EXAMPLE 8.24

WEATHER ON THE WEB

Setting Up an Upper-Tail Test of the Mean

Are more people using the web for weather information? The National Weather Service website (*http://www.nws.fsu.edu/stats/#Browser*) gives the number of "hits" the site gets. In April 1999 the average number of hits per day was 57,622. Suppose the average for April 2000 was 50,000 hits per day. To test whether the average increased you set up this hypothesis test:

$$H_0: \quad \mu \leq 50{,}000 \text{ hits per day}$$
$$H_A: \quad \mu > 50{,}000 \text{ hits per day}$$

Notice that the theory we wish to "prove" becomes the alternative hypothesis. The inequality with the equal sign (in this case, $\leq$) is again placed in the null hypothesis.

✔ TRY IT NOW!

Age of First Marriage
Setting Up an Upper-Tail Test of the Mean

Americans are saying their "I do's" later. The mean age at first marriage for men rose from 23.2 in 1970 to 26.8 in 2000. The increase was even greater for women, with the mean age at first marriage rising from 20.8 in 1970 to 25.1 three decades later. Set up the null and alternative hypotheses to test the theory that the average marrying age for women has increased from 1970.

Summary

Two-tail test

- Is used to test whether the parameter has shifted away from a certain number in either direction, increased or decreased
- Must be set up with the = in the null hypothesis
- Is used when the problem statement contains the key word *changed* or *different*

Lower-tail test

- Is used to test whether the parameter has shifted to a number less than a certain number
- Must be set up with the = as part of the null hypothesis

ANS. $H_0: \mu \leq 20.8$ YEARS; $H_A: \mu > 20.8$ YEARS

- Is used when the problem statement contains the key words *decreased, reduced,* or *less than*
- Is set up with the theory we wish to "prove" as the alternative hypothesis

Upper-tail test

- Is used to test whether the parameter has shifted to a number greater than a certain number
- Must be set up with the = as part of the null hypothesis
- Is used when the problem statement contains the key words *increased* or *greater than*
- Is set up with the theory we wish to "prove" as the alternative hypothesis

8.8.2 WHAT VIEW REQUIRES NO ACTION?

As you saw in the preceding section, there is really no choice about how to set up the null and alternative hypotheses after we decide to use a two-tail test. The next approach is particularly useful when we have decided to use a one-tail test. It considers the question, What view requires that I take no action? Typically, this is the view that the population is behaving as it should be or as someone claims it should be. This view becomes the null hypothesis. Sometimes people call this the *status quo*.

EXAMPLE 8.25

THE VENDING MACHINE

Using the Norm As the Null Hypothesis

The view that requires no action (or adjustments) is that the vending machine is working properly. The specification is that the mean *amount of soda* is greater than or equal to 12 ounces. This becomes the null hypothesis. Thus, H_0: $\mu \geq 12$ ounces. Once we have the null hypothesis, it is easy to construct the alternative hypothesis because it has to cover all the other cases. So H_A: $\mu < 12$ ounces. Here are the hypotheses:

$$H_0: \quad \mu \geq 12 \text{ ounces}$$
$$H_A: \quad \mu < 12 \text{ ounces}$$

Understand the problem.

You should ask, Why do it this way? The reason is that the hypothesis-testing procedure is conservative. It behaves like a conservative person who takes action only when he or she is very sure that action needs to be taken.

Think about one's belief in Santa Claus. Many children believe in Santa Claus until they are 7 or 8 years old. In this case, the null hypothesis is H_0: Santa Claus exists, and the alternative hypothesis is H_A: Santa Claus does not exist. Children start off believing in Santa Claus (it is never proven to them) and they continue to believe until there is overwhelming evidence to the contrary.

In a similar manner, the hypothesis-testing procedure tells us to believe the null hypothesis unless the evidence in the sample data overwhelmingly contradicts the null hypothesis. In other words, the status quo, or the view that implies that no action be taken, is placed in the null hypothesis. It is assumed to be correct until the data in the sample are *really* incompatible with it.

This approach is just like our judicial system. A person is assumed innocent until the evidence is so strong that it is impossible to continue to believe that the person is innocent. Jurors are always instructed to believe "beyond a reasonable doubt" that the person is guilty. This means that, even if most of the evidence indicates that the person is guilty, if there is still some reasonable doubt, the jurors must find the person not guilty.

✔ *TRY IT NOW!*

Judicial System
Setting Up the Null and Alternative Hypotheses

If you think about the judicial system in terms of a hypothesis test, how would you set up the null and alternative hypotheses?

Consider another example.

Understand the problem.

EXAMPLE 8.26

GRADE INFLATION

Illustrating the Status Quo Approach

The president of a university is very worried about all those A's. He wants to find out if the average grade point average has changed from the longstanding average of 2.7. Using the approach described in this section, he asks, What view requires that no action be taken?

Since the view that requires no action is that the true mean GPA is 2.7, this statement is placed in the null hypothesis. The hypotheses are

$$H_0: \quad \mu = 2.7$$
$$H_A: \quad \mu \neq 2.7$$

✔ *TRY IT NOW!*

VCR Manufacturer
Setting Up the Null and Alternative Hypotheses

Suppose a manufacturer of VCRs claims that their average lifetime is at least 3 years. You have a VCR made by this company and you have had problems with it, so you question this claim. Set up the hypotheses to investigate the manufacturer's claim.

8.8.3 EXERCISES—LEARNING IT!

8.16 Administrators at a small college are concerned that part-time evening students may not be familiar with all the services of the college. They wish to offer an orientation program to these students but recognize that most of them work during the day and are very busy. They do not want to prepare an elaborate presentation if only a handful of part-time students will attend. They decide to conduct the orientation if more than 25% of the part-time students are interested in attending. Set up the null and alternative hypotheses to be used to decide whether the administrators should offer the program.

8.17 A company is thinking about setting up an on-site day-care program for its employees. The CEO has stated that she will do so only if more than 80% of the employees favor such a decision. Set up the null and alternative hypotheses to be tested.

8.18 A human resource manager thinks that men use e-mail less than women. Set up the null and alternative hypotheses that should be tested to determine whether the manager is correct.

8.19 You wish to investigate the claim that people who are religious heal faster or cope with illness more effectively than do the nondevout. Suppose that the average healing time for all people with a particular illness is 1 week. Set up the null and alternative hypotheses to test the theory that prayerful people heal faster.

8.20 You are a connoisseur of chocolate chip cookies and you do not think that Nabisco's claim that every bag of Chips Ahoy cookies has 1000 chocolate morsels is correct. Set up the null and alternative hypotheses to test this claim.

8.21 Antilock brake systems (ABS) have been hailed as a revolutionary safety feature. A study by the National Highway Traffic Safety Administration looked at fatal accidents. The claim is that cars with ABS are involved in fewer fatal crashes than those without ABS. Set up the null and alternative hypotheses to test this claim.

8.22 A college placement office wonders whether the average entry-level salary this year is differ- ent from last year's value of $25,000. Set up the null and alternative hypotheses to test this question.

8.23 The college placement office also wonders if there is a difference between the average salaries of criminal justice graduates and social work graduates. Set up the null and alternative hypotheses to test whether these averages are different.

8.24 Your new television has a 1-year warranty. You are given the option to buy a 3-year warranty and you wonder whether it is worth it. You wish to test the hypothesis that the average time before a problem occurs is more than 3 years. Set up the null and alternative hypotheses to test this belief.

8.25 It seems like you spend more money on groceries during the summer months when you eat more ice cream and drink more fluids. You know that you spend an average of $25 a week on groceries during the winter months. Set up the null and alternative hypotheses to decide whether, on the average, you spend more than $25 per week during the summer.

8.26 M&M/Mars claims that at least 20% of the M&M's in each package are the new blue color. Set up the null and alternative hypotheses to test this claim.

8.27 The computer center at a university claims that the average amount of time that students spend online has increased from last year's average of 2 hours per day. Set up the null and alter- native hypotheses to test this claim.

8.9 One-Tail Tests of the Mean: Large Sample

In Section 8.6 you learned how to do a two-tail hypothesis test of the mean. Now we adapt that procedure to do a one-tail hypothesis test of the mean, μ. The procedure applies when we know the standard deviation of the population or we have a suffi- ciently large sample, $n > 30$. In the latter case, the only change in the procedure is to use the sample standard deviation, s, instead of σ. Therefore, we again use the label "large sample" to describe these tests.

8.9.1 LOWER-TAIL TESTS OF THE MEAN

We are often interested in whether the mean has shifted in *one* direction. In the airline arrivals example you are interested in whether the true average *Arrival Delay* is less than 30 minutes. Thus, a lower-tail test makes more sense. We can use this example to illustrate the procedure for a one-tail hypothesis test of μ when the popu- lation standard deviation is known.

Fortunately, the steps for conducting a one-tail test of the mean when the popula- tion standard deviation is known are similar to those we used in earlier sections. In fact, the only thing that changes is the procedure for finding the rejection region in Step 2. Recall that we have been using the following five-step hypothesis-testing procedure:

Step 1: Set up the null and alternative hypotheses.

Step 2: Define the test procedure.

Step 3: Collect the data and calculate the test statistic and the p value.

Steps for any hypothesis test

Step 4: Decide whether to reject the null hypothesis.

Step 5: Interpret the statistical decision in terms of the stated problem.

Reconsider the problem of evaluating the airlines as a one-tail test.

> **Understand the problem.**

EXAMPLE 8.27

AIRLINE ARRIVALS

Step 1: Set Up the Null and Alternative Hypotheses

Since you are checking to see whether the true mean is *less than* 30 minutes, this option becomes the alternative hypothesis. You have this setup:

$$H_0: \quad \mu \geq 30 \text{ minutes}$$
$$H_A: \quad \mu < 30 \text{ minutes}$$

We called this type of test a *lower-tail test*. The fact that the $<$ sign is in the alternative hypothesis makes it a lower-tail test. You will see why it is called a lower-tail test when we find the rejection region.

The second step is to define the test procedure. This includes selecting the appropriate test statistic and setting the value for α, the maximum probability of a Type I error you wish to tolerate. Since we are still doing a large-sample test of the mean, the appropriate test statistic is Z:

Large-sample tests are used whenever we know σ, or $n > 30$.

$$Z = \frac{\overline{X} - \mu}{\sigma/\sqrt{n}} \qquad \text{when } \sigma \text{ is known}$$

and

$$Z \approx \frac{\overline{X} - \mu}{s/\sqrt{n}} \qquad \text{when } \sigma \text{ is unknown and } n > 30$$

Let's do the second step for the airline arrival example.

> **Understand the problem.**

EXAMPLE 8.28

AIRLINE ARRIVALS

Step 2: Define the Test Procedure

Since this is a test of the mean and σ is known, the test statistic is $Z = \frac{\overline{X} - \mu}{\sigma/\sqrt{n}}$. Suppose we set $\alpha = 0.05$.

The third step in the hypothesis-testing procedure is to capture the information in the sample into a single number called the test statistic and calculate the p value.

The calculation of the Z test statistic is done the same way it was done for the two-tail test of the mean. However, there needs to be a slight adjustment to the procedure for finding the p value in the case of a lower-tail test of the mean. We are looking for the chance that we incorrectly reject H_0 when it is true. What Z values should lead us to reject the null hypothesis? We are interested in detecting whether the true mean is less than 30 minutes. So, values of $\overline{X}$ that are much lower than 30 will lead us to be suspicious of the null hypothesis. Following the same logic that we used earlier, if $\overline{X}$ is

Remember! When we talk about "a lot lower" we consider the difference in terms of the size of the standard error.

"a lot lower" than 30, then $(\overline{X} - \mu)$ will be negative and Z will be a negative number. Clearly, "large" negative Z values will then lead us to reject the null hypothesis.

How about large positive Z values? To get a positive Z value, $\overline{X}$ must be greater than 30. If this is the case, would we ever want to reject the null hypothesis? No! An $\overline{X}$ value of, say, 40, would certainly give us a large positive Z value, but a sample mean of 50 is consistent with the null hypothesis. Remember that we reject the null hypothesis

FIGURE 8.7 *p* **value for lower-tail test**

only when the sample evidence is inconsistent with the null hypothesis. Thus, any $\overline{X}$ value greater than 30 will lead us to continue to believe the null hypothesis. In fact, we wouldn't even have to do the hypothesis test in such cases!

So, we will reject H_0 only when Z is "too small." The *p* value is the probability of observing the value of the test statistic or a more extreme value. In the case of a lower-tail test of the mean, the words "more extreme" only refer to values of Z lower than the observed test statistic. The *p* value is the area of the shaded region in Figure 8.7.

The steps are summarized here. To find the *p* value for a large-sample lower-tail test of the mean:

1. Calculate the test statistic, Z.

2. Find the area to the left of Z. That is, find P(Z < test statistic).

Steps for finding the p value for a lower-tail test of the mean

Let's see how this test statistic and *p* value are calculated for the airline arrival example.

EXAMPLE 8.29

AIRLINE ARRIVALS

Collect and analyze the data.

Step 3: Collect the Data, Calculate the Test Statistic and the *p* Value

The data are collected as described in the beginning of the chapter. Recall that the population standard deviation, σ, for the arrival delays is 50 minutes. Suppose that the sample mean, $\overline{X}$, is observed to be 12.2 minutes based on the first 100 flights into LaGuardia in December 2000. The Z statistic is then

$$Z = \frac{\overline{X} - \mu}{\sigma/\sqrt{n}} = \frac{12.2 - 30}{50/\sqrt{100}} = \frac{-17.8}{5} = -3.56$$

$$p \text{ value} = P[Z < -3.56] = 0.0002$$

The fourth step of the hypothesis-testing procedure is do decide whether to reject the null hypothesis. All of your data are evidence and are now summarized in the test statistic. There are two ways to decide—using the *p* value method or using the rejection region method.

Using the p value method, we will reject the null hypothesis if the *p* value is less than the value of α.

Using the rejection region approach, we will reject the null hypothesis if the test statistic falls in the rejection region. If the test statistic does not fall in the rejection region then we fail to reject the null hypothesis.

Combining the ideas that we will reject H_0 when the Z value is too large in the negative direction and that we will reject H_0 incorrectly only a certain percentage, α, of the time, we arrive at the graph in Figure 8.8 on page 396. The shaded region is the rejection region. The rejection region is the *lower tail* of the normal distribution—hence, the name *lower-tail test*. Remember that the rejection region consists of those values of Z, the test statistic, that lead to rejection of the null hypothesis. Thus, a Z value less than $-Z_\alpha$ (or a "large" negative Z value) will lead us to reject the null hypothesis.

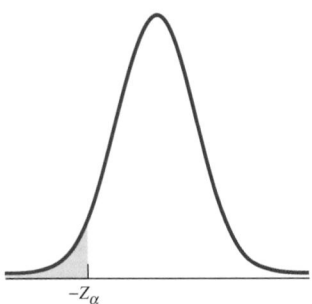

FIGURE 8.8 **Rejection region for lower-tail test of μ – large sample**

By using the standard normal table we can find this critical Z value so that the probability that we are in the rejection region is α.

In general, the procedure for finding the rejection region for a lower-tail test is as follows:

Steps for finding the rejection region for a lower-tail test

Step 1: Look up the value of α in the body of the Z table (or find the value closest to it).

Step 2: Read off the corresponding value; call it $-Z_\alpha$.

Step 3: The rejection region contains all values of Z that are less than $-Z_\alpha$.

When $\alpha = 0.05$, the value of $-Z_\alpha$ is -1.645. You may remember that for the two-tail test, when we set $\alpha = 0.05$, the rejection region is $Z < -1.96$ or $Z > 1.96$. Why is the negative critical Z value different for the two-tail test? Remember that for the two-tail test, we split the value of $\alpha = 0.05$ in half before looking it up in the table. Thus, the critical value of -1.96 has an area of only 0.025 compared with the total value of 0.05 used to get the critical value of -1.645.

TRY IT NOW!

Airline Arrivals
Finding the Rejection Region

Try to predict what will happen to the $-Z_\alpha$ value for the one-tail test if $\alpha = 0.025$. Now find the value to confirm your guess.

Now we can finish the airline arrival evaluation.

Draw conclusions and make recommendations.

EXAMPLE 8.30

AIRLINE ARRIVALS

Step 4: Make the Statistical Decision
Step 5: Interpret the Statistical Decision

The p value is less than α. Since $\alpha = 0.05$, the rejection region for this one-tail, lower-tail hypothesis test is $Z < -1.645$. Clearly the calculated Z value of -3.56 falls in the rejection region. Thus, we reject H_0 and conclude that the true mean arrival delay is less than 30 minutes.

The output from Minitab is shown here.

One-Sample Z: Arrival Delay

Test of mu = 30 vs mu < 30
The assumed sigma = 50

Variable	N	Mean	StDev	SE Mean
Arrival Dela	100	12.20	29.22	5.00

Variable	95.0% Upper Bound	Z	P
Arrival Dela	20.42	−3.56	0.000

Notice that the p value is labeled P and is displayed to only three decimal places.

Let's look at another example.

EXAMPLE 8.31

CRIME RATE

Understand the problem.

Conducting a Lower-Tail Hypothesis Test

Politicians love to tell their constituency that they are making things better for the people. A recent headline in a Massachusetts newspaper claims "State Crime Rate Drops Again." The numbers of murders, rapes, robberies, aggravated assaults, burglaries, cases of larceny, and motor vehicle thefts were reported for a sample of towns in Massachusetts. A portion of the data are shown here:

Collect and analyze the data.

Draw conclusions and make recommendations.

Town	Murder 1999	Murder 2000	Rape 1999	Rape 2000	Robbery 1999	Robbery 2000	Aggravated assault 1999	Aggravated assault 2000
Agawam	2	0	3	11	10	4	247	189
Amherst	0	0	6	6	6	1	22	47
Belchertown	0	1	0	2	0	0	0	34
Bernardston	0	0	0	0	0	0	3	2
Buckland	0	0	0	0	0	0	1	0
Charlemont	0	0	0	1	0	0	0	0
Cheshire	0	0	0	0	0	0	2	0

Town	Burglary 1999	Burglary 2000	Larceny 1999	Larceny 2000	Motor vehicle theft 1999	Motor vehicle theft 2000
Agawam	154	137	325	305	76	71
Amherst	75	129	225	369	43	65
Belchertown	30	21	77	61	6	10
Bernardston	21	16	24	25	1	1
Buckland	9	6	9	8	4	2
Charlemont	4	3	10	19	0	0
Cheshire	2	4	9	2	0	2

There are 41 towns in this sample. We want to determine whether there is enough evidence to conclude that the average number of robberies has decreased from 18.3 in 1999. Use $\alpha = 0.05$.

Step 1: *Set up the null and alternative hypotheses.* Since we are trying to show that the average number of robberies has decreased, that must be the alternative hypothesis. So, we have these null and alternative hypotheses:

$$H_0: \quad \mu \geq 18.3 \text{ robberies}$$
$$H_A: \quad \mu < 18.3 \text{ robberies}$$

Step 2: *Define the test procedure.* This is a large-sample test of the mean since $n > 30$ and σ is unknown. The appropriate test statistic is $Z = \dfrac{\overline{X} - \mu}{s/\sqrt{n}}$. We set $\alpha = 0.05$.

Step 3: *Collect the data, and calculate the test statistic and the p value.* The values of $\overline{X}$ and s must be calculated from the data. They are $\overline{X} = 18.3$ and $s = 78.5$.

$$Z = \frac{\overline{X} - \mu}{s/\sqrt{n}} = \frac{16.7 - 18.3}{78.5/\sqrt{41}} = -0.13$$

$$p \text{ value} = P[Z < -0.13] = 0.4483$$

The output from KaddStat confirms these calculations.

One Sample Test for μ

p-value =	0.4484
Null Hypothesis: μ =	18.3
Alternative Hypothesis:	Less than
Sample Size: n =	41
Sample Mean: X =	16.7000
Sample Std. Dev.: S =	78.5000
Standard Error: S_X =	12.2596

Step 4: *Make the statistical decision.* We look up 0.05 in the body of the standard normal table and find $Z_\alpha = -1.645$. Clearly, $Z = -0.13$ is not in the rejection region and the p value is greater than α. We fail to reject H_0.

Step 5: *Interpret the statistical decision in terms of the stated problem.* The data indicate that the average number of robberies has not decreased.

✔ TRY IT NOW!

Supermarket Survey
Conducting a Lower-Tail Test of the Mean

Jake Bramhall can identify the make, model, and number of cylinders of any passing car, but he can't tell the difference between stewed tomatoes and tomato paste. Although more men are pushing shopping carts these days, many like Mr. Bramhall show little aptitude in the supermarket and display markedly different purchasing behavior than women. A study done by Consumer Network Inc. found that the average amount of money spent by 100 single men on facial tissues was $7.38. On the basis of these data can you conclude that men spend less money on facial tissues than the average $8.19 spent by women? Use a population standard deviation of $3.50 and $\alpha = 0.05$.

Step 1: Set up the null and alternative hypotheses.

Step 2: Define the test procedure.

Step 3: Collect the data, and calculate the test statistic and the p value.

Step 4: Make the statistical decision.

Step 5: Interpret the statistical decision in terms of the stated problem.

Are the results different if you use $\alpha = 0.01$?

8.9.2 UPPER-TAIL TESTS OF THE MEAN

So far all of the one-tail tests we have performed have been lower-tail tests; that is, the $<$ symbol is used in the alternative hypothesis. Now we consider how the procedure will differ when we need to do an upper-tail test—one with the $>$ symbol in the alternative hypothesis. Let's look at an example.

EXAMPLE 8.32

SUPERMARKET SURVEY

Understand the problem.

Step 1: Set Up the Null and Alternative Hypotheses

In the study of supermarket behavior, the average amount spent on frozen dinner/entrees by men was found to be \$41.48. There were 100 men in the study, and the population standard deviation can be assumed to be \$10. Is there any evidence to indicate that men spend more than \$40.71, the national average for women?

The steps of the hypothesis-testing procedure are the same. Since we are interested in showing that the men spend *more* on frozen dinner/entrees, we use a one-tail test with the $>$ symbol in the alternative hypothesis. That gives the following null and alternative hypotheses:

$$H_0: \quad \mu \leq \$40.71$$
$$H_A: \quad \mu > \$40.71$$

The second step is to define the test procedure. This includes selecting the appropriate test statistic and setting the value for α, the maximum probability of a Type I error you wish to tolerate. Since we are still doing a large-sample test of the mean, the appropriate test statistic is Z:

$$Z = \frac{\overline{X} - \mu}{\sigma/\sqrt{n}} \qquad \text{when } \sigma \text{ is known}$$

and

$$Z \approx \frac{\overline{X} - \mu}{s/\sqrt{n}} \qquad \text{when } \sigma \text{ is unknown and } n > 30$$

Let's do the second step for the supermarket survey example.

EXAMPLE 8.33

SUPERMARKET SURVEY

Understand the problem.

Step 2: Define the Test Procedure

This is a large-sample test of the mean because we know the population standard deviation. So the appropriate test statistic is

$$Z = \frac{\overline{X} - \mu}{\sigma/\sqrt{n}}$$

Set α to be 0.05.

The third step in the hypothesis-testing procedure is to calculate the test statistic and the p value. The calculation of the Z test statistic is done the same way it was done for the lower-tail test of the mean. However, there needs to be a slight adjustment to the procedure for finding the p value in the case of an upper-tail test of the mean. We are looking for the chance that we incorrectly reject H_0 when it is true. If you look back at the logic that led us to the rejection region for the lower-tail test, you will see that in this case, values of $\overline{X}$ that are much greater than 40.71 will lead us to reject H_0. These values of $\overline{X}$ will yield large positive Z statistics. So we will reject H_0 only when Z is "too big." The words "more extreme" refer to values of Z greater than the observed test statistic in the case of an upper-tail test of the mean. The p value is the area shaded in Figure 8.9.

Steps for finding the p value for an upper-tail test of the mean

The steps are summarized here. To find the p value for a large-sample upper-tail test of the mean:

Remember: *To find a greater than Z probability you need to find the less than probability in the Z table and subtract it from 1.*

1. Calculate the test statistic, Z.

2. Find the area to the right of Z. That is, find P(Z > test statistic).

Let's see how this test statistic and p value are calculated for the supermarket survey example.

Collect and analyze the data.

EXAMPLE 8.34

SUPERMARKET SURVEY

Steps 3: Collect the Data, Calculate the Test Statistic and the *p* Value

The data have been collected: $n = 100$, $\overline{X} = \$41.48$, and $\sigma = \$10$. For Step 3 we calculate the Z test statistic and the p value:

$$Z = \frac{\overline{X} - \mu}{\sigma/\sqrt{n}} = \frac{41.48 - 40.71}{10/\sqrt{100}} = 0.77$$

$$p \text{ value } = P[Z > 0.77] = 1 - P[Z < 0.77] = 1 - 0.7794 = 0.2206$$

The results from KaddStat confirm these calculations.

One Sample Test for μ

p-value =	0.2206

Null Hypothesis: μ =	40.71
Alternative Hypothesis:	Greater than
Sample Size: n =	100
Sample Mean: X =	41.4800
Sample Std. Dev.: S =	10.0000
Standard Error: S_X =	1.0000

The fourth step of the hypothesis-testing procedure is to decide whether to reject the null hypothesis. All of your data are evidence and are now summarized in the

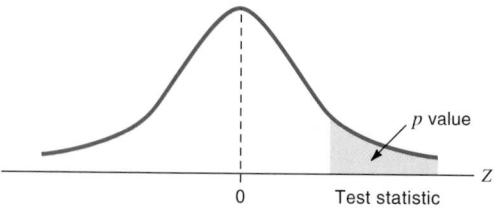

FIGURE 8.9 *p* **value for an upper-tail test**

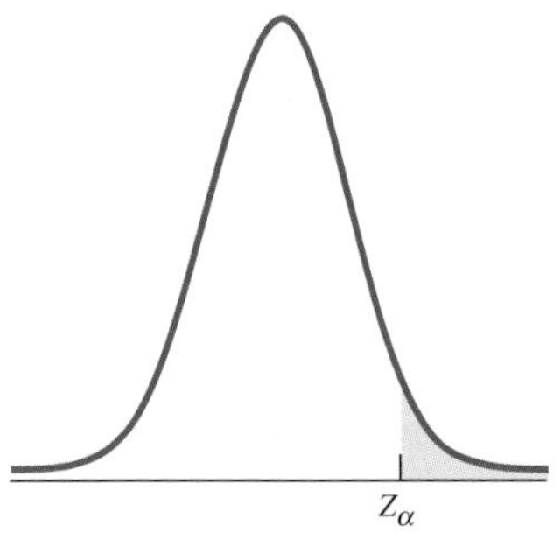

FIGURE 8.10 Rejection region for upper-tail test of μ – large sample

test statistic. There are two ways to decide—using the p value method or using the rejection region method.

> *Using the p value method,* we will reject the null hypothesis if the p value is less than the value of α.

> *Using the rejection region approach,* we will reject the null hypothesis if the test statistic falls in the rejection region. If the test statistic does not fall in the rejection region, then we fail to reject the null hypothesis.

In Figure 8.10, the shaded region has an area of α. We will reject H_0 only when Z is greater than Z_α, thus falling into the rejection region.

Consider finding the value of Z_α when $\alpha = 0.05$. Remember that the normal distribution table gives only lower-tail areas as the probabilities. But, because the distribution is symmetric, we can look up 0.05 and simply drop the negative sign to get Z_α. The value for Z_α is found by dropping the negative sign in front of either -1.64 or -1.65. Hence, we use either 1.64 or 1.65 or their midpoint, 1.645. The rejection region is $Z > 1.645$.

This example illustrates the steps for finding the rejection region for an upper-tail test:

Step 1: Look up the value of α in the body of the Z table (or find the value closest to it).

Step 2: Read off the corresponding value and drop the negative sign. Call it Z_α.

Step 3: The rejection region consists of all values of Z that are greater than Z_α.

We are now ready to complete this example.

Steps for finding the rejection region for an upper-tail test

---

EXAMPLE 8.35

SUPERMARKET SURVEY

Step 4: Make the Statistical Decision
Step 5: Interpret the Statistical Decision

Using the p value method, we see that the p value of 0.2206 is not less than $\alpha = 0.05$.

Using the steps listed above, we find the rejection region is $Z > 1.645$. The Z value of 0.77 does not fall in the rejection region.

Therefore, using both methods, we fail to reject the null hypothesis and conclude that men do not spend more than \$40.71 on average for frozen dinners/entrees.

Draw conclusions and make recommendations.

TRY IT NOW!

Supermarket Survey
Conducting an Upper-Tail Test of the Mean

From the same supermarket survey it is found that the 100 men spent, on the average, $19.98 on low-calorie soft drinks. Is there enough evidence to conclude that men spend more than women, who, on the average, spend $18.86? Assume that the population standard deviation is $10 and use $\alpha = 0.05$.

8.9.3 EXERCISES—LEARNING IT!

8.28 Nike, a major manufacturer of running shoes, thinks it has found a way to make shoes that last longer than the current average of 750 miles. Significant signs of wear and abnormal aches indicate that it's time to purchase a new pair of shoes. Nike wants to see whether its new design yields an average shoe lifetime longer than 750 miles. A sample of 30 runners are selected to wear the new design and the shoes last an average of 815 miles before failing. The failure time is normally distributed with a standard deviation of 25 miles.

(a) Set up the null and alternative hypotheses to test whether the average time to failure is greater than 750 miles.

(b) Complete the remaining hypothesis-testing steps using $\alpha = 0.05$.

(c) Find the p value.

(d) Based on the p value, what can you conclude about the average time to failure for the new running shoes?

8.29 Recent medical research indicates that people who have skin cancer and receive a new medication to treat it live longer than those who do not get the medication. The average length of life prior to the development of this medication was 18 months. The medical community wishes to test the claim made by the developers of this drug. A sample of 35 patients who received the medication lived an average of 21 months. The standard deviation is 5 months.

(a) Set up the null and alternative hypotheses to test whether the average length of life has increased from 18 months.

(b) Complete the remaining hypothesis-testing steps using $\alpha = 0.10$.

(c) Find the p value.

(d) Based on the p value, what can you conclude about the average length of life for patients who receive the medication?

8.30 An automobile company thinks that with new designs, its cars will last longer before having a problem. For this reason, the company wishes to extend the warranty that comes with the vehicle in hopes of attracting more customers. Before making this change, the company needs to test its idea. Prior to the design changes, the cars lasted an average of 43 months before having a major problem. A sample consisting of 50 cars was tested. The cars lasted an average of 44 months before having a major problem. The standard deviation is 2 months.

(a) Set up the null and alternative hypotheses to test whether the average time before a car has a major problem is longer than 43 months.

(b) Complete the remaining hypothesis-testing steps using $\alpha = 0.05$.

(c) Find the p value.

(d) Based on the p value, what can you conclude about the average time before a car has a major problem?

8.31 A manufacturer of top-of-the-line tennis rackets claims that its Smack Em racket will change a player's game. A tennis pro currently serves the ball at an average speed of 115 mph with a standard deviation of 2.5 mph. The speeds are normally distributed. The tennis pro decides to test the company's claim and records the speeds of his serve for 15 balls using the Smack Em racket. The speeds are listed here:

| 117.3 | 112.9 | 115.4 | 113.8 | 114.2 | 115.9 | 120.8 | 116.9 |
| 114.4 | 116.0 | 116.2 | 115.1 | 115.2 | 113.0 | 115.0 | |

(a) Set up the null and alternative hypotheses to test whether the average service speed has increased with the new racket.

(b) Complete the remaining hypothesis-testing steps using $\alpha = 0.01$.

(c) Find the p value.

(d) Based on the p value, should the tennis pro invest in the new racket?

8.32 In an attempt to improve quality, many manufacturers are forming partnerships with their suppliers. A local fast-food burger outfit has partnered with its supplier of potatoes. The burger outfit buys potatoes in bags that weigh 20 pounds. It does not want to accept underweight bags of potatoes. A sample of 40 bags shows an average weight of 19.95 pounds with a standard deviation of 0.1 pound.

(a) Set up the null and alternative hypotheses to test whether the average bag weighs at least 20 pounds.

(b) Complete the remaining hypothesis-testing steps using $\alpha = 0.05$.

(c) Find the p value.

(d) Based on the p value, should the burger outfit accept the shipment of potatoes?

GET IT IN WRITING

The Airline Company

TO: **Travel Manager**
FROM: **Erica Q. Analyst**
RE: **Evaluating American Airlines**

As you know, our company has been evaluating airlines in order to choose one to be the main carrier for our business travel. In this effort I have downloaded all data on American Airlines flights into LaGuardia Airport for the month of December 2000. I am in the process of analyzing these data.

Our company standard is that the airline arrival delays should not be, on average, more than 30 minutes. My team and I have decided that the tool of hypothesis testing should be used to determine whether American Airlines can meet this requirement. If so, then we will recommend that the company use American Airlines as its primary carrier because of the large number of flights it has into LaGuardia.

We have decided to use a series of hypothesis tests to investigate this problem. For each variable, we will use the December data as sample data to test the population mean, μ, against the requirement. We know that we can also use the sample data to test the population variance, σ^2 (recall that the variance is the standard deviation squared), but we must read a little bit more about the details of how to do this.

For the tests of the population mean, we are proposing to use lower-tail tests to see whether the actual population mean arrival delay is less than 30 minutes. We ran a two-tail test of the mean using the first 200 flights into LaGuardia in December 2000 and found that the average delay differed from 30 minutes. However, we are more interested in knowing if the population mean is less than 30 minutes rather than simply different from 30 minutes.

Since our sample data contain information about the origin airport as well as the date and time of day the flight was scheduled to arrive, we will do some analysis of arrival delays by origin airport and day of the week. Perhaps American Airlines will be our airline of choice for weekend travel or for travel from a particular part of the country into New York. Another thing we plan to do is examine a different month of data. December may be the worst-case scenario with weather problems as well as holiday travel.

We will conduct these tests and report the *p* value of each of the tests to you. This will give you complete information about the tests and allow you to decide what level of significance, α, is appropriate. Keep in mind that by using a very small value for α, you protect the company against choosing American Airlines if it cannot meet our requirements, but you increase the chance that we miss out on using American Airlines when in fact it can meet our requirement. The costs of each of these errors should be evaluated before deciding on a value for α.

My team will be completing the hypothesis tests of the mean of each variable in the next few days. Please contact me if you have any questions about how we are proceeding.

8.10   Using Technology in Hypothesis Testing

All of the Excel, Minitab, and TI-83 instructions for hypothesis testing are presented at the end of Chapter 9.

CHAPTER 8 SUMMARY

In this chapter you have learned the key steps involved in doing any hypothesis test. You first formulate two opposing viewpoints called the null and alternative hypotheses. These hypotheses are typically theories or ideas about the value of one or more population parameters. The technique of hypothesis testing helps you decide between these opposing hypotheses using the sample data as the evidence upon which to base your decision. You can make two possible errors in any hypothesis test. These are called Type I and Type II errors, and the probabilities of making these errors are labeled α and β, respectively. It is desirable to make both of these probabilities small, but there is a tradeoff.

You also learned the procedure for doing large-sample tests of the mean. This procedure applies whenever you know the population standard deviation or you have a sufficiently large sample size, $n > 30$. Thus, the tests are called large-sample tests.

KEY TERMS		
Term	**Definition**	**Page reference**
Alpha (α)	The probability of making a Type I error is labeled α. It is the area of the rejection region.	384
Alternative hypothesis	The **alternative hypothesis** is a statement about a parameter of the population(s) that is opposite to the null hypothesis. It is labeled H_A.	363
Beta (β)	The probability of making a Type II error is labeled β.	384
Hypothesis	A **hypothesis** is an idea, an assumption, or a theory about the characteristics of one or more variables in one or more populations.	358

Term	Definition	Page reference
Hypothesis test	A **hypothesis test** is a statistical procedure that involves formulating a hypothesis and using sample data to decide on the validity of the hypothesis.	358
Large-sample test of the mean	A **large-sample test of the mean** is conducted when the characteristic of interest is the population mean, μ, and either of these situations exists: (a) the population standard deviation is known (regardless of the sample size), OR (b) the population standard deviation is unknown but the sample size, n, is greater than 30.	369
Lower-tail hypothesis test	A **lower-tail test** of the population mean has these null and alternative hypotheses: H_0: $\mu \geq$ [a specific number] H_A: $\mu <$ [a specific number]	389
Null hypothesis	The **null hypothesis** is a statement about a parameter of the population(s). It is labeled H_0.	363
p value	The **p value** is the actual probability of rejecting a true null hypothesis based on the evidence in the test statistic. It is also the smallest value of α for which we can reject H_0. It is the probability of obtaining the observed value of the test statistic or a value more extreme if the null hypothesis is true.	366
Rejection region	The **rejection region** is the range of values of the test statistic that will lead us to reject the null hypothesis. It is defined by the **critical value(s).** The area of the rejection region is α, the significance level.	367
Significance level	The **significance level**, α, is the maximum probability tolerated for rejecting a true null hypothesis.	366
Small-sample test of the mean	A **small-sample test of the mean** is conducted when the characteristic of interest is the population mean, μ, and the population standard deviation is unknown but the sample size, n, is less than or equal to 30.	369
Test statistic	A **test statistic** is a number that captures the information in the sample data. It is used to decide between the null and alternative hypotheses.	366
Two-tail hypothesis test	A **two-tail test** of the population mean has these null and alternative hypotheses: H_0: $\mu =$ [a specific number] H_A: $\mu \neq$ [a specific number]	370
Type I error	A **Type I error** is made when we reject the null hypothesis and the null hypothesis is actually true. In other words, we incorrectly reject a true null hypothesis.	383
Type II error	A **Type II error** is made when we fail to reject a false null hypothesis. In other words, we fail to reject H_0 when we should have rejected it.	383
Upper-tail hypothesis test	An **upper-tail test** of the population mean has these null and alternative hypotheses: H_0: $\mu \leq$ [a specific number] H_A: $\mu >$ [a specific number]	390

KEY FORMULAS

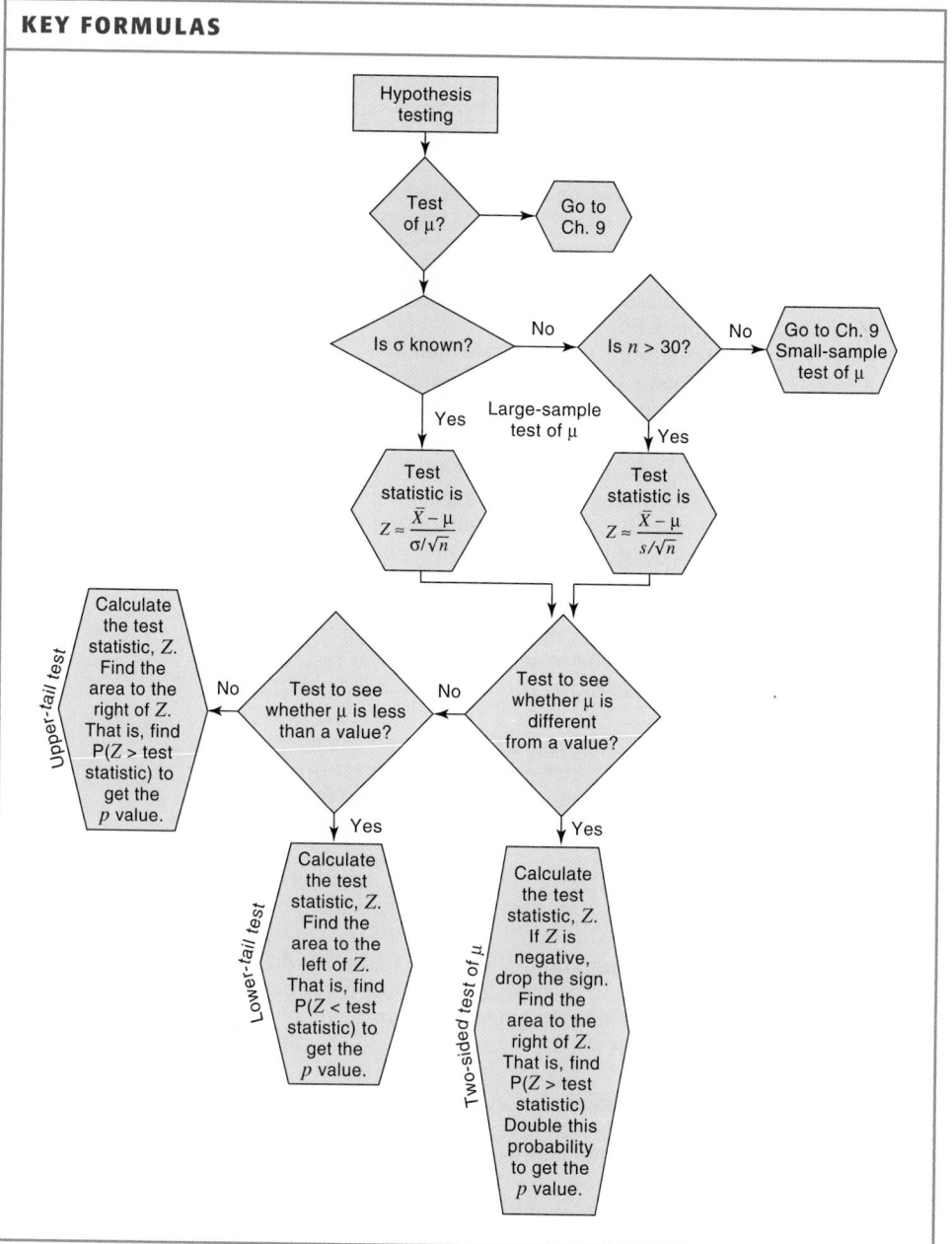

CHAPTER 8 EXERCISES

Learning It!

8.33 The manufacturer of an over-the-counter pain reliever claims that its product brings pain relief to headache sufferers in an average of 3.5 minutes. To be able to make this claim in its television advertisements, the manufacturer is required by a particular television network to present statistical evidence in support of the claim.

(a) Is this a one-tail test or a two-tail test?

(b) Set up the null and alternative hypotheses.

(c) What is a Type I error?

(d) What is a Type II error?

(e) A sample of 40 headache sufferers is used. They took an average of 3.3 minutes to get some pain relief. If the standard deviation is 0.5 minute, perform the hypothesis test using $\alpha = 0.02$.

8.34 Pharmaceutical companies spend millions of dollars annually on research and development of new drugs. After a new drug is formulated, the pharmaceutical company must subject it to lengthy and involved testing before receiving the necessary permission from the Food and Drug Administration (FDA) to market the drug. The pharmaceutical company must provide substantial evidence that a new drug is safe before receiving FDA approval, so that the FDA can confidently certify the safety of the drug.

(a) Set up the null and alternative hypotheses.

(b) What is a Type I error?

(c) What is a Type II error?

8.35 Does raising or lowering the speed limit change driver behavior? This is an age-old question. Suppose you wish to investigate the effect on the actual speed traveled when you lower the speed limit by 5 mph.

(a) Is this a one-tail or a two-tail test?

(b) Set up the null and alternative hypotheses.

(c) What is a Type I error?

(d) What is a Type II error?

(e) A sample of 300 drivers were observed after the speed limit was lowered from 65 to 60 mph. The average speed was found to be 66 mph with a standard deviation of 2 mph. Before the speed limit was lowered, the mean was 68 mph. Test to see whether lowering the speed limit reduced the average speed from 68 mph.

(f) Find the p value. What conclusion can you draw about the effect of lowering the speed limit by 5 mph.

8.36 The LEGO Group, an international company, makes the LEGO blocks that many of us have played with at some time. Many of its products require that the production process perform according to specifications. One of the products is Little People. The neck diameter of each of the Little People must be 0.5 inch so that it can be attached to the head properly. LEGO is interested in testing to see whether this process is performing according to specifications.

(a) Is this a one-tail test or a two-tail test?

(b) Set up the null and alternative hypotheses.

(c) What is a Type I error?

(d) What is a Type II error?

(e) A sample of 30 blocks is tested. The average diameter of the sampled necks is 0.48 inch. If the standard deviation is 0.05 inch, perform the hypothesis test using $\alpha = 0.02$.

(f) Find the p value and make a recommendation to LEGO.

8.37 How long do heart transplant patients have to wait for a heart to become available? The national average wait is 55 days, and the hospital in your city has an average wait of 65 days based on a sample of 30 patients with a standard deviation of 7 days. Is the wait at this hospital atypical?

(a) Is this a one-tail test or a two-tail test?

(b) Set up the null and alternative hypotheses.

(c) What is a Type I error?

(d) What is a Type II error?

(e) Perform the hypothesis test using $\alpha = 0.05$.

(f) Find the p value and decide whether you would like to investigate another hospital for your ailing relative.

8.38 An up-and-coming restaurant chain is trying to decide whether to locate in the town of Longmortgage. It will locate there only if the average number of days of the week that people eat out is three or more.

(a) Is this a one-tail test or a two-tail test?

(b) Set up the null and alternative hypotheses.

(c) What is a Type I error?

(d) What is a Type II error?

Thinking About It!

Requires Exercise 8.2

8.39 Is it possible to switch the null and the alternative hypotheses for the ETS problem? Explain why or why not.

Requires Exercise 8.28

8.40 For Nike which of the possible errors is more costly? Consider the consequences of making a Type I error and the consequences of making a Type II error.

Requires Exercise 8.29

8.41 Consider the case of medical research for skin cancer treatment.

(a) What position are you taking if you make the null hypothesis $H_0: \mu \geq 18$?

(b) What factors other than the medication might influence the length of time a person with skin cancer lives?

Requires Exercise 8.30

8.42 What are the implications of the car company setting the warranty too long?

Requires Exercise 8.6

8.43 Consider the government's data on tourism.

(a) From the government's perspective, which error is more costly?

(b) From a hotel manager's perspective, which error is more costly?

Requires Exercise 8.33

8.44 For the television network, what are the implications and consequences of a Type I error? A Type II error?

Requires Exercise 8.34

8.45 Rewrite the hypothesis test for the FDA exercise by switching the null and alternative hypotheses.

(a) What is a Type I error for this new setup?

(b) What is a Type II error for this new setup?

(c) Compare your answers with those you found in Exercise 8.34. What has happened to the errors?

(d) Which setup do you think is better? Why?

Requires Exercise 8.35

8.46 What else would you like to know about the drivers who were observed after the speed limit was lowered?

Requires Exercise 8.36

8.47 What are the implications for LEGO if the manufacturing process for the Little People is not producing necks with the specified diameter?

Requires Exercise 8.37

8.48 As the waiting recipient of a heart transplant, what other data would you like to know about the hospital in your city?

Requires Exercise 8.38

8.49 From the perspective of the restaurant chain, what are the consequences of a Type I error? A Type II error? Which error is more costly?

Doing It!

Datafile:
NYarrivaldelays.xxx

8.50 Let us return to the evaluation of the airlines presented at the beginning of this chapter. Complete the analysis of the airlines for your boss by analyzing the full data set of all 1421 American Airlines flights into LaGuardia in December 2000. A portion of the data set is shown below:

Carrier code	Date (mm/dd/yyyy)	Flight no.	Origin airport	Scheduled	Actual arrival time (HH:MM)	Arrival delay (min.)
AA	12/01/2000	158	DFW	22:03	22:03	0
AA	12/01/2000	338	ORD	0:54	0:56	2
AA	12/01/2000	340	ORD	23:31	23:45	14
AA	12/01/2000	342	ORD	22:54	23:04	10
AA	12/01/2000	362	ORD	18:06	19:24	78
AA	12/01/2000	370	ORD	16:00	17:07	67
AA	12/01/2000	378	ORD	13:55	14:31	36
AA	12/01/2000	386	ORD	12:01	12:18	17

The variables are as follows:

Carrier Code	Since all the flights in this data file are American Airlines, this will always be designated AA.
Date	The date in December 2000 when the flight arrived.
Flight No.	The flight number
Origin Airport	You will need to decode these using information on the Web. It indicates the airport that the flight originated at
Scheduled	The time (24-hour clock) when the flight was due to land
Actual Arrival Time	The time (24-hour clock) when the flight arrived
Arrival Delay	The number of minutes late (or early) that the flight arrived

(a) Display the data on the variable *Arrival Delay* as a histogram. Describe the histogram in terms of its center, spread, and shape.

(b) Using all flights, test the hypothesis that the *Arrival Delay* is less than 30 minutes. Find the *p* value and draw a conclusion about the average arrival delay of flights into LaGuardia.

(c) Test the hypothesis that *Arrival Delay* is less than 30 minutes on Mondays.

(d) Test the hypothesis that *Arrival Delay* is less than 30 minutes on Fridays.

(e) Sort the data by origination airport. Select one origination airport and test the hypothesis that *Arrival Delay* is less than 30 minutes for flights originating at that airport.

(f) Repeat part (e) for a different origination airport.

(g) Test the hypothesis that *Arrival Delay* is less than 30 minutes for flights scheduled to arrive by noon.

(h) Test the hypothesis that *Arrival Delay* is less than 30 minutes for flights scheduled to arrive after noon.

(i) Write a summary memo to explain your findings and make a recommendation regarding using American Airlines as your carrier.

8.51 After analyzing the American Airlines data for December 2000, you decide to consider some other months. Go to the website *www.bts.gov* and download arrival delay data for a month other than December 2000. Select a summer month. Repeat the analysis you did in Exercise 8.50 for this data set.

8.52 Return to the website *www.bts.gov* and download arrival delay data for December 2000 and for the month you selected in Exercise 8.51, but now choose a different airline. Analyze these data sets the same way. Summarize all of your findings by month and by airline.

Inferences: More One-Population Tests

WASH YOUR HANDS!

It turns out Mom was right: Wash your hands, and you'll avoid illness. A study of elementary school children found that students who used an instant hand sanitizer in the classroom had 20% fewer absentee days due to illness than those who did not. More than 6000 students in four states participated in the 10-month case study. And now school nurses across the country are participating in a program to promote hand washing at school and at home.

"Hand washing is the single most effective hygiene habit that will reduce the spread of illness and shorten illness in an individual," says Jane Tustin, RN, President of the National Association of School Nurses (NASN) in Castle Rock, Colorado, and a coordinator of health services for the school district of Lubbock, Texas.

Source: http://cbshealthwatch.medscape.com/cx/viewarticle/225343

A school nurse decides to see for herself if the average number of absentee days is lower for children who wash their hands regularly. The average number of absentee days is 7.5 a year. Twenty students are randomly selected and taught the proper way to wash hands and their absences are tracked for the school year. A portion of the data set is shown here:

Days missed	Grade	Gender
7	1	M
6	5	F
7	5	M
6	5	F
5	2	F
6	4	M
8	4	F
7	1	F

> Generally accepted rules for hand washing:
> 15 seconds of vigorous rubbing that generates friction. If hands are visibly soiled, rub longer.
> Lather every surface well, especially around the nails. Use the correct type of soap and rinse according to the situation.
> Rinse in a flowing stream of water.
> In the absence of water, use alternative agents like detergent-containing towelettes (for removal of light soil) and alcohol-based handrubs (for reduction of microbial flora).
> Do not use handrubs if hands are soiled.
> Dry hands with paper towels or hand dryers.
>
> *http://cbshealthwatch.medscape.com/cx/viewarticle/225343_2.*

9.1   Chapter Objectives

The school nurse does not know the standard deviation of the number of absentee days and the sample size of 20 is small ($n \leq 30$). Therefore, we cannot use the hypothesis-testing procedure introduced in Chapter 8. Now it is time to examine what to do if the population standard deviation is unknown and the sample size is small. In addition to handling small-sample tests of the mean, we also describe how to perform a hypothesis test for other population parameters.

If we are analyzing a *quantitative variable*, such as the time a child washes his or her hands, then we know that the variable can be described by its distribution. The distribution, in turn, can be described by parameters. Thus, if we are constructing a theory or hypothesis about a *quantitative variable*, it might be a statement about one of these values:

- The mean value, μ, of the variable in one population
- The amount of variability, σ^2, of the variable in one population

If we are analyzing *nominal data*, such as whether an employee is satisfied with the job, the hypothesis might be a statement about:

- The value of the proportion, p, of population members that have a certain characteristic (one of the categories of the nominal variable)

In this chapter we look at the numerical details of the test statistics and rejection regions for hypothesis tests about a single population parameter. We will study only one population. For example, we look at the average age at marriage or the proportion of people who favor the death penalty. In Chapter 10 we extend these results to compare the behaviors of two populations. This might entail comparing the average ages at marriage in different cultures or comparing the views of men and women on the issue of the death penalty.

This chapter covers the following topics:

- Hypothesis test of the population mean, μ—small sample
- Hypothesis test of the population variance, σ^2
- Hypothesis test of the population proportion, p
- The relationship between hypothesis testing and confidence intervals

Remember! Both μ and p describe the behavior of the population, not the sample. They are typically unknown.

9.2 Hypothesis Test of the Mean: Small Sample

Remember! To find the sample variance use

$$s^2 = \frac{n\Sigma x^2 - (\Sigma x)^2}{n(n-1)}$$

In Chapter 8 you learned how to perform a large-sample hypothesis test of the mean. Some of the examples had information about the population standard deviation from either product specifications or some previous study. However, more often than not, the population standard deviation is not known. Then the sample standard deviation, *s*, is used to calculate an estimate of the unknown population standard deviation, σ. If the sample is sufficiently large, $n > 30$, then we can use the Z test statistic from Chapter 8 to do a hypothesis test on the mean. However, very often the sample is small.

Now let us think about the implications of having a small sample on the hypothesis test. The steps in any hypothesis test are repeated here. Notice that Step 3 includes the calculation of the *p* value. The *p* value provides an alternative way to make the decision at Step 4 and gives decision makers more information. However, for small-sample tests of μ and tests of σ^2, we cannot calculate the *p* values by hand with the tables in Appendix A. We need probability tables similar to the normal distribution table. We will follow standard procedure and get our *p* value from software output.

Use the p value from the output of Excel or Minitab.

Steps for any hypothesis test

Step 1: Set up the null and alternative hypotheses.

Step 2: Define the test procedure.

Step 3: Collect the data and calculate the test statistic and the *p* value.

Step 4: Decide whether to reject the null hypothesis.

Step 5: Interpret the statistical decision in terms of the stated problem.

Step 1 is not different because we are just setting up the null and alternative hypotheses. If you guessed that Steps 2 and 3 are affected by the small sample size, then you are right. In Chapter 8 we calculated a Z statistic as the test statistic in Step 3. But the formula for Z uses σ, which is now unknown, and the estimate is based on a small sample. Logic tells us simply to use *s* in the formula instead of σ. However, then the test statistic no longer follows a normal distribution. Instead, it has a *t distribution*.

The Central Limit Theorem does not apply because $n \le 30$.

Remember! A test statistic is a single number based on the sample that allows us to decide between the null and the alternative hypotheses.

t test statistic

We used the *t* distribution in Chapter 7 when we found a confidence interval for the mean of a normally distributed population with unknown σ and a small sample size. We will use our knowledge of the *t* distribution to do a hypothesis test of μ for precisely this situation.

The test statistic is calculated as

$$t = \frac{\overline{X} - \mu}{s/\sqrt{n}}$$

Remember! The population must have a normal distribution to use the t statistic.

Notice that the calculation for the *t* statistic is the same as the formula for Z with σ replaced with *s*.

9.2.1 TWO-TAIL TEST OF THE MEAN: SMALL SAMPLE

Let's first look at a two-tail test of the mean when σ is unknown.

Understand the problem.

EXAMPLE 9.1

WASH YOUR HANDS

Conducting a Two-Tail Test of μ — Small Sample

The nurse wants to test whether the population average number of school days missed is different from 7.5 days. Use $\alpha = 0.05$.

Step 1: *Set up the null and alternative hypotheses.* Because the nurse is interested in detecting a shift from the value of 7.5 days in either direction, this is a two-tail test:

$$H_0: \quad \mu = 7.5 \text{ days}$$

$$H_A: \quad \mu \neq 7.5 \text{ days}$$

Step 2: *Define the test procedure.* From our work in Chapter 8 we know that we will reject H_0 if the calculated t statistic is too large or too small. Here are the numbers of days absent for the 20 students who were observed:

> 7   6   7   6   5   6   8   7   7   6
> 5   8   7   6   7   5   8   5   6   10

We need to use the t table to find the critical values of t. Since this is a two-tail test, we must divide α in half. Thus, we want $0.05/2 = 0.025$ in the upper-tail area and 0.025 in the lower-tail area. Using the column labeled 0.025 and the row for $n - 1 = 20 - 1 = 19$ degrees of freedom, we find $t_{\alpha/2} = 2.093$. Because the t distribution is symmetric, the value for $-t$ is -2.093 and the rejection region is the shaded area in the figure:

Collect and analyze the data.

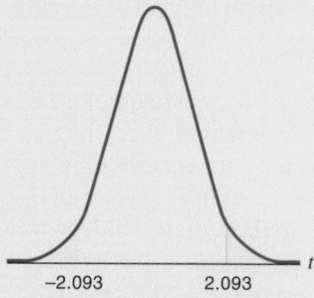

Do you remember what the critical Z value was when $\alpha = 0.05$? It was 1.96. So the critical t value is a bit larger than the corresponding Z value. This will always be the case, and as the degrees of freedom get larger, $t_{\alpha/2}$ gets closer and closer to 1.96. Look down the column for $\alpha = 0.025$ and see that, as the degrees of freedom increase (due to larger sample sizes), the values for t get closer and closer to 1.96.

The t value is always a bit larger than the corresponding Z value.

Step 3: *Collect the data and calculate the test statistic and the p value.* From the data we find $\overline{X} = 6.6$ days and $s = 1.27$ days, and we know $n = 20$. With this information we can calculate the t test statistic:

$$t = \frac{\overline{X} - \mu}{s/\sqrt{n}} = \frac{6.6 - 7.5}{1.27/\sqrt{20}} = -3.16$$

To find the p value we most likely need to use some software support such as Minitab or Excel. This is because the t table is not designed to be used to find probabilities but rather to find the cutoff values for the rejection region. We find

Use the tdist function in Excel to get this probability.

$$p \text{ value} = 2P(t < -3.16) = (2)(0.00258) = 0.00516$$

The output shown here is from Minitab. The t statistic is labeled T. In addition, the p value is calculated by Minitab and shown in the last column labeled P.

One-Sample T: Days Missed

Test of mu = 7.5 vs mu not = 7.5

Variable	N	Mean	StDev	SE Mean
Days Missed	20	6.600	1.273	0.285

Variable	95.0% CI		T	P
Days Missed	(6.004,	7.196)	−3.16	0.005

Step 4: *Decide whether to reject the null hypothesis.* Here we check to see whether the *t* statistic calculated in Step 3 falls in the rejection region. Since the *t* statistic, −3.16, falls in the rejection region, we reject H_0. We can reach the same conclusion by examining the *p* value from Minitab. Since the *p* value is considerably smaller than 0.05, we reject the null hypothesis.

Draw conclusions.

Step 5: *Interpret the statistical decision in terms of the stated problem.* We conclude that the average number of days of school missed is different from 7.5 days.

Let's look at another example.

Understand the problem.

EXAMPLE 9.2

CEREAL BOX WEIGHTS

Conducting a Two-Tail Test of μ When σ Is Unknown

Packaging the correct amount of product is critical to most companies. Underfilling packages clearly results in customer complaints and overfilling packages costs the company money. Suppose you are working for a cereal manufacturer and each box of cereal is supposed to contain 13 ounces. The weight of the cereal boxes is assumed to be normally distributed. Is there evidence that the true mean is not equal to 13 ounces? Use $\alpha = 0.05$.

Step 1: *Set up the null and alternative hypotheses.* Because the company is interested in detecting a difference from the value of 13 ounces in either direction, this is a two-tail test:

$$H_0: \quad \mu = 13 \text{ ounces}$$
$$H_A: \quad \mu \neq 13 \text{ ounces}$$

Collect and analyze the data.

Step 2: *Define the test procedure.* The company has taken a sample of 25 cereal boxes and recorded these weights (in ounces):

12.985	12.976	13.107	13.006	12.910
12.755	12.938	13.139	13.015	13.033
13.029	12.887	13.049	12.823	12.910
13.073	13.024	13.088	13.061	13.111
13.050	13.141	13.008	13.149	12.907

We use the *t* table to find the critical values of *t*. Because this is a two-tail test, we must divide α in half. Thus, we want 0.05/2 = 0.025 in the upper-tail area and 0.025 in the lower-tail area. Using the column labeled 0.025 and the row for $n - 1 = 25 - 1 = 24$ degrees of freedom, we find $t_{\alpha/2}$ to be 2.064. Because the *t* distribution is symmetric, the value for $-t_{\alpha/2}$ is −2.064 and the rejection region is $t < -2.064$ or $t > 2.064$.

Step 3: *Collect the data and calculate the test statistic and the p value.* From the data we calculate $\overline{X} = 13.007$ ounces and $s = 0.1004$ ounce, and we know $n = 25$. With this information we can calculate the *t* test statistic:

$$t = \frac{\overline{X} - \mu}{s/\sqrt{n}} = \frac{13.007 - 13}{0.1004/\sqrt{25}} = 0.35$$

We can use a software package such as Minitab to find the test statistic and the p value. Here is the output from Minitab:

One-Sample T: Cereal Wt

Test of mu = 13 vs mu not = 13

Variable	N	Mean	StDev	SE Mean
Cereal Wt	25	13.0070	0.1004	0.0201

Variable	95.0% CI	T	P
Cereal Wt	(12.9655, 13.0484)	0.35	0.732

Step 4: *Decide whether to reject the null hypothesis.* Here we check to see whether the t statistic calculated in Step 3 falls in the rejection region. Since the t statistic, 0.35, does not fall in the rejection region, we do not reject H_0. Using the p value to make this same decision, we see that the p value is greater than 0.05, leading us to fail to reject H_0.

Step 5: *Interpret the statistical decision in terms of the stated problem.* We conclude that the cereal boxes are being filled properly.

Draw conclusions.

Let's summarize the procedure for finding the rejection region for a two-tail test of μ when σ is unknown:

Step 1: Divide the value of α in half and use that column of the t table.

Step 2: Find the number of degrees of freedom by calculating $n - 1$ and use that row of the t table.

Step 3: Find the value of $t_{\alpha/2}$ at the intersection of the row and column.

Step 4: The value of $-t_{\alpha/2}$ is the same as $t_{\alpha/2}$ with a negative sign.

Step 5: The rejection region is $t < -t_{\alpha/2}$ or $t > t_{\alpha/2}$.

Steps for finding the rejection region for a two-tail test of μ when σ is unknown

✔ TRY IT NOW!

The Vending Machine
Conducting a Test of μ When σ Is Unknown

In Chapter 8 we conducted a hypothesis test to decide whether a vending machine is correctly dispensing 12 ounces of soda. The amount dispensed is assumed to be normally distributed. The machine is not working properly if the bottles are overfilled or underfilled. Is there any evidence to indicate that the machine is not filling the bottles properly? Use $\alpha = 0.05$.

Step 1: Set up the null and alternative hypotheses.

Step 2: Define the test procedure.

Step 3: Collect the data and calculate the test statistic and the p value. You observe the machine filling 30 bottles and record the following data (in ounces):

11.6	12.62	11.4	10.69	11.59	11.49
11.11	10.66	10.9	11.47	11.56	12.74
11.87	11.63	10.83	11.55	12.42	12.18
12.39	12.3	11.26	11.82	11.71	12.94
12.35	11.21	11.36	11.57	11.66	11.42

Step 4: Decide whether to reject the null hypothesis.

Step 5: Interpret the statistical decision in terms of the stated problem.

9.2.2 ONE-TAIL TESTS OF THE MEAN: SMALL SAMPLE

You have learned that the two-tail hypothesis-testing procedure for μ is affected in two major ways by the lack of knowledge about σ. The test statistic is a t test statistic instead of a Z test statistic, and the rejection region cutoff values are found from the t table rather than the Z table. The same can be said about one-tail tests of the mean when σ is unknown.

The only step in the procedure that we need to adjust for one-tail tests is finding the rejection region using the t table. The form of the rejection region is the same as when σ is known. The only difference is in finding the critical values. Remember that we constructed the rejection regions by following a series of logical arguments about what values of $\overline{X}$ will lead us to reject the null hypothesis. These arguments still apply.

For a one-tail test we want to reject H_0 if the calculated t statistic is too small. Thus, we have the rejection region shaded in Figure 9.1. For an upper-tail test we want to reject H_0 if the calculated t statistic is too large. The rejection region for this type of test is shaded in Figure 9.2. The cutoff values are found in a manner similar to that used for the two-tail test except that there is no reason to split the value of α in half. These steps are summarized here:

Steps for finding the rejection region for one-tail tests of μ— small sample

Step 1: Use the column of the t table that corresponds to the value of α you have selected.

Step 2: Find the number of degrees of freedom by calculating $n - 1$ and use that row of the t table.

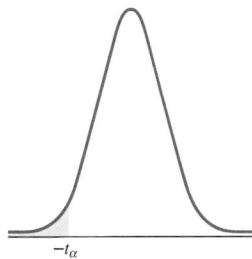

FIGURE 9.1 **Rejection region for a lower-tail test of μ**

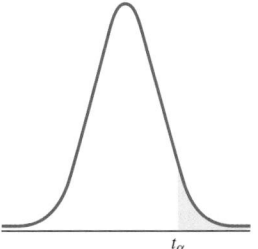

FIGURE 9.2 **Rejection region for an upper-tail test of μ**

Ans. $H_0: \mu = 12$, $H_A: \mu \neq 12$, $t = 2.045$, $t = -2.98$, $p = 0.006$, REJECT H_0

Step 3: For upper-tail tests the desired t_α value is found at the intersection of that row and column. For lower-tail tests, place a negative sign in front of the value to get the value of $-t_\alpha$.

Step 4: For upper-tail tests the rejection region is $t > t_\alpha$. For lower-tail tests the rejection region is $t < -t_\alpha$.

Let's look at some examples.

EXAMPLE 9.3

WASH YOUR HANDS

Conducting a One-Tail Test of μ – Small Sample

The school nurse is interested in testing whether the mean number of absentee days for children who did proper hand washing is less than 7.5 days. This is a one-tail test of μ.

Step 1: *Set up the null and alternative hypotheses.* The nurse is interested in seeing whether the average number of school days missed by those who use proper hand washing is *less than 7.5 days.* This is a lower-tail test with these null and alternative hypotheses:

$$H_0: \quad \mu \geq 7.5 \text{ days}$$
$$H_A: \quad \mu < 7.5 \text{ days}$$

Understand the problem.

Step 2: *Define the test procedure.* The value of α is set at 0.05. The specific value of t_α is found by looking in the t table and using the column labeled 0.05 and the row corresponding to 19 degrees of freedom ($20 - 1 = 19$). The value at the intersection of this column and row is 1.729, the value of t_α. Since this is a lower-tail test, the value of $-t_\alpha$ is needed, or -1.729.

Step 3: *Collect the data and calculate the test statistic and the p value.* The data were collected and are listed in Example 9.1. Because the value of σ is not known, we must calculate both $\overline{X}$ and s from the sample data. These values and the t test statistic were calculated in Example 9.1: $\overline{X} = 6.6$ days, $s = 1.27$ days, and

Analyze the data.

$$t = \frac{\overline{X} - \mu}{s/\sqrt{n}} = \frac{6.6 - 7.5}{1.27/\sqrt{20}} = -3.16$$

The p value is found by calculating the chance of observing $t = -3.16$ or smaller. Notice that since this is a one-tail test, we do not double the p value:

$$p \text{ value} = P(t < -3.16) = 0.003$$

Here is the output from Minitab:

One-Sample T: Days Missed

Test of mu = 7.5 vs mu < 7.5

Variable	N	Mean	StDev	SE Mean
Days Missed	20	6.600	1.273	0.285

Variable	95.0% Upper Bound	T	P
Days Missed	7.092	−3.16	0.003

Step 4: *Decide whether to reject the null hypothesis.* The calculated test statistic of $t = -3.16$ is in the rejection region and the p value is less than 0.05, so we reject H_0.

Draw conclusions.

Step 5: *Interpret the statistical decision in terms of the stated problem.* Based on these data, we conclude that the average number of days missed is less than 7.5 for the group of trained "hand washers."

EXAMPLE 9.4

NEW MARKETING PLAN

Conducting an Upper-Tail Test of μ When σ Is Unknown

A company is trying out a new marketing plan and wishes to evaluate its success. Prior to the new advertising scheme the average store sales per week was $4000. The new method is tried on a random sample of 15 stores. Can the company conclude that the new marketing plan works? The weekly sales are assumed to be normally distributed. Use $\alpha = 0.05$.

Understand the problem.

Step 1: *Set up the null and alternative hypotheses.* Since the company is interested in deciding whether average sales have *increased*, this is an upper-tail test with these null and alternative hypotheses:

$$H_0: \quad \mu \leq \$4000$$
$$H_A: \quad \mu > \$4000$$

Step 2: *Define the test procedure.* The value of α is set at 0.05. The specific value of t_α is found by looking in the t table. We use the column labeled 0.05 and the row corresponding to 14 degrees of freedom $(15 - 1 = 14)$. The value at the intersection of this column and row is 1.761, the value of t. The rejection region is $t > 1.761$.

Collect and analyze the data.

Step 3: *Collect the data and calculate the test statistic and the p value.* The following sales data (in dollars) were collected:

4128	4148	4028	4190	4088	4132	4157	4146
4054	4069	4163	4039	4174	4181	4099	

Since the value of σ is not known, we must calculate both $\overline{X}$ and s from the sample data. We get $\overline{X} = \$4119.70$, $s = \$53.40$, and

$$t = \frac{\overline{X} - \mu}{s/\sqrt{n}} = \frac{4119.7 - 4000}{53.4/\sqrt{15}} = 8.69$$

Minitab can be used to provide the following output:

T-TEST OF THE MEAN

Test of mu = 4000.0 vs. mu > 4000.0

Variable	n	Mean	StDev	SE Mean	T	P
sales	15	4119.7	53.4	13.8	8.69	0.0000

Step 4: *Decide whether to reject the null hypothesis.* The calculated test statistic of $t = 8.69$ is in the rejection region, so we reject H_0. The p value of 0.0000 also tells us to reject H_0.

Draw conclusions.

Step 5: *Interpret the statistical decision in terms of the stated problem.* Based on these data, we conclude that the advertising scheme has indeed increased sales.

✓ TRY IT NOW!

Public Health Around the Globe
Conducting a Lower-Tail Test of μ — Small Sample

In her recently published book *Betrayal of Trust: The Collapse of Global Public Health* (Hyperion Books, 2001), Laurie Garrett traces public health history around the world, making the point that the health of rich and poor alike is in jeopardy because of crumbling or nonexistent public health infrastructures. One of the variables she looks at is the average life expectancy in countries around the world. Test the hypothesis that the average life expectancy in third-world countries is lower than the U.S. average of 77 years. Use $\alpha = 0.05$.

Step 1: Set up the null and alternative hypotheses.

Step 2: Define the test procedure.

Step 3: Collect the data and calculate the test statistic and the p value. The data on life expectancies (in years) for nine third-world countries are listed here:

Angola	47	Kyrgystan	68
Chile	75	Sierra Leone	37
China	70	Somalia	49
El Salvador	69	Vietnam	67
Kenya	54		

Step 4: Decide whether to reject the null hypothesis.

Step 5: Interpret the statistical decision in terms of the stated problem.

9.2.3 SUMMARY OF TESTS OF THE MEAN: SMALL SAMPLE

You have now seen more examples of the five-step hypothesis-testing procedure. In each case, regardless of whether the test is two-tail or one-tail, the same five steps are used. For unknown σ there are two major differences from the tests of μ when σ is known: (1) a t test statistic is used instead of Z and (2) the cutoff values for the rejection region are found from the t table instead of the Z table. The rejection regions are summarized in Table 9.1.

TABLE 9.1

REJECTION REGIONS FOR HYPOTHESIS TESTS OF μ WHEN σ IS UNKNOWN

Type of test	Rejection region
Two-tail test of μ $H_0: \mu =$ [a specific number] $H_A: \mu \neq$ [a specific number]	Reject H_0 if $t < -t_{\alpha/2}$ or $t > t_{\alpha/2}$
Lower-tail test of μ $H_0: \mu \geq$ [a specific number] $H_A: \mu <$ [a specific number]	Reject H_0 if $t < -t_{\alpha}$
Upper-tail test of μ $H_0: \mu \leq$ [a specific number] $H_A: \mu >$ [a specific number]	Reject H_0 if $t > t_{\alpha}$

ANS. $t = -4.00$, p VALUE $= 0.002$, REJECT THE NULL HYPOTHESIS. CONCLUDE THAT LIFE EXPECTANCY IN OTHER COUNTRIES IS LESS THAN 77 YEARS.

9.2.4 EXERCISES—LEARNING IT!

9.1 The costs of common goods and services in five cities are listed in the table (source: *USA Today*):

City	Aspirin (100)	Fast food (hamburger, fries, soft drink)	Woman's haircut/blow dry	Toothpaste (6.4 oz)
Los Angeles	$7.69	$4.15	$20.11	$2.42
Tokyo	35.93	7.62	76.24	4.24
London	9.69	5.80	44.35	3.63
Sydney	7.43	4.53	29.93	2.08
Mexico City	1.16	3.63	17.94	1.08

(a) You have just returned from a business trip. You lost your receipt for the aspirin you purchased but would like to be reimbursed by your company (since you had to take the aspirin after a stressful business meeting!). You guess a cost of $10.00. Your boss claims that the average cost of aspirin is less than $10.00. Using the data, can you "prove" your boss wrong? Conduct the necessary hypothesis test. Assume that all costs are normally distributed.

(b) Based on these data, is there enough evidence to support a cost of $10.00 for the fast-food meal on your trip?

(c) If you eliminate Tokyo from the data set, do your answers to parts (a) and (b) change? What does this tell you about the effect of outliers on the hypothesis test of μ when you have a small sample?

9.2 The marketing manager for a New England ski resort advertises that the resort can make snow whenever the temperature is 32°F or below. To demonstrate how often this happens their brochure includes a line graph of the weekly average temperatures:

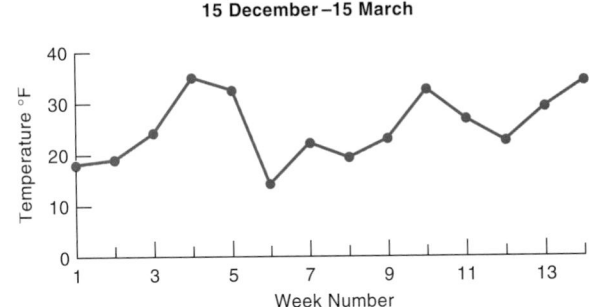

Weekly Average Temperatures
15 December–15 March

These are the data that generated this graph:

Week	1	2	3	4	5	6	7	8	9	10	11	12	13	14
Temperature	18	19	24	35	33	14	22	20	23	33	27	23	30	35

Is there enough evidence for the ski resort to claim that the average weekly temperature is below 32°F? Assume that the average weekly temperature is normally distributed.

9.3 Although job prospects for nurses were once good, many nurses now face an uncertain future as hospitals cut staff and train unlicensed workers to do some of their jobs. The change is reflected in enrollment at several area nursing schools. The following table lists the enrollment at nursing schools in western Massachusetts:

School	1995	1996
Holyoke Community College	60	45
Berkshire Community College	96	96
Greenfield Community College	90	79

Springfield Technical Community College	146	135
American International College	189	180
Elms College	298	300
U. Mass—Amherst	411	429
Baystate Medical Center	150	130

Using the 1996 data, test the hypothesis that the average enrollment in nursing programs has decreased from the average of 180 in 1995. Assume that enrollment in nursing programs is normally distributed.

9.4 If you like shopping for the best deal on long-distance phone service, then you'll enjoy sorting through offers from ten different marketers vying to be your energy supplier. Residents of 16 communities will be the first in Massachusetts to participate in a nationwide experiment in deregulation of the natural gas industry. The average customer uses 1232 therms of natural gas, for which the average cost has been $520.24. The table lists the proposed costs of ten competitors to deliver 1232 therms of natural gas:

All Energy Marketing Co.	$478.66
Broad Street/Energy One	450.24
Global Energy Services	468.16
Green Mountain Energy Partners	471.24
KBC Energy Services	435.53
Louis Dreyfus Energy Services	472.24
National Fuel Resources	468.22
NorAm Energy	442.20
WEPCO Gas	443.52
Western Gas Resources	457.81

Is there enough evidence to conclude that the average cost to the customer from the competitors is less than $520.24? Assume that the costs are normally distributed.

9.5 Computer centers at universities and colleges are aware of the increased number of web surfers. To begin to understand the demands that will be made on the computer center's resources, one school studied 25 children in grades 7 to 12. The numbers of hours that these children spent on the Internet in one week are shown here:

5.0	4.4	5.7	5.6	5.5
5.2	5.0	4.8	3.6	4.1
4.6	4.9	4.0	6.7	5.5
5.4	6.7	5.8	5.4	4.8
5.9	5.1	3.8	4.1	6.7

Is there enough evidence to indicate that children spend more than an average of 5 hours per week web surfing? Assume that the time spent web surfing is normally distributed.

9.3 Hypothesis Test of a Single Variance

Hypothesis tests of the population variance, σ^2, follow the same basic steps that we used to conduct a hypothesis test of the population mean. You might ask why we would be interested in testing a hypothesis about the variance when we have been using the standard deviation as our measure of variability. If this question has occurred to you, then you have correctly remembered that the variance has units of measure equal to the original units of measure squared, such as dollars^2 and minutes^2. These quantities are difficult to interpret, and for this reason the standard deviation is typically the preferred measure of spread. It is expressed in units equal to the original units of the data. However, remember that we cannot get the standard deviation without first finding the variance. Likewise we cannot do a hypothesis test directly on the

Remember! The variance is the standard deviation squared.

standard deviation; we must do a variance test instead. This is not a big problem as you will see in what follows.

Remember that the standard deviation is a measure of the spread in the data. A small standard deviation indicates that the data values are very similar and very consistent, whereas a larger standard deviation indicates that the values vary and are not so consistent. If we are interested in testing how consistent the values of a certain variable are, then we use a variance test. As with tests of the mean, we can do a two-tail or a one-tail test of the variance. To decide which type of test we need, we must consider what we are trying to learn. The key words and guidelines were presented in Chapter 8.

9.3.1 TWO-TAIL TEST OF THE VARIANCE

To show how to do a two-tail test of the variance, we illustrate the five steps of the hypothesis-testing procedure for the lengths of stay of patients in a hospital after having a certain medical procedure.

Understand the problem.

EXAMPLE 9.5

HOSPITAL STAYS

Doing Step 1

Physicians have been experimenting with a new procedure for lung cancer surgery. It involves using a glue to help prevent air leaks in the lungs after the surgery. It is hoped that this new surgery shortens the lengths of stay of patients in the hospital. However, a test of the mean length of stay for 39 patients who had the new procedure did not indicate that this is the case. Now the doctors are looking at the lengths of stay of patients who had the new procedure to see if perhaps the lengths of stay are more predictable, more consistent, and less variable than stays after traditional lung cancer surgery.

In Step 1 of the hypothesis-testing procedure, we set up the null and alternative hypotheses for a test of the population variance. The variable *Length of Stay* for the traditional surgery has a standard deviation of 5 days. So the population variance is $5^2 = 25$ (days)^2. To see if the new procedure has a standard deviation different from 25 days^2, we can use these null and alternative hypotheses:

$$H_0: \quad \sigma^2 = 25 \text{ (days)}^2$$
$$H_A: \quad \sigma^2 \neq 25 \text{ (days)}^2$$

The next step in the hypothesis-testing procedure is to define the test procedure. To do this we must decide what test statistic to use. This is the main difference between a hypothesis test of the population mean, μ, and a hypothesis test of the population variance, σ^2. In testing means, we used the sample mean as the basis for our decision to reject or fail to reject the null hypothesis. Because the Central Limit Theorem states that $\overline{X}$ has an approximately normal distribution for sufficiently large sample sizes, the appropriate test statistic for a large-sample test of the mean is a Z statistic and thus the rejection region is determined by Z values. In the small-sample case, $\overline{X}$ follows a t distribution for normally distributed populations.

Use s^2 as the basis for deciding between H_0 and H_A.

When we are testing the population variance, the sample variance, s^2, is used as the basis for deciding between H_0 and H_A. Relying once again on the mathematical statisticians to do the theoretical work, we learn that the sampling distribution associated with the sample variance, s^2, is called the chi-square (χ^2) distribution. The rejection region is determined by critical values from the chi-square distribution. An example of a chi-square distribution is shown in Figure 9.3.

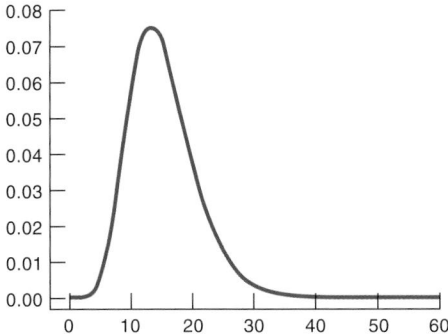

FIGURE 9.3 **A chi-square distribution**

The chi-square distribution, like any distribution, describes how the random variable behaves. Recall from Chapter 6 that a distribution illustrates the most likely values of the random variable (where there is the most area under the curve) and the least likely values of the random variable (where there is little to no area under the curve). The chi-square distribution is discussed again in Chapter 15.

If the variable being studied is assumed to be normally distributed, then the statistic to test whether the population variance is equal to a particular value is calculated as follows:

$$\chi^2 = \frac{(n-1)\,s^2}{\sigma^2}$$

The chi-square test assumes that the underlying population distribution is normal.

Chi-square test statistic for variances

where

n = sample size

s^2 = sample variance

σ^2 = hypothesized value of the population variance under the null hypothesis

Notice that this test statistic compares the variability contained in the sample and reflected in the sample variance with the value of the population variance that is being tested. In this case, the comparison takes the form of a ratio. When we test means, we compare the sample evidence, reflected in $\overline{X}$, with the value of the population mean being tested by means of a subtraction. In both cases the idea is the same: Is the sample evidence consistent with the null hypothesis?

This test statistic has been labeled with the Greek letter χ (chi) and a squared symbol—hence, the name *chi-square test statistic*. Like the t distribution, the shape of the chi-square distribution is determined by the number of degrees of freedom. This test statistic has $n-1$ degrees of freedom. Unlike the Z and t distributions, the χ^2 distribution is not symmetric. In particular, we can never get a negative χ^2 value because $(n-1)$, s^2, and σ^2 are all always positive.

The χ^2 distribution is not symmetric.

Recall that the rejection region is the set of those values of the test statistic that will lead us to reject the null hypothesis. Let's think a minute about what values of the chi-square test statistic will lead us to reject the null hypothesis. Clearly, if the sample variance exactly equals the population variance, then the ratio of s^2 to σ^2 will be 1 and the test statistic will be equal to $n-1$. In the hospital stay example, we get $39 - 1 = 38$. The sample evidence is clearly consistent with the null hypothesis and we fail to reject H_0.

If the sample variance is quite different from the population variance being tested, then it is either greater or less than σ^2. If the sample variance is a great deal less than the population variance being tested, then the ratio of s^2 to σ^2 will be a fraction and the test statistic will be some value less than $n-1$. If the sample variance is much greater than the population variance being tested, then the ratio of s^2 to σ^2 will be greater than 1 and the test statistic will be larger than $n-1$. These situations will

Remember! The total area of the rejection region is α.

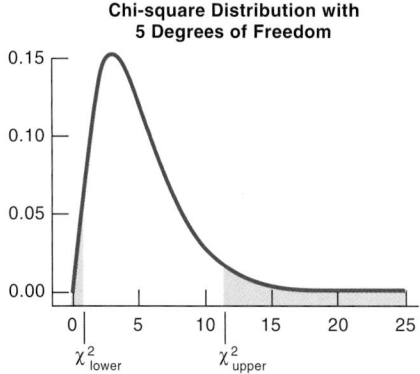

FIGURE 9.4 **Rejection region for a two-tail test of the variance**

Note: We cannot use +/− the same critical value because the distribution is not symmetric.

lead us to reject H_0. This gives us the basic structure for the rejection region. We will reject H_0 if the value of the test statistic is much less than $n - 1$ or much greater than $n - 1$. Since this is a two-tail rejection region, the area in each tail must be $\alpha/2$. A typical rejection region is shaded in Figure 9.4. It is defined by values greater than χ^2_{upper} or less than χ^2_{lower}.

The specific values for χ^2_{lower} and χ^2_{upper} must be found from the chi-square table. A portion of this table is shown in Table 9.2. The complete table is found in Table 5 in Appendix A.

Now let's use Table 9.2 to complete Step 2 of the hypothesis-testing procedure for the hospital stays.

TABLE 9.2

A Portion of the Chi-square Table

Degrees of freedom	Upper-tail areas											
	0.995	0.99	0.975	0.95	0.9	0.75	0.25	0.1	0.05	0.025	0.01	0.005
32	15.134	16.362	18.291	20.072	22.271	26.304	36.973	42.585	46.194	49.480	53.486	56.328
33	15.815	17.073	19.047	20.867	23.110	27.219	38.058	43.745	47.400	50.725	54.775	57.648
34	16.501	17.789	19.806	21.664	23.952	28.136	39.141	44.903	48.602	51.966	56.061	58.964
35	17.192	18.509	20.569	22.465	24.797	29.054	40.223	46.059	49.802	53.203	57.342	60.275
40	20.707	22.164	24.433	26.509	29.051	33.660	45.616	51.805	55.758	59.342	63.691	66.766

Collect and analyze the data.

EXAMPLE 9.6

HOSPITAL STAYS

Doing Step 2

In Step 2 we define the test procedure. The lengths of stay for 39 patients who received the new lung cancer procedure were collected. Here are the summary statistics:

Descriptive Statistics: Length of Stay

Variable	N	Mean	Median	TrMean	StDev	SE Mean
PODINIT	39	4.564	4.000	4.086	2.836	0.454

Variable	Minimum	Maximum	Q1	Q3
PODINIT	3.000	19.000	3.000	5.000

The doctors have set $\alpha = 0.05$. Since the sample size is $n = 39$, we must use the row in the table for $n - 1 = 38$ degrees of freedom. To get the value for χ^2_{upper}, we use the column labeled 0.025 in the upper tail. In Table 9.2 there is no row labeled 38. In this case we use the row corresponding to the degrees of freedom closest to 38 but greater than 38. So we use the row with 40 degrees of freedom. The value at the intersection of the correct row and column is 59.342. You may have noticed that the column values correspond to upper-tail areas. There is no need to have another table for lower-tail areas, however. To have the area of the lower tail be 0.025, the upper tail must be 0.975. To get the value for χ^2_{lower} we use the column labeled 0.975 in the upper tail. The value at the intersection of the row with 40 degrees of freedom and 0.975 in the upper tail is 24.433. Therefore the rejection region is $\chi^2 < 24.433$ or $\chi^2 > 59.342$.

Using a row with more degrees of freedom pushes χ^2_{upper} up, making it harder to reject H_0 and thereby guaranteeing that α is at most 0.05.

Now we can complete Step 3 of the hypothesis-testing procedure. Remember that at this step we calculate the test statistic and the p value.

EXAMPLE 9.7

HOSPITAL STAYS

Collect and analyze the data.

Doing Step 3

A sample of 39 patients' hospital stays has a sample standard deviation length of stay of 2.84. The test statistic is

$$\chi^2 = \frac{(n-1)\,s^2}{\sigma^2} = \frac{(39-1)(2.84)^2}{25} = 12.26$$

The p value must be obtained from the output of software.

The fourth step of the hypothesis-testing procedure is to decide whether to reject the null hypothesis on the basis of the test statistic. The fifth step is to interpret this decision in terms of the original problem statement.

EXAMPLE 9.8

HOSPITAL STAYS

Draw conclusions.

Doing Steps 4 and 5

Since the value of 12.26 does fall in the rejection region, we reject the null hypothesis. The new surgical procedure does have a variance significantly different from that of the traditional procedure.

TRY IT NOW!

Cereal Box Weights
Conducting a Test of σ^2

A cereal manufacturer wishes to test whether the population variance of the weight of the cereal boxes is equal to 0.0500 ounce^2.

Step 1: Set up the null and alternative hypotheses.

Step 2: Define the test procedure. A random sample of 20 boxes has a standard deviation of $s = 0.25$ ounce. Use $\alpha = 0.05$.

Step 3: Collect the data and calculate the test statistic and the p value.

Step 4: Decide whether to reject the null hypothesis.

Step 5: Interpret the statistical decision in terms of the stated problem.

9.3.2 ONE-TAIL TESTS OF THE VARIANCE

When we are testing the population variance, we are usually interested in doing a two-tail test. However, it is possible to do a one-tail test of the variance. The only change in the hypothesis-testing procedure needed to complete a one-tail test of the variance is in Step 2. A one-tail rejection region is used in this case. The two possible rejection regions are shown in Figure 9.5:

$$H_0: \quad \sigma^2 \geq \text{[a specific number]} \qquad H_0: \quad \sigma^2 \leq \text{[a specific number]}$$
$$H_A: \quad \sigma^2 < \text{[a specific number]} \qquad H_A: \quad \sigma^2 > \text{[a specific number]}$$

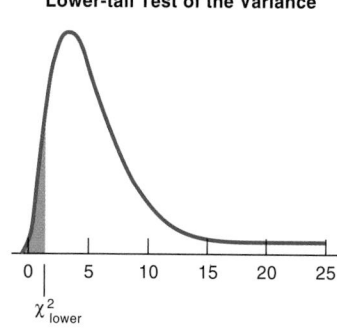
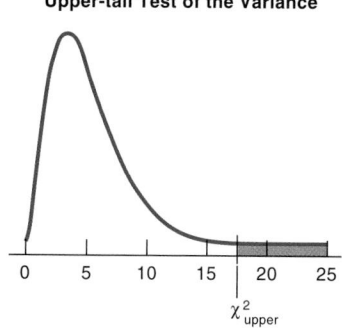

FIGURE 9.5 Rejection regions for one-tail tests of the variance

EXAMPLE 9.9

HOSPITAL STAYS

Understand the problem.

Conducting a One-Tail Test of σ^2

Let's reconsider the variability of *Length of Stay* for the hospital example. Suppose that the hospital wishes to test whether the variance of hospital stays with the new procedure is less than with the traditional procedure. The key words "less than" indicate that we use a lower-tail test of the variance.

Step 1: *Set up the null and alternative hypotheses*:

$$H_0: \quad \sigma^2 \geq 25 \text{ days}^2$$
$$H_A: \quad \sigma^2 < 25 \text{ days}^2$$

Step 2: *Define the test procedure.* Set $\alpha = 0.05$. We use the chi-square table with 38 degrees of freedom to find the rejection region. Again, since no row corresponds to 38 degrees of freedom, we use the row corresponding to 40 degrees of freedom. The value at the intersection of the row for 40 degrees of freedom and the column with 0.95 in the upper-tail area is 26.509. Note that we do not divide α in half because this is a one-tail test with a one-tail rejection region.

Remember! To get χ^2_{lower} we must find χ^2_{upper} for $(1 - \alpha)$.

Step 3: *Collect the data and calculate the test statistic and p value.* The data were collected and the summary statistics were shown in Example 9.6. The test statistic is calculated here:

$$\chi^2 = \frac{(n-1)s^2}{\sigma^2} = \frac{(39-1)(2.84)^2}{25} = 12.26$$

Analyze the data.

Step 4: *Decide whether to reject the null hypothesis.* The test statistic 12.26 falls in the rejection region, so we reject the null hypothesis.

Step 5: *Interpret the statistical decision in terms of the stated problem.* We conclude that the population variance with the new procedure is less than 25 days^2 and therefore the lengths of patient stays are more consistent, less variable, and thus more predictable than with the traditional procedure. This predictability is helpful in managing health care costs.

Draw conclusions.

9.3.3 SUMMARY OF TESTS OF THE VARIANCE

In this section all the tests use the same test statistic in Steps 2 and 3. The purpose of the test statistic is always the same: It is a single number that summarizes the sample information and allows us to decide between the null and alternative hypotheses. The test statistic for tests of variances is

$$\chi^2 = \frac{(n-1)s^2}{\sigma^2}$$

In Step 2 the rejection region is different depending on what type of test we are doing. These differences are summarized in Table 9.3 on page 428.

9.3.4 EXERCISES — LEARNING IT!

9.6 A company that sells mail-order computer systems has been planning inventory and staffing based on the assumption that the variance of its weekly sales is $\$180,000^2$. The weekly

TABLE 9.3

REJECTION REGIONS FOR HYPOTHESIS TESTS OF σ^2

Type of test	Rejection region
Two-tail test of σ^2 $H_0: \sigma^2 =$ [a specific number] $H_A: \sigma^2 \neq$ [a specific number]	Reject H_0 if $\chi^2 < \chi^2_{\text{lower}}$ or $\chi^2 > \chi^2_{\text{upper}}$
Lower-tail test of σ^2 $H_0: \sigma^2 \geq$ [a specific number] $H_A: \sigma^2 <$ [a specific number]	Reject H_0 if $\chi^2 < \chi^2_{\text{lower}}$
Upper-tail test of σ^2 $H_0: \sigma^2 \leq$ [a specific number] $H_A: \sigma^2 >$ [a specific number]	Reject H_0 if $\chi^2 > \chi^2_{\text{upper}}$

sales are normally distributed. The company selects 15 weeks at random from the past year and obtains these weekly data (in thousands of dollars):

$$191 \quad 222 \quad 222 \quad 223 \quad 223 \quad 225 \quad 227 \quad 228$$
$$229 \quad 232 \quad 234 \quad 234 \quad 236 \quad 244 \quad 253$$

(a) What is the sample variance for these data?

(b) Set up the hypotheses to test whether the population variance is different from 180.

(c) At the 0.05 level of significance, what can you conclude about the company's assumption?

9.7 Educators are looking at new and different testing methods to get more consistent results. If test scores vary widely, that indicates either that confusing questions are causing the variability or that the students' abilities are widely varying. One new test was given to ten students and the following scores resulted:

$$78 \quad 73 \quad 81 \quad 86 \quad 85 \quad 88 \quad 69 \quad 78 \quad 85 \quad 74$$

(a) What is the sample variance for these data?

(b) Set up the hypotheses to test whether the actual variance is less than the variance from the standard test, which is 10^2. Assume the scores are normally distributed.

(c) At the 0.05 level of significance, what can you conclude?

9.8 Consistency in the delivery time for e-mail or catalog orders is important to most customers. An online company has received numerous complaints about the inconsistency of delivery times—sometimes a few days and other times a few weeks with no explanation provided. It has been assumed that delivery times are normally distributed with a standard deviation of 2 days. The company decides to look at the last 20 orders placed and finds these delivery times (in days):

$$1 \quad 4 \quad 4 \quad 4 \quad 4 \quad 4 \quad 5 \quad 5 \quad 6 \quad 6$$
$$6 \quad 6 \quad 7 \quad 7 \quad 7 \quad 7 \quad 8 \quad 8 \quad 10 \quad 11$$

(a) What is the sample variance for these data?

(b) Set up the hypotheses to test whether the actual variance is greater than the value the company has been assuming. Be sure you use the *variance* here.

(c) At the 0.05 level of significance, what can you conclude?

9.9 In an effort to understand the cost overruns in a prison, data were collected on the amounts of the overruns (in dollars) on 24 different days:

87.3	89.9	91.5	93.6	93.7	94.9	96.5	96.7	96.8	97.0	97.1	97.3
98.4	99.6	100.3	100.4	100.9	101.3	105.7	107.7	107.8	109.7	111.5	114.2

(a) Calculate the sample variance for the daily cost overrun.

(b) At the 0.05 level of significance, do the data agree with the assumption that the variance of the daily overrun is $\$50^2$? Assume that the cost overruns are normally distributed.

9.4 Hypothesis Test of a Single Proportion

So far we have been conducting tests on quantitative data. However, very often it is not the average or variance that we are interested in but rather some proportion of the population that behaves in a certain manner. If we are analyzing *nominal data*, the hypothesis might be this type of statement:

- The value of the proportion, *p*, of population members that have a certain characteristic (one of the categories of the nominal variable)

For example, the hypothesis might be a statement about one of these proportions:

- Residents of Swingfield who are in favor of a casino
- Students who are interested in graduate school
- Vaccinated patients who remain cancer-free
- CEOs who use computers as a major tool
- People who are unemployed

In these cases, the parameter of interest is a proportion. In Chapter 7 you learned that an estimate for the true unknown proportion, *p*, is the corresponding sample proportion, $\hat{p}$ (*p*-hat). For example, if 100 people are surveyed and 58 are in favor of a casino, then the sample proportion, $\hat{p}$, is 58/100, or 0.58.

Remember! We are using p to represent the population proportion.

Like $\overline{X}$, the sample proportion, $\hat{p}$, rarely if ever actually equals the true population proportion, *p*. Remember that this is not because we made an error but rather because we are looking at only a piece of the population and not the entire population. This is what we call sampling error, and it is a fact of life in statistics. We cannot eliminate sampling error unless we examine the entire population.

Because the sample proportion is not likely to equal the population proportion, the estimate alone is inadequate for making decisions. Suppose, for example, that more than 60% of the voters must be in favor of a casino and the sample proportion comes out to be 0.58. Does this automatically mean that the casino plan should be killed? Is it possible that more than 60% of the population favor the casino but only 58% of those sampled favor the casino? Possibly. How possible again depends on the standard error. So we need to develop the hypothesis-testing procedure for tests of proportions. We start by looking at the two-tail test of proportions.

9.4.1 TWO-TAIL TEST OF PROPORTIONS

Fortunately we can easily adapt much of what we have learned about hypothesis testing to tests of proportions. In fact, we use the same five steps that we have been following for any hypothesis test.

EXAMPLE 9.10

TEEN SPENDING

Conducting a Two-Tail Test of *p*

Nearly one-third of teens polled by Teenage Research Limited said they have been personally affected by the recession. Where do teens get their money? We wish to do a hypothesis test to see whether half or 50% of all teens get some money from their parents.

Understand the problem.

Step 1: *Set up the null and alternative hypotheses.* The same general guidelines from Chapter 8 apply to setting up the null and alternative hypotheses for a test of proportions. We are interested in seeing whether the true proportion *differs* from 50% in either direction, so we use a two-tail test. The null and alternative hypotheses are:

$$H_0: \quad p = 0.50$$
$$H_A: \quad p \neq 0.50$$

Collect and analyze the data.

Step 2: *Define the test procedure.* A survey of 1000 teens found that 47% of them get some money from their parents. In Chapter 7 you learned that the sampling distribution for $\hat{p}$ is approximately normal. Therefore, the test statistic for proportions is a Z statistic. The form of the rejection region is exactly the same as the rejection region for a two-tail test of the mean when the standard deviation is known. If we set $\alpha = 0.05$, then the rejection region is $Z > 1.96$ or $Z < -1.96$.

Step 3: *Collect the data and calculate the test statistic and the p value.* Now we must consider what information in the sample will help us decide between the null and the alternative hypotheses. Clearly, we must use the sample proportion, $\hat{p}$. But we cannot simply use the value of $\hat{p}$ alone. We must take into account the fact that $\hat{p}$ varies from sample to sample. As we have repeatedly shown, variability is taken into account by dividing by the standard error of the estimator. The appropriate test statistic is

Z test statistic for proportions

$$Z = \frac{\hat{p} - p}{\sqrt{p(1-p)/n}}$$

where p is the hypothesized value of the population proportion. This test statistic has the same basic form as the Z statistic we used in testing μ:

$$Z = \frac{\overline{X} - \mu}{\sigma/\sqrt{n}}$$

In the numerator we are calculating the difference between the value of the population proportion, p, and the sample proportion, $\hat{p}$. This is equivalent to the calculation in the numerator of the Z test statistic for μ. In the denominator of both Z statistics we use the standard error of the estimator. In the case of proportions, this is $\sqrt{p(1-p)/n}$. For this example the Z test statistic is

$$Z = \frac{0.47 - 0.50}{\sqrt{0.50(1-0.50)/100}} = -1.90$$

To complete the hypothesis test, we should calculate the p value. The p value for a test of proportions is calculated the same way as it is for the test of means. This is a two-tail test, so we must double the tail area probability. The Z statistic for this example was found to be -1.90. So the p value is

$$p \text{ value} = 2P[Z > 1.90] = 2P[Z < -1.90] = 2(0.0287) = 0.0574$$

The output from Minitab is shown here.

Test and CI for One Proportion

Test of p = 0.5 vs p not = 0.5

Sample	X	N	Sample p	95.0% CI	Exact P-Value
1	470	1000	0.470000	(0.438693, 0.501484)	0.062

Notice that Minitab produces output that looks more like a binomial random variable. There is no information about Z, and the p value doesn't match our calculation. This is because a test of proportions technically involves the binomial distribution, which is typically approximated by the normal distribution, Z. Therefore, the p value you obtain by using $2P[Z > 1.90]$ will not match the exact p value given by Minitab. Since the exact p value is given by Minitab, we should use it.

Step 4: *Decide whether to reject the null hypothesis.* Since the value of $Z = -1.90$ is not in the rejection region and the p value is greater than 0.05, we do not reject H_0.

Step 5: *Interpret the statistical decision in terms of the stated problem.* We conclude that the true proportion of teenagers who receive spending money from their parents is not different from 0.50.

Draw conclusions.

✔ TRY IT NOW!

Poll of Americans
Conducting a Test of p

She did it again. For the zillionth time, your mother casually asked you when you are going to get married and you've been seething ever since. How do you get it off your chest? These days, you might e-mail her. The Pew Internet and American Life Project, based on a telephone survey of 3533 randomly selected adults who use e-mail, recently (July 2000) found that 31% find e-mail an easier way to say frank or unpleasant things to their relatives. Is there evidence that more than 30% of those who send family e-mail use it to communicate unpleasant things? Use $\alpha = 0.05$.

Step 1: Set up the null and alternative hypotheses.

Step 2: Define the test procedure.

Step 3: Collect the data and calculate the test statistic and the p value.

Step 4: Decide whether to reject the null hypothesis.

Step 5: Interpret the statistical decision in terms of the stated problem.

9.4.2 ONE-TAIL TESTS OF PROPORTIONS

Finally, we consider one-tail tests of proportions. The five-step hypothesis-testing procedure in this case is identical to the one we have been using. The test statistic is the same as the test statistic for a two-tail test of proportions, and the rejection regions are the same as those used for one-tail tests of the mean.

Step 1: *Set up the null and alternative hypotheses.* There are two possible ways to set up a one-tail test of proportions. We have already looked at the issues that should be addressed in deciding how to set up the hypothesis test. Here we simply present the two forms again using the terms *upper-tail* and *lower-tail* test.

ANS. $H_0: p \le 0.30$, $H_A: p > 0.30$; CRITICAL VALUE = ± 1.96; $Z = -1.30$; FAIL TO REJECT H_0; p VALUE = 0.0968

Remember! The specific number must be between 0 and 1.

Upper-tail test	**Lower-tail test**
H_0: $p \leq$ [a specific number]	H_0: $p \geq$ [a specific number]
H_A: $p >$ [a specific number]	H_A: $p <$ [a specific number]

Step 2: *Define the test procedure.* Because the test statistic is Z, the rejection regions for the one-tail tests are the same as the ones we used in testing means.

Step 3: *Collect the data and calculate the test statistic and the p value.* The appropriate test statistic is

$$Z = \frac{\hat{p} - p}{\sqrt{p(1 - p)/n}}$$

For an upper-tail test, the p value is the probability of observing a Z value greater than this test statistic value. For a lower-tail test, the p value is the probability of observing a Z value less than this test statistic value. Note that the p value is not doubled for a one-tail test as it is for a two-tail test.

Steps 4 and 5 remain the same. Now let's consider an example.

Understand the problem.

EXAMPLE 9.11

MANAGING YOUR TIME

Conducting Upper-Tail Test of *p*

What do college students need to succeed in school? They need to know their professors; they need to manage their time well; they need to study in groups. That's what Richard J. Light, a professor in Harvard University's Graduate School of Education, says. In a pilot study of students, Light asked them to track how they spend their time for 2 weeks. Fifty-five percent of the students said that it was "the single most valuable thing they did in their entire first year." You are wondering if this exercise would be helpful to students at your school. You decide to do your own study and you wonder if you can conclude that more than 50% of the students found it valuable. Test using $\alpha = 0.05$.

Step 1: *Set up the null and alternative hypotheses.* Because you are interested in seeing whether more than 50% of the students found the tracking valuable, this is an upper-tail test with the following hypotheses:

$$H_0: \quad p \leq 0.50$$
$$H_A: \quad p > 0.50$$

Collect and analyze the data.

Step 2: *Define the test procedure.* You take a sample of 100 students; 60 of them found the exercise helpful. Since $\alpha = 0.05$, the rejection region is $Z > 1.645$.

Step 3: *Collect the data and calculate the test statistic and the p value.* The sample proportion is found to be $\hat{p} = 60/100 = 0.60$ and the sample size is $n = 100$. Using this information, you calculate the test statistic:

$$Z = \frac{\hat{p} - p}{\sqrt{p(1 - p)/n}} = \frac{0.60 - 0.50}{\sqrt{0.50(1 - 0.50)/100}} = \frac{0.10}{0.05} = 2$$

Notice that 0.50 is used as the value for p because it is the hypothesized value of the population proportion under H_0. Here is the output from Minitab:

Test and CI for One Proportion

Test of p = 0.5 vs p > 0.5

Sample	X	N	Sample p	95.0% Lower Bound	Exact P-Value
1	60	100	0.600000	0.512976	0.028

Step 4: *Decide whether to reject the null hypothesis.* Since $Z = 2$ does fall in the rejection region, you reject H_0. The p value is 0.028, which is less than $\alpha = 0.05$, so you reject H_0.

Step 5: *Interpret the statistical decision in terms of the stated problem.* You can conclude that more than 50% of the students at your school would find tracking their time for 2 weeks to be a valuable exercise. Therefore you plan to recommend this to the dean of freshmen.

Draw conclusions.

✓ TRY IT NOW!

The Tobacco Industry
Conducting a One-Tail Test of p

"The multibillion-dollar national tobacco pact has failed to end a torrent of cigarette advertising targeted at teen-age magazine readers" (Associated Press, August 16, 2001). A study published in the *New England Journal of Medicine* says that cigarette makers have maintained a high level of spending for magazine ads targeted at middle- and high-school-age children. In 1999 such ads reached 88% of the teenage readers. Recently a sample of 200 teenage readers found that 82% of them were reached by these ads. Is there enough evidence to conclude that the proportion of teenagers being reached has decreased from 88%? Use $\alpha = 0.05$.

Step 1: Set up the null and alternative hypotheses.

Step 2: Define the test procedure.

Step 3: Collect the data and calculate the test statistic and the p value.

Step 4: Decide whether to reject the null hypothesis.

Step 5: Interpret the statistical decision in terms of the stated problem.

9.4.3 SUMMARY OF TESTS OF PROPORTIONS

In this section all the tests use the same test statistic in Steps 2 and 3. The purpose of the test statistic is always the same: It is a single number that summarizes the sample information and allows us to decide between the null and alternative hypotheses. The test statistic for tests of proportions is

$$Z = \frac{\hat{p} - p}{\sqrt{p(1 - p)/n}}$$

ANS. H_0: $p \geq 0.88$; H_A: $p < 0.88$; CRITICAL VALUE $= -1.645$; $Z = 2.61$; p VALUE $= 0.009$; REJECT H_0; THE INDUSTRY HAS REDUCED THE ADVERTISING REACH.

TABLE 9.4

REJECTION REGIONS FOR HYPOTHESIS TESTS OF P

Type of test	Rejection region
Two-tail test of p $H_0: p =$ [a specific number between 0 and 1] $H_A: p \neq$ [a specific number between 0 and 1]	Reject H_0 if $Z < -Z_{\alpha/2}$ or $Z > Z_{\alpha/2}$
Lower-tail test of p $H_0: p \geq$ [a specific number between 0 and 1] $H_A: p <$ [a specific number between 0 and 1]	Reject H_0 if $Z < -Z_{\alpha}$
Upper-tail test of p $H_0: p \leq$ [a specific number between 0 and 1] $H_A: p >$ [a specific number between 0 and 1]	Reject H_0 if $Z > Z_{\alpha}$

In Step 2 the rejection region is different depending on what type of test we are performing. The differences are summarized in Table 9.4.

9.4.4 EXERCISES—LEARNING IT!

9.10 Companies are increasingly concerned about employees playing video games at work. In addition to reducing productivity, this habit slows down networks and uses valuable computer storage space. A recent article stated that 80% of all employees play video games at work at least once a week. A large company that employs many engineers wonders whether its employees are as bad as the article claims. If they are, the company will install software that detects and removes video games from the network. The company surveys (anonymously) 100 employees and finds that 85 had played video games at work in the past week.

(a) Set up the null and alternative hypotheses to test whether the proportion of the company's employees who play video games is greater than the proportion stated in the article.

(b) At the 0.05 level of significance, test the hypotheses.

(c) What do you recommend that the company do?

9.11 An alumni office is interested in serving alumni better to encourage more donations to the college. A survey of 200 alumni was conducted to determine whether half-day training sessions offered on the campus were of interest. If more than 75% of the alumni were interested, the college would start a program. The survey showed that 160 of the alumni surveyed were interested in such a program.

(a) Set up the null and alternative hypotheses to test whether the college should implement the program.

(b) At the 0.05 level of significance, test the hypotheses.

(c) What do you recommend that the college do?

9.12 The October 19–21, 2001, Gallup Poll found that of 1011 adults surveyed, 52% believe the recent U.S. incidents of anthrax-laden letters signal the beginning of a sustained anthrax campaign against the country.

(a) Set up the null and alternative hypotheses to test whether the proportion of Americans who believe the anthrax letters signal the beginning of a sustained attack is greater than 50%.

(b) At the 0.05 level of significance, test the hypotheses.

9.13 A university in the Northeast claims in its brochures that it has an acceptance rate of 60%. From a sample of 300 high school seniors who applied to this university, 148 were accepted.

(a) Set up the null and alternative hypotheses to test whether the acceptance rate is what the university claims.

(b) At the 0.05 level of significance, test the hypotheses.

(c) Is there a need for the university to change its literature?

9.14 "Computer Jobs Increase by 6%" is the headline in a national newspaper. A study of the classified job ads in that same newspaper indicates that out of 2202 advertisements in a typical Sunday edition, 502 were for computer-related jobs. Statistics released by the Bureau of Labor Statistics for the preceding 12 months show that 15% of available jobs were computer-related.

(a) Set up the null and alternative hypotheses to test whether the headline is correct.

(b) At the 0.05 level of significance, test the hypotheses.

(c) What conclusion can you reach?

9.5 Summary of One-Population Hypothesis Tests

We have now finished with the tests of parameters of one-population variables. In the next chapter you will learn how to extend these tests to compare the parameters from two populations. Before leaving this chapter, however, we summarize all the one-population tests from this chapter and from Chapter 8.

We have said repeatedly that the steps for *any* hypothesis test are the same and the only thing that changes is the test statistic, which affects the determination of the rejection region and the *p* value calculation. However, it is easy to get lost in the details of the particulars for each test. To help you, the accompanying flowchart suggests questions that you can ask to help you decide which test to use. The diamond symbols indicate a decision point.

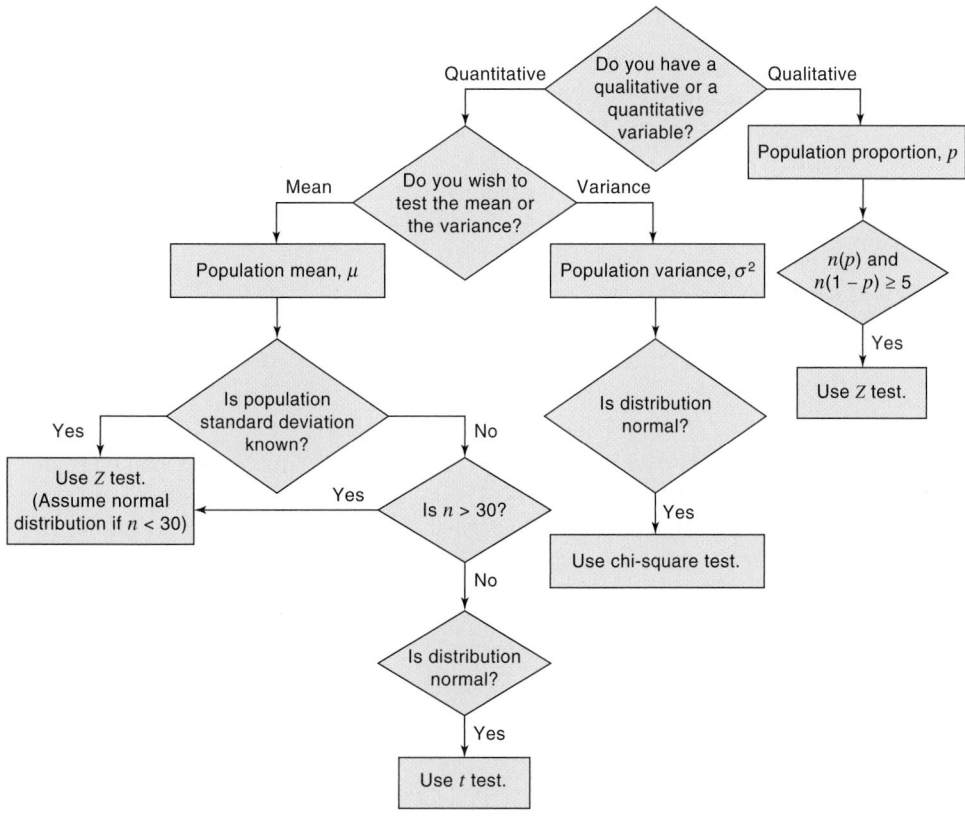

Flowchart for one-population hypothesis tests

Notice that many of the tests require that the underlying population distribution is normal. Most of the time you will not know this for sure. There is a hypothesis test for testing whether the underlying distribution is normal. It is another chi-square test, which is covered in Chapter 15. For now, you should get used to displaying the data as a histogram or a boxplot when you have a quantitative variable. Visually check to see whether the histogram is "reasonably" bell-shaped.

Also notice that there is an assumption to be checked when you are testing a population proportion. Technically, you should use a binomial distribution to test a hypothesis about p. However, when $n(p)$ and $n(1 - p)$ are greater than or equal to 5, the binomial distribution can be approximated by the normal or the Z distribution. Be sure to check this condition before using the Z test for proportions.

Discovery Exercise 9.1

THE CONNECTION BETWEEN CONFIDENCE INTERVALS AND HYPOTHESIS TESTING

Part I

A recent survey is offering the first evidence that PCs are replacing TVs as the primary source of home recreation, information, and entertainment. The survey was conducted among 1200 homes nationwide. The average computer user spends 9.5 hours per week in front of the PC but only 8 hours per week watching prime-time TV. Assume that the standard deviation of hours spent in front of a PC is 3 hours per week.

1. Test the hypothesis that the average number of hours per week a computer user spends in front of a PC is different from 8 hours. Use $\alpha = 0.05$.

2. Using the same data, construct a 95% confidence interval for μ, the population average time spent in front of a PC.

3. Is the value of 8 in the confidence interval?

4. Do you reject the null hypothesis?

Part II

The Casual Businesswear Employee Survey was conducted to assess the attitudes and behavior of white-collar employees whose companies allow casual dress. The study was national in scope and the sample size was 752 people. Of those in the sample, 609 agree that allowing casual dress improves morale.

1. Test the hypothesis that the proportion of white-collar employees who agree that allowing casual dress improves morale is different from 80%. Use $\alpha = 0.05$.

2. Construct a 95% confidence interval for p.

3. Is the value of 0.80 in the confidence interval?

4. Do you fail to reject the null hypothesis?

Part III

Based on these two situations, complete the following statements:

1. If the value of the parameter being tested (the one in the null hypothesis) is not in the confidence interval, then I will ____ the null hypothesis.

2. If the value of the parameter being tested (the one in the null hypothesis) is in the confidence interval, then I will ____ the null hypothesis.

9.6 Connection Between Hypothesis Testing and Confidence Intervals

Now you have learned about both confidence intervals and hypothesis testing. These two techniques are actually closely related even though they are used for different purposes. You can see the relationship between these techniques easily if we reconsider one of the hypothesis-testing examples from this chapter and calculate the corresponding confidence interval. Let's look at the hand washing data again.

EXAMPLE 9.12

WASH YOUR HANDS

Relating Confidence Intervals and Hypothesis Tests

From a sample of 20 students, the average *Days Missed* was 6.6 with a standard deviation of 1.273 days. In Example 9.1 we tested the hypothesis that the population mean *Days Missed* was 7.5 days with these hypotheses:

$$H_0: \quad \mu = 7.5 \text{ days}$$
$$H_A: \quad \mu \neq 7.5 \text{ days}$$

The value of the test statistic was $t = -3.16$. Based on this t value and $\alpha = 0.05$, the null hypothesis was rejected, leading us to conclude that the hand washers had an average number of *Days Missed* different from 7.5 days.

Now we can construct a 95% confidence interval for μ using these data:

$$\text{Upper bound} = \overline{X} + t(s/\sqrt{n}) = 6.6 + 2.093(1.273/\sqrt{20}) = 7.196$$

$$\text{Lower bound} = \overline{X} - t(s/\sqrt{n}) = 6.6 - 2.093(1.273/\sqrt{20}) = 6.004$$

Note: $t_{0.025,19} = 2.093$ is the appropriate t value from the table with $n - 1$ degrees of freedom. The value of μ of 7.5 is not in the confidence interval. This means that 7.5 is not a likely value of μ. Using the same data, we reject the null hypothesis and conclude that μ is not 7.5 days. The results from the confidence interval are consistent with the conclusions drawn from the hypothesis test.

Understand the problem.

Analyze the data.

Draw conclusions.

This result can be stated in general for any hypothesis test, whether it be a test of μ, σ^2, or p. If the value of the parameter being tested (the one in the null hypothesis) is in the confidence interval, then we will fail to reject the null hypothesis of a two-tail test. If the value of the parameter being tested is not in the confidence interval, then we will reject the null hypothesis. Of course, the same value of α must be used for both the confidence interval construction and the hypothesis test, and the hypothesis test must be a two-tail test.

GET IT IN WRITING

Wash Your Hands

TO: **Superintendent of Schools**
FROM: **Erica Q. Nurse**
RE: **Hand washing**

After reading an article about the effect of proper hand washing on student absenteeism, I decided to investigate this in our school. I trained 20 students in the proper hand washing technique, and I collected data on the number of school days these students missed during the year following the training. I also kept track of which grade these students were in and their gender.

Here are the histogram of the number of days missed for these students and the summary statistics:

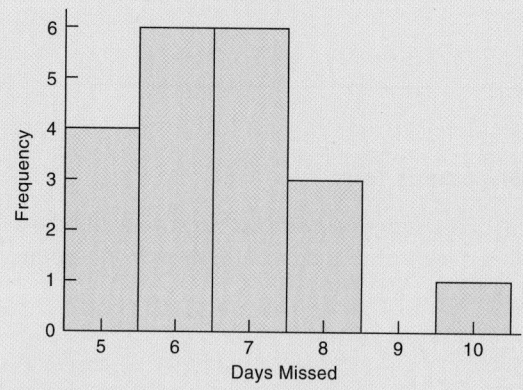

Descriptive Statistics: Days Missed

Variable	N	Mean	Median	TrMean	StDev	SE Mean
Days Missed	20	6.600	6.500	6.500	1.273	0.285

Variable	Minimum	Maximum	Q1	Q3
Days Missed	5.000	10.000	6.000	7.000

For our school the average number of days a student misses is 7.5 days. The 20 students in this experiment missed an average of 6.6 days. Since I had a small sample ($n < 30$) I knew that I needed to use a t test to decide whether the average was statistically less than 7.5 days. The t test requires the assumption that the days missed are normally distributed. The histogram indicates that this is a reasonable assumption.

As you know, the number of days of school a student misses is directly related to performance. Therefore, I recommend a larger study with the likely result that we include a training session on hand washing at the start of each school year. I would also like to continue my analysis of these 20 students to see if there are any differences by gender or grade level.

9.7 Using Technology for Hypothesis Testing

You learned in Chapter 7 that all confidence intervals have the same basic structure. In this chapter you learned the same thing about hypothesis testing. When using Excel or Minitab to do hypothesis testing, the most important decision you will make is which test to use. Doing hypothesis tests using the TI-83 is very similar to doing confidence intervals. All of the tests allow you to input either raw data or summary statistics. KADDSTAT provides statistical functions that allow you to perform hypothesis tests for both means and proportions. The basic steps are the same for each. The only differences are the input that the user must supply. Following, we will give detailed instructions for the small-sample hypothesis test for the mean. Once you understand this procedure, the others are very similar.

9.7.1 SMALL-SAMPLE TEST OF THE POPULATION MEAN USING EXCEL

From the KADD menu choose **Hypothesis Testing > One Sample.** You will see a list of the one-sample hypothesis tests that KADDSTAT will perform. Choose **Population Mean using t** and the dialog box shown in Figure 9.6 will open.

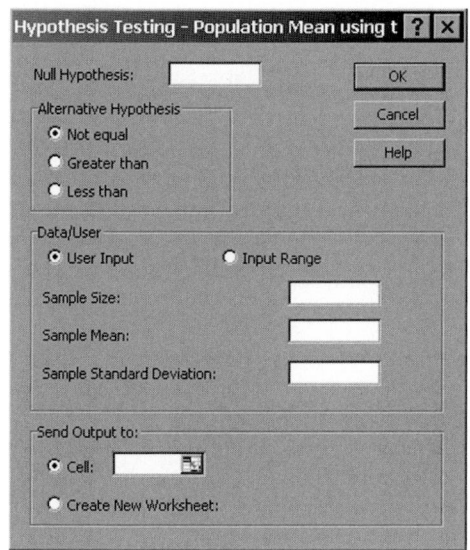

FIGURE 9.6 **Dialog box for small-sample test of** μ

The dialog boxes for hypothesis testing have three main parts: Null Hypothesis, Alternative Hypothesis, and Data/User.

Suppose that we are looking at the hand washing data and we want to perform a test to see whether the mean time spent washing hands is different from 5 seconds. Since this is a small sample, 28 employees, and we do not have the population standard deviation, the t test is appropriate. We will use a level of significance of 0.05.

1. In the box labeled Null Hypothesis, input the value of the hypothesized mean, 5 seconds.

2. Next we must identify the form of the alternative hypothesis. Since we want to know whether the mean has *changed*, this is a two-sided test. Click on the radio button next to **Not Equal.**

3. Now we must indicate whether we are doing the test from raw data or from a set of summary statistics. If you have summary statistics only, you would input the sample size, the sample mean, and the sample standard deviation. In this case we have the data so click on the button labeled **Input Range.** The dialog box will change to allow you to input the range in the Excel worksheet that contains the data. Position the cursor in the box labeled **Input Range** and highlight the data in the Excel worksheet. Click on **Header Row Included** if your data range is labeled with a variable name.

4. Last indicate where you want the results of the test to be located.

The completed dialog box should look like the one in Figure 9.7. Click **OK** and the output shown in Figure 9.8 will appear in the location you specified.

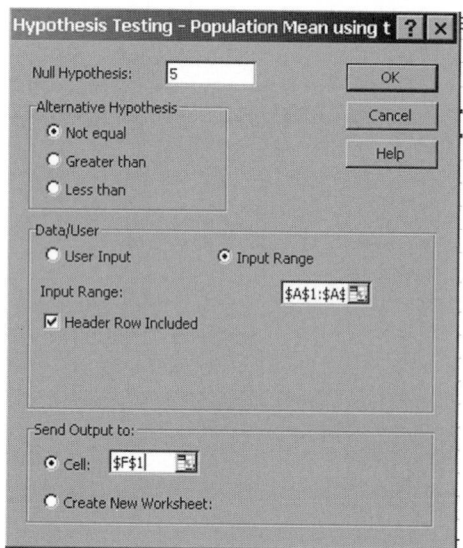

F	G
One Sample Test for μ	
p-value =	0.0015
Null Hypothesis: μ =	5
Alternative Hypothesis:	Not Equal
Sample Size: n =	28
Sample Mean: $\bar{X}$ =	2.5357
Sample Std. Dev.: S =	3.6866
Standard Error: S_X =	0.6967

FIGURE 9.8 **Output from the *t* test for the hand washing data**

FIGURE 9.7 **Completed dialog box for hand washing data**

From the output section, we see that the *p* value of the test is 0.0015. This means that unless our chosen level of significance is less than 0.0015, we will reject H_0 and conclude that the hand washing time has changed.

9.7.2 SMALL-SAMPLE TEST OF THE POPULATION MEAN USING MINITAB

Now we'll perform a hypothesis test about the hand washing data using Minitab. What we'd like to see is whether the mean time spent washing hands is different from 5 seconds. Since this is a small-sample test, $n = 28$, the *t* test is the appropriate test procedure. A portion of the data in a Minitab worksheet is shown in Figure 9.9.

↓	C1-T	C2-T	C3
	Observation	**Unit**	**Time 1**
1	1	CCU	3
2	2	CCU	2
3	3	CCU	0
4	4	CCU	5
5	5	CCU	2
6	6	CCU	0
7	7	CCU	2

FIGURE 9.9 **Hand washing data in Minitab**

To perform the test:

1. From the menu select **Stat** > **Basic Statistics** > **1-sample t** . . . . The dialog box shown in Figure 9.10 will open.

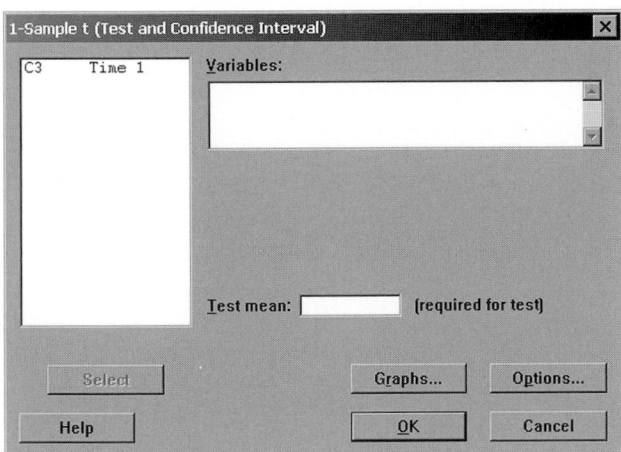

FIGURE 9.10 **1-Sample t test dialog box**

2. Minitab expects to use real data for the problem, not summary statistics. Position the cursor in the box labeled **Variables:** and select **Time 1** from the list of variables at the left.

3. The next input that Minitab expects is the hypothesized value for the mean. In this case we are interested in knowing whether the mean time is different from 5, so enter 5 in the box labeled **Test mean:.**

4. Click on the box labeled **Options** . . . and the dialog box shown in Figure 9.11 at the top of page 442 opens.

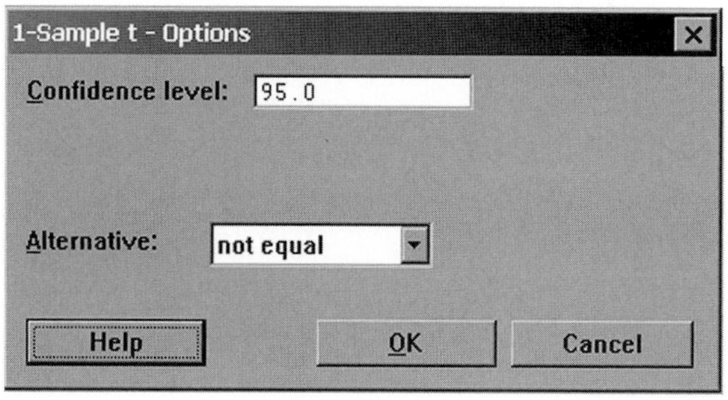

FIGURE 9.11 **Options box for 1-Sample t test**

Minitab uses this dialog box to allow you to specify the alternative hypothesis, H_A. That is, you can tell Minitab whether the test is two-tailed or one-tailed, and if it is one-tailed you can specify an upper- or lower-tail test. You also use this box to indicate the level of significance for the test. Be careful here. Minitab expects you to enter $1 - \alpha$, the confidence level, rather than α, the significance level.

5. Enter 95 in the box labeled **Confidence level:** and select **not equal** from the drop down menu labeled **Alternative:.** Press **OK** to return to the main dialog box.

6. Click **OK.** The output will appear in the Session window as shown in Figure 9.12.

One-Sample T: Time 1

Test of mu = 5 vs mu not = 5

Variable	N	Mean	StDev	SE Mean
Time 1	28	2.536	3.687	0.697

Variable	95.0% CI	T	P
Time 1	(1.106, 3.965)	−3.54	0.001

FIGURE 9.12 **Output for One-Sample T test**

From the output we see that the value of the test statistic is −3.54 and the p value of the test is 0.001. Thus, we reject H_0 and conclude that the mean time is different from 5 seconds.

9.7.3 SMALL-SAMPLE TEST WITH THE TI-83

Doing hypothesis tests using the TI-83 is very similar to doing confidence intervals. All of the tests allow you to input either raw data or summary statistics.

Suppose that we would like to perform a hypothesis test for the hand washing data. We would like to test whether the mean time for hand washing is different from 5 seconds. Since there were 28 times recorded, the appropriate test is the t test. We will do the test with both raw data and summary statistics.

For this example, we will assume that the data are entered into L_1 in the TI-83.

1. Select [STAT] and then use the [▶] key to move over to **TESTS.** From the list shown in Figure 9.13 select option **2: T-Test . . . .**

FIGURE 9.13 **TI-83 Stat Test menu**

2. The screen shown in Figure 9.14 should appear. Since we have the raw data, we are using the **Data** option.

If the screen in Figure 9.14 does not appear, move the cursor over to Data and hit [ENTER].

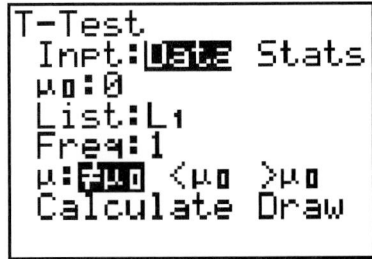

FIGURE 9.14 **T-Test screen for raw data**

3. The inputs required are the value of the hypothesized mean (μ_0), the location of the data **(List),** and the form for the alternative hypothesis (μ:). Enter 5, L_1 (or whichever list you used) and $\neq \mu_0$, respectively, as shown in Figure 9.15.

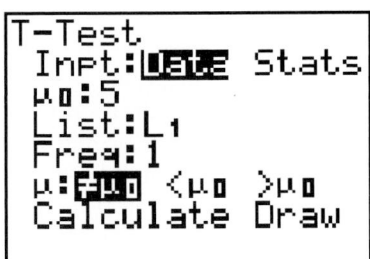

FIGURE 9.15 **Completed T-Test screen**

4. There are two different ways to have the test output reported. The **Calculate** option will produce output similar to that from Excel and Minitab. This is shown in Figure 9.16.

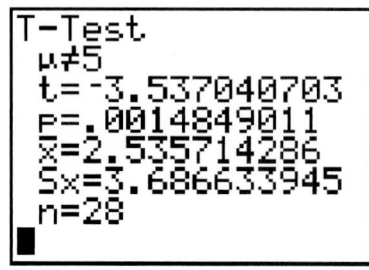

FIGURE 9.16 **Calculate Output from T-Test**

The **Draw** option provides a graphical display of the test statistic as well as the value of the test statistic and the p value. This output is shown in Figure 9.17.

Note: Be sure you don't have any other plots turned on when you use the Draw option.

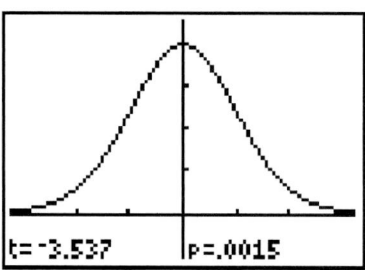

FIGURE 9.17 **Draw Output from T-Test**

CHAPTER 9 SUMMARY

In this chapter you learned how to do a hypothesis test of the population mean, population variance, and population proportion. The five-step hypothesis-testing procedure is the same for all hypothesis tests. The differences in the tests are in the rejection regions and the calculations of the test statistic and the p value. For a test of the population mean, you use a Z test if the population standard deviation is known or if $n > 30$; otherwise, you use a t test. For a hypothesis test of the population variance, use a chi-square test; and for a test of proportions, use a Z test.

In any hypothesis test the rejection region and thus the final decision depend on the value of α. You can alter the outcome of the test by adjusting the value of α. To handle the potential unethical use of hypothesis testing and to provide decision makers with more information than simply whether the null hypothesis is rejected, you can use p values. The p value frees you from specifying a value of α for the test. By reporting the p value for the test, you put the decision to reject or fail to reject the null hypothesis in the hands of decision makers. They must weigh the costs of Type I and Type II errors and make a decision in light of these costs and the p value. The final section of this chapter ties the results of a hypothesis test to the corresponding confidence interval calculation.

KEY FORMULAS

Term	Formula	Page reference
t test statistic	$t = \dfrac{\bar{X} - \mu}{s/\sqrt{n}}$	412
Chi-square test statistic	$\chi^2 = \dfrac{(n-1)\,s^2}{\sigma^2}$	423
Z test statistic	$Z = \dfrac{\hat{p} - p}{\sqrt{p(1-p)/n}}$	430

CHAPTER 9 EXERCISES

Learning It!

9.15 Most traffic lights are set so that there is enough time for pedestrians to cross the road safely. A recent study indicates that a large number of elderly cannot get across the road in the usual 15 seconds allowed. To determine the average amount of time it takes senior citizens to cross the street, a study was done of 25 seniors. On the average, it took them 19.5 seconds to cross the street, with a sample standard deviation of 5 seconds. Assume the time to cross the road has a normal distribution.

(a) Set up the null and alternative hypotheses to see whether the data show that it does indeed take seniors longer than 15 seconds to cross the street.

(b) In terms of traffic flow, what are the implications of a Type I error?

(c) What are the implications of a Type II error?

(d) Find the value of the test statistic.

(e) If $\alpha = 0.05$, what is the rejection region?

(f) What is your recommendation to the city?

9.16 Kids think moms are doing a great job! According to a recent poll conducted by Massachusetts Mutual Insurance Company, 90% of children think their mothers spend enough time with them. A similar survey was conducted in New York to see whether children there felt the same way. Of 1000 children surveyed in New York, 880 thought their mothers spend enough time. Do these data indicate that fewer than 90% of the New York children think their mothers spend enough time with them?

(a) Set up the null and alternative hypotheses.

(b) What is the value of the test statistic?

(c) If $\alpha = 0.10$, what is the rejection region?

(d) What do you conclude about the children of New York?

(e) What might be causing this difference?

9.17 In a new advertising campaign, Coca-Cola pokes fun at the decision to revive its famous contoured bottle. Atlanta-based Coke faced significant technical hurdles in bringing out the well-known bottle in a new material: plastic. Coke conducted extensive marketing tests to make sure that one of the world's best-known packages, seldom seen since the 1970s, would be a hit in the 1990s. Coca-Cola wants to be sure that at least 60% of consumers prefer the contoured bottle. Of 3000 consumers surveyed nationwide, 1900 prefer the new bottle. Is this sufficient evidence to give the new bottle the go-ahead?

(a) Set up the null and alternative hypotheses for Coca-Cola.

(b) In terms of Coke's decision to introduce the new bottle or not, what is a Type I error?

(c) In terms of Coke's decision to introduce the new bottle or not, what is a Type II error?

(d) What is the value of the test statistic?

(e) If $\alpha = 0.05$, what is the rejection region? What is your decision?

9.18   You purchased your home in 1987 for $165,000, which was the average price at the time. Now you are thinking of selling your home. You look at a sample of ten homes sold in your neighborhood during the past 2 months and find these sales prices (in dollars):

130,000	135,500	136,000	140,000	160,000
167,000	168,000	174,500	177,400	180,000

(a) Based on these data, conduct a hypothesis test to see whether the average selling price has increased from $165,000. Assume that the selling price of homes in this town is normally distributed.

(b) What should you decide about selling your home at this time? Explain your answer in terms of the results of the hypothesis test.

9.19   The Charlotte Sting is a team in the Women's NBA. The manager is concerned about attendance during the 2001 season and has collected the following data through August 15, 2001:

Opponent	Home dates	Home att.	Road dates	Road att.
Cleveland	1	6,077	2	21,600
Detroit	2	9,576	1	5,780
Houston	1	11,057	0	0
Indiana	2	10,426	1	8,024
Los Angeles	1	12,843	1	8,823
Miami	1	7,064	2	17,919
Minnesota	0	0	1	7,494
New York	2	15,782	1	13,150
Orlando	2	9,635	1	7,211
Phoenix	0	0	1	7,008
Portland	0	0	1	8,616
Sacramento	1	4,077	1	5,866
Seattle	1	7,007	1	4,016
Utah	1	4,085	0	0
Washington	1	7,896	2	28,533

(a) Calculate the average and standard deviation of attendance at home games.

(b) Calculate the average and standard deviation of attendance at away games.

(c) Suppose that last year the average home game attendance was 5000. Test this year's home game attendance against 5000 to see whether the average attendance is different from last year. Assume the attendance figures are normally distributed.

(d) Suppose that last year the average away game attendance was 8000. Test this year's away game attendance against 8000 to see whether the average attendance is different from last year. Assume the attendance figures are normally distributed.

(e) What additional data do you think the manager should be tracking with regard to attendance? Explain your answer.

9.20   In the wake of a large number of repeat crime offenders, many states are considering the death penalty. A sample of 100 citizens of a New England state was surveyed, and 56 people favored the death penalty.

(a) Legislatures in favor of the death penalty are using the results of this survey to say that the majority of the people in the state are in favor of the death penalty. Conduct the appropriate hypothesis test to determine whether these legislatures are justified in their remark.

(b) What conclusion can you reach from this survey about the preference of citizens of this state for the death penalty?

Thinking About it!

9.21 Coca-Cola has asked for your recommendation about the new bottle. Basically, the company wants to know whether you are secure in your decision before it makes any drastic changes. You decide to test how much the procedure you chose affected your decision.

Requires Exercise 9.17

(a) Suppose you use $\alpha = 0.10$. What is your decision now?

(b) Use $\alpha = 0.01$. What is your decision now?

(c) Find the p value for this problem.

(d) Based on this new information, what will you tell Coca-Cola to do?

9.22 Look again at your recommendation to the city about the length of its walk light signal. In writing your report, you start to think about whether the study you did was adequate.

Requires Exercise 9.15

(a) What other factors might be useful in determining the length of time to allow people to cross the street safely?

(b) What are the drawbacks of making decisions such as this based on the mean?

(c) What statistic might be better to use in this case?

9.23 Many states are considering allowing casino gambling in an attempt to revitalize urban areas. Opponents say gambling will be a magnet for crime and will shift spending away from entertainment and other industries. A recent poll conducted by the Boston *Sunday Herald* found that 46% of the Bay State voters are opposed to gambling, whereas 36% support the proposal. Eighteen percent of those surveyed had no opinion or were neutral. The poll of 450 registered voters across the state was conducted between a Tuesday and a Friday. Legislatures will endorse the casinos if fewer than 50% of the voters oppose it. Do the data indicate that this is the case?

(a) Set up the null and alternative hypotheses.

(b) What should you do with the 18% who had no opinion or were neutral?

(c) Find the value of the test statistic.

(d) If the members of the legislature want to be conservative, how should they set α?

(e) If the members of the legislature want to be more aggressive on this issue, how should they set α?

(f) Using the value of α set in part (d), find the rejection region.

(g) What is your recommendation to your local representative?

(h) Using the value of α set in part (e), find the rejection region.

(i) What is your recommendation now to your local representative?

9.24 Remember when you were trying to convince your boss that your expense report for the aspirin purchase was justified? Your boss asks you for the data you used to justify the expense. He points out that the price of aspirin in Tokyo seems high and asks you to reduce your estimate. The data are repeated here:

Requires Exercise 9.1

City	Aspirin (100)	Fast food (hamburger, fries, soft drink)	Woman's haircut/blow dry	Toothpaste (6.4 oz)
Los Angeles	$7.69	$4.15	$20.11	$2.42
Tokyo	35.93	7.62	76.24	4.24
London	9.69	5.80	44.35	3.63
Sydney	7.43	4.53	29.93	2.08
Mexico City	1.16	3.63	17.94	1.08

(a) If you remove Tokyo from the data set should you change your expense report?

(b) What does this exercise tell you about the effect of outliers on the hypothesis test of μ when the sample is small?

Doing It!

Datafile:
WASHHANDS.XXX

9.25 The data that the school nurse collected are given here:

Days missed	Grade	Gender
7	1	M
6	5	F
7	5	M
6	5	F
5	2	F
6	4	M
8	4	F
7	1	F
7	3	M
6	3	F
5	K	F
8	4	M
7	5	M
6	5	F
7	3	M
5	5	F
8	3	M
5	3	F
6	2	M
10	2	M

You can see that information was also collected on the grade and gender for each student in the experiment. Assume that the days missed are normally distributed.

(a) Find the sample means and sample standard deviations for the days missed by gender and by grade.

(b) Test whether the average number of days missed for the boys was less than 7.5 days.

(c) Test whether the average number of days missed for the girls was less than 7.5 days.

(d) Test the hypothesis that the variance of the days missed for the hand washers is different from the school variance of 2 days^2.

(e) Find the sample proportion of student hand washers who missed more than 7.5 days of school.

(f) Test the hypothesis that the proportion of students who missed more than 7.5 days of school is less than the schoolwide percentage of 25%.

(g) Based on your analysis, write a memo to the superintendent of schools summarizing your results and making a recommendation.

Datafile:
DEATHPENALTY.
XXX

9.26 The death penalty is an ongoing social issue in the United States. A data file containing information on all of the executions in the United States from 1977 to 2001 has been constructed by Dr. David Russell at Western New England College. Here are the first lines of this file:

Number since 1977	Date of execution	Date of conviction	Date of crime	Date of birth	First name(s)	Last name	Suffix	State	Gender of defendant	Gender of victim	Race of defendant	Race of victims	Number of victims
1	17-Jan-77				Gary	Gilmore		UT	Male		White	White	1
2	25-May-79				John	Spenkelink		FL	Male		White	White	1
3	22-Oct-79				Jesse	Bishop		NV	Male		White	White	1
4	9-Mar-81				Steven	Judy		IN	Male		White	White	3
5	10-Aug-82				Frank	Coppola		VA	Male		White	White	1
6	7-Dec-82				Charlie	Brooks		TX	Male		Black	White	1
7	22-Apr-83				John	Evans		AL	Male		White	White	1
8	2-Sep-83				Jimmy	Gray		MS	Male		White	White	1

(a) Find the percentage of those executed by race; that is, what percentage were Asian? Black? Latino/a? Native American? White?

(b) Given that the percentage of Blacks in the United States is 13.3%, test whether the percentage of Black executions is different from 13.3%.

(c) The data file gives the state. Find the percentages of executions that took place in the South, Midwest, West, and Northeast.

(d) Given the 35.1% of the U.S. population lives in the South, test whether the percentage of executions in the South is greater than 35.1%.

(e) Test whether more than 50% of those executed were men.

(f) Find the percentage of same-race defendant and victims. Test whether this percentage is greater than 50%.

(g) Summarize your findings in a memo to your congressperson.

10

Comparing Two Populations

DO ONLINE COURSES REALLY WORK?

Recently colleges and universities have been moving to provide online courses to students. Many courses use the World Wide Web for part of their content, but whole courses and entire degree programs are now available online. Some educators are skeptical about the amount of learning that can be done in an online course compared with a traditional classroom-based course. It would be inconceivable that professors might not be necessary to learning—wouldn't it?

A university professor performed an experiment to examine this issue: Do students in online courses really learn? How does the learning that takes place online compare with learning in a traditional on-ground course? The professor taught four sections of an introductory statistics course in a single semester. Two sections were offered entirely online and two were traditional on-ground courses. Initially 25 students were enrolled in each of the sections, although some students in both types of course dropped out before the end of the semester. Students were given a statistics pretest at the beginning of the semester and a similar posttest at the end of the semester. A sample of the data follows:

Student	Type of class	Gender	Pretest	Posttest	Course grade
1	On ground	M	51	85	P
2	On ground	M	52	65	F
3	On ground	M	71	76	P
4	On ground	M	82	78	P
5	On ground	M	63	87	P
6	On ground	M	56	81	P

In Chapter 9 you learned how to do hypothesis tests for many different population parameters. In each case you tested hypotheses about whether the sample data could have come from a population with a parameter equal to a specific value. Although these tests provide valuable information to decision makers, they are not the only questions you might want answered. For example, suppose you want to know the answer to these questions:

- Who uses e-mail more frequently, men or women?
- Is it better to promote a person from within the organization or to hire from outside the organization?
- Do men and women behave the same way in the supermarket?
- Does one design of a golf ball travel farther than another?
- Do college graduates actually earn more money than high school graduates?

Although these questions cover widely different topics, they do have a common theme. In each case a comparison is being made. In this chapter we extend the hypothesis-testing procedure from Chapter 9 to handle the comparison of two population means, two population proportions, and two population variances. The basic five-step hypothesis-testing procedure is the same; the only difference is the test statistic to be used.

Chapter 9 presented hypothesis tests of a mean, proportion, or variance from a single population.

This chapter covers the following topics:

- Collecting data from two populations
- Large-sample test of the difference in two population means
- Small-sample test of the difference in two population means
- Test of the difference in two population means—paired data
- Test of the difference in two population proportions
- Test of the difference in two population variances

10.2   Collecting Data from Two Populations

When we compare the characteristics of two different populations, we must have a sample from each of the populations. These samples are usually selected independent of each other. In other words, the selection of one sample does not have any effect on the selection of the second sample. We will label all the parameters of one population with a subscript 1 and all the parameters of the second population with a subscript 2. It does not matter which population is labeled 1 or 2. The populations and samples are illustrated in Figure 10.1 on page 452.

Consider the question about whether men or women spend more money on frozen foods. We can label the population of men population 1 and the population of women population 2. Then the parameters and statistics that correspond to the male population are identified with a subscript 1 and those that correspond to the female population carry a subscript 2. The variables are listed in Table 10.1 on page 452.

We can select a sample of men shoppers and a separate sample of women shoppers. We could ask all members of the sample how much money they spent on

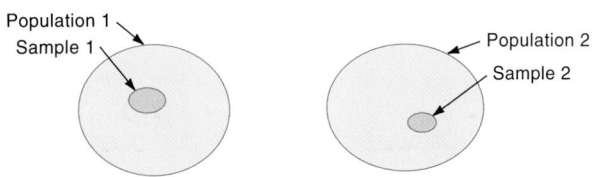

FIGURE 10.1 **Two populations and two samples**

TABLE 10.1		
PARAMETERS AND STATISTICS FOR TWO POPULATIONS		
	Population 1: men	**Population 2: women**
Size of population	N_1	N_2
Population mean	μ_1	μ_2
Population standard deviation	σ_1	σ_2
Sample size	n_1	n_2
Sample mean	$\overline{X}_1$	$\overline{X}_2$
Sample standard deviation	s_1	s_2

frozen foods in the past week. It is not necessary that the two sample sizes be equal, but if possible both sample sizes (n_1, n_2) should be greater than or equal to 30. The reason is that the Central Limit Theorem generally applies when the sample size is 30 or greater. Remember that the Z test statistic is based on the knowledge that the sample mean, $\overline{X}$, has a normal distribution.

Qualitative variables are often used to identify two populations for comparison.

Often we may select a single sample and use a qualitative variable to identify two populations for comparison. For the food shopping example, we might select one sample of shoppers and then record the gender of each respondent as part of the data. The data can then be divided into the two comparison populations after they have been collected. If we also collect data on each shopper's age as "under 40" or "40 and over," then we can compare the average frozen food expenditures for younger and older buyers. Clearly, spending differences that are identified by gender or age could be of great assistance in developing a marketing strategy.

10.3 Hypothesis Test of the Difference in Two Population Means—Overview

Population standard deviations are known or the sample sizes are large.

There are several different cases to consider in testing a hypothesis about the difference between two population means. To determine which case we have, we must first see if we have information on the population standard deviation of both populations. As discussed in Chapter 9, sometimes we know the value of the population standard deviation but most of the time we do not. If we know the value of the standard deviation for each of the two populations we are comparing, then we use a Z test statistic. In this case, it does not matter how large the samples are, but the underlying populations must be normally distributed. This case is covered in Section 10.4.

Remember! When $n \geq 30$ the Central Limit Theorem generally applies.

If we do not know the value of the standard deviation for the two populations, then we must check the size of the sample. If each of the samples has 30 or more observations, then we are in what is commonly referred to as the "large-sample" case. This case is very similar to the situation when we know the standard deviations.

If we do not know the population standard deviations and if one or both of the samples has fewer than 30 observations, then we are in the "small-sample" case. We must decide whether the variances in each of the two populations are the same, even though they are unknown, or different. Section 10.5 covers the case when we do not know the value of the standard deviations but we know that the populations are normally distributed.

Population standard deviations are unknown and sample sizes are small.

Finally, sometimes the data are collected in such a way that the samples are not independent. This happens often when we compare two different medical procedures or treatments. We want to be sure that the differences we observe are not simply due to the fact that we have different patients with different medical histories. In this case we might use the same patients in both samples. If the samples are dependent, then we use what is known as a paired *t* test to compare the population means. This is the subject of Section 10.7.

Paired data must be analyzed differently.

In summary, three sections of this chapter are devoted to hypothesis testing of the difference in means. Section 10.4 covers tests when the standard deviation is known, Section 10.5 covers the cases when the standard deviation is not known, and Section 10.7 covers the test for paired data. Regardless of the particular case, each hypothesis test follows the five-step procedure explained in Chapters 8 and 9:

Step 1: Set up the null and alternative hypotheses.

Step 2: Define the test procedure.

Step 3: Collect the data and calculate the test statistic and the *p* value.

Step 4: Make a statistical decision (reject or fail to reject the null hypothesis).

Step 5: Interpret the statistical decision in terms of the stated problem.

Steps for any hypothesis test

Because you are already familiar with hypothesis testing, in this chapter we often combine the first two steps, which define the test procedure.

10.4 Large-Sample Tests of the Difference in Two Population Means

10.4.1 LARGE-SAMPLE TESTS OF TWO MEANS WITH KNOWN STANDARD DEVIATIONS

In the basic test of two population means we want to know whether the two samples come from populations that have equal means and we assume that the population standard deviations are known. Although this may not seem reasonable, there are cases where a historical or specified standard deviation is appropriate.

EXAMPLE 10.1

TISSUE STRENGTH

Identifying the Two Populations

A tissue company that is looking at customer complaints about tissues tearing wants to reexamine the problem. Because tissue strength seems to vary, your boss has asked you to find out whether there is any difference in the strengths of the tissues made on different days. The company uses two measures of tissue strength: machine-directional strength (MDStrength) measured in pounds per

Understand the problem.

Collect the data.

ream and cross-directional strength (CDStrength) also measured in pounds per ream. You will look at MDStrength first.

The product specifications for MDStrength state that the mean should be 1000 lb/ream with standard deviation of 50 lb/ream. You have collected data on 75 sheets for each of three different days. For each sheet you recorded the MD-Strength and the CDStrength.

The first step is to identify the two populations to be compared. The tissue manufacturer wants to compare the strength of tissues made on day 1 with the strength of tissues made on day 2. Population 1 is all tissue sheets made on day 1 and we have a sample of size $n_1 = 75$ from this population. All tissue sheets made on day 2 make up population 2 and we have a sample of $n_2 = 75$ from this population.

Use the techniques of Chapter 9 to test the variance for each of the populations. It is rare to have the population standard deviations known, really.

In this example we use the specified values of the population variances to perform the test. Of course, this is not something we do blindly. Just because specifications exist does not mean that a company actually complies with them! Still, for this example we have established that the specified values of the population standard deviations are valid.

Once we can establish that the values of the population standard deviations for both populations are known, we know what case we are dealing with and we are ready to proceed with the first step of the hypothesis-testing procedure: to construct the null and alternative hypotheses. As with tests of a single mean, there are three different ways to set up the hypothesis test. The guidelines for determining which of these tests to use were presented in Chapter 9. Here are the three possible setups:

Remember! The key words "different," "less than," and "greater than" indicate which test to use.

Two-tail test

H_0: $\mu_1 = \mu_2$ or H_0: $\mu_1 - \mu_2 = 0$
H_A: $\mu_1 \neq \mu_2$ H_A: $\mu_1 - \mu_2 \neq 0$

Use this test to decide whether the mean of population 1 is different from the mean of population 2.

Lower-tail test

H_0: $\mu_1 \geq \mu_2$ or H_0: $\mu_1 - \mu_2 \geq 0$
H_A: $\mu_1 < \mu_2$ H_A: $\mu_1 - \mu_2 < 0$

Use this test to decide whether the mean of population 1 is less than the mean of population 2.

Upper-tail test

H_0: $\mu_1 \leq \mu_2$ or H_0: $\mu_1 - \mu_2 \leq 0$
H_A: $\mu_1 > \mu_2$ H_A: $\mu_1 - \mu_2 > 0$

Use this test to decide whether the mean of population 1 is greater than the mean of population 2.

Note: Never have $\overline{X}$'s in the statement of the hypotheses. There is no need to formulate a hypothesis about the sample means because we know the values. We do not know the population means.

Notice that the = sign is always part of the null hypothesis. Also notice that the hypotheses are statements about the relationship between the sizes of the mean of population 1 and the mean of population 2. The tests do not give information about the value of the means, only about how the value of μ_1 compares with the value of μ_2.

There are two ways to state each of the hypotheses. For each setup, the first way of writing the test makes a statement about the relative values of μ_1 and μ_2. The second way of writing the same test makes a statement about the value of the difference, $\mu_1 - \mu_2$. The statements are equivalent. Look at the two-tail test. Clearly, if $\mu_1 = \mu_2$, then the difference between them must be 0. It is also possible to test for differences other than 0. For example, we could test that the difference between the two means is 10 or any other value. The procedure is basically the same with only a slight difference in the test statistic formula.

EXAMPLE 10.2

TISSUE STRENGTH

Analyze the data.

Setting Up the Hypotheses and the Test Procedure

The tissue manufacturer wishes to see whether the mean tissue strengths are different for tissues made on day 1 and day 2. The variable to be examined is MD-Strength. The null and alternative hypotheses are

$$H_0: \quad \mu_1 = \mu_2$$
$$H_A: \quad \mu_1 \neq \mu_2$$

The test is also defined by a particular level of significance, α. Suppose for this test we choose $\alpha = 0.05$. The test statistic to be calculated in Step 3 is a Z statistic, so we know how to find the rejection region for the two-tail test. The total area of the rejection region must be $\alpha = 0.05$, so we split α in half and require the area of each tail of the normal distribution to be 0.025. From the standard normal tables we find the critical values for the test to be ± 1.96.

The rejection regions are found using the procedures from Chapter 8.

Once the test has been defined, the data are collected and processed. This allows us to proceed with Step 3 of the hypothesis-testing procedure: calculating the value of the test statistic and the *p* value.

Remember that the test statistic is calculated from sample data and is used to decide between the null and alternative hypotheses. If we are trying to decide whether $\mu_1 = \mu_2$, then it makes sense to look at the size of the difference between $\overline{X}_1$ and $\overline{X}_2$. We know that even if the two population means are exactly the same, we almost never get two equal sample means. So we don't expect the difference $\overline{X}_1 - \overline{X}_2$ to be 0 even if the null hypothesis is true. However, if the null hypothesis is true, then the difference $\overline{X}_1 - \overline{X}_2$ should be small relative to the size of the standard error. Clearly, large differences in the sample means will lead us to be suspicious of the null hypothesis.

The appropriate test statistic is the Z statistic:

$$Z = \frac{(\overline{X}_1 - \overline{X}_2) - 0}{\sqrt{\sigma_1^2/n_1 + \sigma_2^2/n_2}}$$

Test statistic for the difference in two population means, variances known

The numerator of the test statistic simply compares the evidence, $\overline{X}_1 - \overline{X}_2$, with the difference between the population means if H_0 is true. For the hypotheses we are using, if H_0 is true then the true difference is 0, so the comparison is between $\overline{X}_1 - \overline{X}_2$ and 0. As you saw in Chapter 9, the size of the difference is always measured in terms of the standard error. Thus, the denominator is the standard error of the point estimator. In this case we are trying to estimate $\mu_1 - \mu_2$ and the natural point estimator is $\overline{X}_1 - \overline{X}_2$. The standard error of $\overline{X}_1 - \overline{X}_2$ is a natural extension of the standard error $\overline{X}$, $\sqrt{\sigma^2/n}$, which we often wrote as $\sigma/\sqrt{n}$.

When we are testing for a difference of 0, the last part of the numerator is often left out entirely. This makes sense, but it is important that you understand what we are really testing. When we need to test for a difference equal to some value other than 0, we put that value in the formula for the test statistic in place of the 0. For example, if we wanted to test whether the difference in the population means was equal to 10, then we would change the null and alternative hypotheses as follows:

Remember! A point estimator is calculated from sample information only and is used to estimate an unknown population parameter.

$$H_0: \quad \mu_1 - \mu_2 = 10$$
$$H_A: \quad \mu_1 - \mu_2 \neq 10$$

We would then use the value of 10 in the formula for the test statistic instead of 0. This is called the hypothesized difference and is labeled d.

EXAMPLE 10.3

TISSUE STRENGTH

Calculating the Test Statistic and the p Value

Now we are ready to calculate the test statistic for the tissue manufacturer. Here are the relevant data:

Population 1: day 1 tissues	Population 2: day 2 tissues
$n_1 = 75$	$n_2 = 75$
$\overline{X}_1 = 990.8$	$\overline{X}_2 = 977.0$
$\sigma_1 = 50$ lb/ream	$\sigma_2 = 50$ lb/ream

Analyze the data.

We calculate the value of the Z test statistic:

$$Z = \frac{(990.8 - 977.0) - 0}{\sqrt{50^2/75 + 50^2/75}} = 1.69$$

Using the standard normal table, we can find the p value of the test by looking up the test statistic of 1.69. We find that the area to the right of $Z = 1.69$ is 0.0455. Since this is a two-tail test, we double that and find that the P value is 0.09.

Although much of the work is done at this point, the test procedure is not completed. Step 4 of the hypothesis-testing procedure requires that we either reject or fail to reject the null hypothesis on the basis of the test statistic. In Step 5 of the procedure we interpret this decision in terms of the stated problem. Let's finish the tissue strength problem now.

EXAMPLE 10.4

TISSUE STRENGTH

Finishing the Hypothesis Test

The test statistic for the comparison of tissue strengths is 1.69. The rejection region for the test has critical values of ± 1.96. Since 1.69 is not outside the critical values—that is, it is not in the rejection region of the test—we fail to reject the null hypothesis. Our conclusion is that there is no evidence that the mean MDStrength of tissues made on day 1 is different from the mean of tissues made on day 2.

Analyze the data.

From previous experience in hypothesis testing, we know that we should also examine the p value for these data. The p value is less than 0.10. This means that if we set α greater than the p value—say, at 0.10—then we would reject the null hypothesis of equal means. The decision, in this case, depends very much on the choice of α. We might say that the decision is "on the fence" with regard to these two hypotheses. Perhaps we should compare day 2 with day 3 to get a better understanding of what is happening.

✓ TRY IT NOW!

Tissue Strength
Comparing Two Different Means

The tissue manufacturer also recorded values of MDStrength for 75 tissues made on day 3. The sample mean MDStrength for that day was found to be 1000.32 lb/ream. Is there any evidence that the average MDStrength is different on day 2 than on day 3? Use $\alpha = 0.05$. Identify the two populations and conduct a hypothesis test of the difference in the means.

Step 1:

Step 2:

Step 3:

Step 4:

Step 5:

Once the conclusion of the test is reached, the real work of decision making begins. The statistical analysis simply confirms or fails to confirm a hypothesis. It does not help us make decisions about the problem we are trying to solve.

EXAMPLE 10.5

TISSUE STRENGTH

Interpreting and Using the Results of a Test

As a result of our tests, we concluded that the average MDStrength of the tissues made on day 1 and day 2 are equal but the average MDStrength of the tissues made on day 2 and day 3 are different. Clearly, something is not right with this process. How can we interpret this? What should we do next?

What we know as a result of this test is that the tissue strength is not behaving entirely as it is supposed to. It is not clear how the mean strength differs, or whether it is always different or just sometimes different. It is also not clear on which, if any, of the three days the MDStrength is correct.

This is a good time to graph the data and see what is happening and look for any patterns in the data.

Draw conclusions.

To answer the original question, we want to compare all three days. That is, we want to do a hypothesis test that looks like this:

$$H_0: \quad \mu_1 = \mu_2 = \mu_3$$
$$H_A: \quad \text{At least one mean is different.}$$

This is a test of more than two means and we cannot do such a test using the techniques of this chapter. We need a technique called analysis of variance, which is covered in Chapter 14.

ANS. $Z = -2.86$; p VALUE $= 0.0042$. REJECT H_0 AND CONCLUDE THAT THE AVERAGE MDSTRENGTH IS DIFFERENT ON DAY 2 THAN ON DAY 3.

10.4.2 LARGE-SAMPLE TESTS OF TWO MEANS WITH UNKNOWN STANDARD DEVIATIONS

The large-sample test for the difference between two population means requires that both of the sample sizes be greater than 30. Remember that the only differences in the various hypothesis tests are in the calculation of the test statistic.

The test statistic for this case is similar to the case when we know the standard deviations. Because the sample sizes are large, each individual sample standard deviation is a good estimate of the corresponding unknown population standard deviation. So, we simply use each of the sample standard deviations in the formula instead of the corresponding values of σ. The test statistic becomes

Test statistic for the difference in two population means, variances unknown, $n_1, n_2 \geq 30$

$$Z = \frac{(\overline{X}_1 - \overline{X}_2) - 0}{\sqrt{s_1^2/n_1 + s_2^2/n_2}}$$

Finding the rejection region is the same as it is for any hypothesis test involving the Z distribution—it depends on α and whether the test is one-tail or two-tail.

Understand the problem.

EXAMPLE 10.6

ONLINE COURSES

Setting Up the Hypotheses

The university professor is interested in comparing how much students learn in online versus traditional (on-ground) courses. The students in each type of course took a test at the end of the course. The professor wants to know whether students in the on-ground course achieved better results. In this case the two populations are:

- Population 1: Students in traditional on-ground course
- Population 2: Students in online course

Since the professor is interested in whether one group is better than the other, the test is one-tail and the hypotheses are

$$H_0: \quad \mu_1 \leq \mu_2$$
$$H_A: \quad \mu_1 > \mu_2$$

Once the hypotheses are set up and we know the population parameters we are testing, we need to make a decision about which specific type of test to use. We need to assemble the data and look at the standard deviation and sample size. For tests about the mean, the first question is whether the population standard deviation is known. If it is, then we perform the Z test defined in the previous section. If it is not, then we look at the sample sizes to make the next decision. When both sample sizes n_1 and n_2 are large (≥ 30), then the test is also a Z test with the test statistic just described.

Analyze the data.

EXAMPLE 10.7

ONLINE COURSES

Defining the Test Precedure

The professor assembled and processed the data as shown here:

	Population 1: on-ground	Population 2: online
Sample size	$n_1 = 47$	$n_2 = 46$
Sample mean	$\overline{X}_1 = 84.34$	$\overline{X}_2 = 79.78$
Sample standard deviation	$s_1 = 9.08$	$s_2 = 10.06$

After some students dropped the course, 47 students in the on-ground course took the posttest and 46 people in the online course took the posttest. The study was a new one, so the professor had only sample measures of variation. She is able to use a Z test, however, because of the large sample sizes.

Once the professor knows she will use a Z test, she can find the rejection region. In this case there is no reason to choose a level of significance other than $\alpha = 0.05$, so that is what is used.

The one-tail rejection region is found using the techniques of Chapter 9. The value of Z is 1.645; that is, the rejection region is to the right of the value 1.645.

Steps 1 and 2 of the hypothesis test define the test procedure to be used. The only other computational portion of the test is to calculate the test statistic. This is completely determined by the test procedure and so it is a matter of using the correct formula.

EXAMPLE 10.8

ONLINE COURSES

Calculating the Test Statistic and the p Value

The professor studying the different types of courses knows that she is using a Z test for the data. The test statistic is calculated as follows:

$$Z = \frac{(84.34 - 79.78) - 0}{\sqrt{9.08^2/47 + 10.06^2/46}} = \frac{4.56}{1.98} = 2.30$$

To find the p value of the test, we look up the Z score of 2.30 on the standard normal table and find that the p value is 0.0107.

Analyze the data.

The remainder of the hypothesis-testing procedure is the same for this test as it has been for all the others we have done. So let's complete the test by doing Steps 4 and 5.

EXAMPLE 10.9

ONLINE COURSES

Drawing Conclusions and Making an Interpretation

That last two steps involve making decisions. First, we make a statistical decision. Since $Z = 2.30$ is in the rejection region, we reject the null hypothesis. Second, we must make a real decision: What does the test tell us and what might it mean? Because we reject the null hypothesis, we conclude that the mean posttest score for the traditional on-ground course is higher than that for online course.

Draw conclusions.

You are probably not surprised at this conclusion because the average for the on-ground course was higher than the average for the online course. However, this is not sufficient reason to conclude that they are different. You cannot simply look at the two $\overline{X}$ values and conclude that students in the on-ground course did better on the posttest than students in the online course. Yes, $\overline{X}_1$ is higher than $\overline{X}_2$ if you look only at the observed difference. But it is possible that such a difference might occur due to the random nature of the sampling and testing procedure. The only way to tell whether there is a "significant" difference is to compare the observed difference in the sample means with the size of the standard error. This is precisely what the test statistic does for you.

Remember! The standard error is the standard deviation of the point estimate. It is the yardstick by which we judge all differences.

✔ TRY IT NOW!

Online Courses
Comparing Two Different Means

The professor looking at differences in learning for on-ground and online courses is not convinced that she can simply say that on-ground courses are better. She wants to look at how the two groups compared on the pretest. Here are the relevant data:

	Population 1: on-ground	Population 2: online
Sample size	$n_1 = 47$	$n_2 = 46$
Sample mean	$\overline{X}_1 = 58.87$	$\overline{X}_2 = 53.70$
Sample standard deviation	$s_1 = 11.18$	$s_2 = 13.77$

Perform a hypothesis test to see whether there is a difference between the mean pretest scores. Use $\alpha = 0.05$.

Step 1:

Step 2:

Step 3:

Step 4:

Step 5:

10.4.3 EXERCISES—LEARNING IT!

10.1 Many studies have been done comparing the consumer behavior of men and women. One ongoing study concerns take-out food. In particular, the study focuses on whether there is a difference in the mean numbers of times per month that men and women buy take-out food for dinner. The most recent study results are shown here:

	Men	Women
Sample size	$n_1 = 34$	$n_2 = 28$
Sample mean	$\overline{X}_1 = 25.6$	$\overline{X}_2 = 21.2$
Population standard deviation	$\sigma_1 = 4.2$	$\sigma_2 = 3.8$

Because there is so much historical data, the population standard deviations are known.

(a) Set up the hypotheses to test whether there is a difference in the mean numbers of times per month that men and women buy take-out food for dinner.

(b) Use the Z test with known population variances to set up and perform the test. Use a level of significance of 0.05.

(c) Find the p value for the test.

(d) Do the data provide evidence that the mean number of times per month for men differs from that for women?

(e) Does the choice of α in this case affect the decision?

10.2 Professional employees who work for large corporations often contend that the mean salary paid by a company differs by location in the United States. To test that claim, data were collected on teachers working in New England and in the upper Midwest. From an extensive history of salary data, the population standard deviations are available. The study found the following results:

	New England	**Upper Midwest**
Sample size	$n_1 = 25$	$n_2 = 18$
Sample mean	$X_1 = \$38,348$	$X_2 = \$36,782$
Population standard deviation	$\sigma_1 = \$2336$	$\sigma_2 = \$2258$

(a) Set up the appropriate hypotheses to test whether the teachers in New England were paid more, on the average, than those working in the upper Midwest.

(b) Use the Z test with known population variances to set up and perform the test. Use a level of significance of 0.05.

(c) Find the p value for the test.

(d) Do the data support the contention that the mean pay for teachers in New England is higher than the pay for teachers in the upper Midwest?

10.3 A study at a university in New England focused on binge drinking by students. The people who administered the study wanted to determine whether students who live on campus have more episodes of binge drinking per semester on the average than those who live at home. They surveyed all students who were enrolled in a required health course and obtained the following data:

	On campus	**At home**
Sample size	$n_1 = 220$	$n_2 = 196$
Sample mean	$X_1 = 37.3$	$X_2 = 35.6$
Sample standard deviation	$s_1 = 8.6$	$s_2 = 10.1$

(a) Set up the appropriate hypotheses to test whether students who live at home have fewer episodes of binge drinking per semester, on the average, than those who live on campus.

(b) Use the Z test with known population variances to set up and perform the test. Use a level of significance of 0.05.

(c) Find the p value for the test.

(d) Do the data indicate that students who live at home binge less on the average?

10.4 Jiffy Copy uses two suppliers to provide paper for the copy machines. The company has experienced an excessive number of paper jams and wonders whether there is a difference in the paper provided by the two suppliers. The company collects data on the number of jams per ream of paper (500 sheets) over a period of 2 months. Here are the summary data:

	Supplier 1	**Supplier 2**
Sample size	$n_1 = 45$	$n_2 = 52$
Sample mean	$X_1 = 8.1$	$X_2 = 9.3$
Sample standard deviation	$s_1 = 1.1$	$s_2 = 1.2$

(a) Set up the appropriate hypotheses to test whether the paper from the two suppliers results in the same average number of jams per ream.

(b) Use the Z test with known population variances to set up and perform the test. Use a level of significance of 0.05.

(c) Find the p value for the test.

(d) Do the data indicate that there is a difference between suppliers on average?

10.5 You are considering selling your house and need to choose between two real estate agencies. You collect data on the numbers of weeks that a house is on the market for both agencies. The summary statistics are shown here:

	Agency A	**Agency B**
Sample size	$n_1 = 25$	$n_2 = 20$
Sample mean	$X_1 = 22.3$	$X_2 = 18.5$
Population standard deviation	$\sigma_1 = 1.5$	$\sigma_2 = 2.2$

(a) At the 0.05 level of significance, test to see whether houses listed with agency A are on the market on the average longer than those listed with agency B.

(b) Find the *p* value for the test.

(c) Which agency would you use and why?

10.5 Small-Sample Tests of the Difference in Two Population Means

When the population standard deviation is unknown and the sample size is large, we use the Z test. If the sample size is not large, then we use a t test, just as we did for the one-population tests.

10.5.1 SMALL-SAMPLE TESTS OF TWO MEANS WITH UNKNOWN BUT EQUAL STANDARD DEVIATIONS

The reason 30 is used as the cutoff point is that the Central Limit Theorem applies for sample sizes of 30 or more.

Unfortunately, we do not always have the luxury of large samples. Remember that data collection is expensive, so often the sample size for one or both of the samples is less than 30. This is called the "small-sample" case.

Just as in the one-population case, the small-sample test for two population means involve the t distribution. The t test carries with it the same assumption that the populations involved in the test are normally distributed. In addition, though, the two-population case requires that we know whether the two population variances can be considered equal. In this section, we look at both cases.

Understand the problem.

EXAMPLE 10.10

ONLINE LEARNING AND GENDER

Setting Up the Hypotheses

In addition to looking at the test results of students in on-ground versus online courses, the professor wondered if gender plays a role in online learning. She took the group of people in the online course and separated them according to gender. She wanted to know whether there was any difference in the posttest scores for men and women. We can call the women population 1 and the men population 2. Since the professor is interested only in whether the groups are different, she can use a two-tail test with these hypotheses:

$$H_0: \quad \mu_1 = \mu_2$$
$$H_A: \quad \mu_1 \neq \mu_2$$

The first approach is to assume that even though we don't know the variability in the two populations, we know enough to believe that there is the same amount of variability in each population. At first this may seem a bit odd, but actually it is not such a ridiculous assumption. Remember that we are comparing the means of two populations. Because we are measuring the same variable in both populations, it is quite possible that the two populations have a normal distribution with the same shape but with different means. Figure 10.2 illustrates this situation. There is, of course, a way to tell by doing a hypothesis test to compare two variances. This is the subject of Section 10.9.

For now, we will assume that equal variances have been confirmed through a statistical procedure. We have two estimates of this variance, one from each sample; that is, s_1^2 and s_2^2 are two different estimates of the same number. Instead of using two

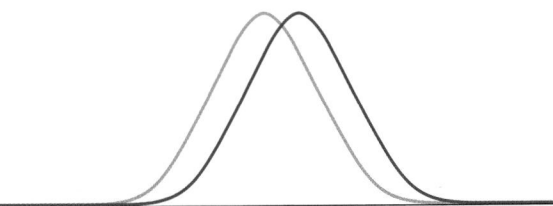

FIGURE 10.2 Two normally distributed populations with the same variance but different means

different estimates of the unknown but common population variance, we can combine the data and find one better estimate of the variance. Instead of using s_1^2 and s_2^2 in the formula for the test statistic, we "pool" the data from the samples and get one estimate of the common variance. This is called the **pooled variance.**

One way to estimate the pooled variance is simply to put all the data together from both samples and calculate the sample variance of this "big" sample. There are two reasons for not doing that. One reason is that if indeed the two population means are different, then the consolidated data will have a mean that is not from either population. We will have a distorted measure of variance. A second reason is that at this point most people have already calculated the sample variance for each sample individually. Thus, we find the pooled variance by building from the individual sample variances. Your first thought might be to average the two sample variances. This is the right idea, but instead we use a weighted average. If, for example, one of the samples sizes is 10 and the other one is 25, then the sample variance from the sample based on 25 observations is more accurate and should have more influence on the pooled variance. The formula for the pooled variance, s_p^2, is shown here:

$$s_p^2 = \frac{(n_1 - 1)s_1^2 + (n_2 - 1)s_2^2}{n_1 + n_2 - 2}$$

The pooled variance

This formula weights the sample variances by 1 less than the sample size. If the sample sizes are the same, this formula is a simple average of the two sample variances.

EXAMPLE 10.11

ONLINE LEARNING AND GENDER

Analyze the data.

Calculating the Pooled Variance

The data on the posttests for the online course are given here:

	Population 1: women in online course	Population 2: men in online course
Sample size	$n_1 = 19$	$n_2 = 27$
Sample average	$\overline{X}_1 = 80.05$	$\overline{X}_2 = 79.59$
Sample standard deviation	$s_1 = 11.43$	$s_2 = 9.20$

The calculation of the pooled variance of the online posttest scores is then

$$s_p^2 = \frac{(19 - 1)(11.43^2) + (27 - 1)(9.20^2)}{19 + 27 - 2}$$

$$= \frac{2351.6082 + 2200.6400}{44} = \frac{4552.2482}{44} = 103.4602$$

The pooled standard deviation is

$$s_p = \sqrt{103.4602} = 10.17$$

As a double check, be sure the number you get for the pooled standard deviation is between the two sample standard deviations. If it is not, recheck your work; you have made a mistake.

Now we are ready to consider the test statistic for the small-sample case when the population variances are equal. The test statistic in this case has a t distribution with $(n_1 + n_2 - 2)$ degrees of freedom. Remember that for simpler notation we can write the t value as $t_{\alpha, n-1}$. The test statistic is calculated as follows:

Test statistic for the difference in two population means, pooled variances

$$t = \frac{\overline{X}_1 - \overline{X}_2}{\sqrt{s_p^2/n_1 + s_p^2/n_2}}$$

or equivalently

$$t = \frac{\overline{X}_1 - \overline{X}_2}{s_p\sqrt{1/n_1 + 1/n_2}}$$

Knowing the degrees of freedom and the formula for the test statistic, we can complete the hypothesis test.

Analyze the data.

EXAMPLE 10.12

ONLINE LEARNING AND GENDER

Conducting the Hypothesis Test

Now the professor who is looking at whether gender makes a difference in online learning can find the rejection region for the test and calculate the test statistic and the p value. She is doing a two-tail test, so the cutoff values are the t statistic with 0.025 in the tail area and 44 degrees of freedom, $t_{0.025,44}$. In the t table the closest critical values are ± 2.021 because the table goes from 40 to 60 degrees of freedom. From statistical software like Minitab the exact critical values are ± 2.015. The test statistic is calculated as

$$t = \frac{80.05 - 79.59}{10.17\sqrt{\frac{1}{19} + \frac{1}{27}}} = \frac{0.46}{10.17\sqrt{0.0897}} = \frac{0.46}{(10.17)(0.2994)} = 0.1510$$

The professor uses a statistical software package to look up the t statistic and finds that the area outside the test statistic is 0.4403. Since this is a two-tail test, she doubles that to find p value = 0.8806. The output of the statistical package for the test follows:

Two-Sample T-Test and CI: Post Test, Gender

Two-sample T for Post Test

Gender	N	Mean	StDev	SE Mean
F	19	80.1	11.4	2.6
M	27	79.59	9.20	1.8

Difference = mu (F) − mu (M)
Estimate for difference: 0.46
95% CI for difference: (−5.68, 6.60)
T-Test of difference = 0 (vs not =): T-Value = 0.15   P-Value = 0.881
DF = 44
Both use Pooled StDev = 10.2

The results of the test show that since 0.1510 is not in the rejection region of the test, the professor cannot reject H_0. There is not enough evidence to say that the mean posttest score for the male online group is different from the mean posttest score for the female online group.

✔ *TRY IT NOW!*

Test Scores and Gender
Comparing Two Different Means

The professor also wants to look at gender differences in the test scores for the students in the on-ground course. She assembles the relevant data:

	Population 1: women in on-ground course	Population 2: men in on-ground course
Sample size	$n_1 = 20$	$n_2 = 27$
Sample mean	$\overline{X}_1 = 85.15$	$\overline{X}_2 = 83.74$
Sample standard deviation	$s_1 = 8.99$	$s_2 = 9.27$

Find the pooled variance for the data.

Perform a hypothesis test to see whether there is evidence that the mean posttest score for the women in the on-ground course is different from the mean posttest score for the men in the on-ground course. Use a level of significance of 0.05.

Step 1:

Step 2:

Step 3:

Step 4:

Step 5:

10.5.2 SMALL-SAMPLE TESTS OF TWO MEANS WITH UNKNOWN AND UNEQUAL STANDARD DEVIATIONS

If we are testing for equality of variances and find that the population variances are not equal, then we cannot pool the data. This is known as the *Behrens–Fisher problem*. Figure 10.3 illustrates the situation in which both populations have a normal distribution with different variances.

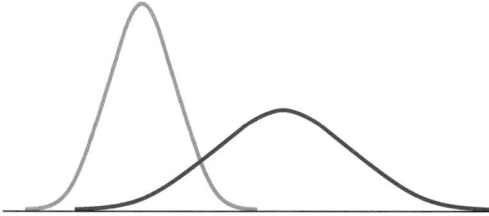

FIGURE 10.3 **Two normally distributed populations with unequal variances**

The test statistic is called a separate-variance t test. As the name indicates, the individual sample variances are used in the test statistic. In fact, the formula for calculating the test statistic looks exactly like the Z statistic for the large-sample case when the variances are unknown. It is calculated as follows:

Test statistic for the diffence in two population means, variances unknown, $n_1, n_2 < 30$

$$t = \frac{(\overline{X}_1 - \overline{X}_2) - d}{\sqrt{s_1^2/n_1 + s_1^2/n_2}}$$

The Z distribution does not apply because the estimates of the unknown variances are based on small samples. We know from our work in Chapter 9 that the appropriate distribution for small-sample situations is the t distribution. This is the case here as well, but the number of degrees of freedom is found from a more complicated formula than you have seen thus far. This test statistic can be approximated by a t distribution with ν (Greek letter *nu*) degrees of freedom. The formula for ν follows:

Degrees of freedom for small samples and unequal variances

$$\nu = \frac{\left(\dfrac{s_1^2}{n_1} + \dfrac{s_2^2}{n_2}\right)^2}{\dfrac{(s_1^2/n_1)^2}{n_1 - 1} + \dfrac{(s_2^2/n_2)^2}{n_2 - 1}}$$

Note: There are several equivalent formulas for calculating the degrees of freedom in this case.

When we calculate ν, we most likely get a fractional result, but we know that degrees of freedom are integer values. We can simply drop the fractional portion of the result to get an integer value for the degrees of freedom. You may wonder why we do this. The reason is that the number that results for the degrees of freedom in this case is smaller than the one used in the pooled variance case. The number of degrees of freedom in the test defines the test's ability to detect significant differences in the population means. That is, when we assume that the population variances are not equal, the hypothesis test becomes less powerful and the chances of a Type II error for a given value of α increase. By not pooling variances we are more likely to miss true differences in the population means.

EXAMPLE 10.13

ONLINE LEARNING AND GENDER

Conducting a Small-Sample Test with Unequal Variances

Suppose we redo the test on gender in the online classes, this time assuming that the population variances are not equal. How does this change the analysis? Here are the relevant data:

	Population 1: women in online course	**Population 2:** men in online course
Sample size	$n_1 = 19$	$n_2 = 27$
Sample average	$\overline{X}_1 = 80.05$	$\overline{X}_2 = 79.59$
Sample standard deviation	$s_1 = 11.43$	$s_2 = 9.20$

First we construct the null and alternative hypotheses:

$$H_0: \quad \mu_1 = \mu_2$$
$$H_A: \quad \mu_1 \neq \mu_2$$

Then we select α and find the rejection region. To do this we need to find the degrees of freedom. Substituting in the formula, we get

$$\nu = \frac{\left(\dfrac{11.43^2}{19} + \dfrac{9.20^2}{27}\right)^2}{\dfrac{(11.43^2/19)^2}{19 - 1} + \dfrac{(9.20^2/27)^2}{27 - 1}}$$

$$= \frac{100.21736}{2.97716} = 33.66207$$

Taking the integer portion of ν gives us 33 degrees of freedom. The value $t_{0.025,36}$ is found from a computer software package to be ± 2.0345. The result is that the critical values are slightly farther out and the rejection region is smaller than we found in Example 10.12. The decision does not change, though. We still do not reject the null hypothesis and we conclude that there is no difference due to gender.

Many statistical packages offer the option of doing the t test assuming equal variances or assuming unequal variances. To make an informed choice about whether to pool the data, you must do a test for the equality of the variances. This is the subject of Section 10.9.

10.5.3 EXERCISES – LEARNING IT!

10.6 The Board of Realtors for Greater Bridgeport, Connecticut, wants to know whether the average price of a single-family home in the area increased in the past. They take a random sample of homes sold in 1995 and 1996 and calculate the sample statistics:

	1995 sales	**1996 sales**
Sample size	$n_1 = 25$	$n_2 = 25$
Sample mean	$\overline{X}_1 = \$151{,}166$	$\overline{X}_2 = \$160{,}669$
Sample standard deviation	$s_1 = \$5332$	$s_2 = \$6468$

(a) Set up the appropriate hypotheses to test whether the mean selling price of a home in Greater Bridgeport increased.

(b) Calculate the pooled variance for the data.

(c) Assuming that the data are normally distributed, use the small-sample test with equal variances to test the hypotheses. Use a level of significance of 0.10.

(d) Do the data provide evidence that the average price of a home in 1996 was higher than in 1995?

10.7 The speed of shipping is of great concern to mail-order businesses. Traditionally, mail-order companies have used United Parcel Services (UPS) for their shipping. The U.S. Postal Service has been promoting its parcel post service as a competitor for UPS. To determine whether parcel post-is faster than UPS ground, a mail-order company sent 20 packages via UPS ground and 20 packages via parcel post and tracked the number of working days for the packages to reach their intended destinations. A summary of the data is given here:

	UPS ground	**U.S. Postal Service**
Sample size	$n_1 = 20$	$n_2 = 20$
Sample mean	$\overline{X}_1 = 7.3$	$\overline{X}_2 = 7.1$
Sample standard deviation	$s_1 = 0.9$	$s_2 = 2.3$

(a) Set up the appropriate hypotheses to test whether the U.S. Postal Service is faster than UPS ground.

(b) Assuming that the data are normally distributed, use the small-sample test with unequal variances to test the hypotheses. Use a level of significance of 0.05.

(c) Do the data provide evidence that the average delivery time for parcel post is less than the time for UPS ground?

10.8 After months of working overtime, you have saved some money for a set of new golf clubs and you want to make sure that you are buying the best. You can get a really good deal on brand X clubs, but you are willing to make the sacrifice to buy brand Z if the clubs really improve your game. The salesperson allows you to take the number 3 wood from each brand and use them to hit balls on a driving range. The data (in yards) are summarized here:

	Brand X	Brand Z
Sample size	$n_X = 15$	$n_Z = 15$
Sample mean	$\bar{X}_X = 255$	$\bar{X}_Z = 271$
Sample standard deviation	$s_X = 8.7$	$s_Z = 9.1$

(a) Set up the appropriate hypotheses to test whether brand Z clubs are better than brand X.

(b) Calculate the pooled variance for the data.

(c) Assuming that the data are normally distributed, use the small-sample test with equal variances to test the hypotheses. Use a level of significance of 0.05.

(d) Should you spend the extra cash or save it for golf balls?

10.9 Few things are more frustrating than needing information and having to wait for it. The time spent on the phone waiting for technical support for software is one of the largest drains on productivity that service industries like libraries experience. Before a large city library system will consider switching over to new software for managing its collection, administrators have to be convinced that they will get better service (that is spend less time on hold) with the new software company. They decide to perform a test to see if this is the case. They place ten calls to both their current software vendor and a competitor and record the amount of time (in minutes) they spent on hold. The summary data follow:

	Current vendor	Competing vendor
Sample size	$n_1 = 10$	$n_2 = 10$
Sample mean	$\bar{X}_1 = 11.2$	$\bar{X}_2 = 9.7$
Sample standard deviation	$s_1 = 3.4$	$s_2 = 1.6$

(a) Assuming that the data are normally distributed, use the small-sample test with unequal variances to test the hypotheses that the library will spend less time on hold per call, on the average, with the competitor's software company.

(b) Should the library switch software?

10.10 Are all discount mail-order companies the same? In an attempt to answer this question, data were collected on the prices of the top ten business software packages reported by *PC Magazine* in August 1997. Two well-known mail-order companies were asked for their prices on each piece of software. The data are shown here:

Top ten business software packages	Computability price	PC Connection price
Norton Anti-Virus 2000 v 6.0	$29	$32
Microsoft W98 Second Edition Upgrade	95	90
Norton System Works 2000 v 3.0	59	60
VirusScan 5.0	29	24
QuickBooks 2000 Pro	200	200
Norton Internet Security 2000	58	48
QuickBooks 2000	120	120
Microsoft W98 Second Edition	180	179
Microsoft Office 2000 Upgrade	220	230
VirusScan 5.0 Deluxe	38	33

(a) For each vendor, calculate the average price and the standard deviation of the prices.

(b) Based on these statistics, do you think the mail-order companies are the same or different? Why?

(c) Set up the hypothesis test to determine whether the average price for software is different at the two mail-order companies.

(d) Assume that the variances are equal and calculate the pooled variance.

(e) Assuming that the data are normally distributed, use the t test with equal variances to perform the test at the 0.05 level of significance.

(f) What can you conclude about software prices at the two mail-order companies?

(g) What assumption was violated in this example?

10.6 Summary of Tests of Two Population Means: Independent Samples

The preceding two sections have presented different cases of testing the difference in two population means. This is a good place to review what you have learned so far. Table 10.2 on page 470 summarizes the tests.

For all but one case the population must be normally distributed. It is unlikely that you will know about the distribution of the population, so you must use the sample data to decide whether this assumption is reasonable. To determine whether the data are normally distributed, you should use a statistical test such as a chi-square goodness of fit test (see Chapter 15) or a normal probability plot (see Chapter 11). Now you can assess the assumption of normality by using the descriptive techniques you learned in Chapters 2 and 3. You can look at boxplots, histograms, or dotplots of the data. Since the assumption of normality is an issue in small-sample tests, histograms often do not work well as visual tools. Although these graphs are certainly not rigorous tests, unless there are large departures from normality (very skewed or non-mound-shaped distributions) the test will be valid. You could also apply the empirical rule for normality, which was covered in Chapter 3.

EXAMPLE 10.14

ONLINE COURSES

Checking for Normality of Data

The professor who is looking at gender issues in learning knows that she needs to check the data to see whether she can reasonably assume that the population posttest scores for men and women are normally distributed. She makes these boxplots of the data:

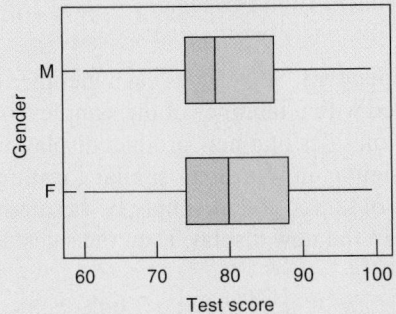

Posttest Scores for Online Course

Draw conclusions.

The data appear to be reasonably symmetric: The medians are near the centers of the boxes and the whiskers are approximately the same lengths. It appears that normality is a reasonable assumption.

TABLE 10.2

SUMMARY OF TESTS OF POPULATION MEANS

Populations	Variances	Sample sizes	Test
Independent and normal	Known	No restrictions	Z test
Independent and nonnormal	Unknown	n_1 and $n_2 \geqslant 30$	Z test
Independent and normal	Unknown but assumed equal	n_1 and $n_2 < 30$	t test with a pooled variance
Independent and normal	Unknown and assumed not equal	n_1 and $n_2 < 30$	Modified t test (Behrens–Fisher problem)

A quick look at Table 10.2 indicates that all of these tests are for independent samples. This means that when we select a sample from population 1, it has no impact on the selection of the sample from population 2. This is typically done, although sometimes it is necessary or desirable to use dependent samples instead. That is the subject of the next section.

10.7   Tests of the Difference in Two Population Means: Dependent Samples

10.7.1   TESTS WITH DEPENDENT POPULATIONS

In many cases, such as the example about online courses, data are collected before and after some experiment is performed. This type of data is known as **pretest** and **posttest data.** When such data are collected, the samples come not from independent populations but from dependent populations. We are interested in knowing whether the experimental procedure has an effect on the variable being studied.

> *Pretest* and *posttest* conditions exist when data are collected on the same sample elements before and after some experiment is performed.

This is not the only way samples can be **dependent.** Sometimes each member of the sample from population 1 is matched or paired with a member of the sample from population 2. For example, in studying the effect on sales of a new product display, we would want to compare stores that are either similar in size or in similar locations. Otherwise, the apparent increase in sales that is observed might simply be because a larger, busier store was randomly chosen to try out the new display. Then the sales did not really increase due to the new display.

> *Dependent samples* are related to each other. The members of one sample are identical to or matched or paired with the members of the other sample according to some characteristic.

Discovery Exercise 10.1

INTRODUCTION TO EXPERIMENTAL DESIGN

Did you ever wonder why the restrooms in restaurants have a sign instructing all employees to wash their hands before returning to work? The owner of a large restaurant chain wants to increase the amount of time that employees wash their hands after reading about the health implications of washing for only a few seconds. A random sample of ten employees was selected. The amount of time each employee washed was recorded. The data were collected in such a way that the employees did not know they were being observed.

These employees were then educated on the benefits of hand washing. They watched a health video that detailed the benefits of increasing the amount of time they washed. One week after the training, employee hand washing was timed again. The times (in seconds) are shown in the table:

Employee	Before training	After training
1	3	3
2	3	3
3	2	4
4	4	4
5	3	5
6	3	4
7	4	4
8	3	4
9	5	6
10	2	3

1. Use a two-population paired t test to test the hypothesis that no learning has occurred. Use a pooled variance.

$$H_0: \quad \mu_1 = \mu_2$$
$$H_A: \quad \mu_1 \neq \mu_2$$

2. What is your conclusion?

3. Why is this surprising? (*Hint:* Of the ten employees, how many washed at least as long after the training?)

4. Explain how these data are different from most of the two-sample data sets you have seen up to this point in the chapter. Why might this be important?

5. Suppose ten different employees were used for the second timing. Would this make a difference? Explain why or why not.

In the situation described at the beginning of this section, one of the characteristics that could be used to pair or match the sample stores is size. A second is the location of the store. This is the first departure we have seen from random sampling. In fact, we are trying to match the items that are sampled in an attempt to control other variables that influence the data. Clearly, the size of the store will influence sales. In a sense, we match sample elements to keep the comparison "fair." Similarly, it is not fair to compare the average miles per gallon of two cars driven by two people with different driving styles. We would want either the same person testing both cars or people with matched driving styles. We may want to use more than one characteristic to match the members of the two samples. This discussion is really the beginning of a topic known as *experimental design*. In designing a statistical experiment, we try to get the most information for the money. Paired data are the simplest case of a designed experiment. This topic fits naturally here because often paired data are collected to compare two population means and yet the data are analyzed incorrectly using the methods discussed in the last section.

You should not analyze these paired or matched data using the small-sample *t* test. If you do, you might miss differences that exist but that are masked by improper analysis. In this section we describe how to analyze dependent or paired data.

10.7.2 THE PAIRED SAMPLE TEST

The problem with using the tests for independent populations is that if there is a large amount of variation among the sample elements, then the standard errors of the estimates are high. Because we divide by the standard error to obtain the test statistic, the test statistic is often smaller than it should be. Real differences in the populations are hidden by the amount of variation among the sample elements.

How can we fix this problem? Well, if you think about it, the real question is whether the difference between the pretest and posttest measurements is significantly different from 0. We can look at the differences as our measurements and perform a test about them. These are now the hypotheses for this test:

$$H_0: \quad \mu_d = 0 \qquad H_0: \quad \mu_d \leq 0 \qquad H_0: \quad \mu_d \geq 0$$
$$H_A: \quad \mu_d \neq 0 \qquad H_A: \quad \mu_d > 0 \qquad H_A: \quad \mu_d < 0$$

Table 10.3 shows how the data are set up for such a test. Each observation in the sample is a row in the table. The second and third columns are the values of the pretest (population 1) and posttest (population 2). For example, in the second column, the entry x_{13} is the pretest value for the third sample element. The fourth column, labeled Difference, is calculated by subtracting the pretest score from the posttest score. These differences are labeled $d_1, d_2, \ldots, d_n$. Finally, the average of these differences is found and labeled $\bar{d}$; that is,

Average difference

$$\bar{d} = \frac{\sum_{i=1}^{n} d_i}{n}$$

TABLE 10.3

DATA SETUP FOR PAIRED DIFFERENCE TEST

Observation	Pretest, X_1	Posttest, X_2	Difference, $d = X_2 - X_1$
1	x_{11}	x_{21}	d_1
2	x_{12}	x_{22}	d_2
3	x_{13}	x_{23}	d_3
$\vdots$	$\vdots$	$\vdots$	$\vdots$
n	x_{1n}	x_{2n}	d_n

EXAMPLE 10.15

ONLINE COURSES

Understand the problem.

Setting Up Hypotheses for a Paired Difference Test

The professor who is looking at the test results of two different types of courses now knows that the end result (the score on the posttest) is higher for students in the on-ground course. She also knows that this does not mean that the online course is not effective. She decides to analyze the online course data further and look at the differences between the pretest and posttest scores for the people in the online courses. She wants to know whether people learned when they took the online course—that is, whether the posttest scores for the students are indeed higher than the pretest scores. She must use a paired difference test because the two populations are dependent. Since she wants to know whether the test scores increase, she uses a one-tail test. The question is, In what direction do each of the hypotheses go?

If the online course does improve test scores, then the differences (posttest − pretest) will be positive. In this case, the test should be an upper-tail test on these hypotheses:

$$H_0: \quad \mu_d \leq 0$$
$$H_A: \quad \mu_d > 0$$

The direction of the test depends on what the experimenter thinks the effect of the experiment will be and the way the pretest and posttest values are subtracted. Most paired difference tests are one-tail.

EXAMPLE 10.16

SLEEP APNEA

Understand the problem.

Setting Up the Hypotheses

Because of skyrocketing health-care costs, many hospital administrators are working to contain costs. Studies are being done to see whether some conditions can be diagnosed and treated at home at a significantly reduced cost. One area of study is the diagnosis and treatment of obstructive sleep apnea syndrome (OSAS), which affects 2%–5% of the adult male population and is characterized by extremely heavy snoring. As a result of OSAS, breathing is suspended either partially or entirely, which can lead to suffocation. After OSAS is diagnosed, the traditional treatment is hospitalization to begin nasal continuous positive airway pressure (NCPAP). In an effort to reduce costs, a study was done in which NCPAP was initiated at home.

One variable observed for each patient before and after treatment was the number of obstructions. If the treatment works, then the average number of obstructions posttreatment should be less than the pretreatment average. The hospital running the study wants to know whether home treatment of OSAS is effective.

Since the number of obstructions should decrease after NCPAP if treatment is effective, the analyst knows that the test should be one-tail and the differences (posttest − pretest) should be negative. This means that the test should be a lower-tail test of these hypotheses:

$$H_0: \quad \mu_d \geq 0$$
$$H_A: \quad \mu_d < 0$$

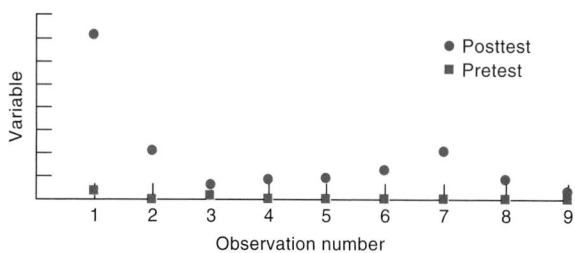

FIGURE 10.4 **Pretest and posttest data for similar observations**

The hypothesis test is now essentially a test of a single mean, and we know how to do this test from Chapter 9. The paired difference test is considered to be a small-sample test, so we will use the t test statistic. The difference is that the mean we are testing is the mean difference between two populations rather than the mean of a single population. Otherwise, the procedure is the same.

We need to define the test statistic for the test. For a paired difference test, the test statistic has a t distribution with $n - 1$ degrees of freedom. Remember that in this test n is the number of pairs and not the number of data points. Here is t:

Test statistic for paired difference test

$$t = \frac{\overline{d}}{s_d/\sqrt{n}}$$

where s_d is the standard deviation of the differences and is found in the usual way. Since the test is a t test, we must, as always, have normally distributed populations.

Figure 10.4 shows typical data for a paired sample test. Because the data points are plotted by observation, you can clearly see that the values after treatment are higher than the values before treatment for every observation in the sample. The statistical test will determine whether the differences are statistically significant.

If we used the small, independent-sample t test we would average all of the values in the first sample and all of the values in the second sample. But in doing so we would lose some of the information contained in the sample. To see this effect, let's look at some different data. In Figure 10.5, it is not obvious that the posttest data have a higher average if you do not view the data in pairs. The variation from sample element to sample element hides the differences between the pretest and posttest data. This is what happens when we simply average all of the pretreatment values and all of the posttreatment values and compare the sample means using a t test.

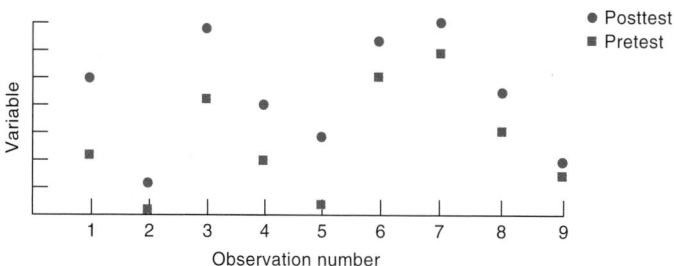

FIGURE 10.5 **Pretest and posttest data for highly variable observations**

Analyze the data.

EXAMPLE 10.17

ONLINE COURSES

Calculating the Test Statistic and the *p* Value

The professor wants to know whether students in an online course learn the material. There were 46 students who completed the online course. The professor analyzes the data and finds the following summary statistics:

$$n = 46$$
$$\overline{d} = 20.09$$
$$s_d = 16.90$$

She decides to test with $\alpha = 0.05$. Since there are 46 pairs of data, the t test has $n - 1 = 45$ degrees of freedom and the critical value is $t_{0.05,45} = 1.679$, obtained from statistical software. The test statistic is calculated to be

$$t = \frac{20.09}{16.90/\sqrt{46}} = 10.47$$

Clearly, the p value is 0.0000.

Analyze the data.

EXAMPLE 10.18

SLEEP APNEA

Calculating the Test Statistic and the *p* Value

Let's complete the OSAS example using a level of significance of 0.05. If the treatment is effective, then the average difference should be significantly less than 0. If the treatment is not effective, then the pretest and posttest values will be close and the differences will be small, resulting in an average difference close to 0. The numbers of obstructions for the $n = 9$ patients are given in the table:

Patient	Pretreatment	Posttreatment	Difference
1	365	22	−343
2	107	1	−106
3	28	12	−16
4	40	8	−32
5	48	3	−45
6	64	8	−56
7	109	0	−109
8	55	4	−51
9	20	0	−20
$\overline{d}$			−86.44
s_d			101.83

The test statistic follows a t distribution with $n - 1$ degrees of freedom. The value of n is equal to the number of pairs or the number of differences. In this case, $n = 9$ so there are 8 degrees of freedom. We need to find $t_{0.05,8}$. From the t table, the critical value is -1.860. The test statistic looks just like the t statistic for a single population mean:

$$t = \frac{-86.44}{101.83/\sqrt{9}} = -2.546$$

The p value of the test is found by looking up the test statistic in the t distribution with 8 degrees of freedom. The p value is 0.0172.

After the testing procedure, we must make a decision and then interpret it in light of the given problem.

EXAMPLE 10.19

ONLINE COURSES

Deciding What the Test Results Mean

The professor who is testing the effectiveness of online courses looks at the results of the test. She sees that the test statistic is in the rejection region, so she will reject H_0. She concludes that the average difference in the scores before and after the training program is greater than 0. This means that the students in the online course improved from the posttest to the pretest; that is, they learned.

EXAMPLE 10.20

SLEEP APNEA

Reaching Conclusions

In the study of the effectiveness of home treatment of sleep apnea, the test statistic of -2.546 is in the rejection region, so we reject H_0. The hospital administration can conclude that the average difference between the number of obstructions after and before home treatment is less than 0; that is, the number has decreased. Therefore, the home treatment is effective.

TRY IT NOW!

Sleep Apnea
Comparing Two Other Differences

In addition to measuring the number of obstructions before and after home treatment in the OSAS study, hospital administrators recorded the number of hypopnea (reduction in airflow) both pretreatment and posttreatment for each patient:

Patient	Pretreatment	Posttreatment	Difference
1	208	22	
2	297	0	
3	126	17	
4	150	13	
5	133	0	
6	201	10	
7	272	7	
8	137	1	
9	310	2	
$\overline{d}$			
s_d			

Complete the table and conduct a hypothesis test. Is there evidence that the treatment has reduced the number of hypopnea at the 0.05 level of significance?

Step 1:

Step 2:

Step 3:

Step 4:

Step 5:

10.7.3 EXERCISES—LEARNING IT!

10.11 Having learned about the paired t test, you realize that you really should have used that test for the software price comparisons. The data are repeated here:

Requires Exercise 10.10

Top ten business software packages	Computability price	PC Connection price
Norton Anti-Virus 2000 v 6.0	$29	$32
Microsoft W98 Second Edition Upgrade	95	90
Norton System Works 2000 v 3.0	59	60
VirusScan 5.0	29	24
QuickBooks 2000 Pro	200	200
Norton Internet Security 2000	58	48
QuickBooks 2000	120	120
Microsoft W98 Second Edition	180	179
Microsoft Office 2000 Upgrade	220	230
VirusScan 5.0 Deluxe	38	33

(a) Calculate the differences between the prices for each software package. Just looking at the differences, do you think that one company charges more than the other? Why or why not?

(b) Calculate the average difference and the standard deviation of the differences.

(c) Set up the hypotheses to test whether the mean difference in price between the two mail-order companies is 0.

(d) Assuming that the data are normally distributed, at the 0.05 level of significance, is there a difference in the mean prices of software for the two mail-order companies?

(e) Did these results differ from your analysis in Exercise 10.10? Why do you think this happened?

10.12 A hospital administrator is concerned about the length of time the nursing staff washes their hands. A recent study in health care showed that longer hand washing greatly reduces the spread of germs. The hospital observed the amount of time that a sample of nine nurses in the cardiac care unit (CCU) washed their hands. The data were collected in such a way that the nurses did not know they were being observed. The hospital then showed the nurses an educational video on the negative effects of short hand washing times. After the video, the researchers again timed the group of nurses washing their hands. Here are the data (with times in seconds):

Observation	Unit	Time 1	Time 2
1	CCU	3	16
2	CCU	2	7
3	CCU	0	5
4	CCU	5	8
5	CCU	2	15
6	CCU	0	15
7	CCU	2	20
8	CCU	3	16
9	CCU	0	18

ANS. $\bar{d} = 195.8$; $s_d = 76.0$; $H_0: \mu_1 = \mu_2$; $H_A: \mu_1 > \mu_2$; CRITICAL VALUE $= 1.860$; $t = 7.73$; REJECT H_0. THE MEAN NUMBER HAS GONE DOWN.

(a) Calculate the difference between the times for each nurse. Just looking at the differences, do you think that, on the average, the nurses washed their hands longer the second time? Why or why not?

(b) Calculate the average difference and the standard deviation of the differences.

(c) Set up the hypotheses to test whether there was an increase in the average amount of time spent washing hands.

(d) Assuming that the data are normally distributed, at the 0.05 level of significance, what can you conclude?

(e) Can you conclude that the video caused the nurses to wash their hands longer? Why or why not?

10.13 The way that advertising affects us influences how we buy things. Clearly advertising affects sales, but advertising is also expensive and companies want to advertise in ways that have the greatest benefit for the amount of money spent. A company that sells snack food designed two different advertising strategies, one focusing on print media and the other on radio. It ran the campaigns in 32 different cities, paired on population size, and measured the sales in the week following the beginning of the campaign. Here are the sales (in thousands):

Print sales	Radio sales	Print sales	Radio sales
$28.3	$22.1	$24.3	$25.5
24.6	19.1	25.2	22.6
23.1	20.3	23.3	24.9
21.0	24.4	25.3	29.7
25.7	22.4	22.2	22.2
22.5	19.2	23.4	28.5
32.0	22.8	23.9	28.2
23.5	20.3	25.7	21.6

(a) Calculate the differences in sales for each pair of cities.

(b) Based on these differences, do you think there is a difference in the mean sales for the two types of advertising campaigns? Why or why not?

(c) Calculate the average difference and the standard deviation of the differences.

(d) Set up the hypotheses to test whether there is a difference in mean sales due to type of advertising.

(e) Assuming that the data are normally distributed, at the 0.01 level of significance, what do you conclude?

10.14 A study was done on the effectiveness with which a person uses a computer mouse. Researchers selected ten people, matched on computer skills, and measured the speed with which they moved a mouse at the beginning of a long session of computer use and then after 2 hours of use. Their speeds (in hundredths of a second) are shown here:

67	57
64	53
69	71
88	61
72	73
80	50
85	53
116	80
77	63
78	41

(a) Calculate the difference between the times for each person. Just looking at the differences, do you think that the average speed changed after 2 hours? Why or why not?

(b) Calculate the average difference and the standard deviation of the differences.

(c) Set up the hypotheses to test whether there was a change in the mean speed with which the people moved the mouse.

(d) Assuming that the data are normally distributed, at the 0.05 level of significance, what can you conclude?

10.15 Does gender affect e-mail use? A large company studied the use of e-mail by counting the number of business-related e-mails generated in a day by ten men and ten women matched by job position:

E-mails by men	E-mails by women
82	48
77	61
78	56
83	59
82	58
78	56
81	60
74	64
86	59
76	63

(a) Calculate the difference in the number of business-related e-mails for each man/woman pair. Just looking at the differences, do you think men use e-mail more? Why or why not?

(b) Calculate the average difference and the standard deviation of the differences.

(c) Set up the hypotheses to test whether the average number of business-related e-mails generated by men is greater than that generated by women.

(d) Assuming that the data are normally distributed, at the 0.05 level of significance, what can you conclude?

10.8   Test of the Difference in Two Population Proportions

10.8.1   TWO-POPULATION TESTS WITH QUALITATIVE DATA

You should realize by now that different types of data require different statistical techniques. So far, all of the two-population tests that we have looked at involve quantitative data. In this section we look at the two-population tests for a kind of qualitative data: population proportions or percentages.

A lot of data are available in the form of proportions or percentages. Here are some examples taken from the newspaper:

- A study of Americans in the 1980s found that 71% said the government should take care of people who can't take care of themselves. In the 1990s the percentage was 57%. Is there evidence that Americans have become more cynical and less compassionate?

- A recent nationwide poll of 1225 adults showed that Blacks and Whites differ in their opinions of the causes and solutions of society's problems. But do they really? Seventy percent of Blacks think progress has been made in easing racial tension in the past decade compared with 65% of Whites. Based on these data, do Blacks and Whites really feel differently about this issue?

- Researchers in San Francisco used the diseased cells of patients with melanoma to develop a vaccine that they say dramatically reduces the recurrence of the deadliest form of skin cancer. After 3 years 70% of those vaccinated remained cancer-free compared with 20% of patients treated with surgery alone.

- A study of 1049 men and women aged 18 to 65 shows that a greater percentage of women (86%) find it difficult to have sex without emotional involvement compared with men (71%).

- A study conducted by an online service found that 30% of respondents under age 45 drove sports cars compared with 17% of the 45-or-over population.

Each of these studies has a structure similar to those we have already looked at in this chapter. Two populations are being compared and we have taken a sample from each population. What has changed is that the parameter being analyzed is no longer the mean but the population proportion. For each sample, the percentage of the sample that has a certain characteristic is found. These percentages then need to be compared.

10.8.2 THE TEST FOR TWO POPULATION PROPORTIONS

We know that even if the percentages of two populations that have a certain characteristic were exactly the same, we would almost never get exactly the same percentages in two samples from the populations. This is because of sampling error. The question is then, How large a difference in the percentages is large enough to declare the difference statistically significant? That is, how large does the difference have to be for us to be reasonably sure that there really is a difference in the behavior of the two populations, that something more than just sampling error is causing the difference? This is the same question we addressed in comparing means. This time the population parameters being compared are proportions, however, and the evidence from the sample is summarized by the sample proportions. The notation in Table 10.4 follows the pattern established for the tests of means.

As in the tests for comparing two population means, there are two different but equivalent ways of stating the null and alternative hypotheses. We can also test for a difference in proportions other than 0. Here are the null and alternative hypotheses for the upper-tail and lower-tail tests of differences in proportions:

Lower-tail test

$H_0: p_1 = p_2$ or $H_0: p_1 - p_2 \neq 0$ Use this test to decide whether the proportion of popu-
$H_A: p_1 < p_2$ $H_A: p_1 - p_2 < 0$ lation 1 is less than the proportion of population 2.

Upper-tail test

$H_0: p_1 \leq p_2$ or $H_0: p_1 - p_2 \leq 0$ Use this test to decide whether the proportion of popu-
$H_A: p_1 > p_2$ $H_A: p_1 - p_2 > 0$ lation 1 is greater than the proportion of population 2.

TABLE 10.4

VARIABLES FOR TWO POPULATION PROPORTIONS

	Population 1	Population 2
Population size	N_1	N_2
Sample size	n_1	n_2
Population proportion	p_1	p_1
Number in sample that have the characteristic being studied	x_1	x_2
Sample proportion	$\hat{p}_1 = \dfrac{x_1}{n_1}$	$\hat{p}_2 = \dfrac{x_2}{n_2}$

EXAMPLE 10.21

ONLINE COURSES

Understand the problem.

Setting Up the Hypotheses

The professor who is looking at the effectiveness of online courses has learned that the average score on the posttest is higher for the on-ground classes than for the online classes. However, she knows that more is involved in course performance than a single test and that just because a student does well on an exam does not mean that he or she will pass the course. She decides to look at whether the percentage of students who actually pass the course is also higher for the on-ground classes.

The professor looks at the final grades for each student and classifies them as pass (D or higher) or fail (F). The data for the two groups follow:

Collect the data.

	Population 1: on-ground	Population 2: online
Sample size	47	46
Population proportion	p_1	p_2
Number of passes	42	35
Sample proportion	0.89 (89%)	0.76 (76%)

Because the professor wants to know whether the proportion who pass the course for the on-ground group is higher than that for the online group, she will do a one-tail test with these hypotheses:

$$H_0: \quad p_1 \le p_2 \quad \text{or} \quad H_0: \quad p_1 - p_2 \le 0$$
$$H_A: \quad p_1 > p_2 \qquad\qquad H_A: \quad p_1 - p_2 > 0$$

After you become proficient at hypothesis testing you might want to change from the standard subscripts of 1 and 2 to denote populations. It is often easier to use subscripts that reflect the two populations, such as M and F for comparisons of males and females or O and I for outside and in-house. In fact, we do that in the next example.

It really doesn't matter which is population 1 and which is population 2 as long as we are consistent.

EXAMPLE 10.22

E-MAIL USE

Setting Up the Hypotheses

Although most companies use some form of e-mail to increase the frequency and timeliness of information, very little is known about the impact e-mail has had on organizational behavior. A study of five different companies was conducted to learn how e-mail is used at different levels of the organization. A total of 70 people responded to the survey. The respondents provided information on their title (VP/Director, Manager, Staff/Administrator), their gender, and the degree to which they use e-mail for various purposes. All companies in this study had used e-mail for at least 1 year. The study focused on the differences in the proportions of men and women who used e-mail to communicate.

	Population 1: men	Population 2: women
Sample size	35	35
Population proportion	p_M	p_W
Number that use e-mail	13	27
Sample proportion	37.1%	77.1%

We want to decide whether there is evidence of different use of e-mail by men and women. To detect differences in the two population proportions we use a two-tail test with these null and alternative hypotheses:

$$H_0: \quad p_M = p_W \quad \text{or} \quad H_0: \quad p_M - p_W = 0$$
$$H_A: \quad p_M \neq p_W \qquad H_A: \quad p_M - p_W \neq 0$$

Remember! Sample proportions must always be between 0 and 1, or between 0 and 100 if expressed as a percentage.

Recall that the test statistic for a single-population proportion is a Z statistic. Comparing two-population proportions also uses a Z test statistic. Therefore, we already know how to find the rejection region.

The test statistic for this test is constructed using logic similar to what we have used before. The two sample proportions are compared by subtracting one from the other. To determine whether this difference is "large," it is compared to the standard error of the estimate. In this case, the estimate of the true difference between the population proportions, $p_1 - p_2$, is the difference between the sample proportions, $\hat{p}_1 - \hat{p}_2$. The standard error of this estimate is similar to the standard error for a single-sample proportion. The test statistic is then given by the formula

Test statistic for comparing two population proportions

$$Z = \frac{\hat{p}_1 - \hat{p}_2}{\sqrt{\bar{p}(1 - \bar{p})\left(\dfrac{1}{n_1} + \dfrac{1}{n_2}\right)}}$$

Remember! The test statistic is calculated under the assumption that the null hypothesis is true.

Notice that $\bar{p}$ is used in the calculation of the standard error. The value of $\bar{p}$ is similar to the pooled variance in the sense that it combines all of the sample data. If the null hypothesis is true and the population proportions are equal, then the best estimate of the common but unknown population proportion is found by pooling the data. Here is the formula for $\bar{p}$:

Common population proportion

$$\bar{p} = \frac{x_1 + x_2}{n_1 + n_2}$$

EXAMPLE 10.23

ONLINE COURSES

Calculating the Test Statistic

After thinking about the problem, the professor who is testing to see whether the proportion of passes in the on-ground course is higher than that in the online course decides to perform the test at the 0.05 level of significance. The critical value for the test is $Z = 1.645$. She calculates the value of $\bar{p}$ to be

$$\frac{42 + 35}{47 + 46} = 0.828, \text{ or } 83\%$$

The value of the test statistic is then

$$Z = \frac{0.89 - 0.76}{\sqrt{0.83 \times 0.17 \times \left(\dfrac{1}{47} + \dfrac{1}{46}\right)}} = 1.669$$

EXAMPLE 10.24

E-MAIL USE

Analyze the data.

Calculating the Test Statistic

For the study of e-mail use, there is no reason to test at any significance level other than 0.05. The rejection region for a two-tail Z test is defined by the critical values ± 1.96. The value of $\bar{p}$ is calculated as follows:

$$\bar{p} = \frac{13 + 27}{35 + 35} = 0.571$$

and the test statistic is found to be

$$Z = \frac{0.371 - 0.771}{\sqrt{(0.571)(0.429)\left(\frac{1}{35} + \frac{1}{35}\right)}} = -3.39$$

As in any hypothesis test, we are not finished until we make a decision about H_0 and interpret that decision in terms of the problem we are trying to solve.

EXAMPLE 10.25

ONLINE COURSES

Reaching a Conclusion

The professor who is comparing on-ground and online courses found that the test statistic for the data was 1.669. Since 1.669 is in the rejection region, she rejects H_0 and concludes that the proportion of students who pass the course is higher for the on-ground courses than the online courses.

Does this end the issue? Does it mean that online courses are not as effective and therefore should not be offered? Not by a long shot. Many other factors influence how students learn. Perhaps online courses are not for every student. Online courses take a lot of self-discipline or students fall quickly behind. Maybe data should be collected about the students themselves, which would give some insight into this question. Also, online courses are new and there are many questions about the best ways to assess students enrolled in them. The bottom line is that students in the online courses learned material. There are still many studies to be done here.

EXAMPLE 10.26

E-MAIL USE

Reaching a Conclusion

For the question about using e-mail, the test statistic falls in the rejection region, so we reject the null hypothesis. We conclude that different proportions of men and women use e-mail to communicate. Although this tells the company something, the results of the test do not indicate whether men or women use e-mail more. If this is what the company wanted to know, then it should have done a one-tail test.

As you can see, testing the difference in two population proportions is not very different from testing the difference in two population means. This test has been developed parallel to the procedure for testing the difference in two population means. It has also been developed as a natural extension of the test of a single-population proportion.

✓ TRY IT NOW!

E-mail Use
Comparing Two Proportions

The company that investigated e-mail use is also interested in the use of e-mail for personal messages. After seeing the results of the first test, the company decides that it really wants to know whether a higher proportion of women than men use e-mail to send personal messages. Here are the data:

	Population 1: men	Population 2: women
Sample size	35	35
Population proportion	p_1	p_2
Number who use e-mail	15	18
Sample proportion		

Finish the table and then perform the hypothesis test.

Step 1:

Step 2:

Step 3:

Step 4:

Step 5:

10.8.3 EXERCISES—LEARNING IT!

10.16 Does television match reality? A recent study looked at the percentage of television characters who survived cardiopulmonary resuscitation (CPR) compared with those who survive in real life. The study looked at 34 instances where CPR was applied to a television character and found that 25 survived, whereas data provided by emergency medical technicians found that in 40 real cases, only 16 survived.

(a) Calculate the sample proportions of people who survive CPR on television and in reality.

(b) Set up the hypotheses to test whether there is a difference in the proportions of people who survive CPR in the two groups.

(c) At the 0.05 level of significance, does television reflect reality?

10.17 Women who smoke have an increased risk of dying of breast cancer, according to a recently published study. In a study of 319,000 women who never smoked, there were 468 deaths from breast cancer, whereas out of 120,000 smokers, there were 187 deaths.

(a) Calculate the sample proportions of women who died of breast cancer for smokers and nonsmokers.

(b) Set up the hypotheses to test whether the proportion of women who die of breast cancer is higher for smokers than for nonsmokers.

(c) At the 0.05 level of significance, can you conclude that smoking causes breast cancer? If not, what can you conclude?

10.18 Selling personal computers is big business and consumers are becoming increasingly aware of vendor reputation. A recent study of two vendors of desktop personal computers reports on the units that need repair for Dell Computer and Gateway 2000. Of 1584 computers manufactured by Dell Computer, 427 needed repair, whereas for Gateway 2000, 825 of 2662 computers needed repair.

(a) Calculate the sample proportion of computers needing repair for each company.

(b) Set up the hypotheses to test whether the proportions of computers needing repairs are different for the two companies.

(c) At the 0.05 level of significance, what can you conclude?

10.19 How punctual are Amtrak trains? One reason passenger trains are late is that almost all of the tracks are owned and maintained by freight companies that control the schedules for use. Amtrak's definition of on time is based on trip length. After customer complaints about late trains, Amtrak has tried to improve and claims that it has. Of 100 runs of the Metroliner between New York and Washington, DC, in March 1993, 87 were on time, whereas in the same period in 1994, 90 were on time.

(a) Calculate the sample proportion of on-time trips for each time period.

(b) Can Amtrak really claim it has improved ($\alpha = 0.05$)? Why or why not?

10.9 Test of the Difference in Two Population Variances

10.9.1 THE F TEST FOR COMPARING POPULATION VARIANCES

Remember the test to compare population means when the samples are small and the standard deviations are unknown? In working with small samples, we had to decide whether the populations had a common but unknown variance. If they did, then we pooled the data and calculated a pooled estimate of the variance, s_p^2. We did not, however, discuss how to decide whether the populations had a common variance. Now it is time to learn how to determine this. A hypothesis test of variances follows the same five steps that we have used repeatedly. Again, the only difference is the test statistic.

To decide whether we should pool the data, we need to test whether the two population variances are equal. Thus, we use a two-tail test with these null and alternative hypotheses:

Two-tail test

$H_0: s_1^2 = s_2^2$ or $H_0: s_1^2/s_2^2 = 1$ Use this test to decide whether the variance of population 1
$H_A: s_1^2 \neq s_2^2$ $H_A: s_1^2/s_2^2 \neq 1$ is different from the variance of population 2.

Remember! The key words "different," "less than," and "greater than" indicate which test to use.

As with the tests to compare two population means or proportions, it is also possible to do one-tail tests.

Lower-tail test

$H_0: s_1^2 \geq s_2^2$ or $H_0: s_1^2/s_2^2 \geq 1$ Use this test to decide whether the variance of population 1
$H_A: s_1^2 < s_2^2$ $H_A: s_1^2/s_2^2 < 1$ is less than the variance of population 2.

Because of the difficulty in finding the rejection region for a lower-tail test, most one-tail tests are conveniently done as upper-tail tests.

Upper-tail test

$H_0: s_1^2 \leq s_2^2$ or $H_0: s_1^2/s_2^2 \leq 1$

$H_A: s_1^2 > s_2^2$ $H_A: s_1^2/s_2^2 > 1$

Use this test to decide whether the variance of population 1 is greater than the variance of population 2.

Since we are trying to decide how two population variances compare, it makes sense to compare the sample variances. You have seen two ways of comparing numbers. One way is to subtract one number from the other and see how close the difference is to 0. If the two numbers are estimates of the same unknown population parameter, then their difference will be close to 0. The other way to compare two numbers is to divide one number by the other and see how close the ratio is to 1. If the two numbers are estimates of the same unknown population parameter, then their ratio will be close to 1. Notice that the second way of writing each of the variance tests shown previously uses a ratio and a comparison to 1. When comparing variances, we always use a ratio to compare them and not a difference.

The test statistic for comparing variances is an F statistic.

You know from Chapter 7 that the best estimate of the population variance is the sample variance. Extending this idea to two populations, we find that the point estimate for the ratio of the population variances is the ratio of the sample variances. This is also the test statistic, shown here:

Test statistic for comparing two variances

$$F = \frac{s_1^2}{s_2^2}$$

The F distribution is determined by two sets of degrees of freedom

Notice that this ratio is labeled F. This means that the test statistic follows an F distribution if the two original populations are normally distributed. The F distribution was briefly introduced in Chapter 7. It is named after the famous statistician R. A. Fisher. Like the χ^2 distribution, which we used to test a single-population variance, the specific shape of the F distribution is determined by its degrees of freedom. But the F distribution has not one but two values that determine its shape. One value is called the degrees of freedom in the numerator and is equal to 1 less than the sample size on which s_1^2 is based, $n_1 - 1$. The other value is called the degrees of freedom in the denominator and is equal to 1 less than the sample size on which s_2^2 is based, $n_2 - 1$.

The procedure for finding the rejection region is similar to the one we have used for other tests. If we are a doing a two-tail test, then the rejection region is two-tail; if we are doing a one-tail test, then the rejection region is one-tail. The rejection regions are shown in Figure 10.6.

The symbols df_1 and df_2 stand for the number of degrees of freedom in the numerator and the denominator, respectively.

The critical values that define the rejection region are labeled $F_{\text{upper,df1,df2}}$ and $F_{\text{lower,df1,df2}}$. To find the values for $F_{\text{upper,df1,df2}}$ and $F_{\text{lower,df1,df2}}$ we need to notice that the shape of the F distribution is not symmetric and the distribution is not centered at 0. Therefore, the absolute values of F_{upper} and F_{lower} are not the same and they are always greater than 0. It happens that there is a relationship between the upper and lower F values. In particular, $F_{\text{lower,df1,df2}}$ can be found from an upper-tail value as follows:

Lower critical value for an F test

$$F_{\text{lower, df1, df2}} = \frac{1}{F_{\text{upper, df2,df1}}}$$

That is, the lower critical value is the reciprocal of the upper critical value with the degrees of freedom reversed. Therefore, we need table values for only F_{upper}.

Since the F distribution is determined by two sets of degrees of freedom, labeled df_1 and df_2, we need an entire table to specify the values for F_{upper} that cut off a certain amount of probability in the upper tail of the distribution. Unlike the other tables we have used, each upper-tail area probability requires a separate table. For instance, Table 6a in Appendix A shows the values for F_{upper} that cut off an area of 0.01 in the upper tail. The row and column we use depend on the number of degrees of freedom in the numerator and the denominator. However, if we need 0.05 in the upper-tail

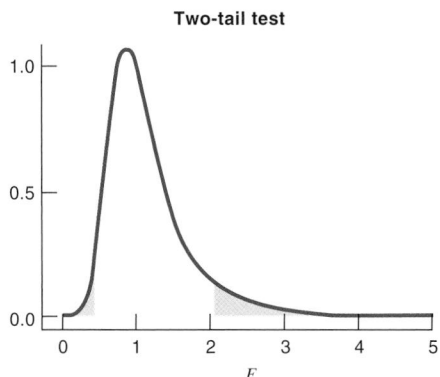

Two-tail test

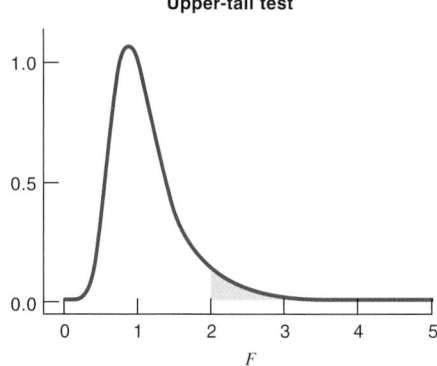

Upper-tail test

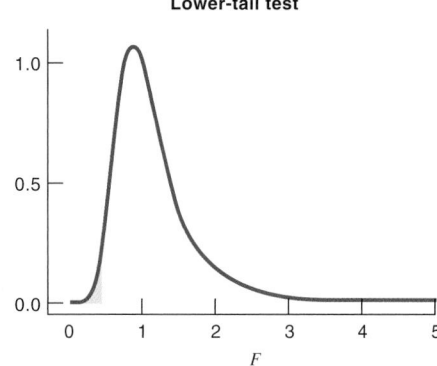

Lower-tail test

FIGURE 10.6 **Rejection regions for test of variances**

area, then we use Table 6b in Appendix A. This is the first time we have encountered a different table for each value of α. When we used the t or the χ^2 distribution, each column corresponded to a different value of α.

EXAMPLE 10.27

ONLINE COURSES

Setting Up the Test

When the professor who is studying online courses wanted to know whether there was a difference in the posttest scores for men and women, she made an assumption about whether the population variances were equal. The relevant data from the problem follow:

	Population 1: women in online course	**Population 2:** men in online course
Sample size	$n_1 = 19$	$n_2 = 27$
Sample average	$\overline{X}_1 = 80.05$	$\overline{X}_2 = 79.59$
Sample standard deviation	$s_1 = 11.43$	$s_2 = 9.20$

To determine whether this assumption is reasonable we need to test the hypotheses:

$$H_0: \quad s_1^2 = s_2^2$$
$$H_A: \quad s_1^2 \neq s_2^2$$

If we reject H_0, then we will conclude that the variances are different and we will know that we cannot pool the variances. For the moment, we assume that

Understand the problem.

Collect the data.

Analyze the data.

the populations are normally distributed and we will do the test at the 0.10 level of significance. This means that we need to use the F table with the upper-tail area of 0.05 in Table 6b. Here is a portion of that table:

Denominator degrees of freedom (df_2)	Numerator degrees of freedom (df_1)									
	18	**19**	**20**	**21**	**22**	**23**	**24**	**25**	**26**	**27**
18	2.217	2.203	2.191	2.179	2.168	2.159	2.150	2.141	2.134	2.126
19	2.182	2.168	2.155	2.144	2.133	2.123	2.114	2.106	2.098	2.090
20	2.151	2.137	2.124	2.112	2.102	2.092	2.082	2.074	2.066	2.059
21	2.123	2.109	2.096	2.084	2.073	2.063	2.054	2.045	2.037	2.030
22	2.098	2.084	2.071	2.059	2.048	2.038	2.028	2.020	2.012	2.004
23	2.075	2.061	2.048	2.036	2.025	2.014	2.005	1.996	1.988	1.981
24	2.054	2.040	2.027	2.015	2.003	1.993	1.984	1.975	1.967	1.959
25	2.035	2.021	2.007	1.995	1.984	1.974	1.964	1.955	1.947	1.939
26	2.018	2.003	1.990	1.978	1.966	1.956	1.946	1.938	1.929	1.921
27	2.002	1.987	1.974	1.961	1.950	1.940	1.930	1.921	1.913	1.905

To find the upper critical value we use the numerator degrees of freedom as $19 - 1 = 18$ and the denominator degrees of freedom as $27 - 1 = 26$. The value of $F_{upper,18,26}$ is 2.018. Now, the lower critical value that we want is $F_{lower,18,26}$, so using the relationship

To find $F_{lower,df1,df2}$ reverse the degrees of freedom, find the corresponding F_{upper} value, and take the reciprocal.

$$F_{lower,18,26} = \frac{1}{F_{upper,26,18}}$$

we find that

$$F_{lower,18,26} = \frac{1}{2.134} = 0.469$$

This example illustrates the procedure for finding the rejection region for a two-tail test. Now we summarize the steps for finding a two-tail rejection region and note the differences for a one-tail rejection region.

Two-tail rejection region

Steps for finding the rejection regions for the F distribution

Step 1: Divide the value of α in half.

Step 2: Find the F table that corresponds to those values that cut off $\alpha/2$ in the upper tail of the distribution.

Step 3: Find $F_{upper,df1,df2}$ at the intersection of the column corresponding to $df_1 = n_1 - 1$ and the row corresponding to $df_2 = n_2 - 1$.

Step 4: Find F_{lower} by first finding $F_{upper,df2,df1}$ with $n_2 - 1$ degrees of freedom in the numerator and $n_1 - 1$ degrees of freedom in the denominator. Then take the reciprocal of this number.

Step 5: Reject H_0 if the F statistic is greater than F_{upper} or less than F_{lower}.

If you are using a one-tail test of the variances, the main difference in finding the rejection region is that you do not split α in half. In addition, you need to find only one of the critical values: either F_{upper} or F_{lower}.

Upper-tail rejection region

Step 1: Find the F table that corresponds to those values that cut off α in the upper tail of the distribution.

Step 2: Find $F_{\text{upper,df1,df2}}$ at the intersection of the column corresponding to $df_1 = n_1 - 1$ and the row corresponding to $df_2 = n_2 - 1$.

Step 3: Reject H_0 if the F statistic is greater than F_{upper}.

Lower-tail rejection region

Step 1: Find the F table that corresponds to those values that cut off α in the upper tail of the distribution.

Step 2: Find F_{lower} by first finding $F_{\text{upper,df2,df1}}$ with $n_2 - 1$ degrees of freedom in the numerator and $n_1 - 1$ degrees of freedom in the denominator. Then take the reciprocal of this number.

Step 3: Reject H_0 if the F statistic is less than F_{lower}.

To avoid the tedious problem of finding the lower-tail rejection region for a one-tail test, just set the test up as an upper-tail test by defining population 1 as the one with the larger of the two sample variances.

The remainder of the hypothesis-testing procedure involves calculating the test statistic and making a decision in terms of both the null and alternative hypotheses and the original problem.

EXAMPLE 10.28

ONLINE COURSES

Concluding the Test

From the data we can calculate the test statistic as

$$F = \frac{11.43^2}{9.20^2} = 1.544$$

Since the critical values for the test were 0.469 and 2.018, we cannot reject H_0. This means there is not enough evidence to say that the variances are different, and so the assumption that they are equal is reasonable.

Analyze the data.

Draw conclusions.

✓ TRY IT NOW!

Online Courses

Testing for Equality of Variances

In Section 10.5.1 we used the pooled variance to compare men and women on the posttest scores. That is, we assumed the population variances were equal. The relevant data are listed here:

	Population 1: women in on-ground course	**Population 2:** men in on-ground course
Sample size	$n_1 = 20$	$n_2 = 27$
Sample mean	$\overline{X}_1 = 85.15$	$\overline{X}_2 = 83.74$
Sample standard deviation	$s_1 = 8.99$	$s_2 = 9.27$

Assume that the data are normally distributed and perform a hypothesis test to determine whether the assumption of equal variances is reasonable. Use a level of significance of 0.10.

Step 1:

Step 2:

Step 3:

Step 4:

Step 5:

10.9.2 EXERCISES – LEARNING IT!

Requires Exercise 10.6

10.20 The Board of Realtors for Greater Bridgeport, Connecticut, is looking at the average selling prices of homes. The data are given again:

	1995 sales	1996 sales
Sample size	$n_1 = 25$	$n_2 = 25$
Sample mean	$\overline{X}_1 = \$151{,}166$	$\overline{X}_2 = \$160{,}669$
Sample standard deviation	$s_1 = \$5332$	$s_2 = \$6468$

(a) Assuming that the populations are normally distributed, set up the hypotheses to test whether the population variances are equal at the 0.10 level of significance.

(b) Was the decision to test using the pooled variance justified?

Requires Example 10.8

10.21 In your quest for the perfect golf clubs you made an assumption about the population variances when you tested the hypotheses. Here are the data (in yards) you collected by hitting brand X and brand Z balls on a driving range:

	Brand X	Brand Z
Sample size	$n_X = 15$	$n_Z = 15$
Sample mean	$\overline{X}_X = 255$	$\overline{X}_Z = 271$
Sample standard deviation	$s_X = 8.7$	$s_Z = 9.1$

(a) Set up the appropriate hypotheses to test whether the variance for brand Z clubs is the same as the variance for brand X clubs.

(b) Assuming that the populations are normally distributed, at the 0.10 level of significance, was your decision to pool the variances reasonable?

(c) In general, would a difference in variation between the clubs be a factor in your purchase decision?

10.22 The members of the Chamber of Commerce of a small city in Fairfield County, Connecticut, are looking at the amount of vacant office space in their city compared to the space in a similar city. The data (in square feet) are given here:

	Our city	Their city
Sample size	$n_1 = 12$	$n_2 = 12$
Sample mean	$\overline{X}_1 = 217{,}000$	$\overline{X}_2 = 167{,}607$
Sample standard deviation	$s_1 = 2200$	$s_2 = 2100$

Given that the populations are normally distributed, at the 0.02 level of significance, should they pool the variances?

ANS. $H_0: \sigma_1^2 = \sigma_2^2$; $H_A: \sigma_1^2 \neq \sigma_2^2$; $F_{lower,19,26} = 0.476$, $F_{upper,19,26} = 2.003$; $F = 0.941$; DO NOT REJECT H_0: THE ASSUMPTION OF EQUAL VARIANCES IS REASONABLE.

10.23 A company has two different production lines that make the plastic cards used for credit cards and ATM cards. Both lines use $\overline{X}$ control charts to make sure that they run to the target specification, and both have been in control for the past 6 weeks. Recently, however, the quality manager has noticed that one of the machines (machine A) has many more items being rejected for the measurement on the width of the card. Since both machines are running to target, he decides the problem must be with the variability and decides to run a test. He samples 40 items from each production line and calculates the following summary statistics:

	Machine A	Machine B
Sample size	40	40
Sample standard deviation	1.1 mm	0.62 mm

(a) Set up the hypotheses to test whether the variance of machine A is greater than the variance of machine B.

(b) Assuming that the populations are normally distributed, perform the test at the 0.10 level of significance.

(c) Is the quality manager correct in his perception that machine A is more variable?

10.24 A large utility company is considering two sites for locating a large-scale wind energy conversion system (windmill). Both the average speed and the variation in speed at a site are important: The more consistent the wind speeds, the more efficient the energy conversion. The variability for the two sites is measured using a sample of 30 wind speed observations at each site:

	Site 1	Site 2
Sample size	30	30
Sample standard deviation	1.80 mph	2.62 mph

(a) Set up the hypotheses to test whether the variability in wind speed at the two sites is the same.

(b) Assuming that the populations are normally distributed, perform the test at the 0.10 level of significance.

(c) What can you tell the utility company about the variability at the two sites?

GET IT IN WRITING

Online Courses

TO: **Dean of Faculty**
FROM: **Professor X Bar**
RE: **Learning in Online Courses**

I recently conducted a study to look at the effectiveness of online courses compared with the traditional on-ground courses. This semester I taught elementary statistics to two sections of each type of course. Students in each course were given a pretest on the first day of class to assess their current knowledge of the subject. At the end of the semester they were given a similar examination. Initial enrollment in each section was 25 students, but after students dropped the course a total of 47 students were in the traditional on-ground course and 46 in the online course. The results were that students in the on-ground course scored higher on the posttest, on average, than students in the online course ($p < 0.05$) and that a higher proportion of students in the on-ground course passed the course. The table summarizes some of the results.

	On-ground	Online
Posttest score mean	89.34	79.78
Standard deviation	9.08	10.06
Percent passing	89% (42/47)	76% (35/50)
Sample size	47	46

The students in the on-ground course scored, on average, ten points higher on the posttest. The variability was not significantly different in the scores of the two groups. To be certain that differences in the posttest scores were real and not a result of inherent differences in the two groups, I also used a t test to see whether the pretest scores of the two groups differed. The results of the test found that the on-ground students scored significantly higher on the pretest ($p < 0.05$) although the difference was only five points. This difference could be a factor in posttest performance.

Although the data provide evidence that students in an on-ground course do better than those in an online course, this does not mean that students in an online course do not learn. The average score for students in the online course went from 53.7 on the pretest to 79.78 on the posttest, a difference of 26 points. Clearly, learning was taking place!

Another issue is that online courses are relatively new. We have not completely learned how to assess students in this environment. The test procedure used may favor students in the on-ground course for some reason. Also, although the on-ground students performed better, it is possible that the online course is better suited to certain kinds of students. Age might be a factor in this.

I suggest that we consider collecting additional data to continue the study. We should consider several demographic variables on the students themselves, such as age, gender, and previous online courses. If you will arrange my teaching schedule to permit it, I will be happy to continue the study.

10.10 Technology and Two-Population Hypothesis Tests

Excel, Minitab, and the TI-83 graphing calculator have tools to help you in two-population hypothesis tests. **Excel** has built-in data analysis tools for two-population hypothesis tests. Once you know which test to use, the procedure is not very different from test to test. **Minitab** has built-in routines that will allow you to do almost all of the two-population tests that you learned about in this chapter. Just as in the one-population tests, the procedures are all similar. All that really changes is the data that you need to input. The **TI-83** will allow you to do almost all of the population tests you learned in this chapter. As in the one-population tests, the TI-83 will allow you to use either raw data or summary statistics.

Here, we will give instructions for large-sample tests of two means using the various software packages. Please refer to the manual associated with the software that you are using for more detailed instructions on small-sample tests and other two-population tests.

10.10.1 LARGE-SAMPLE TESTS OF TWO MEANS IN EXCEL

For a large-sample test of two population means, either the population standard deviations must be known or the sample size from each population must be large — that is, greater than 30. Suppose that we want to consider the data from the professor who was looking at on-ground classes versus online classes. We want to know whether the results obtained in the on-ground classes are better than those of the online classes. This is a one-tail test, and the hypotheses are

$$H_0: \quad \mu_1 = \mu_2$$
$$H_A: \quad \mu_1 > \mu_2$$

To use Excel to perform the hypothesis test, the data must be in a spreadsheet. If you do not know the population standard deviations and are using the sample standard deviations, you must first calculate them using either the Descriptive Statistics tool or the **STDEV** or **VAR** functions. The value you will actually need is the variance, so if you calculate the standard deviation, you will have to square it to find the variance.

In our example, the population standard deviations are not known, but the sample sizes are $n_1 = 47$ and $n_2 = 46$, respectively. To perform the test, open the **Tools > Data Analysis** menu and from the list choose **z Test: Two Sample for Means.** This will open the dialog box for the test as shown in Figure 10.7.

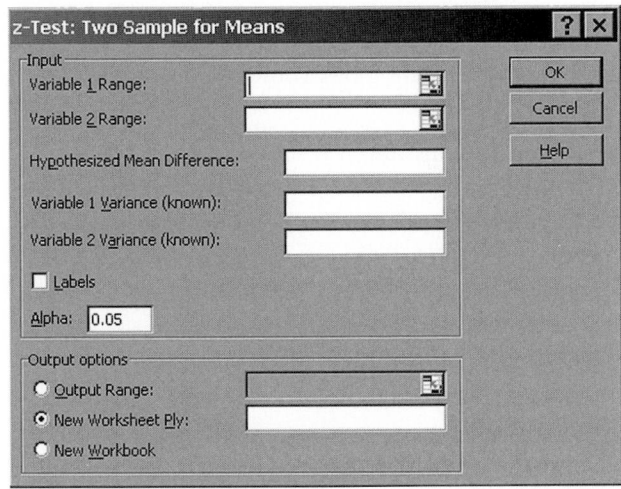

FIGURE 10.7 **Dialog box for *z* test for two means**

Use the following procedure to perform the test:

1. Position the cursor in the text box next to **Variable 1 Range:** and highlight the range that contains the posttest data for the first population — in this case, the group trained by the outside trainer.

2. With the cursor in the text box for **Variable 2 Range:** highlight the location of the data for the second population, the in-house group.

3. Position the cursor in the text box for **Hypothesized Mean Difference** and enter **0.** Since we want to know whether $\mu_1 > \mu_2$, this is the same as $\mu_1 - \mu_2 > 0$.

4. Now, put the cursor in the text box for **Variable 1 Variance (known)** and enter the value of the variance for population 1. For the training example, the variance in posttest scores for the outside group is $(9.08)^2 = 82.4464$.

5. Repeat this for **Variable 2 Variance (known).**

6. If there were labels in any of the data ranges you highlighted, check **Labels.**

7. Enter the level of significance for the test. In this case, we will use the default value of $\alpha = 0.05$.

8. Finally, indicate where you want the output from the test to appear.

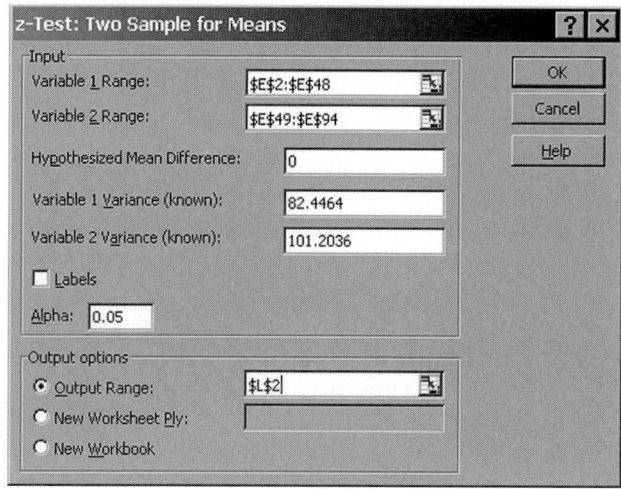

FIGURE 10.8 *z* **test dialog box**

When you have entered all the data, the dialog box should look like the one in Figure 10.8.

9. Hit Enter to perform the test. The resulting output is shown in Figure 10.9.

z-Test: Two Sample for Means		
	Variable 1	*Variable 2*
Mean	84.34042553	79.7826087
Known Variance	82.4464	101.2036
Observations	47	46
Hypothesized Mean Difference	0	
z	2.292051772	
P(Z<=z) one-tail	0.010951297	
z Critical one-tail	1.644853	
P(Z<=z) two-tail	0.021902595	
z Critical two-tail	1.959961082	

FIGURE 10.9 **Test output**

From the output, we see that the value of the test statistic, labeled **z** in the output, is 2.292051772 (which is the same as the 2.30 we obtained by hand). Since the critical value for a one-tailed test is 1.644853, we can reject H_0 and conclude that the mean for the outside group is higher than the mean for the in-house group.

The procedure gives the p value for the test, but does not label it as such. The value labeled **P(Z <= z) one-tail** is the p value for the one-tailed test, and **P(Z <= z) two-tail** is the p value for the two-tailed test.

10.10.2 LARGE-SAMPLE TEST OF TWO MEANS IN MINITAB

Minitab does not actually do a large-sample test (Z) for two population means if you want to use known population standard deviations. This is because, as you have learned, it would be rare for this situation to occur with real data. However, you can still do a large-sample test with population standard deviations unknown by following the procedure for small-sample tests. The only problem you will have is that the p value will be based on the t distribution rather than the Z distribution. For large degrees of freedom (which a large-sample test will have), this will make almost no difference.

10.10.3 LARGE-SAMPLE TEST OF POPULATION MEANS WITH TI-83

Suppose that we would like to do the hypothesis test for the online versus on-ground class data. The professor wants to know whether the results obtained in the on-ground classes are better than those of the online classes. This is a one-tail test, and the hypotheses are

$$H_0: \mu_1 = \mu_2$$
$$H_A: \mu_1 > \mu_2$$

For a large-sample test of two population means, either the population standard deviations must be known or the sample size from each population must be large—that is, greater than 30.

To perform the test:

1. Press $\boxed{\text{STAT}}$ and select **TESTS**.
2. From the **TESTS** menu select option 3, **2-Samp Z Test. . . .**
3. We will do this test using summary data, so position the cursor on **Stats** and press $\boxed{\text{ENTER}}$.
4. Figure 10.10 shows the screen with the summary data from Example 10.7 entered.

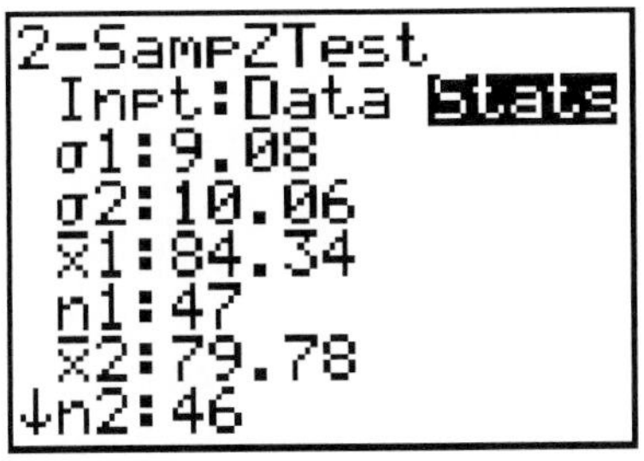

```
2-SampZTest
Inpt:Data Stats
σ1:9.08
σ2:10.06
x̄1:84.34
n1:47
x̄2:79.78
↓n2:46
```

FIGURE 10.10 **Screen for 2-sample Z test**

Be careful. The data are entered in an odd order—the standard deviations first.

5. Since there is a ↓ next to **n2**, you will need to use the $\boxed{\blacktriangledown}$ key or press $\boxed{\text{ENTER}}$ after you input **n2** to scroll down to the rest of the input.
6. Press **Calculate** and the output will appear on the screen as shown in Figure 10.11.

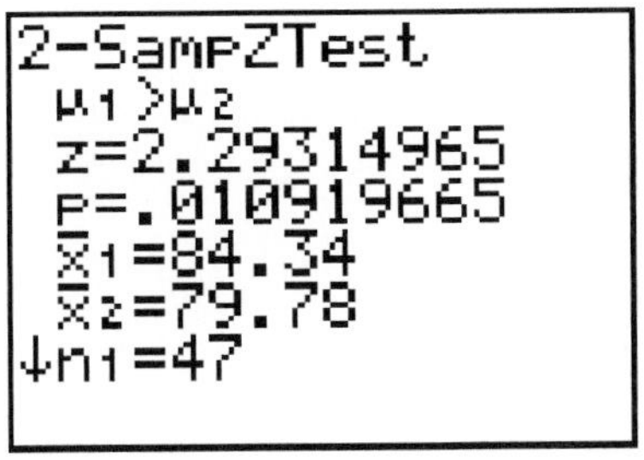

```
2-SampZTest
μ1>μ2
z=2.29314965
p=.010919665
x̄1=84.34
x̄2=79.78
↓n1=47
```

FIGURE 10.11 **Output from 2-Sample Z test**

CHAPTER 10 SUMMARY

KEY TERMS

Term	Definition	Page reference
Dependent samples	**Dependent samples** are related to each other. The members of one sample are identical to or matched or paired with the members of the other sample according to some characteristic.	470
Pretest and posttest	**Pretest** and **posttest** conditions exist when data are collected on the same sample elements before and after some experiment is performed.	470

KEY FORMULAS

Term	Formula	Page reference
Average difference	$\bar{d} = \dfrac{\sum\limits_{i=1}^{n} d_i}{n}$	472
Common population proportion	$\bar{p} = \dfrac{x_1 + x_2}{n_1 + n_2}$	482
Degrees of freedom for small-sample, unequal variances test	$\nu = \dfrac{\left(\dfrac{s_1^2}{n_1} + \dfrac{s_2^2}{n_2}\right)^2}{\dfrac{(s_1^2/n_1)^2}{n_1 - 1} + \dfrac{(s_2^2/n_2)^2}{n_2 - 1}}$	466
Lower critical value for F test	$F_{\text{lower,df1,df2}} = \dfrac{1}{F_{\text{upper,df2,df1}}}$	486
Pooled variance	$s_p^2 = \dfrac{(n_1 - 1)s_1^2 + (n_2 - 1)s_2^2}{n_1 + n_2 - 2}$	463
Test for paired difference	$t = \dfrac{\bar{d}}{s_d/\sqrt{n}}$	474
Tests for two population means:		
• Variances known	$Z = \dfrac{(\bar{X}_1 - \bar{X}_2) - 0}{\sqrt{\sigma_1^2/n_1 + \sigma_2^2/n_2}}$	455
• Variances unknown, $n_1, n_2 \geqslant 30$	$Z = \dfrac{(\bar{X}_1 - \bar{X}_2) - 0}{\sqrt{s_1^2/n_1 + s_2^2/n_2}}$	458
• Variances unknown and not equal, $n_1, n_2 < 30$; degrees of freedom adjusted	$t = \dfrac{(\bar{X}_1 - \bar{X}_2) - d}{\sqrt{s_1^2/n_1 + s_2^2/n_2}}$	466
• Variances unknown but equal, $n_1, n_2 < 30$	$t = \dfrac{\bar{X}_1 - \bar{X}_2}{s_p\sqrt{1/n_1 + 1/n_2}}$	464
Test for two population proportions	$Z = \dfrac{\hat{p}_1 - \hat{p}_2}{\sqrt{\bar{p}(1 - \bar{p})\left(\dfrac{1}{n_1} + \dfrac{1}{n_2}\right)}}$	482
Test for two population variances	$F = \dfrac{s_1^2}{s_2^2}$	486

CHAPTER 10 EXERCISES

Learning It!

10.25 The members of the Chamber of Commerce of a small city in Fairfield County, Connecticut, are wondering whether they need to worry about the amount of vacant office space in the city. They will consider lobbying for tax incentives for businesses if they find that their city has more vacant office space than a comparable city. They gather weekly data on the number of square feet of vacant office space in the second quarter of 1996 for both cities:

Requires Exercise 10.22

	Our city	Their city
Sample size	$n_1 = 12$	$n_2 = 12$
Sample mean	$\overline{X}_1 = 210{,}700$	$\overline{X}_2 = 167{,}607$
Sample standard deviation	$s_1 = 2200$	$s_2 = 2100$

(a) Calculate the pooled variance for the data.

(b) Assuming that the data are normally distributed, at the 0.01 level of significance, should they lobby for tax incentives?

10.26 A recent study of consumer behavior focused on the amount of money spent monthly on frozen foods. Researchers wanted to determine whether there was a difference in the average amounts of money spent by men and women. Data were collected on samples of 50 men and 50 women:

	Men	Women
Sample size	$n_M = 50$	$n_W = 50$
Sample mean	$\overline{X}_M = \$72.24$	$\overline{X}_W = \$67.44$
Sample standard deviation	$s_M = \$8.23$	$s_W = \$8.12$

(a) Set up the appropriate hypotheses to test whether men spend more per month, on the average, for frozen foods than women do.

(b) Use the large-sample test with unknown variances to test the hypotheses. Use a level of significance of 0.10.

(c) Do the data provide evidence that the average amount spent per month on frozen food by men is greater than the amount spent by women?

(d) What is the *p* value of the test?

10.27 The nurses who were part of the hand washing experiment are still not convinced that the length of time spent washing hands makes that much difference. They design their own study and decide to have each nurse in the CCU wash his or her hands twice, once for 2.5 seconds and once for 15 seconds. After each washing, they do a bacteria culture and measure the number of bacteria that remain on the person's hands. The data are shown in the table:

Observation	Culture 1: 2.5 seconds	Culture 2: 15 seconds
1	66	78
2	132	115
3	120	93
4	187	48
5	190	77
6	17	3
7	33	12
8	92	12
9	1000	146

(a) Calculate the difference between the numbers of bacteria for each nurse. Just looking at the differences, do you think that washing longer lowered the number of bacteria? Why or why not?

(b) Calculate the average difference and the standard deviation of the differences.

(c) Set up the hypotheses to test whether the average number of bacteria dropped after washing longer.

(d) Assuming that the data are normally distributed, at the 0.05 level of significance, what can you conclude?

(e) Suppose you were told that the second episode of hand washing was done right after the first. Would that change your interpretation of the results of the study?

Thinking About It!

10.28   Psychologists have done many studies on the effects of shift rotation on employee behaviors. A manufacturing company with many plants rotates its employees in the New England plant so that every time they switch shifts they go to an earlier shift. That is, they go from working the day shift (7 A.M. to 3 P.M.) to the graveyard shift (11 P.M. to 7 A.M.) to the swing shift (3 to 11 P.M.). This is contrary to the practice at the other plants. Human resources personnel think that the employees involved in this backward rotation take more sick days than employees in plants that use forward rotation. They decide to sample 25 employees from the New England plant and 25 employees from a plant in the South. Here are the data on the numbers of sick days used by each employee in the past 12 months:

South: forward rotation					New England: backward rotation				
1	3	5	5	6	0	5	6	7	9
1	4	5	6	7	3	5	7	7	9
2	4	5	6	8	4	6	7	7	10
2	4	5	6	8	4	6	7	8	11
3	4	5	6	8	5	6	7	8	11

At the 0.05 level of significance, do the data provide enough evidence to conclude that employees who are on a forward shift rotation use, on the average, fewer sick days than those who are on a backward rotation? Be sure to justify any assumptions you made in selecting a test procedure.

10.29   A large hospital recently did a study on the treatment of patients who were admitted with pancreatitis. The hospital wanted to know, among other things, how long patients spend in the hospital. Because patients were admitted to either the medical or surgical service, researchers decided to look at the data for both services separately. Here are the numbers of days patients spent in the hospital:

Medical				Surgical		
2	4	7	10	3	6	9
2	4	7	12	4	6	10
2	5	7	13	4	6	11
3	5	7	15	5	7	13
3	6	8		5	8	

The hospital wants to know if patients admitted to the surgical service have a shorter stay than those in the medical service. If the surgical service has a shorter stay, more patients who have pancreatitis may be admitted there.

(a) What level of significance do you suggest the hospital use? Justify your choice.

(b) Assuming that the data are normally distributed, should the hospital start admitting more patients with pancreatitis to the surgical service? Use the level of significance you chose in part (a).

(c) What impact does your choice of α have on the decision?

(d) Do you think the assumption of normality is reasonable? Why or why not?

Requires Exercises 10.10, 10.11

10.30   You are still wondering about the results of the test on the difference in software prices. You wonder why the two tests came to different conclusions and figure that it must be the amount of variability in the software prices for the packages chosen. Look at the data again.

Top ten business software packages	Computability price	PC Connection price
Norton Anti-Virus 2000 v 6.0	$29	$32
Microsoft W98 Second Edition Upgrade	95	90
Norton System Works 2000 v 3.0	59	60
VirusScan 5.0	29	24
QuickBooks 2000 Pro	200	200
Norton Internet Security 2000	58	48
QuickBooks 2000	120	120
Microsoft W98 Second Edition	180	179
Microsoft Office 2000 Upgrade	220	230
VirusScan 5.0 Deluxe	38	33

(a) Do any of the software packages have prices that are *very* different from the others? If so, which ones?

(b) Drop the data for the most unusual observations and perform the hypothesis test again using the test for independent samples.

(c) Does anything change from the last time you did the test? If so, what?

(d) Does dropping the observations change the decision?

(e) Was this the right test to use? Why or why not?

10.31 A study was recently completed by an insurance company concerning a particular surgical procedure. The study looked at the hospital records of 40 patients at two different hospitals and compared the lengths of stay. The data were analyzed using Minitab. This output shows the descriptive statistics for the data:

DESCRIPTIVE STATISTICS

Length of	StayN	Mean	Median	Trim Mean	St. Dev.	SE Mean
Hospital 1	40	7.725	7.500	7.667	2.562	0.405
Hospital 2	40	10.350	10.000	10.222	3.340	0.528

Length of StayMin		Max	Q1	Q3
Hospital 1	2.000	14.000	6.000	9.000
Hospital 2	5.000	18.000	8.000	13.500

(a) Set up the hypotheses to test whether the patients at hospital 1 had, on the average, a shorter stay than those at hospital 2.

(b) What type of test do you use to make this decision? Why?

(c) Perform the appropriate hypothesis test. Use a level of significance of 0.05.

(d) Do the data provide evidence that the mean stay at hospital 1 is shorter than at hospital 2?

10.32 After looking at the results of the data analysis, the director of human resources at the company looking at sick days and shift rotation writes a memo to the vice president of human resources suggesting that the company change the shift rotation in the New England plant to a forward rotation. She cites the results of the test and states that the data "provide evidence that a forward shift rotation causes people to take fewer sick days." Since you did the analysis, she gives you the memo to read before she sends it.

Requires Exercise 10.28

(a) Do you agree with the director of human resources? Why or why not?

(b) Write a memo to the director of human resources explaining your reaction. Include plans for further study if you think it is warranted.

10.33 Reconsider the study of the amount of money spent monthly on frozen foods. Data were collected on samples of 50 men and 50 women:

Requires Exercise 10.26

	Men	Women
Sample size	$n_M = 50$	$n_W = 50$
Sample mean	$\overline{X}_M = \$72.24$	$\overline{X}_W = \$67.44$
Sample standard deviation	$s_M = \$8.23$	$s_W = \$8.12$

(a) Was your decision sensitive to the value chosen for α? Why or why not?

(b) Suppose you are interested only in whether the average spent by men was different than the average spent by women. How does this change the setup of the test? Does it change the conclusion?

Requires Exercises
10.10, 10.11, 10.30

10.34 After looking at the software price data again, and based on the results of the paired t test in Exercise 10.11, you are considering buying your software from Computability. You decide to check the ads for each company one more time to see whether there are any hidden catches and you notice that shipping charges for PC Connection are $5 whereas for Computability they are $16.95. At the 0.05 level of significance, who will you buy your software from?

10.35 Since the data are available, the nursing supervisor wants to test whether there is a difference in the average amounts of time that nurses from two different departments spend washing their hands. She is not sure whether to pool the variances, and Minitab does not do an F test on variances, so she decides to run the t test both ways. The Minitab output is shown here.

Two Sample T-Test and Confidence Interval

Two sample T for C9

C8	N	Mean	StDev	SE Mean
IMCU	10	1.00	1.89	0.60
N4	8	4.87	5.79	2.0

95% CI for mu (IMCU) − mu (N4): (−8.79, 1.0)
T-Test mu (IMCU) = mu (N4) (vs not =): T = −1.82 P = 0.11
DF = 8

Two Sample T-Test and Confidence Interval

Two sample T for C9

C8	N	Mean	StDev	SE Mean
IMCU	10	1.00	1.89	0.60
N4	8	4.87	5.79	2.0

95% CI for mu (IMCU) − mu (N4): (−7.98, 0.2)
T-Test mu (IMCU) = mu (N4) (vs not =): T = −2.00 P = 0.063
DF = 16

(a) Interpret the results of the output for the first test, without pooling the variances. How many degrees of freedom are there for this test? If you use a level of significance of 0.10, what do you conclude about the nurses in the two departments?

(b) Interpret the results of the output for the second test, pooling the variances. How many degrees of freedom are there for this test? If you use a level of significance of 0.10, what do you conclude about the nurses in the two departments?

(c) How does this exercise confirm what you learned about the effects of pooling the variances?

Doing It!

Datafile: LUNG.XXX

10.36 A large hospital in the Northeast wants to study the effects of using fibrin glue in operations for lung cancer in reducing the incidence and length of chest air leaks. As patients who needed lung surgery were admitted, they were randomly assigned to either the experimental group (using glue) or the control group (no glue). Data were collected on variables concerning patient history as well as several outcome variables such as length of stay, number of postoperative days that the patient had a chest tube inserted, and whether or not there were complications. A portion of the data are presented here:

Index	Group	Age	Gender	Number of wedges	Diabetic?	Taking steroids?	Type of fissure	Staple	Smoker history?	Smoker now?
1	C	77	M	0	N	Y	Complete	3	Y	N
2	C	79	F	0	N	N	Complete	4	Y	N
3	C	67	M	2	N	N	Complete	3	N	N
4	C	20	M	0	N	N	Incomplete	3	N	N
5	C	70	M	0	Y	N	Complete	2	Y	N
6	C	78	F	0	N	N	Incomplete	3	N	N
7	C	61	F	0	N	N	Complete	1.5	N	N

Index	Immediate air leak?	Four-hour air leak?	Days tube in	Air leak duration	Complications?	Persistent air leak?	Length of stay
1	Y	Y	3	0.167	1	No	6
2	N	Y	3	1	1	No	4
3	Y	N	2	0	0	No	3
4	Y	Y	6	5	1	No	6
5	Y	Y	2	0.167	1	No	5
6	N	Y	2	1	0	No	3
7	N	N	2	0	0	No	3

(a) Look at the data on length of stay for the control and the experimental groups. Create a plot of the data for each group. Based on the graphs do you think the data are normally distributed? Why or why not? Are outliers affecting your answer?

(b) Perform the appropriate hypothesis test to determine whether the variances in lengths of stay for the two groups are equal. Based on this test, can you assume equal variances?

(c) Based on the results of your answers to parts (a) and (b) select the appropriate test procedure to decide whether the mean length of stay for the fibrin glue group (experimental) is shorter than the mean length of stay for the control group.

(d) Perform the test at the 0.05 level of significance. What is your conclusion? Do you think a 0.05 level of significance is a good choice here? Why or why not?

(e) Repeat the procedure you used to answer parts (a)–(d) to determine whether the mean duration of the air leak for the fibrin glue group is less.

(f) What can you conclude about the proportion of patients in each group who had complications from the surgery?

(g) A persistent air leak is defined as one that lasts longer than 7 days. What can you conclude about the proportion of patients in each group who have persistent air leaks?

(h) Is the proportion of patients who had an immediate air leak less for the experimental group?

(i) Consider the experimental group only. Look at some patient factors and determine whether certain patients do better with fibrin glue than others. Perform any tests and create any graphical displays you think are necessary.

(j) Write a memo summarizing your analysis. Include the answers to parts (a)–(h) as appendix items in your report.

11

Regression Analysis

THE REAL ESTATE MARKET

What makes one house cost more or less than another house that is right around the corner? Many factors influence the cost of a house. Location is one variable. In the next chapter we will look at a broad question, What characteristics of a city or town are related to the selling prices of houses in that town? Here we look at the question of housing prices in a single town. What makes the selling price of one house so different from the price of another? A local real estate company collected data on the listing prices of houses in the town that were on the market in a 1-month period. The company also collected data on the size of each house, the number of bedrooms, and the number of bathrooms. Here is a portion of the data:

House number	Price	Size (square feet)	Bedrooms	Bathrooms
1	$689,500	4000	4	3.5
2	385,000	3400	5	3
3	449,900	3269	4	2.5
4	949,900	5300	5	4
5	848,000	5575	4	3.5
6	559,900	3687	4	3.5

In Chapter 4 you learned about relationships between quantitative variables. In particular, you learned that there are different types of relationships between two variables and that when the relationship is linear, we can use the method of least squares to find the equation that describes the relationship. For example, a company might want to predict the sales of a particular product. The company knows that many different variables can affect sales, but it does not know which variable(s) are most important or exactly how they relate to sales. The company needs to find a model that allows it to predict sales as a function of the other variables. In *regression analysis*, this model is an *equation*.

In this chapter we look at *simple linear regression*, which predicts the value of Y as a linear function of a single independent variable X. After you have learned the basics of regression analysis in the simple linear case, in Chapter 12 we introduce *multiple regression models*, which predict the value of Y as a function of a set of independent variables.

In this chapter you will learn how to do these tasks:

- Find the linear regression equation for a dependent variable Y as a function of a single independent variable X.
- Determine whether a relationship between X and Y exists.
- Analyze the results of a regression analysis to determine whether the simple linear model is appropriate.

11.2 The Simple Linear Regression Model

11.2.1 DETERMINISTIC AND STATISTICAL RELATIONSHIPS

In some cases in which two variables, x and y, are related, the relationship is *deterministic*, or *functional*. This means that when a value of x is selected, the value of y is uniquely determined. For example, if we are interested in the relationship between the total cost of an order, y, and the number of items ordered, x, we can describe this relationship by an equation such as

$$y = \$50 + \$1.20x$$

where $50 is the ordering cost and $1.20 is the cost per item ordered. Figure 11.1 on page 504 is a graph of this type of relationship.

If a person orders $x = 100$ items, then the corresponding cost is $y = \$50 + (\$1.20)(100) = \$170$. Every person who orders 100 items will incur the cost of $170; that is, the value of y is *unique* for a given value of x.

Although many real-world problems are described by this type of relationship, we are interested in a different situation when we study *linear regression*. When we look at the relationship between two variables, X and Y, we are interested in situations where the value of Y varies for a given value of X. That is, we are interested in the *statistical relationship* between two variables.

Suppose we are looking at the relationship between dollars spent on advertising and revenues from sales. Clearly, we expect the two variables to be related, but we do not expect that every time a company spends x on advertising it will have y in revenues. We know that other *factors*, or *variables*, such as the type of

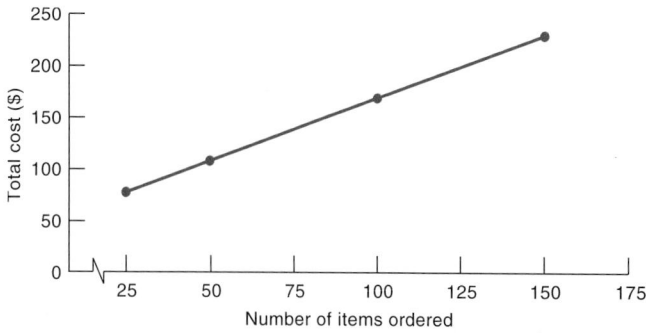

FIGURE 11.1 Deterministic relationship between total cost of order and number of items ordered

product, location, and economic factors, will affect the value of Y for a given value of X.

When we collect data, we are collecting *pairs* of observations on the two variables, X and Y. Thus, we have a set of n data pairs:

$$(x_1, y_1), (x_2, y_2), \ldots, (x_n, y_n)$$

A plot of the data might look like the one in Figure 11.2. You can see that the two variables are related, but a particular value of advertising expenditures can result in more than one value for revenues. This type of plot, a scatter plot, is of primary importance in exploring relationships between variables and should be made *before* any type of statistical analysis is performed.

Understand the problem.

EXAMPLE 11.1

THE REAL ESTATE MARKET

Plotting the Data to Look for a Linear Relationship

The company that is looking at housing prices thinks the variable that is probably most closely related to the listing price of a house is the size. They create a scatter plot of their data to see if this is reasonable.

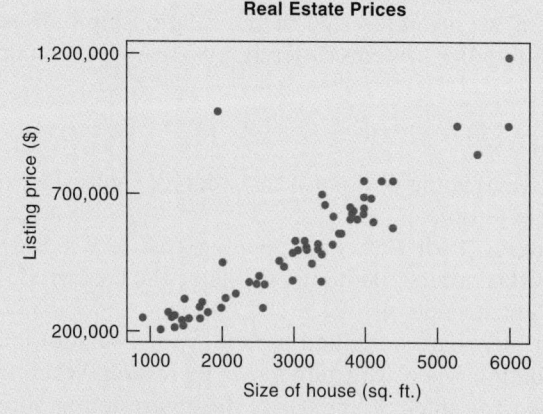

From the plot it looks like a linear relationship is a reasonable model, and so the company decides to proceed with a simple linear regression analysis.

In simple linear regression, we want to find an equation, or model, that allows us to predict a value for a variable, Y. Ideally, since the value of Y depends on many

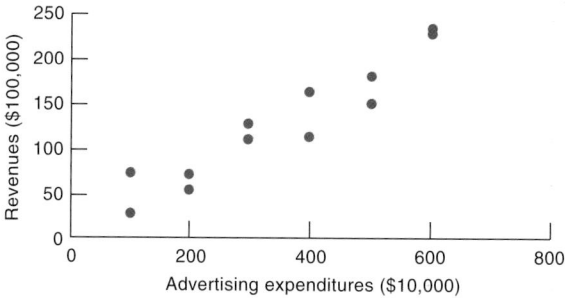

FIGURE 11.2   Statistical relationship between revenues and advertising expenditures

different factors, we would like to find some variable *X* that does a good job of predicting *Y* with a linear equation. By a "good" job of predicting, we mean that the prediction we obtain is *useful* for purposes of planning or problem solving. In regression analysis, the *X* variable is assumed to be controlled or at least controllable. This means that its values can be fixed by the person who is collecting the data.

EXAMPLE 11.2

HMO HEALTH

Understand the problem.

Looking at the Relationship Between Variables

As approaches to health-care coverage change, health maintenance organizations (HMOs) are growing in popularity. Some business analysts wonder how the increase in popularity has affected the financial health of HMOs. They collected these data on revenues and numbers of members for ten HMOs:

Collect the data.

HMO	Members (millions)	Revenues (billions)
United HealthCare	4.24	$5.49
Humana	3.19	4.63
FHP International	1.83	3.86
PacifiCare Health Systems	1.62	3.60
U.S. Healthcare	2.07	3.43
WellPoint Health Network	2.30	2.91
Health Systems International	1.83	2.74
Foundation Health	2.15	2.40
Oxford Health Plans	0.97	1.71
Physician Corp. of America	0.89	1.20

SOURCE: The Economist, April 6, 1996

In this case the dependent variable is revenues, measured in billions of dollars, and the independent variable is number of members, in millions. The analysts look at the scatter plot of the data to see whether a relationship exists:

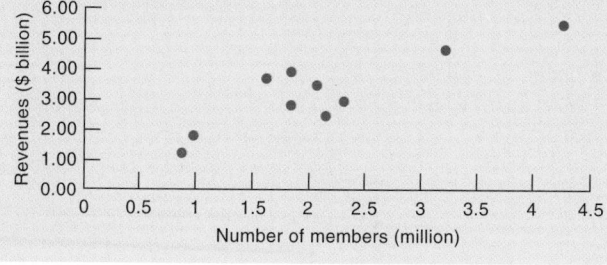

Analyze the data.

From the plot it appears that the relationship between revenues and number of members is linear. This means that a straight line provides a good model for predicting the revenues for an HMO as a function of the number of members of the HMO.

As you learn additional and more sophisticated statistical tools, you should not lose sight of the descriptive tools you learned first. Plotting data is essential to any statistical analysis and should be one of the first things you do to determine what analysis, if any, is appropriate.

✔ TRY IT NOW!

Hearing Difficulties
Plotting Data to Look at the Relationship

As people age they begin to experience hearing loss. A study was done to determine the "comfort level" of sound for people of different ages. The data are given here:

Age (years)	Sound level (decibels)
15	56
25	57
35	64
45	64
55	68
65	74
75	78
85	85

Use the grid to create a scatter plot of the data.

Do you think that a linear model is appropriate?

(See Try It Now! answer at the bottom of page 507.)

11.2.2 DEFINING THE SIMPLE LINEAR REGRESSION MODEL

The objective of simple linear regression is to find a linear equation that describes the existing relationship between two variables, X and Y. The equation we find will be based on data taken from some population. Just as in any problems that involve sampling from a population, we are trying to find an *estimate* for the true **regression model** between the two variables.

> The true relationship between the variables X and Y, the **simple linear regression model,** is described by the equation
>
> $$y = \beta_0 + \beta_1 x + \varepsilon$$

This equation says that for a given value of the variable $X = x$, the actual value of Y is determined by the expression $\beta_0 + \beta_1 x$, plus some random variation, ε, caused by other, unmeasured, factors. Thus, if we know the values of β_0, the true population intercept, and β_1, the true population slope, we can predict the value of Y to within some random error, ε. Figure 11.3 shows the population model for a linear regression.

*We are assuming that a plot of the data has determined that a **linear model** is appropriate.*

For a given value of $X = x_1$, the values of Y vary around the regression line. This variation is measured by the error term, ε. One of the assumptions of regression analysis is that the ε terms are normally distributed with a mean of 0 and a standard deviation of σ. We discuss this assumption along with others later in the chapter.

How can we find estimates for the population values β_0 and β_1? We want to find values that do the best job of describing the relationship between the variables. If you look at the data on advertising and revenues in Figure 11.2, you can probably imagine

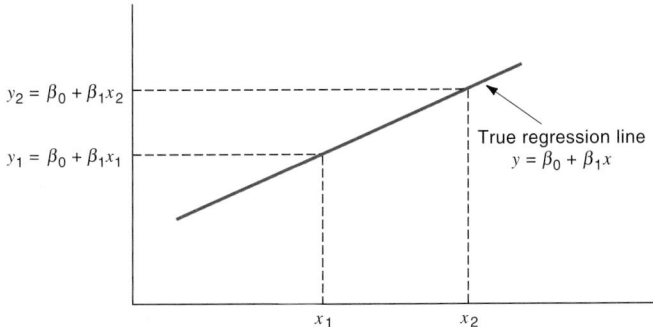

FIGURE 11.3 **The true regression model showing how *Y* varies for a given value of *X***

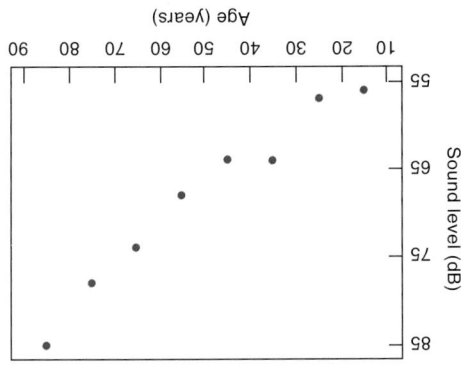

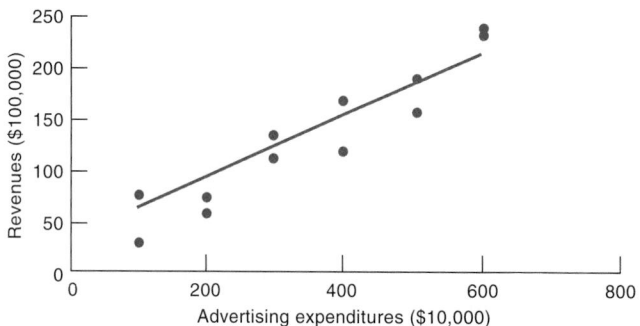

FIGURE 11.4 **Straight line approximating the relationship between advertising expenditures and revenues**

a straight line that captures the relationship. Figure 11.4 shows such a line along with the data. The equation of the line is

$$\hat{y} = b_0 + b_1 x$$

where $\hat{y}$ is the predicted value of Y for a particular value of $X = x$. The quantities b_0 and b_1 are the *estimates* of the population values β_0 and β_1. This line is called the *regression line of y on x* or the estimate of the *simple regression model*.

The problem is that we might draw many different lines depending on what our criterion for "best" is. Here are some possible criteria:

- Hit as many points as possible.
- Have an equal number of points above and below the line.
- Pick two representative points and connect them.
- Connect the first and the last points.

The trouble with these criteria is that they do not produce a unique line. That is, many lines hit three, four, or five points, and many lines have an equal number of points above and below them. (The last criterion actually produces a unique line, but it is not a very good criterion.) Figure 11.5 illustrates the problem. In Figure 11.5(*a*) both of the lines drawn have six points above and six points below. The three lines drawn in Figure 11.5(*b*) all go through exactly three of the data points.

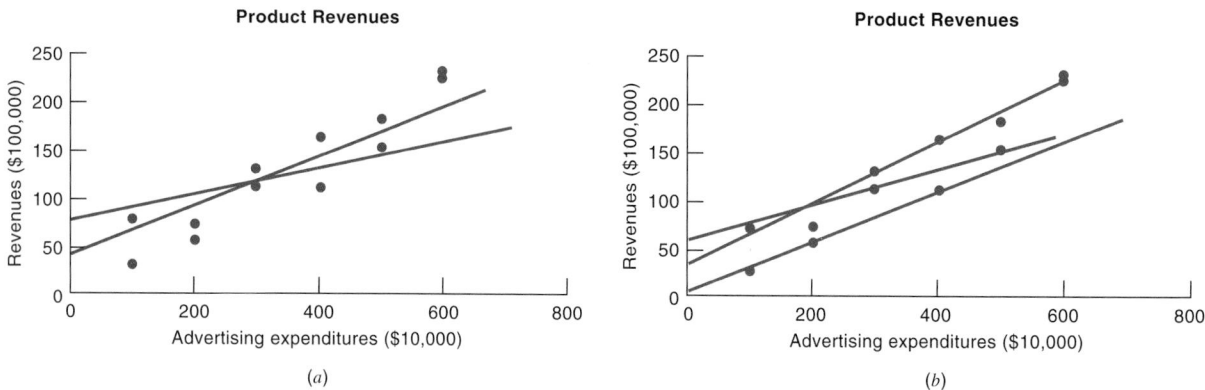

FIGURE 11.5 **A single criterion can produce many different lines.**

You may wonder why it is important that we find a unique line to fit the data. Remember that we want to use the model (the equation) to predict values for Y, the dependent variable, for different possible values of X, the independent variable. One important feature of a good model is that it is consistent. If the technique we use to find the line can produce many different models, then how will the user know which model or prediction to use? If everyone is allowed to choose the model he or she likes best, then the technique is, for all practical purposes, useless. We might as well just pick a number out of a hat.

11.2.3 THE LEAST-SQUARES LINE

At this point, we are sure about two things. We want the method we use for finding the equation of the line to produce a line that is unique, and we want the line to be a *good* representation of reality. Although there are certainly many ideas about what is good, we can agree that the line should be close to as many of the data points as is possible. Figure 11.6 shows a set of data points and a line drawn to represent the relationship between the variables. Although the line does not actually go through any of the data points, it is very close to most of them. The distance from each data point to the line is shown. These distances are called the **deviations** or **errors** of the line. We want to find a line that somehow minimizes the overall deviation of the data points from the line.

> The distance between the predicted value of Y, $\hat{y}$, and the actual value of Y, y, is called the ***deviation*** or ***error***.

When we introduced the variance in Chapter 3, you learned that when deviations are both positive and negative (in this case because the data points fall both above and below the line), we cannot simply look at the *sum* or *total* deviation because it will be 0. For the variance, we solved the problem by squaring the deviations and then adding them together. We use a similar approach to solving this problem.

The technique that we use to find the line that best fits the data is the **least-squares method.**

> The technique that finds the equation of the line that minimizes the total or sum of the squared deviations between the actual data points and the line is called the ***least-squares method.***

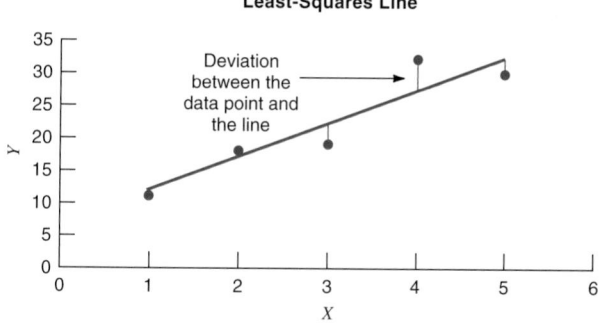

FIGURE 11.6   Deviations between the data points and the line

Note: Sometimes, when a data point is a true outlier, this leads to problems with the model.

You may wonder why we always square negative quantities to make them positive instead of just taking the absolute value. One reason is that mathematically the square is an easier quantity to work with than the absolute value. The other reason, in the case of fitting a line to a set of data, is that using the square of the error makes the line try to fit *all* of the points. This is because the penalty for avoiding certain points is much larger. Suppose that the actual distance from a data point to the line is -10 units. If we simply wanted to minimize the *absolute* deviation, then the penalty attached to missing the point is $|-10| = 10$. When we minimize the square of the distance, the penalty for missing that point becomes $(-10)^2 = 100$ and so the line tries to get closer to that point.

The least-squares method finds the equation of the line

$$\hat{y} = b_0 + b_1 x$$

that minimizes

$$\sum_{i=1}^{n} (y_i - \hat{y}_i)^2$$

or the total of the squared deviations from the data points to the line. The values for b_0 (the intercept of the line) and b_1 (the slope of the line) are found with these equations:

$$b_1 = \frac{n \sum_{i=1}^{n} x_i y_i - \sum_{i=1}^{n} x_i \sum_{i=1}^{n} y_i}{n \sum_{i=1}^{n} x_i^2 - \left(\sum_{i=1}^{n} x_i \right)^2} \quad \text{or} \quad \frac{n \sum xy - \sum x \sum y}{n \sum x^2 - \left(\sum x \right)^2}$$

Slope and intercept of a least-squares line

and

$$b_0 = \frac{\sum_{i=1}^{n} y_i}{n} - b_1 \frac{\sum_{i=1}^{n} x_i}{n} \quad \text{or} \quad b_0 = \bar{y} - b_1 \bar{x}$$

Although these equations may seem complex, only five quantities have to be calculated. Four of the quantities involve the data values for X and Y, and the fifth is simply the number of observations in the sample, n. For example, the quantity $\sum xy$ is simply the sum of the products of the x and y values for each data point. The easiest way to look at what is involved in the calculations is to make a table with a column for each sum needed; see Table 11.1.

TABLE 11.1

TABLE FOR CALCULATING THE LEAST-SQUARES LINE

Observation number	x	y	xy	x^2
1	x_1	y_1	$x_1 y_1$	$x_1 x_1$
1	x_2	y_2	$x_2 y_2$	$x_2 x_2$
.	.	.	.	.
.	.	.	.	.
.	.	.	.	.
n	x_n	y_n	$x_n y_n$	$x_n x_n$
Total	$\sum x$	$\sum y$	$\sum xy$	$\sum x^2$

EXAMPLE 11.3

THE REAL ESTATE MARKET

Finding the Least-Squares Line

The real estate company that is trying to determine the relationship between house size and price has decided from a plot of the data that a linear relationship is reasonable. To make the numbers manageable they decide to use the listing price in thousands of dollars (that is, a price of $665,000 is recorded as 665) and the size in thousands of square feet (that is, a 2350-square-foot house is recorded as 2.35). To find the equation of the least-squares line, they have assembled the following information:

$$\sum x = 199.2 \qquad \sum y = 33{,}177.0 \qquad \sum xy = 113{,}999.80 \qquad \sum x^2 = 692.033 \qquad n = 66$$

They first calculate b_1, the estimate of the slope:

$$b_1 = \frac{(66)(113{,}999.80) - (199.2)(33{,}177.0)}{(66)(692.033) - (199.2)^2} = \frac{915{,}128.4}{5993.538} = 152.6858427$$

and then use that estimate to find b_0, the estimate of the y intercept of the line:

$$b_0 = \frac{33{,}177.0}{66} - (152.6858427)\left(\frac{199.2}{66}\right) = 41.84818385$$

The equation of the regression line that relates the listing price and the house size is

$$\hat{y} = 41.848 + 152.685x$$

Or, translated back to the original numbers,

$$\hat{y} = 41{,}848 + 153x$$

This equation indicates that there is a positive relationship between the size of the house and the listing price; that is, as the size increases so does the price. In fact, the equation tells the real estate people that for every thousand-square-foot change in the size of a house, the listing price increases by 152.685 thousand dollars. This is equivalent to saying that for every square-foot change in house size, the price increases by $153. This change rate is the definition of the slope of a line.

Although the positive relationship is encouraging, the people looking at the equation wonder whether it really means anything. They wonder whether the increase in price in relation to the house size is *significant*. The equation of the line alone cannot tell that. They need to do further analysis.

Understand the problem.

Collect the data.

Do not round until you have done the calculations.

Analyze the data.

Although we would never use the estimate of b_1 to this many decimal places, do not round until after the value is used in the equation for b_0. In the final equation, the values for b_0 and b_1 should be rounded to reflect the precision of the original data.

Although you divided all the data by 1000, don't just automatically multiply it back in the end. THINK about what the numbers mean.

Draw conclusions.

The equation of the least-squares regression line gives information about the relationship between the independent variable Y and the dependent variable X. The *sign* of the slope estimate tells whether the relationship is positive or negative. The *value* of the slope gives the change that will occur in Y when X is changed by one unit.

It is more difficult to explain the interpretation of the intercept. By definition, the intercept of a line is the value of Y when $X = 0$. In some situations, this number can be thought of as the value of the dependent variable that is related to *other factors* that are not considered in this model. For example, when we look at the relationship between sales and advertising, it is possible that for $0 spent on advertising, there will still be sales of a product. In many cases, however, it does not make sense for the value of Y to be nonzero when $X = 0$. In the real estate example, if the house size is 0 square feet, *there will be no selling price*—that is, Y must be 0. We discuss this problem in more detail as we proceed.

EXAMPLE 11.4

Analyze the
data.

HMO HEALTH

Calculating the Least-Squares Line

After they plotted the data and looked at the graph, the analysts who are interested in the relationship between HMO revenues, Y, and number of members, X, decide to use the least-squares method to find the equation of the regression line relating the two variables. They assembled the data listed in the table:

Although finding the equation of the regression line by hand is not difficult, it is tedious, even for small data sets. Many calculators can find the equation of the least-squares line. If you have such a calculator, you should learn how to use it now.

Observation number	Members, X (millions)	Revenues, Y (billions)	XY	X²
1	4.24	$5.49	23.2776	17.9776
2	3.19	4.63	14.7697	10.1761
3	1.83	3.86	7.0638	3.3489
4	1.62	3.60	5.8320	2.6244
5	2.07	3.43	7.1001	4.2849
6	2.30	2.91	6.6930	5.29
7	1.83	2.74	5.0142	3.3489
8	2.15	2.40	5.1600	4.6225
9	0.97	1.71	1.6587	0.9409
10	0.89	1.20	1.0680	0.7921
Total	21.09	31.97	77.6371	53.4063

Substituting the values into the equations, the analysts get

$$b_1 = \frac{(10)(77.6371) - (21.09)(31.97)}{(10)(53.4063) - (21.09)^2} = \frac{102.1237}{89.2749} = 1.143924$$

$$b_0 = \frac{31.97}{10} - (1.143924)\left(\frac{21.09}{10}\right) = 0.7844643$$

Thus, the regression equation of HMO revenues on number of members is

$$\hat{y} = 0.78 + 1.14x$$

The equation means that for an increase of 1 million members, revenues will increase by $1.14 billion. The intercept of the line is 0.78, which means that when an HMO has no members, it will still generate $0.78 billion in revenues. In this case our interpretation of the intercept does not make sense.

Because least-squares analysis is mechanical, the equation alone cannot tell whether the relationship is real or the two variables are in fact unrelated. We can find a least-squares line that relates any two variables, but that certainly does not mean that the two variables are really related. To make that decision, other statistical tools are needed.

Most spreadsheet programs and all statistical software also perform regression analysis. It is not necessary to use the computer to find the equation of the regression line, but further analysis can be done only by using a computer package. Although it is not necessary now, the remainder of this chapter assumes that you have access to some statistical software package. At the end of this section we look at regression output from different software packages. Throughout this chapter we use the output extensively to make decisions about the problems we are trying to solve.

✓ TRY IT NOW!

Hearing Difficulties
Finding the Equation of the Least-Squares Regression Line

In the study about sound levels and age a linear relationship seems appropriate. Fill in the table values or use some other means to find the equation of the least-squares line:

Observation number	Age, X (years)	Sound level, Y (decibels)	XY	X²
1	15	56		
2	25	57		
3	35	64		
4	45	64		
5	55	68		
6	65	74		
7	75	78		
8	85	85		
Total				

Interpret the meaning of the estimate of the slope of the line. Does the *y* intercept make sense for these data?

11.2.4 USING THE COMPUTER TO DO REGRESSION ANALYSIS

Any statistical software is capable of performing regression analysis. In addition, most spreadsheet packages do regression. The output from the analyses may look different, but they have common elements. We start by identifying the estimates for the parameters of the regression equation in the output from several software packages. As we progress in the chapter we continue to use the output in conjunction with other analyses.

Figure 11.7 on page 514 gives the regression output from two different computer packages for the real estate problem. The output from any software package contains three sections: (1) general information about the regression line, (2) information about the regression coefficients, and (3) analysis of variance (ANOVA) information. In each output, the coefficients of the regression equation are highlighted.

As you can see from the output, the equation is identified in a slightly different way for each package. You can also see that there is a lot more to the output of a regression analysis than just the equation. It is important that you understand the language of regression so that you can identify the output from whatever software package you are using.

11.2.5 USING THE REGRESSION EQUATION TO MAKE PREDICTIONS

When we know the equation of the regression line, we can use it to predict values of the dependent variable, Y, for different values of the independent variable, X. We substitute different values of X into the regression equation and calculate the corresponding values of $\hat{y}$. Remember Figure 11.3 showed that the values of Y vary around the true regression line.

> The value of $\hat{y}$ that we find is really a prediction of the mean value of Y for a given value of X.

ANS: $\hat{y} = 47.9 + 0.41x$. HEARING COMFORT LEVEL INCREASES 0.41 DECIBEL FOR EVERY YEAR INCREASE IN AGE. NOT REALLY.

REGRESSION ANALYSIS: PRICE VERSUS SQ. FT.

The regression equation is

Price = 41240 + 153 Sq. Ft.

Predictor	Coef	SE Coef	T	P
Constant	41240	36478	1.13	0.262
Sq. Ft.	152.86	11.27	13.57	0.000

S = 107220   R – Sq = 74.2%   R – Sq(adj) = 73.8%

ANALYSIS OF VARIANCE

Source	DF	SS	MS	F	P
Regression	1	2.11665E+12	2.11665E+12	184.12	0.00
Residual Error	64	7.35756E+11	11496184916		
Total	65	2.85240E+12			

Minitab

SUMMARY OUTPUT

Regression Statistics	
Multiple R	0.861427567
R Square	0.742057454
Adjusted R Square	0.738027101
Standard Error	107220.2636
Observations	66

ANOVA

	df	SS	MS	F	Significance F
Regression	1	2.11665E+12	2.11665E+12	184.1172684	1.69204E-20
Residual	64	7.35756E+11	11496184916		
Total	65	2.8524E+12			

	Coefficients	Standard Error	t Stat	P-valve	Lower 95%	Upper 95%	Lower 95.0%
Intercept	41240.33213	36478.26655	1.130545282	0.262465536	–31633.31686	114113.9811	–31633.31686
Sq. Ft.	152.8586913	11.26530295	13.56898185	1.69204E-20	130.3536819	175.3637008	130.3536819

Excel

FIGURE 11.7   **Regression output for Minitab and Excel**

EXAMPLE 11.5

THE REAL ESTATE MARKET

Using the Regression Equation to Predict

The company that is looking at real estate prices found this regression equation for its sample data:

$$\hat{y} = 41{,}848 + 153x$$

A developer who is thinking about building homes in the area wants to know what price he can expect for a 3500-square-foot home. He substitutes 3500 into the equation and finds

$$\hat{y} = 41{,}848 + 153(3500) = \$577{,}348$$

He also wants to know what the price would be if the size is increased to 4500 square feet because this is the size of many new homes in the area. He finds

$$\hat{y} = 41{,}848 + 153(4500) = \$730{,}348$$

It looks like the house price will increase with size.

Analyze the data.

When we first looked at the regression equation and talked about what the parameters mean, we noted that the slope of the line does not always make sense. What happens when we use the regression equation to predict Y when $X = 0$?

THE REAL ESTATE MARKET

Using the Regression Equation When $X = 0$

One of the people in the real estate company thinks something is wrong with the model. He uses the equation to predict the listing price for a house size of 0 square feet and finds

$$\hat{y} = 41{,}848 + 153(0) = \$41{,}848$$

If this is true, he could be rich very quickly! He is not so sure about using the model to predict prices if it makes errors like this.

Analyze the data.

Draw conclusions.

When using a regression model, we must remember that the model is constructed only from sample data. It is relevant only *over the range of observed values.* The equation provides information about the relationship between the dependent and independent variables, only in this range. We can make two kinds of predictions: **interpolation** and **extrapolation.**

Using the equation to predict values of Y within the range of the X data is called *interpolation.* Predicting values of Y for values of X outside the observed range is called *extrapolation.*

Extrapolation is risky and should almost never be done. Without actual data we have no idea what the relationship is like beyond the "boundaries." Over a larger range of the dependent variable, the relationship might change shape and be nonlinear. It might even change direction, making the predictions totally inappropriate.

✔ TRY IT NOW!

Hearing Difficulties
Using the Regression Equation to Predict y

Use the equation of the least-squares line found earlier to predict the sound level for each of the observed values of X, age. Add the calculated values to the table.

Observation number	Age, X (years)	Sound level, Y (decibels)	$\hat{y} = 47.9 + 0.41x$
1	15	56	
2	25	57	
3	35	64	
4	45	64	
5	55	68	
6	65	74	
7	75	78	
8	85	85	
Total			

11.2.6 CALCULATING RESIDUALS

We used the method of least squares to find the line for regression analysis because we wanted the line to be close to the actual data points. How can we assess how well the regression line accomplishes this?

If we plot the regression line on the same plot with the data, we can get a *visual* or *graphical* idea of the connection between the model and the data. The proximity of the data points to the line gives an overall picture of how well the regression line describes the relationship between the variables.

EXAMPLE 11.7

Draw conclusions.

THE REAL ESTATE MARKET

Plotting the Regression Line

The company that is looking at real estate prices decides to plot its data and the regression equation together to get a visual idea of the least-squares line:

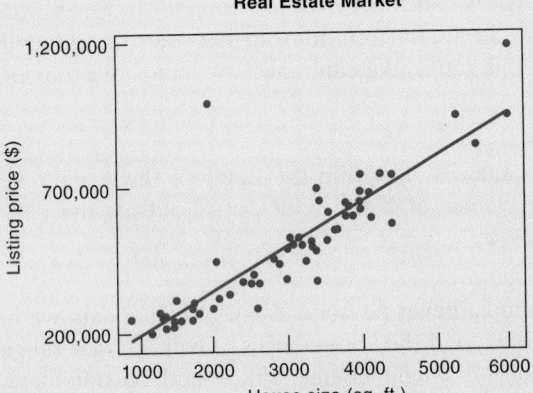

The graph shows that most of the data points are very close to the line, although some (one in particular) are quite a bit off. Still, the company thinks the line does a pretty good job of describing the relationship between listing price and house size.

EXAMPLE 11.8

HMO HEALTH

Plotting the Regression Line

The analysts who are looking at the relationship between revenues and number of members of HMOs plot the data and the regression equation:

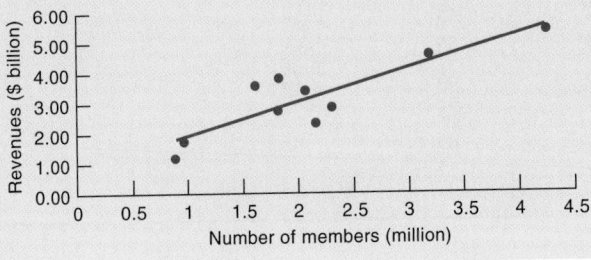

They see that some of the points are very close to the line, but others are farther away. Still, the line does a pretty good job of describing the relationship between revenues and number of members. The analysts wonder whether they can get a numerical measure of the differences between the predicted values and the actual data.

Draw conclusions.

A *numerical* measure of the agreement between the line and the data points is obtained by examining the differences between the observed and expected values of Y for each value of X. In regression analysis, these differences are known as the **residuals.**

> The difference between the observed value of Y, y_i, and the value of Y predicted from the regression equation, $\hat{y}$, for a value of $X = x_i$ is called the ith **residual, e_i.**

The *sign* of the residual, positive or negative, tells whether the prediction is lower or higher than the actual data value. The *size* of the residual gives an idea of how much the actual data vary around the line.

EXAMPLE 11.9

THE REAL ESTATE MARKET

Calculating Residuals

To obtain a better idea of how well the regression line agrees with their data, the real estate company decides to calculate the residuals for the two values that the developer is interested in: 3500 and 4500 square feet. They have actual data for house sizes of 3453 and 4400 square feet, which are close to the 3500 and 4500 sizes. They can use the regression equation to find the predicted values for $X = 3453$:

$$\hat{y} = 41{,}848 + 153(3453) = \$570{,}157$$

and for $X = 4400$:

$$\hat{y} = 41{,}848 + 153(4400) = \$715{,}048$$

Then they calculate the residuals, $y - \hat{y}$, for $X = 3453$:

$$y - \hat{y} = \$659{,}900 - \$570{,}157 = \$89{,}743$$

and for $X = 4400$:

$$y - \hat{y} = \$579{,}900 - \$715{,}048 = -\$135{,}148$$

Analyze the data.

In one case, the difference between the observed data point and the point on the least-squares line is positive; in the other, negative. The magnitude of the deviation is about $100,000$, which is sizeable.

The residuals play an important role in regression analysis. As you just saw, they give a numerical value for the difference between the model and the actual data for each of the data points in the sample. We have pointed out that, for a statistical relationship, the actual values of Y vary for a fixed value of X. The residuals also provide a means to measure the overall variation in Y for any value of X. In addition, as we discuss later in this chapter, the residuals are used to determine the appropriateness of a linear model.

Analyze the data.

EXAMPLE 11.10

HMO HEALTH

Calculating Residuals

For the HMO problem, the analysts want to use the model to predict revenues for different sizes of HMOs. They are interested in predicting Y for $X = 1.0$ (1 million members) and $X = 2.0$ (2 million members). To have a way to compare the predicted values with actual data, they use the data points with $X = 0.97$ and $X = 2.07$ because these points are closest to the values of interest. The analysts calculate the predicted values using the equation $\hat{y} = 0.78 + 1.14x$ for each value and then calculate the residuals. Here are the results:

Observation number	x_i	y_i	$\hat{y}_i = 0.78 + 1.14x_i$	$e_i = y_i - \hat{y}_i$
9	0.97	1.71	$\hat{y}_9 = 0.78 + 1.14(0.97) = 1.89$	$e_9 = 1.71 - 1.89 = -0.18$
5	2.07	3.43	$\hat{y}_5 = 0.78 + 1.14(2.07) = 3.14$	$e_5 = 3.43 - 3.14 = 0.29$

The differences between the predicted revenues and the actual revenues are $180 million and $290 million, or about 10%.

Although residuals are important, they do not tell us everything about how good the model is for predicting values of Y. Just because the residual for a certain data point is large, it does not mean that the regression equation does a poor job of predicting Y for values of X near that point. Perhaps the data point used for the residual is unusually far from the regression line and the prediction is actually more representative of the population. For this reason it is helpful to look at the residuals in context. Graphing the regression line with the data points provides additional insight about individual residuals.

Analyze the data.

EXAMPLE 11.11

HMO HEALTH

Looking at the Residuals Graphically

When the HMO analysts look at the following plot of the data and the regression line, they see that the data points they used to calculate the residuals in Example 11.10 are actually very close to the regression line.

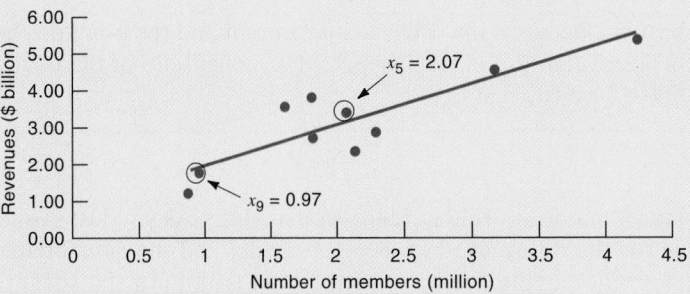

They think the errors for the numbers of members they were interested in are not unusually large.

TRY IT NOW!

Hearing Difficulties

Calculating the Residuals

The researchers who are looking at the relationship between age and comfortable sound levels want to get a better idea of how the regression line relates to the actual data. They decide to calculate the residual for each observed value of X, age. Find the residuals and fill in the table:

Observation number	Age, X (years)	Sound level, Y (decibels)	$\hat{y} = 47.9 + 0.41x$	$e_i = y_i - \hat{y}_i$
1	15	56	54	
2	25	57	58	
3	35	64	62	
4	45	64	66	
5	55	68	70	
6	65	74	74	
7	75	78	78	
8	85	85	83	
Total				

To get a picture of how the residuals and the regression line fit together, the researchers also decide to graph the regression line on a plot of the data. Graph the regression line on the data plot. How well do you think the line represents the data?

Comfort Hearing Levels of Different Subjects

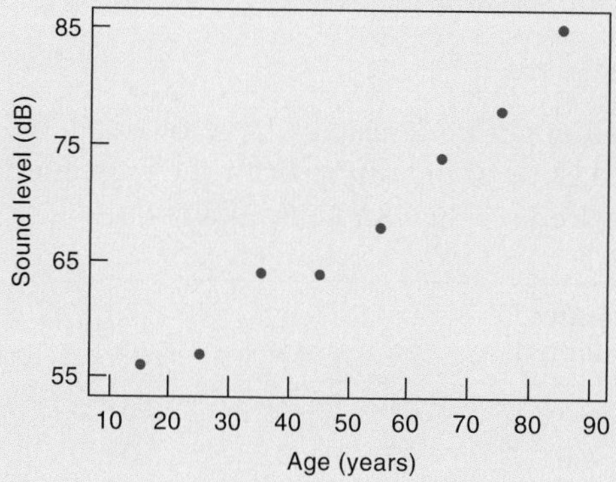

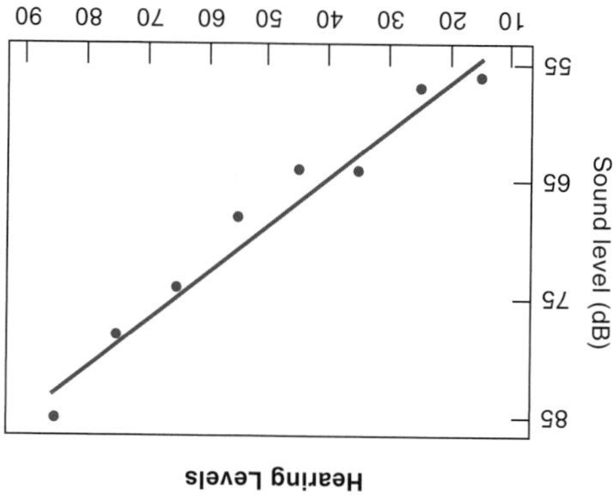

ANS. RESIDUALS: 2, −1, 2, −2, −2, 0, 0, 2. THE LINE DOES A PRETTY GOOD JOB.

11.2.7 THE STANDARD ERROR OF THE ESTIMATE

Since each residual corresponds to a single data point, no single residual indicates how much the data differ from the model. Still, we would like to have some way to measure how much the data points vary around the regression line. This measure is known as the **standard error of the estimate, $s_{y|x}$.**

> *The standard error of the estimate, $s_{y|x}$,* is a measure of how much the data vary around the regression line.

The quantity that we are trying to measure is the *overall* or *average* variation of the data around the line. Remember that the residuals measure the *individual* variation of each data point from the line. We can obtain an overall measure of this variability by using a formula similar to the one we used to measure the variability of the data from its mean in Chapter 3. In fact,

Standard error of the estimate

$$s_{y|x} = \sqrt{\frac{\Sigma(y - \hat{y})^2}{n - 2}} = \sqrt{\frac{\Sigma e^2}{n - 2}}$$

Several different computational formulas may be used to find the standard error of the estimates. The one shown here makes it easiest to understand what the quantity really measures. It is most convenient if you have already calculated the residuals. We will not look at any other formulas because, as we have said, beyond finding the least-squares equation, regression analysis should be done using a computer.

EXAMPLE 11.12

HMO HEALTH

Calculating the Standard Error of the Estimate

Since they knew that any single residual might be unusually large or small, the analysts decide to calculate the standard error of the estimate for the regression line. The table gives the residual and squared residual for each data point:

Observation number	Members, X (millions)	Revenues, Y (billions)	$\hat{y}$	$e = y - \hat{y}$	e^2
1	4.24	$5.49	5.61	−0.12	0.01
2	3.19	4.63	4.42	0.21	0.04
3	1.83	3.86	2.87	0.99	0.98
4	1.62	3.60	2.63	0.97	0.94
5	2.07	3.43	3.14	0.29	0.08
6	2.30	2.91	3.40	−0.49	0.24
7	1.83	2.74	2.87	−0.13	0.02
8	2.15	2.40	3.23	−0.83	0.69
9	0.97	1.71	1.89	−0.18	0.03
10	0.89	1.20	1.79	−0.59	0.35
Total					**3.38**

The standard error of the estimate is

$$s_{y|x} = \sqrt{\frac{3.38}{8}} = \sqrt{0.423} = 0.65$$

This means that the typical deviation or distance of each data point from the line is 0.65, or $650 million.

SUMMARY OUTPUT

Regression Statistics	
Multiple R	0.879773847
R Square	0.774002022
Adjusted R Square	0.745752275
Standard Error	0.65297731
Observations	10

The regression equation is
Rev $bn = 0.784 + 1.14 Members (m)

Predictor	Coef	StDev	T	P
Constant	0.7845	0.5050	1.55	0.159
Members	1.1439	0.2185	5.23	0.000

S = 0.6530   R – Sq = 77.4%   R – Sq(adj) = 74.6%

Excel   **Minitab**

FIGURE 11.8 **Computer output showing the standard error of the estimate**

The standard error of the estimate is analogous to the standard deviation of data from the mean and to the standard error of the mean, which you have studied before. From our knowledge of what these quantities measure and the empirical rule, we can get an intuitive sense of what the quantity $s_{y|x}$ means. We can get a *mental image* of how the data are spread out around the line. Later in the chapter we develop some more formal ways to use the standard error of the estimate to look at predictions.

Figure 11.7 showed the computer output for the real estate problem from two different software packages. Figure 11.8 shows only the portion of the output that contains the standard error of the estimate for the HMO data. Notice that in the Excel output it is called the "Standard Error," whereas in the Minitab output it is simply labeled "S."

11.2.8 EXERCISES—LEARNING IT!

11.1 In a study of postage tariffs and revenue, the European Economic Community wanted to look at the relationship between the number of postal employees in a country and the amount of domestic mail that was processed. They collected data for six different countries:

Country	Postal employees	Domestic mail (billions of pieces)
Germany	342,413	18.32
France	289,156	23.87
Italy	221,534	6.62
Britain	189,000	16.75
Spain	65,355	4.06
Sweden	52,251	4.21

Source: *The Economist,* June 15, 1996

(a) Create a scatter plot of the data and determine whether a linear relationship is appropriate.

(b) Find the regression equation for the data.

(c) The Netherlands employs 53,560 postal workers. Use the regression equation to predict the amount of domestic mail for the Netherlands.

(d) Calculate the residuals for France and Germany.

11.2 To study the effects of shopping center expansion, the U.S. Department of Commerce decided to look at the relationship between the number of shopping centers and the retail sales for different states in the same region. It collected the following data for the North Central states:

State	Shopping centers	Retail sales (billions)
Illinois	2096	$41.8
Indiana	905	21.4
Michigan	1018	25.3
Minnesota	471	13.9
Ohio	1704	41.6
Iowa	308	7.5

(continued)

Missouri	887	22.7
Wisconsin	625	14.6
South Dakota	58	1.3
North Dakota	87	2.1
Nebraska	264	5.7
Kansas	481	11.6

Source: *Statistical Abstract of the United States 1999*

(a) Create a scatter plot of the data.

(b) Find the regression equation that relates retail sales and numbers of shopping centers.

(c) Plot the regression line on the same plot as the data. Does the line fit the data well? Why or why not?

(d) Use the regression line to predict retail sales for each state.

(e) Calculate the residual for each state. Which state has the largest residual? Which state has the smallest? Do the residuals support your answer to part (d)?

(f) Find the standard error of the estimate.

11.3 As part of an international study on energy consumption, data were collected from 12 countries on the number of cars in a country and the total travel in kilometers:

Country	Cars (millions)	Travel (billions of kilometers)
United States	142.35	3140.29
Finland	1.82	34.66
Denmark	1.66	30.76
Britain	21.32	352.76
Australia	8.53	138.22
Sweden	3.32	53.21
Netherlands	5.53	83.69
France	23.27	348.20
Norway	1.59	23.54
Italy	26.12	367.85
Germany	43.75	608.52
Japan	40.25	439.30

Source: *The Economist,* June 22, 1996

(a) Create a scatter plot of the data. Do you think there is a linear relationship between the number of kilometers traveled and the number of cars?

(b) Find the least-squares regression line for the data. Interpret the value of the slope.

(c) Does the intercept make sense for these data? Why or why not?

(d) Plot the regression line on the same plot with the data. Does the line make you feel confident about predicting travel as a function of the number of cars?

(e) Use the regression line to predict the number of kilometers traveled for Sweden and Japan. How well do the predictions agree with the original data?

Datafile:
KIDSCOUNT1998.
XXX

11.4 The Annie E. Casey Foundation is a private charitable organization dedicated to helping disadvantaged children. Through a program called Kids Count they conduct yearly surveys of the status of children in the United States. As part of the survey they record for each state the percentage of babies born with low birthweights and the infant mortality rate (deaths per 1000 live births).

(a) Create a scatter plot of the data. Does it appear that the infant mortality rate is related to the percentage of low-birthweight babies?

(b) Find the equation of the regression line for the data.

(c) Plot the regression line on the same plot as the data. Does the line do a good job of predicting infant mortality rates from the percentages of low-birthweight babies? Why or why not?

(d) From the plot, which state do you think will have the largest residual? Which will have the smallest?

(e) Use the line to predict the infant mortality rate for 10% low-birthweight babies.

(f) Find the residual for each state. Do the values support your answer to part (d)?

(g) Calculate the standard error of the estimate for the regression line.

11.5 As part of the anticrime bill passed in 1995, the U.S. government granted money to cities to hire new beat-patrol officers. Nine cities in New Jersey were given grants and hired officers. To assess the program, government analysts used these data to look at the relationship between the numbers of officers hired and the amounts of the grant:

Grant	Number of officers
$150,000	2
375,000	5
471,125	6
70,967	1
450,000	6
525,000	7
375,370	7
750,000	10
1,000,000	12

Source: U.S. Justice Department

(a) Create a scatter plot of the data.

(b) Find the equation of the regression line.

(c) Plot the regression line on the same plot as the data.

(d) How well does the line fit the data?

(e) Use the regression equation to predict the number of officers hired for each city.

(f) Calculate the residuals and the standard error of the estimate.

11.6 For many countries tourism accounts for a large part of revenues. To predict tourism revenues, one of the independent variables that is considered important is the number of foreign visitors to the country. Data for six different countries were collected:

Country	Visitors (millions)	Tourism receipts (billions)
France	60	$27.3
Spain	48	25.1
United States	45	58.4
Italy	30	27.1
Britain	23	17.5
Germany	15	11.9

Source: The Economist, July 1996

(a) Create a scatter plot of the data and find the equation of the regression line.

(b) Use the regression line to predict tourism revenues for Italy and the United States.

(c) For which country does the regression line do a better job of predicting tourism receipts?

(d) Calculate the residuals and the standard error of the estimate.

11.7 The *Chronicle for Higher Education* publishes data on graduation rates for athletes at colleges and universities in the United States. The data for football players in the Big East Conference and for students entering in 1994–1995 are given in the table:

Institution	Graduation rate (%)	
	Male students	Male athletes
Boston College	0.84	0.77
Georgetown University	0.91	0.86
Providence College	0.82	0.78
Rutgers University—New Brunswick	0.71	0.55
Seton Hall University	0.56	0.54
St. John's University	0.62	0.57
Syracuse University	0.69	0.66

(*continued*)

University of Connecticut	0.64	0.56
University of Miami	0.57	0.45
University of Notre Dame	0.93	0.82
University of Pittsburgh	0.59	0.59
Villanova University	0.82	0.75
Virginia Tech	0.69	0.49
West Virginia University	0.51	0.53

(a) Create a scatter plot of the data. Do you think there is a relationship between the graduation rates of all males and male athletes?

(b) Find the regression line for the data.

(c) Use the regression line to predict the graduation rates for Virginia Tech and Providence College. Which prediction is better?

(d) Leave out the data point for Virginia Tech and recalculate the regression line. Does this line do a better job of predicting athletes' graduation rates? Why or why not?

11.3 Inferences About The Linear Regression Model

The tools you have learned for finding the simple linear regression equation are descriptive. The method of least squares allows you to find the equation of the line that best describes the sample data. Looking at the residuals gives you some idea about how well the line fits the data. Residuals are useful to some degree but, as in most cases, we would really like to be able to make *inferences* about the population based on our sample data. In particular, we would like to answer these questions:

- Is the relationship described by the regression equation *meaningful?*

- How well does the regression equation enable us to predict values of the dependent variable?

- How useful are the predictions for decision making?

Now we need to learn some inferential methods related to regression analysis.

11.3.1 HYPOTHESIS TESTING ABOUT THE SLOPE, b_1

Saying that there is a linear relationship between X and Y means that as X changes, Y changes in some predictable, corresponding way. The parameter that describes the magnitude of the relationship is the *slope* of the line. Remember that the slope is the change in Y for a unit change in X. Thus, if the variables X and Y are related, then the slope of the line is some number. If there is no relationship between X and Y, then the slope of the line is 0; that is, as X changes, Y *does not change* in a related way. See Figure 11.9. Thus, one of the ways we can decide whether the relationship between the two variables is real is to determine whether the slope of the regression line is 0.

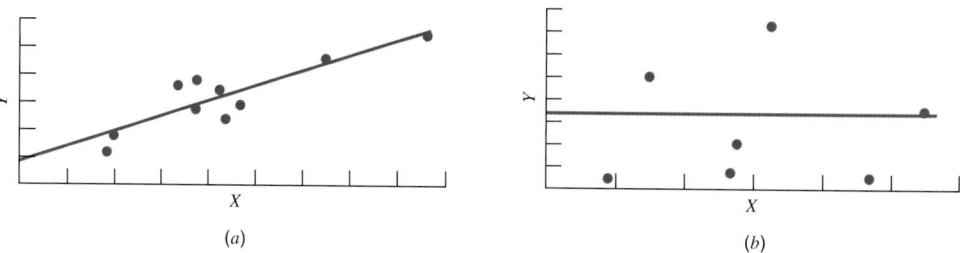

FIGURE 11.9 **(a) Line with nonzero slope; (b) line with zero slope**

Although looking at data gives us an idea about what might be true in the population, we cannot simply look at the numbers to make these decisions. If the data are small numbers, then the slope and intercept values, b_1 and b_0, are also small. In the same way, if the data are large numbers, then the corresponding slope and intercepts might or might not be large. We must consider the *standard error* of the statistic in question to make decisions involving the values. We will use hypothesis testing to decide whether the slope of the regression line is *significantly different from 0*.

The first step in testing a hypothesis is to set up the null and alternative hypotheses. In this case we want to test

$$H_0: \quad \beta_1 = 0$$
$$H_A: \quad \beta_1 \neq 0$$

If the test leads us to reject the null hypothesis, then we will conclude that the slope of the regression line is not equal to 0 and the relationship between the X and Y variables is real.

Our estimate of β_1 is b_1, and to proceed with the steps of the hypothesis test we need to know about the sampling distribution of b_1. It turns out that the sampling distribution associated with the least-squares estimate of the slope is the Student's t distribution. The test statistic for the hypothesis test is therefore

$$t = \frac{b_1 - \beta_1}{s_{b_1}}$$

If you took a different sample, the regression line would not have the same slope and intercept values. Thus, these estimates are sample statistics and have sampling distributions.

Test statistic for the slope

which has a t distribution with $n - 2$ degrees of freedom. In the formula, s_{b_1} is the **standard error of the slope** b_1 and is calculated by

$$s_{b_1} = \frac{s_{y|x}}{\sqrt{\Sigma x^2 - (\Sigma x)^2/n}}$$

Standard error of the slope

Although the equation may look complicated, we have said that for any purposes other than finding the regression line, we will use statistical software to find the values we need. Because the test is two-tail, after the significance level of the test, α, is chosen, the critical values of the test are $\pm t_{\alpha/2, n-2}$. We have now set up the test. All that remains is to perform it and make a decision.

This does not mean that the values cannot be calculated by hand. Once you find the regression coefficients and the standard error, you have all the pieces you need to find the test statistic.

EXAMPLE 11.13

THE REAL ESTATE MARKET

Conducting a Hypothesis Test About the Slope Coefficient

The real estate company that is looking at housing prices has determined that the equation of the regression line relating price to house size is

$$\hat{y} = 41{,}848 + 153x$$

When the regression line and the data are graphed, it appears that the regression line does a good job of representing the data. Still, the company looks at the value of the slope, 153, and wonders whether it really means anything. The company decides to perform a hypothesis test at the 0.05 level of significance to see whether the relationship is real.

The hypotheses are

$$H_0: \quad \beta_1 = 0$$
$$H_A: \quad \beta_1 \neq 0$$

Analyze the data.

Since the value of α is 0.05 and there are 66 data points, the critical values of the test are $\pm t_{0.025, 64}$ or ± 1.998. From previous analyses the company knows that $\Sigma x = 199.2$, $\Sigma x^2 = 692.033$, and $s_{y|x} = 107.8$. The value of the test statistic is

Remember that the hypothesized value for β_1 is 0.

Draw conclusions.

$$t = \frac{153 - 0}{(107.8)/\sqrt{692.033 - (199.2)^2/66}} = \frac{153}{11.312} = 13.53$$

Comparing the value of the test statistic with the critical values, the company sees that 13.53 is definitely beyond the critical values and so it rejects H_0 and concludes that the slope of the regression line is not equal to 0. There is a significant linear relationship between house price and size.

Computer output usually has three sections. We are interested in the section with information about the regression coefficients. Figure 11.10 shows the relevant sections of the computer output for two different statistical packages. The output provides all of the information we need to perform the hypothesis test about the slope coefficient. We can either look up the critical value for the level of significance we have chosen and compare it with the t statistic, or use the p value of the test, which is also included in the output. In all cases, the p value of the test is 0.000. Since this is less than our chosen α, we reject H_0.

Analyze the data.

EXAMPLE 11.14

HMO HEALTH

Conducting a Hypothesis Test About the Slope of the Regression Line

The analysts who are looking at the relationship between HMO revenues and numbers of members used statistical software to perform the regression analysis and to determine whether the relationship is significant. They want to test, at the 0.05 level of significance, these hypotheses:

$$H_0: \quad \beta_1 = 0$$
$$H_A: \quad \beta_1 \neq 0$$

At the 0.05 level of significance with $10 - 2 = 8$ degrees of freedom, the critical values for the test are ± 2.306.

The analysts obtained this output:

Regression Analysis

The regression equation is
Rev $bn = 0.784 + 1.14 Members (m)

Predictor	Coef	St Dev	T	P
Constant	0.7845	0.5050	1.55	0.159
Members	1.1439	0.2185	5.23	0.000

From the output line that corresponds to the slope estimate, they see that the coefficient is 1.1439 and the standard error of the slope is 0.2185. The corresponding t statistic is 5.23. Since the critical value for the test is 2.306, they reject H_0 and conclude that the slope of the regression line is not 0. There is a significant linear relationship between revenues and numbers of members.

Draw conclusions.

The analysts can also use the p value of the test to make the decision. From the output, the p value of the test is 0.000. Since that is less than the specified α of 0.05, they know that they should reject H_0.

Predictor	Coef	SE Coef	T	P
Constant	40.80	36.76	1.11	0.271
Sq. Ft.	152.57	11.32	13.47	0.000

Minitab

	Coefficients	Standard Error	t Stat	P-value
Intercept	41.24033213	36.47826655	1.130545282	0.262465536
Sq. Ft.	152.8586913	11.26530295	13.56898185	1.69204E-20

Excel

FIGURE 11.10 *t*-test portion of computer output

The computer output also contains information about b_0, the intercept of the regression line. It is tempting to test hypotheses about β_0, the intercept, in particular whether it is equal to 0. *Do not do this!* Remember that the regression line is valid only in the range of X that has been observed in the data. If we do not have X data for $X = 0$, then we cannot make a determination about whether the intercept should remain in the equation. Removing the intercept from the equation when the slope is not equal to 0 means that the regression line will move downward, parallel to itself, and will no longer be the line that best fits the data.

✔ TRY IT NOW!

Hearing Difficulties
Testing for Significance of the Regression Model

The researchers who are doing the study on comfortable sound levels and age want to determine whether the relationship between the two factors is significant. Write down the hypotheses they must test.

The researchers decide to use a 0.01 level of significance for the test. Find the critical values for the test.

Remember to find the degrees of freedom—in this case, n − 2.

A computer software package was used to obtain this output:

Predictor	Coef	SE Coef	T	P
Constant	47.893	1.742	27.49	0.000
Age in Y	0.40714	0.03168	12.85	0.000

From the output find the slope of the regression line, the standard error of the slope, and the value of the *t* statistic.

Perform the hypothesis test and make a decision about the regression line.

Find the *p* value of the test from the output and explain how to use the *p* value to reach the same decision.

ANS. $H_0: \beta_1 = 0$; $H_A: \beta_1 \neq 0$; ± 2.447; 0.40714, 0.03168, 12.85; REJECT H_0: THE REGRESSION IS SIGNIFICANT, SINCE 0.01 > 0.000, WE REJECT H_0.

Once we have determined that the relationship between X and Y is significant, we can perform some additional analyses to see whether the predictions we obtain are useful to make decisions and to determine the strength of the relationship.

11.3.2 PARTITIONING THE VARIANCE IN LINEAR REGRESSION

In statistics, we are always interested in looking at the *variation* in data. In Chapter 3 you learned about the standard deviation and how that can be used to measure the variation of data around its mean value. When we are looking at the relationship between two variables, Y (the dependent variable) and X (the independent variable), we are interested in how Y *varies* considering the values of the variable X.

When two variables, X and Y, are linearly related, the variation of the y values from the overall mean of Y can be attributed to two different factors: Y's relationship with X and pure chance. In fact, the variation in the y values from the overall mean of Y can be partitioned, or divided, into parts called **sums of squares:**

> **SST:** the total variation in the y values around the mean $\bar{y}$
>
> **SSR:** the variation in Y that is caused by Y's relationship with X
>
> **SSE:** the variation in Y that remains unexplained

Notice that SSE is related to the residuals you learned about previously.

The quantities **SST, SSR,** and **SSE** are known as the **sums of squares:** SST is the total sum of squares, SSR is the regression sum of squares, and SSE is the error sum of squares. Figure 11.11 illustrates how these quantities relate to the data and to the regression line. You see that SST represents the distance of the data value y_i from the overall mean $\bar{y}$; SSR represents the distance of the predicted value $\hat{y}$ from the overall mean $\bar{y}$; and SSE represents the distance of the data value y_i from the predicted value $\hat{y}$. These formulas are used for calculating the sums:

Sums of squares

$$SST = \Sigma(y - \bar{y})^2 \qquad SSR = \Sigma(\hat{y} - \bar{y})^2 \qquad SSE = \Sigma(y - \hat{y})^2$$

In addition you see that

$$SST = SSR + SSE$$

When X and Y are related, SSR is a large part of the total variation. This implies that Y varies so much *because* it is related to X. When this is true, the SSE component of the variation is small and is the variation in Y that happens "naturally" or entirely due to chance.

When X and Y are not related, the regression line is horizontal ($\beta_1 = 0$) and the SSR component of the variation disappears. The SSE part of the variation becomes

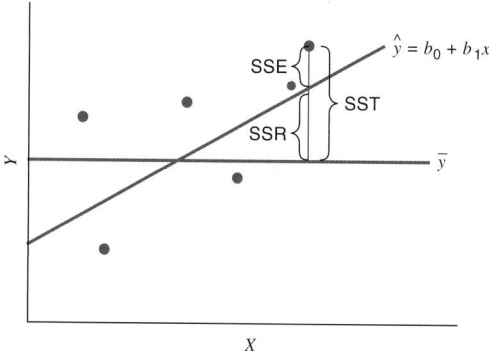

FIGURE 11.11 **Components of the variation in y values**

dominant and we say that we cannot really explain the variation in Y using the linear model with X.

We can use these ideas to create a formal test for the significance of the linear regression model. The test is called an *analysis of variance (ANOVA)* test because it looks at the *variation* in the Y variable. We test these hypotheses:

H_0: The linear regression model is significant.

H_A: The linear regression model is not significant.

The test statistic for this test uses the **mean squares,** which are obtained by dividing the sums of squares, SSR and SSE, by their respective degrees of freedom:

$$MSR = \frac{SSR}{1} \quad \text{and} \quad MSE = \frac{SSE}{n-2}$$

Mean squares

The ANOVA test looks at the ratio of the regression mean square (MSR) to the error mean square (MSE). As we explained, when the regression model is really significant (that is, when X and Y are really linearly related), SSR (and therefore MSR) is much larger than SSE (and therefore MSE). The ratio of the two mean squares has an F distribution with 1 and $n-2$ degrees of freedom. If the ratio is larger than the critical F value, then the model is significant. Thus, ANOVA is a one-tail test.

You saw the F distribution when you looked at the test for comparing the variances of two normal populations in Chapter 10.

Calculation of SST, SSR, and SSE is not difficult, but for even small data sets it can be computationally exhausting. If we have any two of the components, we can obtain the third by subtraction. The formulas show that SSE is really the sum of the squared residuals that we used to find $s_{y|x}$, and SST is really the numerator portion of the variance of Y.

If you have been doing the analyses by hand or using a calculator, you probably have already calculated most of the quantities you need. However, if you are doing this much analysis, you really should be using a statistical software package to do the job. The output from a statistical analysis program includes an ANOVA section. Figure 11.12 shows the ANOVA output from Excel and Minitab.

You might wonder why we need another method for testing the significance of a linear regression model. In reality we need only one method because both tests (the t test and the F test) yield the same decision. The difference is, to some degree, a matter of preference. Some statistical analyses use the ANOVA approach and some people prefer that test. In addition, the ANOVA approach is used in multiple regression and model building, as you will see in the next chapter.

ANOVA

	df	SS	MS	F	Significance F
Regression	1	37.7124744	37.7124744	182.6413924	1.9909E-12
Residual	23	4.749125596	0.206483722		
Total	24	42.4616			

Excel

Analysis of Variance

Source	DF	SS	MS	F	P
Regression	1	37.712	37.712	182.64	0.000
Error	23	4.749	0.206		
Total	24	42.462			

Minitab

FIGURE 11.12 Computer ANOVA output for regression analysis

The quantities SST, SSR, and SSE generated during the ANOVA procedure yield additional information about how good the linear regression model is. One reason we like to know whether two variables are related is so we can use the information to predict one given the other. Another reason is so we can control or reduce the variation in one quantity by controlling the variation in the other. The regression sum of squares (SSR) measures the amount of the variation in the Y variable that can be accounted for or *explained by* Y's relationship with X. If we look at SSR as a portion of SST, then we can determine the amount of the variability in Y that can be explained or accounted for. This value is called the **coefficient of determination, R^2:**

Coefficient of determination

$$R^2 = \frac{\text{SSR}}{\text{SST}} \times 100\%$$

R^2 is usually given in the output from statistical packages.

The coefficient of determination is a measure of how much the variation in Y could be reduced if X were controlled to a single value. This is a way of measuring how useful a model is for planning purposes. Even if a linear regression model is significant, it may not explain enough of the variation in the Y variable to be useful for planning. When this is the case, you may want to use a regression model with more than one independent or *explanatory* variables. These models are called multiple regression models and are discussed in Chapter 12.

11.3.3 EXERCISES–LEARNING IT!

Requires Exercise 11.1

11.8 The data on the number of postal employees and the amount of domestic mail processed were analyzed using Minitab. Here is a portion of the output:

Predictor	Coef	StDev	T	P
Constant	0.955	4.608	0.21	0.846
_Staff	0.00005872	0.00002087	2.81	0.048

S = 5.461   R = Sq = 66.4%   R-Sq(adj) = 58.1%

(a) Set up the hypotheses to test whether the slope is equal to 0.

(b) The sample had six observations. At the 0.05 level of significance, what are the critical values of the test?

(c) From the output, find the value of the slope coefficient, the standard error of the slope, and the value of the t statistic.

(d) Compare the t statistic with the critical value and determine whether the relationship between the number of postal employees and the amount of mail processed is significant.

Requires Exercise 11.4

11.9 For the relationship between countries' infant mortality rate and percentage of low-birthweight babies, the regression model is $\hat{y} = 1.36 + 0.808x$ with these values:

$$s_{y|x} = 0.937 \qquad \Sigma x = 396.2 \qquad \Sigma x^2 = 3118.04$$

(a) Find the standard error of the slope.

(b) Set up the null and alternative hypotheses and find the value of the test statistic.

(c) At the 0.10 level of significance, test to see whether the slope coefficient is significantly different from 0.

Requires Exercise 11.6

11.10 For the relationship between the number of visitors to a country, X, and the receipts from tourism, Y, the regression equation is $\hat{y} = 9.95 + 0.487x$ with these values:

$$s_{y|x} = 15.50 \qquad \Sigma x = 221 \qquad \Sigma x^2 = 9583$$

(a) Calculate the standard error of the slope.

(b) Set up the hypotheses to determine whether the regression equation is significant at the 0.05 level.

(c) Find the critical values of the test and calculate the test statistic.

(d) Is the relationship between the number of visitors and tourism receipts significant?

11.11 Consider the data on the number of kilometers of travel, Y, and the number of cars, X, in 12 countries. The results of a regression analysis using Microsoft Excel are given here: *Requires Exercise 11.3*

	Coefficients	Standard Error	t Stat	P-value
Intercept	−106.212984	55.14747208	−1.92598102	0.082984278
Total Cars	2.15816E-05	1.19E-06	18.07496921	5.76E-09

(a) From the output, find the slope coefficient, the standard error of the slope, the t statistic, and the p value of the test.

(b) At the 0.01 level of significance, use the p value to determine whether the relationship between the number of kilometers traveled and the number of cars is significant. (The p value in Excel is a one-tail p value.)

(c) The data set had 12 observations. Find the critical values for the test and verify your answer to part (b).

11.12 Consider the relationship between the amount of money, X, that a city was granted and the number of new police officers hired, Y, for nine cities in New Jersey. The Minitab output from the regression analysis is shown here: *Requires Exercise 11.5*

Predictor	Coef	SE Coef	T
Constant	0.7077	0.5360	1.32
Grant ($)	0.00001191	0.00000100	11.88

(a) Use the output to find the regression equation relating X and Y.

(b) Use the output to test whether the relationship between the amount of the grant and the number of new officers is significant.

11.13 Data on graduation rates at Big East Conference schools were analyzed using Microsoft Excel and the output is given here: *Requires Exercies 11.7*

	Coefficients	Standard Error	t Stat	P-value
Intercept	0.125681427	0.083709811	1.50139423	0.159101295
Percent of Football Players	0.912607626	0.128777793	7.086684788	1.27098E-05

Use the output to determine whether there is a significant relationship between the graduation rates for all male students and for all male athletes.

11.14 Data on the number of shopping centers and retail sales for the 12 North Central states were analyzed with Minitab and the output is shown here: *Requires Exercise 11.2*

Predictor	Coef	StDev	T	P
Constant	1.331	1.006	1.32	0.215
No. Shop	0.019021	0.0011291	6.84	0.000

At the 0.01 level of significance, is the relationship between the number of shopping centers and retail sales significant?

11.4 Confidence Intervals and Prediction Intervals

We have discussed that in a *statistical* relationship the values of the dependent variable vary for any single value of the independent variable. No matter how strong the relationship between X and Y is, some inherent variation exists in Y for any given value of X. This natural variation is identified in the population regression model $y = \beta_0 + \beta_1 x + \varepsilon$ as the quantity ε.

When the relationship between X and Y is strong, the variation in Y for a given value of X is small and can be thought of simply as *natural* variation. On a plot of the data and the regression line, the points will be very close to the line. That is, the variation that exists is entirely due to chance and we should not waste time trying to figure out why it happens. The predictions from the estimate of the regression line will be fairly precise and can be used for decision making.

When the relationship between X and Y is not so strong, the variation in Y for a given value of X occurs because the linear model is *not adequate*. The data points will not be close to the regression line, the relationship might not be linear, or other factors may cause Y to vary. Even if X and Y have a statistically significant relationship, the predictions obtained from the regression model might not be precise enough for decision making.

In Section 11.2 you learned that the standard error of the estimate, $s_{y|x}$, is a measure of the overall variation in Y for any value of X. In this section we put that value in context and see how it can be used to evaluate the usefulness of a regression model.

11.4.1 CONFIDENCE INTERVALS FOR REGRESSION ANALYSIS

The value of $\hat{y}$ obtained from the regression model is an estimate of the *mean value of Y for a given value of X*; that is, the value $\hat{y}$ is an *estimate* of the true mean, $\mu_{y|x}$. If the variable X represents advertising expenditures, then the regression model that relates X to sales of a product, Y, predicts the *average* sales for a given level of advertising. Because the estimate is based on sample data, there is some error in that estimate.

In Chapter 9 you learned how to find confidence intervals for the mean of a population, μ, based on a sample estimate, $\overline{X}$. We can use the same ideas to create a confidence interval for $\mu_{y|x}$ based on our sample estimate, $\hat{y}$.

To find a confidence interval for a population parameter we need to know the sampling distribution of the estimate and the standard error of the sample statistic. The estimate $\hat{y}$ has a Student's t distribution with $n - 2$ degrees of freedom. The standard error is related to the standard error of the estimate, $s_{y|x}$. The formula for a $(1 - \alpha)100\%$ **confidence interval** for the mean value of Y for a given value of $X = x_i$, $\mu_{y|x_i}$, is

Confidence interval

$$\hat{y}_i - t_{\alpha/2, n-2} s_{y|x} \sqrt{\frac{1}{n} + \frac{(x_i - \overline{x})^2}{\Sigma x^2 - (\Sigma x)^2/n}} \leq \mu_{y|x_i} \leq \hat{y}_i + t_{\alpha/2, n-2} s_{y|x} \sqrt{\frac{1}{n} + \frac{(x_i - \overline{x})^2}{\Sigma x^2 - (\Sigma x)^2/n}}$$

> A **confidence interval** provides an estimate for the mean value of Y ($\mu_{y|x}$) at a particular value of X.

Again, the expression looks formidable, but it does not contain anything we have not already dealt with!

Earlier, when we found confidence intervals, the width of the interval was related to the level of confidence that we chose. This is also true for the confidence in-

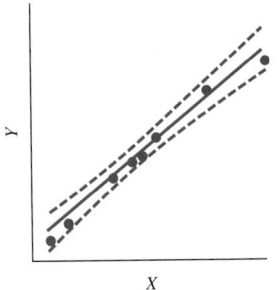

FIGURE 11.13 Confidence intervals for the mean estimates

terval for $\mu_{y|x_i}$, but the width of this interval also depends on how far the selected value of X is from the average value of X. The farther the value of interest is from the average X value, the wider (and less precise) the confidence interval becomes! Figure 11.13 illustrates how the confidence interval is related to the values of the X variable.

From the formula for the confidence interval, we see that when $X = \overline{X}$, the part of the formula in the square root reduces to $1/\sqrt{n}$ and the formula looks exactly like the confidence interval for the mean presented earlier.

EXAMPLE 11.15

THE REAL ESTATE MARKET

Calculating Confidence Intervals for the Mean Estimates

The real estate company that is looking at house prices and sizes wants to know something about the accuracy of the predictions from the regression model. They decide to find 95% confidence intervals for the two different size houses that are being built in town, 3500 and 5000 square feet. The regression model is

$$\hat{y} = 41{,}848 + 153x$$

Analyze the data.

From previous work they know that the predicted value for 3500 square feet is $577,348. They find that for 5000 square feet the predicted value is

$$\hat{y} = 41{,}848 + 153(5000) = \$806{,}848$$

They also know that $\Sigma x = 199.2$, $\Sigma x^2 = 692.033$, and $s_{y|x} = 107.8$. They calculate that $\overline{X} = 3.018$ (or 3108 square feet).

For a 95% confidence interval the t values are $\pm t_{0.025,64} = \pm 1.998$. For a house of 3500 square feet, the company calculates

$$t_{\alpha/2, n-2} s_{y|x} \sqrt{\frac{1}{n} + \frac{(x_i - \overline{x})^2}{\Sigma x^2 - (\Sigma x)^2/n}} = (1.998)(107.8) \sqrt{\frac{1}{66} + \frac{(3.500 - 3.108)^2}{692.033 - (199.2)^2/66}}$$

$$= (1.998)(107.8) \sqrt{0.016843\ldots} = 27.957$$

To actually calculate the interval it is probably easiest to calculate the ± part and then add and subtract that value from the prediction, $\hat{y}$.

The confidence limits are

$$577{,}348 \pm 27{,}957 = (549{,}391, 605{,}305)$$

or

$$\$549{,}391 \leqslant \mu_{y|3500} \leqslant \$605{,}305$$

The real estate company sees that the mean listing price is somewhere between $549,391 and $605,305. The width of the interval is

$$\$605{,}305 - \$549{,}391 = \$55{,}914$$

which is really not that wide.

The company does a similar calculation for a house of 5000 square feet and obtains:

The only change in the calculation is substituting 5.000 for 3.500 in the square root part of the formula.

$$(1.998)(107.8)\sqrt{\frac{1}{66} + \frac{(5.000 - 3.108)^2}{(692.033) - (199.2)^2/66}} = (1.998)(107.8)\sqrt{0.054570\ldots} = 50.312$$

So the 95% confidence interval is

$$806{,}848 \pm 50{,}312 = (756{,}536, 857{,}160)$$

or

$$\$756{,}536 \leq \mu_{y|5000} \leq \$857{,}160$$

For a house of 5000 square feet, the average listing price will be between $756,536 and $857,160. This interval is much wider.

It is not difficult to calculate confidence intervals for one or two estimates, but if you want intervals for more estimates, using a computer is much easier.

EXAMPLE 11.16

HMO HEALTH

Finding Confidence Intervals for the Mean Estimates

The analysts who are looking at the relationship between HMO revenues and number of members wants to know how accurate the estimates from the regression model are. They use Minitab to find 95% confidence intervals for the mean for each value of X in the data set. The output (edited) from the analysis is shown here:

Analyze the data.

X Value	Fit	95.0% CI
0.89	1.803	(1.025, 2.580)
0.97	1.894	(1.148, 2.640)
1.62	2.638	(2.101, 3.174)
1.83	2.878	(2.381, 3.374)
1.83	2.878	(2.381, 3.374)
2.07	3.152	(2.676, 3.629)
2.15	3.244	(2.767, 3.721)
2.30	3.415	(2.930, 3.901)
3.19	4.434	(3.710, 5.157)
4.24	5.635	(4.460, 6.809)

The analysts see that the uncertainty in the estimated revenue (the width of the interval) goes from a low of $0.953 billion to a high of $2.349 billion. For the middle X values (close to the mean) the width of the interval is about $1 billion. The confidence intervals are very wide, especially considering the size of the actual estimates. It might not be practical to use this model to predict mean revenues.

Most statistical software packages calculate the confidence intervals for either a single X value or for an entire set of X values. In a spreadsheet program, entering the formula is not a difficult task.

TRY IT NOW!

Hearing Difficulties
Finding Confidence Intervals for the Mean Estimates

After finding the regression model and deciding that it is significant, the researchers who are doing the study on age and sound levels want to know about the accuracy of the estimates from the model. They decide to calculate 95% confidence intervals for ages 35 and 75. They know from previous work that for the set of eight observations in the model, $s_{y|x} = 2.053$, $\Sigma x = 400$, and $\Sigma x^2 = 24{,}200$. Find 95% confidence intervals for the mean estimates.

Do you think these estimates are useful? Why or why not?

11.4.2   PREDICTION INTERVALS FOR REGRESSION ANALYSIS

A confidence interval for the mean estimate from the regression model gives the decision maker an idea of how accurate the estimate is *over a long period of time*; that is, it is an interval estimate for the mean. This is certainly important for decisions that involve long time horizons, but what about models of predictions in a single instance?

   If a company is looking at the relationship between sales and advertising, a confidence interval indicates that, for a specific level of advertising, *average* sales will vary between these two amounts. This is certainly useful information, but some additional information might also be helpful. The company might be interested in the accuracy of a *single observation*. For example, it might like to have an interval estimate for next month's sales for some specific advertising outlay. In a problem related to cash flow, this information is more important than an overall mean. Intervals for individual estimates are called **prediction intervals** in regression analysis.

> A *prediction interval* gives an estimate for an individual value of Y at a particular value of X.

   The formula used to calculate a $(1 - \alpha)100\%$ prediction interval for a regression model is almost identical to the formula for the confidence interval:

$$\hat{y}_i - t_{\alpha/2,\,n-2}s_{y|x}\sqrt{1+\frac{1}{n}+\frac{(x_i - \bar{x})^2}{\Sigma x^2 - (\Sigma x)^2/n}} \leq y \leq \hat{y}_i + t_{\alpha/2,\,n-2}s_{y|x}\sqrt{1+\frac{1}{n}+\frac{(x_i - \bar{x})^2}{\Sigma x^2 - (\Sigma x)^2/n}} \qquad \textit{Prediction interval}$$

   The only difference is the 1 in the square root part of the formula. The impact of this change is that the prediction intervals for a specific value of X are *wider* than the corresponding confidence intervals. If you think about it, a wider interval for prediction makes sense. You know from the ideas of sampling and the Central Limit Theorem that the distribution of individual observations is always more variable than the distribution of the sample means.

EXAMPLE 11.17

THE REAL ESTATE MARKET

Finding Prediction Intervals for Regression Estimates

A builder who is working with the real estate company thinks that the estimates of the mean price are useful, but he would really like to know how much variability there is in the prices of individual homes of the sizes he is interested in building. He asks the real estate company to find 95% prediction intervals for the estimates for home sizes of 3500 and 5000 square feet. The relevant values for the 25 observations are $\hat{y}_{3500} = \$577{,}348$, $\hat{y}_{5000} = \$806{,}848$, $\Sigma x = 199.2$, $\Sigma x^2 = 692.033$, $s_{y|x} = 107.8$, and $\overline{X} = 3.018$ (or 3108 square feet). The t value has not changed from the confidence interval calculations in Example 11.15 and so the analysts calculate the interval as follows:

Analyze the data.

$$(1.998)(107.8)\sqrt{1 + \frac{1}{66} + \frac{(3.500 - 3.108)^2}{692.033 - (199.2)^2/66}} = (1.998)(107.8)\sqrt{1.016843\ldots} = 217.191$$

The 95% prediction interval is then

$$577{,}348 \pm 217{,}191 = (360{,}157, 794{,}539)$$

or

$$\$360{,}157 \leq y \leq \$794{,}539$$

So, any given house of 3500 square feet will have a listing price between \$360,157 and \$794,539. Similarly for the house of 5000 square feet they find

$$(1.998)(107.8)\sqrt{1 + \frac{1}{66} + \frac{(5.000 - 3.108)^2}{(692.033) - (199.2)^2/66}}$$

$$= (1.998)(107.8)\sqrt{1.054570\ldots} = 221{,}183$$

The 95% prediction interval is

$$806{,}848 \pm 221{,}183 = (585{,}665, 1{,}028{,}031)$$

or

$$\$585{,}665 \leq y \leq \$1{,}028{,}031$$

At 5000 square feet, the listing price will vary from \$585,665 to \$1,028,031.

The variability for an individual house price is greater than the variability for the average price. The width of the intervals might make planning difficult.

Prediction intervals are important when the individual values of the Y variable are critical to the decision process. If, for example, the home builder is interested in estimating cash flow, then an average price does not help much.

TRY IT NOW!

Hearing Difficulties

Calculating Prediction Intervals for Regression Estimates

The researchers doing the study on sound levels and age decide to calculate 95% prediction intervals for the two X values of 35 and 75. They know from previous work that for the set of eight observations in the model, $s_{y|x} = 2.053$, $\Sigma x = 400$, and $\Sigma x^2 = 24{,}200$. Find 95% prediction intervals for the comfortable sound levels for ages 35 and 75.

Do you think confidence intervals or prediction intervals are more useful to the research?

ANS. (56.688, 67.598) AND (72.758, 84.099). FOR GENERAL PURPOSES, CONFIDENCE INTERVALS. FOR DESIGNING THINGS INVOLVING SOUND, PROBABLY PREDICTION INTERVALS.

11.4.3 EXERCISES—LEARNING IT!

11.15 For the data on federal grant money and police officers hired the relevant values are:

$$\Sigma x = 4{,}167{,}462 \qquad \Sigma x^2 = 2{,}571{,}647{,}717{,}614 \qquad s_{y|x} = 0.8033 \qquad n = 9$$

Requires Exercises 11.5, 11.12

(a) Find a 95% confidence interval for the mean number of officers hired when the grant is $350,000.

(b) Find a 95% prediction interval for the number of officers hired when the grant is $350,000.

11.16 For the data on the number of postal workers in a country and the amount of domestic mail processed, the relevant values are:

$$\Sigma x = 1{,}159{,}709 \qquad \Sigma x^2 = 292{,}657{,}611{,}087 \qquad s_{y|x} = 5.4614 \qquad n = 6$$

Requires Exercises 11.1, 11.8

(a) Find 95% confidence intervals for the mean amount of mail processed when the number of postal workers is 50,000 and 250,000.

(b) Find 95% prediction intervals for the same two values of X.

11.17 For the data on the number of shopping centers and retail sales for the North Central states, the relevant values are:

$$\Sigma x = 8308 \qquad \Sigma x^2 = 9{,}517{,}766 \qquad s_{y|x} = 2.1918 \qquad n = 12$$

Requires Exercises 11.2, 11.14

(a) Find 98% confidence intervals for the mean retail sales when the number of shopping centers is 1000 and 1500.

(b) Which interval in part (a) is wider? Why?

(c) Find 98% prediction intervals for the same values of X.

(d) Do you think that prediction or confidence intervals are more appropriate for these data?

11.18 For the data on the number of cars in a country and the number of kilometers traveled, the relevant values are:

$$\Sigma x = 319.51 \qquad \Sigma x^2 = 25{,}598.8975 \qquad s_{y|x} = 156.0981 \qquad n = 12$$

Requires Exercises 11.3, 11.11

(a) Find 99% confidence intervals for the number of kilometers traveled for $X = 1.5$ and $X = 10$ (million) cars.

(b) Find 99% prediction intervals for the same two values of X.

11.5 Correlation Analysis

When people talk about regression analysis, they also often talk about *correlation analysis*. The two topics are related but not interchangeable. Both regression and correlation analyses deal with bivariate quantitative data and the relationship between the two variables. The main purpose of regression analysis is to find an equation or model that allows the decision maker to predict the value of the dependent variable. *Correlation analysis* simply *measures the strength* of the linear relationship between two quantitative variables. The output of the analysis is a single number. In correlation analysis, there is no need to identify which variable is dependent and which is independent because prediction is not the end result.

We have talked about the types of linear relationships that can exist between two variables. Figure 11.14 on page 538 illustrates three types of relationships: perfect negative, none, and perfect positive.

If we simply want to measure the strength of a relationship between two variables, we use the **correlation coefficient.**

> The **correlation coefficient** is a measure of the strength of a linear relationship. A correlation of -1 corresponds to a perfect negative relationship, a correlation of 0 corresponds to no relationship, and a correlation of $+1$ corresponds to a perfect positive relationship.

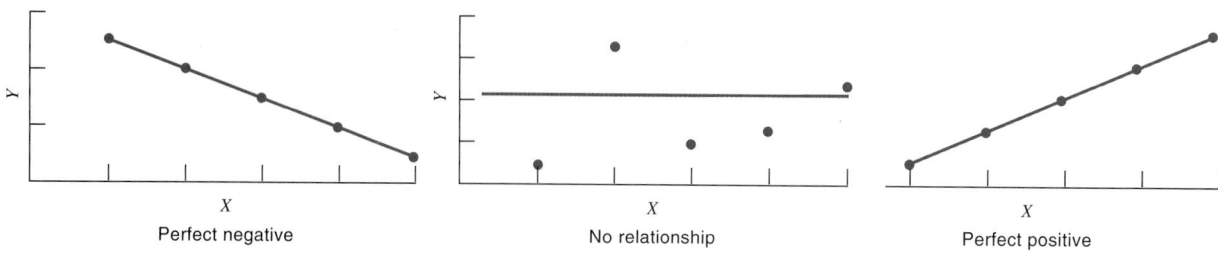

FIGURE 11.14 **Three types of relationships: perfect negative, no relationship, and perfect positive**

The correlation coefficient, *r*, is calculated using the formula

Correlation coefficient

$$r = \frac{\Sigma xy - (\Sigma x)(\Sigma y)/n}{\sqrt{\Sigma x^2 - (\Sigma x)^2/n}\sqrt{\Sigma y^2 - (\Sigma y)^2/n}}$$

Again, we have a complicated formula, but it uses all of the familiar elements. Any scientific calculator that does regression will output the correlation coefficient. Software that performs regression analysis usually does not output the correlation coefficient directly because correlation and regression are not interchangeable. In the next section we introduce a statistic that is part of regression and can be used to obtain the correlation coefficient.

In many cases, the correlation coefficient is used with regression analysis as another measure of how good the regression model is. It is important to recognize that the correlation coefficient can be used as a statistic in its own right when prediction of one variable as a function of the other is not appropriate or necessary.

For example, suppose that a company is interested in knowing whether there is a relationship between the score on an aptitude test and the number of months that a person remains in an entry-level position. The company does not necessarily want to *predict* the number of months in the entry-level position; it simply wants to know whether the test score is related to that variable. In this case the company could calculate the correlation coefficient between test score and number of months in the entry-level job.

EXAMPLE 11.18

THE REAL ESTATE MARKET

Calculating the Correlation Coefficient

The real estate company wonders what the correlation coefficient is for the relationship between price and house size. They decide to calculate the correlation coefficient from the relevant numbers:

$\Sigma x = 199.2$ $\Sigma y = 33,177.0$ $\Sigma xy = 113,999.80$ $\Sigma x^2 = 692.033$
$\Sigma y^2 = 19,529,846$ $n = 66$

Analyze the data.

The calculation is

$$r = \frac{113,999.80 - (199.2)(33,177.0)/66}{\sqrt{692.033 - (199.2)^2/66}\sqrt{19,529,846 - (33,177.0)^2/66}}$$

$$= \frac{13,865.58182}{(9.5294...)(1688.896...)} = 0.8615$$

The correlation coefficient is very close to +1, which indicates a strong positive relationship between listing price and house size.

You may wonder what values indicate good correlation and what values indicate that there is not really a relationship between two variables. The answer is, It depends. When data are collected under uncontrolled circumstances, there is usually a lot of variation in the data from other uncontrolled, unmeasured variables. When this happens, the correlation between two variables might not appear to be strong because of the additional factors. This is also reflected in the widths of the confidence and prediction intervals. Sometimes a correlation coefficient in the range of 0.6 to 0.7 (or -0.6 to -0.7), which might not provide *useful* information, is an indication that further investigation is appropriate.

EXAMPLE 11.19

HMO HEALTH

Calculating the Correlation Coefficient

The analysts looking at the relationship between revenues and number of members for HMOs are pretty certain that the relationship between the two variables will be a strong positive one, so they decide to calculate the correlation coefficient to check their assumption. They use these numbers to perform the calculation:

$$\Sigma x = 21.09 \quad \Sigma y = 31.97 \quad \Sigma xy = 77.6371 \quad \Sigma x^2 = 53.4063$$
$$\Sigma y^2 = 117.3013 \quad n = 10$$

Analyze the data.

The calculation is

$$r = \frac{77.6371 - (21.09)(31.97)/10}{\sqrt{53.4063 - (21.09)^2/10}\sqrt{117.3013 - (31.97)^2/10}}$$

$$= \frac{10.2124}{11.60795\ldots} = 0.8798$$

The analysts think that the number they found indicates the strong positive correlation they expected.

✔ TRY IT NOW!

Hearing Difficulties
Calculating the Correlation Coefficient

The relevant data to calculate the correlation coefficient for the sound level and age problem are $\Sigma x = 400$, $\Sigma y = 546$, $\Sigma xy = 29{,}010$, $\Sigma x^2 = 24{,}200$, $\Sigma y^2 = 37{,}986$, and $n = 8$. Find the correlation coefficient for the data.

The correlation coefficient is also related to one of the quantities that we looked at in regression analysis, the coefficient of determination, R^2. The value of r is equal to the square root of R^2. The sign of r is the same as the sign of the slope coefficient. We have

$$r = \sqrt{R^2} \quad \text{with the appropriate sign}$$

11.6 Regression Assumptions and Residual Analysis

Although regression is certainly one of the most powerful and frequently used tools in statistical analysis, it is often misused. The model is built on certain *assumptions* and people are often unaware of the assumptions or choose to ignore them. As you have learned before in this book, when we ignore the *assumptions* behind a statistical tool, the results of our analysis might be incorrect. In this section we look at the assumptions of the simple linear regression model and the ramifications of violating those assumptions. We also look at some simple techniques for determining whether the assumptions of the model are appropriate for a particular set of data.

Another reason the simple linear model is misused is that people do not realize that the model is inappropriate to use with certain data. This can be true even when the results of the analysis indicate that the simple linear model is appropriate. We also look at ways to determine when the simple linear model is not appropriate.

11.6.1 ASSUMPTIONS AND PROBLEMS IN THE REGRESSION MODEL

Remember that the simple linear model is

$$y = \beta_0 + \beta_1 x + \varepsilon$$

and the ε represents the random error in the model. Most of the assumptions that are built into the simple linear regression model involve this error term.

Here are the basic assumptions about the error, ε:

1. It has a mean value of 0 ($\mu_\varepsilon = 0$).
2. For every value of X, the standard deviation, σ, of ε is the same.
3. The distribution of ε is normal.
4. The error terms for different observations are not correlated with one another.

What happens if these assumptions are not valid? Obviously, the simple linear model is not correct, but what does that really mean? One possible result of violating the assumptions is that you may decide that a model is significant when it is not, or vice versa. Another result is that although you may be correct in your decision about the significance of the model, the equation you obtain is completely wrong. That is, although the model may give reasonable predictions for Y, the coefficients in the model, b_0 and b_1, cannot be interpreted correctly. When this is the case, using the model for planning or control purposes is risky.

In addition to violating the assumptions of the regression model, other factors make the linear model inappropriate. Here are two of these factors:

- Fitting a linear model when the true relationship is nonlinear
- Ignoring the influence of outliers on the regression equation

Sometimes data that appear to have a linear relationship in a scatter plot may actually have a nonlinear relationship. It is often hard to see curves or nonlinear patterns on scatter plots. The results of the analysis might indicate that the linear model is appropriate when in fact a nonlinear model would be better. Also, sometimes certain observations in the data set have a large influence on the regression equation. This often happens when the X value for one observation is far away from the other values. When these data points are removed, the entire analysis may change.

11.6.2 RESIDUAL ANALYSIS

How can we tell if we are violating the assumptions or using the linear model incorrectly? One tool that answers many of these questions is called *residual analysis*. You have already learned how to calculate residuals. Now we look at them as a diagnostic tool.

According to the assumptions of linear regression, for all values of the independent variable, the residuals of the model should be normally distributed random variables with a mean of 0 and a standard deviation of σ; that is, the ε values are $N(0, \sigma)$. If we plot the residuals from the regression model, the e_i's, as a function of X, they should be randomly distributed around a value of 0. Furthermore, the amount of variation around 0 should be the same for all values of X. When the plot exhibits departures from this, it indicates that certain assumptions of the model are not valid or that a linear model was not appropriate.

Figure 11.15 shows four plots of the residuals versus the values of the independent variable, X. Figure 11.15(a) shows what the residual plot should look like when the assumptions of the linear regression model are met and when the simple linear model is appropriate. The residuals are randomly distributed around the mean value of 0 and the plot has no apparent pattern. The others plots show different patterns that can result when the assumptions are violated or when the model is not appropriate. Figure 11.15(b) shows a typical plot that results when the linear model is used for data that have a non-linear relationship. The residuals have a systematic pattern. Figures 11.15(c) and (d) show plots that can result when the assumption of equal variances is violated.

Some statistical packages plot the residuals against the fitted values ($\hat{y}$) instead of X. When the regression model is significant, X and Y are related and so the plots

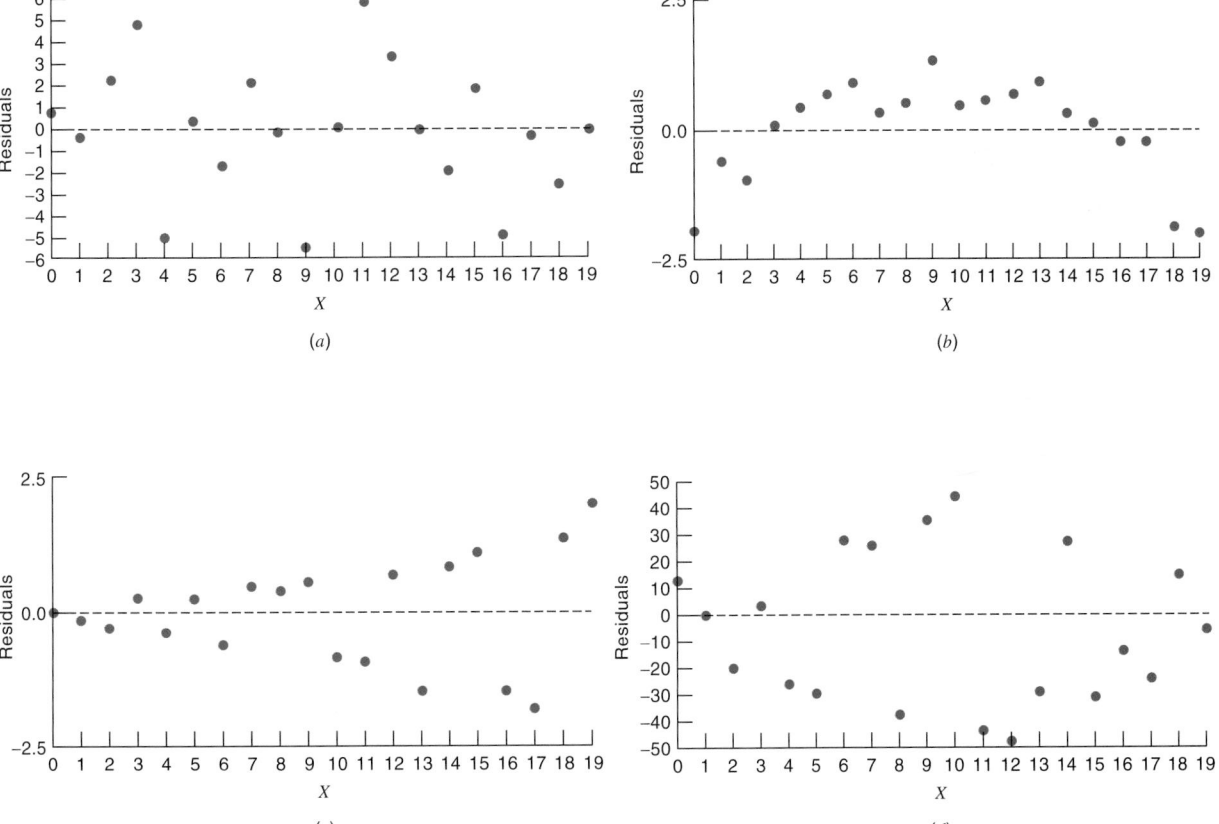

FIGURE 11.15 **Examples of residual plots**

will not have different shapes. Also, some packages use "standardized residuals" (the residuals divided by the standard error of the residuals) instead of the residuals. Again, this does not produce a difference in the plots.

In addition to checking that the residuals have a mean of 0 and are randomly dispersed around this value, we must check the shape of the distribution to see if it is normal. This was also an assumption for some of the hypothesis tests described in Chapters 9 and 10. How can we check to see whether data come from a normal distribution? There are several ways to accomplish this. We will look at one *informal* method, a histogram, and one *formal* method, a normal probability plot, available in many statistical software packages.

Note: For small data sets histograms may be misleading. You might want to use dotplots or normal plots.

If you look at a data distribution to see whether it is normally distributed, it should be approximately bell-shaped—that is, unimodal and symmetric. Figure 11.16 shows histograms of the residuals from the plots shown in Figure 11.15. In each case, a normal curve with a mean of 0 is overlaid on the histogram. Even in part (*a*), where the residuals looked randomly distributed, the residuals do not appear to come from a normal distribution. It may be that the residuals are not normally distributed, or perhaps there are just not enough data to make a really good histogram. This is one of the problems with such an informal method.

Another more formal method for determining whether data are normally distributed is a **normal probability plot.**

A *normal probability plot* is a plot of the ordered data against their expected values under a normal distribution. When data are normally distributed, the plot is a straight line.

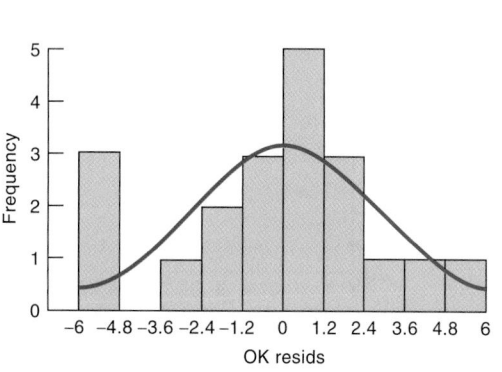

(*a*)

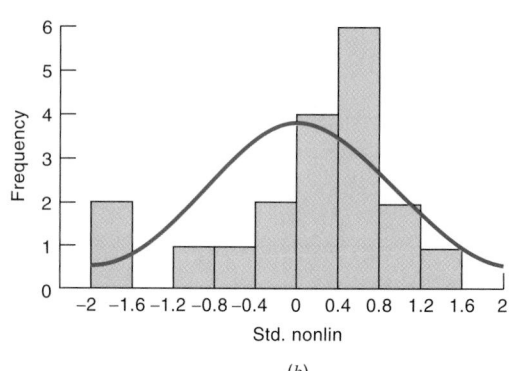

(*b*)

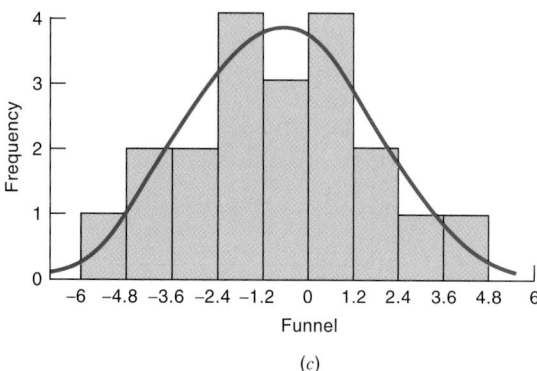

(*c*)

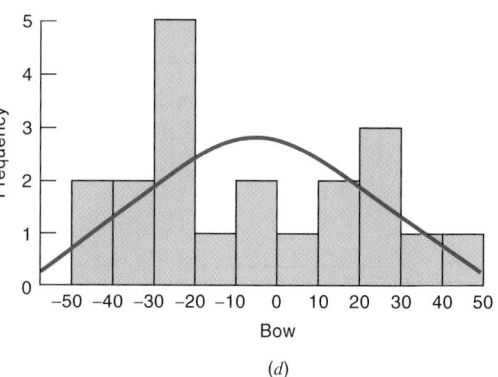

(*d*)

FIGURE 11.16 **Histograms of residuals**

The methods for creating a normal plot are well beyond the scope of this book, but the plots are created easily with statistical software packages. In fact, most statistical software make normal plots as part of their output from regression analysis. Figure 11.17 shows a set of residual plots from Minitab, including a normal plot, for each of the sets of residuals we have been looking at.

Another factor that can cause problems in using the linear regression model is the effect of observations that have excessive influence on the equation of the line. This can happen as a result of data pairs that are outliers and data points with X values that are very far removed from the rest of the data. Both of these factors can cause the regression model to shift toward the suspect data point and skew the resulting regression line.

There are ways to determine whether a data set has such an extreme observation. One visual method, as seen in the Minitab output in Figure 11.17, is to create a control chart for the residuals and see whether any of the values indicate an "out of control" situation. Another visual method is to make a boxplot of the residuals or the X data values and see whether any are determined to be outliers.

It is also possible to determine whether a data point has a large influence on the model by using some statistical techniques. Many statistical software packages use these techniques and report the problem as part of the output. Figure 11.18 on page 544 shows this part of the output for Minitab.

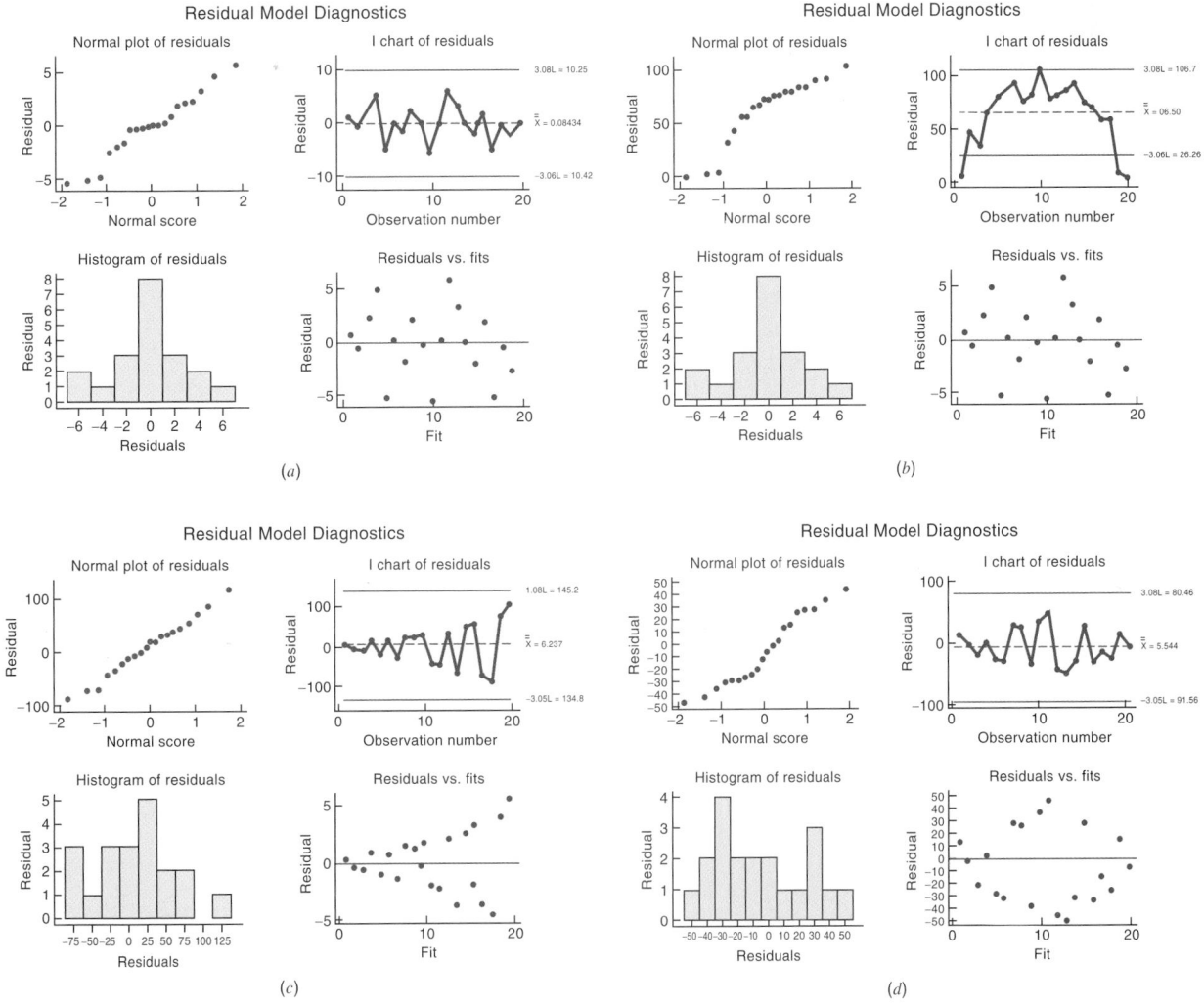

FIGURE 11.17 **Regression diagnostic plots from Minitab**

Obs	Members	HMORev $	Fit	StDev Fit	Residual	St Resid
1	4.24	5.486	5.631	0.509	−0.146	−0.36 X
2	3.19	4.629	4.432	0.313	0.197	0.34
3	1.83	3.857	2.878	0.215	0.979	1.59
4	1.62	3.600	2.639	0.232	0.961	1.58
5	2.07	3.429	3.153	0.206	0.276	0.45
6	2.30	2.914	3.415	0.210	−0.501	−0.81
7	1.83	2.743	2.878	0.215	−0.136	−0.22
8	2.15	2.400	3.244	0.206	−0.844	−1.37
9	0.97	1.714	1.896	0.323	−0.182	−0.32
10	0.89	1.200	1.805	0.336	−0.605	−1.08

X denotes an observation whose X value gives it large influence.

Obs	of Sta	RadioRev	Fit	StDev Fit	Residual	St Resid
1	83	1.0500	0.4628	0.1191	0.5872	2.75 P
2	57	0.3143	0.3446	0.0783	−0.0303	−0.13
3	104	0.3143	0.5583	0.1720	−0.2440	−1.40
4	35	0.3048	0.2446	0.0942	0.0602	0.27
5	21	0.2857	0.1809	0.1232	0.1048	0.50
6	63	0.2286	0.3719	0.0831	−0.1433	−0.62
7	67	0.2190	0.3901	0.0882	−0.1710	−0.75
8	41	0.2095	0.2718	0.0852	−0.0623	−1.27
9	38	0.2095	0.2582	0.0894	−0.0487	−0.21
10	20	0.1238	0.1764	0.1256	−0.0526	−0.25

FIGURE 11.18 **Warning output from Minitab**

Analyze the data.

EXAMPLE 11.20

THE REAL ESTATE MARKET

Checking Regression Assumptions Using Residuals

The real estate company that is using linear regression to determine whether listing prices are related to house size wants to make sure that what it has done so far is correct. Since the linear regression model is significant, the analysts want to make sure that it is valid and that they can use it for decision-making purposes. They use Minitab to analyze the data and look at the output of residual plots. The output is shown here:

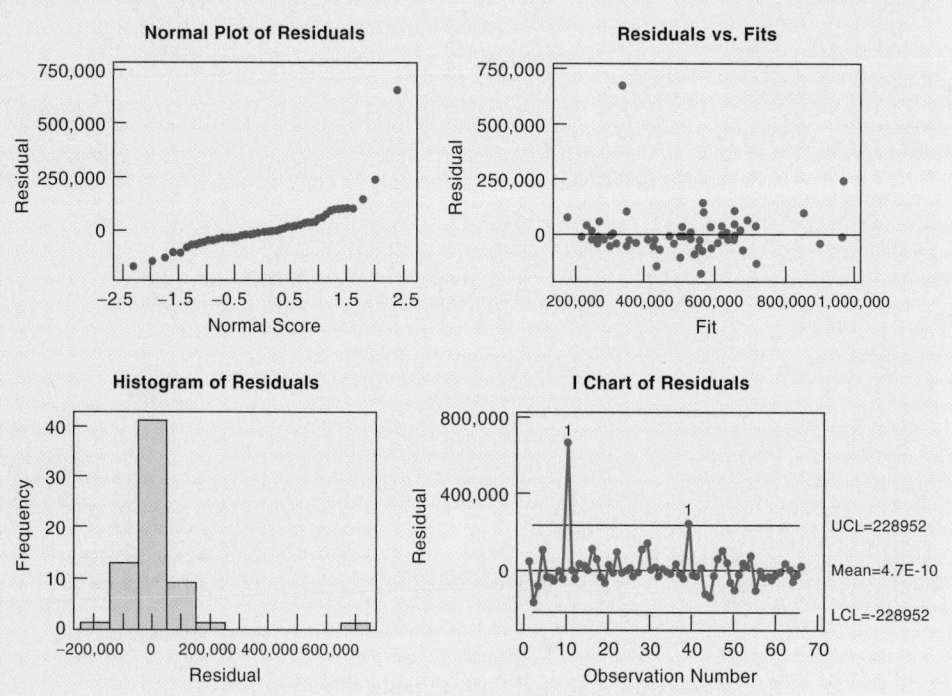

The histogram and normal probability plots suggest that there might be a problem with the assumption of normality. There are some outliers rather than skewness. The plot of the residuals versus the y values appears to be random, although there is an impression that the data actually might have a nonlinear relationship. The control chart indicates one or two extreme values.

Before they decide to use this model, the analysts should look more carefully at the data and see whether they can explain the outliers or perhaps use a nonlinear model.

11.6.3 EXERCISES—LEARNING IT!

11.19 Look at the regression model you developed for the data on federal grant money and police officers hired.

Requires Exercises 11.5, 11.12

(a) Make a plot of the residuals versus the values of the independent variable.

(b) From the plot, does it appear that a linear model is appropriate for these data? Why or why not?

(c) From the same residual plot, does the assumption of equal variances appear reasonable? Why or why not?

(d) Use an appropriate graphical technique to display the distribution of the residuals. Be sure to consider how many residuals you have in selecting the technique. Does the assumption of normality appear to be reasonable?

(e) Create a normal probability plot of the residuals. What does the plot suggest about the normality assumption?

(f) Considering your answers to parts (a)–(e), do you think the linear regression model is a good one for these data? Be sure to cite specifics.

11.20 Consider the regression model you developed for the data on the number of postal workers in a country and the amount of domestic mail processed.

Requires Exercises 11.1, 11.8

(a) Plot the residuals versus the values of the independent variable.

(b) What does the plot lead you to believe about the appropriateness of the linear model?

(c) Now plot the residuals versus the predicted values of y.

(d) How do the graphs in parts (a) and (c) compare? Why do you think this happened?

(e) Make a normal probability plot of the data. Do you think the assumption of normality is reasonable?

(f) Considering your answers to parts (a)–(e), do you think that the linear regression model is appropriate for these data. Why or why not?

11.21 Look at the model you developed for the data on the number of shopping centers and retail sales for the North Central states.

Requires Exercises 11.2, 11.14

(a) Make a plot of the residuals versus the independent variable.

(b) From the plot, does it appear that a linear model is appropriate?

(c) From the residual plot, do you think the assumption of equality of variances is violated?

(d) Make a graph that shows the distribution of the residuals.

(e) Does it appear that the residuals are normally distributed?

(f) Considering your answers to parts (a)–(e), do you think the linear regression model is appropriate for these data? Why or why not?

11.22 Look at the model you developed for the data on the number of cars in a country and the number of kilometers traveled.

Requires Exercises 11.3, 11.11

(a) Make a residual plot of the data versus the independent variable.

(b) What does the residual plot tell you about the assumption of a linear relationship? The assumption of equal variances?

(c) Using the residual plot, the computer output, or a plot of the values of the independent variable, do you think that any observation had a lot of influence on the model? If so, which point(s) are they?

(d) Do you think the assumption of normality is reasonable?

(e) Considering your answers to parts (a)–(d), do you think the linear model is appropriate for these data? Why or why not?

GET IT IN WRITING

Real Estate Market

TO: Board of Realtors
FROM: Acme Realty Company
RE: Housing Price Study

As part of the grant that you gave us to look at the current real estate market in our town, we have done a study to determine whether there is a relationship between the listing price and the size of a house. We studied all of the houses that have been put on the market in the past month since we did not want outside economic factors to affect the data.

We used linear regression to determine whether a relationship exists, what the exact relationship is, and whether the model is useful for predicting house prices. The accompanying figure is a plot of listing prices versus house sizes with the least-squares line drawn in. As you can see, with a few exceptions, the data points fit the line very well.

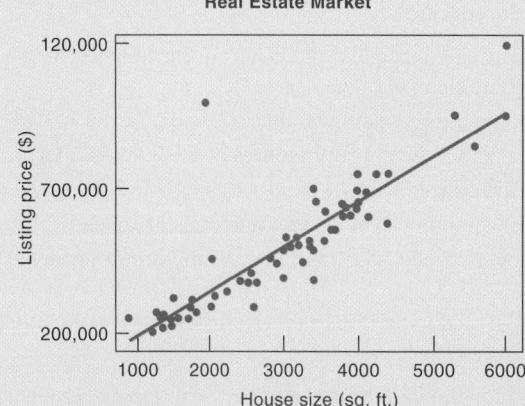

Real Estate Market

The relationship between listing price and house size is given by the equation

$$\text{Listing price} = \$41{,}848 + 153 \times \text{House size}$$

This means that for every increase in house size of 1 square foot, the listing price increases by $153. This relationship holds only for the house sizes covered in the study. We have no idea what the relationship is for house sizes beyond the minimum (880 square feet) and maximum (6000 square feet), but that range covers almost every house in town.

The predictions are fairly accurate for house sizes close to the average, which is 3108 square feet. The farther the house size is from the average, the less accurate the predictions.

It is possible that other variables affect the listing price of a house. For example, one house listed at $995,000 is only 1920 square feet. Clearly something else, perhaps lot size, is driving the price of this house. During the next 2 weeks, we will be looking at additional variables, and we will keep you informed of our progress.

11.7   Using Technology for Simple Linear Regression

In Chapter 4 you learned how to find and plot the least-squares line for a set of bivariate data. You learned in this chapter that the least-squares line is just the beginning of regression analysis. This section covers the tools of regression analysis using Excel, Minitab, and the TI-83 graphing calculator.

11.7.1   THE LINEAR REGRESSION MODEL IN EXCEL

Let's look at the problem of predicting HMO revenues based on the number of members. The data should be located in an Excel worksheet similar to the one shown in Figure 11.19.

	A	B	C
1	HMOs	Members (m)	Rev $bn
2	United HealthCare	4.24	5.49
3	Humana	3.19	4.63
4	FHP International	1.83	3.86
5	PacifiCare Health systems	1.62	3.60
6	U.S. Healthcare	2.07	3.43
7	WellPoint Health Network	2.3	2.91
8	Health Systems International	1.83	2.74
9	Foundation Health	2.15	2.40
10	Oxford Health Plans	0.97	1.71
11	Physician Corp. of America	0.89	1.20

FIGURE 11.19   The HMO data in Excel

To find the simple linear regression model for a set of data, follow these steps:

1. From the list of data analysis tools, select **Regression.** The dialog box shown in Figure 11.20 opens. The dialog box has two sections, Input and Output. The

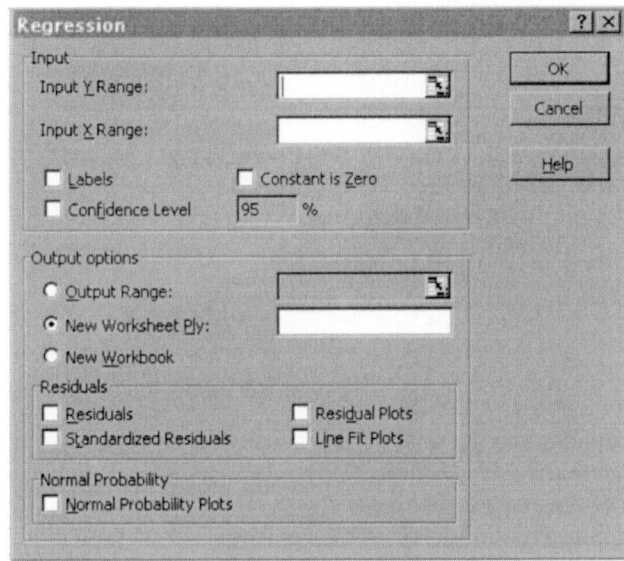

FIGURE 11.20   The Regression dialog box

Input section gives Excel information about the data. The Output section lets you make decisions about specific types of extra output that are available.

2. Position the cursor in the text box labeled **Input Y Range:** and highlight the data range for the *Y* variable—in this case, Revenues.

3. Move the cursor to the text box for **Input X Range:** and highlight the data range of the *X* variable—in this case, Members.

4. If the data ranges contain labels, click on the **Labels** check box. If you want confidence intervals for the regression estimates, click the check box for **Confidence Level.**

5. Specify the location where you want the output to appear, either in the current sheet, in a new worksheet, or in a new workbook.

In addition to standard regression output, Excel will create some of the diagnostic plots that we discussed in this chapter. You can choose to have residuals, standardized residuals, or both as part of the output. You can also have residual plots, fitted line plots, and normal probability plots. We recommend that you not create some of these plots, since the values are calculated incorrectly.

6. Click the check box for **Residuals.** Do *not* check the **Standardized Residuals** check box. Excel does not calculate these values correctly.

7. Click the check boxes for **Residual Plots** and **Line Fit Plots.** Do *not* click the check box for **Normal Probability Plot.** The plot is not created correctly. The completed dialog box should resemble the one in Figure 11.21.

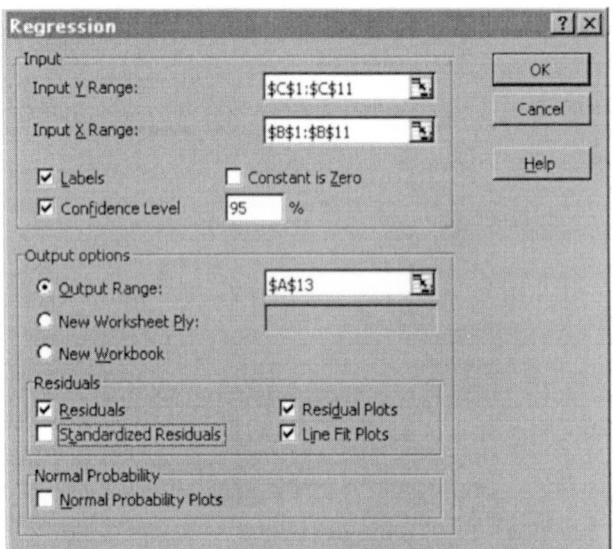

FIGURE 11.21 **Completed Regression dialog box**

8. Click on **OK.** The output will appear in the location you specified.

The output from the regression analysis is quite extensive. Although the whole output is still highlighted, you should select **Format > Column > AutoFit Selection** so that the columns adjust to fit the various output sections.

The first section of the output is the Summary Output. This section gives information about the regression model, including the coefficients, the value of R^2, the standard error, and the ANOVA output. This output is shown in Figure 11.22.

13	SUMMARY OUTPUT						
14							
15	*Regression Statistics*						
16	Multiple R	0.879773847					
17	R Square	0.774002022					
18	Adjusted R Square	0.745752275					
19	Standard Error	0.65297731					
20	Observations	10					
21							
22	ANOVA						
23		*df*	*SS*	*MS*	*F*	*Significance F*	
24	Regression	1	11.68217506	11.68217506	27.3985468	0.000788678	
25	Residual	8	3.411034941	0.426379368			
26	Total	9	15.09321				
27							
28		*Coefficients*	*Standard Error*	*t Stat*	*P-value*	*Lower 95%*	*Upper 95%*
29	Intercept	0.784464301	0.505044505	1.55325777	0.158965828	-0.380171168	1.94909977
30	Members (m)	1.143923992	0.218541241	5.23436212	0.000788678	0.639966661	1.647881323

FIGURE 11.22   Summary section of regression output

The coefficients for the regression equation are at the end of the section. You see that the equation is $\hat{y} = 0.78 + 1.1439x$. In the same section, you see the information for testing whether the regression model is significant. You can perform this test using the ANOVA output just above. In both cases, we see that the regression model is significant.

The top section of the output, Regression Statistics, includes the value for R^2 and the standard error. Excel does not include the correlation coefficient specifically, but you can find it by taking the square root of the R^2 value.

The second section of the output you created is the Residual Output, as shown in Figure 11.23. This section contains the values of $\hat{y}$ and the residuals.

34	RESIDUAL OUTPUT		
35			
36	*Observation*	*Predicted Rev $bn*	*Residuals*
37	1	5.634702027	-0.144702027
38	2	4.433581835	0.196418165
39	3	2.877845206	0.98215494
40	4	2.637621168	0.962378832
41	5	3.152386964	0.277613036
42	6	3.415489482	-0.505489482
43	7	2.877845206	-0.137845206
44	8	3.243900884	-0.843900884
45	9	1.894070573	-0.184070573
46	10	1.802556654	-0.602556654

FIGURE 11.23   Residual output

The plots that are created as part of the output are stacked on top of each other. Figure 11.24 (page 550) shows the plots after they have been moved in the worksheet.

The Residual Plot is a plot of the residuals versus the X variable. From this plot, you can determine whether the assumption of equal variances is violated. You can also get information about whether a linear model is appropriate or whether any observation exerts a lot of influence on the model.

The Line Fit Plot shows the predicted and fitted values of y on the same plot. This gives you information about how well the model fits the data. If the predictions were perfect, the two sets of symbols would coincide exactly.

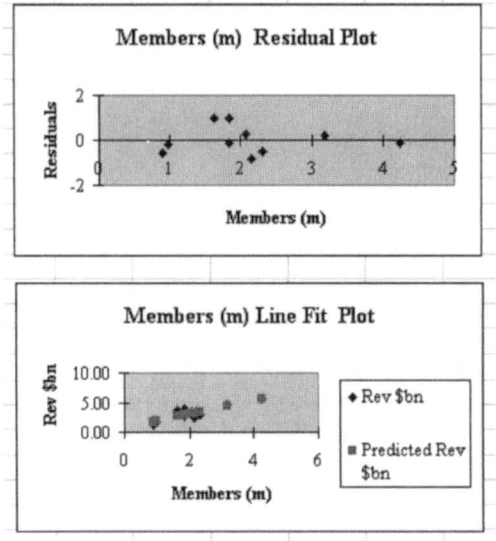

FIGURE 11.24 **Plots from regression analysis**

11.7.2 CONFIDENCE AND PREDICTION INTERVALS USING EXCEL

Although Excel does not provide an option to output confidence intervals or prediction intervals as part of the regression routine, it is not hard to calculate these intervals. All the values that are needed are included in the regression output. You can use formulas to find the upper and lower values for the intervals.

Recall that the formulas for the $(1 - \alpha)100\%$ confidence interval for the mean value of Y at a particular value of X are given by

$$\hat{y}_i - t_{\alpha/2,n-2}s_{y|x}\sqrt{\frac{1}{n} + \frac{(x_i - \bar{x})^2}{\Sigma x^2 - (\Sigma x)^2/n}} \leq \mu_{y|x_i}$$

$$\leq \hat{y}_i + t_{\alpha/2,n-2}s_{y|x}\sqrt{\frac{1}{n} + \frac{(x_i - \bar{x})^2}{\Sigma x^2 - (\Sigma x)^2/n}}$$

and the formulas for the corresponding prediction intervals are given by

$$\hat{y}_i - t_{\alpha/2,n-2}s_{y|x}\sqrt{1 + \frac{1}{n} + \frac{(x_i - \bar{x})^2}{\Sigma x^2 - (\Sigma x)^2/n}} \leq \mu_{y|x_i}$$

$$\leq \hat{y}_i + t_{\alpha/2,n-2}s_{y|x}\sqrt{1 + \frac{1}{n} + \frac{(x_i - \bar{x})^2}{\Sigma x^2 - (\Sigma x)^2/n}}$$

Excel allows you to obtain the predicted values $\hat{y}$ as part of the output; all other parts of the formula either are part of the output or can be calculated using other Excel functions. Unless you are adept at manipulating formulas, entering a formula this complicated in Excel is best accomplished by breaking it into smaller pieces and then combining the final results. For example, you should calculate the quantity under the square root separately and take the square root in a separate step. This minimizes possible problems with using parentheses.

11.7.3 USING KADDSTAT FOR REGRESSION ANALYSIS

Although Excel does perform linear regression, KADDSTAT can also be used for the analysis. The basic input is the same, although KADDSTAT has slightly different output.

From the **KADD** menu select **Regression and Correlation > Single/Multiple.**
The dialog box shown in Figure 11.25 opens. KADD allows you to specify both the
dependent and independent variables by name rather than by location. It also provides histograms of the residuals and standardized residuals in addition to the standard residual plot.

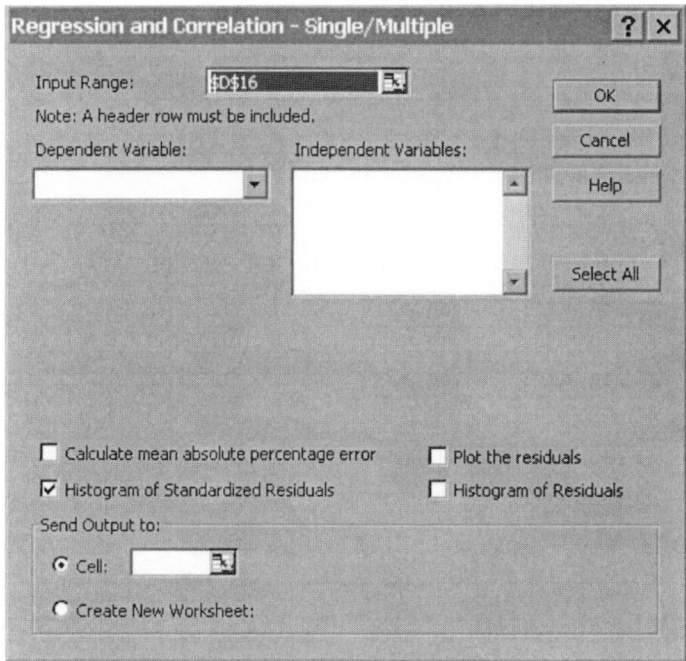

FIGURE 11.25   Regression dialog box

1. Position the cursor in the box labeled **Input Range** and highlight your data in
the Excel worksheet. Although nothing changes immediately, if you click on the
drop-down arrow in the box labeled **Dependent Variable,** all of the variable
names in the Input Range appear in the boxes for **Dependent Variable** and **Independent Variables,** as shown in Figure 11.26.

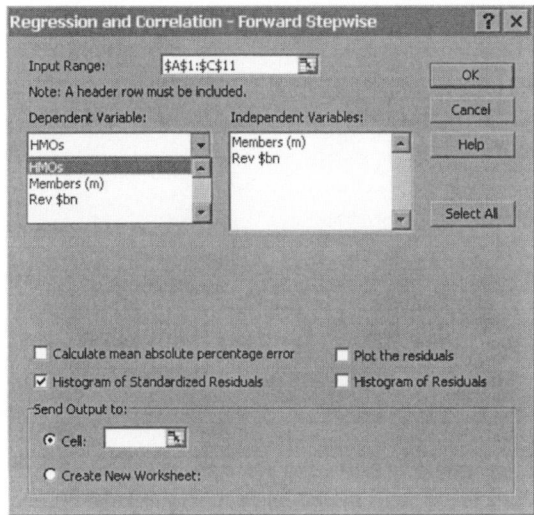

**FIGURE 11.26   Variable lists for regression
analysis**

2. From the drop-down list, select **Rev $bn** for the Dependent Variable.

3. Move the cursor over to the box labeled **Independent Variables** and from the list, click on the variable that you want to use for the independent variable—in this case, **Members (m).**

4. In the bottom part of the dialog box indicate which plots you want included in the output.

5. Indicate where you want the output to appear and click OK.

The output is extensive. The main portion of the output shown in Figure 11.27 contains the same information that is in the linear regression output from the Data Analysis Tools. The remainder of the output consists of the graphs requested and the residuals and standardized residuals, as shown in Figure 11.28.

Rather than inputting the equation of the regression line as a formula, KADD calculates the predicted values for the data points or for any other x values that you want to use. In the main portion of the output, click on the box labeled **Forecast** and

G	H	I	J	K	L	M	N	O
Regression and Correlation								
Observations	10			ANOVA				
R Square	0.7740		df	SS	MS	F	p value	
Standard Error	0.6530	Regression	1	11.6822	11.6822	27.3985	0.0008	
Adjusted R Square	0.7458	Residual	8	3.4110	0.4264			
Multiple R	0.8798	Total	9	15.09321022				
	Coefficients	Standard Error	t value	p value				
Intercept	0.7845	0.5050	1.5533	0.1590				
Members (m)	1.1439	0.2185	5.2344	0.0008			Forecast?	
Caution: If any independent variable can be perfectly predicted using one or								
more of the other independent variables, then the regression analysis may								
give an error message, or it may yield results which are not reasonable.								

FIGURE 11.27   Main output from regression analysis

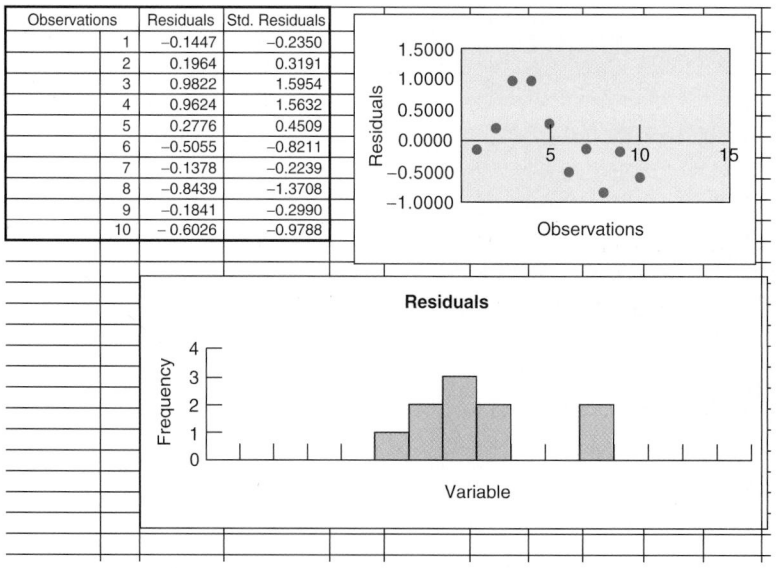

Observations	Residuals	Std. Residuals
1	−0.1447	−0.2350
2	0.1964	0.3191
3	0.9822	1.5954
4	0.9624	1.5632
5	0.2776	0.4509
6	−0.5055	−0.8211
7	−0.1378	−0.2239
8	−0.8439	−1.3708
9	−0.1841	−0.2990
10	−0.6026	−0.9788

FIGURE 11.28   Additional regression output

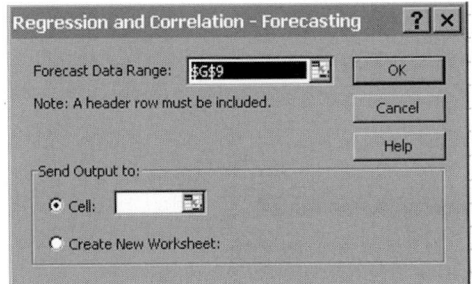

FIGURE 11.29   **Forecasting dialog box**

T	U	V	W
Regression and Correlation - Forecasting			
	Intercept	Members (m)	
Coefficient	0.7845	1.1439	
	Observations		Forecasted Value
	1	4.24	5.6347
	2	3.19	4.4336
	3	1.83	2.8778
	4	1.62	2.6376
	5	2.07	3.1524
	6	2.3	3.4155
	7	1.83	2.8778
	8	2.15	3.2439
	9	0.97	1.8941
	10	0.89	1.8026

FIGURE 11.30   **Predictions for the values of the independent variable**

the dialog box shown in Figure 11.29 opens. Place the cursor in the **Forecast Data Range:** box and highlight the location of the values of the independent variable for which you want predictions. Then indicate where you want the output located and click **OK.** The output shown in Figure 11.30 appears.

11.7.4   THE LINEAR REGRESSION MODEL IN MINITAB

You can find the equation for the least-squares line using the **Stat > Regression > Fitted Line Plot** option in Minitab. You will also get the basic information you need to do regression analysis. However, Minitab does not have all the available options. In this section we use **Stat > Regression > Regression** to do the analysis for the HMO revenue data. A portion of the data is shown in Figure 11.31.

↓	C1-T	C2	C3
	HMOs	Members (m)	Rev $bn
1	United HealthCare	4.24	5.49
2	Humana	3.19	4.63
3	FHP International	1.83	3.86
4	PacifiCare Health systems	1.62	3.60
5	U.S. Healthcare	2.07	3.43
6	WellPoint Health Network	2.30	2.91

FIGURE 11.31   **HMO Data in Minitab**

To perform a regression analysis use the following procedure:

1. Select **Stat** > **Regression** > **Regression** from the menu. The dialog box shown in Figure 11.32 opens.

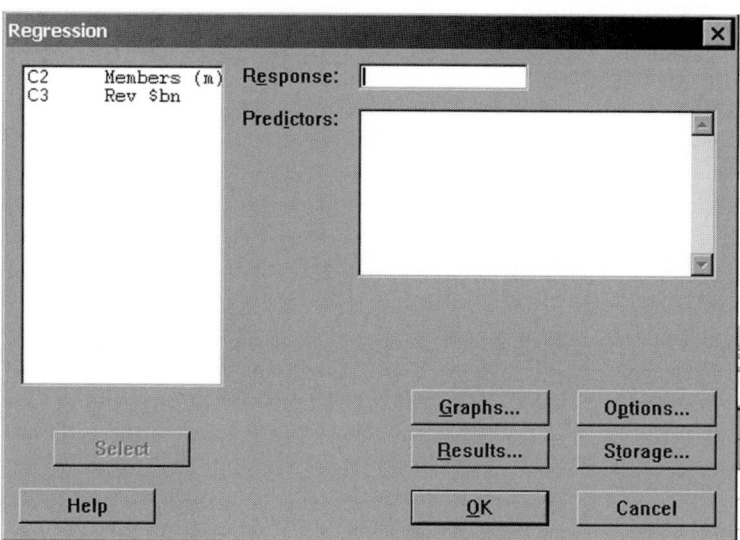

FIGURE 11.32 **Regression dialog box**

2. Position the cursor in the box labeled **Response:** and select the Y (dependent) variable from the list on the left—in this case, Revenues.

3. Position the cursor in the box labeled **Predictors:** and select the X (independent) variable from the list at the left—in this case, Members.

4. For the basic results, click **OK.** The output will appear in the Session window as shown in Figure 11.33.

Regression Analysis: Rev $bn versus Members (m)

The regression equation is
Rev $bn = 0.784 + 1.14 Members (m)

Predictor	Coef	SE Coef	T	P
Constant	0.7845	0.5050	1.55	0.159
Members	1.1439	0.2185	5.23	0.001

S = 0.6530   R-Sq = 77.4%   R-Sq(adj) = 74.6%

Analysis of Variance

Source	DF	SS	MS	F	P
Regression	1	11.682	11.682	27.40	0.001
Residual Error	8	3.411	0.426		

FIGURE 11.33 **Regression output for HMO data**

The first section of the output gives the regression equation. The second section gives all the information needed to do a t test for the coefficients as well as the values for $s_{y|x}$ (labeled "S" in the output) and R^2. The third section gives the information for performing the test using the ANOVA method.

11.7.5 PREDICTED VALUES AND RESIDUALS WITH MINITAB

In addition to the basic information Minitab allows you to calculate the predicted values (called "Fits") and the residuals. You can either print these values out or store them in the worksheet for further analysis.

To obtain the predicted values and residuals, from the Regression dialog box, click on the **Storage** button. The dialog box shown in Figure 11.34 opens. Check the boxes for **Residuals** and **Fits** and click **OK.** Click **OK** again to get the output.

Be careful! If you have these boxes checked and you run the regression more than once, Minitab will keep storing them in two new columns. Don't do this until you have everything else correct.

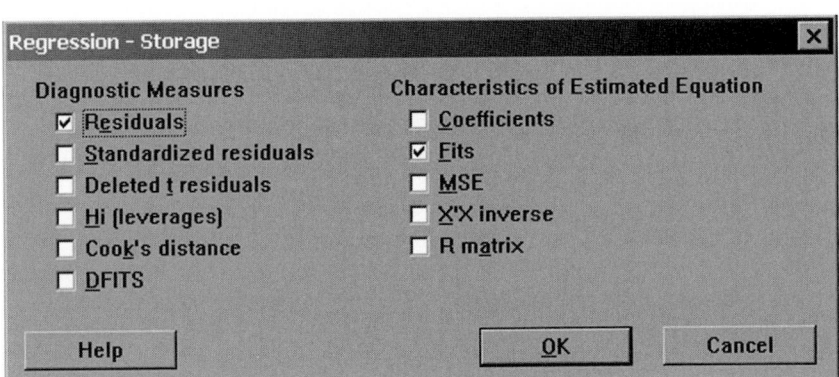

FIGURE 11.34 **Regression–Storage dialog box**

The values of the residuals and fits are stored in the next two empty columns of the worksheet as shown in Figure 11.35.

↓	C1-T	C2	C3	C4	C5
	HMOs	Members (m)	Rev $bn	RESI1	FITS1
1	United HealthCare	4.24	5.49	-0.144702	5.63470
2	Humana	3.19	4.63	0.196418	4.43358
3	FHP International	1.83	3.86	0.982155	2.87785
4	PacifiCare Health systems	1.62	3.60	0.962379	2.63762
5	U.S. Healthcare	2.07	3.43	0.277613	3.15239

FIGURE 11.35 **Residuals and fitted values**

You can create plots of the residuals as you learned in this chapter by using the **Graph > Plot** menu selection. Minitab has a built-in routine that does an excellent graphical analysis of residuals.

Once the residuals and fitted values are stored in the worksheet, select **Stat > Regression > Residual Plots** from the menu. The dialog box shown in Figure 11.36 on page 556 opens. You need to enter the columns that contain the fitted values and the residuals. You can also give your plots a title. The output is shown in Figure 11.37.

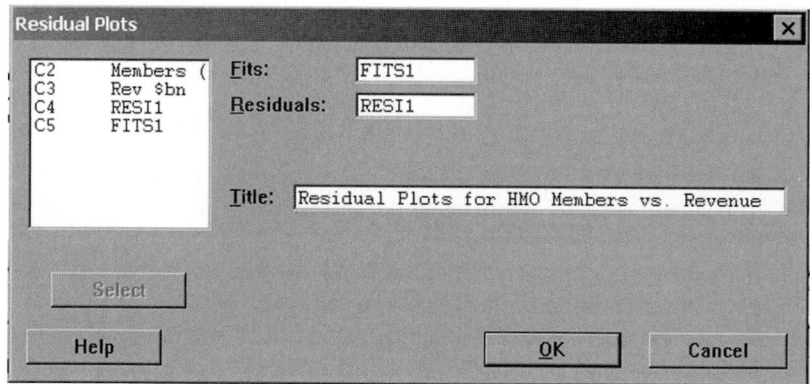

FIGURE 11.36 **Residual Plots dialog box**

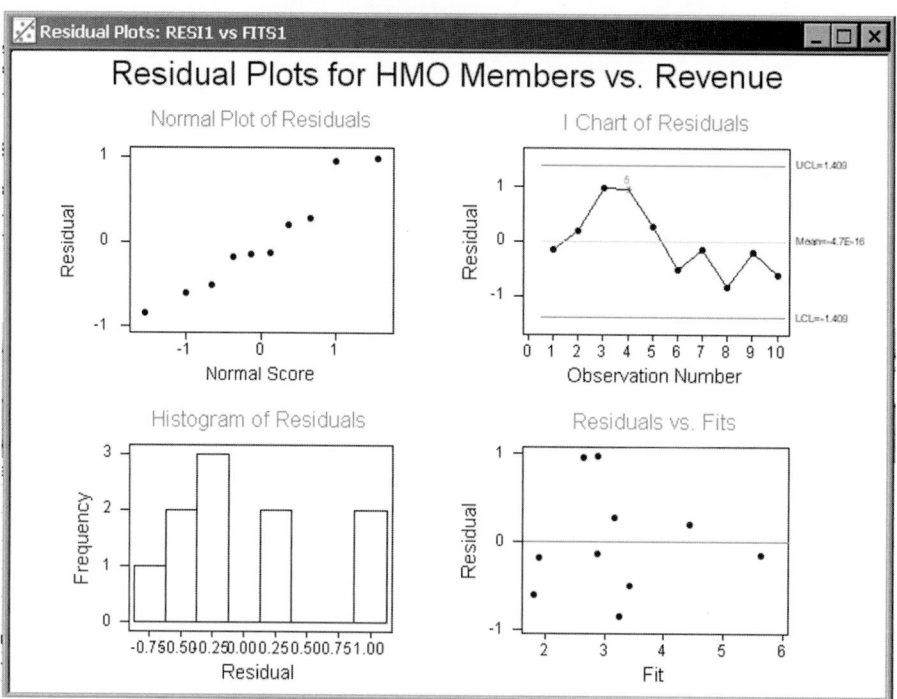

FIGURE 11.37 **Residual Plots output from Minitab**

The plots on the left are a normal probability plot and a histogram of the residual values. Both of these plots allow you to check the assumption that the residuals are normally distributed. If they are normally distributed, then the normal plot should look like a straight line. The plot on the right allows you to determine whether any of the residuals (and therefore the data values) are larger than you should expect. If the residuals fall outside the lines labeled UCL and LCL, then they are considered to be unusually large. The last plot lets you assess the assumptions of randomness and equal variances.

11.7.6 PREDICTION AND CONFIDENCE INTERVALS WITH MINITAB

From the Regression dialog box you can tell Minitab that you want to calculate prediction and/or confidence intervals—either for all the data points or for specific data points. To do this, click on the box labeled **Options** in the Regression dialog box. The box shown in Figure 11.38 opens.

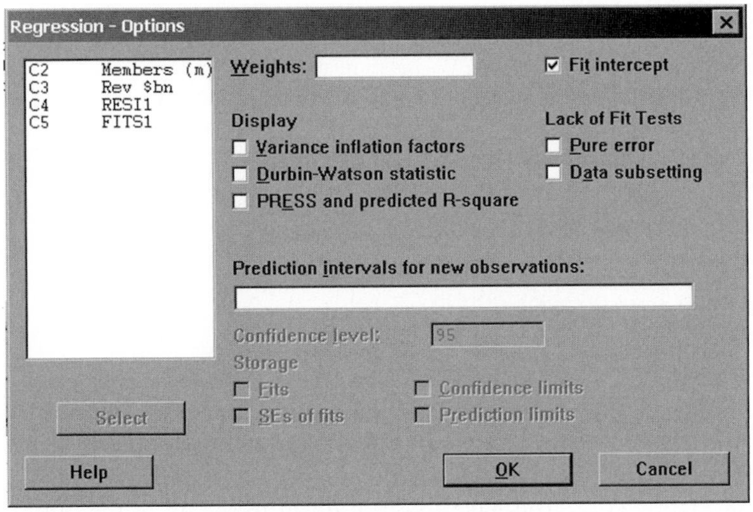

FIGURE 11.38   **Regression Options dialog box**

In the section labeled **Prediction intervals for new observations:** you can enter either the column with the original X values or a single value for which you would like prediction and confidence intervals. If you want the prediction and confidence limits added to the spreadsheet, click on the boxes for **Confidence limits** and **Prediction limits** under **Storage.** The output appears in the Session window (and in the worksheet if you specified it) as shown in Figure 11.39. Note that you also receive information on any data points with unusual X values that might be affecting the results.

Predicted Values for New Observations

New Obs	Fit	SE Fit	95.0% CI	95.0% PI
1	5.635	0.509	(4.460, 6.809)	(3.725, 7.545) X
2	4.434	0.314	(3.710, 5.157)	(2.763, 6.104)
3	2.878	0.215	(2.381, 3.374)	(1.292, 4.463)
4	2.638	0.233	(2.101, 3.174)	(1.039, 4.236)
5	3.152	0.207	(2.676, 3.629)	(1.573, 4.732)
6	3.415	0.211	(2.930, 3.901)	(1.833, 4.998)
7	2.878	0.215	(2.381, 3.374)	(1.292, 4.463)
8	3.244	0.207	(2.767, 3.721)	(1.665, 4.823)
9	1.894	0.323	(1.148, 2.640)	(0.214, 3.574)
10	1.803	0.337	(1.025, 2.580)	(0.108, 3.497)

X denotes a row with X values away from the center

FIGURE 11.39   **Prediction and confidence intervals**

11.7.7 REGRESSION ANALYSIS WITH THE TI-83 GRAPHING CALCULATOR

In Chapter 4 you learned how to calculate the equation of the least-squares line for a set of data and how to plot the data and the line using the TI-83. You can also use the TI-83 to perform the t test for a linear regression analysis.

1. Press $\boxed{\text{STAT}}$ and then select **TESTS.** Choose Option E: LinRegTTEST from the list. The screen shown in Figure 11.40 on page 558 opens.

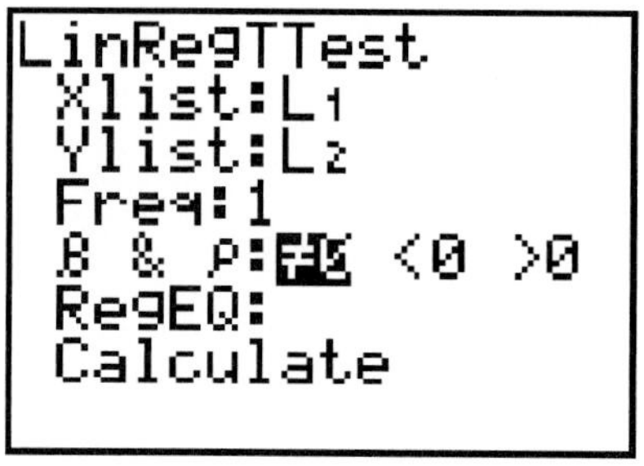

```
LinRegTTest
Xlist:L₁
Ylist:L₂
Freq:1
β & ρ:≠0 <0 >0
RegEQ:
Calculate
```

FIGURE 11.40 **Linear regression *t* test screen**

2. Enter the lists that contain the X and Y data and select **Calculate.** The output screen is shown in Figure 11.41. You have to scroll down to see all of the output shown.

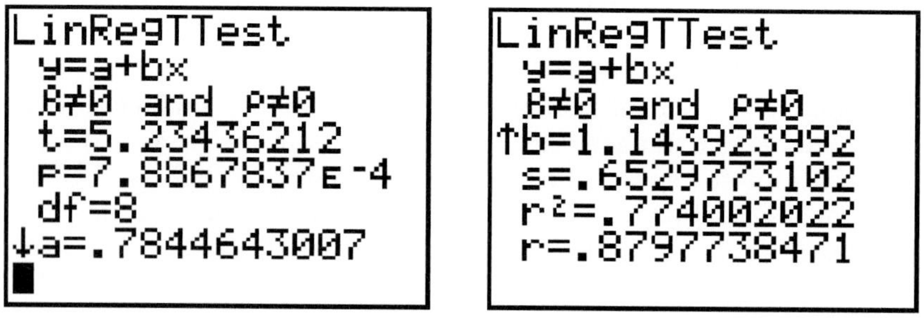

```
LinRegTTest
y=a+bx
β≠0 and ρ≠0
t=5.23436212
p=7.8867837ᴇ-4
df=8
↓a=.7844643007
■
```

```
LinRegTTest
y=a+bx
β≠0 and ρ≠0
↑b=1.143923992
 s=.6529773102
 r²=.774002022
 r=.8797738471
```

FIGURE 11.41 **Linear regression *t* test output**

The TI-83 does the test to see whether the slope coefficient β_1 is different from 0 as well as the test to see whether the correlation coefficient ρ is different from 0. They are essentially the same test.

CHAPTER 11 SUMMARY

Linear regression analysis is a powerful tool for determining how two variables are related. The regression equation can be used for *description, control*, and *prediction*. Description is important when the user is simply trying to understand the way that two variables are related. Control describes when the model is used to set standards or reduce variability. Prediction is when the model is used to determine what the resulting Y value should be when X takes on certain values. Although finding the simple linear model itself is not numerically difficult, performing a complete regression analysis requires the use of computer software.

You have also seen that even though the simple linear model may be significant, it might not be correct. It is necessary to test the *assumptions* of the linear model to see whether the model you obtain is appropriate.

KEY TERMS

Term	Definition	Page reference
Confidence interval	A **confidence interval** provides an estimate for the mean value of Y ($\mu_{y|x}$) at a particular value of X.	532
Correlation coefficient	The **correlation coefficient** is a measure of the strength of a linear relationship. A correlation of -1 corresponds to a perfect negative relationship, a correlation of 0 corresponds to no relationship, and a correlation of $+1$ corresponds to a perfect positive relationship.	537
Deviation	The distance between the predicted value of Y, $\hat{y}$, and the actual value of Y, y, is called the **deviation** or **error.**	509
Interpolation and extrapolation	Using the equation to predict values of Y within the range of the X data is called **interpolation.** Predicting values of Y for values of X outside the observed range is called **extrapolation.**	515
Least-squares method	The technique that finds the equation of the line that minimizes the total or sum of the squared deviations between the actual data points and the line is called the **least-squares method.**	509
Normal probability plot	A **normal probability plot** is a plot of the ordered data against their expected values under a normal distribution. When data are normally distributed, the plot is a straight line.	542
Prediction interval	A **prediction interval** gives an estimate for an individual value of Y at a particular value of X.	535
Residual, e_i	The difference between the observed value of Y, y_i, and the value of Y predicted from the regression equation, $\hat{y}_i$, for a value of $X = x_i$ is called the ith **residual, e_i.**	517
Simple linear regression model	The true relationship between the variables X and Y, the **simple linear regression model,** can be described by $$y = \beta_0 + \beta_1 x + \varepsilon$$	507
Standard error of the estimate, $s_{y|x}$	The **standard error of the estimate, $s_{y|x}$,** is a measure of how much the data vary around the regression line.	520
Sums of squares	**SST:** the total variation in the y values around the mean $\bar{y}$ **SSR:** the variation in Y that is caused by Y's relationship with X **SSE:** the variation in Y that remains unexplained	528
$\hat{y}$ (y−hat)	The value of $\hat{y}$ that we find is really a prediction of the mean value of Y for a given value of X.	513

KEY FORMULAS

Term	Formula	Page reference
b_1, Slope of least-squares line	$b_1 = \dfrac{n\sum_{i=1}^{n} x_i y_i - \sum_{i=1}^{n} x_i \sum_{i=1}^{n} y_i}{n\sum_{i=1}^{n} x_i^2 - \left(\sum_{i=1}^{n} x_i\right)^2}$ or $\dfrac{n\sum xy - \sum x \sum y}{n\sum x^2 - \left(\sum x\right)^2}$	510

(continued)

KEY FORMULAS

Term	Formula	Page reference
b_0, Intercept of least-squares line	$b_0 = \dfrac{\sum_{i=1}^{n} y_i}{n} - b_1 \dfrac{\sum_{i=1}^{n} x_i}{n}$ or $b_0 = \bar{y} - b_1\bar{x}$	510
Standard error of the estimate, $s_{y\vert x}$	$s_{y\vert x} = \sqrt{\dfrac{\Sigma(y - \hat{y})^2}{n - 2}} = \sqrt{\dfrac{\Sigma e^2}{n - 2}}$	520
Standard error of the slope	$s_{b_1} = \dfrac{s_{y\vert x}}{\sqrt{\Sigma x^2 - (\Sigma x)^2/n}}$	525
Test statistic for the slope	$t = \dfrac{b_1 - \beta_1}{s_{b_1}}$	525
Sums of squares	$SST = \Sigma(y - \bar{y})^2$ $SSR = \Sigma(\hat{y} - \bar{y})^2$ $SSE = \Sigma(y - \hat{y})^2$	528
Mean squares	$MSR = \dfrac{SSR}{1}$ and $MSE = \dfrac{SSE}{n - 2}$	529
Coefficient of determination, R^2	$R^2 = \dfrac{SSR}{SST} 100\%$	530
Confidence interval	$\hat{y}_i - t_{\alpha/2,n-2}\, s_{y\vert x} \sqrt{\dfrac{1}{n} + \dfrac{(x_i - \bar{x})^2}{\Sigma x^2 - (\Sigma x)^2/n}} \leq \mu_{y\vert x_i}$ $\leq \hat{y}_i + t_{\alpha/2,n-2}\, s_{y\vert x} \sqrt{\dfrac{1}{n} + \dfrac{(x_i - \bar{x})^2}{\Sigma x^2 - (\Sigma x)^2/n}}$	532
Prediction interval	$\hat{y}_i - t_{\alpha/2,n-2}\, s_{y\vert x} \sqrt{1 + \dfrac{1}{n} + \dfrac{(x_i - \bar{x})^2}{\Sigma x^2 - (\Sigma x)^2/n}} \leq y$ $\leq \hat{y}_i + t_{\alpha/2,n-2}\, s_{y\vert x} \sqrt{1 + \dfrac{1}{n} + \dfrac{(x_i - \bar{x})^2}{\Sigma x^2 - (\Sigma x)^2/n}}$	535
Correlation coefficient, r	$r = \dfrac{\Sigma xy - (\Sigma x)(\Sigma y)/n}{\sqrt{\Sigma x^2 - (\Sigma x)^2/n}\sqrt{\Sigma y^2 - (\Sigma y)^2/n}}$	538

CHAPTER 11 EXERCISES

Learning It!

11.23 How much does advertising affect market penetration? To assess the impact of advertising in the tobacco industry, a study looked at the amount of money spent on advertising a particular brand of cigarettes and brand preference among adolescents and adults. Here are the data:

Brand	Advertising (millions)	Brand preference (%) Adolescents	Adults
Marlboro	$75	60.0	23.5
Camel	43	13.3	6.7
Newport	35	12.7	4.8
Kool	21	1.2	3.9
Winston	17	1.2	3.9
Benson & Hedges	4	1.0	3.0
Salem	3	0.3	2.5

Source: Centers for Disease Control website

(a) Look at the data on brand preference for adolescents and the amount spent on advertising. Which is the dependent variable? Which is the independent variable?

(b) Create a scatter plot of advertising expenditures and adolescent brand preferences. Do you think there is a linear relationship between the two variables? Why or why not?

(c) Now create another scatter plot using adult brand preferences. How does this plot compare with the one in part (b)? From the plots, do you think that adolescent or adult brand preferences are more strongly related to advertising expenditures? Why?

(d) Find the least-squares line for adolescent brand preferences and advertising expenditures.

(e) Interpret the meaning of the slope and intercept for the model in part (d). Do they make sense?

(f) Use the model to predict adolescent brand preference for each brand studied. How well do the predicted values agree with the actual data?

(g) With $\alpha = 0.05$, is the model significant?

11.24 Retention of students is one of the greatest problems facing colleges and universities today. Loss of students has a negative impact on revenues from not only tuition but also state and federal funding programs. If a university can understand what factors influence student retention, it can form strategic plans aimed at keeping students enrolled. Academic performance is often cited as one reason students leave. Data were collected from 33 colleges in the Midwest on the freshman retention rates (percentage of freshmen who stay for a second year) and the 25th percentile scores on the American College Testing (ACT) examination. (The ACT is a college entrance examination similar to the SAT.) The data are listed here:

Freshman retention rate	ACT 25th percentile	Freshman retention rate	ACT 25th percentile
0.84	22	0.82	20
0.82	22	0.81	18
0.86	22	0.77	21
0.88	23	0.80	22
0.83	23	0.78	21
0.87	21	0.73	18
0.87	22	0.75	18
0.85	25	0.73	20
0.84	22	0.78	18
0.85	23	0.77	24
0.84	21	0.73	21
0.81	22	0.78	19
0.81	22	0.72	21
0.83	21	0.77	21
0.75	20	0.74	19
0.86	20	0.72	21
0.79	20		

(a) Which is the independent variable? Which is the dependent variable?

(b) Create a scatter plot of the data. Does it appear that the freshman retention rate is related to the 25th percentile score on the ACT?

(c) Find the equation of the regression line for the data.

(d) Plot the regression line on the same plot as the data. Does the line do a good job of predicting the freshman retention rate at a college or university? Why or why not?

(e) Calculate the standard error of the estimate, $s_{y|x}$, for the regression line.

(f) At the $\alpha = 0.05$ level, is the model significant?

11.25 The British Bankers' Association wants to understand the relationship between a bank's deposits (in billions of £) and number of customers. Analysts collected data on six large banks:

Bank	Deposits (£ billions)	Customers (millions)
Abbey National	101.7	13.6
Barclays	108.2	10.0
Lloyds	96.9	15.0
National Westminster	113.8	7.5
Woolrich	27.5	4.0
Halifax	77.1	7.6

(a) Which is the independent variable? Which is the dependent variable?

(b) Create a scatter plot of the data. Does it appear that deposits are related to the number of customers?

(c) Find the equation of the regression line for the data.

(d) Plot the regression line on the same plot as the data. Does the line do a good job of predicting the amount of deposits? Why or why not?

(e) Calculate the standard error of the estimate, $s_{y|x}$, for the regression line.

(f) At the 0.05 level, is the model significant?

Datafile:
POVERTY.XXX

11.26 An education task force is studying poverty levels in the United States. It has collected data for each state and the District of Columbia on the total number of people below the poverty level and the number of adults over the age of 25 who did not graduate from high school for the year 1993.

State	Not HS graduates	People below the poverty level (thousands)
Alabama	922,624	609
Alaska	57,717	60
Arizona	845,022	812
Arkansas	588,886	377
California	6,500,643	5,118
Colorado	412,981	363
Connecticut	533,673	310
Delaware	110,053	80
District of Columbia	84,746	114
Florida	2,699,792	1,923
Georgia	1,528,441	1,034
Hawaii	183,722	131
Idaho	212,562	165
Illinois	1,903,162	1,234
Indiana	973,367	547
Iowa	352,081	257
Kansas	283,939	250
Kentucky	869,966	521
Louisiana	934,959	821
Maine	165,485	131
Maryland	785,626	359
Massachusetts	885,187	528
Michigan	1,433,317	1,099
Minnesota	500,894	498
Mississippi	624,725	486
Missouri	929,994	531
Montana	95,969	153
Nebraska	204,514	211
Nevada	190,412	195
New Hampshire	189,608	119
New Jersey	1,095,526	693

New Mexico	354,334	371
New York	3,362,431	3,068
North Carolina	1,403,648	1,039
North Dakota	100,204	97
Ohio	1,546,910	1,253
Oklahoma	515,394	458
Oregon	475,886	503
Pennsylvania	1,908,231	1,338
Rhode Island	190,777	112
South Carolina	820,896	527
South Dakota	101,129	77
Tennessee	1,254,473	749
Texas	4,287,836	2,994
Utah	224,674	190
Vermont	78,587	58
Virginia	1,181,694	589
Washington	455,141	512
West Virginia	427,433	312
Wisconsin	626,820	449
Wyoming	48,091	51

(a) The task force wants to find a model that will predict the number of people living below the poverty level from the number of adults who are not high school graduates. Which is the dependent variable? Which is the independent variable?

(b) Create a scatter plot of the data. Do you think that a linear relationship exists between the two variables?

(c) Find the linear regression model for the data.

(d) Interpret the meaning of the slope and the y intercept of the model. Do you think the y intercept makes sense for these data?

(e) Use the model to predict the number of people who are below the poverty level in Wyoming, Connecticut, and North Carolina. What are the residuals for these three observations?

(f) At the 0.05 level of significance, is the model significant?

Thinking About It!

11.27 The U.S. Department of Commerce has data available on the numbers of shopping centers and retail sales for the South Central states.   *Requires Exercise 11.2*

State	Shopping centers	Retail sales (billions)
Kentucky	616	$13.9
Alabama	630	15.5
Tennessee	1,200	23.0
Arkansas	370	7.5
Mississippi	430	8.2
Louisiana	700	18.7
Oklahoma	568	13.2
Texas	2,976	87.3

Source: Statistical Abstract of the United States 1999

(a) Find the linear regression model for the South Central states.

(b) How does the equation for the South Central states compare with the one you found for the North Central states?

(c) Would you expect the equations to be exactly the same? Why or why not?

(d) What similarities would you expect the equations to have? What differences?

(e) Combine the data from the North Central and South Central states and find the linear regression model for the combined data.

(f) Do you think the individual models are better or worse than the combined model? On what criteria do you base your conclusion?

Requires Exercise 11.23

11.28 Look at the data on advertising expenditures and brand preferences for cigarettes again.

(a) Find the simple linear regression model that relates adult brand preferences to advertising expenditures.

(b) At the 0.05 level, is the model significant?

(c) What is R^2 for this model? What does it mean? What is the correlation coefficient?

(d) Compare the model for adult brand preferences with the one you found for adolescent brand preferences. Which variable do you think is more strongly related to advertising expenditures? Why?

(e) Do these data help support the claims against the tobacco industry that its advertising targets teenagers? Why or why not?

Requires Exercise 11.26

11.29 Consider the model you developed relating poverty levels and education.

(a) Find 95% confidence intervals for the mean number of people who live below the poverty level when the number of adults who are not high school graduates is 500,000; 1,000,000; and 2,000,000.

(b) Find 95% prediction intervals for the number of people who live below the poverty level when the number of adults who are not high school graduates is 500,000; 1,000,000; and 2,000,000.

(c) Do you think that the model is useful for predicting the number of people below the poverty level? Why or why not?

(d) The task force contends that if it can institute programs to reduce the number of adults who have not finished high school by 10% in each state, that will have a significant impact on the number of people who are living below the poverty level. Do you agree or disagree with this statement? Why?

Requires Exercise 11.25

11.30 Consider the data on the British banks.

(a) What is R^2 for this model? What does it mean? What is the correlation coefficient?

(b) Find 95% confidence intervals for the mean amounts of deposits when the number of customers is 12 million; 15 million; and 7 million.

(c) Find 95% prediction intervals for the amounts of deposits when the number of customers is 12 million; 15 million; and 7 million.

(d) Do you think the model is useful for predicting the amounts of deposits? Why or why not?

11.31 The British Bankers' Association decided to look at another variable to predict deposits, the number of branches for each bank. The data for 1996 are given here:

Bank	Deposits (£ billions)	Branches
Abbey National	101.7	867
Barclays	108.2	1997
Lloyds	96.9	2797
National Westminster	113.8	1920
Woolrich	27.5	430
Halifax	77.1	938

(a) Find the simple linear regression model that relates deposits to number of branches.

(b) At the 0.05 level, is the model significant?

(c) What is R^2 for this model? What does it mean? What is the correlation coefficient?

(d) Compare the model for predicting deposits using the number of branches with the one you found for number of customers. Which variable do you think is more strongly related to deposits? Why?

11.32   Look at the model you developed for adolescent brand preferences for cigarettes and the amount of money spent on advertising.

Requires Exercise 11.23

(a) Make a plot of the residuals versus the values of the independent variables.

(b) What does the plot lead you to believe about the appropriateness of the linear model?

(c) Now plot the residuals versus the predicted values of *y*.

(d) How do the graphs in parts (a) and (c) compare? Why do you think this happened?

(e) Make a normal probability plot of the data. Do you think the assumption of normality is reasonable?

(f) Considering your answers to parts (a)–(e), do you think the linear regression model is appropriate for these data? Why or why not?

11.33   Look at the model you developed for shopping centers and retail sales for the South Central states.

Requires Exercise 11.27

(a) Make a plot of the residuals versus the values of the independent variables.

(b) What does the plot lead you to believe about the appropriateness of the linear model?

(c) Now plot the residuals versus the predicted values of *y*.

(d) How do the graphs in parts (a) and (c) compare? Why do you think this happened?

(e) Make a normal probability plot of the data. Do you think the assumption of normality is reasonable?

(f) Considering your answers to parts (a)–(e), do you think the linear regression model is appropriate for these data? Why or why not?

11.34   Consider the model you developed for poverty levels and education.

Requires Exercise 11.26

(a) Make a plot of the residuals versus the values of the independent variables.

(b) What does the plot lead you to believe about the appropriateness of the linear model?

(c) Now plot the residuals versus the predicted values of *y*.

(d) How do the graphs in parts (a) and (c) compare? Why do you think this happened?

(e) Make a normal probability plot of the data. Do you think the assumption of normality is reasonable?

(f) Considering your answers to parts (a)–(e), do you think the linear regression model is appropriate for these data? Why or why not?

Doing It!

11.35   The Children's Defense Fund publishes annual data on the condition of children and their families by state. One philanthropic organization is interested in funding projects that reduce infant mortality. They decide to try to determine which family and household characteristics are most closely related to infant mortality. The data collected are for 13 different variables from 1998. A portion of that data is presented here:

Datafile: CHILDRENS DEFENSE FUND.XXX

Number of grandparents living with own grandchildren under 18	Number of grandparents responsible for raising grandchildren	Percent in poverty	Percent poor under 18	Percent poor 18+	Percent poor 0–4
87,059	50,159	20.5	21.4	14.2	26.3
9,946	5,052	17.2	12.7	7.2	15.4
109,577	49,274	16.1	22.5	13.1	25.3
55,573	32,286	26.4	25.5	14.7	31
866,415	289,546	16	19.9	11.7	21.8
63,871	24,473	9.4	9.8	8.5	11.2

(continued)

State	Per pupil expenditures	Percentage of children in poverty	Percentage of low-birthweight babies	Infant death rate per 1000 births	Percentage immunized	Percentage of early prenatal care	Percentage under 19 without health care
AL	$4849	23.8	9.3	10.2	78.4	82.4	15
AK	8271	16.2	6	5.9	80.1	81.4	15.2
AZ	4595	23.2	6.8	7.5	72.4	75.1	25.4
AR	4708	25	8.9	8.9	77.1	77.8	19.9
CA	5644	24.6	6.2	5.8	75.3	82.4	19.4
CO	5656	14.6	8.6	6.7	75.8	82.2	14.3

(a) Make scatter plots of infant mortality versus each of the other variables.

(b) Which variables do you think are most closely related to infant mortality? Pick the three that you think are best. Do any of these variables have a linear relationship? If so, which ones?

(c) Find the simple linear regression model with infant mortality as the dependent variable and the variables that you think have the best linear relationship as the independent variables.

(d) At the 0.05 level, is there a significant linear relationship between infant mortality and any of these variables? If so, which ones?

(e) Find the value of R^2 for each model. What do the values mean?

(f) For the model you think is best, find the prediction and confidence intervals for each state. Is the model useful for predicting infant mortality? Why or why not?

(g) Calculate the predicted values and the residuals for the model that you think is best. Use the residuals to check the assumptions of the linear regression model.

(h) For the model that you think is the best, are there any outliers? If so, delete them from the model and redo the analysis. What, if anything, changes?

(i) Perform any additional analyses you think might be helpful to the philanthropic organization.

(j) Write a report with your conclusions.

Datafile:
RACING.XXX

11.36 A competitive runner who runs races "almost every Saturday" in a road racing series has kept careful records of his performance for the past 15 years. Data include the year, week in the series, distance of the race, place, pace (minutes per mile), and overall time (in minutes). Here is a portion of the data:

Year	Week	Place	Distance	Pace	Time
2001	1	5	3.55	6.11	21.70
2001	2	4	4.8	6.27	30.08
2001	3	2	3.95	6.07	23.98
2001	4	4	3.95	6.14	24.25
2001	5	4	4.8	6.24	29.93
2001	6	1	4.3	6.36	27.33

(a) Create a scatter plot of pace versus distance for the data as a whole. Does it appear that there is a linear relationship between distance and pace?

(b) Redo part (a) for each year separately. Does this make the relationship between pace and distance better or worse?

(c) Find linear regression models for the data as a whole and for each year separately.

(d) At the 0.05 level of significance, is there a significant relationship between pace and distance overall? For each year separately?

(e) Compare the slopes and intercepts of the models for each of the individual years. What can you learn from these data?

(f) Find all of the 3.95-mile races that he ran over the 15-year period. He was 35 in 1986 when he started recording the data. Create a plot of pace versus age for the 3.95-mile races.

(g) Does there appear to be a linear relationship between pace and age for this distance?

(h) Find the linear regression model for the data. Is there a significant relationship between pace and age?

(i) Do the same thing for the 2.4-mile races. What can you conclude?

(j) Combine the 4.3-, 4.4-, and 4.5-mile races and repeat part (i).

(k) Perform any other analyses that will help you explain to this runner how his performance has been over the past 15 years.

(l) Write the runner a letter with your findings.

12

Multiple Regression Models

WHERE SHOULD WE LIVE?

One of the problems that face city planners and real estate developers is the price of homes in an area. What factors influence the prices of homes? On a microlevel, real estate developers can look at factors related directly to the individual dwelling, such as square footage, number of bedrooms, and number of bathrooms. For large-scale planning purposes, a micro-model is not appropriate. Many other global factors affect the price of a home, such as location, school systems, and access to public transportation.

City planners in Dallas and Fort Worth, Texas, wanted to look at the "commuter" suburbs of the city that have grown up over the years and find a model to help them understand what factors play a role in the prices of homes in these areas. They decided to look at 41 communities regarded as "commuter" suburbs of Dallas/Fort Worth, and they collected data on five different variables that may affect the average selling price of a home: average household income, total instructional expenditures, mean SAT scores, population diversity, and the violent crime rate. Here is a portion of the data:

Suburb	1994 average value of home sold	Average household income	Total instructional expenditures per pupil	Mean SAT score	Population diversity (% nonwhite)	1993/94 violent crime (per 1000)
Addison	$204,800	$49,803	$2,600	980	32.40	4.5
Allen	99,000	60,031	2,185	955	9.30	1.9
Arlington	97,613	50,188	2,154	929	14.70	8.3
Balch Springs	48,300	37,111	1,947	935	24.00	14.6
Bedford	173,073	58,982	2,473	951	10.10	2.3
Carrollton	108,700	58,551	2,600	980	22.10	2.5
Cedar Hill	84,500	55,768	2,174	901	24.10	1.1
Cockrell Hill	58,200	30,434	2,714	775	70.10	8.3
Colleyville	235,427	121,962	2,471	952	4.90	0.4
Coppell	149,500	78,917	2,472	925	13.80	0.5
Desoto	109,300	62,575	2,306	885	27.20	3.1

The planners wanted to find a model that explains the average value of homes in a location based on these variables. They think this model will help them make decisions about current as well as future communities.

12.1 Chapter Objectives

The purposes of regression models are to *predict*, to *explain*, and to *control*. Many real-world phenomena are too complex for the simple linear model that we looked at in Chapter 11. Some variables that we are interested in depend on *more than one* independent variable. When this is the case, we use *multiple regression models* to describe the relationship between the independent variable, Y, and the set of dependent variables, $X_1, X_2, ..., X_k$.

In this chapter you will learn how to do these tasks:

- Find the regression equation for a dependent variable Y as a function of a set of independent variables $X_1, X_2, ..., X_k$.
- Determine whether the relationship is significant.
- Determine which variables contribute to the model and which do not.
- Analyze the results of a regression analysis to determine whether the model is appropriate.

12.2 The Multiple Regression Model

12.2.1 A DESCRIPTION OF THE MULTIPLE REGRESSION MODEL

When two quantitative variables are related in a linear manner, we can use a simple linear regression model to describe the relationship. In many cases, the simple linear model does a good job of predicting or describing the way one variable behaves relative to another. In some cases, though, the simple linear model is not appropriate because the relationship between the variables is not linear or because one or more of the assumptions of the simple linear model is violated. In other cases, the simple linear model may be appropriate and even statistically significant, but the results may not be *useful* for making decisions.

One of the quantities that we look at in simple linear regression is the value of R^2, the coefficient of determination. You remember that R^2 is a measure of the amount of variation in the dependent variable, Y, that can be explained by the linear model involving the independent variable X. Sometimes even when a linear model is significant, however, the independent variable, X, does not do a good enough job of explaining how Y behaves. One possible solution to this problem is to find another, better independent variable to work with, but there may not be a single variable that will do the job. Another solution, the one we look at in this chapter, is to find a *set* of independent variables, known as **input variables,** that together provide a good model for predicting the dependent variable known as the **output variable.**

> The dependent variable, Y, is often referred to as the **output variable,** whereas the independent variables, $X_1, X_2, ..., X_k$, are referred to as the **input variables.**

Models that describe this type of relationship are known as **multiple regression models.**

> The true relationship between the independent variable Y and the set of independent variables $X_1, X_2, ..., X_k$, the **multiple regression model,** can be described by the equation
>
> $$y = \beta_0 + \beta_1 x_1 + \beta_2 x_2 + \cdots + \beta_k x_k + \varepsilon$$

This equation says that the actual value of the variable Y is determined by the equation $\beta_0 + \beta_1 x_1 + \cdots + \beta_k x_k$ plus some random variation, ε, caused by other unmeasured factors.

The coefficients $\beta_1, \beta_2, ..., \beta_k$ are similar to the slope coefficient β_1 in the simple linear model, $y = \beta_0 + \beta_1 x_1 + \varepsilon$, with just a small difference. In the simple linear model, β_1 represents the slope of the line, or the change in the dependent variable, Y, for a unit change in the independent variable, X. In the multiple regression model, the parameters $\beta_1, \beta_2, ..., \beta_k$ are really not the slope because we are not talking about a line, but they have a similar interpretation. Each of the β_i coefficients represents the change that occurs in the independent variable, Y, if the variable associated with the coefficient of interest, X_i, is changed by one unit *and all other variables in the model are held constant.* In the case of only two independent variables, the multiple regression equation has a physical interpretation: The equation is the *plane* that describes the relationship, similar to the line in the one-variable model.

12.2.2 THE LEAST-SQUARES MULTIPLE REGRESSION MODEL

Our objective in the multiple regression model is to find the estimates for the population values $\beta_1, \beta_2, ..., \beta_k$ that do the "best" job of describing the real relationship between Y and the set of X variables that we have chosen.

In the simple linear model, we used the method of least squares to find the equation of the line that minimizes the sum of the squared differences between the actual values of Y and the values predicted by the model. In the simple linear model we have equations to find the values of b_0 and b_1, although most of the analysis in simple linear regression is best done using an appropriate computer software package, such as Minitab or Excel.

Least squares can also be extended to models with more than one independent variable. We want to find the equation

$$\hat{y} = b_0 + b_1 x_1 + b_2 x_2 + \cdots + b_k x_k$$

that minimizes the total squared deviation between the actual data and the values obtained from the regression model. Remember that $\hat{y}$ is the value of the dependent variable as predicted by the model *for a specific set of values of the independent variables*, $X_1 = x_1$, $X_2 = x_2$, ..., $X_k = x_k$. Finding the values of b_0, b_1, b_2, ..., b_k involves solving a system of k simultaneous equations and is too tedious to do by hand. For such an analysis, an appropriate software package is essential for finding the regression model.

EXAMPLE 12.1

TEXAS REAL ESTATE

Understand the problem.

Finding the Multiple Regression Model

The city planners who are looking at the prices of homes in the suburbs of Dallas/Fort Worth, Texas, are trying to find the relationship between the average price of a home (Y) and a set of five independent variables: *average household income* (X_1), *total instructional expenditures* (X_2), *mean SAT scores* (X_3), *population diversity* (*% nonwhite*) (X_4), and *violent crime rate* (per 1000) (X_5). Their equation is

$$\hat{y} = b_0 + b_1 x_1 + b_2 x_2 + b_3 x_3 + b_4 x_4 + b_5 x_5$$

The analysts use Minitab to find the equation of the multiple regression model that best fits the data. The results follow:

Regression Analysis

The regression equation is
1994 Avg. Value _of Home Sold = −140352 + 1.64 Avg. Household Income
  + 39.2 Total Instructional Expenditure + 90 Mean SAT Score
  − 47820 Diversity − 1132 1993/94 Avg. Violent Crime

Analyze the data.

The computer output gives the values of the coefficients as $b_0 = -140{,}352$, $b_1 = 1.64$, $b_2 = 39.2$, $b_3 = 90$, $b_4 = -47{,}820$, and $b_5 = -1132$.

The analysts interpret the coefficients as the change in the average selling price of a home in dollars for a unit change in each of the variables. That is, if the average household income in a location increases by \$1, the average selling price of a home in that location will increase by \$1.64, *as long as all other variables remain constant*. Similarly, if the mean SAT score increases by one point, the average price of a home will increase by \$90. Again, this assumes that nothing else in the model has changed.

From the coefficients the analysts see that for the most part the model makes sense. That is, as income, total instructional expenditures, and SAT scores go up, so do the selling prices of homes. Similarly, as the violent crime rate increases, the price of the homes goes down. The analysts are not pleased about the coefficient of X_4, but they know that it reflects reality.

Draw conclusions.

The output from statistical software for multiple regression models is almost identical to the output for the simple linear model. Figure 12.1 on page 572 shows the output from Excel and Minitab. The sections of the output that contain the coefficients of the multiple regression model are highlighted.

You see that the basic sections of the output are the same: general information about the model, information about the coefficients, and the ANOVA information. We look at the output in more detail as the chapter progresses.

SUMMARY OUTPUT

Regression Statistics	
Multiple R	0.866778185
R Square	0.751304423
Adjusted R Square	0.715776483
Standard Error	37957.7302
Observation	41

ANOVA

	df	SS	MS	F	Significance F
Regression	5	1.52341E+11	30468171569	21.14686162	1.08463E-09
Residual	35	50427624859	1440789282		
Total	40	2.02768E+11			

	Coefficient	StandardError	tStat	P-value	Lower 95%	Upper 95%	Lower 95.0%
Intercept	−140351.6689	112957.8302	−1.242513853	0.222307657	−369668.5358	88965.198	−369668.5358
Avg. Household Income	1.637671454	0.312118144	5.246960128	7.59702E-06	1.004037162	2.271305746	1.004037162
Total Instructional Expenditures per pupil	39.15871129	21.78140875	1.797804345	0.080839566	−5.059953319	83.3773759	−5.059953319
Mean SAT	89.63285763	113.8778516	0.787096493	0.436523067	−141.5517541	320.8174694	−141.5517541
Population Diversity (% non-white)	−47820.09741	62143.05497	−0.769516359	0.446749366	−173977.3601	78337.1653	−173977.3601
1993/94 Avg. Violent Crime (per 1000)	−1132.378569	2146.787097	−0.527475952	0.601191161	−5490.5934	3225.836262	−5490.5934

Excel

Note: The software packages do not highlight these areas.

Regression Analysis

The regression equation is

1994 Avg. Value _of Home Sold = − 140352 + 1.64 Avg. Household_ Income
+ 39.2 Total Instructional_ Expenditur + 90 Mean SAT_ Score
− 47820 Population Diversity_ (% non-white
− 1132 1993/94 Avg. Violent _Crime (per 1000)

Predictor	Coef	StDev	T	P
Constant	−140352	112958	−1.24	0.222
Avg. Hou	1.6377	0.3121	5.25	0.000
Total In	39.16	21.78	1.80	0.081
Mean SAT	89.6	113.9	0.79	0.437
Populati	−47820	62143	−0.77	0.447
1993/94	−1132	2147	−0.53	0.601

S = 37958   R-Sq = 75.1%   R-Sq (adj) = 71.6%

Analysis of Variance

Source	DF	SS	MS	F	P
Regression	5	1.52341E+11	30468171569	21.15	0.000
Error	35	50427624859	1440789282		
Total	40	2.02768E+11			

Source	DF	Seq SS
Avg. Hou	1	1.46103E+11
Total In	1	3261836377
Mean SAT	1	1202680560
Populati	1	1372121063
1993/94	1	400872070

Unusual Observations

Obs	Avg. Hou	1994 Avg	Fit	StDev Fit	Residual	St Resid
1	49803	204800	110273	13150	94527	2.65R
8	30434	58200	42311	28127	15889	0.62 X
21	196697	361100	397689	30411	−36589	−1.61 X
38	84007	269926	178710	9319	91216	2.48R
39	87854	269926	178242	13518	91684	2.58R

Minitab

FIGURE 12.1 **Computer output from Excel and Minitab**

PHILLIPS PETROLEUM

EXAMPLE 12.2

Finding the Multiple Regression Model

Understand
the problem.

Financial analysts would like to understand the relationship among common stockholders' equity, total revenue, and total assets. Data on these variables for Phillips Petroleum from 1986 to 1996 are shown in the table:

Year	Equity, Y (per share)	Revenue, X_1 (billions)	Assets, X_2 (billions)
1986	$7.55	$10.018	$9.186
1987	7.08	10.917	8.772
1988	8.69	11.490	8.417
1989	8.74	12.492	7.832
1990	10.51	13.975	8.301
1991	10.61	13.259	8.298
1992	10.37	12.140	8.489
1993	10.28	12.545	7.961
1994	11.29	12.367	8.042
1995	12.16	13.521	8.493
1996	16.15	15.807	9.120

The analysts decide to use a multiple regression model to find a model with equity as the dependent variable and revenue and assets as the independent variables. A portion of the computer output for their analysis follows:

SUMMARY OUTPUT

Regression Statistics	
Multiple R	0.923228385
R Square	0.852350651
Adjusted R Square	0.815438314
Standard Error	1.067662859
Observation	11

ANOVA

	df	SS	MS	F	Significance F
Regression	2	52.64353179	26.32176589	23.09121324	0.000475254
Residual	8	9.119231848	1.139903981		
Total	10	61.76276364			

	Coefficients	Standard Error	tStat	P-value	Lower 95%
Intercept	−16.71847039	7.179010344	−2.328798761	0.048252087	−33.27330864
Revenue	1.452643014	0.216118353	6.72516208	0.00014936	0.954272876
Assets	1.034291795	0.769247372	1.344550312	0.215646057	−0.739596974

The model is $\hat{y} = -16.72 + 1.45x_1 + 1.03x_2$.

Draw
conclusions.

From this model, the analysts see that an increase in revenue of $1 billion will increase equity by $1.45/share for a fixed value of assets, whereas an increase in assets of $1 billion will increase equity by $1.03/share for a fixed revenue. The constant for the model is −16.72, which indicates that with revenue and assets equal to 0, shareholder equity is −$16.72 billion. This does not make much sense, but it would be extrapolating anyway.

TRY IT NOW!

Order Filling
Finding the Multiple Regression Model

A mail-order catalog company is investigating the time it takes to prepare an order for shipping. In particular, the company is looking for the amount of time that is spent collecting the items ordered and packing them. In this operation, an employee (a checker) is given an order to fill. Items are located in bins in one of six different sections of the warehouse. The checkers move around the warehouse retrieving the items and packing them into the shipping cartons. The company has looked at the operation in some detail and believes that three major variables are involved in the process: the number of items ordered, the number of different locations (sections of the warehouse) in which the items are located, and the experience (in months) of the checker. Data are collected on 45 orders. A portion of the data is shown here:

Time (minutes)	Items	Locations	Experience (months)
9.3	1	6	8
4.4	1	3	13
4.4	5	2	3
5.6	3	3	5
4.9	11	1	4
8.8	14	3	5

The company wants to know how the time it takes to fill an order is related to the other three variables, so it decides to use a multiple regression model. The output from the computer software is given here:

SUMMARY OUTPUT

Regression Statistics	
Multiple R	0.915838993
R Square	0.838761061
Adjusted R Square	0.82696309
Standard Error	1.021396967
Observations	45

ANOVA

	df	SS	MS	F	Significance F
Regression	3	222.5057888	74.16859627	71.09366969	2.71039E-16
Residual	41	42.7733223	1.043251763		
Total	44	265.2791111			

	Coefficients	Standard Error	tStat	P-value	Lower 95%
Intercept	0.215224599	0.55856837	0.385314691	0.701996141	−0.912827729
Number of Items	0.281391801	0.029191616	9.639473395	4.26275E-12	0.222438105
Locations	1.259340956	0103175681	12.20579253	3.10059E-15	1.050973327
Experience	0.019087782	0.047667374	0.400437032	0.690913642	−0.0077178487

Write the equation of the regression model.

Interpret the value of each of the coefficients of the model.

ANS. $\hat{y} = 0.21 + 0.28x_1 + 1.26x_2 + 0.02x_3$; ASSUMING THAT ALL OTHER FACTORS REMAIN CONSTANT, IF YOU INCREASE THE NUMBER OF ITEMS BY 1, THEN THE TIME INCREASES BY 0.28 MINUTE; IF YOU INCREASE THE NUMBER OF LOCATIONS BY 1, THEN THE TIME INCREASES BY 1.26 MINUTES; AND IF YOU INCREASE EXPERIENCE BY 1 MONTH, THEN THE TIME INCREASES BY 0.02 MINUTE.

12.2.3 USING THE MULTIPLE REGRESSION MODEL FOR PREDICTION

Like the simple linear model, the multiple regression equation can be used to predict the value of the dependent variable for a specific set of values of the independent variables. We must ask the same question: How useful are these predictions?

You remember that one way to determine the usefulness of the simple linear model is to look at the *residuals*, the differences between the actual values of the dependent variable and the values predicted by the model for a specific set of input variables. The value of the residual for the *i*th observation is

$$e_i = y_i - \hat{y}_i$$

The size of the residual gives a measure of how good a particular model is at predicting the value of the dependent variable.

EXAMPLE 12.3

TEXAS REAL ESTATE

Using the Model to Make Predictions

The planners who are looking at average selling prices of homes in Dallas/Fort Worth suburbs decide to use the model to predict the selling prices in two of the communities in the data set to see how well the values predicted by the model agree with the data. To decide which locations to pick, the analysts make a boxplot of all the data:

Analyze the data.

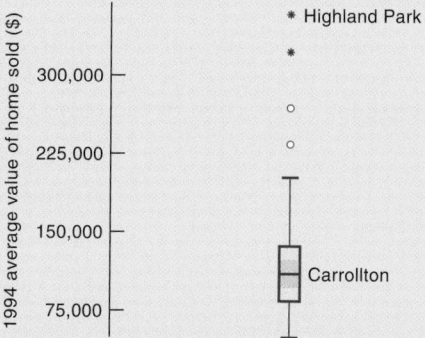

After looking at the boxplot, they choose Highland Park because it appears to be an outlier and Carrollton because it is at the center of the data. They use this regression model

> 1994 average value of home sold = −140,352 + 1.64 average household income
> + 39.2 total instructional expenditures per pupil
> + 90 mean SAT score
> −47,820 population diversity
> −1132 1993/94 average violent crime rate

For Carrollton they calculate

> $\hat{y} = 140,352 + 1.64(58,551) + 39.2(2600) + 90(980) - 47,820(0.221) - 1132(2.5)$
> $= \$132,393$

and for Highland Park

> $\hat{y} = -140,352 + 1.64(196,697) + 39.2(3159) + 90(1070) - 47,820(0.037) - 1132(1.7)$
> $= \$398,670$

They compare these values with the actual data of $361,100 for Highland Park and $108,700 for Carrollton and find the deviations:

Carrollton: $108,700 − $132,393 = −$23,693   % error = ($23,693/$108,700)(100)
                                                            = 21.8%

Highland Park: $361,100 − $398,670 = −$37,570   % error = ($37,570/$361,100)(100)
                                                            = 10.4%

They observe that whereas the actual deviation is greater for Highland Park, the percent error is greater for Carrollton.

As you see in the example, it is not difficult to use the multiple regression model to predict the values of the independent variable. However, as models become more complex, it can get tedious. In fact, if we want to compare the predicted and actual values for every observation in the sample, it will certainly be time-consuming. For this reason, almost all software used for regression analysis allows us to find predicted values of Y either for all of the actual sets of X data or for some specific data of interest as well as the residuals.

EXAMPLE 12.4

PHILLIPS PETROLEUM

Using the Model for Prediction

The analysts who are looking at a model to predict equity using revenue and assets wonder how well the multiple regression model does the job. They analyze the data, and as part of the output they have the software calculate, for each set of input data, the predicted equity and the residual. A portion of the output is shown here:

Residual Output

Observation	Predicted Equity	Residuals
1	7.335111748	0.214888252
2	8.212841015	−1.132841015
3	8.678031875	0.011968125
4	9.528519475	0.788519475
5	12.16787192	−1.657871917
6	11.12467664	−0.514676643
7	9.696718843	0.673281157
8	9.738933196	0.541066804
9	9.564140375	1.725859625
10	11.70695601	0.453043987
11	15.6761989	0.473801102

Analyze the data.

They see that the largest deviation is about $1.73/share, whereas the smallest is $0.01/share.

Another way to look at how well the model predicts the values of the independent variable is to create a scatter plot of the predicted values. If the model is a perfect fit, then we expect the predicted values to be equal to the actual data—that is, $\hat{y} = y$. If we plot the predicted values versus the actual values, the results should form a straight line that intercepts the origin if the fit is good.

EXAMPLE 12.5

PHILLIPS PETROLEUM

Plotting the Predicted Versus the Actual Values

The analysts who hope to predict equity want to look at how well the model predicts equity graphically. They decide to plot the predicted values ($\hat{y}$) versus the actual values (y). They also plot the line $\hat{y} = y$ for reference. They see that the points fall very close to the line. It appears that the model provides a good fit.

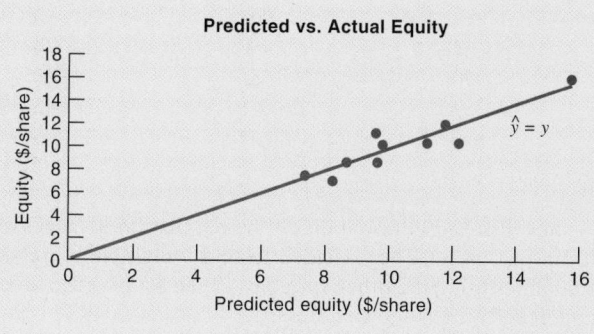

Predicted vs. Actual Equity

Analyze the data.

✓ **TRY IT NOW!**

Order Filling
Finding Predicted Values

The mail-order catalog company that is looking at the time it takes to prepare an order for shipping wants to see how well the model it has found predicts the time to fill an order. Here are the data for six of the observations:

Time (minutes)	Items	Locations	Experience (months)
9.3	6	1	8
4.4	3	1	13
4.4	5	2	3
5.6	3	3	5
4.9	11	1	4
8.8	14	3	5

Use the model to find the predicted time to fill an order for each of the six sets of input data.

Compare the predicted results with the actual data. Does this model do a good job of predicting the time to fill an order? Why or why not?

12.2.4 EXERCISES—LEARNING IT!

Datafile:
TRAFFAT.XXX

12.1 A group of legislators wants to look at factors that affect the number of traffic fatalities. They collected 1994 data from the National Transportation Safety Board on the number of fatalities in 50 states and the District of Columbia, the number of licensed drivers, the number of registered vehicles, and the number of vehicle miles traveled. A portion of the data is shown here:

State	Fatalities	Population	Licensed drivers, X_1 (thousands)	Registered vehicles, X_2 (thousands)	Vehicle miles, X_3 (millions)
AL	1,083	4,219	3,043	3,422	48,956
AK	85	606	443	508	4,150
AZ	903	4,075	2,654	2,980	38,774
AR	610	2,453	1,770	1,560	24,948
CA	4,226	31,431	20,359	23,518	271,943
CO	585	3,656	2,620	3,144	33,705

Source: *Statistical Abstract of the United States 1996*

As you can see from the data table, they also had information on the population of each state, but they did not think that it would prove useful. They decided to find a multiple regression model that predicts the number of fatalities using the other variables as the set of independent variables. The output from the computer software is shown here:

SUMMARY OUTPUT

Regression Statistics	
Multiple R	0.982548538
R Square	0.96540163
Adjusted R Square	0.963193224
Standard Error	154.5407481
Observations	51

ANOVA

	df	SS	MS	F	Significance F
Regression	3	31321046.9	10440348.97	437.1485021	2.54274E-34
Residual	47	1122493.613	23882.84282		
Total	50	32443540.51			

	Coefficients	Standard Error	tStat	P-value	Lower 95%
Intercept	51.7481659	30.43306219	1.700392999	0.095666076	−9.475200509
X Variable 1	0.06294764	0.048829545	1.289130172	0.20366155	−0.035284642
X Variable 2	−0.211896991	0.055989427	−3.784589385	0.000435566	−0.324533083
X Variable 3	0.029349954	0.003525079	8.326041167	8.34285E-11	0.022258416

(a) How many independent variables are in the proposed model? What are they?

(b) Use the computer output to write the regression model.

(c) Interpret the coefficients of the model.

(d) Use the model to predict the number of traffic fatalities for the states shown in the data table.

(e) Compare the values predicted from the model with the actual values. Based on the plot, does the model do a good job of predicting the number of traffic fatalities?

Datafile:
SEATBELT.XXX

12.2 In a study on mandatory seatbelt laws, a group of lobbyists wanted to determine whether there was a relationship among the percentage of motorists who comply with the law, the size of the fine for violation of the law, and the amount of time the law has been in effect. They decided to look at data on 48 states that have mandatory seatbelt laws. A portion of the data they had available is shown here:

State	Percent compliance	Fine	Years in effect
AL	55	$25	5.0
AK	69	15	6.8
AZ	60	10	6.5

AR	51	30	6.0
CA	83	20	11.5
CO	54	15	10.0

Note: ME and NH do not have mandatory seatbelt laws.
Data are not available for Wyoming.
Source: Bureau of Transportation

(a) Which variable do you think is the dependent variable and which are the independent variables? Why?

(b) The output from a multiple regression model is shown here. Does this model agree with your answer to part (a)?

Regression Analysis

The regression equation
Percent Compliance = 0.416 + 0.00187 Fine + 0.0194 Years In Effect

47 cases used 4 cases contain missing values

Predictor	Coef	StDev	T	P
Constant	0.41558	0.05153	8.06	0.000
Fine	0.001874	0.001050	1.78	0.081
Years In	0.019397	0.004672	4.15	0.000

S = 0.09682   R-Sq = 30.8%   R-Sq(adj) = 27.7%

Analysis of Variance

Source	DF	SS	MS	F	P
Regression	2	0.183552	0.091776	9.79	0.000
Error	44	0.412431	0.009373		
Total	46	0.595983			

Source	DF	Seq SS
Fine	1	0.021952
Years In	1	0.161600

(c) Interpret the values of the coefficients in the model.

(d) Use the model to predict the percent compliance with the law for the six states shown in the data table.

(e) Make a plot of the values of $\hat{y}$ versus y for these states.

(f) From the plot, do you think that the model does a good job of predicting compliance with seatbelt laws? Why or why not?

12.3 A committee studying crime in large cities wants to know whether it is possible to find a relationship between the number of serious crimes in the city and several other crime/sociological variables. The committee members decided to focus on the number of police officers in the city, the percentage of arrested men who tested positive for drugs, the number of families in the city below the poverty level, the number of people between 16 and 19 who were not enrolled in school and were not high school graduates, and the number of single-female-parent households. They were focusing on 23 large cities, but all of the data were not available for all of the cities in the study. A portion of the data is shown here:

Datafile:
CRIME.XXX

City	Police officers	Nonenrolled/ not grad	Familes below poverty level	Serious crimes	Single-female-parent households	Male arrests for drugs (%)
Atlanta, GA	1,533	3,247	21,686	76,398	36,577	69
Birmingham, AL	724	1,891	14,075	33,895	22,391	64
Chicago, IL	12,132	27,838	116,645	NA	197,631	69
Cleveland, OH	1,682	5,262	31,340	45,610	45,455	64
Dallas, TX	2,857	10,702	35,085	154,929	54,434	59
Denver, CO	1,361	3,312	14,417	36,558	24,215	60
Detroit, MI	3,954	13,455	71,673	127,080	113,553	58
Fort Lauderdale, FL	463	1,212	4,545	24,334	7,346	64

Source: County and City Data Book 1994

(a) Which variable is the dependent variable? Which are the independent variables?

(b) Look at the set of input variables. Is it reasonable that there is a relationship between the number of serious crimes in a city and these variables? Why or why not? Can you think of other variables that the committee should consider?

(c) Use a computer software package to find the multiple regression model for the data. Interpret the coefficients in the model.

(d) Use the software package to find the predicted value of the number of serious crimes for each of the cities in the study and calculate the residuals. Which city has the best fit? Which has the worst?

(e) Use the model to predict the number of serious crimes for the three cities that were missing those data. How do these predictions fit into the overall data set?

(f) Make a plot of the predicted versus actual data. Overall, how well do you think the model predicts the number of serious crimes in a city?

Datafile:
INTERNET.XXX

12.4 A survey done by a local Internet provider in a medium-size city focused on the amount of time per week that the computer was used for Internet access. Analysts surveyed 35 households and collected data on the number of hours of Internet use per week, the number of computers in the household, the number of children in the household, household income, and the education level of the head of the household (in years of education).

(a) Decide which variable in this study is the dependent variable and which are the independent variables.

(b) Which two independent variables do you think are most likely related to the number of hours of Internet use per week?

(c) Use a computer software package to find the model that predicts hours of use using the two variables that you identified in part (b). Interpret the coefficients of the model.

(d) Use the computer software to find predicted values and residuals for each of the 35 households in the survey. Make a plot of the predicted values versus the actual data.

(e) Based on the plot do you think that the model you found does a good job of predicting the number of hours of Internet use per week? Why or why not?

12.3   Assessing the Multiple Regression Model

In the preceding section we presented the multiple regression model and what it means. In addition to finding the multiple regression model we investigated the *quality* of the model. This investigation, looking at the size of the residuals, is subjective in that no formal statistical test is attached to it. In this section we look at formal, *statistical* methods for assessing the multiple regression model we obtain.

12.3.1   TESTING THE SIGNIFICANCE OF THE MODEL

Once we have the regression equation, we want to know whether that relationship is *statistically significant*. That is, we want to know whether the model, defined by the set of input or independent variables, does a good job of explaining or accounting for the variation in the output or dependent variable.

For the simple linear model, we have two ways to test its significance. The *t* test tests whether the slope coefficient, β_1, is nonzero. If β_1 is not 0, then there is a significant relationship between the dependent and independent variables. The *F* test, the ANOVA test, tests whether the amount of variation in the dependent variable that is explained by the linear model is greater than the amount left unexplained. In multiple regression analysis we test the significance of the *model as a whole*. The null and alternative hypotheses are

$$H_0: \quad \beta_1 = \beta_2 = \cdots = \beta_k = 0$$
$$H_A: \quad \text{At least one of the } \beta_i\text{'s} \neq 0.$$

To do this, we use the F test. Remember that the variation in the y values from the overall mean of Y can be partitioned, or divided, into parts called sums of squares:

SST: the total variation in the y values around the mean $\bar{y}$

SSR: the variation in Y that is caused by Y's relationship with X

SSE: the variation in Y that remains unexplained

You remember from the simple linear model that SST = SSR + SSE and that SST is known as the total sum of squares and represents the distance of the data value y_i from the overall mean $\bar{y}$. SSR is the regression sum of squares and represents the distance of the predicted value $\hat{y}$ from the overall mean $\bar{y}$. SSE is the error sum of squares and represents the distance of a particular data value y_i from the predicted value $\hat{y}$. Here are the formulas for calculating these quantities:

$$\text{SST} = \Sigma(y - \bar{y})^2 \qquad \text{SSR} = \Sigma(\hat{y} - \bar{y})^2 \qquad \text{SSE} = \Sigma(y - \hat{y})^2$$

When the dependent variable, Y, is related to the set of independent variables included in the model, then SSR is large and SSE is small. When Y is not related to the set of X variables, then SSE dominates. The test statistic, F, is calculated from the mean squares, MSR and MSE, which are found by dividing the sums of squares by their respective degrees of freedom:

$$F = \frac{\text{MSR}}{\text{MSE}}$$

F test statistic

How many degrees of freedom does each sum of squares have? The entire model with n observations has $n - 1$ degrees of freedom, so SST has $n - 1$ degrees of freedom. Since the model has k independent variables, SSR has k degrees of freedom, and by subtraction SSE has $(n - 1) - k$ or $n - k - 1$ degrees of freedom. Table 12.1 is a general ANOVA table for a multiple regression model.

TABLE 12.1

ANOVA TABLE FOR THE MULTIPLE REGRESSION MODEL

Source of variation	Degrees of freedom	Sum of squares	Mean square	F value
Regression	k	SSR	$\text{MSR} = \dfrac{\text{SSR}}{k}$	$\dfrac{\text{MSR}}{\text{MSE}}$
Error	$n - k - 1$	SSE	$\text{MSE} = \dfrac{\text{SSE}}{n - k - 1}$	
Total	$n - 1$	SST		

EXAMPLE 12.6

TEXAS REAL ESTATE

Using the ANOVA Table

The city planners who are interested in predicting the average selling price of a home in the Dallas/Fort Worth area used Minitab to perform the regression analysis and to get the multiple regression model. They looked at the residuals to see how well the model was able to predict the price, but now they need to see whether the model is significant. They want to test these hypotheses:

$$H_0: \quad \beta_1 = \beta_2 = \beta_3 = \beta_4 = \beta_5 = 0$$

$$H_A: \quad \text{At least one of the coefficients is not equal to 0.}$$

To do this they look at the output from the Minitab analysis again (page 582).

Analyze the data.

Regression Analysis

The regression equation is
1994 Avg. Value of Home Sold = − 140352 + 1.64 Avg. Household Income
    + 39.2 Total Instructional Expenditure + 90 Mean SAT Score
    − 47820 Population Diversity − 1132 1993/94 Avg. Violent Crime

Predictor	Coef	StDev	T	P
Constant	− 140352	112958	− 1.24	0.222
Avg. Hou	1.6377	0.3121	5.25	0.000
Total In	39.16	21.78	1.80	0.081
Mean SAT	89.6	113.9	0.79	0.437
Populati	− 47820	62143	− 0.77	0.447
1993/94	− 1132	2147	− 0.53	0.601

S = 37958   R-Sq = 75.1%   R-Sq(adj) = 71.6%

Analysis of Variance

Source	DF	SS	MS	F	P
Regression	5	1.52341E+11	30468171569	21.15	0.000
Error	35	50427624859	1440789282		
Total	40	2.02768E+11			

In the section of the output labeled Analysis of Variance, they see that the value of the test statistic, F, is

$$\frac{\text{MSR}}{\text{MSE}} = \frac{30,468,171,569}{1,440,789,282} = 21.15$$

Now they must determine what this means.

In an ANOVA test we are looking at the ratio of the explained variation to the unexplained variation. If it is high enough (if the explained variation is significantly larger than the unexplained part), then we can conclude that the model is significant. The F test is then a one-tail test.

You remember that an F statistic has two parameters, the numerator degrees of freedom and the denominator degrees of freedom. In this case the numerator, MSR, has k (the number of variables in the model) degrees of freedom and the denominator, MSE, has $n - k - 1$ degrees of freedom. For a test at the α level of significance, the F critical value is

F critical value

$$F_{\alpha, k, n-k-1}$$

EXAMPLE 12.7

TEXAS REAL ESTATE

Testing the Significance of the Multiple Regression Model

Analyze the data.

The planners who are looking at the average selling price of houses want to perform the test at the 0.05 level of significance. Since they are looking at 41 locations and their model has 5 independent variables, they need an F critical value with 5 degrees of freedom in the numerator and $41 - 5 - 1 = 35$ degrees of freedom in the denominator. From the tables of the F distribution, they find that the critical value is $F_{0.05,5,35} = 2.4852$. The F test statistic is 21.15, which is definitely outside the critical value. They reject H_0 and conclude that the model is significant.

In the example we showed the calculation for the test statistic and found the critical value using the F distribution tables. When we use the computer to perform the analysis this is not always necessary. Most statistical software packages print out the value of the test statistic. Unless the software asks for the level of significance of the test it will not print out the critical value, but many packages do print out the p value of the test. Remember that you can use the p value to determine whether to reject H_0 by comparing it with the level of significance you choose. If the p value is less than the chosen level of significance, then you reject H_0; if not, then you fail to reject.

EXAMPLE 12.8

PHILLIPS PETROLEUM

Testing the Significance of the Multiple Regression Model

The analysts who are looking at predicting shareholders' equity using total revenue and assets need to know whether their model is significant. They want to test the hypotheses

$$H_0: \quad \beta_1 = \beta_2 = 0$$

$$H_A: \quad \text{At least one coefficient is not equal to 0.}$$

at the 0.05 level of significance. They look again at the output from the Excel computer analysis:

SUMMARY OUTPUT

Regression Statistics

Multiple R	0.923228385
R Square	0.852350651
Adjusted R Square	0.815438314
Standard Error	1.067662859
Observations	11

Analyze the data.

ANOVA

	df	SS	MS	F	Significance F
Regression	2	52.64353179	26.32176589	23.09121324	0.000475254
Residual	8	9.119231848	1.139903981		
Total	10	61.76276364			

	Coefficients	Standard Error	tStat	P-value	Lower 95%
Intercept	−16.71847039	7.179010344	−2.328798761	0.048252087	−33.27330864
Revenue	1.452643014	0.216118353	6.721516208	0.00014936	0.954272876
Assets	1.034291795	0.769247372	1.344550312	0.215646057	−0.739596974

They see that the F statistic is 23.09 and the p value is 0.0005. Since 0.0005 is less than their specified level of significance, 0.05, they reject H_0 and conclude that the model is significant.

TRY IT NOW!

Order Filling
Testing the Significance of the Model

The mail-order company that is looking at factors related to the time to fill an order wants to know whether its model is significant. Write down the hypotheses that it must test.

The Excel output from the company's analysis is shown here:

SUMMARY OUTPUT

Regression Statistics	
Multiple R	0.915838993
R Square	0.838761061
Adjusted R Square	0.82696309
Standard Error	1.021396967
Observations	45

ANOVA

	df	SS	MS	F	Significance F
Regression	3	222.5057888	74.16859627	71.09366969	2.71039E-16
Residual	41	42.7733223	1.043251763		
Total	44	265.2791111			

	Coefficients	Standard Error	tStat	P-value	Lower 95%
Intercept	0.215224599	0.55856837	0.385314691	0.701996141	−0.912827729
Number of Items	0.281391801	0.029191616	9.639473395	4.26275E-12	0.222438105
Locations	1.259340956	0.103175681	12.20579253	3.10059E-15	1.050973327
Assets	0.019087782	0.047667374	0.400437032	0.690913642	−0.077178487

Using the output locate the values for MSR and MSE and their degrees of freedom.

Use the values of MSR and MSE to calculate the value of the F statistic and compare it with the value in the table.

At the 0.05 level of significance, what is the critical value for the test?

What can the mail-order company conclude as a result of its test? Verify your answer by using the p value from the printout.

At this point you should realize that just because the result of a statistical analysis is *statistically significant* does not mean that it will be useful or meaningful for decision making. If you have a lot of data, they might result in significant but essentially useless results.

The results of the significance test for a multiple regression model indicate that the *model* as a whole is significant, specifically that at least one of the coefficients is not 0. The F test tells that the amount of variation in the dependent variable that is explained by the model is significantly greater than the amount left unexplained. These results lead to some questions—in particular, Which coefficients are nonzero? and How much of the variation in the dependent variable is accounted for by the model? We can answer these questions by looking at some additional results of the analysis.

ANS. MSR = 74.17, DF = 3, MSE = 1.04, DF = 41, F = 74.17/1.04. CRITICAL VALUE IS 2.8327. SINCE 71.09 > 2.8327, THE MODEL IS SIGNIFICANT; AT LEAST ONE OF THE COEFFICIENTS IN THE MODEL IS NOT 0. SINCE p IS LESS THAN 0.05, THE COMPANY REJECTS H_0.

12.3.2 THE COEFFICIENT OF MULTIPLE DETERMINATION

In the simple linear regression model you learned that it is sometimes useful to look at the amount of the variation in Y that can be explained by its association with X (SSR) as a percentage of the total variation in Y (SST). This measure is called the coefficient of determination, R^2, which can be interpreted as the amount by which the variation in Y *could be reduced* if the value of X is held constant. This type of interpretation is particularly useful in manufacturing control applications where the objective is to reduce the variation in the output (product) by controlling the variation in the input (raw materials, machine settings).

In multiple regression models there is an analogous measure called the **coefficient of multiple determination, R^2.**

> The *coefficient of multiple determination, R^2,* is a measure of the percentage of the variation in the dependent variable, Y, that can be accounted for by the complete set of independent variables, $X_1, X_2, \ldots, X_k$, in the model.

The coefficient of multiple determination is calculated in the same way as the coefficient of determination from the simple linear model:

$$R^2 = \frac{\text{SSR}}{\text{SST}} \, 100\%$$

Coefficient of multiple determination

EXAMPLE 12.9

TEXAS REAL ESTATE

Looking at the Coefficient of Multiple Determination

The city planners who are looking at the selling prices of homes have found that their regression model with five independent variables is statistically significant. They would like a better idea of how much of the variation in selling price can be accounted for by the set of variables they chose. They look at the output from the model for the value of the coefficient of multiple determination:

Understand the problem.

Regression Analysis

The regression equation is
1994 Avg. Value of Home Sold = −140352 + 1.64 Avg. Household Income
  + 39.2 Total Instructional Expenditure + 90 Mean SAT Score
  − 47820 Population Diversity −1132 1993/94 Avg. Violent Crime

Predictor	Coef	StDev	T	P
Constant	−140352	112958	−1.24	0.222
Avg. Hou	1.6377	0.3121	5.25	0.000
Total In	39.16	21.78	1.80	0.081
Mean SAT	89.6	113.9	0.79	0.437
Populati	−47820	62143	−0.77	0.447
1993/94	−1132	2147	−0.53	0.601

S = 37958   R-Sq = 75.1%   R-Sq(adj) = 71.6%

Draw conclusions.

They see that the value of R^2 is 75.1%, which means that the set of five variables they chose accounts for 75.1% in the variation of the average selling price of a house. The value makes them feel good about the set of variables that they chose, but at the same time they wonder if they could find a better set of variables.

You may be thinking that "more is better" applies to the multiple regression model. Certainly, the more variables we add to the model, the larger SSR will get and therefore R^2 will increase. This is true in an absolute sense—numerically adding variables will never decrease SSE, so that as we add variables, R^2 will increase—but that does not mean the model is better.

One objective of modeling in general, and the multiple regression model in particular, is to create a model that does a good job of predicting the value of the dependent variable, *with as small a set of input variables as possible*. There are several reasons for this. As the number of variables in a model increases, so does the complexity of the model. Complex models are harder to maintain and harder to explain to the people who want to use them. Also, as we increase the number of independent variables in a model, we increase the number of ways in which these variables may interact *with each other*. This results in unstable models—models that are sensitive to small changes in the inputs. For these reasons, more is not necessarily better.

Another measure that is used to evaluate multiple regression models is called the **adjusted R^2.**

> The *adjusted R^2* is the value of the coefficient of multiple determination adjusted to reflect the number of variables in the model.

Although it is true that as we increase the number of variables, SSR will always increase, increasing the number of variables also *decreases* the number of degrees of freedom in the model. It is possible that adding a variable will increase SSR, and thereby decrease SSE, in such a way that the MSE actually increases! This happens because although the value we are dividing into (SSE) decreases, the value we are dividing by (degrees of freedom) decreases even more. The adjusted R^2 may actually decrease as variables are added to the model! We talk about the adjusted R^2 further when we discuss model building.

EXAMPLE 12.10

PHILLIPS PETROLEUM

Looking at the Coefficient of Multiple Determination

The financial analysts who created a model for shareholders' equity want to see how the variables they selected account for the variation in shareholders' equity. They look at the output from their computer analysis:

Analyze the data.

SUMMARY OUTPUT

Regression Statistics	
Multiple R	0.923228385
R Square	0.852350651
Adjusted R Square	0.815438314
Standard Error	1.067662859
Observations	11

They see that the two variables in the model account for about 85% of the variation in shareholders' equity. They notice that the value of the adjusted R^2 is about 82%, but they are not sure how this number should be interpreted.

TRY IT NOW!

Order Filling
Looking at the Coefficient of Multiple Determination

The mail-order company that is looking at factors related to the time to fill an order wants to know whether the set of independent variables that it selected does a good job of accounting for the variation in the times to fill an order. The relevant output from the model is shown here:

SUMMARY OUTPUT

Regression Statistics	
Multiple R	0.915838993
R Square	0.838761061
Adjusted R Square	0.82696309
Standard Error	1.021396967
Observations	45

Using the output, find the coefficient of multiple determination.

If you were the manager of the company, would you be satisfied with this model? Why or why not?

Look at the value of adjusted R^2. What might this value be telling the company?

12.3.3 TESTING THE INDIVIDUAL COEFFICIENTS

One way to test whether the simple linear regression model is significant is to perform a t test on the slope coefficient. This tests the hypotheses

$$H_0: \quad \beta_1 = 0$$
$$H_A: \quad \beta_1 \neq 0$$

In the simple linear model, the t test for β_1 and the ANOVA test are equivalent. In the multiple regression model, the ANOVA tests all of the coefficients simultaneously. The result of rejecting the null hypothesis in this test of significance is that we know that at least one of the coefficients is nonzero. Although this tells us, to some extent, whether we are on the right track, it does not tell us whether all of the variables in the model are useful or necessary to predict the value of the dependent variable. It is certainly possible that in a model with three independent variables, only one of the coefficients is nonzero and that the other two variables are not adding any information to the model. If this were the case, it would be nice to be able to identify which variables contribute to the model and which variables do not.

This process is actually the beginning of a complex process called "model building," which we discuss later in the chapter. Now we simply look at how to do a simple assessment of the value of each of the individual variables in the model.

We can test whether individual coefficients in a multiple regression model are equal to 0 using a t test like the one we used in the simple linear model. The major difference in this test is that we are testing whether the coefficient of the particular variable is equal to 0 *assuming that all other variables remain in the model*. This assumption is critical because it does not allow us to simply toss out all of the variables that do not prove to be significant. To test whether the coefficient of some variable X_i in the model is different from 0, we test the hypotheses

$$H_0: \quad \beta_1 = 0$$
$$H_A: \quad \beta_1 \neq 0$$

using a *t* test. The test statistic is given by

t test statistic for individual coefficients

$$t = \frac{b_i}{s_{b_i}}$$

Most computer packages give the information necessary to perform this test as part of their standard output. It is important that you understand exactly what it tells you so that you do not make errors in formulating your model.

EXAMPLE 12.11

TEXAS REAL ESTATE

Performing *t* Tests for Individual Coefficients

The city planners who are looking at the average selling prices of homes in the Dallas/Fort Worth area are pleased with the results of their model, but they wonder whether all of the variables in the model really contribute to the final result. They decide to look at the computer output again and see whether they can obtain any additional information about the model. The relevant portion of the output is shown here:

Analyze the data.

Regression Analysis

The regression equation is
1994 Avg. Value of Home Sold = −140352 + 1.64 Avg. Household Income
    + 39.2 Total Instructional Expenditure + 90 Mean SAT Score
    − 47820 Population Diversity − 1132 1993/94 Avg. Violent Crime

Predictor	Coef	StDev	T	P
Constant	−140352	112958	−1.24	0.222
Avg. Hou	1.6377	0.3121	5.25	0.000
Total In	39.16	21.78	1.80	0.081
Mean SAT	89.6	113.9	0.79	0.437
Populati	−47820	62143	−0.77	0.447
1993/94	−1132	2147	−0.53	0.601

S = 37958   R-Sq = 75.1%   R-Sq(adj) = 71.6%

The planners want to perform five individual *t* tests to see whether the individual coefficients are equal to 0. They decide to do this at the 0.05 level of significance and look at the *p* values from the output. They see that only one variable, *average household income*, has a *p* value less than 0.05. They reason that this means that the coefficient of X_1 is not 0. Using the same approach, they see that the result of the other four tests is that the coefficients of those variables, β_2 through β_5, are not nonzero.

What does this mean? Can the planners simply drop the other four variables from the model? What would happen if they did?

EXAMPLE 12.12

PHILLIPS PETROLEUM

Looking at Individual Coefficients

Although the model for predicting shareholders' equity had only two variables in it, the analysts still wonder whether both variables are necessary. They decide to look at the relevant output again:

SUMMARY OUTPUT

Regression Statistics	
Multiple R	0.923228385
R Square	0.852350651
Adjusted R Square	0.815438314
Standard Error	1.067662859
Observations	11

ANOVA

	df	SS	MS	F	Significance F
Regression	2	52.64353179	26.32176589	23.09121324	0.000475254
Residual	8	9.119231848	1.139903981		
Total	10	61.76276364			

	Coefficients	Standard Error	tStat	P-value	Lower 95%
Intercept	−16.71847039	7.179010344	−2.328798761	0.048252087	−33.27330864
Revenue	1.452643014	0.216118353	6.721516208	0.00014936	0.954272876
Assets	1.034291795	0.769247372	1.344550312	0.215646057	−0.739596974

They see that the t statistic for *revenue* is 6.72 and the p value is essentially 0, indicating that the coefficient of that variable is nonzero. They also see that the t statistic for *assets* is 1.34 and the p value for that test is 0.22. This leads to the conclusion that the coefficient of the *assets* variable is 0.

Should the analysts drop the variable? At this point they are unsure what to do.

Analyze the data.

What do we learn from testing the individual coefficients in a model? We learn something about the utility of the variables. Clearly, some of the variables are important and should stay in the model, but what about the others? It seems that we cannot throw them all out because of the assumption, but what happens if we do? If we don't throw them all out, which ones should we discard?

EXAMPLE 12.13

TEXAS REAL ESTATE

Adjusting the Model

The city planners who are looking at the real estate model decide to drop all of the variables with low t statistics from the model. They are pretty sure this is not the right thing to do, but they want to see what happens if they take the most drastic approach.

Regression Analysis

The regression equation is
1994 Avg. Value of Home Sold = −1167 + 2.09 Avg. Household Income

(continued)

Analyze the data.

```
Predictor      Coef      StDev        T        P
Constant      −1167      14281      −0.08    0.935
Avg. Hou     2.0894     0.2084      10.03    0.000

S = 38118      R-Sq = 72.1%      R-Sq(adj) = 71.3%

Analysis of Variance

Source        DF            SS              MS          F        P
Regression     1      1.46103E+11      1.46103E+11    100.56    0.000
Error         39       56665134929      1452952178
```

The first thing they notice is that the model itself has changed. The intercept value has decreased from −$140,352 to −$1167. This is quite a change. In addition, the coefficient of the variable *average household income* has increased from 1.64 to 2.09.

It is interesting that the value of R^2 has not changed much—from 75.1% to 72.1%. It certainly appears that the variables that were dropped were not of any use, but the planners are still not convinced they should drop them all.

The answer to the question about dropping variables is not a simple yes or no. The t statistics and/or p values for the individual coefficients give a hierarchy for the variables in the model. With all variables in the model, we can consider the variable with the smallest t value (or largest p value) to be contributing the least to the model and we can eliminate that variable. But the simple approach stops there. If we drop a variable, then *the entire model changes*. It is necessary to rerun the analysis and get a new model and look at those results in the same way we looked at the original model.

EXAMPLE 12.14

PHILLIPS PETROLEUM

Looking at Individual Coefficients

The financial analysts from Phillips Petroleum decide to look at the effect of dropping assets from the model, so they rerun the model as a simple linear model. The results of that analysis are shown here:

Analyze the data.

SUMMARY OUTPUT

Regression Statistics	
Multiple R	0.904978067
R Square	0.818985302
Adjusted R Square	0.798872558
Standard Error	1.114548637
Observations	11

ANOVA

	df	SS	MS	F	Significance F
Regression	1	50.58279565	50.58279565	40.71971956	0.000128074
Residual	9	11.17996798	1.242218665		
Total	10	61.76276364			

	Coefficients	Standard Error	tStat	P-value	Lower 95%
Intercept	−7.795059065	2.857364461	−2.728059081	0.023298434	−14.25887147
Revenue	1.437769522	0.225313329	6.381200479	0.000128074	0.928074973

The first thing they look at is the model itself. Without assets included as a variable, the new model is

$$\hat{y} = -7.80 + 1.44x$$

The intercept term has changed considerably, but the coefficient for the revenue variable is essentially the same.

The analysts also notice that the R^2 value has changed. Without assets in the model, the model accounts for only 81.9% of the variation in shareholders' equity, whereas the two-variable model had an R^2 of 85.2%. The analysts also notice that the new R^2 is close to the value of the adjusted R^2 for the two-variable model. They wonder how to interpret this.

At this point they cannot decide what to do. They like the model with the larger R^2 value, but it seems pointless to leave the variable in if it is not contributing significantly to the model. They decide to try to find a model that has a higher R^2 by looking for other independent variables that might prove to be significant.

You can see that there are no easy answers here. In the two examples you have seen, dropping the variables with low t values had different impacts. In one case the model changed but the value of R^2 did not, whereas in the other case the R^2 value decreased but the model coefficients remained essentially the same. In other cases both the model and the R^2 values can change.

TRY IT NOW!

Order Filling
Testing Individual Regression Coefficients

The mail-order company that is looking at factors related to the time to fill an order wants to know how the individual variables contribute to the model. It decides to look at the computer analysis again:

SUMMARY OUTPUT

Regression Statistics	
Multiple R	0.915838993
R Square	0.838761061
Adjusted R Square	0.82696309
Standard Error	1.021396967
Observations	45

ANOVA

	df	SS	MS	F	Significance F
Regression	3	222.5057888	74.16859627	71.09366969	2.71039E-16
Residual	41	42.7733223	1.043251763		
Total	44	265.2791111			

	Coefficients	Standard Error	tStat	P-value	Lower 95%
Intercept	0.215224599	0.55856837	0.385314691	0.701996141	−0.912827729
Number of Items	0.281391801	0.029191616	9.639473395	4.26275E-12	0.222438105
Locations	1.259340956	0.103175681	12.20579253	3.10059E-15	1.050973327
Experience	0.019087782	0.047667374	0.400437032	0.690913642	−0.077178487

Look at the coefficients of each of the three variables in the model and perform the appropriate hypothesis tests.

Which variables have nonzero coefficients? Which variables have coefficients that are equal to 0?

As a result of these tests, what recommendation would you make?

The questions raised by looking at the individual model coefficients do not have simple answers. They give rise to a group of techniques called model-building techniques, which try to find the "best" model from a set of input variables. By "best" we generally mean the model with the smallest number of variables that explains the largest portion of the variation in the dependent variable.

12.3.4 EXERCISES—LEARNING IT!

Requires Exercise 12.1   **12.5** Consider the model on traffic fatalities by state. The variable X_1 represents the number of licensed drivers, X_2 is the number of registered vehicles, and X_3 is the number of vehicle miles driven annually. The output from the computer analysis is shown here:

SUMMARY OUTPUT

Regression Statistics	
Multiple R	0.982548538
R Square	0.96540163
Adjusted R Square	0.963193224
Standard Error	154.5407481
Observations	51

ANOVA

	df	SS	MS	F	Significance F
Regression	3	31321046.9	10440348.97	437.1485021	2.54274E-34
Residual	47	1122493.613	23882.84282		
Total	50	32443540.51			

	Coefficients	Standard Error	tStat	P-value	Lower 95%
Intercept	51.7481659	30.43306219	1.700392999	0.095666076	−9.475200509
X Variable 1	0.06294764	0.048829545	1.289130172	0.20366155	−0.035284642
X Variable 2	−0.211896991	0.055989427	−3.784589385	0.000435566	−0.324533083
X Variable 3	0.029349954	0.003525079	8.326041167	8.34285E-11	0.022258416

(a) Set up the hypotheses to test whether the model as a whole is significant.

(b) Use the output to test the hypotheses at the 0.05 level of significance.

(c) What is the coefficient of multiple determination for this model?

(d) Does the model do a good job of explaining the variation in the number of traffic fatalities in a state? Why or why not?

(e) Set up the hypotheses to test each regression coefficient individually.

(f) Perform the tests at the 0.05 level of significance.

(g) What can you conclude about the coefficient of variable X_2?

(h) What are your conclusions about X_1? About X_3?

(i) Would you recommend dropping any of these variables from the model? Why or why not?

Requires Exercise 12.2   **12.6** The analysts who are looking at the model on compliance with mandatory seatbelt laws are wondering whether their model is statistically significant. They look at the output from the model:

ANS. FOR NUMBER OF ITEMS, REJECT H_0; FOR LOCATION, REJECT H_0; FOR EXPERIENCE, FAIL TO REJECT H_0; THE COEFFICIENTS OF NUMBER OF ITEMS AND LOCATION ARE NOT 0; THE COEFFICIENT OF EXPERIENCE IS 0; TRY RUNNING THE MODEL WITHOUT EXPERIENCE.

Regression Analysis

The regression equation
Percent Compliance = 0.416 + 0.00187 Fine + 0.0194 Years In Effect

47 cases used 4 cases contain missing values

Predictor	Coef	StDev	T	P
Constant	0.41558	0.05153	8.06	0.000
Fine	0.001874	0.001050	1.78	0.081
Years In	0.019397	0.004672	4.15	0.000

S = 0.09682   R-Sq = 30.8%   R-Sq(adj) = 27.7%

Analysis of Variance

Source	DF	SS	MS	F	P
Regression	2	0.183552	0.091776	9.79	0.000
Error	44	0.412431	0.009373		
Total	46	0.595983			

Source	DF	Seq SS
Fine	1	0.021952
Years In	1	0.161600

(a) Set up the hypotheses to test whether the model is significant.

(b) At the 0.01 level of significance, is the model they tested significant? What does this mean?

(c) What is the value of R^2 for this model? Does the model do a good job of explaining the variation in compliance with the seatbelt laws? Why or why not?

(d) Set up the hypotheses to test for each of the regression coefficients individually and perform the test at the 0.05 level of significance.

(e) What are your conclusions from the tests on individual coefficients? What recommendations would you make about including the variables currently under consideration?

12.7 Consider the committee that is studying serious crimes in some large cities. *Requires Exercise 12.3*

(a) Set up the hypotheses to test whether the multiple regression model is significant. *Datafile:*

(b) Use the output from your analysis to perform the test at the 0.05 level of significance. What is your conclusion? *CRIME.XXX*

(c) What percentage of the variation in the number of serious crimes in a city is accounted for by the set of variables in the model? Based on this number, do you think the set of variables chosen by the committee does a good job of explaining the number of serious crimes in a large city?

(d) Set up the hypotheses to test the individual regression coefficients.

(e) At the 0.01 level of significance, what are your conclusions? What recommendations would you make about the set of variables currently in the model?

12.8 Look at the data on Internet usage again and consider the two-variable model that you *Requires Exercise 12.4* developed.

(a) At the 0.05 level of significance, what can you conclude about the significance of your model? *Datafile:*

(b) What percentage of the variation in hours of Internet usage is explained by the set of variables in your model? Based on this percentage, do you think your model is a good one? Why or why not? *INTERNET.XXX*

(c) Set up the hypotheses and test the individual regression coefficients in your model at the 0.05 level of significance. What are your conclusions?

(d) Based on the results of the tests on individual coefficients, what would you do next? Why?

12.9 In the problem about the number of traffic fatalities the model was rerun, dropping the *Requires Exercises* data on number of licensed drivers, which had the lowest *t* statistic. The output is shown at the *12.1, 12.5* top of page 594.

SUMMARY OUTPUT

Regression Statistics

Multiple R	0.981925801
R Square	0.964178279
Adjusted R Square	0.962685707
Standard Error	155.6025568
Observations	51

ANOVA

	df	SS	MS	F	Significance F
Regression	2	31281357.04	15640678.52	645.9845512	1.99303E-35
Residual	48	1162183.473	24212.15568		
Total	50	32443540.51			

	Coefficients	Standard Error	tStat	P-Value	Lower 95%
Intercept	46.03605608	30.31563375	1.518558262	0.135433408	−14.91757512
X Variable 2	−0.162799336	0.041321977	−3.939776022	0.000263617	−0.245882691
X Variable 3	0.029996183	0.003513227	8.538070709	3.42795E-11	0.02293237

(a) Write the equation of the new two-variable model.

(b) Compare the new model to the model with three variables. How much does the model change when the number of licensed drivers is dropped?

(c) Compare the value of R^2 for both models. What does this make you think about the decision to drop the number of licensed drivers from the model?

(d) Do you consider the two-variable model a good model? Why or why not?

(e) Based on the value of R^2 would you be satisfied with this model or would you want to consider other variables?

Requires Exercises 12.2, 12.6

12.10 The lobbyists who are looking at compliance with seatbelt laws decided to drop the variable associated with the amount of the fine for noncompliance from the model. The output from the modified model is shown here:

Regression Analysis

The regression equation is
Percent Compliance = 0.462 + 0.0188 Years In

48 cases used 3 cases contain missing values

Predictor	Coef	StDev	T	P
Constant	0.46203	0.04443	10.40	0.000
Years In	0.018846	0.004673	4.03	0.000

S = 0.09806   R-Sq = 26.1%   R-Sq(adj) = 24.5%

Analysis of Variance

Source	DF	SS	MS	F	P
Regression	1	0.15642	0.15642	16.27	0.000
Error	46	0.44231	0.00962		
Total	47	0.59873			

(a) Write the equation for the new model and compare it to the model with two independent variables. Do you think that the coefficients have changed significantly? Why or why not?

(b) Test the significance of the new model at the 0.05 level of significance.

(c) Compare the value of R^2 from the new model with R^2 from the old model. What does this make you think about the decision to drop the variable associated with the fine?

(d) If you were among the group of lobbyists would you be satisfied with this model? Why or why not? What other variables might be considered?

Discovery Exercise 12.1

FINDING THE BEST MODEL

The Office of Institutional Planning at a university in the West is interested in understanding what factors influence the graduation rate—that is, the percentage of entering freshmen who actually graduate from the university. The university planners have collected data on 46 universities with traits similar to theirs and have selected three variables that they think are related to the graduation rate: 25th percentile combined SAT scores of accepted students, the acceptance rate (percentage of students who apply that are accepted by the university), and education expenditures per student by the university. A sample of the data is shown here:

School	1995 actual graduation rate (%)	SAT 25th	Acceptance rate (%)	Education expenditures per student
Indiana University—Bloomington	59	NA	80	$9,713
SUNY—Binghamton	74	NA	40	9,080
Allegheny University	52	786	61	33,270
Univ. of California—Riverside	56	910	78	13,403
Oregon State University	53	949	89	10,182
University of Hawaii—Manoa	55	960	65	13,360
New Jersey Inst. of Technology	63	970	67	14,030

The entire data set is in the datafile GRADRATE.XXX.

1. Which variable do you think will have the greatest effect on the graduation rate? Explain why you chose this variable.

2. Find a simple linear model that predicts the graduation rate using the variable that you think is most important. What is the value of R^2 for your model? Is the model significant?

3. Redo part 2 using the other two independent variables, and fill in the table.

Independent variable	R^2	p Value

4. Which one-variable model do you think is the best? Why?

5. How many different models with two independent variables could you find? List them all.

6. Find the multiple regression models for all of the two-variable models, and fill in the table.

Independent variables	R^2	p Value

7. Which two-variable model do you think is best? Why? Does the two-variable model you think is best contain the variable from the best one-variable model? Would you expect it to?

(continued)

8. Find the multiple regression model that predicts the graduation rate using all three independent variables. What is R^2 for this model? Is it significant?

9. Now, fill in the table for each of the "best" models you have found.

Number of independent variables	Independent variables	R^2	p Value
1			
2			
3			

10. If you were going to use a model to predict the graduation rate, which of these three models would you choose? Why?

12.4 Building A Multiple Regression Model

The problem of deciding which variables to include in the multiple regression model is not an easy one. The objective is to find the "best" model, but that is not well defined. Two somewhat conflicting criteria are involved in the definition: The model should do an acceptable job of predicting or explaining the dependent variable, and the model should use the smallest possible set of independent variables.

The task of deciding which variables should be included and excluded from the model must be done in a systematic way. If variables are dropped from a model, then the entire analysis must be redone. To keep the analysis from being chaotic and unproductive, several methods known as **model-building techniques** have been developed.

> *Model-building techniques* are methods used for identifying the best multiple regression model from a set of independent variables. These methods include forward selection, backward elimination, stepwise regression, and all possible regressions.

This section focuses on the issues of choosing the set of independent variables to be considered in a model as well as some of the model-building techniques available in most statistical software packages. We will not go into the technical details of most of the techniques, since they can be complex, but rather we discuss the ideas behind the techniques and how to interpret the results.

12.4.1 CHOOSING A SET OF INDEPENDENT VARIABLES

After reading the examples in this chapter and doing some of the exercises presented so far, you may be wondering how and why the analysts chose the variables they did. In many cases, we might have chosen an entirely different set of variables! The choice of the dependent variable is usually straightforward. The problem to be solved or decision to be made involves the variable in question. The choice for the set of independent variables is not quite so easy. Many factors are involved in the decision.

One major influence on the decision process is the experience and intuition of the "experts" involved. It is difficult, if not impossible, to solve a problem or make a decision if you know nothing about the subject. Often analysts will brainstorm to find a list of variables that might be related to the dependent variable. Once this list is established and prioritized, other factors influence the final choice.

A factor that must be considered in the decision process is whether the data for the study will come from existing data or will be collected especially for this analysis. When the data are to come from existing sources, availability and cost are two important considerations. Sometimes variables that the group considers important might be too expensive to obtain or might not even be available. In this case, similar variables that are less desirable might have to be included. Occasionally, the set of available data is overwhelming and decisions have to be made about which variables to include.

EXAMPLE 12.15

TEXAS REAL ESTATE

Choosing the Set of Input Variables

The city planners who are looking at the selling prices of homes in different locations in the Dallas/Fort Worth area had 56 different variables on their list of possible variables. They were able to obtain data on all the locations of interest for 20 of the 56 variables. They thought that 20 variables were too many to work with, so after some discussion they decided to try to find a set of five that they felt were most important. The question was, How should they choose the five variables?

Of course, everyone in the group had opinions about which variables to include and there was some overlap, but in the end the planners decided on a more objective approach. They knew that the correlation coefficient measures the strength of the relationship between two variables, and so they decided to calculate the correlation coefficient for each of the 20 variables with the average selling price of a home. Then they would pick the five variables with the highest correlations. The correlations are listed here:

Understand the problem.

Average household income	0.84885	Square miles	−0.25142
Percent passing AA	0.63989	Population density	0.20480
Percent taking SAT	0.51699	Property tax rate	−0.20097
Mean SAT score	0.50495	1994 population	−0.19425
Total instructional exp.	0.48006	Distance to Dallas/Fort Worth	−0.12979
Population diversity	−0.44283	Annual insurance	−0.10825
1993/94 violent crime	−0.42840	Percent owner occupied	0.09989
Hazardous waste sites	−0.28200	Median age	−0.09370
Appreciation of sale price	0.27094	Park land	−0.05087
1993/94 nonviolent crime	−0.26457	Annual growth rate	0.01887

They looked at the five variables that were most highly correlated with the average selling price of a home and decided not to use them all. The reason was that four out of the five variables were related to education, and they wanted some diversity in the variable set. They decided to take two of the education-related variables and agreed on mean SAT scores and total instructional expenditures.

When data come from existing sources, it is often tempting to include every bit of data available. This approach can be counterproductive and can lead to overly complex and unstable models. We talk more about these and other problems with models in the next section. If a model from a subset of available data is not acceptable, it is always possible to add more variables to the set under consideration.

When data are going to be collected as part of an analysis, it is wise to include as many possible variables as you can afford. If you collect data by means of a survey or an experiment, then usually the only way to add variables to the study is to repeat the entire data collection procedure. This can be time-consuming and expensive, if not impossible.

TRY IT NOW!

GPA
Choosing the Independent Variables

Many studies have been done on what factors are related to a college student's grade point average (GPA). These studies often focus on precollege factors, such as high school performance, SAT or ACT scores, and general socioeconomic factors such as family income and race. Students know that once they are in college, many other factors influence their GPA. You are going to try to find a model relating current factors to GPA. Make a list of all variables that you think are related to a college student's GPA.

From this list, identify what you think are the five most important variables.

How would you go about gathering the data you need to do your study?

Once the set of variables is chosen and the analysis begins, statistical software is used to actually find the "best" model from the set of variables under consideration.

12.4.2 FORWARD SELECTION AND BACKWARD ELIMINATION

The simplest model-building techniques are forward selection and backward elimination. The techniques are very well described by their names. In forward selection, a model is built by adding variables one at a time, until adding variables does not cause the model to improve. In backward elimination, all variables are included at first and then dropped one by one until the model begins to degrade. There are several different ways to define improvement and degradation of a model. We focus on the most commonly used methods.

Forward selection starts by running all possible one-variable models and selecting the "best" one. There are several equivalent measures of "best": largest R^2, smallest MSE, largest F statistic, and largest t statistic. After the best one-variable model is found, each of the remaining unchosen variables is combined with the current model. Again, the best of these is chosen. Most computer packages define "best" in terms of something called the partial F statistic, which is equivalent to the increase in MSR, the decrease in MSE, or the highest t statistic. The technique continues adding variables until the change in the model is not significant at some specified level.

Backward elimination works the same way as forward selection, except that it starts with the full model and drops the "worst" (using the same criteria) until dropping a variable causes the model to become significantly worse.

EXAMPLE 12.16

TEXAS REAL ESTATE

Using Forward Selection

The city planners decide to use forward selection to find the "best" model for their set of five variables. The computer output from the procedure is enlarged here:

****MULTIPLE REGRESSION****

Listwise Deletion of Missing Data

Equation Number 1   Dependent Variable..   1994 Avg.   Value of Home

Block Number 1.   Method: Forward Criterion PIN .0500   AVG. HOU TOTAL IN MEAN SAT 1993 94 POPULATION

Variable(s) Entered on Step Number 1. AVG.HOU. Avg. Household Income

Multiple R	.84885	Analysis of Variance			
R Square	.72054		DF	Sum of Squares	Mean Square
Adjusted R Square	.71338	Regression	1	146103347775.79370	146103347775.794
Standard Error	38117.60981	Residual	39	56665134929.23070	1452952177.67258

F = 100.55620 Signif F = .0000

-------------------------------Variables not in the Equation-------------------------------

Variable	B	SE B	Beta	T	Sig T
AVG. HOU	2.089442	.208365	.848848	10.28	.0000
(Constant)	−1166.758282	14280.72621		−.082	.9353

----------------------------Variables not in the Equation----------------------------

Variable	Beta In	Partial	Min Toler	T	Sig T
TOTAL IN	.140535	.239924	.814503	1.523	.1359
MEAN SAT	.072320	.115670	.714894	.718	.4772
1993 94	−.066667	−.113157	.805119	−.702	.4869
POPULATI	−.049250	−.081890	.772608	−.507	.6154

Analyze the data.

-------------------------------Variables not in the Equation-------------------------------

Variable	B	SE B	Beta	T	Sig T
AVG. HOU	2.089442	.208365	.848848	10.028	.0000
(Constant)	−1166.758282	14280.72621		0.082	.9353

----------------------------Variables not in the Equation----------------------------

Variable	Beta In	Partial	Min Toler	T	Sig T
TOTAL IN	.140535	.239924	.814503	1.523	.1359
MEAN_SAT	.072320	.115670	.714894	.718	.4772
1993_94_	−.066667	−.113157	.805119	−.702	.4869
POPULATI	−.049250	−.081890	.772608	−.507	.6154

From the output they see that only one variable is included in the final model. They notice that the next variable for consideration is total instructional expenditures, which had the second highest t statistic from the original model.

The model is $\hat{y} = -1167 + 2.09$ (Average household income).

TRY IT NOW!

Order Filling

Using Forward Selection

The mail-order company wants to use a standard method to find the best model from its set of three variables. It decides on forward selection because this method is the easiest to understand. The relevant portions of the computer output are shown here:

Method: Forward Criterion PIN .0500 NUMBER_O LOCATION EXPERIEN

Variable(s) Entered on Step Number 1.. LOCATION Locations

Multiple R	.68766	Analysis of Variance			
R Square	.47287		DF	Sum of Squares	Mean Square
Adjusted R Square	.46061	Regression	1	125.44303	125.44303
Standard Error	1.80333	Residual	43	139.83608	3.25200

F = 38.57409 Signif F = .0000

```
-----------------------Variables in the Equation------------------------

Variable              B          SE B       Beta        T        Sig T

LOCATION       1.116792     .179814     .687657     6.211      .0000
(Constant)     2.650287     .704624                 3.761      .0005
```

Variable(s) Entered on Step Number 2.. NUMBER_O Number of Items

		Analysis of Variance			
Multiple R	.91549				
R Square	.83813		DF	Sum of Squares	Mean Square
Adjusted R Square	.83042	Regression	2	222.33850	111.16925
Standard Error	1.01114	Residual	42	42.94061	1.02240

F = 108.73411   Signif F = .0000

```
-------------------------Variables in the Equation-------------------------

Variable              B          SE B       Beta        T        Sig T

Number_O        .281325     .028898     .610427     9.735      .0000
Location       1.256149     .101834     .773465    12.335      .0000
(Constant)      .338773     .460945                  .735      .4665
```

END PIN = .050 Limits reached.

What variable is added in the first step of the procedure? What is its coefficient? What is the value of R^2 for the first model considered?

What variable is added in the second step? What is its coefficient? What does R^2 change to after this step?

Does the coefficient of the first variable change when the second variable is added? If so, what is the new value?

Write down the final model.

Forward selection and backward elimination are straightforward techniques, but there is no guarantee that they will find the "best" model. In fact, as you can see, many combinations of variables are not even considered as part of the process. For this reason, a technique called stepwise regression, a variation on these methods, is often used.

12.4.3 STEPWISE REGRESSION

Stepwise regression is actually a combination of the forward selection and backward elimination methods. It starts the same way as forward selection, by running all of the one-variable models and selecting the variable with the largest F or t statistic. It then adds variables one at a time to the current model using the same type of criteria as forward selection. The difference occurs when three variables have been included in the model. Stepwise regression includes a routine that considers dropping previously added variables from the model using criteria similar to the ones used by backward elimination. In this way the method considers combinations of variables that were never considered in either of the other methods.

For example, suppose that five variables are being considered for addition in the model and that the first variable chosen is X_2, the second is X_1, and the third is X_5. Up to this point the stepwise procedure is the same as forward selection. However, at this point the stepwise regression procedure considers dropping either X_1 or X_2 from the model, looking at the possible combination of X_1 and X_5 or X_2 and X_5. Although the second of these combinations, X_2 and X_5, was considered in the second iteration, the combination X_1 and X_5 was not. In this way, stepwise regression considers many more combinations of variables than the other two methods.

If this sounds complicated, it is. Many people make errors in using the stepwise regression technique because they do not understand it well enough. The method is very sensitive to the significance levels chosen for adding and dropping variables. Small changes in these values can lead to completely different models being picked as the "best" model. If people are not familiar with the technique or with the software package they are using, they can get easily confused. Minitab requires that the user specify the criteria for adding and dropping variables from the model in terms of an F statistic and defaults to the same value, 4.00 for each. Other packages may allow the user to choose between significance levels or F statistics.

Clearly, the defaults and options are different among the packages and in many cases will produce different results. A powerful mainframe statistical package called SAS defaults to 0.15 for both significance levels. Without knowing what they are doing, users can get very different results!

EXAMPLE 12.17

TEXAS REAL ESTATE

Using Stepwise Regression

Knowing that forward selection does not consider many different combinations of variables, the city planners decided to use stepwise regression to see whether it results in a different model. The output from the Minitab procedure is given here. The parts in **_bold italics_** indicate responses made by the user to questions from the procedure.

Stepwise Regression

F-to-Enter: 4.00   F-to-Remove: 4.00

Response is 1994 Avg on 5 predictors, with N = 41

Step	1
Constant	−1167
Avg. Hou	2.09
T-Value	10.03
S	38118
R-sq	72.05

More? (Yes, No, Subcommand, or Help)
SUBC>**_y_**

No variables entered or removed

More? (Yes, No, Subcommand, or Help)
SUBC>**_no_**

Analyze the data.

The output is not very helpful, but after looking at it, the planners find that one variable, average household income, was added to the model and that its coefficient is 2.09. The R^2 from the model is 72.05.

In this case, the model from stepwise regression is the same as the model from forward selection.

The output from the stepwise regression procedure varies among software packages. The output from Minitab is interactive, requiring the user to make decisions at several points, and very lengthy because it gives the results of each iteration. The output from KADDSTAT is slightly less involved, with only the end results output. The major portion of the output contains information on the variables included in the model and those considered but not included. If we want the additional information available with standard regression output, such as predicted values and residuals, we need to rerun the model with the common regression routine.

✔ TRY IT NOW!

Order Filling
Testing Individual Regression Coefficients

The mail-order company that is looking for a model for order filling decides to use stepwise regression to see whether it finds a model that is different from the one found using forward selection. Here are the Minitab results (with the interactive parts edited out):

Stepwise Regression

F-to-Enter: 4.00   F-to-Remove: 4.00

Response is   Time   on 3 predictors, with N = 45

Step	1
Constant	2.650

| Location | 1.12 |
| T-Value | 6.21 |

| S | 1.80 |
| R-Sq | 47.29 |

| Step | 2 |
| Constant | 0.3388 |

| Location | 1.26 |
| T-Value | 12.34 |

| Number o | 0.281 |
| T-Value | 9.74 |

| S | 1.01 |
| R-Sq | 83.81 |

No variables entered or removed

From the output, how many iterations did the procedure take?

Which variable was added to the model first? What is its coefficient?

Which variable was added second? What is its coefficient?

Did the coefficient for the first variable change? If so, what is its coefficient after the second variable is added?

What is the final model from the stepwise procedure?

ANS. TWO; LOCATION, 1.12; NUMBER OF ITEMS, 0.281; YES; FROM 1.12 TO 1.26; $\hat{y} = 0.3388 + 1.26$LOCATION + 0.281NUMBER OF ITEMS

12.4.4 ALL POSSIBLE REGRESSIONS

One method of finding the "best" regression model that has been largely overlooked in the past is known as all possible regressions or all subsets. Practitioners have hesitated to use this technique because of the number of calculations it requires. In the past, the technique has proved to be too computer-intensive for most situations. With the advent of powerful personal computers, however, this is no longer an issue and the technique is becoming widely used.

The method does exactly what its name states. Given a set of candidate variables, it runs regression models for all possible combinations of variables and then identifies the "best" model for each different size. For example, if five variables are under consideration for a model, the all subsets method runs all models with one variable, all possible combinations of two variables, all three-variable combinations, and so on. It then chooses, according to one or several criteria, the best one-variable model, the best two-variable model, and so on. The user is then free to choose among the models presented based on other decision criteria, such as cost, ease of access to the data, and personal preferences.

The reason this technique is so computer-intensive is the way the number of possible combinations grows with respect to the number of variables. If k variables are under consideration, then there are $2^k - 1$ combinations to consider. Table 12.2 gives you an idea of how large this number gets.

The number of possible models for 5 variables is not unreasonable; even for 10 it is manageable, but doubling the number of variables from 10 to 20 multiplies the result by a factor of 1000!

TABLE 12.2

NUMBER OF SUBSETS

Number of variables	Number of subsets	Result
1	$2^1 - 1$	1
2	$2^2 - 1$	3
3	$2^3 - 1$	7
4	$2^4 - 1$	15
5	$2^5 - 1$	31
⋮	⋮	⋮
10	$2^{10} - 1$	1,023
⋮	⋮	⋮
20	$2^{20} - 1$	1,048,575

The method of all possible regressions usually uses several criteria to define "best" for each level of model. You are already familiar with two of the criteria: R^2 and adjusted R^2. The third criterion is called the C_p **statistic,** which is a measure of the total squared error. The quantity p is the number of *terms* in the model being evaluated, including the intercept:

$$C_p = \frac{\text{SSE}_p}{\text{MSE}_{\text{ALL}}} (n - 2p)$$

C_p statistic

The C_p statistic uses the SSE from the model being evaluated and the MSE from the "full" model, the model that contains all of the potential variables. When we use C_p, the "best model" with p terms results in $C_p = p$. For this model the estimates of the coefficients are unbiased.

EXAMPLE 12.18

TEXAS REAL ESTATE

Using All Possible Regressions

Although the forward selection and stepwise regression procedures both came up with the same "best" model, the planners are not sure they want a one-variable model. They decide to use the all possible regressions method. We show the Minitab output from the procedure. In this case, the output is the best two models for each size considered.

Best Subsets Regression

Response is 1994 Avg

Vars	R-Sq	R-Sq (adj)	C-p	S	A v g . H o u i	P o p 9 1 H a t i	M e 9 9 4 / S A T	T o t a n a 1 S A l n
1	72.1	71.3	2.3	38118	X			
1	25.5	23.6	67.9	62238			X	
2	73.7	72.3	2.1	37488	X			X
2	72.4	71.0	3.8	38357	X		X	
3	74.5	72.5	2.8	37351	X	X		X
3	74.3	72.2	3.2	37561	X		X	X
4	74.9	72.1	4.3	37575	X	X		X X
4	74.7	71.9	4.6	37742	X		X X X
5	75.1	71.6	6.0	37958	X	X X X X

Analyze the data.

The planners found the results of this model helpful. The best one-variable model is the same one they found by the forward and stepwise procedures. From the value of adjusted R^2, they see that in going from a three-variable model to a four-variable model, the value actually decreases from 72.5 to 72.1, indicating that adding the variable mean SAT scores provides no additional information. There is a two-variable model that's the best on all criteria and that according to the C_p statistic is better than the one-variable model. This model has both average household income and total instructional expenditures included.

One benefit of the all possible regressions procedure is that, for reasonably sized sets of input variables, it can be done with any software package that does regression analysis. The evaluation criteria R^2 and adjusted R^2 are included with normal regression output; the exception is C_p, which is often omitted but easy to calculate from the information available. The user simply needs to carefully list and run all of the combinations of independent variables.

✔ TRY IT NOW!

Order Filling
Using All Possible Regressions

The mail-order company is pretty sure that it has identified the best model, but it decides to use the all possible regressions method to make sure it is not missing something. The output from the analysis follows:

Best Subsets Regression

Response is Time

Vars	R-Sq	R-Sq (adj)	C-p	S	Number	Location	Experience
1	47.3	46.1	93.0	1.8033		X	
1	25.2	23.4	149.3	2.1486	X		
2	83.8	83.0	2.2	1.0111	X	X	
2	47.3	44.8	94.9	1.8239		X	X
3	83.9	82.7	4.0	1.0214	X	X	X

Is there a one-variable model that is best on all criteria? If so, what is it?

What is the best two-variable model?

Is there any benefit from moving to a three-variable model? Why or why not?

What model do you recommend the company use? Why?

The relative ease with which all possible regressions can be done has made the forward, backward, and stepwise procedures less popular. These latter techniques are still used, however, to reduce an extremely large set of input variables down to a smaller set so that all subsets can be used.

12.5 Checking Model Adequacy

Just as in the simple linear model, some assumptions in the model must be checked to ensure that the model is appropriate. A model may appear to give good predictions and have a large value of R^2 and still not be a good multiple regression model. In this section we discuss some of the problems that can occur in a multiple regression model.

12.5.1 RESIDUAL ANALYSIS

In the simple linear model some of the assumptions involved the error term. Remember that the multiple regression model is given by

$$y = \beta_0 + \beta_1 x_1 + \beta_2 x_2 + \cdots + \beta_k x_k + \varepsilon$$

and that ε is the error term. The basic assumptions about the error term ε follow:

1. It has a mean value of 0 ($\mu_\varepsilon = 0$).
2. For every value of X, the standard deviation, σ, of ε is the same.
3. The distribution of ε is normal.
4. The error terms for different observations are not correlated with one another.

These assumptions also hold for the multiple regression model. In the simple linear model we used residual plots to tell us whether the assumptions were violated. These same plots help in the multiple regression model.

A plot of the residuals versus the predicted values gives information about whether the relationship is linear and whether the error terms have equal variances. If the plot of the residuals versus the $\hat{y}$ values shows a pattern, then the relationship between Y and the set of predictor variables might be nonlinear. Similarly, if the amount of scatter in the residuals varies as the level of Y varies, then the assumption of equal variances may be violated.

In addition to these plots, in multiple regression models it is useful to plot the residuals against each of the independent variables to obtain additional information. If any problems are indicated by the plot of the residuals versus the predicted values, we have information about a problem with the model as a whole. Plots against the individual variables can pinpoint whether the problem exists with all or just specific variables. If the problem exists with only one or two variables, the analyst can look at alternative models that do not contain these variables. Although the residual plots can be made by any package that creates scatter plots, Excel and Minitab include the plots as part of the output options in their regression routines.

To decide whether the assumption that the residuals are normally distributed is valid, we use a normal probability plot or a histogram, just as we did in the simple linear model.

EXAMPLE 12.19

TEXAS REAL ESTATE

Looking at Residual Plots

The city planners have decided to use the two-variable model identified by all possible regressions. They plot the residuals from the model against each of the independent variables:

Analyze the data.

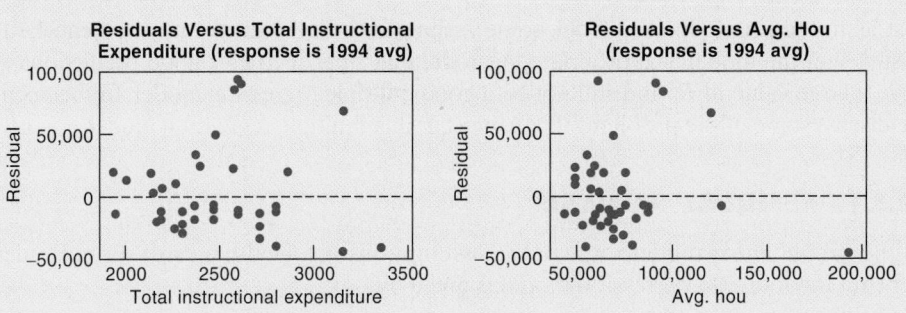

Both plots exhibit more variation on the high side than on the low side. In the plot against the total instructional expenditures, it appears that the variance of the residuals increases as the variable increases. In the second plot against average household income, the problem is not so severe, although it is more difficult to evaluate because there are some isolated values at the high end of the variable range.

A normal plot of the residuals was also done, as shown here:

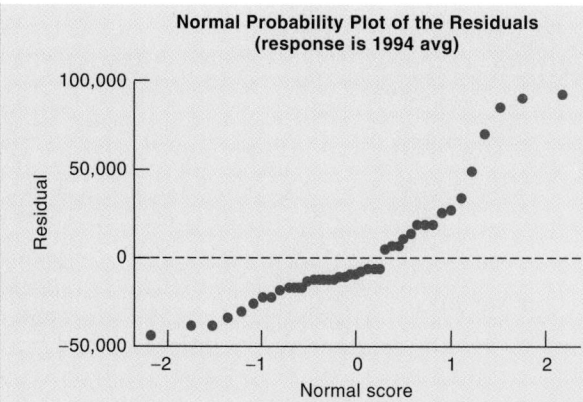

Normal Probability Plot of the Residuals
(response is 1994 avg)

It appears that the normality assumption is not valid because the points do not fall in a straight line.

From the diagnostic plots, it appears that the planners' "best" model violates some of the regression model assumptions.

Draw conclusions.

At this point you might be thinking that you would be really upset if, after doing all that work and finding the "best" model, there is a problem with it. You might wonder why you didn't check these things before all that work. In fact, modeling is not a sequential procedure. It is an iterative procedure from which a solution gradually evolves. Checking earlier would involve a lot more checks than might be necessary with the final, candidate model.

✓ TRY IT NOW!

Order Filling
Using All Possible Regressions

Now that it has a model it likes, the mail-order company wants to make sure that the model does not violate any assumptions of the multiple regression model. The company prints out a set of residual plots:

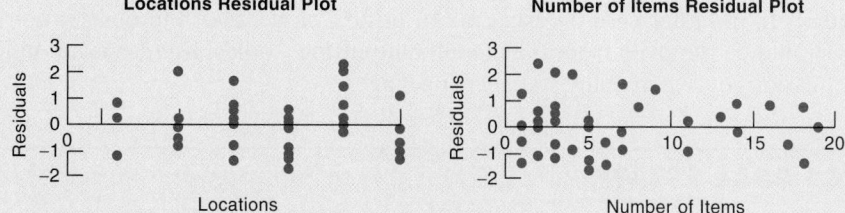

Look at the plot of residuals versus the locations. Does it appear that there is a problem with the assumption of equal variances?

Look at the plot of residuals versus the number of items. Does it appear that there is a problem with the assumption of equal variances?

The company also created a normal probability plot of the residuals:

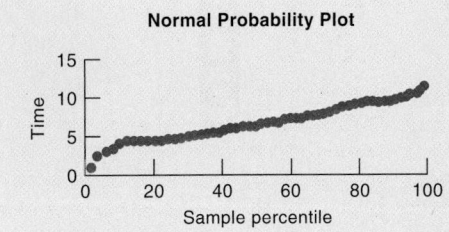

From this plot, what can you say about the assumption of normality?

12.5.2 VALUES WITH LARGE INFLUENCE

Another problem that we encountered in the simple linear model was data points that exert a large influence on the model. This can happen when the value of the dependent variable does not agree with the model or when an observation has an X value distant from the other data points. Since there was only one independent variable, we could use plots such as scatter plots and boxplots to help identify these observations.

The problem of observations exerting a large influence can also happen in the multiple regression model, but it is more difficult to detect. The influence might come from a combination of values of the independent variables, and we lose the element of visualization that we had in the one-variable model. In multiple regression it is necessary to rely on other measures to identify observations with a large influence. As we saw with the linear model, as part of the output Minitab generates a warning about observations that exert a large influence on the model either because of a large residual or because of X values that are significantly larger.

One method that has emerged as a way to detect influential values uses a statistic called **Cook's distance, D.**

> The **Cook's distance** method compares the values of the regression coefficients with all observation to the values when the ith observation is removed from the model.

If the ith observation is influencing the model, then the difference in coefficients between the two models is large and the corresponding value of Cook's distance is also large. A suggested definition of "large" is to compare D with $F_{0.5, k+1, n-k-1}$, an F statistic of a tail area of 0.50 and with $k + 1$ degrees of freedom in the numerator and $n - k - 1$ in the denominator. A rougher rule of thumb is to compare Cook's distance with 1. If, in either case, D exceeds the critical value, then the observation in question should be removed from the model. Outputting Cook's distance is an option of the regression routine in many statistical packages.

EXAMPLE 12.20

TEXAS REAL ESTATE

Looking at Residual Plots

Looking at their data, the city planners wonder whether one of the locations might be exerting a large influence on the model. In particular, they remember that when they made a boxplot of average home values, Highland Park was an outlier. They decide to look at the regression model again and to calculate Cook's distance for their model and see whether they should eliminate those data. They specify for Minitab to calculate Cook's distances and also to output warnings about the data. The warning output is shown here:

Unusual Observations

Obs	Avg. Hou	1994 Avg	Fit	Stdev Fit	Residual	St Resid
1	49803	204800	108753	7458	96047	2.61R
21	196697	361100	410877	28165	−49777	−2.01RX
25	42483	71800	117684	20993	−45884	−1.48X
38	84007	269926	175521	7392	94405	2.57R
39	87854	269926	181703	7868	88223	2.41R
40	115368	324990	253062	15077	71928	2.10R

R denotes an observation with a large standardized residual
X denotes an observation whose X value gives it large influence

The planners notice that observation 21 has both a large residual and a large X value influence. They find that it is indeed Highland Park. Minitab outputs Cook's distance for Highland Park to be 1.749 and compares it with $F_{0.5,3,38} = 0.803$. Since Cook's distance is so much larger than the F value, the planners have to consider redoing the analysis without that bit of data. None of the other Cook's distances exceeds the critical value.

Analyze the data.

12.5.3 MULTICOLLINEARITY

One problem that occurs in the multiple regression model that is not a factor in the simple linear model is known as **multicollinearity.**

> A multiple regression model has *multicollinearity* when variables in the set of independent variables are correlated with each other.

It is not unusual to find that some of the variables in the set of independent variables are related to each other. This is especially likely to happen when the data used for a model come from existing sources. When data come from experimentation, it is possible to control elements that might cause multicollinearity. When data come from existing sources, this is not easy.

Suppose we are trying to find a model to predict the amount of money that a family spends on food, and your set of independent variables includes family income, house value, number of children, family size, and level of education of the head of the household. It seems almost certain that some of these variables are related to each other. It is also possible that variables in the set are related to other variables that are not included in the model. When this happens, the results of a multiple regression analysis may be affected.

The main effect of multicollinearity on the model is instability of the regression coefficients. When two variables in the set of independent variables are highly correlated, then the coefficients of the model are not stable. This means that when we remove one of the variables and rerun the model, the coefficients will change—sometimes even changing sign from negative to positive! If this is the case, then interpretation of the coefficients is not meaningful. We know that the coefficient in the model represents the change in the dependent variable when the variable of interest is changed *and all other variables are held constant*. If two variables are highly correlated, then it may not be possible to vary one variable without also having the other variables change.

Another problem that occurs when multicollinearity is present in a model is that the individual coefficients have large standard errors and therefore small t statistics. It is possible that a model might test significant as a whole but that no individual coefficient is significantly different from 0. This is certainly a contradiction and an indication that something is wrong!

Several complex statistical procedures are used to find out whether a model has multicollinearity, but some simple procedures work well too. One method for detecting multicollinearity is to calculate the correlations between every pair of independent variables and look for large values near 1 or −1. If it appears that some variables are highly correlated, then the analyst may wish to concentrate on models that do not include both variables. Some statistical packages, such as Minitab, warn the user when variables are highly correlated and, in cases where the problem is severe, refuse to include a variable in the model.

EXAMPLE 12.21

Analyze the data.

TEXAS REAL ESTATE

Checking for Multicollinearity

The city planners know that their data came from existing sources and so might exhibit multicollinearity. They decide to run correlations for the set of five variables that they are considering:

Correlations (Pearson)

	Avg. Hou	Populati	1993/94	Mean SAT
Populati	−0.477			
1993/94	−0.441	0.436		
Mean SAT	0.534	−0.400	−0.208	
Total In	0.431	0.073	−0.163	0.138

From the correlation matrix the planners see that the highest correlation is 0.534, which is not high at all. The correlation between the two variables used in the model is 0.434. Hence, they know that they do not have to worry about multicollinearity.

A way to examine the effects of multicollinearity in a model is to run the model with and without one of the variables. If the coefficients change to an unacceptable degree—for example, a change in sign from positive to negative—then another analysis method should be used. One method that produces coefficients that are not affected by multicollinearity is called ridge regression.

✔ ## TRY IT NOW!

Order Filling
Checking for Multicollinearity

The mail-order company wants to check whether the model it is thinking about using has any problems with multicollinearity. It calculates the correlation between the two variables that are in the final model and gets −0.141. Based on this information do you think multicollinearity is a problem in their model? Why or why not?

Although multicollinearity causes big problems with the regression coefficients, the overall ability of the model to predict values of the dependent variable is not affected. If the users of the model are not looking at the actual equation but only at the predictors, the model may still be useful.

12.5.4 USING THE MODEL

We have discussed some of the problems that can occur in a multiple regression model. Although none of the problems make the model unusable, we must understand their impact on the decision-making process. The biggest problem in terms of the usefulness of the model is instability of the regression coefficients. When the coefficients are unstable, they really do not represent the change in the dependent variable that results from a corresponding change in one of the independent variables. If we try to use the model to bring about such a change, the result may not even resemble what we expect.

Problems with coefficient instability or error result from multicollinearity, overspecified models, and influential observations. Unless the multicollinearity is severe (variables correlated with a correlation coefficient of 0.9999 and above), instability is not much of a problem.

Overspecification of a model occurs when the "more is better" approach is taken to modeling. If we add enough variables to a model, we can get an almost perfect fit to the set of input data. The problem is that the perfect fit is just for that particular set of data. If we vary the data set at all, the model varies wildly, which makes it almost useless for "what if" analysis and scenario management. Overspecification can be avoided by using one of the techniques designed to find the "best" regression model.

Observations that exert a large influence on the model also cause the coefficients of the model to be inaccurate. Often dropping the observations that are erroneous will solve the problem.

It is useful to check the stability of the model. One simple way to do this is to run the regression with the selected independent variables on subsets of the original data. If we have a large enough data set, then we can divide it into two randomly chosen subsets and run the model on both subsets. If the resulting models are not similar, then the model may have problems with stability. If we do not have a large enough data set to run two separate analyses, then we can choose two or more random samples of a fixed size from the set of data and run the regression on each subset. Again, if the models do not agree, then there is a problem we must be aware of before using the model to make decisions.

GET IT IN WRITING

Housing Prices

TO: **City Councils, Cities of Dallas and Fort Worth**
FROM: **City Planners**
RE: **Housing Price Study**

We recently conducted a study to try to determine what factors, if any, can be used to predict the selling prices of houses in the commuter suburbs of Dallas/Fort Worth. In this study we considered a macromodel, concentrating on characteristics of the towns rather than characteristics of the houses themselves.

After some deliberation we decided to use five different variables in our model: average household income, total instructional expenditures per pupil, mean SAT score, population diversity (% nonwhite), and number of violent crimes (per 1000 population). Many other variables might be considered, but we wanted to keep the analysis reasonable.

Using multiple regression models, we found two models that might be considered "best." Each was statistically significant and no one model was clearly better than the other. The result of the first model is

Average selling price = −1167 + 2.09(Average household income)

This says that, for every $1000 increase in household income, the selling price will increase by $2090. The intercept term is essentially 0, which makes sense. This model was considered best using two different methods of analysis.

We also looked at a model that considered an additional variable, total instructional expenditures per pupil. The result of this model is

Average selling price = −67,346 + 1.94(Average household income)
+ 30.6 (Total instructional expenditures per pupil)

From a statistical perspective the two models are not very different. The coefficients in the one-variable model make more sense, but that model does not tell us anything about what characteristics of the town are related to selling prices. In the two-variable model, the intercept is very large and might be misleading. It is our conclusion that neither model is satisfactory.

We would like to continue the analysis by considering a different set of variables. In particular we would like to replace average household income with town characteristics to see whether we can find a model that will help us in our planning. We will keep you informed of our progress.

12.6 Using Technology for Multiple Regression Models

The basic tools for finding and testing the multiple regression models in Excel and Minitab are the same as those for the simple linear regression model.

In **Excel,** the only difference is that the data range for the X variables covers more than one column. It is *very* important, however, that the X variable columns be adjacent to one another. If they are not, you have to rearrange the worksheet before you do the analysis.

The output section includes information about each of the regression coefficients. If you choose the residual plots option, you will get a plot of residuals versus each independent variable. This is also true for the line fit plots. These plots allow you to look at the assumptions of the multiple regression model and decide whether the model is appropriate.

Excel does not have any tools for finding the "best" regression model. These are sophisticated statistical tools, and Excel is not really a statistics software package. If you want to find the best model using Excel, you will have to use the all possible regressions approach.

KADDSTAT makes it easier to do multiple regression using Excel because it does not require that the data be in adjacent columns, unlike the Data Analysis Tools routine. Using KADDSTAT for multiple regression is almost identical to using it for simple linear regression. The only difference is that you will select more than one variable from the list of independent variables. To accomplish this, press the control (Ctrl) key while you click on each variable that you want to include in the analysis. If you want to include the entire list, you can just click on the **Select All** button.

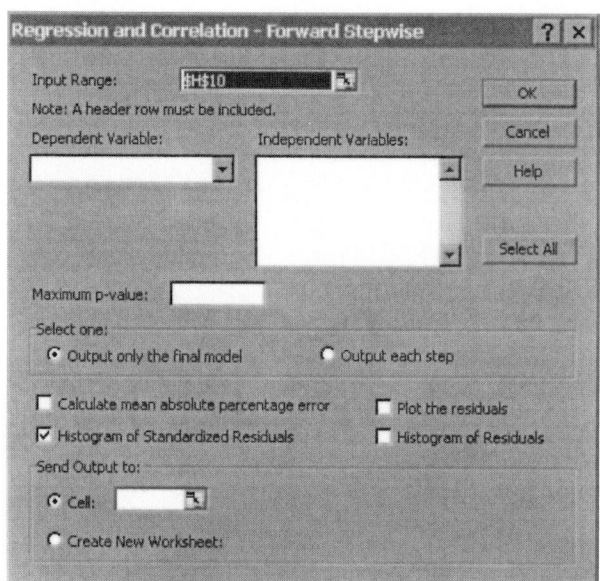

FIGURE 12.2 **Dialog box for forward stepwise regression**

The basic tools for finding and testing the multiple regression model in **Minitab** are the same as those for the simple linear regression model. Minitab also has tools that allow you to do stepwise regression and best subset analysis.

12.6.1 STEPWISE REGRESSION WITH KADDSTAT

KADDSTAT has a routine that will allow you to perform stepwise regression, both forward and backward, in Excel. From the **KADD** menu select **Regression and Correlation > Forward Stepwise.** The dialog box shown in Figure 12.2 opens. The dialog box is identical to the one for Single/Multiple regression with two additional inputs.

Just above the area where you indicate which graphical output you want is a section labeled **Select one:.** This allows you to indicate whether you want to have only the final model output or whether you want each step output. In addition, you need to specify the p value to be used for adding and dropping variables from the model. KADDSTAT uses the same p value for both.

12.6.2 STEPWISE REGRESSION IN MINITAB

Suppose you want to use stepwise regression to find the "best" model for the Texas real estate data. Follow this procedure:

1. From the main menu select **Stat > Regression > Stepwise.** The dialog box shown in Figure 12.3 on page 614 opens.

2. Position the cursor in the box labeled **Response:** and select the Y variable from the list at the left.

3. Position the cursor in the box labeled **Predictors:** and select all the potential X variables from the list at the left. If, for some reason, there are any independent variables that *must* appear in the final model, enter them under **Predictors to include in every model:.**

4. Click **OK.** The output shown in Figure 12.4 appears in the Session window.

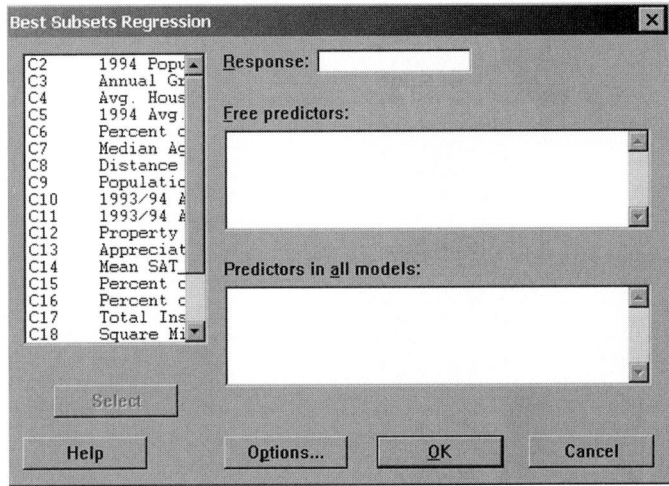

FIGURE 12.3 **Stepwise Regression dialog box**

Stepwise.Regression: 1994 Avg. Va versus Avg. Househo, Total Instru, . . .

Alpha-to-Enter: 0.15 Alpha-to-Remove: 0.15
Response is 1994 Avg on 5 predictors, with N = 41

Step	1	2
Constant	−1167	−67346
Avg. Hou	2.09	1.94
T-Value	10.03	8.55
P-Value	0.000	0.000
Total In	31	
T-Value	1.52	
P-Value	0.136	
S	38118	37488
R-Sq	72.05	73.66
R-Sq(adj)	71.34	72.28
C-p	2.3	2.1

More? (Yes, No, Subcommand, or Help)
SUBC > Yes

No variables entered or removed

More? (Yes, No, Subcommand, or Help)
SUBC > No

Note: The stepwise regression routine is interactive. You must reply No to the last SUBC> in order to keep using Minitab for other analyses.

FIGURE 12.4 **Output from stepwise regression**

Note: Older versions of Minitab specify the F value, not the alpha value.

From the output you can see that first variable to enter was average household income and the second was total instructional expenditures per pupil. The default values for *F* to enter and remove variables from the model are 0.15. You can change these by clicking on the **Methods** button in the Stepwise dialog box. You can also specify Forward Selection or Backward Elimination instead of the Stepwise method.

12.6.3 ALL POSSIBLE REGRESSIONS IN MINITAB

You can also use Minitab to do the all possible regressions analysis. From the menu select **Stat** > **Regression** > **Best Subsets.** The dialog box shown in Figure 12.5 opens.

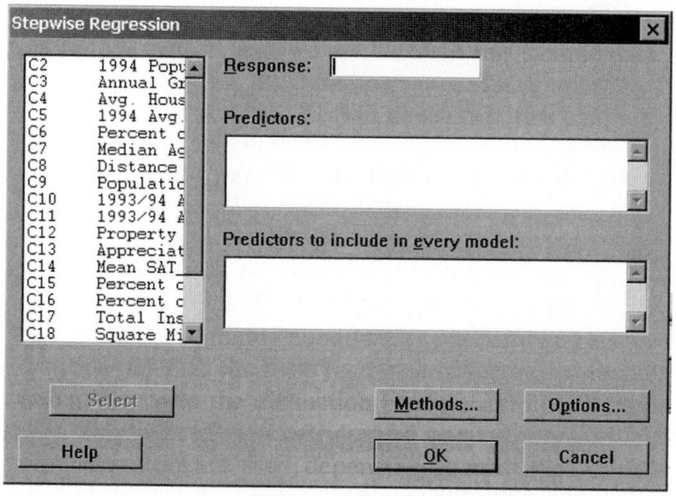

FIGURE 12.5 **Best Subsets Regression dialog box**

You can see that the input is identical to that for stepwise regression. After the variables are entered, the output appears in the Session window as shown in Figure 12.6.

Best Subsets Regression: 1994 Avg. Va versus Avg. Househo. Total Instru. . . .

Response is 1994 Avg

					A	T	M	P	
					v	o	e	o	1
					g	t	a	p	9
					.	a	n	u	9
					1			1	3
					H		S	a	/
					o	l	A	t	9
Vars	R-Sq	R-Sq(adj)	C-p	S	u	n	T	i	4
1	72.1	71.3	2.3	38118	X				
1	25.5	23.6	67.9	62238			X		
2	73.7	72.3	2.1	37488	X	X			
2	72.4	71.0	3.8	38357	X		X		
3	74.5	72.5	2.8	37351	X	X		X	
3	74.3	72.2	3.2	37561	X	X	X		
4	74.9	72.1	4.3	37575	X	X	X	X	
4	74.7	71.9	4.6	37742	X	X	X		X
5	75.1	71.6	6.0	37958	X	X	X	X	X

FIGURE 12.6 **Output from best subsets**

The output includes R^2, the adjusted R^2, and the C_p statistic. The default has Minitab output the best two models of each size. If you want to see more than two, then you can change this using the **Options** button.

12.6.4 MULTIPLE REGRESSION ANALYSIS IN MINITAB

In addition to the output that you learned about in Chapter 11, you can have Minitab calculate and output other diagnostics for the model. To obtain the Cook's distance values as part of the output, click on **Storage** from the Regression dialog box and click on the **Cook's Distance** box.

CHAPTER 12 **SUMMARY**

Finding a multiple regression model is easy. Finding a multiple regression model that is useful for decision making is an art. Modeling is an iterative process that has no single correct answer. The answer you choose depends on what you know and understand about the problem to be solved and how you intend to use the model.

There are several steps to the modeling process: Identifying potential independent variables, collecting data, and finding a potential model are the beginning steps. Once a potential model is found, the process of *model building* takes place to find a "best" model. The objective of this process is to find a model that does an acceptable job of explaining or predicting the dependent variable with as few independent variables as possible. Some of the model-building techniques used are *forward selection, backward elimination, stepwise regression*, and *all possible regressions*. Finally, once the "best" model is identified, it must be checked for problems such as *violation of assumptions, influential observations*, and *multicollinearity* before it can be used for decision making.

KEY TERMS		
Term	**Definition**	**Page reference**
Adjusted R^2	The **adjusted R^2** is the value of the coefficient of multiple determination adjusted to reflect the number of variables in the model.	586
Coefficient of multiple determination, R^2	The **coefficient of multiple determination, R^2**, is a measure of the percentage of the variation in the dependent variable, Y, that can be accounted for by the complete set of independent variables, $X_1, X_2, ..., X_k$ in the model.	585
Cook's distance	The **Cook's distance** method compares the values of the regression coefficients with all observations to the values when the ith observation is removed from the model.	608
Input variable	The set of independent variables, $X_1, X_2, ..., X_k$, is referred to as the **input variables.**	570
Model-building techniques	**Model-building techniques** are methods used for identifying the best multiple regression model from a set of independent variables. These methods include forward selection, backward elimination, stepwise regression, and all possible regressions.	596
Multicollinearity	A multiple regression model has **multicollinearity** when variables in the set of independent variables are correlated with each other.	609
Multiple regression model	The true relationship between the independent variable Y and the set of independent variables $X_1, X_2, ..., X_k$, the **multiple regression model,** can be described by the equation $$y = \beta_0 + \beta_1 x_1 + \beta_2 x_2 + \cdots + \beta_k x_k + \varepsilon$$	570
Output variable	The dependent variable, Y, is often referred to as the **output variable.**	570

KEY FORMULAS		
Term	**Formula**	**Page reference**
F **test statistic**	$F = \dfrac{MSR}{MSE}$	581
F **critical value**	$F_{\alpha,k,n-k-1}$	582
R²	$R^2 = \dfrac{SSR}{SST} \, 100\%$	585
t **test statistic**	$t = \dfrac{b_i}{s_{b_i}}$	588

CHAPTER 12 EXERCISES

Learning It!

12.11 The financial analysts who are looking at the model for shareholders' equity at Phillips Petroleum decided that the model they had was not adequate for their needs. They decide to try the model with a different set of independent variables: debt and number of employees. The data and output of the regression analysis are given here:

Year	Equity (per share)	Debt (billions)	Employees (thousands)
1986	$7.55	$6.2	21.8
1987	7.08	5.8	22.5
1988	8.69	4.9	21.0
1989	8.74	4.0	21.8
1990	10.51	3.9	22.4
1991	10.61	4.0	22.7
1992	10.37	3.8	21.4
1993	10.28	3.2	19.4
1994	11.29	3.1	18.4
1995	12.16	3.1	17.4
1996	16.15	3.0	17.2

SUMMARY OUTPUT

Regression Statistics	
Multiple R	0.848198989
R Square	0.719441525
Adjusted R Square	0.649301907
Standard Error	1.471736507
Observations	11

ANOVA

	df	SS	MS	F	Significance F
Regression	2	44.43469688	22.21734844	10.2572774	0.006195745
Residual	8	17.32806676	2.166008345		
Total	10	61.76276364			

	Coefficients	Standard Error	tStat	P-value	Lower 95%
Intercept	25.01669053	4.991469246	5.011889145	0.001037315	13.50633437
Debt	−1.187745989	0.571806762	−2.077180733	0.07142751	−2.506335599
Employees	−0.47922578	0.303483518	−1.579083382	0.152970523	−1.179060481

(a) Write the equation of the regression model and interpret the coefficients.

(b) Use the regression equation to predict shareholders' equity for each year from 1986 through 1996.

(c) Calculate the residuals and compare the actual data with the predicted values. Which year has the largest residual? Which has the smallest?

(d) Does the model do a good job of predicting shareholders' equity? Why or why not?

(e) Set up the hypotheses to test the significance of the overall model.

(f) At the 0.05 level of significance, what is your conclusion about the model?

(g) What is the value of R^2 for this model? What does it mean?

(h) Compare this model with the model using revenue and assets developed in Example 12.2. Which model do you think is better? Why?

Datafile:
SALES.XXX

12.12 Analysts at a company that produces small appliances are looking at sales of 24 food preparation products in a medium-size city in the Midwest. They have noticed that sales in this city have not been meeting forecast values for several months, and they want to look at the problem in more detail. They have collected data on monthly sales, advertising expenditures, number of competing products available, number of discount opportunities (sales, coupons, etc.) offered during the month, and the warranty period of the item. A portion of the data is shown here:

Sales	Advertising	Competitors	Discounts	Warranty (years)
$4565	$459	1	1	2.00
4896	545	0	0	0.25
4480	472	2	2	1.00
4300	482	3	3	2.00
3502	435	3	3	0.25
4413	499	3	3	1.00
5868	604	0	0	1.00
4527	501	1	1	1.00
3849	370	3	3	1.00

(a) Consider the problem that must be solved. What do you think is the dependent variable? Which are the independent variables?

(b) As advertising expenditures increase, what do you think the effect on sales will be? Will they increase or decrease? How do you think each of the other independent variables will affect monthly sales?

(c) Use a computer package to find the multiple regression model you identified in part (a).

(d) Write the regression equation and interpret the coefficients in the model. How do the actual coefficients agree with your expectation of their effect?

(e) Use the software package to calculate the predicted sales for each of the products in the data set and compare them with the actual sales.

(f) Make a plot of predicted versus actual sales.

(g) How well do you think the model predicts monthly sales?

(h) Set up the hypotheses and test the significance of the model using $\alpha = 0.05$.

(i) What is the value of R^2 for this model and what does it mean?

Datafile:
NATLPARK.XXX

12.13 The U.S. Department of the Interior is looking at state parks in the United States. It is wondering what variables might affect revenues and has assembled data for each state on the number of acres, number of day visitors, number of overnight visitors, and operating expenses for 1992. A portion of the data is shown here:

State	Acreage (thousands)	Day visitors (thousands)	Overnight visitors (thousands)	Total revenues (thousands)	Operating expenses (thousands)
Alabama	50	4,740	1,175	$25,724	$28,279
Alaska	3,250	3,567	703	1,957	5,294
Arizona	46	1,721	580	4,114	11,392
Arkansas	51	6,931	683	12,805	23,364
California	1,345	64,765	6,187	63,689	175,189
Colorado	347	11,973	727	10,708	15,771
Connecticut	176	7,672	357	3,571	8,875
Delaware	17	2,795	207	5,013	8,829

Source: Statistical Abstract of the United States 1999

(a) Find a multiple regression model that predicts revenues as a function of the number of acres of parkland, the number of day visitors, the number of overnight visitors, and operating expenses.

(b) Interpret the coefficients of the model.

(c) Use the model to predict the revenues for each state.

(d) Calculate the residuals and compare them with the actual data.

(e) Which state has the largest residual? Which state has the smallest?

(f) What is the value of R^2 for this model? Does the model do a good job of predicting revenues for state parks? Why or why not?

12.14 A national chain of women's clothing stores with locations in large shopping malls thinks that it can do a better job of planning store renovations and expansions if it understands what variables affect sales. It plans a small pilot study on stores in 24 different mall locations. The data it collects consist of monthly sales, store size, number of linear feet of window display, number of competitors in the mall, size of the mall, and distance to nearest competitor. A sample of the data is shown here:

Datafile:
MALL.XXX

Store	Monthly sales	Size (square feet)	Windows (linear feet)	Competitors	Mall size (square feet)	Nearest competitor (feet)
1	$4453	3860	39	12	943,700	227
2	4770	4150	41	15	532,500	142
3	4821	3880	39	15	390,500	263
4	4912	4000	39	13	545,500	219
5	4774	4140	40	10	329,600	232
6	4638	4370	48	14	802,600	257

(a) What is the dependent variable in this model? Which variables are the independent variables?

(b) How do you think the size of a store will affect sales? How do you think the other independent variables will affect sales?

(c) Use a computer software package to find a multiple regression model for the data.

(d) Interpret the values of the coefficients in the model. Do they agree with how you expected them to behave?

(e) Set up the hypotheses to test whether the model as a whole is significant. At the 0.05 level of significance, what is your conclusion?

(f) Use the model to predict monthly sales for each store in the study. Calculate the residuals.

(g) Which store has the largest residual? Which has the smallest?

(h) Plot the residuals versus the actual values. Does the model do a good job of predicting monthly sales? Why or why not?

Thinking About It!

12.15 The financial analysts who are looking at a model for shareholders' equity at Phillips Petroleum have obtained data from Mobil Oil. They would like to construct models for Mobil similar to the two they have constructed for Phillips. The data are not available for all the years of interest. The data for Mobil are shown here:

*Requires Example
12.2, Exercise 12.11*

Datafile:
MOBIL.XXX

Year	Assets (millions)	Revenues (millions)	Shareholder equity (per share)	Total debt (millions)	Employees (thousands)
1986	NA	$44,936	$32.86	NA	70.9
1987	NA	51,678	36.46	NA	68.1
1988	NA	54,740	38.19	NA	69.4
1989	NA	56,388	39.84	NA	67.9
1990	NA	64,774	42.44	NA	67.3
1991	$25,464	63,311	43.74	$8229	67.5
1992	25,075	64,456	41.06	8520	63.7
1993	25,037	63,975	42.74	8027	61.9
1994	25,503	67,383	42.61	7727	58.5
1995	24,850	75,370	44.71	6756	50.4
1996	27,479	81,503	47.62	6581	43.0

Source: Mobil Oil Annual Report, World Wide Web site: *www.mobiloil.com*

(a) Using the data that are available, create a model for Mobil that relates shareholders' equity to assets and revenues.

(b) In what ways do you expect the model for Mobil to be similar to the one for Phillips? In what ways might they be different?

(c) Compare the model for Mobil with the one found in Example 12.2 for Phillips Petroleum. Do the models agree with your answer to part (b)?

(d) Find a model for Mobil that relates shareholders' equity to total debt and number of employees.

(e) Compare the model for Mobil with the one for Phillips from Exercise 12.11. How are they similar? How are they different?

(f) Which of the two models for Mobil does a better job of predicting shareholders' equity?

Datafile:
MILES.XXX

12.16 Automobile salespeople want to understand the amount of driving done by a household so that they can plan advertising campaigns. They propose to use data from a survey sponsored by the U.S. Department of Transportation to help them understand what factors affect the annual number of miles driven by a household.

(a) What variables might influence the number of miles driven annually by a household? Why did you choose these variables?

As part of the study, data were extracted from a database containing the survey information. The data consisted of 85 households with two or more people for a small city in the Northeast. The variables in the study were total miles driven in the previous 12 months, number of drivers in the household, number of vehicles in the household, number of children ages 5 to 21 in the household, total number of trips taken in the 24-hour period prior to the survey, and total number of miles driven in the same 24-hour period.

(b) Do you see any problems with these data? If so, what are they? Suggest any additional variables that would help to eliminate or clarify the problems.

(c) A portion of the data available from the U.S. Department of Transportation is shown here:

HHVMILES	DRVRCNT	VEHCOUNT	NUM_KIDS	DTVCNT_H	DTVMILH
0	2	0	2	3	3
20,975	2	2	1	6	35
23,000	3	2	0	7	99
36,000	1	1	0	2	70
6,327	2	2	2	2	10
27,423	2	2	0	5	58
7,058	1	2	0	0	0
24,365	1	3	1	4	10
101,313	1	2	1	4	13
2,519	2	2	0	2	6

The six variables are defined as follows:

   HHVMILES—total annual miles for all vehicles in household

   DRVRCNT—number of drivers in household

   VEHCOUNT—number of vehicles in household

   NUM_KIDS—number of children in household aged 5–21

   DTVCNT_H—number of travel day vehicle trips for the household

   DTVMILH—travel day vehicle miles for household

The travel day is the 24-hour period prior to the survey.

(d) Use a computer software package to find the multiple regression model that predicts total annual vehicle miles using the other five variables as the set of independent variables.

(e) Write down the model and interpret the coefficients. Do they make sense?

(f) At the 0.05 level of significance, what can you say about the model?

(g) What is the value of R^2 for this model? Do you think the model will be useful for predicting total annual vehicle miles for a household? Why or why not?

12.17 The planners who are looking at store characteristics and sales want to know more about the model they have found. *Requires Exercise 12.14*

(a) Find and interpret the value of R^2 for this model.

(b) Do you think this model will be useful in helping the planners? Why or why not?

(c) Set up the hypotheses to test the individual regression coefficients. At the 0.05 level of significance, what are your conclusions?

(d) If you were going to drop just one variable from the model, which one would you choose? Why?

12.18 The analysts who are looking at financial data from Phillips Petroleum are not satisfied with either of the two models they have developed. They want to combine the two models and determine if a better model is available. *Requires Exercise 12.11*
Datafile:
PHILLIPS.XXX

(a) Based on the results of the previous models, which two variables do you think will be most helpful in predicting shareholders' equity? Why?

(b) Use the all possible regressions method to find the best one-, two-, three-, and four-variable models.

(c) From the list of the best models for each size, pick the model that you think should be used to predict shareholders' equity.

(d) Write a memo to the chief financial officer of the company explaining your model and justifying your choice.

12.19 Look at the output from the analysis of the monthly sales data for the small appliances manufacturer. Four variables were in the original analysis, but the company wants to know whether they are all important in the model. *Requires Exercise 12.12*

(a) Set up the hypotheses to test whether the individual coefficients are different from 0 and perform the tests at the 0.05 level of significance.

(b) What are your conclusions from this test?

(c) Consider the variables that have coefficients that are 0. Which of these do you recommend dropping from the model? Why?

(d) Rerun the analysis, dropping the variable you identified in part (c). How does this change the model as a whole?

(e) Does dropping this variable change the contribution of the other variables? If so, how?

(f) Would you consider dropping another variable? If so, which one?

(g) If you answered yes to part (f), then rerun the analysis dropping that variable.

(h) Based on all of your analyses, what model do you think the company should use?

(i) Does this approach to finding a model resemble any of the model-building techniques you learned about? If so, which one?

12.20 The store planners for the women's clothing chain want to find the best model for understanding what store characteristics affect monthly sales. *Requires Exercises 12.14, 12.17*

(a) Use stepwise regression or all possible regressions to find the best model for the data.

(b) Analyze the model you have identified to determine whether it has any problems.

(c) Write a memo reporting your findings to your boss. Identify the strengths and weaknesses of the model you have chosen.

12.21 Look again at the data on monthly sales of small appliances. *Requires Exercises 12.12, 12.19*

(a) Use stepwise regression or all possible regressions to find the best model from the set of four variables.

(b) Compare the results of the formal procedure with the model you chose in Exercise 12.19. Are the two models the same? Would you expect them to be?

(c) Consider the two "best" models you have. If they are different, which one would you choose as your model? Why?

12.22 Look at the results of the analysis on the state park data collected by the U.S. Department of the Interior. *Requires Exercise 12.13*

(a) Set up the hypotheses for testing individual regression coefficients. At the 0.01 level of significance, what are your conclusions?

(b) Drop any of the variables with coefficients that tested to be 0 from the model and rerun the analysis.

(c) How does the model change?

(d) Are you satisfied with this model or do you think that additional analysis should be done to find the best model?

(e) Perform any additional analyses you think are necessary to find the best model for predicting revenues.

Requires Exercise 12.16

12.23 Look again at the model for predicting the total annual miles driven by a household.

(a) At the 0.05 level of significance, what can you say about the individual regression coefficients?

(b) Which variables would you consider dropping from the model? Why?

(c) Consider what the model is trying to predict. Do you think that households with no vehicles should be included in these data? Why or why not?

(d) Identify the observations that have no vehicles and drop them from the data. Rerun the analysis. Does this change anything?

(e) Again, think about what the model is trying to predict. What other data values do not represent the population of interest? Why?

(f) Drop any observations that correspond to your answer to part (e) and rerun the model. How do the results compare with the other two models? Do you feel more confident with this model? Why or why not?

Requires Exercises 12.12, 12.19, 12.21

12.24 Look at the model that you decided on for the monthly small appliance sales.

(a) Prepare a set of residual plots for the model. Does the model appear to violate any of the assumptions of the multiple regression model? If so, what are the problems?

(b) Calculate the correlations between each pair of independent variables. Does there appear to be multicollinearity in the model? If so, what would you recommend doing?

(c) Divide the data set in half using some random selection process and run the model on each subset. Compare the models. Do the coefficients of the model appear stable? Why or why not?

(d) Use either Cook's distance or some other measure available in your software to determine whether any of the observations are exerting a strong influence on the model.

(e) Drop the observations identified in step (d) from the data and rerun the analysis. How does this change the model?

(f) Prepare a memo describing your findings and suggesting a model to use for predicting revenue. Be sure to include any limitations of the model you suggest.

Requires Exercises 12.16, 12.23

12.25 Consider the model that you think is best for predicting total annual miles for a household.

(a) Prepare a set of residual plots for the model. Does the model appear to violate any of the assumptions of the multiple regression model? If so, what are the problems?

(b) Calculate the correlations between each pair of independent variables. Does there appear to be multicollinearity in the model? If so, what would you recommend doing?

(c) Use either Cook's distance or some other measure available in your software to determine whether any of the observations are exerting a strong influence on the model.

(d) Drop the observations identified in step (c) from the data and rerun the analysis. How does this change the model?

(e) Prepare a memo describing your findings and suggesting a model to use for predicting revenues. Be sure to include any limitations of the model you suggest.

Doing It!

12.26 As you saw in the example about Texas real estate, many factors affect home prices. A group looking at the city of Boston collected data on these variables believed to be predictors of home prices:

CRIM—per capita crime rate by town

ZN—proportion of residential land zoned for lots larger than 25,000 square feet

INDUS—proportion of nonretail business acres per town

CHAS—Charles River dummy variable (1 if tract bounds river, 0 otherwise)

NOX—nitric oxides concentration (parts per 10 million)

RM—average number of rooms per dwelling

AGE—proportion of owner-occupied units built prior to 1940

DIS—weighted distances to five Boston employment centers

RAD—index of accessibility to radial highways

TAX—full-value property-tax rate per $10,000

PTRATIO—pupil–teacher ratio by town

LSTAT—percent lower status of population

MEDV—median value of owner-occupied homes (in $thousands)

CRIM	ZN	INDUS	CHAS	NOX	RM	AGE	DIS	RAD	TAX	PTRATIO	LSTAT	MEDV
0.00632	18.0	2.31	0	0.538	6.575	65.2	4.0900	1	296	15.3	4.98	24.0
0.02731	0.0	7.07	0	0.469	6.421	78.9	4.9671	2	242	17.8	9.14	21.6
0.02729	0.0	7.07	0	0.469	7.185	61.1	4.9671	2	242	17.8	4.03	34.7
0.03237	0.0	2.18	0	0.458	6.998	45.8	6.0622	3	222	18.7	2.94	33.4
0.06905	0.0	2.18	0	0.458	7.147	54.2	6.0622	3	222	18.7	5.33	36.2
0.02985	0.0	2.18	0	0.458	6.430	58.7	6.0622	3	222	18.7	5.21	28.7
0.08829	12.5	7.87	0	0.524	6.012	66.6	5.5605	5	311	15.2	12.43	22.9

(a) Calculate the correlation coefficients of each variable with the median value of owner-occupied homes (MEDV).

(b) Select a set of no more than five variables to use as the independent variables.

(c) Using the variables you selected, find the best model for predicting the median value of homes.

(d) Test to see whether your model is appropriate.

(e) Summarize your findings in a memo to the group.

Experimental Design and ANOVA

THE TISSUE PROBLEM

Have you ever pulled the first tissue out of the box and had it tear? Have you ever opened a box of tissues and, in trying to get one tissue out, ended up with several tissues? The cause of both of these annoying problems is that the box does not have enough airspace. Do you ever think about calling the tissue company to complain? Next time you purchase a box of tissues, look at the bottom of the box. You will find an 800 number to call with customer complaints. In fact, many people do call to complain and the problems described here are frequently mentioned by the callers.

Airspace is the amount of space between the top of the tissues and the top of the box. It is measured in millimeters and it should be at least 9 mm. Even if there is 9 mm of airspace when the box is manufactured, there many not be enough airspace by the time the customer opens the box. This is caused by a phenomenon called "growback." As the box sits in the warehouse or on the supermarket shelf or in your cupboard at home, the tissues, which were heavily compressed when they were put into the box, begin to expand or "grow back." Thus, the airspace is reduced.

The tissue company needs to estimate how much growback will occur in order to determine how much airspace should be left in the box at the time of manufacture. An experiment was done to try to understand this growback phenomenon. Here is a portion of the data:

Position	Time	Airspace (mm)
1	1	23
1	1	25
1	1	23
1	1	23
1	1	23

13.1 Chapter Objectives

Throughout most of this book we have assumed that samples have been selected at random from the population we are studying. These types of samples are called *simple random samples*, defined in Chapter 1. The reasons we have studied primarily simple random samples are twofold. First, very often real-world data are collected this way. Second, data analysis is simplest when the samples are simple random samples, so it makes sense to use these samples first when we present the data analysis techniques.

However, we are now at a point where it makes sense to think about *how* the sample is selected to get the most information for our money. Recall that a statistical experiment is any action with outcomes that are recordable data. Now it is time to learn how to set up or design that statistical experiment. In doing so we will specify how the data are to be collected. You may recall that in Chapter 1 we recognized that the amount of resources we have available does constrain the size of our sample. However, the formulas for determining the sample size that we presented in Chapter 7 do not explicitly take this into account. So, it seems that we have, in effect, ignored the fact that we have limited resource. This chapter addresses that issue.

By now, you know that the techniques used to analyze data must take into account the type of data being analyzed and how the data were collected. Thus, this chapter presents a technique to analyze data that result from a designed experiment. This technique is called *analysis of variance*, abbreviated *ANOVA*. You have seen this technique used in conjunction with regression in Chapter 11.

In particular, this chapter covers the following topics:

- Motivation for using a designed experiment
- Analysis of data from one-way designs
- Assumptions of ANOVA
- Analysis of data from blocked designs
- Analysis of data from two-way designs
- Other types of experimental designs

A simple random sample is a sample that has been selected in such a way that all members of the population have an equal chance of being selected and every sample of size n has the same chance of becoming the sample.

13.2 Motivation for Using a Designed Experiment

To help the tissue company, you might suggest that the company take a random sample of tissue boxes and measure the amount of airspace in each box. But when should they measure the airspace? If you suggested measuring the airspace at the time the tissue box is manufactured, then you are partway to the right answer. Such measurements are called *in-process* data because they are taken at the time the box is manufactured.

Suppose an average of 12 mm of airspace is in the tissue boxes right after they are manufactured. Is this enough to ensure that, even after growback, there is an average of 9 mm of airspace when the customer opens the tissue box? It looks like we need to check the airspace in the box at some future time—say, 2 weeks after the box is made. Can we use the same boxes and measure the airspace 2 weeks later? Clearly, we cannot, since the boxes must be opened to measure the airspace right after they are manufactured. This is an example of destructive inspection. The product is, in a sense, destroyed once we open it and measure the airspace. We certainly cannot sell opened tissue boxes! We need to use another set of boxes that have been sitting on the shelf for 2 weeks.

Initially, it looks like we could use the two-population tests of Chapter 10 to compare the mean airspaces of two populations: the population of boxes right after they have been manufactured and the population of boxes that have been sitting on the shelf for 2 weeks. If these samples are selected randomly from the two populations, then this plan would work. But what might happen if we use this approach? Well, the boxes that we check 2 weeks after manufacturing might have been manufactured on a different day, under different conditions than those we check right after manufacturing. How can we tell if the differences in airspace are due to different manufacturing conditions or to the growback during the 2-week time delay? If we decide that the boxes to be checked after 2 weeks should be selected at the same time as the ones checked immediately, then we have addressed this issue. Then we have samples that are, in fact, comparable. They are matched in terms of the manufacturing conditions. This is an example of the simplest type of experimental design.

Remember! In dependent samples the members of one sample are matched with members of the other sample.

Actually, you were introduced to the idea of a designed experiment in Section 10.7. We explained that in some situations, other characteristics influence the variable being observed. One example we looked at was the effect on sales of a new product display. Clearly, the size of the store influences sales, so to do a fair comparison we compared stores of similar size. The stores were matched on the size characteristic. These types of data are called paired data or dependent samples. The technique used to analyze these types of data is a paired *t* test. We presented it in Chapter 10 when you were learning how to draw conclusions about two population parameters.

If we compared only the average airspace in process with the average airspace 2 weeks later, then we could use a *t* test. But how do we know that 2 weeks is the right time to check the airspace? Perhaps we should examine the airspace at several different times. This is precisely what was done at the tissue company. A total of four cases were sampled, one case every hour for 4 hours. One case was used to obtain in-process measurements. The other three cases were saved for observations taken after 24 hours, 2 weeks, and 4 weeks. Each case contained 24 cartons of tissue boxes, or 120 tissue boxes.

A portion of the data is shown in Figure 13.1. The variable *time* indicates when the measurement was taken and is coded as follows: 1 = in process, 2 = after 24 hours, 3 = after 2 weeks, and 4 = after 4 weeks. The variable *airspace* is the measurement of the airspace in millimeters. Table 13.1 gives the sample mean and sample standard deviation for each of the four time periods.

As you can see, we are now comparing the mean airspaces of four different samples to draw a conclusion about four different populations: the population of boxes at the time of manufacturing (in process), the population of boxes that have been sitting on the shelf for 24 hours, the population of boxes that have been sitting on the shelf for 2 weeks, and the population of boxes that have been sitting on the shelf for 4 weeks. The average amount of airspace seems to be decreasing the longer the tissue box sits on the shelf. But we know from our work in previous chapters that just because the sample means are different does not automatically mean that the

TABLE 13.1

SUMMARY STATISTICS FOR THE TISSUE COMPANY

Time	Sample mean	Sample standard deviation
1	21.01	1.53
2	17.46	1.48
3	15.78	2.08
4	15.02	2.32

	A	B
1	Time	Airspace
2	1	23
3	1	25
4	1	23
5	1	23
6	1	23

FIGURE 13.1 A portion of the tissue company data set

population means are different. We would like to be able to tell the tissue company whether the mean airspace is the same for all four of these populations. Thus, the null and alternative hypotheses are as follows:

H_0: $\mu_1 = \mu_2 = \mu_3 = \mu_4$

H_A: At least one of the population means is different from the others.

We need a technique that extends the work we did in Chapter 10 beyond two populations. If the population means test different—that is, if we reject the null hypothesis—then we would like a model to explain how the airspace decreases as the box sits on the shelf for a longer period of time. Both of these issues are addressed in the next section.

13.3 Analysis of Data from One-Way Designs

13.3.1 ONE-WAY DESIGNS: THE BASICS

The tissue company is interested in comparing the characteristics of four populations that differ on the basis of one **factor:** the amount of time that has elapsed since the box was manufactured. In this experiment the factor of interest has four different **levels** or possible values.

> A *factor* is a variable that can be used to differentiate one group of population from another. It is a variable that may be related to the variable of interest. A *level* is one of several possible values or settings that the factor can assume.

The variable that is being studied is called the **response variable.** This is just a description of the term *variable* that you have been using throughout this text. The descriptor *response* in front of the word *variable* indicates that what we are measuring or observing may respond to the experimental conditions—that is, the setting of the factor(s).

> The *response variable* is a quantitative variable that we are measuring or observing.

Many situations involve populations that differ on the basis of one factor. In fact, all of our work in Chapter 10 involved comparing populations that were different on the basis of one factor and that factor had only two levels. In Chapter 1 we said that a qualitative variable is often used to divide a large population into two smaller groups for comparisons. A commonly used qualitative variable is gender, which clearly has two levels: male and female. Thus, using the techniques of Chapter 10, we could compare the average salaries of males and females. The factor is gender and we are comparing average male salaries with average female salaries.

The natural extension of this situation is to handle a factor that has more than two levels. For example, a teacher might wish to study the differences in student performances when different teaching methods are used. Here, the factor is the teaching method. If the teacher is considering four different methods, then there are four levels of this factor. The response variable is student learning. A university career office might wish to study differences in starting salaries for different majors. Here, the factor is the academic major and the response variable is starting salary. If the career office is studying computer science, nursing, and education majors, then there are three levels of this factor. A soft drink manufacturer might be interested in conducting a taste test of four different versions of a new drink. In this case, the factor is the version of the drink and there are four levels. The response variable is the taste test score. A medical researcher may wish to study the amount of time it takes for five different experimental drugs to work. In this case, the factor is the drug and there are five different levels. The response variable is the time it takes for the drug to work. A manufacturer may wish to study the life of products made at three different plants. The plant location is the factor and there are three levels. The response variable is the life of the product.

Notice that, in each of these situations, the response variable is a quantitative variable (airspace, learning, salary, tastiness, time to work, product life) and the factor takes on one of several possible values. These are all examples of **one-way** or **completely randomized designs.**

> An experiment has a ***one-way*** or ***completely randomized design*** if several different levels of one factor are being studied and the objects or people being observed or measured are randomly assigned to one of the levels of the factor.

The term *one-way* refers to the fact that the groups differ with regard to the one factor being studied. The term *completely randomized* means that individual observations are assigned to the groups in a random manner. For example, in the case of the soft drink manufacturer, people are assigned to sample one of the four versions of the drink in a random manner.

✔ *TRY IT NOW!*

One-Way Designs
Designing a Simple Study

Select a population that you might be interested in studying and identify a quantitative variable that you might wish to analyze.

Now, specify a factor that might be of interest. This should be some characteristic that you think might influence the variable you are analyzing.

Indicate at least three levels of the factor.

13.3.2 UNDERSTANDING THE TOTAL VARIATION

When data are the result of a one-way design and certain assumptions are met, the proper tool to analyze the data is called **analysis of variance (ANOVA).**

> *Analysis of variance (ANOVA)* is the technique used to analyze the variation in the data to determine whether more than two population means are equal.

R. A. Fisher originated the technique of ANOVA in England in connection with agricultural experiments. He compared the yields of crops when the farmland was treated differently. In this case the factor was the treatment of the farmland and the response variable was the yield of the crop. The word **treatment** has remained a part of the language of designed experiments. It need not refer to treatment of the farmland but has a more general meaning.

> A *treatment* is a particular setting or combination of settings of the factor(s) being studied.

Your initial reaction is most likely that *analysis of variance* is misnamed. After all, we are really interested in determining whether the groups, which correspond to differing levels of the factor of interest, have different *means*. What does the variance have to do with this technique? Why are we analyzing the variance and not the means? This is completely normal thinking at this point. In fact, it is through an analysis and breakdown of what is causing the variation we see in the data that we reach conclusions about the group means! Let's see how this works.

Suppose we think about *all* of the data that are collected for the tissue company— that is, 120 observations per time period times the four time periods, or 480 observations of airspace. We want to decide whether the airspace changes, so we might ask the question, How different are these airspace measurements from one another? We know from our work in descriptive statistics that we measure how spread out the data values are by using a measure of spread or variation. Thus, we could calculate the variance of the entire data set. The variance is calculated using the formula for the sample variance:

Remember! The units of measure in a variance calculation are the original units squared.

$$s^2 = \frac{\sum\limits_{i=1}^{n}(x_i - \bar{x})^2}{n-1}$$

Sample variance

In this case $n = 480$ and the value of $\bar{x}$, the overall average for the data set, is found to be 17.32 mm. Since the data were collected in four different groups corresponding to the four levels of the factor time, it is convenient to label each observation with a double subscript rather than just a single subscript ranging from 1 to 480. We use the letter c for the number of different levels of the factor. In this case $c = 4$. Each observation can be written as x_{ij}, where the first subscript, i, tells how the observation is numbered within the group and the second subscript, j, tells what group the observation is in.

EXAMPLE 13.1

TISSUE AIRSPACE

Understand the problem.

Setting Up the Notation

In the case of the tissue company, x_{11} is the first observation in the first group, which is the in-process data. Similarly, x_{14} is the first observation in the fourth group, the data collected 4 weeks after the time of manufacturing.

TRY IT NOW!

Tissue Airspace
Getting Used to ANOVA Notation

Using the information in Figure 13.1, what is the value of x_{31}?

What notation refers to the 120th observation of the data taken 2 weeks after the time of manufacturing?

What is the range of values for the first subscript for group 1?

What is the range of values for the second subscript?

Carrying this notation further, we notice that the number of observations in each group does not have to be the same. Thus, n, which is the total number of observations in the experiment, can be written as follows:

$$n = n_1 + n_2 + \cdots + n_j + \cdots + n_c$$

where each n_j is the sample size of group j and c is the number of different groups or levels of the factor.

Understand the problem.

EXAMPLE 13.2

TISSUE AIRSPACE

Calculating the Sample Size

There are four time periods or four levels of the factor, so $c = 4$. There are 120 observations at each time period, so $n_1 = n_2 = n_3 = n_4 = 120$ and

$$n = 120 + 120 + 120 + 120 = 480$$

Finally the **overall mean** is relabeled $\bar{\bar{x}}$, read as "x double bar." For the tissue company $\bar{\bar{x}}$ is 17.32 mm. It is based on all 480 of the observations and therefore is sometimes called the **grand mean**.

> The **grand mean** or the **overall mean** is the sample average of all the observation in the experiment. It is labeled $\bar{\bar{x}}$.

We can now rewrite the variance calculation as follows:

Sample variance, with double subscript notation

$$s^2 = \frac{\sum\limits_{j=1}^{c} \sum\limits_{i=1}^{n_j} (x_{ij} - \bar{\bar{x}})^2}{n - 1}$$

This formula yields the same result as the formula for the variance presented earlier. The numerator is simply rewritten to reflect the fact that each observation belongs to some level of the factor.

The numerator of this formula is called the **total variation** or **sum of squares total (SST)**.

The **total variation** or **sum of squares total (SST)** is a measure of the variability in the entire data set considered as a whole.

SST is calculated as follows:

$$\sum_{j=1}^{c} \sum_{i=1}^{n_j} (x_{ij} - \bar{\bar{x}})^2$$

Sum of squares total (SST)

The technique of ANOVA breaks down this total variation into several components and examines the contribution each component makes to the total variation. So instead of analyzing the sample variance, we will actually be analyzing the total variation, which is the numerator of the sample variance.

You could calculate SST by hand, but it can be a long and tedious calculation, even for small data sets. The next example shows the first few terms of the calculation of SST for the tissue company.

Always use software to calculate SST.

EXAMPLE 13.3

TISSUE AIRSPACE

Analyze the data.

Calculating SST

Using the information in Figure 13.1 and the fact that $\bar{\bar{x}} = 17.32$ mm, we can set up the first five terms of the calculation of SST for the tissue company:

Time	Airspace, x_{ij}	$(x_{ij} - \bar{x})^2$
1	23	$(23 - 17.32)^2 = 32.3$
1	25	$(25 - 17.32)^2 = 59.0$
1	23	$(23 - 17.32)^2 = 32.3$
1	23	$(23 - 17.32)^2 = 32.3$
1	23	$(23 - 17.32)^2 = 32.3$

The calculation shown in the last column 3 would be done for all 480 observations and then the values totaled. This gives the values SST = 4249.87.

Any statistical package will calculate SST and produce what is known as an ANOVA table. You were introduced to SST and ANOVA tables in Chapter 11 in relation to regression analysis. The formula for SST given in Chapter 11 is written like the numerator of the variance formula presented previously, before we introduced the double subscript notation. It measures the same thing as the formula for SST here. The ANOVA tables produced by Excel and Minitab are shown in the next example.

EXAMPLE 13.4

TISSUE AIRSPACE

Analyze the data.

Looking at Output from Excel and Minitab

ANOVA tables from Excel and Minitab for the tissue company follow:

ANOVA

Source of Variation	SS	df	MS	F	P-value	F crit
Between Groups	2554.75	3	851.58333	239.13025	1.342E-94	2.6236364
Within Groups	1695.116667	476	3.5611695			
Total	4249.866667	479				

Excel

In Excel SST is shown in the line labeled Total and the column *SS*.

Analysis of Variance for Airspace

Source	DF	SS	MS	F	P
Time	3	2554.75	851.58	239.13	0.000
Error	476	1695.12	3.56		
Total	**479**	**4249.87**			

Minitab

In Minitab SST is shown in the line labeled Total and the column SS.

Other values are printed in the ANOVA table besides SST. The next section explains what the other numbers mean and how to use them.

13.3.3 COMPONENTS OF THE TOTAL VARIATION

The ANOVA tables in Example 13.4 have a column labeled SS in both the Excel and the Minitab outputs. The SS label stands for sum of squares. So far we have seen what the SST or total sum of squares term means and how it is calculated. A small amount of detective work indicates that the two numbers above SST add up to SST. This is exactly correct. The technique of ANOVA focuses on the relative sizes of these components of the total variation.

Notice that the between groups variation is labeled "Time" in the Minitab output. This is because the label "Time" was used when the data were entered into Minitab to indicate that time was the factor being studied.

To start, we use the Excel output terminology and then relate the discussion to the Minitab output. In the first column of the ANOVA table, you can see that one of the components of SST is called the **between groups variation** and the other component is called the **within groups variation.** The between groups variation is also called the **sum of squares between** or the **sum of squares among,** and it measures how much of the total variation comes from actual differences in the treatments. The dotplot shown in Figure 13.2 displays the sample average for each of the four time treatments. These are called **treatment means.**

A *treatment mean* is the average of the response variable for a particular treatment.

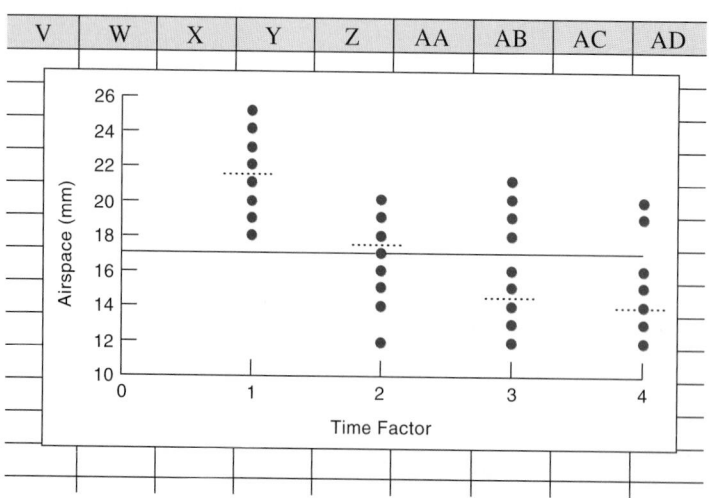

FIGURE 13.2   Dotplot of airspace by time for the tissue company

By examining the graph we can see that the treatment means are "pretty different" from the grand mean of 17.32 mm. Thus, we might expect that a substantial portion of the total variation is due to the fact that the airspace measurements are taken at four different times. This is what the between groups variation component measures.

Note: The term "pretty different" will be quantified once you get a handle on the concept.

In computing this component, we look at differences between the treatment means ($\bar{x}_j$, shown as dotted lines) and the overall mean ($\bar{\bar{x}}$, shown as a solid line). Those differences are weighted by the sample size of each group (n_j).

> **Between groups variation** measures how different the individual treatments are from the overall grand mean. It is often called the **sum of squares between** or the **sum of squares among (SSA)**.

The formula for the sum of squares among (SSA) is

$$\text{SSA} = \Sigma n_j(\bar{x}_j - \bar{\bar{x}})^2$$

Sum of squares among (SSA)

Let's verify the calculation of SSA for the tissue company.

EXAMPLE 13.5

TISSUE AIRSPACE

Calculating SSA

The calculation of SSA for the tissue company is shown in the table:

Analyze the data.

Column A	Column B	Column C	Column D
Time	Sample mean	(Sample mean − Grand Mean)²	Column C × Group sample size
1	21.01	13.62	1634.40
2	17.46	0.02	2.40
3	15.78	2.37	284.40
4	15.02	5.29	634.80
		SSA	2556.00

Remember $n = 120$ for each group.

The calculation of SSA is not so long and tedious as that of SST because there are only c terms, but SSA is always part of an ANOVA table regardless of what statistical package you are using. There is really little value in calculating it by hand. Since you are learning the technique of ANOVA for the first time, it may help you to understand what the number SSA measures. For this reason it is time to try your hand at a small example.

Do not calculate SSA by hand either.

✔ TRY IT NOW!

Career Office
Calculating SSA

The career office is interested in studying the starting salaries for three different majors: computer science, nursing, and education. The overall average starting salary is $\bar{\bar{x}} = \$28,200$. There are 30 students in each group and the averages are shown here. Find SSA by completing the table.

Column A Major	Column B Sample mean	Column C (Sample mean − Grand mean)^2	Column D Column C × Group sample size
Computer science	$28,100		
Nursing	25,600		
Education	20,900		
Total			

The second component of the total variation is labeled Within groups variation in the Excel output and Error in the Minitab output. These are the most common labels used to identify this component of variation. This component is also called experimental error, which explains why the word *error* is used in the Minitab output. Once again, *error* does not indicate that you have made a mistake but rather that you are studying only a sample from the population.

If you take another look at the dotplot in Figure 13.2 you can see the within groups variation there as well. Concentrate for the moment on time period 1 or one group. Clearly, the depths of airspace in the boxes checked at the time of manufacturing differ from each other and from the treatment mean for that time period ($\bar{x}_1 =$ 21.01 mm). It is this variation that is captured in the within groups component of the total variation. The difference between each observation and the mean of the group it is in is squared and these squared differences are accumulated into the **sum of squares within,** which is also called the **sum of squares error (SSE).**

> *Within groups variation* measures the variability in the measurements within the groups. It is often called *sum of squares within* or the *sum of squares error (SSE).*

The sum of squares error is calculated as follows:

Sum of squares error (SSE)

$$SSE = \sum_{j=1}^{c} \sum_{i=1}^{n_j} (x_{ij} - \bar{x}_j)^2$$

Let's set up the first few terms for the calculation of SSE for the tissue company. To show the detailed calculation of SSE for the tissue company we would have to show all 480 squared differences. It is not particularly helpful to do this.

Analyze the data.

EXAMPLE 13.6

TISSUE AIRSPACE

Finding the First Five Terms of SSE

Using the information in Figure 13.1 and the fact that $\bar{x}_1 = 21.01$ mm, we can set up the first five terms of the calculation of SSE:

$$SSE = (23 - 21.01)^2 + (2.5 - 21.01)^2 + (2.3 - 21.01)^2 + (23 - 21.01)^2 + (23 - 21.01)^2 + \cdots$$

We know from the computer output that SSE = 1695.12.

Because SSA and SSE are the only two components of SST and we know two of these numbers, we can find SSE by simple subtraction:

$$SSE = SST - SSA$$

This formula shows that SSE measures whatever variation is not a result of the fact that the data were collected in several different groups.

EXAMPLE 13.7

TISSUE AIRSPACE

Analyze the data.

Finding SSE by Subtraction

We can find SSE for the airspace data by subtraction:

$$SSE = SST - SSA = 4249.87 - 2554.75 = 1695.12$$

Let's summarize what we know so far about the airspace data set. From the dotplot and the relative sizes of SSA and SSE, much of the total variation in the data is due to the time factor being studied. To draw the inference that the means of the four populations are in fact different, we must perform a hypothesis test using the information in the ANOVA table. This is discussed in the next section.

Before moving on, we present one more example. This example was introduced in Chapter 8 and reconsidered in Section 10.4 when you learned how to compare the means of two populations. However, there are really three populations in this situation. Now you have the right tool to do the complete analysis.

EXAMPLE 13.8

TISSUE STRENGTH

Analyze the data.

Finding SST, SSA, and SSE

Customer complaints about tissues tearing led the company to look at the strength of the tissues. Two tissue strength measurements were studied: machine-directional strength (*MDStrength*) and cross-directional strength (*CDStrength*). They are both measured in pounds per ream. We look at only *MDStrength* in this example. The data were collected over three days and the company would like to know if the average *MDStrength* is the same for all three days.

The complete data set contains 75 observations per day. In this example we will look at only five observations per day so you can see the computational details. In the last exercise at the end of the chapter you will have the opportunity to do a complete analysis of this data set.

Five values of *MDStrength* for each of the three days are listed here:

	A	B	C	D
1		Day 1	Day 2	Day 3
2		1006	951	993
3		994	994	1093
4		1032	1017	939
5		875	965	966
6		1043	966	992
7	Averages	990	978.6	996.6
8	Grand Mean	988.4		

Here is a scatter plot of the data created in Excel:

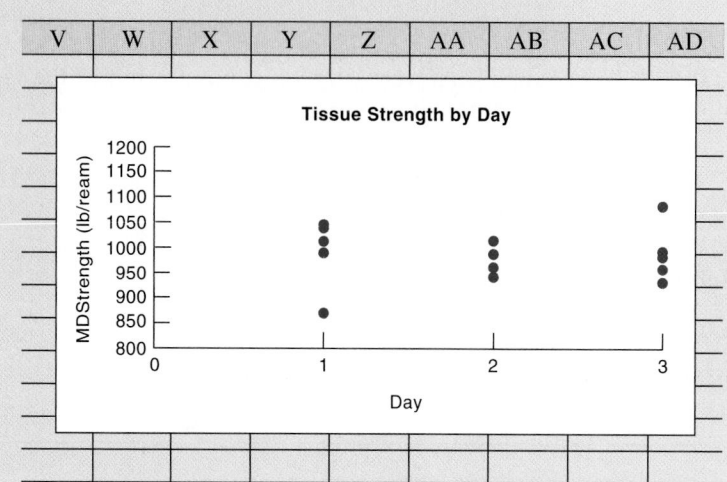

There are $c = 3$ levels of the factor "day" and $n_1 = n_2 = n_3 = 5$. The treatment means have been calculated as well as the grand mean and are shown in the table: $\bar{x}_1 = 990$ lb/ream, $\bar{x}_2 = 978.6$ lb/ream, $\bar{x}_3 = 996.6$ lb/ream, and $\bar{\bar{x}} = 988.4$ lb/ream.

Recall that the formula for the sum of squares total is

$$\text{SST} = \sum_{j=1}^{c} \sum_{i=1}^{n_j} (x_{ij} - \bar{\bar{x}})^2$$

The following Excel spreadsheet displays the 15 squared differences, which added together give SST.

	F	G	H	I
	Day	**MDStrength**	**(MDStrength – Grand Mean)**	**(MDStrength – Grand Mean) squared**
1				
2	1	1006	17.6	309.76
3	1	994	5.6	31.36
4	1	1032	43.6	1900.96
5	1	875	–113.4	12859.56
6	1	1043	54.6	2981.16
7	2	951	–37.4	1398.76
8	2	994	5.6	31.36
9	2	1017	28.6	817.96
10	2	965	–23.4	547.56
11	2	966	–22.4	501.76
12	3	993	4.6	21.16
13	3	1093	104.6	10941.16
14	3	939	–49.4	2440.36
15	3	966	–22.4	501.76
16	3	992	3.6	12.96
17			**SST**	**35,297.60**

Recalling that the formula for the sum of squares among is

$$\text{SSA} = \Sigma n_j (\bar{x}_j - \bar{\bar{x}})^2$$

we can easily compute this component as follows:

$$\text{SSA} = (5)(990 - 988.4)^2 + (5)(978.6 - 988.4)^2 + (5)(996.6 - 988.4)^2 = 829.2$$

Finally, the sum of squares error is found by subtracting SSA from SST:

$$SSE = 35{,}297.6 - 829.2 = 34{,}468.4$$

Alternatively, SSE can be calculated using the formula

$$SSE = \sum_{j=1}^{c} \sum_{i=1}^{n_j} (x_{ij} - \bar{x}_j)^2$$

The details of the calculation using the formula for SSE are shown here in an Excel spreadsheet:

	A	B	C	D	E	F	G
1		Day 1	SSE Contribution	Day 2	SSE Contribution	Day 3	SSE Contribution
2		1006	256	951	761.76	993	12.96
3		994	16	994	237.16	1093	9292.96
4		1032	1764	1017	1474.56	939	3317.76
5		875	13225	965	184.96	966	936.36
6		1043	2809	966	158.76	992	21.16
7	Day Average	990		978.6		996.6	
8	TOTAL		18,070.0		2,817.2		13,581.2
9							
10						SSE	34,468.4

All of our computations are confirmed in the following ANOVA table:

	A	B	C	D	E	F	G
27	Anova: Single Factor						
28							
29	SUMMARY						
30	Groups	Count	Sum	Average	Variance		
31	Day 1	5	4950	990	4517.5		
32	Day 2	5	4893	978.6	704.3		
33	Day 3	5	4983	996.6	3395.3		
34							
35							
36	ANOVA						
37	Source of Variation	SS	df	MS	F	P-value	F crit
38	Between Groups	829.2	2	414.6	0.144341	0.867073	3.885290312
39	Within Groups	34468.4	12	2872.367			
40							
41	Total	35297.6	14				
42							

To use Excel to generate the ANOVA table, you must have the data organized in columns as displayed at the beginning of this example.

Clearly, you want to use a statistics package or a spreadsheet package that has some statistical tools to find SST, SSA, and SSE. Even this small example ($n = 15$) took two pages to perform by hand.

The next sections explain what the other numbers in the ANOVA table mean and how to do the hypothesis test to determine whether any of the population means are different from the others.

13.3.4 THE MEAN SQUARE TERMS IN THE ANOVA TABLE

Remember that the technique of ANOVA analyzes the variances to make an inference about the equality of the population means. What we have found in the previous sections is the numerator of three variances. To turn these sums of squares into variances we must divide by a number that is typically one less than the number of observations in the sample. This number is called the degrees of freedom. It is typically labeled "df" or "DF." You have used this term in conjunction with the t distribution and the F distribution.

The most obvious formula is the degrees of freedom associated with SST. Since this term measures the total variability in the data set and uses all n observations, it has $n - 1$ degrees of freedom associated with it. Following this line of thinking, we know that c levels of the factor are being compared and therefore c terms are added together to find SSA. Thus, SSA has $c - 1$ degrees of freedom. This leaves us with $n - c$ degrees of freedom for SSE. There are two ways to think about why this is correct. First, just like the SS column, the degrees of freedom column must add up correctly. If the column labeled degrees of freedom must total $n - 1$ and we have used $c - 1$ degrees of freedom for SSA, then that leaves us with $(n - 1) - (c - 1)$ or $n - c$ degrees of freedom for SSE. Another way to think about this is to realize that each of the c levels contributes $n_j - 1$ degrees of freedom to SSE. Summing these, we get the right number for the degrees of freedom for SSE:

$$\sum_{j=1}^{c} (n_j - 1) = n - c$$

If we divide each of the sums of squares (SS) terms by the appropriate degrees of freedom, we get three variances or **mean square terms.**

> The *mean square among* is labeled **MSA.** The *mean square error* is labeled **MSE.** The *mean square total* is labeled **MST.**

The formulas for the mean squares are

Mean squares

$$\text{MSA} = \frac{\text{SSA}}{c - 1} \qquad \text{MSE} = \frac{\text{SSE}}{n - c} \qquad \text{MST} = \frac{\text{SST}}{n - 1}$$

In the output from Excel and Minitab these values are found in the column labeled MS. Notice that typically only MSA and MSE are printed.

Analyze the data.

EXAMPLE 13.9

TISSUE AIRSPACE

Calculating MSA and MSE

For the airspace data the mean square calculations are

$$\text{MSA} = \frac{2554.75}{3} = 851.58$$

$$\text{MSE} = \frac{1695.12}{476} = 3.56$$

✓ **TRY IT NOW!**

Tissue Strength
Finding MSA and MSE

Find MSA and MSE for the tissue strength data shown in Example 13.8. You should do the calculations using the formula and then find those values in the computer output.

Ans. MSA = 414.6; MSE = 2872.367

13.3.5 TESTING THE HYPOTHESIS OF EQUAL MEANS

So far we have examined the first four columns of the ANOVA table. Typically, the table has two more columns, labeled F and P or p value. Excel provides one additional column labeled F crit, which stands for the critical value of the F distribution. These columns are based on the mean square column and allow us to do the hypothesis test that we set out to do—that is, test whether at least one of the population means is different.

In general, these are the null and alternative hypotheses for a one-way designed experiment:

H_0: $\mu_1 = \mu_2 = \mu_3 = \cdots = \mu_c$

H_A: At least one of the population means is different from the others.

EXAMPLE 13.10

TISSUE AIRSPACE

Setting Up the Null and Alternative Hypotheses

For the airspace data there are $c = 4$ levels of the factor, so these are the null and alternative hypotheses:

H_0: $\mu_1 = \mu_2 = \mu_3 = \mu_4$

H_A: At least one of the population means is different from the others.

Analyze the data.

Think about what it means if the null hypothesis is true. For the airspace data, if the null hypothesis is true, then the amount of time the box sits on the shelf does not affect the average airspace in the tissue box. In this case was say that there is no *treatment effect.*

If there is no treatment effect, then we conclude that the factor does not affect the variable being studied. Even in this case, we would not expect every observation in the data set to be identical because some natural variation is inherent in the process. Remember that SSE measures the variation within the groups and so MSE is certainly an estimate of this unknown population variability, σ^2.

If the null hypothesis is true and there is no treatment effect, then not only is MSE an estimate of σ^2 but all three of the mean square values, MSA, MSE, and MST, are estimates of the natural variability in the data. We have three estimates of the same parameter, σ^2. We known from our work in Section 10.9 how to test for equality of variances of two normally distributed populations: We use an F test. We use this same F test here to see whether MSA and MSE are actually two different estimates of the same parameter. If MSA and MSE test equal, then we can conclude that there is no treatment effect and the population means are equal. We will fail to reject the null hypothesis.

The formula for the F test statistic from Section 10.9 is the ratio of the two sample variances: $F = s_1^2/s_2^2$. In ANOVA, MSA and MSE are the two sample variances. So the F statistic is calculated as

$$F = \frac{\text{MSA}}{\text{MSE}}$$

F test statistic for ANOVA

The hypothesis test is easily done by determining whether the F value is in the rejection region. Excel provides the critical F value as part of its ANOVA

The value for F critical is obtained from the F table and the procedure for reading this table was explained in Section 10.9.

output. It is labeled F crit. If the F test statistic is "too large"—that is, greater than the cutoff value shown in F critical—then we conclude that MSA and MSE are not estimates of the same number and there is a treatment effect. If your software does not provide the critical F value, you can look it up in the F table in the Appendix. The F statistic has $c - 1$ degrees of freedom in the numerator (from the MSA term) and $n - c$ degrees of freedom in the denominator (from the MSE term).

An easier way to perform the hypothesis test is to use the p value. Remember that we have used p values to do all of the hypotheses tests in this book. Both Minitab and Excel provide a p value. Most statistics packages typically output both the F test statistic and the corresponding p value as part of the ANOVA table.

Now we can complete the airspace data analysis.

Analyze the data.

EXAMPLE 13.11

TISSUE AIRSPACE

Conducting the Hypothesis Test

The F test statistic for the airspace data is

$$F = \frac{\text{MSA}}{\text{MSE}} = \frac{851.58}{3.56} = 239.13$$

In the ANOVA table from Excel or Minitab shown in Example 13.4, this F value was printed under the label F. This means that the ratio of the two mean squares is about 240:1, or MSA is 240 times larger than MSE. Clearly, the ratio is not even close to 1 and these two values are not likely to be estimates of the same number.

The p value of 0.000 shown in the Minitab output and the p value of 1.342×10^{-94} (equivalent to a decimal point followed by 93 zeros and then 1342) in Excel confirm this line of thinking and tell us to reject the null hypothesis.

Draw conclusions.

Thus, we conclude that at least one of the population means is different from the others and there is a treatment effect: The time the box sits on the shelf does affect the amount of airspace in the tissue box.

That's it! We have looked at all of the numbers in the ANOVA table and learned what they measure, how to calculate them, and how to use them to draw an inference about the population means. You should have noticed that the technique of ANOVA ties together several concepts and techniques that have been developed in earlier chapters. For this reason, ANOVA is in a sense a "capstone" technique for this book. It is a very commonly used tool and it is an interesting technique because it analyzes the variability in the data to see whether there are any differences in the population means. We have glossed over some assumptions up to this point and it is now time to take a look at them. People tend to use ANOVA without checking to see whether it is actually appropriate, which can lead to erroneous conclusions. Section 14.4 discusses the assumptions necessary to use ANOVA.

Before moving on you should complete the analysis for the tissue strength data that was started in Example 13.8.

TRY IT NOW!

Tissue Strength
Completing the Analysis

Set up the null and the alternative hypotheses for the tissue strength example.

Use the ANOVA table shown in Example 13.8 to find the F statistic and decide whether you should reject the null hypothesis or fail to reject it.

What do you conclude about the variable *MDStrength*? What is your recommendation to the company?

13.3.6   A SUMMARY OF ONE-WAY DESIGNS

You have learned how to examine a set of data that has resulted from a one-way de-signed experiment. There is a variable of interest and you wish to know whether a particular factor affects the average level of this variable. So, you have one factor, which can be "set" to various values or levels. The variable of interest is repeatedly observed or measured with the factor set at each of the levels.

Once you have the data, you quickly see that not all the values of the variable are the same! No surprise here. The technique of ANOVA allows you to decide what is causing the variation that you see. Is the variation largely due to the fact that you have treated the observations differently by setting the factor at different levels? If so, you will conclude there is a treatment effect. Or is the variation largely due to natural or inherent variation in the variable being studied? If so, you will conclude that there is no treatment effect. You divide the total variability in the data into two components and then examine the relative sizes of these two esti-mates of variability.

All of the computation necessary to carry out a one-way ANOVA are summa-rized in the ANOVA table in Table 13.2.

You can calculate each of the values in the ANOVA table by hand, but it is much more likely that you will use a statistical software package or a spreadsheet package that has some built-in data analysis tools. The calculations are done from left to right across the columns of the table, with the p value being the final value

TABLE 13.2

SUMMARY TABLE FOR ONE-WAY ANOVA

Source of variation	SS	df	MS	F	p Value
Between groups	SSA	$c - 1$	$MSA = \dfrac{SSA}{c - 1}$	$F = \dfrac{MSA}{MSE}$	
Within groups	SSE	$n - c$	$MSE = \dfrac{SSE}{n - c}$		
Total	SST	$n - 1$			

calculated. Based on the *p* value you make a decision to reject or fail to reject the null hypothesis. If you fail to reject the null hypothesis, then you conclude that the factor you studied does *not* affect the average value of the variable. If you reject the null hypothesis, then you conclude that there is a treatment effect and at least one of the population means is different from the others. In this case, although you cannot tell at this point which population mean(s) are different, you do know that the factor you have investigated is important to the response variable. This is one of the major focuses of the technique of ANOVA. It tells you whether anything interesting is going on with respect to this factor. It tells you whether you need to investigate this factor further or it can simply be ignored because it does not affect the average level of the variable being studied.

13.3.7 BUILDING THE MODEL FOR THE RESPONSE VARIABLE

Remember that we want to be able to tell the tissue company whether the mean airspace differs when the box is left on the shelf for different amounts of time. If the population means are different, then we want to be able to tell the tissue company how the airspace changes as the box sits on the shelf. Now that we know there is a treatment effect, let's consider how to estimate the airspace.

Just as the overall variability in the data can be split into component parts, so can a single observation be divided into its component parts. Each observation can be thought of like this:

$$\text{Response} = \text{Grand mean} + \text{Treatment effect} + \text{Error}$$

Notice that if we concluded that there was no treatment effect (by failing to reject H_0), then each response would be equal to the overall mean plus some random variation called error.

We must use the data to estimate the grand mean and the treatment effect. As you will see in Section 13.4, the error term is assumed to have an average of 0. The grand mean can be estimated by $\bar{\bar{x}}$, and the treatment effect is the adjustment that must be made to the grand mean to predict a response. The treatment effect can be estimated by taking the difference between the treatment mean and the overall mean, $\bar{x}_j - \bar{\bar{x}}$. Thus, the estimate of the response is

$$\text{Response} = \bar{\bar{x}} + (\bar{x}_j - \bar{\bar{x}}) + \text{Error} = \bar{x}_j + \text{Error}$$

or just Treatment mean + Error.

Analyze the data.

EXAMPLE 13.12

TISSUE AIRSPACE

Using the Prediction Model

For the airspace data consider $x_{11} = 23$ mm. Writing this in terms of its components gives us

$$\text{Response} = \text{Grand mean} + \text{Treatment effect} + \text{Error}$$

$$23 = 17.32 + (21.01 - 17.32) + \text{Error}$$

$$23 = 17.32 + 3.69 + 1.99$$

Our model predicts an average of 21.01 mm of airspace in process. In this case we are in error by 1.99 mm.

✓ TRY IT NOW!

Tissue Strength
Using the Prediction Model

For the tissue strength data, write x_{11} in terms of its components.

What is the error for this particular observation?

13.3.8 THE NEXT STEP: MULTIPLE COMPARISONS

If we reject the null hypothesis of equal population means, as in the case of the airspace data, we can say that there is sufficient evidence in the data to state that not all the population means are the same. Clearly, this is only the first step. To decide what type of experiment to run next to further investigate the effect this factor has on the variable of interest, we should learn a little bit more from this data set.

Remember! The goal is to understand how the airspace changes as the tissue box sits on the shelf.

Let's see what tools we have that might help us with our detective work. We started off by simply looking at the sample mean for each level of the factor. But we know from our work in Chapter 7 that the sample mean is a point estimate and is never likely to actually "hit" the population mean right on the money. To get an idea of how far off our point estimate was likely to be, we constructed confidence intervals for the population mean. This might be a useful tool at this point. In fact, most statistical software packages calculate the individual sample means and the corresponding confidence intervals when they run ANOVA. This portion of the Minitab output for the airspace data is shown in Figure 13.3. Excel ANOVA output does not include confidence intervals.

By this point in the text, you know that you can learn a great deal by examining graphs and noticing patterns. From these confidence intervals you can see that some of them overlap and some do not. For example, the confidence interval for the mean airspace in process (population or level 1) seems particularly different from the confidence interval for the mean airspace after 4 weeks (level 4). However, the confidence intervals for the mean airspaces for levels 3 (2 weeks) and 4 (4 weeks) overlap a bit.

Visual analysis for pairwise comparison gets you started.

Analysis of Variance

Source	DF	SS	MS	F	P
Time	3	2554.75	851.58	239.13	0.000
Error	476	1695.12	3.56		
Total	479	4249.87			

```
                                  Individual 95% CIs For Mean
                                  Based on Pooled StDev
Level    N    Mean    StDev    -------- + -------- + -------- + --------
  1     120   21.008   1.526                              (-+-)
  2     120   17.458   1.483                  (*-)
  3     120   15.783   2.083     (-*-)
  4     120   15.017   2.319   (-*-)
                               -------- + -------- + -------- + --------
Pooled StDev =  1.887            16.0     18.0     20.0
```

FIGURE 13.3 **Confidence intervals for airspace data from Minitab**

Ans. $1006 = 988.4 + (900 - 988.4) + 16$; error $= 16$ LB/REAM

TABLE 13.3

DIFFERENCE IN SAMPLE MEANS OF AIRSPACE DATA

Level	1	2	3
1			
2	3.55		
3	5.23	1.68	
4	5.99	2.44	0.76

Although this is not a formal statistical test, we can certainly intelligently speculate about whether one population mean is causing the F value in ANOVA to be large (leading to a rejection of the null hypothesis) or whether all of the means are different. In the case of the airspace data there is a great difference between the in-process data and the 24-hour data. Perhaps it would be a good idea to look at the differences in the sample means. Table 13.3 shows the amount of growback (in mm) in the tissues from the time of manufacturing until 4 weeks after manufacturing. These values are easily found by subtracting pairs of sample means. For example, the difference between $\bar{x}_1$ and $\bar{x}_2$ is $(12.01 - 17.46 \text{ mm}) = 3.55$ mm.

Table 13.3 tells us that the airspace decreases 3.55 mm between the time of manufacturing (level 1) and 24 hours later (level 2), it decreases an additional 1.68 mm between the 24-hour mark and 2 weeks later, and it decreases only an additional 0.76 mm between 2 weeks and 4 weeks. In total, the tissues "grow back" about 6 mm from the time of manufacturing until the customer opens the box 4 weeks later. Clearly, our idea that most of the growback occurs during the first 24 hours after manufacturing seems to be correct. In fact, the tissues are greatly compressed at the time of manufacturing to get them into the box, so perhaps focusing on the first 24 hours is a good approach for the next experiment.

This analysis also points out how much easier it is to do data analysis when we are actually involved in the manufacturing process or the situation that generated the data. If we understand the situation, we are much more likely to be able to make sense of the data and make relevant recommendations.

The intuitive, visual approach we have taken here works well at pointing us in the right direction and will not differ much from the conclusions we could draw from formal techniques if the sample sizes are equal, as they were in the airspace data. In fact, the most common formal statistical test done at this point uses the difference between two of the sample means as the center of a confidence interval, which estimates the difference in the corresponding population means. Several techniques could be used to do this. Alternatively, a hypothesis test of the difference in the two population means can be done. It might seem that a t test or a paired t test from Chapter 10 would work here. It is natural to think so, since we now wish to compare two population means. Although the setup of the null and alternative hypotheses is the same here, unfortunately we cannot use either a t test or a paired t test to do the test. The reason for this is that we are making multiple pairwise comparisons from a single data set and we are doing the comparison after the data have been analyzed.

Techniques for doing pairwise comparisons sometimes give conflicting conclusions.

Several techniques can be used and most statistical software packages have several options for doing the pairwise comparisons. For example, Minitab can use any one of four techniques to do the pairwise comparisons. Sometimes these techniques lead to conflicting conclusions. Many issues need to be addressed to explain these techniques properly and they will not be addressed here. However, by examining the confidence intervals for the mean of each level of the factor, looking at how they overlap and how much the sample means change as we change levels of the factor, we can detect a good deal about how the factor influences the variable we are studying. This will guide the design of the next experiment.

13.3.9 EXERCISES—LEARNING IT!

13.1 The number of days that a patient must have a breathing tube after surgery is an indicator of the speed of the patient's recovery. Three different surgical methods have been studied on 35 patients each. The numbers of days the breathing tube was left in are recorded here:

Method 1	Method 2	Method 3	Method 1	Method 2	Method 3
3	4	5	2	3	5
3	2	4	2	3	3
2	2	5	15	3	7
6	7	6	8	2	6
2	4	3	12	4	6
2	3	3	16	10	6
2	5	9	5	2	6
3	2	6	3	2	6
4	4	8	2	2	3
3	3	8	2	3	8
2	5	6	2	3	7
3	3	6	17	4	6
18	5	5	2	3	5
2	2	4	4	3	4
2	4	5	3	2	7
3	3	4	3	2	7
2	2	6	5	3	9
3	2	6			

(a) What is the response variable? What factor is being studied?

(b) How many levels of the factor are being studied?

(c) Is there any difference in the average numbers of days from the three different approaches? If so, which ones are different?

(d) What conclusions can you draw from these data?

13.2 Many schools are experimenting with different teaching approaches in order to increase student performance but not increase costs. A traditional statistics course was taught in a small classroom (30 students), a large lecture hall (300 students), and online (no face-to-face meetings). The final exam scores for a sample of 15 students from each setting are listed here:

Small classroom	Large lecture hall	Online
72	73	82
66	70	91
76	72	76
83	66	72
83	64	83
87	69	76
59	67	76
73	75	72
82	58	79
67	81	95
70	66	79
63	72	65
62	66	72
68	54	74
69	70	79

(a) What is the response variable? What factor is being studied?

(b) How many levels of the factor are being studied?

(c) Is there any difference in the average final exam scores in the three environments?

(d) Considering that online courses are more flexible and require fewer physical campus resources, what is your recommendation about how to teach this material?

Datafile:
HOMEWORK.XXX

13.3 Grading homework is a real problem. It takes an enormous amount of time. Many students do not do a good job on their homework or they just copy answers from other students or the back of the book. A teacher of elementary statistics decided to conduct a study to determine what effect grading homework had on her students' exam scores. She taught three sections of elementary statistics and randomly assigned each class to one of three conditions: (1) no homework given, (2) homework given but not collected, and (3) homework given, collected, and graded. After the first exam, she collected these data on exam scores:

No homework	Homework, not collected	Homework, collected
69	73	83
69	63	97
92	68	72
84	79	79
79	57	84
84	68	76
76	72	91
63	74	76
76	49	83
82	84	88
89	79	91
72	71	96
72	80	68
65	74	99
73	71	89
47	63	80
92	88	79
71	83	91
83	89	83
81	82	83
92	69	76
80	92	90
64	79	79
72	81	67
84	76	86
79	81	86
74	81	82
81	75	84

(a) What is the response variable? What is the factor?

(b) How many levels of the factor are being studied?

(c) Is there any difference in the average exam scores with the three different approaches to homework? If so, which ones are different?

(d) What is your recommendation to the teacher and why?

13.4 If we are going to rely on e-mail as a means of communication, then people must check their e-mail. A study examined the numbers of times a week that people in four age groups check their e-mail. Here are the data:

Ages 10–18	Ages 18–25	Ages 25–35	Over age 35
22	20	16	15
25	21	17	11
24	20	17	11
20	21	16	14
21	23	15	16
25	20	16	13

27	22	16	9
26	22	18	12
23	21	15	12
24	21	13	16
20	21	16	14
23	19	19	15
25	26	19	14
22	23	17	15
21	22	20	17
17	22	21	11
26	21	18	13
22	16	15	12
23	22	16	15
23	23	17	10

(a) Display the data for each age group as a dotplot.

(b) Find the grand mean and the treatment means.

(c) Is there any difference in the average numbers of times per week that people in the four age groups check their e-mail?

(d) Considering that much of our communication today is via e-mail, what conclusions can you reach from these data?

13.5 The sports industry is a large and competitive market. A manufacturer of golf balls is considering four new ball designs. A sample of 36 balls from each model is tested, and the distances the balls carry are recorded. The balls are hit by a machine. The distances (in yards) follow:

Datafile:
BALLDESN.XXX

M1 model	M2 model	M3 model	M4 model
257	256	244	250
255	255	243	255
256	258	241	251
255	257	243	249
255	257	240	250
256	257	249	251
255	258	244	251
258	258	249	248
252	256	248	247
256	258	243	253
253	258	242	250
254	256	245	255
260	259	253	250
258	255	250	256
258	259	251	260
257	257	249	259
259	256	249	252
257	255	249	253
255	255	255	257
256	257	249	255
259	260	251	251
261	255	250	255
260	261	248	253
257	262	250	256
260	258	258	262
262	260	258	258
258	260	257	263
260	256	258	259
258	255	254	256
258	256	256	261
258	257	255	258
256	251	254	258

(continued)

M1 model	M2 model	M3 model	M4 model
260	253	255	256
260	254	260	263
261	254	254	264
257	255	253	257

(a) What is the response variable? What is the factor?

(b) How many levels of the factor are being studied?

(c) Is there any difference in the average distances the ball carries?

(d) What is your recommendation to the golf manufacturer and why?

13.4 Assumptions of ANOVA

We have mentioned that ANOVA is a commonly used technique. Unfortunately, many times ANOVA is used when it is not the appropriate technique. This may be partly because it is relatively easy to run ANOVA using a statistical package and most software tools simply do the calculations but do not check that it is the appropriate technique. This is your job as the data analyst. As with many of the tools presented in this book, the tool of ANOVA gives valid conclusions only if the data meet certain assumptions. If the data do not meet these assumptions and we use ANOVA, we can easily draw the wrong conclusions. Thus, you should always check that the three major assumptions of ANOVA are met *before* you use this technique.

Here are the three major assumptions of ANOVA:

1. The errors are random and independent of one another.

2. Each population has a normal distribution.

3. All of the populations have the same variance.

13.4.1 ASSUMPTION ABOUT THE ERRORS

The first assumption concerns the behavior of the errors. What are these errors? The term *error* refers to the difference between any observed value and the sample mean of its group. We looked at these differences when we developed the formula for SSE in Section 13.3.3. Remember that SSE is found by adding the squared differences of the form $x_{ij} - \bar{x}_j$. Each term in the sum for SSE is considered an "error." Thus, the first assumption says that these differences should be random and the error for one observation should not influence the error for another observation. If we randomly select a sample from each of the populations, then typically this assumption is met.

Data observed over time may violate this assumption.

The most common situation in which this assumption is violated is with data observed over time. The observation at time period t may influence the observation at time period $t + 1$. If the observations are dependent, then the errors will be dependent. A violation of this assumption could seriously affect the conclusions reached using ANOVA. If you are suspicious of a bias in the sampling process or if you have time-dependent data, then you should consult a more advanced textbook.

13.4.2 ASSUMPTIONS ABOUT THE UNDERLYING POPULATIONS

The second and third assumptions of ANOVA refer to the behavior of the variable being collected. The second assumption says that the variable should have a normal distribution in each of the populations (each level of the factor). We can check this assumption formally using a chi-square goodness of fit test, which is the subject of the next chapter. At this point we can visually check the shape of the data in each sample by looking at a histogram of the data.

EXAMPLE 13.13

TISSUE AIRSPACE

Analyze the data.

Looking at Histograms for Normality

Histograms for all four levels of the factor *time* for the airspace data are shown here:

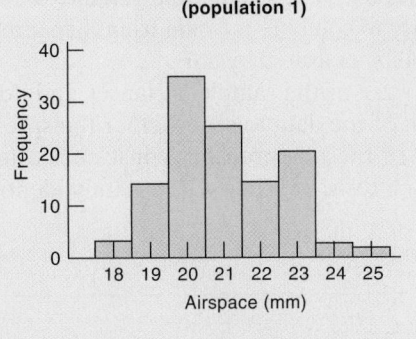

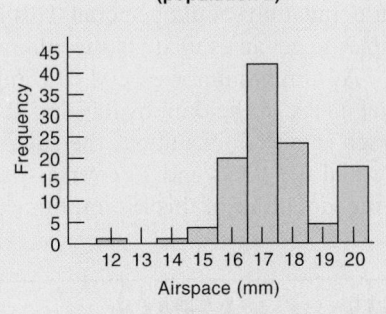

Note: If you use Excel's Histogram command, change the gap width to zero.

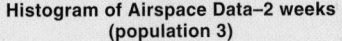

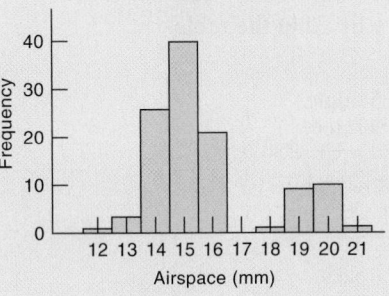

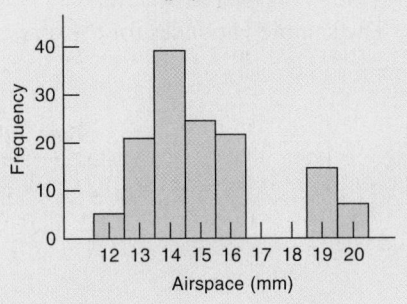

These graphs show that we may be violating the assumption of normality for at least one of the populations.

Another way to test the assumption of normality is to use a normal probability plot. This plot was introduced in the discussion of the assumptions of simple linear regression in Section 11.6.1. A normal probability plot generated by Minitab for the in-process data is shown in Figure 13.4.

Remember! When the data are normally distributed, the plot is a straight line.

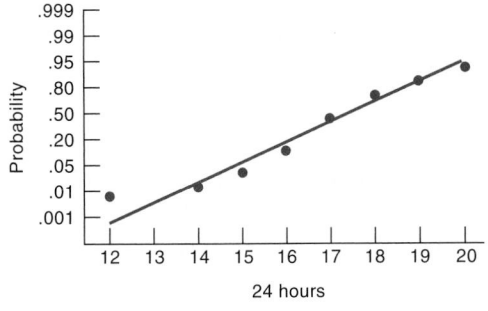

Average: 17.4583
StDev: 1.48322
N: 120

Anderson-Darling Normality Test
A. Squared: 4.758
P. Value: 0.000

The normal probability plot in Excel should not be used because it is not correct.

FIGURE 13.4 Normal probability plot of 24-hour airspace data

Fortunately, the technique of ANOVA is not particularly sensitive to violations of the normality assumption. If the data follow a distribution that is not extremely different from a normal distribution, then the conclusions from ANOVA are probably fine, especially for a large sample size.

Remember! You are using ANOVA to test equality of means.

The third assumption of ANOVA also has to do with the behavior of the distribution of the variable in each of the populations. Once we know that the second assumption has been met, or at least is not severely violated, then we must check to see that the amount of variability in each of the populations is the same. That is, the variance within each population should be equal. This is necessary to combine the data from the various samples to get an estimate of the inherent variability measured by SSE.

At a minimum, we need to compare the sizes of the sample variances and do a visual check of the data by looking at a boxplot of the data to see whether the spread in each sample looks about the same. Neither of these approaches constitutes a test for equal variances and, of course, there are such tests, but they will certainly identify glaring violations of this assumption.

Analyze the data.

EXAMPLE 13.14

TISSUE AIRSPACE

Checking for Equal Variances Using a Boxplot

The sample variances for the airspace data are listed in the table:

Time	Sample standard deviation	Sample variance
1	1.53	2.34
2	1.48	2.19
3	2.08	4.33
4	2.32	5.38

Here are the boxplots of airspace by time:

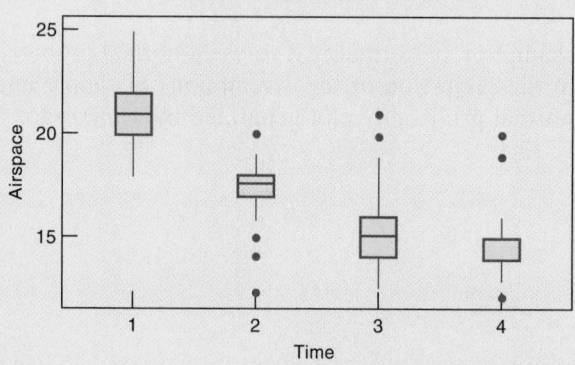

Boxplots for Airspace by Time
(means are indicated by solid circles)

The largest sample variance (5.38) is almost 2.5 times greater than the smallest sample variance (2.19). This represents a large difference in the variances given the size of this sample. These data may violate the third assumption of ANOVA.

Transformations can be used to stabilize the variance.

There are formal statistical procedures to test the assumption of equal variances. Statistical software has more than one way to test this assumption. If the data are in violation of the third assumption, you will need to use an appropriate data transformation to normalize the data and stabilize the variance. Alternatively, you could use a nonparametric technique.

13.4.3 EXERCISES—LEARNING IT!

13.6   Check the ANOVA assumptions for the medical researcher interested in the numbers of days that breathing tubes are left in after surgery.

Requires Exercise 13.1

(a) Is there any reason to believe that the errors are not independent?

(b) Construct histograms of the numbers of days for each of the three surgical methods. Do they look normally distributed?

(c) Calculate and compare the size of the sample variances and do a visual check of the data by looking at boxplots of the data to see whether the spread in each sample looks about the same.

(d) Comment on the validity of the assumptions of ANOVA for these data.

13.7   Check the ANOVA assumptions for the final exam scores in three teaching settings.

Requires Exercise 13.2

(a) Is there any reason to believe that the errors are not independent?

(b) Construct histograms of the final exam scores for each of the three settings. Do they look normally distributed?

(c) Calculate and compare the sizes of the sample variances and do a visual check of the data by looking at boxplots of the data to see whether the spread in each sample looks about the same.

(d) Comment on the validity of the assumptions of ANOVA for these data.

13.8   Check the ANOVA assumptions for the statistics teacher investigating the effect of homework on exam scores.

Requires Exercise 13.3

(a) Is there any reason to believe that the errors are not independent?

(b) Construct histograms of the exam scores for each of the three methods. Do they look normally distributed?

(c) Calculate and compare the sizes of the sample variances and do a visual check of the data by looking at boxplots of the data to see whether the spread in each sample looks about the same.

(d) Comment on the validity of the assumptions of ANOVA for these data.

13.9   Check the ANOVA assumptions for the golf ball manufacturer.

Requires Exercise 13.5

(a) Is there any reason to believe that the errors are not independent?

(b) Construct histograms of the distances the balls carried for the four ball designs. Do they look normally distributed?

(c) Calculate and compare the sizes of the sample variances and do a visual check of the data by looking at boxplots of the data to see whether the spread in each sample looks about the same.

(d) Comment on the validity of the assumptions of ANOVA for these data.

13.5   Analysis of Data from Blocked Designs

13.5.1   MOTIVATION FOR BLOCK DESIGNS

We have looked at two issues for the tissue company: the effect of time on the air-space in the box and the strength of the tissues as measured by the variable *MD-Strength*. In both situations there is one factor of interest and the company is able to collect a sample of *comparable* observations from each population.

Often there is one factor of interest but the observations in the sample may not be comparable due to some other factor. Consider the effect on sales of product display. One factor is being studied here: the product display. Suppose we are considering three different product displays; then there are three levels for this factor. To use a one-way ANOVA we need to collect sales data from a sample of stores using display method 1, a sample of stores using display method 2, and a sample of stores using display method 3. This is not a problem, but one question arises: Are the sales from

Use a block design when it is difficult to get comparable samples.

these stores different for some reason other than the display method being used? It is likely that the answer to this question is yes. The stores are probably different in size and geographic location. These factors would not matter if they did not affect the variable we were studying: sales. If they did not affect sales, then we could happily use a one-way ANOVA, randomly assigning stores to one of the three product display methods. But these factors most likely do affect sales. We are not going to find a sample of like size stores in the same location! So we cannot get a sample of *comparable* stores to sample.

In this example, it seems as though we actually have three factors: product display method, size of the store, and location of the store. But really we are interested only in the effect of the product display method on sales. The other two factors are in a sense "nuisance factors." They need to be taken into account, but we are not directly interested in their effect on sales. The factors size and location need to be blocked out so that we can see the effect of the product display on sales. Otherwise, there is no way to tell whether the differences in sales are a result of the differing product displays or the differing size and/or location of the store.

There are two possible ways to handle the data collection for this example. We could find three stores that are similar in size and location (large city, small city, rural, etc.) and randomly assign a product display method to each store. Then we would need to find another three stores that are similar in size and location and do the same thing. Each set of three stores becomes what is called a **block.** We would use as many blocks as feasible given the amount of time, the amount of money, and the availability of stores.

Alternatively, we could observe a set of 20 stores for 3 weeks. During the first week each store would randomly be assigned to use one of the three product display methods, during the second week the store would use a second product display method, and during the third week the store would use the remaining product display method. In this case each store is a **block.**

> A **block** is a group of objects or people that have been matched. An object or person can be matched with itself, meaning that repeated observations are taken on that same object or person and these observations form a block.

If the realities of data collection lead you to use blocks, then you must take this into account in your analysis. This experimental design is called a **randomized block design.** Instead of using a one-way ANOVA, you use a block ANOVA.

> An experiment has a **randomized block design** if several different levels of one factor are being studied and the objects or people being observed or measured have been matched. Each object or person is randomly assigned to one of the c levels of the factor.

13.5.2 PARTITIONING THE TOTAL VARIATION IN A BLOCK DESIGN

The analysis approach is conceptually the same as in a one-way ANOVA.

The analysis of data from a block design is conceptually the same as that from a one-way design. We essentially still have just one factor of interest, and we are interested in deciding whether there is a treatment effect. That is, we are interested in seeing if the population means are all equal or if at least one is different. The null and alternative hypotheses remain the same:

$$H_0: \quad \mu_1 = \mu_2 = \mu_3 = \cdots = \mu_c$$
$$H_A: \quad \text{At least one of the means is different.}$$

Like the approach we used with data from a one-way design, the idea is to take the total variability as measured by SST and break it down into its components. With a block design there is one additional component: the variability between the blocks. It is called the **sum of squares blocks** and labeled **SSBL.**

> The *sum of squares blocks* measures the variability between the blocks. It is labeled *SSBL.*

The difference between each observation and the mean of its block is squared, and these squared differences are accumulated into the sum of squares blocks. This component measures the portion of the variation that is attributable to the fact that the data were collected in blocks. Since we use blocks because we think there are some "interfering factors," we expect this component to be of significant size. If it is not, then we are probably worried about factors that do not, in fact, affect the variable of interest.

The block factor is probably significant but you must test it.

For a block design, the variation in the data is caused by one of three things: the level of the factor, the block, or the error. Thus, the total variation is divided into three components:

$$SST = SSA + SSBL + SSE$$

SST in terms of other components in a block design

The terms SSA and SSE have the same formulas and definition as for a one-way ANOVA. The effect of partitioning the total variation into another component is to reduce the error term, SSE. Recall that SSE is used to find MSE, which becomes the denominator of the F test statistic. If we can reduce SSE by the use of blocks, then the MSE term will be smaller, making the F test statistic larger. Since "big" F values lead us to reject the null hypothesis of equal means, the use of blocks allows us to see differences in the treatment means that we might not have seen otherwise.

SSA and SSE have the same formulas as in a one-way ANOVA.

13.5.3 USING THE ANOVA TABLE IN A BLOCK DESIGN

Consider a randomized block design with r blocks and c groups and a total of $n = rc$ observations. The ANOVA table for such a block design looks just like the ANOVA table for a one-way design with an additional row. It is shown in Table 13.4.

The last two columns of the ANOVA table are used to do the hypothesis test of the equality of the population means. The hypothesis test is easily done by determining if the F value is in the rejection region. If the F test statistic is "too large"—that is, greater than the critical F value from the table—then we conclude that MSA and MSE are not, in fact, estimates of the same number and there is a treatment effect. If your software does not provide the critical F value, you can look it up in the F table in Appendix A. The F statistic has $(c - 1)$ degrees of freedom in the numerator (from the MSA term) and $(r - 1)(c - 1)$ degrees of freedom in the denominator (from the MSE term).

Use the F test statistic and p value to test the hypothesis of equal means.

TABLE 13.4

SUMMARY TABLE FOR ANOVA WITH A BLOCK DESIGN

Source of variation	SS	df	MS	F	p Value
Between groups	SSA	$c - 1$	$MSA = \dfrac{SSA}{c - 1}$	$F = \dfrac{MSA}{MSE}$	
Between blocks	SSBL	$r - 1$	$MSBL = \dfrac{SSBL}{r - 1}$		
Within groups	SSE	$(r - 1)(c - 1)$	$MSE = \dfrac{SSE}{(r - 1)(c - 1)}$		
Total	SST	$rc - 1$			

Let's see how this works by following the sales and product display example.

Collect and analyze the data.

EXAMPLE 13.15

PRODUCT DISPLAY LOCATION

Conducting a Hypothesis Test with a Block Design

A national chain of stores is investigating the effect of product display on sales. The analysts are interested in three product display locations: on the shelf (group 1), at the end of the aisle (group 2), and at the entrance to the store (group 3). They have matched stores with respect to size and location and have created ten blocks of stores. The stores in the blocks were randomly assigned a product display location and the monthly sales were recorded. The data are shown here:

What do you notice about the "front of store" sales?

	A	B	C	D
1	**Block**	**Shelf**	**End of Aisle**	**Front of Store**
2	1	4457	4500	5800
3	2	4400	4370	5290
4	3	4310	4300	5000
5	4	4600	4400	5600
6	5	5000	5000	6000
7	6	4500	4500	5300
8	7	4700	5100	5900
9	8	4590	4280	5460
10	9	8510	8670	10600
11	10	6470	6500	8410
12				
13	**Average**	**5153.7**	**5162**	**6336**

The data are displayed by location in the graph:

Sales by Location of Product Display

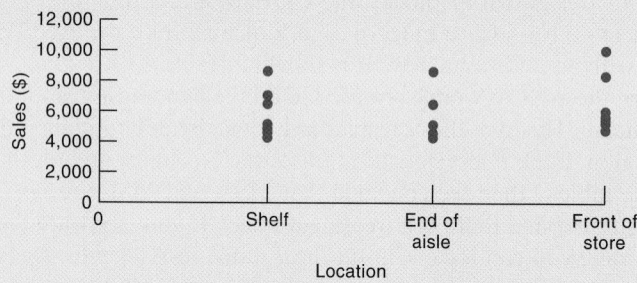

Here are the ANOVA table and the group sample means and confidence intervals generated by Minitab:

Analysis of Variance of Sales

Source	DF	SS	MS	F	P
Location	2	9253927	4626964	60.70	0.000
Block	9	60738440	6748716	88.54	0.000
Error	18	1371972	76221		
Total	29	71364339			

```
                              Individual 95% CI

Location        Mean     ------ + ------ + ------ + ------ + ------
End of Aisle     5162    (----*----)
Front of Store   6336                              (---*---)
Shelf            5154    (----*---)
                         ------ + ------ + ------ + ------ + ------
                          5200    5600    6000    6400
```

From the sample means and the confidence intervals for the individual population means, we can see that locating the product at the front of the store generates considerably more sales.

To complete the hypothesis test we need to find the F value and then use the F table to find the rejection region. Minitab does not calculate the F or p values for a block design. The null and alternative hypotheses are

$$H_0: \quad \mu_1 = \mu_2 = \mu_3$$
$$H_A: \quad \text{At least one population mean is different.}$$

The F statistic is MSA/MSE $= 4{,}626{,}964/76{,}221 = 60.7$. The critical value for the rejection region is found from the F table with 2 $(c - 1)$ and 18 $(r - 1)(c - 1)$ degrees of freedom. If $\alpha = 0.05$, then the critical F value is 3.55. Clearly, $F = 60.7$ is greater than the critical F value and we reject H_0.

We conclude that the location of the product display affects the average sales. The confidence interval plot indicates that it is probably group 3 (front of store) that has the different mean because the other two sample means are similar and their confidence intervals overlap substantially.

Draw conclusions.

All the assumptions of one-way ANOVA pertain to the block design. Thus, you should be sure to check these assumptions for your data set.

Be sure to check your assumptions!

✔ TRY IT NOW!

Airline Arrivals
Using the Block Design

In Chapter 8 we considered the average arrival delays into New York's LaGuardia Airport, but we considered only one airline. Now we want to see whether there is a difference in the mean arrival delays for three different carriers: TWA, United Airlines (UA), and American Airlines (AA). To compare the arrival delays we block on the date. We collect data on 90 flights into LaGuardia on August 1–6, 2001. On each date we record the arrival delays for five flights for each airline. Use the Minitab output shown here to determine whether the mean arrival delays differ for the three airlines.

```
Analysis of Variance for Arrival
Source      DF      SS       MS       F        P
Carrier      2     135       68      0.10     0.904
Block        5    9496     1899      2.84     0.021
Error       82   54914      670
Total       89   64546


                     Individual 95% CI
Carrier    Mean    ------+---------+---------+---------+-----
TWA        11.9    (---------------*---------------)
AA         14.3      (---------------*---------------)
UA         14.7       (---------------*---------------)
                    ------+---------+---------+---------+-----
                        6.0      12.0      18.0      24.0
```

13.5.4 BENEFITS OF BLOCKING

Is it worth it to use a block design?

The purpose of blocking is to allow us to see a treatment effect with a smaller sample size. However, it is often difficult to set up the blocks and so it is interesting to examine whether the effort to set up the blocks is worth it. The following formula gives a measure of the relative efficiency of the block design compared with a one-way design:

Relative efficiency of block design

$$\text{Relative efficiency} = \frac{(r - 1)\text{MSBL} + r(c - 1)\text{MSA}}{(n - 1)\text{MSE}}$$

The terms in this formula are all found in the ANOVA table from a block design. The value of n is the total sample size. The relative efficiency value tells us by what factor the sample size would have to be increased to see the treatment effect without using a block design. Clearly, if the relative efficiency is a low number such as 1.5, then perhaps it is not worth the effort to set up the blocks. Let's see how this formula works for the product display location.

Analyze the data.

Draw conclusions.

EXAMPLE 13.16

PRODUCT DISPLAY LOCATION

Finding the Relative Efficiency

The ANOVA table for the product display location data is shown here:

Source	DF	SS	MS
Display	2	9,253,927	4,626,964
Block	9	60,738,440	6,748,716
Error	18	1,371,972	76,221
Total	**29**	**71,364,339**	

The relative efficiency of this design is calculated as follows:

$$\text{Relative efficiency} = \frac{(10 - 1)6,748,716 + 10(3 - 1)4,626,964}{(30 - 1)76,221} = 69.3$$

This number tells us that we would have needed 69 times more observations to see the treatment effect if we did not use a block design. Clearly, the blocking was worth it in this case.

✓ TRY IT NOW!

Airline Arrivals

Finding the Relative Efficiency

Calculate the relative efficiency of the block design for the data on airline arrival delays. Comment on the usefulness of the blocking.

Ans. Relative efficiency = 0.17. Blocking does not seem to have mattered.

13.5.5 EXERCISES – LEARNING IT!

Datafile: AUCTION1.XXX

13.10 You have become a regular user of the electronic auctions on eBay. You are trying to decide on the "best" strategy for bidding on an item. You wonder if the number of bids differs for each passing day of the auction. You decide to collect data on the auctions of three items: a grandfather clock, a computer, and a set of knives. For each auction you record the numbers of bids on each of the seven days the item is on auction. Here are the data:

Item	Day	Number of bids
Grandfather clock	1	24
Grandfather clock	2	16
Grandfather clock	3	12
Grandfather clock	4	10
Grandfather clock	5	10
Grandfather clock	6	22
Grandfather clock	7	35
Computer	1	32
Computer	2	28
Computer	3	29
Computer	4	21
Computer	5	33
Computer	6	42
Computer	7	45
Knives	1	37
Knives	2	34
Knives	3	21
Knives	4	16
Knives	5	25
Knives	6	31
Knives	7	40

(a) What is the factor? How many levels are in this design?

(b) What is the blocking factor?

(c) Is there a significant difference in the numbers of bids over the life of the auction?

(d) Calculate the relative efficiency of the blocking. Is the blocking worth it?

Datafile: ALGORTH.XXX

13.11 Many techniques for finding the "best" or optimal solution to a problem require that you provide the computer software with a starting solution. There are often different ways to get this starting solution. Four algorithms for generating starting solutions are run on 30 different problems. The numbers of iterations required to move from the starting solution to the optimal solution were recorded. A portion of the data is shown here:

Problem	Method	Iterations
1	1	5
2	1	4
3	1	4
4	1	4
5	1	2
6	1	2

(a) Is there any difference in the numbers of iterations required to find the optimal solution that is due to the difference in the method used to generate the starting solution?

(b) Which method would you recommend and why?

13.12 No. 1 Foods contracts with colleges and universities to provide their food services. It has three different cafeteria layouts, which it is evaluating in terms of the time (in minutes) it takes for a "typical" student to get his or her food. To do a fair comparison No. 1 Foods has blocked on the size of the school. Is there a difference in the average times spent getting food for the three different layouts? If so, which layout is best? The data are shown here:

	Layout 1	Layout 2	Layout 3
1	6.5	2.9	6.9
2	5.8	4.5	6.1
3	5.8	4.1	7.0
4	6.7	4.2	7.4
5	6.4	5.1	6.9
6	5.8	4.7	7.7
7	6.2	5.2	7.8
8	5.7	4.9	8.1
9	6.9	4.8	7.8
10	5.0	4.3	8.2
11	5.8	4.2	7.1
12	4.8	4.1	7.4
13	6.7	4.7	7.2
14	6.6	4.3	7.6
15	4.7	4.4	7.5

13.13 Should you listen to classical music, watch TV, or just have silence when you study? To answer this question a group of ten undergraduate students were randomly selected from a large university. For one semester they listened to classical music while they studied, for the next semester they watched TV while they studied, and for the third semester they studied in the library. Their GPAs for each semester were recorded:

Student	Music	TV	Library
1	3.62	3.35	3.23
2	3.38	3.23	3.07
3	3.09	2.24	2.70
4	2.69	1.88	2.41
5	3.29	2.31	3.01
6	2.97	2.75	2.82
7	2.60	1.98	2.14
8	3.56	3.26	3.07
9	3.50	2.74	3.23
10	3.74	3.19	3.55

(a) What is the response variable?

(b) What is the factor? How many levels are there?

(c) What is the blocking factor?

(d) Use block ANOVA to determine whether there are any differences in the average GPAs for the three treatment groups.

(e) What other factors would you suggest be included in a subsequent study?

Datafile:
GRADSAL.XXX

13.14 What can I expect as a starting salary if I major in nursing? Computer science? Education? The career office at a university decided to see whether there were any differences in the starting salaries of these three majors. Analysts selected 50 students from last year's graduating class and recorded the starting salary for each student. To do a fair comparison, the students were matched on the basis of the overall grade point average. The data are shown here:

Nursing	Computer science	Education
$27,220	$35,685	$24,476
26,204	33,500	24,246
28,065	32,955	24,612
28,778	32,878	25,057
28,498	32,802	24,821
29,150	33,389	25,588
25,389	33,875	24,562
27,310	32,435	24,390
28,501	33,131	23,689
26,433	33,556	24,832

26,557	32,996	24,698
25,707	33,545	25,214
25,297	33,705	24,222
26,636	35,324	24,632
26,624	31,834	24,661
25,447	33,309	23,711
26,943	34,988	24,688
27,064	33,941	24,732
27,666	33,628	24,092
27,150	32,888	24,916
27,230	32,950	24,800
27,042	33,735	24,617
28,949	34,607	24,218
27,240	33,921	24,191
27,512	34,573	24,726
27,031	32,849	24,417
29,402	33,686	24,226
28,487	33,839	24,424
29,841	33,770	24,411
26,408	33,017	23,915
29,049	34,669	24,333
26,191	35,282	24,638
28,137	33,676	25,528
28,347	34,547	25,034
29,584	33,065	25,062
27,569	33,938	24,434
26,976	33,686	24,853
27,960	33,094	24,459
26,835	33,494	24,609
28,228	33,414	23,643
25,731	32,811	25,267
26,554	34,200	23,586
25,770	32,487	24,578
27,307	34,261	24,880
27,498	33,211	24,783
27,235	33,320	24,144
27,227	33,385	25,203
29,677	33,736	24,538
25,758	34,803	24,359
27,073	33,694	24,166

(a) What is the response variable?

(b) What is the factor? How many levels are there?

(c) What is the blocking factor?

(d) Use block ANOVA to determine whether there are any differences in the average starting salaries for the three groups.

(e) What other factors would you suggest be included in a subsequent study?

13.15 Surely you have heard of the "freshman ten," or the average weight gain of 10 pounds by students during the freshman year. Does the same phenomenon occur to graduate students? A sample of 15 students was taken and their weights were recorded in September, February of the following year, July of the following year, and then July 2 years later. Here are the weights (in pounds):

Datafile:
WEIGHT.XXX

Person	September	February	July	July, 2 years later
1	110	115	115	111
2	140	144	146	141
3	165	169	171	165

(continued)

Person	September	February	July	July, 2 years later
4	123	128	131	124
5	158	167	169	159
6	180	187	187	181
7	173	179	179	173
8	178	185	190	180
9	195	203	207	196
10	145	154	158	145
11	136	137	137	137
12	170	179	180	170
13	166	168	171	168
14	180	184	185	181
15	130	136	138	130

A new student of statistics decided to use a one-way ANOVA to analyze these data. The results from Minitab follow:

Analysis of Variance for Weight

Source	DF	SS	MS	F	P
Month	3	634	211	0.36	0.784
Error	56	33108	591		
Total	59	33742			

(a) Using this ANOVA table, what do you conclude about the average weights at the four times the data were collected?

(b) Explain why this analysis is incorrect and why it must be analyzed as a block design.

(c) What are the blocks?

(d) The Minitab output for the block ANOVA is shown here. What do you conclude about the average weights at the four times?

Analysis of Variance for Weight

Source	DF	SS	MS
Person	14	32990.73	2356.48
Month	3	633.87	211.29
Error	42	117.13	2.79
Total	59	33741.73	

Discovery Exercise 13.1

THE BENEFITS OF BLOCKING

In manufacturing electronic products such as loudspeakers, it is important that connections are strong and hold so that they do not disconnect. The same is true for commercial heaters. One of the customers of a heater manufacturer complained that the pull poundage was too low. Thus, the customer did not have confidence in the terminal connection. New wires with the terminals attached were made and a pull test was done on a sample of 25. The terminals were connected and pulled apart. The pull poundages at which the terminals disconnected from each other were recorded. Then the same terminals were reconnected and the test was repeated a total of six times. The object was to determine if the pull poundages changed over time. The minimum pull poundage required by the customer was 5 pounds. The data are shown here:

Trial 1	Trial 2	Trial 3	Trial 4	Trial 5	Trial 6
11.0	12.3	9.8	12.3	10.8	8.5
8.5	10.8	7.5	11.0	8.0	8.0
9.5	10.0	9.0	9.0	9.0	7.8
10.0	7.5	8.0	9.8	10.5	8.5
10.5	7.5	10.3	9.0	9.0	9.8
11.8	11.8	10.0	9.0	9.5	10.3
11.5	10.0	14.0	11.3	14.0	13.0
10.0	9.5	10.0	8.0	8.3	8.0
11.3	12.0	11.0	11.3	10.5	10.3
10.3	10.5	9.8	9.8	11.8	10.5
11.0	15.0	13.5	11.5	9.8	10.0
11.3	10.8	9.8	10.3	10.0	9.5
10.5	10.0	10.5	10.5	10.0	9.5
11.8	10.8	10.0	11.8	9.8	9.5
10.0	10.0	9.5	11.8	9.0	9.3
11.5	11.0	11.0	9.8	9.0	9.8
10.5	9.0	12.5	11.3	11.0	11.8
11.3	11.0	9.5	9.0	10.0	8.0
11.0	11.3	11.3	10.0	11.3	9.5
11.0	8.0	10.5	7.0	7.0	12.0
11.8	10.3	12.3	11.8	12.5	11.0
12.0	12.3	11.5	12.0	12.3	11.0
11.5	12.0	11.0	10.0	11.5	11.5
12.0	9.5	12.5	9.5	9.0	9.3
10.0	7.5	10.5	10.3	9.5	9.0

Datafile:
HEATERS.XXX

Part I: One-Way ANOVA

Use a one-way ANOVA to analyze these data.

1. What is the response variable? What is the factor?
2. How many levels of the factor are being studied?
3. Is there any difference in the average pull poundage among the trials? If so, which ones are different?
4. What is your recommendation to the company and why?

Part II: Block Design

Since the same terminals were connected and pulled apart six times, each row is actually a block. Reanalyze these data using a block design.

1. Is there any difference in the average pull poundage among the trials? If so, which ones are different?
2. Explain why the answer is different when you analyze the data as a one-way design.
3. Calculate the sample size needed to see the treatment effect using a randomized one-way design.
4. Now, what is your recommendation to the company?

13.6 Analysis of Data from Two-Way Designs

After learning about block designs and thinking about the results of the airspace analysis, you wonder whether some other factor might be influencing the airspace in the box (something other than the length of time the box sits on the shelf). You return to ask the people who work in manufacturing about how the tissue boxes are made.

13.6.1 MOTIVATION FOR A FACTORIAL DESIGN MODEL

Figure 13.5 is a picture of a hardroll, which is a large roll of raw paper stock. Each box of 250 tissues is made from 25 hardrolls, each of which has 10 different slit positions. Four such slit positions are shown in the diagram. One tissue is taken from each slit and they are pressed into a tissue box and sealed. A hardroll lasts for 4 hours in production.

The manufacturing personnel think that the mean airspace might differ by position in the hardroll from which the tissues are made. This hypothesis was proposed because of two facts:

1. As the hardroll sits in the warehouse, the outside collects moisture. Thus, boxes made from tissues taken from the outside of the roll might have less growback.

2. The core (very center) of the roll is very compressed (a hardroll weighs 1000 pounds). Thus, boxes of tissues made from the core have the "life stretched out of them."

As a result of this discussion you have identified another potential factor, the position in the hardroll from which the box is produced. This is not a blocking factor because you are interested in knowing whether this factor affects the airspace in the box. Hence, it is not a nuisance factor. You return to look at your data and realize that a total of four cases of tissue boxes were sampled, one case every hour for 4 hours. One case was used to obtain in-process measurements. The other three cases were saved for observations taken after 24 hours, 2 weeks, and 4 weeks. Each case contained 24 cartons of tissue boxes or 120 tissue boxes. This means that the tissues that were allowed to sit on the shelf for 24 hours were made from a different part of the hardroll than the ones that sat on the shelf for 2 weeks or 4 weeks. You cannot tell from this data set if the position in the hardroll is indeed a significant factor because you did not account for this factor in your experimental design.

Based on this new information you suggest the following sampling plan:

Revised sampling plan for the tissue company

1. Sample five times through the hardroll. Since a hardroll lasts for 4 hours, the five observations are taken at the beginning and once an hour thereafter. This results in observations at the top, the core, and three in the middle of the hardroll.

2. Each sample consists of a case of tissues (24 cartons or 120 tissue boxes).

3. A total of four cases are sampled every hour. One is used to obtain in-process measurements. The other three cases are saved for observations taken after 24 hours, 2 weeks, and 4 weeks.

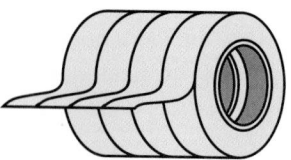

FIGURE 13.5 **Schematic of a hardroll**

Each airspace measurement is taken on a tissue box made from a certain position in the hardroll and allowed to sit on the shelf a certain amount of time before measurement. Thus, we have two factors: position and time. The position indicates from what part of the hardroll the sample is taken. The values for position are 1, 2, 3, 4, and 5. Position 1 indicates that it is taken from the outside of the roll and position 5 indicates it is taken from the core. The factor time indicates when the measurement is taken: 1 = in process, 2 = after 24 hours, 3 = after 2 weeks, and 4 = after 4 weeks.

This experimental design is an example of a **factorial design with two factors.** The two factors do not need to have the same number of levels, but we assume that the sample size is the same for each combination of the levels. For the airspace data this means that the number of tissue boxes sampled from position 1 and tested in process is the same as the number of tissue boxes sampled from position 3 and allowed to sit 2 weeks on the shelf, and this is the same as the number of tissue boxes sampled from each of the five positions and allowed to sit for each of four different time possibilities. In this design, a sample of one case (24 cartons or 120 boxes) is selected from each of the $5 \times 4 = 20$ populations. This is referred to as **equal replication.** Each observation within a population is referred to as a **replicate.** If you have unequal sample sizes, you must refer to a more advanced text.

> An experimental design is called a **factorial design with two factors** if several different levels of two factors are being studied. The first factor is called **factor A** and there are r levels of factor A. The second factor is called **factor B** and there are c levels of factor B.

EXAMPLE 13.17

TISSUE AIRSPACE

Setting Up the Two-Factor Design

For the airspace data, we can label the position factor as factor A and the time factor as factor B. There are $r = 5$ levels of factor A: positions 1, 2, 3, 4, and 5. Position 1 refers to tissues made from the very outside of the hardroll and position 5 refers to tissues made from the core of the hardroll. There are $c = 4$ levels of the factor B, as we had before.

Understand the problem.

> The design is said to have **equal replication** if the same number of objects or people being observed or measured are randomly selected from each population. The population is described by a specific level for each of the two factors. Each observation is called a **replicate.** There are n' observations or replicates observed from each population. There are $n = n'rc$ observations in total.

EXAMPLE 13.18

TISSUE AIRSPACE

Calculating the Number of Replicates

For the airspace data, one case or 120 tissue boxes are selected for each combination of position and time. There are 20 combinations of position and time ($rc = (5)(4) = 20$). Thus, n' is 120 and there are $(2)(120) = 2400$ observations in total.

Collect the data.

The tissue company wishes to know whether there is a difference in airspace due to position in the hardroll and whether there is a difference in airspace by time tested. To answer these questions we must extend our analysis technique to handle this second factor.

13.6.2 PARTITIONING THE VARIATION

Break down SST into four components.

The underlying concept in analyzing data from a two-factor design is the same as the approach we took in analyzing data from a one-way design. Remember that we looked at the relative sizes of the components of variation. For a two-factor design the variation in the data is due to one of four things: factor A, factor B, the interaction between factors A and B, and error. The idea is to take the total variability as measured by SST and break it down into four component sum of squares.

The names and labels for these components are **sum of squares due to factor A (SSA), sum of squares due to factor B (SSB), sum of squares due to the interacting effect of A and B (SSAB),** and **sum of squares error (SSE).** Although the names are different from those we used in a one-way ANOVA, they are an extension of the same concept.

The *sum of squares due to factor A* is labeled *SSA.* It measures the squared differences between the mean of each level of factor A and the grand mean.

The *sum of squares due to factor B* is labeled *SSB.* It measures the squared differences between the mean of each level of factor B and the grand mean.

The *sum of squares due to the interacting effect of A and B* is labeled *SSAB.* It measures the effect of combining factor A and factor B.

The *sum of squares error* is labeled *SSE.* It measures the variability in the measurements within the groups.

SST in terms of other components in a two-way design

Thus, the total variation is divided into four components:

$$SST = SSA + SSB + SSAB + SSE$$

The terms SSA and SSE have the same formulas and definitions as for a one-way ANOVA. The SSB and SSAB terms are calculated in a similar fashion. If you are using a two-way experimental design, you need to use software to do the calculations. The formulas for the sum of squares terms are long and tedious and any statistical software will generate the ANOVA table for a two-way design.

13.6.3 USING THE ANOVA TABLE IN A TWO-WAY DESIGN

Consider a factorial design with r levels of factor A and c levels of factor B with n' replicates in each combination of factors A and B. There are a total of $n = n'rc$ observations. The ANOVA table for such a design looks just like the ANOVA table for a one-way design with two additional rows. It is shown in Table 13.5.

TABLE 13.5

SUMMARY TABLE FOR ANOVA WITH A TWO-WAY DESIGN

Source of variation	SS	df	MS	F	p Value
Factor A	SSA	$r - 1$	$MSA = \dfrac{SSA}{r - 1}$	$F = \dfrac{MSA}{MSE}$	
Factor B	SSB	$c - 1$	$MSBL = \dfrac{SSA}{c - 1}$	$F = \dfrac{MSB}{MSE}$	
Interaction of A and B	SSAB	$(r - 1)(c - 1)$	$MSAB = \dfrac{SSAB}{(r - 1)(c - 1)}$	$F = \dfrac{MSAB}{MSE}$	
Error	SSE	$rc(n' - 1)$	$MSE = \dfrac{SSE}{rc(n' - 1)}$		
Total	**SST**	$rcn' - 1$			

The last two columns of the ANOVA table are used to do the hypothesis tests. In a two-way ANOVA, three hypothesis tests should be done.

1. To test the hypothesis of no difference due to factor A we have these null and alternative hypotheses:

   H_0:   There is no difference in the population means due to factor A.

   H_A: There is a difference in the population means due to factor A.

   Hypothesis test for factor A

   This hypothesis test is easily done by determining whether the F value in the row of the ANOVA table corresponding to factor A is in the rejection region. If the F test statistic is "too large"—that is, greater than the critical F value from the table—then we conclude that factor A is significant. If our software does not provide the critical F value, we can look it up in the table of F distribution values. The F statistic has $(r - 1)$ degrees of freedom in the numerator (from the MSA term) and $rc(n' - 1)$ degrees of freedom in the denominator (from the MSE term).

2. To test the hypothesis of no difference due to factor B we have these null and alternative hypotheses:

   H_0:   There is no difference in the population means due to factor B.

   H_A: There is a difference in the population means due to factor B.

   Hypothesis test for factor B

   This hypothesis test is easily done by determining whether the F value in the row of the ANOVA table corresponding to factor B is in the rejection region. If the F test statistic is "too large"—that is, greater than the critical F value from the table—then we conclude that factor B is significant. If our software does not provide the critical F value, we can look it up in the F distribution table. The F statistic has $c - 1$ degrees of freedom in the numerator (from the MSB term) and $rc(n' - 1)$ degrees of freedom in the denominator (from the MSE term).

3. To test the hypothesis of no difference due to the interaction of factors A and B we have these null and alternative hypotheses:

   H_0:   There is no difference in the population means due to the interaction of factors A and B.

   H_A: There is a difference in the population means due to the interaction of factors A and B.

   Hypothesis test for interaction of factors A and B

This hypothesis test is easily done by determining whether the F value in the row of the ANOVA table corresponding to the interaction of factors A and B is in the rejection region. If the F test statistic is "too large"—that is, greater than the critical F value from the table—then we conclude that the interaction is significant. If our software does not provide the critical F value, we can look it up in the F distribution table. The F statistic has $(r-1)(c-1)$ degrees of freedom in the numerator (from the MSAB term) and $rc(n'-1)$ degrees of freedom in the denominator (from the MSE term). The interaction effect is discussed further in the next section.

Analyze the data.

EXAMPLE 13.19

TISSUE AIRSPACE

Conducting a Hypothesis Test with Two-Way ANOVA

The output from Minitab for the airspace data is shown here. Factor A is position and factor B is time.

Two-way Analysis of Variance

Analysis of Variance for Airspace

Source	DF	SS	MS
Position	4	5942.98	1485.74
Time	3	12773.75	4257.92
Interaction	12	865.52	72.13
Error	2380	1667.08	0.70
Total	2399	21249.33	

To test the hypothesis of no difference due to factor A, we use these null and alternative hypotheses:

H_0: There is no difference in the population means due to position.

H_A: There is a difference in the population means due to position.

The F value is MSA/MSE = 1485.74/0.70 = 2122.5. With such a large F value we really don't need an F critical to tell us that our F value is greater than F critical. However, for completeness the table F value is found for $\alpha = 0.05$ with 4 degrees of freedom in the numerator and 2380 degrees of freedom in the denominator. F critical = 2.37, so clearly position is a significant factor.

To test the hypothesis of no difference due to factor B, we use these null and alternative hypotheses:

H_0: There is no difference in the population means due to time.

H_A: There is a difference in the population means due to time.

The F value is MSB/MSE = 4257.92/0.70 = 6082.7. With such a large F value we really don't need an F critical to tell us that our F value is greater than F critical. However, for completeness the table F value is found for $\alpha = 0.05$ with 3 degrees of freedom in the numerator and 2380 degrees of freedom in the denominator. F critical = 2.60, so clearly time is a significant factor.

To test the hypothesis of no difference due to the interaction of factors A and B we use these null and alternative hypotheses:

H_0: There is no difference in the population means due to the interaction of position and time.

H_A: There is a difference in the population means due to the interaction of position and time.

The F value is MSAB/MSE = 72.13/0.70 = 103.0. With such a large F value we really don't need an F critical to tell us that our F value is greater than F critical. However, for completeness the table F value is found for $\alpha = 0.05$ with 12 degrees of freedom in the numerator and 2380 degrees of freedom in the denominator. F critical = 1.75, so clearly the interaction is significant.

At this point we know from the analysis that both the position in the hardroll from which the tissue box is made and the length of time the box sits on the shelf affect the average airspace in the box. We also know that these two factors interact.

Draw conclusions.

✔ TRY IT NOW!

Airline Arrivals
Conducting a Hypothesis Test with a Two-Way Design

In an earlier Try It Now exercise we looked at the arrival delays (in minutes) for flights into LaGuardia Airport for three airlines: TWA, American Airlines (AA), and United Airlines (UA). Now we are wondering if the city is a factor. Using the data we had for August 1–6, 2001, for flights into LaGuardia in New York, we also consider flights into Boston Logan Airport and Chicago O'Hare Airport. Use this Minitab output to determine whether there are any differences in the mean arrival delays by airline or by airport and whether there is any interaction effect.

Two-Way ANOVA: Arrival Delay (Min.) versus Carrier Code, Airport

Analysis of Variance for Arrival

Source	DF	SS	MS	F	P
Carrier	2	5182	2591	2.01	0.136
Airport	2	93	47	0.04	0.964
Interaction	4	6929	1732	1.35	0.253
Error	261	336034	1287		
Total	269	348238			

```
                         Individual 95% CI
Carrier      Mean    ----------+---------+---------+---------+-
TWA          7.7     (----------*----------)
AA           12.5               (----------*----------)
UA           18.4                         (---------*----------)
                     ----------+---------+---------+---------+-
                          7.0      14.0      21.0      28.0
```

```
                         Individual 95% CI
Airport           Mean   ---------+---------+---------+---------+--
NY LaGuardia      13.7            (----------------*-----------------)
Boston Logan      12.2      (-----------------*-----------------)
Chicago O'Hare    12.8      (------------------*------------------)
                         ---------+---------+---------+---------+--
                             8.0      12.0      16.0      20.0
```

Ans: $F = 2.10$; $F_{\text{CRITICAL}} = 3.072$ using 2 and 120 degrees of freedom and $\alpha = 0.05$. Fail to reject H_0. There are no differences in the mean arrival delays for the three airlines; $F = 0.04$; $F_{\text{CRITICAL}} = 3.072$. Fail to reject H_0. There are no differences in mean arrival delays for the three airports. $F = 1.35$; $F_{\text{CRITICAL}} = 2.447$ using 4 and 120 degrees of freedom and $\alpha = 0.05$. Fail to reject H_0. There is no significant interaction effect.

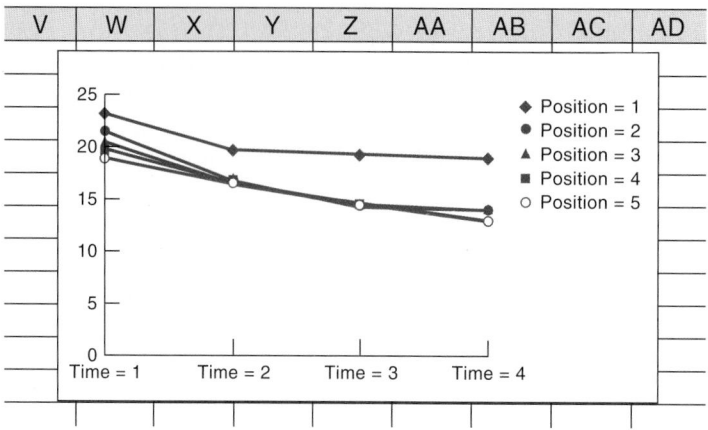

FIGURE 13.6 **Average airspace based on time on shelf for different positions**

13.6.4 UNDERSTANDING THE INTERACTION EFFECT

In the previous section we performed a hypothesis test to determine whether the two factors had a significant interaction effect. The easiest way to understand this effect is to look at a graph of the sample averages for each of the possible combinations of the two factors. The line graph shown in Figure 13.6 displays the 20 sample means for airspace.

From this graph you can see that the mean airspace decreases as the box sits on the shelf longer, regardless of from what position in the hardroll the box was made. However, the line connecting the sample means for those made from position 2 crosses the line connecting the sample means for those made from position 5. This indicates that the amount of growback in the tissue box between the 24-hour mark and the 2-week mark depends on where in the hardroll the tissue came from. Thus, the airspace behavior is affected by the interaction of the time on the shelf and the position in the hardroll from which it was made. This is what we expected based on our hypothesis test.

The information needed to look at the interaction effect graphically is the average of the observed variable for each group. Such information can be generated by a statistical software package. Sometimes it is part of the output of ANOVA and sometimes it is a separate command.

Analyze the data.

EXAMPLE 13.20

TISSUE AIRSPACE

Looking at the Sample Means for All Combinations of Position and Time Tested

In Minitab the analysis of means command generates output as shown here for the airspace data:

Rows: Position   Columns: Time

	1	2	3	4	All
1	120	120	120	120	480
	23.00	19.75	19.58	19.33	20.42
	0.648	0.523	0.705	0.473	1.613
2	120	120	120	120	480
	20.83	16.88	15.08	14.13	16.73
	0.803	1.171	0.762	0.668	2.713

3	120	120	120	120	480
	20.25	17.08	14.54	14.08	16.49
	0.664	1.120	0.916	0.705	2.605
4	120	120	120	120	480
	19.29	16.67	15.04	13.29	16.07
	0.679	0.853	0.938	0.938	2.371
5	120	120	120	120	480
	21.67	16.92	14.67	14.25	16.88
	1.286	0.816	0.690	0.882	3.096
All	600	600	600	600	2400
	21.01	17.46	15.78	15.02	17.32
	1.521	1.478	2.076	2.311	2.976

Cell Contents-

Airspace: N

Mean

StDev

If there were no interaction effect, the lines connecting the sample means would be parallel. This would tell us that the growback behavior of the tissues in the box is the same no matter from what position in the hardroll the tissues are made. The graph might look like the one shown in Figure 13.7.

13.6.5 EXERCISES—LEARNING IT!

13.16 After having lived in several different regions of the United States, you realize that you should purchase a car battery that is best suited for the temperatures of the region. You and your friends decide to study the lifetimes of three different car batteries exposed to three different temperatures: $-30°F$, $50°F$, and $100°F$. For each combination of battery and temperature, four batteries are tested. Here are the lifetimes (in hours):

	Temperature		
	−30°F	50°F	100°F
Battery A	130,155,74,180	34,40,80,75	20,70,82,58
Battery B	150,188,159,126	126,122,106,115	25,70,58,45
Battery C	138,110,168,160	174,120,150,139	96,104,82,60

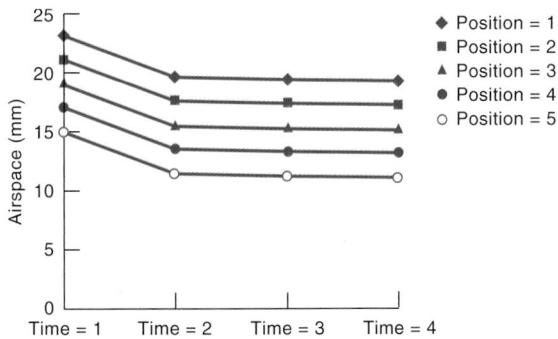

FIGURE 13.7 **Hypothetical average airspace if there were no interaction effect**

(a) Calculate the average life for each of the three battery types.

(b) Calculate the average life for each of the three temperatures.

(c) Calculate the average life for each of the nine treatment groups.

(d) Plot the nine treatment means on a graph with temperature on the x axis and the life of the battery (in hours) on the y axis. Use a different color for each of the three batteries and connect the averages for those of the same battery. What do you speculate about the interaction effect based on the graph?

(e) Confirm your suspicions by doing a two-way ANOVA and testing to see if there is a significant interaction effect.

(f) What battery do you recommend and why?

Requires Exercise 13.5

13.17 The golf ball data that you analyzed as a one-way ANOVA were actually collected at three different times of day. To see if the time of day is a factor, reanalyze the data using a two-way ANOVA. The first 12 observations of each model were taken early in the morning, the second 12 observations were taken mid-day, and the last 12 observations were taken late in the afternoon.

(a) Is there any difference in the average distances the ball carries by design?

(b) Is there any difference in the average distances the ball carries by time of day?

(c) Is there any interaction effect? If so, interpret this in terms of the application.

(d) Check the assumptions of ANOVA for these data.

Requires Exercise 13.10

13.18 After investigating the numbers of bids on eBay by day of the auction, you want to find out whether the number of bids is affected by the type of auction: single unit (one unit for sale) or multiple unit (more than one unit of the same item for sale in the same auction). Each auction lasts seven days. New data are collected, and a portion of the data are shown here:

Datafile:
AUCTION2.XXX

Item	Type of auction	Day	Number of bids
Grandfather clock	Single	1	25
Grandfather clock	Single	2	28
Grandfather clock	Single	3	22
Grandfather clock	Single	4	15
Grandfather clock	Single	5	24
Grandfather clock	Single	6	27
Grandfather clock	Single	7	28
Grandfather clock	Multiple	1	22
Grandfather clock	Multiple	2	20
Grandfather clock	Multiple	3	13
Grandfather clock	Multiple	4	13

(a) With the additional information about the type of auction, this problem can now be considered a two-factor design. What is factor A? What is factor B? How may replicates are there?

(b) Is there any difference in the numbers of bids by day of the auction?

(c) Is there any difference in the numbers of bids by type of auction?

(d) Is there any interaction effect? If so, interpret what it means in terms of the application.

13.19 You and your friends spend a great deal of your spare time playing computer games and you wonder if this is having any effect on the amount of time it takes you to complete tasks. You wish to study students' efficiency as measured by how long they take to complete a particular task (in seconds). You and your friends in the Internet Association decide to do an experiment. Twenty students are timed at the beginning of the day and at 2, 4, and 6 hours after they have been playing intensive computer games. In each case the student uses his or her right hand to complete the task 20 times and the left hand 20 times. The data are shown here:

Beginning of day		After 2 hours of intensive work		After 4 hours of intensive work		After 6 hours of intensive work	
Right	Left	Right	Left	Right	Left	Right	Left
67	162	84	63	52	89	57	114
64	86	78	71	53	53	53	61
69	88	74	86	56	122	71	61
88	99	91	111	66	88	61	61
72	83	70	99	59	93	73	53
80	88	73	88	77	116	50	43
85	113	86	121	64	104	53	56
116	69	71	66	62	53	80	111
77	88	76	56	54	58	63	38
78	101	76	56	65	119	41	40
68	73	61	48	71	96	63	58
51	61	62	51	92	104	41	41
54	76	94	123	71	146	53	84
75	53	63	56	50	81	63	58
71	71	70	83	71	53	61	136
64	61	63	63	58	58	46	93
86	63	66	86	77	73	68	71
98	101	71	63	53	56	64	68
103	86	53	61	81	136	49	73
91	71	81	171	70	83	70	74

(a) Is there any difference in the average times to complete the task after differing amounts of intensive computer work?

(b) Is there any difference in the average times to complete the task for the right hand versus the left hand?

(c) Is there a significant interaction effect?

(d) Display the treatment means to get some additional insight into the interaction effect. What does the graph tell you that the hypothesis test does not tell you?

(e) What is your recommendation?

13.20 You are in the market for a new adhesive product for home use. You are trying to decide between two different varieties of *Always Stick Glue*: regular and gel. You suspect that the amount of time for the glue to take hold also depends on whether or not the item has been broken before in the same spot. So you decide to test both varieties of glue on three items that have experienced a break for the first time and three items that you have already repaired once before. The times (in seconds) for the glue to take hold are shown here:

Type of break	Regular	Gel
First-time repair	175, 100, 175	43, 43, 44
Already repaired once	95, 115, 85	95, 105, 116

(a) Does the glue behave differently if the product has already been repaired once?

(b) Do the regular glue and the gel behave differently?

(c) Is there a significant interaction effect?

(d) Which variety of glue should you buy if you are a student with a low budget who is constantly breaking things? Explain why.

13.7 Other Types of Experimental Designs

Often we wish to investigate more than two factors. The technique of ANOVA extends directly to what is known as a *factorial design* with more than two factors. The interaction terms become more complicated and the required sample size increases as

Two-way ANOVA extends to factorial design.

we consider additional factors. However, the basic approach that we have taken works and the concepts you have learned extend directly to handle more than two factors.

The number of observations that we need increases very quickly as we increase the number of factors. For the airspace data we ended up with 20 groups from two factors. This means the total sample size increases. Suppose, for example, that the tissue company wants to consider a third factor that has three levels. We would have (20)(3) or 60 groups to study. If we select a sample of five from each of these populations, we would need a total sample size of 300. We often do not have the resources to collect this many data.

With the addition of a third factor we can perform more hypothesis tests. We would have three hypothesis tests for the effect of the individual factors, three hypothesis tests to test for interactions between two of the variables (factor A and factor B, factor B and factor C, factor A and factor C), and one hypothesis test to check for interaction among all three variables. Although there are more tests to be performed, each one is done using an *F* test just as we did in this chapter.

Higher order interactions are often difficult to interpret.

The two-way interaction terms are often relatively easy to interpret, as we have seen in the previous section. However, the higher order interactions are typically difficult to interpret and often do not tell us much. Thus, instead of using a full factorial design as we have done with two factors, we use a technique that keeps the sample size down while still giving information about the main effect of the factors on the variable we are studying. These are called *fractional factorial designs*. In the case of the tissue company faced with 60 groups to study, the fractional design may call for a sample to be selected from only 30 groups. Thus, half the number of groups are studied—hence, the term fractional design. We can select the 30 groups to be studied judiciously so that we get information on the three factors and some of the important interaction effects. We lose information on some of the higher order interactions, but at least we have information on the major factors individually and some of the two-way interactions. This is better than not doing the experiment at all because of cost reasons. If you are in a situation where you wish to study more than two factors, you should consult an advanced text on experimental design.

Fractional designs help reduce cost.

GET IT IN WRITING

The Airspace Problem

TO: **Tissue Manufacturing Manager**
FROM: **Erica Q. Analyst**
RE: **Analysis of tissue airspace**

This report addresses the customer complaints that have been received regarding tissues tearing when a new box of tissues is opened. As you know this is caused because there is not enough airspace at the top of the box. Although there is sufficient airspace in the tissue box at the time of manufacturing, as the tissues sit on the shelf they "grow back," leaving less airspace in the box.

To understand this "grow back," phenomenon, we designed an experiment, collected the data, and analyzed the data. This memo will summarize the experiment and the results.

The airspace in 120 tissue boxes (24 cartons) was measured at four different time periods: in process, 24 hours after manufacturing, 2 weeks after manufacturing, and 4 weeks after manufacturing. Clearly, this was destructive sampling because we could not sell the tissue boxes after they had been opened and the airspace measured! Also, we were looking at four populations, so we could not use a *t* test to compare the means. In this case the correct technique to use is ANOVA. Because we were interested in only one factor, time, a one-way ANOVA analysis was done. The output from Excel is shown here:

	J	K	L	M	N	O	P
3	Anova: Single Factor						
4							
5	SUMMARY						
6	*Groups*	*Count*	*Sum*	*Average*	*Variance*		
7	In Process	120	2521	21.00833	2.327661		
8	24 hours	120	2095	17.45833	2.19993		
9	2 weeks	120	1894	15.78333	4.339216		
10	4 weeks	120	1802	15.01667	5.377871		
11							
12							
13	ANOVA						
14	*Source of Variation*	*SS*	*df*	*MS*	*F*	*P-value*	*F crit*
15	Between Groups	2554.75	3	851.5833	239.1302	1.34E-94	2.623636
16	Within Groups	1695.117	476	3.561169			
17							
18	Total	4249.867	479				
19							

From the summary section of this output you can see that the average amount of airspace decreases from 21 mm to just over 15 mm as the boxes sit on the shelf. The *F* value for the between groups variation ($F = 239.1302$) is greater than the *F* critical value (labeled *F* crit) (*F* crit $= 2.62$). The *F* value and the tiny *p* value (1.34×10^{-94}) indicate that the mean airspace is different for at least one of the time periods. The one most likely to be different is the in-process time period. Within 4 weeks of the time of manufacture the airspace decreases on the average about 6 mm.

A word of caution should be added at this point regarding the assumptions of ANOVA. The major assumptions of this statistical technique were checked, and it is possible that the assumption of equal variance for the four time periods may not hold. Further testing of this assumption should be done and if necessary a transformation should be used to stabilize the variance.

After doing this analysis we talked to your employees and learned that the mean airspace might differ by the position in the hardroll from which the tissues were made. A new experiment was designed and data were collected to investigate this possibility. Two factors were used in this experiment: time (with the same four levels as the first experiment) and position in the hardroll (five levels ranging from the outside of the roll to the core of the roll). A two-way ANOVA was done on these data. We conclude that the mean airspace differs with time and with position in the hardroll and that these two factors interact with each other.

Ideally, we would like to specify a target amount of airspace to be left at the time of manufacturing so that by the time the customer opens the box 4 weeks later there will be sufficient airspace left in the box to keep the tissues from tearing. The original estimate of a 6-mm change in airspace due to the time factor must now be adjusted to reflect the position in the hardroll from which the tissues are made. It is possible that different target values should be specified for different parts of the hardroll. We will continue to investigate this problem.

13.8 Using Technology for Analysis of Variance

Excel and Minitab can be used for performing the basics of analysis of variance. Excel has some built-in tools for this purpose. With Minitab you can do all of the analyses in this chapter as well as much more complicated designs. Here we take a look at how to use Excel and Minitab for one- and two-way designs.

13.8.1 ONE-WAY ANOVA IN EXCEL

To perform a one-factor ANOVA using Excel, the data must be in a worksheet with the data for the factor levels in separate, adjacent columns. A portion of the airspace data is shown in Figure 13.8.

	A	B	C	D
1	Time 1	Time 2	Time 3	Time 4
2	23	20	19	19
3	25	19	19	19
4	23	20	20	19
5	23	20	20	19
6	23	20	19	19
7	23	19	20	19
8	23	20	20	19
9	23	20	19	19
10	22	20	19	19
11	22	20	20	20
12	23	20	19	20
13	23	19	19	19
14	22	20	19	19
15	22	20	21	19
16	24	20	20	19
17	23	20	18	20
18	23	20	19	20

FIGURE 13.8 **Airspace data in Excel**

To perform the analysis to see whether there is a difference in the mean airspace at the different times, we use the **ANOVA: Single Factor** tool from the **Tools > Data Analysis** menu. The dialog box shown in Figure 13.9 opens.

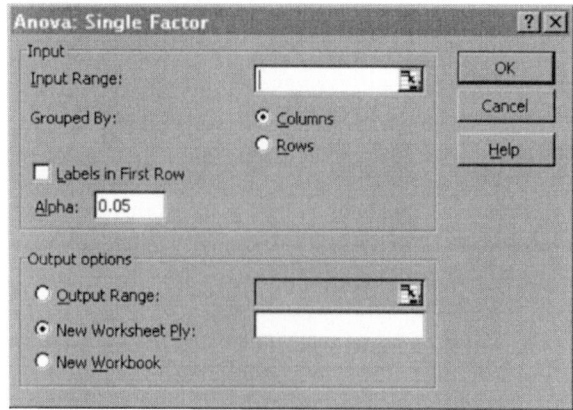

FIGURE 13.9 **Anova: Single Factor dialog box**

The input is similar to most Excel data analysis tools.

1. Enter the range that contains the data in the **Input Range:** text box. If your data range contains labels, click on the text box for **Labels in First Row**.

2. Since each factor is in a different column, click the radio button for **Grouped By: Columns.** This is the default value.

3. In the text box for **Alpha,** enter the level of significance that you want to use for the test.

4. Indicate where you want Excel to put the output from the analysis.

5. Click **OK;** the output appears in the location specified. The output for the air-space data is shown in Figure 13.10. As with the output from the regression analysis, you might want to select **Format > Column > AutoFit Selection** while the output is still highlighted.

	F	G	H	I	J	K	L
Anova: Single Factor							
SUMMARY							
	Groups	Count	Sum	Average	Variance		
	Time 1	120	2521	21.00833333	2.327661064		
	Time 2	120	2095	17.45833333	2.199929972		
	Time 3	120	1894	15.78333333	4.339215686		
	Time 4	120	1802	15.01666667	5.377871148		
ANOVA							
	Source of Variation	SS	df	MS	F	P-value	F crit
	Between Groups	2554.75	3	851.5833333	239.1302467	1.34182E-94	2.623636419
	Within Groups	1695.116667	476	3.561169468			
	Total	4249.866667	479				

FIGURE 13.10 **Output from ANOVA on airspace data**

Excel provides some summary statistics for each factor and the ANOVA table as output. From the ANOVA table, you see that the F value for Between Groups is 239.13, compared with the critical value of 2.6236. The p value of the test is 0 for all practical purposes. Thus, the conclusion is that there is a difference in the mean air-space values for the different time periods.

Excel does not provide any tools for doing multiple comparisons to determine which levels are different, nor does it provide any diagnostic plots for checking assumptions. Since the fitted values are not provided, there is no way to find the residuals and perform the analysis yourself.

13.8.2 TWO-WAY ANOVA IN EXCEL

The data analysis tools in Excel allow you to analyze two-way designs for both situations where there are no replications and situations where there are replications. The no-replication tool is similar to the method you would use to analyze a blocked design, whereas the replications tool allows you to analyze a full two-way design.

To analyze a blocked design, we look at the data on the effects of display type on sales. The data should be in a spreadsheet with the factors (display type) in columns and the blocks (stores) in rows, as shown in Figure 13.11 on page 676.

1. From the analysis tools list, select **ANOVA: Two-Factor Without Replication** and the dialog box shown in Figure 13.12 opens.

2. Enter the worksheet range for the data and fill in the other text boxes with the appropriate information.

3. Click **OK;** the output is placed in the location specified. The output for the store display data is shown in Figure 13.13.

	A	B	C	D
1	Store	Shelf	End of Aisle	Front of Store
2	1	4457	4500	5800
3	2	4400	4370	5290
4	3	4310	4300	5000
5	4	4600	4400	5600
6	5	5000	5000	6000
7	6	4500	4500	5300
8	7	4700	5100	5900
9	8	4590	4280	5460
10	9	8510	8670	10600
11	10	6470	6500	8410

FIGURE 13.11   **Display/sales data**

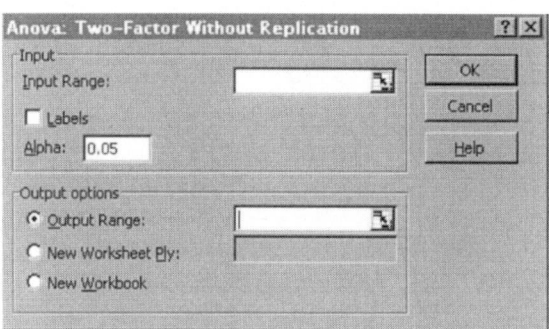

FIGURE 13.12   **Two-Factor Without Replication dialog box**

F	G	H	I	J	K	L
Anova: Two-Factor Without Replication						
SUMMARY	Count	Sum	Average	Variance		
1	3	14757	4919	582583		
2	3	14060	4686.667	273233.3		
3	3	13610	4536.667	161033.3		
4	3	14600	4866.667	413333.3		
5	3	16000	5333.333	333333.3		
6	3	14300	4766.667	213333.3		
7	3	15700	5233.333	373333.3		
8	3	14330	4776.667	374233.3		
9	3	27780	9260	1353100		
10	3	21380	7126.667	1235433		
Shelf	10	51537	5153.7	1782646		
End of Aisle	10	51620	5162	1970196		
Front of Store	10	63360	6336	3148316		
ANOVA						
Source of Variation	SS	df	MS	F	P-value	F crit
Rows	60738440.03	9	6748716	88.5418	4.73E-13	2.456282
Columns	9253927.267	2	4626964	60.70484	9.97E-09	3.554561
Error	1371972.067	18	76220.67			
Total	71364339.37	29				

FIGURE 13.13   **Output for blocked design**

The output includes summary information for each factor level and block in addition to the ANOVA table. We see from the p values that both the blocks (located in the rows) and the display types (located in the columns) are significant.

The method for analyzing a full two-way design with replications is very similar to the method for the one-way and two-way without replications. The only difference is in the way Excel expects the data to be stored in the worksheet.

Excel expects to find the data in table format, with the levels of one factor in columns and the levels of the other factors in rows. Each replicate (observation) for a pair of levels is placed in a cell under the previous one. An explanation of this data setup from the Excel documentation is shown in Figure 13.14.

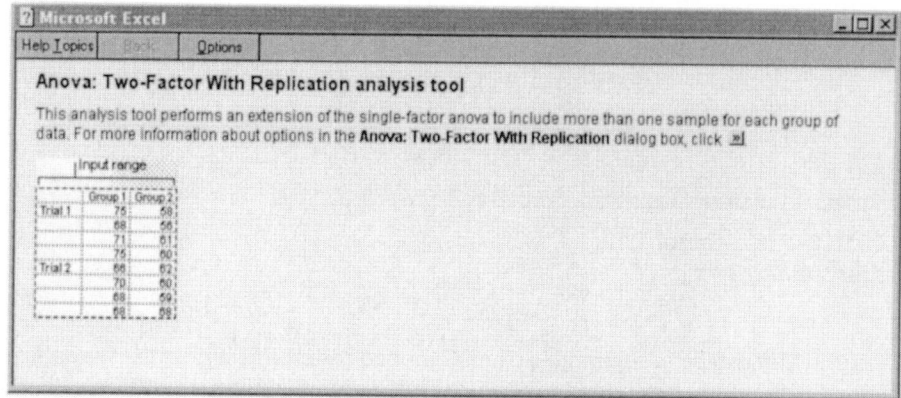

FIGURE 13.14   Data setup for two-way with replications

A subset of the airspace data is shown in Figure 13.15. In this example there are 24 boxes taken from each of five positions. The airspace is measured at four different times.

	A	B	C	D	E	F	
1			Time1	Time2	Time 3	Time 4	
2		Position 1	23	20	19	19	
3			25	19	19	19	
4			23	20	20	19	
5			23	20	20	19	
6			23	20	19	19	
7			23	19	20	19	
8			23	20	20	19	
9			23	20	19	19	
10			22	20	19	19	
11			22	20	20	20	
12			23	20	19	20	
13			23	19	19	19	
14			22	20	19	19	
15			22	20	21	19	
16			24	20	20	19	
17			23	20	18	20	
18			23	20	19	20	
19			23	20	20	19	
20			23	18	20	20	
21			23	20	20	19	
22			23	20	20	19	
23			23	20	21	20	
24			24	20	20	20	
25			23	19	19	20	
26		Position 2	20	18	14	14	
27			20	17	15	14	
28			21	17	15	14	
29			20	16	16	14	
30			21	17	14	15	
31			21	18	15	14	
32			20	17	15	14	

FIGURE 13.15   Airspace data for two-way analysis

The completed dialog box for the two-way with replications analysis tool is shown in Figure 13.16.

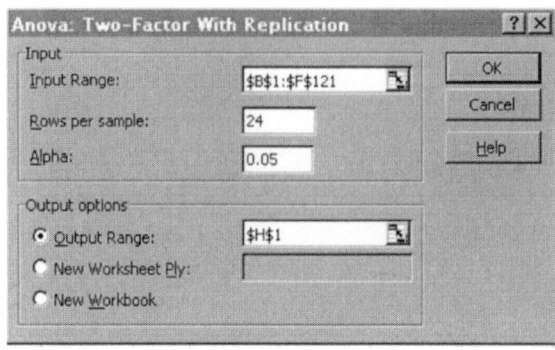

FIGURE 13.16   Two-Factor With Replications dialog box

The only difference in the input is that you need to tell Excel how many replications there are by filling in the text box labeled **Rows per sample:.** In this case there were 24. You *must* highlight the row with the first factor labels and the column with the second factor labels. Excel does not give you a choice here. The output is shown in Figure 13.17. It is similar to the output from the other ANOVA tools.

H	I	J	K	L	M	N
Anova: Two-Factor With Replication						
SUMMARY	Time1	Time2	Time 3	Time 4	Total	
Position 1						
Count	24	24	24	24	96	
Sum	552	474	470	464	1960	
Average	23	19.75	19.5833333	19.3333333	20.4166667	
Variance	0.43478261	0.2826087	0.51449275	0.23188406	2.6245614	
Position 2						
Count	24	24	24	24	96	
Sum	500	405	362	339	1606	
Average	20.8333333	16.875	15.0833333	14.125	16.7291667	
Variance	0.66666667	1.41847826	0.60144928	0.46195652	7.42061404	
Position 3						
Count	24	24	24	24	96	
Sum	486	410	349	338	1583	
Average	20.25	17.0833333	14.5416667	14.0833333	16.4895833	
Variance	0.45652174	1.29710145	0.86775362	0.51449275	6.84199561	
Position 4						
Count	24	24	24	24	96	
Sum	463	400	361	319	1543	
Average	19.2916667	16.6666667	15.0416667	13.2916667	16.0729167	
Variance	0.47644928	0.75362319	0.91123188	0.91123188	5.6683114	
Position 5						
Count	24	24	24	24	96	
Sum	520	406	352	342	1620	
Average	21.6666667	16.9166667	14.6666667	14.25	16.875	
Variance	1.71014493	0.6884058	0.49275362	0.80434783	9.66842105	
Total						
Count	120	120	120	120		
Sum	2521	2095	1894	1802		
Average	21.0083333	17.4583333	15.7833333	15.0166667		
Variance	2.32766106	2.19992997	4.33921569	5.37787115		
ANOVA						
Source of Variation	*SS*	*df*	*MS*	*F*	*P-value*	*F crit*
Sample	1188.59593	4	297.148958	409.963072	3.869E-150	2.39132447
Columns	2554.75	3	851.583333	1174.89128	3.56E-215	2.62429012
Interaction	173.104167	12	14.4253472	19.9020037	4.7221E-35	1.77324111
Within	333.416667	460	0.72481884			
Total	4249.86667	479				

FIGURE 13.17   Output from airspace two-way design

13.8.3 ONE-WAY ANOVA IN MINITAB

There are two routines that do one-way ANOVA in Minitab. They differ only in the way they expect the input data to appear. If the data for each level of the factor are in different columns, then select **Stat > ANOVA > One-way (Unstacked).** If the data values are all in the same column and another column is used to specify which factor level is assigned to each data value, then select **Stat > ANOVA > One-way.** In this section we use the option that has the data values all in one column because that is the way the data are set up for more complicated designs. A portion of the airspace data is shown in Figure 13.18.

↓	C1	C2	C3
	Position	**Time**	**Airspace**
1	1	1	23
2	1	1	25
3	1	1	23
4	1	1	23
5	1	1	23
6	1	1	23
7	1	1	23

FIGURE 13.18 **Portion of airspace data in Minitab worksheet**

To perform the analysis, follow this procedure:

1. Select **Stat > ANOVA > One-way. . .** from the main Minitab menu. The dialog box shown in Figure 13.19 opens.

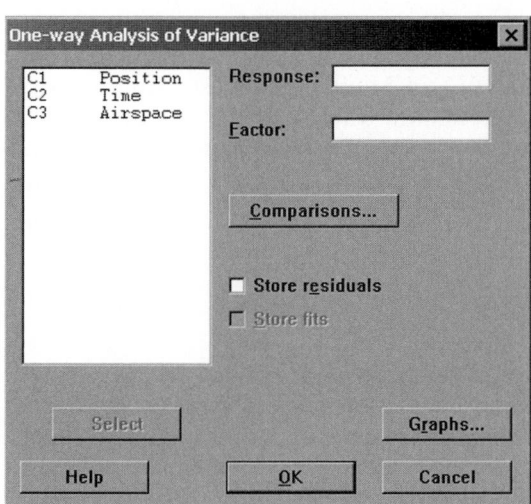

FIGURE 13.19 **One-way ANOVA dialog box**

2. Position the cursor in the box labeled **Response:** and select the variable that contains the data values from the list on the left.

3. Position the cursor in the box labeled **Factor:** and select the variable that contains the factor levels from the list on the left.

4. Click **OK.** The output shown in Figure 13.20 should appear in the Session window.

One-way ANOVA: Airspace versus Time

```
Analysis of Variance for Airspace
Source     DF       SS        MS        F       P
Time        3  2554.75    851.58   239.13   0.000
Error     476  1695.12      3.56
Total     479  4249.87
                                  Individual 95% CIs For Mean
                                  Based on Pooled StDev
Level       N     Mean     StDev   -------+---------+---------+---------
1         120   21.008     1.526                                  (-*-)
2         120   17.458     1.483                     (*-)
3         120   15.783     2.083         (-*-)
4         120   15.017     2.319   (-*-)
                                  -------+---------+---------+---------
Pooled StDev =    1.887            16.0      18.0      20.0
```

FIGURE 13.20   Output from one-way ANOVA

From the output you can see that the value of the test statistic is 239.13 and the p value of the test is 0.000. Clearly we can reject H_0 and conclude that there is a difference in airspace due to time of measurement. As part of the output, Minitab also includes a confidence level for the mean for each level. The graphical display is particularly useful for investigating which of the means differ. From this display it appears that the only means that might be equal are those from time periods 3 and 4, and then just barely.

Minitab also allows you to do formal statistical comparisons of the means. You can access this option by clicking on **Comparisons** in the ANOVA dialog box. However, unless you know something about the methods used, you should be very careful. Each method makes different assumptions and they often have conflicting results.

13.8.4 TWO-WAY ANOVA IN MINITAB

The procedure for doing a two-way ANOVA in Minitab is almost identical to the method for the one-way ANOVA. Suppose we now want to look at whether position also is a factor in airspace. Thus, we have two factors: time and position. Follow these steps:

1. Select **Stat > ANOVA > Two-way. . .** from the main Minitab menu. The dialog box shown in Figure 13.21 opens.

2. Position the cursor in the box labeled **Response:** and select the variable that contains the data values from the list on the left.

3. Position the cursor in the box labeled **Row Factor:** and select the variable that contains one of the factor levels from the list on the left. Click on the box labeled **Display means** to get the graphical display of confidence intervals for the first factor.

4. Position the cursor in the box labeled **Column factor:** and select the variable that contains one of the factor levels from the list on the left. Click on the box labeled **Display means** to get the graphical display of confidence intervals for the second factor.

Note: To do a blocked design, click on the Fit additive model box.

5. Since this is a true two-way design and not a blocked ANOVA, do *not* click on the box labeled **Fit additive model** in order to include the interaction term.

6. Click **OK,** and the output shown in Figure 13.22 should appear in the Session window.

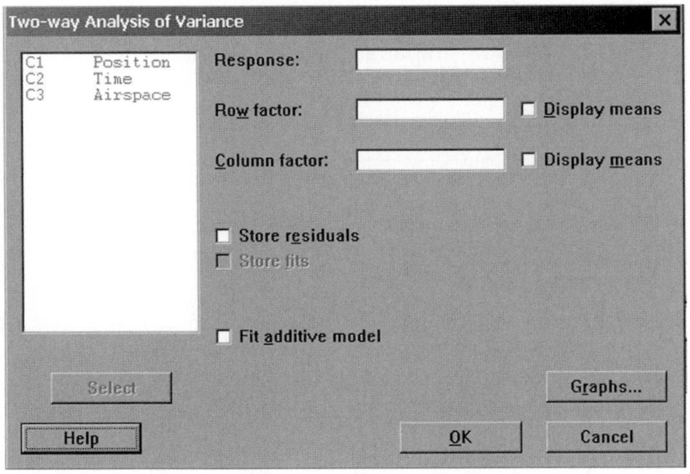

FIGURE 13.21   Two-way ANOVA dialog box

Two-way ANOVA: Airspace versus Time, Position

```
Analysis of Variance for Airspace
Source        DF        SS        MS        F        P
Time           3   2554.750   851.583  1174.89    0.000
Position       4   1188.596   297.149   409.96    0.000
Interaction   12    173.104    14.425    19.90    0.000
Error        460    333.417     0.725
Total        479   4249.867

                        Individual 95% CI
Time          Mean    -+---------+---------+---------+---------+
1           21.008                                           (*)
2           17.458                          (*)
3           15.783             (*)
4           15.017      (*)
                      -+---------+---------+---------+---------+
                  15.000     16.500    18.000    19.500    21.000

                        Individual 95% CI
Position      Mean    --------+---------+---------+---------+---
1           20.417                                         (*-)
2           16.729            (*-)
3           16.490           (*-)
4           16.073      (*)
5           16.875             (-*)
                      --------+---------+---------+---------+---
                          16.800    18.000    19.200    20.400
```

FIGURE 13.22   Output from two-way ANOVA

The output is very similar to the one from the one-way ANOVA.
Minitab has some routines that make very nice plots of the means for each of the
main factors and for the interactions. To obtain these plots select **Stat > ANOVA >
Main effects plot** or **Stat > ANOVA > Interactions plot** from the menu. The dialog
boxes ask for the response variables and the factors. The outputs from these two routines are shown in Figures 13.23 and 13.24 on page 682.

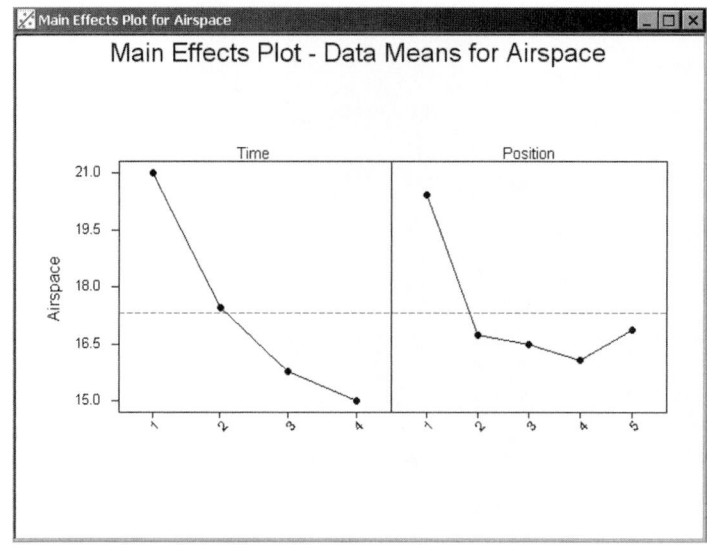

FIGURE 13.23 **Main effects plot**

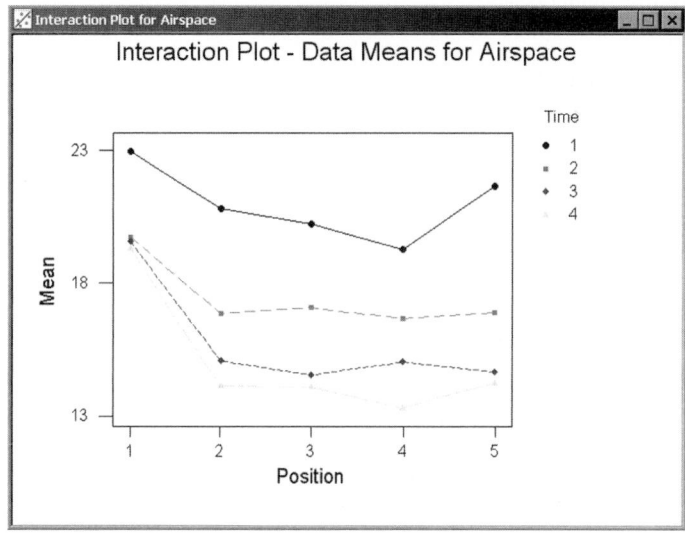

FIGURE 13.24 **Interaction plot**

CHAPTER 13 SUMMARY

In this chapter you have learned the tool of analysis of variance (ANOVA), which is used to draw inferences about more than two population means. It extends directly the two-population tests to compare means, which you learned in Chapter 10. You typically compare more than two populations when you are trying to determine what factors influence the variable you are studying.

Regardless of the number of factors you are investigating, the technique of ANOVA partitions the variation in the data into components and compares the relative sizes of these components. So it is through an analysis of the variation in the data that we learn something about the means. It is a clever approach that draws on many of the basic concepts and tools that you have learned in earlier chapters.

The results of hypothesis tests guide your design of subsequent experiments as you try to understand the factors that cause the variable of interest to vary. Ultimately you are interested in building a model to predict or control this variation.

KEY TERMS

Term	Definition	Page reference
ANOVA	**Analysis of variance (ANOVA)** is the technique used to analyze the variation in the data to determine whether more than two population means are equal.	629
Block	A **block** is a group of objects or people that have been matched. An object or person can be matched with itself, meaning that repeated observations are taken on that same object or person and these observations form a block.	652
Factor	A **factor** is a variable that can be used to differentiate one group or population from another.	627
Factorial design with two factors	An experimental design is called a **factorial design with two factors** if several different levels of two factors are being studied. The first factor is called **factor A** and there are r levels of factor A. The second factor is called **factor B** and there are c levels of factor **B**.	663
Grand mean	The **grand mean** or the **overall mean** is the sample average of all the observations in the experiment. It is labeled $\bar{\bar{x}}$.	630
Level	A **level** is one of several possible values or settings that the factor can assume.	627
Mean squares	The **mean square among** is labeled **MSA**. The **mean square error** is labeled **MSE**. The **mean square total** is labeled **MST**.	638
One-way or completely randomized design	An experiment has a **one-way** or **completely randomized design** if several different levels of one factor are being studied and the objects or people being observed or measured are randomly assigned to one of the c levels of the factor.	628
Randomized block design	An experiment has a **randomized block design** if several different levels of one factor are being studied and the objects or people being observed or measured have been matched. Each object or person is randomly assigned to one of the c levels of the factor.	652
Replicate	The design is said to have **equal replication** if the same number of objects or people being observed or measured are randomly selected from each population. The population is described by a specific level for each of the two factors. Each observation is called a **replicate**. There are n' observations or replicates observed from each population. There are $n = n'rc$ observations in total.	663
Response variable	The **response variable** is a quantitative variable that we are measuring or observing.	627
Sum of squares among (SSA)	**Between groups variation** measures how different the individual treatment means are from the overall grand mean. It is often called the **sum of squares between** or the **sum of squares among (SSA)**.	633

(continued)

Term	Definition	Page reference
Sum of squares blocks (SSBL)	The **sum of squares blocks** measures the variability between the blocks. It is labeled **SSBL.**	653
Sum of squares error (SSE)	**Within groups variation** measures the variability in the measurements within the groups. It is often called the **sum of squares within** or the **sum of squares error (SSE).**	634
Sum of squares due to factor A (SSA)	The **sum of squares due to factor A** is labeled **SSA.** It measures the squared differences between the mean of each level of factor A and the grand mean.	664
Sum of squares due to factor B (SSB)	The **sum of squares due to factor B** is labeled **SSB.** It measures the squared differences between the mean of each level of factor B and the grand mean.	664
Sum of squares due to the interacting effect of A and B (SSAB)	The **sum of squares due to the interacting effect of A and B** is labeled **SSAB.** It measures the effect of combining factor A and factor B.	664
Sum of squares total (SST)	The **total variation** or **sum of squares total (SST)** is a measure of the variability in the entire data set considered as a whole.	631
Treatment	A **treatment** is a particular setting or combination of settings of the factor(s) being studied.	629
Treatment mean	A **treatment mean** is the average of the response variable for a particular treatment.	632

KEY FORMULAS

Term	Formula	Page reference
Sample variance	$s^2 = \dfrac{\sum\limits_{i=1}^{n}(x_i - \bar{x})^2}{n - 1}$	629
SST, one-way design	$\sum\limits_{j=1}^{c} \sum\limits_{i=1}^{n_j}(x_{ij} - \bar{\bar{x}})^2$	631
SSA, one-way design	$SSA = \Sigma n_j(\bar{x}_j - \bar{\bar{x}})^2$	633
SSE, one-way design	$\sum\limits_{j=1}^{c} \sum\limits_{i=1}^{n_j}(x_{ij} - \bar{x}_j)^2$	634
MSA, one-way design	$\dfrac{SSA}{c - 1}$	638
MSE, one-way design	$\dfrac{SSE}{n - c}$	638
MST, one-way design	$\dfrac{SST}{n - 1}$	638
***F* statistic, one-way design**	$\dfrac{MSA}{MSE}$	639
SST, block design	$SST = SSA + SSBL + SSE$	653
Relative efficiency of block design	$\dfrac{(r - 1)MSBL + r(c - 1)MSA}{(n - 1)MSE}$	656
SST, two-way design	$SST = SSA + SSB + SSAB + SSE$	664

CHAPTER 13 EXERCISES

Learning It!

13.21 With the increased use of the Internet people are "demanding" faster and more reliable connections to the Internet from home. To decide among a DSL, cable, and phone dial-up you analyze data on the percentages of time people tried to connect but couldn't on each of ten days for each type of connection. The failure rates are listed here:

	DSL	Cable	Phone dial-up
Day 1	0.020	0.037	0.036
Day 2	0.016	0.032	0.032
Day 3	0.018	0.028	0.037
Day 4	0.021	0.029	0.033
Day 5	0.020	0.031	0.038
Day 6	0.022	0.034	0.035
Day 7	0.019	0.028	0.034
Day 8	0.019	0.032	0.027
Day 9	0.019	0.031	0.036
Day 10	0.021	0.030	0.034

(a) What type of design is this?

(b) Is there any difference in the failure rates by type of connection? If so, which one(s)?

(c) Check the assumptions of ANOVA for these data.

13.22 Once connected to the Internet, many people download MP3 files. Being a student of time management, you wonder if you can save some time by downloading at certain times of the day; that is, you decide to see whether download time depends on the time of day. Data are collected on the amounts of time to download an MP3 file at 6 A.M., noon, 3 P.M., and midnight. Thirty songs are downloaded at each of these four times of day. The download times (in seconds) follow:

Sample	6 A.M.	Noon	3 P.M.	Midnight
1	14.5	14.4	14.3	14.63
2	14.8	14.8	14.7	14.98
3	14.8	14.8	14.7	14.93
4	14.6	14.6	14.5	14.80
5	15.8	15.7	15.6	15.95
6	14.6	14.6	14.5	14.80
7	14.8	14.8	14.7	14.99
8	14.4	14.4	14.3	14.62
9	14.6	14.5	14.4	14.71
10	14.3	14.2	14.2	14.43
11	14.3	14.3	14.2	14.49
12	14.6	14.6	14.5	14.75
13	14.3	14.2	14.2	14.44
14	15.0	14.9	14.8	15.11
15	14.5	14.5	14.4	14.68
16	14.7	14.7	14.5	14.86
17	14.6	14.6	14.5	14.76
18	14.7	14.6	14.5	14.81
19	14.7	14.7	14.6	14.85
20	14.6	14.5	14.4	14.71
21	14.5	14.4	14.3	14.58
22	14.4	14.3	14.3	14.50
23	15.3	15.3	15.2	15.44
24	14.4	14.4	14.3	14.54
25	14.5	14.4	14.3	14.58

(continued)

Sample	6 A.M.	Noon	3 P.M.	Midnight
26	14.8	14.7	14.6	14.86
27	14.6	14.5	14.4	14.72
28	14.5	14.4	14.3	14.57
29	14.4	14.3	14.2	14.50
30	14.6	14.6	14.5	14.75

(a) What type of design is this?

(b) Is there any difference in the measurements taken by the different times of day? If so, which one(s)?

(c) Check the assumptions of ANOVA for these data.

13.23 Unfortunately after you do the experiment in Exercise 13.22 you realize that none of the times you tested corresponds to a time when you would typically be doing downloads. So you collect data during one more time slot, 10 P.M. The download times (in seconds) for the same MP3 files at this new time of day are shown here:

Samples 1–5	14.62	15.00	14.95	14.78	15.93
Samples 6–10	14.79	15.01	14.61	14.70	14.45
Samples 11–15	14.48	14.76	14.45	15.09	14.67
Samples 16–20	14.86	14.77	14.79	14.84	14.71
Samples 21–25	14.58	14.50	15.45	14.55	14.59
Samples 26–30	14.85	14.72	14.58	14.51	14.76

(a) Rerun the ANOVA including these data as a fifth treatment.

(b) Do any of your conclusions change?

Datafile:
ONHOLD.XXX

13.24 Many online service companies have a technical support telephone service to assist users who are having difficulty getting connected or using the service. Recently there has been a concern about the amount of time that a customer has to wait "on hold" before hearing a human voice. In an effort to decide which online service to purchase, a study was conducted. Thirty calls were made to four online services during three times of day: early morning (6–8 A.M.), midday (noon to 2 P.M.), and late afternoon (4–6 P.M.). The amounts of time spent on hold (in minutes) were recorded. A portion of the data is shown in the table:

	Company 1	Company 2	Company 3	Company 4
Early A.M.	11	11	7	3
	8	11	9	8
	7	11	7	6
	8	9	9	9
	8	9	8	11
	7	6	8	9
	8	9	5	9

(a) Is there any difference in the average times on hold among the four companies?

(b) Is there any difference in the average amounts of time on hold by time of day?

(c) Is there any interaction between the time of day and the four companies?

(d) If you could use more than one of the companies, which company would you use during each of the three time periods? Explain your reasoning.

(e) If you cannot use more than one company, which company would you select? Explain your reasoning.

(f) What other factors might influence the time on hold?

13.25 Everything changed, utterly changed, on September 11, 2001. Our lives will never be the same and we are still adjusting to the changes. Changes in our travel patterns have been the most clear and obvious. More people traveled by car during the fall of 2001 than ever before. Also, there have been substantial changes in U.S. border security. Coupling the increased car travel with the heightened border security, customs agents and travelers saw increases in the waiting times to cross the border from Canada into the United States. Some times (in minutes) to cross the border before and after September 11 are shown here:

Before 9-11-01	After 9-11-01
17, 9, 13	26, 29, 22

(a) Based on these data, what can you conclude about the average wait times at the border before and after September 11, 2001?

(b) What else would you like to know about these data to be certain that the increase in wait time at the border is a result of 9/11 actions?

13.26 There has been increasing concern about the declining reading and math abilities of American students. A study was conducted to determine the effect of income on four response variables: behavior problems index, reading score, math score, and vocabulary score. All three of the scores range from 0 to 100. The behavior index ranges from 0 to 100, with low scores indicating fewer behavior problems. One hundred children aged 5 to 7 were tested from families with varying income levels. Three income levels were used: an annual income of $15,000 or less, an annual income of $25,000, and an annual income of $40,000.

Datafile: SCORES.XXX

(a) For each of the response variables, test to see whether there is a difference in the average scores for the three levels of the income factor.

(b) Have any of the assumptions of ANOVA been violated?

(c) What recommendations can you make based on your analysis?

13.27 The student from Exercise 13.19 is also concerned about the accuracy of her friends after many hours of intensive computer work. At the same time the data on times to complete the task were taken, a measure of accuracy (% correct) was also recorded. These data are shown here:

Requires Exercise 13.19
Datafile: MOUSE2.XXX

Beginning of day		After 2 hours of intensive work		After 4 hours of intensive work		After 6 hours of intensive work	
Right	**Left**	**Right**	**Left**	**Right**	**Left**	**Right**	**Left**
95.88	89.70	93.60	90.78	87.96	82.54	79.88	92.38
91.94	90.00	98.00	94.17	96.39	87.96	84.70	86.58
92.79	85.68	90.10	94.17	98.59	95.76	93.68	94.17
95.76	95.53	97.00	89.70	96.00	91.00	93.92	95.00
91.46	81.97	87.96	96.84	84.35	83.24	97.76	96.84
92.72	90.15	95.53	90.78	96.39	95.00	74.68	84.48
93.60	92.72	96.84	92.72	94.90	94.61	86.40	89.23
94.61	83.45	80.76	91.00	94.61	87.96	99.00	90.00
99.00	98.00	95.53	92.19	79.38	79.60	94.61	86.11
97.76	88.30	92.00	88.82	89.70	85.00	76.65	82.91
100.00	93.60	96.39	97.17	93.68	96.84	98.59	94.17
89.18	89.23	80.90	85.58	98.59	90.00	81.13	90.00
87.47	97.00	85.68	94.61	92.93	91.75	86.96	95.88
94.17	96.39	93.29	91.94	85.68	94.90	92.72	83.00
91.40	91.94	96.00	84.97	93.60	87.27	91.94	94.90
95.53	84.19	94.90	90.15	90.57	89.70	80.69	94.61
93.68	96.39	94.34	89.00	96.39	84.70	91.75	87.79
94.90	84.44	92.19	87.79	87.96	95.53	95.00	95.00
97.76	94.17	95.00	91.40	94.90	93.92	85.58	82.97
97.00	93.29	98.59	91.40	95.53	90.57	96.84	88.69

(a) Is there any difference in the average accuracy levels after differing amounts of intensive computer work?

(b) Is there any difference in the average accuracy levels for the right hand versus the left hand?

(c) Is there a significant interaction effect?

(d) What is your recommendation to this student?

13.28 In an effort to reduce crime, several cities have implemented a plan that encourages members of the police force to live in the city they patrol. These police officers have been given the opportunity to purchase homes at half price, with the other half of the cost funded by the city budget. To see whether this program has had the desired effect, data were recorded on the yearly number of murders in five cities the year before the program was initiated, 1 year after the program started, and 2 years after the program was initiated. The data are shown here:

City	Year before program started	1 Year after program started	2 Years after program started
Los Angeles	44	43	43
Chicago	35	34	33
Detroit	41	39	38
Springfield	21	20	19
Philadelphia	29	28	27

(a) What type of experimental design was used in this situation?

(b) Has the program been effective? Why or why not?

(c) Have any of the assumptions of ANOVA been violated?

(d) A government publication states that the average number of murders has dropped from 34 to 31.8 for the cities that adopted this program. The publication uses this statement to propose expanding this program to other cities. Do you agree with this analysis? Why or why not? Should this program be adopted by other cities or should it be scrapped?

Requires Exercise 13.20

13.29 After doing your analysis in Exercise 13.20 your friend tells you about a new company called Really Stick It! This company also has regular and gel varieties of glue. You decide to run the same experiment with the Really Stick It glues. Here are the times (in seconds) for the glue to take hold:

Type of break	Regular	Gel
First-time repair	437, 437, 450	9, 10, 6
Already repaired once	115, 115, 50	105, 87, 105

(a) Does the Really Stick It glue behave differently if the item has already been repaired once?

(b) Do the Really Stick It regular glue and the gel behave differently?

(c) Is there a significant interaction effect?

(d) Summarize your findings about the two glue companies' products in terms of your needs as a student on a low budget who is constantly breaking things.

Thinking About It!

Requires Exercises 13.22, 13.23

13.30 After further thought about the download times for the MP3 files, you decide to compare just two of the columns of data: noon and midnight.

(a) Use a paired *t* test to make this comparison. Is there any difference in the download times?

(b) Run the appropriate ANOVA using just the noon and midnight times as the data set. Does the ANOVA indicate any difference in the treatment means?

(c) Should you have expected the same answer using the paired *t* test and the ANOVA? Explain why or why not.

(d) Take the value of the *t* statistic from the paired test and square it. Compare it with the *F* statistic from the ANOVA. What do you notice? Do you think this will always happen or is this just a coincidence for this data set?

Requires Exercise 13.3

13.31 Reconsider the teacher of statistics who is wondering whether to assign and grade homework. The teacher used three different classes to investigate the three different ways of dealing with homework. The teacher used a one-way design thinking that students randomly select sections of a course to take and so they should be comparable. Is there any reason to think that a block design should have been used in this situation instead of the one-way design? Explain why or why not.

Requires Exercise 13.1

13.32 When the doctor investigated the number of days that the breathing tube was left in after the three different surgical methods, the experiment was set up as a one-way design. Can you think of any reason a block design might be better? If so, what would you consider as the blocking factor?

13.33 In any ANOVA you test the null hypothesis of equal means using an F statistic. If you fail to reject the null hypothesis and you construct a confidence interval for the difference in two of the treatment means, what number should you expect to find in the confidence interval? Explain why.

13.34 In Exercise 13.18 you analyzed the number of bids on eBay auctions using a two-way design with factors of time (days into the auction) and type of auction. Since then we have experienced the tragedies of September 11, 2001, and you wonder if they have had any effect on the numbers of people who are bidding on eBay. Unfortunately this time the auctions you collect data on run for ten days, not seven.

Requires Exercise 13.18

(a) Explain why you cannot simply add another three levels to the time factor and reanalyze the entire data set using ANOVA.

(b) What analysis would you suggest be used with these additional data?

13.35 You have decided to use your newly acquired data analysis skills in your coaching of a 6th-grade girls' basketball team. You want to see whether there is any difference in the average numbers of points your team scores when you use different defense strategies. You suggest the following design: use "man-to-man" defense strategy for the first third of the season and record the points scored in each game. Switch to full-court press strategy for the second third of the season and record the points scored in each game. During the last third of the season use a mixture of defense strategies.

(a) Will you be able to tell which strategy is best from the data collected from this experiment?

(b) Suggest an alternative experimental design.

13.36 In Exercise 13.15 you concluded that the average weights for the students were different at the four different times the data were collected. Suppose you now learned that these students were enrolled in a 1-year weekend MBA program starting in September of one year and ending a year later and that they were all working at full-time jobs during this time.

Requires Exercise 13.15

(a) What might be causing the weight differences you observed?

(b) Support your theory by taking another look at this data set and analyzing it differently.

13.37 In Exercises 13.19 and 13.27 you analyzed the amounts of time to complete a task and the accuracy levels after varying amounts of time of intensive computer work. What other factors should be studied in a subsequent experiment on this subject?

Requires Exercises 13.19, 13.27

Doing It!

13.38 Customers complain about many things. If you have ever been in the middle of cooking something particularly messy and been in desperate need of a paper towel, you may have been frustrated by your inability to get the roll to turn because it is "too wide around." In your frustration you call the paper company to complain!

The problem appears to be that the diameters of the towel rolls are too large. Many factors in the manufacturing process appear to affect roll diameter, but the engineers involved are not sure exactly how the different machine settings really affect roll diameter. In fact, they are not convinced that all of the factors make a difference! The towel machine team designed a study to determine the effect that different machine settings have on roll diameter. The specification for the diameter of the roll of towels is 5.35 inches. Several different people on the team express a concern that some settings that result in a good roll diameter might have an adverse impact on another towel roll characteristic, roll firmness.

If a towel is not firm its wrapping will be affected, and if it is too firm consumers will still react negatively. Fixing the roll diameter problem at the expense of firmness is not an option.

The team decides to look at two machine factors:

1. Embosser roll gap: the mechanism that puts the pattern in the towel

2. Speed: the speed at which the machine winds the rolls of towels

The machine was run at two different settings for speed and two different sizes of the embosser roll gap. The response variables are diameter of the roll of paper towels in inches and firmness measured on a specialized scale. Lower numbers indicate softer rolls, whereas larger numbers indicate firmer rolls. A portion of the data set is shown here:

Speed	Embosser	Diameter	Firmness
0	0	5.43307	0.271667
0	0	5.39370	0.336667
0	0	5.47244	0.260333
0	0	5.39370	0.261333
0	0	5.43307	0.297000
1	0	5.39370	0.312000
1	0	5.47244	0.294667
1	0	5.43307	0.342333
1	0	5.47244	0.306000
1	0	5.43307	0.365667

Datafile:
TOWEL.XXX

The variable *Speed* is a 0–1 variable that indicates the machine speed. A 0 indicates that the slower machine speed was used and a 1 indicates that the faster machine speed was used. The variable *Embosser* is a 0–1 variable that indicates the size of the embosser roll gap. A 0 indicates that the smaller gap measurement was used, whereas a 1 indicates that the larger gap measurement was used.

(a) Is there a difference in the diameters of the paper towel roll due to the speed at which the machine is run?

(b) Is there a difference in the diameters of the paper towel roll due to the size of the embosser roll gap?

(c) Is there any interaction of the two factors with respect to the diameter of the paper towel roll?

(d) Is there a difference in the firmness of the paper towel roll due to the speed at which the machine is run?

(e) Is there a difference in the firmness of the paper towel roll due to the size of the embosser roll gap?

(f) Is there any interaction of the two factors with respect to the firmness of the paper towel roll?

(g) Check the ANOVA assumptions for both response variables.

(h) Prepare a report explaining the effects that each of the machine settings have on *Diameter* and *Firmness*. Indicate which settings result in acceptable values for roll diameter. Make a recommendation on machine settings, if you can.

13.39 We have looked at a portion, of a data set measuring tissue strength. The management of the tissue company is concerned about complaints involving sheets tearing on removal. One possibility is too little airspace in the box. We studied that data set extensively in this chapter. Another possible explanation is that the tissues are not strong enough. Two tissue strength measurements are studied: machine-directional strength (*MDStrength*) and cross-directional strength (*CDStrength*). They are both measured in lb/ream. The data were collected over three days and the company would like to know if the average *MDStrength* and the average *CDStrength* are the same for all three days. There are 75 observations per day. A portion of the data set is shown here:

Datafile:
TISSUE.XXX

Day	MDStrength	CDStrength	Day	MDStrength	CDStrength
1	1006	422	1	962	464
1	994	448	1	973	472
1	1032	423	1	1036	489
1	875	435	1	1084	440
1	1043	445			

The specifications indicate that *MDStrength* should have a mean of 1000 lb/ream and a standard deviation of 50 lb/ream. The specifications indicate that *CDStrength* should be normally distributed with a mean of 450 and a standard deviation of 25 lb/ream.

(a) Is there any difference in the average *MDStrength* for the three days? If so, which day(s) are different?

(b) Is the *MDStrength* running to specification?

(c) Is there any difference in the average *CDStrength* for the three days? If so, which day(s) are different?

(d) Is the *CDStrength* running to specification?

(e) Check the assumptions of ANOVA for both response variables.

(f) Using the results of your analysis, prepare a report telling management about the current tissue manufacturing process. Make recommendations on whether the process must be adjusted or whether it is running to target. Remember, if it is running to target, management will consider making changes to the specifications to reduce customer complaints about dispensing.

14

The Analysis of Qualitative Data

COLLEGE DRINKING

For many students consumption of alcohol is part of the "college experience." Some colleges and universities attract students based on their reputations as "party schools." Students do not often think about the ramifications of their drinking until they experience problems related to it, such as failing classes or flunking out entirely, unwanted pregnancies, or encounters with law enforcement. University administrators are often unaware of or in denial about the problems and have no programs in place to help students address the problems of drinking and alcohol-related behaviors.

The problem of excessive alcohol consumption in college has been documented many times in many different ways. A national survey of 140 colleges and universities, conducted by Harvard University, looked at the problems connected with binge drinking, which is defined as consuming five or more drinks during one episode of drinking for men and consuming four or more drinks during one episode for women. The study found that as the number of binge drinking episodes for an individual increased, the number of alcohol-related problems experienced increased as well. The types of problems included having a hangover, missing class, having unplanned sex, and driving while intoxicated.

A senior public health major at a university in the Northeast conducted a survey similar to the Harvard survey to examine the problem of binge drinking on the officially "alcohol-free" campus. The survey was administered to students enrolled in a public health course entitled "Wellness." Although the class was not a random sample of the student population, the course was required of all students at the university. The survey was limited to students 25 years of age or younger and participation was voluntary.

The study resulted in 221 usable responses. Data were collected in four different areas: information about drinking habits (4 questions), problems experienced related to the students' drinking (15 questions), problems experienced because of other students' drinking (8 questions), and some demographic information (9 questions).

Here is a sample of the data:

Five	Four	Three	Last	Binger	Hangover	Miss	Behind
0	0	0	2	1	1	1	1
0	0	0	4	1	1	1	1
0	1	1	3	2	1	1	1
0	0	0	3	1	1	1	1
0	0	0	3	1	1	1	1
0	0	0	4	1	1	1	1

The variable names are abbreviated and are explained in the text as needed.

14.1 Chapter Objectives

You have learned that although qualitative data are important in understanding the results of statistical analyses, not many statistical tools deal specifically with these types of data. You learned some graphical techniques for displaying qualitative data in Chapters 3 and 5, and you learned how to calculate probabilities for qualitative variables in Chapter 6. In Chapters 9 and 10 you learned some techniques for testing hypotheses about population proportions, which arise from qualitative data.

This chapter deals primarily with the analysis of qualitative data. Almost all the techniques used to analyze qualitative data are known as chi-square tests. This chapter focuses on these uses for chi-square tests:

- Testing whether a particular probability model fits a set of data (goodness of fit test)
- Testing equality of proportions from more than two populations
- Testing whether two qualitative variables are dependent or independent

14.2 Test for Goodness of Fit

In Chapters 9 and 10 you learned several different hypothesis tests that assume data come from a population that is normally distributed. At the time you may have wondered how you could know whether that assumption is valid for a particular set of data. In Chapter 11 you used a normal probability plot to determine informally whether the residuals are normally distributed.

There are some other, informal, ways to check whether the assumption of normality is valid, such as creating a histogram, dotplot, or boxplot of the data and seeing whether the shape resembles that of the normal distribution. This is not a bad idea, but whenever we can substantiate an eyeball test with a formal statistical technique, we are more secure in the decisions we make.

One method of testing whether data come from a population with a certain distribution is to perform a test called a **chi-square goodness of fit test.**

> The *chi-square goodness of fit test* determines how well a set of data fits the model for a particular probability distribution.

14.2.1 THE CHI-SQUARE TEST

You know by now that every hypothesis test has these components:

- A set of hypotheses, H_0 and H_A
- A test procedure that defines how the data are collected and processed
- A test statistic that has a certain sampling distribution
- One or more critical values
- A decision or conclusion that is reached from the test results

The chi-square goodness of fit test is always a one-tail test.

For a chi-square goodness of fit test, these are the hypotheses:

H_0: The data come from a population with a specific probability distribution (normal, binomial, etc.).

H_A: The data do not come from a population with the specified distribution.

The test procedure is to collect a set of sample data and create a frequency histogram for the data. The test then compares the **observed** frequency distribution of the data with the frequency distribution that would be **expected** if the null hypothesis were true.

> The *observed frequencies* are the actual number of observations that fall into each class in a frequency distribution or histogram.

> The *expected frequencies* are the number of observations that should fall into each class in a frequency distribution under the hypothesized probability distribution.

EXAMPLE 14.1

COLLEGE DRINKING

Identifying the Hypothesized Distribution

Understand the problem.

Because the data collected in the survey on binge drinking are not a random sample, the researcher wants to know how well the sample represents the actual student population. The researcher thinks the distribution of students among the four classes—freshman, sophomore, junior, and senior—is approximately uniform. That is, he expects the probability distribution shown in the table:

Analyze the data.

Class	Freshman	Sophomore	Junior	Senior
Probability	25%	25%	25%	25%

This type of distribution is known as a **uniform distribution.**

> A *uniform distribution* is one in which each outcome or class of outcomes is equally likely to occur.

EXAMPLE 14.2

COLLEGE DRINKING

Setting Up the Goodness of Fit Test

We wish to test these hypotheses:

H_0:   The distribution of students is uniform over the classes.

H_A: The distribution of students is not uniform.

Because of a mixup in the early administration of the survey, the question about class was left off some of the surveys. As a result there were only 171 responses to this question. The data are listed here:

Class	Observed frequency
Freshman	86
Sophomore	36
Junior	30
Senior	19
Total (n)	**171**

From simply looking at the data we may reasonably wonder whether the distribution is uniform because the number of freshmen is larger than that of the other classes combined.

Analyze the data.

Draw Conclusions.

Remember that we often test hypotheses when observed data seem unusual or inconsistent with what we expect.

EXAMPLE 14.3

QUALITY PROBLEMS

Setting Up the Goodness of Fit Test

A company that manufactures CD jewel cases contracts with various computer software manufacturers to provide a product that is 10% defective. One of the software manufacturers is wondering about the quality of the product it receives and decides to check the quality of the cases.

Since the company has just received a shipment of product, it decides to sample 1000 boxes of the jewel cases and take 5 jewel cases from each. The reason for checking 5 jewel cases is that this is consistent with the company's current acceptance sampling procedures. The company will inspect and count the number of defective cases. If the vendor is conforming to the contract, then the data collected on the number of defective cases should have a binomial distribution with $p = 0.10$, or 10%.

The company wishes to test these hypotheses:

H_0:   The distribution of defective cases is binomial with $n = 5$ and $p = 0.10$.

H_A: The distribution of defective cases is not binomial with $n = 5$ and $p = 0.10$.

The company decides to perform the test at the 0.05 level of significance.

You might wonder why we would want to do this test because we already know how to do hypothesis tests about proportions. This is a good question. The reason is tied to the information we are trying to obtain. If the software manufacturer tests the hypothesis $p = 0.10$ (against $p < 0.10$ or $p > 0.10$), then the conclusion is about only the percent defective over the 5000 jewel cases in the sample. Depending on the length of time over which the product was manufactured, an average percent defective may not be a good indication of what is happening.

The chi-square goodness of fit test gives information about the distribution of the number of defective cases in a sample of size 5, which may enable the software manufacturer to obtain additional information about when or how the population changes. Remember that for the binomial probability distribution, one of the assumptions is that p, the proportion of successes in the population, remains constant over time. In a manufacturing process this is not always a safe assumption; the process may shift over time. If the process changes but averages 0.10 over time, the hypothesis test about proportions might not be able to identify the problem. If we test the distribution as well as the value of p, we may be able to identify a problem. You will see what we mean as the example progresses.

In a chi-square goodness of fit test, we are always trying to decide whether the data we observe fit a particular probability distribution model. We can use this model to determine how many data points we expect to fall into each class of the frequency distribution because we know the theoretical probabilities of observing any of the possible values.

✔ **TRY IT NOW!**

Seatbelt Use

Setting Up the Goodness of Fit Test

Analysts for insurance companies assume that the number of drivers who wear seatbelts is a binomial random variable with $p = 0.70$. To test this assumption they decide to set up checkpoints and sample ten drivers every 2 hours. Set up the null and alternative hypotheses to perform an appropriate chi-square goodness of fit test.

EXAMPLE 14.4

COLLEGE DRINKING

Finding the Expected Frequencies

If the distribution of students over the classes is uniform, then we expect the same number of students in each class; that is,

$$(25\%)(171) = 42.75 \text{ students in each class}$$

If we add an expected frequency column to the table, we see that indeed the observed frequency does differ from what we expected to happen.

Analyze the data.

Class	Observed frequency	Expected frequency
Freshman	86	42.75
Sophomore	36	42.75
Junior	30	42.75
Senior	19	42.75
Total (n)	**171**	**171**

We can see the deviation from what is expected by looking at the frequency histograms for both the observed and expected distributions.

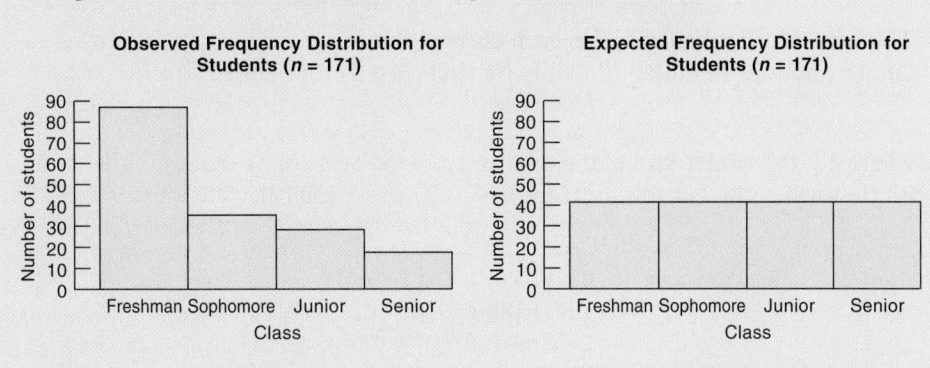

In the case of the uniform distribution, not much calculation is involved in finding the expected frequencies of the probability distribution. For other probability distributions, we need to do some additional calculations to find the expected frequencies.

EXAMPLE 14.5

QUALITY PROBLEMS

Calculating the Expected Frequencies

The software manufacturer who is wondering about the quality of the CD jewel cases it purchases from an outside vendor has collected the data and obtained a frequency distribution and histogram:

Number of defective cases	Frequency
0	485
1	340
2	127
3	48
4	0
5	0
Total	**1000**

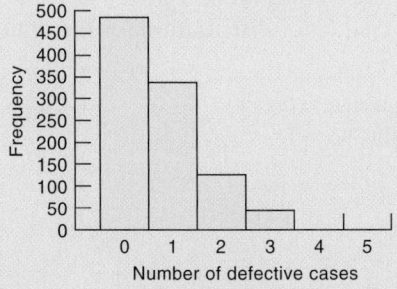

The data have the right shape for a binomial distribution with a small value of p but there seem to be a high number of defective cases.

The next step is to determine the expected frequencies for the binomial distribution with $n = 5$ and $p = 0.10$. From the binomial tables, the manufacturer finds this probability distribution:

Collect the data.

Analyze the data.

x	0	1	2	3	4	5
$p(x)$	0.590	0.328	0.073	0.008	0.000	0.000

The expected frequency, e_i, for each class of the frequency distribution is calculated by multiplying the probability for that class by the sample size:

Formula for expected frequency for goodness of fit

$$e_i = np_i$$

where n is the sample size of the data and p_i is the probability that the value in the ith class will occur. For this problem, $n = 1000$, so we can calculate these values:

x	np_i
0	$(1000)(0.590) = 590$
1	$(1000)(0.328) = 328$
2	$(1000)(0.073) = 73$
3	$(1000)(0.008) = 8$
4	$(1000)(0.000) = 0$
5	$(1000)(0.000) = 0$

Putting all the data together in one table, we see that there is definitely a difference between the observed and expected frequencies.

The reason for the discrepancy in the totals is rounding. If the binomial tables had four decimal places, you could find the expected frequencies to one decimal place and avoid the problem.

Number of defective cases	Observed frequency	Expected frequency
0	485	590
1	340	328
2	127	73
3	48	8
4	0	0
5	0	0
Total	**1000**	**999**

TRY IT NOW!

Seatbelt Use

Calculating the Expected Frequencies

The insurance analysts collect data for 1000 samples of 10 drivers and obtain the frequency distribution shown in the table. Complete the table by finding the expected frequency distribution for the data if the distribution is really binomial with $n = 10$ and $p = 0.70$.

Number wearing seatbelts, x	Observed frequency, $p(x)$	Expected frequency
0	0	
1	0	
2	1	
3	6	
4	33	
5	116	
6	213	
7	275	
8	216	
9	119	
10	21	
Total	**1000**	

Don't worry about rounding if the expected frequency column does not sum to 1.

Create frequency histograms for both the observed and the expected frequency distributions. Does it appear that the observed data conform to the binomial distribution with $n = 10$ and $p = 0.70$? Why or why not?

Once we have collected the data, formed a hypothesis, and found the expected frequencies for the hypothesized distribution, we need to find a way to quantify the results of the test.

14.2.2 THE CHI-SQUARE STATISTIC

Often departures from what we expect to happen lead us to perform a hypothesis test. But how can we quantify the "deviation" from what is expected? How can we know when what we observe deviates by more than it should or more than is likely by pure chance? To answer these questions we need to define a test statistic that measures the deviation of interest and has a sampling distribution with a behavior that we understand. This test statistic is known as the **chi-square statistic** and is calculated as

$$\chi^2 = \sum_{i=1}^{k} \frac{(o_i - e_i)^2}{e_i}$$

Chi-square statistic

where

o_i = observed frequency in the *i*th class of the frequency distribution

e_i = expected frequency in the *i*th class of the frequency distribution

k = number of classes in the frequency distribution

This test statistic has a chi-square distribution with $k - p - 1$ degrees of freedom, where p is the number of parameters of the theoretical distribution that are estimated from the data.

Don't worry about the value of p right now. It is explained in detail later on.

The chi-square statistic is a little different from those you have encountered before. We are trying to measure the amount by which the observed frequency distribution deviates from the expected frequency distribution. Clearly, looking at the

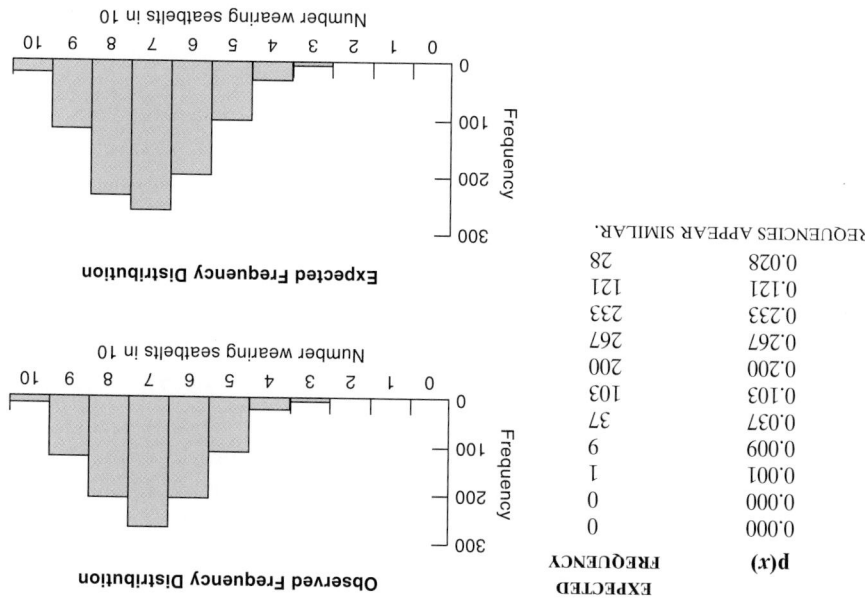

p(x)	EXPECTED FREQUENCY
0.000	0
0.000	0
0.001	1
0.009	6
0.037	37
0.103	103
0.200	200
0.267	267
0.233	233
0.121	121
0.028	28

THE FREQUENCIES APPEAR SIMILAR.

ANS.

difference between the two quantities, $o_i - e_i$, is a good start. Just as clearly, though, if we simply add the deviations, they will cancel each other out because some will be positive and some will be negative. So, just as we did when we learned about the variance and the standard deviation, we square the quantities so that they are all positive. Adding the squared deviations together, we get a measure of total deviation. Dividing by e_i is similar to calculating the percent error instead of the absolute error.

EXAMPLE 14.6

COLLEGE DRINKING

Calculating the Chi-Square Statistic

For the example about the distribution of students, we can calculate the chi-square statistic by extending the frequency table:

Class	Observed frequency, o_i	Expected frequency, e_i	$o_i - e_i$	$\dfrac{(o_i - e_i)^2}{e_i}$
Freshman	86	42.75	43.25	43.76
Sophomore	36	42.75	−6.75	1.07
Junior	30	42.75	−12.75	3.80
Senior	19	42.75	−23.75	13.19
Total (*n*)	171	171		61.82

Analyze the data.

The value of the chi-square test statistic is 61.82. This is a measure of the **total** deviation of the observed frequencies from the expected frequencies. The question now is, What does this tell us about the distribution of students?

EXAMPLE 14.7

QUALITY PROBLEMS

Calculating the Chi-Square Statistic

From the data that were collected it appears that there are definitely differences between the observed data and what would be expected if indeed the population is binomial with $p = 0.10$ and $n = 5$. To quantify the total difference we need to calculate the chi-square test statistic:

To avoid large errors due to rounding, the probabilities are calculated with a computer package and the expected frequencies are calculated to one decimal place.

Number of defective cases	Observed frequency, o_i	Expected frequency, e_i	$o_i - e_i$	$\dfrac{(o_i - e_i)^2}{e_i}$
0	485	590.5	−105.5	18.85
1	340	328.1	11.9	0.43
2	127	73.0	54.0	39.95
3	48	8.1	39.9	196.54
4	0	0.5	−0.5	0.50
5	0	0.0	0.0	0.00
Total	1000	1000.2		256.27

Analyze the data.

We see that the total deviation between the expected and observed frequencies is 256.27. We also see that the classes in the frequency distribution for values of 2 and 3 contribute the most to the total.

✔ TRY IT NOW!

Seatbelt Use
Calculating the Chi-Square Statistic

Complete the table to calculate the value of the chi-square statistic for the data obtained by the insurance analysts.

Number wearing seatbelts	Observed frequency, o_i	p(x)	Expected frequency, e_i	$o_i - e_i$	$\dfrac{(o_i - e_i)^2}{e_i}$
0	0	0.000	0		
1	0	0.000	0		
2	1	0.001	1		
3	6	0.009	9		
4	33	0.037	37		
5	116	0.103	103		
6	213	0.200	200		
7	275	0.267	267		
8	216	0.233	233		
9	119	0.121	121		
10	21	0.028	28		
Total	**1000**	**0.999**	**999**		

Once we quantify the difference between the observed and expected frequencies, we need to determine whether the difference is significant.

14.2.3   THE CRITICAL VALUE AND THE DECISION RULE

We now have a way to measure the deviation of the observed data from what is expected, the chi-square statistic, and we know that this test statistic has a chi-square distribution with $k - 1 - p$ degrees of freedom. All that remains of the test is to find the critical value and make a decision about the hypotheses.

To find the appropriate critical value we need a level of significance for the test, α. Chi-square goodness of fit tests are always upper-tail tests. This is because we are looking at the total deviation and trying to see whether it exceeds some reasonable level. To find the critical value we look up $\chi^2_{\alpha, k-p-1}$ for the appropriate values of α and $k - p - 1$, as shown in Figure 14.1 on page 702.

7.18		
1.75	-7	
0.03	-2	
1.24	-17	
0.24	8	
0.85	13	
1.64	13	
0.43	-4	
1.00	-3	
0.00	0	
—	0	
—	0	
$\dfrac{(o_i - e_i)^2}{e_i}$	$o_i - e_i$	ANS.

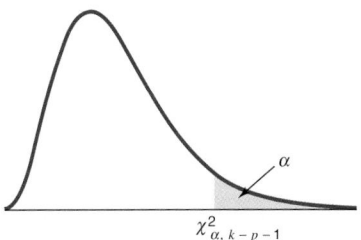

FIGURE 14.1 **The upper tail of a chi-square distribution**

To make a decision about the hypotheses we compare the value of the test statistic with the critical value. Since it is an upper-tail test, if the chi-square test statistic is greater than the χ^2 value from the table, we reject H_0 and conclude that the data did not come from the hypothesized distribution. If the chi-square test statistic is less than the critical value, we fail to reject H_0 and conclude that there is no evidence that the hypothesized distribution is not appropriate.

We must know two things to find the critical value for the test: the level of significance of the test, α, and the degrees of freedom, $k - p - 1$.

EXAMPLE 14.8

Analyze the data.

COLLEGE DRINKING

Performing the Chi-Square Test

Suppose we would like to perform the test for the student distribution at the 0.05 level of significance. This means there is a 5% chance that the test will conclude that the data collected did not come from the uniform distribution when in fact it did.

To determine the number of degrees of freedom, we first look at the number of classes in the frequency distribution, k. In this case, $k = 4$. The value p represents the number of population parameters we estimated from the data. In this case we did not use the data to estimate any parameters. Thus, the number of degrees of freedom for the test is $4 - 0 - 1 = 3$. From the chi-square table, we look up $\chi^2_{0.05,3}$ and find that the critical value is 7.82.

Draw conclusions.

To make a decision, we compare the chi-square test statistic, 61.82, with the critical value of 7.82. Since 61.82 is greater than 7.82 (the test statistic falls beyond or outside the critical value), we reject H_0. The test concludes that the data do not come from a population that has a uniform distribution. This indicates that the survey was not evenly spread out over students from the four classes.

EXAMPLE 14.9

QUALITY PROBLEMS

Performing the Chi-Square Test

The software company found the value of 256.27 for the chi-square statistic for the data on the defective CD jewel cases. To decide what this means, it needs to find the critical value for the test, compare the test statistic with the critical value, and make a decision.

Analyze the data.

To find the critical value the company analysts must know α and the degrees of freedom for the test. They have already decided to test at the 0.05

level of significance. The number of degrees of freedom for the test is calculated as

$$k - p - 1 = 6 - 0 - 1 = 5$$

Remember that p is the number of parameters of the hypothesized distribution that are estimated from the data. The binomial distribution has two parameters, n and p, and both were given or assumed, not estimated.

This critical value, $\chi^2_{0.05,5}$, is 11.07. Since 256.27 is clearly beyond the critical value, the software company can reject H_0 and conclude that the number of defective jewel cases in a sample of five does not have a binomial distribution with $p = 0.10$. That is, the data do not fit the assumed probability model.

Draw conclusions.

✔ TRY IT NOW!

Seatbelt Use
Finding the Critical Value and Performing the Test

The insurance analysts decide that they want to test the goodness of fit hypotheses at the 0.01 level of significance. How many degrees of freedom will the critical value for the test have?

Find the critical value for the test.

Based on the chi-square test statistic and the critical value, what can you conclude about the distribution of the number of people in a sample of ten who wear seatbelts?

As we learn more and more statistical techniques, we may lose sight of the fact that we are doing the analyses to help identify problems—that is, to understand why things happen and to make informed decisions. After the chi-square goodness of fit test, what do we know and how can we use the information?

EXAMPLE 14.10

COLLEGE DRINKING

Following Up on the Test

As a result of the chi-square test, the researcher who administered the survey on binge drinking knows that the students in his sample are not uniformly distributed over the four classes. This means that his sample may not be representative of the population he is trying to study and that he must be careful about any conclusions or generalizations he makes.

Now the researcher began to think a little more about his null hypothesis. Does it really make sense that students are uniformly distributed over the four classes? Given the problems of student retention (loss of students from transfer or dropout) and the transfer in of students from other universities and community colleges, it is more likely that the true percentage of students in a class decreases with each class level. The researcher contacted a university

Understand the problem.

ANS. 10 DEGREES OF FREEDOM. CRITICAL VALUE IS 23.21. DO NOT REJECT H_0. THERE IS NO REASON TO BELIEVE THE HYPOTHESIZED DISTRIBUTION IS INCORRECT.

administrator who keeps track of enrollment and found this historical distribution of students:

Collect the data.

Class	Expected distribution (%)	Expected frequency
Freshman	45.2	77.3
Sophomore	18.6	31.8
Junior	18.0	30.8
Senior	18.2	31.1
Total (n)	100.0	171.0

The researcher then decided to compare the actual frequency distribution with the new distribution obtained from the university and redo the chi-square test. The comparison is shown in the table:

Class	Observed frequency, o_i	Expected frequency, e_i	$o_i - e_i$	$\dfrac{(o_i - e_i)^2}{e_i}$
Freshman	86	77.3	8.7	0.98
Sophomore	36	31.8	4.2	0.55
Junior	30	30.8	−0.8	0.02
Senior	19	31.1	−12.1	4.71
Total (n)	171	171.0		6.26

Analyze the data.

Draw conclusions.

Since nothing else changed, the critical value of the test is still $\chi^2_{0.05,3} = 7.82$. The value of the test statistic, 6.26, is not outside the critical value, so there is not sufficient evidence to reject H_0. This means there is no reason to say that the sample did not come from a population with the hypothesized distribution, and so it is reasonable to assume that the sample is representative of the student population of the university as *defined by the university administration.*

One benefit of the chi-square test is that it lets us proceed with confidence in our analysis. Knowing that the sample does indeed represent the population of interest gives much more credibility to any conclusions drawn from the analysis.

EXAMPLE 14.11

QUALITY PROBLEMS

Following Up on the Test

The company that used the outside vendor for CD jewel cases knows as a result of the test that the number of defective cases in a sample of size 5 does not have a binomial distribution with $p = 0.10$. The company does not know whether the distribution is binomial with a different value of p or whether some other factor makes the binomial distribution inappropriate. This is the time to use some common sense, knowledge of how the data were collected, and descriptive and graphical statistics to identify the problem.

Draw conclusions.

The company analysts could do many things, depending on how they collected the data and what variables they recorded, but in this case they decided to plot the data in the order in which the cases were produced. Each sample they took came from a different box, and they recorded the lot numbers and production data along with the data on the number of defective cases. A time plot of the data indicated that the number of defectives in the sample started to increase at about the 750th sample. Since the first plot was very dense, the analysts also decided to plot every fifth sample so that they could see the shift more clearly.

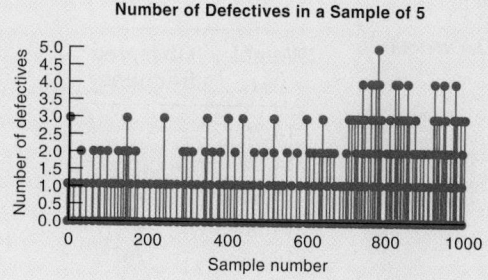

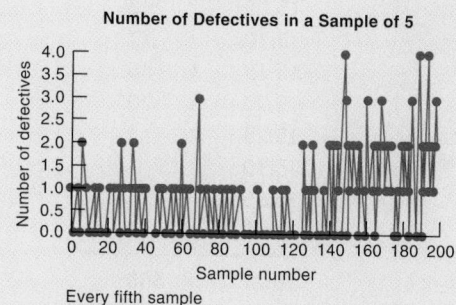

Every fifth sample

As a result of the chi-square test and the plot of the data, the analysts learned that the percent defective apparently changed during production. This is why the binomial distribution was not correct.

14.2.4 TESTING FOR NORMALITY AND OTHER CONSIDERATIONS

The chi-square goodness of fit test is similar to the method you learned for testing whether the variances of two populations are equal. The test does not result in a definitive answer that the hypothesized distribution is correct, just as the hypothesis test to see whether two variances are equal does not tell that the variances are equal. Rather, if we reject H_0, it tells us when the hypothesized distribution is not correct. When we use a chi-square test to test the assumption of normality, we decide either that the assumption of normality is not appropriate or that, since there is no evidence to the contrary, the assumption is a reasonable one.

EXAMPLE 14.12

JAR WEIGHTS

Testing for Normality

The quality specifications for a 15.0-ounce jar of peanut butter state that the weights of the jars should be normally distributed with a mean of 15.1 ounces. The quality focus group selects a random sample of 500 jars to determine whether the specifications are being met. It decides to test at the 0.05 level of significance. These are the hypotheses for the test:

H_0: The jar weights are normally distributed with $\mu = 15.1$ ounces.

H_A: The jar weights are not normally distributed with $\mu = 15.1$ ounces.

Understand the problem.

The group collects the data and uses a computer package to create a frequency distribution and histogram:

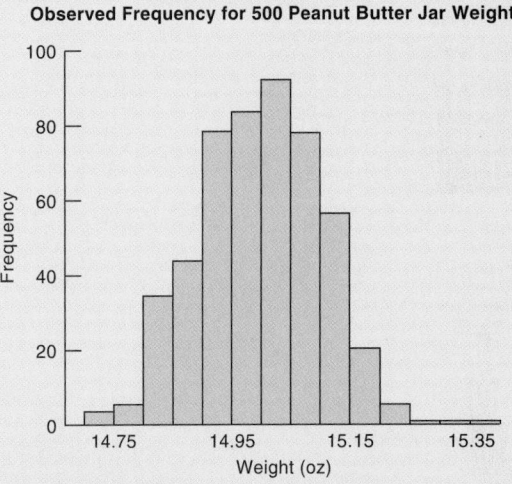

Weight (≤)	Observed frequency
14.75	3
14.80	5
14.85	34
14.90	43
14.95	79
15.00	83
15.05	92
15.10	77
15.15	56
15.20	20
15.25	5
15.30	1
15.35	1
15.40	1
Total	**500**

Observed Frequency for 500 Peanut Butter Jar Weights

Collect the data.

The data are symmetric, although this does not mean they are definitely normally distributed.

To calculate the expected frequency for each class, the group needs to calculate the probability that a randomly selected jar will fall into the weight range defined by each class. For the first class they need to find $P(14.70 < X \le 14.75)$. Remember from Chapter 6 that we need to turn the two values into standard normal variables (Z's) and look them up. But to do that we need μ and σ! The specifications for the jars give only a value for μ, so the analysts have to use the data to estimate σ. Using the same software package, the group finds that the sample standard deviation, s, is 0.102 ounce. So, to find the probability of interest they use:

Analyze the data.

$$\text{For } 14.70: \quad Z = \frac{14.70 - 15.10}{0.102} = -3.92$$

$$\text{For } 14.75: \quad Z = \frac{14.75 - 15.10}{0.102} = -3.43$$

Looking up the Z values in the table and subtracting the results, they find that $P(14.70 < X \le 14.75) = 0.0003 - 0.0000 = 0.0003$. When they were doing the calculations, they noticed that there was a discrepancy between the total for the expected frequency and the total of 500 observations. Looking at the table, they realized that this was probably because a few additional classes on the high side had nonzero probabilities. They expanded the table to include these classes. To find the expected number of jars out of the 500 that would weigh between 14.70 and 14.75 ounces, the group multiplied the probability by 500 and obtained $(0.0003)(500) = 0.15$. They found the probabilities and expected frequencies for all of the classes:

Weight (<)	Observed frequency, o_i	$P(a < X < b)$	Expected frequency, e_i	$\dfrac{(o_i - e_i)^2}{e_i}$
14.75	3	0.0003	0.1	84.10
14.80	5	0.0013	0.7	28.13
14.85	34	0.0055	2.7	355.99
14.90	43	0.0179	8.9	130.33
14.95	79	0.0458	22.9	137.71
15.00	83	0.0927	46.4	28.93
15.05	92	0.1486	74.3	4.23
15.10	77	0.1879	94.0	3.07
15.15	56	0.1879	94.0	15.36
15.20	20	0.1486	74.3	39.66
15.25	5	0.0927	46.4	36.91
15.30	1	0.0458	22.9	20.92
15.35	1	0.0179	8.9	7.03
15.40	1	0.0055	2.7	1.11
15.45	0	0.0013	0.7	0.67
15.50	0	0.0003	0.1	0.13
Total	**500**	**1.0000**	**500.0**	**894.28**

The chi-square statistic indicates some kind of problem. The data may not be exactly normally distributed, but they are not that far off. Why is the value of the test statistic so large?

Before we proceed with the example we need to discuss some problems with the chi-square goodness of fit test. The problems relate to the number of observations that are used to create the frequency distribution. You may remember from Chapter 3 that a fairly large sample size is required to obtain a good frequency distribution. If a frequency histogram is highly variable due to lack of data, the frequencies in some classes will have greater differences from the expected frequencies.

Another related problem is that when the total sample size is small, the expected frequencies for each class will also be small. Since the chi-square statistic divides by the expected frequency, small values can inflate the value of the statistic artificially and lead to rejection of H_0 when, in fact, it is true. For this reason, it is recommended that the chi-square test not be used if the expected frequency in any cell is less than 5. It is possible to combine adjacent cells to get the expected frequencies above 5, but this results in a loss of degrees of freedom.

Remember! Dividing by numbers close to 0 causes the result to be large.

EXAMPLE 14.13

JAR WEIGHTS

Collapsing Classes for Small Expected Frequencies

After looking at the table of the observed and expected frequencies, the quality focus group for peanut butter called in a statistician. The statistician explained that too many classes with very small expected frequencies were inflating the chi-square statistic. The group decided to collapse the cells on both ends of the distribution to get the expected frequencies above 5. The table was adjusted and the chi-square statistic recalculated:

Analyze the data.

Weight (<)	Observed frequency, o_i		P(a < X < b)	Expected frequency, e_i		$\dfrac{(o_i - e_i)^2}{e_i}$
14.75	3		0.0003	0.1		
14.80	5	85	0.0013	0.7	12.4	423.89
14.85	34		0.0055	2.7		
14.90	43		0.0179	8.9		
14.95	79		0.0458	22.9		137.71
15.00	83		0.0927	46.4		28.93
15.05	92		0.1486	74.3		4.23
15.10	77		0.1880	94.0		3.07
15.15	56		0.1880	94.0		15.36
15.20	20		0.1486	74.3		39.66
15.25	5		0.0927	46.4		36.91
15.30	1		0.0458	22.9		20.92
15.35	1		0.0179	8.9		
15.40	1	2	0.0055	2.7	12.4	8.78
15.45	0		0.0013	0.7		
15.50	0		0.0003	0.1		
Total	**500**		**1.0000**	**500.0**		**719.46**

After collapsing, only ten classes are left. Also, the normal distribution has two parameters, μ and σ, and the data were used to estimate one of them, σ.

Clearly, the chi-square statistic is still large, but it has been considerably reduced.

To complete the test, the quality focus group found the critical value for the test. It calculated the degrees of freedom as

$$k - p - 1 = 10 - 1 - 1 = 8$$

Since it wanted to test at $\alpha = 0.05$, it used $\chi^2_{0.05,8} = 15.51$.

The value of the test statistic, 719.46, is beyond the critical value of 15.51, so the group rejects H_0 and concludes that the jar weights are not normally distributed with a mean of 15.1 ounces and a standard deviation of 0.102 ounce.

Again, we need to look at the conclusion of the chi-square goodness of fit test and realize that, in fact, it is a package deal. In Example 14.13 the conclusion is that the data do not come from a normal distribution with a mean of 15.1 ounces. It is not necessarily true that the data are not normally distributed; they might be normally distributed with a different mean than the one hypothesized. It is also not necessarily true that the mean is not 15.1 ounces; the mean could be 15.1 but the distribution might not be normal. To really know what is going on here, the quality focus group might want to perform a hypothesis test about the value of the mean to see whether that is the problem, or rerun the chi-square test using the data to estimate μ.

14.2.5 EXERCISES—LEARNING IT!

14.1 The administration of a university has been using the following distribution to classify the ages of students:

Age	Estimated percent of student population
Younger than 18	2.7
18–19	29.9
20–24	53.4
Older than 24	14.0

A recent student survey provided the following data on the ages of students:

Age	Frequency
Younger than 18	6
18–19	118
21–24	102
Older than 24	26

(a) Set up a table that compares the expected and observed frequencies for each group.

(b) Based on the table, do you think the data represent the estimated distribution?

(c) Set up the hypotheses for the chi-square goodness of fit test.

(d) Perform the goodness of fit test at the 0.05 level of significance.

(e) Based on the chi-square test, is the estimated age distribution that the university is using correct?

14.2 As part of a survey on the use of portable music devices such as MP3 players, a public opinion research company wanted to know whether its population was uniformly distributed over the following age groups: under 25, 25 to 44, 45 and up. The company looked at the data it had collected so far and found the following distribution:

Age	Number of respondents
Under 25	73
25 to 44	61
45 and up	66
Total	**200**

(a) Based on the data, do you think the respondents are uniformly distributed over the age categories?

(b) Set up the hypotheses to test whether the data are uniformly distributed over the age categories.

(c) Find the expected frequency distribution and perform the chi-square goodness of fit test.

(d) At the 0.05 level of significance, would you say that the respondents are uniformly distributed over the age groups?

14.3 The transit authority in a large city estimates that 80% of commuters get a seat for their entire commute. It decided to take a random sample of its subway system over a 12-week period to see whether its estimate is correct. The authority set up an exit poll at the most common destination station and asked 15 randomly selected commuters whether they got a seat for the entire commute. Here are the data the transit authority obtained:

Number of commuters in 15 who got seats	Count
0	0
1	0
2	0
3	0
4	0
5	0
6	0
7	0
8	1
9	3
10	10
11	21
12	31
13	20
14	9
15	5
Total	**100**

(a) Why is the number of commuters in 15 who got seats a binomial random variable?

(b) Set up the hypotheses to test whether the sample data come from a binomial distribution with $n = 15$ and $p = 0.80$.

(c) Find the expected frequencies for the hypothesized distribution.

(d) Perform the chi-square goodness of fit test at the 0.01 level of significance.

(e) Is it reasonable to assume that the data come from a binomial distribution with $n = 15$ and $p = 0.80$?

14.4 Did you ever wonder where all your money goes? One of the things people notice when they start to track their expenditures is that they spend a lot of money dining away from home. A recent consumer expenditure study done by a state agency looked at the amount of money spent per week on dining out by a family of four. Here are the amounts of money from a sample of size 50:

17	18	22	24	25	26	27	28	31	34
35	37	38	39	40	40	41	42	42	44
44	44	45	45	45	45	45	46	47	49
49	50	50	52	54	58	60	61	63	64
66	68	69	70	75	76	79	80	83	86

(a) Make a frequency distribution and a histogram for the data. Do not use more than five classes.

(b) From the histogram, do you believe that the data are normally distributed?

(c) Set up the hypotheses to test that the data come from a normal distribution with a mean of $50 and a standard deviation of $15.

(d) Find the expected frequency distribution of the data.

(e) At the 0.05 level of significance, does it appear reasonable to assume that the data are normally distributed?

14.5 A health agency believes that 80% of women who are prescribed a popular drug for osteoporosis stop taking the medicine because the drug has such complicated directions for taking it. Researchers decide to take a random sample of ten women who filled prescriptions for the drug from a large retail pharmacy chain every day for 3 months. One month after the prescription was filled, they call each person to see if she is still taking the drug and record the number in the sample who are no longer taking it. The following data are obtained:

Number of women in ten who stopped taking the drug	Frequency
4	1
5	5
6	11
7	19
8	27
9	18
10	7
Total	**88**

(a) Set up the necessary hypotheses to test whether the data come from a binomial distribution with $n = 10$ and $p = 0.80$.

(b) Find the expected frequency distribution for the data.

(c) At the 0.05 level of significance, is it reasonable to assume that the number of women who stopped taking the drug has a binomial distribution with $p = 0.80$?

14.3 Test for Equality of Proportions

In Chapter 10 you learned how to compare parameters from two different populations. In particular, you learned that you can use a Z test to determine whether the population proportions for two populations are equal. What if we are interested in comparing the population proportions for more than two populations?

With a chi-square test, it is possible to compare proportions for more than two populations. Although it would be possible to compare a set of c populations by using the Z test you learned in Chapter 10 to test all of the possible pairs of populations, this is not a good idea. The reason for doing the single test is that each hypothesis test has a Type I error probability associated with it. When we do multiple tests, we increase the chances that we will decide that two population proportions are different when in fact they are not. The single test controls the probability of making this mistake.

14.3.1 TESTING PROPORTIONS FOR MORE THAN TWO POPULATIONS

In general, some characteristic of importance (a success) is defined for several different populations. We want to determine whether the proportion of successes in each population is the same. We wish to test this general set of hypotheses:

$$H_0: \quad p_1 = p_2 = \cdots = p_c$$
$$H_A: \quad \text{At least one } p_i \text{ is different.}$$

To test these hypotheses for two populations we take a sample from each population and count the numbers of times that the characteristic of interest occurs. The data collection for more than two populations is the same.

In Chapter 5 you learned how to summarize data that involve two qualitative variables—the contingency table. For this test the two variables are the population that the sample item comes from ($i = 1, 2, \ldots, c$) and whether the item is a success (the characteristic is present) or a failure (the characteristic is not present). Table 14.1

TABLE 14.1

CONTINGENCY TABLE FOR TESTING EQUALITY OF PROPORTIONS

	Population				
Result	**1**	**2**	**...**	**c**	**Total**
Success	s_1	s_2	...	s_c	s
Failure	f_2	f_2	...	f_c	f
Total	n_1	n_2	...	n_c	n

is an example of the contingency table. The proportion of successes for each sample can be calculated as

$$\hat{p}_i = \frac{s_i}{n_i}$$

EXAMPLE 14.14

COLLEGE DRINKING

Setting Up a Chi-Square Test of Proportions

Understand the problem.

In the study on binge drinking, students were classified as nonbinge drinkers, infrequent binge drinkers, and frequent binge drinkers. One question of interest to the researcher was whether there was a difference in academic responsibility among the three different categories of drinkers. He decided to look at whether the proportion of people who missed class because of drinking is the same for each of the three groups. He created this contingency table:

	Type of drinker			
Miss class	**Nonbinge**	**Infrequent**	**Frequent**	**Total**
Never	77	39	31	**147**
Once or more	7	19	39	**65**
Total	**84**	**58**	**70**	**212**

Analyze the data.

The populations in this case are the different types of drinkers and the characteristic of interest is whether the student missed class. Here are the hypotheses:

$$H_0: \quad p_N = p_I = p_F$$
$$H_A: \quad \text{At least one } p_i \text{ is different.}$$

Draw conclusions.

From the data we have $\hat{p}_N = 0.083$, $\hat{p}_I = 0.328$, and $\hat{p}_F = 0.557$. The proportions certainly appear to be different, but because of sampling error a statistical test is necessary.

The hypotheses are equivalent to asking the question, Is missing class related to the amount a student drinks? This is similar to the question we asked when looking at regression models. The difference is that these variables are qualitative.

EXAMPLE 14.15

CHANGING LIVES

Setting Up a Chi-Square Test of Proportions

After the tragic events of September 11, 2001, many people vowed to "do things differently." People said they would spend less time working and more time with their families. A survey done 2 months after the tragedy asked people if they had done anything to spend more time with their families. The people surveyed were also asked their marital status and were classified as (1) married with children, (2) unmarried with children, or (3) no children. Suppose we want to know whether the proportion of people who are not spending more time with their families is the same for all marital groups. We wish to test these hypotheses:

Understand the problem.

$$H_0: \quad p_1 = p_2 = p_3$$
$$H_A: \quad \text{At least one } p_i \text{ is different.}$$

The statistical tool to test these types of hypotheses is a chi-square test similar to the one used to test goodness of fit.

Collect the data.

The survey selected 580 people at random and assembled a contingency table:

| Spend more time with family? | Marital status | | | |
	Married with children	Unmarried with children	No children	Total
No	12	22	14	**48**
Yes	188	148	196	**532**
Total	**200**	**170**	**210**	**580**

From the data we can estimate the proportion who spend more time with family for each marital status: $\hat{p}_1 = 0.06$, $\hat{p}_2 = 0.13$, and $\hat{p}_3 = 0.07$. Clearly, the numbers are different, but we know enough about sampling error to realize that we cannot be sure without a statistical test.

Contingency tables like this one are tedious to do by hand for large data sets, but they are done easily by computer software packages like Minitab or Excel.

 **TRY IT NOW!**

Customer Service

Setting Up the Chi-Square Test for Proportions

A mail-order company has set up customer service phone centers in three different locations. The support representatives keep a log for each call to customer service, and as part of that log they record whether the problem was resolved successfully. Company analysts are interested in knowing whether the percentage of calls that are successfully resolved is the same for each location. They randomly select logs from each location and collect data on the numbers of calls that result in a successful resolution of the problem. The data are summarized in the table:

	Location			
Calls	1	2	3	Total
Successful	257	264	283	**804**
Unsuccessful	43	86	97	**226**
Total	**300**	**350**	**380**	**1030**

Set up the hypotheses for the test.

Calculate the proportion of successfully resolved calls for each location. Based solely on these numbers, do you think the proportion of successfully resolved calls for all three locations is the same?

14.3.2 PERFORMING THE CHI-SQUARE TEST FOR PROPORTIONS

In Section 14.2 you learned that a chi-square test compares expected and observed frequencies. The observed frequencies are the raw data that are collected, and the expected frequencies are the frequencies that would be predicted if the null hypothesis were true. The data in Table 14.1 can be analyzed using exactly this method.

If the null hypothesis, H_0, is true, then the data for all of the populations being sampled can be combined to give one overall estimate of the true proportion of successes, p. This overall estimate, $\hat{p}$, can then be used to make predictions. In general, the estimate for p is calculated by counting all of the successes observed and dividing by the total number of objects sampled. That is, the **overall proportion of successes** is given by

Overall proportion of successes

$$\hat{p} = \frac{s_1 + s_2 + \cdots + s_c}{n_1 + n_2 + \cdots + n_c} = \frac{s}{n}$$

and the **expected number of successes** in the sample from population i is

Expected number of successes

$$e_i = \frac{s}{n} n_i = \frac{sn_i}{n} = \hat{p} n_i$$

These are equivalent ways of calculating the expected frequency. For the chi-square test of proportions, the last expression is probably easiest to use. For other chi-square tests, people find the second expression simplest.

Since we know the sample size for each population and the expected number of successes in the sample, the easiest way to find the expected number of failures is by subtraction.

EXAMPLE 14.16

COLLEGE DRINKING

Calculating the Expected Frequencies

For the college drinking study the researcher wants to estimate p, the overall proportion of students who miss class because of drinking. The value of p is estimated assuming that the null hypothesis is true. The overall proportion of students who miss class because of drinking is

ANS. $H_0: p_1 = p_2 = p_3$, H_A: AT LEAST ONE p_i IS DIFFERENT; $p_1 = 0.86$, $p_2 = 0.75$, $p_3 = 0.74$, NO.

$$\hat{p} = \frac{7 + 19 + 39}{84 + 58 + 70} = \frac{65}{212} = 0.307$$

The expected number for each group is calculated as

$$e_N = (0.307)(84) = 25.8$$
$$e_1 = (0.307)(58) = 17.8$$
$$e_F = (0.307)(70) = 21.5$$

Analyze the data.

The next step is to calculate the number of students who did not miss class for each group:

Nonbinge drinkers:	$84 - 25.8 = 58.2$	
Infrequent binge drinkers:	$58 - 17.8 = 40.2$	
Frequent binge drinkers:	$70 - 21.5 = 48.5$	

To get a better picture, the observed and expected frequencies are put into a single table as shown here. The expected frequencies are given in parentheses below the observed frequencies.

	Type of drinker			
Miss class	**Nonbinge**	**Infrequent**	**Frequent**	**Total**
Never	77	39	31	**147**
	(58.2)	(40.2)	(48.5)	
Once or more	7	19	39	**65**
	(25.8)	(17.8)	(21.5)	
Total	**84**	**58**	**70**	**212**

CHANGING LIVES

EXAMPLE 14.17

Calculating the Expected Frequencies

We want to know whether the proportions of people who are spending more time with their families are the same for different marital groups. The overall proportion defective is found to be

Analyze the data.

$$\hat{p} = \frac{12 + 22 + 14}{200 + 170 + 210} = \frac{48}{580} = 0.083$$

So the expected numbers who are not spending more time with their families for each group are listed here:

$$e_1 = (0.083)(200) = 16.6$$
$$e_2 = (0.083)(170) = 14.1$$
$$e_3 = (0.083)(210) = 17.4$$

The following table shows the observed and expected frequencies for each group. The expected frequencies are given in parentheses below the observed frequencies.

Spend more time with family?	Marital status			Total
	Married with children	Unmarried with children	No children	
No	12 (16.6)	22 (14.1)	14 (17.4)	48
Yes	188 (183.4)	148 (155.9)	196 (192.6)	532
Total	200	170	210	580

✔ TRY IT NOW!

Customer Service
Calculating the Expected Frequencies

The data for the mail-order company interested in its customer service locations are presented in the table:

Calls	Location			Total
	1	2	3	
Successful	257	264	283	804
Unsuccessful	43	86	97	226
Total	300	350	380	1030

As with the goodness of fit test, the expected frequencies might not add up to the total because of rounding.

Estimate p, the percentage of calls that are resolved successfully, assuming that the three locations are the same.

Use the overall proportion of successful calls to find the expected frequency of successful calls for each location.

14.3.3 PERFORMING THE TEST

Once we have the expected frequency for each population, all that remains to be done is to calculate the chi-square test statistic, find the critical value for the test; and make a decision.

We have already defined the chi-square test statistic in general and have shown how it is used for the goodness of fit test. The application of the chi-square statistic to testing proportions is exactly the same:

Numbering the cells is not critical, although generally we number cells in a table row by row.

$$\chi^2 = \sum_{\text{all cells}} \frac{(o_i - e_i)^2}{e_i}$$

where

o_i = observed frequency in the ith cell of the table

e_i = expected frequency in the ith cell of the table

The degrees of freedom for a chi-square test involving a table are given by

$$(r - 1)(c - 1)$$

where r is the number of rows in the table and c is the number of columns.

ANS. $\hat{p} = 0.781 (78\%)$; 234, 273, 297

EXAMPLE 14.18

COLLEGE DRINKING

Performing a Chi-Square Test for Equal Proportions

The data on missing class and drinking are shown again here:

	Type of drinker			
Miss class	Nonbinge	Infrequent	Frequent	Total
Never	77 (58.2)	39 (40.2)	31 (48.5)	147
Once or more	7 (25.8)	19 (17.8)	39 (21.5)	65
Total	84	58	70	212

To perform the chi-square test at the 0.05 level of significance, the researcher must calculate the test statistic and find the critical value for comparison. The test statistic is calculated as shown:

$$\frac{(77 - 58.2)^2}{58.2} + \frac{(39 - 40.2)^2}{40.2} + \frac{(31 - 48.5)^2}{48.5} + \frac{(7 - 25.8)^2}{25.8} + \frac{(19 - 17.8)^2}{17.8}$$

$$+ \frac{(39 - 21.5)^2}{21.5} = 40.45$$

Analyze the data.

To find the critical value, he must first find the degrees of freedom of the test:

$$(r - 1)(c - 1) \overset{\cdot}{=} (2 - 1)(3 - 1) = 2$$

The critical value is then $\chi^2_{0.05,2} = 5.99$.

To finish the test he compares the test statistic of 40.45 with the critical value of 5.99. Because 40.45 is outside the critical value, the decision is to reject H_0 and conclude that the proportion of students who miss class is not the same for each population. That is, missing class is related to the amount that a student drinks.

Draw conclusions.

Software packages make it easy to perform the calculations for the expected frequencies, and some software displays these values as part of their output, as shown in Figure 14.2.

	Non-Binge	Infrequent	Frequent	All
Never	77 58.25	39 40.22	31 48.54	147 147.0
Once or more	7 25.75	19 17.78	39 21.46	65 65.0
All	84 84.00	58 58.00	70 70.00	212 212.0

Chi-Square = 40.484,   DF = 2,   P-Value = 0.00

Cell Contents—

      Count
      Exp Freq

FIGURE 14.2 Minitab output from chi-square analysis

Analyze the data.

EXAMPLE 14.19

CHANGING LIVES

Performing a Chi-Square Test for Equal Proportions

For the survey about spending more time with family, we have collected and summarized the data and calculated the expected number of people who do not spend more time with their families for each group:

Spend more time with family?	Marital status			
	Married with children	Unmarried with children	No children	Total
No	12 (16.6)	22 (14.1)	14 (17.4)	48
Yes	188 (183.4)	148 (155.9)	196 (192.6)	532
Total	200	170	210	580

We want to perform the test at the 0.05 level of significance. We calculate the chi-square statistic:

$$\frac{(12 - 16.6)^2}{16.6} + \frac{(22 - 14.1)^2}{14.1} + \frac{(14 - 17.4)^2}{17.4} + \frac{(188 - 183.4)^2}{183.4} + \frac{(148 - 155.9)^2}{155.9}$$

$$+ \frac{(196 - 192.6)^2}{192.6} = 6.94$$

Since the level of significance is 0.05 and the degrees of freedom is $(2 - 1) \times (3 - 1) = 2$, the critical value for the test is $\chi^2_{0.05,2} = 5.99$.

Comparing the test statistic of 6.94 with the critical value of 5.99, we reject H_0 and conclude that the proportions for the three marital groups are not the same.

TRY IT NOW!

Customer Service

Performing the Chi-Square Test for Equal Proportions

The mail-order company with three different customer service locations wants to complete the test to determine whether the percentages of successfully resolved calls are the same at all three locations. It wants to test at the 0.01 level of significance. Calculate the value of the chi-square test statistic and complete the test.

Is the proportion of successfully resolved calls the same at each location?

14.3.4 USING THE RESULTS OF THE CHI-SQUARE TEST FOR PROPORTIONS

You might be wondering (and with good reason) how we can use the chi-square test for equality of proportions to gain useful information or to make decisions. The results of the chi-square test are similar to the results of analysis of variance that you

ANS. $\chi^2 = 14.65$, CRITICAL VALUE = 9.21; REJECT H_0, AT LEAST ONE OF THE LOCATIONS IS DIFFERENT.

learned in Chapter 14. When the test does not lead to rejection of the null hypothesis, there is really nothing else to do. However, when the test leads to rejection of the null hypothesis, we conclude that at least one of the populations is different from the others. Does it mean that one of the populations is different from all of the others? Does it mean that each population is different from every other population? How useful is the conclusion?

Remember that statistical analyses are not answers in themselves but rather tools that can be used to identify problems or aid in decision making. The key words here are *identify* and *aid*. If we conduct a chi-square test for proportions, then we probably suspect that something is going on that is causing the populations to be different. The results of the test verify that our suspicions are justified. The test result indicates that further investigation is warranted.

How do we go about determining exactly which of the populations are the same and which are different? Certainly, we know that the two samples that are farthest apart are probably different, but how about the ones in between? How do they compare? For ANOVA there are formal techniques for testing which means are different. For chi-square tests, the techniques are much less formal.

Often descriptive and graphical techniques can help. We might simply plot the sample proportions for the populations on the same graph and see how they compare.

EXAMPLE 14.20

COLLEGE DRINKING

Using the Chi-Square Test Results

As a result of the chi-square test, the researcher knows that at least one of the populations is different from the others with respect to missing classes. The problem is that that is all he knows. The chi-square test does not tell whether all three populations are different or whether the two binge drinking categories are similar but different from the nonbingers.

He decides to plot the proportion for each sample on a graph. It appears that all three groups are different. The chi-square test verifies his original ideas, but if he wants more definitive conclusions he might need to use a designed experiment.

Draw conclusions.

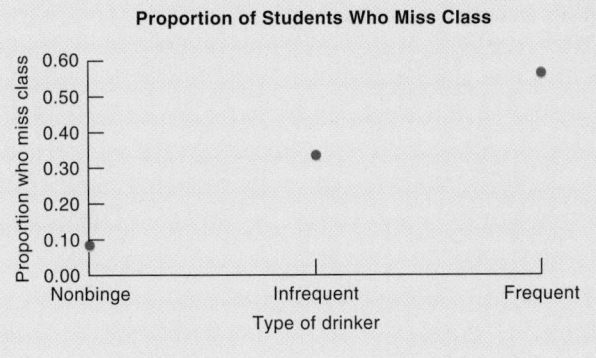

As you can see, the answer is not often obvious from a plot. If the sample sizes are relatively the same, you could find confidence intervals for each of the samples and compare them.

EXAMPLE 14.21

CHANGING LIVES

Using the Chi-Square Test Results

Now that we know that at least one of the marital groups is different from the others, we want to identify which one(s) is different so we can do some further analysis. We can plot the proportion who did not spend more time with their families for each group:

**Percent of People Who Did Not
Spend More Time with Family**

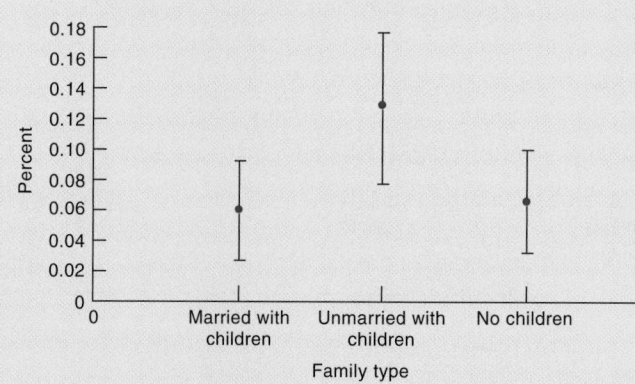

From the plot it certainly appears that the married with children and no children groups are comparable, whereas unmarried with children has a much higher proportion. To see if they are really different, we can calculate and plot 95% confidence intervals for the proportions:

Note: There has been some rounding in these calculations.

Married with children:
$$0.06 \pm 1.96\sqrt{\frac{(0.06)(0.94)}{200}} = 0.06 \pm 0.0329$$
$$= (0.0271, 0.0929)$$

Unmarried with children:
$$0.13 \pm 1.96\sqrt{\frac{(0.13)(0.87)}{170}} = 0.13 \pm 0.0504$$
$$= (0.0796, 0.1804)$$

No children:
$$0.07 \pm 1.96\sqrt{\frac{(0.07)(0.93)}{210}} = 0.07 \pm 0.0337$$
$$= (0.0363, 0.1037)$$

**Confidence Intervals For Percent Who Did Not Spend More
Time With Family Since 9/11**

From the plot we see that the confidence intervals for the married with children and no children groups almost completely overlap, whereas the confidence interval for the unmarried with children group barely overlaps the other two intervals. Thus, we can be pretty sure that the unmarried with children group differs from the other two and that it has a higher proportion of people who did not spend more time with their families after September 11, 2001.

Draw conclusions.

An important thing to remember when doing the chi-square test for proportions, or any statistical test, is that the test is not the answer or solution to the problem. It is simply verification that something is causing the differences we observed. Once the statistical analysis is done, it always requires knowledge of the situation to identify the causes and correct the problem if that is needed.

14.3.5 EXERCISES – LEARNING IT!

14.6 In an experiment to study the attitudes of voters concerning term limitations in Congress, voters in Indiana, Ohio, and Kentucky were polled with the following results:

Opinion	Indiana	Kentucky	Ohio
Support	82	107	93
Do not support	97	66	74

(a) Set up the hypotheses to test whether the proportions of voters who support congressional term limits are the same for all three states.

(b) Calculate the proportion of voters who support congressional term limits for each state individually. Based on these values, do you think there is a difference in the proportions?

(c) Calculate the overall proportion of voters who support term limits for Congress.

(d) Calculate the expected frequency for each cell and find the value of the chi-square test statistic.

(e) At the 0.05 level of significance, is there a difference in the proportions of voters who support congressional term limits in the three states?

14.7 In a survey about satisfaction with local phone service, respondents who rated their current service as Excellent and those who rated it as Poor – Very Poor were asked to classify their current local service provider. The results are given in the table:

Current service rating	Type of service				
	Long distance	Local phone	Power	Cable TV	Cellular phone
Excellent	264	444	131	215	198
Poor – Very Poor	1394	1318	485	431	572

(a) Set up the hypotheses to test whether the proportion of people who rated their company as Excellent is the same for each type of company.

(b) Calculate the overall proportion of people who rate their current phone service as Excellent.

(c) Find the expected frequency for each cell and calculate the chi-square test statistic.

(d) If you want to perform the test at the 0.05 level of significance, what is the critical value of the test?

(e) At the 0.05 level of significance, is there a difference in the proportions of people who rate their local phone service as Excellent among the different types of service?

14.8 The National Center for Health Statistics conducts the Immunization Survey. In the survey done in 1999, respondents were asked whether or not they had immunization records for the child in the survey and whether or not the parents were present in the family. A random sample of 3000 of the surveys provided these data:

Have shot records?	Parents present in family				Total
	Mother, no father	Father, no mother	Mother and father	Neither mother nor father	
Yes	189	32	564	19	804
No	614	119	1386	77	2196
Total	803	151	1950	96	3000

(a) Suppose you are interested in determining whether the proportion who had shot records is the same for each family group. Set up the hypotheses to test this.

(b) Calculate the overall proportion of people who had shot records.

(c) Use the value from part (b) to find the expected frequency for each cell.

(d) Calculate the value of the chi-square test statistic.

(e) At the 0.01 level of significance, is the proportion of respondents who had shot records the same for each family group?

14.9 In a survey on the use of online price comparison sites, 201 people who responded to the survey were asked about their experience with such sites. The people administering the survey were interested in determining whether women had different experiences with such sites than men. The data they compiled are shown in the table:

Gender	Experience with price comparison sites			
	Use it frequently	Use infrequently	Heard about it	No experience
Female	46	3	3	4
Male	118	12	5	10

(a) Set up the hypotheses to determine whether experience with online price comparison sites is related to gender.

(b) Calculate the overall proportion of female respondents in the survey.

(c) Use the overall proportion to calculate the expected frequency for each cell.

(d) If you want to do this test at the 0.10 level of significance, what is the critical value?

(e) At the 0.10 level of significance, are gender and experience with online price comparison sites related?

14.4   Chi-Square Test for Independence

In Chapter 5 you looked at bivariate data. For qualitative data you learned how to summarize the data using a contingency table, and for quantitative data you learned how to look at relationships between two variables. You learned that two quantitative variables can be dependent, meaning there is a relationship between the variables, or they can be independent, meaning there is no relationship between the variables. You also learned how to identify the type of relationship (linear or nonlinear) and to quantify a linear relationship in the form of an equation.

Now you know that it is also possible for there to be a relationship between two qualitative variables. For example, suppose the human resources department of a large company collected data on employees' job levels and the types of medical coverage that the employees chose. Both of these variables, job level and type of medical coverage, are qualitative, but it makes sense that there might be a relationship between them. People with higher level, higher paying jobs might have better health coverage. How can we determine whether such a relationship exists? That is, how can we determine whether two qualitative variables are dependent or independent?

The chi-square test for proportions involved a contingency table that had two rows, successes and failures, and a column for each population. This test can be extended to look at a table of two qualitative variables, where each row corresponds to a different value of the first variable and each column corresponds to a different value of the second variable.

14.4.1 PROBABILITY AND INDEPENDENCE

As you know, we rarely know the populations that we are studying, and instead we use data to estimate or approximate different characteristics of these populations. You learned in Chapter 6 that data collected over time are often used as estimates of the probabilities that different events will occur. These probabilities are known as *empirical* probabilities.

When we collect data on two qualitative variables, we can display the data in a contingency table. The rows of the table represent the possible categories of one of the variables, and the columns of the table represent the categories of the other variable. We can use the frequencies in the table to estimate different probabilities about the elements of the population. We can estimate whether an element will have a certain characteristic (event A), a particular pair of characteristics (event A AND B), or one or more of the characteristics (event A OR B).

EXAMPLE 14.22

HEALTH-CARE PLANS

Calculating Probabilities

Suppose the human resources department mentioned earlier looked at the records of 300 employees and collected data on two characteristics: job classification and health coverage. The data are summarized here:

Collect the data.

Job classification	Health coverage Physician network	HMO	No coverage	Total
Salaried professional	35	12	3	**50**
Salaried clerical	21	67	12	**100**
Hourly	6	112	32	**150**
Total	**62**	**191**	**47**	**300**

If we define A as the event that a person selects the HMO for medical coverage and B as the event that a person is an hourly worker, then we have the probabilities:

Analyze the data.

$$P(A) = \frac{191}{300} \qquad P(B) = \frac{150}{300} \qquad P(A \text{ AND } B) = \frac{112}{300}$$

14.4.2 PERFORMING THE CHI-SQUARE TEST FOR INDEPENDENCE

We want to determine whether two qualitative variables are related. We test these hypotheses:

H_0: The two variables are independent of each other.

H_A: The two variables are not independent.

TABLE 14.2

**CONTINGENCY TABLE FOR TEST
FOR INDEPENDENCE**

Variable 1	Variable 2				Total
	1	**2**	...	**c**	
1	o_{11}	o_{12}	...	o_{1c}	r_1
2	o_{21}	o_{22}	...	o_{2c}	r_2
$\vdots$	$\vdots$	$\vdots$	$\vdots$	$\vdots$	
r	o_{r1}	o_{r2}	...	o_{rc}	r_r
Total	c_1	c_2	...	c_c	**n**

When we collect data on two qualitative variables and summarize them in a contingency table, the number in each cell of the table is the observed frequency associated with one particular category of the first variable and one particular category of the second variable. A general contingency table for two variables is shown in Table 14.2.

Variable 1 has r possible categories and variable 2 has c possible categories. The value o_{12} is the number of sample elements in category 1 of variable 1 and category 2 of variable 2.

EXAMPLE 14.23

Understand the problem.

Collect the data.

Analyze the data.

COLLEGE DRINKING

Setting Up the Contingency Table

In addition to questions used to classify drinking habits, the survey on drinking asked students about where they lived: residence halls or dormitories, fraternity or sorority houses, other university housing, or off-campus houses or apartments. In analyzing the results of the survey, the researcher asked whether the amount of drinking by students depends on where they live. Here are the hypotheses:

H_0: The amount a student drinks is independent of where he or she lives.

H_A: The amount a student drinks depends on where he or she lives.

The researcher compiled this contingency table of the data:

Residence	Type of drinker			Total
	Nonbinge	Infrequent	Frequent	
Residence hall or dormitory	35	25	46	**106**
Fraternity or sorority house	0	1	0	**1**
Other university housing	0	2	1	**3**
Off-campus house or apartment	49	30	24	**103**
Total	**84**	**58**	**71**	**213**

To do a chi-square test, we must find the expected frequency for each cell in the table and compare it with the observed frequency. As in the other chi-square tests we have done, we find the expected frequencies by assuming that H_0 is true and calculating the probability for each cell under that assumption. For the test of independence, the null hypothesis is that the two variables are independent. To find the probability of each value of a variable, we use the row or column totals the same way we did in Chapter 6 on probability. That is, for row i,

$$P(i) = \frac{r_i}{n}$$

and for column j,

$$P(j) = \frac{c_j}{n}$$

Remember that the definition of independence works two ways. If we know that two events are independent, then we can multiply their probabilities together to find the probability that they will both occur. That is, when we have independence for each cell in the contingency table, the probability that variable 1 will have category i and variable 2 will have category j is

$$P(i \text{ AND } j) = \frac{r_i}{n} \frac{c_j}{n}$$

To find the **expected frequency** for any cell we multiply the probability by the sample size, so

$$e_{ij} = \left(\frac{r_i}{n} \frac{c_j}{n} \right) n = \frac{r_i c_j}{n}$$

Expected frequency in a cell

EXAMPLE 14.24

COLLEGE DRINKING

Finding Expected Frequencies

The data on residence and type of drinker are shown again here:

Residence	Type of drinker			
	Nonbinge	Infrequent	Frequent	Total
Residence hall or dormitory	35	25	46	106
Fraternity or sorority house	0	1	0	1
Other university housing	0	2	1	3
Off-campus house or apartment	49	30	24	103
Total	**84**	**58**	**71**	**213**

Here are the calculations for the expected frequency for each cell:

Analyze the data.

$$e_{11} = \frac{(106)(84)}{213} = 41.80 \quad e_{12} = \frac{(106)(58)}{213} = 28.86 \quad e_{13} = \frac{(106)(71)}{213} = 35.33$$

$$e_{21} = \frac{(1)(84)}{213} = 0.39 \quad e_{22} = \frac{(1)(58)}{213} = 0.27 \quad e_{23} = \frac{(1)(71)}{213} = 0.33$$

$$e_{31} = \frac{(3)(84)}{213} = 1.18 \quad e_{32} = \frac{(3)(58)}{213} = 0.82 \quad e_{33} = \frac{(3)(71)}{213} = 1.00$$

$$e_{41} = \frac{(103)(84)}{213} = 40.62 \quad e_{42} = \frac{(103)(58)}{213} = 28.05 \quad e_{43} = \frac{(103)(71)}{213} = 34.33$$

The expected frequencies are shown (in parentheses) under the observed frequencies in the table:

| | Type of drinker | | | |
Residence	Nonbinge	Infrequent	Frequent	Total
Residence hall or dormitory	35 (41.80)	25 (28.86)	46 (35.33)	106
Fraternity or sorority house	0 (0.39)	1 (0.27)	0 (0.33)	1
Other university housing	0 (1.18)	2 (0.82)	1 (1.00)	3
Off-campus house or apartment	49 (40.62)	30 (28.05)	24 (34.33)	103
Total	**84**	**58**	**71**	**213**

Remember that in any chi-square test the expected frequency for each cell or class should be at least 5. When this is not the case, the categories must be collapsed or joined. For the goodness of fit test, the data are quantitative, so we always join classes that are adjacent. In the test for proportions or the test for independence, this is not the case. The categories should be collapsed on a logical basis—that is, one that preserves the intent of the question.

EXAMPLE 14.25

COLLEGE DRINKING

Adjusting the Contingency Table

When he looked at the contingency table on type of drinker and residence, the researcher saw that two of the categories for residence did not have expected frequencies of at least 5. He knew he should collapse some of the categories. His first thought was to combine the two low-frequency categories (Fraternity or sorority house and Other university housing) into a single category designated Other university housing, but he realized that this would still not raise the expected values to above 5.

His final decision was to collapse all of the on-campus housing into a single category to meet the requirements of the chi-square test. The modified contingency table is shown here:

| | Type of drinker | | | |
Residence	Nonbinge	Infrequent	Frequent	Total
On campus	35 (43.38)	28 (29.95)	47 (36.67)	110
Off campus	49 (40.62)	30 (28.05)	24 (34.33)	103
Total	**84**	**58**	**71**	**213**

Understand the problem.

Analyze the data.

EXAMPLE 14.26

HEALTH-CARE PLANS

Setting Up the Contingency Table

For the human resources department, these are the hypotheses to be tested:

H_0: Choice of health-care plan and job classification are independent.

H_A: Choice of health-care plan and job classification are not independent.

For the health coverage data the expected frequencies are calculated as follows:

$$e_{11} = \frac{(50)(62)}{300} = 10.3 \quad e_{12} = \frac{(50)(191)}{300} = 31.8 \quad e_{13} = \frac{(50)(47)}{300} = 7.8$$

$$e_{21} = \frac{(100)(62)}{300} = 20.7 \quad e_{22} = \frac{(100)(191)}{300} = 63.7 \quad e_{23} = \frac{(100)(47)}{300} = 15.7$$

$$e_{31} = \frac{(150)(62)}{300} = 31.0 \quad e_{32} = \frac{(150)(191)}{300} = 95.5 \quad e_{33} = \frac{(150)(47)}{300} = 23.5$$

The expected frequencies together with the original data are given in the next table. There are clearly discrepancies between the observed and expected frequencies, but it is not clear whether they are simply the result of random variation or evidence that H_0 is false.

> **Analyze the data.**

Job classification	Health coverage			
	Physician network	HMO	No coverage	Total
Salaried professional	35 (10.3)	12 (31.8)	3 (7.8)	50
Salaried clerical	21 (20.7)	67 (63.7)	12 (15.7)	100
Hourly	6 (31.0)	112 (95.5)	32 (23.5)	150
Total	62	191	47	300

As with the goodness of fit test and the test for proportions, it is important to remember that the expected frequency in each cell should be at least 5. If not, then categories for one or both of the variables need to be combined. Since the categories are not numerical, the categories should be grouped on similarities.

✔ **TRY IT NOW!**

College Drinking
Setting Up the Contingency Table

The public health student who did the study on drinking also collected data on the number of times students drove while intoxicated in the last 2 weeks. The contingency table for the usable responses is given here:

| | **Drove while intoxicated** | | | |
Class	Not at all	Once	Twice or more	Total
Freshman	72	5	9	86
Sophomore	19	8	9	36
Junior	16	8	6	30
Senior	8	4	7	19
Total	**115**	**25**	**31**	**171**

The university is interested in whether the number of times a student drove while intoxicated is related to his or her class. This information will help the school target student audiences for programs on drinking and driving. Set up the hypotheses that the university should test.

Calculate the expected frequencies for each cell and put them in the appropriate locations in the table.

Are any of the expected frequencies less than 5? If so, suggest a logical way to combine categories to avoid this problem.

14.4.3 PERFORMING THE TEST FOR INDEPENDENCE

Once the expected and observed frequencies for each cell are recorded in the table, nothing new needs to be learned to perform the test. The chi-square test statistic is calculated in the same way as for the test of proportions:

$$\chi^2 = \sum_{\text{all cells}} \frac{(o_{ij} - e_{ij})^2}{e_{ij}}$$

and the degrees of freedom are calculated as

$$(r - 1)(c - 1)$$

EXAMPLE 14.27

COLLEGE DRINKING

Performing the Chi-Square Test for Independence

The researcher who is looking at whether a student's residence is related to binge drinking calculated the chi-square statistic for the revised contingency table:

| | **Type of drinker** | | | |
Residence	Nonbinge	Infrequent	Frequent	Total
On campus	35	28	47	110
	(43.38)	(29.95)	(36.67)	
Off campus	49	30	24	103
	(40.62)	(28.05)	(34.33)	
Total	**84**	**58**	**71**	**213**

Analyze the data.

$$\chi^2 = \frac{(35 - 43.38)^2}{43.38} + \frac{(28 - 29.95)^2}{29.95} + \cdots + \frac{(30 - 28.05)^2}{28.05} + \frac{(24 - 34.33)^2}{34.33}$$

$$= 1.62 + 0.13 + \cdots + 0.14 + 3.11 = 9.63$$

He decided to do the test at the 0.05 level of significance and calculated the degrees of freedom of the test to be $(2 - 1)(3 - 1) = 2$. The critical value of the test is $\chi^2_{0.05,2} = 5.99$. Because the test statistic of 9.63 is greater than the critical value, he rejects H_0 and concludes that the variables residence and type of drinker are not independent. That is, the amount of drinking that students do is related to where they live.

Draw conclusions.

EXAMPLE 14.28

HEALTH-CARE PLANS

Performing the Chi-Square Test for Independence

The human resources department that is looking at the relationship between employee job classification and type of health coverage calculates the chi-square statistic for the data:

$$\chi^2 = \frac{(35 - 10.3)^2}{103} + \frac{(12 - 31.8)^2}{31.8} + \cdots + \frac{(112 - 95.5)^2}{95.5} + \frac{(32 - 23.5)^2}{23.5}$$

$$= 58.88 + 12.36 + \cdots + 2.85 + 3.07 = 101.35$$

Analyze the data.

The analysts decide to perform the test with $\alpha = 0.05$. The degrees of freedom for the test are $(3 - 1)(3 - 1) = 4$, and they find the critical value to be 9.49. Since the chi-square test statistic is definitely outside the critical value, they reject H_0 and conclude that job classification and health coverage are not independent; that is, they are related.

Draw conclusions.

✔ **TRY IT NOW!**

College Drinking
Performing the Chi-Square Test for Independence

The university that is looking at the relationship between class and driving while intoxicated wants to perform the test at the 0.05 level of significance. Calculate the value of the chi-square statistic for the data.

Find the critical value and perform the test.

Are class and driving while intoxicated independent?

Don't forget to collapse the categories so that no cells have expected values less than 5.

ANS. 22.11 (COLLAPSING THE DRINKING CATEGORIES), 21.9 (COLLAPSING JR./SR.); 7.82, 9.49; NO

14.4.4 EXERCISES—LEARNING IT!

14.10 A report by the U.S. Department of Justice on rape victims reports on interviews with 3721 victims. The attacks were classified by the age of the victim and the relationship of the victim to the rapist. The results of the study are shown here:

	Relationship of rapist		
Age of victim	Family	Acquaintance or friend	Stranger
Under 12	153	167	13
12 to 17	230	746	172
Over 17	269	1232	739

(a) Set up the hypotheses to test whether age of victim and relationship of rapist are independent.

(b) Calculate the expected frequency for each cell.

(c) How many degrees of freedom does the chi-square test for independence have? Using a 0.01 level of significance, what is the critical value for the test?

(d) Calculate the value of the chi-square test statistic.

(e) Is the age of the victim independent of the relationship of the rapist?

14.11 A company that manufactures cardboard boxes for packaging cereals wants to determine whether the type of defect that a particular box has is related to the shift on which it was produced. It compiles the following data. In each case, if a box had multiple defects, the most serious defect was recorded.

	Type of defect		
Shift	Printing	Rips/tears	Size
1	55	60	85
2	58	63	79
3	89	63	48

(a) Set up the appropriate hypotheses for the test.

(b) Calculate the expected frequency for each cell and calculate the value of the chi-square test statistic.

(c) How many degrees of freedom does the test have?

(d) At the 0.05 level of significance, are defect type and shift related?

14.12 A company that depends heavily on advertising for selling its products wants to know if its various advertising media have different effectiveness relative to the age of the customer. The company uses its warranty return cards to collect the following information:

Type of advertisement	Age of customer				Total
	21–30	31–40	41–50	Over 50	
Store display	21	28	8	6	**63**
Catalog	8	5	1	1	**15**
Magazine	1	23	8	1	**33**
Newspaper	18	14	2	4	**38**
Total	**48**	**70**	**19**	**12**	**149**

(a) Set up the hypotheses to test whether type of advertisement and age of customer are independent.

(b) What population will the test apply to?

(c) Calculate the expected frequency for each cell.

(d) How many degrees of freedom does the test have?

(e) Without collapsing any cells, calculate the value of the chi-square test statistic.

(f) At the 0.05 level of significance, are type of advertisement and age of customer independent?

(g) Do you think categories should have been collapsed? Why or why not?

(h) What categories do you recommend collapsing to fix the problem?

14.13 The people who did the survey on online price comparison sites were also interested in whether experience with such sites is related to age. They created a contingency table to look at this question:

	Age		
Experience	**Under 25**	**25 to 44**	**45 and up**
Frequent use	11	51	101
Infrequent use	2	2	11
Heard about it	2	2	4
No experience	2	3	10

(a) Set up the hypotheses to test whether experience with online price comparison sites is independent of age.

(b) How many degrees of freedom does the chi-square test have?

(c) Calculate the expected frequency for each cell.

(d) Without collapsing any categories, calculate the value of the chi-square test statistic.

(e) At the 0.05 level of significance, is experience with online price comparison sites independent of age?

(f) Are the cells with frequencies less than 5 a factor in this case? Why or why not? Do you think you should redo the test after collapsing categories?

GET IT IN WRITING

College Drinking

TO: **Dean of Students**
FROM: **Phil Student**
RE: **Survey of Binge Drinking on Campus**

As part of my honor's thesis in public health and because I am the president of one of the fraternities on campus, I decided to address the issue of binge drinking among students. The Wexler survey that I used was developed at Harvard and is used widely at colleges across the country. It is accepted as a valid survey instrument. I thought that you might be interested in some preliminary results of the study.

The survey was given to students enrolled in a public health course that is required of all students; this way the sample would be representative of the student population at our university. The study yielded 221 usable responses, although not every student answered every question. The breakdown across class years is shown in the table:

Class	Observed frequency	Percentage
Freshman	86	50.3
Sophomore	36	21.1
Junior	30	17.5
Senior	19	11.1
Total (*n*)	**171**	**100.0**

Although this distribution is not uniform over the four years, it is not statistically different from the distribution of students at the university as reported by the university administration (45.2%, 18.6%, 18.0%, and 18.2%).

The study examined the impact of drinking on academic responsibility. For the purpose of the study students are classified as nondrinkers, nonbinge drinkers, infrequent binge drinkers, and binge drinkers. Binge drinkers are those who consume five or more drinks during one episode of drinking for men and four or more drinks during one episode of drinking for women.

One question on the study asked students how many times they had missed class because of drinking. The table shows the results:

| | Type of drinker | | | |
Miss class	Nonbinge	Infrequent	Frequent	Total
Never	77	39	31	147
Once or more	7	19	39	65
Total	**84**	**58**	**70**	**212**

It is critical to note that of the 221 responses, only 9 (4%) were classified as nondrinkers. A chi-square statistical test was used to determine whether the proportion of students who missed class because of drinking was different for the different types of drinkers. The test found a significant difference in the proportions ($p < 0.05$). The following graph depicts the differences quite well:

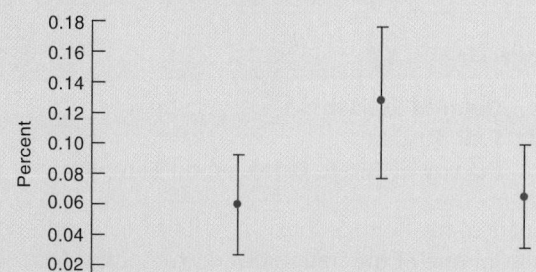

Confidence Intervals For Percent Who Did Not Spend Time With Family Since 9/11

It is clear that the proportion of drinkers who miss class increases as the amount of drinking increases, from 3.3% for nonbinge drinkers to 18.4% for frequent binge drinkers. This is an alarming result.

Another portion of the study looked at what situational variables might be related to binge drinking. One question in particular examined where the student lived and what type of drinker he or she was. The table shows these data:

| | Type of drinker | | | |
Residence	Nonbinge	Infrequent	Frequent	Total
Residence hall or dormitory	35	25	46	**106**
Fraternity or sorority house	0	1	0	**1**
Other university housing	0	2	1	**3**
Off-campus house or apartment	49	30	24	**103**
Total	**84**	**58**	**71**	**213**

Because the data set was not very large for the purpose of statistical testing, residence was reclassified as either on or off campus by combining the first three categories. That data are shown here:

	Type of drinker			
Residence	Nonbinge	Infrequent	Frequent	Total
On campus	35	28	47	**110**
Off campus	49	30	24	**103**
Total	**84**	**58**	**71**	**213**

The results indicate that the amount of drinking done by students does depend on where they live ($p < 0.05$). Determining the exact nature of the relationship will take further analysis.

I intend to perform further analysis of the data and will share the results with you as soon as they are available.

14.5 Using Technology for the Chi-Square Test

Excel and Minitab both provide tools to do a chi-square test. The TI-83 graphing calculator, however, does not. You learned the basic tool for doing the chi-square test in Chapter 5, when you learned how to create a contingency table. Here we look at how to add the results of a chi-square analysis to that tool.

14.5.1 CREATING THE CONTINGENCY TABLE IN EXCEL

In Chapter 5 you learned how to create a contingency table for two qualitative variables in Excel. This contingency table provides the backbone for the chi-square test for independence in Excel. We will look at the data from the binge drinking survey and test whether where a student lives is independent of the type of binge drinker. A small portion of the data in an Excel worksheet is shown in Figure 14.3.

	A	B	C
1	BINGER	MISS	LIVE
2	1	1	4
3	1	1	4
4	2	1	1
5	1	1	4
6	1	1	4
7	1	1	1
8	1	1	4
9	3	1	4
10	1	1	4
11	1	1	4
12	2	2	1
13	3	2	4
14	0	1	4
15	3	1	1
16	3	3	1
17	3	2	4
18	2	1	1

FIGURE 14.3 **Binge drinking data**

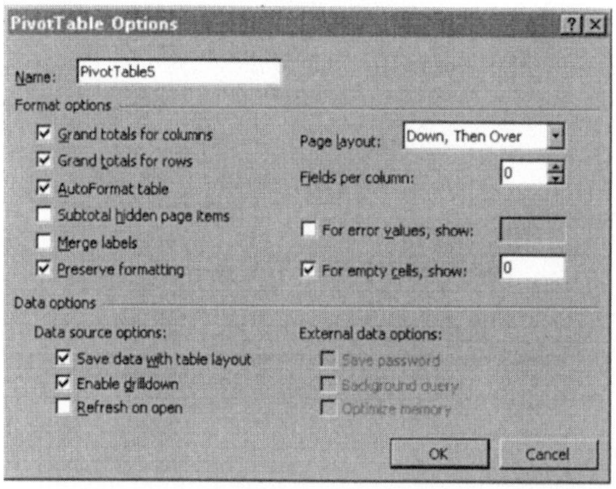

FIGURE 14.4 **Pivot Table Options dialog box**

The first step is to create a contingency table in Excel using the **Pivot table** tool. From the **Data** menu, select **Pivot Table Report** and follow the steps of the pivot table wizard to create the table. In this example we put the variable for type of binge drinker (BINGER) in the columns and the variable for where a student lives (LIVE) in the rows. At step 4 of 4, after you indicate where you want the table to be placed, do not select **Finish.** Instead, click **Options** . . . and the dialog box shown in Figure 14.4 opens.

Make sure that the check box next to **For empty cells, show** is checked, and in the text box next to it, type "0." To perform a chi-square test, Excel will not accept empty cells in the contingency table. Click **OK** and then **Finish.** The contingency table for the data is shown in Figure 14.5.

Since we are interested only in students who are drinkers, we do not want the Binger column for 0, which indicates that the student does not drink. To eliminate this column, double click on the field button BINGER and the pivot table field dialog box shown in Figure 14.6 opens.

O	P	Q	R	S	T
Count of BINGER	BINGER				
LIVE	0	1	2	3	Grand Total
1	11	35	25	46	117
2	0	0	1	0	1
3	0	0	2	1	3
4	28	49	30	24	131
Grand Total	39	84	58	71	252

FIGURE 14.5 **Contingency table for binge drinking data**

Click on the drop-down arrow next to the field button for Binger and uncheck the box labeled 0, as shown in Figure 14.6.

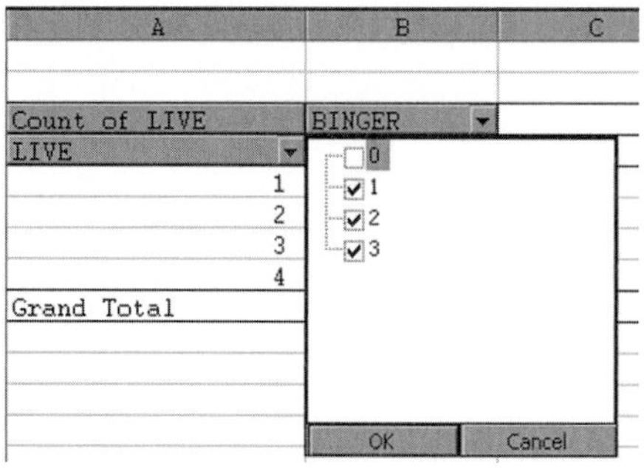

FIGURE 14.6 **Hiding the nondrinkers**

14.5.2 PERFORMING THE CHI-SQUARE TEST

KADDSTAT provides a statistical function for performing the chi-square test. The function calculates the expected values and the chi-square test statistic as well as the *p* value of the test.

To perform the chi-square test, select **Hypothesis Testing > Chi Square Test** from the KADD menu. The dialog box shown in Figure 14.7 opens.

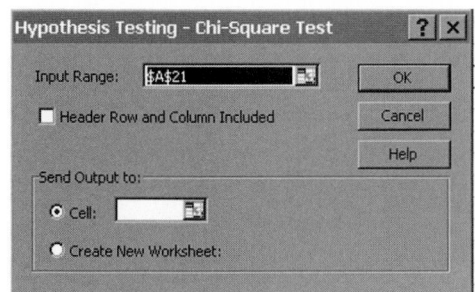

FIGURE 14.7 Dialog box for chi-square test

The input is very simple. The input range should be the location of the contingency table for the data. Make sure that you do not select the column and row that contain the totals for the table. If you have labels for your columns and rows (as in Figure 14.3), then check the box labeled **Header Row and Column Included.**

Indicate where you want the output located and then click **OK.** The output for the binge drinking data is shown in Figure 14.8.

Chi-square test statistic = 15.3884
p-value = 0.0174

Number of:
rows = 4
columns = 3

Actual frequencies

		Variable B			
		Non-binge	Infrequent	Frequent	Totals
Variable A	Residence Hall or Dormitory	35	25	46	106
	Fraternity or Sorority	0	1	0	1
	Other University Housing	0	2	1	3
	Off Campus House or Apartment	49	30	24	103
	Totals	84	58	71	213

Chi-square calculations

		Variable B		
		Non-binge	Infrequent	Frequent
Variable A	Residence Hall or Dormitory	1.1071	0.5172	3.2201
	Fraternity or Sorority	0.3944	1.9447	0.3333
	Other University Housing	1.1831	1.7135	0.0000
	Off Campus House or Apartment	1.7289	0.1360	3.1100

Expected frequencies

		Variable B			
		Non-binge	Infrequent	Frequent	Totals
Variable A	Residence Hall or Dormitory	41.8028	28.8638	35.3333	106
	Fraternity or Sorority	0.3944	0.2723	0.3333	1
	Other University Housing	1.1831	0.8169	1.0000	3
	Off Campus House or Apartment	40.6197	28.0469	34.3333	103
	Totals	84	58	71	213

NOTE: Expected frequencies should not be less than 5.0

FIGURE 14.8 KADDSTAT output from chi-square test

14.5.3 SUBSETTING THE DATA IN MINITAB

Look at the data from the survey about binge drinking. Suppose we want to know, for the drinkers, whether type of drinker is independent of where the person lives. A very small portion of the data is shown in Figure 14.9.

↓	C1 BINGER	C2 MISS	C3 LIVE
2	1	1	4
3	2	1	1
4	1	1	4
5	1	1	4
6	1	1	1
7	1	1	4

Note: We are interested only in drinkers, so nondrinkers were eliminated by using the **Manip > Subset worksheet** *command.*

FIGURE 14.9 **Portion of binge drinking survey data**

In Minitab you can use the following procedure to create the contingency table and do the chi-square analysis all at once:

1. Select **Stat > Tables > Cross Tabulation** from the menu. The dialog box shown in Figure 14.10 opens.

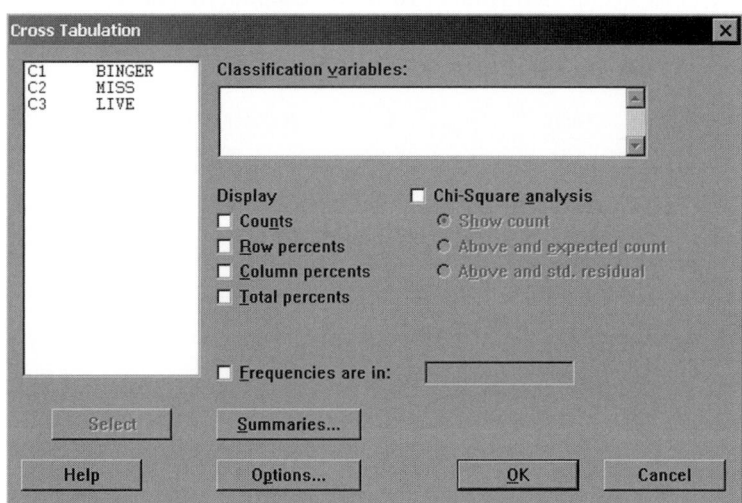

FIGURE 14.10 **Cross Tabulation dialog box**

2. Position the cursor in the box labeled **Classification variables:** and select the two variables that you are interested in from the list on the left—in this case, Miss and Live.

3. In the section marked **Display** indicate which statistics you want Minitab to display in the table. For this example we include Count only.

4. Check the box labeled **Chi-Square analysis.** When you click this box, the options below become active. If you want to include the expected counts too, click on **Above and expected count.** This will include both the counts (above) and the expected counts. If you do this, *uncheck* the counts under **Display** or you will get them twice!

5. Click **OK** and the output shown in Figure 14.11 appears in the Session window.

Tabulated Statistics: BINGER, LIVE

Rows: BINGER   **Columns: LIVE**

	1	2	3	4	All
1	35	0	0	49	84
	41.80	0.39	1.18	40.62	84.00
2	25	1	2	30	58
	28.86	0.27	0.82	28.05	58.00
3	46	0	1	24	71
	35.33	0.33	1.00	34.33	71.00
All	106	1	3	103	213
	106.00	1.00	3.00	103.00	213.00

Chi-Square = 15.388, DF = 6
* WARNING * 4 cells with expected counts less than 1.0
        * Chi-Square approximation probably invalid
6 cells with expected counts less than 5.0

Cell Contents --
        Count
        Exp Freq

FIGURE 14.11 **Output from chi-square test**

The output includes the value of the test statistic, 15.388, and the degrees of freedom, 6. You will need to look up the critical value either with the software or in a table. Minitab also gives warnings about cells with very low expected values and indicates that the test is probably invalid. You should collapse cells if this happens.

CHAPTER 14 SUMMARY

The chi-square test involves comparing observed and expected frequencies for different classes of data. The test is versatile and can be used to test *goodness of fit, equality of proportions for more than two populations*, and *independence of qualitative variables*.

As is true with any statistical tool, the results of a chi-square test do not *solve* a problem or make a decision for the user. They simply point out when further action is indicated and when it is not. Solving the real problem requires knowledge of the situation and sometimes additional data collection and analysis.

KEY TERMS		
Term	**Definition**	**Page reference**
Chi-square goodness of fit test	The **chi-square goodness of fit test** determines how well a set of data fits the model for a particular probability distribution.	694
Expected frequencies, e_i	The **expected frequencies** are the number of observations that should fall into each class in a frequency distribution under the hypothesized probability distribution.	694
Observed frequencies, o_i	The **observed frequencies** are the actual number of observations that fall into each class in a frequency distribution or histogram.	694
Uniform distribution	A **uniform distribution** is one in which each outcome or class of outcomes is equally likely to occur.	694

KEY FORMULAS

Term	Formula	Page reference
Chi-square critical value	$\chi^2_{\alpha, k-p-1}$	701
Chi-square statistic	$\chi^2 = \dfrac{\sum\limits_{i=1}^{k}(o_i - e_i)^2}{e_i}$	699
Degrees of freedom for goodness of fit test	$k - p - 1$	699
Degrees of freedom for test for independence	$(r - 1)(c - 1)$	728
Expected frequency for test for independence	$e_{ij} = \left(\dfrac{r_i}{n}\dfrac{c_j}{n}\right)n = \dfrac{r_i c_j}{n}$	725
Expected frequency for test for proportions	$e_i = \dfrac{s}{n}n_i = \dfrac{sn_i}{n} = pn_i$	714
Expected frequency for goodness of fit test	$e_i = np_i$	698
Overall proportion of success	$p = \dfrac{s_1 + s_2 + \cdots + s_c}{n_1 + n_2 + \cdots + n_c} = \dfrac{s}{n}$	714

CHAPTER 14 EXERCISES

Learning It!

14.14 A random sample of 69 Porsche drivers were asked how many miles they had driven their vehicle in the past calendar year. Here is a frequency table for the data:

Miles	Observed frequency
$0 < x \leqslant 4000$	20
$4000 < x \leqslant 8000$	22
$8000 < x \leqslant 12{,}000$	14
$12{,}000 < x \leqslant 16{,}000$	6
$16{,}000 < x \leqslant 20{,}000$	1
$20{,}000 < x \leqslant 24{,}000$	4
$24{,}000 < x \leqslant 28{,}000$	0
$28{,}000 < x \leqslant 32{,}000$	2

(a) Set up the hypotheses to test whether the number of miles driven by Porsche drivers is normally distributed with a mean of 7500 miles and a standard deviation of 6500 miles.

(b) Find the expected frequency for each cell.

(c) Calculate the chi-square test statistic. If necessary, collapse cells so that the expected frequency for each category is at least 5.

(d) At the 0.05 level of significance, what can you conclude about the number of miles driven annually by Porsche drivers?

14.15 The average yearly salary of entry-level social workers in the United States is $32,500 with a standard deviation of $930. A state agency collects data on recent hires and asks you to analyze the data and show how its social workers' salaries compare with others in the country. Here are the salaries for 30 social workers:

26,633	26,764	27,306	27,619	28,445	28,827
29,045	29,453	29,620	29,690	29,864	29,974
30,192	30,260	30,269	30,346	30,400	30,532
30,628	30,829	31,270	31,489	32,731	33,190
33,397	33,553	33,685	34,466	34,944	35,751

(a) Create a frequency distribution for the data using six classes and display the data graphically.

(b) Do you think the assumption that the data are normally distributed is reasonable? Why or why not?

(c) Set up the hypotheses to test whether the data come from a normal population that has a mean of $32,500 and a standard deviation of $930.

(d) Perform the chi-square goodness of fit test with a level of significance of 0.05.

(e) What does the test tell you about the data?

14.16 Workplace violence is a growing problem. A survey of 600 full-time American workers on workplace violence concentrated on those respondents who were victims of harassment, threat of physical violence, and actual physical violence. In follow-up interviews the victims were asked to identify the major effect that the violence had on them. The data are shown here:

Major effect or worker	Type of violence		
	Harassment	Threat	Physical
Psychological	56	28	15
Disrupted work life	39	13	12
Physical injury or sickness	15	5	10
No negative effect	5	7	6

(a) Suppose the group that conducted the study is interested in whether the type of violence that a person experienced is related to the effect of the violence on the worker. Set up the hypotheses for this test.

(b) Find the expected frequency for each cell.

(c) Without collapsing any categories, how many degrees of freedom does the test have?

(d) At the 0.05 level of significance, are type of violence and effect on a worker independent?

14.17 Throughout the 1990s the number of people who took the drug Prozac increased dramatically. As part of the General Social Survey done in 1998, information was collected on whether respondents had ever taken Prozac and the region of the country in which they lived. The data follow:

	Region								
Taken Prozac?	New England	Middle Atlantic	East North Central	West North Central	South Atlantic	East South Atlantic	West South Central	Mountain	Pacific
Yes	8	11	17	10	25	8	6	7	5
No	52	148	188	84	202	67	108	81	148

(a) Set up the hypotheses to test whether the proportion of people who took Prozac was the same for all regions of the United States.

(b) Calculate the overall proportion of people who took Prozac.

(c) Find the expected frequency for each cell.

(d) Calculate the chi-square statistic.

(e) At the 0.05 level of significance, do the data indicate that the proportion of people who took Prozac was the same for all regions?

14.18 In a survey of elementary school teachers and technology, teachers were asked whether they were comfortable installing software on classroom computers or whether they relied on the school district's computer personnel to do the job. The responses, by grade level taught, are shown here:

Install software?	Grade level taught		
	Pre-K to 2	Grades 3 to 5	Grades 6 to 8
Yes	18	48	108
No	20	55	12

(a) Set up the hypotheses to test whether the proportions of teachers who install their own classroom software are the same for all three grade levels.

(b) Use the data to calculate the overall proportion of teachers who install their own classroom software.

(c) Find the expected frequency for each cell and calculate the chi-square statistic.

(d) At the 0.05 level of significance, what can you conclude about the proportion of teachers who install their own classroom software?

14.19 A software company that was looking at the time to failure of the diskettes it uses decides to look at its two suppliers of the product. The specifications for the product state that the lifetime of the diskette should be normally distributed with a mean of 500 hours and a standard deviation of 5 hours. The diskette lifetimes (in hours) for each of the suppliers are shown here:

Supplier A					Supplier B				
474	492	498	504	511	487	492	495	497	499
486	492	500	505	511	488	492	495	497	499
489	494	501	506	512	489	492	495	497	499
490	494	501	507	513	489	492	495	497	501
490	494	501	508	514	489	493	495	498	502
490	495	502	508	515	491	493	496	498	503
491	496	502	509	517	491	494	496	498	503
491	498	504	509	519	491	494	496	498	505
491	498	504	510	528	492	494	496	499	506

(a) Create a frequency distribution and a histogram for each supplier.

(b) From the histogram, does it appear that the data for each supplier are normally distributed?

(c) Perform a chi-square goodness of fit test for each supplier to see if it is reasonable to say that the diskettes meet the specifications. Use a level of significance of 0.05.

14.20 As part of an annual survey by the U.S. Department of Transportation, randomly selected households are asked about the number of vehicles in the household and the availability of public transportation. The results for people who lived in urban areas are tabulated as follows:

Public transportation?	Number of vehicles					
	0	**1**	**2**	**3**	**4**	**5+**
Yes	974	2592	2427	687	194	59
No	98	378	540	199	58	23

(a) Set up the hypotheses to determine whether the number of vehicles in a household and the availability of public transportation are independent.

(b) How many degrees of freedom does the critical value for the test have?

(c) Calculate the expected frequencies and compute the test statistic.

(d) At the 0.05 level of significance, what can you conclude about the number of vehicles in a household and the availability of public transportation?

14.21 In a poll sponsored by NBC News and the Wall Street Journal in December 1999, people were asked how much of their holiday shopping they intended to do over the Internet. Here are the results by gender of the respondent:

	Holiday shopping over the Internet				
Gender	**All of it**	**Most of it**	**Some of it**	**None of it**	**Not sure**
Male	4	17	94	346	3
Female	2	14	85	431	7

(a) Set up the hypotheses to test whether gender and the amount of holiday shopping done on the Internet are independent.

(b) Find the expected frequency for each cell.

(c) Perform the chi-square test for independence.

(d) At the 0.05 level of significance, is the amount of holiday shopping done over the Internet independent of gender?

14.22 In the same poll sponsored by NBC News and the Wall Street Journal in December 1999, 1003 people were asked whether or not they intended to do any of their holiday shopping over the Internet. They were also asked their annual household income. The results are given in the table:

	Household income							
Shop on Internet?	Less than $10,000	Between $10,000 and $20,000	Between $20,000 and $30,000	Between $30,000 and $40,000	Between $40,000 and $50,000	Between $50,000 and $75,000	Between $75,000 and $100,000	More than $100,000
No	31	57	101	113	90	12	54	61
Yes	2	5	9	20	19	3	49	44

Note: Not all 1003 respondents chose to answer both questions.

(a) Set up the hypotheses to test whether the proportion of people who intend to do some holiday shopping over the Internet is the same for each income level.

(b) Calculate the overall proportion of people who intend to do some holiday shopping over the Internet.

(c) Calculate the expected frequency for each cell.

(d) At the 0.10 level of significance, what can you say about the proportion of people who intend to do some holiday shopping over the Internet for the different income levels?

Thinking About It!

14.23 One of the big controversies during the Vietnam War era was whether the draft lottery was truly random. Many people thought that the lottery was biased against people who were born in certain months of the year. The accompanying data show the number of birth dates in each month that were chosen in the first half of the draft lottery (there are 366 possible birth dates, so these are the first 183 birth dates that were chosen).

Month	Number of birth dates in the first half of the draft	Month	Number of birth dates in the first half of the draft
January	13	July	14
February	12	August	18
March	9	September	19
April	11	October	13
May	14	November	21
June	14	December	25

(a) Just by looking at the data, do you think that the claim of bias is worth investigating? Why or why not?

(b) At the 0.05 level of significance, do the data indicate that the distribution of the first 183 draft dates was something other than random? (*Hint:* Think about what *random* would mean in this situation.)

14.24 Consider the company that is looking at the relationship between type of advertising and age of customer. The data are shown again here:

Requires Exercise 14.12

Type of advertisement	Age of customer				
	21–30	31–40	41–50	Over 50	Total
Store display	21	28	8	6	**63**
Catalog	8	5	1	1	**15**
Magazine	1	23	8	1	**33**
Newspaper	18	14	2	4	**38**
Total	**48**	**70**	**19**	**12**	**149**

(a) Create a revised contingency table that collapses categories and fixes the problem of cells with expected frequencies less than 5.

(b) Perform the chi-square test for independence again.

(c) Compare the results of this test with the results of the test where categories were not collapsed.

Requires Exercise 14.19

14.25 Consider the data on the diskette suppliers.

(a) Why is doing a chi-square goodness of fit test for each supplier to see if they meet specifications different from doing a test to see whether the mean for each supplier is 500 hours and a test to see if the variance for each supplier is 25 hours^2?

(b) Just because you reject the hypothesis that the data are normally distributed with a mean of 500 hours and a standard deviation of 5 hours, does that mean that the data are not normally distributed?

(c) Calculate the sample mean and standard deviation for each supplier from the data provided.

(d) Do a chi-square test for each supplier to see if the data are normally distributed using the sample means and standard deviations.

(e) How do the changes in part (d) change the test procedure itself?

(f) How do the results of the new tests in part (d) compare with the results of the previous tests?

Requires Exercise 14.17

14.26 Look at the results of the chi-square test on proportions for the data on Prozac use.

(a) Prepare a plot of the proportion of people who took Prozac by region.

(b) When you reject H_0, you decide that at least one population (in this case region) is different. Use your plot from part (a) to come to a more informative conclusion.

Requires Exercise 14.14

14.27 Look at the data on the number of miles driven by the sample of Porsche drivers.

(a) Make a boxplot of the data.

(b) Do the data look normally distributed?

(c) Delete the outliers and replot the data. Did this change your opinion about the normality of the data? Why or why not?

(d) Calculate the mean and standard deviation of the data without the outliers.

(e) Perform a chi-square goodness of fit test to determine whether the data come from a normally distributed population with the calculated mean and standard deviation.

(f) What are your conclusions?

Doing It!

Datafile: BINGE.XXX

14.28 The public health student who was looking at the issue of college drinking administered a survey that asked questions on four different areas: drinking habits, problems related to drinking, problems related to other people's drinking, and living habits. A portion of the data from the survey and an explanation of the variables are given here:

Five	Four	Three	Last	Binger	Hangover	Miss	Behind	Regret	Forget
0	0	0	2	1	1	1	1	1	1
0	0	0	4	1	1	1	1	1	1
0	1	1	3	2	1	1	1	1	1
0	0	0	3	1	1	1	1	1	1
0	0	0	3	1	1	1	1	1	1
0	0	0	4	1	1	1	1	1	1

Argue	Engage	Protect	Damage	Police	Injured	Medical	Drive	DWI	Ride
1	1	1	1	1	1	1	1	1	1
1	1	1	1	1	1	1	1	1	1
1	1	1	1	1	1	1	1	1	1
1	1	1	1	1	1	1	1	1	1
1	1	1	1	1	1	1	3	1	3
1	1	1	1	1	1	1	1	1	1

Age	Gender	Fulltime	Greek	Live	Roommate	Race	Class
18	1	Y	N	4	P	2	1
62	2	Y	N	4	S	1	
18	2	Y	N	1	R	1	1
31	1	Y	N	4	R	1	
19	2	Y	N	4	P	1	2
20	2	Y	N	1	R	1	2

For the purposes of this survey a "drink" means any of the following:

12-ounce can or bottle of beer

4-ounce glass of wine

12-ounce bottle or can of wine cooler

1-ounce (shot) of liquor straight or in a mixed drink

Binge drinking is defined as:

The consumption of five or more drinks in one episode of drinking for males.

The consumption of four or more drinks in one episode of drinking for females.

The variables **Five, Four,** and **Three** refer to the 2-week period just before the survey was administered and answer the questions indicated. They are coded as follows:

0—none

1—once

2—twice

3—three to five times

4—six to nine times

5—ten or more times

- Five: How many times have you had five or more drinks in a row?
- Four: How many times have you had four drinks in a row (but no more than that)?
- Three: How many times have you had three drinks in a row (but no more than that)?

The variable **Last** answers the question, When did you have your last drink? The answers are coded as follows:

0—I never had a drink

1—not in the past year

2—more than 30 days ago, but less than a year ago

3—more than one week ago, but less than 30 days ago

4—within the last week

The variable **Binger** is defined as the type of drinker that a student is based on his or her answer to the first four questions.

0—nondrinker

1—nonbinge drinker

2—infrequent binge drinker

3—frequent binge drinker

The next 15 variables answer questions that start with the statement "Since the beginning of the school year, how often has your drinking caused you to . . . ." The answers are coded as follows:

0—not at all

1—once

2—twice or more

- **Hangover**   ". . . have a hangover?"
- **Miss**   ". . . miss a class?"
- **Behind**   ". . . get behind in school work?"
- **Regret**   ". . . do something you later regret?"
- **Forget**   ". . . forget where you were or what you did?"
- **Argue**   ". . . argue with friends?"
- **Engage**   ". . . engage in unplanned sexual activity?"
- **Protect**   ". . . not use protection when you had sex?"
- **Damage**   ". . . damage property?"
- **Police**   ". . . get into trouble with campus or local police?"
- **Injured**   ". . . get hurt or injured?"
- **Medical**   ". . . require medical treatment for an alcohol overdose?"
- **Drive**   ". . . drive after drinking alcohol?"
- **DWI**   ". . . drive after having five or more drinks?"
- **Ride**   ". . . ride with a driver who was high or drunk?"

The variable **Age** is the age of the student. The variable **Gender** is the gender of the student and is coded as

1—male

2—female

The variable **Fulltime** is Y if the student is full-time and N if the student is part-time. The variable **Greek** is Y if the student is a member of a fraternity or sorority and N if not. The variable **Live** records where the student lives and is coded as follows:

1—residence hall or dormitory

2—fraternity or sorority house

3—other university housing

4—off-campus house or apartment

The variable **Roommate** describes with whom the student lives and is coded as follows:

A—alone

R—roommate(s) or housemate(s)

S—spouse

P—parent(s) or other relative(s)

O—significant other

C—children

The variable **Race** describes the race or ethnic category of the student and is coded as

Blank—Hispanic (separate question)

1—White

2—Black/African American

3—Asian/Pacific Islander

4—Native American/Native Alaskan

5—other

The variable **Class** describes the student's year at the university:

1—freshman

2—sophomore

3—junior

4—senior

(a) Look at the variable *Drive*, which refers to whether a student drove after drinking. Create a contingency table for this variable and the variable *Binger*.

(b) At the 0.05 level of significance, is whether a student drove after drinking independent of the type of drinker he or she is?

(c) If you exclude nondrinkers and look at only those students who drink, does the answer to part (b) change?

(d) From the variables *Live* and *Roommate* derive a way to classify students as residents/commuters and to separate commuters living at home from commuters living with nonrelatives.

(e) Create contingency tables that allow you to look at the relationship between how often a student missed class (*Miss*) and resident/commuter students. Do you expect these two variables to be related? Why or why not?

(f) Do a chi-square test to determine whether the two variables in part (e) are related.

(g) Look at the variables *Miss* and *Gender* and do the same analysis.

(h) Look at the variables *Binger* and *Age*. Is there a relationship between them?

(i) From the variable *Age* create a new variable that classifies students as above and below the legal drinking age (assume that it is age 21 in the state of interest). How is this new variable related to the type of drinker that a student is? To *Drive*? To *Miss*?

(j) Is there a relationship between *Race* and *Binger*?

(k) Consider all of the other variables related to behavior after drinking. Determine whether these variables are related to the type of drinker that a student is.

(l) Consider any other variables and relationships that you think are important and investigate them. Prepare a report for the university administration that describes the drinking behaviors of the students at the university.

14.29 The Chamber of Commerce of a large city is studying the credit problems of small businesses. It asked the businesses three questions to classify their business and seven questions related to the issue of credit problems. A portion of the datafile and an explanation of each variable follow:

Datafile:
CREDIT.XXX

Size	Employees	Nature	Problem	Understd	Concern	Call	Loan	Collateral	Access
2	2	1	1	2	1	2	2	0	2
1	2	3	1	2	2	0	2	2	1
4	3	1	2	1	2	2	2	2	0
1	1	1	2	1	2	2	2	2	2
1	2	1	2	0	0	0	0	0	0
3	1	5	2	0	2	2	2	2	1
3	3	2	2	1	1	2	2	2	1
2	2	3	2	1	2	2	2	2	2
3	5	2	2	1	2	2	2	1	2

The variable **Size** refers to the annual sales of the company and is coded as follows:

1—under \$1 million
2—\$1–\$5 million
3—\$6–\$10 million
4—\$11–\$20 million
5—over \$20 million

The variable **Employees** refers to the number of employees that the company currently employs. This variable is coded as:

1—0–5 employees
2—6–10 employees
3—11–50 employees
4—51–150 employees
5—151–250 employees
6—over 250 employees

The variable **Nature** refers to the type of business and is coded as:

1—manufacturing
2—retail
3—service
4—real estate
5—other

The next seven variables contain the responses to the questions or statements indicated and are coded as follows:

1—yes
2—no

- **Problem**—"Are you experiencing credit-related problems?"
- **Understd**—"The bank understands my problems."
- **Concern**—"I am concerned that my note might be recalled."
- **Call**—"The bank is planning to recall my loan."
- **Loan**—"The bank has called my loan."
- **Collateral**—"The bank has demanded more collateral."
- **Access**—"Access to credit is affecting my business."

(a) Look at the variable *Nature*, which refers to the type of business. Is the sample of businesses that were surveyed by the Chamber of Commerce uniformly distributed over the different types of companies?

(b) Answer the question posed in part (a) for *Size* and *Employees*.

(c) Create a contingency table for the variables *Problem* and *Nature*. Are the proportions of companies that are having credit-related problems different for the different types of companies?

(d) Answer the question in part (c) for the variable *Nature* and each of the variables *Understd* and *Access*.

(e) Investigate the relationship between the variables *Size* and *Access* and between the variables *Size* and *Problem*.

(f) Investigate the relationship between the variables *Employees* and *Access* and the variables *Employees* and *Problem*.

(g) Is there a relationship between the size of the company and the type of business that the company does?

(h) Is there a relationship between the size of the company and the number of employees in the company?

(i) Look at the variables related to loan recall: *Concern*, *Call*, and *Loan*. Investigate whether these variables are related to *Size*, *Employees*, and *Nature*.

(j) Look at any other relationships that you think might be useful to the Chamber of Commerce.

(k) Prepare a report for the Chamber of Commerce about your findings.

15

Nonparametric Statistics

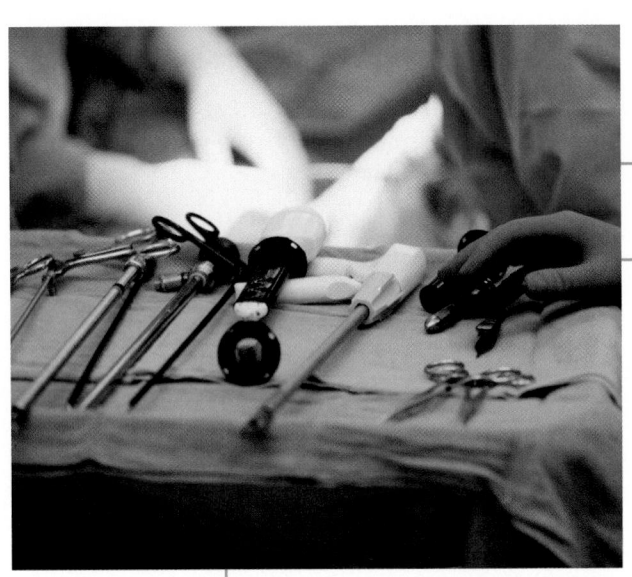

SURGICAL LEADERSHIP DEVELOPMENT PROGRAM

Administrators in the surgical department of a large hospital were dissatisfied with the way their current training program for surgical residents was working. In particular they were not happy with the leadership skills developed and demonstrated by the residents and the chief resident. In turn, the residents were not satisfied with what the program provided them in the way of support. The administrators hired a management consulting firm to develop a new training program. The new program was designed to reorganize the resident teams and the way the teams, the chief resident, and the surgical attending physicians interacted.

To assess the impact of the new program over time hospital administrators developed a set of questionnaires, which were given out to personnel in each department that interacted with the residents and to the residents themselves. The groups were the surgical residents (SR), surgical attending physicians (SA), surgical nursing staff (SNS), emergency department physicians (EDP), and emergency department nursing staff (EDN). The survey had a series of statements and used a five-point Likert scale (1 = Strongly Disagree to 5 = Strongly Agree) to assess the respondents' agreement with each statement. A portion of the data is given here:

Category	Time	Q1	Q2	Q3	Q4	Q5	Q6
SNS	June 1998	3	3	2	4	2	4
EDP	June 1998	4	2	4	3	4	3
SR	June 1998	5	4	5	1	2	5
SR	June 1998	4	4	3	4	4	4
SR	June 1998	4	4	5	5	2	4

SR	June 1998	4	4	4	2	4	2
SR	June 1998	4	4	4	4	3	4
SR	June 1998	4	4	5	5	3	4
SR	June 1998	3	3	4	2	1	2
SR	June 1998	4	4	4	4	4	4
SR	June 1998	5	4	5	4	4	4
SR	June 1998	4	5	5	5	4	4

The survey was given to all groups at the start of the new program, to selected groups after 6 months, and to the entire group again at the end of the year. The administrators were interested in finding out how perceptions compared for the different groups and whether and how perceptions changed over time.

15.1 Chapter Objectives

In hypothesis testing you learned techniques for testing inferences about *quantitative* data. In particular you learned that when you have large sample sizes, you can assume that the sampling distribution of the test statistic is normally distributed. When the sample size is small or when you are testing population parameters such as variances, the test procedures rely on the fact that the data collected come from populations that are normally distributed. What about data for which this assumption is not true? What can we do to make inferences about these populations?

The problem arises in two cases. First, the data collected are simply not normally distributed; that is, they come from populations that have skewed distributions. Second, the data are not truly quantitative; for example, the data represent ranks or ratings. Remember that this type of data is called ordinal data. Even though these data can be numerical, they are not really quantitative.

Procedures for working with these types of data are called *nonparametric*, or *distribution-free*, statistical methods.

> **Nonparametric** or **distribution-free statistical methods** are used when the data collected are not normally distributed.

Many of the techniques are analogous to ones that you learned in earlier chapters. The basic five-step hypothesis-testing procedure still applies. The five steps are listed below as a reminder.

Step 1: Set up the **null and alternative hypotheses**.

Step 2: Define the test procedure. This includes selecting the right **test**, picking the value of α, and finding the **rejection region.**

Step 3: Collect the data and calculate the **test statistic.**

Step 4: Make a statistical decision. Decide **whether to reject the null hypothesis.**

Step 5: **Interpret** and communicate the statistical decision in terms of the stated problem.

Steps for any hypothesis test

This chapter describes two nonparametric tests:

- Wilcoxon rank sum test for comparing the centers of two populations
- Kruskal–Wallis test, a nonparametric alternative to ANOVA

15.2 Method for Comparing Two Populations

15.2.1 THE WILCOXON RANK SUM TEST

The Wilcoxon rank sum test is sometimes called the Mann-Whitney test because the tests appeared simultaneously in the literature.

The **Wilcoxon rank sum test** is used to determine whether or not two independent populations have the same distribution. It can also be used to determine whether one population has values that are significantly higher or lower than the values of another population. If the populations that are sampled have the same shape (not necessarily normal) and variability, then the Wilcoxon rank sum test is equivalent to comparing the centers (means) of the two populations.

> The *Wilcoxon rank sum test* is used to compare the distributions of two independent populations when the samples are small and the data are not normally distributed or when the data are not quantitative.

The Wilcoxon rank sum test combines the two populations, ranks the observations, and then looks at the sum of the ranks for each population. If the two populations are the same, then the sums should be approximately equal. If one of the populations is lower or higher than the other, then the sums of the ranks should differ accordingly.

By convention, the Wilcoxon test calls the population with the smaller sample size population 1 and the one with the larger sample size population 2. Let's look at an example.

EXAMPLE 15.1

SURGICAL RESIDENTS

Identifying the Populations

Understand the problem.

The administrators of the surgery department who are about to begin the new training programs want to know how the surgical residents are viewed by different departments. In particular, they want to know if there is a difference in the way surgical residents are perceived by nurses in two different departments. They think that nurses outside the surgical department are more satisfied with the surgical residents. They decide to compare the responses from the emergency department (EDN) nurses and the surgical department (SNS) nurses to the statement: "Members of the resident teams appear to work well together." The survey administered to the two departments had seven responses from the EDN nurses and ten responses from the SNS nurses. Thus the emergency department is designated population 1 and the surgical department is population 2. The data (ratings on the five-point scale) follow:

Collect the data.

Emergency department	Surgical department
4	2
4	2
4	4
4	3
5	4
4	4
3	2
	2
	4
	4

The Wilcoxon rank sum test is also used when the data are quantitative but the samples are small and the normality assumption does not hold.

EXAMPLE 15.2

EMPLOYEE BREAKS

Identifying the Populations

The employees of a large retail company have asked to have their current break policies changed. They want to take breaks as they are needed rather than at scheduled times. The workers think the increased freedom will increase productivity, and so they convince management to do a pilot study to collect some data. A small group of employees is randomly assigned to two groups. One group remains on the existing break schedule and the other is allowed to take breaks whenever they want. During this time, productivity (in numbers of items produced) is monitored for each employee. Six employees are assigned to the fixed break group and five are assigned to the free break group. Here are the data:

Free	Fixed
351	357
316	347
480	380
446	259
470	342
	282

Because the free break group is smaller, it is designated as population 1.

Understand the problem.

Collect the data.

The first step in any hypothesis test is to set up the null and alternative hypotheses. The null and alternative hypotheses for the Wilcoxon test differ slightly depending on the assumptions we make. If we assume that the two populations have similar shapes and variability, then the hypotheses are about differences in the means. If we cannot assume this, then the hypotheses are about the population distributions.

Step 1 of hypothesis-test procedure

EXAMPLE 15.3

SURGICAL RESIDENTS

Setting Up the Hypotheses

To see whether the populations are similar in shape and variability the hospital administrators looked at these dotplots of the data:

Analyze the data.

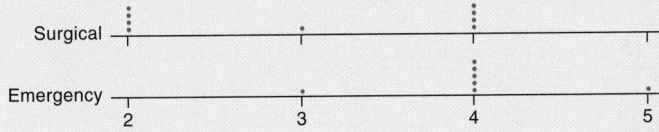

From the dotplots, it appears that the variability is similar but the shapes are very different. Since the administrators think that emergency department nurses are more satisfied (higher ranks) than surgical department nurses, these are the hypotheses for this statement:

H_0: Responses from the EDN nurses are not higher than those from the SNS nurses.

H_A: Responses from the EDN nurses are higher than those from the SNS nurses.

EXAMPLE 15.4

Analyze the data.

EMPLOYEE BREAKS

Setting Up the Hypotheses

To set up the hypotheses for the study on employee breaks, management creates dotplots of the data:

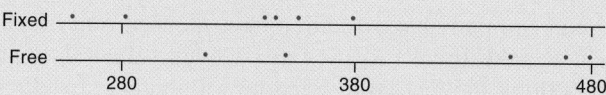

Since the variability and shapes look similar, the hypotheses are about the mean of each population:

H_0:   The mean productivity for the free break group is not higher than for the fixed break group.

H_A: The mean productivity for the free break group is higher than for the fixed break group.

Step 2 of hypothesis-test procedure

   The second step of the hypothesis-test procedure is to define the test procedure. The Wilcoxon rank sum test combines the observations from both populations and puts them in order from lowest to highest. Each observation is assigned a rank, which is the position of that observation in the ordered list. When there is a tie, each observation is assigned the average of the rank that would have been assigned if there had not been a tie. Then the populations are separated, and the ranks for each population are added to obtain the sum.

   If there are n_1 observations in one population and n_2 observations in the second population, then the ranks go from 1 to n, where $n = n_1 + n_2$. If the two populations are not different, then the ranks should be evenly distributed between the two samples. That is, each sample should account for about half of the rank total. If the sum of the ranks for one population is disproportionately large or small, then we conclude that the populations are not the same.

   The Wilcoxon test looks at the sum of the ranks. A mathematical formula may be used to calculate the sum of the integers from 1 to n as

Sum of integers from 1 to n

$$\frac{n(n + 1)}{2}$$

For example, the sum of the integers from 1 to 5 is

$$\frac{(5)(5 + 1)}{2} = \frac{(5)(6)}{2} = \frac{30}{2} = 15$$

This formula saves a lot of time in summing the ranks. Imagine adding the numbers from 1 to 100 by hand!

   The critical values for the hypothesis test are found in Table A13 for sample sizes up to ten. The values given are the lower and upper critical values T_L and T_U, for tail areas of 0.025 and 0.05. The 0.025 values are given in Table 15.1. You can see that the critical value for a lower-tail test with $\alpha = 0.025$ and $n_1 = 4$ and $n_2 = 7$ is 13. Thus you will reject the null hypothesis if the test statistic is less than 13.

   For larger sample sizes the sampling distribution that applies is the normal probability distribution and the standard normal table is used.

TABLE 15.1

LOWER AND UPPER CRITICAL VALUES *T* OF WILCOXON RANK SUM TEST

One-tail $\alpha =$ 0.025	Two-tail $\alpha =$ 0.05														
	T_L	T_U	T_L	T_U	T_L	T_U	T_L	T_U	T_L	T_U	T_L	T_U	T_L	T_U	T_L	T_U
n_2 n_1	3		4		5		6		7		8		9		10	
4	6	18	11	25												
5	6	21	12	28	18	37										
6	7	23	12	32	19	41	26	52								
7	7	26	13	35	20	45	28	56	37	68						
8	8	28	14	38	21	49	29	61	39	73	49	87				
9	8	31	15	41	22	53	31	65	41	78	51	93	63	108		
10	9	33	16	44	24	56	32	70	43	83	54	98	66	114	79	131

EXAMPLE 15.5

SURGICAL RESIDENTS

Finding the Critical Value

The hospital administrators decide to perform the test at the 0.05 level of significance. With $n_1 = 7$ and $n_2 = 10$, the critical value for the upper-tail test is 80.

Analyze the data.

EXAMPLE 15.6

EMPLOYEE BREAKS

Finding the Critical Value

The management of the retail company looks at the results of the analysis. For $n_1 = 5$ and $n_2 = 6$ at the 0.05 level of significance, the critical value of the test is 41.

Analyze the data.

The third step in the hypothesis-testing procedure is to determine the value of the test statistic. The data from the two samples are combined and put in order from smallest to largest. Each value is assigned a rank from 1 to *n*. If values are repeated in the data—that is, there are ties in the ranks—then the rank assigned is the average of the ranks for the tied data.

Step 3 of hypothesis-test procedure

EXAMPLE 15.7

SURGICAL RESIDENTS

Determining the Ranks

To find the ranks both samples need to be combined and ordered. Since the emergency department had fewer observations, it was designated population 1 and $n_1 = 7$ and $n_2 = 10$. For the data on surgical residents the ordered data and ranks are shown in the table:

Analyze the data.

Emergency department	Surgical department	Rank	Position
	2	2.5	1
	2	2.5	2
	2	2.5	3
	2	2.5	4
3		5.5	5
	3	5.5	6
4		11.5	7
4		11.5	8
4		11.5	9
4		11.5	10
4		11.5	11
	4	11.5	12
	4	11.5	13
	4	11.5	14
	4	11.5	15
	4	11.5	16
5		17	17

At first glance assigning the ranks may seem complicated, but it really is not. The first rating, 2, occurs four times, so the average of the positions involved is assigned. In this case,

$$\text{Rank} = \frac{1 + 2 + 3 + 4}{4} = \frac{10}{4} = 2.5$$

The next data value, 3, occurs two times, and so the average of those positions (5–7) is assigned:

$$\text{Rank} = \frac{5 + 6}{2} = \frac{11}{2} = 5.5$$

The next data value, 4, occurs ten times, in positions 8–17, and so the rank is:

$$\text{Rank} = \frac{7 + 8 + \cdots + 15 + 16}{10} = \frac{115}{10} = 11.5$$

Finally, the last rating, 5, has rank 17.

Finding the ranks is really not difficult if you are organized about the way you do it. It is important to keep track of which population each value is in so that you can separate them again later.

EXAMPLE 15.8

EMPLOYEE BREAKS

Finding the Ranks

Analyze the data.

The management of the retail company looks at the data they collected for the week of the pilot study. They calculate the rank for each data value. Since none of the data values repeats, each rank is the same as the position of the data value in the list.

Free break	Fixed break	Rank	Position
	259	1	1
	282	2	2
316		3	3
	342	4	4
	347	5	5
351		6	6
	357	7	7
	380	8	8
446		9	9
470		10	10
480		11	11

✔ TRY IT NOW!

Restaurant Specials
Setting Up the Hypothesis Test

A large restaurant is interested in finding out whether offering "specials" during the week will increase business. The owner decides to run a trial for 2 months, offering the specials once a week. She will compare weekly sales in these 2 months with the previous 2 months' records and see if they differ. Because one of the weeks in the special menu period included a holiday, she has only seven observations for that period. Here are data (in numbers of customers):

Regular menu	Specials menu
44	47
45	49
46	49
48	52
42	53
47	46
48	53
44	

Which group is population 1 and which is population 2?

Create dotplots of the data. Can you assume that the shape and variability for both populations are the same?

Set up the hypotheses for the test.

At the 0.05 level of significance, what is the critical value for the test?

$N_1 = 7$ AND $N_2 = 8$ THE CRITICAL VALUE IS 71.
HIGHER THAN FOR THE REGULAR MENU; H_A: THE MEAN FOR THE SPECIALS MENU IS HIGHER THAN FOR THE REGULAR MENU. WITH
IT SEEMS REASONABLE TO ASSUME THAT THE SHAPE AND VARIABILITY ARE SIMILAR. H_0: THE MEAN FOR THE SPECIALS MENU IS NOT

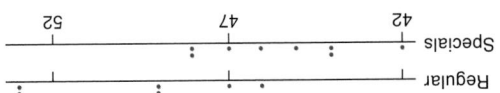

ANS. SPECIALS IS POPULATION 1.

By convention, the value of the test statistic is the sum of the ranks in the smaller population. When the sample sizes are large, the test statistic is approximately normally distributed with mean

Mean of the Wilcoxon rank sum test statistic

$$\mu = \frac{n_1(n_1 + n_2 + 1)}{2}$$

and standard deviation

Standard deviation of the Wilcoxon rank sum test statistic

$$\sigma = \sqrt{\frac{n_1 n_2(n_1 + n_2 + 1)}{12}}$$

For example, if $n_1 = 20$ and $n_2 = 25$, then the test statistic T, the sum of the ranks, is normally distributed with mean

$$\mu = \frac{20(20 + 25 + 1)}{2} = \frac{(20)(46)}{2} = 460$$

and standard deviation

$$\sigma = \sqrt{\frac{(20)(25)(46)}{12}} = \sqrt{23{,}000} = 151.66$$

To understand what these formulas mean, remember that for samples of 20 and 25 we have a total of 45 ranks. The sum of the numbers from 1 to 45 is $(45)(46)/2 = 1035$. Since population 1 is smaller than population 2, we expect the sum of the ranks, if the populations are the same size, to be a little less than half of the total, which is 517.5. That is what the mean, 460, represents.

EXAMPLE 15.9

Analyze the data.

SURGICAL RESIDENTS

Summing the Ranks

The ranks for the emergency department nurses, population 1, are listed here:

Rating	Rank
3	5.5
4	11.5
4	11.5
4	11.5
4	11.5
4	11.5
5	17
Sum	80

The test statistic, the rank sum, is 80.

EXAMPLE 15.10

Analyze the data.

EMPLOYEE BREAKS

Summing the Ranks

From the data table in Example 15.8 the retail company management finds that the sum of the ranks for the free break group is $3 + 6 + 9 + 10 + 11 = 39$.

TRY IT NOW!

Restaurant Specials
Finding the Rank Sum

Consider the restaurant data given earlier. Complete the table to find the rank for each data value.

Specials menu	Regular menu	Rank	Position
			1
			2
			3
			4
			5
			6
			7
			8
			9
			10
			11
			12
			13
			14
			15

Find the sum of the ranks for the specials menu sample.

The last steps of the hypothesis-testing process are to compare the test statistic with the critical value and make a decision. The results of the Wilcoxon rank sum hypothesis test are the same as those for any hypothesis-testing procedure. We decide to either reject H_0 or fail to reject H_0.

Steps 4 and 5 of hypothesis-test procedure

Ans.
Specials	Regular	Rank	Position
42		1	1
44		2.5	2
44		2.5	3
45		4	4
	46	5.5	5
46		5.5	6
	47	7.5	7
47		7.5	8
48		9.5	9
48		9.5	10
	49	11.5	11
	49	11.5	12
	52	13	13
	53	14.5	14
	53	14.5	15

THE SUM OF THE RANKS IS 78.

EXAMPLE 15.11

Draw conclusions.

SURGICAL RESIDENTS

Concluding the Test

Since the rank sum is 80 and the critical value for the test is 80, the hospital administrators decide to reject H_0. The data indicate that the emergency department nurses have more favorable attitudes toward the surgical residents than the surgical department nurses do. From this the administrators decide to look at other, more specific statements to try to determine in what areas the differences exist.

Remember that simply rejecting or not rejecting the null hypothesis is not enough. We must use the information from the test to make an informed decision about the problem.

EXAMPLE 15.12

Draw conclusions.

EMPLOYEE BREAKS

Concluding the Test

Since the rank sum for the data is 39 and the critical value of the test is 41, the management of the retail company concludes that they cannot reject H_0. The data do not provide evidence that the change in break policy will increase productivity. Management tells the employees that there is no justification for changing the policy but that they will continue to study the issue and see what else can be done.

✔ TRY IT NOW!

Restaurant Specials
Drawing Conclusions

Compare the rank sum for the restaurant data with the critical value for the test. What is your conclusion?

What action should the owner of this restaurant take? Should she consider any other factors before making this decision? If so what should she do and why?

15.2.2 EXERCISES—LEARNING IT!

15.1 A school department is using two different vendors to supply the packing cases used to store supplies in the warehouse. Recently employees have been noticing that more cases are being rejected for printing defects. They believe that the majority of the rejected cases come from Vendor X. They decide to run some tests and collect data by looking at ten cases from each of five lots for both vendors. They count the number of defective cases in each sample and record the data:

ANS. RANK SUM = 78, CRITICAL VALUE = 71, REJECT H_0. THE MEAN FOR THE SPECIALS IS GREATER THAN FOR THE REGULAR MENU. OWNER SHOULD RUN WEEKLY SPECIALS BUT DO A COST ANALYSIS TO SEE IF THE REVENUE LOST TO LOWER PRICES IS OFFSET BY MORE CUSTOMERS.

Vendor X	Vendor Y
1	2
4	0
2	1
3	2
2	3

(a) Create a dotplot for each vendor. What assumptions do you think are reasonable?

(b) Set up the hypotheses to see whether the number of defective cases from Vendor X is greater than that for Vendor Y.

(c) Find the rank for each data value and the rank sum for each sample.

(d) At the 0.05 level of significance, what can you conclude?

15.2 An Internet service provider is asking customers to rate their current service. The provider is interested in knowing whether people who have been customers for longer than 1 year are less satisfied than new customers (less than 6 months). They use a five-point scale where 1 = Extremely Satisfied and 5 = Extremely Dissatisfied. Here are the data (already sorted):

Old customers	New customers
1	1
2	1
3	1
3	1
3	1
3	2
3	2
4	3
5	4
5	4

(a) Create a dotplot for each sample. What assumptions do you think are reasonable?

(b) Set up the hypotheses to see whether the old customers are less satisfied than the new customers.

(c) Find the rank for each data value and the rank sum for each sample.

(d) At the 0.05 level of significance, what can you conclude?

15.3 A business with an Internet presence wonders about the effectiveness of banner ads. They decide to create two different websites for their company. One of the sites uses a banner ad and the other does not. You decide to count the number of daily hits for each site for a 1-week period. The data are listed here:

Banner	No banner
19	12
19	13
19	15
21	16
21	19
25	20
28	32

(a) Create a dotplot for each sample. What assumptions do you think are reasonable?

(b) Set up the hypotheses to see whether banner ads produce more hits for the website.

(c) Find the rank for each data value and the rank sum for each sample.

(d) At the 0.05 level of significance, what can you conclude?

15.4 A golf ball manufacturer wants to compare two different ball designs. He decides to take ten balls of each type and hit them in a controlled environment. He measures the distance that each balls travels to where it first hits the ground. Here are the data (in yards):

Ball design 1	Ball design 2
212	201
231	208
232	210
233	217
240	220
241	227
266	232
311	237
313	238
325	265

(a) Make dotplots of the two data sets and decide whether the shapes and variability are similar.

(b) Set up the hypotheses to test whether there is a difference in the amounts of travel for the two different designs.

(c) Carry out the Wilcoxon rank sum test at the 0.05 level of significance.

(d) What can you conclude about the distances for the two different designs?

15.3   Method for Comparing More Than Two Populations

15.3.1 THE KRUSKAL-WALLIS TEST

You know that ANOVA is the statistical technique used to compare the means of more than two populations. ANOVA relies on the assumption that the populations are normally distributed and have equal variances. If this is not true, then we must either transform the data or use an analogous nonparametric statistical test.

The nonparametric technique is the **Kruskal-Wallis test.** The idea behind the test is very similar to the idea for the Wilcoxon rank sum test. Just as in the Wilcoxon test, if the populations have similar shapes and variation, then the hypotheses are about the means of the populations; otherwise, the hypotheses are about the distributions of the populations.

> The *Kruskal-Wallis test* is a nonparametric method for comparing the data from more than two populations.

EXAMPLE 15.13

Understand the problem.

SURGICAL RESIDENTS

Setting Up the Hypothesis Test

The hospital administrators are also interested in finding out whether the perceptions of the surgical residents changed during the training program. This information reflects the effectiveness of the program. They administered the same questionnaire to certain groups at three different times: at the beginning of the training, 6 months into the program, and at the end of the program. They decide to look at the surveys for the surgical attending physicians, the people who work most closely with the surgical residents. The table lists the ratings on the five-point scale for the statement "Members of the resident teams appear to work well together":

June 1998	December 1998	June 1999
4	4	2
4	4	2
2	4	4
3	4	3
4	4	4
4	2	4
3	3	5
2	4	4
4	4	4
4	3	4
4	4	5
4	4	4
4	2	
4	4	
	2	
	3	
	4	
	4	
	4	
	4	

Collect the data.

The administrators made plots of the data to see what assumptions they can make. Except for one observation in June 1998, the assumptions of equal shape and variation seem reasonable.

Surgical Attendings Ratings

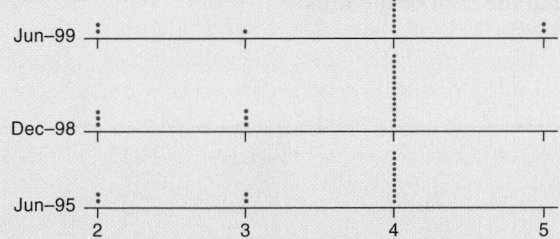

The null and alternative hypotheses are:

H_0: The mean rating for each time period is the same.

H_A: At least one mean is different.

EXAMPLE 15.14

TAX PREPARATION SOFTWARE

Setting Up the Hypothesis Test

A company that creates tax preparation software for the general public is considering changing the software they currently use to process returns. They wonder how the three alternatives that they are considering compare in ease of use. They decide to take the last 21 tax returns they have done and randomly assign one return to each of the three software packages. The tax return data are input into the respective software and the time until a usable, correct return is obtained is measured (in hours). The data and plots are presented here:

Understand the problem.

Collect the data.

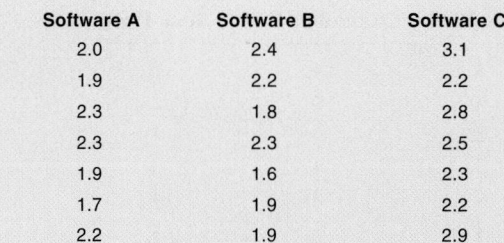

	Software A	Software B	Software C
	2.0	2.4	3.1
	1.9	2.2	2.2
	2.3	1.8	2.8
	2.3	2.3	2.5
	1.9	1.6	2.3
	1.7	1.9	2.2
	2.2	1.9	2.9

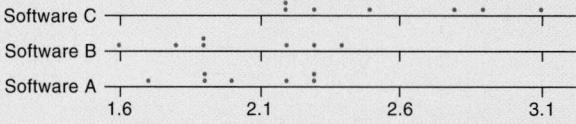

Plots of the data indicate that the assumptions of equal shape and variability are reasonable, and so they use these hypotheses:

H_0:   The mean time for each software package is the same.

H_A: At least one mean is different.

The Kruskal-Wallis test proceeds much like the Wilcoxon rank sum test. The data are combined, ranked, and then separated back into the original samples. The idea is that if the populations are similar, then the ranks should be about equally distributed among the populations. The Wilcoxon rank sum test looks at the mean rank for each group, however, rather than the sum of the ranks.

EXAMPLE 15.15

SURGICAL RESIDENTS

Calculating the Mean Ranks

Analyze the data.

For the surgical residents the data, sorted and ranked, are listed in the table:

June 1998	Rank	December 1998	Rank	June 1999	Rank
2	4	2	4	2	4
2	4	2	4	2	4
3	10.5	2	4	3	10.5
3	10.5	3	10.5	4	29
4	29	3	10.5	4	29
4	29	3	10.5	4	29
4	29	4	29	4	29
4	29	4	29	4	29
4	29	4	29	4	29
4	29	4	29	4	29
4	29	4	29	5	45.5
4	29	4	29	5	45.5
4	29	4	29		
4	29	4	29		
		4	29		

	4	29			
	4	29			
	4	29			
	4	29			
	4	29			

| **Average rank** | **22.8** | | **22.5** | | **26.0** |

The average of all the ranks is 23.5.

EXAMPLE 15.16

TAX PREPARATION SOFTWARE

Finding the Average Ranks

The tax preparers assembled their data, ranked them, and calculated the average ranks:

Analyze the data.

Software A	Rank	Software B	Rank	Software C	Rank
1.7	2	1.6	1	2.2	10.5
1.9	5.5	1.8	3	2.2	10.5
1.9	5.5	1.9	5.5	2.3	14.5
2.0	8	1.9	5.5	2.5	18
2.2	10.5	2.2	10.5	2.8	19
2.3	14.5	2.3	14.5	2.9	20
2.3	14.5	2.4	17	3.1	21
Average rank	**8.6**		**8.1**		**16.2**

In ANOVA, if the means of the populations are not different, then the variation among the means of the populations should be similar to the variation within each population. The Kruskal-Wallis test uses the same idea, but it tests the variation among the mean ranks rather than the means of the data values themselves. In addition, the Kruskal-Wallis test compares the variation among the populations with the total variation in ranks rather than the variation within each population.

The test statistic for the Kruskal-Wallis test is

$$H = \frac{\text{SSA}}{\text{SST}/(n-1)}$$

Kruskal-Wallis test statistic

SSA, the variation among the populations, is calculated in the same way as for ANOVA; that is, for the Kruskal-Wallis test,

$$\text{SSA} = \sum_{j=1}^{c} n_j (\bar{r}_j - \bar{\bar{r}})^2$$

SSA for the Kruskal-Wallis test

where

n_j = number of observations in population j

c = total number of populations

$\bar{r}_j$ = average rank of population j

$\bar{\bar{r}}$ = average of all the ranks

SST, the total variation in the ranks, is calculated as

*SST for the
Kruskal-Wallis test*

$$\sum_{j=1}^{c} \sum_{i=1}^{n_j} (r_{ij} - \bar{\bar{r}})^2$$

When there are no (or few) ties in the data, there is a simpler way to calculate the test statistic H:

*Approximate
Kruskal-Wallis test
statistic*

$$H = \frac{12}{n(n+1)} \sum_{j=1}^{c} \frac{R_j^2}{n_j} - 3(n+1)$$

where

R_j = sum of the ranks for population j

c = number of population

n = total number of observations ($n = n_1 + n_2 + \cdots + n_c$)

There are tables of exact critical values for the Kruskal-Wallis test. However, if all the sample sizes are greater than 5 and if the shapes of the underlying distributions are approximately the same, then the sampling distribution of H is the chi-square distribution with $c - 1$ degrees of freedom. Because these assumptions are not very restrictive, the chi-square distribution is almost always used.

EXAMPLE 15.17

SURGICAL RESIDENTS

Calculating the Test Statistic

*Analyze the
data.*

The data for the surgical resident ratings have many ties, so we use the exact formula to find the value of the test statistic:

$$SSA = \sum_{j=1}^{c} n_j(\bar{r}_j - \bar{\bar{r}})^2$$

$$= 14(22.8 - 23.5)^2 + 20(22.5 - 23.5)^2 + 12(26.0 - 23.5)^2 = 104.27$$

*When the data are this
complicated, it is
probably easiest to run
an ANOVA using the
ranks as the data to
find H.*

Similarly, SST is calculated to be 5581.5 as follows:

$$SST = (4 - 23.5)^2 + (10.5 - 23.5)^2 + 10(29 - 23.5)^2$$
$$+ 3(4 - 23.5)^2 + 3(10.5 - 23.5)^2 + 14(29 - 23.5)^2$$
$$+ 2(4 - 23.5)^2 + (10.5 - 23.5)^2 + 7(29 - 23.5)^2 + 2(45.5 - 23.5)^2$$
$$= 7(4 - 23.5)^2 + 6(10.5 - 23.5)^2 + 31(29 - 23.5)^2 + 2(45.5 - 23.5)^2$$
$$= 5581.5$$

Thus, the test statistic, H, is

$$H = \frac{SSA}{SST/(n-1)} = \frac{104.27}{5581.5/45} = 0.84$$

Using a level of significance of 0.05, we find the critical value of the test to be $\chi_{0.05,2}^2 = 5.991$. Because the test statistic does not fall outside the critical value of the test, there is not enough evidence to say that the means for the three time periods are different.

From this analysis, the hospital administrators conclude that the training program did not change the attitudes of the surgical attending physicians toward the surgical residents.

EXAMPLE 15.18

TAX PREPARATION SOFTWARE

Calculating the Test Statistic

The data for the tax preparation software did not have many ties, so we can use the approximation to find the value of the test statistic. The sums of the ranks for the three populations are 60.5, 57, and 113.5. The test statistic is calculated as follows:

$$H = \frac{12}{n(n+1)} \sum_{j=1}^{c} \frac{R_j^2}{n_j} - 3(n+1)$$

$$H = \frac{12}{(21)(22)} \left(\frac{60.5^2}{7} + \frac{57^2}{7} + \frac{113.5^2}{7} \right) - 3(22) = 7.44$$

The tax preparers decide to use a level of significance of 0.05 and so the critical value of the test is $\chi_{0.05,2}^2 = 5.991$. Because the test statistic is greater than the critical value of the test, the tax preparers conclude that the mean time for at least one of the software packages is different. Looking at the data, they think the first two packages, A and B, are similar but package C takes considerably longer. This information can help them decide which package to use.

Analyze the data.

✔ TRY IT NOW!

Travel Expenses
Conducting the Kruskal-Wallis Test

The mayor's office wants to look at the differences in the amounts of money charged for travel reimbursement in three different city departments. Each department uses a different method for booking travel arrangements, so the mayor thinks one of the methods may lead to significant savings. The mayor looks at the travel expenses submitted in the past 3 weeks in each department and finds these amounts:

Police	Fire	Education
$529	$604	$567
451	633	409
546	568	504
272	457	605
426	520	682
651	509	477
	551	480

Make dotplots for the data. Is it reasonable to assume similar shapes and variability?

Write down the hypotheses for the test.

Calculate the rank for each department, the sums of the ranks, and the average ranks.

Which formula for the test statistic is best to use for these data?

Calculate the value of the Kruskal-Wallis test statistic.

At the 0.05 level of significance, what can the mayor conclude about the departments' travel expenses?

Is the mayor's conclusion correct? What other factors should have been considered in collecting the data?

15.3.2 EXERCISES – LEARNING IT!

Requires Exercise 15.2

15.5 The Internet service provider who is interested in how customers rate their service decides to include another group in the study. The new group consists of people who have been customers between 6 months and a year. The company conducts a new survey using the same scale (1 = Extremely Satisfied and 5 = Extremely Dissatisfied) and gets these results:

More than a year	Six months to one year	Less than 6 months
1	1	1
2	1	1
3	2	1
3	3	1
3	3	1
3	3	2
3	4	2
4	5	3
5	5	4
5	5	4

(a) Create dotplots for all three samples. What assumptions do you think are reasonable?

(b) Set up the hypotheses to see whether there is a difference in satisfaction among the three groups.

(c) Find the rank for each data value and the rank sum and rank average for each sample.

(d) Perform the Kruskal-Wallis test at the 0.05 level of significance. What can you conclude?

15.6 A psychologist is interested in starting an online shopping service and wants to find out whether or not there are differences in how women shop online. He is interested in people who are already connected to the Internet, and so he runs a web-based survey. He asks respondents how many purchases they have made online in the last 3 months. In addition he asks

ANS.

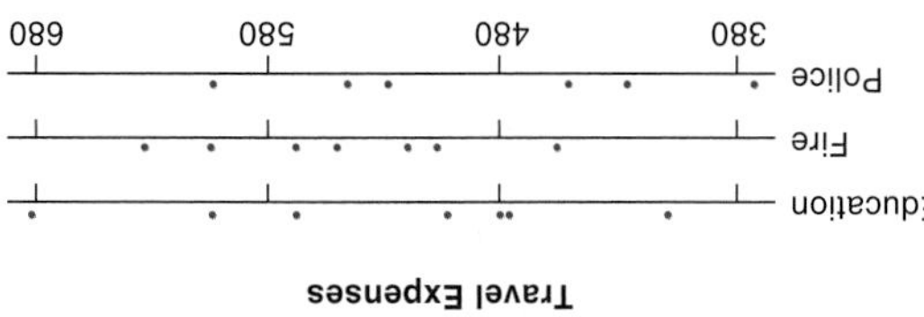

Police	Rank	Fire	Rank	Education	Rank
372	1	457	5	409	2
426	3	509	9	477	6
451	4	520	10	480	7
529	11	551	13	504	8
546	12	568	15	567	14
603	16	604	17	605	18
		633	19	682	20
Sum	47		88		75
Average	7.833333		12.6		10.71429

H_0: THE MEAN TRAVEL EXPENSES ARE THE SAME FOR ALL 3 DEPARTMENTS.

H_A: AT LEAST ONE OF THE DEPARTMENTS HAS A DIFFERENT MEAN.

IT IS REASONABLE TO ASSUME SIMILAR SHAPES AND VARIABILITY. SINCE THERE ARE NO TIES, USE THE APPROXIMATE TEST STATISTIC; $H = 2.09$; $\chi^2_{0.05,2} = 5.991$. H IS NOT OUTSIDE THE CRITICAL VALUE, SO DO NOT REJECT H_0. THERE IS NO DIFFERENCE IN THE MEANS FOR THE THREE DEPARTMENTS. NO; THEY SHOULD CONSIDER LENGTH OF TRIPS, NUMBER OF TRIPS, NUMBER OF DAYS INVOLVED, AND SO ON.

demographic questions about gender, age, and level of education. The numbers of purchases by women respondents in three age categories are shown in the table:

Ages 21–30	Ages 30–45	Ages 45–60
2	4	3
2	5	3
3	6	4
4	6	5
4	8	6
5	9	6
8	9	9

(a) Create dotplots for all three samples. What assumptions do you think are reasonable?

(b) Set up the hypotheses to see whether the number of times women shop online differs among the three groups.

(c) Find the rank for each data value and the rank sum and rank average for each sample.

(d) Perform the Kruskal-Wallis test at the 0.05 level of significance. What can you conclude?

15.7 The regional planning board for a metropolitan area is wondering how often people use public transportation. They want to use these data with other information they have collected on housing starts to plan for future expansion of services. They divide the area up into three different geographical regions and conduct a survey of households in each. The respondents are asked how many times the head of the household used public transportation on weekdays in a 2-week period in February. The results follow:

Region A	Region B	Region C
3	4	5
3	5	6
4	5	6
4	5	6
4	5	7
5	5	7
5	5	

(a) Create dotplots for all three samples. What assumptions do you think are reasonable?

(b) Set up the hypotheses to see whether there is a difference in the use of public transportation for the three regions.

(c) Find the rank for each data value and the rank sum and rank average for each sample.

(d) Perform the Kruskal-Wallis test at the 0.05 level of significance. What can you conclude?

15.8 The food services manager at a large university is looking at the types of hotdogs that are available. She wonders if there is really a difference in the numbers of calories in different kinds of hotdogs. She decides to look at types of hotdogs and randomly select brands from each type. The numbers of calories are listed in the table:

Beef	Meat	Poultry
111	147	152
132	153	152
139	179	129
152	182	129
181	190	113
181		102

(a) Create dotplots for all three samples. Is it reasonable to assume that the shapes and variability are similar?

(b) Set up the hypotheses to see whether the number of calories differs for the three types of hotdogs.

(c) Find the rank for each data value and the rank sum and rank average for each sample.

(d) Perform the Kruskal-Wallis test at the 0.05 level of significance. What can you conclude?

15.4 Using Nonparametric Tests

Now you might be asking these questions: Will I ever figure out which test to use? What happens if I use the wrong test? How good are these nonparametric tests anyway?

Determining which test to use is important. What is basic to this skill is knowing what assumptions each test makes and *carefully* checking those assumptions. This is why it is so important to really look at your data with descriptive methods *before* you start any analysis.

For the most part, the variable that is most affected by using a statistical test incorrectly is the probability of making a Type II error. For example, if you use a small-sample *t* test for the population mean when the underlying population is not normally distributed, then the chance that you will make a Type II error is higher than if the populations had been normally distributed. That is, your test is not as *powerful* at detecting false hypotheses.

If you have any doubts about which test to use, do them both. If they agree, your decision is probably correct. If they disagree, look at the data more carefully.

When data are normally distributed, the classic statistical tests you learned first are always more powerful than the nonparametric tests, although in many cases not overwhelmingly so. When the data are not normally distributed, the nonparametric tests are the more powerful test.

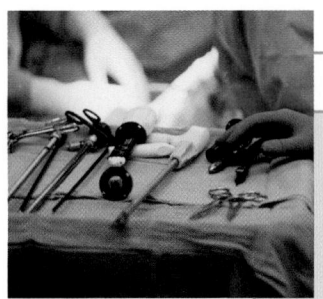

GET IT IN WRITING

The Hospital

TO: **Administrators in Surgical Department**
FROM: Erica Q. Analyst
RE: **New Training Program**

In order to assess the impact of the new training program we have developed a set of questionnaires that were given out to personnel in each department that interacted with the residents and to the residents themselves. The survey was given to all groups at the start of the new program, to selected groups after 6 months, and to the entire group again at the end of the year.

Since you indicated an interest in knowing about how the surgical residents were viewed by different departments, we have analyzed the data to investigate that. In particular, we looked to see if there is a difference in the way surgical residents were perceived by nurses in two different departments. Since your feeling was that the nurses outside the surgical department were more satisfied with the surgical residents, we decided to compare the responses from the emergency department (ED) nurses and the surgical department (SD) nurses to the statement: "Members of the resident teams appear to work well together." The survey administered in the two departments had seven responses from the ED nurses and ten responses from the SD nurses.

The data indicated that the ED nurses had more favorable attitudes toward the surgical residents than the SD nurses did. From this we decided to look at other, more specific statements to try to determine in what areas of leadership the differences exist.

The hospital administration was also interested in whether perceptions of the surgical residents changed during the new training program. They thought that this would be an indication of whether or not the program was effective. We decided to look at the surveys from the surgical attending physicians, the people who work most closely with the surgical residents. Plots of the data were made to see what assumptions could be made. Except for one observation in June 1998, the assumptions of equal shape and variation seem reasonable.

However, since the test statistic did not fall outside the critical value of the test, there was not enough evidence to say that the means for the three time periods were different. Thus, the training program did not change the attitudes of the surgical attending physicians toward the surgical residents.

Considerably more information can be learned by further analyzing the survey data. We plan to continue the analysis, looking for differences in perceptions by department and over time. We will report back to you in 2 weeks with additional results.

15.5 Using Technology for Nonparametric Tests

Minitab has routines that perform a variety of nonparametric tests. Excel does not have any built-in routines for nonparametric tests and neither does the TI-83 graphing calculator. Here we show some of the Minitab routines. All of the tests are selected by choosing **Stat** > **Nonparametrics** and then the test of interest.

15.5.1 THE WILCOXON RANK SUM (MANN-WHITNEY) TEST

Suppose we want to perform the Wilcoxon rank sum test for the data about breaks. The data are shown in a Minitab worksheet in Figure 15.1.

↓	C1	C2
	Free	Fixed
1	351	357
2	316	347
3	480	380
4	446	259
5	470	342
6		282

FIGURE 15.1 Break data in Minitab

To perform the test follow this procedure:

1. Select **Stat** > **Nonparametrics** > **Mann-Whitney** from the menu. The dialog box shown in Figure 15.2 on page 770 opens.

2. Position the cursor in the box labeled **First Sample:** and select the variable Free.

3. Position the cursor in the box labeled **Second Sample:** and select the variable Fixed.

4. Since we want to know whether the production for the group with free breaks is higher than for the fixed breaks, select **greater than** from the drop-down menu **Alternative:**.

5. Click **OK**, and the output appears in the Session window as shown in Figure 15.3.

Minitab refers to the test as the Mann-Whitney test.

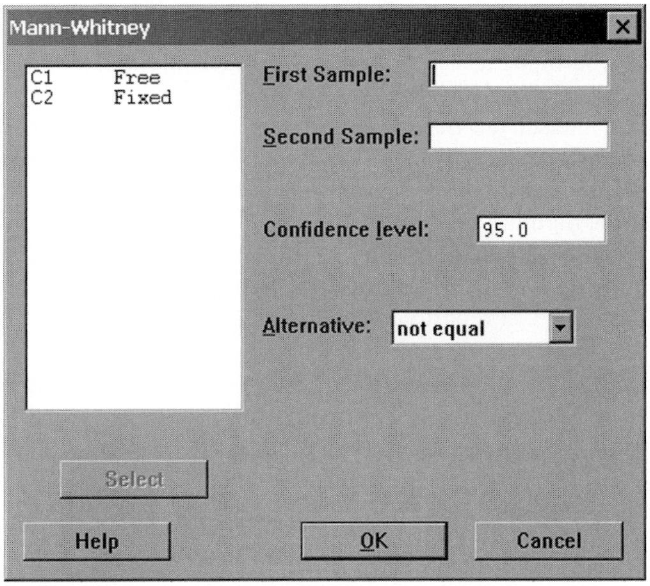

FIGURE 15.2 **Mann-Whitney dialog box**

Mann-Whitney Test and CI: Free, Fixed

Free   N = 5   Median = 446.0
Fixed   N = 6   Median = 344.5
Point estimate for ETAl-ETA2 is 95.5
96.4 Percent CI for ETA1-ETA2 is (−29.0, 188.0)
W = 39.0
Test of ETA1 = ETA2 vs ETA1 > ETA2 is significant at 0.0603

Cannot reject at alpha = 0.05

FIGURE 15.3 **Output from the Mann-Whitney test**

The value of the rank sum is labeled **W.** The output also includes the p value of the test and the conclusion (cannot reject).

15.5.2 THE KRUSKAL-WALLIS TEST

For the Kruskal-Wallis test procedure, Minitab expects to find the data values in a single column with a second column designating the population from which the data value was obtained. A sample of the data for the tax preparation software example is shown in Figure 15.4.

C7	C8-T
Time	Software
1.7	Software A
1.9	Software A
1.9	Software A
2.0	Software A
2.2	Software A
2.3	Software A

FIGURE 15.4
Software data

To perform the test follow this procedure:

1. Select **Stat > Nonparametrics > Kruskal-Wallis** from the menu. The dialog box shown in Figure 15.5 opens.

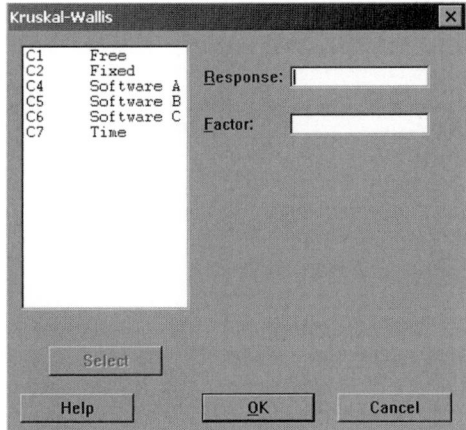

FIGURE 15.5   Kruskal-Wallis dialog box

2. Position the cursor in the box labeled **Response:** and select the variable that contains the data values from the box on the left.
3. Position the cursor in the box labeled **Factor:** and select the variable that contains the population designations from the box on the left.
4. Click **OK** and the output appears in the Session window as shown in Figure 15.6.

Kruskal-Wallis Test: Time versus Software

Kruskal-Wallis Test on Time

Software	N	Median	Ave Rank	Z
Software	7	2.000	8.6	−1.23
Software	7	1.900	8.1	−1.49
Software	7	2.500	16.2	2.72
Overall	21		11.0	

H = 7.44   DF = 2   P = 0.024
H = 7.59   DF = 2   P = 0.023 (adjusted for ties)

FIGURE 15.6   Output from the Kruskal-Wallis test

From the output we see that the value of the test statistic, H, is 7.44 and the p value of the test is 0.024.

CHAPTER 15 SUMMARY

In this chapter you learned techniques for testing inferences about *quantitative* data when the sample size is small and the assumption of normality is violated. This problem arises in two cases. The first occurs when the data collected are simply not normally distributed; that is, they come from populations with skewed distributions. The second occurs when the data are not truly quantitative—for example, when the data represent ranks or ratings. Remember that this type of data is called ordinal data. Even though these data can be numerical, they are not really quantitative.

Procedures for working with these types of data are called nonparametric, or distribution-free, statistical methods.

KEY TERMS

Term	Definition	Page reference
Nonparametric methods	**Nonparametric methods** are used when the data collected are not normally distributed.	749
Kruskal-Wallis test	The **Kruskal-Wallis test** is a nonparametric method for comparing the data from more than two populations.	760
Wilcoxon rank sum test	The **Wilcoxon rank sum test** is used to compare the distributions of two independent populations when the samples are small and the data are not normally distributed or when the data are not quantitative.	750

KEY FORMULAS

Term	Formula	Page reference
Sum of the integers from 1 to n	$\dfrac{(n)(n + 1)}{2}$	752
Mean of the Wilcoxon rank sum test statistic	$\mu = \dfrac{n_1(n_1 + n_2 + 1)}{2}$	756
Standard deviation of the Wilcoxon rank sum test statistic	$\sigma = \sqrt{\dfrac{n_1 n_2(n_1 + n_2 + 1)}{12}}$	756
Test statistic for the Kruskal-Wallis test	$H = \dfrac{SSA}{SST/(n - 1)}$	763
SSA for the Kruskal-Wallis test	$SSA = \displaystyle\sum_{j=1}^{c} n_j (\bar{r}_j - \bar{\bar{r}})^2$ where n_j = number of observations in population j c = total number of populations $\bar{r}_j$ = average rank of population j $\bar{\bar{r}}$ = average of all the ranks	763
SST for the Kruskal-Wallis test	$\displaystyle\sum_{j=1}^{c} \sum_{i=1}^{n_j}(r_{ij} - \bar{\bar{r}})^2$	764
Approximate Kruskal-Wallis test statistic	$H = \dfrac{12}{n(n + 1)} \displaystyle\sum_{j=1}^{c} \dfrac{R_j^{\,2}}{n_j} - 3(n + 1)$ where R_j = sum of the ranks for population j c = number of populations n = total number of observations $(n = n_1 + n_2 + \cdots + n_c)$	764

CHAPTER 15 EXERCISES

Learning It!

15.9 A large food company is considering changing the preparation directions on packets of instant oatmeal since customer complaints are increasing. They run a palatability study on a test group of consumers. For one group they use a high level of liquid in the oatmeal and for the other a low level of liquid. The data are ratings based on a questionnaire:

Low	High
16	21
35	24
39	39
77	60
84	64
97	65
104	86
129	94

(a) Create a dotplot for each sample. What assumptions do you think are reasonable?

(b) Set up the hypotheses to see whether there is a difference in palatability for the two levels of liquid.

(c) Find the rank for each data value and the rank sum for each sample.

(d) Perform the Wilcoxon rank sum test at the 0.05 level of significance. What can you conclude about the effect of the liquid level on palatability?

15.10 A major retailer of stereos is interested in finding out whether type of advertising affects sales. They pick two geographically similar areas and use print advertising (newspapers, magazines) in one area and Internet advertising (banner ads, website ads) in the other. The data are the numbers of sales per week in an 8-week period:

Internet	Print
10	18
13	20
15	21
15	22
16	25
19	25
20	26
21	26

(a) Create a dotplot for each sample. What assumptions do you think are reasonable?

(b) Set up the hypotheses to see whether print advertising leads to more sales than Internet advertising.

(c) Find the rank for each data value and the rank sum for each sample.

(d) Perform the Wilcoxon rank sum test at the 0.05 level of significance. What can you conclude about sales and type of advertising?

15.11 A group of consumers in an urban area are given the opportunity to test three different types of high-speed Internet service: DSL (direct subscriber line), cable modem, and satellite. A service type is randomly assigned to each consumer, and each is asked to use the service for 1 month. At the end of the month the consumers are asked to rate the service on a five-point scale, where 1 = Extremely Satisfied and 5 = Extremely Dissatisfied. Here are the ratings:

DSL	Cable	Satellite
1	1	1
1	1	1
1	2	1
1	2	2

(continued)

1	2	3
1	3	3
1	3	4
2	4	4
2	4	5
3	4	5

(a) Create dotplots for all three samples. Can you assume that the shapes and variability are the same for the three groups?

(b) Set up the hypotheses to see whether satisfaction differs for the three high-speed Internet services.

(c) Find the rank for each data value and the rank sum and rank average for each sample.

(d) Perform the Kruskal-Wallis test at the 0.05 level of significance. What can you conclude?

15.12 The psychologist who is trying to understand online shopping is interested in how often people use the Internet. The survey asks respondents how many times per week they access the Internet for purposes other than business or e-mail. The table lists the numbers of times for women in three different household income categories:

Under $25,000	$25,000 to $50,000	Over $50,000
1	2	8
3	5	12
3	6	13
3	6	13
5	6	17
10	10	18

(a) Create dotplots for all three samples. Can you assume the shapes and variability are the same for the three groups?

(b) Set up the hypotheses to see whether there is a difference in the numbers of times women use the Internet for the three different income levels.

(c) Find the rank for each data value and the rank sum and rank average for each sample.

(d) Perform the Kruskal-Wallis test at the 0.05 level of significance. What can you conclude?

Thinking About It!

15.13 It is a widely held belief that switching to participative management increases employees' "buy-in" to the company. One of the benefits should be a reduction in the number of sick days that employees use. A hospital that has made the switch in some departments wonders about the results. They decide to sample 25 employees from each of two hospital departments. The first has been using a participative management style for almost 2 years and the second is still using a traditional management style. Here are the numbers of sick days used by employees in the past 12 months:

Participative					Traditional				
1	3	5	5	6	0	5	6	7	9
1	4	5	6	7	3	5	7	7	9
2	4	4	6	8	4	6	7	7	10
2	4	5	6	8	4	6	7	8	11
3	4	5	6	8	5	6	7	8	11

(a) Create dotplots of the data. Do you think they are normally distributed?

(b) Use the appropriate nonparametric test to decide if the data provide enough evidence to say that employees who use participative management take fewer sick days than those who use a traditional management style?

(c) Redo the test using a small-sample (Student's t) test. What are the results of this test?

(d) Which test do you think should be used? Why?

15.14 A software company that was looking at the time to failure of diskettes decides to look at an alternative supplier of the product. The times to failure (in hours) for their current supplier and for the new supplier are listed here:

Current supplier				Alternative supplier			
486	494	502	508	489	492	495	498
490	496	504	510	489	492	496	499
491	498	505	514	491	493	497	502
491	498	506	515	492	493	497	503
494	498	507	527	492	494	497	505

(a) Create dotplots of the data. Do you think they are normally distributed?

(b) Use the appropriate nonparametric test to decide whether the data provide enough evidence to indicate that the alternative supplier has better diskettes than the current one.

(c) Redo the test using a small-sample (Student's t) test. What are the results of this test?

(d) Do you think the nonparametric test is the better test here? Why or why not?

15.15 A diaper company is considering three different filler materials for its disposable diapers. Eight diapers were tested with each of the three filler materials and 24 toddlers were randomly given a diaper to wear. As the child played, fluid was injected into the diaper every 10 minutes until the product failed (leaked). The amount of fluid (in grams) at the time of failure was recorded for each diaper:

Material 1	Material 2	Material 3
791	809	828
789	818	814
796	803	855
802	781	844
810	813	847
790	808	848
800	805	836
790	811	873

(a) Create dotplots of the data. Do you think the three populations are normally distributed?

(b) Use the appropriate nonparametric test to decide whether the data provide enough evidence to indicate that the absorptions of the three materials are different.

(c) Redo the test as a one-way ANOVA. What are the results of this test?

(d) Do you think the nonparametric test is the better alternative here? Why or why not?

15.16 Look again at the data from the golf ball company.   *Requires Exercise 15.4*

(a) Is it reasonable to assume that the data are normally distributed? If so, how would this have changed your approach to solving the problem?

(b) If the normality assumption is not justified, what is the impact of using a small-sample hypothesis test procedure?

15.17 The consumers who conducted the survey on high-speed Internet service want to tell   *Requires Exercise 15.11*
other consumers which type of service received the highest ratings.

(a) Based on your conclusions, what would you recommend they do?

(b) As a consultant, you are concerned with the way the study was done. Write a memo to the consumers explaining how their study was faulty and suggest changes for a new study. Be sure to include what other factors you think should be considered and what other types of questions should be asked.

15.18 The stereo retailer who is looking at the effects of advertising type on sales decides to   *Requires Exercise 15.10*
look at one more type of advertising. They use broadcast advertising (radio, television) in another similar area for a different 8-week period. The new data, along with the old, are listed here:

Internet	Print	Broadcast
10	18	15
13	20	16
15	21	17
15	22	19
16	25	23
19	25	23
20	26	26
21	26	27

(a) Perform the Kruskal-Wallis test at the 0.05 level of significance to determine whether there is a difference in weekly sales due to advertising type.

(b) Do you think that what the retailer did was reasonable? Why or why not?

(c) Write a memo explaining why the study was incorrect and suggest a better method to use.

Doing It!

Datafile:
CHURCH1.XXX

15.19 After September 11, 2001, many people have returned to church in an attempt to understand and process this tragedy. One New England church wishes to do a needs assessment in an attempt to serve its members better. Below are some of the questions included in this needs assessment:

Please indicate your satisfaction with the performance of the following:

	Highest			**Lowest**	
Q1: Pastors	1	2	3	4	5
Q2: Office staff	1	2	3	4	5
Q3: Music staff	1	2	3	4	5
Q4: Council	1	2	3	4	5

Please indicate your satisfaction with the following components of our worship experience:

	Highest			**Lowest**	
Q5: Congregational singing	1	2	3	4	5
Q6: Choir	1	2	3	4	5
Q7: Special music	1	2	3	4	5
Q8: Sermon	1	2	3	4	5
Q9: Contemporary music	1	2	3	4	5
Q10: Traditional music	1	2	3	4	5
Q11: Announcements	1	2	3	4	5
Q12: Sound system	1	2	3	4	5
Q13: Worship bulletin	1	2	3	4	5

Q14: What is your age group?

Under 21

Over 21 to 34

Over 34 to 49

Over 49 to 65

Over 65

Q15: What is your individual personal income?

Under $15,000

$15,000 to under $30,000

$30,000 to under $45,000

$45,000 to under \$60,000

$60,000 to under \$75,000

$75,000 or over

Q16: Gender: Male or Female

Q17: Marital status: Married or Single

(a) Analyze each question to see if the mean rating differs by age group.

(b) Analyze each question to see if the mean rating differs by income group.

(c) Analyze each question to see if the mean rating differs by gender.

(d) Analyze each question to see if the mean rating differs by marital status.

(e) Write a memo to the church leadership summarizing your findings and making some recommendations of areas that need attention.

16

Making Your Case

THE GREAT DIVIDE

The earnings gap between executives at the very top of corporate America and middle managers and workers has stretched into a vast chasm.

—Wall Street Journal, April 11, 1996

A *Wall Street Journal* article reports that in 1995 the heads of about 30 major companies received compensation that was 212 times higher than the pay of the average American employee. The widening gulf has ignited a political firestorm and CEOs are scurrying to explain their pay increases. Compensation of chief executives "has reached Marie Antoinette proportions," says Nell Minow, a principal at Lens Inc., an activist investment fund in Washington, DC. "People are getting disgusted with it."

William M. Mercer Inc., New York compensation consultants, analyzed 350 proxy statements of the biggest U.S. businesses. The Securities and Exchange Commission (SEC) rules for proxy statements require disclosure of the compensation of the chief executive officer plus four or more other highly compensated executives. For each business the following data were collected:

Company name
Type of business
Standard Industrial Classification Code (SIC)
Number of employees — 1995

Company revenue—1995

CEO's name

1995 salary—base salary earned in 1995

1995 bonus—annual bonus earned in 1995

Percent change from 1994—change in salary and bonus from 1994 to 1995. Percent change data have been excluded for new CEOs and in cases where salary and bonus data for 1995 or 1994 are not valid.

Long-term compensation: options gain—gains from the exercise of stock options and/or stock appreciation rights during 1995

Long-term compensation: other—includes 1995 (1) value of shares of restricted stock or restricted stock units at grant, and (2) value of payouts under any other long-term incentive compensation plan.

The furor over the great pay divide is creating pressure for higher tax rates on the rich and new federal limits on executive compensation. One idea is to give tax breaks for businesses that treat their workers well—for example, by capping the highest-paid executive's pay at perhaps 50 times the lowest-paid employee's.

16.1   Chapter Objectives

This is the final chapter of the book. It is a good time to look back on the techniques that have been covered and use them to "make your case"! We have been building up to this point throughout the entire book.

Each of the chapters in this book started with a problem. That problem was used to explain the specific techniques of the chapter. Many of the worked examples in the chapter pertained to this opening business problem. Finally, each chapter ended with a memo that included the analysis and resulting conclusions based on the work done on that business problem in the examples in that chapter. Sometimes the memo was more of an in-progress report, and the remaining analysis was part of the exercises at the end of the chapter. In completing the analysis, you were completing a structured case. You received some guidance in the analysis by the nature of the questions that were asked, and you knew to apply the techniques that had been presented in the chapter.

Now it is time to complete the course and to solve some cases without any guidance. These are called Making Your Case exercises. This is, after all, what you will need to be able to do as you enter the workforce. You must use all that you have learned about data analysis to make some sense out of these data sets, see the information in the data, draw some conclusions, and make your case. We hope you have completed the paradigm shift that we talked about in Chapter 1, and you now see the world through a "statistical thinking" lens. It is this vision that will make you an important member of any team in the workplace.

Specifically, this chapter looks at these topics:

- Summary of the techniques covered in this book
- Guidance in how to analyze the "great divide" data set

16.2   Summary of the Techniques Covered in This Book

As you look back over the material covered in this book, try not to see the small numerical details of each of the techniques; instead, look for the unifying themes that underlie what you have learned. Take a global look at the book and use this summary to identify the major threads that tie the book together.

This book has been divided into 16 chapters. We can take the chapters and group them into six major parts.

Part	Chapters	Topic
Part I	Chapters 1 and 2 Main idea: Sample as a snapshot of the population	Getting started
Part II	Chapters 3–5 Main ideas: Summarize the data in picture and numerical form Look for the center, the spread, and trends	Exploratory data analysis
Part III	Chapters 6 and 7 Main ideas: Distributions and parameters Link between probability and inferential statistics	The keys that unlock the door to inferential statistics
Part IV	Chapters 8–10 Main idea: Use the sample to draw conclusions about the population	Inferential statistics
Part V	Chapters 11 and 12 Main idea: Use sample data to build a model to explain the relationship between two or more variables	Model building
Part VI	Chapters 13–15	Additional topics

Now let's see if there are some major themes that tie these ideas together.

1. Early in the text you learned that even if we do our sampling perfectly, we will still undoubtedly have incomplete information. This is not because we made any mistakes but because of what is called *sampling error*. Sampling error is the difference between what the sample predicts and what is true for the population. It is due to the fact that the sample is only a piece of the population. No matter what we do in statistics, unless we take a census (examine the entire population), we will have some sampling error. You have learned that even if the sample data are "a bit" inconsistent with the hypothesized theory, this doesn't automatically mean the theory is wrong. It is possible and maybe even likely to find that 52% of a sample of employees feel the company has empowered them when in fact less than half of all of the employees feel this way.

2. Without a doubt, the concept of *variability* has been a common thread from the beginning of this book. One of the great quality gurus, W. Edwards Deming, said

that variation is our greatest enemy. We have indeed seen this ourselves. In fact, we observed that if there were no variability in the population, we could take a sample of size $n = 1$ and have perfect information. The *standard deviation*, which is the square root of the variance, is the yardstick by which we evaluate all differences when analyzing data.

3. Another theme that you may have noticed throughout the text is the *relationships* between variables. Right from the beginning of descriptive statistics we compared graphs of variables and used scatter plots to look for relationships between variables. This theme was extended when you learned to do hypothesis tests to compare two or more population means or proportions. In one of the later chapters you learned how to build models that would allow you to exploit the relationship between two or more variables for prediction purposes.

4. *Decision making in any form should be based on a proper analysis of data.* To remain competitive in today's fast-changing business environment, more data than ever before are being collected. However, the data alone are useless. They must be properly analyzed to reveal the information they contain. The entire quality movement depends on the proper use of data to make informed business decisions.

So now we have reduced the entire book to four fundamental themes. Despite the many different techniques that have been covered and even some that have not been covered, there are really only a few fundamental concepts that you need to learn in statistics.

16.3 How to Analyze the "Great Divide" Data Set

The "story" presented at the beginning of this chapter is a real-life case. Your job is to analyze this data set to reveal the information it contains. This section provides some ideas for your analysis. Before we look at the particulars of this data set, let's see if we can identify some basic steps that should be followed anytime we begin to analyze data to solve a problem.

Datafile:
DIVIDE.XXX

The material in the text has been presented in the order that you would use the techniques.

Step 1: Explore the data visually using the graphical techniques of descriptive statistics. These include bar charts, pie charts, and histograms.

Step 2: Summarize the data numerically. This includes finding the mean, median, quartiles, range, and standard deviation (at a minimum).

Step 3: Look for relationships in the data visually and numerically. Visually you should examine scatter plots. Numerically, you should use cross-tabulation tables to examine averages and standard deviations for subsets of the data.

Step 4: On the basis of the results of Steps 1–3, formulate some hypotheses that might help you answer your questions and solve the problem.

Step 5: Test the hypotheses, being sure to check that the assumptions of the test are not violated. For example, if the variables must be normally distributed for the test to be valid, then first run a chi-square goodness of fit test. If the technique requires you to assume a common variance, use an F test to be sure this is a reasonable assumption.

Step 6: If there appears to be a pattern in the data over time, use a time series technique. (Note that time series techniques have not been presented in this text.)

Step 7: If appropriate, estimate the parameters of a multiple regression model for the variable you are trying to predict. Be sure that you check the assumptions of the model. Determine whether your model is useful by testing for the significance of each of the variables and the significance of the regression model.

Step 8: Draw some conclusions from your analysis and make some recommendations.

First, examine the data visually using the graphical techniques of descriptive statistics.

Let's look at the details of implementing these eight steps for the "great divide" data set. At Step 1, you should visually explore the data. Consider using pie charts and bar charts for the qualitative variables such as type of business and SIC code. Consider using histograms for the quantitative variables such as 1995 salary, 1995 bonus, number of employees, and annual revenue.

Next, calculate numerical descriptors for the quantitative variables.

At Step 2, for each of the quantitative variables you should calculate summary statistics including the mean, median, quartiles, range, and standard deviation.

Look for relationships in the data.

At Step 3, you should start to look for patterns, trends, and relationships among the variables. Calculate some summary statistics for subsets of the data. For example, you might wish to calculate the average CEO salary for all "Energy" companies. For these data you might construct scatter plots to answer these questions:

- Is salary and/or bonus related to type of business?
- Is salary and/or bonus related to annual revenue of the company?
- Is salary and/or bonus related to size of the company as measured by the number of employees?

Formulate hypotheses based on your analysis of the sample.

Step 4 is the point at which you summarize what you have learned from your exploratory data analysis. Remember that the tools you have used thus far in the procedure simply describe the sample. At this step, you formulate some hypotheses in an attempt to use the sample to answer questions about the population related to the problem you are trying to solve.

Use hypothesis testing and ANOVA to determine if the differences in the sample are significant.

At Step 5, you should consider using t tests to test for differences in the average salary. For example, you could divide the companies into two groups: companies with less than 500 employees and those with more than 500 employees. Then you could test to see if the average salary is different for the two groups. To see whether salary is related to the type of business, you would use ANOVA to see whether the average CEO salaries are equal across the various types of businesses. If they are not, determine which ones are different.

Use time series analysis if there appears to be a pattern in a variable over time.

Step 6 proposes that you examine the data over time when appropriate. For this data set, you essentially have 2 years worth of data because you have the percent change in salary and bonus from 1994 to 1995. This is not really enough data to do any time series analysis but you could look for patterns in the percent changes by type of business, SIC code, or size of the company.

Use regression to develop a prediction model.

At Step 7, you should try to make some predictions. For this data set, you should consider trying to predict the salary of the CEO based on the type of business, size of the company, SIC code, and annual revenue of the company. To do this you would use a multiple regression model and determine which factors are significant.

Draw some conclusions.

At Step 8, you should reach some conclusions and make some recommendations on the basis of your analysis. Perhaps the "great divide" is most pronounced in large companies or in a particular type of business. This type of conclusion might influence the type of legislation that would help solve the problem. Your data analysis must support your case but it must do so ethically.

CHAPTER 16 **SUMMARY**

In this chapter we have taken a global look at the techniques you learned in this book. Despite the large number of different techniques and many formulas presented in this book, we looked at the "bigger picture." We have grouped the chapters into parts and found four themes that tie all of our work in this book together.

You have learned the basic tools to allow you to see the world through a statistical paradigm. It is this lens that will help you advance in your career.

At whatever level you find yourself in an organization, you may be called upon to analyze data. You have learned how to do this, and remember that this analysis must always be done ethically and fairly. This has been a constant point throughout the discussions in this book. Lives have been lost and saved by fair data analysis. You must use your statistical paradigm responsibly.

CHAPTER 16 **EXERCISES**

Thinking About It!

16.1 After you received your master's degree in statistics last year, you took the only statistical job you could find near your fiancée. It is with a large pharmaceutical corporation, where you are now the only statistician. You spend most of your time supporting some randomized clinical trials initiated during the tenure of your predecessor, who left on very short notice and for reasons referred to rather vaguely in terms of "mutual agreement." The company has been losing ground to the competition, and management is now counting on new product development to "save the company."

The protocol for the largest of the ongoing trials, now coming to a close, clearly states the null hypothesis to be that cancer patients treated by a new procedure will have the same 2-year survival rate as those treated by the standard regimen X, but no statistical method for the comparison is specified. Your analysis produces a two-tail p value of 0.08. Your boss says that switching the hypothesis to a one-tail test yields a p value of 0.05 and the data clearly support the alternative hypothesis. He is prepared to recommend that the CEO approve a large budget (including your salary) to further develop this new procedure.

What do you do now? OK, what do you *really* do now?

16.2 You have been hired by an attorney to "prove" that Company ZZZ does not practice age discrimination. The specifics of the case are that Mr. Oldstone, who had worked for the company for over 20 years, lost his job a year ago. Mr. Oldstone is suing Company ZZZ for $1 million, claiming age discrimination. The company claims that it reorganized and that Mr. Oldstone's position was eliminated in the reorganization. Since that time Ms. Youngsville has been hired to perform tasks similar in nature to those formerly performed by Mr. Oldstone. Ms. Youngsville is a recent college graduate and is paid $30,000 less than Mr. Oldstone's salary. You have access to the company's human resource records. An initial review of the data seems to support Mr. Oldstone's claim. However, you are being paid to disprove this claim. You are a poor, starving graduate student and you and your spouse are expecting your first child in 3 months. You also know that several other companies have similar suits against them and they are watching the results of this trial. You know that you can probably manipulate the highly disorganized human resource data to support the company's claim that it does not practice age discrimination.

What do you do now? OK, what do you *really* do now?

Doing It!

16.3 Complete the analysis of the "great divide" data set by following the suggestions outlined for each of the eight steps in Section 16.3.

CASE STUDIES

The variables in these case studies are explained in Appendix B.

MAKING YOUR CASE

Advising the President: Exploratory Data Analysis

You have taken a job at the White House on the staff of the President's Council of Economic Advisors. As the new kid on the block, you have been given the assignment of providing the President with a view of the country. The President wants you to capture the country with statistics, so that he understands how the different demographic and economic factors vary around the country.

You have data available to you from the U.S. Census Bureau by state and region of the country. The Census Bureau tracks 63 different variables covering nine main categories: population, health, education, income, crime, employment, banking, transportation, and housing. A very small portion of the data follows:

Region division, and state	Resident population\1			Resident population\1		
	Total,\2 1994 (1000)	Percent increase,\3 1990–94 (%)	65 years old and over, 1994 (%)	Householder 65 years old and over, 1994\3 (%)	In metro areas, 1992\4 (%)	Net domestic immigration, 1990–94\1
United States	260,341	4.7	12.7	21.8	79.7	0
Northeast	51,396	1.2	14.1	23.7	89.4	(1,439,371)
New England	13,270	0.5	13.9	22.9	84.1	(364,734)
Maine	1,240	1.0	13.9	22.8	35.7	(12,109)
New Hampshire	1,137	2.5	11.9	19.5	59.4	(9,912)
Vermont	580	3.1	12.1	20.1	27.0	2,287
Massachusetts	6,041	0.4	14.1	23.3	96.2	(183,995)
Rhode Island	997	−0.7	15.6	25.6	93.6	(35,219)
Connecticut	3,275	−0.4	14.2	23.1	95.7	(125,786)

The data are found in a file named *CENSUS.XXX*. You notice that the data cover the time period from 1990 to 1994. You wonder about the data in parentheses and remember from your accounting course that parentheses mean negative numbers. You also notice summary lines for geographic regions and realize that you had better remove them before you do any type of analysis by state.

Completely overwhelmed, you decide to approach your boss and ask for some direction. She tells you that the President is looking for summaries and relationships among the variables and regions of the country and that reminds you of the popularity of Ross Perot's charts and graphs. In addition, she tells you that the President wants to be prepared for attacks from Capitol Hill on health care and education, and that you should start your analysis there.

1. Prepare a set of graphical displays and statistics that summarize the availability of health care in the country. Do any of the states or regions exhibit unusual behavior? If so, which ones are they? Is any state or region consistently low or high on all aspects of health care?

2. Of particular interest is the cost of health care. Does there appear to be any relationship between the cost of hospital care and the other variables? If so, how are they related?

Now that you have gotten the idea, you understand what you need to do for each of the rest of the categories in order to have your report on the President's desk by the end of the week. Oh, by the way, the President is wondering where the best place in the country to live is.

MAKING YOUR CASE

What's on the Road? Who's Driving It?: Exploratory Data Analysis

Before you can make decisions about the future, you need to understand the current situation. The transportation industry controls the infrastructure of the country, and the decisions it makes affect everyone.

One method the industry uses to understand current trends is the Nationwide Personal Transportation Survey (NPTS). This survey is done approximately every 7 years under the sponsorship of the U.S. Department of Transportation. The NPTS compiles national data on the nature and characteristics of personal travel. NPTS data may be used to describe current travel patterns and, given projections of demographic change, can provide a valuable tool to forecast future travel demand. One aspect of transportation planning is knowing what is currently on the road. The NPTS data include motor vehicle information such as year, make, model, and other vehicle-related information.

You are asked to describe the current vehicle population of the United States. To assist in this, you have been given a portion of the NPTS data from the 1990 survey, in which data were collected on 26,172 households using computer-assisted telephone interviewing (CATI). Specifically, you have been given data from all survey interviews conducted in April 1990. A portion of the data is shown here:

CMSA	HHFAMINC	HHLOC	HHMSA	HHSIZE	HH_RACE	HOUSEID	LIF_CYC	MAKECODE
5602	05	1	5640	01	01	00038	01	014
5602	04	1	5640	06	01	00039	08	038
5602	04	1	5640	06	01	00039	08	022
5602	04	1	5640	06	01	00039	08	021
5602	98	1	5640	05	01	00041	06	021
5602	98	1	5640	05	01	00041	06	021
5602	98	1	5640	05	01	00041	06	035
5602	14	2	5190	02	01	00045	10	009
5602	14	2	5190	02	01	00045	10	006
5602	99	2	5190	03	01	00047	06	019

The data are found in a datafile called *CARS.XXX*. As is the case with much publicly available data, the variable names are cryptic and the qualitative variables are coded. Clearly, you will need to know what each variable and code mean. This information is detailed in Appendix B.

Of particular interest to automobile manufacturers is information about the ages and types of vehicles on the road.

1. Use graphical and numerical summary tools to describe the distribution of vehicles by make and year.

2. Are the distributions affected by various demographics? Who drives foreign cars? Who drives old cars?

3. Is the distance driven annually affected by where people live, gender, age, or type of vehicle?

Using these questions as a starting point, prepare a report that completely describes the vehicle population of the United States. Be sure to include any limitations of the data.

MAKING YOUR CASE

Who Spends Money? What Do They Buy?: Inferential Statistics

A large marketing research firm has an international client who has business interests in many consumer industries such as food, apparel, alcohol, and automobile fuel. As a person with some statistical experience, you have been hired by this marketing research company. Before you spend any of the client's money on surveying consumers, the client has asked you to report on general trends in consumer spending. Specifically, you are asked which business areas have experienced increases, and what are the characteristics of people who are spending money in these areas. Your report will guide future survey design and eventually new product development and advertising expenditures.

Using your Internet-surfing skills, you find the website for the U.S. Census Bureau and find the results of the Consumer Expenditure Survey conducted periodically. The most recent survey available online is from 1993 and so you decide to use it. The Consumer Expenditure Survey includes information on many variables but you select for study 49 variables that you think would be helpful. A portion of the data is shown here:

AGE2	AGE_REF	ALCBEVCQ	ALCBEVPQ	APPARCQ	APPARPQ	AS_COMP1	AS_COMP2
	46	97	30	0	33	2	1
	78	0	0	0	0	0	1
	55	0	65	0	159	0	1
73	78	0	40	0	145.83	1	1
45	45	85	170	482	845.4	1	1
52	56	54	27	175	840	2	1
36	35	40	20	375.38	120	1	1
	63	0	0	22	10	1	0

Each row in the datafile (*CONSEXP.XXX*) represents one person's (the Reference Person) responses to the survey questions. As is the case with much publicly available data, the variable names are cryptic and the qualitative variables are coded. Clearly, you need to know what each variable means and also what the codes mean. This information is detailed in Appendix B.

One portion of your client's business is the food industry. The company owns restaurants and supermarkets. The first thing you decide to do is examine trends in the amount of money spent on food. You have information on the amount of money spent in the current quarter of the survey year and the amount spent in the previous quarter.

1. Did consumers spend significantly more money in the current quarter of the survey year than in the previous quarter on food? If so, is the difference meaningful?

2. Was there a change in how the money spent on food was distributed between the amount spent on eating out and the amount spent for food eaten at home?

3. Do households with teenagers spend more money on food eaten at home?

4. Do households where both adults are working spend more money on eating out?

Having done the analysis of food expenditures, you decide to take the same approach for the other business areas. In addition to looking for changes in expenditures between quarters, you wonder what other demographic factors affect spending habits. You set about doing the appropriate statistical tests to provide your client with a report by the end of the week.

MAKING YOUR CASE

Advising the President: Inferential Statistics

The President is pleased with your exploratory data analysis. Although he agrees with your impressions, he is hesitant to propose new legislation based on a subjective interpretation of the data. You are smarter now, too, and you recognize the need for the formal statistical tools of hypothesis testing.

Remember the data available to you are from the U.S. Census Bureau and are found in a datafile named *CENSUS.XXX*. A portion of the data follows:

Region division, and state	Resident population\1			Resident population\1		
	Total,\2 1994 (1,000)	Percent increase,\3 1990–94 (%)	65 years old and over, 1994 (%)	Householder 65 years old and over, 1994\3 (%)	In metro areas, 1992\4 (%)	Net domestic immigration, 1990–94\1
United States	260,341	4.7	12.7	21.8	79.7	0
Northeast	51,396	1.2	14.1	23.7	89.4	(1,439,371)
New England	13,270	0.5	13.9	22.9	84.1	(364,734)
Maine	1,240	1.0	13.9	22.8	35.7	(12,109)
New Hampshire	1,137	2.5	11.9	19.5	59.4	(9,912)
Vermont	580	3.1	12.1	20.1	27.0	2,287
Massachusetts	6,041	0.4	14.1	23.3	96.2	(183,995)
Rhode Island	997	−0.7	15.6	25.6	93.6	(35,219)
Connecticut	3,275	−0.4	14.2	23.1	95.7	(125,786)

The attacks continue on the health-care front. Specifically, legislators from the South Central region claim that they need additional federal funding to encourage doctors and nurses to practice there. Before the President supports such legislation, he wants answers to these questions:

1. Is the availability of medical professionals in the South Central states really lower than in the rest of the country?

2. Are the South Central states the only area of the country that needs this type of support? If you had to make a recommendation, what area(s) would you support?

Some senators from the Midwest region say that they will support the health-care initiatives for the South Central region only if they are assured that they have equity in the area of education, specifically teachers' salaries.

3. Can the President ignore their lobbying? Why or why not?

You now know that you need to be prepared to advise the President if legislators from other regions are going to try similar tactics in other categories. Do the necessary analyses and prepare your report.

MAKING YOUR CASE

Responding to the Customer: Pulling It All Together

Congratulations! You have completed your introductory statistics course and been hired by a company that is interested in applying for ISO (International Standards Organization) certification. You know that this certification requires the company to document all its processes, including the process by which it obtains and uses customer feedback for continuous improvement.

The company has provided the customers with a fax number and an e-mail address to report complaints or problems. When a complaint comes in, the company must respond to the customer. Your job is to analyze the complaint data collected when customers call to complain about the product.

Your boss is most concerned about the time it takes to respond to a customer complaint. The response time is the number of minutes from receipt of the fax or e-mail until first contact with the customer. Clearly, shorter response times are desirable, with all response times targeted at less than one-half day (240 minutes). The data include the number of staff personnel scheduled for that day, whether that type of problem had been encountered previously (1 = yes; 0 = no), how the issue was received (via fax or e-mail), whether the issue statement was transferred into a written resolution log, and finally the volume (or number) of issues electronically recorded for that day. Information is also provided on the coded volume level (Cvol = 1 if Volume ≤ 25; Cvol = 2 if 26 < Volume ≤ 50; Cvol = 3 for [51, 75]; Cvol = 4 for [76, 100]; Cvol = 5 for > 100). A small portion of the data set is shown here:

Run	Staff	Exper	Type	Log	Volume	Cvol	RespTime
1	5	1	E	Y	114	5	97
2	8	1	E	Y	54	3	39
3	6	1	F	Y	36	2	54
4	9	1	F	Y	66	3	55
5	10	0	E	Y	125	5	125
6	5	1	E	Y	121	5	108
7	8	0	F	Y	69	3	115
8	6	1	F	Y	40	2	60

The data are found in a file named *CUSTOMER.XXX*. You notice that some of the variables are qualitative variables, some are quantitative variables, and some are coded responses. You know that the type of data will certainly influence what techniques you use to analyze the data.

Here are your boss's instructions:

1. Describe the data. This includes graphs and numerical summaries when appropriate.

2. Look for relationships in the data. Check to see whether the response time to the customer is related to any of the other variables and, if so, how they are related.

3. Test to see what factors affect the response time.

4. Look for patterns or trends in the complaints over time.

Prepare a report for your boss addressing these issues and make recommendations based on your findings.

MAKING YOUR CASE

Needs Assessment: Pulling It All Together

With the increased terrorism in the world, many churches have found their pews filled with people who have been away from church for sometime. These people seem to be trying to process world events from a spiritual perspective. In an attempt to meet the needs of these people, a church has decided to do a needs assessment. A survey was designed and it is shown on the next few pages. The survey was mailed to the current members of the church and they were given a data collection date. Either people could fill out the survey at home and bring it with them to church that Sunday or they could fill it out at the end of the worship service that day. The church received 184 completed surveys.

The data are stored in *CHURCH.XXX*. The lines for the first few questions are shown here:

CODE	Q1	Q2	Q3	Q4	Q5	Q6	Q7	Q8	Q9
1	5	5	5	1	4	3	1	1	0
2	4	2	3	1	3	1	1	1	0
3	2	0	0	1	3	1	1	0	0
4	4	5	5	2	3	3	1	2	2
5	3	3	0	3	3	4	2	2	0
6	5	5	0	1	0	0	1	0	0

Notice that each response is given a numeric code. In addition, some of the survey questions have multiple parts and each question is given a separate column so the question numbers in the file do not match up exactly with the survey. The survey shown on the next few pages also includes the correspondence with the data file and the codes used for each possible response.

1. Summarize the responses to each survey question numerically and graphically.

2. Perform bivariate analyses to investigate differences in the responses to the questions by gender, age, marital status, and any other variable you think might be of interest to the church leadership.

3. Use the appropriate nonparametric analysis to compare the various groups (men, women, etc.) in this church.

4. Comment on the bias, if any, in these data.

5. Summarize your analysis and present your findings to the church leadership in the form of recommendations.

Needs Assessment Survey Questions

This survey is being conducted by the Church Council in an effort to better understand who we are as a church body, what we think, and how we should proceed for the future. Findings will be studied, which will lead to a new vision statement. Thank you!

1. How would you define the size of the "community" this church serves? **(Q1)**

    (Choose one)   Coding
    Immediate neighborhood   _____   1
    City of Springfield   _____   2
    15-minute driving radius   _____   3
    One-hour driving radius   _____   4
    Other   _____   5
                                                   0—Blanks

2. How should this size be redefined for the purpose of our church's ministry outreach? **(Q2)**

    (Choose one)   Coding
    Immediate neighborhood   _____   1
    City of Springfield   _____   2
    15-minute driving radius   _____   3
    One-hour driving radius   _____   4
    Other   _____   5
                                                   0—Blanks

3. Indicate how this church meets the spiritual needs of:

    Circle your rating.

	Highest				**Lowest**	Coding—As is with 0 for blanks
Seniors	1	2	3	4	5	**(Q3)**
Families	1	2	3	4	5	**(Q4)**
Singles	1	2	3	4	5	**(Q5)**
Couples	1	2	3	4	5	**(Q6)**
Youth	1	2	3	4	5	**(Q7)**
Children	1	2	3	4	5	**(Q8)**

4. Does this church meet your spiritual needs?   Yes _____   No _____ **(Q9)**

    Comment:

    Coding   0   1   2—Blanks

5. Do you understand how money is designated to the various missions programs of the church? **(Q10)**

    Yes _____   No _____

    Comment:

    Coding   0   1   2—Blanks

5a. Are you satisfied with the present missions program of the church? **(Q11)**

Yes _____   No _____   No opinion _____

Comment:

Coding   0   1   2   3—Blanks

5b. Please respond to the present emphasis placed on each of the following components of our missions program. (Check the appropriate column)

	About Right	**Too Much**	**Too Little**	**Don't Know**	
(Q12) World missions	1	2	3	4	5—Blanks
(Q13) Covenant missions	1	2	3	4	
(Q14) Short-term trips	1	2	3	4	
(Q15) Community ministry	1	2	3	4	

6. How much financial support are you giving *annually* to missions organizations or missionaries outside of your church giving? $_____ **(Q16)**
   number

   Please list persons and/or organizations to whom you give. _____

   _____

   not captured

7. Please indicate your satisfaction with the performance of the following:

Circle your response.

	Highest				**Lowest**	Coding—as is
(Q17) Pastors	1	2	3	4	5	6—Blanks
(Q18) Office staff	1	2	3	4	5	
(Q19) Music staff	1	2	3	4	5	
(Q20) Council leadership	1	2	3	4	5	

8. Please indicate your satisfaction with the following:

Circle your response.

	Highest				**Lowest**	Coding—as is
(Q21) Physical plant	1	2	3	4	5	6—Blanks
(Q22) General maintenance	1	2	3	4	5	

9. Please indicate your satisfaction with the following components of our worship experience.

Circle your response.

	Highest				**Lowest**	Coding—as is
(Q23) Congregational singing	1	2	3	4	5	6—Blanks
(Q24) Choir	1	2	3	4	5	
(Q25) Special music	1	2	3	4	5	
(Q26) Sermon	1	2	3	4	5	
(Q27) Contemporary music	1	2	3	4	5	
(Q28) Traditional music	1	2	3	4	5	
(Q29) Announcements	1	2	3	4	5	
(Q30) Sound system	1	2	3	4	5	
(Q31) Worship bulletin	1	2	3	4	5	

10. Do you attend Sunday School?

     (Q32) Adult Sunday School   Yes <u> 0 </u>   No <u> 1 </u>   Blanks—2

     (Q33) Children's Sunday School   Yes <u> 0 </u>   No <u> 1 </u>   Blanks—2

10a. If yes, please indicate your satisfaction with the Sunday School program

Circle your response.

	Highest				Lowest	Coding—as is
(Q34) Quality of teaching	1	2	3	4	5	Blanks—6
(Q35) Variety of course offerings	1	2	3	4	5	
(Q36) Overall learning experience	1	2	3	4	5	

10b. If no, why not? (Choose the most important one) **(Q37)**

     Course offerings   —— 1

     Time doesn't allow   —— 2

     Not relevant   —— 3

     Other   —— 4

          Blank—5

11. Does the church provide you with ample communication relative to the church activities and ministry? **(Q38)**

     Yes, well informed   —— 0

     No, needs improvement   —— 1

     Comment: Blank—2

12. How well does the church provide for social experience in the various programs and ministries? **(Q39)**

Circle your response.

Highest				Lowest	Coding—as is
1	2	3	4	5	6—Blank

     Comment:

13. How long have you been attending this church? **(Q40)**   Years —— number

14. Do you attend church at least three times per month? **(Q41)**   Yes <u> 0 </u> No <u> 1 </u> 2—Blank

15. How far do you travel from home to church (one way)? **(Q42)**   Miles —— number

16. What is your age group? **(Q43)**

     Under 21   —— 1

     21–34   —— 2

     35–49   —— 3

     50–65   —— 4

     Over 65   —— 5

          Blank— 6

17. What is your individual personal income? (Do not include spouse.) **(Q44)**

Under $15,000 ——— 1
$15,000–$30,000 ——— 2
$30,000–$45,000 ——— 3
$45,000–$60,000 ——— 4
$60,000–$75,000 ——— 5
Over $75,000 ——— 6
                Blank—7

18. Gender: **(Q45)**

Male ——— 0
Female ——— 1
        Blank—2

19. Marital status: **(Q46)**

Married ——— 0
Single ——— 1
        Blank—2

Thank you for your help in completing this survey.

APPENDIX **A**

Statistical Tables

TABLE 1 **Random Number Table** A2
TABLE 2 **Binomial Probability Tables** A4
TABLE 3 **Standard Normal Table** A6
TABLE 4 *t* **Critical Values** A8

TABLE 5 **Chi-Square Distribution Table** A9
TABLE 6 **The *F* Distribution Table** A10
TABLE 7 **Critical Values of the Wilcoxon Rank Sum Test** A14

TABLE 1

RANDOM NUMBER TABLE

Row #	1	2	3	4	5	Column # 6	7	8	9
1	094632795	711501513	537971597	562758635	410398128	182794408	773761503	455139927	132682754
2	033413186	653475420	289063704	485441982	460744361	328703833	289612212	569540556	620271271
3	297556368	658953044	738968017	414437050	296126017	075254187	702140315	467039889	762226273
4	472960570	785645638	574817322	817883255	976076280	843373358	118284363	445336907	327380271
5	256883707	716249997	378236162	467694224	193707682	380141891	605807481	180164558	473854769
6	179451522	878902420	602450872	987686989	686677180	242196303	517640224	691116863	275385608
7	894964682	704841116	241902107	750429362	794778197	693242123	316755091	193593484	913974355
8	738120861	744470405	873393138	758824215	394004646	496696605	006936567	163371217	727267920
9	803156944	653387115	716335974	835667154	066959782	908783760	165946696	735683921	894672507
10	187636922	953598780	481536873	055734541	493193305	566923120	435549770	007706188	839596393
11	102021077	286953643	851411058	132935798	745770831	187026467	363837178	791264282	107184709
12	734254842	133959443	113708008	443989454	786207141	772432741	682053431	048076059	617648837
13	757417865	524596578	240504889	544970942	340054233	417544234	302126745	003333205	250247568
14	227658086	233543943	487116060	577966118	524453480	934483237	367425608	431112250	536516890
15	746794759	557361146	826373105	870360802	412399571	804914923	128067420	659566961	452000520
16	235158119	336776002	728424416	086967212	040966064	335090111	985461873	832921870	461741235
17	944558037	787700710	060386364	635482046	558143223	600009181	448499754	064172342	713707601
18	658344036	271853277	275251035	744269244	877186509	130398637	367142231	846275675	485443650
19	179411521	104680475	020354893	576185422	778014690	380931445	886031872	320231466	062555147
20	865570814	699503925	628956988	683503622	276170341	744494133	081246804	523527226	198219562
21	984902072	022065717	504274676	136174524	195356906	027900159	809382340	381669407	544140648
22	139846732	496390379	582502144	768571665	177715615	830320391	105937107	329901920	618226629
23	492146502	493503513	138813631	479880385	684082619	010963692	268892703	552334849	488002392
24	739727377	916314641	263944162	861966588	286459084	491798049	760316559	837966446	371951811
25	075823838	115491339	547215506	007869049	323138362	193432798	361574944	787418390	016648846

26	915745558	104176259	349828840	546922404	266406684	490531595	336155799	242136076	641061181
27	136001350	309685268	986533618	587428568	052231831	422269870	302793461	564542482	915031158
28	691703257	926306032	988266746	716231516	519662016	986665536	993015206	066999095	731533696
29	154222042	316334873	963715901	044966315	846937935	104409586	768790545	113341348	108261519
30	983823702	641345385	203912928	219869690	208288383	861497163	149918954	160034550	759951927
31	277515731	805241329	549047139	285206828	534033122	130094940	970748730	798208639	485614463
32	566308464	543263173	711363354	339738940	051286779	714375200	698531722	072971345	762369710
33	533605194	994619567	798813607	079914804	405016946	275797011	801942743	814124918	033457635
34	374266385	237398626	014653680	107885763	848594153	093210516	751461171	121583622	385898493
35	739563239	736604709	251789737	480977217	432264262	975146204	639768152	455086460	573742841
36	539277994	536590953	293592699	279474008	525803109	281596944	199856046	646139218	124051051
37	609535944	455877404	251255783	334162937	523110770	461537085	359043224	423641232	223047648
38	254030985	962503045	747584829	988588554	631976076	901087130	961891746	149014870	557453130
39	869881662	489992047	240739861	875737562	237409010	135678267	196964820	343397286	364892121
40	935509137	382168564	659392779	628617853	533473897	569590209	333032014	937163707	460780426
41	727184789	651476235	562537081	259345598	118701307	970343244	678458590	189103174	840164700
42	076139028	276588587	329963439	184510242	904140612	761154973	175127287	520167477	242700207
43	181087637	087629199	028292894	460181436	623140518	207937371	238398056	136009368	581975565
44	514095654	401875869	095986936	976620836	391483115	713574093	679457157	184527765	553593737
45	804743163	129181317	005547120	712455031	948648814	909230622	839276576	726704039	427115217
46	492640056	038991626	749280637	430162677	414656226	603291802	983746203	756647413	333907575
47	105925664	168586200	348119547	829517480	244539107	448715108	154400559	634475954	701530403
48	688908833	510541646	706386776	935447659	798110618	127019341	889979390	455625169	283128630
49	806445894	067701203	566577304	808746117	093933115	198530698	142531634	042491555	776838859
50	249997369	047395403	102944245	987149692	239682871	971259345	193515078	797533485	459099813

TABLE 2

BINOMIAL PROBABILITY TABLES

$$P(X = x)$$

n = 5   *p*

x	0.05	0.10	0.20	0.25	0.30	0.40	0.50	0.60	0.70	0.75	0.80	0.90	0.95
0	0.774	0.590	0.328	0.237	0.168	0.078	0.031	0.010	0.002	0.001	0.000	0.000	0.000
1	0.204	0.328	0.410	0.396	0.360	0.259	0.156	0.077	0.028	0.015	0.006	0.000	0.000
2	0.021	0.073	0.205	0.264	0.309	0.346	0.313	0.230	0.132	0.088	0.051	0.008	0.001
3	0.001	0.008	0.051	0.088	0.132	0.230	0.313	0.346	0.309	0.264	0.205	0.073	0.021
4	0.000	0.000	0.006	0.015	0.028	0.077	0.156	0.259	0.360	0.396	0.410	0.328	0.204
5	0.000	0.000	0.000	0.001	0.002	0.010	0.031	0.078	0.168	0.237	0.328	0.590	0.774

n = 10   *p*

x	0.050	0.100	0.200	0.250	0.300	0.400	0.500	0.600	0.700	0.750	0.800	0.900	0.950
0	0.599	0.349	0.107	0.056	0.028	0.006	0.001	0.000	0.000	0.000	0.000	0.000	0.000
1	0.315	0.387	0.268	0.188	0.121	0.040	0.010	0.002	0.000	0.000	0.000	0.000	0.000
2	0.075	0.194	0.302	0.282	0.233	0.121	0.044	0.011	0.001	0.000	0.000	0.000	0.000
3	0.010	0.057	0.201	0.250	0.267	0.215	0.117	0.042	0.009	0.003	0.001	0.000	0.000
4	0.001	0.011	0.088	0.146	0.200	0.251	0.205	0.111	0.037	0.016	0.006	0.000	0.000
5	0.000	0.001	0.026	0.058	0.103	0.201	0.246	0.201	0.103	0.058	0.026	0.001	0.000
6	0.000	0.000	0.006	0.016	0.037	0.111	0.205	0.251	0.200	0.146	0.088	0.011	0.001
7	0.000	0.000	0.001	0.003	0.009	0.042	0.117	0.215	0.267	0.250	0.201	0.057	0.010
8	0.000	0.000	0.000	0.000	0.001	0.011	0.044	0.121	0.233	0.282	0.302	0.194	0.075
9	0.000	0.000	0.000	0.000	0.000	0.002	0.010	0.040	0.121	0.188	0.268	0.387	0.315
10	0.000	0.000	0.000	0.000	0.000	0.000	0.001	0.006	0.028	0.056	0.107	0.349	0.599

n = 15   *p*

x	0.050	0.100	0.200	0.250	0.300	0.400	0.500	0.600	0.700	0.750	0.800	0.900	0.950
0	0.463	0.206	0.035	0.013	0.005	0.000	0.000	0.000	0.000	0.000	0.000	0.000	0.000
1	0.366	0.343	0.132	0.067	0.031	0.005	0.000	0.000	0.000	0.000	0.000	0.000	0.000
2	0.135	0.267	0.231	0.156	0.092	0.022	0.003	0.000	0.000	0.000	0.000	0.000	0.000
3	0.031	0.129	0.250	0.225	0.170	0.063	0.014	0.002	0.000	0.000	0.000	0.000	0.000
4	0.005	0.043	0.188	0.225	0.219	0.127	0.042	0.007	0.001	0.000	0.000	0.000	0.000
5	0.001	0.010	0.103	0.165	0.206	0.186	0.092	0.024	0.003	0.001	0.000	0.000	0.000
6	0.000	0.002	0.043	0.092	0.147	0.207	0.153	0.061	0.012	0.003	0.001	0.000	0.000
7	0.000	0.000	0.014	0.039	0.081	0.177	0.196	0.118	0.035	0.013	0.003	0.000	0.000
8	0.000	0.000	0.003	0.013	0.035	0.118	0.196	0.177	0.081	0.039	0.014	0.000	0.000
9	0.000	0.000	0.001	0.003	0.012	0.061	0.153	0.207	0.147	0.092	0.043	0.002	0.000
10	0.000	0.000	0.000	0.001	0.003	0.024	0.092	0.186	0.206	0.165	0.103	0.010	0.001
11	0.000	0.000	0.000	0.000	0.001	0.007	0.042	0.127	0.219	0.225	0.188	0.043	0.005
12	0.000	0.000	0.000	0.000	0.000	0.002	0.014	0.063	0.170	0.225	0.250	0.129	0.031
13	0.000	0.000	0.000	0.000	0.000	0.000	0.003	0.022	0.092	0.156	0.231	0.267	0.135
14	0.000	0.000	0.000	0.000	0.000	0.000	0.000	0.005	0.031	0.067	0.132	0.343	0.366
15	0.000	0.000	0.000	0.000	0.000	0.000	0.000	0.000	0.005	0.013	0.035	0.206	0.463

n = 20 　　　　　　　　　　　　　　　　　　　　　　　　　　 *p*

x	0.050	0.100	0.200	0.250	0.300	0.400	0.500	0.600	0.700	0.750	0.800	0.900	0.950
0	0.358	0.122	0.012	0.003	0.001	0.000	0.000	0.000	0.000	0.000	0.000	0.000	0.000
1	0.377	0.270	0.058	0.021	0.007	0.000	0.000	0.000	0.000	0.000	0.000	0.000	0.000
2	0.189	0.285	0.137	0.067	0.028	0.003	0.000	0.000	0.000	0.000	0.000	0.000	0.000
3	0.060	0.190	0.205	0.134	0.072	0.012	0.001	0.000	0.000	0.000	0.000	0.000	0.000
4	0.013	0.090	0.218	0.190	0.130	0.035	0.005	0.000	0.000	0.000	0.000	0.000	0.000
5	0.002	0.032	0.175	0.202	0.179	0.075	0.015	0.001	0.000	0.000	0.000	0.000	0.000
6	0.000	0.009	0.109	0.169	0.192	0.124	0.037	0.005	0.000	0.000	0.000	0.000	0.000
7	0.000	0.002	0.055	0.112	0.164	0.166	0.074	0.015	0.001	0.000	0.000	0.000	0.000
8	0.000	0.000	0.022	0.061	0.114	0.180	0.120	0.035	0.004	0.001	0.000	0.000	0.000
9	0.000	0.000	0.007	0.027	0.065	0.160	0.160	0.071	0.012	0.003	0.000	0.000	0.000
10	0.000	0.000	0.002	0.010	0.031	0.117	0.176	0.117	0.031	0.010	0.002	0.000	0.000
11	0.000	0.000	0.000	0.003	0.012	0.071	0.160	0.160	0.065	0.027	0.007	0.000	0.000
12	0.000	0.000	0.000	0.001	0.004	0.035	0.120	0.180	0.114	0.061	0.022	0.000	0.000
13	0.000	0.000	0.000	0.000	0.001	0.015	0.074	0.166	0.164	0.112	0.055	0.002	0.000
14	0.000	0.000	0.000	0.000	0.000	0.005	0.037	0.124	0.192	0.169	0.109	0.009	0.000
15	0.000	0.000	0.000	0.000	0.000	0.001	0.015	0.075	0.179	0.202	0.175	0.032	0.002
16	0.000	0.000	0.000	0.000	0.000	0.000	0.005	0.035	0.130	0.190	0.218	0.090	0.013
17	0.000	0.000	0.000	0.000	0.000	0.000	0.001	0.012	0.072	0.134	0.205	0.190	0.060
18	0.000	0.000	0.000	0.000	0.000	0.000	0.000	0.003	0.028	0.067	0.137	0.285	0.189
19	0.000	0.000	0.000	0.000	0.000	0.000	0.000	0.000	0.007	0.021	0.058	0.270	0.377
20	0.000	0.000	0.000	0.000	0.000	0.000	0.000	0.000	0.001	0.003	0.012	0.122	0.358

n = 25 　　　　　　　　　　　　　　　　　　　　　　　　　　 *p*

x	0.050	0.100	0.200	0.250	0.300	0.400	0.500	0.600	0.700	0.750	0.800	0.900	0.950
0	0.277	0.072	0.004	0.001	0.000	0.000	0.000	0.000	0.000	0.000	0.000	0.000	0.000
1	0.365	0.199	0.024	0.006	0.001	0.000	0.000	0.000	0.000	0.000	0.000	0.000	0.000
2	0.231	0.266	0.071	0.025	0.007	0.000	0.000	0.000	0.000	0.000	0.000	0.000	0.000
3	0.093	0.226	0.136	0.064	0.024	0.002	0.000	0.000	0.000	0.000	0.000	0.000	0.000
4	0.027	0.138	0.187	0.118	0.057	0.007	0.000	0.000	0.000	0.000	0.000	0.000	0.000
5	0.006	0.065	0.196	0.165	0.103	0.020	0.002	0.000	0.000	0.000	0.000	0.000	0.000
6	0.001	0.024	0.163	0.183	0.147	0.044	0.005	0.000	0.000	0.000	0.000	0.000	0.000
7	0.000	0.007	0.111	0.165	0.171	0.080	0.014	0.001	0.000	0.000	0.000	0.000	0.000
8	0.000	0.002	0.062	0.124	0.165	0.120	0.032	0.003	0.000	0.000	0.000	0.000	0.000
9	0.000	0.000	0.029	0.078	0.134	0.151	0.061	0.009	0.000	0.000	0.000	0.000	0.000
10	0.000	0.000	0.012	0.042	0.092	0.161	0.097	0.021	0.001	0.000	0.000	0.000	0.000
11	0.000	0.000	0.004	0.019	0.054	0.147	0.133	0.043	0.004	0.001	0.000	0.000	0.000
12	0.000	0.000	0.001	0.007	0.027	0.114	0.155	0.076	0.011	0.002	0.000	0.000	0.000
13	0.000	0.000	0.000	0.002	0.011	0.076	0.155	0.114	0.027	0.007	0.001	0.000	0.000
14	0.000	0.000	0.000	0.001	0.004	0.043	0.133	0.147	0.054	0.019	0.004	0.000	0.000
15	0.000	0.000	0.000	0.000	0.001	0.021	0.097	0.161	0.092	0.042	0.012	0.000	0.000
16	0.000	0.000	0.000	0.000	0.000	0.009	0.061	0.151	0.134	0.078	0.029	0.000	0.000
17	0.000	0.000	0.000	0.000	0.000	0.003	0.032	0.120	0.165	0.124	0.062	0.002	0.000
18	0.000	0.000	0.000	0.000	0.000	0.001	0.014	0.080	0.171	0.165	0.111	0.007	0.000
19	0.000	0.000	0.000	0.000	0.000	0.000	0.005	0.044	0.147	0.183	0.163	0.024	0.001
20	0.000	0.000	0.000	0.000	0.000	0.000	0.002	0.020	0.103	0.165	0.196	0.065	0.006
21	0.000	0.000	0.000	0.000	0.000	0.000	0.000	0.007	0.057	0.118	0.187	0.138	0.027
22	0.000	0.000	0.000	0.000	0.000	0.000	0.000	0.002	0.024	0.064	0.136	0.226	0.093
23	0.000	0.000	0.000	0.000	0.000	0.000	0.000	0.000	0.007	0.025	0.071	0.266	0.231
24	0.000	0.000	0.000	0.000	0.000	0.000	0.000	0.000	0.001	0.006	0.024	0.199	0.365
25	0.000	0.000	0.000	0.000	0.000	0.000	0.000	0.000	0.000	0.001	0.004	0.072	0.277

TABLE 3

STANDARD NORMAL TABLE

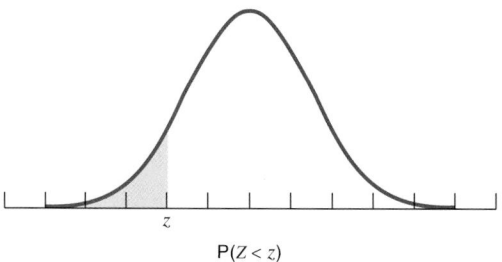

$P(Z < z)$

Second Decimal Place

z	0.00	0.01	0.02	0.03	0.04	0.05	0.06	0.07	0.08	0.09
−3.9	0.0000	0.0000	0.0000	0.0000	0.0000	0.0000	0.0000	0.0000	0.0000	0.0000
−3.8	0.0001	0.0001	0.0001	0.0001	0.0001	0.0001	0.0001	0.0001	0.0001	0.0001
−3.7	0.0001	0.0001	0.0001	0.0001	0.0001	0.0001	0.0001	0.0001	0.0001	0.0001
−3.6	0.0002	0.0002	0.0001	0.0001	0.0001	0.0001	0.0001	0.0001	0.0001	0.0001
−3.5	0.0002	0.0002	0.0002	0.0002	0.0002	0.0002	0.0002	0.0002	0.0002	0.0002
−3.4	0.0003	0.0003	0.0003	0.0003	0.0003	0.0003	0.0003	0.0003	0.0003	0.0002
−3.3	0.0005	0.0005	0.0005	0.0004	0.0004	0.0004	0.0004	0.0004	0.0004	0.0003
−3.2	0.0007	0.0007	0.0006	0.0006	0.0006	0.0006	0.0006	0.0005	0.0005	0.0005
−3.1	0.0010	0.0009	0.0009	0.0009	0.0008	0.0008	0.0008	0.0008	0.0007	0.0007
−3.0	0.0013	0.0013	0.0013	0.0012	0.0012	0.0011	0.0011	0.0011	0.0010	0.0010
−2.9	0.0019	0.0018	0.0018	0.0017	0.0016	0.0016	0.0015	0.0015	0.0014	0.0014
−2.8	0.0026	0.0025	0.0024	0.0023	0.0023	0.0022	0.0021	0.0021	0.0020	0.0019
−2.7	0.0035	0.0034	0.0033	0.0032	0.0031	0.0030	0.0029	0.0028	0.0027	0.0026
−2.6	0.0047	0.0045	0.0044	0.0043	0.0041	0.0040	0.0039	0.0038	0.0037	0.0036
−2.5	0.0062	0.0060	0.0059	0.0057	0.0055	0.0054	0.0052	0.0051	0.0049	0.0048
−2.4	0.0082	0.0080	0.0078	0.0075	0.0073	0.0071	0.0069	0.0068	0.0066	0.0064
−2.3	0.0107	0.0104	0.0102	0.0099	0.0096	0.0094	0.0091	0.0089	0.0087	0.0084
−2.2	0.0139	0.0136	0.0132	0.0129	0.0125	0.0122	0.0119	0.0116	0.0113	0.0110
−2.1	0.0179	0.0174	0.0170	0.0166	0.0162	0.0158	0.0154	0.0150	0.0146	0.0143
−2.0	0.0228	0.0222	0.0217	0.0212	0.0207	0.0202	0.0197	0.0192	0.0188	0.0183
−1.9	0.0287	0.0281	0.0274	0.0268	0.0262	0.0256	0.0250	0.0244	0.0239	0.0233
−1.8	0.0359	0.0351	0.0344	0.0336	0.0329	0.0322	0.0314	0.0307	0.0301	0.0294
−1.7	0.0446	0.0436	0.0427	0.0418	0.0409	0.0401	0.0392	0.0384	0.0375	0.0367
−1.6	0.0548	0.0537	0.0526	0.0516	0.0505	0.0495	0.0485	0.0475	0.0465	0.0455
−1.5	0.0668	0.0655	0.0643	0.0630	0.0618	0.0606	0.0594	0.0582	0.0571	0.0559
−1.4	0.0808	0.0793	0.0778	0.0764	0.0749	0.0735	0.0721	0.0708	0.0694	0.0681
−1.3	0.0968	0.0951	0.0934	0.0918	0.0901	0.0885	0.0869	0.0853	0.0838	0.0823
−1.2	0.1151	0.1131	0.1112	0.1093	0.1075	0.1056	0.1038	0.1020	0.1003	0.0985
−1.1	0.1357	0.1335	0.1314	0.1292	0.1271	0.1251	0.1230	0.1210	0.1190	0.1170
−1.0	0.1587	0.1562	0.1539	0.1515	0.1492	0.1469	0.1446	0.1423	0.1401	0.1379
−0.9	0.1841	0.1814	0.1788	0.1762	0.1736	0.1711	0.1685	0.1660	0.1635	0.1611
−0.8	0.2119	0.2090	0.2061	0.2033	0.2005	0.1977	0.1949	0.1922	0.1894	0.1867
−0.7	0.2420	0.2389	0.2358	0.2327	0.2296	0.2266	0.2236	0.2206	0.2177	0.2148
−0.6	0.2743	0.2709	0.2676	0.2643	0.2611	0.2578	0.2546	0.2514	0.2483	0.2451
−0.5	0.3085	0.3050	0.3015	0.2981	0.2946	0.2912	0.2877	0.2843	0.2810	0.2776
−0.4	0.3446	0.3409	0.3372	0.3336	0.3300	0.3264	0.3228	0.3192	0.3156	0.3121
−0.3	0.3821	0.3783	0.3745	0.3707	0.3669	0.3632	0.3594	0.3557	0.3520	0.3483
−0.2	0.4207	0.4168	0.4129	0.4090	0.4052	0.4013	0.3974	0.3936	0.3897	0.3859
−0.1	0.4602	0.4562	0.4522	0.4483	0.4443	0.4404	0.4364	0.4325	0.4286	0.4247
0.0	0.5000	0.4960	0.4920	0.4880	0.4840	0.4801	0.4761	0.4721	0.4681	0.4641

TABLE 3

(CONTINUED)

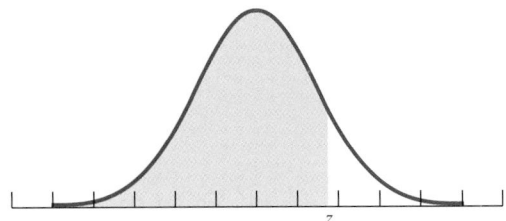

$P(Z < z)$

Second Decimal Place

z	0.00	0.01	0.02	0.03	0.04	0.05	0.06	0.07	0.08	0.09
0.0	0.5000	0.5040	0.5080	0.5120	0.5160	0.5199	0.5239	0.5279	0.5319	0.5359
0.1	0.5398	0.5438	0.5478	0.5517	0.5557	0.5596	0.5636	0.5675	0.5714	0.5753
0.2	0.5793	0.5832	0.5871	0.5910	0.5948	0.5987	0.6026	0.6064	0.6103	0.6141
0.3	0.6179	0.6217	0.6255	0.6293	0.6331	0.6368	0.6406	0.6443	0.6480	0.6517
0.4	0.6554	0.6591	0.6628	0.6664	0.6700	0.6736	0.6772	0.6808	0.6844	0.6879
0.5	0.6915	0.6950	0.6985	0.7019	0.7054	0.7088	0.7123	0.7157	0.7190	0.7224
0.6	0.7257	0.7291	0.7324	0.7357	0.7389	0.7422	0.7454	0.7486	0.7517	0.7549
0.7	0.7580	0.7611	0.7642	0.7673	0.7704	0.7734	0.7764	0.7794	0.7823	0.7852
0.8	0.7881	0.7910	0.7939	0.7967	0.7995	0.8023	0.8051	0.8078	0.8106	0.8133
0.9	0.8159	0.8186	0.8212	0.8238	0.8264	0.8289	0.8315	0.8340	0.8365	0.8389
1.0	0.8413	0.8438	0.8461	0.8485	0.8508	0.8531	0.8554	0.8577	0.8599	0.8621
1.1	0.8643	0.8665	0.8686	0.8708	0.8729	0.8749	0.8770	0.8790	0.8810	0.8830
1.2	0.8849	0.8869	0.8888	0.8907	0.8925	0.8944	0.8962	0.8980	0.8997	0.9015
1.3	0.9032	0.9049	0.9066	0.9082	0.9099	0.9115	0.9131	0.9147	0.9162	0.9177
1.4	0.9192	0.9207	0.9222	0.9236	0.9251	0.9265	0.9279	0.9292	0.9306	0.9319
1.5	0.9332	0.9345	0.9357	0.9370	0.9382	0.9394	0.9406	0.9418	0.9429	0.9441
1.6	0.9452	0.9463	0.9474	0.9484	0.9495	0.9505	0.9515	0.9525	0.9535	0.9545
1.7	0.9554	0.9564	0.9573	0.9582	0.9591	0.9599	0.9608	0.9616	0.9625	0.9633
1.8	0.9641	0.9649	0.9656	0.9664	0.9671	0.9678	0.9686	0.9693	0.9699	0.9706
1.9	0.9713	0.9719	0.9726	0.9732	0.9738	0.9744	0.9750	0.9756	0.9761	0.9767
2.0	0.9772	0.9778	0.9783	0.9788	0.9793	0.9798	0.9803	0.9808	0.9812	0.9817
2.1	0.9821	0.9826	0.9830	0.9834	0.9838	0.9842	0.9846	0.9850	0.9854	0.9857
2.2	0.9861	0.9864	0.9868	0.9871	0.9875	0.9878	0.9881	0.9884	0.9887	0.9890
2.3	0.9893	0.9896	0.9898	0.9901	0.9904	0.9906	0.9909	0.9911	0.9913	0.9916
2.4	0.9918	0.9920	0.9922	0.9925	0.9927	0.9929	0.9931	0.9932	0.9934	0.9936
2.5	0.9938	0.9940	0.9941	0.9943	0.9945	0.9946	0.9948	0.9949	0.9951	0.9952
2.6	0.9953	0.9955	0.9956	0.9957	0.9959	0.9960	0.9961	0.9962	0.9963	0.9964
2.7	0.9965	0.9966	0.9967	0.9968	0.9969	0.9970	0.9971	0.9972	0.9973	0.9974
2.8	0.9974	0.9975	0.9976	0.9977	0.9977	0.9978	0.9979	0.9979	0.9980	0.9981
2.9	0.9981	0.9982	0.9982	0.9983	0.9984	0.9984	0.9985	0.9985	0.9986	0.9986
3.0	0.9987	0.9987	0.9987	0.9988	0.9988	0.9989	0.9989	0.9989	0.9990	0.9990
3.1	0.9990	0.9991	0.9991	0.9991	0.9992	0.9992	0.9992	0.9992	0.9993	0.9993
3.2	0.9993	0.9993	0.9994	0.9994	0.9994	0.9994	0.9994	0.9995	0.9995	0.9995
3.3	0.9995	0.9995	0.9995	0.9996	0.9996	0.9996	0.9996	0.9996	0.9996	0.9997
3.4	0.9997	0.9997	0.9997	0.9997	0.9997	0.9997	0.9997	0.9997	0.9997	0.9998
3.5	0.9998	0.9998	0.9998	0.9998	0.9998	0.9998	0.9998	0.9998	0.9998	0.9998
3.6	0.9998	0.9998	0.9999	0.9999	0.9999	0.9999	0.9999	0.9999	0.9999	0.9999
3.7	0.9999	0.9999	0.9999	0.9999	0.9999	0.9999	0.9999	0.9999	0.9999	0.9999
3.8	0.9999	0.9999	0.9999	0.9999	0.9999	0.9999	0.9999	0.9999	0.9999	0.9999
3.9	1.0000	1.0000	1.0000	1.0000	1.0000	1.0000	1.0000	1.0000	1.0000	1.0000

TABLE 4

t CRITICAL VALUES

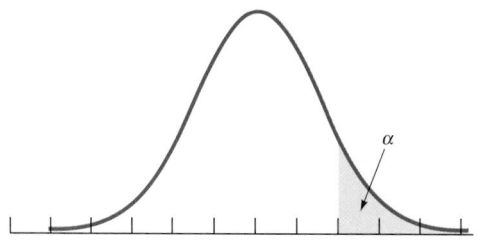

Degrees of Freedom	Upper Tail Probability (α)								
	0.15	0.10	0.05	0.025	0.015	0.01	0.005	0.001	0.0005
1	1.963	3.078	6.314	12.706	21.205	31.821	63.657	318.309	1273.155
2	1.386	1.886	2.920	4.303	5.643	6.965	9.925	22.327	44.703
3	1.250	1.638	2.353	3.182	3.896	4.541	5.841	10.215	16.326
4	1.190	1.533	2.132	2.776	3.298	3.747	4.604	7.173	10.305
5	1.156	1.476	2.015	2.571	3.003	3.365	4.032	5.893	7.976
6	1.134	1.440	1.943	2.447	2.829	3.143	3.707	5.208	6.788
7	1.119	1.415	1.895	2.365	2.715	2.998	3.499	4.785	6.082
8	1.108	1.397	1.860	2.306	2.634	2.896	3.355	4.501	5.617
9	1.100	1.383	1.833	2.262	2.574	2.821	3.250	4.297	5.291
10	1.093	1.372	1.812	2.228	2.527	2.764	3.169	4.144	5.049
11	1.088	1.363	1.796	2.201	2.491	2.718	3.106	4.025	4.863
12	1.083	1.356	1.782	2.179	2.461	2.681	3.055	3.930	4.717
13	1.079	1.350	1.771	2.160	2.436	2.650	3.012	3.852	4.597
14	1.076	1.345	1.761	2.145	2.415	2.625	2.977	3.787	4.499
15	1.074	1.341	1.753	2.131	2.397	2.602	2.947	3.733	4.417
16	1.071	1.337	1.746	2.120	2.382	2.583	2.921	3.686	4.346
17	1.069	1.333	1.740	2.110	2.368	2.567	2.898	3.646	4.286
18	1.067	1.330	1.734	2.101	2.356	2.552	2.878	3.611	4.233
19	1.066	1.328	1.729	2.093	2.346	2.539	2.861	3.579	4.187
20	1.064	1.325	1.725	2.086	2.336	2.528	2.845	3.552	4.146
21	1.063	1.323	1.721	2.080	2.328	2.518	2.831	3.527	4.109
22	1.061	1.321	1.717	2.074	2.320	2.508	2.819	3.505	4.077
23	1.060	1.319	1.714	2.069	2.313	2.500	2.807	3.485	4.047
24	1.059	1.318	1.711	2.064	2.307	2.492	2.797	3.467	4.021
25	1.058	1.316	1.708	2.060	2.301	2.485	2.787	3.450	3.997
26	1.058	1.315	1.706	2.056	2.296	2.479	2.779	3.435	3.974
27	1.057	1.314	1.703	2.052	2.291	2.473	2.771	3.421	3.954
28	1.056	1.313	1.701	2.048	2.286	2.467	2.763	3.408	3.935
29	1.055	1.311	1.699	2.045	2.282	2.462	2.756	3.396	3.918
30	1.055	1.310	1.697	2.042	2.278	2.457	2.750	3.385	3.902
40	1.050	1.303	1.684	2.021	2.250	2.423	2.704	3.307	3.788
50	1.047	1.299	1.676	2.009	2.234	2.403	2.678	3.261	3.723
60	1.045	1.296	1.671	2.000	2.223	2.390	2.660	3.232	3.681
120	1.041	1.289	1.658	1.980	2.196	2.358	2.617	3.160	3.578
Z critical value	1.036	1.282	1.645	1.960	2.170	2.326	2.576	3.090	3.290
Level of Significance for a one-tailed test	0.15	0.10	0.05	0.025	0.015	0.01	0.005	0.001	0.0005
Level of Significance for a two-tailed test	0.30	0.20	0.10	0.05	0.03	0.02	0.01	0.002	0.001

TABLE 5

CHI-SQUARE DISTRIBUTION TABLE

The entries in this table give the critical values of χ^2 for the specified number of degrees of freedom and areas in the right tail.

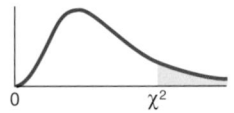

Degrees of Freedom	Upper Tail Areas											
	0.995	0.99	0.975	0.95	0.9	0.75	0.25	0.1	0.05	0.025	0.01	0.005
1	0.000	0.000	0.001	0.004	0.016	0.102	1.323	2.706	3.841	5.024	6.635	7.879
2	0.010	0.020	0.051	0.103	0.211	0.575	2.773	4.605	5.991	7.378	9.210	10.597
3	0.072	0.115	0.216	0.352	0.584	1.213	4.108	6.251	7.815	9.348	11.345	12.838
4	0.207	0.297	0.484	0.711	1.064	1.923	5.385	7.779	9.488	11.143	13.277	14.860
5	0.412	0.554	0.831	1.145	1.610	2.675	6.626	9.236	11.070	12.832	15.086	16.750
6	0.676	0.872	1.237	1.635	2.204	3.455	7.841	10.645	12.592	14.449	16.812	18.548
7	0.989	1.239	1.690	2.167	2.833	4.255	9.037	12.017	14.067	16.013	18.475	20.278
8	1.344	1.647	2.180	2.733	3.490	5.071	10.219	13.362	15.507	17.535	20.090	21.955
9	1.735	2.088	2.700	3.325	4.168	5.899	11.389	14.684	16.919	19.023	21.666	23.589
10	2.156	2.558	3.247	3.940	4.865	6.737	12.549	15.987	18.307	20.483	23.209	25.188
11	2.603	3.053	3.816	4.575	5.578	7.584	13.701	17.275	19.675	21.920	24.725	26.757
12	3.074	3.571	4.404	5.226	6.304	8.438	14.845	18.549	21.026	23.337	26.217	28.300
13	3.565	4.107	5.009	5.892	7.041	9.299	15.984	19.812	22.362	24.736	27.688	29.819
14	4.075	4.660	5.629	6.571	7.790	10.165	17.117	21.064	23.685	26.119	29.141	31.319
15	4.601	5.229	6.262	7.261	8.547	11.037	18.245	22.307	24.996	27.488	30.578	32.801
16	5.142	5.812	6.908	7.962	9.312	11.912	19.369	23.542	26.296	28.845	32.000	34.267
17	5.697	6.408	7.564	8.672	10.085	12.792	20.489	24.769	27.587	30.191	33.409	35.718
18	6.265	7.015	8.231	9.390	10.865	13.675	21.605	25.989	28.869	31.526	34.805	37.156
19	6.844	7.633	8.907	10.117	11.651	14.562	22.718	27.204	30.144	32.852	36.191	38.582
20	7.434	8.260	9.591	10.851	12.443	15.452	23.828	28.412	31.410	34.170	37.566	39.997
21	8.034	8.897	10.283	11.591	13.240	16.344	24.935	29.615	32.671	35.479	38.932	41.401
22	8.643	9.542	10.982	12.338	14.041	17.240	26.039	30.813	33.924	36.781	40.289	42.796
23	9.260	10.196	11.689	13.091	14.848	18.137	27.141	32.007	35.172	38.076	41.638	44.181
24	9.886	10.856	12.401	13.848	15.659	19.037	28.241	33.196	36.415	39.364	42.980	45.558
25	10.520	11.524	13.120	14.611	16.473	19.939	29.339	34.382	37.652	40.646	44.314	46.928
26	11.160	12.198	13.844	15.379	17.292	20.843	30.435	35.563	38.885	41.923	45.642	48.290
27	11.808	12.878	14.573	16.151	18.114	21.749	31.528	36.741	40.113	43.195	46.963	49.645
28	12.461	13.565	15.308	16.928	18.939	22.657	32.620	37.916	41.337	44.461	48.278	50.994
29	13.121	14.256	16.047	17.708	19.768	23.567	33.711	39.087	42.557	45.722	49.588	52.335
30	13.787	14.953	16.791	18.493	20.599	24.478	34.800	40.256	43.773	46.979	50.892	53.672
31	14.458	15.655	17.539	19.281	21.434	25.390	35.887	41.422	44.985	48.232	52.191	55.002
32	15.134	16.362	18.291	20.072	22.271	26.304	36.973	42.585	46.194	49.480	53.486	56.328
33	15.815	17.073	19.047	20.867	23.110	27.219	38.058	43.745	47.400	50.725	54.775	57.648
34	16.501	17.789	19.806	21.664	23.952	28.136	39.141	44.903	48.602	51.966	56.061	58.964
35	17.192	18.509	20.569	22.465	24.797	29.054	40.223	46.059	49.802	53.203	57.342	60.275
40	20.707	22.164	24.433	26.509	29.051	33.660	45.616	51.805	55.758	59.342	63.691	66.766
60	35.534	37.485	40.482	43.188	46.459	52.294	66.981	74.397	79.082	83.298	88.379	91.952
120	83.852	86.923	91.573	95.705	100.624	109.220	130.055	140.233	146.567	152.211	158.950	163.648

TABLE 6

THE F DISTRIBUTION TABLE

a. Area in the Right Tail under the F Distribution Curve = 0.01

Degrees of Freedom Denominator	Degrees of Freedom – Numerator																
	1	2	3	4	5	6	7	8	9	10	11	12	13	14	15	16	17
1	4052.185	4999.340	5403.534	5624.257	5763.955	5858.950	5928.334	5980.954	6022.397	6055.925	6083.399	6106.682	6125.774	6143.004	6156.974	6170.012	6181.188
2	98.502	99.000	99.164	99.251	99.302	99.331	99.357	99.375	99.390	99.397	99.408	99.419	99.422	99.426	99.433	99.437	99.441
3	34.116	30.816	29.457	28.710	28.237	27.911	27.671	27.489	27.345	27.228	27.132	27.052	26.983	26.924	26.872	26.826	26.786
4	21.198	18.000	16.694	15.977	15.522	15.207	14.976	14.799	14.659	14.546	14.452	14.374	14.306	14.249	14.198	14.154	14.114
5	16.258	13.274	12.060	11.392	10.967	10.672	10.456	10.289	10.158	10.051	9.963	9.888	9.825	9.770	9.722	9.680	9.643
6	13.745	10.925	9.780	9.148	8.746	8.466	8.260	8.102	7.976	7.874	7.790	7.718	7.657	7.605	7.559	7.519	7.483
7	12.246	9.547	8.451	7.847	7.460	7.191	6.993	6.840	6.719	6.620	6.538	6.469	6.410	6.359	6.314	6.275	6.240
8	11.259	8.649	7.591	7.006	6.632	6.371	6.178	6.029	5.911	5.814	5.734	5.667	5.609	5.559	5.515	5.477	5.442
9	10.562	8.022	6.992	6.422	6.057	5.802	5.613	5.467	5.351	5.257	5.178	5.111	5.055	5.005	4.962	4.924	4.890
10	10.044	7.559	6.552	5.994	5.636	5.386	5.200	5.057	4.942	4.849	4.772	4.706	4.650	4.601	4.558	4.520	4.487
11	9.646	7.206	6.217	5.668	5.316	5.069	4.886	4.744	4.632	4.539	4.462	4.397	4.342	4.293	4.251	4.213	4.180
12	9.330	6.927	5.953	5.412	5.064	4.821	4.640	4.499	4.388	4.296	4.220	4.155	4.100	4.052	4.010	3.972	3.939
13	9.074	6.701	5.739	5.205	4.862	4.620	4.441	4.302	4.191	4.100	4.025	3.960	3.905	3.857	3.815	3.778	3.745
14	8.862	6.515	5.564	5.035	4.695	4.456	4.278	4.140	4.030	3.939	3.864	3.800	3.745	3.698	3.656	3.619	3.586
15	8.683	6.359	5.417	4.893	4.556	4.318	4.142	4.004	3.895	3.805	3.730	3.666	3.612	3.564	3.522	3.485	3.452
16	8.531	6.226	5.292	4.773	4.437	4.202	4.026	3.890	3.780	3.691	3.616	3.553	3.498	3.451	3.409	3.372	3.339
17	8.400	6.112	5.185	4.669	4.336	4.101	3.927	3.791	3.682	3.593	3.518	3.455	3.401	3.353	3.312	3.275	3.242
18	8.285	6.013	5.092	4.579	4.248	4.015	3.841	3.705	3.597	3.508	3.434	3.371	3.316	3.269	3.227	3.190	3.158
19	8.185	5.926	5.010	4.500	4.171	3.939	3.765	3.631	3.523	3.434	3.360	3.297	3.242	3.195	3.153	3.116	3.084
20	8.096	5.849	4.938	4.431	4.103	3.871	3.699	3.564	3.457	3.368	3.294	3.231	3.177	3.130	3.088	3.051	3.018
21	8.017	5.780	4.874	4.369	4.042	3.812	3.640	3.506	3.398	3.310	3.236	3.173	3.119	3.072	3.030	2.993	2.960
22	7.945	5.719	4.817	4.313	3.988	3.758	3.587	3.453	3.346	3.258	3.184	3.121	3.067	3.019	2.978	2.941	2.908
23	7.881	5.664	4.765	4.264	3.939	3.710	3.539	3.406	3.299	3.211	3.137	3.074	3.020	2.973	2.931	2.894	2.861
24	7.823	5.614	4.718	4.218	3.895	3.667	3.496	3.363	3.256	3.168	3.094	3.032	2.977	2.930	2.889	2.852	2.819
25	7.770	5.568	4.675	4.177	3.855	3.627	3.457	3.324	3.217	3.129	3.056	2.993	2.939	2.892	2.850	2.813	2.780
26	7.721	5.526	4.637	4.140	3.818	3.591	3.421	3.288	3.182	3.094	3.021	2.958	2.904	2.857	2.815	2.778	2.745
27	7.677	5.488	4.601	4.106	3.785	3.558	3.388	3.256	3.149	3.062	2.988	2.926	2.872	2.824	2.783	2.746	2.713
28	7.636	5.453	4.568	4.074	3.754	3.528	3.358	3.226	3.120	3.032	2.959	2.896	2.842	2.795	2.753	2.716	2.683
29	7.598	5.420	4.538	4.045	3.725	3.499	3.330	3.198	3.092	3.005	2.931	2.868	2.814	2.767	2.726	2.689	2.656
30	7.562	5.390	4.510	4.018	3.699	3.473	3.305	3.173	3.067	2.979	2.906	2.843	2.789	2.742	2.700	2.663	2.630
40	7.314	5.178	4.313	3.828	3.514	3.291	3.124	2.993	2.888	2.801	2.727	2.665	2.611	2.563	2.522	2.484	2.451
60	7.077	4.977	4.126	3.649	3.339	3.119	2.953	2.823	2.718	2.632	2.559	2.496	2.442	2.394	2.352	2.315	2.281
120	6.851	4.787	3.949	3.480	3.174	2.956	2.792	2.663	2.559	2.472	2.399	2.336	2.282	2.234	2.191	2.154	2.119

Degrees of Freedom—Numerator

Degrees of Freedom Denominator	18	19	20	21	22	23	24	25	26	27	28	29	30	40	60	120
1	6191.432	6200.746	6208.662	6216.113	6223.097	6228.685	6234.273	6239.861	6244.518	6249.174	6252.900	6257.091	6260.350	6286.427	6312.970	6339.513
2	99.444	99.448	99.448	99.451	99.455	99.455	99.455	99.459	99.462	99.462	99.462	99.462	99.466	99.477	99.484	99.491
3	26.751	26.719	26.690	26.664	26.639	26.617	26.597	26.579	26.562	26.546	26.531	26.517	26.504	26.411	26.316	26.221
4	14.079	14.048	14.019	13.994	13.970	13.949	13.929	13.911	13.894	13.878	13.864	13.850	13.838	13.745	13.652	13.558
5	9.609	9.580	9.553	9.528	9.506	9.485	9.466	9.449	9.433	9.418	9.404	9.391	9.379	9.291	9.202	9.112
6	7.451	7.422	7.396	7.372	7.351	7.331	7.313	7.296	7.281	7.266	7.253	7.240	7.229	7.143	7.057	6.969
7	6.209	6.181	6.155	6.132	6.111	6.092	6.074	6.058	6.043	6.029	6.016	6.003	5.992	5.908	5.824	5.737
8	5.412	5.384	5.359	5.336	5.316	5.297	5.279	5.263	5.248	5.234	5.221	5.209	5.198	5.116	5.032	4.946
9	4.860	4.833	4.808	4.786	4.765	4.746	4.729	4.713	4.698	4.684	4.672	4.660	4.649	4.567	4.483	4.398
10	4.457	4.430	4.405	4.383	4.363	4.344	4.327	4.311	4.296	4.283	4.270	4.258	4.247	4.165	4.082	3.996
11	4.150	4.123	4.099	4.077	4.057	4.038	4.021	4.005	3.990	3.977	3.964	3.952	3.941	3.860	3.776	3.690
12	3.910	3.883	3.858	3.836	3.816	3.798	3.780	3.765	3.750	3.736	3.724	3.712	3.701	3.619	3.535	3.449
13	3.716	3.689	3.665	3.643	3.622	3.604	3.587	3.571	3.556	3.543	3.530	3.518	3.507	3.425	3.341	3.255
14	3.556	3.529	3.505	3.483	3.463	3.444	3.427	3.412	3.397	3.383	3.371	3.359	3.348	3.266	3.181	3.094
15	3.423	3.396	3.372	3.350	3.330	3.311	3.294	3.278	3.264	3.250	3.237	3.225	3.214	3.132	3.047	2.959
16	3.310	3.283	3.259	3.237	3.216	3.198	3.181	3.165	3.150	3.137	3.124	3.112	3.101	3.018	2.933	2.845
17	3.212	3.186	3.162	3.139	3.119	3.101	3.083	3.068	3.053	3.039	3.026	3.014	3.003	2.920	2.835	2.746
18	3.128	3.101	3.077	3.055	3.035	3.016	2.999	2.983	2.968	2.955	2.942	2.930	2.919	2.835	2.749	2.660
19	3.054	3.027	3.003	2.981	2.961	2.942	2.925	2.909	2.894	2.880	2.868	2.855	2.844	2.761	2.674	2.584
20	2.989	2.962	2.938	2.916	2.895	2.877	2.859	2.843	2.829	2.815	2.802	2.790	2.778	2.695	2.608	2.517
21	2.931	2.904	2.880	2.857	2.837	2.818	2.801	2.785	2.770	2.756	2.743	2.731	2.720	2.636	2.548	2.457
22	2.879	2.852	2.827	2.805	2.785	2.766	2.749	2.733	2.718	2.704	2.691	2.679	2.667	2.583	2.495	2.403
23	2.832	2.805	2.780	2.758	2.738	2.719	2.702	2.686	2.671	2.657	2.644	2.632	2.620	2.536	2.447	2.354
24	2.789	2.762	2.738	2.716	2.695	2.676	2.659	2.643	2.628	2.614	2.601	2.589	2.577	2.492	2.403	2.310
25	2.751	2.724	2.699	2.677	2.657	2.638	2.620	2.604	2.589	2.575	2.562	2.550	2.538	2.453	2.364	2.270
26	2.715	2.688	2.664	2.642	2.621	2.602	2.585	2.569	2.554	2.540	2.526	2.514	2.503	2.417	2.327	2.233
27	2.683	2.656	2.632	2.609	2.589	2.570	2.552	2.536	2.521	2.507	2.494	2.481	2.470	2.384	2.294	2.198
28	2.653	2.626	2.602	2.579	2.559	2.540	2.522	2.506	2.491	2.477	2.464	2.451	2.440	2.354	2.263	2.167
29	2.626	2.599	2.574	2.552	2.531	2.512	2.495	2.478	2.463	2.449	2.436	2.423	2.412	2.325	2.234	2.138
30	2.600	2.573	2.549	2.526	2.506	2.487	2.469	2.453	2.437	2.423	2.410	2.398	2.386	2.299	2.208	2.111
40	2.421	2.394	2.369	2.346	2.325	2.306	2.288	2.271	2.256	2.241	2.228	2.215	2.203	2.114	2.019	1.917
60	2.251	2.223	2.198	2.175	2.153	2.134	2.115	2.098	2.083	2.068	2.054	2.041	2.028	1.936	1.836	1.726
120	2.089	2.060	2.035	2.011	1.989	1.969	1.950	1.932	1.916	1.901	1.886	1.873	1.860	1.763	1.656	1.533

TABLE 6

(CONTINUED)

b. Area in the Right Tail under the F Distribution Curve = 0.05

Degrees of Freedom Denominator	Degrees of Freedom—Numerator																
	1	2	3	4	5	6	7	8	9	10	11	12	13	14	15	16	17
1	161.446	199.499	215.707	224.583	230.160	233.988	236.767	238.884	240.543	241.882	242.981	243.905	244.690	245.363	245.949	246.466	246.917
2	18.513	19.000	19.164	19.247	19.296	19.329	19.353	19.371	19.385	19.396	19.405	19.412	19.419	19.424	19.429	19.433	19.437
3	10.128	9.552	9.277	9.117	9.013	8.941	8.887	8.845	8.812	8.785	8.763	8.745	8.729	8.715	8.703	8.692	8.683
4	7.709	6.944	6.591	6.388	6.256	6.163	6.094	6.041	5.999	5.964	5.936	5.912	5.891	5.873	5.858	5.844	5.832
5	6.608	5.786	5.409	5.192	5.050	4.950	4.876	4.818	4.772	4.735	4.704	4.678	4.655	4.636	4.619	4.604	4.590
6	5.987	5.143	4.757	4.534	4.387	4.284	4.207	4.147	4.099	4.060	4.027	4.000	3.976	3.956	3.938	3.922	3.908
7	5.591	4.737	4.347	4.120	3.972	3.866	3.787	3.726	3.677	3.637	3.603	3.575	3.550	3.529	3.511	3.494	3.480
8	5.318	4.459	4.066	3.838	3.688	3.581	3.500	3.438	3.388	3.347	3.313	3.284	3.259	3.237	3.218	3.202	3.187
9	5.117	4.256	3.863	3.633	3.482	3.374	3.293	3.230	3.179	3.137	3.102	3.073	3.048	3.025	3.006	2.989	2.974
10	4.965	4.103	3.708	3.478	3.326	3.217	3.135	3.072	3.020	2.978	2.943	2.913	2.887	2.865	2.845	2.828	2.812
11	4.844	3.982	3.587	3.357	3.204	3.095	3.012	2.948	2.896	2.854	2.818	2.788	2.761	2.739	2.719	2.701	2.685
12	4.747	3.885	3.490	3.259	3.106	2.996	2.913	2.849	2.796	2.753	2.717	2.687	2.660	2.637	2.617	2.599	2.583
13	4.667	3.806	3.411	3.179	3.025	2.915	2.832	2.767	2.714	2.671	2.635	2.604	2.577	2.554	2.533	2.515	2.499
14	4.600	3.739	3.344	3.112	2.958	2.848	2.764	2.699	2.646	2.602	2.565	2.534	2.507	2.484	2.463	2.445	2.428
15	4.543	3.682	3.287	3.056	2.901	2.790	2.707	2.641	2.588	2.544	2.507	2.475	2.448	2.424	2.403	2.385	2.368
16	4.494	3.634	3.239	3.007	2.852	2.741	2.657	2.591	2.538	2.494	2.456	2.425	2.397	2.373	2.352	2.333	2.317
17	4.451	3.592	3.197	2.965	2.810	2.699	2.614	2.548	2.494	2.450	2.413	2.381	2.353	2.329	2.308	2.289	2.272
18	4.414	3.555	3.160	2.928	2.773	2.661	2.577	2.510	2.456	2.412	2.374	2.342	2.314	2.290	2.269	2.250	2.233
19	4.381	3.522	3.127	2.895	2.740	2.628	2.544	2.477	2.423	2.378	2.340	2.308	2.280	2.256	2.234	2.215	2.198
20	4.351	3.493	3.098	2.866	2.711	2.599	2.514	2.447	2.393	2.348	2.310	2.278	2.250	2.225	2.203	2.184	2.167
21	4.325	3.467	3.072	2.840	2.685	2.573	2.488	2.420	2.366	2.321	2.283	2.250	2.222	2.197	2.176	2.156	2.139
22	4.301	3.443	3.049	2.817	2.661	2.549	2.464	2.397	2.342	2.297	2.259	2.226	2.198	2.173	2.151	2.131	2.114
23	4.279	3.422	3.028	2.796	2.640	2.528	2.442	2.375	2.320	2.275	2.236	2.204	2.175	2.150	2.128	2.109	2.091
24	4.260	3.403	3.009	2.776	2.621	2.508	2.423	2.355	2.300	2.255	2.216	2.183	2.155	2.130	2.108	2.088	2.070
25	4.242	3.385	2.991	2.759	2.603	2.490	2.405	2.337	2.282	2.236	2.198	2.165	2.136	2.111	2.089	2.069	2.051
26	4.225	3.369	2.975	2.743	2.587	2.474	2.388	2.321	2.265	2.220	2.181	2.148	2.119	2.094	2.072	2.052	2.034
27	4.210	3.354	2.960	2.728	2.572	2.459	2.373	2.305	2.250	2.204	2.166	2.132	2.103	2.078	2.056	2.036	2.018
28	4.196	3.340	2.947	2.714	2.558	2.445	2.359	2.291	2.236	2.190	2.151	2.118	2.089	2.064	2.041	2.021	2.003
29	4.183	3.328	2.934	2.701	2.545	2.432	2.346	2.278	2.223	2.177	2.138	2.104	2.075	2.050	2.027	2.007	1.989
30	4.171	3.316	2.922	2.690	2.534	2.421	2.334	2.266	2.211	2.165	2.126	2.092	2.063	2.037	2.015	1.995	1.976
40	4.085	3.232	2.839	2.606	2.449	2.336	2.249	2.180	2.124	2.077	2.038	2.003	1.974	1.948	1.924	1.904	1.885
60	4.001	3.150	2.758	2.525	2.368	2.254	2.167	2.097	2.040	1.993	1.952	1.917	1.887	1.860	1.836	1.815	1.796
120	3.920	3.072	2.680	2.447	2.290	2.175	2.087	2.016	1.959	1.910	1.869	1.834	1.803	1.775	1.750	1.728	1.709

Degrees of Freedom—Numerator

Degrees of Freedom Denominator	18	19	20	21	22	23	24	25	26	27	28	29	30	40	60	120
1	247.324	247.688	248.016	248.307	248.579	248.823	249.052	249.260	249.453	249.631	249.798	249.951	250.096	251.144	252.196	253.254
2	19.440	19.443	19.446	19.448	19.450	19.452	19.454	19.456	19.457	19.459	19.460	19.461	19.463	19.471	19.479	19.487
3	8.675	8.667	8.660	8.654	8.648	8.643	8.638	8.634	8.630	8.626	8.623	8.620	8.617	8.594	8.572	8.549
4	5.821	5.811	5.803	5.795	5.787	5.781	5.774	5.769	5.763	5.759	5.754	5.750	5.746	5.717	5.688	5.658
5	4.579	4.568	4.558	4.549	4.541	4.534	4.527	4.521	4.515	4.510	4.505	4.500	4.496	4.464	4.431	4.398
6	3.896	3.884	3.874	3.865	3.856	3.849	3.841	3.835	3.829	3.823	3.818	3.813	3.808	3.774	3.740	3.705
7	3.467	3.455	3.445	3.435	3.426	3.418	3.410	3.404	3.397	3.391	3.386	3.381	3.376	3.340	3.304	3.267
8	3.173	3.161	3.150	3.140	3.131	3.123	3.115	3.108	3.102	3.095	3.090	3.084	3.079	3.043	3.005	2.967
9	2.960	2.948	2.936	2.926	2.917	2.908	2.900	2.893	2.886	2.880	2.874	2.869	2.864	2.826	2.787	2.748
10	2.798	2.785	2.774	2.764	2.754	2.745	2.737	2.730	2.723	2.716	2.710	2.705	2.700	2.661	2.621	2.580
11	2.671	2.658	2.646	2.636	2.626	2.617	2.609	2.601	2.594	2.588	2.582	2.576	2.570	2.531	2.490	2.448
12	2.568	2.555	2.544	2.533	2.523	2.514	2.505	2.498	2.491	2.484	2.478	2.472	2.466	2.426	2.384	2.341
13	2.484	2.471	2.459	2.448	2.438	2.429	2.420	2.412	2.405	2.398	2.392	2.386	2.380	2.339	2.297	2.252
14	2.413	2.400	2.388	2.377	2.367	2.357	2.349	2.341	2.333	2.326	2.320	2.314	2.308	2.266	2.223	2.178
15	2.353	2.340	2.328	2.316	2.306	2.297	2.288	2.280	2.272	2.265	2.259	2.253	2.247	2.204	2.160	2.114
16	2.302	2.288	2.276	2.264	2.254	2.244	2.235	2.227	2.220	2.212	2.206	2.200	2.194	2.151	2.106	2.059
17	2.257	2.243	2.230	2.219	2.208	2.199	2.190	2.181	2.174	2.167	2.160	2.154	2.148	2.104	2.058	2.011
18	2.217	2.203	2.191	2.179	2.168	2.159	2.150	2.141	2.134	2.126	2.119	2.113	2.107	2.063	2.017	1.968
19	2.182	2.168	2.155	2.144	2.133	2.123	2.114	2.106	2.098	2.090	2.084	2.077	2.071	2.026	1.980	1.930
20	2.151	2.137	2.124	2.112	2.102	2.092	2.082	2.074	2.066	2.059	2.052	2.045	2.039	1.994	1.946	1.896
21	2.123	2.109	2.096	2.084	2.073	2.063	2.054	2.045	2.037	2.030	2.023	2.016	2.010	1.965	1.916	1.866
22	2.098	2.084	2.071	2.059	2.048	2.038	2.028	2.020	2.012	2.004	1.997	1.990	1.984	1.938	1.889	1.838
23	2.075	2.061	2.048	2.036	2.025	2.014	2.005	1.996	1.988	1.981	1.973	1.967	1.961	1.914	1.865	1.813
24	2.054	2.040	2.027	2.015	2.003	1.993	1.984	1.975	1.967	1.959	1.952	1.945	1.939	1.892	1.842	1.790
25	2.035	2.021	2.007	1.995	1.984	1.974	1.964	1.955	1.947	1.939	1.932	1.926	1.919	1.872	1.822	1.768
26	2.018	2.003	1.990	1.978	1.966	1.956	1.946	1.938	1.929	1.921	1.914	1.907	1.901	1.853	1.803	1.749
27	2.002	1.987	1.974	1.961	1.950	1.940	1.930	1.921	1.913	1.905	1.898	1.891	1.884	1.836	1.785	1.731
28	1.987	1.972	1.959	1.946	1.935	1.924	1.915	1.906	1.897	1.889	1.882	1.875	1.869	1.820	1.769	1.714
29	1.973	1.958	1.945	1.932	1.921	1.910	1.901	1.891	1.883	1.875	1.868	1.861	1.854	1.806	1.754	1.698
30	1.960	1.945	1.932	1.919	1.908	1.897	1.887	1.878	1.870	1.862	1.854	1.847	1.841	1.792	1.740	1.683
40	1.868	1.853	1.839	1.826	1.814	1.803	1.793	1.783	1.775	1.766	1.759	1.751	1.744	1.693	1.637	1.577
60	1.778	1.763	1.748	1.735	1.722	1.711	1.700	1.690	1.681	1.672	1.664	1.656	1.649	1.594	1.534	1.467
120	1.690	1.674	1.659	1.645	1.632	1.620	1.608	1.598	1.588	1.579	1.570	1.562	1.554	1.495	1.429	1.352

TABLE 7

CRITICAL VALUES OF THE WILCOXON RANK SUM TEST

One-tail $\alpha = 0.025$; Two-tail $\alpha = 0.05$

		T_L	T_U	T_L	T_U	T_L	T_U	T_L	T_U	T_L	T_U	T_L	T_U	T_L	T_U	T_L	T_U
n_2	n_1	3		4		5		6		7		8		9		10	
4		6	18	11	25												
5		6	21	12	28	18	37										
6		7	23	12	32	19	41	26	52								
7		7	26	13	35	20	45	28	56	37	68						
8		8	28	14	38	21	49	29	61	39	73	49	87				
9		8	31	15	41	22	53	31	65	41	78	51	93	63	108		
10		9	33	16	44	24	56	32	70	43	83	54	98	66	114	79	131

One-tail $\alpha = 0.05$; Two-tail $\alpha = 0.10$

		T_L	T_U	T_L	T_U	T_L	T_U	T_L	T_U	T_L	T_U	T_L	T_U	T_L	T_U	T_L	T_U
n_2	n_1	3		4		5		6		7		8		9		10	
3		6	15														
4		7	17	12	24												
5		7	20	13	27	19	36										
6		8	22	14	30	20	40	28	50								
7		9	24	15	33	22	43	30	54	39	66						
8		9	27	16	36	24	46	32	58	41	71	52	84				
9		10	29	17	39	25	50	33	63	43	76	54	90	66	105		
10		11	31	18	42	26	54	35	67	46	80	57	95	69	111	83	127

Explanation of Large Data Sets

The Information That Follows Will Help You Understand the Data File CARS.XXX.

EXPLANATION OF THE VARIABLE NAMES IN CARS

CMSA	Household location—CMSA
HHFAMINC	Household family income category
HHLOC	MSA status
HHMSA	Household location—MSA
HHSIZE	Total number of persons in household
HH_RACE	Race of household reference person
HOUSEID	Household identification number
LIF_CYC	Family life cycle
MAKECODE	NASS Code for vehicle make
MSASIZE	Size of MSA/CMSA of household
MSTR_MON	Date of household master interview—Month
POPDNSTY	Population density of household ZIP code area
POVERTY	Household below, near, or above poverty level
REF_AGE	Age of household reference person
REF_EDUC	Education of household reference person
REF_SEX	Sex of household reference person
URBAN	Urbanized area indicator
URBNAREA	Urbanized area status
URBNSIZE	Size of urbanized area
VEH12MNT	Vehicle received in last 12 months
VEHHHOWN	Vehicle owned by household member
VEHMILES	Reported vehicle mileage previous 12 months
VEHNEW	Vehicle new or used when received
VEHTYPE	Vehicle type
VEHYEAR	Model year of vehicle

EXPLANATION OF CODED VALUES FOR VARIABLES IN CARS

CMSA Household location—CMSA

Value	Label
1122	Boston-Lawrence-Salem, MA-NH
1282	Buffalo-Niagara Falls, NY
1602	Chicago-Gary-Lake County, IL-IN-WI
1642	Cincinnati-Hamilton, OH-KY-IN
1692	Cleveland-Akron-Lorain, OH
1922	Dallas-Fort Worth, TX
2082	Denver-Boulder, CO
2162	Detroit-Ann Arbor, MI
3282	Hartford-New Britain-Middletown, CT
3362	Houston-Galveston-Brazoria, TX
4472	Los Angeles-Anaheim-Riverside, CA
4992	Miami-Fort Lauderdale, FL
5082	Milwaukee-Racine, WI
5602	New York-Northern New Jersey-Long Island, NY-NJ-CT
6162	Philadelphia-Wilmington-Trenton, PA-NJ-DE-MD
6282	Pittsburgh-Beaver Valley, PA
6442	Portland-Vancouver, OR-WA
6482	Providence-Pawtucket-Fall River, RI-MA
7362	San Francisco-Oakland-San Jose, CA
7602	Seattle-Tacoma, WA

HHFAMINC Household family income category

Value	Label
01	Less than $5,000
02	$5,000 to $9,999
03	$10,000 to $14,999
04	$15,000 to $19,999
05	$20,000 to $24,999
06	$25,000 to $29,999
07	$30,000 to $34,999
08	$35,000 to $39,999
09	$40,000 to $44,999
10	$45,000 to $49,999
11	$50,000 to $54,999
12	$55,000 to $59,999
13	$60,000 to $64,999
14	$65,000 to $69,999
15	$70,000 to $74,999
16	$75,000 to $79,999
17	$80,000 or over
98	Not Ascertained
99	Refused

HHLOC MSA status

Value	Label
1	In MSA central city
2	In MSA, not in central city
3	Not in MSA

HHSIZE Total number of persons in household

Value	Label
01	One person in household
02	Two people in household
03	Three people in household
04	Four people in household
05	Five people in household
06	Six people in household
07	Seven people in household
08	Eight people in household
09	Nine people in household
10	Ten people in household

HH_RACE Race of household reference person

Value	Label
01	White
02	Black
03	Other
98	Not Ascertained
99	Refused

HOUSEID Household identification number

(This is just an index number that is used to keep track of each interview.)

LIF_CYC Family life cycle

Value	Label
01	Single adult, no children
02	Two or more adults, no children
03	Single adult, youngest child age 0–5
04	Two or more adults, youngest child age 0–5
05	Single adult, youngest child age 6–15
06	Two or more adults, youngest child age 6–15
07	Single adult, youngest child age 16–21
08	Two or more adults, youngest child age 16–21
09	Single adult, retired, no children
10	Two or more adults, retired, no children
98	Not Ascertained

MAKECODE NASS code for vehicle make

Value	Label	Value	Label	Value	Label
001	American Motors	030	Volkswagen	048	Subaru
002	Jeep (includes Kaiser-Jeep)	031	Alfa Romeo	049	Toyota
003	AM General	032	Audi	050	Triumph
006	Chrysler	033	Austin/Austin Healey	051	Volvo
007	Dodge	034	BMW	052	Mitsubishi
008	Imperial	035	Nissan/Datsun	053	Suzuki
009	Plymouth	036	Fiat	054	Acura
010	Eagle	037	Honda	055	Hyundai
012	Ford	038	Isuzu	056	Merkur
013	Lincoln	039	Jaguar	057	Yugo
014	Mercury	040	Lancia	058	Infiniti
018	Buick	041	Mazda	059	Lexus
019	Cadillac	042	Mercedes-Benz	060	Daihatsu
020	Chevrolet	043	MG	069	Other foreign
021	Oldsmobile	044	Peugeot	084	International Harvester/Navistar
022	Pontiac	045	Porsche	089	Other medium/heavy trucks and buses
023	GMC	046	Renault	099	Unknown
024	Saturn	047	Saab	994	Legitimate skip
029	Other domestic				

MSASIZE Size of MSA/CMSA of household

Value	Label
01	Less than 250,000
02	250,000 to 499,999
03	500,000 to 999,999
04	1,000,000 to 2,999,999
05	3,000,000 or more
94	Not in MSA

MSTR_MON Date of household master interview—Month

Value	Label
01	January
02	February
03	March
04	April
05	May
06	June
07	July
08	August
09	September
10	October
11	November
12	December
98	Not ascertained

POPDNSTY Population density of household ZIP code area

Value	Label
01	0 to 99
02	100 to 249
03	250 to 499
04	500 to 749
05	750 to 999
06	1,000 to 1,999 and in MSA
07	2,000 to 2,999 and in MSA
08	3,000 to 3,999 and in MSA
09	4,000 to 4,999 and in MSA
10	5,000 to 7,499 and in MSA
11	7,500 to 9,999 and in MSA
12	10,000 to 49,999 and in MSA
13	50,000 or more and in MSA
14	1,000 or more and not in MSA

POVERTY Household below, near, or above poverty level

Value	Label
01	Below poverty level
02	Near poverty level
03	Above poverty level
98	Not Ascertained
99	Refused

REF_AGE Age of household reference person

Number except the following:

Value	Label
077	Reference person age 76 through 79
082	Reference person age 80 through 84
088	Reference person age 85 and over
998	Not Ascertained
999	Refused

REF_SEX Sex of household reference person

Value	Label
01	Male
02	Female
98	Not Ascertained
99	Refused

REF_EDUC Education of household reference person

Value	Label
01	First grade
02	Second grade
03	Third grade
04	Fourth grade
05	Fifth grade
06	Sixth grade
07	Seventh grade
08	Eighth grade
09	Ninth grade
10	Tenth grade
11	Eleventh grade
12	Twelth grade
13	Technical school after high school
21	1st (Freshman) year of college or equivalent
22	2nd (Sophomore) year of college or equivalent
23	3rd (Junior) year of college or equivalent
24	4th (Senior) year of college or equivalent
31	One year of graduate school
32	Two or more years of graduate school
98	Not Ascertained
99	Refused

The Following Information Will Help You Understand the Data File CONSEXP.XXX

EXPLANATION OF VARIABLE NAMES USED IN THE CONSUMER EXPENDITURES SURVEY

AGE2	Age of Spouse
AGE_REF	Age of Ref. Person
ALCBEVCQ	Alcoholic Beverages This Quarter
ALCBEVPQ	Alcoholic Beverages Last Quarter
APPARCQ	Apparel and Services This Quarter
APPARPQ	Apparel and Services Last Quarter
AS_COMP1	No. of Males Age 16 and Over in Cu
AS_COMP2	No. of Females Age 16 and Over
AS_COMP3	No. of Males Age 2 Through 15
AS_COMP4	No. of Females Age 2 Through 15
AS_COMP5	No. of Members Under Age 2 in Cu
BLS_URBN	Urban/rural
CUTENURE	Housing Tenure
EARNINCX	Cu Earned Inc. Before Taxes
EDUCA2	Education of Spouse
EDUC_REF	Education of Ref. Person
ENTERTCQ	Entertainment This Quarter
FAMTFEDX	Fed Inc. Tax Deducted From Last Pay
FAM_SIZE	No. of Members in Cu
FDAWAYCQ	Food Away From Home This Quarter
FDAWAYPQ	Food Away From Home Last Quarter
FDHOMECQ	Food At Home This Quarter
FDHOMEPQ	Food At Home Last Quarter
FINCATAX	Cu Inc. After Taxes
FINDRETX	Money Placed in Individual Retirement Account (IRA)

FOODCQ	Total Food This Quarter
FOODPQ	Total Food Last Quarter
FRRETIRX	Soc. Sec. and Railroad Income
FSALARYX	Cu Wage and Salary Inc. Before Taxes
GASMOCQ	Gasoline and Motor Oil This Quarter
GASMOPQ	Gasoline and Motor Oil Last Quarter
HEALTHCQ	Health Care This Quarter
HEALTHPQ	Health Care Last Quarter
HOUSEQCQ	House furnishings, Equip. This Quarter
HOUSEQPQ	House furnishings, Equip. Last Quarter
INC_HRS1	Wkly. Hrs. Reference Person Usually Worked
INC_HRS2	Weekly Hrs. Spouse Usually Worked
NEWID	Cu Identification Number
NUM_AUTO	No. of Automobiles
PUBTRACQ	Pub. Transportation This Quarter
PUBTRAPQ	Pub. Transportation Last Quarter
RENTEQVX	Approx. Mthly. Rental Value of Home
SAVACCTX	Savings Accts. Balances At Banks
SECESTX	Est. Market Value of Securities
TRANSCQ	Transportation This Quarter
TRANSPQ	Transportation Last Quarter
UTILCQ	Utilities, Fuels, Public Services This Quarter
UTILPQ	Utilities, Fuels, Public Services Last Quarter
VEHQ	Number of Vehicles

Explanation of Code Values for Variables in Consumers Expenditure Survey

CUTENURE Type of ownership of house

Value	Label
1	Owned With Mortgage
2	Owned Without Mortgage
3	Owned Mort. Not Reported
4	Rented
5	Occ. Without Payment of Cash Rent
6	Stud. Housing

BLS_URBN—Whether the location of the household is urban or rural

Value	Label
1	Urban
2	Rural

EDUCA2 and EDUC_REF Education level of head of household

Value	Label
1	Elementary 1–8 Yrs.
2	High School, Less Than H.S. Grad.
3	High School Graduate
4	Coll., Less Than Coll. Grad.
5	Coll. Graduate
6	Graduate School
7	Never Att. School

URBAN Urbanized area indicator

Value	Label
1	Household in urbanized area
2	Household not in urbanized area

URBNAREA Urbanized area status

Value	Label
1	Urbanized, in MSA central city
2	Urbanized, not in MSA central city
3	Not in urbanized area

URBNSIZE Size of urbanized area

Value	Label
01	50,000 to 199,999
02	200,000 to 499,999
03	500,000 to 999,999
04	1,000,000 or more without subway/rail
05	1,000,000 or more with subway/rail
94	Not in urbanized area

VEHYEAR Model year of vehicle

Value	Label
055	1919–1959
063	1960–1964
065	1965
066	1966
067	1967
068	1968
069	1969
070	1970
071	1971
072	1972
073	1973
074	1974
075	1975
076	1976
077	1977
078	1978
079	1979
080	1980
081	1981
082	1982
083	1983
084	1984
085	1985
086	1986
087	1987
088	1988
089	1989
090	1990
091	1991
994	Legitimate skip
998	Not Ascertained
999	Refused

VEH12MNT Vehicle received in last 12 months

Value	Label
01	Yes
02	No
98	Not ascertained
99	Refused

VEHHHOWN Vehicle owned by household member

Value	Label
01	Yes
02	No
98	Not ascertained
99	Refused

VEHMILES Reported vehicle mileage previous 12 months

Value	Label
999994	Legitimate skip
999998	Not ascertained

VEHTYPE Vehicle type

Value	Label
01	Automobile (includes station wagon)
02	Passenger van
03	Cargo van
04	Pickup truck (including pickup with camper)
05	Other truck
06	RV or motor home
07	Motorcycle
08	Moped
09	Other
98	Not Ascertained
99	Refused

VEHNEW Vehicle new or used when received

Value	Label
01	New
02	Used
98	Not ascertained
99	Refused

Conditional Probability and Bayes' Rule

You know that the formula for finding the conditional probability $P(A|B)$ is

$$P(A|B) = \frac{P(A \text{ AND } B)}{P(B)}$$

and you have used this formula to solve conditional probability problems. In this section, we will use this formula to solve problems that are more complex.

Look at the data in the table below on homeschooling.

Number of Parents in Household	Percent of Students	Percent of Homeschooled Students
Two parents	66%	2.1
One parent	31%	0.9
Nonparental guardians	3%	1.4

Source: National Center for Education Statistics, 1999.

If you chose a child at random there would be two variables of interest about that child—the number of parents in their household and whether or not they were homeschooled. The data in the column labeled Percent of Students represent the percent of students in the United States for each of the different categories of Number of Parents in Household. In other words, the probability that a student had one parent living at home is 0.31 or 31%. The numbers in the column labeled Percent of Homeschooled Students represent the percent of students *in each category* that are homeschooled. They are conditional probabilities. That is, the 2.1% represents the probability that a student is homeschooled *given that there are two parents living at home.*

We will call *Number of Parents in Household* event A with each possible outcome (two parents, etc.) represented as A_1, A_2, and A_3, and we will call *Homeschooled* event B. If you look again at the table of probabilities you can see that the probabilities in the second column are $P(A_1)$, $P(A_2)$, and $P(A_3)$, and those in the third column are $P(B|A_1)$, $P(B|A_2)$, and $P(B|A_3)$. It is easier to visualize the problem using a tree diagram as shown at the top of the next page.

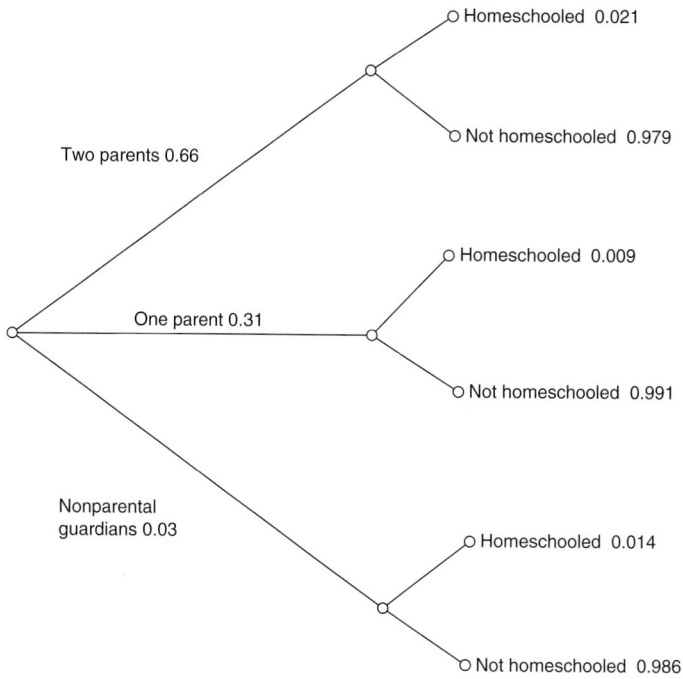

The first set of branches represents the event A. At the end of each branch of A, we have the event B, which either occurs or does not occur. The values of the probabilities for B *depend on* the branch of A that they follow. Notice also that, for each junction in the tree, the sum of the probabilities of the branches that come from there is equal to 1. This structure is the basis for a type of problem that is solved using a rule of conditional probability called *Bayes' Theorem* or *Bayes' Rule*.

Suppose that for the homeschooling data we wanted to know the following: If a child is selected at random in the United States and we find that they are homeschooled, what is the probability that they came from a household with two parents? We know that event B has occurred (the child is homeschooled) and we want to find $P(A_1|B)$. This requires working backward from the original information that we were given. Using the formula for conditional probability, we know that

$$P(A_1|B) = \frac{P(A_1 \text{ AND } B)}{P(B)}$$

We will start with the denominator. (You will see why we do this later in the problem.) How do we find $P(B)$, the probability that a child selected at random is homeschooled? This event can happen in more than one way—the child could come from a two-parent household and be homeschooled, or from a one-parent household and be homeschooled, or from a nonparental guardian household and be homeschooled. We need to find the probabilities of each of these possible outcomes and then add them together to find $P(B)$. This might sound difficult, but we will work with one part of the problem and then apply that solution to the other parts.

Look at the probability of the event that a child is homeschooled and comes from a two-parent household, $P(B \text{ AND } A_1)$. We know that if the events were independent, then we could just multiply the probabilities together, but clearly these events are *not independent*. In fact, we have already said that the probability for B *depends on* which branch of A it follows. However, we can use the conditional probability formula in another way to find $P(A_1 \text{ AND } B)$. Since $P(B|A_1) = P(B \text{ AND } A_1)/P(A_1)$, we can rewrite that as $P(B \text{ AND } A_1) = P(B|A_1) \times P(A_1)$. We know both of those values from the table! Thus,

$$P(B \text{ AND } A_1) = 0.021 \times 0.66 = 0.0139$$

We can find the other two probabilities in the same way:

$$P(B \text{ AND } A_2) = 0.009 \times 0.31 = 0.0028 \quad \text{and} \quad P(B \text{ AND } A_3) = 0.014 \times 0.03 = 0.0004$$

Putting these together, we have

$$P(B) = 0.0139 + 0.0028 + 0.0004 = 0.0171$$

You might notice that, although this value is not any of the values in the third column of the table, it is a *weighted* average of the three.

Now there is only one step left. We have found the denominator of the formula for $P(A_1|B)$. Looking at the formula, though, you see that the numerator is actually one of the parts of the denominator. We already know that $P(A_1 \text{ AND } B) = 0.0139$. This will be true in *every* Bayes' Rule problem: The numerator will be one of the components of the denominator.

Putting it all together, we have

$$P(A_1|B) = \frac{0.0139}{0.0171} = 0.813$$

Therefore, given that a randomly selected student is homeschooled, the probability that he or she is from a two-parent household is 0.813.

When you are attacking problems of this type, the tree diagram is very useful. Look at the endpoints of the diagram. The first endpoint represents the event that the child comes from a two-parent household and is homeschooled. If you multiply the probabilities on those two branches, you get $0.66 \times 0.021 = 0.0139$, which is the first component of $P(B)$. In fact, if you multiply across the branches for each of the endpoints and then add those endpoints that correspond to the event that the child is homeschooled, you will get 0.0171, which is the denominator of the problem! Using the tree diagram avoids the use of the notation, which can get very messy and sometimes confusing.

The Poisson Probability Distribution

The *Poisson probability distribution* was originally defined as an approximation to the binomial distribution under certain circumstances. The Poisson distribution made calculating the binomial probabilities less computationally difficult. With the availability of computers and even calculators, this use is no longer as important. However, the Poisson distribution still plays a very important role as a distribution on its own.

We have learned about a binomial random variable, which describes the number of successes in a random sample of size *n*, or in *n* independent trials of an experiment. We might be interested, however, in the number of successes that occur when we are observing successes in a time interval or in an area. For example, we might be interested in collecting data on the number of people who arrive at a hospital emergency room in a 1-hour interval, the number of calls to a consumer hotline in a 5-minute period, or the number of a certain type of insect that can be found in a 1-square-foot area of farmland. The random variable that describes this type of behavior is known as a **Poisson random variable**.

Suppose that we were observing the number of arrivals at the drive-in window of a fast-food restaurant. From past experience, we know that the average number of customers who arrive at the window in a 5-minute interval is 6. We would like to find the probability that there might be 10 customers arriving in a 5-minute interval. This is an example of a Poisson probability.

To use the Poisson distribution as the probability distribution for the number of successes in a time interval or area, certain conditions must be met. The mathematics behind these conditions is well beyond the scope of an introductory text, but the ideas are not difficult. The following are some of the conditions:

- The numbers of successes occurring in non-overlapping intervals must be independent.
- The probability of a success in an interval must remain constant over the length of the interval.
- In a very short interval, no more than one success can occur.

The Poisson probability distribution relies on one parameter, the average number of successes in the time interval—that is, the *rate* of successes per unit interval.

The formula for the Poisson probability distribution is given by

$$P(X = x) = \frac{e^{-\mu} \mu^x}{x!}$$

where e is the familiar constant 2.718..., and μ is the average number or rate of successes in the time interval of interest.

This might not seem to be computationally any better than the binomial (it really is) but, as is the case with the binomial distribution, there are tables available; in addition, most statistical software packages calculate these probabilities very easily.

For the fast-food drive-in example, μ is 6 customers per 5-minute interval. To find $P(X = 10)$, we can calculate (without too much difficulty)

$$P(X = 10) = \frac{e^{-6} 6^{10}}{10!} = 0.0413$$

When using the Poisson distribution to solve problems, it is important to remember that the rate of successes you use must match the time interval of the problem. For example, if we were interested in knowing the probability that 15 customers arrive at the drive-in window in a 10-minute period, we would use $\mu = 12$ customers per 10-minute interval.

Exercises Appendix

Chapter 1 Exercises

Section 1.2

1. What is the relationship between a population of interest and a sample? When are both the same?

2. The U.S. government conducts a census every 10 years. The purpose is to collect information on the number of people living in the United States. The entire population of the United States is questioned. Why do you think this is not done more frequently?

3. A social worker wishes to study the day-care expenses of 12,000 working mothers in her area. A sample of 200 working mothers reveals that the average cost of day care per child is $110 per week. However, unknown to the social worker, the average cost of day care per child of all working mothers in the area is only $90 per week.
a. What is the population?
b. What is the sample?
c. What type of error was made?

4. A weight-loss company claims that its clients lose an average of 2 pounds a week when following its diet program. The company thinks that results may differ depending on gender and amount of weight one needs to lose.
a. What is the population?
b. What variables should be measured?

5. A pediatrician is studying the characteristics of newborn babies she delivers.
a. What is the population?
b. What are some variables she should measure?
c. Will a sample or a population be used to address this issue? Why?

6. A dietitian is studying the eating habits of American families. She selects 2000 families and sends them a survey requesting information about their eating habits.
a. Are the 2000 families a sample or a population?
b. What variables might be measured?

7. It is estimated that the religion of 76% of all people in the United States is Christianity. Was this number calculated from a sample or a population? Why?

8. A parent of a young child wants to decide whether to homeschool her child or send him to public school.
a. Would the parent gather information from a sample or a population? Why?
b. What variables could be used to address this question?

9. *Gallup Poll vault: Opinion of the FBI*   In November 1965, 84% of Americans said they had a very favorable opinion of the Federal Bureau of Investigation. By July 1973, this percentage had dropped to 52%. It fell further, to 37% in November 1975, and to 27% in July 2001. In a May 2002 poll, 24% of Americans said they had a very favorable opinion of the FBI.
a. What is the population of interest in these Gallup polls?
b. Are these numbers calculated from a population or a sample? Why?

10. *Gallup Poll survey*   The following results are based on telephone interviews with a randomly selected national sample of 1002 adults, 18 years and older, conducted May 20–22, 2002.

Do you plan to take a vacation this summer?   Yes: 55%;   No: 43%;   No opinion: 2%

Are the figures calculated from a sample or a population? Why?

Section 1.3

1. *Gallup Poll survey*   These results are based on telephone interviews with a randomly selected national sample of 1002 adults, 18 years and older, conducted May 20–22, 2002.

Do you plan to take a vacation this summer?   Yes: 55%;   No: 43%;   No opinion: 2%

a. What is the parameter of interest in this survey?
b. Are the figures reported statistics or parameters? Why?

2. *Gallup Poll vault: Opinion of the FBI*   In November 1965, 84% of Americans said they had a very favorable opinion of the Federal Bureau of Investigation. By July 1973, this percentage had dropped to 52%. It fell further, to 37% in November 1975, and to 27% in July 2001. In a May 2002 poll, 24% of Americans said they had a very favorable opinion of the FBI.
a. What is the parameter of interest in these surveys?
b. Are the figures reported statistics or parameters? Why?

3. According to the Core Institute, an organization that surveys college drinking practices, almost one-third of college students admit to having missed at least one class because of their alcohol or drug use, and nearly one-quarter of students report bombing a test or project because of the after-effects of drinking or doing drugs.
a. What is the parameter of interest in this survey?
b. Is the reported "one-third" a parameter or a statistic? Why?

4. A social worker wishes to study the day-care expenses of 12,000 working mothers in her area. A sample of 200 working mothers reveals that the average cost of day care per child is $110 per week. However, unknown to the social worker, the average cost of day care per child of all working mothers in the area is only $90 per week.
a. Is $110 per week a parameter or a statistic?
b. Is $90 per week a parameter or a statistic?
c. Explain why the parameter and the statistic are not equal.

5. A hospital administrator is interested in the average birth weight of babies born at the hospital. A sample of 68 births last month indicated that the average weight was 7.25 lb. However, unknown to the administrator, the average birth weight of all babies ever born at this hospital is 7.10 lb.
a. Is 7.25 lb a parameter or a statistic?
b. Is 7.10 lb a parameter or a statistic?
c. Explain why the parameter and the statistic are not equal.

6. A university bookstore is interested in the average amount that students pay for textbooks each semester. A sample of 200 students indicates that the average cost is $325 per semester. However, unknown to the bookstore, the average cost of textbooks for all 25,000 students at the university is $375 per semester.
a. Is $325 per semester a parameter or a statistic?
b. Is $375 per semester a parameter or a statistic?
c. Explain why the parameter and the statistic are not equal.

7. A doctor is interested in the average number of colds that his patients catch each year. He calls a sample of 40 patients and finds that they had an average of 3.7 colds this year. However, unknown to the doctor, the average number of colds this year for all 270 patients is 4.2.
a. Is 3.7 colds a parameter or a statistic?
b. Is 4.2 colds a parameter or a statistic?
c. Explain why the parameter and the statistic are not equal.

8. A credit card company is interested in studying unpaid balances of its credit card holders. A sample of 500 credit card accounts is selected from 50,000 accounts. The average unpaid balance of the 500 credit card accounts is $2438. The average unpaid balance of all 50,000 accounts is $2776.
a. Is $2438 a parameter or a statistic?
b. Is $2776 a parameter or a statistic?
c. Explain why the parameter and the statistic are not equal.

9. A weight-loss company claims that people who follow the plan lose an average of 2 lb per week. A skeptical potential customer selects a sample of 500 people who follow the plan. Their average weight loss is 1.9 lb per week.
a. Is 2 lb a parameter or a statistic?
b. Is 1.9 lb a parameter or a statistic?

10. In the months prior to a national election, a poll of 1500 people reveals that 55% plan to vote for the Republican candidate. After the election, the Republican candidate wins with 58% of the votes.
a. Is 55% a parameter or a statistic?
b. Is 58% a parameter or a statistic?
c. Explain why the parameter and the statistic are not equal.

Section 1.4

1. Name two factors that can affect sampling errors.

2. List three factors that influence the sample size required to obtain information about a population.

3. A social worker wishes to study the day-care expenses of 12,000 working mothers in her area. A sample of 200 working mothers reveals that the average cost of day care per child is $110 per week. However, unknown to the social worker, the average cost of day care per child of all working mothers in the area is only $90 per week.
a. What is the size of the population, N?
b. What is the size of the sample, n?

4. A university bookstore is interested in the average amount that students pay for textbooks each semester. A sample of 200 students indicates that the average cost is $325 per semester. However, unknown to the bookstore, the average cost of textbooks for all 25,000 students at the university is $375 per semester.
a. What is the size of the population, N?
b. What is the size of the sample, n?

5. A doctor is interested in the average number of colds that his patients catch each year. He calls a sample of 40 patients and finds that they had an average of 3.7 colds this year. However, unknown to the doctor, the average number of colds this year for all 270 patients is 4.2.
a. What is the size of the population, N?
b. What is the size of the sample, n?

6. A credit card company is interested in studying unpaid balances of its credit card holders. A sample of 500 credit card accounts is selected from 50,000 accounts. The average unpaid balance of the 500 credit card accounts is $2438. The average unpaid balance of all 50,000 accounts is $2776.
a. What is the size of the population, N?
b. What is the size of the sample, n?

7. *Gallup Poll survey* These results are based on telephone interviews with a randomly selected national sample of 1002 adults, 18 years and older, conducted May 20–22, 2002. Assume there were 200 million adults, 18 years and older, living in the United States at the time of this survey.

Do you plan to take a vacation this summer?   Yes: 55%;   No: 43%;   No opinion: 2%

a. What is the size of the population, N?
b. What is the size of the sample, n?

8. A drug company is testing an antibiotic to treat ear infections. It offers $5 back to patients who decide to participate in the study. It would like to test the antibiotic on as many patients as possible, but can afford to pay only 300 patients the $5 incentive. What factor influenced the sample size?

9. A credit card company is interested in studying unpaid balances of its credit card holders. Some pay their cards off each month, while others carry a balance close to the maximum each month. The company, therefore, needs a large sample in order to properly study the balances. What factor influenced the sample size?

10. The Gallup organization conducts opinion polls from a national sample of 1000–1500 people. The organization wants to obtain results that are no more than 3% away from the percent of how the whole United States would respond. What factor influenced the sample size?

Section 1.5

1. A large university has 25,000 students, made up of both graduate and undergraduate students. A survey about parking is sent to a sample of 500 undergraduate students. Is this sample biased? Explain.

2. A teacher of introductory statistics wishes to determine what percent of students who take this course are liberal arts majors. There are 10 sections of the course that meet at different times during the day. She uses her own section, which meets at 10 A.M. on MWF, as a sample. Explain how this sample is biased.

3. Fox News posts questions of the day to which viewers can call in or e-mail answers. Explain why the samples obtained are biased.

4. Suppose a sample is taken and all the samples of size n from the population have an equal chance of being selected. What type of sampling is described?

5. Explain how a table of random digits is compiled.

The following is a list of random numbers taken from a random number table:

094632795   711501534   537971597   410398128   558723092

Use this list to select a simple random sample for each of the following situations.

6. There are 46 students enrolled in an introductory statistics class. Each student is assigned a random number, 1 to 46. Select a simple random sample of 5 students from this class. List the numbers for the 5 selected students.

7. There are 90 parking spaces in a parking lot. Each space is assigned a number, 1 to 90. Select a simple random sample of 10 cars from this parking lot. List the numbers for the 10 parking spaces selected.

8. There are 500 employees in a large company. Each employee is given a number, 1 to 500. Select a simple random sample of 5 employees. List the numbers for the 5 employees selected.

9. There are 9 faculty members in a small department. The department chair wishes to select 3 faculty members to serve on a committee. Select a simple random sample of 3 faculty members. List the numbers of the faculty members selected.

10. A store receives a shipment of 30 watermelons. The produce manager will select 5 for a quality inspection before accepting the shipment. Select a simple random sample of 5 watermelons. List the numbers for the watermelons to be selected.

Section 1.6

1. A weight-loss company claims that its clients lose an average of 2 pounds a week when following its diet program. The company thinks that results may differ depending on gender and amount of weight one needs to lose. Tell whether each variable is quantitative or qualitative.
a. Weight at beginning of diet program
b. Gender
c. Amount of weight to lose

2. An airline collects data about its flights. Decide whether the following variables are qualitative or quantitative. If qualitative, tell whether it is nominal or ordinal. If quantitative, tell whether it is discrete or continuous.
a. The number of seats available on a flight
b. The number of miles between cities that the airplane flies
c. The class in which the passengers are seated (first class or coach)
d. The name of the airline (American, Delta, United, etc.)

3. The following student data are collected. Decide whether the following variables are qualitative or quantitative. If qualitative, tell whether it is nominal or ordinal. If quantitative, tell whether it is discrete or continuous.

a. Height
b. Gender
c. Major
d. Class (Freshman, Sophomore, Junior, Senior)
e. Number of siblings

4. Teacher evaluation surveys include a question on the availability of the teacher outside of class. Students rate their satisfaction as follows:

Very dissatisfied   Dissatisfied   Neutral   Satisfied   Very satisfied

What type of variable is measured here? Be specific.

5. Psychology questionnaires often include questions that describe a scenario and ask for level of agreement on the following scale:

Strongly disagree   Disagree   Neutral   Agree   Strongly agree

What type of variable is measured here? Be specific.

Identify each of the following variables as nominal qualitative, ordinal qualitative, discrete quantitative, or continuous quantitative.

6. Rating of the effectiveness of a new cold remedy (Not effective, Somewhat effective, Very effective)

7. Amount of time spent assembling a five-shelf bookcase

8. Number of children in a beginning swimming class

9. University where a student is enrolled

10. Distance of current residence from place of birth

Section 1.10

1. The following data set lists the number of minutes that 10 Internet subscribers spent on the Internet during their most recent session.

$$21, 7, 37, 30, 44, 22, 17, 33, 41, 36$$

Write the sigma notation to represent the sum of these 10 data values.

2. The values below represent the number of car crashes in Chatauqua County for each of the last nine months.

$$22, 8, 17, 11, 15, 14, 14, 9, 13$$

Write the sigma notation to represent the sum of these nine data values.

3. The amounts (in dollars) spent on groceries in a week for 12 families were recorded and appear below.

$$120, 200, 135, 143, 177, 347, 137, 143, 126, 128, 138, 205$$

Write the sigma notation to represent the sum of these 12 data values.

4. The number of home runs (HRs) for Mark McGwire's first 12 seasons (1987–1998) are displayed in the table.

Year	1987	1988	1989	1990	1991	1992	1993	1994	1995	1996	1997	1998
HRs	49	32	33	39	22	42	9	9	39	52	58	70

Write the sigma notation to represent the sum of Mark McGwire's home runs in these 12 seasons.

5. The following data are the sodium content (grams) in 11 types of cheese.

$$4, 8, 13, 22, 42, 24, 18, 4, 31, 9, 10$$

Write the sigma notation to represent the sum of these 11 data values.

6. The number of points scored by the USC Gamecock football team for the 11 games in the 2001 season are displayed below.

$$32, 14, 16, 37, 42, 7, 46, 10, 38, 17, 20$$

Write the sigma notation to represent the sum of these 11 data values.

7. Eight men who were at least 60 years old were selected at random and their serum cholesterol level was measured. Here are their levels, in mg/dL.

$$177, 197, 190, 185, 163, 181, 205, 222$$

Write the sigma notation to represent the sum of these 8 data values.

8. A sample of 15 women with breast cancer was taken. Here are their ages at the time of first detection.

$$33, 42, 45, 50, 53, 56, 57, 57, 59, 60, 60, 62, 65, 68, 70$$

Write the sigma notation to represent the sum of these 15 data values.

9. Eight participants in a bike race had the following finishing times, in minutes.

$$28, 22, 26, 35, 21, 23, 37, 24$$

Write the sigma notation to represent the sum of these 8 data values.

10. A study in Switzerland studied the number of cesarean sections (surgical deliveries of babies) performed in a year by doctors. The data for 10 male doctors are:

$$27, 50, 33, 25, 86, 25, 59, 31, 37, 44$$

Write the sigma notation to represent the sum of these 10 data values.

Chapter 2 Exercises

Section 2.2

1. A survey of size 1000 asked teens to give the most important invention of the last century. The results of the survey were as follows:

Personal computer	320
Pacemaker	260
Wireless communication	180
Television	240

Express the relative frequency of each as a fraction, decimal, and percentage.

2. The question "How often do you experience stress in your daily life" is asked of a class of students. The results were as follows. Create a frequency table and express each class as a frequency as well as a relative frequency.

Frequently	Frequently	Rarely	Sometimes	Frequently
Never	Rarely	Frequently	Sometimes	Sometimes
Sometimes	Frequently	Frequently	Frequently	Sometimes
Rarely	Sometimes	Sometimes	Rarely	Never
Frequently	Sometimes	Frequently	Frequently	Frequently

3. Tea has been linked to survival after a heart attack. One hundred heart attack survivors were asked to give the number of cups of tea they consumed per day, on the average, before they had a heart attack. The results of the survey were as follows:

```
3  1  4  4  2  3  2  2  4  3
2  1  2  2  1  2  4  3  1  2
4  2  3  2  4  2  4  3  4  4
2  3  2  1  3  1  1  2  2  4
2  3  3  2  2  2  3  2  2  3
2  4  2  4  2  2  4  2  4  3
3  2  0  4  2  1  2  4  4  0
4  3  4  0  4  0  4  4  2  3
3  2  4  4  4  4  2  4  4  2
0  1  4  2  0  3  2  2  2  1
```

Create a relative frequency table for the data. Give the relative frequencies as percents.

4. The number of cups of tea consumed per day by 100 heart attack survivors prior to the heart attack had the following relative frequency distribution. Form the cumulative relative frequency distribution for the tea consumption.

Cups	Percent
0	6
1	10
2	37
3	18
4	29

5. The number of e-mails sent per week by a company is of interest. The 100 employees of the company are asked to give an average figure for the number of e-mails sent. The data are shown below.

```
18  17  16  21  13   9  16  12   5  18
21  11  11   9  20  12  14  13  24   7
16  15  24  15  15  18  20  15   8  11
17  13  14  20  12  18  23  14  13  22
19  13   8  14  14  12  25  13  23  13
12  19  17   8  11  16  13  15  16  19
22  12  17  20  18  18  13  10  20  15
12  17   9  13  13  15   9  15  17  12
14  12  18  17  12  15   7  13  11  21
21  15  20  13  19  14  11  15  10  18
```

a. How many classes do you recommend?
b. How wide should each class be?
c. What classes do you recommend? Use classes that do not include the upper boundary.
d. Group the data into the classes.

6. On June 19, 2002, nineteen large fires were burning in 11 states. The fires and their names are given in the table. The fires are named for the areas where they started.

Name	Acres	Name	Acres
Sudden Ranch	7,161	Garfield	50
Cannon	15,000	BMG	350
Borel	3,430	Roybal	2,700
Bluecut	5,500	Hayman	113,000
Sanford	60,000	Hensel	2,415
Walker	16,369	Elgin Complex	3,100
West Dome	930	Blackjack Bay	123,239
Miracle Ranch	3,951	Marbleyard Complex	2,991
Missionary Ridge	37,200	Legends	1,658
Coal Seam	12,105		

a. Using the square root rule, into how many classes would you group the data?
b. What should the class width be?
c. The class boundaries are adjusted to those given at the top of the next page. Fill in the frequencies.

Acres	Frequency
$0 \le x < 31{,}000$	
$31{,}000 \le x < 62{,}000$	
$62{,}000 \le x < 93{,}000$	
$93{,}000 \le x < 124{,}000$	

7. Consider the following miles per gallon data obtained by testing 50 Mazda Proteges.

Miles per gallon	Frequency
$20 \le x < 21$	2
$21 \le x < 22$	7
$22 \le x < 23$	9
$23 \le x < 24$	15
$24 \le x < 25$	8
$25 \le x < 26$	7
$26 \le x < 27$	2

a. Which class contained the most values?
b. Which class contained the smallest number of values?
c. In what class is the value so that half the data are larger and half the data are smaller than that value?

8. The question "What is your primary source of news?" was asked of 50 people. The answers obtained were as follows. Form a relative frequency table.

Radio	Newspaper	Internet	TV	Other
Internet	Newspaper	Other	Radio	Newspaper
Newspaper	TV	Radio	Newspaper	Internet
Radio	Newspaper	Internet	Radio	Other
Radio	Other	TV	Newspaper	Internet
TV	Newspaper	Internet	TV	Internet
Internet	TV	Internet	Internet	Radio
Newspaper	Internet	TV	TV	Newspaper
Newspaper	TV	TV	Radio	Radio
Internet	Newspaper	Newspaper	TV	TV

9. A group of diabetics were classified according to the number of insulin injections they took per day. The data were given as follows.

1	2	1	1	1	3	3	1	1	0
0	0	0	2	1	1	1	2	1	1
1	1	0	0	0	0	2	0	1	1
1	1	1	2	0	0	0	1	0	1
0	1	1	1	2	1	1	1	2	1
1	2	1	0	2	2	1	0	2	1
1	1	1	1	0	1	0	0	0	1
0	1	1	3	1	1	2	2	1	2
0	0	4	0	1	2	1	0	2	0
2	1	3	0	3	2	1	0	2	2

Form a frequency distribution for these data. Express as a percent distribution. Diabetics who took more than 2 injections per day were to be part of a national study. How many from this group were eligible for the national study?

10. The heights of 50 college female basketball players were recorded. The results were as follows.

69.6	69.6	72.5	71.4	72.6
71.8	67.8	70.5	68.0	70.2
73.6	72.2	71.9	68.8	72.3
69.8	71.7	73.4	72.8	69.4
71.4	71.4	72.7	72.5	70.6
69.9	68.3	70.3	71.3	72.2
72.7	72.2	69.8	70.7	69.4

(*continued*)

69.0	69.4	68.3	71.9	69.0
70.9	72.1	73.2	70.3	71.4
71.9	73.0	72.3	71.5	72.0

a. How many classes should be used to form a frequency distribution?
b. Approximately how wide should each class be?
c. Group the data into the proper number of classes as determined in part **a**.

Consider the following information for Exercises 11–14.

Doug Larch, the manager of a popular local barbershop, has decided to evaluate the time required by his customers to obtain a haircut, pay, and depart. Doug borrowed the clock punch from one of the offices in the shopping center and asked the next 36 customers to punch in when they arrived at his shop and punch out after paying for their haircut. The following data were obtained (in minutes):

23	36	41	33	24	19	17	40	32
32	25	28	46	31	22	24	51	19
15	39	42	53	20	19	26	39	43
27	33	44	28	21	37	52	45	27

Doug chose the following classes for the data (left endpoint included, right endpoint excluded):

14–21
21–28
28–35
35–42
42–49
49–56

11. Doug's frequency table contains _____ class intervals.
a. 3
b. 4
c. 5
d. 6

12. The width of each class interval in Doug's frequency table is _____ minutes.
a. 4
b. 5
c. 6
d. 7

13. How many customers are contained in the lowest (shortest time) class interval?
a. 7
b. 6
c. 5
d. 4

14. Which class interval contains the most customers?
a. 14–21 minutes.
b. 21–28 minutes.
c. 28–35 minutes.
d. 35–42 minutes.

Section 2.3

1. Construct a bar chart for the following frequency distribution. This represents a survey where teens were asked to name the most important invention of the twentieth century.

Personal computer	320
Pacemaker	260
Wireless communication	180
Television	240

2. Construct a pie chart for the following frequency distribution. This represents a survey where teens were asked to name the most important invention of the twentieth century.

Personal computer	320
Pacemaker	260
Wireless communication	180
Television	240

Give the frequency and percent for each class.

3. A medical study involved the study of aphasia, the loss of the faculty of understanding spoken or written language. Three type of aphasia have been identified. They are broca, anomic, and conduction. Thirty patients were classified as follows:

Conduction	Anomic	Broca
Anomic	Anomic	Broca
Conduction	Anomic	Broca
Broca	Anomic	Conduction
Anomic	Conduction	Broca
Broca	Conduction	Anomic
Anomic	Anomic	Broca
Conduction	Conduction	Anomic
Anomic	Broca	Broca
Broca	Anomic	Conduction

Construct a bar chart for these data.

4. A medical study involved the study of aphasia, the loss of the faculty of understanding spoken or written language. Three type of aphasia have been identified. They are broca, anomic, and conduction. Fifty patients were classified as follows:

Broca	Conduction	Broca	Anomic	Broca
Conduction	Anomic	Conduction	Anomic	Broca
Conduction	Broca	Conduction	Conduction	Conduction
Broca	Conduction	Conduction	Conduction	Conduction
Anomic	Anomic	Conduction	Broca	Conduction
Broca	Conduction	Broca	Anomic	Conduction
Broca	Conduction	Anomic	Conduction	Broca
Broca	Conduction	Broca	Broca	Anomic
Conduction	Broca	Conduction	Conduction	Conduction
Broca	Broca	Broca	Conduction	Conduction

Construct a pie chart for these data and identify the condition that occurs the most often.

5. The frequency distribution for the type of aphasia of 50 patients is given below. Determine the angles of the three pieces of the pie.

Type of Aphasia	Frequency
Conduction	24
Broca	18
Anomic	8

6. The number of vacations of five nights or more that Americans take in a typical year was determined for a sample of 100 families. The results were as follows.

1	1	1	1	1	1	1	1	1	2
1	1	1	1	3	2	1	1	1	2
1	1	1	1	1	1	1	2	2	1
1	1	1	1	1	1	0	1	1	2
2	1	1	1	1	2	1	1	1	1
1	1	1	1	0	1	1	1	1	1
1	1	1	1	1	1	1	1	2	0
3	1	1	1	1	1	1	1	1	3
2	1	0	1	1	1	2	1	1	1
1	2	1	2	1	0	1	3	1	1

Construct a histogram for these integer data.

```
165   111   167   151   128
121   115   151   144   122
116   123   123   125   133
162   154   163   174   156
142   151   131   130   141
```

3. Graph the following mixture of male and female weights. Where is the center of the distribution? What is the shape of the distribution?

```
178   154   174   173   178
174   190   184   196   170
199   181   170   186   182
182   181   177   188   185
201   164   181   163   182
132   112   147   129   131
146   122   138   147   134
148   124   120   143   119
132   130   130   135   136
124   141   126   135   141
```

4. The following dotplot represents the yearly salaries (in thousands of dollars) at ABC Software. Can you explain the distribution?

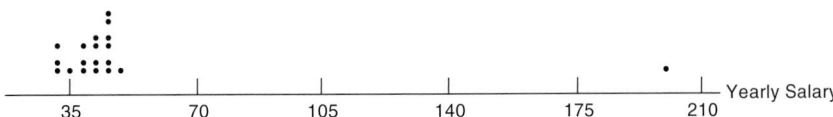

5. The following is a histogram of groom ages taken from the newspaper. Explain the shape of the histogram.

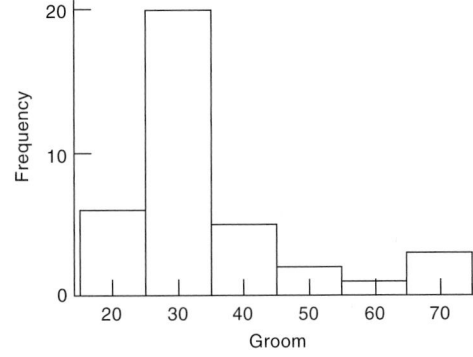

6.

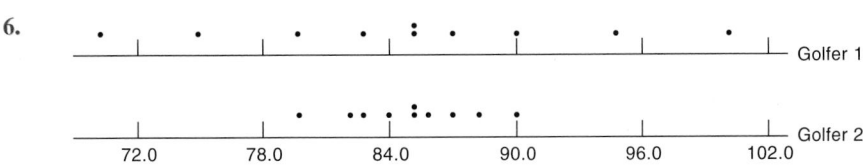

Both of these golfers shoot 85 on the average. Which is the more consistent golfer (that is, which one has the lesser amount of variability)?

7.

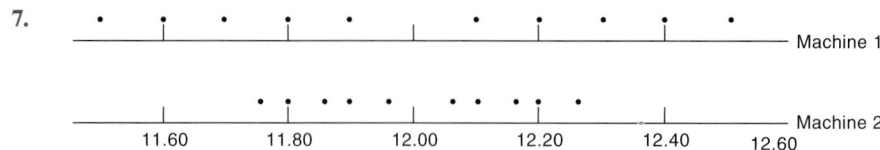

Two filling machines are compared with respect to their consistency in filling 12-ounce containers. Each fills with 12 ounces on the average. Ten 12-ounce soft drink containers are filled by machine 1 and 10 are filled by machine 2. The results are shown above. Which machine has the smaller variability?

8. Fifty lifetimes for males are selected from the obituary pages. The data are as follows.

82	89	85	13	82
69	91	55	72	80
99	80	89	77	35
82	69	25	65	76
78	82	70	79	79
77	94	67	88	87
85	72	88	99	81
68	103	94	90	89
44	87	55	77	64
85	75	79	93	70

Construct a histogram and describe its shape.

9. Twenty-five hospital records are randomly selected and the length of stay (LOS) is one of the variables noted. A plot of the data follows. Describe the shape of the distribution.

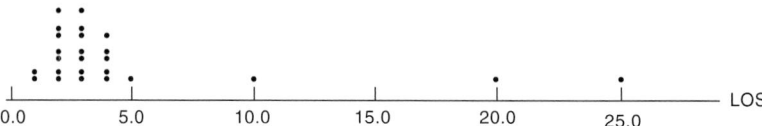

10. Three methods for controlling hypertension are compared. Thirty hypertensive patients are randomly divided and 10 are put on an exercise program, 10 are put on a diet, and 10 are put on a combination of diet and exercise program. The change in diastolic blood pressure is recorded after 6 weeks for all 30 patients and plotted. Interpret the following dotplot.

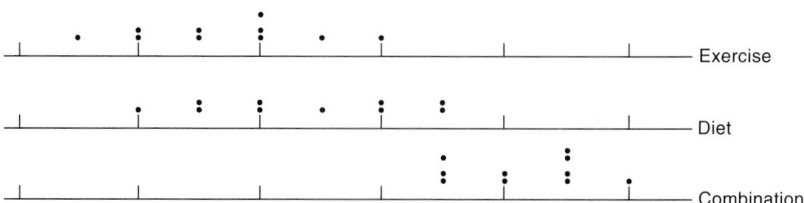

11. If you compared a histogram of home prices in San Francisco to a histogram of home prices in Omaha, Nebraska, on the same scale, what would you notice about the two histograms?

12. Suppose you compared a dotplot of fasting blood sugar readings for a group of diabetic patients on the same scale with fasting blood sugar readings of a group of nondiabetics. What would you notice about the relationship of the two dotplots?

13. How would you describe the following histogram?

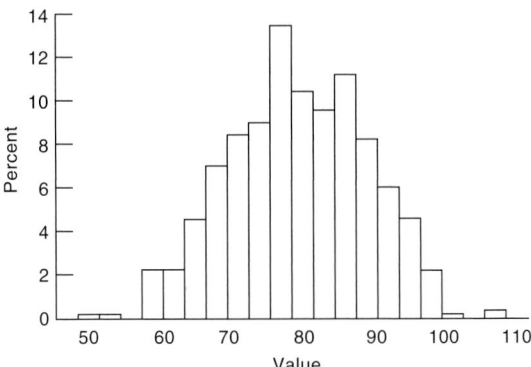

14. A machine is designed to cut table legs to 3.5 feet in length. Fifty table legs are selected from production and measured in inches. Forty-two inches is subtracted from each measurement and a histogram is formed for the difference measurements. Comment on the distribution.

0.434	− 0.240	0.254	0.442	0.038
0.084	0.286	− 0.213	− 0.488	− 0.257
0.486	− 0.151	0.289	− 0.053	− 0.471
0.153	− 0.296	− 0.395	0.186	− 0.181
− 0.379	− 0.223	− 0.347	0.374	− 0.361
0.044	− 0.335	− 0.487	0.164	− 0.098
0.142	− 0.412	− 0.107	− 0.497	0.046
− 0.233	0.058	0.201	0.270	0.272
− 0.500	− 0.429	0.210	0.001	− 0.138
0.159	0.038	− 0.413	− 0.269	0.268

15. Samples of 50 women college basketball players and 50 men college basketball players are selected. Their heights (in inches) are determined and are recorded in the following table. Plot a histogram and comment on its shape.

72	71	71	70	72	73	75	74	74	75
73	72	71	70	72	77	75	76	75	75
72	71	71	72	71	75	74	76	75	75
72	71	70	73	71	74	74	76	75	76
70	71	72	72	72	76	75	74	75	75
71	71	70	70	71	75	77	74	74	76
71	71	71	71	72	74	75	74	73	76
72	71	73	70	71	75	76	74	77	77
71	72	72	72	71	75	74	74	75	75
71	69	71	71	72	76	76	74	75	75

Chapter 2 End-of-Chapter Exercises

1. A survey was conducted to determine the interest that exists in the United States for the game of soccer. In answer to the question "How interested are you in soccer?" one of five responses could be given: Very, Somewhat, Not too, Not at all, and No opinion. The results of a preliminary survey of 50 people are shown here. Put the results into a frequency table.

Very	Not at all	No opinion	Not too	Not at all
No opinion	Somewhat	No opinion	Not at all	No opinion
No opinion	No opinion	Not at all	Somewhat	Not at all
Somewhat	Somewhat	No opinion	Somewhat	Somewhat
Not too	Not at all	Very	No opinion	Not too
No opinion	Not at all	Very	Not at all	Not too
Not too	Somewhat	No opinion	Not too	Very

(*continued*)

Very	Not too	Very	Very	Somewhat
Very	Somewhat	Not at all	Very	Not too
Not at all	Very	Not at all	Somewhat	Somewhat

2. To determine whether a die was balanced, it was rolled 100 times and a relative frequency of the outcomes was formed. Form a relative frequency of the outcomes and comment on the fairness of the die. The outcomes are given in the following table.

5	5	3	6	6	6	4	1	6	6
5	6	6	1	5	2	2	5	6	5
1	4	2	3	1	2	4	2	3	3
6	5	4	4	2	2	5	3	2	2
2	3	4	1	3	5	3	3	2	6
3	4	6	1	6	1	3	4	6	6
5	1	1	6	2	3	2	4	3	3
1	3	1	6	4	1	4	2	1	5
6	2	1	1	5	6	1	1	3	2
1	4	2	2	3	5	2	4	3	2

3. The scores of the winning team in the 36 Super Bowls are as follows. Form a frequency distribution consisting of 6 classes. About how wide should each class be?

35	14	35	38	20	35
33	24	31	46	37	31
16	16	27	39	52	34
23	21	26	42	30	23
16	32	27	20	49	34
24	27	38	55	27	20

Group the data into the following classes.

$$14 \leq \text{winning score} < 21$$
$$21 \leq \text{winning score} < 28$$
$$28 \leq \text{winning score} < 35$$
$$35 \leq \text{winning score} < 42$$
$$42 \leq \text{winning score} < 49$$
$$49 \leq \text{winning score} < 56$$

4. A survey of Americans' feelings toward soccer was taken. The answers to the survey question "How interested are you in soccer?" are summarized here. Construct a pie chart showing the results.

Response	Percent
No opinion	10
Not at all	11
Not too	28
Somewhat	41
Very	10

5. The frequency distribution of blood type for a group of African-Americans is given below. A pie chart is to be constructed for this distribution. Determine the angle that goes with each class.

Blood type	Percent
O	49
A	27
B	20
AB	4

6. A survey asked 60 individuals to list the legal drug that was taken most often for pain. The results were as follows. Construct a bar chart for these data.

OxyContin	Aspirin	Advil	Valium	Valium	Aspirin
Valium	Valium	Advil	Aspirin	Aspirin	Aspirin

(*continued*)

Aspirin	Aspirin	OxyContin	Advil	Aspirin	Tylenol
Advil	Advil	Valium	Advil	Advil	Aspirin
Tylenol	Aspirin	Aspirin	Advil	Advil	Aspirin
Aspirin	Advil	Aspirin	Tylenol	Aspirin	Aspirin
Aspirin	Aspirin	Tylenol	Aspirin	Advil	OxyContin
Advil	Aspirin	Aspirin	Tylenol	Advil	Aspirin
Advil	Advil	Aspirin	Aspirin	OxyContin	Tylenol
Aspirin	Advil	OxyContin	Aspirin	Valium	Tylenol

7. Thirty individuals were asked to give the time in minutes that they spent in planned exercise the past week. Construct a dotplot for these data.

300	420	360	360	360
360	360	210	75	300
360	300	0	360	300
0	0	360	0	360
360	210	210	360	210
0	360	0	360	210

8. Two different teaching methods were used in an experiment. Method 2 used software and the other did not. The results on a common final exam are shown below in the two dotplots. Which method produced the better results?

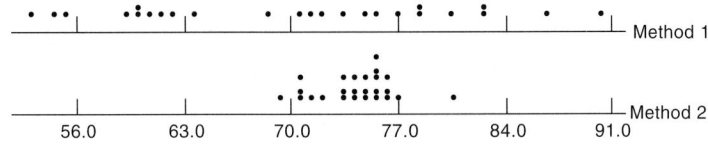

9. A frequency distribution of summer daily high temperatures for New Orleans and Seattle are graphed for a 5-year period. What would be the relationship between the two distributions?

10. Suppose you were to construct a histogram of the yearly salaries of male and female professional basketball players combined. What type of distribution would you expect to see?

Chapter 3 Exercises

Section 3.3

1. Form a dotplot and on that dotplot locate the mean of the following data. Comment on the location of the mean.

$$8, 10, 8, 7, 86, 5, 10, 10, 25, 6, 9, 9, 11, 9, 12, 15$$

2. Form a dotplot and on that dotplot locate the mean of the following data. Comment on the location of the mean.

$$1, 2, 3, 18, 23, 25, 26, 25, 26, 27, 21, 23, 24, 26, 30$$

3. Form a dotplot and on that dotplot locate the mean of the following data. Comment on the location of the mean.

$$1, 2, 2, 3, 3, 3, 4, 4, 4, 4, 5, 5, 5, 6, 6, 7$$

4. To give some idea of the time required to complete a Schedule C tax form, 15 people were randomly selected and timed. The times, in minutes, required to complete the form were as follows. Make a dotplot of the data. Use the dotplot to describe the amount of time required to fill out the form.

$$39, 40, 42, 35, 42, 42, 51, 39, 45, 42, 37, 41, 33, 45, 41$$

5. The mean credit hours taken by students at Midwestern University was estimated by taking a sample of 100 students and using the sample mean to estimate the mean for the university. The total number of credit hours taken by the 100 students was 1340. What is the estimate of the university mean for the present semester?

6. A software company consists of nine programmers and the owner of the company. The salaries, in thousands, of the company employees and the owner are given below. Construct a dotplot of the salaries. Determine the mean and median, and indicate which is more representative of the company salaries.

$$40, 45, 47, 43, 44, 39, 37, 40, 41, 250$$

7. Construct a dotplot of the following data. Find the mean, median, and mode. Comment on what you observe.

$$10, 20, 20, 30, 30, 30, 40, 40, 40, 40, 50, 50, 50, 60, 60, 70$$

8. In the case of home prices, which of the three measures of central tendency is preferred and why?

9. Calculate the mean and the 10% trimmed mean, and compare the two for the following data.

0.33	3.30	0.57	1.47
0.87	0.46	10.07	1.09
0.72	2.42	0.49	0.19
0.22	1.40	4.01	1.23
1.12	1.08	0.83	0.28
0.71	12.02	0.65	0.55
1.67	0.95	1.01	1.34
3.80	0.51	2.24	0.77
0.16	0.68	1.77	1.19
1.23	2.89	2.55	2.13

10. Draw a histogram for the following data.

0.33	3.30	0.57	1.47
0.87	0.46	10.07	1.09
0.72	2.42	0.49	0.19
0.22	1.40	4.01	1.23
1.12	1.08	0.83	0.28
0.71	12.02	0.65	0.55
1.67	0.95	1.01	1.34
3.80	0.51	2.24	0.77
0.16	0.68	1.77	1.19
1.23	2.89	2.55	2.13

a. Does the histogram lead you to believe that the data are symmetric or skewed? If they are skewed, in which direction?

b. Calculate the mean and the median of the data and compare them. What does this indicate about the skewness of the data?

11. The number of e-mails sent during the past week by the employees at a mail-order company were as follows.

e-mails	Frequency
28	11
30	14
32	18
34	20
36	22
38	18
40	14
44	12
48	10

What is the mean number of e-mails sent?

12. Refer to the distribution in Exercise 11. Find the median.

13. Refer to the distribution in Exercise 11. Find the mode.

14. The numbers of faxes sent during the past week by the employees at a mail-order company were as follows. What is the mean number of faxes sent per employee?

32	19	20	23
12	21	22	27
28	23	22	22
22	27	19	24
20	21	19	26
25	21	26	23
22	22		

15. Refer to the data in Exercise 14. What is the median number of faxes sent per employee?

16. Refer to the data in Exercise 14. What is the modal number of faxes sent per employee?

Section 3.4

1. Calculate the range of the following data. Is the range a reliable measure of how spread out the data are in this case?

$$6.5, 3.4, 4.3, 5.1, 5.4, 5.7, 6.0, 5.7, 4.7, 4.9$$

2. Calculate the range of the following data. Is the range a reliable measure of how spread out the data are in this case?

$$10.2, 11.1, 5.6, 5.8, 6.2, 2.9, 8.2, 13.4, 9.1, 15.2$$

3. Using the definition of variance, calculate the variance and standard deviation for the following data.

$$1.5, 3.4, 5.4, 4.2, 2.5$$

4. Using the shortcut formula for variance, calculate the variance and standard deviation for the following data.

$$1.5, 3.4, 5.4, 4.2, 2.5$$

5. The following are the service times (in seconds) of McDonald's drive-through customers. Find the variance and standard deviation.

79.1	69.3	101.4	77.3	79.6	123.0	51.5	76.6	73.4	59.2
92.9	80.4	94.5	70.7	80.4	77.0	54.6	64.1	79.3	49.6

6. The following are the weights of 10-pound bags of potatoes. Find the mean and standard deviation as well as $\bar{x} \pm s$, $\bar{x} \pm 2s$, and $\bar{x} \pm 3s$. Determine the number of data values within the three intervals and compare with the empirical rule.

$$10.2, 9.7, 10.2, 9.9, 9.9, 10.2, 10.3, 10.2, 9.9, 10.0$$

7. The following are the service times (in seconds) of McDonald's drive-through customers. Find the mean and standard deviation as well as $\bar{x} \pm s$, $\bar{x} \pm 2s$, and $\bar{x} \pm 3s$. Determine the number of data values within the three intervals and compare with the empirical rule.

79.1	69.3	101.4	77.3	79.6	123.0	51.5	76.6	73.4	59.2
92.9	80.4	94.5	70.7	80.4	77.0	54.6	64.1	79.3	49.6

8. The numbers of collision claims reported monthly to an insurance office during the past year were as follows:

$$66, 62, 53, 45, 53, 42, 57, 55, 70, 53, 45, 125$$

Give the z-value for the December reading, 125. Is it an outlier? Why or why not?

9. Two machines were compared with respect to the number of defectives reported per ten thousand produced. Eleven values were reported over the same reporting periods. Comment on the defectives produced.

Machine 1:	8	12	7	11	11	10	10	10	11	12	9
Machine 2:	11	10	9	10	9	9	11	11	10	9	10

10. Compute the z-value for each value in the following data set, and comment on the z-values.

10	32	24	27	23	25	19	28	24	29
24	30	34	50	21	17	21	21	20	31

11. A statistician is not located near her statistical software and has to calculate the variance of 41 numbers the hard way, by using a calculator that does not have a built-in routine to calculate variance. She found the sum of the observations to be 320 and the sum of the squares of the observations to be 4200. What is the sample variance?

12. The Hope Index is derived from the answers to a battery of questions to people in different nations about whether they expect to be better off after 1 year, after 10 years, and whether their children will be better off than they are. The maximum possible score is 100. Recent results, published in the *Economist*, are shown in the table for the 29 nations that were sampled. Calculate the range, variance, and the standard deviation for the Hope Index for the data.

Country	Hope Index	Country	Hope Index
Argentina	32	Japan	12
Australia	39	Malaysia	66
Belgium	29	Netherlands	30
Brazil	64	Norway	23
Britain	48	Poland	33
Canada	45	Russia	20
China	58	South Africa	42
Colombia	66	South Korea	62
Czech Republic	26	Spain	34
Denmark	25	Sweden	33
France	17	Taiwan	50
Germany	18	Thailand	60
Greece	33	Turkey	44
Israel	42	United States	62
Italy	34		

13. A study of the life expectancy of 100 lightbulbs found that the mean time they lasted before burning out was 1900 hours. If the standard deviation was 150 hours, then according to the empirical rule, approximately 68 of the bulbs burned out between _____ and _____ hours.

14. Margaret Morris just correctly computed the standard deviation to be 6.0 from a sample of 50 observations. What is the variance of the sample?

15. A student scored 96 on her last biology unit test. The mean score of the 36 students in her class was 78 and the standard deviation was 12. What was her z-score?

Section 3.5

1. The following is a set of 100 test scores made on a test.

21	38	43	49	54	58	62	67	71	78
27	38	44	50	54	58	62	67	72	78
27	39	45	51	55	58	63	68	73	80
29	39	46	51	55	58	63	68	73	86
32	39	47	51	56	58	64	69	73	86

(*continued*)

32	39	47	52	56	58	64	69	74	88
33	41	47	52	56	58	64	70	75	89
34	41	48	53	57	59	64	70	75	91
34	42	49	53	57	60	65	70	76	93
34	42	49	53	57	61	65	71	77	96

Give the percentile rank for the score 58.

2. For the test scores in Exercise 1, find the first and third quartiles.

3. Find the inner fences. Locate the ends of the two whiskers.

4. Construct the boxplot.

5. Find the outer fences.

6. The following table gives the number of defectives on 100 days produced by a machine. The data are ordered.

0	5	6	7	9	10	11	12	13	14
4	6	6	7	9	10	11	12	13	15
4	6	6	7	9	10	11	12	13	15
5	6	7	8	9	11	11	12	13	15
5	6	7	8	9	11	11	12	14	16
5	6	7	8	9	11	11	12	14	17
5	6	7	8	9	11	11	13	14	18
5	6	7	8	9	11	12	13	14	18
5	6	7	8	9	11	12	13	14	25
5	6	7	9	9	11	12	13	14	35

Give the percentile rank for the score 11.

7. For the numbers of defectives in Exercise 6 of this section, find the first and third quartiles.

8. Find the inner fences. Locate the ends of the two whiskers.

9. Find the outer fences.

10. Construct the boxplot.

11. The following table gives the Monday closing values for the 15 most widely held stocks. All fractions have been converted to decimal equivalents.

Stock	Monday Close	Stock	Monday Close
AT&T	52.9375	IBM	115.5
Bell Atlantic	60.875	Johnson & Johnson	90.625
Cisco Systems	107.8125	Lucent	58.3125
Compaq	22.0	Merck	66.25
Disney	29.0	Microsoft	77.5625
Exxon	81.125	Pfizer	95.125
GE	103.75	SBC Comm.	54.625
Intel	54.375		

Give the value of the first and third quartiles.

12. Find the inner fences. Locate the ends of the two whiskers.

13. Find the outer fences.

14. Construct the boxplot.

15. The table shows the seating capacity of the home stadiums of the 30 major league baseball teams (in thousands, rounded to the nearest thousand) as of the 1998 season.

Team	Capacity	Team	Capacity
Diamondbacks	49	Angels	65
Braves	51	Orioles	48
Cubs	39	Red Sox	34
Reds	53	White Sox	44
Rockies	50	Indians	42
Marlins	48	Tigers	52
Astros	54	Royals	41
Dodgers	56	Brewers	53
Expos	47	Twins	57
Mets	56	Yankees	58
Phillies	63	A's	43
Pirates	57	Mariners	60
Cardinals	57	Devil Rays	45
Padres	47	Rangers	49
Giants	63	Blue Jays	51

Give the first and third quartiles.

16. Find the inner fences. Locate the ends of the two whiskers.

17. Find the outer fences.

18. Construct the boxplot.

19. The following data show the number of deaths caused by major tornadoes for 33 recent years. Find the first and third quartiles.

57	17	25
61	10	23
33	12	26
71	57	52
32	15	4
110	17	23
47	75	4
315	29	7
22	18	4
60	9	26
4	13	27

20. Find the inner fences. Locate the ends of the two whiskers.

21. Find the outer fences.

22. Construct the boxplot.

Chapter 3 End-of-Chapter Exercises

The following table gives the distances (in feet) of the 73 home runs that Barry Bonds hit during the 2001 baseball season. (This set the record for home runs during a season.) Use these data in Exercises 1–9.

420	370	415	410	320	410	488	442
417	420	436	450	360	380	361	404
440	400	430	320	375	430	394	385
410	360	410	430	370	415	410	
390	410	400	380	440	380	411	
417	420	390	375	400	375	365	
420	391	420	375	405	400	360	
410	416	410	347	430	435	440	
380	440	420	380	350	420	435	
430	410	410	429	396	420	454	

1. Find the mean, median, and mode for these data.

2. Construct a dotplot for the data.

3. Find the range, variance, and standard deviation of the home run distances.

4. Find the mean and standard deviation as well as $\bar{x} \pm s$, $\bar{x} \pm 2s$, and $\bar{x} \pm 3s$. Determine the number of data values within the three intervals and compare with the empirical rule.

5. Give the percentile rank for the distance 440.

6. Find the first and third quartiles.

7. Find the inner fences. Locate the ends of the two whiskers.

8. Find the outer fences.

9. Construct the boxplot.

Chapter 4 Exercises

Section 4.2

1. During the 2000 Presidential race, polls were conducted to see how each candidate was doing. The following poll was conducted in four large counties prior to the election. The candidates were classified as Bush, Gore, and Other.

	Midwest	Easton	Weston	Southern
Bush	100	50	175	125
Gore	50	110	125	100
Other	25	15	50	35

Form a clustered bar chart where the counties are displayed along the horizontal direction. In what county does Bush have the greatest percentage lead? In what county does he have the smallest percent?

2. Refer to the data in Exercise 1. Form a clustered bar chart where the candidates are displayed along the horizontal direction. What conclusion can be reached?

3. Refer to the data in Exercise 1. Form a stacked bar chart where the counties are displayed along the horizontal direction. What conclusion can be reached?

4. Refer to the data in Exercise 1. Form a stacked bar chart where the candidates are displayed along the horizontal direction. What conclusion can be reached?

5. The following contingency table was partially completed. What are the values for A, B, C, and D?

	Male	Female	Total
Undergraduate	15	A	55
Graduate	B	15	45
Total	C	55	D

6. The following table shows the living arrangements for graduates and undergraduates at Midwestern University. Form a clustered bar chart with living arrangement on the horizontal axis. What conclusion can be reached?

	Dorm	**Apartment**	**Other**
Graduate	50	250	30
Undergraduate	2000	500	50

7. Refer to the data of Exercise 6 of this section. Form a clustered bar chart with undergraduate–graduate status on the horizontal axis. What conclusion can be reached?

8. Your office associate, Julie Brown, has been building a stacked bar chart to analyze three variables that contain three categories each. Julie is called away on business and leaves you with the task of finishing the table needed to build the chart. She assures you in an e-mail that the raw data need not be known. The variable of interest is the column variable. The following table of percentages has thus far materialized from her work.

	Category A (%)	Category B (%)	Category C (%)
Category 1	19	24	30
Category 2	35	A	20
Category 3	B	32	50
Total	C	D	E

Find the correct values for *A* and *B*.

9. Refer to Exercise 8. Find the correct values for *C*, *D*, and *E*.

10. Refer to Exercise 8 of this section. Recall that when the data were originally collected, 200 observations for the variable of interest fell into category 2. Therefore, you conclude that the number of these observations that also fell into the row variable's category 2 is equal to what?

11. An auto tire manufacturer conducted a survey of 300 automobile owners who were using a specific set of original equipment tires on a new car they had purchased from a dealer. Three different brands of vehicles were isolated, and each owner was asked to rate the overall performance of their tires after driving the car for 1 year. The following contingency table summarizes the results of the study:

Tire Rating	Car A	Car B	Car C	Total
Excellent	20	40	50	110
Good	30	20	30	80
Fair	40	20	10	70
Poor	10	20	10	40
Total	100	100	100	300

Construct a clustered bar chart with tire rating as the horizontal axis. Which car got the best rankings?

12. Refer to Exercise 11 of this section. Construct a clustered bar chart with car type as the horizontal axis. Which car got the best rankings?

13. Refer to Exercise 11 of this section. Construct a stacked bar chart with tire ratings as the horizontal axis.

14. Refer to Exercise 11 of this section. Construct a stacked bar chart with car type as the horizontal axis.

Section 4.3

1. The height and weight of 10 adult males are plotted on a scatter plot. The data are as follows. What relationship is apparent from the plot?

Height	Weight
64	143
67	159
68	164
69	167
71	169
71	171
73	174

(*continued*)

73	192
73	201
74	205

2. Find the equation of the least-squares line. Predict the weight of an adult male who is 70 inches tall.

3. Ten plots are given differing amounts of moisture and the weight of tomatoes on the plot is determined. The data are as follows. What relationship is apparent from the plot?

Moisture	Tomatoes
1	5.5
2	6.1
3	6.5
4	6.5
5	6.7
6	7.0
7	6.5
8	6.3
9	5.5
10	4.0

4. Ten individuals were chosen and two variables were measured: the number of letters in the person's last name and the number of e-mails the person sent per day. The data are as follows. What relationship is apparent from the plot?

e-mail	Letters
8	8
8	15
12	11
15	12
8	12
5	12
11	12
8	8
11	7
5	9

5. The weights of autos (in thousand of pounds) and their miles per gallon (mpg) ratings are determined for a wide variety of cars and trucks in Europe. The data are as follows.

Weight	mpg
2.6	44
3.5	38
3.6	34
4.5	34
4.6	33
4.9	31
6.1	30
6.2	29
6.5	24
9.8	15

What relation is apparent from a plot of the data?

6. Refer to Exercise 5. Find the equation of the least-squares line. Predict the mpg of a car that weighs 5000 pounds.

7. A study records the hours spent watching TV per week and the GPA of eighth grade students. The data were as follows.

Hours	20	25	30	32	34	36	38	41	43
GPA	3.4	3.2	3.1	3.0	2.9	2.8	2.6	2.5	2.3

What would you conclude about the relationship between the two variables?

8. Refer to Exercise 7 of this section. Suppose the equation connecting GPA and Hours is GPA = 4.41 – 0.0465Hours. What would you predict the GPA to be for an eighth grade student who watches 25 hours of TV per week?

9. A study records the high school GPA of 15 female seniors and the number of interscholastic high school sports they participated in during their high school years. The results are as follows.

GPA	2.3	2.3	2.4	2.5	2.8	2.5	2.8	3.0	3.4	3.2	3.3	3.4	2.9	3.9	3.7
Sports	0	2	2	3	3	5	7	7	8	9	9	10	11	12	12

What relation is apparent from the scatter plot of the data?

10. Find the equation of the least squares line.

11. Plot the following data. What relationship is apparent from the scatter plot of the data?

x	y
5	35
7	30
10	28
15	20
25	25

12. Refer to Exercise 11 of this section. What is the equation of the least-squares line?

13. Refer to Exercises 11 and 12 of this section. Make a table showing the deviations, e, and show that they sum to 0.

14. Refer to Exercises 11, 12, and 13 of this section. The equation of the least-squares line is $\hat{y} = 33.5 - 0.475x$. Plot the least-squares line and the data points on the same graph.

Chapter 4 End-of-Chapter Exercises

1. A survey asked males and females to give their primary news source. The results of the survey are given in the table below.

	Newspaper	Radio	TV	Other
Male	50	250	150	50
Female	100	150	200	50

Construct a clustered bar chart with gender on the horizontal axis. What is the primary source for both genders?

2. Refer to Exercise 1 of this section. Construct a clustered bar chart with news source on the horizontal axis. Which gender is primary for each source?

3. Refer to Exercise 1 of this section. Construct a stacked bar chart with gender on the horizontal axis. What percent of the males choose radio as their primary source, and what percent of the females choose TV as their primary source?

4. Refer to Exercise 1 of this section. Construct a stacked bar chart with news source on the horizontal axis. Which gender is primary for each source?

5. Change the table given in Exercise 1 of this section to a percent table.

6. Refer to the data in Exercise 1 of this section. Fill in the following table with percents.

	Newspaper	Radio	TV	Other
Male				
Female				
Total (%)	100	100	100	100

7. Refer to the data in Exercise 1 of this section. Fill in the following table with percents.

	Newspaper	Radio	TV	Other	Total (%)
Male					100
Female					100

8. The selling price (in thousands) and the area (in square feet) for five houses are given below. Make a scatter plot for the data and comment on the pattern of the data.

Price	Area
125	1250
150	1750
325	2500
300	2750
375	3500

9. Refer to Exercise 8 of this section. Find the equation of the least-squares line. Predict the cost of a house that has an area of 3000 square feet.

10. A study records the average yearly rainfall (in inches) and the number of forest fires for a particular location in the western United States for a 5-year period. The data are as follows. Make a scatter plot for the data and comment on the pattern of the data.

Rainfall	Fires
10.1	12
15.2	10
13.2	11
17.5	8
8.7	15

11. Refer to Exercise 10 of this section. Find the equation of the least-squares line. Predict the number of fires for 12 inches of rainfall.

12. A study is performed in which the reaction time, y, and the amount of drug, x, are measured for 20 patients. The amount of drug varies between 1 and 5 units for all individuals in the study. The least-squares line is found to be $\hat{y} = -0.1 + 0.7x$. Find the estimated reaction time for $x = 6$ units of drug. What is wrong with this estimate?

13. A study is performed in which the reaction time, y, and the amount of drug, x, are measured for 20 patients. The amount of drug varies between 1 and 5 units for all individuals in the study. The least-squares line is found to be $\hat{y} = -0.1 + 0.7x$. Find the estimated reaction time for $x = 3$ units of drug. Will this estimate be a good one?

14. A least-squares line was fit to the following data. The equation of the least-squares line is $\hat{y} = 1.30 + 0.900x$. Find the deviations, $e = \hat{y} - y$, and show that they sum to 0.

x	y
1	2
2	3
3	5
4	4
5	6

15. A least-squares line was fit to the following data. The equation of the least-squares line is $\hat{y} = 21.6 + 0.42x$. Plot the least-squares line and the data points on the same graph.

x	y
10	25
15	30
20	28
25	33
30	34

Chapter 5 Exercises

Section 5.2

1. Consider the experiment of rolling a pair of dice one time. Give a representation of the sample space for this experiment.

2. Consider the experiment of tossing a coin three times and observing the outcome on the three tosses. Give a representation of the sample space for this experiment.

3. An experiment consists of selecting one card from a standard deck of 52. Give a representation of the sample space for this experiment.

4. An experiment consists of rolling a die followed by flipping a coin. Give a representation of the sample space for this experiment.

5. A pair of dice is rolled one time. Find the probability that the sum of the two faces is not equal to 2.

6. A coin is tossed three times. The sample space for this experiment is given in Exercise 2 of this section. Let A be the event that exactly one head is obtained. Find A and A$'$, and show that P(A) + P(A$'$) = 1.

7. A survey asked couples to give the primary contraception method they used. The survey size was 5000. The results of the survey were as follows.

Birth Control Method	%
Tubal sterilization	26
Pill	25
Male condom	19
Partner's vasectomy	10
Other	20

a. If one couple is selected at random from the survey, what is the probability that they specified tubal sterilization or pill as their birth control method?

b. If one couple is selected at random from the survey, what is the probability that they did not specify male condom as their birth control method?

8. A sampling of Black and White Americans was obtained with respect to blood type. The observed frequencies of the four blood types and race are given in the table. One person is chosen at random.

	Blood type			
	O	A	B	AB
Black	85	44	36	7
White	445	404	110	26

a. What is the probability the person selected does not have blood type AB?

b. What is the probability that a person with blood type O is chosen?

9. U.S. mothers had a median weight gain of 30.5 pounds during their pregnancies in 2002. A breakdown of pregnancy weight gains is shown in the table. If a woman is selected at random, what is the probability she gained more than 25 pounds? Solve by using the complement of the event and also by not using the complement of the event.

Weight Gain in Pounds	Percent
Under 16	12
16–25	25
26–35	32
36–45	20
46+	11

10. Most crashes occur within 2–5 miles of home. The following table gives the percent of crashes as a function of distance (in miles) from home.

Distance	1 or less	2–5	6–10	11–15	16–20	>20
Percent	23	29	17	8	6	17

Find the percent of crashes that occur within 5 miles of home.

11. Find the probability of each of the sums, 2 through 12, when a pair of dice is rolled.

12. An experiment consists of rolling three dice. There are 216 outcomes ranging from the triple $(1, 1, 1)$ to the triple $(6, 6, 6)$. Find the probability that the same number occurs on each of the three dice.

Section 5.3

1. One thousand people, half male and half female, were asked to give their primary source of the news. The results of the survey were as follows. One of the people surveyed is selected at random.

	Radio	TV	Newspaper	Other
Male	257	135	58	50
Female	189	201	68	42

a. What is the probability that the person was a female or responded Radio?
b. What is the probability that the person did not respond Newspaper?
c. What is the probability that the person responded TV?

2. A survey comparing males and females with respect to time on-line was conducted. The observed frequencies were as follows.

	Less Than 10 Hours	10–20 Hours	More Than 20 Hours
Male	24	76	33
Female	17	95	27

a. What is the probability of spending more than 20 hours on-line?
b. What is the probability of spending less than 10 hours and being female?

3. The number of vehicles per household was determined for 1990 and 2000. The numbers of households (in millions) were as follows.

	No Vehicles	One Vehicle	2 Vehicles	3 or More
1990	10.6	31	34.4	15.9
2000	10.9	36.1	40.5	18

a. What percent of households had two or more vehicles in 2000?
b. What is the percent difference with 3 or more vehicles in 2000 than in 1990?

4. A study recorded two pieces of information about the people in a survey: the number of letters in their last name (either 8 or not 8) and party affiliation (Democrat, Republican, or Other). The results of the survey were as follows.

	Democrat	Republican	Other
8 letters in last name	85	105	30
Other than 8 letters in last name	255	315	90

a. Find the percent of Democrats, the percent of Republicans, and the percent of Others in the survey.
b. Find the percent who were Democrats and had 8 letters in the last name, find the percent who were Republicans and had 8 letters in the last name, and find the percent who were Other and had 8 letters in the last name.

5. Survey respondents $(n = 377)$ were classified according to two criteria: White or African American, and response to the question "How often do you experience stress in your life?"

How often do you experience stress in your daily life?

	Frequently	Sometimes	Rarely	Never
White	52	132	43	12
African American	33	76	23	6

An individual is selected at random.

a. What is the probability that the individual did not respond that he or she frequently experiences stress?

b. What is the probability that the individual was White or responded that he or she sometimes experiences stress?

6. A survey was made to determine whether individuals had been in a chat room on the Internet. The respondents were also classified according to their age.

	Age		
Chat Room?	19 or Less	20 to 50	More Than 50
Yes	134	75	34
No	34	84	125

a. What percent of the survey respondents had been in a chat room?

b. What percent were more than 50 and had been in a chat room?

c. What percent had been in a chat room or were 19 or less?

7. One card is drawn from a deck of 52. Event A is that the card is a club, and event B is that the card is a heart. Find the probability of the event A OR B.

8. One card is drawn from a deck of 52. Event A is that the card is a club, and event B is that the card is a face card (jack, queen, or king). Find the probability of the event A OR B.

9. A die is rolled, followed by a coin flip. Find the probability of an odd number or a head occurring.

10. Find the probability of a 7 or an 11 when a pair of dice is rolled.

11. One card is drawn from a deck of 52. Let A be the event that the card is black, and let B be the event that the card is a face card. Find the probability that the card is black and a face card.

12. One of the Barry Bonds home runs, listed in the following table, is selected at random. Find the probability that the Giants won that game (Outcome = W) or the home run was hit in the first inning (Inn = 1).

Number	Inn	Dist.	Field	Outcome	Number	Inn	Dist.	Field	Outcome
1	5	420	CF	W	19	7	440	RF	W
2	4	417	LCF	L	20	8	410	CF	W
3	1	440	RCF	W	21	1	415	RCF	L
4	5	410	RF	L	22	7	436	CF	L
5	8	390	LCF	L	23	4	430	CF	L
6	8	417	RF	W	24	9	410	LCF	L
7	7	420	RF	W	25	3	400	RF	L
8	4	410	CF	W	26	1	390	RF	W
9	3	380	RCF	L	27	2	420	RF	L
10	8	430	CF	L	28	6	410	CF	L
11	4	370	RF	L	29	3	420	RF	W
12	5	420	CF	W	30	4	410	CF	W
13	1	400	RCF	L	31	3	410	CF	W
14	6	360	RF	W	32	7	450	CF	L
15	4	410	CF	W	33	1	320	RF	W
16	3	420	CF	L	34	6	430	RCF	W
17	8	391	RF	L	35	1	380	LCF	W
18	3	416	RCF	W	36	6	375	RF	W

(*continued*)

37	5	375	RCF	L	56	5	375	RF	W
38	8	347	RF	W	57	8	400	RF	L
39	1	380	RF	L	58	4	435	CF	L
40	1	429	CF	L	59	7	420	CF	W
41	4	320	RF	W	60	2	420	RCF	W
42	5	360	LF	W	61	1	488	CF	W
43	4	375	RF	W	62	5	361	RF	W
44	5	370	LF	W	63	11	394	RF	W
45	4	440	RCF	W	64	5	410	CF	L
46	1	400	RCF	W	65	2	411	CF	W
47	6	405	RF	L	66	4	365	LF	W
48	11	430	RCF	W	67	7	360	RF	W
49	3	350	RF	W	68	2	440	RCF	W
50	2	396	CF	W	69	6	435	RF	W
51	6	410	RF	W	70	9	454	RCF	W
52	4	380	RCF	W	71	1	442	RCF	L
53	8	430	CF	W	72	3	404	LCF	L
54	8	415	RCF	L	73	1	385	RF	W
55	9	380	RF	W					

13. Refer to Exercise 12. Find the probability that the home run was hit to left field (Field = LF) or it came in the fifth inning (Inn = 5).

Section 5.4

1. A card is drawn from a deck of 52 cards. A is the event that the card is a face card, and B is the event that the card is a club. Find $P(A|B)$ and $P(B|A)$.

2. A card is drawn from a deck of 52 cards. A is the event that the card is red, and B is the event that the card is a diamond. Find $P(A|B)$ and $P(B|A)$.

3. Consider the experiment of tossing a coin three times and observing the outcome on the three tosses. A is the event that more heads than tails were obtained, and B is the event that two heads were obtained. Find $P(B|A)$.

4. An experiment consists of rolling a die followed by flipping a coin. A is the event that a head occurred when the experiment was performed. B is the event that an odd number occurred when the experiment was performed. Determine whether A and B are independent.

5. Consider the experiment of rolling a pair of dice one time. Event A is the event that a 1 occurred on die 1. Event B is the event that the total on the two dice was less than 7. Are A and B independent?

6. One thousand people, half male and half female, were asked to give their primary source of the news. The results of the survey were as follows.

	Radio	TV	Newspaper	Other
Male	257	135	58	50
Female	189	201	68	42

Find the probability that the person's news source is radio given that the person is a female.

7. A survey was made to determine whether individuals had been in a chat room on the Internet. The respondents were also classified according to their age.

	Age		
Chat Room	19 or Less	20 to 50	More Than 50
Yes	134	75	34
No	34	84	125

Find the probability that a person is 19 or less given that they have been in a chat room.

8. A study recorded two pieces of information about the people in a survey: the number of letters in their last name (either 8 or not 8) and party affiliation (Democrat, Republican, or Other). The results of the survey were as follows. A is the event that the person has 8 letters in their last name, and B is the event that the person is a Republican. Show that A and B are independent.

	Democrat	Republican	Other
8 letters in last name	85	105	30
Other than 8 letters in last name	255	315	90

9. Given that $P(A) = 0.4$, $P(B) = 0.3$, and A and B are independent, find $P(A \text{ OR } B)$.

10. Given that $P(A) = 0.5$, $P(B) = 0.4$, and $P(A \text{ OR } B) = 0.7$, are A and B independent?

Chapter 5 End-of-Chapter Exercises

1. A game consists of flipping a coin followed by the drawing of a card. How many outcomes are there in the sample space? Give a general description of the sample space.

2. A game consists of flipping a coin followed by rolling a pair of dice. How many outcomes are there in the sample space? Give a general description of the sample space.

3. In Exercise 1 of this section, what is the probability of a head followed by an ace?

4. Refer to Exercise 2 of this section. Event A is the event that a sum of 3 or greater occurred on the die roll. List the outcomes in the complement of A, and give the probability of A.

5. In Exercise 1 of this section, the experiment of flipping a coin followed by the drawing of a card is described. Let event A be that a king was obtained, and let B be the event that a queen was obtained. Find the probability of the event A OR B.

6. In Exercise 1 of this section, the experiment of flipping a coin followed by the drawing of a card is described. Let A be the event that a tail followed by a face card was obtained, and let B be the event that a tail followed by a spade was obtained. Find the probability of the event A OR B.

7. Office workers were asked how long it generally takes them to respond to e-mail. Their responses are given in the following table.

Response	Percent
As soon as I return to my desk	36%
Within an hour or two	35%
Before the end of the business day	24%
When I can	5%

What percent did not respond, "When I can"?

8. Two cards are drawn without replacement from a deck of 52. Find the probability that both are red.

9. A cancer study reported the following results from smoking cigars.

Cigars	Died from cancer		
	Yes	No	Totals
Never smoked	780	120,000	120,780
Former smoker	90	8,000	8,090
Current smoker	140	8,000	8,140
Totals	1010	136,000	137,010

Given that the person was a former or current cigar smoker, what is the probability that he or she died from cancer?

10. A container has 7 red marbles and 3 white marbles. Three are selected with replacement.
a. What is the probability that all 3 are red?
b. What is the probability that all 3 are white?

Chapter 6 Exercises

Section 6.2

1. Determine which of the following are probability distributions for random variables.

a.

x	0	1	2	3
$p(x)$	0.3	0.4	0.2	0.1

b.

x	0	1	2	3	4
$p(x)$	0.5	−0.2	−0.3	0.6	0.4

c.

x	1	3	5	7	9	11	13
$p(x)$	0.1	0.2	0.3	0.2	0.1	0.1	0.1

2. Let X = # of heads in three tosses of a fair coin. Derive the probability distribution for X.

3. Draw a probability distribution histogram for the variable in Exercise 2 of this section.

4. Four-feet by eight-feet sheets of plywood are sold by House Depot, a large superstore. The following gives the distribution of the number of defects per sheet of plywood.

x	0	1	2	3	4	5	6
$p(x)$	0.8	0.05	0.05	0.025	0.025	0.025	0.025

a. Find the probability that a sheet will have at least 1 defect.
b. Find the probability of more than 2 defects per sheet.
c. Find the probability of at most 3 defects per sheet.
d. Find the probability of fewer than 3 defects per sheet.
e. Find the probability of between 3 and 5 defects per sheet.
f. Find the probability of between 3 and 5 defects, inclusive, per sheet.

5. Find the distribution for the random variable X = the sum to appear when a pair of dice are rolled.

6. Draw a probability distribution histogram for the variable in Exercise 5 of this section.

7. A random variable takes on the values x = 1, 2, 3, . . . with probabilities 1/2, 1/4, 1/8, 1/16, ... Such a random variable is said to take on a countably infinite number of values. Show that the probabilities add to 1.

8. Show that the following is a probability distribution.

x	1	2	3	4	5
$p(x)$	0.5	0.25	0.125	0.0625	0.0625

9. Draw the probability histogram for the random variable in Exercise 8.

10. Show that the following is a probability distribution.

x	1	2	3	4	5	6
$p(x)$	0.03125	0.03125	0.0625	0.125	0.25	0.5

11. Draw the probability histogram for the random variable in Exercise 10.

12. Show that the following is a probability distribution.

x	1	2	3	4	5	6	7
$p(x)$	0.05	0.1	0.2	0.3	0.2	0.1	0.05

13. Draw the probability histogram for the random variable in Exercise 12.

14. The number of defectives per lot has the following probability distribution.

x	0	1	2	3	4
$p(x)$	0.9	0.07	0.01	0.01	0.01

Ten thousand lots are produced over a year. About how many will be defect-free?

15. The number of peanuts per shell follows the following probability distribution.

x	0	1	2	3
$p(x)$	0.03	0.3	0.6	0.07

What is the most likely number of peanuts in a shell?

Section 6.3

1. A sample of 15 individuals is chosen and the number of left-handers in the 15 is determined. Show that X = the number of left-handers in the 15 is a binomial random variable.

2. About 15% of the population is left-handed. Fifteen individuals are randomly selected. What is the probability that:
a. 5 or more are left-handed?
b. 3 or fewer are left-handed?

3. What are the mean and standard deviation of the number of left-handers to be found in a group of 15 individuals if 15% of the population is left-handed? Is it unusual to find 10 left-handers in the 15?

4. It is estimated that 10% of the population have a personal digital assistant (PDA). A sample of 10 is selected. What is the probability of finding more than 2 in the sample who have PDAs?

5. It is estimated that 10% of the population have a personal digital assistant (PDA). A sample of 10 is selected. Suppose 4 in the 10 are found to have a PDA. Calculate the mean, the standard deviation, and the z-score for 4.

6. Suppose binomial tables are not available and neither is a software package. Calculate the probability of 4 successes in 10 trials when $p = 0.25$.

7. Plot the binomial distribution with $n = 10, p = 0.1$.

8. Plot the binomial distribution with $n = 10, p = 0.5$.

9. Plot the binomial distribution with $n = 10, p = 0.8$.

10. A coin is flipped 3 times. X is the number of heads to occur. Derive the binomial distribution by the use of and/or rules from probability.

11. Refer to Exercise 10. Use the binomial formula to derive the binomial distribution for $n = 3$ and $p = 0.5$.

12. Seventy-five percent of college students have a home computer. A sample of 20 college students is selected. Find the probability that all 20 have a home computer.

13. Seventy-five percent of college students have a home computer. A sample of 20 college students is selected. Find the mean and standard deviation of the number in the 20 who have a home computer.

14. A coin is flipped 100 times. Find the mean and standard deviation. If 80 heads are obtained, find the z-score and comment on it.

15. A pair of dice are rolled 20 times. Five 7s are rolled. Find the mean and standard deviation, and comment on obtaining five 7s.

Section 6.4

1. Give a probability statement to describe the shaded area under the following curve.

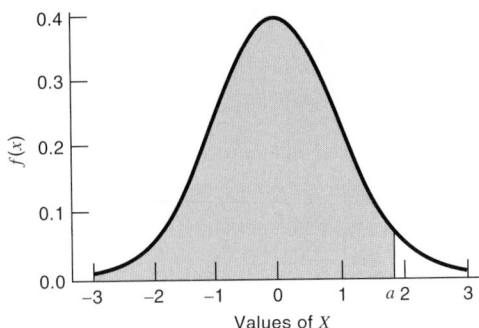

2. Give a probability statement to describe the shaded area under the following curve.

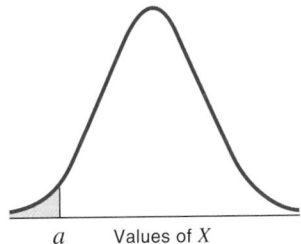

3. Give a probability statement to describe the shaded area under the following curve.

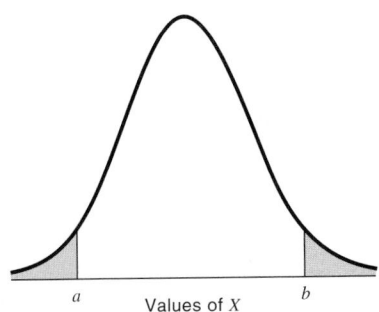

4. Give a probability statement to describe the shaded area under the following curve.

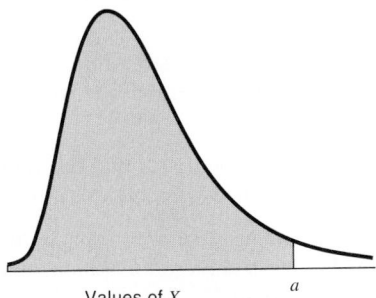

5. Give a probability statement to describe the shaded area under the following curve.

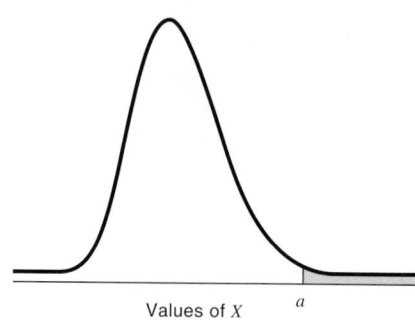

Values of X a

Section 6.5

1. Find the following area under the standard normal curve: $P(Z < 1.23)$.

2. Find the following area under the standard normal curve: $P(Z > 2.34)$.

3. Find the following area under the standard normal curve: $P(-1.56 < Z < 1.87)$.

4. Find the following areas under the standard normal curve.

$$P(-1.00 < Z < 1.00) \qquad P(-2.00 < Z < 2.00) \qquad P(-3.00 < Z < 3.00)$$

5. The age for the onset of menopause is normally distributed with a mean of 47.5 years and a standard deviation of 2.5 years. What percent of women have an onset before 42.5 years?

6. Adult male heights are normally distributed with a mean equal to 70 inches and a standard deviation equal to 3 inches. What percent are taller than 75 inches?

7. For those who use chat rooms, the mean time spent per week is normally distributed with a mean equal to 5 hours and a standard deviation equal to 1.2 hours. What percent of chat room users spend between 3 hours and 8 hours?

8. The time of play for a minor league baseball game is normally distributed with mean equal to 2 hours and 40 minutes (160 minutes) and standard deviation equal to 25 minutes. What percent of the games are shorter than 2 hours?

9. The number of e-mails sent per day by a mail-order company is normally distributed with a mean equal to 550 and a standard deviation equal to 75. On what percent of the days are more than 700 e-mails sent?

10. Standardized test scores are normally distributed with mean equal to 500 and a standard deviation equal to 100. Find the cutoff for the lower 10% of the scores.

11. Standardized test scores are normally distributed with mean equal to 500 and a standard deviation equal to 100. Find the 90th percentile of the test scores.

12. Times to assemble a lawn mower from scratch are normally distributed with mean equal to 120 minutes with standard deviation equal to 10 minutes. Find the 80th percentile for assembly times.

13. If you were to approximate the binomial distribution having $n = 100$ and $p = 0.5$ with a normal curve, where would you center the curve and what standard deviation would the curve have?

14. Adult male heights are normally distributed with a mean equal to 70 inches and a standard deviation equal to 3 inches. What is the 95th percentile of male heights?

15. A population has cholesterol readings that are normally distributed with mean equal to 180 and standard deviation equal to 25. A cholesterol reading of 200 needs to be treated. What percent of the population needs to be treated?

Chapter 6 End-of-Chapter Exercises

1. Let X be the number of girls in families of 4 children. Give the binomial distribution for X.

2. Refer to Exercise 1. What is the probability of the event that all four children are of the same sex?

3. Two cards are drawn from a deck of 52 without replacement. Let X = the number of aces drawn. Give the distribution for X.

4. Plot the binomial distribution $n = 25, p = 0.5$.

5. Refer to Exercise 4 of this section. Note how much the binomial distribution resembles a normal curve. If a normal curve were to be overlaid on the binomial distribution, where would the center be? What would the standard deviation be?

6. Give a probability statement to describe the shaded area under the following curve.

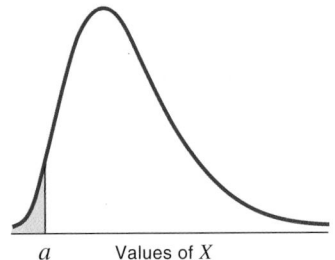

a Values of X

7. The amount spent on Valentine's Day is normally distributed with mean per household equal to $94.50 and standard deviation equal to $15.50. What is the percent of households that spend between $75 and $125?

8. Find the 95th percentile of the amount spent in Exercise 7 of this section.

9. A drug company claims a drug controls cholesterol in 75% of those who take it. The drug is administered to 25 individuals with elevated cholesterol. Twelve of the 25 have their cholesterol controlled. What is the probability that 12 or fewer will have their cholesterol controlled if the 75% figure is true for the population?

10. A drug company claims that with a zinc nasal spray, the time to be cured of a cold is normally distributed with a mean of 3 days and a standard deviation of 0.75 day. It takes you 6 days to be cured of the cold even though you use the zinc nasal spray. What is the probability that, if the claim is correct, it would take 6 or more days to be cured?

Chapter 7 Exercises

Section 7.2

1. A number used to estimate a parameter of the population is called a(n) _____.

2. The formula or rule used to calculate the point estimate is called a(n) _____.

3. Which point estimator(s) could be used to estimate the population mean?

4. *Fortune* magazine publishes the list of billionaires annually. The 1992 list included 233 individuals or families. Their wealth, age, and geographic location are reported in the sample data set. The summary statistics below are on age (in years) of the billionaires.

Variable	n	Mean	Variance	Std. Dev.	Median
age	225	64.03111	183.08385	13.5308485	65

Variable	Range	Min	Max	Q1	Q3
age	95	7	102	56	72

a. What is the point estimate of the population mean using the sample mean?
b. What is the point estimate of the population mean using the sample median?

5. *Fortune* magazine publishes the list of billionaires annually. The 1992 list included 233 individuals or families. Their wealth, age, and geographic location are reported in the sample data set. The summary statistics below are on wealth (in billion dollars).

Variable	n	Mean	Variance	Std. Dev.	Median
wealth	233	2.681545	11.014701	3.3188403	1.8

Variable	Range	Min	Max	Q1	Q3
wealth	36	1	37	1.3	3

a. What is the point estimate of the population mean using the sample mean?
b. What is the point estimate of the population mean using the sample median?

6. The number of calories per serving was recorded for 77 breakfast cereals. Summary statistics appear below.

Variable	n	Mean	Variance	Std. Dev.	Median
calories	77	106.88312	379.6309	19.48412	110

Variable	Range	Min	Max	Q1	Q3
calories	110	50	160	100	110

a. What is the point estimate of the population mean using the sample mean?
b. What is the point estimate of the population mean using the sample median?

7. The amount of sodium per serving, in milligrams, was recorded for 77 breakfast cereals. Summary statistics appear below.

Variable	n	Mean	Variance	Std. Dev.	Median
sodium	77	159.67532	7027.8535	83.8323	180

Variable	Range	Min	Max	Q1	Q3
sodium	320	0	320	130	210

a. What is the point estimate of the population mean using the sample mean?
b. What is the point estimate of the population mean using the sample median?

Section 7.3

1. The birth weights (in pounds) of 19 sets of twins are listed below. TwinA represents the weight of the firstborn twin and TwinB represents the weight of the secondborn twin.

TwinA	TwinB	Type	TwinA	TwinB	Type
5.9375	5.5625	BG	6.125	7.5625	BB
5.1875	5.6875	GB	5.6875	5.4375	BG
6.4375	5.8125	BB	4.375	4.9375	GG
6.4375	7.5625	BB	5.25	5.1875	BB
4.3125	5.5	BB	3.5	3.9375	GB
5.75	5.8125	GG	4.8125	5.125	GB
7.375	7.5625	BB	6.125	6.625	BB
5.5625	4.4375	BG	6.375	6.125	GG
5.875	4.875	GB	4.5	5.3125	BB
4.6875	4.6875	BB			

a. What is your estimate for the average birth weight of the firstborn twin?
b. What is your estimate for the average birth weight of the secondborn twin?
c. What is the difference in the two estimates you found in parts **a** and **b**?
d. Suppose these samples were both selected from a population of birth weights of all twins that has population mean birth weight $\mu = 5.5$ lb. Explain why they have different sample averages.

2. Refer to the data set on birth weights (in pounds) of 19 sets of twins, listed in Exercise 1 of this section.
a. What is your estimate for the standard deviation of birth weights of firstborn twins?
b. What is your estimate for the standard deviation of birth weights of secondborn twins?
c. Suppose these samples were both selected from a population of birth weights of all twins that has population standard deviation $\sigma = 1.0$ lb. Explain why they have different sample standard deviations.

3. An experiment was conducted to test whether directed reading activities in the classroom help elementary school students improve aspects of their reading ability. A treatment class of 21 third-grade students participated in these activities for eight weeks, and a control class of 23 third-graders followed the same curriculum without the activities. After the eight-week period, students in both classes took a Degree of Reading Power (DRP) test, which measures the aspects of reading ability that the treatment is designed to improve. [*Reference*: Moore, David S., and George P. McCabe (1989). *Introduction to the Practice of Statistics. Original source*: Schmitt, Maribeth C. "The Effects on an Elaborated Directed Reading Activity on the Metacomprehension Skills of Third Graders," Ph.D. dissertation, Purdue University, 1987.]

Control		Treatment	
42	46	24	56
43	10	43	59
55	17	58	52
26	60	71	62
62	53	43	54
37	42	49	57
33	37	61	33
41	42	44	46
19	55	67	43
54	28	49	57
20	48	53	
85			

a. What is your estimate of the average DRP score for the control group, no activities?
b. What is your estimate of the average DRP score for the treatment group, with activities?
c. What is your estimate of the average difference in score between the control group and the treatment group?
d. Do you think it is likely that the population mean difference between the control group and the treatment group, $\mu_1 - \mu_2$, is 0? Explain.

4. Refer to the data set of test scores for the control and treatment groups in a reading experiment, listed in Exercise 3 of this section.
a. What is your estimate of the standard deviation of DRP score for the control group, no activities?
b. What is your estimate of the standard deviation of DRP score for the treatment group, with activities?
c. What is your estimate of the ratio of standard deviation for the control group to the treatment group?
d. Do you think it is likely that the ratio of standard deviation for the control group to the treatment group, σ_1/σ_2, is 1? Explain.

5. During presidential election years, we are interested in the proportion of all voters who plan to vote for each candidate. A survey of 1200 registered voters yielded 540 who plan to vote for the Republican candidate.
a. What is the parameter we wish to estimate from this sample?
b. What is the value of the statistic that estimates the parameter described in part **a**?

6. A researcher wishes to determine the remission rate for a new treatment for a particular form of cancer. Suppose that 26 out of 100 patients using the new method go into remission.
a. What is the parameter we wish to estimate from this sample?
b. What is the value of the statistic that estimates the parameter described in part **a**?

7. A taste-test experiment is conducted to determine whether a popular name-brand soft drink tastes better than a less expensive store-brand soft drink. Of the 200 tasters, 79 preferred the store-brand soft drink, and 121 preferred the name-brand soft drink.
a. What are the parameters we wish to estimate from this sample (there are two)?
b. What are the values of the statistics that estimate the parameters described in part **a**?

Section 7.4

1. State in statistical terms the three properties that are desirable for a point estimator to have.

2. State in nonstatistical terms what it means for a point estimator to be "unbiased."

3. State in nonstatistical terms what it means for a point estimator to be "consistent."

4. State in nonstatistical terms what it means for a point estimator to be "efficient."

5. The average weight loss following a particular diet program is 2 pounds per week. Five samples of five dieters were selected from the population of all dieters following the program. Their average weight losses for 1 week appear below.

Sample 1	1.2	1.9	2.1	1.6	2.0
Sample 2	2.3	2.0	1.9	2.4	2.0
Sample 3	1.9	2.0	2.0	2.3	1.5
Sample 4	1.5	1.7	1.3	1.9	3.0
Sample 5	1.8	2.4	1.7	2.0	1.9

a. For each sample, find the sample mean and the sample median.
b. Find the average (mean) and the standard deviation of the five sample means.
c. Find the average (mean) and the standard deviation of the five sample medians.
d. Which point estimator is less biased?
e. Which point estimator has less variability (more consistent)?

6. The average birth weight of newborn babies is 7 lb. Five samples of five babies are selected from five hospitals. The birth weights of these babies (in pounds) appear below.

Hospital 1	7.2	6.9	7.4	6.5	5.8
Hospital 2	6.4	7.8	8.0	6.9	7.2
Hospital 3	7.1	7.6	6.5	8.4	7.0
Hospital 4	6.5	7.2	8.2	7.2	7.5
Hospital 5	7.6	7.2	6.6	6.4	7.3

a. For each sample, find the sample mean and the sample median.
b. Find the average and the standard deviation of the five sample means.
c. Find the average and the standard deviation of the five sample medians.
d. Which point estimator is less biased?
e. Which point estimator has less variability (more consistent)?

Section 7.5

1. The probability distribution of a point estimator or a sample statistic is called a(n)

_____.

2. Finish the statement of the Central Limit Theorem: In random sampling from a population with mean μ and standard deviation σ, when n is large enough, the distribution of $\overline{X}$ is
a. approximately _____ with
b. mean _____ equal to _____ and
c. standard deviation _____ equal to _____.

3. Replacement times for TV sets are normally distributed with a mean of 8.2 years and a standard deviation of 1.1 years. A sample of 36 TV sets is selected.
a. What is the approximate distribution of the sample mean replacement time for 36 TV sets?
b. What is the mean of the sampling distribution for the sample mean replacement time for 36 TV sets?
c. What is the standard error of the sampling distribution for the sample mean replacement time for 36 TV sets?

4. The heights of adult women are normally distributed with a mean of 63.6 inches and a standard deviation of 2.5 inches. A sample of 49 adult women is selected.
a. What is the approximate distribution of the sample mean height for 49 women?
b. What is the mean of the sampling distribution for the sample mean height for 49 women?
c. What is the standard error of the sampling distribution for the sample mean height for 49 women?

5. The time adults, age 18–24, spend reading the newspaper each day is normally distributed with a mean of 9 minutes and a standard deviation of 1.5 minutes. A sample of 40 adults in this age range is selected.
a. What is the approximate distribution of the sample mean time spent reading the newspaper each day for 40 adults in this age group?
b. What is the mean of the sampling distribution for the sample mean time spent reading the newspaper each day for 40 adults in this age group?
c. What is the standard error of the sampling distribution for the sample mean time spent reading the newspaper each day for 40 adults in this age group?

6. Assume that the average income of adult women is $481 per week and the standard deviation is $85 per week. A sample of 45 women is selected.
a. What is the approximate distribution of the sample mean weekly income of 45 adult women?
b. What is the mean of the sampling distribution for the sample mean weekly income of 45 adult women?
c. What is the standard error of the sampling distribution for the sample mean weekly income of 45 adult women?

7. Birth weights of newborn infants at a particular hospital have a normal distribution with a mean of 7.3 lb and a standard deviation of 0.60 lb. A sample of 36 newborn infants is selected.
a. What is the approximate distribution of the sample mean birth weight for 36 newborn infants?
b. What is the mean of the sampling distribution for the sample mean birth weight for 36 newborn infants?
c. What is the standard error of the sampling distribution for the sample mean birth weight for 36 newborn infants?

8. The weights of adult males are normally distributed with mean 170 lb and standard deviation 15 lb. A sample of 25 adult males is selected.
a. What is the approximate distribution of the sample mean weight for 25 adult males?
b. What is the mean of the sampling distribution for the sample mean weight for 25 adult males?
c. What is the standard error of the sampling distribution for the sample mean weight for 25 adult males?

Section 7.6

1. According to the CLT, when will the distribution of sample means be approximately normal?

2. Suppose a histogram of a sample is approximately bell shaped. Is this proof that the population is also bell shaped? Explain.

3. Replacement times for TV sets are normally distributed with a mean of 8.2 years and a standard deviation of 1.1 years.
a. Find the standard error for a sample of 9 TV sets.
b. Find the standard error for a sample of 36 TV sets.
c. Find the standard error for a sample of 100 TV sets.
d. What pattern do you notice from parts **a–c**?

4. Assume that the average income of adult women is $481 per week and the standard deviation is $85 per week.
a. Find the standard error for a sample of 16 women.
b. Find the standard error for a sample of 36 women.
c. Find the standard error for a sample of 81 women.
d. What pattern do you notice from parts **a–c**?

5. The heights of adult women are normally distributed with a mean of 63.6 inches and a standard deviation of 2.5 inches.
a. Find the standard error for a sample of 16 adult women.
b. Find the standard error for a sample of 49 women.
c. Find the standard error for a sample of 144 women.
d. What pattern do you notice from parts **a–c**?

6. Birth weights of newborn infants at hospital A have a normal distribution with a mean of 7.3 lb and a standard deviation of 0.60 lb. Birth weights of newborn infants at hospital B have a normal distribution with a mean of 7.3 lb and a standard deviation of 0.75 lb. A sample of 36 newborn infants is selected at each hospital.

a. Find the mean and standard error for the average of samples of 36 infants at hospital A.

b. Find the mean and standard error for the average of samples of 36 infants at hospital B.

c. What pattern do you notice from parts **a** and **b**?

7. The time adults, age 18–24, spend reading the newspaper each day is normally distributed with a mean of 9 minutes and a standard deviation of 1.5 minutes. The time adults, age 25–31, spend reading the newspaper each day is normally distributed with a mean of 9 minutes and a standard deviation of 1.2 minutes. A sample of 40 adults in this age range is selected from each age group.

a. Find the mean and standard error for the average of samples of 40 adults, age 18–24.

b. Find the mean and standard error for the average of samples of 40 adults, age 25–31.

c. What pattern do you notice from parts **a** and **b**?

8. Two large community bowling leagues are being compared. League 1 bowlers have a mean score of 150 and a standard deviation of 25. League 2 had a mean score of 150 and a standard deviation of 15. Samples of 36 bowlers from each league are selected.

a. Find the mean and standard error for the average of samples of 36 bowlers in league 1.

b. Find the mean and standard error for the average of samples of 36 bowlers in league 2.

c. What pattern do you notice from parts **a** and **b**?

Section 7.7

1. Replacement times for TV sets are normally distributed with a mean of 8.2 years and a standard deviation of 1.1 years. A store sells 36 TV sets during a one-month period. There are many complaints that these TV sets were defective and did not last as long as they should have. The sample mean replacement time for these 36 TV sets was 7.3 years.

a. Between what two values do the middle 99.7% of average TV replacement times fall?

b. Find the Z-score for a sample mean replacement time of 7.3 years.

c. Based on the Z-score, does it seem that these TV sets were in fact defective? Explain.

2. The heights of adult women are normally distributed with a mean of 63.6 inches and a standard deviation of 2.5 inches. A sample of 49 adult women is selected and the sample mean height is 66 inches.

a. Between what two values do the middle 95% of average adult women's heights fall?

b. Find the Z-score for a sample mean height of 66 inches.

c. Based on the Z-score, does it seem that this sample of women is unusually tall? Explain.

3. The time adults, age 18–24, spend reading the newspaper each day is normally distributed with a mean of 9 minutes and a standard deviation of 1.5 minutes. A sample of 40 adults in this age range is selected and the mean time they spend reading the newspaper each day is 8.6 minutes.

a. Between what two values do the middle 95% of average times spent reading the newspaper fall for this age group?

b. Find the Z-score for a sample mean time spent reading the newspaper of 8.6 minutes.

c. Based on the Z-score, does it seem that this sample spends significantly less time reading the newspaper than what is claimed about the population? Explain.

4. Assume the average income of adult women is $481 per week and the standard deviation is $85 per week. A sample of 45 women is selected and their mean income is $495 per week.

a. Between what two values do the middle 99.7% of average weekly incomes of women fall?

b. Find the Z-score for a sample mean income of $495 per week.

c. Based on the Z-score, does this sample of 45 women seem to make significantly more than average? Explain.

5. Birth weights of newborn infants at a particular hospital have a normal distribution with a mean of 7.3 lb and a standard deviation of 0.60 lb. A sample of 36 newborn infants has a mean weight of 7.5 lb.

a. Between what two values do the middle 99.7% of birth weights fall?

b. Find the Z-score for a sample mean birth weight of 7.5 lb.

c. Based on the Z-score, does it seem that this sample of infants is unusually heavy? Explain.

6. The mean weight of adult males is normally distributed with mean 170 lb and standard deviation 15 lb. A sample of 25 adult males is selected and their mean weight is 180 lb.
a. Between what two values do the middle 95% of average male weights fall?
b. Find the Z-score for a sample mean weight of 180 lb.
c. Based on the Z-score, does it seem that this sample of men is overweight? Explain.

7. An instructor gives the same final exam each semester. The mean score over the years is 75 and the standard deviation is 7. Consider this semester's class of 46 students as a sample. This class has a mean score of 80 on the final exam.
a. Between what two values do the middle 99.7% of average final exam scores fall?
b. Find the Z-score for this class's mean final exam score of 80.
c. Based on the Z-score, does it seem that this class performed particularly well? Explain.

8. The mean amount of potato chips in a "16-ounce" bag is 16 ounces and the standard deviation is 0.5 ounce. A consumer protection agency is concerned that consumers are getting less than 16 ounces of chips per bag. A sample of 36 bags is selected and the mean weight is 15.9 ounces.
a. Between what two values do the middle 95% of average weights of "16-ounce" potato chip bags fall?
b. Find the Z-score for a sample mean weight of 15.9 ounces.
c. Based on the Z-score, does it seem that customers are getting significantly less potato chips than they should? Explain.

Section 7.8

1. The lengths of human pregnancies from conception to birth are approximately normally distributed. A hospital administrator wishes to estimate the mean length of pregnancies in his community. He looks at the records of 36 pregnancies over the past year and calculates the mean length of these pregnancies to be 268 days. Assume the standard deviation is 16 days. Find a 95% confidence interval for the mean length of pregnancies in this community.

2. A magazine editor wants to determine the average age of his readers. A survey of a random sample of 45 readers produced an average age of 29 years. Assume the population standard deviation is 4.2 years. Find a 90% confidence interval for the mean age of all readers of this magazine.

3. A dietician wishes to determine the mean weight of adult males in her community. A sample of 40 adult males from her community have a mean weight of 172 lb. Assume the standard deviation is 10 lb. Assuming the distribution of weights is normal, find a 90% confidence interval for the mean weight of all adult males in this community.

4. A restaurant owner is interested in the mean time it takes customers to receive their food once an order is placed. The time between placing an order and receiving food was recorded for a random sample of 36 customers. The mean for the sample was 14.3 minutes. Assume the population standard deviation is 3 minutes. Assuming the distribution of time to receive food is normal, find a 90% confidence interval for the mean time it takes all customers to receive their food.

5. We are interested in estimating the weight of paper discarded by households each week in a community. A sample of 50 families is selected and the weight of paper discarded is recorded for each family. The sample mean is 8.2 lb. Assume the standard deviation is 4.0 lb. Assuming the distribution of weights is normal, find a 95% confidence interval for the mean weight of paper discarded by all families in this community.

6. A pharmaceutical company is interested in how long its new brand of pain reliever controls muscle pain. In a clinical study, 55 subjects with muscle pain take the drug and record the length of time it remains effective. The 55 subjects reported a mean time of effectiveness of 8.6 hours. Assume the standard deviation is 1.5 hours. Assuming the distribution of pain relief times is normal, find a 95% confidence interval for the mean time this pain reliever will control muscle pain for all people.

7. *Fortune* magazine collected data on 225 billionaires in 1992. The mean age was 64.03 years. Assume the standard deviation is 13.53. Assuming the distribution of ages is normal, find a 95% confidence interval for the mean age of all billionaires.

8. The average lifetime of a brand of battery is being studied. A sample of 60 batteries last a mean time of 490 minutes. Assume the standard deviation is 60 minutes. Assuming the distribution of lifetimes is normal, find a 90% confidence interval for the mean lifetime of all batteries.

Section 7.9

1. What is the difference between the Z-score and the t-score?

2. Explain why the t-score has greater variability than the Z-score.

3. How are the sample size and degrees of freedom (df) related?

4. When is the sample size, n, large enough for the graph of the t distribution to be indistinguishable from the graph of the standard normal distribution?

5. Let $t_{\alpha/2, n-1}$ be the value from the t distribution with $n - 1$ degrees of freedom such that the area in the upper tail is $\alpha/2$. Find the following:
   a. $t_{0.01,24}$
   b. $t_{0.05,20}$
   c. $t_{0.025,26}$
   d. $t_{0.005,21}$

Section 7.10

1. Eight men who were at least 60 years old were selected at random and their serum cholesterol level was measured. Here are their levels, in mg/dL:

$$177, 197, 190, 185, 163, 181, 205, 222$$

Assuming the distribution is normal, find a 95% confidence interval for the mean serum cholesterol level for all men 60 years of age and older.

2. A sample of 15 women with breast cancer was taken. Here are their ages at the time of first detection.

$$33, 42, 45, 50, 53, 56, 57, 57, 59, 60, 60, 62, 65, 68, 70$$

Assuming the distribution is normal, find a 95% confidence interval for the mean age of all women at time of first detection of breast cancer.

3. Here are the prices (in dollars) for 12 concert tickets:

$$37, 38, 27, 31, 36, 34, 41, 48, 25, 36, 40, 28$$

Assuming the distribution is normal, find a 90% confidence interval for the mean price of all concert tickets.

4. The following data are the amounts of sugar (in grams) contained in a sample of 13 breakfast cereals. The data are presented in order from least to greatest.

$$1, 5, 5, 6, 6, 7, 8, 8, 9, 10, 12, 13, 14$$

Assuming the distribution is normal, find a 90% confidence interval for the mean amount of sugar contained in all breakfast cereals.

5. Eight participants in a bike race had the following finishing times, in minutes.

$$28, 22, 26, 35, 21, 23, 37, 24$$

Assuming the distribution is normal, find a 90% confidence interval for the mean finishing time of all race participants.

6. The following list gives the duration (in minutes) of 12 power failures in a particular area:

$$18, 125, 44, 96, 31, 53, 26, 80, 49, 125, 63, 58$$

Assuming the distribution is normal, find a 90% confidence interval for the mean duration of all power failures in this area.

7. A sample of eight doctors in a community was asked how many flu shots they had given to patients this fall. The numbers of flu shots were 6, 3, 5, 14, 2, 6, 0, and 8. [Note: $\Sigma x = 44$, $\Sigma x^2 =$

370.] Assuming the distribution is normal, find a 95% confidence interval for the mean number of flu shots all doctors in this community give.

8. A study in Switzerland studied the number of cesarean sections (surgical deliveries of babies) performed in a year by doctors. The data for 10 male doctors are:

$$27, 50, 33, 25, 86, 25, 59, 31, 37, 44$$

Assuming the distribution is normal, find a 95% confidence interval for the mean number of cesarean sections that all male doctors perform yearly.

Section 7.11

1. We are interested in estimating the percentage of families in a large community living below the poverty level. A random sample of 400 families in this community is studied and 70 are found to be living below the poverty level. Calculate a 95% confidence interval for the percentage of all families in this community who live below the poverty level.

2. A statistics class collects data on colors of M&Ms chocolate candies. In all, 1000 M&Ms are inspected, of which 205 are red.
a. Calculate a 95% confidence interval for the proportion of all M&Ms chocolate candies that are red.
b. The makers of M&Ms candies claim the proportion of all M&Ms that are red is 0.20. Based on your answer to part **a**, is there any reason to doubt this claim? Explain.

3. You read that the percent of college students who have used illegal drugs in the past month is 20%. You decide to investigate this percentage at your own college. You ask a sample of 500 students whether they have used illegal drugs in the past month. Of these, 55 say "yes."
a. Calculate a 90% confidence interval for the percentage of all students at your college who have used illegal drugs in the past month.
b. Based on your answer to part **a** and assuming the students at your college are typical of all college students, is there any reason to doubt the claim in the study? Explain.
c. Illegal drug use is a sensitive issue and sometimes people will lie. Explain what effect this would have on the confidence interval.

4. A movie theater is interested in the proportion of films that are R-rated. A random sample of 60 films are considered, of which 35 are R-rated. Calculate a 90% confidence interval for the proportion of all films that are R-rated.

5. A police officer wishes to determine the proportion of traffic fatalities in his jurisdiction that are alcohol-related. In a study of 200 traffic fatalities in the police officer's jurisdiction, 37 of the fatalities were alcohol-related. Calculate a 95% confidence interval for the proportion of all traffic fatalities in the officer's jurisdiction that are alcohol-related.

6. A community program to reduce smoking is initiated. In the past, 25% of the community's residents were smokers. After a year, a sample of 500 people in the community is selected, of whom 105 indicated that they were smokers.
a. Calculate a 95% confidence interval on the percent of all community residents who are smokers.
b. Based on your answer to part **a**, does the program seem to be effective? Explain.

7. It is claimed that 10% of the population is left-handed. In a random sample of 750 people, 92 are left-handed.
a. Calculate a 90% confidence interval for the percent of all people who are left-handed.
b. Based on your answer to part **a**, does the claim about the population seem to be correct? Explain.

8. We wish to estimate the proportion of all students at our college who work at a job. In a sample of 200 students, 65 say they work at a job.
a. Calculate a 90% confidence interval for the proportion of all students who work at a job.
b. Suppose the sample included only daytime students, and the college has some students who take only evening classes. Can we trust the interval from part **a**? Explain.

9. A researcher wishes to determine the remission rate for a new treatment for a particular form of cancer. Suppose that 26 out of 100 patients using the new method go into remission.

Calculate a 95% confidence interval for the remission rate of all cancer patients who use the new treatment.

Section 7.12

1. The lengths of human pregnancies from conception to birth are approximately normally distributed. A hospital administrator wishes to estimate the mean length of pregnancies in his community. Assume the standard deviation is 16 days. How many pregnancies need to be observed to be 95% confident that the error in estimating the mean length is at most 5 days?

2. A dietician wishes to determine the mean weight of adult males in her community. Assume the distribution of weights is normal and the standard deviation is 10 lb. How many males need to be sampled to be 90% confident that the error in estimating the mean weight is at most 5 lb?

3. A pharmaceutical company is interested in how long its new brand of pain reliever controls muscle pain. A clinical study is to be conducted. Assume the standard deviation is 1.5 hours and the distribution of pain relief time is normal. How many subjects are needed in the clinical study to be 95% confident that the error in estimating the mean time is at most $\frac{1}{2}$ hour?

4. The average lifetime of a brand of battery is being studied. Assume the standard deviation is 60 minutes and the distribution of lifetimes is normal. How many batteries are needed to be 90% confident that the error in estimating the mean lifetime is at most 30 minutes?

5. We are interested in estimating the mean weight of paper discarded by households each week in a community. Assume the distribution of weights is normal and the standard deviation is 4.0 lb. How many households need to be sampled to be 90% confident that the error in estimating the mean weight is at most 1.5 lb?

6. We are interested in estimating the percentage of families in a large community living below the poverty level. How many families need to be sampled to be 95% confident that the error in estimating this percentage is at most 5%?

7. A police officer wishes to determine the proportion of traffic fatalities in his jurisdiction that are alcohol-related. How many fatalities need to be observed to be 90% confident that the error in estimating this proportion is at most 0.04?

8. A researcher wishes to determine the remission rate for a new treatment for a particular form of cancer. How many cancer patients need to be included in a study to be 90% confident that the error in estimating this rate is at most 0.05?

9. A movie theater is interested in the proportion of films that are R-rated. How many movies need to be sampled to be 95% confident that the error in estimating this proportion is at most 0.07?

10. We wish to estimate the proportion of all students at our college who work at a job. How many students need to be sampled to be 90% confident that the error in estimating this proportion is at most 0.07?

Chapter 8 Exercises

Section 8.3

Which type of hypothesis test should be used in the following situations? Choose from the following:
   A. The shape of the distribution in one population
   B. The mean value, μ, of the variable in one population
   C. How the mean value of one population compares with the mean value of the variable in the second population
   D. The equality of the mean values of more than two populations
   E. The amount of variability, σ^2, of the variable in one population
   F. How the amount of variability compares with the amount of variability of the variable in a second population

G. The value of the proportion, p, of the population members that have a certain characteristic

H. How the population proportion that have a certain characteristic in one population compares with the corresponding proportion in the second population

1. The mean weight of adult males is greater than 160 lb.

2. The percent of college students who have used illegal drugs in the past month is 20%.

3. The mean lifetime of General Electric lightbulbs is greater than the mean lifetime of Sylvania lightbulbs.

4. The distribution of breaking strength of facial tissues is normal.

5. The percent of red M&Ms that are manufactured is 20%.

6. The variability in the weights of females is larger than the variability in the weights of males.

7. The mean score on the SAT verbal section differs from the mean score on the SAT mathematics section.

8. The mean amount of water absorbed by Bounty paper towels is greater than the mean amount of water absorbed by Wal-Mart paper towels.

9. The percent of plants that germinate using seed coating is higher than the percent of plants that germinate when no coating is used.

10. The mean score on the SAT for freshmen at a particular university is greater than 1000.

Section 8.4

1. List the five steps for a hypothesis test.

For Exercises 2–6, state the correct null and alternative hypotheses.

2. A nurse is studying weights of adult males in her community. She suspects that the mean weight of adult males is greater than 160 lb.

3. A university official is interested in studying illegal drug use among his students. A report indicates that the proportion of college students who have used illegal drugs in the past month is 20%. The university official thinks the percent at his college is lower.

4. The manufacturer of M&Ms candies claims that the percent of red M&Ms that are manufactured is 20%. You plan to test whether this claim is correct.

5. The manufacturer of a popular brand of facial tissue desires the breaking strength of its tissue to be 6 oz. If the breaking strength is lower than this, then the tissues will tear too easily. If the breaking strength is higher than this, then the tissue will be too rough on the face.

6. The mean number of calories per serving of breakfast cereals is thought to be 110. A dietician suspects that the true mean value may be lower.

7. Suppose the significance level for a test is set at 10%. For which p-values will the null hypothesis be rejected?

8. Suppose the significance level for a test is set at 1%. For which p-values will the null hypothesis be rejected?

9. Suppose the significance level for a test is set at 5%. For which p-values will the null hypothesis be rejected?

10. What is the name for the number that is used to decide between the null and alternative hypotheses?

11. The probability of obtaining the value of the test statistic or something more extreme is known as the _____.

12. The range of values of the test statistic that will lead us to the decision to reject the null hypothesis is called the _____.

Section 8.5

Which test statistic, the t or the Z, is more appropriate for the following situations?

1. Test of the population mean when $n > 30$ and the standard deviation is known.

2. Test of the population mean when $n > 30$ and the standard deviation is unknown.

3. Test of the population mean when $n < 30$ and the standard deviation is known.

4. Test of the population mean when $n < 30$ and the standard deviation is unknown.

Section 8.6

1. The lengths of human pregnancies from conception to birth are approximately normally distributed with a mean of 266 days and a standard deviation of 4 days. A hospital administrator wishes to determine whether pregnancies in his community differ from this mean. He looks at the records of 36 pregnancies over the past year and calculates the mean length of these pregnancies to be 268 days.
   a. Set up the null and alternative hypotheses to test whether pregnancy lengths at this hospital differ from the overall average.
   b. Complete the remaining hypothesis-testing steps using $\alpha = 0.05$.
   c. Find the p-value.
   d. Based on the p-value, what can the hospital administrator conclude about the mean length of pregnancies at his hospital?

2. A magazine editor wants to determine whether the average age of his readers differs from 30 years of age. A survey of a random sample of 45 readers produced an average age of 29 years. Assume the population standard deviation is 4.2 years.
   a. Set up the null and alternative hypotheses to test the magazine editor's hypothesis.
   b. Complete the remaining hypothesis-testing steps using $\alpha = 0.05$.
   c. Find the p-value.
   d. Based on the p-value, what can the magazine editor conclude about the average age of his readers?

3. It is known that the birth weights of newborn babies in the United States have a mean of 7 pounds and a standard deviation of 1.6 pounds. A State Health Department director wishes to determine whether the mean weight in his state differs from the national average. He randomly samples 64 birth certificates from the State Health Department and records the birth weights of these babies. The sample mean birth weight is 6.8 pounds. Use the standard deviation of 1.6 pounds.
   a. Set up the null and alternative hypotheses to test the State Health Department director's hypothesis.
   b. Complete the remaining hypothesis-testing steps using $\alpha = 0.10$.
   c. Find the p-value.
   d. Based on the p-value, what can the State Health Department director conclude about the mean weight of newborn babies in his state?

4. The mean number of calories per serving in a particular brand of breakfast cereal is claimed to be 110 calories. A consumer protection agency wishes to test whether this claim is correct. A sample of 50 servings of this cereal is selected for analysis. The mean number of calories is 106.88. Assume the population standard deviation is 5 calories.
   a. Set up the null and alternative hypotheses to test the consumer protection agency's hypothesis.
   b. Complete the remaining hypothesis-testing steps using $\alpha = 0.05$.
   c. Find the p-value.
   d. Based on the p-value, what can the consumer protection agency conclude about the mean number of calories per serving in this cereal?

5. The mean weight of adult males is claimed to be 170 lb. A dietician wishes to test whether this claim is correct. A sample of 40 adult males have mean weight 175 lb. Assume the standard deviation is 10 lb.
   a. Set up the null and alternative hypotheses to test the dietician's hypothesis.
   b. Complete the remaining hypothesis-testing steps using $\alpha = 0.05$.

c. Find the *p*-value.

d. Based on the *p*-value, what can the dietician conclude about the mean weight of adult males?

6. A 1990 census report revealed that the mean distance that U.S. citizens live from their birthplace is 50 miles and the standard deviation is 15 miles. A random sample of 49 residents in your city yields a mean of 56 miles.

a. Set up the null and alternative hypotheses to test whether the mean distance for people living away from their birthplace in your city differs from the national average.

b. Complete the remaining hypothesis-testing steps using $\alpha = 0.01$.

c. Find the *p*-value.

d. Based on the *p*-value, what can you conclude about the mean distance for people living away from their birthplace in your city?

7. An owner of a fast-food franchise reported to corporate headquarters that the average bill paid by his customers in the last quarter was $6.20 and the standard deviation was $1.90. Not knowing exactly what the effect would be, headquarters launched a promotional campaign featuring a discount for a multi-sandwich purchase. The stubs from 81 purchases at the owner's franchise after the campaign was launched averaged $6.65.

a. Set up the appropriate null and alternative hypotheses.

b. Complete the remaining hypothesis-testing steps using $\alpha = 0.05$.

c. Find the *p*-value.

d. Based on the *p*-value, what can you conclude about the mean sales of the franchise after launching the multi-sandwich campaign?

8. A national golf magazine reports that the average weekend golfer carries a handicap of 15 strokes and the standard deviation is 4 strokes. A local men's church league has 64 players and just tallied the end-of-season totals. The church league finished the year with an average handicap of 14 strokes.

a. Set up the null and alternative hypotheses to test whether the mean handicap for the local church league differs from the national average reported in the magazine.

b. Complete the remaining hypothesis-testing steps using $\alpha = 0.02$.

c. Find the *p*-value.

d. Based on the *p*-value, what can you conclude about the mean handicap for men in the local church league?

9. A grocery store owner is interested in determining whether the average weight of a package of ground beef sold in the store is 1 pound. He selects 49 packages of ground beef and weighs them. The sample mean weight is 0.97 lb. Use a standard deviation of 0.10 lb.

a. Set up the null and alternative hypotheses to test whether the average weight of packages of ground beef differs from 1 pound.

b. Complete the remaining hypothesis-testing steps using $\alpha = 0.01$.

c. Find the *p*-value.

d. Based on the *p*-value, what can you conclude about the average weight of all packages of ground beef?

10. A restaurant manager thinks that the average bill paid by his customers is $25. To test his hypothesis, the next 50 tabs are tallied. The sample mean bill paid by these customers is $27. Use a standard deviation of $5.

a. Set up the null and alternative hypotheses to test whether the average bill paid by his customers differs from $25.

b. Complete the remaining hypothesis-testing steps using $\alpha = 0.05$.

c. Find the *p*-value.

d. Based on the *p*-value, what can you conclude about the average bill paid by his customers?

Section 8.7

1. Two drugs are being compared, a new drug and a standard drug. If the new drug is found to be more effective than the standard drug, then the drug company will market the new drug in place of the standard drug. For each statement, identify whether a Type I, Type II, or no error has occurred.

a. The sample evidence leads you to the decision that the new drug is no more effective than the old drug. In reality, the new drug is truly as effective as the old drug.

b. The sample evidence leads you to the decision that the new drug is more effective than the old drug. In reality, the new drug actually is as effective as the old drug.

2. Two drugs are being compared, a new drug and a standard drug. If the new drug is found to be more effective than the standard drug, then the drug company will market the new drug in place of the standard drug.
a. State the consequences of a Type I error.
b. State the consequences of a Type II error.
c. Suggest a value for α and justify your choice.

3. Quality control officers test bottles of ketchup to see whether the filling machines are putting the proper amount in each bottle. They do not want to shut down production unless there is strong evidence indicating that the machines are not functioning properly. For each statement, identify whether a Type I, Type II, or no error has occurred.
a. After testing a sample of bottles, a quality control officer decides to allow the filling machines to continue operating. Actually, however, the filling machines are not operating properly.
b. After testing a sample of bottles, a quality control officer decides to shut down the filling machines. Actually, however, the filling machines are operating properly.

4. Quality control officers test bottles of ketchup to see whether the filling machines are putting the proper amount in each bottle. They do not want to shut down production unless there is strong evidence indicating that the machines are not functioning properly.
a. State the consequences of a Type I error.
b. State the consequences of a Type II error.
c. Suggest a value for α and justify your choice.

5. We are interested in determining whether the mean number of hours of sleep per day for children aged 3–5 is more than 8 hours per day. We will conduct a hypothesis test. For each statement, identify whether a Type I, Type II, or no error has occurred.
a. The results of our test indicate there is no evidence to conclude that children aged 3–5 sleep an average of more than 8 hours per day. In truth, children aged 3–5 sleep an average of 10 hours per day.
b. The results of our test indicate there is evidence to conclude that children aged 3–5 sleep an average of more than 8 hours per day. In truth, children aged 3–5 sleep an average of 8 hours per day.

6. A mail-order catalog company claims customers will receive their product within 4 days of ordering. A competitor believes that this claim is an underestimate; that is, the competitor thinks it will take longer to receive the product. The competitor will conduct a hypothesis test. If the competitor finds that it takes longer than 4 days to receive products ordered by this company, its marketing department plans to use this information in advertisements. For each statement, identify whether a Type I, Type II, or no error has occurred.
a. The competitor concludes that the mean time to receive products ordered from this company is longer than 4 days. However, the actual mean time to receive products ordered from this company is 3 days.
b. The competitor concludes that the mean time to receive products ordered from this company is longer than 4 days. The actual mean time to receive products ordered from this company is 5 days.

7. A mail-order catalog company claims customers will receive their product within 4 days of ordering. A competitor believes that this claim is an underestimate; that is, the competitor thinks it will take longer to receive the product. The competitor will conduct a hypothesis test. If the competitor finds that it takes longer than 4 days to receive products ordered by this company, its marketing department plans to use this information in advertisements.
a. State the consequences of a Type I error.
b. State the consequences of a Type II error.
c. Suggest a value for α and justify your choice.

8. A dog food manufacturer sells "50-pound" bags of dog food. A hypothesis test is conducted to decide whether the mean weight of all bags of this dog food differs from the advertised weight of 50 pounds.
a. State the consequences of a Type I error.
b. State the consequences of a Type II error.
c. Suggest a value for α and justify your choice.

9. The mean stopping distance of a particular type of car is claimed to be 25 ft. A consumer magazine wishes to test whether the mean stopping distance for this car is actually longer than 25 ft.

a. State the consequences of a Type I error.

b. State the consequences of a Type II error.

c. Suggest a value for α and justify your choice.

10. The mean weight loss for the first week on a particular diet program is claimed to be 5 lb. A potential customer is skeptical and wishes to test whether the mean weight loss is less than 5 lb.

a. State the consequences of a Type I error.

b. State the consequences of a Type II error.

c. Suggest a value for α and justify your choice.

Section 8.8

1. A consumer protection agency wants to prove that packages of a popular cereal average less than 24 oz. Set up the null and alternative hypotheses to be tested.

2. A mail-order catalog company claims customers will receive their product within 4 days of ordering. A competitor believes that this claim is an underestimate; that is, the competitor thinks it will take longer to receive the product. Set up the null and alternative hypotheses to be tested.

3. The lengths of human pregnancies from conception to birth are approximately normally distributed with a mean of 266 days. A hospital administrator wishes to determine whether pregnancies in his community differ from this mean. Set up the null and alternative hypotheses to be tested.

4. A restaurant owner is interested in the mean time it takes customers to receive their food once an order is placed. He would like to see his customers receive their food in less than 15 minutes. Set up the null and alternative hypotheses to be tested.

5. A dog food manufacturer sells "50-pound" bags of dog food. A hypothesis test is conducted to decide whether the mean weight of all bags of this dog food differs from the advertised weight of 50 pounds.

6. The mean stopping distance of a particular type of car is claimed to be 25 ft. A consumer magazine wishes to test whether the mean stopping distance for this car is actually longer than 25 ft. Set up the null and alternative hypotheses to be tested.

7. The mean weight loss for the first week on a particular diet program is claimed to be 5 lb. A potential customer is skeptical and wishes to test whether the mean weight loss is less than 5 lb. Set up the null and alternative hypotheses to be tested.

8. In a community, the weights of paper discarded by households each week are normally distributed with a mean of 9.4 lb and a standard deviation of 4.0 lb. A recycling program is started to help decrease this amount. Set up the null and alternative hypotheses to be tested.

9. A magazine editor wants to determine whether the average age of his readers differs from 30 years of age. Set up the null and alternative hypotheses to be tested.

10. The average lifetime of a brand of battery is being studied. The manufacturer claims the batteries will last, on average, at least 500 hours. We wish to test whether this claim is correct. Set up the null and alternative hypotheses to be tested.

Section 8.9

1. A consumer protection agency wants to prove that packages of a popular cereal average less than 24 oz. A random sample of 101 boxes is chosen. The 101 boxes are weighed; the mean weight is 23.94 oz. Assume the population standard deviation is 0.13 oz.

a. Set up the null and alternative hypotheses to test the agency's hypothesis.

b. Complete the remaining hypothesis-testing steps using $\alpha = 0.10$.

c. Find the p-value.

d. Based on the p-value, what can the consumer protection agency conclude about the mean weight of the cereal?

2. A restaurant owner is interested in the mean time it takes customers to receive their food once an order is placed. He would like to see his customers receive their food in less than 15 minutes. The time between placing an order and receiving food was recorded for a random sample of 36 customers. The mean for the sample was 14.3 minutes. Assume the population standard deviation is 2 minutes.

a. Set up the null and alternative hypotheses to test the restaurant owner's hypothesis.
b. Complete the remaining hypothesis-testing steps using $\alpha = 0.01$.
c. Find the p-value.
d. Based on the p-value, what can the restaurant owner conclude about the mean time it takes his customers to get their meals?

3. In a community, the weights of paper discarded by households each week are normally distributed with a mean of 9.4 lb. and a standard deviation of 4.0 lb. A recycling program is started to help decrease this amount. After implementing the program for six months, a sample of 50 families is selected and the weight of paper discarded is recorded for each family. The sample mean is 8.2 lb. Use the standard deviation of 4.0 lb.

a. Set up the null and alternative hypotheses to test the success of the recycling program.
b. Complete the remaining hypothesis-testing steps using $\alpha = 0.05$.
c. Find the p-value.
d. Based on the p-value, what conclusion can be made about the success of the recycling program?

4. A student group at the University of South Carolina (USC) is concerned that USC students pay too much money for textbooks. The national average cost of textbooks for college students is $350 per semester and the standard deviation is $80. A random sample of 46 USC students spent a mean of $365.20 on books this semester. Use the standard deviation of $80.

a. Set up the null and alternative hypotheses to test the student group's hypothesis.
b. Complete the remaining hypothesis-testing steps using $\alpha = 0.05$.
c. Find the p-value.
d. Based on the p-value, what can the student group conclude about the amount of money USC students pay for textbooks?

5. A pharmaceutical company claims that its new brand of pain reliever controls muscle pain for over 8 hours. In a clinical study, 55 subjects with muscle pain took the drug and recorded the length of time it remained effective. The 55 subjects reported a mean time of effectiveness of 8.6 hours. Assume the standard deviation is 1.5 hours.

a. Set up the null and alternative hypotheses to test the pharmaceutical company's hypothesis.
b. Complete the remaining hypothesis-testing steps using $\alpha = 0.05$.
c. Find the p-value.
d. Based on the p-value, what can the pharmaceutical company conclude about the length of time the pain medication remains effective?

6. The mean cost for a home nationwide is reported to be $90,000. To test the mean cost for a home in Nebraska is lower than the national mean, 36 homes for sale are randomly selected. The sample mean selling price is $86,500. Assume the standard deviation is $11,500.

a. Set up the null and alternative hypotheses to test the hypothesis of interest.
b. Complete the remaining hypothesis-testing steps using $\alpha = 0.05$.
c. Find the p-value.
d. Based on the p-value, what can we conclude about the mean selling price of homes in Nebraska?

7. A pediatrician is testing the claim that the mean time spent sleeping per day by children, age 3–5, is greater than 480 minutes (8 hours). The pediatrician sampled 36 children, age 3–5, and recorded the length of time that they slept the day before. The mean time slept was 527.5 minutes. Use a standard deviation of 90 minutes.

a. Set up the null and alternative hypotheses to test the pediatrician's hypothesis.
b. Complete the remaining hypothesis-testing steps using $\alpha = 0.02$.
c. Find the p-value.
d. Based on the p-value, what can we conclude about the mean time spent sleeping per day by children, age 3–5?

8. The average lifetime of a brand of battery is being studied. The manufacturer claims the batteries will last, on average, at least 500 hours. We wish to test whether this claim is correct. A

sample of 60 batteries last a mean time of 490 minutes. Assume the standard deviation is 60 minutes.
a. Set up the null and alternative hypotheses to test the hypothesis of interest.
b. Complete the remaining hypothesis-testing steps using $\alpha = 0.05$.
c. Find the p-value.
d. Based on the p-value, what can we conclude about the mean lifetime of this brand of battery?

9. It is believed that the average age of billionaires is over 60 years of age. *Fortune* magazine collected data on 225 billionaires in 1992. The mean age was 64.03 years. Assume the standard deviation is 13.53.
a. Set up the null and alternative hypotheses to test the hypothesis of interest.
b. Complete the remaining hypothesis-testing steps using $\alpha = 0.05$.
c. Find the p-value.
d. Based on the p-value, what can we conclude about the mean age of billionaires?

10. A 1999 salary survey of 500 subscribers to the *Visual Basic Programmer's Journal* yielded a mean annual salary of $64,000 and a standard deviation of $17,000. Readers between the ages of 25 and 29 earned a mean of $55,000. Suppose 49 readers were included in this group.
a. Set up the null and alternative hypotheses to test whether the mean salary of the 25-to-29 age group is less than the mean for all readers.
b. Complete the remaining hypothesis-testing steps using $\alpha = 0.01$.
c. Find the p-value.
d. Based on the p-value, what can we conclude about the mean salary of the 25-to-29 age group?

Chapter 9 Exercises

Section 9.2

1. A nurse suspects that the average serum cholesterol level for men 60 and over is higher than 180. Eight men who were at least 60 years old were selected at random and their serum cholesterol level was measured. Here are their levels, in mg/dL.

$$177, 197, 190, 185, 163, 181, 205, 222$$

Assume the distribution of cholesterol levels is normal.
a. State the appropriate null and alternative hypotheses.
b. Calculate the test statistic and the p-value.
c. At $\alpha = 0.01$, should you reject the null hypothesis?
d. What conclusion can you draw?

2. A doctor believes that the average age of women at the time of first detection of breast cancer is over 50. A sample of 15 women with breast cancer was taken. Here are their ages at the time of first detection.

$$33, 42, 45, 50, 53, 56, 57, 57, 59, 60, 60, 62, 65, 68, 70$$

Assume that the distribution of age at first detection of breast cancer is normal.
a. State the appropriate null and alternative hypotheses.
b. Calculate the test statistic and the p-value.
c. At $\alpha = 0.05$, should you reject the null hypothesis?
d. What conclusion can you draw?

3. A ticket agent thinks the average cost of concert tickets is $40. To test this claim, a sample of 12 concert tickets is selected. Here are the prices for 12 concert tickets:

$$37, 38, 27, 31, 36, 34, 41, 48, 25, 36, 40, 28$$

Assume that the distribution of concert ticket prices is normal.
a. State the appropriate null and alternative hypotheses.
b. Calculate the test statistic and the p-value.
c. At $\alpha = 0.05$, should you reject the null hypothesis?
d. What conclusion can you draw?

4. The mean amount of sugar contained in breakfast cereals is claimed to be 9 grams. A dietician wishes to test whether this is true. Data were collected on the amount of sugar contained in a sample of 13 breakfast cereals. The data are presented in order from least to greatest.

$$1, 5, 5, 6, 6, 7, 8, 8, 9, 10, 12, 13, 14$$

Assume that the distribution of amount of sugar contained in breakfast cereals is normal.
a. State the appropriate null and alternative hypotheses.
b. Calculate the test statistic and the p-value.
c. At $\alpha = 0.10$, should you reject the null hypothesis?
d. What conclusion can you draw?

5. In the past, the mean time for participants to finish a particular bike race was 30 minutes. It is suspected that racers are becoming faster and finishing in less time. Eight participants in a bike race had the following finishing times, in minutes.

$$28, 22, 26, 35, 21, 23, 37, 24$$

Assume the distribution of finishing times is normal.
a. State the appropriate null and alternative hypotheses.
b. Calculate the test statistic and the p-value.
c. At $\alpha = 0.05$, should you reject the null hypothesis?
d. What conclusion can you draw?

6. A power company works to keep power failures as short as possible. In the past, the mean duration of power failures was 65 minutes. The following list gives the duration, in minutes, of 12 power failures in a particular area:

$$18, 125, 44, 96, 31, 53, 26, 80, 49, 125, 63, 58$$

Assume that the distribution of duration of power failures in this area is normal.
a. State the appropriate null and alternative hypotheses.
b. Calculate the test statistic and the p-value.
c. At $\alpha = 0.10$, should you reject the null hypothesis?
d. What conclusion can you draw?

7. The national average number of flu shots given by doctors is 7. We wish to determine whether the average number of flu shots given by doctors in a particular community differs from the national average. A sample of eight doctors in a community were asked how many flu shots they had given to patients this fall. The numbers of flu shots were 6, 3, 5, 14, 2, 6, 0, and 8. Assume that the distribution of flu shots given by doctors in this community is normal.
a. State the appropriate null and alternative hypotheses.
b. Calculate the test statistic and the p-value.
c. At $\alpha = 0.05$, should you reject the null hypothesis?
d. What conclusion can you draw?

8. The average number of cesarean sections (surgical deliveries of babies) performed annually by doctors in the past has been 30. A researcher suspects that more cesarean sections are being performed today. A recent study in Switzerland studied the number of cesarean sections performed in a year by doctors. The data for 10 male doctors are:

$$27, 50, 33, 25, 86, 25, 59, 31, 37, 44$$

Assume that the distribution of number of cesarean sections performed is normal.
a. State the appropriate null and alternative hypotheses.
b. Calculate the test statistic and the p-value.
c. At $\alpha = 0.05$, should you reject the null hypothesis?
d. What conclusion can you draw?

9. The employees of Jones Construction Co. are allowed a 30-minute lunch break. The owner wants to know whether the employees are actually taking 30-minute breaks. The foreman decides to record the times taken for the next 25 breaks taken at the site. The data are as follows:

$$25\ 34\ 28\ 27\ 33\ 40\ 23\ 22\ 26\ 33\ 38\ 25\ 22$$
$$27\ 30\ 32\ 31\ 24\ 22\ 28\ 32\ 34\ 26\ 24\ 35$$

a. State the appropriate null and alternative hypotheses.
b. Calculate the test statistic and the *p*-value.
c. At $\alpha = 0.05$, should the foreman reject the null hypothesis?
d. What conclusion can the foreman present to the owner?

10. Mary Martinez loves to weigh things with her new electronic scale that is accurate to the nearest gram. Mary has always wondered whether the cereal inside the boxes of her favorite brand actually weighs 300 grams as shown on the carton, so she decides to weigh the cereal inside the next 21 boxes that she buys. Results are as follows:

297 303 289 314 302 301 300 302 288 294 296
297 300 299 391 287 286 288 295 298 302

a. State the appropriate null and alternative hypotheses.
b. Calculate the test statistic and the *p*-value.
c. At $\alpha = 0.05$, should Mary reject the null hypothesis?
d. What conclusion can Mary draw?

11. The players on last year's football team at State College were able to bench press a mean of 312 lb. Coach Juarez made it clear to the players during spring training that the team's average best lift had to improve. A special weight-training program was launched, and all the players participated. In an effort to measure the team's progress, the coach recorded the heaviest lifts of the starting offensive and defensive lineups at the start of this season. Results are as follows:

346 412 332 285 396 461 321 275 246 315 298
347 430 419 406 311 319 385 377 365 385 400

a. State the appropriate null and alternative hypotheses.
b. Calculate the test statistic.
c. At $\alpha = 0.01$, should Coach Juarez reject the null hypothesis?
d. Assuming the starting lineup is a representative sample, what conclusion can the coach draw?

Section 9.3

1. The average birth weight of premature infants is known to be less than that of full-term infants. An obstetrician wished to determine whether there is also greater variability in the weights of premature infants. Suppose the variance for full-term infants is 0.36 lb². Data on 30 premature infants are collected and the variance is 0.49 lb². Assume the distribution of infant birth weights is normal.
a. State the appropriate null and alternative hypotheses.
b. Calculate the test statistic.
c. At $\alpha = 0.05$, should you reject the null hypothesis?
d. What conclusion can you draw?

2. A soft-drink dispensing machine distributes 8-oz soft drinks in cups. A random sample of 16 cups has a variance of 0.42. Assume the distribution of fill amounts is normal. Test the hypothesis that the variance is less than 0.50.
a. State the appropriate null and alternative hypotheses.
b. Calculate the test statistic.
c. At $\alpha = 0.01$, should you reject the null hypothesis?
d. What conclusion can you draw?

3. The waiting time in a doctor's office has had a standard deviation of 8.6 minutes. The staff at the office hears complaints about the wait time being too long. In an effort to reduce the wait time, a reduced number of patients are scheduled for each time slot. The standard deviation of a sample of 27 patients scheduled on a day after the new scheduling system was implemented is 6.4 min. Assume the distribution of wait times is normal.
a. State the appropriate null and alternative hypotheses.
b. Calculate the test statistic.
c. At $\alpha = 0.05$, should you reject the null hypothesis?
d. What conclusion can you draw?

4. A farmer grows cantaloupe melon each year. The diameter of last year's crop had a standard deviation of 2.4 inches. The farmer selects a sample of 20 cantaloupes from this year's crop and calculates the standard deviation of the diameters to be 2.6 inches. Assume the distribution of cantaloupe diameters is normal. Test the hypothesis that the standard deviation has changed.

a. State the appropriate null and alternative hypotheses.
b. Calculate the test statistic.
c. At $\alpha = 0.05$, should you reject the null hypothesis?
d. What conclusion can you draw?

5. Dieters who follow a certain diet plan lose an average of 2 pounds a week. A sample of 25 dieters selected one week produce a variance of 0.49. Assume the distribution of weight loss for these dieters is normal. Test the hypothesis that the variance differs from 0.50.

a. State the appropriate null and alternative hypotheses.
b. Calculate the test statistic.
c. At $\alpha = 0.01$, should you reject the null hypothesis?
d. What conclusion can you draw?

6. Bubble Bottling Co. is operating an old machine that fills each 1-liter plastic bottle of soda prior to capping. The average fill seems satisfactory, but the operator has reported lately that too many bottles seem either too full or too empty coming off the line. The problem, of course, could be the variation in the capacity of the bottles rather than with the machine's ability to fill them. The filling machine is rated at a standard deviation of 0.01 liter. In an effort to get to the bottom of the situation, you decide to open up 30 filled bottles and measure the volume of the contents. The sample variance is computed at 0.00012 $(\text{liters})^2$.

a. State the appropriate null and alternative hypotheses.
b. Calculate the test statistic.
c. At $\alpha = 0.05$, should you reject the null hypothesis?
d. What conclusion can you draw?

7. A nurse suspects that the average serum cholesterol level for men 60 and over has a standard deviation higher than 10 mg/dL. Eight men who were at least 60 years old were selected at random and their serum cholesterol level was measured. Here are their levels, in mg/dL.

$$177, 197, 190, 185, 163, 181, 205, 222$$

Assume the distribution of cholesterol levels is normal.
a. State the appropriate null and alternative hypotheses.
b. Calculate the test statistic.
c. At $\alpha = 0.05$, should you reject the null hypothesis?
d. What conclusion can you draw?

8. A doctor believes that the standard deviation of the age of women at the time of first detection of breast cancer is less than 10 years. A sample of 15 women with breast cancer was taken. Here are their ages at the time of first detection.

$$33, 42, 45, 50, 53, 56, 57, 57, 59, 60, 60, 62, 65, 68, 70$$

Assume that the distribution of time of first detection of breast cancer is normal.
a. State the appropriate null and alternative hypotheses.
b. Calculate the test statistic.
c. At $\alpha = 0.01$, should you reject the null hypothesis?
d. What conclusion can you draw?

9. The variance of the sugar content in breakfast cereals is claimed to be $4(\text{grams})^2$. A dietician wishes to test whether this is true. Data were collected on the amount of sugar contained in a sample of 13 breakfast cereals. The data are presented in order from least to greatest.

$$1, 5, 5, 6, 6, 7, 8, 8, 9, 10, 12, 13, 14$$

Assume that the distribution of sugar contained in breakfast cereals is normal.

a. State the appropriate null and alternative hypotheses.
b. Calculate the test statistic.
c. At $\alpha = 0.10$, should you reject the null hypothesis?
d. What conclusion can you draw?

10. A manufacturer claims the lifetimes of its batteries have a standard deviation of 60 minutes. We wish to test whether this claim is correct. A sample of 60 batteries have lifetimes with a standard deviation of 70 minutes. Assume the distribution of battery lifetimes is normal.
a. State the appropriate null and alternative hypotheses.
b. Calculate the test statistic.
c. At $\alpha = 0.10$, should you reject the null hypothesis?
d. What conclusion can you draw?

Section 9.4

1. We are interested in estimating the percentage of families in a large community living below the poverty level. A random sample of 400 families in this community is studied and 70 are found to be living below the poverty level. Test the hypothesis that the percentage of all families in this community who live below the poverty level is less than 20%.
a. State the appropriate null and alternative hypotheses.
b. Calculate the sample proportion, test statistic, and p-value.
c. At $\alpha = 0.05$, should you reject the null hypothesis?
d. What conclusion can you draw?

2. The makers of M&Ms candies claim that the proportion of all M&Ms that are red is 0.20. A statistics class collects data on colors of M&Ms chocolate candies. In all, 1000 M&Ms are inspected, of which 205 are red.
a. State the appropriate null and alternative hypotheses.
b. Calculate the sample proportion, test statistic, and p-value.
c. At $\alpha = 0.05$, should you reject the null hypothesis?
d. What conclusion can you draw?

3. You read that the percent of college students who have used illegal drugs in the past month is 20%. You decide to investigate this percentage at your own college. You ask a sample of 500 students whether they have used illegal drugs in the past month. Of these, 55 say "yes."
a. State the appropriate null and alternative hypotheses.
b. Calculate the sample proportion, test statistic, and p-value.
c. At $\alpha \approx 0.05$, should you reject the null hypothesis?
d. What conclusion can you draw?
e. Illegal drug use is a sensitive issue and sometimes people will lie. Explain what effect this would have on the results.

4. A movie theater is interested in the proportion of films that are R-rated. A random sample of 60 films are considered, of which 35 are R-rated. Test the hypothesis that the proportion of all films that are R-rated is greater than 0.50.
a. State the appropriate null and alternative hypotheses.
b. Calculate the sample proportion, test statistic, and p-value.
c. At $\alpha = 0.10$, should you reject the null hypothesis?
d. What conclusion can you draw?

5. A police officer wishes to study the proportion of traffic fatalities in his jurisdiction that are alcohol-related. In a study of 200 traffic fatalities in the police officer's jurisdiction, 37 of the fatalities were alcohol-related. Test the hypothesis that the percentage of all traffic fatalities in the officer's jurisdiction that are alcohol-related is less than 25% using the 0.05 level of significance.
a. State the appropriate null and alternative hypotheses.
b. Calculate the sample proportion, test statistic, and p-value.
c. At $\alpha = 0.05$, should you reject the null hypothesis?
d. What conclusion can you draw?

6. A community program to reduce smoking is initiated. In the past, 25% of the community's residents were smokers. After a year, a sample of 500 people in the community is selected, of whom 105 indicated that they were smokers.
a. State the appropriate null and alternative hypotheses.
b. Calculate the sample proportion, test statistic, and p-value.
c. At $\alpha = 0.01$, should you reject the null hypothesis?
d. What conclusion can you draw?

7. It is claimed that 10% of the population is left-handed. In a random sample of 750 people, 92 are left-handed.
a. State the appropriate null and alternative hypotheses.
b. Calculate the sample proportion, test statistic, and p-value.
c. At $\alpha = 0.05$, should you reject the null hypothesis?
d. What conclusion can you draw?

8. A researcher wishes to determine the remission rate for a new treatment designed to treat a particular form of cancer. Suppose 26 out of 100 patients using the new method go into remission. The treatment used in the past produced remission in 20% of all patients. Test the hypothesis that the new treatment produces remission in a higher percentage of patients using the 0.05 level of significance.
a. State the appropriate null and alternative hypotheses.
b. Calculate the sample proportion, test statistic, and p-value.
c. At $\alpha = 0.01$, should you reject the null hypothesis?
d. What conclusion can you draw?

9. Nielsen Media Research reported that in 1998, 70% of the households in the nation owned at least one working television set. Suppose that you conduct a survey of 150 households in your city and discover that 73% of the households owned a working television set. Does your city differ from the rest of the nation?
a. State the appropriate null and alternative hypotheses.
b. Calculate the test statistic and find the p-value.
c. At $\alpha = 0.05$, do you reject the null hypothesis?
d. What conclusion can you draw?

10. A feature article in *Newsweek* reported in the summer of 1999 that 68.5% of those using tobacco products were either trying to cut down their consumption or quit altogether. Suppose the management of a tobacco company firmly believes that the figure is inflated. To prove its point, the company hires an independent research firm to conduct a survey of 250 customers, asking exactly the same question. The tobacco company's study reveals that 64.2% are trying to cut back or quit.
a. State the appropriate null and alternative hypotheses.
b. Calculate the test statistic and find the p-value.
c. At $\alpha = 0.05$, should the tobacco company reject the null hypothesis?
d. What conclusion can the tobacco company's management draw?
e. What is the largest proportion reported by the company's survey that would have reversed the decision?

Chapter 10 Exercises

Section 10.4

1. A supermarket chain just added two new stores located on opposite sides of a city. Management is interested in determining whether customers are spending equal times shopping in the two stores. The research team temporarily installed a clock punch in each store and asked 100 customers at each location to collect a time card and punch in when they arrived and punch out as they left with their sacks full of groceries. Results from the time cards are summarized in the table.

	Store 1	Store 2
Mean shopping time (min)	89.3	83.5
Sample standard deviation (s, min)	6.6	7.1

Test the hypothesis that the two means are the same at $\alpha = 0.05$ versus the alternative hypothesis that the two means are not the same.

2. George Harris is trying to decide between two on-line investment companies to handle his stock transactions. George has decided that he will open an account with the company that can process his transaction requests more quickly. Data measuring the average transaction time for the past 42 days of trading are shown below for each company.

	Company 1	Company 2
Mean transaction time (min)	4.35	4.85
Sample standard deviation (s, min)	1.6	1.1

Test the hypothesis that the two means are the same at $\alpha = 0.05$ versus the alternative hypothesis that the two means are not the same.

3. Silver Gym's aerobics class grew so large that the class had to be split in half and offered on two different nights. Silver also hired a new aerobics leader to handle the Tuesday night class, whereas an existing staff member continued to handle the Thursday night class. Concerned about customer satisfaction with the new leader, Silver Gym's management decided to secure a leader-approval rating from each class after four weeks. Results are shown in the table.

	Tuesday	Thursday
Class size	38	41
Mean approval rating	87.3	89.7
Sample standard deviation (s)	7.4	6.3

Did the new leader of the Tuesday night class receive a lower approval rating than the leader of the Thursday night class? Set up and test your hypotheses at $\alpha = 0.05$.

4. Data were collected from 117 homes that sold in Albuquerque, NM. A realtor thinks that homes located in the northeast part of town sell for more, on average, than those located in other areas of town. Summary statistics appear in the table.

	Northeast Area	Other Areas
Number of homes	78	39
Mean selling price ($100)	1107.7	972.8
Sample standard deviation (s)	401.5	320.4

Test the appropriate hypotheses using the 0.05 level of significance.

5. Data were collected from 117 homes that sold in Albuquerque, NM. A realtor thinks that homes that were custom-built sell for more, on average, than those that were built on speculation. Summary statistics appear in the table.

	Custom-Built	Speculation
Number of homes	27	90
Mean selling price ($100)	1446.8	947.5
Sample standard deviation (s)	475.8	253.7

Test the appropriate hypotheses using the 0.05 level of significance.

6. Wait times between eruptions of the Old Faithful Geyser in Yellowstone National Park were recorded from August 1 to August 10, 1985. A scientist wishes to determine whether the wait time between eruptions during the first 5 days differs from the wait time between eruptions during the second 5 days. Summary statistics appear in the table.

	First 5 Days (8/1–8/5)	Second 5 Days (8/6–8/10)
Number of eruptions	100	100
Mean wait time	71.62	72.76
Sample standard deviation (s)	1.42	1.32

Test the appropriate hypotheses using the 0.01 level of significance.

7. Williams Software purchases its PCs from two different manufacturers. Most of the bugs are worked out a few weeks after a new machine is installed, but only after several phone calls are placed to the manufacturer's technical support staff. Frank Williams, the company's PC buyer, has asked the programmers to keep a log of the number of minutes spent on the phone with each company for three weeks after installation. Results stemming from last year's activity with new machines are shown in the table.

	Company A	Company B
Installed computers	38	36
Minutes per machine	57.3	69.7
Sample standard deviation (s)	6.1	9.2

Set up the hypotheses to test whether the time spent with the technical staff of the two manufacturers is the same. Use the 0.01 level of significance.

8. The Y2K syndrome concerned many companies, particularly those involved heavily with data-processing activities. The following question was written for a small sample survey: "How many hours per employee were devoted to preparation for the year 2000?" The results were divided between data-processing-intensive companies and all others. Results are given in the table.

	Data-Processing Companies	All Other Companies
Sample size	43	45
Hours per employee	147.3	108.7
Sample standard deviation (s)	16.1	12.4

Do the results support the belief that data-processing companies devoted more time to being Y2K-compliant than other companies? State and test your hypotheses at level of significance 0.01.

9. A driving machine was employed to hit a sample of 41 each of two different brands of golf balls using the same force each time. The research team then measured the distance between the point of impact and where the ball eventually came to rest on the ground. Results are shown in the table.

	Brand A	Brand B
Mean distance (yd)	285	281
Sample standard deviation (s)	17.3	11.6

At the 0.05 level of significance, state and test the appropriate hypotheses needed to determine whether the mean distance traveled by brand A differs from that of brand B.

10. A university admissions officer is interested in whether there is a difference in SAT scores for males and females. Summary statistics appear in the table.

	Males	Females
Number of students	46	56
Mean SAT score	1072	1068
Sample standard deviation (s)	70	67

Test the appropriate hypotheses using the 0.05 level of significance.

Section 10.5

1. Michelson developed methods to measure the speed of light. He first measured the speed of light in 1879 and then again in 1882. We wish to determine whether there is a difference in the mean measurements for the two methods. Summary statistics appear in the table.

	Measurements in 1879	Measurements in 1882
Number of measurements	100	23
Mean speed	852.4	756.2
Sample standard deviation (s)	7.9	22.3

Test the appropriate hypotheses using the 0.01 level of significance. Do not assume the population variances are equal.

2. After the Hyde Park area of Chicago experienced several burglaries, a citizen-police program was designed to reduce crime in this area. Data were collected on the number of monthly burglaries before and after the program was implemented. Summary statistics appear in the table. Conduct the appropriate test to determine whether the program is successful in reducing the number of burglaries.

	Before Program	After Program
Number of months	41	17
Mean number of burglaries	64.32	60.65
Sample standard deviation (s)	16.81	15.93

Test the appropriate hypotheses using the 0.05 level of significance. Assume the population variances are equal.

3. Does the average salary of professors at the "Big Ten" schools differ from that of those at other universities? The "Big Ten" schools are universities that are in the Committee on Institutional Cooperation. (There are actually 12 universities in the "Big Ten.") Average salaries were calculated from 50 schools, the 12 in the "Big Ten" and 38 others. Summary statistics appear below.

	"Big Ten" Universities	Other Universities
Number of schools	12	38
Mean salary	56.20	58.83
Sample standard deviation (s)	4.37	7.44

Test the appropriate hypotheses using the 0.10 level of significance. Do not assume the population variances are equal.

4. A football analyst wishes to determine whether gate attendance has increased from the time the San Francisco Giants baseball team moved to Oakland (and later became the Oakland As). Gate attendance figures (thousands) from the 1958–1967 seasons, when the team was in San Francisco, and for 1968–1978, when the team was in Oakland, appear in the table.

Attendance 1958–1967 (thousands)	Attendance 1968–1978 (thousands)
1273	1674
1422	1652
1795	1519
1391	2021
1593	1569
1571	1835
1504	1366
1546	1601
1657	1408
1242	1196
	2267

Test the appropriate hypotheses using the 0.10 level of significance. Do not assume the population variances are equal.

5. A patient receives medication for which the effect on blood pressure is unknown. A nurse monitors the patient's blood pressure each month to determine whether there is a change as a result of taking the medication. The table below summarizes the results.

	Monthly BP Before Medication	Monthly BP While on Medication
Number of months	10	31
Mean blood pressure	86.400	81.355
Sample standard deviation (s)	4.700	4.903

Test the appropriate hypotheses using the 0.01 level of significance. Assume the population variances are equal.

6. A teacher experiences disruptions (students talking) while teaching her education class. The teacher decides to try remedial interventions to reduce such disruptions. The numbers of daily disruptions before and after the teacher begins use of remedial interventions are recorded. Summary statistics appear below. Conduct the appropriate hypothesis test to determine whether the mean number of disruptions is reduced as a result of using remedial interventions.

	Before Remedial Interventions	After Remedial Interventions
Number of days	20	20
Mean number of disruptions	19.35	4.85
Sample standard deviation (s)	2.11	2.39

Test the appropriate hypotheses using the 0.01 level of significance. Assume the population variances are equal.

7. Gasoline service stations P and T obtain their gas for resale from two different terminals. An independent testing agency decides to test the octane level of the top-grade, premium fuel being pumped at each station. The agency secured 21 one-pint samples of fuel from station P and 31 one-pint samples from station T. Each one-pint sample was extracted 10 days apart. Results are shown in the table.

	Station P	Station T
Mean octane level	91.7	91.4
Sample standard deviation (s)	1.6	1.1

At the 0.10 level of significance, state and test the appropriate hypotheses needed to determine whether the mean octane level of station P's premium fuel differs from that of station T's premium fuel. Assume the population variances are equal.

8. Is the average height of male college students greater than the average height of female college students? Data were collected in an introductory college statistics class. Summary statistics of the heights of the students by gender are displayed in the table.

	Males	Females
Number of students	21	22
Mean height	69.238	65.364
Sample standard deviation (s)	4.170	2.122

Test the appropriate hypotheses using the 0.05 level of significance. Assume the population variances are equal.

9. Is there a difference in the average time that students spend studying on weeknights between male and female college students? Data were collected in an introductory college statistics class. Summary statistics of the time spent studying on weeknights (in minutes) by gender are displayed in the table.

	Males	Females
Number of students	21	22
Mean time (min)	103.10	97.73
Sample standard deviation (s)	75.70	38.81

Test the appropriate hypotheses using the 0.05 level of significance. Do not assume the population variances are equal.

10. Is the average time that students spend studying on weeknights lower for college students who work 20 hours or more per week than for those who work less than 20 hours per week? Data were collected in an introductory college statistics class. Summary statistics of the time spent studying on weeknights (in minutes) by work hours are displayed in the table.

	Less Than 20 Hours per Week	20 Hours or More per Week
Number of students	33	10
Mean time (min)	102.1	94.5
Sample standard deviation (s)	63.5	43.6

Test the appropriate hypotheses using the 0.05 level of significance. Do not assume the population variances are equal.

Section 10.7

1. Jones Real Estate School trains real estate salespeople and brokers and prepares them for their state certification examination. A new software program has been constructed for the students so that they can learn what they need to know using their own computer as an aid. The school selected 25 students before enrolling in the course and measured their "real estate IQ" before using the software. After two weeks of computer drill and no other form of instruction, real estate IQs of the same group were measured again using a similar but different test composed of questions rated at the same level of difficulty by the experts. The mean difference in the two scores was 9.5, and the standard deviation was 12.8.

a. State the appropriate hypotheses for a one-tail test and compute the test statistic.

b. At the 0.05 level of significance, what can Jones Real Estate School conclude?

2. An eye clinic is using a new laser surgery to correct the vision of nearsighted patients. The clinic measured the vision of 21 highly nearsighted patients while wearing contact lenses before they agreed to have the surgery performed. Six weeks following the surgery, the vision of the same patients without wearing any corrective lenses was measured again. On a standardized scale, the mean difference in the patients' abilities to read a Snellen chart was 0.5, and the standard deviation was 1.1.

a. State the appropriate hypotheses for a two-tail test and compute the test statistic.

b. At the 0.01 level of significance, what can the clinic conclude?

3. An auto mechanic wishes to compare two methods of measuring treadwear of tires. A sample of 16 tires is selected and measured for treadwear using two methods, the first based on weight loss and the second based on groove wear. The data appear in the table. Conduct a hypothesis test to determine whether the two methods differ using the 0.05 level of significance.

Weight Loss	Groove Wear	Weight Loss	Groove Wear
45.9	35.7	30.4	23.1
41.9	39.2	27.3	23.7
37.5	31.1	20.4	20.9
33.4	28.1	24.5	16.1
31.0	24.0	20.9	19.9
30.5	28.7	18.9	15.2
30.9	25.9	13.7	11.5
31.9	23.3	11.4	11.2

4. Gosset compared the corn yield from regular seeds to the yield from seeds that were kiln-dried. Each type of seed (regular and kiln-dried) was planted in adjacent plots. There were 11 "split" plots and the corn yield, in lb/acre, was recorded for each half-plot. The data appear below.

Regular	Kiln-dried
1903	2009
1935	1915
1910	2011
2496	2463
2108	2180
1961	1925
2060	2122
1444	1482
1612	1542
1316	1443
1511	1535

Conduct the appropriate test to determine whether the kiln-dried seeds produce a greater yield. Use the 0.05 level of significance.

5. Data were collected on labor force participation rates of women in 1968 and in 1972 in 19 major cities across the United States. We wish to determine whether the mean participation rate of women in the labor force increased during this time period. The data appear in the table. Conduct the appropriate hypothesis test using the 0.01 level of significance.

City	Women in 1972	Women in 1968	City	Women in 1972	Women in 1968
New York	0.45	0.42	Washington, D.C.	0.52	0.42
Los Angeles	0.50	0.50	Cincinnati	0.53	0.51
Chicago	0.52	0.52	Baltimore	0.57	0.49
Philadelphia	0.45	0.45	Newark	0.53	0.54
Detroit	0.46	0.43	Minn./St. Paul	0.59	0.50
San Francisco	0.55	0.55	Buffalo	0.64	0.58
Boston	0.60	0.45	Houston	0.50	0.49
Pittsburgh	0.49	0.34	Patterson	0.57	0.56
St. Louis	0.35	0.45	Dallas	0.64	0.63
Hartford	0.55	0.54			

6. An association called Acorn (acronym for Association of Community Organizations for Reform Now) investigates possible discrimination in banking between minorities and whites. Refusal rates in mortgage lending were collected for 20 banks in major cities. The data are displayed below.

BANK = Name of bank

MIN = Overall refusal rate for minority applicants

WHITE = Overall refusal rate for white applicants

HIMIN = Refusal rate for high-income minority applicants

HIWHITE = Refusal rate for high-income white applicants

BANK	MIN	WHITE	HIMIN	HIWHITE
HARRISTRUST	20.90	3.7	21.4	2.2
NCNBTEXAS	23.23	5.5	8.0	8.0
CRESTAR	23.10	6.7	11.3	3.6
MERCANTILE	30.40	9.0	17.3	5.5
FIRSTNBCOMMERC	42.70	13.9	38.0	7.6
TEXASCOMMERCE	62.20	20.6	33.3	10.3
COMERICA	39.50	13.4	33.6	9.4
FIRSTOFAMERICA	38.40	13.2	29.5	7.3
BOATMANSNATL	26.20	9.3	21.7	7.4
1STCOMML	55.90	21.0	39.1	15.8
PROVIDENTNATL	49.70	20.1	36.6	15.3
WORTHEN	44.60	19.1	28.6	10.1
HIBERNIANATL	36.40	16.0	32.9	9.2
SOVRON	32.00	16.0	21.0	13.0
BELLFEDERAL	10.60	5.6	5.8	4.2
SECPACAZ	34.30	18.4	24.2	14.1
CORESTATES	42.30	23.3	38.3	15.0
CITIBANKAZ	26.50	15.6	27.3	16.1
MFERSHANOVER	51.50	32.4	41.3	25.1
CHEMICAL	47.20	29.7	41.1	26.8

Conduct the appropriate hypothesis test to determine whether the overall refusal rate for minority applicants is higher than the overall refusal rate for white applicants. Use the 0.01 level of significance.

7. An association called Acorn (acronym for Association of Community Organizations for Reform Now) investigates possible discrimination in banking between minorities and whites. Refusal rates in mortgage lending were collected for 20 banks in major cities. The data are displayed in Exercise 6 above. Conduct the appropriate hypothesis test to determine whether the refusal rate for high-income minority applicants is higher than the refusal rate for high-income white applicants. Use the 0.01 level of significance.

8. Is there a difference in the birth weight between the firstborn twin and the secondborn twin? Data collected on 19 sets of twins appear below, where Twin A = the birth weight of the firstborn twin and Twin B = the birth weight of the secondborn twin.

Twin A	Twin B	Twin A	Twin B
5.9375	5.5625	6.1250	7.5625
5.1875	5.6875	5.6875	5.4375
6.4375	5.8125	4.3750	4.9375
6.4375	7.5625	5.2500	5.1875
4.3125	5.5000	3.5000	3.9375
5.7500	5.8125	4.8125	5.1250
7.3750	7.5625	6.1250	6.6250
5.5625	4.4375	6.3750	6.1250
5.8750	4.8750	4.5000	5.3125
4.6875	4.6875		

Conduct the appropriate hypothesis test to determine whether the birth weights differ. Use the 0.10 level of significance.

9. Data were collected on the percentages of the labor force employed in agriculture (AGR), industry (IND), and services (SER) for 20 countries in 1960. The data appear below.

Country	AGR	IND	SER
Canada	13	43	45
Sweden	14	53	33
Switzerland	11	56	33
Luxembourg	15	51	34
U. Kingdom	4	56	40
Denmark	18	45	37
W. Germany	15	60	25
France	20	44	36
Belgium	6	52	42
Norway	20	49	32
Iceland	25	47	29
Netherlands	11	49	40
Austria	23	47	30
Ireland	36	30	34
Italy	27	46	28
Japan	33	35	32
Greece	56	24	20
Spain	42	37	21
Portugal	44	33	23
Turkey	79	12	9

Conduct the appropriate hypothesis test to determine whether the percent employed in agriculture was lower than the percent employed in industry worldwide in 1960. Use the 0.05 level of significance.

10. Data were collected on the percentages of the labor force employed in agriculture, industry, and services for 20 countries in 1960. The data appear in Exercise 9 above. Conduct the appropriate hypothesis test to determine whether the percent employed in industry was higher than the percent employed in service worldwide in 1960. Use the 0.05 level of significance.

Section 10.8

1. Prior to installing a new computerized baggage system, 680 of a sample of 2000 passengers checked their bags rather than carry them on board. Six months after installing the new system, 950 of a sample of 2500 checked their bags. State and test the appropriate hypotheses to determine whether the proportion of the passengers checking bags increased after the new system was installed. Use $\alpha = 0.01$, and draw a conclusion for the airport.

2. Media Disk, Inc., duplicates over a million 3.5″ floppy disks each year by copying masters to stacks of blank disks. The company buys its blank stock from two different suppliers, A and B. The manager has decided to check each supplier's stock by counting the rejected disks in the next run from 5000 that just arrived from supplier A and 4500 that just arrived from supplier B. During the run, the disk duplicators rejected 73 of A's disks and 56 of B's. State the appropriate hypotheses to test whether the proportions of defective disks from the two suppliers are the same. At $\alpha = 0.05$, what can Media Disk conclude?

3. A recent Gallup poll asked respondents to indicate their health status (excellent, good, fair, or poor) and whether or not they were overweight. Of the 472 people who rated their health status as "good," 212 indicated they were overweight. Of the 221 people who rated their health status as either "fair" or "poor," 119 indicated they were overweight. Conduct the appropriate hypothesis test to determine whether the proportion of overweight people in "good" health is lower than the proportion of overweight people in "fair" or "poor" health. Use the 0.05 level of significance.

4. In a clinical study of a new medication, 500 people received the experimental drug and 500 people received a placebo (a pill with no active ingredient). Of the 500 people taking the experimental drug, 127 reported that they experienced severe headaches while taking the drug. Of the 500 people taking the placebo, 56 reported that they experienced severe headaches. Conduct the appropriate hypothesis test to determine whether the proportion of people who experience severe headaches is lower for those taking the placebo than for those taking the experimental drug. Use the 0.10 level of significance.

5. In a clinical study of a new medication, 500 people received the experimental drug and 500 people received a placebo (a pill with no active ingredient). Of the 500 people taking the experimental drug, 83 reported that they experienced drowsiness while taking the drug. Of the 500 people taking the placebo, 76 reported that they experienced drowsiness. Conduct the appropriate hypothesis test to determine whether the proportion of people who experience drowsiness is lower for those taking the placebo than for those taking the experimental drug. Use the 0.10 level of significance.

6. According to a survey conducted by the Bureau of Labor Statistics in 1998, 5.1% of women compared to 2.7% of men were absent from their job during the average work week. This means that they worked less than 35 hours during the week because of injury, illness, or a variety of other reasons. Assume that 12,000 women and 15,000 men were included in this survey. Conduct the appropriate hypothesis test to determine whether the proportion of women who are absent from their job is greater than the proportion of men who are absent from their job. Use the 0.05 level of significance.

7. Is the poverty rate higher for women who are single women with children under 18 than for single men with children under 18? The results of the Current Population Survey conducted by the Bureau of Labor Statistics in 1999 indicated that 1,651 of 6,920 single women with children under 18 lived below poverty. The survey also indicated that 211 of 1,965 single men with children under 18 lived below poverty. Conduct the appropriate hypothesis test to answer this question using the 0.05 level of significance.

8. An exercise science student is interested in comparing the proportion of males and females who can do 40 or more push-ups. Of the 44 males he observes, 34 can do 40 or more push-ups. Of the 52 females he observes, 22 can do 40 or more push-ups. Conduct the appropriate hypothesis test to determine whether the proportion who can do 40 or more push-ups is greater for males than females. Use the 0.10 level of significance.

9. A taste test comparing a name-brand cola and a store-brand cola is conducted. Suppose we are interested in comparing the percent of males who prefer the name brand to the percent of females who prefer the name brand. Suppose 50 males and 50 females taste the products. The results indicate that 41 males prefer the name-brand cola while 36 females prefer the name-brand cola. Conduct the appropriate hypothesis test to determine whether there is a difference in the percent of males and females who prefer the name-brand cola. Use the 0.10 level of significance.

10. A politician is running for office and is planning his campaign strategy on the issue of gun control. He wants to know whether there is a significant difference in the percentage of voters who favor the issue in district 1 and district 2. In a poll of residents in district 1, 89 out of 135 voters are in favor of gun control. In a poll of residents in district 2, 65 out of 110 voters are in favor of gun control. Conduct the appropriate hypothesis test using the 0.05 level of significance.

Section 10.9

1. A driving machine was employed to hit a sample of 41 each of two different brands of golf balls using the same force each time. The research team then measured the distance between the point of impact and where the ball eventually came to rest on the ground. Results are shown in the table.

	Brand A	Brand B
Sample standard deviation (s)	17.3	11.6

At the 0.05 level of significance, state and test the appropriate hypotheses needed to determine whether the variance of the distance traveled by brand A exceeds that of brand B.

2. Gasoline service stations P and T obtain their gas for resale from two different terminals. An independent testing agency decides to test the octane level of the top-grade, premium fuel being pumped at each station. The agency secured 21 one-pint samples of fuel from station P and 31 one-pint samples from station T. Each one-pint sample was extracted 10 days apart. Results are shown in the table.

	Station P	Station T
Sample standard deviation (s)	1.6	1.1

At the 0.01 level of significance, state and test the appropriate hypotheses needed to determine whether the variance of the octane level of station P's premium fuel is greater than that of station T's premium fuel.

3. Michelson developed methods to measure the speed of light. He first measured the speed of light in 1879 and then again in 1882. We wish to determine whether there is a difference in the variance for the two methods. Summary statistics appear in the table.

	Measurements in 1879	Measurements in 1882
Number of measurements	100	23
Sample standard deviation (s)	7.9	22.3

Test the appropriate hypotheses using the 0.01 level of significance.

4. Does the average salary of professors at the "Big Ten" schools have less variability than that of those at other universities? The "Big Ten" schools are universities that are in the Committee on Institutional Cooperation. (There are actually 12 universities in the "Big Ten.") Average salaries and standard deviations were calculated from 50 schools, the 12 in the "Big Ten" and 38 others. Summary statistics appear below.

	Big Ten Universities	Other Universities
Number of schools	12	38
Sample standard deviation (s)	4.37	7.44

Test the appropriate hypotheses using the 0.10 level of significance.

5. A football analyst wishes to determine whether the standard deviation of gate attendance has increased from the time the San Francisco Giants baseball team moved to Oakland (and later became the Oakland As). Gate attendance (thousands) figures from the 1958–1967 seasons, when the team was in San Francisco, and for 1968–1978, when the team was in Oakland, appear below.

Attendance 1958–1967 (thousands)	Attendance 1968–1978 (thousands)
1273	1674
1422	1652
1795	1519
1391	2021
1593	1569
1571	1835
1504	1366

(*continued*)

1546	1601
1657	1408
1242	1196
	2267

Test the appropriate hypotheses using the 0.05 level of significance.

6. A patient receives medication for which the effect on blood pressure is unknown. A nurse monitors the patient's blood pressure each month to determine whether there is a change in the variability of blood pressure as a result of taking the medication. The table summarizes the results.

	Monthly BP Before Medication	Monthly BP While on Medication
Number of months	10	31
Sample standard deviation (s)	4.700	4.903

Test the appropriate hypotheses using the 0.05 level of significance.

7. A teacher experiences disruptions (students talking) while teaching her education class. The teacher decides to try remedial interventions to reduce such disruptions. The number of daily disruptions before and after the teacher begins use of remedial interventions are recorded. Summary statistics appear below. Conduct the appropriate hypothesis test to determine whether the standard deviation in number of disruptions changes as a result of using remedial interventions.

	Before Remedial Interventions	After Remedial Interventions
Number of days	20	20
Sample standard deviation (s)	2.11	2.39

Test the appropriate hypotheses using the 0.05 level of significance.

8. Do the heights of male college students show more variability than the heights of female college students? Data were collected in an introductory college statistics class. Summary statistics of the heights of the students by gender are displayed in the table.

	Males	Females
Number of students	21	22
Sample standard deviation (s)	4.170	2.122

Test the appropriate hypotheses using the 0.05 level of significance.

9. Is there a difference in variability of the time that students spend studying on weeknights between male and female college students? Data were collected in an introductory college statistics class. Summary statistics of the time spent studying on weeknights (in minutes) by gender are displayed in the table.

	Males	Females
Number of students	21	22
Sample standard deviation (s)	75.70	38.81

Test the appropriate hypotheses using the 0.05 level of significance.

10. Is the variability in time that students spend studying on weeknights greater for college students who work 20 hours or more per week than for those who work less than 20 hours per week? Data were collected in an introductory college statistics class. Summary statistics of the time spent studying on weeknights (in minutes) by work hours are displayed in the table.

	Less than 20 hours per week	20 hours or more per week
Number of students	33	10
Sample standard deviation (s)	63.5	43.6

Test the appropriate hypotheses using the 0.05 level of significance.

Chapter 11

Section 11.2

1. Does a heavier bowler knock down more pins? Jim Samuels, manager of Maple Lanes, asked 15 men's league players to report their body weight (lb) and their league average, rounded to the nearest pin. Results are shown in the table.

Weight, X	Average, Y	Weight, X	Average, Y
154	156	215	159
159	145	226	152
165	178	237	132
172	201	245	178
178	205	262	196
192	191	280	145
205	187	289	168
211	146		

a. Construct a scatter plot of the data.
b. Find the equation for the least-squares line of best fit.
c. Draw the least-squares line on the scatter plot.
d. Predict the league average for an adult male bowler who weighs 190 lb. What can you conclude?
e. Find the standard error of the estimate.

2. The expense ratio of a mutual fund measures the percentage of the fund's assets used to pay for annual administrative overhead. Funds with higher ratios spend more of the fund's return to operate the fund. Does this hurt or help the fund's return? A random selection of 16 stock mutual funds in 1998 revealed the following data:

Expense Ratio, X	Total Return, Y	Expense Ratio, X	Total Return, Y
0.62	34.3	0.75	33.8
1.21	33.0	0.92	27.9
1.03	31.4	0.57	32.6
0.67	30.8	0.95	24.3
0.40	32.9	0.20	36.3
1.12	27.6	1.00	31.7
0.78	30.3	1.25	29.9
0.47	31.9	1.03	34.7

a. Construct a scatter plot of the data.
b. Find the equation for the least-squares line of best fit.
c. Draw the least-squares line on the scatter plot.
d. Predict the total return for a stock fund with an expense ratio of 0.81%. What can you conclude?
e. Find the standard error of the estimate.

3. Is the size of a home related to selling price? Use SQFT (square feet of living space in the home) as the explanatory variable and PRICE (selling price in hundreds of dollars) as the response variable to conduct the analysis.

PRICE	SQFT	AGE	PRICE	SQFT	AGE	PRICE	SQFT	AGE
2050	2650	13	700	1215	*	1295	3750	*
2080	2600	*	720	1121	46	975	1500	7
2150	2664	6	720	1050	*	939	1428	40
2150	2921	3	749	1733	43	820	1375	*
1999	2580	4	731	1299	*	780	1080	*

(continued)

1900	2580	4	725	1140	*	770	900	*
1800	2774	2	670	1181	*	700	1505	*
1560	1920	1	2150	2848	4	620	1480	*
1450	2150	*	1599	2440	*	540	1142	*
1449	1710	1	1350	2253	23	1070	1464	*
1375	1837	4	1299	2743	25	2100	2116	25
1270	1880	8	1250	2180	17	725	1280	*
1250	2150	15	1239	1706	14	660	1159	*
1235	1894	14	1200	1948	*	600	1198	*
1170	1928	18	1125	1710	16	580	1051	15
1180	1830	*	1100	1657	*	1844	2250	40
1155	1767	16	1080	2200	26	1580	2563	*
1110	1630	15	1050	1680	13	699	1400	45
1139	1680	17	1049	1900	34	1330	1850	5
995	1725	*	955	1565	*	1160	1720	5
995	1500	15	934	1543	20	1109	1740	4
975	1430	*	875	1173	6	1129	1700	6
975	1360	*	889	1549	*	1050	1620	6
900	1400	16	855	1900	*	1045	1630	6
960	1573	17	835	1560	*	1050	1920	8
860	1385	*	810	1365	*	1020	1606	5
1695	2931	28	805	1258	7	1000	1535	7
1553	2200	28	799	1314	*	1030	1540	6
1250	2277	*	750	1338	*	975	1739	13
1300	2000	*	759	997	4	950	1715	*
1020	1478	53	755	1275	*	940	1305	5
1020	1713	30	750	1030	*	920	1415	7
922	1326	*	730	1027	*	945	1580	9
925	1050	*	729	1007	19	874	1236	3
899	1464	*	710	1083	22	872	1229	6
850	1190	41	773	1320	*	870	1273	4
876	1156	*	690	1348	15	869	1165	7
890	1746	*	670	1350	*	766	1200	7
870	1280	*	619	837	*	739	970	4

a. Construct a scatter plot of the data.
b. Find the equation for the least-squares line of best fit.
c. Draw the least-squares line on the scatter plot.
d. Predict the selling price of a home that has 2200 square feet of living space. What can you conclude?
e. There is one prominent outlier apparent in the scatter plot. Identify this outlier in the data. Give the square footage and the selling price of this home.

4. Is the number of calories in a serving of breakfast cereal related to the amount of fat? Use cal (calories) as the explanatory variable and fat (fat, in grams) as the response variable to conduct the analysis.

cal	fat	carbo	sugar	cal	fat	carbo	sugar
70	1	5.0	6	140	1	20.0	9
120	5	8.0	8	110	1	21.0	3
70	1	7.0	5	100	2	12.0	6
50	0	8.0	0	110	1	12.0	12
110	2	14.0	8	100	1	16.0	3
110	2	10.5	10	150	3	16.0	11
110	0	11.0	14	150	3	16.0	11
130	2	18.0	8	160	2	17.0	13
90	1	15.0	6	100	1	15.0	6
90	0	13.0	5	120	1	15.0	9
120	2	12.0	12	140	2	21.0	7
110	2	17.0	1	90	0	18.0	2
120	3	13.0	9	130	2	13.5	10

(*continued*)

110	2	13.0	7	120	1	11.0	14
110	1	12.0	13	100	0	20.0	3
110	0	22.0	3	50	0	13.0	0
100	0	21.0	2	50	0	10.0	0
110	0	13.0	12	100	1	14.0	6
110	1	12.0	13	100	2	*	*
110	3	10.0	7	120	1	14.0	12
100	0	21.0	0	100	2	10.5	8
110	0	21.0	3	90	0	15.0	6
100	1	11.0	10	110	0	23.0	2
100	0	18.0	5	110	0	22.0	3
110	1	11.0	13	80	0	16.0	0
110	0	14.0	11	90	0	19.0	0
100	0	14.0	7	90	0	20.0	0
120	2	12.0	10	110	1	9.0	15
120	0	14.0	12	110	0	16.0	3
110	1	13.0	12	90	0	15.0	5
100	0	11.0	15	110	1	21.0	3
110	1	15.0	9	140	1	15.0	14
100	1	15.0	5	100	1	16.0	3
110	0	17.0	3	110	1	21.0	3
120	3	13.0	4	110	1	13.0	12
120	2	12.0	11	100	1	17.0	3
110	1	11.5	10	100	1	17.0	3
110	0	14.0	11	110	1	16.0	8
110	1	17.0	6				

a. Construct a scatter plot of the data.

b. Find the equation for the least-squares line of best fit.

c. Draw the least-squares line on the scatter plot.

d. Predict the amount of sugar in a breakfast cereal with 120 calories per serving. What can you conclude?

e. There is one prominent outlier apparent in the scatter plot. Identify this outlier in the data. Give the number of calories and fat grams for this cereal.

5. Is the amount of carbohydrates in a serving of breakfast cereal related to the amount of sugar? Refer to the data given in Exercise 4 of this section. Use carbo (grams of carbohydrates per serving) as the explanatory variable and sugar (grams of sugar per serving) as the response variable to conduct the following analysis.

a. Construct a scatter plot of the data.

b. Find the equation for the least-squares line of best fit.

c. Draw the least-squares line on the scatter plot.

d. Interpret the slope.

e. Predict the amount of sugar in a breakfast cereal with 16 grams of carbohydrates per serving. What can you conclude?

6. Is the weight of a firstborn twin an indicator of the weight of the secondborn twin? The accompanying data give the weights of the firstborn (Twin A) and secondborn (Twin B) infants for 19 sets of twins. Use Twin A as the explanatory variable and Twin B as the response variable for the following analysis.

Twin A	Twin B	Twin A	Twin B
5.9375	5.5625	6.1250	7.5625
5.1875	5.6875	5.6875	5.4375
6.4375	5.8125	4.3750	4.9375
6.4375	7.5625	5.2500	5.1875
4.3125	5.5000	3.5000	3.9375
5.7500	5.8125	4.8125	5.1250
7.3750	7.5625	6.1250	6.6250
5.5625	4.4375	6.3750	6.1250
5.8750	4.8750	4.5000	5.3125
4.6875	4.6875		

a. Construct a scatter plot of the data.
b. Find the equation for the least-squares line of best fit.
c. Draw the least-squares line on the scatter plot.
d. Interpret the slope.
e. Predict the weight of the secondborn twin when the firstborn twin weighs 4.5 lb. What can you conclude?
f. Compute the residual for the last set of twins.

7. A baseball analyst is interested in the relationship between number of hits and number of strikeouts for baseball pitchers. The following data are from pitchers who played from 1920 to 1950. Use number of hits (HITS) as the explanatory variable and number of strikeouts (SO) as the response variable for the following analysis.

PITCHER	HITS	SO
C. Hubbell	25.1	11.2
D. Dean	25.3	14.4
L. Grove	25.4	13.6
G. Alexander	27.3	6.4
D. Vance	25.4	16.8
D. Leonard	26.5	8.8
B. Walters	25.4	8.8
W. Johnson	25.6	12.0
L. Gomez	24.3	13.6
P. Derringer	27.2	9.6
F. Fitzsimmons	27.2	6.4
T. Lyons	27.5	6.4

a. Construct a scatter plot of the data.
b. Find the equation for the least-squares line of best fit.
c. Draw the least-squares line on the scatter plot.
d. Predict the number of strikeouts for a pitcher who has 26 hits. What can you conclude?
e. Compute the residuals for P. Derringer and F. Fitzsimmons.

8. The accompanying data are from eruptions of the Old Faithful Geyser in Yellowstone National Park. Duration is the length of time the eruption lasted and interval is the length of time from the previous eruption. We wish to determine whether there is a relationship between the length of time from the previous eruption and the duration of the eruption. Use interval as the explanatory variable and duration as the response variable for the following analysis.

day	interval	duration	day	interval	duration	day	interval	duration
1	78	4.4	2	91	4.1	4	73	3.7
1	74	3.9	2	51	1.8	4	67	3.7
1	68	4.0	2	79	3.2	4	68	4.3
1	76	4.0	2	53	1.9	4	86	3.6
1	80	3.5	2	82	4.6	4	72	3.8
1	84	4.1	2	51	2.0	4	75	3.8
1	50	2.3	3	76	4.5	4	75	3.8
1	93	4.7	3	82	3.9	4	66	2.5
1	55	1.7	3	84	4.4	4	84	4.5
1	76	4.9	3	53	2.3	4	70	4.1
1	58	1.7	3	86	3.8	4	79	3.7
1	74	4.6	3	51	1.9	4	60	3.8
1	75	3.4	3	85	4.6	4	86	3.4
2	80	4.3	3	45	1.8	5	71	4.0
2	56	1.7	3	88	4.7	5	67	2.3
2	80	3.9	3	51	1.8	5	81	4.4
2	69	3.7	3	80	4.6	5	76	4.1
2	57	3.1	3	49	1.9	5	83	4.3
2	90	4.0	3	82	3.5	5	76	3.3
2	42	1.8	4	75	4.0	5	55	2.0

(*continued*)

day	interval	duration	day	interval	duration	day	interval	duration
5	73	4.3	6	54	1.8	7	86	4.5
5	56	2.9	6	83	4.4	7	48	2.0
5	83	4.6	6	51	1.9	8	77	4.2
5	57	1.9	6	80	4.6	8	73	4.4
5	71	3.6	6	78	2.9	8	70	4.1
5	72	3.7	7	81	3.5	8	88	4.1
5	77	3.7	7	53	2.0	8	75	4.0
6	55	1.8	7	89	4.3	8	83	4.1
6	75	4.6	7	44	1.8	8	61	2.3
6	73	3.5	7	78	4.1	8	78	4.6
6	70	4.0	7	61	1.8	8	61	1.9
6	83	3.7	7	73	4.7	8	81	4.5
6	50	1.7	7	75	4.2	8	51	2.0
6	95	4.6	7	73	3.9	8	80	4.8
6	51	1.7	7	76	4.3	8	79	4.1
6	82	4.0	7	55	1.8			

a. Construct a scatter plot of the data.
b. Find the equation for the least-squares line of best fit.
c. Draw the least-squares line on the scatter plot.
d. Predict the duration of an eruption that occurs 80 min after the previous eruption. What can you conclude?
e. Find the standard error of the estimate.

9. The data given here appeared in the *Wall Street Journal*. The advertisements were selected by an annual survey conducted by Video Board Tests, Inc., a New York ad-testing company, based on interviews with 20,000 adults who were asked to name the most outstanding TV commercial they had seen, noticed, and liked. The retained impressions were based on a survey of 4000 adults, in which regular product users were asked to cite a commercial they had seen for that product category in the past week. Use SPEND as the explanatory variable and MILIMP as the response variable for the following analysis.

FIRM: Firm name
SPEND: TV advertising budget, 1983 ($ millions)
MILIMP: Millions of retained impressions per week

FIRM	SPEND	MILIMP
Miller Lite	50.1	32.1
Pepsi	74.1	99.6
Stroh's	19.3	11.7
Fed'l Express	22.9	21.9
Burger King	82.4	60.8
Coca-Cola	40.1	78.6
McDonald's	185.9	92.4
MCI	26.9	50.7
Diet Coke	20.4	21.4
Ford	166.2	40.1
Levi's	27.0	40.8
Bud Lite	45.6	10.4
ATT/Bell	154.9	88.9
Calvin Klein	5.0	12.0
Wendy's	49.7	29.2
Polaroid	26.9	38.0
Shasta	5.7	10.0
Meow Mix	7.6	12.3
Oscar Meyer	9.2	23.4
Crest	32.4	71.1
Kibbles 'n Bits	6.1	4.4

a. Construct a scatter plot of the data.

b. Find the equation for the least-squares line of best fit.

c. Draw the least-squares line on the scatter plot.

d. Predict the millions of retained impressions per week for a TV advertising budget of $50 million. What can you conclude?

e. Compute the residuals for Pepsi and Ford.

10. We wish to use the accompanying data to predict the top speed (SP) of a car in miles per hour based on its engine horsepower (HP).

MAKE/MODEL	HP	MPG	SP	MAKE/MODEL	HP	MPG	SP
GM/Geo Metro XF1	49	65.4	96.0	Chevrolet Beretta	95	32.2	106.0
GM/Geo Metro	55	56.0	97.0	Toyota Corolla	102	32.2	109.0
GM/Geo Metro LSI	55	55.9	97.0	Pontiac Sunbird Conv.	95	32.2	106.0
Suzuki Swift	70	49.0	105.0	Dodge Shadow	93	31.5	105.0
Daihatsu Charade	53	46.5	96.0	Dodge Daytona	100	31.5	108.0
GM/Geo Sprint Turbo	70	46.2	105.0	Eagle Spirit	100	31.4	108.0
GM/Geo Sprint	55	45.4	97.0	Ford Tempo	98	31.4	107.0
Honda Civic CRXHF	62	59.2	98.0	Toyota Celica	130	31.2	120.0
Honda Civic CRXHF	62	53.3	98.0	Toyota Camry	115	33.7	109.0
Daihatsu Charade	80	43.4	107.0	Toyota Camry	115	32.6	109.0
Subaru Justy	73	41.1	103.0	Toyota Camry	115	31.3	109.0
Honda Civic CRX	92	40.9	113.0	Toyota Camry Wagon	115	31.3	109.0
Honda Civic	92	40.9	113.0	Olds Cutlass Sup.	180	30.4	133.0
Subaru Justy	73	40.4	103.0	Olds Cutlass Sup.	160	28.9	125.0
Subaru Justy	66	39.6	100.0	Saab 9000	130	28.0	115.0
Subaru Justy4wd	73	39.3	103.0	Ford Mustang	96	28.0	102.0
Toyota Tercel	78	38.9	106.0	Toyota Camry	115	28.0	109.0
Honda Civic CRX	92	38.8	113.0	Chrysler LeBaron Conv.	100	28.0	104.0
Toyota Tercel	78	38.2	106.0	Dodge Dynasty	100	28.0	105.0
Ford Escort	90	42.2	109.0	Volvo 740	145	27.7	120.0
Honda Civic	92	40.9	110.0	Ford Thunderbird	120	25.6	107.0
Pontiac LeMans	74	40.7	101.0	Chevrolet Caprice	140	25.3	114.0
Isuzu Stylus	95	40.0	111.0	Lincoln Continental	140	23.9	114.0
Dodge Colt	81	39.3	105.0	Chrysler New Yorker	150	23.6	117.0
GM/Geo Storm	95	38.8	111.0	Buick Reatta	165	23.6	122.0
Honda Civic CRX	92	38.4	110.0	OldsTrof/Toronado	165	23.6	122.0
Honda Civic	92	38.4	110.0	Oldsmobile 98	165	23.6	122.0
Subaru Loyale	90	29.5	109.0	Pontiac Bonneville	165	23.6	122.0
Volks Jetta Diesel	52	46.9	90.0	Lexus LS400	245	23.5	148.0
Mazda 323 Protege	103	36.3	112.0	Nissan 300ZX	280	23.4	160.0
Ford Escort Wagon	84	36.1	103.0	Volvo 760 Wagon	162	23.4	121.0
Ford Escort	84	36.1	103.0	Audi 200 Quatro Wagon	162	23.1	121.0
GM/Geo Prism	102	35.4	111.0	Buick Electra Wagon	140	22.9	110.0
Toyota Corolla	102	35.3	111.0	Cadillac Brougham	140	22.9	110.0
Eagle Summit	81	35.1	102.0	Cadillac Brougham	175	19.5	121.0
Nissan Sentra Coupe	90	35.1	106.0	Mercedes 500SL	322	18.1	165.0
Nissan Sentra Wagon	90	35.0	106.0	Mercedes 560SEL	238	17.2	140.0
Toyota Celica	102	33.2	109.0	Jaguar XJS Convert.	263	17.0	147.0
Toyota Celica	102	32.9	109.0	BMW 750IL	295	16.7	157.0
Toyota Corolla	130	32.3	120.0	Rolls-Royce Various	236	13.2	130.0
Chevrolet Corsica	95	32.2	106.0				

a. Construct a scatter plot of the data.

b. Find the equation for the least-squares line of best fit.

c. Draw the least-squares line on the scatter plot.

d. Interpret the slope.

e. Predict the top speed of a car that has a 200-horsepower engine.

11. Refer to the data given in Exercise 10 of this section. We wish to predict the gas mileage of a car in miles per gallon (MPG) based on its engine horsepower (HP).
a. Construct a scatter plot of the data.
b. Find the equation for the least-squares line of best fit.
c. Draw the least-squares line on the scatter plot.
d. Interpret the slope.
e. Predict the gas mileage of a car that has a 200-horsepower engine.

12. A baseball team's batting success could be evaluated theoretically by its ability to wear down the opposing teams' pitchers. Players call this "working the count." As the game progresses, the opposing pitchers are often required to throw more pitches and tend to get tired. Meanwhile, the batters get time to study the pitchers and improve their timing. All of this additional work can make the pitcher more vulnerable to an offensive attack as the innings go by. The following data from *Sports Illustrated* show the number of pitches faced per game in 1998 and the winning percentages of the teams in the American League:

Team	Pitches Faced per Game	Winning Percentage	Team	Pitches Faced per Game	Winning Percentage
Yankees	152.6	70.3	Orioles	145.3	48.8
Athletics	151.7	45.7	Royals	143.6	44.4
Indians	150.7	54.9	Devil Rays	143.0	38.9
Rangers	149.4	54.3	White Sox	142.8	49.4
Mariners	148.4	46.9	Tigers	142.6	40.1
Red Sox	145.9	56.8	Angels	142.5	52.5
Blue Jays	145.7	54.3	Twins	140.2	43.2

a. Construct a scatter plot of the data.
b. Find the equation for the least-squares line of best fit.
c. Draw the least-squares line on the scatter plot.
d. Predict the winning percentage for a team that can force its opposing pitchers to throw an average of 152 pitches per game. What can you conclude?
e. Calculate the standard error of the estimate.

Section 11.3

1. Does a heavier bowler knock down more pins? Jim Samuels, manager of Maple Lanes, asked 15 men's league players to report their body weight (lb) and their league average, rounded to the nearest pin. The data appear in Exercise 1 of Section 11.2. Here is a portion of the output:

Predictor	Coef	SE Coef	T	P
Constant	190.90	31.64	6.03	0.000
Weight	−0.1017	0.1459	−0.70	0.498

$S = 23.78$ R-Sq = 3.6%   R-Sq(adj) = 0.0%

a. Set up the hypotheses to test whether the slope is 0.
b. The sample had 15 observations. What is the critical value at $\alpha = 0.05$?
c. From the output, find the value of the slope coefficient, the standard error of the slope, and the value of the t statistic.
d. Compare the t statistic with the critical value and determine whether the relationship between the two variables is significant.

2. The expense ratio of a mutual fund measures the percentage of the fund's assets used to pay for annual administrative overhead. Funds with higher ratios spend more of the fund's return to operate the fund. Does this hurt or help the fund's return? A random selection of 16 stock mutual funds in 1998 was selected. The data appear in Exercise 2 of Section 11.2. Here is a portion of the output:

Predictor	Coef	SE Coef	T	P
Constant	35.221	2.021	17.43	0.000
ExpRatio	−4.636	2.343	−1.98	0.068

$S = 2.760$ R-Sq = 21.9%   R-Sq(adj) = 16.3%

a. Set up the hypotheses to test whether the slope is 0.
b. The sample had 16 observations. What is the critical value at $\alpha = 0.05$?

c. From the output, find the value of the slope coefficient, the standard error of the slope, and the value of the *t* statistic.

d. Compare the *t* statistic with the critical value and determine whether the relationship between the two variables is significant.

3. Is the size of a home related to selling price? Use the home prices data given in Exercise 3 of Section 11.2. Use SQFT (square feet of living space in the home) as the explanatory variable and PRICE (selling price in hundreds of dollars) as the response variable to conduct the analysis. Here is a portion of the output:

Predictor	Coef	SE Coef	T	P
Constant	47.82	62.85	0.76	0.448
SQFT	0.61367	0.03625	16.93	0.000

$S = 204.5$ R-Sq = 71.4%   R-Sq(adj) = 71.1%

a. Set up the hypotheses to test whether the slope is 0.

b. The sample had 117 observations. What is the critical value at $\alpha = 0.01$?

c. From the output, find the value of the slope coefficient, the standard error of the slope, and the value of the *t* statistic.

d. Compare the *t* statistic with the critical value and determine whether the relationship between the two variables is significant.

4. Is the age of a home related to selling price? Use the home prices data given in Exercise 3 of Section 11.2. Use AGE as the explanatory variable and PRICE as the response variable to conduct the analysis. Here is a portion of the output:

Predictor	Coef	SE Coef	T	P
Constant	1244.27	75.46	16.49	0.000
AGE	−5.366	3.860	−1.39	0.169

$S = 400.2$ R-Sq = 2.8%   R-Sq(adj) = 1.4%

a. Set up the hypotheses to test whether the slope is 0.

b. The sample had 68 observations. What is the critical value at $\alpha = 0.10$?

c. From the output, find the value of the slope coefficient, the standard error of the slope, and the value of the *t* statistic.

d. Compare the *t* statistic with the critical value and determine whether the relationship between the two variables is significant.

5. Is the number of calories in a serving of breakfast cereal related to the amount of fat? Use the cereals data given in Exercise 4 of Section 11.2. Use cal (calories) as the explanatory variable and fat as the response variable to conduct the following analysis. Here is a portion of the output:

Predictor	Coef	SE Coef	T	P
Constant	−1.7399	0.5616	−3.10	0.003
cal	0.025756	0.005170	4.98	0.000

$S = 0.8782$ R-Sq = 24.9%   R-Sq(adj) = 23.9%

a. Set up the hypotheses to test whether the slope is 0.

b. The sample had 77 observations. What is the critical value at $\alpha = 0.02$?

c. From the output, find the value of the slope coefficient, the standard error of the slope, and the value of the *t* statistic.

d. Compare the *t* statistic with the critical value and determine whether the relationship between the two variables is significant.

6. Is the amount of carbohydrates in a serving of breakfast cereal related to the amount of sugar? Use the cereals data given in Exercise 4 of Section 11.2. Use carbo (grams of carbohydrates per serving) as the explanatory variable and sugar (grams of sugar per serving) as the response variable to conduct the analysis. Here is a portion of the output:

Predictor	Coef	SE Coef	T	P
Constant	14.842	1.758	8.44	0.000
carbo	−0.5280	0.1149	−4.60	0.000

$S = 3.888$ R-Sq = 22.2%   R-Sq(adj) = 21.2%

a. Set up the hypotheses to test whether the slope is 0.

b. The sample had 76 observations. What is the critical value at $\alpha = 0.05$?

c. From the output, find the value of the slope coefficient, the standard error of the slope, and the value of the *t* statistic.

d. Compare the t statistic with the critical value and determine whether the relationship between the two variables is significant.

7. Is the weight of a firstborn twin an indicator of the weight of the secondborn twin? The data in Exercise 6 of Section 11.2 give the weights of the firstborn (Twin A) and secondborn (Twin B) infants for 19 sets of twins. Use Twin A as the explanatory variable and Twin B as the response variable for the following analysis. Here is a portion of the output:

Predictor	Coef	SE Coef	T	P
Constant	1.1557	0.9682	1.19	0.249
TwinA	0.8224	0.1739	4.73	0.000

S = 0.7004   R-Sq = 56.8%   R-Sq (adj) = 54.3%

a. Set up the hypotheses to test whether the slope is 0.
b. The sample had 19 observations. What is the critical value at $\alpha = 0.05$?
c. From the output, find the value of the slope coefficient, the standard error of the slope, and the value of the t statistic.
d. Compare the t statistic with the critical value and determine whether the relationship between the two variables is significant.

8. A baseball analyst is interested in the relationship between number of hits and number of strikeouts for baseball pitchers. The data in Exercise 7 of Section 11.2 are from pitchers who played from 1920 to 1950. Use number of hits (HITS) as the explanatory variable and number of strikeouts (SO) as the response variable for the following analysis. Here is a portion of the output:

Predictor	Coef	SE Coef	T	P
Constant	77.73	16.55	4.70	0.001
HITS	−2.5778	0.6357	−4.05	0.002

S = 2.247   R-Sq = 62.2%   R-Sq (adj) = 58.4%

a. Set up the hypotheses to test whether the slope is 0.
b. The sample had 12 observations. What is the critical value at $\alpha = 0.02$?
c. From the output, find the value of the slope coefficient, the standard error of the slope, and the value of the t statistic.
d. Compare the t statistic with the critical value and determine whether the relationship between the two variables is significant.

9. The Old Faithful data set appears in Exercise 8 of Section 11.2. These data are from eruptions of the Old Faithful Geyser in Yellowstone National Park. Duration is the length of time the eruption lasted and interval is the length of time from the previous eruption. We wish to determine whether there is a relationship between the length of time from the previous eruption and the duration of the eruption. Use interval as the explanatory variable and duration as the response variable for the following analysis. Here is a portion of the output:

Predictor	Coef	SE Coef	T	P
Constant	−1.4342	0.2897	−4.95	0.000
interval	0.068904	0.004015	17.16	0.000

S = 0.5360   R-Sq = 73.7%   R-Sq (adj) = 73.5%

a. Set up the hypotheses to test whether the slope is 0.
b. The sample had 107 observations. What is the critical value at $\alpha = 0.01$?
c. From the output, find the value of the slope coefficient, the standard error of the slope, and the value of the t statistic.
d. Compare the t statistic with the critical value and determine whether the relationship between the two variables is significant.

10. The data in Exercise 9 of Section 11.2 appeared in the *Wall Street Journal*. The advertisements were selected by an annual survey conducted by Video Board Tests, Inc., a New York ad-testing company, based on interviews with 20,000 adults who were asked to name the most outstanding TV commercial they had seen, noticed, and liked. The retained impressions were based on a survey of 4000 adults, in which regular product users were asked to cite a commercial they had seen for that product category in the past week. Use SPEND as the explanatory variable and MILIMP as the response variable for the following analysis.

FIRM: Firm name
SPEND: TV advertising budget, 1983 ($ millions)
MILIMP: Millions of retained impressions per week

Here is a portion of the output:

Predictor	Coef	SE Coef	T	P
Constant	22.163	7.089	3.13	0.006
SPEND	0.36317	0.09712	3.74	0.001

S = 23.50   R-Sq = 42.4%   R-Sq (adj) = 39.4%

a. Set up the hypotheses to test whether the slope is 0.
b. The sample had 21 observations. What is the critical value at $\alpha = 0.10$?
c. From the output, find the value of the slope coefficient, the standard error of the slope, and the value of the t statistic.
d. Compare the t statistic with the critical value and determine whether the relationship between the two variables is significant.

11. We wish to predict the top speed of a car (SP) based on its engine horsepower (HP). Use the passenger cars data given in Exercise 10 of Section 11.2 for the following analysis. Here is a portion of the output:

Predictor	Coef	SE Coef	T	P
Constant	84.4541	0.9210	91.70	0.000
HP	0.238705	0.007082	33.70	0.000

S = 3.623   R-Sq = 93.4%   R-Sq (adj) = 93.3%

a. Set up the hypotheses to test whether the slope is 0.
b. The sample had 81 observations. What is the critical value at $\alpha = 0.05$?
c. From the output, find the value of the slope coefficient, the standard error of the slope, and the value of the t statistic.
d. Compare the t statistic with the critical value and determine whether the relationship between the two variables is significant.

12. We wish to predict the gas mileage of a car (MPG) based on its engine horsepower (HP). Use the passenger cars data given in Exercise 10 of Section 11.2 for the following analysis. Here is a portion of the output:

Predictor	Coef	SE Coef	T	P
Constant	50.066	1.569	31.90	0.000
HP	−0.13902	0.01207	−11.52	0.000

S = 6.174   R-Sq = 62.4%   R-Sq (adj) = 61.9%

a. Set up the hypotheses to test whether the slope is 0.
b. The sample had 81 observations. What is the critical value at $\alpha = 0.01$?
c. From the output, find the value of the slope coefficient, the standard error of the slope, and the value of the t statistic.
d. Compare the t statistic with the critical value and determine whether the relationship between the two variables is significant.

13. A baseball team's batting success could be evaluated theoretically by its ability to wear down the opposing teams' pitchers. Players call this "working the count." As the game progresses, the opposing pitchers are often required to throw more pitches and tend to get tired. Meanwhile, the batters get time to study the pitchers and improve their timing. All of this additional work can make the pitcher more vulnerable to an offensive attack as the innings go by. The data in Exercise 12 of Section 11.2 are from *Sports Illustrated*. The data show the number of pitches faced per game in 1998 and the winning percentages of the teams in the American League. Here is a portion of the output:

Predictor	Coef	SE Coef	T	P
Constant	−135.03	69.26	−1.95	0.075
Pitches	1.2673	0.4742	2.67	0.020

S = 6.680   R-Sq = 37.3%   R-Sq (adj) = 32.1%

a. Set up the hypotheses to test whether the slope is 0.
b. The sample had 14 observations. What is the critical value at $\alpha = 0.05$?
c. From the output, find the value of the slope coefficient, the standard error of the slope, and the value of the t statistic.
d. Compare the t statistic with the critical value and determine whether the relationship between the two variables is significant.

14. The pupil per teacher ratio is commonly used as an indication of academic excellence. The lower the ratio, the smaller the enrollments in classrooms, and the more time teachers can devote to individual attention for their students. Does this also mean that SAT scores will be higher? The following table is a random selection of combined SAT scores (verbal and math) and pupil/teacher ratios from 20 states.

Pupil/Teacher Ratio	SAT Score	Pupil/Teacher Ratio	SAT Score
16.9	1116	13.9	1011
19.6	1045	14.6	1016
24.0	1010	17.8	1174
14.4	1016	15.4	1135
18.9	998	14.5	1126
17.1	1140	15.7	1039
17.8	995	17.0	1099
15.5	1190	16.2	978
16.9	1094	17.1	1071
17.0	993	23.8	1146

Here is a portion of the output:

Predictor	Coef	SE Coef	T	P
Constant	1062.0	101.4	10.47	0.000
P/TRatio	0.443	5.824	0.08	0.940

S = 69.30   R-Sq = 0.0%   R-Sq (adj) = 0.0%

a. Set up the hypotheses to test whether the slope is 0.
b. The sample had 20 observations. What is the critical value at $\alpha = 0.05$?
c. From the output, find the value of the slope coefficient, the standard error of the slope, and the value of the t statistic.
d. Compare the t statistic with the critical value and determine whether the relationship between the two variables is significant.

Section 11.4

1. The expense ratio of a mutual fund measures the percentage of the fund's assets used to pay for annual administrative overhead. Funds with higher ratios spend more of the fund's return to operate the fund. Does this hurt or help the fund's return? A random selection of 16 stock mutual funds in 1998 revealed the data that appear in Exercise 2 of Section 11.2. Let X = Expense ratio and Y = Total return. Relevant values are shown below:

$$\Sigma x = 12.97 \quad \Sigma x^2 = 11.9017 \quad s_{y|x} = 2.760 \quad n = 16$$

a. Compute a 95% confidence interval for the mean total return of mutual funds with expense ratios of 0.75.
b. Compute a 95% prediction interval for the total return of a mutual fund with an expense ratio of 0.75.

2. Is the size of a home related to selling price? Refer to the home prices data given in Exercise 3 of Section 11.2. Use SQFT (square feet of living space in the home) as the explanatory variable and PRICE (selling price in hundreds of dollars) as the response variable to conduct the following analysis. Relevant values are shown below:

$$\Sigma x = 193,501 \quad \Sigma x^2 = 351,839,665 \quad s_{y|x} = 204.5 \quad n = 117$$

a. Compute a 98% confidence interval for the mean selling price of homes that have 2200 square feet of living space.
b. Compute a 98% prediction interval for the selling price of a home that has 2200 square feet of living space.

3. Is the number of calories in a serving of breakfast cereal related to the amount of fat? Refer to the cereals data given in Exercise 4 of Section 11.2. Use cal (calories) as the explanatory variable and fat (in grams) as the response variable to conduct the following analysis. Relevant values are shown below:

$$\Sigma x = 8230 \quad \Sigma x^2 = 908,500 \quad s_{y|x} = 0.8782 \quad n = 77$$

a. Compute a 99% confidence interval for the mean amount of fat in cereals that have 120 calories per serving.

b. Compute a 99% prediction interval for the amount of fat in a cereal that has 120 calories per serving.

4. Is the amount of carbohydrates in a serving of breakfast cereal related to the amount of sugar? Refer to the cereals data given in Exercise 4 of Section 11.2. Use carbo (grams of carbohydrates per serving) as the explanatory variable and sugar (grams of sugar per serving) as the response variable to conduct the following analysis. Relevant values are shown below:

$$\Sigma x = 1125 \quad \Sigma x^2 = 17{,}798 \quad s_{y|x} = 3.888 \quad n = 76$$

a. Compute a 99% confidence interval for the mean amount of sugar in cereals that have 15 grams of carbohydrates per serving.

b. Compute a 99% prediction interval for the amount of sugar in a cereal that has 15 grams of carbohydrates per serving.

5. Is the weight of a firstborn twin an indicator of the weight of the secondborn twin? The data in Exercise 6 of Section 11.2 give the weights of the firstborn (Twin A) and secondborn (Twin B) infants for 19 sets of twins. Use Twin A as the explanatory variable and Twin B as the response variable for the following analysis. Relevant values are shown below:

$$\Sigma x = 104.31 \quad \Sigma x^2 = 588.91 \quad s_{y|x} = 0.7004 \quad n = 19$$

a. Compute a 95% confidence interval for the mean weight of secondborn twins when the firstborn twin weighs 5.5 lb.

b. Compute a 95% confidence interval for the weight of a secondborn twin when the firstborn twin weighs 5.5 lb.

6. A baseball analyst is interested in the relationship between number of hits and number of strikeouts for baseball pitchers. The data in Exercise 7 of Section 11.2 are from pitchers who played from 1920 to 1950. Use number of hits (HITS) as the explanatory variable and number of strikeouts (SO) as the response variable for the following analysis. Relevant values are shown below:

$$\Sigma x = 312.20 \quad \Sigma x^2 = 8134.9 \quad s_{y|x} = 2.247 \quad n = 12$$

a. Compute a 90% confidence interval for the mean number of strikeouts when the number of hits is 26.

b. Compute a 90% prediction interval for the number of strikeouts when the number of hits is 26.

7. The Old Faithful data set appears in Exercise 8 of Section 11.2. These data are from eruptions of the Old Faithful Geyser in Yellowstone National Park. Duration is the length of time the eruption lasted and interval is the length of time from the previous eruption. We wish to determine whether there is a relationship between the length of time from the previous eruption and the duration of the eruption. Use interval as the explanatory variable and duration as the response variable for the following analysis. Relevant values are shown below:

$$\Sigma x = 7597 \quad \Sigma x^2 = 557{,}209 \quad s_{y|x} = 0.5360 \quad n = 107$$

a. Compute a 98% confidence interval for the mean duration of eruptions in which the interval between eruptions is 75 minutes.

b. Compute a 98% prediction interval for the duration of an eruption in which the interval between eruptions is 75 minutes.

8. The data in Exercise 9 of Section 11.2 appeared in the *Wall Street Journal*. The advertisements were selected by an annual survey conducted by Video Board Tests, Inc., a New York ad-testing company, based on interviews with 20,000 adults who were asked to name the most outstanding TV commercial they had seen, noticed, and liked. The retained impressions were based on a survey of 4000 adults, in which regular product users were asked to cite a commercial they had seen for that product category in the past week. Use SPEND as the explanatory variable and MILIMP as the response variable for the following analysis.

> FIRM: Firm name
> SPEND: TV advertising budget, 1983 ($ millions)
> MILIMP: Millions of retained impressions per week

Relevant values are shown at the top of the next page.

$$\Sigma x = 1058.4 \quad \Sigma x^2 = 111,899 \quad s_{y|x} = 23.50 \quad n = 21$$

a. Compute a 98% confidence interval for the mean number of retained impressions per week for TV advertisements with a budget of $50 million.

b. Compute a 98% prediction interval for the number of retained impressions per week for a TV advertisement with a budget of $50 million.

9. We wish to predict the top speed of a car (SP) based on its engine horsepower (HP). Use the passenger cars data given in Exercise 10 of Section 11.2 for the following analysis. Relevant values are shown below:

$$\Sigma x = 9605 \quad \Sigma x^2 = 1,386,775 \quad s_{y|x} = 3.623 \quad n = 81$$

a. Compute a 99% confidence interval for the mean top speed of cars with 150-horsepower engines.

b. Compute a 99% confidence interval for the top speed of a car with a 150-horsepower engine.

10. We wish to predict the gas mileage of a car (MPG) based on its engine horsepower (HP). Use the passenger cars data given in Exercise 10 of Section 11.2 for the following analysis. Relevant values are shown below:

$$\Sigma x = 9605 \quad \Sigma x^2 = 1,386,775 \quad s_{y|x} = 6.174 \quad n = 81$$

a. Compute a 99% confidence interval for the mean gas mileage (miles per gallon) of cars with 150-horsepower engines.

b. Compute a 99% confidence interval for the gas mileage (miles per gallon) of a car with a 150-horsepower engine.

11. A baseball team's batting success could be evaluated theoretically by its ability to wear down the opposing teams' pitchers. Players call this "working the count." As the game progresses, the opposing pitchers are often required to throw more pitches and tend to get tired. Meanwhile, the batters get time to study the pitchers and improve their timing. All of this additional work can make the pitcher more vulnerable to an offensive attack as the innings go by. The data appear in Exercise 12 of Section 11.2. These data from *Sports Illustrated* show the number of pitches faced per game in 1998 and the winning percentages of the teams in the American League. Relevant values are shown below:

$$\Sigma x = 2044.4 \quad \Sigma x^2 = 298,739 \quad s_{y|x} = 6.68 \quad n = 14$$

a. Compute a 90% confidence interval for the mean winning percentage when 146 pitches are faced per game.

b. Compute a 90% prediction interval for the winning percentage when 146 pitches are faced per game.

Section 11.5

1. Does a heavier bowler knock down more pins? Jim Samuels, manager of Maple Lanes, asked 15 men's league players to report their body weight (lb) and their league average, rounded to the nearest pin. Results are shown in Exercise 1 of Section 11.2. Let X = body weight and Y = league average. Relevant values are shown below:

$$\Sigma x = 3190 \quad \Sigma y = 2539 \quad \Sigma x^2 = 704,960 \quad \Sigma y^2 = 437,395 \quad \Sigma xy = 537,259 \quad n = 15$$

a. Find the value of the correlation coefficient.

b. Comment on the strength and direction of the relationship between these two variables.

c. Find the value of the coefficient of determination.

2. The expense ratio of a mutual fund measures the percentage of the fund's assets used to pay for annual administrative overhead. Funds with higher ratios spend more of the fund's return to operate the fund. Does this hurt or help the fund's return? A random selection of 16 stock mutual funds in 1998 revealed the data that appear in Exercise 2 of Section 11.2. Let X = Expense ratio and Y = Total return. Relevant values are shown below:

$$\Sigma x = 12.97 \quad \Sigma y = 503.40 \quad \Sigma x^2 = 11.9017 \quad \Sigma y^2 = 15,974.7 \quad \Sigma xy = 401.634 \quad n = 16$$

a. Find the value of the correlation coefficient.

b. Comment on the strength and direction of the relationship between these two variables.

c. Find the value of the coefficient of determination.

3. Is the size of a home related to selling price? Refer to the home prices data given in Exercise 3 of Section 11.2. Use SQFT (square feet of living space in the home) as the explanatory variable and PRICE (selling price in hundreds of dollars) as the response variable to conduct the following analysis. Relevant values are shown below:

$$\Sigma x = 193{,}501 \quad \Sigma y = 124{,}340 \quad \Sigma x^2 = 351{,}839{,}665 \quad \Sigma y^2 = 148{,}929{,}422$$
$$\Sigma xy = 225{,}165{,}408 \quad n = 117$$

a. Find the value of the correlation coefficient.
b. Comment on the strength and direction of the relationship between these two variables.
c. Find the value of the coefficient of determination.

4. Is the age of a home related to selling price? Refer to the home prices data given in Exercise 3 of Section 11.2. Use AGE as the explanatory variable and PRICE as the response variable to conduct the following analysis. Relevant values are shown below:

$$\Sigma x = 1018 \quad \Sigma y = 79{,}148 \quad \Sigma x^2 = 25{,}994 \quad \Sigma y^2 = 103{,}006{,}012 \quad \Sigma xy = 1{,}127{,}188 \quad n = 68$$

a. Find the value of the correlation coefficient.
b. Comment on the strength and direction of the relationship between these two variables.
c. Find the value of the coefficient of determination.

5. Is the number of calories in a serving of breakfast cereal related to the amount of fat? Refer to the cereals data given in Exercise 4 of Section 11.2. Use cal (calories) as the explanatory variable and fat as the response variable to conduct the following analysis. Relevant values are shown below:

$$\Sigma x = 8230 \quad \Sigma y = 78 \quad \Sigma x^2 = 908{,}500 \quad \Sigma y^2 = 156 \quad \Sigma xy = 9080 \quad n = 77$$

a. Find the value of the correlation coefficient.
b. Comment on the strength and direction of the relationship between these two variables.
c. Find the value of the coefficient of determination.

6. Is the amount of carbohydrates in a serving of breakfast cereal related to the amount of sugar? Refer to the cereals data given in Exercise 4 of Section 11.2. Use carbo as the explanatory variable and sugar as the response variable to conduct the following analysis. Relevant values are shown below:

$$\Sigma x = 1125 \quad \Sigma y = 534 \quad \Sigma x^2 = 17{,}798 \quad \Sigma y^2 = 5190 \quad \Sigma xy = 7300 \quad n = 76$$

a. Find the value of the correlation coefficient.
b. Comment on the strength and direction of the relationship between these two variables.
c. Find the value of the coefficient of determination.

7. Is the weight of a firstborn twin an indicator of the weight of the secondborn twin? The data in Exercise 6 in Section 11.2 give the weights of the firstborn (Twin A) and secondborn (Twin B) infants for 19 sets of twins. Use Twin A as the explanatory variable and Twin B as the response variable for the following analysis. Relevant values are shown below:

$$\Sigma x = 104.31 \quad \Sigma y = 107.75 \quad \Sigma x^2 = 588.91 \quad \Sigma y^2 = 630.367 \quad \Sigma xy = 604.902 \quad n = 19$$

a. Find the value of the correlation coefficient.
b. Comment on the strength and direction of the relationship between these two variables.
c. Find the value of the coefficient of determination.

8. A baseball analyst is interested in the relationship between number of hits and number of strikeouts for baseball pitchers. The data in Exercise 7 of Section 11.2 are from pitchers who played from 1920 to 1950. Use number of hits (HITS)) as the explanatory variable and number of strikeouts (SO) as the response variable for the following analysis. Relevant values are shown below:

$$\Sigma x = 312.20 \quad \Sigma y = 128 \quad \Sigma x^2 = 8134.9 \quad \Sigma y^2 = 1498.88 \quad \Sigma xy = 3297.2 \quad n = 12$$

a. Find the value of the correlation coefficient.
b. Find the value of the coefficient of determination.

9. The Old Faithful data set appears in Exercise 8 of Section 11.2. These data are from eruptions of the Old Faithful Geyser in Yellowstone National Park. Duration is the length of time the eruption lasted and interval is the length of time from the previous eruption. We wish to determine whether there is a relationship between the length of time from the previous erup-

tion and the duration of the eruption. Use interval as the explanatory variable and duration as the response variable for the following analysis. Relevant values are shown below:

$\Sigma x = 7597$ $\Sigma y = 370$ $\Sigma x^2 = 557{,}209$ $\Sigma y^2 = 1394.22$ $\Sigma xy = 27{,}498$ $n = 107$

a. Find the value of the correlation coefficient.
b. Comment on the strength and direction of the relationship between these two variables.
c. Find the value of the coefficient of determination.

10. The data in Exercise 9 of Section 11.2 appeared in the *Wall Street Journal*. The advertisements were selected by an annual survey conducted by Video Board Tests, Inc., a New York ad-testing company, based on interviews with 20,000 adults who were asked to name the most outstanding TV commercial they had seen, noticed, and liked. The retained impressions were based on a survey of 4000 adults, in which regular product users were asked to cite a commercial they had seen for that product category in the past week. Use SPEND as the explanatory variable and MILIMP as the response variable for the following analysis.

> FIRM: Firm name
> SPEND: TV advertising budget, 1983 ($ millions)
> MILIMP: Millions of retained impressions per week

Relevant values are shown below:

$\Sigma x = 1058.4$ $\Sigma y = 849.80$ $\Sigma x^2 = 111{,}899$ $\Sigma y^2 = 52{,}606$ $\Sigma xy = 64{,}096$ $n = 21$

a. Find the value of the correlation coefficient.
b. Comment on the strength and direction of the relationship between these two variables.
c. Find the value of the coefficient of determination.

11. We wish to predict the top speed of a car (SP) based on its engine horsepower (HP). Use the passenger cars data given in Exercise 10 of Section 11.2 for the following analysis. Relevant values are shown below:

$\Sigma x = 9605$ $\Sigma y = 9218$ $\Sigma x^2 = 1{,}386{,}775$ $\Sigma y^2 = 1{,}052{,}200$ $\Sigma xy = 1{,}142{,}212$ $n = 81$

a. Find the value of the correlation coefficient.
b. Comment on the strength and direction of the relationship between these two variables.
c. Find the value of the coefficient of determination.

12. We wish to predict the gas mileage of a car (MPG) based on its engine horsepower (HP). Use the passenger cars data given in Exercise 10 of Section 11.2 for the following analysis. Relevant values are shown below:

$\Sigma x = 9605$ $\Sigma y = 2770$ $\Sigma x^2 = 1{,}386{,}775$ $\Sigma y^2 = 101{,}686$ $\Sigma xy = 288{,}091$ $n = 81$

a. Find the value of the correlation coefficient.
b. Comment on the strength and direction of the relationship between these two variables.
c. Find the value of the coefficient of determination.

13. A baseball team's batting success could be evaluated theoretically by its ability to wear down the opposing teams' pitchers. Players call this "working the count." As the game progresses, the opposing pitchers are often required to throw more pitches and tend to get tired. Meanwhile, the batters get time to study the pitchers and improve their timing. All of this additional work can make the pitcher more vulnerable to an offensive attack as the innings go by. The data appear in Exercise 12 of Section 11.2. These data from *Sports Illustrated* show the number of pitches faced per game in 1998 and the winning percentages of the teams in the American League. Relevant values are shown below:

$\Sigma x = 2044.4$ $\Sigma y = 700.50$ $\Sigma x^2 = 298{,}739$ $\Sigma y^2 = 35{,}904.3$ $\Sigma xy = 102{,}545$ $n = 14$

a. Find the value of the correlation coefficient.
b. Comment on the strength and direction of the relationship between these two variables.
c. Find the value of the coefficient of determination.

14. The pupil per teacher ratio is commonly used as an indication of academic excellence. The lower the ratio, the smaller the enrollments in classrooms, and the more time teachers can devote to individual attention for their students. Does this also mean that SAT scores will be higher? The data from a random selection of combined SAT scores (verbal and math) and pupil/teacher ratios from 20 states appear in Exercise 14 of Section 11.3. Relevant values are shown at the top of the next page.

$\Sigma x = 344.10$ $\Sigma y = 21{,}392$ $\Sigma x^2 = 6061.85$ $\Sigma y^2 = 22{,}967{,}368$ $\Sigma xy = 368{,}112$ $n = 20$

a. Find the value of the correlation coefficient.
b. Comment on the strength and direction of the relationship between these two variables.
c. Find the value of the coefficient of determination.

Section 11.6

1. Does a heavier bowler knock down more pins? Jim Samuels, manager of Maple Lanes, asked 15 men's league players to report their body weight (lb) and their league average, rounded to the nearest pin. Results are shown in Exercise 1 of Section 11.2. Let X = body weight and Y = league average.
a. Make a plot of the residuals versus the values of the independent variable.
b. From the plot, does it appear that a linear model is appropriate for these data? Why or why not?
c. From the sample residual plot, does the assumption of equal variances appear reasonable? Why or why not?
d. Create a graph to display the distribution of the residuals. Does the assumption of normality seem reasonable?
e. Considering your answers to parts **a–d**, do you think the linear regression model is appropriate for these data? Why or why not?

2. The expense ratio of a mutual fund measures the percentage of the fund's assets used to pay for annual administrative overhead. Funds with higher ratios spend more of the fund's return to operate the fund. Does this hurt or help the fund's return? A random selection of 16 stock mutual funds in 1998 revealed the data that appear in Exercise 2 of Section 11.2. Let X = Expense ratio and Y = Total return.
a. Make a plot of the residuals versus the values of the independent variable.
b. From the plot, does it appear that a linear model is appropriate for these data? Why or why not?
c. From the sample residual plot, does the assumption of equal variances appear reasonable? Why or why not?
d. Create a graph to display the distribution of the residuals. Does the assumption of normality seem reasonable?
e. Considering your answers to parts **a–d** do you think the linear regression model is appropriate for these data? Why or why not?

3. Is the size of a home related to selling price? Refer to the home prices data given in Exercise 3 of Section 11.2. Use SQFT (square feet of living space in the home) as the explanatory variable and PRICE (selling price in hundreds of dollars) as the response variable to conduct the following analysis.
a. Make a plot of the residuals versus the values of the independent variable.
b. From the plot, does it appear that a linear model is appropriate for these data? Why or why not?
c. From the sample residual plot, does the assumption of equal variances appear reasonable? Why or why not?
d. Create a graph to display the distribution of the residuals. Does the assumption of normality seem reasonable?
e. Considering your answers to parts **a–d**, do you think the linear regression model is appropriate for these data? Why or why not?

4. Is the age of a home related to selling price? Refer to the home prices data given in Exercise 3 of Section 11.2. Use AGE as the explanatory variable and PRICE as the response variable to conduct the following analysis.
a. Make a plot of the residuals versus the values of the independent variable.
b. From the plot, does it appear that a linear model is appropriate for these data? Why or why not?
c. From the sample residual plot, does the assumption of equal variances appear reasonable? Why or why not?
d. Create a graph to display the distribution of the residuals. Does the assumption of normality seem reasonable?
e. Considering your answers to parts **a–d**, do you think the linear regression model is appropriate for these data? Why or why not?

5. Is the number of calories in a serving of breakfast cereal related to the amount of fat? Refer to the cereals data given in Exercise 4 of Section 11.2. Use cal (calories) as the explanatory variable and fat as the response variable to conduct the following analysis.

a. Make a plot of the residuals versus the values of the independent variable.

b. From the plot, does it appear that a linear model is appropriate for these data? Why or why not?

c. From the sample residual plot, does the assumption of equal variances appear reasonable? Why or why not?

d. Create a graph to display the distribution of the residuals. Does the assumption of normality seem reasonable?

e. Considering your answers to parts **a–d**, do you think the linear regression model is appropriate for these data? Why or why not?

6. Is the amount of carbohydrates in a serving of breakfast cereal related to the amount of sugar? Refer to the cereals data given in Exercise 4 of Section 11.2. Use carbo as the explanatory variable and sugar as the response variable to conduct the following analysis.

a. Make a plot of the residuals versus the values of the independent variable.

b. From the plot, does it appear that a linear model is appropriate for these data? Why or why not?

c. From the sample residual plot, does the assumption of equal variances appear reasonable? Why or why not?

d. Create a graph to display the distribution of the residuals. Does the assumption of normality seem reasonable?

e. Considering your answers to parts **a–d**, do you think the linear regression model is appropriate for these data? Why or why not?

7. Is the weight of a firstborn twin an indicator of the weight of the secondborn twin? The data in Exercise 6 of Section 11.2 give the weights of the firstborn (Twin A) and secondborn (Twin B) infants for 19 sets of twins. Use Twin A as the explanatory variable and Twin B as the response variable for the following analysis.

a. Make a plot of the residuals versus the values of the independent variable.

b. From the plot, does it appear that a linear model is appropriate for these data? Why or why not?

c. From the sample residual plot, does the assumption of equal variances appear reasonable? Why or why not?

d. Create a graph to display the distribution of the residuals. Does the assumption of normality seem reasonable?

e. Considering your answers to parts **a–d**, do you think the linear regression model is appropriate for these data? Why or why not?

8. A baseball analyst is interested in the relationship between number of hits and number of strikeouts for baseball pitchers. The data in Exercise 7 of Section 11.2 are from pitchers who played from 1920 to 1950. Use number of hits (HITS) as the explanatory variable and number of strikeouts (SO) as the response variable for the following analysis.

a. Make a plot of the residuals versus the values of the independent variable.

b. From the plot, does it appear that a linear model is appropriate for these data? Why or why not?

c. From the sample residual plot, does the assumption of equal variances appear reasonable? Why or why not?

d. Create a graph to display the distribution of the residuals. Does the assumption of normality seem reasonable?

e. Considering your answers to parts **a–d**, do you think the linear regression model is appropriate for these data? Why or why not?

9. The Old Faithful data set appears in Exercise 8 of Section 11.2. These data are from eruptions of the Old Faithful Geyser in Yellowstone National Park. Duration is the length of time the eruption lasted and interval is the length of time from the previous eruption. We wish to determine whether there is a relationship between the length of time from the previous eruption and the duration of the eruption. Use interval as the explanatory variable and duration as the response variable for the following analysis.

a. Make a plot of the residuals versus the values of the independent variable.

b. From the plot, does it appear that a linear model is appropriate for these data? Why or why not?

c. From the sample residual plot, does the assumption of equal variances appear reasonable? Why or why not?

d. Create a graph to display the distribution of the residuals. Does the assumption of normality seem reasonable?

e. Considering your answers to parts **a–d**, do you think the linear regression model is appropriate for these data? Why or why not?

10. The data in Exercise 9 of Section 11.2 appeared in the *Wall Street Journal*. The advertisements were selected by an annual survey conducted by Video Board Tests, Inc., a New York ad-testing company, based on interviews with 20,000 adults who were asked to name the most outstanding TV commercial they had seen, noticed, and liked. The retained impressions were based on a survey of 4000 adults, in which regular product users were asked to cite a commercial they had seen for that product category in the past week. Use SPEND as the explanatory variable and MILIMP as the response variable for the following analysis.

a. Make a plot of the residuals versus the values of the independent variable.

b. From the plot, does it appear that a linear model is appropriate for these data? Why or why not?

c. From the sample residual plot, does the assumption of equal variances appear reasonable? Why or why not?

d. Create a graph to display the distribution of the residuals. Does the assumption of normality seem reasonable?

e. Considering your answers to parts **a–d**, do you think the linear regression model is appropriate for these data? Why or why not?

11. We wish to predict the top speed of a car (SP) based on its engine horsepower (HP). Use the passenger cars data given in Exercise 10 of Section 11.2 for the following analysis.

a. Make a plot of the residuals versus the values of the independent variable.

b. From the plot, does it appear that a linear model is appropriate for these data? Why or why not?

c. From the sample residual plot, does the assumption of equal variances appear reasonable? Why or why not?

d. Create a graph to display the distribution of the residuals. Does the assumption of normality seem reasonable?

e. Considering your answers to parts **a–d**, do you think the linear regression model is appropriate for these data? Why or why not?

12. We wish to predict the gas mileage of a car (MPG) based on its engine horsepower (HP). Use the passenger cars data given in Exercise 10 of Section 11.2 for the following analysis.

a. Make a plot of the residuals versus the values of the independent variable.

b. From the plot, does it appear that a linear model is appropriate for these data? Why or why not?

c. From the sample residual plot, does the assumption of equal variances appear reasonable? Why or why not?

d. Create a graph to display the distribution of the residuals. Does the assumption of normality seem reasonable?

e. Considering your answers to parts **a–d**, do you think the linear regression model is appropriate for these data? Why or why not?

13. A baseball team's batting success could be evaluated theoretically by its ability to wear down the opposing teams' pitchers. Players call this "working the count." As the game progresses, the opposing pitchers are often required to throw more pitches and tend to get tired. Meanwhile, the batters get time to study the pitchers and improve their timing. All of this additional work can make the pitcher more vulnerable to an offensive attack as the innings go by. The data appear in Exercise 12 of Section 11.2. These data from *Sports Illustrated* show the number of pitches faced per game in 1998 and the winning percentages of the teams in the American League.

a. Make a plot of the residuals versus the values of the independent variable.

b. From the plot, does it appear that a linear model is appropriate for these data? Why or why not?

c. From the sample residual plot, does the assumption of equal variances appear reasonable? Why or why not?

d. Create a graph to display the distribution of the residuals. Does the assumption of normality seem reasonable.

e. Considering your answers to parts **a–d**, do you think the linear regression model is appropriate for these data? Why or why not?

14. The pupil per teacher ratio is commonly used as an indication of academic excellence. The lower the ratio, the smaller the enrollments in classrooms, and more time teachers can devote to individual attention for their students. Does this also mean that SAT scores will be higher?

The data from a random selection of combined SAT scores (verbal and math) and pupil/teacher ratios from 20 states appear in Exercise 14 of Section 11.3.

a. Make a plot of the residuals versus the values of the independent variable.

b. From the plot, does it appear that a linear model is appropriate for these data? Why or why not?

c. From the sample residual plot, does the assumption of equal variances appear reasonable? Why or why not?

d. Create a graph to display the distribution of the residuals. Does the assumption of normality seem reasonable?

e. Considering your answers to parts **a–d**, do you think the linear regression model is appropriate for these data? Why or why not?

Chapter 12

Section 12.2 Exercises

1. Do heavy users of products tend to look for bargains before they shop? A bar soap manufacturer was interested in predicting the number of bars purchased by 15 consumers over a 6-month period. They also counted the number of coupons they redeemed and the proportion of their purchases bought when the item was on sale (price markdown). Results are summarized in the table.

Number of Bars of Soap	Coupons Redeemed	Proportion on Sale
31	22	0.87
24	15	0.79
17	9	0.41
14	3	0.36
13	4	0.38
26	15	0.69
10	8	0.10
33	24	0.76
17	10	0.47
28	21	0.82
33	27	0.85
42	31	0.95
11	1	0.09
18	9	0.44
12	4	0.17

a. Which variable do you think is the dependent variable and which are the independent variables in this model?

b. Construct a multiple regression model for the data.

c. Interpret the coefficients of the model.

d. Use the model to predict the number of bars of soap a consumer would buy if he or she redeemed 20 coupons in 6 months and bought 65% of his or her purchases on sale.

e. Does this value seem reasonable? Do you think you are interpolating or extrapolating? Explain.

f. Use the model to predict the number of bars of soap purchased for each of the consumers in the data.

g. Compare the predicted values to the actual values. Do you think that the model does a good job of predicting the number of purchases?

2. When a basketball player steals the ball from an opposing player, it may very well be a break-away steal that leads to an easy lay-up and two points. On the other hand, when an assist is made, one of the other players on the team gets credit for the score. Both steals and assists are offensive weapons, and scorekeepers track both of them. The table shows the average points per game (scoring average) that Michael Jordan scored in his career, together with his average number of assists and steals per game during the years that he played in the NBA for the first time.

Season	Points per Game	Assists per Game	Steals per Game
1984–85	28.2	5.9	2.4
1985–86	22.7	2.9	2.0
1986–87	37.1	4.6	2.9
1987–88	35.0	5.9	3.2
1988–89	32.5	8.0	2.9
1989–90	33.6	6.3	2.8
1990–91	31.5	5.5	2.7
1991–92	30.1	6.1	2.3
1992–93	32.6	5.5	2.8
1994–95	26.9	5.3	1.8
1995–96	30.4	4.3	2.2
1996–97	29.6	4.3	1.7
1997–98	28.7	3.5	1.7

a. Construct a multiple regression model to predict the number of points per game.
b. Interpret the coefficients in the model. Does the number of assists increase or decrease the number of points per game? What about the number of steals?
c. Use the regression equation to predict how many points per game Jordan would have scored after completing a season with his career mean number of assists (5.4) and steals (2.5) per game. *Note:* The number of games Jordan played per season was not the same from one year to the next. His career scoring average was 31.5 points per game.

Section 12.3

1. Last week you left your colleague, Jerry Martinez, to fill in the missing values in a bivariate ANOVA table you were working on so that you could go on vacation. Well, it is payback time because Jerry just called you from the airport terminal before boarding his plane and said, "Hi, partner! I may have stumbled onto a dandy multiple regression model for predicting next month's required purchases. The equation uses three input variables, and the residual plot contained 33 unique points. Can you figure out the missing values in the attached ANOVA table the same way I did last week for that little bivariate model you cranked out? The boys at the top are putting the heat on us. Let me know how this thing flies. See you next week."

Source	df	SS	MS	F
Regression	A	C	91.43	F
Residual	B	65.78	E	
Total		D		

a. Fill in the missing values for the ANOVA table.
b. Set up the hypotheses to test whether the overall model is significant.
c. Perform the test at the 0.05 level of significance. What can you conclude?
d. Calculate the coefficient of determination.

2. Consider the problem of predicting the number of soap purchases in Exercise 1 of Section 12.2.
a. Set up the hypotheses to test whether the multiple regression model is significant.
b. At the 0.05 level of significance, what can you conclude about the model?
c. What is the coefficient of determination for this model? Based on the value, how well do you think this set of independent variables does in predicting the number of bars of soap purchased?
d. Set up the hypotheses to test each of the individual regression coefficients.
e. At the 0.01 level of significance, what can you conclude?

3. Look again at the problem on predicting Michael Jordan's average points per game for a season, in Exercise 2 of Section 12.2.
a. At the 0.05 level of significance, what can you conclude about the model as a whole?
b. What percent of the variation in average points per game can be explained by the number of assists and the number of steals?
c. Do you think this model is a good one? Why or why not?

d. Test each of the coefficients in the model separately. What are your conclusions?

e. Based on your results, what would be your recommendation about a model to use?

Chapter 13 Exercises

Section 13.3

1. Four samples of sizes 8, 9, 10, and 11 have means 21.2, 22.5, 23.6, and 24.5, respectively. Find the grand mean of all four samples combined.

2. Five samples of sizes 12, 14, 17, 19, and 22 have means 111.6, 113.8, 115.8, 113.2, and 105.6, respectively. Find the grand mean of all five samples combined.

3. Four samples, each of size 20, have been selected from four populations. Their means are shown in the following table:

Sample	Mean
1	11.3
2	16.2
3	9.5
4	17.1

Find the grand mean and sum of squares between (SSA).

4. Four samples have been selected from four populations and their sample sizes and means are shown in the following table:

Sample	Sample Size	Mean
1	25	21.5
2	28	22.0
3	29	22.5
4	24	21.0

Find the grand mean and sum of squares between (SSA).

5. An incomplete ANOVA table is shown below:

Source	SS	df	MS	F
Between groups	30	A	B	E
Within groups	C	15	D	
Total	120	20		

Complete the table with the correct numerical values for A, B, C, D, and E.

6. Johnson's Service Center has devised three potential options available to preferred customers who redeem coupons and buy at least 10 gallons of fuel when they stop in. A is a flat 3 cents off each gallon. Option B is a combination of 2 cents off plus another $1 discount on the regular price of a $5 deluxe car wash. Option C is a $2 discount on the same $5 deluxe car wash but no reduction in the fuel purchase. The owner, Harold Johnson, ran each option on three different 2-week trial periods and tracked daily sales receipts from those customers who redeemed their coupons. Results are shown in the table.

Option A	Option B	Option C
$453	$492	$467
507	514	525
513	536	516
521	511	500
511	528	435
615	678	462

(*continued*)

601	611	411
552	653	674
551	596	512
505	516	559
515	534	624
512	543	711
476	498	512
427	437	416

Harold elected to conduct a one-way ANOVA for his single-factor experiment.
a. What is the total sum of squares (SST)?
b. What is the sum of squares between (SSA)?
c. What is the sum of squares error (SSE)?
d. What are the values of the F statistic and the critical value at the 0.05 level of significance?
e. What conclusion can Harold make based on the results of the ANOVA?

7. Three samples of size 10 were used to compare three population means. If the total sum of squares (SST) is 337.51 and the mean square error (MSE) is 6.23, what is the correct value for the mean square between (MSA)?

8. Four samples of size 11 were used to compare four population means. If the total sum of squares (SST) is 221.32 and the mean square between (MSA) is 31.29, what is the correct value for the mean square error (MSE)?

9. Three samples of size 13 were used to compare three population means. If the total sum of squares (SST) is 463.7 and the sum of squares between (SSA) is 223.6, what is the correct value for the F statistic?

10. Manny Hernandez has collected five samples of equal size and is using them to compare five population means. If the mean square error (MSE) is 18.68 and the sum of squares error (SSE) is 373.6, what is the size of each of Manny's samples?

11. The 1-year mean percentage returns of five types of bond funds in 1998 are listed in the table.

U.S. Government	Investment-Grade Corporate	High-Yield Corporate	Tax-Exempt	World Income
15.2	13.9	2.6	4.1	2.7
20.2	2.0	0.0	6.8	10.8
16.2	11.7	5.0	7.0	11.0
15.5	5.0	3.3	10.0	10.8
13.2	3.9	1.9	4.8	12.0

a. Use Excel to conduct a one-way ANOVA of these data. Show both the summary table and the ANOVA table of results.
b. State the null hypothesis to test for equal population means.
c. Build a scatter plot of the data.
d. At the 0.05 level of significance, do the sample data indicate that at least one of the five population mean percentage returns is different from the others?

Section 13.4

1. Refer to Exercise 6 in Section 13.3. Check the ANOVA assumptions for Harold Johnson's Service Center.
a. Is there any reason to believe that the errors are not independent?
b. Construct the histogram of the daily sales for each of the three options. Do they look normally distributed?
c. Calculate and compare the sizes of the sample variances and do a visual check of the data by looking at boxplots to see whether the spread in each sample looks about the same.
d. Comment on the validity of the assumptions of ANOVA for these data.

2. Refer to Exercise 11 in Section 13.3. Check the ANOVA assumptions for the returns on the bond funds.
a. Is there any reason to believe that the errors are not independent?

b. Construct the histogram of the 1-year mean percentage returns for each of the five bond funds. Do they look normally distributed?

c. Calculate and compare the sizes of the sample variances and do a visual check of the data by looking at boxplots to see whether the spread in each sample looks about the same.

d. Comment on the validity of the assumptions of ANOVA for these data.

Section 13.5

1. Refer to the data shown in Exercise 11 of Section 13.3. Assume that each row of observations presented in the table represents funds offered from the same investing firm and that five funds are being considered.

a. Find the block means and the sum of squares blocks (SSBL), and rebuild the ANOVA table.

b. Compute the relative efficiency of the block design.

c. Comment on the benefits of blocking in this case.

Section 13.6

1. Major League baseball teams are divided into two leagues, American and National. Each league is divided into three divisions: East, Central, and West. The following table lists all the teams' batting averages in 1998:

American			National		
East	Central	West	East	Central	West
0.288	0.272	0.272	0.272	0.280	0.252
0.266	0.264	0.289	0.259	0.258	0.274
0.273	0.271	0.276	0.249	0.262	0.291
0.280	0.263	0.257	0.264	0.264	0.246
0.261	0.266	0.260	0.248	0.254	0.253

a. Use Excel to conduct a two-way ANOVA for these data using league as factor A and batting average as factor B.

b. State and test the two main effects and the interaction effect hypotheses at the 0.05 level of significance.

c. Plot the six means with batting average as the vertical axis and division as the horizontal axis.

2. Stock mutual funds are often divided for analysis on the basis of the average size of the companies in which they invest. Large-cap funds focus on companies with a market capitalization of more than $5 billion; mid-cap funds, between $1 billion and $5 billion; and small-cap funds, generally less than $1 billion. The fund can be further subdivided on the basis of investment style: growth, value, and blend of both growth and value. The growth funds concentrate on companies with faster earnings growth, and the value funds seek investments in companies with stock prices that are undervalued for one reason or another. Are the total returns different for each type of fund? The following percent total return data were collected in 1998:

Large-Cap			Mid-Cap			Small-Cap		
Growth	Value	Blend	Growth	Value	Blend	Growth	Value	Blend
26.3	33.0	31.0	20.9	28.3	33.6	24.5	33.3	36.2
27.9	36.1	33.0	22.7	26.0	28.7	15.0	26.1	21.5
34.7	34.3	33.0	15.9	32.5	37.0	32.1	33.3	33.0
29.7	30.4	32.6	18.3	29.2	37.0	21.3	34.9	50.0
29.9	16.0	27.9	21.1	29.3	29.7	22.5	42.4	28.8
35.4	31.2	30.2	21.6	29.3	36.8	24.3	23.9	27.6
31.7	27.5	32.8	20.1	38.9	35.7	21.7	27.8	35.9

a. Use Excel to conduct a two-way ANOVA for these data using size of company as factor A and investment style as factor B.

b. State and test the two main effects and the interaction effect hypotheses at the 0.05 level of significance.

c. Plot the nine means with percent total return as the vertical axis and investment style as the horizontal axis.

Chapter 14 Exercises

Section 14.2

1. The following table shows the average number of daily e-mail messages that Acme Company received over a 6-week measurement period.

Weekday	M	T	W	TH	F
Messages	57	53	52	45	38

George Warren, general manager, has always claimed that the day of the week has no bearing on the number of messages received; that is, he thinks that the number of calls should be uniformly distributed over the weekdays.
a. Set up a table that compares the expected and observed frequencies for each day.
b. Based only on the table, do you think that Mr. Warren's claim is accurate?
c. Set up the hypotheses for testing the claim.
d. Perform the goodness-of-fit test at the 0.05 level of significance.
e. Based on the chi-square test, is the claim valid?

2. A die that is supposed to be fair was rolled 96 times. The table shows the frequency of the number of dots showing face up after each toss:

Dots Showing	1	2	3	4	5	6
Frequency	11	16	20	21	15	13

a. Build a table of observed and expected frequencies.
b. Calculate chi-square and test the hypothesis H_0: $p_1 = p_2 = p_3 = p_4 = p_5 = p_6 = 1/6$ at the 0.05 level of significance.
c. Do you think the die is fair or loaded (unfair)? Why or why not?

3. Bill Stark is the general manager of Bubbles, Inc., a soft-drink company that distributes carbonated soda pop in cans to local stores. Bill likes to include a surprise coupon in about 30% of the cases that are delivered to the store. The coupon offers a case for half price the next time the customer buys a Bubbles brand. Recently one of the biggest stores asked Bill to show some evidence that the surprise coupons were actually inside 30% of the cases after they left the plant. In an attempt to satisfy the store manager's request, Bill collected 400 random samples of 5 cases each and tallied the number of coupons that were found in each sample:

Coupons	0	1	2	3	4	5
Frequency	70	150	110	55	10	5

a. Set up the table to test whether the data fit a binomial distribution with $n = 5$ and $p = 0.30$.
b. Calculate the chi-square test statistic for the data.
c. What can Bill report back to the store?

4. Mindy Schwartz just opened a beauty salon that employs five full-time beauticians. Within reason, Mindy wants to be fair to her employees by scheduling an equal number of customers for each beautician to groom. After 4 weeks in operation, Mindy tallied the total number of different customers serviced at each booth:

Beautician	1	2	3	4	5
Customers	152	165	149	153	141

a. Assuming the distribution is uniform, build a table of observed and expected frequencies.
b. Calculate chi-square and test the hypothesis H_0: $p_1 = p_2 = p_3 = p_4 = p_5 = 1/5$ at the 0.05 level of significance.
c. Do you think Mindy is accomplishing her objective?

5. Eugene Field Elementary School has classes with grades K through 6. Sarah Gerbach, principal, was asked by one of the parents at a recent PTA meeting if enrollments were about the same throughout all grades. Sarah responded, "I believe so, but I'll have to check. Some grades have more pupils than others." The following morning the principal checked the class rosters for each grade and counted all the children:

Grade	K	1	2	3	4	5	6
Roster	33	39	38	42	40	38	41

a. If the number of students per grade were really the same, how many students would you expect to find in each grade?

b. Set up the chi-square table to compare these data to the expected data.

c. Set up the hypotheses and calculate the test statistic.

d. At the 0.05 level of significance, what should the principal tell the PTA?

Section 14.3

1. A survey of 203 adults living in the United States was taken. The respondents were asked questions about race and whether or not they thought that elderly people should live with their children. The resulting data are given here.

Should aged live with their children?	Race of Respondent			Total
	White	Black	Other	
A Good Idea	83	26	10	119
A Bad Idea	66	14	4	84
Total	149	40	14	203

a. Set up the hypotheses to test whether the percent who think it is a bad idea for the aged to live with their children is the same for each racial group.

b. Calculate the proportion of respondents who think it is a bad idea for aged to live with their children for each racial group separately. Based on these data, do you think the proportions are different?

c. Find the expected values and calculate the test statistic for the data.

d. At the 0.05 level of significance, is there a difference in the proportion of people who think is is a bad thing for the aged to live with their children for the racial groups reported?

2. The General Social Survey (GSS) for 1998 compiled data on whether or not the respondent said they believed in God and the region of the country in which they lived. The data, for those that responded, are presented here.

Region	Describe Your Beliefs about God		TOTAL
	Don't Believe in God	Believe in God	
New England	5	45	50
Middle Atlantic	15	157	172
East North Central	14	172	186
West North Central	6	72	78
South Atlantic	13	205	218
East South Central	1	89	90
West South Central	8	114	122
Mountain	10	53	63
Pacific	20	122	142
Total	92	1029	1121

a. For each region of the country, calculate the proportion of people who said that they did not believe in God.

b. Calculate the overall proportion of those surveyed who said that they did not believe in God.

c. At the 0.05 level of significance, is there a difference in the proportion of people who say that they do not believe in God for each region? Explain your answer.

3. A random telephone survey of 300 people asked people about different television shows. One question was "Do you watch 'Buffy the Vampire Slayer'?" The results by level of education are given in the table.

	Level of Education					
	< HS	HS Grad	Some College	College Grad	Post College	Total
Yes	20	32	17	65	29	163
No	43	45	16	28	5	137
Total	63	77	33	93	34	300

a. For each level of education, calculate the proportion of people who said that they do watch "Buffy the Vampire Slayer."

b. Calculate the overall proportion of those surveyed who said that they do watch "Buffy the Vampire Slayer."

c. At the 0.05 level of significance, is there a difference in the proportion of people who say that they do watch "Buffy the Vampire Slayer" for each education level? Explain your answer.

Section 14.4

1. A contingency table has been constructed for students in a private college. The table indicates the numbers of students who do or do not own a computer and, if so, what kind(s):

Type of Computer	Freshman	Sophomore	Junior	Senior
Laptop	111	115	117	157
Desktop	187	201	205	167
Both	67	78	82	65
Neither	145	143	156	83

a. Set up the hypotheses to test whether the type of computer a student has is independent of their year in college.

b. Calculate the expected frequencies for each cell.

c. Perform the test of independence at the 0.05 level of significance.

d. What can you conclude from your test?

2. Your colleague, Jerry Martinez, called in sick with the flu today. Your secretary hands you his personal message as you walk in. "Sorry to bother you again, good buddy, but the boss is desperate for this analysis, and I'm under the weather. It contains some missing values, but I think there is enough information for you to complete it in time for his meeting with the officers this afternoon. The data report preferences for the top three brands from 300 of our customers in the three biggest geographical markets we cover. Can you complete the table for me? The expected frequencies should be shown in parentheses. The boss wants to know where we're thick or thin." You mumble to yourself, "No problem, Jerry. Hope you feel better."

	Company Brand		
Market Area	1	2	3
Northeast	30 ( )	( )	40 ( )
Midwest	( )	50 ( )	30 ( )
Southeast	50 ( )	20 ( )	( )
Total	100	110	

a. Fill in the missing values in the table.

b. Set up the hypotheses to test whether market area and brand are independent.

c. Calculate the chi-square statistic.

d. Perform the chi-square test for independence at the 0.01 level of significance.

e. What is your conclusion for the meeting in the afternoon?

3. In the General Social Survey (GSS) for 1998, there is information on whether the respondent thought that marijuana should be made legal. The table shows the responses by race of the respondent.

| | Should Marijuana Be Made Legal? | | |
Race of Respondent	Yes	No	Total
White	425	979	1404
Black	67	197	264
Other	33	87	120
Total	525	1263	1788

a. Set up the hypotheses to test whether attitude toward marijuana legalization and race are independent.

b. Find the expected value for each cell and compute the chi-square statistic.

c. At the 0.05 level of significance, are attitude toward marijuana legalization and race independent?

Chapter 1 Answers

Section 1.2

1. A population is the entire group about which information is desired. The sample is a subset of a population on which measurements are made. They are the same when a census is taken.

3. a. The 12,000 working mothers in the area
   b. The 200 working mothers
   c. Sampling error

5. a. All babies this pediatrician delivers
   b. Weight, gender, race/ethnicity, weeks of gestation, others possible
   c. Sample. The population keeps growing with each baby she delivers. It would not be possible to look at all babies she will ever deliver.

7. It is very difficult to measure the entire U.S. population, so a sample was most likely used to obtain this information.

9. a. All American adults
   b. A sample; the American population is too large and it is not feasible to poll everyone.

Section 1.3

1. a. The percent of all American adults planning a vacation this summer
   b. Statistics; they were calculated from a sample.

3. a. The proportion of all college students who have missed at least one class due to alcohol or drug use
   b. Statistic; it was calculated from a sample.

5. a. Statistic
   b. Parameter
   c. This is due to sampling error; we do not expect a statistic to be exactly equal to the parameter. The value varies from sample to sample.

7. a. Statistic
   b. Parameter
   c. This is due to sampling error; we do not expect a statistic to be exactly equal to the parameter. The value varies from sample to sample.

9. a. Parameter
   b. Statistic

Section 1.4

1. Variability in the populations and size of the sample

3. **a.** $N = 12{,}000$
   b. $n = 200$

5. **a.** $N = 270$
   b. $n = 40$

7. **a.** $N = 200$ million
   b. $n = 1002$

9. Large variability in the population

Section 1.5

1. Yes, only the views of undergraduate students will be represented. Their views may differ from those of graduate students.

3. This is a volunteer response sample. Only people with strong opinions tend to reply.

7. 9, 46, 32, 79, 57, 11, 50, 15, 34, 53

9. 9, 4, 6

Section 1.6

1. **a.** Quantitative
   b. Qualitative
   c. Quantitative

3. **a.** Quantitative; continuous
   b. Qualitative; nominal
   c. Qualitative; nominal
   d. Qualitative; ordinal
   e. Quantitative; discrete

5. Ordinal qualitative

7. Continuous quantitative

9. Nominal qualitative

Section 1.10

1. $\Sigma x = 288$

3. $\Sigma x = 1999$

5. $\Sigma x = 185$

7. $\Sigma x = 1520$

9. $\Sigma x = 216$

Chapter 2 Answers

Section 2.2

1.

Personal computer	320/1000	0.320	32%
Pacemaker	260/1000	0.260	26%
Wireless communication	180/1000	0.180	18%
Television	240/1000	0.240	24%

3.

Cups	Percent
0	6
1	10
2	37
3	18
4	29

5. **a.** 10
   b. 2
   c. 5 to 7, 7 to 9, 9 to 11, 11 to 13, 13 to 15, 15 to 17, 17 to 19, 19 to 21, 21 to 23, 23 to 25, and 25 to 27
   d.

Class	Frequency
5 to 7	1
7 to 9	5
9 to 11	6
11 to 13	16
13 to 15	20
15 to 17	16
17 to 19	15
19 to 21	10
21 to 23	6
23 to 25	4
25 to 27	1

7. **a.** $23 \leq x < 24$
   b. $20 \leq x < 21$ and $26 \leq x < 27$
   c. $23 \leq x < 24$

9.

Injections	Percent
0	28
1	46
2	20
3	5
4	1

Six from this group were eligible for the national study.

11. d

13. a

Section 2.3

1.

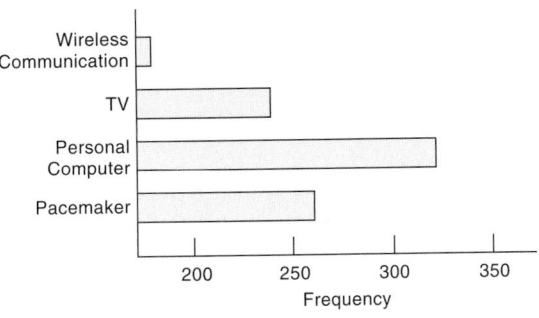

3.

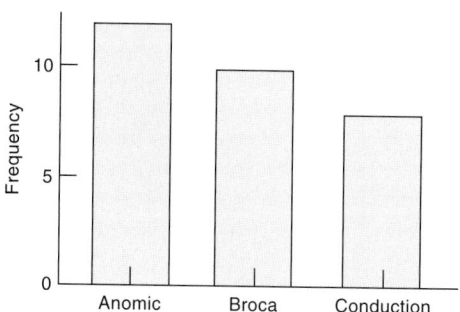

Type of Aphasia for Thirty Patients

5.

Type of Aphasia	Frequency
Conduction	24
Broca	18
Anomic	8

Conduction (24/50)360 degrees = 172.8 degrees
Broca   (18/50)360 degrees = 129.6 degrees
Anomic   (8/50)360 degrees = 57.6 degrees

7.

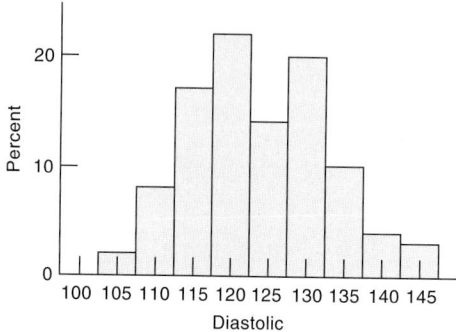

Diastolic Blood Pressures for
Hypertensive Patients

9.

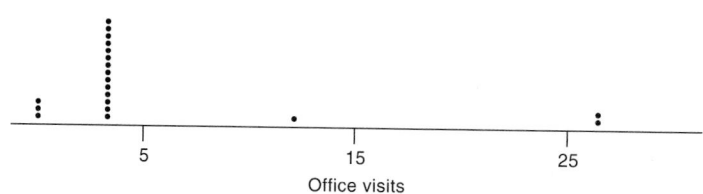

11. b

13. b

15. c

Section 2.4

1.

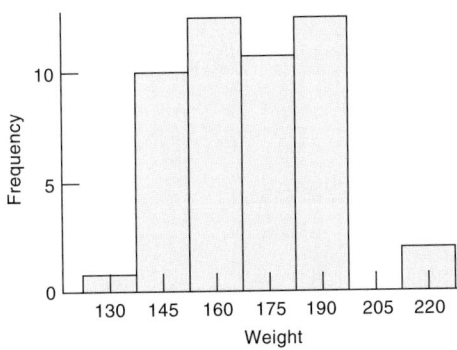

The center is close to 170 pounds. The data are symmetrically spread about the center.

3.

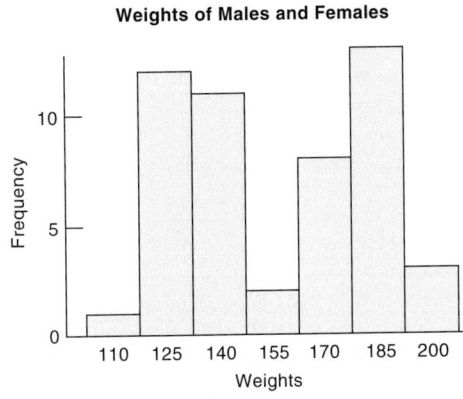

There is a mixture of two distributions here. One center is near 130 pounds and the other is near 180 pounds.

5. Most of the grooms are between 25 and 35 years of age. However, a few are in their forties, fifties, sixties, and seventies. This gives the distribution a skew to the right.

7. Machine 2 has the smaller variability.

9. LOS values are skewed to the right.

11. The histogram of home prices in San Francisco would be shifted to the right of the histogram of home prices in Omaha, Nebraska.

13. The histogram is symmetrical about a center of 80 and has a spread that goes from 50 to 110.

15.

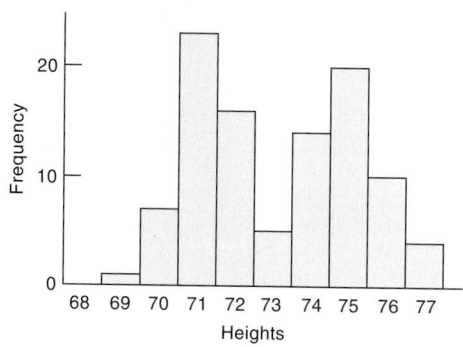

The distribution has two centers. One is at 5 feet 11 inches and the other is at 6 feet 3 inches.

Chapter 2 End-of-Chapter Exercises

1.

Response	Frequency	Percent
No opinion	10	20
Not at all	11	22
Not too	8	16
Somewhat	11	22
Very	10	20

3. Each class should be about 6.83 wide.

	Frequency
$14 \leq$ winning score < 21	7
$21 \leq$ winning score < 28	10
$28 \leq$ winning score < 35	7
$35 \leq$ winning score < 42	7
$42 \leq$ winning score < 49	2
$49 \leq$ winning score < 56	3

5.

Blood type	Percent	Angle
O	49	0.49×360 degrees $= 176.4$ degrees
A	27	0.27×360 degrees $= 97.2$ degrees
B	20	0.20×360 degrees $= 72$ degrees
AB	4	0.04×360 degrees $= 14.4$ degrees

7.

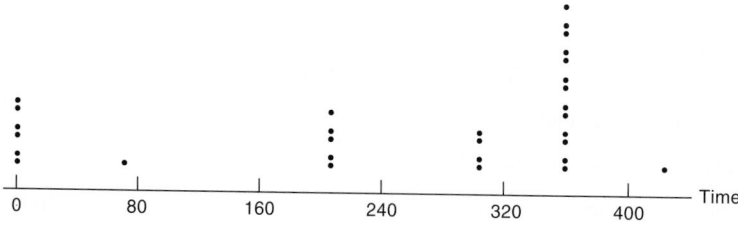

9. The distribution for the New Orleans daily high temperatures would be shifted to the right of the Seattle distribution.

Chapter 3 Answers

Section 3.3

1.

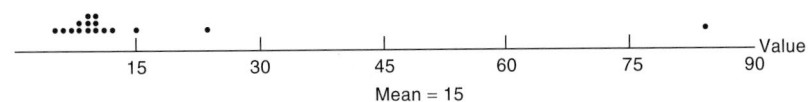

The mean is 15. The two extreme values, 25 and 86, pull the mean to the right of most of the data.

3.

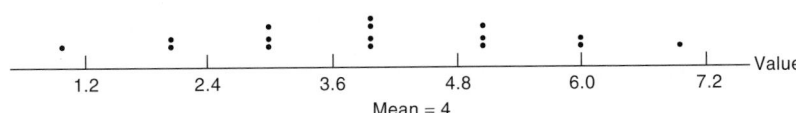

The mean, 4, is located at the center of the data.

5. Sample mean = 1340/100 = 13.4 credit hours

7.

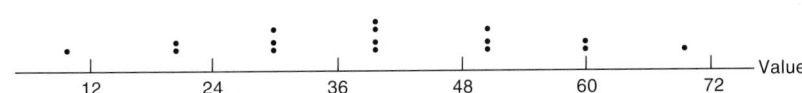

Mean = Median = Mode = 40

In the case of a mound-shaped, symmetrical distribution, the mean, median, and mode are all equal.

9. Mean: sum = 70.98, mean = 70.98/40 = 1.77.

Trimmed mean: sum = 40.22, trimmed mean = 40.22/32 = 1.26

11. $\Sigma xf = 5028$; $\Sigma f = 139$; 5028/139 = 36.2

13. The number 36 occurs most often. The mode is 36.

15. Median = 22

Section 3.4

1. Range = 6.5 − 3.4 = 3.1. Yes; there do not appear to be any unusual values.

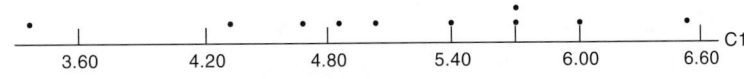

The range gives a good description of the data variability.

3.

x	$(x - \text{mean})$	$(x - \text{mean})^2$
1.5	$1.5 - 3.4 = -1.9$	3.61
3.4	$3.4 - 3.4 = 0.0$	0
5.4	$5.4 - 3.4 = 2.0$	4
4.2	$4.2 - 3.4 = 0.8$	.64
2.5	$2.5 - 3.4 = -0.9$	.81
	$\Sigma(x - \text{mean}) = 0$	$\Sigma(x - \text{mean})^2 = 9.06$

$$s^2 = \Sigma(x - \text{mean})^2/(n - 1) = 9.06/4 = 2.265; s = 1.505$$

5. $\Sigma x = 1533.9; \quad \Sigma x^2 = 123{,}448.8;$

$$s^2 = \frac{n\Sigma x^2 - (\Sigma x)^2}{n(n - 1)} = \frac{20(123{,}448.8) - (1533.9)^2}{20(19)} = \frac{116{,}126.79}{380} = 305.597; s = 17.481$$

7. Mean $= 76.695;$ Standard deviation $= 17.481$

Interval	Number of Data Values	Percentage of Data Values	Empirical Rule
59.214 to 94.176	13	65	68
41.733 to 111.657	19	95	95
24.252 to 129.138	20	100	99

9. The means for the two machines were:

Machine 1: Mean $= 10.1;$ Machine 2: Mean $= 9.9$

The standard deviations were:

Machine 1: $s = 1.6;$ Machine 2: $s = 0.8$

Machine 2 produced a smaller number defective. Also, machine 2 was much more consistent.

11.

$$s^2 = \frac{n\Sigma x^2 - (\Sigma x)^2}{n(n - 1)} = \frac{41(4200) - (320)^2}{41(40)} = \frac{69{,}800}{1640} = 42.56$$

13. $1900 - 150 = 1750$ hours and $1900 + 150 = 2050$ hours

15. $z\text{-score} = (96 - 78)/12 = 18/12 = 1.5$

Section 3.5

1. $b = 50, e = 7,$ and $n = 100.$ $P = (50 + 7/2)/100 = 0.535 = 53.5\%.$ Fifty-three and one-half percent of the test scores were at or below hers.

3. Lower inner fence $= Q_1 - 1.5(\text{IQR}) = 47 - 1.5(22) = 14;$
Upper inner fence $= Q_3 + 1.5(\text{IQR}) = 69 + 1.5(22) = 102;$
The lower whisker ends at 21 and the upper whisker ends at 96.

5. Lower outer fence $= Q_1 - 3(\text{IQR}) = 47 - 3(22) = -19;$
Upper outer fence $= Q_3 + 3(\text{IQR}) = 69 + 3(22) = 135$

7. $Q_1 = (7 + 7)/2 = 7;$ $Q_3 = (12 + 12)/2 = 12$

9. Lower outer fence $= Q_1 - 3(\text{IQR}) = 7 - 3(5) = -8;$
Upper outer fence $= Q_3 + 3(\text{IQR}) = 12 + 3(5) = 27$

11. $Q_1 = 54.375;$ $Q_3 = 95.125$

13. Lower outer fence $= Q_1 - 3(\text{IQR}) = 54.375 - 3(40.75) = -67.875;$
Upper outer fence $= Q_3 + 3(\text{IQR}) = 95.125 + 3(40.75) = 217.375$

15. $Q_1 = 47; Q_3 = 57$

17. Lower outer fence = $Q_1 - 3(\text{IQR}) = 47 - 3(10) = 17$;
Upper outer fence = $Q_3 + 3(\text{IQR}) = 57 + 3(10) = 87$

19. $Q_1 = 12.5$; $Q_3 = 54.5$

21. Lower outer fence = $Q_1 - 3(\text{IQR}) = 12.5 - 3(42) = -113.5$;
Upper outer fence = $Q_3 + 3(\text{IQR}) = 54.5 + 3(42) = 180.5$

Chapter 3 End-of-Chapter Exercises

1. Mean = 403.67 feet;   Median = 410 feet;   Mode = 410 feet (occurred 10 times)

3. Range = $488 - 320 = 168$ feet; $\Sigma x^2 = 11{,}962{,}970$; $\Sigma x = 29{,}468$;

$$s^2 = \frac{n\Sigma x^2 - (\Sigma x)^2}{n(n-1)} = \frac{73(11{,}962{,}970) - (29{,}468)^2}{73(72)} = \frac{4{,}933{,}786}{5256} = 938.69597;\ s = 30.638$$

5. $b = 65, e = 4$, and $n = 73$. $P = (65 + 4/2)/73 = 0.918 = 91.8\%$. 91.8% of the distances were at or below 440.

7. Lower inner fence = $Q_1 - 1.5(\text{IQR}) = 380 - 1.5(40) = 320$;
Upper inner fence = $Q_3 + 1.5(\text{IQR}) = 420 + 1.5(40) = 480$;
The lower whisker ends at 320 and the upper whisker ends at 454.

9.

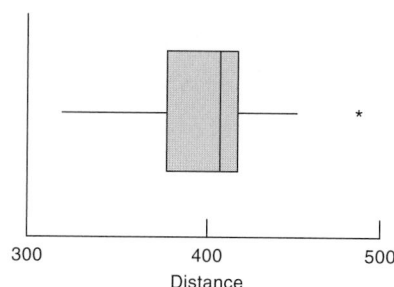

The 488 feet is a possible outlier.

Chapter 4 Answers

Section 4.2

1.

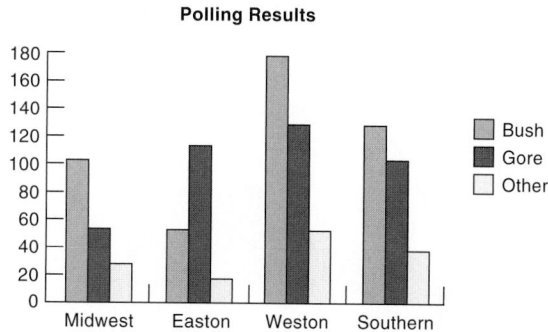

Bush has the greatest percentage lead in Midwest County, 57.1%. He is trailing in Easton County and has 28.57% of the voters.

3.

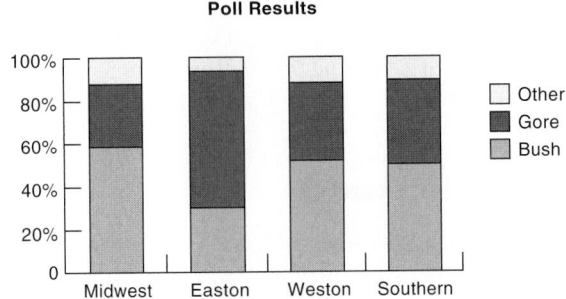

Bush has the majority in three of the four counties.

5. $A = 40; B = 30; C = 45; D = 100$

7.

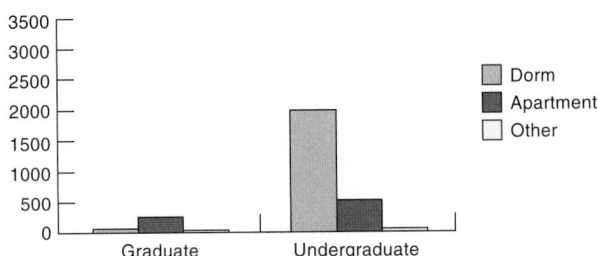

Most of the undergraduates live in the dorm. Most of the graduate students live in apartments.

9. $C = D = E = 100\%$

11.

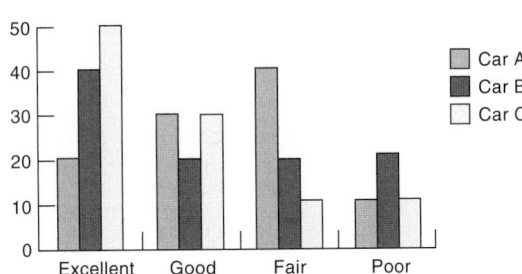

Overall, the tire ratings were best with car C owners.

13.

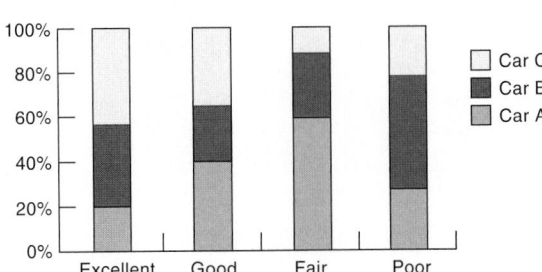

Section 4.3

1.

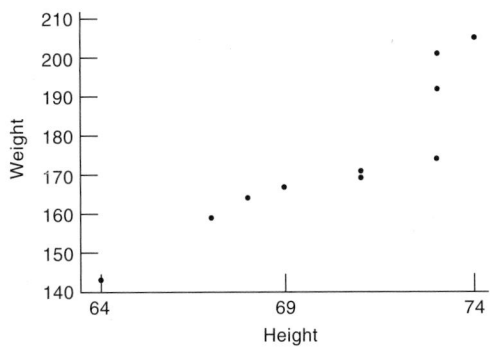

There is a positive linear relationship between height and weight for adult males.

3. There is a nonlinear relationship between the two variables.

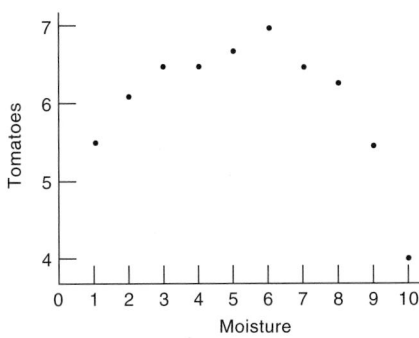

5. A negative linear relationship

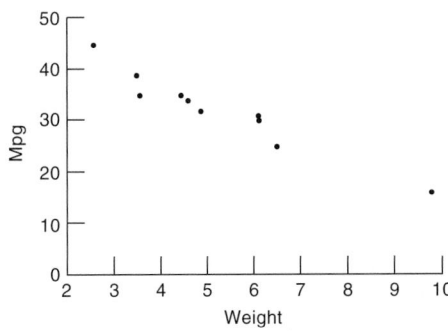

7. I would conclude that there is a negative linear relationship.

9. A direct linear relationship

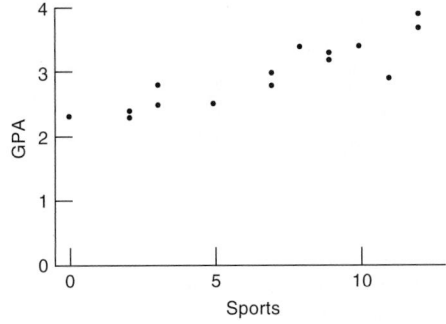

11.

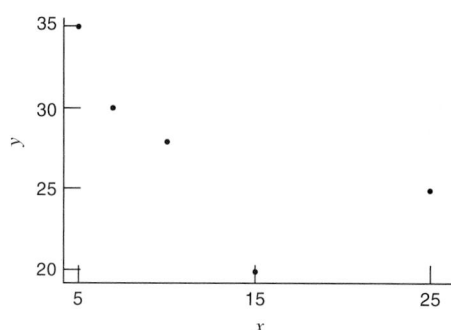

A negative linear relationship connects the two variables.

13. The sum is not 0 because of round-off error.

x	y	$\hat{y} = 33.5 - 0.475\,x$	$e = y - \hat{y}$
5	35	31.125	−3.875
7	30	30.175	0.175
10	28	28.75	0.75
15	20	26.375	6.375
25	25	21.625	−3.375
			Sum = 0.05

Section 4.4

1. Primary source for news for males is radio and for females is TV.

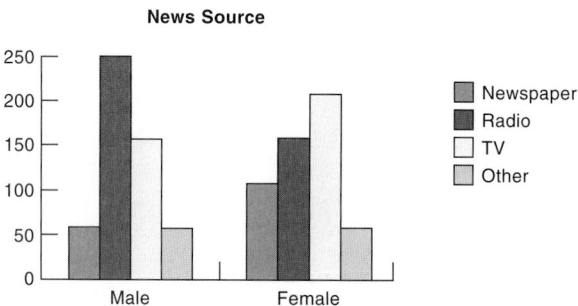

3.

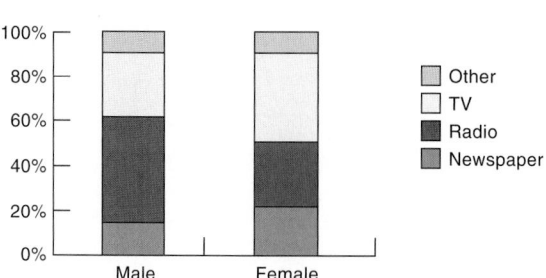

Fifty percent of the males choose radio as their primary source of news, and 40% of females choose TV as their primary source of news.

5.

	Newspaper	Radio	TV	Other	Total
Male	5%	25%	15%	5%	50%
Female	10%	15%	20%	5%	50%
Total	15%	40%	35%	10%	100%

7.

	Newspaper	Radio	TV	Other	Total (%)
Male	10%	50%	30%	10%	100%
Female	20%	30%	40%	10%	100%

9. $b = \dfrac{n\Sigma XY - \Sigma X \Sigma Y}{n\Sigma X^2 - (\Sigma X)^2} = \dfrac{5(3,368,750) - (11,750)(1275)}{5(30,687,500) - (11,750)^2} = \dfrac{1,862,500}{15,375,000} = 0.121;$

$a = \dfrac{\Sigma Y}{n} - b\dfrac{\Sigma X}{n} = 255 - 0.121(2350) = -29.35;$

$\hat{y} = -29.35 + 0.121(3000) = \$333,650$

11. $b = \dfrac{n\Sigma XY - \Sigma X \Sigma Y}{n\Sigma X^2 - (\Sigma X)^2} = \dfrac{5(688.9) - (64.7)(56)}{5(889.23) - (64.7)^2} = \dfrac{-178.7}{260.06} = -0.687;$

$a = \dfrac{\Sigma Y}{n} - b\dfrac{\Sigma X}{n} = 11.2 + 0.687(12.94) = 20.09;$

$\hat{y} = 20.09 - 0.687(12) = 11.846;$ that is, 12 fires.

13. $\hat{y} = -0.1 + 0.7(3) = 2.0.$ This is an example of interpolation. If the linear relationship is a strong one, the estimate will be a good one.

15.

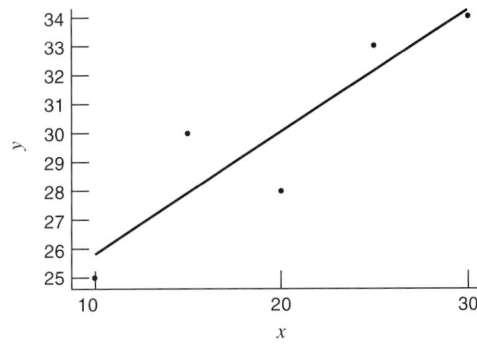

Chapter 5 Answers

Section 5.2

1. The sample space is represented as the 36 points in the following figure. Each outcome in the sample space is equally likely to occur.

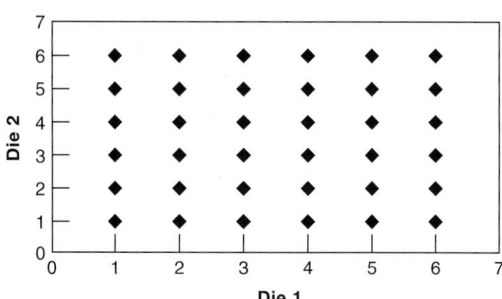

Sample Space for Rolling a Pair of Dice

3. The sample space consists of one of the following outcomes. Each outcome in the sample space is equally likely to occur.

A♣	2♣	3♣	4♣	5♣	6♣	7♣	8♣	9♣	10♣	J♣	Q♣	K♣
A♦	2♦	3♦	4♦	5♦	6♦	7♦	8♦	9♦	10♦	J♦	Q♦	K♦
A♥	2♥	3♥	4♥	5♥	6♥	7♥	8♥	9♥	10♥	J♥	Q♥	K♥
A♠	2♠	3♠	4♠	5♠	6♠	7♠	8♠	9♠	10♠	J♠	Q♠	K♠

5. Referring to the sample space for Exercise 1 of this section, we see that there are 36 equally likely outcomes possible for this experiment. Let A be the event that the sum of the faces is not equal to 2. Then A′ is the event that the sum of the faces is equal to 2. Since A′ consists of only one outcome, (1, 1), we know that $P(A') = 1/36$. Therefore $P(A) = 1 - P(A') = 35/36$.

7. **a.** $0.26 + 0.25 = 0.51$ **b.** $1 - 0.19 = 0.81$

9. Not using the complement: $0.32 + 0.20 + 0.11 = 0.63$;

Using the complement: $1 - 0.12 - 0.25 = 1 - 0.37 = 0.63$

11.

Sum	Outcome	Probability
2	(1, 1)	1/36
3	(1, 2), (2, 1)	2/36
4	(1, 3), (2, 2), (3, 1)	3/36
5	(1, 4), (2, 3), (3, 2), (4, 1)	4/36
6	(1, 5), (2, 4), (3, 3), (4, 2), (5, 1)	5/36
7	(1, 6), (2, 5), (3, 4), (4, 3), (5, 2), (6, 1)	6/36
8	(2, 6), (3, 5), (4, 4), (5, 3), (6, 2)	5/36
9	(3, 6), (4, 5), (5, 4), (6, 3)	4/36
10	(4, 6), (5, 5), (6, 4)	3/36
11	(5, 6), (6, 5)	2/36
12	(6, 6)	1/36

Section 5.3

1. **a.** Using the general addition rule, we obtain

$$P(A \text{ OR } B) = P(A) + P(B) - P(A \text{ AND } B)$$
$$= 500/1000 + 446/1000 - 189/1000 = 757/1000$$

b. Using the complement rule, we obtain $1 - 126/1000 = 874/1000$.
c. 336/1000

3. a. $(40.5 + 18)/(10.9 + 36.1 + 40.5 + 18) = 58.5/105.5 = 55.5\%$
   b. $18/(10.9 + 36.1 + 40.5 + 18) - 15.9/(10.6 + 31 + 34.4 + 15.9) = 18/105.5 - 15.9/91.9$
   $= 0.171 - 0.173 = -0.002 = -0.2\%$.

The percent decreased from 1990 to 2000.

5. a. Using the complementary property, $1 - 85/377 = 292/377 = 0.775$.
   b. Using the general addition rule, $239/377 + 208/377 - 132/377 = 315/377 = 0.836$.

7. The event A consists of the 13 clubs and the event B consists of the 13 hearts. Since A and B are mutually exclusive, the simple addition rule applies:

$$P(A \text{ OR } B) = P(A) + P(B) = 13/52 + 13/52 = 26/52$$

9. The probability of an odd number is 6/12, since an odd number occurs for the outcomes 1H, 3H, 5H, 1T, 3T, 5T. The probability of a head occurring is 6/12 since a head occurs for the outcomes 1H, 2H, 3H, 4H, 5H, 6H. Now an odd number and a head occur for the outcomes 1H, 3H, 5H. Using the general addition rule, the probability is $6/12 + 6/12 - 3/12 = 9/12$.

11. The cards that are black and face cards are the following. The probability of the event A and B is 6/52.

J♣	Q♣	K♣
J♠	Q♠	K♠

13. Let A be the event that it was hit to left field and B be the event that it was hit in the fifth inning.

$$P(A \text{ OR } B) = P(A) + P(B) - P(A \text{ AND } B) = 3/73 + 9/73 - 2/73 = 10/73$$

Section 5.4

1. Event A consists of 12 face cards and event B consists of the 13 clubs.

$$P(A|B) = P(A \text{ AND } B)/P(B) = (3/52)/(13/52) = 3/13$$
$$P(B|A) = P(A \text{ AND } B)/P(A) = (3/52)/(12/52) = 3/12$$

3. $A = \{HHT, HTH, THH, HHH\}$ and $B = \{HTH, THH, HHT\}$;

$$P(B|A) = P(A \text{ AND } B)/P(A) = (3/8)/(4/8) = 3/4$$

5. $A = \{(1,1), (1,2), (1,3), (1,4), (1,5), (1,6)\}$;
$B = \{(1,1), (1,2), (2,1), (1,3), (2,2), (3,1), (1,4), (2,2), (3,1), (4,1), (1,5), (2,4), (3,3), (4,2), (5,1)\}$;
$A \text{ AND } B = \{(1,1), (1,2), (1,3), (1,4), (1,5)\}$;

$$P(A) = 6/36 = 1/6; \quad P(A|B) = P(A \text{ AND } B)/P(B) = (5/36)/(15/36) = 5/15 = 1/3$$

Therefore, A and B are dependent.

7.

Chat Room	19 or Less	20 to 50	More Than 50	Total
Yes	134	75	34	243
No	34	84	125	243
Total	168	159	159	486

$P(19 \text{ or less} \mid \text{Have been in chat room}) = (134/486)/(243/486) = 134/243 = 0.551$

9. A and B independent means $P(A \text{ AND } B) = P(A)P(B) = (0.4)(0.3) = 0.12$. Using the general addition rule of probability, we have

$$P(A \text{ OR } B) = P(A) + P(B) - P(A \text{ AND } B) = 0.4 + 0.3 - 0.12 = 0.58$$

Chapter 5 End-of-Chapter Exercises

1. There are 104 outcomes in the sample space. The sample space may be visualized as follows:

Head with each of the following

A♣	2♣	3♣	4♣	5♣	6♣	7♣	8♣	9♣	10♣	J♣	Q♣	K♣
A♦	2♦	3♦	4♦	5♦	6♦	7♦	8♦	9♦	10♦	J♦	Q♦	K♦
A♥	2♥	3♥	4♥	5♥	6♥	7♥	8♥	9♥	10♥	J♥	Q♥	K♥
A♠	2♠	3♠	4♠	5♠	6♠	7♠	8♠	9♠	10♠	J♠	Q♠	K♠

or

Tail with each of the following

A♣	2♣	3♣	4♣	5♣	6♣	7♣	8♣	9♣	10♣	J♣	Q♣	K♣
A♦	2♦	3♦	4♦	5♦	6♦	7♦	8♦	9♦	10♦	J♦	Q♦	K♦
A♥	2♥	3♥	4♥	5♥	6♥	7♥	8♥	9♥	10♥	J♥	Q♥	K♥
A♠	2♠	3♠	4♠	5♠	6♠	7♠	8♠	9♠	10♠	J♠	Q♠	K♠

3. There are 4 favorable outcomes and 104 outcomes in the sample space. The probability is 4/104. The four favorable outcomes are:

H,A♣
H,A♦
H,A♥
H,A♠

5. Event A is as follows:

H,K♣	T,K♣
H,K♦	T,K♦
H,K♥	T,K♥
H,K♠	T,K♠

Event B is as follows:

H,Q♣	T,Q♣
H,Q♦	T,Q♦
H,Q♥	T,Q♥
H,Q♠	T,Q♠

Since A and B are mutually exclusive, we use the simple addition rule:

$$P(A \text{ OR } B) = P(A) + P(B) = 8/104 + 8/104 = 16/104$$

7. Using the complementary rule, $100\% - 5\% = 95\%$.

9. $230/16230 = 0.014$

Chapter 6 Answers

Section 6.2

1. **a.** Yes; $0 \leq p(x) \leq 1$, and $\Sigma\, p(x) = 1$
   b. No; not all $0 \leq p(x)$.
   c. No; $\Sigma\, p(x) > 1$

3.

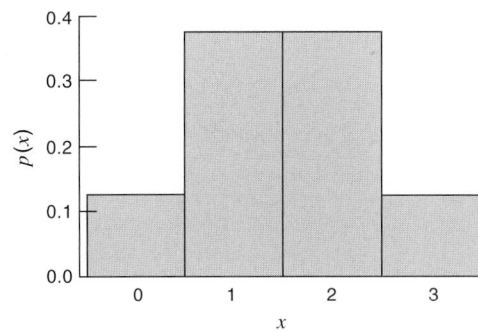

5.

x	Outcome	Probability
2	$(1,1)$	1/36
3	$(1,2),(2,1)$	2/36
4	$(1,3),(2,2),(3,1)$	3/36
5	$(1,4),(2,3),(3,2),(4,1)$	4/36
6	$(1,5),(2,4),(3,3),(4,2),(5,1)$	5/36
7	$(1,6),(2,5),(3,4),(4,3),(5,2),(6,1)$	6/36
8	$(2,6),(3,5),(4,4),(5,3),(6,2)$	5/36
9	$(3,6),(4,5),(5,4),(6,3)$	4/36
10	$(4,6),(5,5),(6,4)$	3/36
11	$(5,6),(6,5)$	2/36
12	$(6,6)$	1/36

x	2	3	4	5	6	7	8	9	10	11	12
p(x)	0.028	0.056	0.083	0.111	0.139	0.167	0.139	0.111	0.083	0.056	0.028

7. $1/2 + 1/4 + 1/8 + 1/16 + \cdots$ is an infinite geometric series with first term $a = 1/2$ and common ratio $r = 1/2$. The sum of such a series is $a/(1 - r) = (1/2)/(1 - 1/2) = 1$.

9.

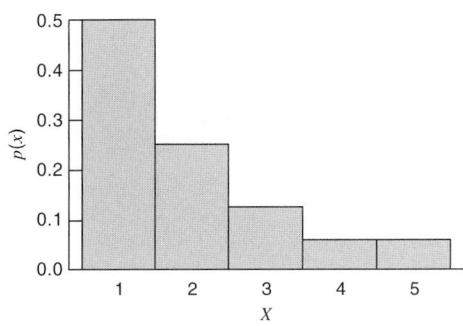

Such a distribution is said to be skewed to the right.

11.

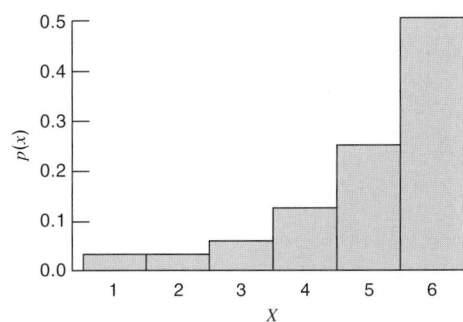

Such a distribution is said to be skewed to the left.

13.

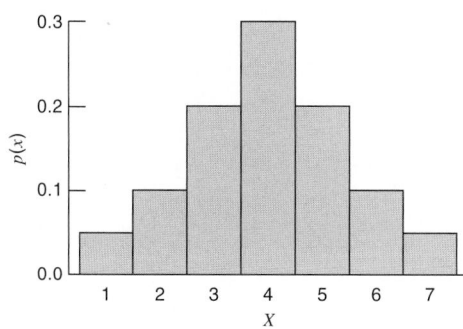

Such a distribution is said to be mound- or bell-shaped.

15. Two

Section 6.3

1. (i) Each experiment has a fixed number of trials, 15. (ii) Success is being left-handed; failure is not being left-handed. (iii) p and q remain the same from trial to trial. (iv) If the 15 are randomly chosen, the trials are independent. (v) The random variable is the number of left-handers among the 15.

3. Mean $= np = 15(0.15) = 2.25$; standard deviation $= \sqrt{npq} = \sqrt{15(0.15)(0.85)} = 1.383$. The z-score is $(10 - 2.25)/1.383 = 5.6$. Ten may be unusual.

5. Mean $= np = 10(0.10) = 1$; standard deviation $= \sqrt{npq} = \sqrt{10(0.1)(0.9)} = 0.949$. The z-score is $(4 - 1)/0.949 = 3.16$. If the 10% figure is correct, they may have an unusual sample.

7. Skewed to the right.

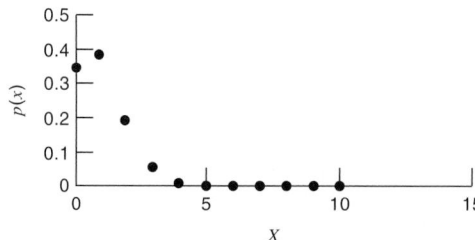

9. Skewed to the left.

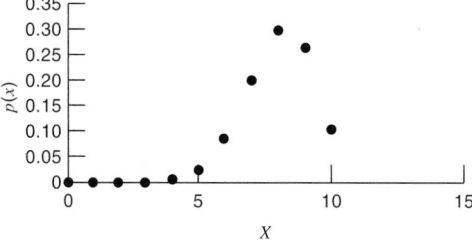

11. $p(0) = \dfrac{3!}{0!3!}(0.5)^0(0.5)^3 = 0.125$; $\qquad p(1) = \dfrac{3!}{1!2!}(0.5)^1(0.5)^2 = 0.375$;

$\quad\;\; p(2) = \dfrac{3!}{2!1!}(0.5)^2(0.5)^1 = 0.375$; $\qquad p(3) = \dfrac{3!}{3!0!}(0.5)^3(0.5)^0 = 0.125$

13. Mean $= np = 20(0.75) = 15$; standard deviation $= \sqrt{npq} = \sqrt{20(0.75)(0.25)} = 1.936$

15. Mean $= np = 20(1/6) = 3.333$; standard deviation $= \sqrt{npq} = \sqrt{20(1/6)(5/6)} = 1.667$

$$z = (5 - 3.333)/1.667 = 1.00$$

Not too unusual to get a z-score of 1.

Section 6.4

1. $P(X < a)$

3. $P(X < a \text{ OR } X > b)$

5. $P(X > a)$

Section 6.5

1. $P(Z < 1.23) = 0.8907$

3. $P(-1.56 < Z < 1.87) = P(Z < 1.87) - P(Z < -1.56) = 0.9693 - 0.0594 = 0.9099$

5. We are looking for $P(X < 42.5)$. The Z-value that corresponds to 42.5 is $Z = (42.5 - 47.5)/2.5 = -2.00$. $P(X < 42.5)$ is equal to $P(Z < -2.00)$. From the tables, this is equal to 0.0228. Thus, 2.28% of the women will have onset before 42.5.

7. We are looking for $P(3 < X < 8)$. The Z-value corresponding to $X = 3$ is $(3 - 5)/1.2 = -1.67$ and the Z-value corresponding to $X = 8$ is $(8 - 5)/1.2 = 2.50$.

$$P(3 < X < 8) = P(-1.67 < Z < 2.50)$$
$$= P(Z < 2.50) - P(Z < -1.67) = 0.9938 - 0.0475 = 0.9463$$

94.63% spend between 3 and 8 hours in the chat room per week.

9. We are looking for $P(X > 700)$. The Z-value corresponding to $X = 700$ is $(700 - 550)/75 = 2.00$. $P(X > 700) = P(Z > 2.00) = 1 - P(Z < 2.00) = 1 - 0.9772 = 0.0228$. On 2.28% on the days the number of e-mails exceeds 700.

11. The Z-value with 90% of the area to the left of it is 1.28.

$$1.28 = (x - 500)/100$$

Solving for x, we find 628.

13. The binomial has a mean equal to $np = 50$ a standard deviation equal to $\sqrt{npq} = \sqrt{100(0.5)(0.5)} = 5$. Center the normal curve at 50 and choose the one that has a standard deviation equal to 5.

15. We are looking for $P(X > 200)$. The Z-value that corresponds to 200 is $Z = (200 - 180)/25 = 0.80$. $P(X > 200)$ is equal to $P(Z > 0.80)$. From the tables, this is equal to $1 - P(Z < 0.80) = 1 - 0.7881 = 0.2119$. Thus, 21.2% need to be treated.

Chapter 6 End-of-Chapter Exercises

1.

x	0	1	2	3	4
$p(x)$	0.0625	0.2500	0.3750	0.2500	0.0625

3. $P(X = 0) = (48/52)(47/51) = 2256/2652 = 0.8507$
$P(X = 1) = (4/52)(48/51) + (48/52)(4/51) = 192/2652 + 192/2652 = 384/2652 = 0.1448$
$P(X = 2) = (4/52)(3/51) = 12/2652 = 0.0045$

x	0	1	2
$p(x)$	0.8507	0.1448	0.0045

5. The center of the normal curve would be at $\mu = 25(0.5) = 12.5$. The standard deviation would be $\sqrt{25(0.5)(0.5)} = 2.5$.

7. We are looking for P(75 < X < 125). The Z-value corresponding to X = 75 is Z = (75 − 94.50)/15.50 = −1.26 and the Z-value corresponding to X = 125 is (125 − 94.50)/15.50 = 1.97.

$$P(75 < X < 125) = P(-1.26 < Z < 1.97) = 0.9756 - 0.1038 = 0.8718$$

9. Assume p = 0.75 and n = 25. Find p(0) + p(1) + ⋯ + p(12) = 0.0034. This casts doubt on the 75% figure. Also check the Z-value for X = 12, with mean = 25(0.75) = 18.75 and standard deviation = $\sqrt{25(0.75)(0.25)}$ = 2.165. Z = (12 − 18.75)/2.165 = −3.12. The number 12 seems to be unusually small if p = 0.75 is correct.

Chapter 7 Answers

Section 7.2

1. Point estimate

3. Mean, median, trimmed mean, or mode

5. **a.** 2.68
   b. 1.8

7. **a.** 159.675
   b. 180

Section 7.3

Output for Exercises 1 and 2

Summary Statistics:

Variable	n	Mean	Variance	Std. Dev.	Median
Twin A	19	5.4901314	0.9011559	0.9492923	5.6875
Twin B	19	5.6710525	1.0728481	1.0357838	5.5

Variable	Range	Min	Max	Q1	Q3
Twin A	3.875	3.5	7.375	4.6875	6.125
Twin B	3.625	3.9375	7.5625	4.9375	6.125

1. **a.** 5.49 lb
   b. 5.67 lb
   c. −0.18 lb
   d. The difference is due to taking a sample rather than looking at the entire population. Values will vary from sample to sample.

Output for Exercises 3 and 4

Summary Statistics:

Variable	n	Mean	Variance	Std. Dev.	Median
Control	23	41.52174	294.07904	17.148733	42
Treatment	21	51.47619	121.1619	11.007357	53

Variable	Range	Min	Max	Q1	Q3
Control	75	10	85	28	54
Treatment	47	24	71	43.5	58.5

3. **a.** 41.52 points
   b. 51.48 points
   c. −9.96 points
   d. No; the sample difference is very close to −10, which suggests the population difference is not 0.

5. **a.** The proportion of all voters who will vote for the Republican candidate
   b. 0.45

7. **a.** The proportion of all people who prefer the taste of the name-brand soft drink and the proportion of all people who prefer the taste of the store-brand soft drink
   b. Name brand: 0.605; Store brand: 0.395

Section 7.4

1. A point estimator should be unbiased, consistent, and efficient.

3. A point estimator is consistent if it yields a value close to the unknown parameter as the sample size increases.

5. **a.**

	Mean	Median
Sample 1	1.76	1.90
Sample 2	2.12	1.90
Sample 3	1.94	2.00
Sample 4	1.88	1.70
Sample 5	1.96	1.90

   b. Average = 1.932; Std. dev. = 0.1308
   c. Average = 1.88; Std. dev. = 0.1095
   d. The mean
   e. The median

Section 7.5

1. Sampling distribution

3. **a.** Normal
   b. 8.2 years
   c. 0.183 year

5. **a.** Normal
   b. 9 minutes
   c. 0.237 minute

7. **a.** Normal
   b. 7.3 lb
   c. 0.1 lb

Section 7.6

1. If the distribution of the individual values is normal or if the sample size is sufficiently large, 30 or greater

3. **a.** 0.367 year
   b. 0.183 year
   c. 0.110 year
   d. As the sample size increases, the standard error decreases.

5. **a.** 0.625 inch
   b. 0.357 inch
   c. 0.208 inch
   d. As the sample size increases, the standard error decreases.

7. **a.** 9 min; 0.237 min
   b. 9 min; 0.190 min
   c. Both populations have the same mean, but the one with the smaller standard deviation has a smaller standard error.

Section 7.7

1. **a.** 7.65 to 8.75 years
   b. $Z = -4.91$
   c. Yes, the Z-score is very small, which indicates that this sample mean is very unusual for the claimed population mean.

3. **a.** 8.535 to 9.465 min
   b. $Z = -1.687$
   c. No; the Z-score indicates that this sample mean is within two standard deviations of the mean, which is not so unusual.

5. **a.** 7.0 to 7.6 lb
   b. $Z = 2.00$
   c. No; the Z-score indicates that this sample mean is two standard deviations above the mean, which is not so unusual.

7. **a.** 71.937 to 78.063
   b. $Z = 4.84$
   c. Yes; the Z-score is very large, which indicates that this sample mean is very unusual for the claimed population mean.

Section 7.8

1. 262.77 to 273.23 days

3. 169.4 to 174.6 lb

5. 7.1 to 9.3 lb

7. 62.26 to 65.80 years

Section 7.9

1. The Z-score formula involves the population standard deviation, and the t-score formula replaces the population standard deviation with the sample standard deviation.

3. df $= n - 1$

5. **a.** 2.492
   b. 1.725
   c. 2.056
   d. 2.831

Section 7.10

1. 174.84 to 205.16 mg/dL

3. \$31.68 to \$38.49

5. 22.98 to 31.02 min

7. 1.93 to 9.08 shots

Section 7.11

1. 13.8% to 21.2%

3. **a.** 8.7% to 13.3%
   b. Yes, since this 90% confidence interval is entirely below 20%.
   c. If the students in the sample lied, then the true value of the population parameter is actually higher than the estimated values from the confidence interval.

5. 0.13 to 0.24

7. **a.** 10.3% to 14.2%
   b. No, since the 95% confidence interval is entirely above 10%. The true percent is probably higher than 10%.

9. 17.4% to 34.6%

Section 7.12

1. 40 pregnancies

3. 35 subjects

5. 20 households

7. 423 traffic fatalities

9. 196 movies

Chapter 8 Answers

Section 8.3

1. B

3. C

5. G

7. C

9. H

Section 8.4

1. (1) State the null and alternative hypotheses. (2) Decide test procedure, statistic, and α. (3) Calculate the test statistic and p-value. (4) Make a decision. (5) Interpret the results.

3. $H_0: p \geq 0.20; H_A: p < 0.20$

5. $H_0: \mu = 6$ oz; $H_A: \mu \neq 6$ oz

7. When the p-value is less than or equal to 0.10

9. When the p-value is less than or equal to 0.05

11. p-value

Section 8.5

1. Z

3. t

Section 8.6

1. **a.** $H_0: \mu = 266; H_A: \mu \neq 266$
   b. $\alpha = 0.05, Z_{cutoff} = \pm 1.96, Z = (268 - 266)/(4/\sqrt{36}) = 3.0$
   Therefore, the calculated value of Z is not in the critical region.
   c. p-value $= 2(1 - 0.9987) = 0.0026$
   d. We reject the null hypothesis and conclude that the mean length of all pregnancies at this hospital differs from the overall average.

3. **a.** $H_0: \mu = 7; H_A: \mu \neq 7$
   b. $\alpha = 0.10, Z_{cutoff} = \pm 1.645, Z = (6.8 - 7)/(1.6/\sqrt{64}) = -1.00$
   Therefore, the calculated value of Z is not in the critical region.
   c. p-value $= 2(0.1587) = 0.3174$
   d. We fail to reject the null hypothesis and conclude that the mean birth weight in this state does not differ from 7 lb.

5. **a.** $H_0: \mu = 170; H_A: \mu \neq 170$
   b. $\alpha = 0.05, Z_{cutoff} = \pm 1.96, Z = (175 - 170)/(10/\sqrt{40}) = 3.16$
   Therefore, the calculated value of Z is in the critical region.
   c. p-value $= 2(1 - 0.9992) = 0.0016$
   d. We reject the null hypothesis and conclude that the mean weight of adult males differs from 170 lb.

7. **a.** $H_0: \mu = 6.20; H_A: \mu \neq 6.20$
   b. $\alpha = 0.05, Z_{cutoff} = \pm 1.96, Z = (6.65 - 6.20)/(1.90/\sqrt{81}) = 2.13$
   Therefore, the calculated value of Z is in the critical region.
   c. p-value $= 2(1 - 0.9834) = 0.0332$
   d. We reject the null hypothesis and conclude that the mean bill at the owner's franchise is not the same as before the promotional campaign.

9. **a.** $H_0: \mu = 1; H_A: \mu \neq 1$
   b. $\alpha = 0.01, Z_{cutoff} = \pm 2.57, Z = (0.97 - 1)/(0.10/\sqrt{49}) = -2.10$
   Therefore, the calculated value of Z is not in the critical region.
   c. p-value $= 2(1 - 0.9821) = 0.0358$
   d. We fail to reject the null hypothesis and conclude that the mean weight for all packages of ground beef does not differ from 1 pound.

Section 8.7

1. **a.** No error
   b. Type I error

3. **a.** Type II error
   b. Type I error

5. **a.** Type II error
   b. Type I error

7. **a.** The mail-order company's advertisements will be misleading in that its competitor actually does deliver the product within 4 days.
   b. The mail-order company will think its competitor delivers the products as claimed, when it actually does not. This information will not be advertised.
   c. 0.01, since a Type I error is more serious than a Type II error

9. **a.** The consumer magazine will assert that the manufacturer is incorrect about the stopping distance when the stopping distance is actually 25 ft, as claimed.
   b. The consumer magazine will assert that the manufacturer is correct about the stopping distance, when the car actually stops in more than 25 ft.
   c. 0.05; this is the standard value and both errors are about equally as serious.

Section 8.8

1. $H_0: \mu \geq 24$ oz; $H_A: \mu < 24$ oz

3. $H_0: \mu = 266$ days; $H_A: \mu \neq 266$ days

5. $H_0: \mu = 50$ lb; $H_A: \mu \neq 50$ lb

7. $H_0: \mu \geq 5$ lb; $H_A: \mu < 5$ lb

9. $H_0: \mu = 30$ years; $H_A: \mu \neq 30$ years

Section 8.9

1. **a.** $H_0: \mu \geq 24$ oz; $H_A: \mu < 24$ oz
   b. $\alpha = 0.10, Z_{cutoff} = -1.28, Z = (23.94 - 24)/(0.13/\sqrt{101}) = -4.64$
   Therefore, the calculated value of Z is in the critical region.
   c. p-value $= 0.0000017$
   d. We reject the null hypothesis and conclude that the mean amount of cereal in a package is less than 24 oz.

3. **a.** $H_0: \mu \geq 9.4$ lb; $H_A: \mu < 9.4$ lb
   b. $\alpha = 0.05, Z_{cutoff} = -1.645, Z = (8.2 - 9.4)/(4.0/\sqrt{50}) = -2.12$
   Therefore, the calculated value of Z is in the critical region.
   c. p-value $= 0.0170$
   d. We reject the null hypothesis and conclude that the mean weight of paper discarded by households is lower since the recycling program has been initiated.

5. **a.** $H_0: \mu \leq 8$ hours; $H_A: \mu > 8$ hours
   b. $\alpha = 0.05, Z_{cutoff} = 1.645, Z = (8.6 - 8)/(1.5/\sqrt{55}) = 2.97$
   Therefore, the calculated value of Z is in the critical region.
   c. p-value $= 0.0015$
   d. We reject the null hypothesis and conclude that the mean time the pain reliever controls muscle pain is longer than 8 hours.

7. **a.** $H_0: \mu \le 480$ min; $H_A: \mu > 480$ min
    b. $\alpha = 0.02, Z_{cutoff} = 2.05, Z = (527.5 - 480)/(90/\sqrt{36}) = 3.17$
    Therefore, the calculated value of Z is in the critical region.
    c. p-value $= 0.0008$
    d. We reject the null hypothesis and conclude that the mean time spent sleeping by children, age 3–5, is greater than 480 min (8 hours).

9. **a.** $H_0: \mu \le 60$ years; $H_A: \mu > 60$ years
    b. $\alpha = 0.05, Z_{cutoff} = 1.645, Z = (64.03 - 60)/(13.53/\sqrt{225}) = 4.47$
    Therefore, the calculated value of Z is in the critical region.
    c. p-value $= 0.0000$ (0.000004)
    d. We reject the null hypothesis and conclude that the mean age of billionaires is over 60 years of age.

Chapter 9 Answers

Section 9.2

1. **a.** $H_0: \mu \le 180; H_A: \mu > 180$
    b. Sample mean $= 190$, sample standard deviation $= 18.13$;
    $t = (190 - 180)/(18.13/\sqrt{8}) = 1.56, p$-value $= 0.081$
    c. At $\alpha = 0.01, t_{cutoff} = 2.998$ (upper-tail test, df $= 7$); do not reject the null hypothesis.
    d. There is not enough evidence to conclude that the mean cholesterol level of men age 60 and older is higher than 180 mg/dL.

3. **a.** $H_0: \mu = 40; H_A: \mu \ne 40$
    b. Sample mean $= 35.08$, sample standard deviation $= 6.57$;
    $t = (35.08 - 40)/(6.57/\sqrt{12}) = -2.59, p$-value $= 0.025$
    c. At $\alpha = 0.05, t_{cutoff} = 2.201$ (two-tail test, df $= 11$); reject the null hypothesis.
    d. There is enough evidence to conclude that the mean price of a concert ticket differs from $40.

5. **a.** $H_0: \mu \ge 30; H_A: \mu < 30$
    b. Sample mean $= 27$, sample standard deviation $= 6$;
    $t = (27 - 30)/(6/\sqrt{8}) = -1.41, p$-value $= 0.100$
    c. At $\alpha = 0.05, t_{cutoff} = -1.895$ (lower-tail test, df $= 7$); do not reject the null hypothesis.
    d. There is not enough evidence to conclude that the mean time in which racers currently finish is less than 30 min.

7. **a.** $H_0: \mu = 7; H_A: \mu \ne 7$
    b. Sample mean $= 5.5$, sample standard deviation $= 4.28$;
    $t = (5.5 - 7)/(4.28/\sqrt{8}) = -0.99, p$-value $= 0.354$
    c. At $\alpha = 0.05, t_{cutoff} = 2.365$ (two-tail test, df $= 7$); do not reject the null hypothesis.
    d. There is not enough evidence to conclude that the mean number of flu shots given by doctors in this community differs from the national average of 7.

9. **a.** $H_0: \mu = 30$ min; $H_A: \mu \ne 30$ min
    b. Sample mean $= 28.84$ min, sample standard deviation $= 5.145$ min;
    $t = (28.84 - 30.00)/(5.145/\sqrt{25}) = -1.127, p$-value $= 0.2708$
    c. At $\alpha = 0.05, t_{cutoff} = 2.064$ (two-tail test, df $= 24$); do not reject the null hypothesis.
    d. Based on the sample evidence, the foreman can report to the owner of the company that there is no difference between the lunch breaks being taken by the employees and the 30 minutes being allowed.

11. **a.** $H_0: \mu \le 312$ lb;
    $H_A: \mu > 312$ lb
    b. Sample mean $= 355.96$ lb, sample standard deviation $= 56.05$ lb;
    $t = (355.96 - 312)/(56.05/\sqrt{22}) = 3.68, p$-value $= 0.0007$
    c. At $\alpha = 0.01, t_{cutoff} = 2.518$ (upper-tail test, df $= 21$); reject the null hypothesis.
    d. Based on the sample evidence, Coach Juarez can conclude that the average bench press now surpasses the 312-lb mean obtained by last year's team.

Section 9.3

1. **a.** $H_0: \sigma^2 \leq 0.36; H_A: \sigma^2 > 0.36$
    b. Sample variance = 0.49;
    $\chi^2 = [(30 - 1)0.49]/0.36 = 14.21/0.36 = 39.47$
    c. At $\alpha = 0.05, \chi^2_{upper} = 42.6$ (df = 29, upper-tail test); do not reject the null hypothesis.
    d. There is not enough evidence to conclude that the variance in birth weights of premature infants is greater than the variance in birth weights of full-term infants.

3. **a.** $H_0: \sigma^2 \geq 73.96; H_A: \sigma^2 < 73.96$
    b. Sample variance = 40.96;
    $\chi^2 = [(27 - 1)40.96]/73.96 = 1064.96/73.96 = 14.40$
    c. At $\alpha = 0.05, \chi^2_{lower} = 15.38$ (df = 26, lower-tail test); reject the null hypothesis.
    d. There is enough evidence to conclude that the standard deviation in wait time has been reduced to less than 8.6 min.

5. **a.** $H_0: \sigma^2 = 0.50; H_A: \sigma^2 \neq 0.50$
    b. Sample variance = 0.49;
    $\chi^2 = [(25 - 1)0.49]/0.50 = 11.76/0.50 = 23.52$
    c. At $\alpha = 0.01, \chi^2_{upper} = 45.6$ (df = 24, two-tail test); do not reject the null hypothesis.
    d. There is not enough evidence to conclude that the variance in weight loss of these dieters differs from 0.50.

7. **a.** $H_0: \sigma^2 \leq 100; H_A: \sigma^2 > 100$
    b. Sample variance = 328.857;
    $\chi^2 = [(8 - 1)328.857]/100 = 2302/100 = 23.02$
    c. At $\alpha = 0.05, \chi^2_{upper} = 14.07$ (df = 7, upper-tail test); reject the null hypothesis.
    d. There is enough evidence to conclude that the standard deviation in cholesterol levels for men over 60 is higher than 10 mg/dL.

9. **a.** $H_0: \sigma^2 = 4; H_A: \sigma^2 \neq 4$
    b. Sample variance = 13.167;
    $\chi^2 = [(13 - 1)13.167]/4 = 158/4 = 39.5$
    c. At $\alpha = 0.10, \chi^2_{upper} = 21.0$ (df = 12, two-tail test); reject the null hypothesis.
    d. There is enough evidence to conclude that the variance in the amount of sugar contained in breakfast cereals differs from 4.

Section 9.4

1. **a.** $H_0: p \geq 0.20; H_A: p < 0.20$
    b. Sample proportion = 0.175, $Z = -1.25$, p-value = 0.1056
    c. At $\alpha = 0.05, Z_{cutoff} = -1.645$ (lower-tail test); do not reject the null hypothesis.
    d. There is not sufficient evidence to conclude that less than 20% of residents in this community live below poverty.

3. **a.** $H_0: p \geq 0.20; H_A: p < 0.20$
    b. Sample proportion = 0.11, $Z = -5.03$, p-value = 0.0000 (2.44×10^{-7})
    c. At $\alpha = 0.05, Z_{cutoff} = -1.645$ (lower-tail test); reject the null hypothesis.
    d. There is sufficient evidence to conclude that less than 20% of students at this college have used illegal drugs in the past month.
    e. The sample would not be truly representative of the population, which would mean we could not trust the results of this test.

5. **a.** $H_0: p \geq 0.25; H_A: p < 0.25$
    b. Sample proportion = 0.185, $Z = -2.12$, p-value = 0.0169
    c. At $\alpha = 0.05, Z_{cutoff} = -1.645$ (lower-tail test); reject the null hypothesis.
    d. There is sufficient evidence to conclude that the percentage of alcohol-related traffic fatalities in this officer's jurisdiction is less than 25%.

7. **a.** $H_0: p = 0.10; H_A: p \neq 0.10$
    b. Sample proportion = 0.12, $Z = 2.07$, p-value = 0.0385
    c. At $\alpha = 0.05, Z_{cutoff} = 1.96$ (two-tail test); reject the null hypothesis.

d. There is sufficient evidence to conclude that the percent of left-handed people differs from 10%.

9. **a.** $H_0: p = 0.70; H_A: p \neq 0.70$
   b. Sample proportion $= 0.73, Z = 0.802, p\text{-value} = 0.423$
   c. At $\alpha = 0.05, Z_{\text{cutoff}} = 1.96$ (two-tail test); do not reject the null hypothesis.
   d. Based on the sample evidence, you can conclude that there is no difference between the proportion of households owning a working television set in your city and the nation as a whole.

Chapter 10 Answers

Section 10.4

1. $H_0: \mu_1 = \mu_2; H_A: \mu_1 \neq \mu_2; Z = 5.983, p\text{-value} = 0.0000 (2.196 \times 10^{-9})$. At $\alpha = 0.05, Z_{\text{cutoff}} = 1.96$ (two-tail test). Reject the null hypothesis. There is enough evidence to conclude that the mean shopping times at the two stores are not the same.

3. $H_0: \mu_T \geq \mu_{\text{Th}}; H_A: \mu_T < \mu_{\text{Th}}; Z = -1.546, p\text{-value} = 0.0610$. At $\alpha = 0.05, Z_{\text{cutoff}} = -1.645$ (one-tail lower test). Do not reject the null hypothesis. There is not enough evidence to conclude that the mean rating for the Tuesday night class is lower than the mean rating for the Thursday night class.

5. $H_0: \mu_{\text{cost}} \leq \mu_{\text{spec}}; H_A: \mu_{\text{cost}} > \mu_{\text{spec}}; Z = 5.234, p\text{-value} = 0.000$. At $\alpha = 0.05, Z_{\text{cutoff}} = 1.645$ (one-tail upper test). Reject the null hypothesis. There is enough evidence to conclude the mean selling price for custom built homes is higher than the mean selling price for speculation homes.

7. $H_0: \mu_A = \mu_B; H_A: \mu_A \neq \mu_B; Z = -6.795, p\text{-value} = 0.000$. At $\alpha = 0.01, Z_{\text{cutoff}} = 2.56$ (two-tail test). Reject the null hypothesis. There is enough evidence to conclude that the mean time spent with the technical staff differs for company A and company B.

9. $H_0: \mu_A = \mu_B; H_A: \mu_A \neq \mu_B; Z = 1.230, p\text{-value} = 0.2188$. At $\alpha = 0.05, Z_{\text{cutoff}} = 1.96$ (two-tail test). Do not reject the null hypothesis. There is not enough evidence to conclude that the mean distance for the golf balls differs between brand A and brand B.

Section 10.5

1. $H_0: \mu_1 = \mu_2; H_A: \mu_1 \neq \mu_2; t = 20.4, p\text{-value} = 0.000$. At $\alpha = 0.01, df = 23, t_{\text{cutoff}} = 2.807$ (two-tail test). Reject the null hypothesis. There is enough evidence to conclude that the mean measurement of the speed of light differs between the two methods, that used in 1879 and that used in 1882.

3. $H_0: \mu_{\text{BigTen}} = \mu_{\text{Others}}; H_A: \mu_{\text{BigTen}} \neq \mu_{\text{Others}}; t = -1.51, p\text{-value} = 0.142$. At $\alpha = 0.10, df = 32, t_{\text{cutoff}} = 1.694$ (two-tail test). Do not reject the null hypothesis. There is not enough evidence to conclude that the mean salary for professors at the "Big Ten" universities differs from the mean salary for professors at other universities.

5. $H_0: \mu_{\text{before}} = \mu_{\text{after}}; H_A: \mu_{\text{before}} \neq \mu_{\text{after}}; t = 2.86, p\text{-value} = 0.007$. At $\alpha = 0.01, df = 39, t_{\text{cutoff}} = 2.708$ (two-tail test). Reject the null hypothesis. There is enough evidence to conclude that the mean blood pressure for this patient changed while taking the medication.

7. $H_0: \mu_P = \mu_T; H_A: \mu_P \neq \mu_T; t = 0.802, p\text{-value} = 0.426$. At $\alpha = 0.10, df = 50, t_{\text{cutoff}} = 1.676$ (two-tail test). Do not reject the null hypothesis. There is not enough evidence to conclude the mean octane level differs between the two gas stations.

9. $H_0: \mu_{\text{male}} = \mu_{\text{female}}; H_A: \mu_{\text{male}} \neq \mu_{\text{female}}; t = 0.29, p\text{-value} = 0.387$. At $\alpha = 0.05, df = 29, t_{\text{cutoff}} = 2.045$ (two-tail test). Do not reject the null hypothesis. There is not enough evidence to conclude that the mean time spent studying on weeknights differs between males and females.

Section 10.7

1. **a.** $H_0: \mu_A \le \mu_B; H_A: \mu_A > \mu_B; t = 3.71, p\text{-value} = 0.0005.$ At $\alpha = 0.05$, df $= 24, t_{cutoff} = 1.711$ (one-tail upper test).
   b. Reject the null hypothesis. There is enough evidence to conclude the mean score on the real estate IQ test is higher after taking the course.

3. $H_0: \mu_1 = \mu_2; H_A: \mu_1 \ne \mu_2; t = 5.65, p\text{-value} = 0.000.$ At $\alpha = 0.05$, df $= 15, t_{cutoff} = 2.131$ (two-tail test). Reject the null hypothesis. There is enough evidence to conclude that the mean tread-wear measure differs between the two methods.

5. $H_0: \mu_{1972} \le \mu_{1968}; H_A: \mu_{1972} > \mu_{1968}; t = 2.46, p\text{-value} = 0.012.$ At $\alpha = 0.01$, df $= 18, t_{cutoff} = 2.552$ (one-tail upper test). Do not reject the null hypothesis. There is not enough evidence to conclude the mean participation rate of women in the work force was higher in 1972 than in 1968.

7. $H_0: \mu_{himin} \le \mu_{hiwhite}; H_A: \mu_{himin} > \mu_{hiwhite}; t = 8.94, p\text{-value} = 0.000.$ At $\alpha = 0.01$, df $= 19, t_{cutoff} = 2.539$ (one-tail upper test). Reject the null hypothesis. There is enough evidence to conclude the mean refusal rate in mortgage lending is higher for high-income minorities than for high-income whites.

9. $H_0: \mu_{agr} \ge \mu_{ind}; H_A: \mu_{agr} < \mu_{ind}; t = -2.67, p\text{-value} = 0.007.$ At $\alpha = 0.05$, df $= 19, t_{cutoff} = -1.729$ (one-tail lower test). Reject the null hypothesis. There is enough evidence to conclude that, in 1960, the mean percent employed in agriculture was lower than the mean percent employed in industry.

Section 10.8

1. $H_0: p_{before} \ge p_{after}; H_A: p_{before} < p_{after}; Z = -2.77, p\text{-value} = 0.003.$ At $\alpha = 0.01$, $Z_{cutoff} = -2.33$ (one-tail lower test). Reject the null hypothesis. There is sufficient evidence to conclude the proportion of passengers checking bags increased after the new system was installed.

3. $H_0: p_{good} \ge p_{fair/poor}; H_A: p_{good} < p_{fair/poor}; Z = -2.19, p\text{-value} = 0.014.$ At $\alpha = 0.05$, $Z_{cutoff} = -1.645$ (one-tail lower test). Reject the null hypothesis. There is sufficient evidence to conclude the proportion of overweight people in "good" health is lower than the proportion of overweight people in "fair" or "poor" health.

5. $H_0: p_{placebo} \ge p_{drug}; H_A: p_{placebo} < p_{drug}; Z = -0.61, p\text{-value} = 0.272.$ At $\alpha = 0.10$, $Z_{cutoff} = -1.28$ (one-tail lower test). Do not reject the null hypothesis. There is not sufficient evidence to conclude the proportion of people who experience drowsiness is lower for those taking the placebo than for those taking the experimental drug.

7. $H_0: p_{women} \le p_{men}; H_A: p_{women} > p_{men}; Z = 12.61, p\text{-value} = 0.000.$ At $\alpha = 0.05$, $Z_{cutoff} = 1.645$ (one-tail upper test). Reject the null hypothesis. There is sufficient evidence to conclude the proportion of single women with children living in poverty is higher than the proportion of single men with children living in poverty.

9. $H_0: p_{males} = p_{females}; H_A: p_{males} \ne p_{females}; Z = 1.19, p\text{-value} = 0.235.$ At $\alpha = 0.10$, $Z_{cutoff} = 1.645$ (two-tail test). Do not reject the null hypothesis. There is not sufficient evidence to conclude the proportion of males preferring the name-brand cola differs from the proportion of females preferring the name-brand cola.

Section 10.9

1. $H_0: \sigma^2_A \le \sigma^2_B; H_A: \sigma^2_A > \sigma^2_B; F = 2.224, p\text{-value} = 0.0066.$ At $\alpha = 0.05$, $df_{num} = 40$, $df_{den} = 40$, $F_{cutoff} = 1.693$ (one-tail test upper). Reject the null hypothesis. There is enough evidence to conclude that the variance in the distance traveled by brand A golf balls is greater than that of brand B.

3. $H_0: \sigma^2_{1882} = \sigma^2_{1879}; H_A: \sigma^2_{1882} \neq \sigma^2_{1879}; F = 7.968$, p-value $= 0.000$. At $\alpha = 0.01$, $df_{num} = 22$, $df_{den} = 99$, $F_{cutoff} = 2.18$ (two-tail test). Reject the null hypothesis. There is enough evidence to conclude that the variances of the two methods to measure the speed of light differ.

5. $H_0: \sigma^2_{Oakland} \leq \sigma^2_{SanFran}; H_A: \sigma^2_{Oakland} > \sigma^2_{SanFran}; F = 3.159$, p-value $= 0.0490$. At $\alpha = 0.05$, $df_{num} = 10$, $df_{den} = 9$, $F_{cutoff} = 3.14$ (one-tail test upper). Reject the null hypothesis. There is enough evidence to conclude that the variance in attendance increased after the team moved to Oakland.

7. $H_0: \sigma^2_{after} = \sigma^2_{before}; H_A: \sigma^2_{after} \neq \sigma^2_{before}; F = 1.283$, p-value $= 0.5924$. At $\alpha = 0.05$, $df_{num} = 19$, $df_{den} = 19$, $F_{cutoff} = 2.53$ (two-tail test). Do not reject the null hypothesis. There is not enough evidence to conclude that the variance in the number of daily disruptions changed after the teacher began using remedial interventions.

9. $H_0: \sigma^2_{males} = \sigma^2_{females}; H_A: \sigma^2_{males} \neq \sigma^2_{females}; F = 3.805$, p-value $= 0.0037$. At $\alpha = 0.05$, $df_{num} = 20$, $df_{den} = 21$, $F_{cutoff} = 2.42$ (two-tail test). Reject the null hypothesis. There is enough evidence to conclude that the variance in study time on weeknights differs between males and females.

Chapter 11 Answers

Section 11.2

1. a. and **c.**

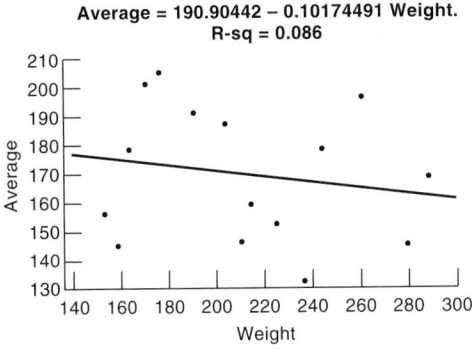

Average = 190.90442 − 0.10174491 Weight.
R-sq = 0.086

b. $\hat{y} = 190.9 - 0.1017x$
d. $190.9 - 0.1017(190) = 171.6$.
The negative slope suggests that heavier bowlers tend to have lower averages. However, due to the scattered appearance of the data on the plot, the relationship does not appear to be very strong.
e. 23.78

3. a. and **c.**

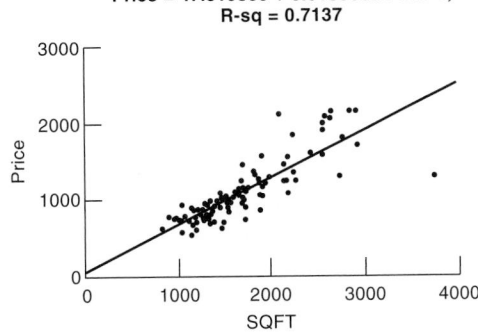

Price = 47.819305 + 0.61366683 SQFT,
R-sq = 0.7137

b. $\hat{y} = 47.8193 + 0.6137x$

d. $47.8193 + 0.6137(2200) = 1397.96$ or \$139,796.

The positive slope indicates that as the square footage increases, the selling price tends to increase. The graph shows a strong linear pattern, with a prominent outlier.

e. 3750 square feet, sold for \$129,500

5. **a.** and **c.**

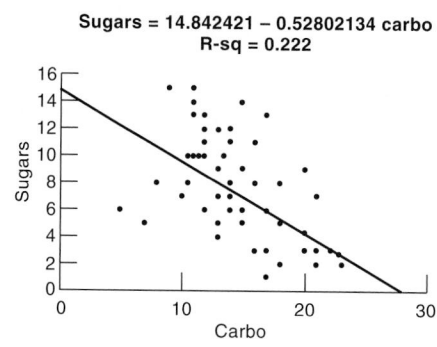

Sugars = 14.842421 − 0.52802134 carbo
R-sq = 0.222

b. $\hat{y} = 14.8424 - 0.5280x$

d. The negative slope suggests that as the number of grams of carbohydrates increases, the number of grams of sugar tends to decrease. The scatter plot suggests that the relationship is moderate.

e. $14.8424 - 0.5280(16) = 6.4$ grams of sugar. As the amount of carbohydrates increases by 1 gram, expect a decrease in the sugar content by 0.53 gram.

7. **a.** and **c.**

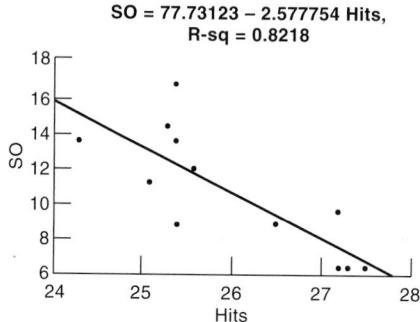

SO = 77.73123 − 2.577754 Hits,
R-sq = 0.8218

b. $\hat{y} = 77.7312 - 2.5778x$

d. $77.7312 - 2.5778(26) = 10.7$ strikeouts.

There is a negative slope, which indicates that as the number of hits increases, the number of strikeouts tends to decrease. The scatter plot shows a moderate to strong relationship.

e. Predicted $= 77.7312 - 2.5778(27.2) = 7.6$ strikeouts; P. Derringer: $9.6 - 7.6 = 2$, F. Fitzsimmons: $6.4 - 7.6 = -1.2$.

9. **a.** and **c.**

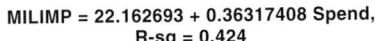

MILIMP = 22.162693 + 0.36317408 Spend,
R-sq = 0.424

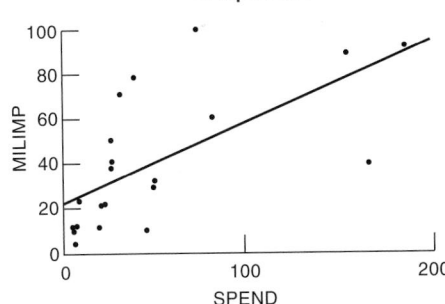

b. $\hat{y} = 22.1627 + 0.3632x$

d. $22.1627 + 0.3632(50) = 40.32$ million. The positive slope indicates that as the TV advertising budget increases, the retention of the advertisement increases. The graph shows a moderate relationship, with three notable outliers.

e. Pepsi: Predicted $= 22.1627 + 0.3632(74.1) = 49.1$, Residual $= 99.6 - 49.1 = 50.5$; Ford: Predicted $= 22.1627 + 0.3632(166.2) = 82.5$, Residual $= 40.1 - 82.5 = -42.4$.

11. **a.** and **c.**

MPG = 50.06608 − 0.13902326 HP,
R-sq = 0.6239

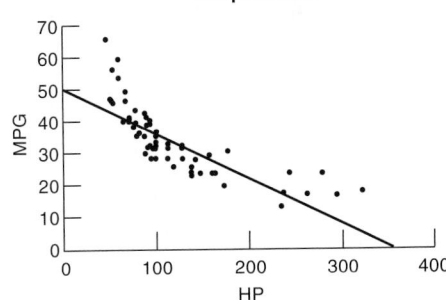

b. $\hat{y} = 50.066 - 0.139x$

d. Expect a decrease of 0.14 mile per gallon in gas mileage, for each additional horsepower of the engine.

e. $50.066 - 0.139(200) = 22.27$ miles per gallon.

The negative slope suggests that as the horsepower of the engine increases, the gas mileage will tend to decrease. The graph shows more of a curved relationship than a linear pattern.

Section 11.3

1. **a.** $H_0: \beta_1 = 0; H_A: \beta_1 \neq 0$

b. At $\alpha = 0.05$ with df $= 13$ for two-tails, $t_{cutoff} = 2.16$.

c. Slope coef. $= -0.1017$, SE slope $= 0.1459, t = -0.70$

d. The relationship between these two variables is not significant.

3. **a.** $H_0: \beta_1 = 0; H_A: \beta_1 \neq 0$
   b. At $\alpha = 0.01$ with df $= 115$ for two-tails, $t_{cutoff} = 2.62$.
   c. Slope coef. $= 0.61367$, SE slope $= 0.03625$, $t = 16.93$
   d. The relationship between these two variables is significant.

5. **a.** $H_0: \beta_1 = 0; H_A: \beta_1 \neq 0$
   b. At $\alpha = 0.02$ with df $= 75$ for two-tails, $t_{cutoff} = 2.38$.
   c. Slope coef. $= 0.025756$, SE slope $= 0.00517$, $t = 4.98$
   d. The relationship between these two variables is significant.

7. **a.** $H_0: \beta_1 = 0; H_A: \beta_1 \neq 0$
   b. At $\alpha = 0.05$ with df $= 17$ for two-tails, $t_{cutoff} = 2.11$.
   c. Slope coef. $= 0.8224$, SE slope $= 0.1739$, $t = 4.73$
   d. The relationship between these two variables is significant.

9. **a.** $H_0: \beta_1 = 0; H_A: \beta_1 \neq 0$
   b. At $\alpha = 0.01$ with df $= 105$ for two-tails, $t_{cutoff} = 2.62$.
   c. Slope coef. $= 0.068904$, SE slope $= 0.004015$, $t = 17.16$
   d. The relationship between these two variables is significant.

11. **a.** $H_0: \beta_1 = 0; H_A: \beta_1 \neq 0$
    b. At $\alpha = 0.05$ with df $= 79$ for two-tails, $t_{cutoff} = 1.99$.
    c. Slope coef. $= 0.238705$, SE slope $= 0.007082$, $t = 33.70$
    d. The relationship between these two variables is significant.

13. **a.** $H_0: \beta_1 = 0; H_A: \beta_1 \neq 0$
    b. At $\alpha = 0.05$ with df $= 12$ for two-tails, $t_{cutoff} = 2.18$.
    c. Slope coef. $= 1.2673$, SE slope $= 0.4742$, $t = 2.67$
    d. The relationship between these two variables is significant.

Section 11.4

1. **a.** 95.0% CI: (30.232, 33.255)   **b.** 95.0% PI: (25.633, 37.854)

3. **a.** 99.0% CI: (1.031, 1.670)   **b.** 99.0% PI: (-0.992, 3.694)

5. **a.** 95.0% CI: (5.340, 6.018)   **b.** 95.0% PI: (4.163, 7.195)

7. **a.** 98.0% CI: (3.6054, 3.8617)   **b.** 98.0% PI: (2.4608, 5.0063)

9. **a.** 99.0% CI: (119.038, 121.481)   **b.** 99.0% PI: (110.622, 129.898)

11. **a.** 90.0% CI: (46.82, 53.18)   **b.** 90.0% PI: (37.68, 62.32)

Section 11.5

1. **a.** $r = -0.1897$
   b. There is a weak negative association between the two variables.
   c. $r^2 = 0.0360$

3. **a.** $r = 0.8448$
   b. There is a fairly strong positive association between the two variables.
   c. $r^2 = 0.7137$

5. **a.** $r = 0.4986$
   b. There is a moderate positive association between the two variables.
   c. $r^2 = 0.2486$

7. **a.** $r = 0.7538$
   b. There is a fairly strong positive association between the two variables.
   c. $r^2 = 0.5682$

9. **a.** $r = 0.8586$
   b. There is a fairly strong positive association between the two variables.
   c. $r^2 = 0.7372$

11. **a.** $r = 0.9665$
    b. There is a very strong positive association between the two variables.
    c. $r^2 = 0.9342$

13. **a.** $r = 0.6109$
    b. There is a moderate positive association between the two variables.
    c. $r^2 = 0.3732$

Section 11.6

1. **a.**

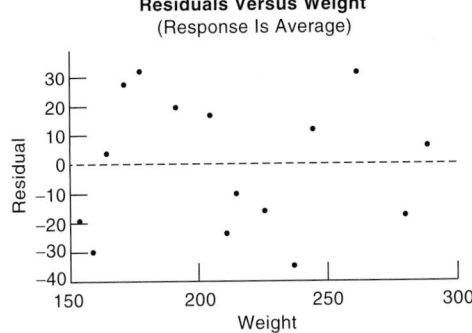

b. and **c.** Yes. The points are scattered.
d.

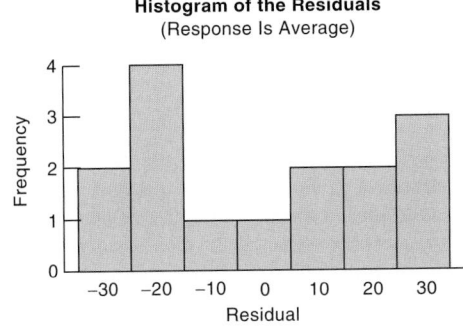

No. The graph shows a bimodal appearance.
e. No, since the residuals are not normal and r is very low.

3. **a.**

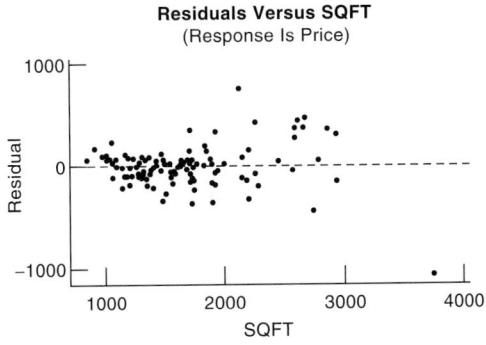

b. and **c.** Yes. The points on the graph are scattered with the exception of an outlier.
d.

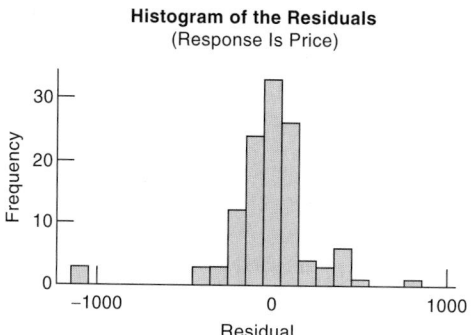

Yes. The graph is bell shaped with the exception of a couple of outliers.
e. Yes, the graphs show that the assumptions are met.

5. **a.**

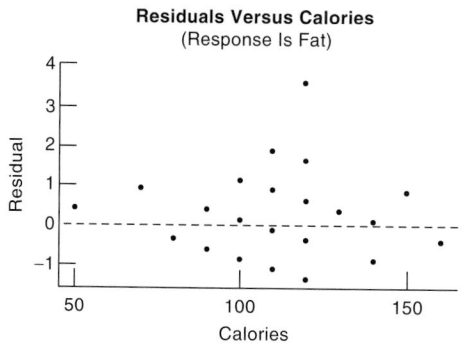

b. and **c.** Yes, the points on the graph are pretty scattered.
d.

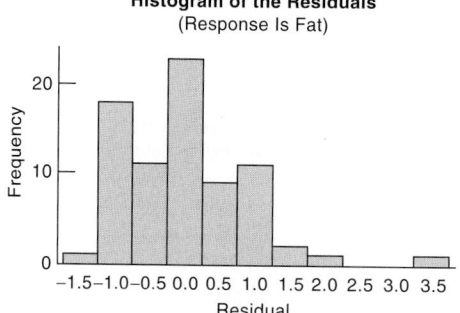

Yes, the curve is reasonably normal. There is a slight skew to the right, but not too bad.
e. Yes, the graphs indicate that the assumptions are met.

7. a.

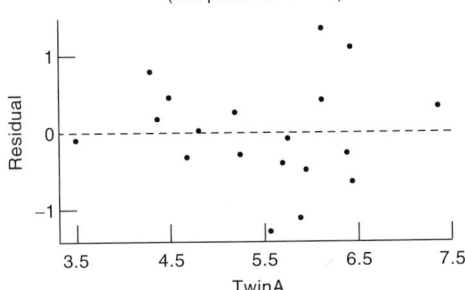

Residuals Versus TwinA
(Response Is TwinB)

b. and c. Yes, the points on the graph are pretty scattered.
d.

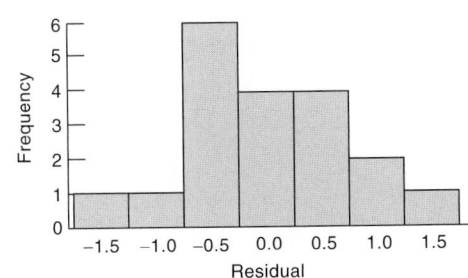

Histogram of the Residuals
(Response Is TwinB)

Yes, the graph is bell shaped.
e. Yes, the graphs indicate that the assumptions are met.

9. a.

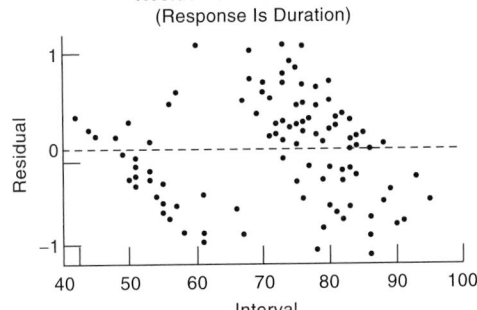

Residuals Versus Interval
(Response Is Duration)

b. and c. Yes, the points on the graph are reasonably scattered.
d.

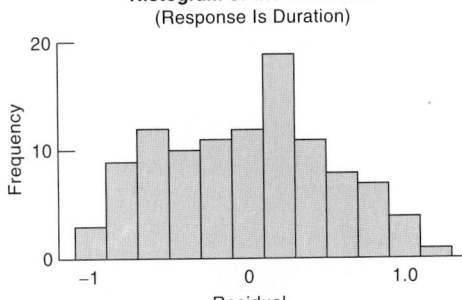

Histogram of the Residuals
(Response Is Duration)

Yes, the graph is fairly bell shaped.
e. Yes, the graphs indicate that the assumptions are met.

11. **a.**

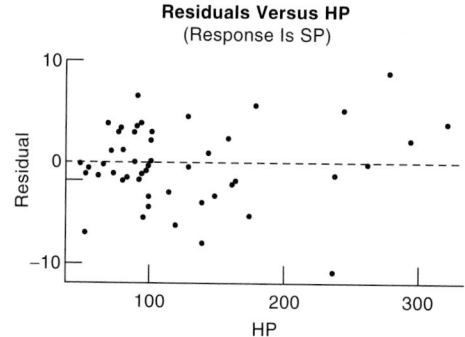

Residuals Versus HP
(Response Is SP)

b. and **c.** Yes, the points on the graph are reasonably scattered.

d.

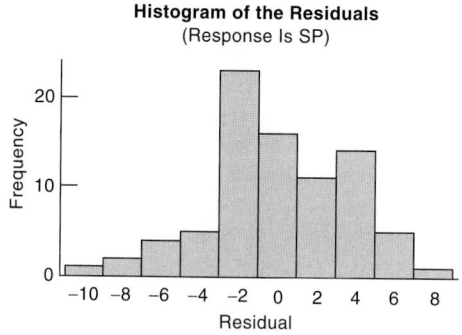

Histogram of the Residuals
(Response Is SP)

Yes, the graph is fairly bell shaped.

e. Yes, the graphs indicate that the assumptions are met.

13. **a.**

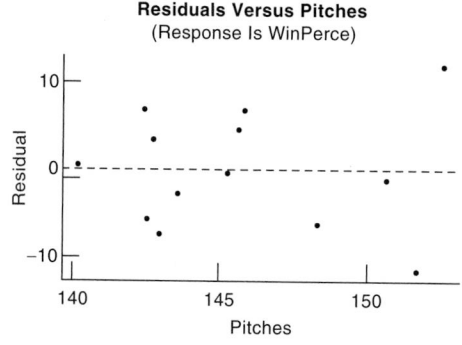

Residuals Versus Pitches
(Response Is WinPerce)

b. and **c.** Yes, the points on the graph are reasonably scattered.

d.

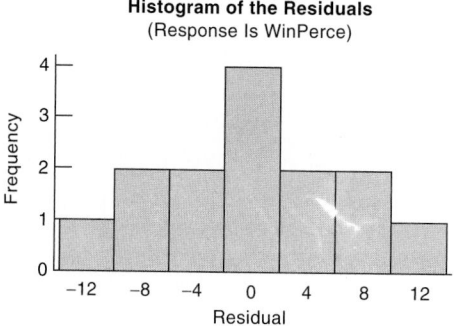

Histogram of the Residuals
(Response Is WinPerce)

Yes, the graph is bell shaped.

e. Yes, the graphs suggest that the assumptions are met.

Chapter 12 Answers

Section 12.2

1. **a.** Dependent: Number of bars of soap; Independent: Coupons redeemed and Proportion on sale

   b. $\hat{y} = 6.524 + 0.7008x_1 + 10.905x_2$

   c. For every extra coupon redeemed, the number of bars increases by 0.7008 and for every increase of 0.01 (1%) in proportion on sale, the number increases by 0.10905.

   d. Expected sales $= 6.524 + (0.7008)(20) + 10.905(0.65)$

   $$= 6.524 + 14.016 + 7.09$$

   $$= 27.6 \approx 28 \text{ bars}$$

   e. This value seems reasonable because both of the values of the independent variables are within the range observed and the y value is within the range, too. You are interpolating.

   f. and **g.**

Observation	Predicted Y	Residuals
1	31.43	−0.429
2	25.65	−1.651
3	17.30	−0.302
4	12.55	1.448
5	13.47	−0.471
6	24.56	1.439
7	13.22	−3.221
8	31.63	1.369
9	18.66	−1.657
10	30.18	−2.183
11	34.72	−1.715
12	38.61	3.391
13	8.21	2.794
14	17.63	0.371
15	11.18	0.819

The model does a decent job of predicting. The largest error is just over 3 bars of soap and the smallest is less than 1 bar.

Section 12.3

1. **a.**

Source	df	SS	MS	F
Regression	3	274.29	91.43	40.3
Residual	29	65.78	2.27	
Total		340.07		

   b. $H_0: \beta_1 = \beta_2 = 0$; H_A: At least one coefficient is not equal to 0.

   c. The critical value for the test is $F_{0.05, 3, 29} = 2.934$. Since 40.3 exceeds this value, we reject H_0 and conclude that the model is significant (at least one coefficient is not 0).

   d. The coefficient of determination is 80.7%.

3. **a.**

Source	df	SS	MS	F
Regression	2	98.58	49.29	7.34
Residual	10	67.11	6.71	
Total	12	165.69		

$F = 7.34$, p-value $= 0.0109$. Reject the null hypothesis. Conclude that at least one of the two coefficients is not equal to zero.

b. The coefficient of determination is 59.5%.

c. The model is not bad, but it does not explain much more than half of the variation.

d.
$$t = -0.0008284/0.7223 = -0.00115, \quad p\text{-value} = 0.991$$

Do not reject the null hypothesis. Conclude that the first coefficient is equal to zero. Jordan's assists did not decrease the number of points he scored per game.

$$t = 5.643/1.889 = 2.988, \quad p\text{-value} = 0.0136$$

Reject the null hypothesis. Conclude that the second coefficient is not equal to zero. Jordan's steals increased the number of points he scored per game.

e. Based on the results, I would suggest looking at other variables to see whether there might be other factors that affected Michael Jordan's average points per game. The R^2 value is not high enough.

Chapter 13 Answers

Section 13.3

1. 23.1

3. 13.525, 821.75

5. $A = 5; B = 6; C = 90; D = 6; E = 1$

7. 84.65

9. 16.76

11. **a.**

Summary

Groups	Count	Sum	Average	Variance
Column 1	5	80.3	16.06	6.598
Column 2	5	36.5	7.30	26.965
Column 3	5	12.8	2.56	3.373
Column 4	5	32.7	6.54	5.308
Column 5	5	47.3	9.46	14.528

ANOVA

Source of Variation	SS	df	MS	F	p-value	F crit
Between groups	492.866	4	123.216	10.852	7.65E-05	2.866081
Within groups	227.088	20	11.354			
Total	719.954	24				

b. $H_0: \mu_1 = \mu_2 = \mu_3 = \mu_4 = \mu_5$

c.

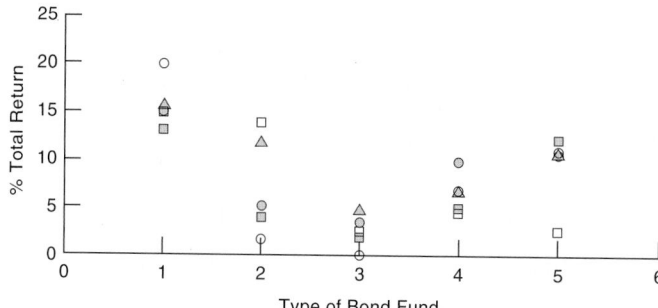

Total Return by Type of Bond Fund

d. Reject the null hypothesis. The evidence supports the claim that at least one of the means is different from the others.

Section 13.4

1. **a.** No, there are no obvious reasons to believe the errors are not independent.
   b.

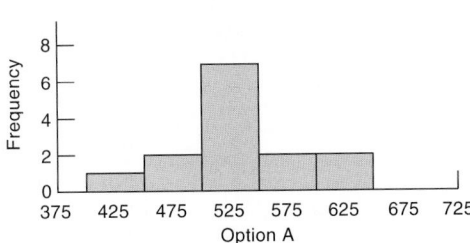

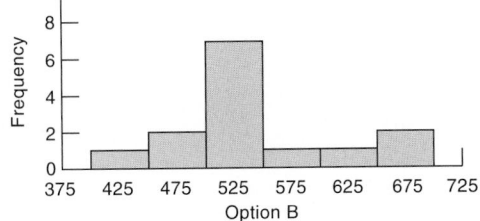

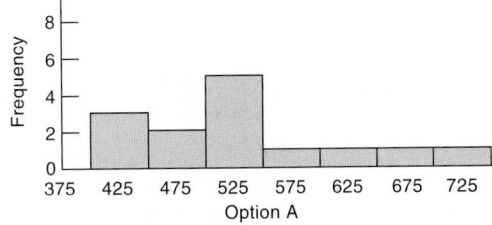

The plot for option C does not appear normally distributed.

c.

Option	Sample Variance
A	2555.96
B	4340.34
C	8389.21

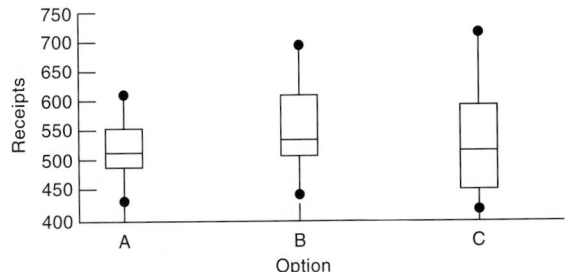

The variances do not appear to be equal.

d. The assumption of independent errors appears to be the only valid assumption.

Section 13.5

1. **a.**

Summary

Groups	Count	Sum	Average	Variance
Row 1	5	38.5	7.70	39.665
Row 2	5	39.8	7.96	64.528
Row 3	5	50.9	10.18	19.042
Row 4	5	44.6	8.92	23.737
Row 5	5	35.8	7.16	25.943
Column 1	5	80.3	16.06	6.598
Column 2	5	36.5	7.30	26.965
Column 3	5	12.8	2.56	3.373
Column 4	5	32.7	6.54	5.308
Column 5	5	47.3	9.46	14.528

ANOVA

Source of Variation	SS	df	MS	F	p-value	F crit
Between Groups	492.866	4	123.216	9.9171	0.0003	3.007
Between Blocks	28.294	4	7.073	0.5693	0.6887	3.007
Within Groups	198.794	16	12.425			
Total	719.954	24				

b.

$$\text{Relative efficiency} = \frac{4(7.073) + 5(4)(123.216)}{24(12.425)} = 8.359$$

c. Although the variance due to the treatment is relatively small, the relative efficiency indicates that 8 times as many observations would have been needed to detect the treatment effect without blocks, so blocking was worth it.

Section 13.6

1. **a.**

Two-Way ANOVA

Source of Variation	SS	df	MS	F	p-value	F crit
League	0.00058080	1	0.00058080	4.0734	0.0548	4.260
Division	0.00001307	2	0.00000653	0.0458	0.9553	3.403
Interaction	0.00017360	2	0.00008680	0.6088	0.5522	3.403
Error	0.00342200	24	0.00014258			
Total	0.00418947	29				

b. H_0: There is no difference in the population means due to league.
H_A: There is a difference in the population means due to league.
$F = 4.0734$; do not reject the null hypothesis. A team's league is not a factor in its batting average.

H_0: There is no difference in the population means due to division.
H_A: There is a difference in the population means due to division.
$F = 0.0458$; do not reject the null hypothesis. A team's division is not a factor in its batting average.

H_0: There is no difference in the population means due to the interaction between league and division.
H_A: There is a difference in the population means due to the interaction between league and division.
$F = 0.6088$; do not reject the null hypothesis. The interaction of league and division is not a factor in batting average.

c.

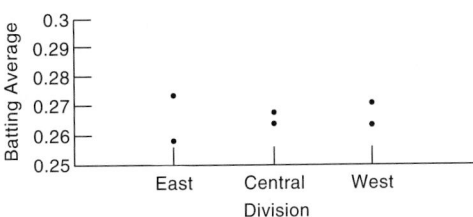

Chapter 14 Answers

Section 14.2

1. a.

Weekday	M	T	W	TH	F
Messages	57	53	52	45	38
Expected	49	49	49	49	49

b. The observed data seem to agree with the expected frequencies, so he appears to be correct.

c. H_0: The messages are uniformly distributed over the weekdays.
H_A: The messages are not uniformly distributed over the weekdays.

d. The chi-square statistic is 4.612. The critical value is 9.488. We cannot reject the null hypothesis.

e. Based on the test, the claim is valid and the messages are uniformly distributed.

3. a.

Coupons Inserted	Number Observed	Expected ($p = 30\%$)
0	70	67.23
1	150	144.06
2	110	123.48
3	55	52.92
4	10	11.34
5	5	0.97

b. Chi-square $= 18.81$

c. Since the critical value with 5 degrees of freedom is 11.070, we reject H_0. The data do not appear to be consistent with a binomial distribution with $p = 0.30$.

5. a. 38.7
b.

Grade	K	1	2	3	4	5	6
Roster	33	39	38	42	40	38	41
Expected	38.7	38.7	38.7	38.7	38.7	38.7	38.7

c. H_0: The distribution of pupils per grade is uniform.

H_A: The distribution of pupils per grade is not uniform.

The value of the test statistic is 1.33.

d. Since the critical value is 12.592, we cannot reject H_0. The principal can tell the PTA that the distribution of pupils per grade is uniform.

Section 14.3

1. **a.** $H_0: p_1 = p_2 = p_3$

H_A: At least one proportion is different.

b. 0.44, 0.35, 0.29. It appears that the proportions might be different.

c. The value of the chi-square statistic is 2.140.

d. The p-value is 0.343, so we cannot reject H_0. It appears that there is no difference in the proportions by racial group.

3. **a.** 0.32, 0.42, 0.52, 0.70, 0.85

b. 0.54

c. Chi-square value is 40.333. Reject H_0. There is a difference in proportions for different education levels.

Section 14.4

1. **a.** H_0: Year and type of computer owned are independent.

H_A: Year and type of computer owned are not independent.

b.

Expected counts are printed below observed counts

	Freshman	Sophomore	Junior	Senior	Total
Laptop	111	115	117	157	500
	122.66	129.15	134.68	113.52	
Desktop	187	201	205	167	760
	186.44	196.31	204.71	172.54	
Both	67	78	82	65	292
	71.63	75.42	78.65	66.29	
Neither	145	143	156	83	527
	129.28	136.12	141.95	119.65	
Total	510	537	560	472	2079

c. The test statistic is 37.357. There are 9 degrees of freedom so the critical value is 16.919. Since 35.357 is greater than 16.919, we reject H_0.

d. We can conclude that year in college and type of computer owned are not independent.

3. **a.** H_0: Race and attitude toward marijuana legalization are independent.

H_A: Race and attitude toward marijuana legalization are not independent.

b. Expected counts are printed below observed counts

	Yes	No	Total
White	425	979	1404
	412.25	991.75	
Black	67	197	264
	77.52	186.48	
Other	33	87	120
	35.23	84.77	
Total	525	1263	1788

c. The value of the chi-square statistic is 2.779. The degrees of freedom are $(3-1)(2-1) = 2$ so the critical value is 5.991. We cannot reject H_0 so we conclude that there is not enough evidence to say that race and attitude toward marijuana legalization are dependent.

Index

A

Adjusted R^2, 586
All possible regressions, 596, 603
  in Minitab, 614
Alpha (α), 384
Analysis of variance (ANOVA), 625, 629
  assumptions, 648, 655
  in Excel, 674–675
  in Minitab, 679–680
  multiple comparisons, 643
  multiple regression, 581
  notation, 629–630
  one-way, 628, 641, 674, 679
  randomized block, 652, 656
  simple linear regression, 529
  two-way, 662–663, 675, 680
Analysis of variance table
  multiple regression, 581
  one-way design, 641
  randomized block design, 653
  two-way design, 665
Arithmetic mean. *See* Sample mean.
Average. *See* Sample mean.
Average difference, 472

B

Backward elimination, 596, 598, 605
Bar chart, 73, 76
  clustered, 171, 194–195
  in Excel, 96, 194–195
  in Minitab, 96, 98, 195
  for qualitative data, 72
  stacked, 171, 173, 195
Bayes' Rule, C2
Beta (β), 384
Biased sample, 27
Bimodal
  data set, 87
  sample, 122
Binomial probabilities, table of, A4
Bivariate data, 167
  contingency table, 167–168, 170
  cross-classification table, 167
  in Excel, 194

Bivariate data (*continued*)
  least-squares line, 185
  in Minitab, 195
  qualitative, 167
  quantitative, 178
  with TI-83, 196
Block, 652
Box and whisker diagram, 145
Boxplot
  creation of, 146
  definition, 145
  in Excel, 153
  fences, 145
  in identification of outliers, 148
  in Minitab, 153
  with TI-83, 153

C

C_p statistic, 603
Census, 14
Central Limit Theorem, 306–307, 452, 462
Central tendency, measures of, 111, 128
  population mean, 112
  sample mean, 112, 117
  sample median, 115, 117
  sample mode, 121
Chi-square distribution table, 424, A9
Chi-square statistic, 423, 427, 699, 716, 728
Chi-square test
  in Excel, 735
  for goodness of fit, 694, 705, 707
  for independence, 722–723, 728, 735
  with KADDSTAT, 735
  in Minitab, 736
  of population variance, 422, 426–427
  for proportions, 711, 716, 718
Class, in frequency table, 57, 64, 699
  lower boundary, 66
  modal, 121
  upper boundary, 66

Class interval, 64, 79
Coefficient of determination, 530, 539
Coefficient of multiple determination, 585
  adjusted, 586
Complement of an event, 211
  probability of, 211
Confidence interval, 320
  in Excel, 345, 550
  interpretation of, 324
  with KADDSTAT, 346
  for the mean, large sample, 320, 322–323
  for the mean, small sample, 333
  for mean value of Y in regression, 532
  in Minitab, 556
  for multiple comparisons, 643
  for the population proportion, 337
  relation to hypothesis testing, 437
  with TI-83, 346
Confidence level, 320
Contingency table, 167–168, 170, 712, 724
  in Excel, 194, 733
  in Minitab, 736
Cook's distance, 608
Correlation coefficient, 537–538
Critical value, 366
  of chi-square distribution, 423, 701
  of F distribution, 486, 502, 582
  of standard normal distribution, 271, A6
  of t distribution, 333, A8
Cross-classification table, 167. *See also* Contingency table.

D

Data, 32
  bivariate, 167
  categorical, 33
  continuous, 37
  description of, 86
  discrete, 36
  measurement, 64

Data (*continued*)
  nominal, 33–34
  ordinal, 34
Data, distribution of
  center, 86
  comparison of, 90
  shape, 86
  skewed, 87
  symmetric, 87, 134
  variability, 86
Data, qualitative, 32–33
  bar chart for, 73
  categorical, 33
  frequency table for, 57
  graphical display of, 73, 76
  Likert scale, 34
  nominal, 33–34
  ordinal, 34
  pie chart for, 76
Data, quantitative, 32, 36
  continuous, 37
  description of, 86
  discrete, 36
  frequency table for, 61
  graphical display of, 77
  measurement, 64
Degrees of freedom
  of F distribution, 486
  of goodness of fit test, 699
  of one-way ANOVA, 638
  of paired difference test, 474
  of regression ANOVA, 581
  of t distribution, 466
  of test of independence, 728
  of test of proportions, 716
  of two-way ANOVA, 665
Deming, W. Edwards, 780
Descriptive statistics, 38–39
Deviation, 185–186, 509
Dispersion, measures of, 128
  population standard deviation,
    131, 453
  population variance, 131
  sample range, 128, 130
  sample standard deviation, 131,
    134
  sample variance, 131–132,
    629–630
Distribution
  binomial, 250, 254, 257, 262, A4
  chi-square, 424, 701, A9
  continuous, 264
  discrete, 244
  F, 486, A10
  normal, 266, 273–274, 277, 542,
    648
  Poisson, D1–D2
  skewed, 87
  standard normal, 268, 273, 277
  symmetric, 87, 134, A6

Distribution (*continued*)
  t, 331, 333, 412, 462, A8
  uniform, 694
Distribution-free methods. *See*
  Nonparametric methods.
Dotplot, 81

E
Empirical probabilities, 213, 723
Empirical rule, 134, 137, 305
  and z-scores, 137
Equal replication, 663
Error, 185–186, 509, 642
Event(s), 209
  A AND B, 217
  A OR B, 217
  complement of, 211
  independent, 229
  intersection of, 220, 223
  mutually exclusive, 218, 221
  probability of, 209, 217
Excel, 8
  analysis of variance (ANOVA),
    674–675
  bar chart, creation of, 96, 194–195
  bivariate data, 194
  boxplot, creation of, 153
  chi-square test for independence,
    735
  confidence interval, calculation of,
    345, 550
  contingency table, creation of, 194,
    733
  frequency distribution table, 96
  graphical displays, creation of, 96
  histogram, creation of, 96–97
  hypothesis testing, 439, 492
  multiple regression, 612
  normal probabilities, calculation
    of, 282
  pie chart, creation of, 77, 96–97
  prediction intervals, creation of,
    550
  probabilities, calculation of, 282
  random data, generation of, 233
  random sample, selection of, 46
  scatter plot, 195
  simple linear regression, 547
  summary statistics, creation of, 153
  t test of population mean, 439
  Z test of population means, 493
Exhaustive hypotheses, 264
Expected number of successes, 714
Experiment, 207, 470
Experimental design, 470, 625
  completely randomized, 628
  factorial, fractional, 672
  factorial, with two factors, 663
  randomized block, 652
Extrapolation, 189, 515

F
F distribution. *See* Distribution, F.
F test
  for comparing population
    variances, 485
  in one-way ANOVA, 639
  in randomized block ANOVA, 653
  for regression, 581
  in two-way ANOVA, 665
Factor, 627, 663
Forward selection, 596, 598, 605
Frequency
  expected, 694, 698–699, 707, 714,
    725
  observed, 694, 699, 714
Frequency distribution, 57. *See also*
  Frequency table.
  in Excel, 96
Frequency table, 57
  class selection, 61, 64
  for continuous data, 64
  for integer data, 61
  for qualitative data, 57

G
Grand mean, 630, 642
Graphical displays, 56, 72
  bar chart, 73, 76, 171, 173
  box and whisker diagram, 145
  boxplot, 145–148
  computer-generated, 82
  contingency table, 170
  dotplot, 81
  in Excel, 96
  frequency distribution, 57
  histogram, 77–79, 96–97
  in Minitab, 96
  pie chart, 76
  for qualitative data, 72
  for quantitative data, 77
  scatter plot, 178, 181

H
Histogram, 77
  class intervals, 89
  for continuous data, 79
  in Excel, 96–97
  for integer data, 78
  in Minitab, 96
  probability, 247
  of skewed data, 88
  with TI-83, 98–99
Hypothesis, 358
  alternative, 363, 391
  design of, 359
  about nominal variables, 361
  null, 363, 390–391
  about quantitative variables,
    359
  summary of, 362

Hypothesis test, 358
    alternative hypothesis, 363, 391
    critical value, 366–367
    lower-tail, 389–390
    null hypothesis, 390–391
    one-tail, 388, 390–391
    outcome, 382
    p value, 366–367, 377, 379–380
    procedure, 362, 749
    rejection region, 366–367, 384, 396
    significance level, 366–367
    test statistic, 366
    two-tail, 370, 388, 390
    upper-tail, 390–391
Hypothesis testing
    chi-square test for goodness of fit, 694, 705, 707
    chi-square test for independence, 722–723, 728
    chi-square test of population variance, 422, 426–427
    chi-square test for proportions, 711, 716, 718
    in Excel, 439, 492
    F test for comparing population variances, 485
    F test in one-way ANOVA, 639
    F test in randomized block ANOVA, 653
    F test in regression, 581
    F test in two-way ANOVA, 665
    individual coefficients in multiple regression, 587
    in Minitab, 440, 492
    relation to confidence intervals, 437
    t test of difference in two population means, 462, 465, 570
    t test of population mean, small sample, 412, 416, 419
    t test of population means, paired samples, 472
    t test for regression coefficients, 588
    t test about the slope, 524
    with TI-83, 442, 492
    Z test of difference in two population means, 453, 458
    Z test of difference in two population proportions, 480
    Z test for one mean, 370, 393, 399
    Z test of proportion, 429, 431, 433–434

I
Independence of events, 229
Inference, 39
Inferential statistics, 39–40
Inner fence, 145

Input variable, 570
Interaction, 668
Intercept, of least-squares line, 510
Interpolation, 189, 515
Interquartile range (IQR), 145
Interval estimate. See Confidence interval.
Interval probability, 271

K
KADDSTAT
    chi-square test for independence, 735
    confidence interval, calculation of, 346
    multiple regression, 612
    regression analysis, 550
    stepwise regression, 613
Kruskal-Wallis test, 760
    compared to ANOVA, 763
    in Minitab, 770
    test statistic, 763–764

L
Law of averages, 216
Law of large numbers, 216
Least-squares line, 185
    calculation of, 186, 510
    deviation, 185–186
    equation of, 510
    extrapolation, 189
    intercept of, 510
    interpolation, 189
    in multiple regression, 571
    predicted value, 186
    in simple linear regression, 509
    slope of, 510
    technique, 185
Least-squares method, 509
Level of a factor, 627
Likert scale, 34

M
Mann-Whitney test, 750, 769
Margin of error, 339
Mean
    of binomial distribution, 257
    of normal distribution, 267
    of sample mean, 308
Mean square among (MSA), 638
Mean square error (MSE), 529, 581
    in one-way ANOVA, 638
    in randomized block design, 653
Mean square regression (MSR), 529, 581
Mean square total (MST), 638
Measures of central tendency. See Central tendency, measures of.
Measures of dispersion. See Dispersion, measures of.

Measures of relative standing. See Relative standing, measures of.
Midpoint, of an interval, 80
Minitab, 8
    all possible regressions, 614
    analysis of variance (ANOVA), 679–680
    bar chart, creation of, 96, 98, 195
    bivariate data, 195
    boxplot, creation of, 153
    chi-square test for independence, 736
    confidence interval, calculation of, 345, 556
    contingency table, creation of, 736
    graphical displays, creation of, 96
    histogram, creation of, 96
    hypothesis testing, 440, 492
    multiple regression, 612, 615
    pie chart, creation of, 96, 98
    predicted values, 555
    prediction interval, creation of, 556
    probabilities, calculation of, 283
    random data, generation of, 234
    random sample, selection of, 47
    residuals, 555
    scatter plot, 196
    simple linear regression, 553
    stepwise regression, 613
    summary statistics, creation of, 154
    t test of population mean, 440
Modal class, 121
Model-building techniques, in multiple regression, 596
Multicollinearity, 609
Multimodal sample, 122
Multiple comparisons, 643
Mutually exclusive
    events, 218, 221
    hypotheses, 364

N
Nonparametric methods, 749, 768
    Kruskal-Wallis test, 760
    Mann-Whitney test, 750, 769
    Wilcoxon rank sum test, 750
Normal distribution. See Distribution, normal.
Normal probability plot, 542, 608, 649

O
Outer fence, 145
Outlier, 137, 145
Output variable, 570
Overall mean. See Grand mean.
Overall proportion of successes, 714

P

p value, 366–367, 377, 379–380
  in ANOVA table, 639–640
  in test of coefficient of multiple
    regression, 590
Paradigm, 2
Parameter, 18, 111, 294
Percentile, 140
Percentile rank, 141
Pie chart, 76
  in Excel, 77, 96–97
  in Minitab, 96, 98
Point estimate, 294, 337
Point estimator, 294
  properties of, 302
  for qualitative variable, 297
  for quantitative variable, 295
  summary of, 298
  unbiased, 302
Poisson distribution. *See*
    Distribution, Poisson.
Poisson random variable, D1
Pooled variance, 463
Population, 11–12
  sampling frame, 28
  size, 24
  variation in, 15, 25
Population mean, 112
  large-sample vs. small-sample
    tests, 369
  small-sample tests in Excel, 439
Population proportion, 297, 482
Population standard deviation, 131
  comparison of two population
    means, *t* test, 462, 465
  comparison of two population
    means, Z test, 453, 458
Population variance, 131
Possible outlier, 148
Posttest conditions, 470
Predicted value, 186, 513
  in Minitab, 555
Prediction interval, 535
  in Excel, 550
  in Minitab, 556
Pretest conditions, 470
Probability, 40, 207
  Bayes' Rule and, C2
  conditional, 227, C1
  empirical, 213
  equally likely outcomes, 209
  of an event, 209, 217
  interval, 245–246, 271
  law of large numbers and, 216
  in Minitab, 283
  notation, 245
  as relative frequency, 213
  sample space, 207–208
  with TI-83, 284
Probability density function, 265

Probability distribution, 244
Probability rules, 207, 244–245
  addition, general, 222
  addition, simple, 218, 220
Probable outlier, 148

Q

Quartile, 142–143

R

R^2, 603
Random numbers
  in Excel, 46, 233
  in Minitab, 47, 234
  table of, 28, A2
  with TI-83, 234
Random variable, 243
  binomial, 250
  continuous, 264
  discrete, 244
  normal, 267
  standard normal, 269
Rank, 752
Regression, multiple, 569–570, 580,
    611
  assumptions, 605
  coefficient of multiple
    determination, 585
  in Excel, 612
  with KADDSTAT, 612
  least-squares model, 570
  in Minitab, 612, 615
  model, 570, 611
  model adequacy, 605
  model building, 596
  multicollinearity, 609
  for prediction, 575
  residual analysis, 605
  testing individual coefficients, 587
  testing for significance, 580
Regression, simple linear, 503, 507
  assumptions, 540
  computer-generated, 513
  confidence intervals for, 532
  in Excel, 547
  inferences about, 524
  with KADDSTAT, 550
  least-squares method, 509
  in Minitab, 553
  model, 507
  prediction intervals for, 535
  with TI-83, 557
  variance, partitioning of, 528
Regression line, 509
Rejection region, 366–367, 384, 396
  one-tail, 416–417, 419, 486–487,
    489
  two-tail, 415, 419, 486–488
Relative efficiency of block design,
    656

Relative frequency, 59
  cumulative, 62
Relative standing, measures of, 140
  first quartile, 142
  interquartile range, 145
  percentile, 140
  third quartile, 143
Replicate, 663
Replication, in experiments, 663
Residual, 516–517, 575
Residual analysis
  in Minitab, 555
  in multiple regression, 605
  in simple linear regression, 541
Residual plot, 541, 606
  in Minitab, 556
Response variable. *See* Variable,
    response.

S

Sample, 13
  biased, 27
  dependent, 470
  selection of, 26–28
  simple random, 27
  unbiased, 26
Sample mean, 112
  comparison with sample median,
    117
Sample median, 115
  comparison to sample mean, 117
  second quartile, 143
Sample mode, 121
Sample range, 128, 130
Sample size, 14, 24, 26
  in binomial distribution, 252
  for comparing two populations, 452
  for estimating mean, 341
  for estimating proportion, 341
  notation, 26
  for range, 130
  relation to standard error, 313
Sample space, 207
  finite, 208
  infinite, 208
Sample standard deviation, 131, 134
Sample variance, 131–132, 629–630
Sampling distribution
  definition, 306
  of sample mean, 305, 307, 331
Sampling error, 14, 25, 296, 339–340
Sampling frame, 28
Sampling with replacement, 250
Scatter plot, 178, 181
  in Minitab, 196
  with TI-83, 196
Sigma notation, 43
Significance level, 702
Simple random sample, 27, 625
Slope, of least-squares line, 510

Standard deviation
    of binomial distribution, 257
    of normal distribution, 267
    relation to standard error, 310
    of sample mean, 309
Standard error, 306, 459
    of the estimate, 520
    relation to sample size, 313
    relation to standard deviation, 310
    of sample mean, 309
    of the slope, 525
Standard normal random variable, 269
Standard normal table, 268–269, 271, 273, A6
Statistic, 19, 111
Statistics, 1–2, 8, 11
Stepwise regression, 596, 600, 605
    with KADDSTAT, 613
    in Minitab, 613
Student's *t* distribution, 331
Sum of squares among (SSA), 632–633, 653, 763
Sum of squares between. *See* Sum of squares among (SSA).
Sum of squares blocks (SSBL), 653
Sum of squares error (SSE), 528, 581, 634–635, 653, 664
Sum of squares regression (SSR), 528, 581
Sum of squares total (SST), 528, 581, 631, 653, 664, 764
Sum of squares within. *See* sum of squares error (SSE).
Summation notation, 43

T

t distribution. *See* Distribution, *t*.
t test
    of difference in two population means, 462, 465, 470

in Excel, 439
in Minitab, 440
of population mean, small sample, 412, 416, 419
of population means, paired samples, 472
for regression coefficients, 587
about the slope, 524
statistic, 412
with TI-83, 442
Test statistic, 366
TI-83 graphing calculator, 8
    binomial probabilities, calculation of, 284
    bivariate data, 196
    boxplot, creation of, 153
    confidence interval, calculation of, 346
    histogram, creation of, 98–99
    hypothesis testing, 442, 492
    normal probabilities, calculation of, 284
    random data, generation of, 234
    regression analysis, 557
    scatter plot, creation of, 196
    summary statistics, creation of, 155
    t test of population means, 442
    Z test of population means, 495
Treatment, experimental, 629
Treatment effect, 639, 641–642
Treatment mean, 632
t-score, 331
Type I error, 366, 383–384
Type II error, 383–384

U

Unbiased
    estimator, 302, 320
    sample, 26
Unimodal data set, 87

V

Variable, 12
    dependent, 178, 511, 596
    independent, 178, 511, 596
    input, 570, 597
    output, 570
    random, 243
    response, 627, 642
Variance
    pooled, 463
    sample, 629
Variation, 15, 25
    between groups, 632–633
    in block design, 652
    within groups, 632, 634
    in linear regression, 528
    total. *See* Sum of squares total (SST).
    in two-factor design, 664

W

Wilcoxon rank sum test, 750
    critical values, table of, A14
    in Minitab, 769
    rank, 752
    test statistic, 753

Y

$\hat{y}$, 186, 513

Z

Z random variable, 269
Z test
    of difference in two population means, 453, 458
    of difference in two population proportions, 480
    in Excel, 493
    of population proportions, 429, 431, 433–434
    with TI-83, 495
Z-score, 136–137, 269, 305, 318